Advances in Modal Logic

Volume 14

Advances in Modal Logic
Volume 14

Edited by

David Fernández-Duque
Alessandra Palmigiano

and

Sophie Pinchinat

ISBN 978-1-84890-413-2

College Publications
Scientific Director: Dov Gabbay
Managing Director: Jane Spurr

http://www.collegepublications.co.uk

Contents

Preface

Advances in Modal Logic (AiML) was initiated in 1995 and aims at providing a venue in which the state of the art of modal logic and its many applications may be regularly presented. It consists of a series of conferences and their respective volumes of proceedings. AiML is the main international forum at which research on all aspects of modal logic is presented. The first installment was held in 1996 in Berlin, Germany, and since then it has been organized biennially, with meetings in 1998 in Uppsala, Sweden; in 2000 in Leipzig, Germany (jointly with ICTL 2000); in 2002 in Toulouse, France; in 2004 in Manchester, UK; in 2006 in Noosa, Australia; in 2008 in Nancy, France; in 2010 in Moscow, Russia; in 2012 in Copenhagen, Denmark; in 2014 in Groningen, The Netherlands; in 2016 in Budapest, Hungary; in 2018 in Bern, Switzerland (Jointly with LATD 2018); and in 2020, organized virtually by the University of Helsinki, Finland, due to the exceptional circumstances of the COVID pandemic. Information about AiML and related events, including conference proceedings, is available at the website `www.aiml.net`.

The fourteenth conference in the series was was organized at the University of Rennes, France, by Sophie Pinchinat (IRISA), with the assistance of Sophie Maupilé, Aurélie Amet, Guillaume Aucher, Dylan Bellier, Frédéric Bouvet, Lénaïg Cornanguer, Catherine Jacques-Orban, Antoine LAzou, Pierre Le Scornet, Hervé Marchand, Nicolas Markey, Alexandre Terefenko and Adrien Thomas. It was held jointly with LAMAS&SR 2022. Due to loosening of the COVID restrictions and improvements in the status of the pandemic, AiML 2022 was held as a mostly in-person conference with assistance for participants who could only attend virtually. The conference website can be found at `https://aiml2022.irisa.fr/`.

This volume contains abstracts of invited talks and contributed papers from the conference. The invited talks were given by

- Willem Conradie (University of the Witwatersrand)
- Wesley Holliday (Univeristy of California, Berkeley)
- Francesca Poggiolesi (CNRS, IHPST)
- Rineke Verbrugge (University of Groningen)

The Programme Committee received 54 regular paper submissions. Of these, 35 were selected for this volume by a reviewing process where every paper received at least three independent expert reviews. The volume includes papers on propositional modal logics, their products, predicate modal logics, temporal and epistemic reasoning, modal logic with non-boolean basis, provability and interpretability logics, inquisitive, dynamic, connexive, intuitionistic, substructural, dependence logics and hybrid logics, and other related logics. The topics include history of modal reasoning, decidability and complexity results, proof theory, model theory, interpolation, as well as other related problems in algebraic logic.

In addition, there were 28 submissions for short presentations at the conference, and 19 were accepted for presentation.

The members of the Programme Committee for the conference were

- Erman Acar, VU Amsterdam
- Bahareh Afshari, University of Amsterdam
- Natasha Alechina, University of Utrecht
- Steve Awodey, Carnegie Mellon
- Philippe Balbiani, CNRS, Toulouse University
- Marta Bilkova, Academy of Sciences of the Czech Republic
- Xavier Caicedo, University of los Andes
- Walter Carnielli, State University of Campinas
- Agata Ciabattoni, TU Wien
- Ivano Ciardelli, University of Munich
- Willem Conradie, University of the Witwatersrand
- Laurent De Rudder, University of Liege
- Tommaso Flaminio, Spanish National Research Council
- Sabine Frittella, INSA Centre Val de Loire
- Nick Galatos, University of Denver
- Sam van Gool, IRIF, Université de Paris
- Giuseppe Greco, VU Amsterdam
- Thomas Icard, Stanford University
- Ramon Jansana, University of Barcelona
- Peter Jipsen, Chapman University
- Joost Joosten, University of Barcelona
- Stanislav Kikot, Sber Automotive Technologies
- Philip Kremer, University of Toronto
- Alexander Kurz, Chapman University
- Roman Kuznets, TU Wien
- Fei Liang, University of Shandong
- Minghui Ma, Sun Yat-Sen University, Guangzhou
- Morteza Moniri, Shahid Beheshti University
- Tommaso Moraschini, University of Barcelona
- Drew Moshier, Chapman University, Orange CA

- Eric Pacuit, University of Maryland
- Fedor Pakhomov, Ghent University
- Sophie Pinchinat, IRISA, University of Rennes I
- Daniele Porello, University of Genova
- Vit Puncochar, Academy of Sciences of the Czech Republic
- Revantha Ramanayake, University of Groningen
- Christian Retoré, University of Montpellier
- Umberto Rivieccio, Universidade Federal do Rio Grande do Norte
- Claudette Robinson, University of Johannesburg
- Gabriel Sandu, University of Helsinki
- Igor Sedlar, Academy of Sciences of the Czech Republic
- Ilya Shapirovsky, New Mexico State University
- Apostolos Tzimoulis, VU Amsterdam
- Sara Uckelman, Durham University
- Jouko Väänänen, University of Helsinki
- Heinrich Wansing, University of Bochum
- Frank Wolter, University of Liverpool

The Programme Committee was chaired by

- David Fernández-Duque (ICS of the Czech Academy of Sciences and Ghent University)
- Alessandra Palmigiano (VU Amsterdam and University of Johannesburg)

The Steering Committee of AiML for 2018–2020 consisted of

- Lev Beklemishev (Steklov Mathematical Institute)
- Guram Bezhanishvili (New Mexico State University)
- Rajeev Goré (Australian National University)
- Giovanna D'Agostino (University of Udine)
- Stéphane Demri (CNRS, France)
- Agi Kurucz (King's College London)
- Sara Negri (University of Helsinki) (local organizer AiML 2020)
- Nicola Olivetti (Aix-Marseille University)
- Rineke Verbrugge (University of Groningen)

Many other people assisted with the reviewing process, including: Dylan Bellier, Luca Carai, Davide Catta, Diana Costa, Andrew Craig, Tiziano Dalmonte, Andrea De Domenico, Fabio De Martin Polo, Martín Diéguez, Ali Farjami, Davide Fazio, Raul Fervari, Damiano Fornasiere, David Gabelaia, Patrick Girard, Maksim Gladyshev, Quentin Gougeon, Gianluca Grilletti, Andreas Herzig, Loan Ho, Emil Jerábek, Alex Kavvos, Kohei Kishida, Daniil

Kozhemiachenko, Andrey Kudinov, Taishi Kurahashi, Timo Lang, Tim Lyon, Krishna Balajirao Manoorkar, Sérgio Marcelino, Brett McLean, Jérôme Mengin, Yoàv Montacute, Emery Neufeld, Mattia Panettiere, Xavier Parent, Pawel Pawlowski, Elaine Pimentel , Adam Prenosil, Renyan Feng, Mehrnoosh Sadrzadeh, Giorgio Sbardolini, Dmitry Shkatov, Daniel Skurt, Gavin St. John, Lutz Strassburger, Michal Stronkowski, Andrew Tedder, Bruno Teheux, Matteo Tesi, Sara Ugolini, Giovanni Varricchione, Kentaro Yamamoto, and Aybüke Özgün. We apologize to anyone whose name was inadvertently left off this list.

We are grateful to the organizers of the conference for bringing it to life despite the challenges that come with the pandemic, including the necessary arrangements for those participants who could not join us physically. We would like to thank the members of the Programme Committee and subreviewers for their time and effort and the thoughtful reports and discussion to ensure the highest scientific standards of the conference and its proceedings. We are also grateful to the authors for their excellent contributions and we thank Jane Spurr, without whom this volume would not have been possible. Special thanks go to Guram Bezhanishvili and Nicola Olivetti, for their invaluable advice throughout the preparation for our conference. We thank the IRISA staff, Sophie Maupilé, Antoine L'Azou, Catherine Jacques-Orban and Agnès Cottais for their help on the administrative, financial and communication tasks of the conference organization. Finally, we are tremendously grateful to Sophie Maupilé for closely assisting Sophie Pinchinat in coordinating the local organization of the conference, and to Pierre Le Scornet who was in charge of both the website of the conference and of the participants registration.

We would like to thank Inria, Université de Rennes 1, Rennes Métropole, Région Bretagne, Fondation Rennes 1, MDPI AG and Mitsubishi Electric R&D Centre Europe for generously sponsoring the conference, the Université de Rennes 1 for help in the organization and online facilities, and Inria for providing the conference rooms.

July 8th, 2022

David Fernández-Duque
Alessandra Palmigiano
Sophie Pinchinat

Abstracts of Invited Talks

On parametric phenomena in correspondence theory

Willem Conradie

School of Mathematics, University of the Witwatersrand, Johannesburg, National Institute for Theoretical and Computational Sciences (NITheCS), South Africa

Abstract

I survey a number of recent results which study the syntactic shape of modal formulas' first-order frame correspondents on different relation semantics and establish systematic links between the correspondents of given modal formula across these different semantic environments. It has been observed that, for instance, in some cases, correspondents remains syntactically unchanged under such semantic shifts; in other cases, correspondents across different semantics can be recovered as parametric instances of the same pattern, with the parameter capturing the embedding between semantic contexts. I discuss some open problems and further directions.

Keywords: Correspondence theory, modal logic, many-valued modal logic, intuitionistic modal logic, non-distributive logics.

Unified correspondence. In recent years, the research programme of unified correspondence (see, e.g., [9] [14]) has developed Sahlqvist correspondence and canonicity results for large families of non-classical logics. In [10] we formulate a general definition of the classes of Sahlqvist and inductive formulas which applies uniformly to any logic algebraically captured by a class of lattice expansions (LE-logics). This definition is based solely on the order-theoretic properties of the algebraic interpretation of the connectives and, when projected onto specific LE-logics, captures or generalises many existing definitions in the literature. A further feature of unified correspondence, is a division of labour which separates the computation of a first-order correspondent for a modal axiom into two phases: a correspondent is first computed and expressed as a conjunction of quasi-inequalities in a extended modal language interpreted on perfect lattice expansions. This is independent of any particular choice of relational semantics for the logic. The second phase of the computation translates this algebraic correspondent by applying the standard translation associated with a particular choice of relational semantics linked with the algebraic semantics via a suitable duality.

Parametric correspondence. This and other recent work in correspondence theory, has focused mostly on the (modal) propositional side of the correspondence theoretic equation, i.e., on identifying classes of (modal) propositional formulas for which first-order correspondents exists. However, beyond the fact of their existence, the properties of the first-order correspondents themselves have received little attention. This situation has changed recently, with a number of results studying the syntactic shape of first-order correspondents on particular relation semantics and, more specifically, comparing the shape of the first-order correspondent of given modal formula across different relational semantics. It has been observed, for instance, that, in some cases, correspondents remains syntactically unchanged under such semantic shifts; in other cases, correspondents across different semantics can be recovered as parametric instances of the same pattern, with the parameter capturing the embedding between semantic contexts. This talk will survey a range of phenomena and results of this type, some of which I discuss below.

Many-valued modal logic. The first result of this type was proven in the context of the correspondence theory of many-valued modal logic (see, e.g., [12,3]). Specifically, considering many-valued modal logics with complete Heyting algebras as truth value spaces, [4] shows that the first-order correspondent of any modal Sahlqvist formula over many-valued Kripke frames is syntactically identical to its first-order correspondent on (crisp) Kripke frames. Of course, the many-valued correspondent should be read as a formula of many-valued predicate logic and therefore its semantic meaning generalizes that of the classical correspondent. Apart from purely theoretical interest, this result has great practical utility as it means that the correspondents of well-known modal axioms can be directly transferred to the many-valued versions of the logics they axiomatize.

Non-distributive modal logics with polarity-based semantics. The next important instance of parametric correspondence which I would like to discuss comes from the study of non-distributive modal logics [11]. These logics can be given semantics based on polarities, or formal contexts, in the sense of formal concept analysis (FCA) [13] and given an epistemic interpretation [8,7] where formulas denote formal categories. In [8,7] it is observed that, over this semantics, the classical S5 axioms of epistemic logic have first-order correspondents which are, in a sense, dual to their well-known correspondents on Kripke frames. Moreover, it is argued that the epistemic meanings of these axioms are preserved in a modified form in this new context. This observation is extensively developed in [6] where the first-order correspondents of modal reduction principles [18] are captured as inclusions of relational compositions, both in the setting of Kripke frames and in that of polarity based frames. It is established that one my pass from the Kripke frame correspondent, thus expressed, to the correspondent on polarity-based frames, roughly speaking, by reversing the inclusion and parameterising the relational compositions with the incidence relation of thc polarity. Mathematically, this is underpinned by the embedding of the class of Kripke frames into the class of enriched polarities.

Non-distributive modal logics with graph-based semantics. Alternatively, non-distributive logics may be given graph-based semantics [11,5] based on Ploščica's [15] representation of general lattices. Here too, the correspondents of modal reduction principles may be obtained by appropriately transforming their correspondents on Kripke frames.

Intuitionistic modal logic. In the case of intuitionistic modal logic with semantics based on [19], the correspondents of inductive (and hence also Sahlqvist) formulas can be transparently obtained from the classical correspondents. This is mainly due to the strong interaction conditions between the pre-order of the intuitionistic Kripke frame and the additional accessibility relation used to interpret modalities. In the case of Fisher-Servi's semantics for intuitionistic modal logic [16,17], the interaction conditions between the relations are weaker, and the relationship between correspondence in the classical case and in this setting is somewhat more involved.

Relativized correspondence. When imposing restrictions on the class of Kripke frames considered, more modal formulas become first order definable, in a phenomenon known as relativized correspondence. For instance, all modal reduction principles are first-order definable over transitive frames [18], while all modal formulas have first-order correspondents over S5 frames [2] and over Euclidean frames [1]. Relativized correspondence has been little studied for modal logics on non-classical propositional base or for many-valued modal logics, but some initial results indicates that an interesting landscape waits to be discovered here. I will conclude by sketching some of these results and directions.

References

[1] Balbiani, P., D. Georgiev and T. Tinchev, *Modal correspondence theory in the class of all euclidean frames*, Journal of Logic and Computation **28** (2018), pp. 119–131.

[2] Balbiani, P. and T. Tinchev, *Definability over the class of all partitions*, Journal of Logic and Computation **16** (2006), pp. 541–557.

[3] Bou, F., F. Esteva, L. Godo and R. O. Rodríguez, *On the minimum many-valued modal logic over a finite residuated lattice*, Journal of Logic and computation **21** (2011), pp. 739–790.

[4] Britz, C., "Correspondence theory in many-valued modal logics," Master's thesis, University of Johannesburg, South Africa (2016).

[5] Conradie, W., A. Craig, A. Palmigiano and N. Wijnberg, *Modelling informational entropy*, in: *Proc. WoLLIC 2019*, Lecture Notes in Computer Science **11541** (2019), pp. 140–160.

[6] Conradie, W., A. De Domenico, K. Manoorkar, A. Palmigiano, M. Panettiere, D. P. Prieto and A. Tzimoulis, *Modal reduction principles across relational semantics*, arXiv preprint arXiv:2202.00899 (2022).

[7] Conradie, W., S. Frittella, A. Palmigiano, M. Piazzai, A. Tzimoulis and N. Wijnberg, *Toward an epistemic-logical theory of categorization*, in: *16th conference on Theoretical Aspects of Rationality and Knowledge (TARK 2017)*, Electronic Proceedings in Theoretical Computer Science **251**, 2017, pp. 170–189.

[8] Conradie, W., S. Frittella, A. Palmigiano, M. Piazzai, A. Tzimoulis and N. M. Wijnberg, *Categories: how I learned to stop worrying and love two sorts*, in: *Proc. WoLLIC 2016*, Lecture Notes in Computer Science **9803**, Springer, 2016, pp. 145–164.

[9] Conradie, W., S. Ghilardi and A. Palmigiano, *Unified Correspondence*, in: A. Baltag and S. Smets, editors, *Johan van Benthem on Logic and Information Dynamics*, Outstanding Contributions to Logic **5**, Springer International Publishing, 2014 pp. 933–975.

[10] Conradie, W. and A. Palmigiano, *Algorithmic correspondence and canonicity for non-distributive logics*, Annals of Pure and Applied Logic **170** (2019), pp. 923–974.

[11] Conradie, W., A. Palmigiano, C. Robinson and N. Wijnberg, *Non-distributive logics: from semantics to meaning*, in: A. Rezus, editor, *Contemporary Logic and Computing*, Landscapes in Logic **1**, College Publications, 2020 pp. 38–86.
URL `arXivpreprintarXiv:2002.04257`

[12] Fitting, M., *Many-valued modal logics*, Fundam. Inform. **15** (1991), pp. 235–254.

[13] Ganter, B. and R. Wille, "Formal concept analysis: mathematical foundations," Springer Science & Business Media, 2012.

[14] Greco, G., M. Ma, A. Palmigiano, A. Tzimoulis and Z. Zhao, *Unified correspondence as a proof-theoretic tool*, Journal of Logic and Computation **28** (2018), pp. 1367–1442.

[15] Ploščica, M., *A natural representation of bounded lattices*, Tatra Mountains Mathematical Publications **4** (1994).

[16] Servi, G. F., *Semantics for a class of intuitionistic modal calculi*, in: *Italian studies in the philosophy of science*, Springer, 1980 pp. 59–72.

[17] Servi, G. F., *Axiomatizations for some intuitionistic modal logics*, Rend. Sem. Mat. Univers. Politecn. Torino **42** (1984), pp. 179–194.

[18] Van Benthem, J., *Modal reduction principles*, The Journal of Symbolic Logic **41** (1976), pp. 301–312.

[19] Wolter, F. and M. Zakharyaschev, *Intuitionistic modal logic*, in: *Logic and foundations of mathematics*, Springer, 1999 pp. 227–238.

Non-classical modal logic for natural language

Wesley H. Holliday

University of California, Berkeley

Abstract

Modality is a central topic in natural language semantics [18,27,26,19,20]. Modal logic has contributed to the study of modality in natural language in important ways, including by providing perspicuous axiomatizations of the consequence relations defined by proposed formal semantics for fragments of modal language (see [13] for an overview). In this talk, based on joint work with Matthew Mandelkern [14], I will explain how attempting to solve one of the most discussed problems in modal semantics in recent years—accounting for the peculiar empirical behavior of *epistemic* modals [28,8,29,1,6,30,21,17,5,4,24,25,23,9,16,2]—led us to a non-classical modal logic for the relevant fragment of language. We have characterized the logic axiomatically as well as semantically, using algebraic semantics based on ortholattices, relational representations of ortholattices [3,22,7], and possibility semantics for modals [15,10,12,11]. Thus, I will discuss how the kinds of tools developed in the AiML community for mathematical modal logic can be brought to bear on a well-known applied problem.

References

[1] Aloni, M., *Conceptual covers in dynamic semantics*, in: L. Cavedon, P. Blackburn, N. Braisby and A. Shimojima, editors, *Logic, Language and Computation, Vol. III*, CSLI, 2000 .

[2] Aloni, M., L. Incurvati and J. J. Schlöder, *Epistemic modals in hypothetical reasoning*, Erkenntnis (Forthcoming).

[3] Birkhoff, G., "Lattice Theory," American Mathematical Society, New York, 1940.

[4] Bledin, J., *Logic informed*, Mind **123** (2014), pp. 277–316.

[5] Dorr, C. and J. Hawthorne, *Embedding epistemic modals*, Mind **122** (2013), pp. 867–913.

[6] Gillies, A., *Epistemic conditionals and conditional epistemics*, Noûs **38** (2004), pp. 585–616.

[7] Goldblatt, R. I., *Semantic analysis of orthologic*, Journal of Philosophical Logic **3** (1974), pp. 19–35.

[8] Groenendijk, J., M. Stokhof and F. Veltman, *Coreference and modality*, in: S. Lappin, editor, *Handbook of Contemporary Semantic Theory*, Oxford, Blackwell, 1996 pp. 179–216.

[9] Hawke, P. and S. Steinert-Threlkeld, *Semantic expressivism for epistemic modals*, Linguistics and Philosophy **44** (2021), pp. 475–511.

[10] Holliday, W. H., *Partiality and adjointness in modal logic*, in: R. Goré, B. Kooi and A. Kurucz, editors, *Advances in Modal Logic, Vol. 10*, College Publications, London, 2014 pp. 313–332.

[11] Holliday, W. H., *Possibility semantics*, in: M. Fitting, editor, *Selected Topics from Contemporary Logics*, Landscapes in Logic, College Publications, London, 2021 pp. 363–476.

[12] Holliday, W. H., *Possibility frames and forcing for modal logic*, The Australasian Journal of Logic (Forthcoming).
URL `https://escholarship.org/uc/item/0tm6b30q`

[13] Holliday, W. H. and T. F. Icard, *Axiomatization in the meaning sciences*, in: D. Ball and B. Rabern, editors, *The Science of Meaning: Essays on the Metatheory of Natural Language Semantics*, Oxford University Press, 2018 pp. 73–97.

[14] Holliday, W. H. and M. Mandelkern, *The orthologic of epistemic modals* (2021), arXiv:2203.02872 [cs.LO].

[15] Humberstone, I. L., *From worlds to possibilities*, Journal of Philosophical Logic **10** (1981), pp. 313–339.

[16] Incurvati, L. and J. J. Schlöder, *Epistemic multilateral logic*, Review of Symbolic Logic **15** (2022), pp. 505–536.

[17] Klinedinst, N. and D. Rothschild, *Connectives without truth-tables*, Natural Language Semantics **20** (2012), pp. 137–175.

[18] Kratzer, A., *Modality*, in: A. von Stechow and D. Wunderlich, editors, *Semantics: An International Handbook of Contemporary Research*, de Gruyter, Berlin, 1991 pp. 639–650.

[19] Kratzer, A., "Modals and Conditionals," Oxford University Press, 2012.

[20] Lassiter, D., "Graded Modality," Oxford University Press, 2017.

[21] MacFarlane, J., *Epistemic modals are assessment sensitive*, in: A. Egan and B. Weatherson, editors, *Epistemic Modality*, Oxford University Press, 2011 pp. 144–177.

[22] MacLaren, M. D., *Atomic orthocomplemented lattices*, Pacific Journal of Mathematics **14** (1964), pp. 597–612.

[23] Mandelkern, M., *Bounded modality*, The Philosophical Review **128** (2019), pp. 1–61.

[24] Moss, S., *On the semantics and pragmatics of epistemic vocabulary*, Semantics and Pragmatics **8** (2015), pp. 1–81.

[25] Ninan, D., *Quantification and epistemic modality*, The Philosophical Review **127** (2018), pp. 433–485.

[26] Portner, P., "Modality," Oxford University Press, 2009.

[27] Swanson, E., *Modality in language*, Philosophy Compass **3** (2008), pp. 1193–1207.

[28] Veltman, F., "Logics for Conditionals," Ph.D. thesis, University of Amsterdam (1985).

[29] Veltman, F., *Defaults in update semantics*, Journal of Philosophical Logic **25** (1996), pp. 221–261.

[30] Yalcin, S., *Epistemic modals*, Mind **116** (2007), pp. 983–1026.

Explanations in Logic

Francesca Poggiolesi [1]

IHPST, UMR 8590
Université Paris 1 Panthéon-Sorbonne

Abstract

To explain phenomena in the world, to answer the question "why" rather than the question "what", is one of the central human activities and one of the main goals of rational inquiry. Causal explanations have been dominant in this field, occupying interest and attention at least from 1940's (see e.g. see [1], [7]). However, in the last decade philosophers have become receptive to another type of explanation, called *non-causal* or *conceptual explanations* (e.g. see [2], [3]). Conceptual explanations do not derive their explanatory power from a network of causal relations, but rather from a network of conceptual relations. Mathematical explanations – that is, mathematical proofs that explain the theorem they prove – are an emblematic example of conceptual explanations. Whilst many have argued that logic has little to contribute to the study of causal explanation (e.g. see [6]), conceptual explanations are *prime facie* a natural object for logical analysis. The main aim of the talk is to propose an account of the logical structure of conceptual explanations. We will do so by using the resources of proof theory and by introducing the novel notion of *formal explanation* (e.g. see [4], [5]). The results we provide not only shed light on conceptual explanations themselves, but also on the role that logic and logical tools might play in the burgeoning field of inquiry concerning explanation.

Keywords: Complexity, derivations, explanations, proofs.

References

[1] Hempel, C., "Aspects of Scientific Explanation and Other Essays in the Philosophy of Science," Free Press, New York, 1965.

[2] Lange, M., "Because Without Cause: Non-causal Explanations in Science and Mathematics," Oxford University Press, Oxford, 2017.

[3] Mancosu, P., *Mathematical explanation: Problems and prospects*, Topoi **20** (2001), pp. 97–117.

[4] Poggiolesi, F., *On defining the notion of complete and immediate formal grounding*, Synthese **193** (2016), pp. 3147–3167.

[5] Poggiolesi, F., *On constructing a logic for the notion of complete and immediate formal grounding*, Synthese **195** (2018), pp. 1231–1254.

[6] Scriven, M., *The logic of cause*, Theory and Decision **2** (1971), pp. 49–66.

---

[1] This work has been developed in the framework of the project IBS (ANR-18-CE27-0012-01), poggiolesi@gmail.com.

[7] Woodward, J., “Making Things Happen: A Theory of Causal Explanation,” Oxford University Press, Oxford, 2004.

Not the sky, but the third floor is the limit: Zero-one laws for provability logic, S4, and K4

Rineke Verbrugge

University of Groningen

Abstract

It has been shown in the late 1960s that each formula of first-order logic without constants and function symbols obeys a zero-one law: As the number of elements of finite models increases, every formula holds either in almost all or in almost no models of that size [1]. For modal logics, limit behavior for models and frames may differ. In 1994, Halpern and Kapron proved zero-one laws for classes of models corresponding to the modal logics K, T, S4, and S5 [2]. They also proposed zero-one laws for the corresponding classes of frames, but their zero-one law for K-frames has since been disproved [5,3], and so has more recently their zero-one law for S4-frames [6].
In this talk, we prove zero-one laws for provability logic with respect to both model and frame validity. Moreover, we axiomatize validity in almost all irreflexive transitive finite models and in almost all irreflexive transitive finite frames, leading to two different axiom systems. In the proofs, we use a combinatorial result by Kleitman and Rothschild about the structure of finite (strict) partial orders: almost all of them consist of only three layers [4]. Finally, we present empirical results in order to give an idea of the number of elements from which onwards a formula's almost sure validity or almost sure invalidity stabilizes in such three-layer Kleitman-Rothschild frames. We also discuss possible extensions of the zero-one laws to the modal logics S4 and K4.

References

[1] Glebskii, Y. V., D. I. Kogan, M. Liogon'kii and V. Talanov, *Range and degree of realizability of formulas in the restricted predicate calculus*, Cybernetics and Systems Analysis **5** (1969), pp. 142–154.

[2] Halpern, J. Y. and B. Kapron, *Zero-one laws for modal logic*, Annals of Pure and Applied Logic **69** (1994), pp. 157–193.

[3] Halpern, J. Y. and B. M. Kapron, *Erratum to "zero-one laws for modal logic" [Ann. Pure Appl. Logic 69 (1994) 157–193]*, Annals of Pure and Applied Logic **121** (2003), pp. 281–283.

[4] Kleitman, D. J. and B. L. Rothschild, *Asymptotic enumeration of partial orders on a finite set*, Transactions of the American Mathematical Society **205** (1975), pp. 205–220.

[5] Le Bars, J.-M., *The 0-1 law fails for frame satisfiability of propositional modal logic*, in: *Proc. Logic in Computer Science, 2002*, 2002, pp. 225–234.

[6] Verbrugge, R., *Zero-one laws for provability logic: Axiomatizing validity in almost all models and almost all frames*, in: *2021 36th Annual ACM/IEEE Symposium on Logic in Computer Science (LICS)*, IEEE, 2021, pp. 1–13.

Contributed Papers

Combinatorial Proofs for Constructive Modal Logic

Matteo Acclavio

Università Roma Tre

Lutz Straßburger

Inria Saclay & Ecole Polytechnique

Abstract

Combinatorial proofs form a syntax-independent presentation of proofs, originally proposed by Hughes for classical propositional logic. In this paper we present a notion of combinatorial proofs for the constructive modal logics CK and CD, we show soundness and completeness of combinatorial proofs by translation from and to sequent calculus proofs, and we discuss the notion of proof equivalence enforced by these translations.

Keywords: combinatorial proofs, proof equivalence, arena nets, constructive modal logic.

1 Introduction

Combinatorial proofs have first been introduced by Hughes in order to give a "syntax-free" presentation of proof in classical propositional logic [19]. Their motivation is to capture the essence of a proof independently from any deductive proof system, such that we can speak about *proof equivalence* for proofs given in different formalisms [4]. Only recently it was possible to extend this idea to richer logics: (classical) modal logics [6], relevant logics [5,8], first order logic [20,21], and intuitionistic propositional logic [17]. In this paper we investigate combinatorial proofs for intuitionistic logic with modalities.

There are many different flavours of "intuitionistic modal logics" (see, e.g., [14,30,29,31,9,12]), depending on which additional variants of the classical k-axiom $\Box(A \supset B) \supset (\Box A \supset \Box B)$ are added. It is necessary to add more than just k, as k does not speak of the diamond modality $\Diamond$, which is in the intuitionistic case no longer the De Morgan dual of the box modality $\Box$.

We take here the minimal approach and only add $\Box(A \supset B) \supset (\Diamond A \supset \Diamond B)$ in addition to the k-axiom, leading to what is now called *constructive modal logics* in the literature [30,9,18,28,13,23]. We chose this setting because (1) we would like to make as few assumptions as possible, (2) these logics have a sequent calculus presentation, which makes it easier to show soundness and

$$\dfrac{\dfrac{\dfrac{\dfrac{\dfrac{\dfrac{}{a \vdash a}\,\mathsf{ax} \quad \dfrac{}{a\vdash a}\,\mathsf{ax}}{a,a\vdash a\wedge a}\,\wedge^{\mathsf{R}}}{a\vdash a\wedge a}\,\mathsf{C}}{c,a\vdash a\wedge a}\,\mathsf{W}}{\Box c,\Box a\vdash \Box(a\wedge a)}\,\mathsf{K}^\Box \quad \dfrac{}{b\vdash b}\,\mathsf{ax}}{\Box(a\wedge a)\supset b,\Box c,\Box a\vdash b}\,\supset^{\mathsf{L}}}{(\Box(a\wedge a)\supset b)\supset(\Box c\supset(\Box a\supset b))}\,3\times\supset^{\mathsf{R}} \quad \rightsquigarrow$$

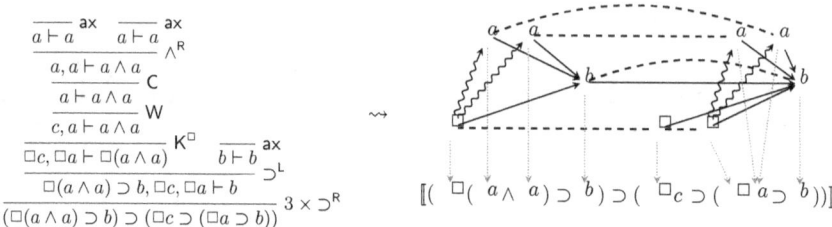

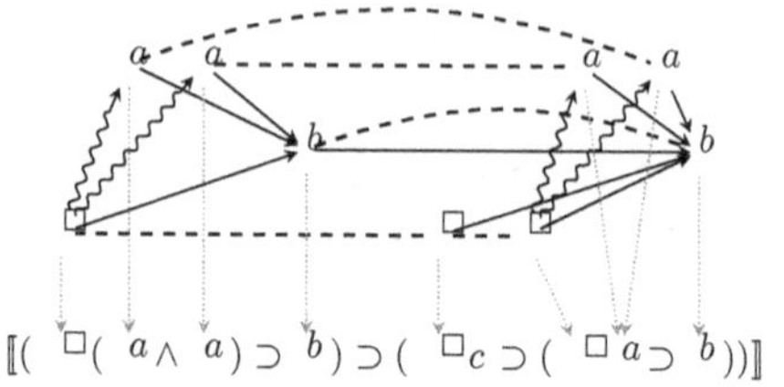

Fig. 1. A sequent calculus derivation of the formula $(\Box(a \wedge a) \supset b, \Box c, \Box a) \supset b$ and its corresponding combinatorial proof

completeness of combinatorial proofs, and (3) there is a close relation to game semantics for modalities [3], extending the work in [17], and lambda-calculus proof terms for constructive modal logics [9].

The main contribution of this paper is the definition of combinatorial proofs for the constructive modal logics CK and CD and to prove their soundness and completeness. We also show that they form a proof system in the sense of Cook and Reckhow [10], that is, checking the correctness of a combinatorial proof can be done in polynomial time in its size.

A combinatorial proof of a formula F is a certain kind of homomorphism $f: \mathcal{G} \to [\![F]\!]$ between two directed graphs. The directed graph $[\![F]\!]$ is a *modal arena* and encodes the formula F. Modal arenas are an extension of the arenas of [17], and are introduced in Section 4 of this paper. The directed graph $\mathcal{G}$ is a *modal arena net* and encodes the "linear part" of the proof. Modal arena nets (introduced in Section 5) are modal arenas equipped with a partition on their vertices, carrying the information of axiom linkings in proof nets. Finally, the homomorphism f is a *skew fibration* [19,17] and encodes the "resource management part" of the proof, i.e., it collects the information carried by the rule instances of contraction and weakening in the sequent calculus. We discuss skew fibrations in Section 6. Figure 1 shows an example of a combinatorial proof for the formula $F = (\Box(a \wedge a) \supset b) \supset (\Box c \supset (\Box a \supset b))$, where the solid and squiggly arrows are the edges of the arena $\mathcal{G}$ and the dashed edges represent the partition of $\mathcal{G}$ (encoding the ax and $\mathsf{K}^\Box$ rules). The dotted downwards directed arrows represent the skew fibration f (encoding the W and C rules).

In order to establish a close correspondence between combinatorial proofs and syntactic proofs in a deductive system, we need to have a *decomposition theorem* which allows to factorize proofs into a linear part, capturing the logic interactions between the components of the proof, and a resource management part, capturing resources duplication or erasing.

The second contribution of this paper is such a decomposition theorem for the logics CK and CD. To obtain this result, we use a combination of the sequent calculus and deep inference. More precisely, we use the cut-free sequent systems given in [24], and we add *deep* rules for contraction and weakening, in a similar way as it has been done in [6].

However, in an intuitionistic setting, we have to distinguish between the left-hand side and the right-hand side of a sequent, where contraction and

$$\frac{}{a \vdash a}\,\mathsf{ax} \qquad \frac{}{\vdash \top}\,\top \qquad \frac{\Gamma, A, A \vdash B}{\Gamma, A \vdash B}\,\mathsf{C} \qquad \frac{\Gamma \vdash B}{\Gamma, A \vdash B}\,\mathsf{W}$$

$$\frac{\Gamma, A \vdash B}{\Gamma \vdash A \supset B}\,\supset^{\mathsf{R}} \qquad \frac{\Gamma \vdash A \quad \Delta, B \vdash C}{\Gamma, \Delta, A \supset B \vdash C}\,\supset^{\mathsf{L}} \qquad \frac{\Gamma \vdash A \quad \Delta \vdash B}{\Gamma, \Delta \vdash A \wedge B}\,\wedge^{\mathsf{R}} \qquad \frac{\Gamma, A, B \vdash C}{\Gamma, A \wedge B \vdash C}\,\wedge^{\mathsf{L}}$$

$$\frac{\Gamma \vdash A}{\Box\Gamma \vdash \Box A}\,\mathsf{K}^{\Box} \qquad \frac{B, \Gamma \vdash A}{\Diamond B, \Box\Gamma \vdash \Diamond A}\,\mathsf{K}^{\Diamond} \qquad \frac{\Gamma \vdash A}{\Box\Gamma \vdash \Diamond A}\,\mathsf{D}$$

$$\begin{aligned} \mathsf{IMLL} &= \{\mathsf{ax}, \top, \supset^{\mathsf{R}}, \supset^{\mathsf{L}}, \wedge^{\mathsf{L}}, \wedge^{\mathsf{R}}\} \\ \mathsf{LI} &= \mathsf{IMLL} \cup \{\mathsf{C}, \mathsf{W}\} \\ \mathsf{LCK} &= \mathsf{LI} \cup \{\mathsf{K}^{\Box}, \mathsf{K}^{\Diamond}\} \\ \mathsf{LCD} &= \mathsf{LCK} \cup \{\mathsf{D}\} \end{aligned}$$

Fig. 2. Sequent rules and sequent systems

weakening apply only to one side. Consequently, the *deep* versions of these rules need to have access to the information on which side a subformula will eventually occur. For this reason, we use polarities, in a similar way as done in [25,32]. The polarized system and the decomposition theorem are given in Section 3.

Finally, we discuss in Section 8 the proof equivalence (in terms of sequent calculus rule permutations) that is induced by our combinatorial proofs and compare it to the one induced by λ-terms [7] and the one induced by winning innocent strategies [3].

2 Preliminaries on Constructive Modal Logics

We consider the *(modal) formulas* generated by a countable set of (atomic) propositional variables $\mathcal{A} = \{a, b, \dots\}$ and the following grammar

$$A, B ::= a \mid \top \mid A \supset B \mid A \wedge B \mid \Box A \mid \Diamond A$$

We say that a formula is *modality-free* if it contains no occurrences of $\Box$ and $\Diamond$. A formula is a $\supset$-formula (resp. a $\wedge$-formula, $\Box$-formula, or $\Diamond$-formula) if it is a formula of the form $A \supset B$ (resp. $A \wedge B$, $\Box A$, or $\Diamond A$).

The constructive modal logic CK is obtained by extending the propositional intuitionistic logic with the *necessitation rule*: if F is provable then so is $\Box F$, and the two modal axiom schemes k_1 and k_2 shown below on the left:

$$\mathsf{k}_1 \colon \Box(A \supset B) \supset (\Box A \supset \Box B) \quad \mathsf{k}_2 \colon \Box(A \supset B) \supset (\Diamond A \supset \Diamond B) \quad \Big| \quad \mathsf{d} \colon \Box A \supset \Diamond A$$

The logic CD is obtained from CK by adding the axiom scheme d on the right above.

We are now recalling the sequent system for these logics. For this, we denote by capital Greek letters Γ or Δ a multiset of formulas, separated by comma. We write $\Box\Gamma$ (resp. $\Diamond\Gamma$) for any such multiset made only of $\Box$-formulas (resp. $\Diamond$-formulas). A *sequent* $\Gamma \vdash A$ is a pair of a multiset of formulas and a formula.

In Figure 2, we show the sequent system LI (e.g., given in [33]) for disjunction-free intuitionistic logic, its linear fragment IMLL [1], and the sequent systems LCK and LCD for the logics CK and CD, respectively, as presented in [24]. The cut rule

$$\frac{\Gamma \vdash A \quad \Delta, A \vdash B}{\Gamma, \Delta \vdash B}\,\mathsf{cut}$$

is admissible for all four systems, and we have the following:

[1] Intuitionistic multiplicative linear logic (see e.g. [26])

Theorem 2.1 *Let* X *for* $\mathsf{X} \in \{\mathsf{CK}, \mathsf{CD}\}$*. The sequent system* LX *is a sound and complete proof system for the disjunction-free fragment of the logic* X.

Proof. Soundness follows from the observation that our systems are the disjunction-free versions of the calculi in [24]. Completeness follows from the cut elimination property. □

If X is a set of rules, we write $F' \overset{\mathsf{X}}{\vdash} F$ if there is a derivation from $\vdash F'$ to $\vdash F$ using rules in X. Moreover, we write $\overset{\mathsf{X}}{\vdash} F$ if there is a proof of F in X, i.e., a derivation with empty premises of $\vdash F$ using rules in X.

Finally, we define the *formula isomorphism* as the equivalence relation $\overset{\mathsf{f}}{\sim}$ over formulas generated by the following relations:

$$\begin{array}{ccc} A \wedge \top \overset{\mathsf{f}}{\sim} A & A \supset \top \overset{\mathsf{f}}{\sim} \top & \top \supset A \overset{\mathsf{f}}{\sim} A \\ A \wedge B \overset{\mathsf{f}}{\sim} B \wedge A & A \wedge (B \wedge C) \overset{\mathsf{f}}{\sim} (A \wedge B) \wedge C & (A \wedge B) \supset C \overset{\mathsf{f}}{\sim} A \supset (B \supset C) \end{array} \tag{1}$$

3 Polarized System and Decomposition

We define the set of *polarized formulas* (or P-*formulas*) as the set generated by $\mathcal{A} = \{a, b, \dots\}$ using the following grammar

$$\begin{array}{l} A^\circ, B^\circ ::= a^\circ \mid \top^\circ \mid A^\circ \wedge B^\circ \mid A^\bullet \supset B^\circ \mid \Box A^\circ \mid \Diamond A^\circ \\ A^\bullet, B^\bullet ::= a^\bullet \mid \top^\bullet \mid A^\bullet \wedge B^\bullet \mid A^\circ \supset B^\bullet \mid \Box A^\bullet \mid \Diamond A^\bullet \end{array} \tag{2}$$

We say that formulas $A^\circ, B^\circ, \dots$ are of *even* polarity and formulas $A^\bullet, B^\bullet, \dots$ are of *odd* polarity. Note that the polarity of a formula determines the polarity of each subformula of that formula. For this reason we will omit the polarity markings for subformulas. A polarized formula is *clean* if it contains no subformulas of the shape $A \supset \top^\bullet$.

A *polarized sequent* is a sequent of P-formulas. We write $\Gamma^\bullet$ for a sequent containing only formulas of odd polarity. Then $\Gamma^\bullet, A^\circ$ is simply the polarized version of a sequent $\Gamma \vdash A$. A *context* is a (polarized) sequent $\Gamma\{\ \}$ in which an atom (or more generally a subformula) has been replaced by an hole $\{\ \}$. Then $\Gamma\{\Delta\}$ stands for the sequent obtained from $\Gamma\{\ \}$ by replacing $\{\ \}$ with Δ.

We can now define the polarized sequent rules given in Figure 3. Observe that the upper part and the two rules $\mathsf{c}^\bullet$ and $\mathsf{w}^\bullet$ on the lower left are just the polarized version of the rules in Figure 2.[2] The two rules $\mathsf{c}^\bullet_\downarrow$ and $\mathsf{w}^\bullet_\downarrow$ on the lower right are the *deep* version of $\mathsf{c}^\bullet$ and $\mathsf{w}^\bullet$, respectively. They can be applied deep inside any formula context. Note that they can only be applied to formulas of odd polarity. Figure 4 lists the various proof systems that are defined with these rules, and that we are using in this paper.

We can define the function $\lfloor\cdot\rfloor$ on polarized formulas that forgets the polarities. This function can be extended to polarized sequents with exactly one

[2] Note that the $\top^\bullet$-rule is just a special case of the $\mathsf{w}^\bullet$-rule. It is introduced here to simplify the presentation of some of the results in this paper.

$$\frac{}{a^\bullet, a^\circ}\,\mathsf{ax} \quad \frac{\Gamma^\bullet, A^\bullet, B^\circ}{\Gamma^\bullet, (A \supset B)^\circ}\,\supset^\circ \quad \frac{\Gamma^\bullet, A^\circ \qquad \Delta^\bullet, B^\bullet, C^\circ}{\Gamma^\bullet, \Delta^\bullet, (A \supset B)^\bullet, C^\circ}\,\supset^\bullet \quad \frac{\Gamma^\bullet, A^\circ \qquad \Delta^\bullet, B^\circ}{\Gamma^\bullet, \Delta^\bullet, (A \wedge B)^\circ}\,\wedge^\circ \quad \frac{\Gamma^\bullet, A^\bullet, B^\bullet, C^\circ}{\Gamma^\bullet, (A \wedge B)^\bullet, C^\circ}\,\wedge^\bullet$$

$$\frac{}{\top^\circ}\,\top^\circ \quad \frac{\Gamma^\bullet, A^\circ}{\top^\bullet, \Gamma^\bullet, A^\circ}\,\top^\bullet \quad \frac{\Gamma^\bullet, A^\circ}{\Box\Gamma^\bullet, \Box A^\circ}\,\mathsf{k}^\Box \quad \frac{A^\bullet, \Gamma^\bullet, B^\circ}{\Diamond A^\bullet, \Box\Gamma^\bullet, \Diamond B^\circ}\,\mathsf{k}^\Diamond \quad \frac{\Gamma^\bullet, A^\circ}{\Box\Gamma^\bullet, \Diamond A^\circ}\,\mathsf{d}$$

$$\frac{\Gamma^\bullet, A^\bullet, A^\bullet, B^\circ}{\Gamma^\bullet, A^\bullet, B^\circ}\,\mathsf{c}^\bullet \quad \frac{\Gamma^\bullet, A^\circ}{\Gamma^\bullet, B^\bullet, A^\circ}\,\mathsf{w}^\bullet \quad \Bigg| \quad \frac{\Gamma\{(A \wedge A)^\bullet\}}{\Gamma\{A^\bullet\}}\,\mathsf{c}^\bullet_\downarrow \quad \frac{\Gamma\{\top^\bullet\}}{\Gamma\{A^\bullet\}}\,\mathsf{w}^\bullet_\downarrow \text{ (for a } A \neq \top)$$

Fig. 3. Sequent rules and deep inference rules on P-formulas

$$\begin{aligned} \mathsf{LI}^{\LEFTcircle} &= \mathsf{LI}^{\LEFTcircle}_\ell \cup \{\mathsf{c}^\bullet, \mathsf{w}^\bullet\} & \mathsf{LI}^{\LEFTcircle}_\ell &= \{\mathsf{ax}, \top^\circ, \top^\bullet, \supset^\circ, \supset^\bullet, \wedge^\bullet, \wedge^\circ\} \\ \mathsf{LCK}^{\LEFTcircle} &= \mathsf{LI}^{\LEFTcircle} \cup \{\mathsf{k}^\Box, \mathsf{k}^\Diamond\} & \mathsf{LCK}^{\LEFTcircle}_\ell &= \mathsf{LI}^{\LEFTcircle}_\ell \cup \{\mathsf{k}^\Box, \mathsf{k}^\Diamond\} \\ \mathsf{LCD}^{\LEFTcircle} &= \mathsf{LI}^{\LEFTcircle} \cup \{\mathsf{k}^\Box, \mathsf{k}^\Diamond, \mathsf{d}\} & \mathsf{LCD}^{\LEFTcircle}_\ell &= \mathsf{LI}_\ell \cup \{\mathsf{k}^\Box, \mathsf{k}^\Diamond, \mathsf{d}\} \end{aligned}$$

Fig. 4. Rules systems for P-formulas

even formula via $\lfloor B_1^\bullet, \ldots, B_n^\bullet, A^\circ \rfloor = \lfloor B_1 \rfloor, \ldots, \lfloor B_n \rfloor \vdash \lfloor A \rfloor$. Since all systems defined in Figure 4 can only prove such sequents, we have an immediate one-to-one correspondence between derivations in the polarized and the corresponding unpolarized systems. However, the motivation for introducing the polarized systems is the following result.

Theorem 3.1 (Decomposition) *Let* $\mathsf{X} \in \{\mathsf{CK}, \mathsf{CD}\}$ *and* H *be a* P*-formula. The following are equivalent:* (i) $\overset{\mathsf{LX}}{\vdash} \lfloor H \rfloor$; (ii) $\overset{\mathsf{LX}^{\LEFTcircle}}{\vdash} H$; (iii) *there is a clean* P*-formula* H' *such that* $\overset{\mathsf{LX}^{\LEFTcircle}_\ell}{\vdash} H' \overset{\{\mathsf{c}^\bullet_\downarrow, \mathsf{w}^\bullet_\downarrow\}}{\vdash} H$.

Proof. (i) $\iff$ (ii) follows from the paragraph above. And (ii) $\iff$ (iii) can be obtained by a simple rule permutation argument and observing that every instance of $\mathsf{c}^\bullet$ can be decomposed into a $\wedge^\bullet$ followed by a $\mathsf{c}^\bullet_\downarrow$, and every instance of $\mathsf{w}^\bullet$ is a $\top^\bullet$ followed by a $\mathsf{w}^\bullet_\downarrow$. If a non-clean formula is introduced by a $\supset^\bullet$, we perform the following transformation.

$$\frac{\overset{\mathsf{LX}^{\LEFTcircle}}{\vdots} \atop \Gamma^\bullet, B^\circ \qquad \dfrac{\overset{\vdots}{\Delta^\bullet, A^\circ}}{\top^\bullet, \Delta^\bullet, A^\circ}\,\top^\bullet}{\Gamma^\bullet, B \supset \top^\bullet, \Delta \vdash A}\,\supset^\bullet \quad \rightsquigarrow \quad \dfrac{\dfrac{\overset{\vdots}{\Delta^\bullet, A^\circ}}{\top^\bullet, \ldots, \top^\bullet, \Delta^\bullet, A^\circ}\,\top^\bullet \times (|\Gamma| + 1)}{\Gamma^\bullet, B \supset \top^\bullet, \Delta^\bullet, A^\circ}\,\mathsf{w}^\bullet \times (|\Gamma| + 1) \qquad (3)$$

We conclude by permuting the rules $\mathsf{c}^\bullet_\downarrow$ and $\mathsf{w}^\bullet_\downarrow$ below all other rules, while applying the transformation above whenever a non-clean formula is introduced. $\Box$

4 Modal Arenas

A *directed graph* $\mathcal{G} = \langle V_\mathcal{G}, \overset{\mathcal{G}}{\to} \rangle$ is given by a set $V_\mathcal{G}$ of *vertices* and a set $\overset{\mathcal{G}}{\to} \subseteq V_\mathcal{G} \times V_\mathcal{G}$ of direct *edges*. A vertex v is a $\overset{\mathcal{G}}{\to}$*-root*, denoted $v \not\to$, if there is no vertex w such that $v \overset{\mathcal{G}}{\to} w$. We denote by $\vec{R}_\mathcal{G}$ the set of $\overset{\mathcal{G}}{\to}$-roots of $\mathcal{G}$. A *path* from v to w of length n is a sequence of vertices $x_0 \ldots x_n$ such that $v = x_0$ and $w = x_n$ and $x_i \overset{\mathcal{G}}{\to} x_{i+1}$ for $i \in \{0, \ldots, n-1\}$. We write $v \overset{\mathcal{G}}{\to}{}^n w$ if there is a path from v to w of length n. A *directed acyclic graph* (or dag for short) is a directed graph such that $v \stackrel{\mathcal{G}}{\to}{}^n v$ implies $n = 0$ for all $v \in V$. A *two-color directed acyclic graph* (or 2-dag for short) $\mathcal{G} = \langle V_\mathcal{G}, \stackrel{\mathcal{G}}{\to}, \stackrel{\mathcal{G}}{\rightsquigarrow} \rangle$ is given by a set of vertices $V_\mathcal{G}$ and two disjoint sets of edges $\stackrel{\mathcal{G}}{\to}$ and $\stackrel{\mathcal{G}}{\rightsquigarrow}$ such that the graph $\langle V_\mathcal{G}, \stackrel{\mathcal{G}}{\to} \cup \stackrel{\mathcal{G}}{\rightsquigarrow} \rangle$ is acyclic. We omit the superscript when clear from context and we denote by $\emptyset$ the empty 2-dag. We write $u \leftrightsquigarrow v$ if $u \rightsquigarrow v$ or $v \rightsquigarrow u$.

If $\mathcal{L}$ is a set, a 2-dag is *$\mathcal{L}$-labeled* if a *label* $\ell(v) \in \mathcal{L}$ is associated to each vertex $v \in V$. In this paper we fix the set of labels to be the set $\mathcal{L} = \mathcal{A} \cup \{\Box, \Diamond\}$, where $\mathcal{A}$ is the set of propositional variables occurring in formulas. We use the notation a, $\Box$ and $\Diamond$ to denote the graphs consisting of a single vertex labeled by a, $\Box$ and $\Diamond$, respectively.

Definition 4.1 Let $\mathcal{G}$, $\mathcal{H}$, $\mathcal{F}$ be 2-dags with $\mathcal{F} \neq \emptyset$. We write $R^\mathcal{G}_\mathcal{F}$ for the set of edges from the $\to$-roots of $\mathcal{G}$ to the $\to$-roots of $\mathcal{F}$, that is $R^\mathcal{G}_\mathcal{F} = \{(u,v) \mid u \in \vec{R}_\mathcal{G}, v \in \vec{R}_\mathcal{F}\}$. We define the following operations on 2-dags:

$$\begin{array}{ll} \mathcal{G} + \mathcal{H} = \langle V_\mathcal{G} \cup V_\mathcal{H}, \stackrel{\mathcal{G}}{\to} \cup \stackrel{\mathcal{H}}{\to}, \stackrel{\mathcal{G}}{\rightsquigarrow} \cup \stackrel{\mathcal{H}}{\rightsquigarrow} \rangle & \\ \mathcal{G} \mathbin{-\!\triangleright} \mathcal{F} = \langle V_\mathcal{G} \cup V_\mathcal{F}, \stackrel{\mathcal{G}}{\to} \cup \stackrel{\mathcal{F}}{\to} \cup R^\mathcal{G}_\mathcal{F}, \stackrel{\mathcal{G}}{\rightsquigarrow} \cup \stackrel{\mathcal{F}}{\rightsquigarrow} \rangle & \text{and} \quad \mathcal{G} \mathbin{-\!\triangleright} \emptyset = \emptyset \\ \mathcal{G} \mathbin{\sim\!\triangleright} \mathcal{H} = \langle V_\mathcal{G} \cup V_\mathcal{H}, \stackrel{\mathcal{G}}{\to} \cup \stackrel{\mathcal{H}}{\to}, \stackrel{\mathcal{G}}{\rightsquigarrow} \cup \stackrel{\mathcal{H}}{\rightsquigarrow} \cup R^\mathcal{G}_\mathcal{H} \rangle & \end{array}$$

We associate to each formula F a $\mathcal{L}$-labeled 2-dag $[\![F]\!]$ as follows:

$$\begin{array}{rclrclrcl} [\![a]\!] & = & a & [\![A \supset B]\!] & = & [\![A]\!] \mathbin{-\!\triangleright} [\![B]\!] & [\![\Box A]\!] & = & \Box \mathbin{\sim\!\triangleright} [\![A]\!] \\ [\![\top]\!] & = & \emptyset & [\![A \wedge B]\!] & = & [\![A]\!] + [\![B]\!] & [\![\Diamond A]\!] & = & \Diamond \mathbin{\sim\!\triangleright} [\![A]\!] \end{array} \tag{4}$$

For a sequent $B_1, \ldots, B_n \vdash A$, we define $[\![B_1, \ldots, B_n \vdash A]\!]$ as $[\![(B_1 \wedge \cdots \wedge B_n) \supset A]\!]$.

Example 4.2 Consider the sequent $\Gamma \vdash A = \Box a \supset \Box(b \wedge c), d \vdash \Diamond(e \supset f)$. We have

$[\![\Box a \supset \Box(b \wedge c), d \vdash \Diamond(e \supset f)]\!] =$

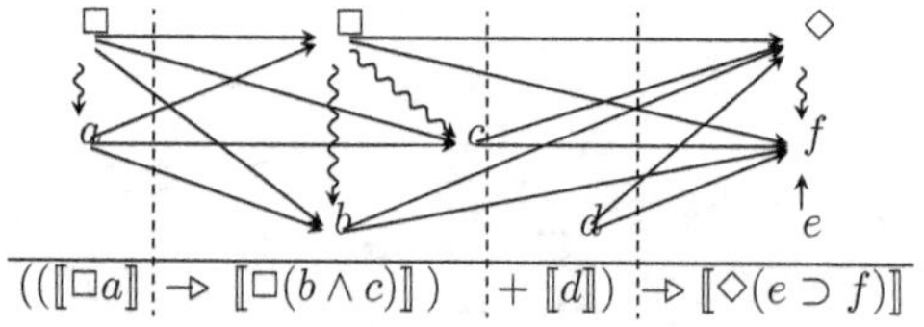

In the following, we give a characterization of those 2-dags that are encodings of formulas.

Definition 4.3 A $\mathcal{L}$-labeled dag $\mathcal{G} = \langle V_\mathcal{G}, \stackrel{\mathcal{G}}{\to} \rangle$ is an *arena* if $\mathcal{G}$ is

- L-free: if $a \to u$ and $a \to w \to v$ then $u \to v$;
- Σ-free: if $a \to v$, $a \to w$, $b \to w$ and $b \to u$ then $a \to u$ or $b \to v$;

A *modal arena* $\mathcal{G} = \langle V_\mathcal{G}, \stackrel{\mathcal{G}}{\to}, \stackrel{\mathcal{G}}{\rightsquigarrow} \rangle$ is an $\mathcal{L}$-labeled 2-dag such that

- $\langle V_\mathcal{G}, \stackrel{\mathcal{G}}{\to} \rangle$ is an arena;
- $\mathcal{G}$ is *properly labeled*: if $v \stackrel{\mathcal{G}}{\rightsquigarrow} w$, then $\ell(v) \in \{\Box, \Diamond\}$;
- $\rightsquigarrow$ is *modal*, that is:

MA.1 if $v \rightsquigarrow w$ and $w \rightsquigarrow u$, then $v \rightsquigarrow u$; **MA**.4 if $v \rightsquigarrow w$ and $u \rightarrow v$, then $u \rightarrow w$;
MA.2 if $v \rightsquigarrow w$ and $u \rightsquigarrow w$, then $u \leftrightsquigarrow v$; **MA**.5 if $v \rightsquigarrow w$ and $v \rightarrow u$, then $w \rightarrow u$;
MA.3 if $v \rightsquigarrow w$ and $v \rightsquigarrow u$, then $u \not\rightarrow w$; **MA**.6 if $v \rightsquigarrow w$ and $w \rightarrow u$, then $v \rightarrow u$.

We write $V_{\mathcal{G}}^{\mathcal{A}}$ (resp. $V_{\mathcal{G}}^{\Box}$, $V_{\mathcal{G}}^{\Diamond}$) for the subsets of vertices of $\mathcal{G}$ with labels in $\mathcal{A}$ (resp. in $\{\Box\}$, in $\{\Diamond\}$). The vertices in $V_{\mathcal{G}}^{\mathcal{A}}$ are called *atomic*, and the vertices in $V_{\mathcal{G}}^{\Box\Diamond} = V_{\mathcal{G}}^{\Box} \cup V_{\mathcal{G}}^{\Diamond}$ are called *modal.*

The relation between arenas and modality-free clean formulas has been established in [17].

Lemma 4.4 ([17]) *In an arena, if $u \rightarrow^n w$ and $v \rightarrow^m w$, then*

$$\text{either} \quad \{x \mid v \rightarrow^n x\} \subseteq \{x \mid w \rightarrow^m x\} \qquad \text{or} \quad \{x \mid w \rightarrow^m x\} \subseteq \{x \mid v \rightarrow^n x\} \quad .$$

Theorem 4.5 ([17]) *An $\mathcal{L}$-labeled* 2-dag *$\mathcal{G}$ is an arena iff there is a modality-free clean formula F such that $\mathcal{G} = [\![F]\!]$.*

We now extend this result to modal formulas.

Lemma 4.6 *Let $\mathcal{G}$ be a modal arena, and let $u, v, w \in V_{\mathcal{G}}$. If $v \rightsquigarrow w$ then:*

(i) *v is a $\rightarrow$-root iff w is a $\rightarrow$-root;*

(ii) *$v \rightarrow^n u$ iff $w \rightarrow^n u$;*

(iii) *if $u \rightarrow^n v$ then $u \rightarrow^n w$.*

Proof. The first statement follows from the fact that in a modal arena, if $v \rightsquigarrow w$, then $v \rightarrow u$ iff $w \rightarrow u$. The second statement is proven using the same argument, proceeding by induction on n making use of Lemma 4.4. The third statement is also proven using Lemma 4.4 and the fact that in a modal arena if $v \rightsquigarrow w$ and $u \rightarrow v$, then $u \rightarrow w$. ☐

Lemma 4.7 *If F is a formula, then the $\mathcal{L}$-labeled* 2-dag *$[\![F]\!]$ is a modal arena.*

Proof. By induction over the number of connectives and modalities of a formula. It suffices to remark that the graph operations $+$ and $\rightarrow\!\!\triangleright$ cannot introduce forbidden modal arena configurations. Similarly, the operation $\rightsquigarrow\!\!\triangleright$ introduces no forbidden configurations whenever $\mathcal{G} = \mathcal{G}_1 \rightsquigarrow\!\!\triangleright \mathcal{G}_2$ with $\mathcal{G}_1$ a single vertex graph of the form $\Box$ or $\Diamond$. ☐

In order to prove the converse, we need the following definitions.

Definition 4.8 Let v be a modal vertex of a modal arena $\mathcal{G}$. The *scope* of v is the set

$$\mathsf{Scope}(v) = \{\ w \in V_{\mathcal{G}} \mid \text{there is a } u \in V_{\mathcal{G}} \text{ s.t. } v \rightsquigarrow u \text{ and } w \rightarrow^* u \text{ and } w \not\rightarrow^* v\ \}$$

Intuitively, the scope of a modal vertex v in $[\![F]\!]$ is the set of vertices corresponding to modalities and atoms in the scope of the corresponding modality in F. To give an example, consider the arena in Example 4.2. There, e is in the scope of the $\Diamond$ while d is not. In fact, despite the existence of f such that $\Diamond \rightsquigarrow f$ and $d \rightarrow f$ and $e \rightarrow f$, we have $e \in \mathsf{Scope}(\Diamond)$ since $e \rightarrow f$ and $e \not\rightarrow f$, while $d \notin \mathsf{Scope}(\Diamond)$ since $d \rightarrow f$ and $d \rightarrow \Diamond$.

Theorem 4.9 *An $\mathcal{L}$-labeled* 2-dag *$\mathcal{G}$ is a modal arena iff $\mathcal{G} = [\![F]\!]$ for some formula F.*

Proof. The "if" direction has been shown in Lemma 4.7. For the "only if" direction, we proceed by induction on the size of $\mathcal{G}$. If $\mathcal{G} = \emptyset$ then $F = \top$. If $|V_\mathcal{G}| = 1$ then if $\ell(v) \in \mathcal{A}$, then $F = a \in \mathcal{A}$, if $\ell(v) = \Diamond$ or $\ell(v) = \Box$ then $F = \Diamond\top$ or $F = \Box\top$, respectively. Otherwise, since $\langle V_\mathcal{G}, \overset{\mathcal{G}}{\rightarrow} \rangle$ is a arena, we conclude by Lemma 4.4 (see [17]) that

(i) either every vertex in $V_\mathcal{G} \setminus \vec{R}_\mathcal{G}$ has a $\rightarrow$-paths to all roots in $\vec{R}_\mathcal{G}$,

(ii) or $\vec{R}_\mathcal{G}$ admits a partition $\vec{R}_\mathcal{G} = R_1 \uplus R_2$ such that any vertex in $\mathcal{G}$ has $\rightarrow$-paths only to roots in one of the two sets.

If (i) holds, then we define $\mathcal{G}_2$ as the modal arena obtained from $\mathcal{G}$ taking the vertices in $V_2 = \vec{R}_\mathcal{G} \cup \left(\bigcup_{v \in \vec{R}_\mathcal{G}} \mathsf{Scope}(v)\right)$ and $\mathcal{G}_1$ as the modal arena over the remaining vertices $V_1 = V_\mathcal{G} \setminus V_2$. Since each vertex in $\mathcal{G}$ has a path to all the roots in $\vec{R}_\mathcal{G}$, then there is a $\rightarrow$ from any root of $\mathcal{G}_1$ to any root of $\mathcal{G}_2$. Since by definition $\vec{R}_{\mathcal{G}_2} = \vec{R}_\mathcal{G}$, then we have that $\mathcal{G} = \mathcal{G}_1 \rightarrowtriangle \mathcal{G}_2$.

If (ii) holds and $\vec{R}_\mathcal{G} = R_1 \uplus R_2$ with R_1 and R_2 non-empty sets. Since $\leadsto$ is modal, we have the following possibilities:

(a) if $R_1 = \{v\}$ and $v \leadsto w$ for all $w \in R_2$, then there is no u such that $u \rightarrow v$. Otherwise $u \rightarrow v$ and $u \rightarrow w$ for all w such that $v \leadsto w$, that is for all $w \in R_2$. This implies that $u \leadsto w$ for all $w \in \vec{R}_\mathcal{G}$, which contradicts (ii). Thus we conclude that $\mathcal{G} = v \leadsto\!\triangleright \mathcal{G}'$ where $\mathcal{G}'$ is the modal arena with vertices $\mathsf{Scope}(v)$;

(b) if there are no $\leadsto$-edges between R_1 and R_2, then $\mathcal{G} = \mathcal{G}_1 + \mathcal{G}_2$ where $\mathcal{G}_1$ and $\mathcal{G}_2$ are the modal arenas with vertices $V_1 = \{v \mid v \rightarrow^* w \text{ for a } w \in R_1\}$ and $V_2 = \{v \mid v \rightarrow^* w \text{ for a } w \in R_2\}$. In fact by definition there are no $\rightarrow$-edges between vertices in V_1 and V_2 otherwise by Lemma 4.4 we should have $R_1 = R_2$. Similarly there are no $\leadsto$-edges between vertices in V_1 and V_2 since there are no $\leadsto$-edges between R_1 and R_2 (by hypothesis) and if there is $v \in V_1 \setminus R_1$ and $w \in V_2$ such that $v \leadsto w$, then by Lemma 4.6 $w \notin R_2$ and we should have again $R_1 = R_2$;

(c) otherwise, we pick a $v \in \vec{R}_\mathcal{G} \cap \overset{\leadsto}{R}_\mathcal{G}$ and define $R_1 = \{v\} \cup \{w \mid v \leadsto w\}$ and $R_2 = \vec{R}_\mathcal{G} \setminus R_1$. If there is no $u \in \vec{R}_\mathcal{G}$ such that $v \not\leadsto u$, then $R_1 = \vec{R}_\mathcal{G}$ and we conclude by (a). If $R_2 \neq \emptyset$, then we define $V_1 = \{v \mid v \rightarrow^* w \text{ for a } w \in R_1\}$ and $V_2 = \{v \mid v \rightarrow^* w \text{ for a } w \in R_2\}$ and we conclude by (b). □

In light of this theorem, we may say that a vertex in $[\![F]\!]$ *corresponds* to an occurrence of an atom or a modality in the formula F.

We conclude this section by remarking that modal arenas identify formulas modulo the formula isomorphism $\overset{\mathsf{f}}{\sim}$ defined by the relations in Equation (1).

Proposition 4.10 *For any formulas F and G we have $F \overset{\mathsf{f}}{\sim} G$ iff $[\![F]\!] = [\![G]\!]$.*

Proof. This follows from the definition of the modal arena operations $+$, $\rightarrowtriangle$ and $\leadsto\!\triangleright$. □

5 Modal Arena Nets

We introduce the notion of CK- and CD-arena nets, which are modal arenas equipped with an equivalence relation over vertices, satisfying certain conditions capturing the idea of "axiom links" in proof nets. We then show the correspondence between these modal arena nets and the linear proofs in $\mathsf{LCK}_\ell^\bullet$ and $\mathsf{LCD}_\ell^\bullet$, respectively.

Definition 5.1 A *partitioned modal arena* $\mathcal{G} = \langle V_\mathcal{G}, \overset{\mathcal{G}}{\to}, \overset{\mathcal{G}}{\rightsquigarrow}, \overset{\mathcal{G}}{\sim} \rangle$ is given by a modal arena $\langle V_\mathcal{G}, \overset{\mathcal{G}}{\to}, \overset{\mathcal{G}}{\rightsquigarrow} \rangle$ together with an equivalence relation $\overset{\mathcal{G}}{\sim}$ over vertices such that:

- if $v \in V_\mathcal{G}^\mathcal{A}$ and $v \overset{\mathcal{G}}{\sim} w$, then $w \in V_\mathcal{G}^\mathcal{A}$ and $\ell(v) = \ell(w)$;
- if $v \in V_\mathcal{G}^\mathcal{A}$, then $v \overset{\mathcal{G}}{\sim} w$ for a unique $w \in V_\mathcal{G}^\mathcal{A}$.

In a partitioned modal arena we represent the equivalence relation $\sim$ by drawing a (dashed non-oriented blue) edge u - w between two distinct vertices in the same $\sim$-class. For better readability, we only represent a minimal subset of these edges relying on the fact that $\sim$ is an equivalence relation. By means of example, if $\{u, v, w\}$ is an $\sim$-class, we may only draw u - v - w omitting the edge between u and w.

We say that a formula (or P-formula) F is *associated* to $\mathcal{G} = \langle V_\mathcal{G}, \overset{\mathcal{G}}{\to}, \overset{\mathcal{G}}{\rightsquigarrow}, \overset{\mathcal{G}}{\sim} \rangle$ if $[\![F]\!] = \langle V_\mathcal{G}, \overset{\mathcal{G}}{\to}, \overset{\mathcal{G}}{\rightsquigarrow} \rangle$, and we denote by $\emptyset$ the empty arena net.

Remark 5.2 If v and w are vertices in a partitioned modal arena $\mathcal{G}$ such that $v \overset{\mathcal{G}}{\sim} w$, then $v \in V_\mathcal{G}^{\Box\Diamond}$ iff $w \in V_\mathcal{G}^{\Box\Diamond}$. If an $\overset{\mathcal{G}}{\sim}$ equivalence class contains more than two vertices then they are all labelled by $\Diamond$ or $\Box$.

If $\mathcal{G}$ is a modal arena and $v \in V_\mathcal{G}$, we define the *depth* of v (denoted $d(v)$) to be the length of the $\to$-paths from v to a $\to$-root $w \in \vec{R}_\mathcal{G}$. This is well-defined as all such paths have the same length (see [17, Lemma 9]). The *parity of a vertex* v is the parity of $d(v)$, which can be either *even* or *odd*. We write v° or $v^\bullet$ if the parity of v is respectively even or odd. Note that if F° is a P-formula, then the parity of the vertices in $[\![F^\circ]\!]$ are the same as the polarity of the corresponding atoms (and modal subformulas) in F°.

The *parity* of an $\to$-edge $v \to w$ is the parity of $d(w)$. We say that an edge $v \to w$ is a *chord* if there is a vertex u such that either $v \to u$ and $u \rightsquigarrow w$; or $u \to w$ and $u \rightsquigarrow v$. By means of example, in the following modal arenas the edges $a \to b$ are chords.

We write by $\overset{\mathcal{G}}{\to}_\bullet$ the set of odd $\to$-edges in $\mathcal{G}$ that are not chords. In the following, we may depict $\to$-edges which are not $\to_\bullet$-edges using dotted edges.

If v is a vertex in a modal arena $\mathcal{G} = [\![F]\!]$, we denote by $\hat{v}$ either v itself if there is no w such that $v \in \mathsf{Scope}(w)$, or the vertex $w = \hat{v}$ such that $v \in \mathsf{Scope}(w)$ and $w \in \mathsf{Scope}(u)$ for all $u \neq w$ with $v \in \mathsf{Scope}(u)$. That is, if $v \neq \hat{v}$, then $\hat{v}$ is the first modal vertex we encounter on in the path from v to the

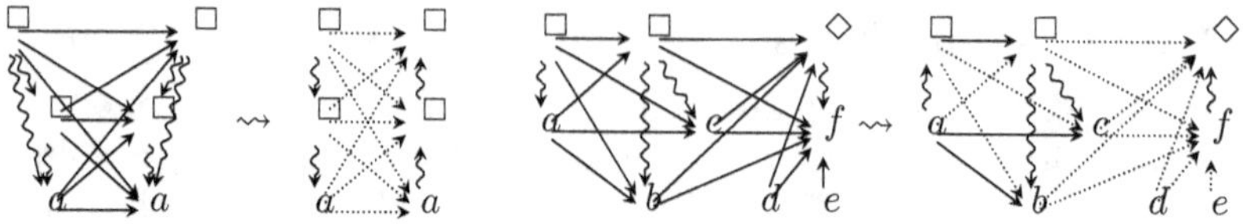

Fig. 5. The arenas of the sequents $\vdash \Box\Box a \supset \Box\Box a$ and $\Box a \supset \Box(b \wedge c), d \vdash \Diamond(e \supset f)$ and the corresponding graphs obtained by replacing the $\rightsquigarrow$-edges with $\rightsquigarrow_\partial$-edges.

root of the formula tree of F. By means of example, in Example 4.2 we have $\hat{f} = \hat{e} = \Diamond = \Diamond$ and $\hat{d} = d$.

We define the edge relation $\overset{\mathcal{G}}{\rightsquigarrow}_\partial$ as follows

$$v \overset{\mathcal{G}}{\rightsquigarrow}_\partial w \qquad \text{if either} \qquad w^\circ \text{ and } w = \hat{v} \neq v, \quad \text{or} \quad v^\bullet \text{ and } v = \hat{w} \neq w$$

Intuitively, the $\rightsquigarrow_\partial$-edges connect a modality to the root of the formula in its scope "one step at the time" (see Figure 5). Note that $v^\bullet \rightsquigarrow_\partial w^\bullet$ implies $v^\bullet \rightsquigarrow w^\bullet$, while $v^\circ \rightsquigarrow_\partial w^\circ$ implies $w^\circ \rightsquigarrow v^\circ$.

Definition 5.3 A partitioned modal arena $\mathcal{G}$ is *linked* if every $\overset{\mathcal{G}}{\sim}$-class is of the form $\{v_1^\bullet, \ldots, v_n^\bullet, w^\circ\}$. This induces the set directed edges $\overset{\mathcal{G}}{\rightharpoonup} = \{(v, w) \mid v^\bullet \overset{\mathcal{G}}{\sim} w^\circ\}$. The *linking graph* $\overset{\frown}{\mathcal{G}}$ of a modal arena is the directed graph with vertices $V_\mathcal{G}$ and edges $\overset{\mathcal{G}}{\to}_\bullet \cup \overset{\mathcal{G}}{\rightsquigarrow}_\partial \cup \overset{\mathcal{G}}{\rightharpoonup}$. We say that path in $\overset{\frown}{\mathcal{G}}$ is *checked* if it ends in a vertex in $\vec{R}_\mathcal{G} \cap \overset{\rightsquigarrow}{R}_\mathcal{G}$ and it contains no edge $v \rightharpoonup w$ with w a modal vertex with $\mathsf{Scope}(w) \neq \emptyset$.

A CK-*arena net* is a linked modal arena which satisfies Conditions (i)–(iv) below:

(i) $\overset{\frown}{\mathcal{G}}$ is acyclic: every checked path is acyclic;

(ii) $\overset{\frown}{\mathcal{G}}$ is *functional*: every checked path in $\overset{\frown}{\mathcal{G}}$ from a vertex $v^\bullet$ to a root includes a vertex w° such that $v \to w$;

(iii) $\mathcal{G}$ is *functorial*: if $v \rightsquigarrow w$ and $w \sim w'$ then there is v' such that $v \sim v'$ and $v' \rightsquigarrow w'$;

(iv) $\mathcal{G}$ is CK-*correct*: if $\{v_1^\bullet, v_2^\bullet, \ldots, v_n^\bullet, w^\circ\}$ is a $\sim$-class of modal vertices, then either $v_1, v_2, \ldots, v_n, w \in V_\mathcal{G}^\Box$ or there is a unique i such that $v_i, w \in V_\mathcal{G}^\Diamond$.

A linked modal arena is a CD-*arena net* if it satisfies Conditions (i)-(iii) above, plus the following:

(v) $\mathcal{G}$ is CD-*correct*: if $\{v_1^\bullet, v_2^\bullet, \ldots, v_n^\bullet, w^\circ\}$ is a $\sim$-class of modal vertices, then either $v_1, v_2, \ldots, v_n, w \in V_\mathcal{G}^\Box$ or $w \in V_\mathcal{G}^\Diamond$ there is at most one $i \in \{1, \ldots, n\}$ such that $v_i \in V_\mathcal{G}^\Diamond$.

A *modal arena net* is either a CK- or a CD-arena net. An *arena net* is a modal arena net with $V^{\Box\Diamond} = \emptyset$. Note that in this case Conditions (iii)-(v) are vacuous.

The intuition for Conditions (iv) and (v) is that $\sim$-classes represent either atoms paired by an ax, or the set of modalities introduced by a same instance

$$\frac{}{a \dashrightarrow a}\,\mathsf{ax} \qquad \frac{\mathcal{F},\mathcal{G} \vdash \mathcal{H}}{\mathcal{F} \vdash \mathcal{G} \mathrel{-\!\!\triangleright} \mathcal{H}}\supset^{\circ} \qquad \frac{\mathcal{F} \vdash \mathcal{G} \quad \mathcal{J},\mathcal{K} \vdash \mathcal{H}}{\mathcal{F},\mathcal{J},\mathcal{G} \mathrel{-\!\!\triangleright} \mathcal{K} \vdash \mathcal{H}}\supset^{\bullet} \qquad \frac{\mathcal{F} \vdash \mathcal{G} \quad \mathcal{I} \vdash \mathcal{K}}{\mathcal{F},\mathcal{I} \vdash \mathcal{G} + \mathcal{K}}\wedge^{\circ} \qquad \frac{\mathcal{F},\mathcal{G},\mathcal{H} \vdash \mathcal{K}}{\mathcal{F},\mathcal{G} + \mathcal{H} \vdash \mathcal{K}}\wedge^{\bullet}$$

$$\frac{\langle \mathcal{G}_1,\dots,\mathcal{G}_n \vdash \mathcal{H} \mid \overset{\mathcal{G}}{\sim} \rangle}{\langle \Box \mathrel{\sim\!\!\triangleright} \mathcal{G}_1,\dots,\Box \mathrel{\sim\!\!\triangleright} \mathcal{G}_n \vdash \Box \mathrel{\sim\!\!\triangleright} \mathcal{H} \mid \overset{\mathcal{G}}{\sim} \cup \overset{\mathsf{k}}{\sim} \rangle}\,\mathsf{k}^{\Box} \qquad \frac{\langle \mathcal{G}_1,\dots,\mathcal{G}_n \vdash \mathcal{H} \mid \overset{\mathcal{G}}{\sim} \rangle}{\langle \Box \mathrel{\sim\!\!\triangleright} \mathcal{G}_1,\dots,\Diamond \mathrel{\sim\!\!\triangleright} \mathcal{G}_i,\dots,\Box \mathrel{\sim\!\!\triangleright} \mathcal{G}_n \vdash \Diamond \mathrel{\sim\!\!\triangleright} \mathcal{H} \mid \overset{\mathcal{G}}{\sim} \cup \overset{\mathsf{k}}{\sim} \rangle}\,\mathsf{k}^{\Diamond}$$

$$\frac{}{\emptyset}\,\top^{\circ} \qquad \frac{\mathcal{F} \vdash \mathcal{G}}{\emptyset,\mathcal{F} \vdash \mathcal{G}}\,\top^{\bullet} \qquad \frac{\langle \mathcal{G}_1,\dots,\mathcal{G}_n \vdash \mathcal{H} \mid \overset{\mathcal{G}}{\sim} \rangle}{\langle \Box \mathrel{\sim\!\!\triangleright} \mathcal{G}_1,\dots,\Box \mathrel{\sim\!\!\triangleright} \mathcal{G}_n \vdash \Diamond \mathrel{\sim\!\!\triangleright} \mathcal{H} \mid \overset{\mathcal{G}}{\sim} \cup \overset{\mathsf{k}}{\sim} \rangle}\,\mathsf{d}$$

Fig. 6. Translation of $\mathsf{LCK}^{\bullet}_{\ell}$ and $\mathsf{LCD}^{\bullet}_{\ell}$ sequent rules in modal arenas rules where $\overset{\mathsf{k}}{\sim}$ is the equivalence class containing all vertices in the conclusion which are not in the premise. Note that $\mathcal{G},\mathcal{H}$ and $\mathcal{K}$ have to be non-empty.

of a $\mathsf{K}^{\Box}$, $\mathsf{K}^{\Diamond}$ or D-rule. Following this intuition, if $\mathsf{c} = \{v_0, v_1, \dots, v_n\} \subseteq V_{\mathcal{G}}^{\Box\Diamond}$ is a $\sim$-class, then the modal arena with vertices $\bigcup_{v\in\mathsf{c}} \mathsf{Scope}(v)$ corresponds to the sub-proof of the premise of any such rule.

Lemma 5.4 *Let $\mathsf{X} \in \{\mathsf{CK}, \mathsf{CD}\}$ and F be a clean P-formula. If $\overset{\mathsf{LX}^{\bullet}_{\ell}}{\vdash} F$, then there is a X-arena net $\mathcal{G} = \langle V_{\mathcal{G}}, \overset{\mathcal{G}}{\to}, \overset{\mathcal{G}}{\rightsquigarrow}, \overset{\mathcal{G}}{\sim} \rangle$ such that $[\![F]\!] = \langle V_{\mathcal{G}}, \overset{\mathcal{G}}{\to}, \overset{\mathcal{G}}{\rightsquigarrow} \rangle$.*

Proof. Let $\mathfrak{D}$ be a derivation of F in $\mathsf{LX}^{\bullet}_{\ell}$. We proceed by induction on $\mathfrak{D}$ translating it into a derivation of the desired modal arena net $\mathcal{G}$ via the rules in Figure 6. By definition, each rule in $\mathsf{LX}^{\bullet}_{\ell}$ preserves X-arena net conditions, that is, if the premises of a rule are X-arena nets, then the conclusion is. Note that Condition (iv) fails for the rule D, while Condition (v) holds. $\Box$

Lemma 5.5 *Let $\mathsf{X} \in \{\mathsf{CK}, \mathsf{CD}\}$ and F be a clean P-formula. If $\mathcal{G}$ is an X-arena net with associated P-formula F, then $\overset{\mathsf{LX}^{\bullet}_{\ell}}{\vdash} F$.*

Proof. We prove the theorem for CK-arena nets since the proof for CD-arena nets is similar by considering also the rule D.

If $\mathcal{G} = \langle \emptyset, \emptyset, \emptyset, \emptyset \rangle$, we conclude since $[\![\top]\!] = \emptyset$ and $\overset{\mathsf{LCK}^{\bullet}_{\ell}}{\vdash} \top^{\circ}$. Otherwise to prove this theorem we define from the CK-arena net $\mathcal{G}$, with associated clean P-formula F, an arena net $\partial(\mathcal{G})$ with associated formula $\partial(F)$. We then use use of the result in [17] on (non-modal) arena nets to produce an $\mathsf{LX}^{\bullet}_{\ell}$-derivation of $\partial(F)$. Then we conclude by showing how to define a $\mathsf{LX}^{\bullet}_{\ell}$-derivation of F using the $\mathsf{LI}^{\bullet}_{\ell}$-derivation of $\partial(F)$.

Step 1: definition of $\partial(\mathcal{G})$. Let $\mathcal{G} = \langle V_{\mathcal{G}}, \overset{\mathcal{G}}{\to}, \overset{\mathcal{G}}{\rightsquigarrow}, \overset{\mathcal{G}}{\sim} \rangle$ be a CK-arena net. We write $v \mathrel{\updownarrow} w$ either if $\hat{v} \sim \hat{w}$, or if $v = \hat{v}$ and $w = \hat{w}$, that is, $v \mathrel{\updownarrow} w$ iff either both v and w are not in the scope of any modality, or both v and w belong to the scope of modalities in the same $\sim$-class.

We define the arena $\partial(\mathcal{G})$ by removing all $\rightsquigarrow$-edges in $\mathcal{G}$ and keeping only the $\to$ between vertices $v, w \in V_{\mathcal{G}}$ such that $v \mathrel{\updownarrow} w$. Then we replace each modal vertex v by a pair of $\rightharpoonup$-linked vertices $v^{\mathsf{in}}, v^{\mathsf{out}}$ in such a way that the vertex v^{in} keeps track of the subformulas of the modality, while v^{out} is a placeholder to keep track of the interaction of the subformulas with the context.

Formally we define $\partial(\mathcal{G}) = \langle \partial(V_{\mathcal{G}})\ ,\ \partial(\overset{\mathcal{G}}{\to} \cup \overset{\mathcal{G}}{\rightsquigarrow})\ ,\ \overset{\partial(\mathcal{G})}{\sim} \rangle$ by:

- $\partial(V_{\mathcal{G}}) = V_{\mathcal{G}}^{\mathcal{A}} \cup \{v^{\mathsf{in}}, v^{\mathsf{out}} \mid v \in V_{\mathcal{G}}^{\Box\Diamond}\}$;

- $\partial(\overset{\mathcal{G}}{\to} \cup \overset{\mathcal{G}}{\rightsquigarrow})$ is the union of the following five sets:

$$\begin{array}{rcl} \{(l^{\mathsf{out}}, r^{\mathsf{out}}) \mid l \rightharpoonup r\} & \bigcup & \{(u,v) \mid u \not\updownarrow v \text{ and } u \to v\} \\ \{(u, r^{\mathsf{in}}) \mid u \rightsquigarrow_{\partial} r\} & \bigcup & \{(l^{\mathsf{in}}, u) \mid l \rightsquigarrow_{\partial} u\} \\ \{(u, m^{\mathsf{out}}) \mid u \to m \text{ and } u \not\updownarrow m\} & \bigcup & \{(m^{\mathsf{out}}, v) \mid m \to v \text{ and } m \not\updownarrow v\} \\ \{(m^{\mathsf{out}}, n^{\mathsf{out}}) \mid m \to n \text{ and } m \not\updownarrow n\} & \bigcup & \{(m^{\mathsf{out}}, n^{\mathsf{out}}) \mid m \to n \text{ and } m \not\updownarrow n\} \end{array}$$

  where we assume $u, v \in V_{\mathcal{G}}^{\mathcal{A}}$ and $l^{\bullet}, r^{\circ}, m, n, p \in V_{\mathcal{G}}^{\square\Diamond}$;

- $\overset{\partial(\mathcal{G})}{\sim}$ is defined as: $v \overset{\partial(\mathcal{G})}{\sim} w$ if $v \overset{\mathcal{G}}{\sim} w$ and as $v^{\mathsf{in}} \overset{\partial(\mathcal{G})}{\sim} v^{\mathsf{out}}$ for each $v \in V_{\mathcal{G}}^{\square\Diamond}$.

See the first line of Figure 8for a running example.

We observe that if $\{v_0^{\circ}, v_1^{\bullet}, \ldots, v_n^{\bullet}\}$ is a $\overset{\mathcal{G}}{\sim}$-class of modal vertices, then a P-formula associated to $\mathcal{G}$ is of the form $H = H\{\ell(v_0)A_0^{\circ}\}\{\ell(v_1)A_1^{\bullet}\}\cdots\{\ell(v_n)A_n^{\bullet}\}$ for an $(n+1)$-ary context $H\{\ \}\cdots\{\ \}$. In this case, a P-formula associated to the arena $\partial(\mathcal{G})$ is of the form $\partial(H) = \partial(H)\{v_0^{\mathsf{out}\circ}\}\{v_1^{\mathsf{out}\bullet}\}\cdots\{v_n^{\mathsf{out}\bullet}\}\{H_{\mathsf{c}}^{\bullet}\}$ with $\partial(H)\{\ \}\cdots\{\ \}$ is an $(n+2)$-ary context, fresh propositional variables $v_i^{\mathsf{in}}, v_i^{\mathsf{out}}$ for all $i \in \{0, \ldots, n\}$ and

$$H_{\mathsf{c}}^{\bullet} = \Big(\Big(\big(v_1^{\mathsf{in}} \supset \partial(A_1^{\bullet}) \wedge \cdots \wedge v_n^{\mathsf{in}} \supset \partial(A_n^{\bullet})\big) \supset \partial(A_0^{\circ})\Big) \supset v_0^{\mathsf{in}}\Big)^{\bullet}$$

Step 2: proof that $\partial(\mathcal{G})$ is an arena net. We observe that, by definition of $\partial(\mathcal{G})$, every path $\partial(\mathsf{p})$ in $\overset{\curvearrowright}{\partial(\mathcal{G})}$ can be constructed from a checked path p in $\overset{\curvearrowright}{\mathcal{G}}$ by induction:

- the empty path is a path in both $\overset{\curvearrowright}{\mathcal{G}}$ and $\overset{\curvearrowright}{\partial(\mathcal{G})}$;
- if $\mathsf{p} = v \cdot \mathsf{p}'$
  - if $v \in V_{\mathcal{G}}^{\mathcal{A}}$, then $\partial(\mathsf{p}) = v \cdot \partial(\mathsf{p})'$;
  - if $v^{\circ} \in V_{\mathcal{G}}^{\square\Diamond}$, then $\partial(\mathsf{p}) = v^{\mathsf{out}} \cdot v^{\mathsf{in}} \cdot \partial(\mathsf{p})'$;
  - if $v^{\bullet} \in V_{\mathcal{G}}^{\square\Diamond}$, then $\partial(\mathsf{p}) = v^{\mathsf{in}} \cdot v^{\mathsf{out}} \cdot \partial(\mathsf{p})'$;

We remark that the parity of atomic vertices is preserved by ∂, while the parity of a modal vertex $v \in V_{\mathcal{G}}$ is the same as the corresponding vertex $v^{\mathsf{out}} \in V_{\partial(\mathcal{G})}$. Since if $v^{\bullet}$ then $v^{\mathsf{out}} \rightharpoonup v^{\mathsf{in}}$, and if v° then $v^{\mathsf{in}} \rightharpoonup v^{\mathsf{out}}$, then we have that in $\overset{\curvearrowright}{\partial(\mathcal{G})}$ an even (odd) vertex may occur only in a even (odd) position in a path in $\overset{\curvearrowright}{\mathcal{G}}$. We conclude since from any path in $\overset{\curvearrowright}{\partial(\mathcal{G})}$ we obtain a path in $\overset{\curvearrowright}{\mathcal{G}}$ by replacing every subpath $v^{\mathsf{out}} \rightharpoonup v^{\mathsf{in}}$ and $v^{\mathsf{in}} \rightharpoonup v^{\mathsf{out}}$ by a the corresponding modal vertex v in $\mathcal{G}$.

By this correspondence between checked paths in $\overset{\curvearrowright}{\mathcal{G}}$ and paths in $\overset{\curvearrowright}{\partial(\mathcal{G})}$ we conclude that $\overset{\curvearrowright}{\partial(\mathcal{G})}$ is acyclic and functional. That is, $\partial(\mathcal{G})$ is an arena net.

Step 3: construct the derivation associated to $\partial(\mathcal{G})$. Since $\partial(\mathcal{G})$ is an arena net, then we apply the result in [17] to produce a derivation in $\mathsf{LI}_{\ell}^{\circ}$ of the formula $\partial(F)$. In such a derivation, by functionality and functoriality of $\mathcal{G}$, whenever v and w are modal vertices such that $v \overset{\mathcal{G}}{\rightharpoonup} w$, then if a path in $\overset{\curvearrowright}{\partial(\mathcal{G})}$ contains v^{out}, then it also contains v^{in}, w^{in}, w^{out}. This means that if $\mathsf{c} = \{v_0^{\circ}, v_1^{\bullet}, \ldots, v_n^{\bullet}\}$

$$\cfrac{\cfrac{A_1^\bullet,\ldots,A_n^\bullet,A_0^\circ}{\Box_1^{\mathsf{out}\bullet},\ldots,\Box_n^{\mathsf{out}\bullet},F_c^\bullet,\Box_0^{\mathsf{out}\circ}}\;\mathsf{CK}_\ell}{\cfrac{\mathcal{D}\{\Box_0^{\mathsf{out}\circ}\}\{\Box_1^{\mathsf{out}\bullet}\}\cdots\{\Box_n^{\mathsf{out}\bullet}\}\{F_c^\bullet\}\,\|\,\mathsf{CK}_\ell}{\partial(F)\{\Box_0^{\mathsf{out}\circ}\}\{\Box_1^{\mathsf{out}\bullet}\}\cdots\{\Box_n^{\mathsf{out}\bullet}\}\{F_c^\bullet\}}\;\partial(\mathcal{D}')\,\|\,\mathsf{LI}_\ell}\quad\leadsto\quad\cfrac{\cfrac{A_1^\bullet,\ldots,A_n^\bullet,A_0^\circ}{\Box A_1^\bullet,\ldots,\Box A_n^\bullet,\Box A_0^\circ}\;\mathsf{k}^{\Box}}{\cfrac{\mathcal{D}\{\Box A_0^\circ\}\{\Box A_1^\bullet\}\cdots\{\Box A_n^\bullet\}\{\emptyset\}\,\|\,\mathsf{CK}_\ell}{F\{\Box_0 A_0^\circ\}\{\Box_1 A_1^\bullet\}\cdots\{\Box_n A_n^\bullet\}}}$$

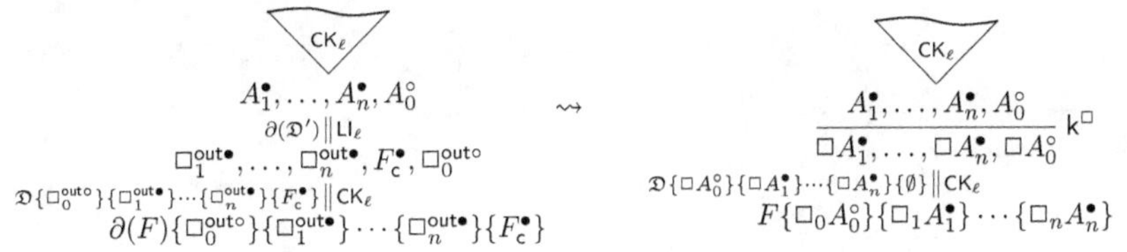

Fig. 7. An example of the construction of the derivation of F from the derivation of $\partial(F)$ assuming that in $\mathcal{G}$ there is only one $\sim$-class of the form $\{\Box_0,\ldots,\Box_n\}$

is an $\overset{\mathcal{G}}{\sim}$-class of vertices in $\overset{\mathcal{G}}{\leadsto}$, then any derivation of $\partial(F)$ in LX_ℓ° contains a subderivation of the sequent $v_1^{\mathsf{out}\bullet},\ldots,v_n^{\mathsf{out}\bullet},H_c^\bullet,v_0^{\mathsf{out}\circ}$ of the following form

$$\cfrac{\cfrac{\cfrac{\cfrac{\cfrac{\overline{v_1^{\mathsf{out}\bullet},v_1^{\mathsf{in}\circ}}\;\mathsf{ax}\quad\cdots\quad\overline{v_n^{\mathsf{out}\bullet},v_n^{\mathsf{in}\circ}}\;\mathsf{ax}\quad\overline{\partial(A_1)^\bullet,\ldots,\partial(A_n)^\bullet,\partial(A_0)^\circ}\;\mathsf{LI}_\ell^\circ}{v_1^{\mathsf{out}\bullet},\ldots,v_n^{\mathsf{out}\bullet},v_1^{\mathsf{in}}\supset\partial(A_1)^\bullet,\ldots,v_n^{\mathsf{in}}\supset\partial(A_n)^\bullet,\partial(A_0)^\circ}\;\supset\mathsf{L}}{v_1^{\mathsf{out}\bullet},\ldots,v_n^{\mathsf{out}\bullet},\bigwedge_{i=1}^n(v_i^{\mathsf{in}}\supset\partial(A_i))^\bullet,\partial(A_0)^\circ}\;\wedge\mathsf{L}}{v_1^{\mathsf{out}\bullet},\ldots,v_n^{\mathsf{out}\bullet},\bigwedge_{i=1}^n(v_i^{\mathsf{in}}\supset\partial(A_i))\supset\partial(A_0)^\circ}\;\supset\mathsf{R}\quad\overline{v_0^{\mathsf{in}\bullet},v_0^{\mathsf{out}\circ}}\;\mathsf{ax}}{v_1^{\mathsf{out}\bullet},\ldots,v_n^{\mathsf{out}\bullet},(\bigwedge_{i=1}^n(v_i^{\mathsf{in}}\supset\partial(A_i))\supset\partial(A_0))\supset v_0^{\mathsf{in}\bullet},v_0^{\mathsf{out}\circ}}\;\supset\mathsf{L}$$

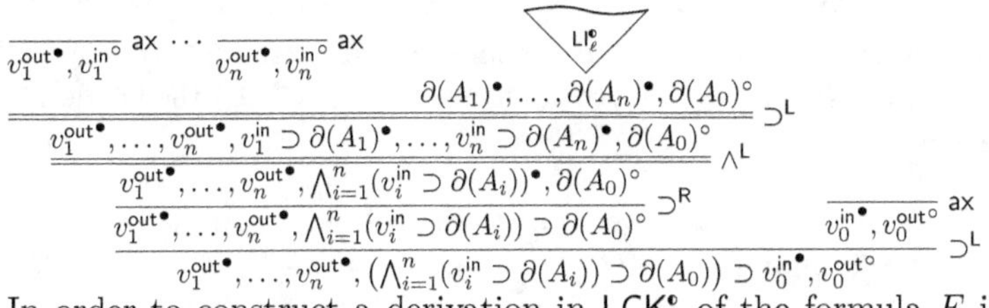

In order to construct a derivation in LCK_ℓ° of the formula F it suffices to proceed by induction over the number of $\overset{\mathcal{G}}{\sim}$-classes of modal vertices. Starting from the top of the derivation, we replace every such subderivation in the derivation of $\partial(F)$ with an application of a $\mathsf{K}^\Box$- or a $\mathsf{K}^\Diamond$-rule, we remove all the occurrences of the formula $H_c^\bullet = \left(\bigwedge_{i=1}^n(v_i^{\mathsf{in}}\supset\partial(A_i))\supset\partial(A_0)\right)\supset v_0^{\mathsf{in}\bullet}$ in the derivation, and we replace for each $i \in \{0,\ldots,n\}$ the atom v_i^{in} with the corresponding formula $\ell(v_i)A_i$ as shown in Figure 7. For a running example, refer to the lower line of Figure 8. □

By Lemma 5.4 and Lemma 5.5 we have the following theorem.

Theorem 5.6 *Let* $\mathsf{X} \in \{\mathsf{CK},\mathsf{CD}\}$ *and* F *be a clean* P*-formula. Then*

$$\vdash^{\mathsf{LX}_\ell^\circ} F \iff \text{there is a X-arena net } \mathcal{G} \text{ with } \mathcal{G} = [\![F]\!]$$

6 Skew Fibrations

After having characterized the linear part of a proof in CK or CD, we will now characterize the maps between modal arenas that characterize derivations built from the deep rules $\mathsf{w}_\downarrow^\bullet$ for weakenning and $\mathsf{c}_\downarrow^\bullet$ for contraction (shown in Figure 3).

Let u, v, and w be vertices in a modal arena. We say that u is a *meeting point* of v and w whenever $v\to^*u$ and $w\to^*u$, and there is no vertex $u' \neq u$ such that $v\to^*u'$ and $w\to^*u'$ and $u'\to^*u$. The *meeting depth* of v and w is the depth of their meeting point, or -1 if no meeting point exists. Note that this is well defined as all meeting points of v and w have the same depth (this follows from [17, Lemma 9]). Two distinct vertices v and w are *conjunct*, denoted $v \curlywedge w$ if their meeting depth is odd (or equal to -1).

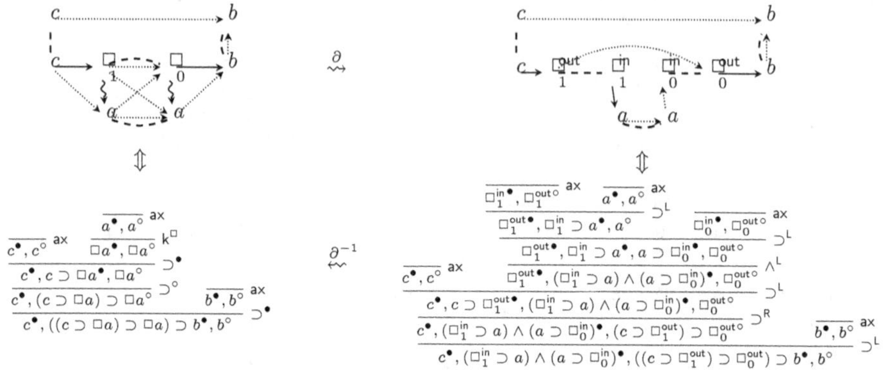

Fig. 8. A K-arena net $\mathcal{G}$ with associated formula $(c \wedge ((c \supset \Box a) \supset \Box a) \supset b) \supset b$, its corresponding arena $\partial(\mathcal{G})$, the $\mathsf{LCK}^{\mathsf{c}}_{\ell}$-derivation associated to $\mathcal{G}$ and the $\mathsf{LI}^{\mathsf{c}}_{\ell}$-derivation associated to $\partial(\mathcal{G})$

Definition 6.1 A *modal arena homomorphism* is either a map $\emptyset_{\mathcal{G}} \colon \emptyset \to \mathcal{G}$ from the empty 2-dag to a modal arena $\mathcal{G}$, or a structure preserving map $f \colon \mathcal{H} \to \mathcal{G}$ between two modal arenas, i.e., its a function $f \colon V_{\mathcal{H}} \to V_{\mathcal{G}}$ that preserves: [3]

$$\begin{array}{llcll} \to : & \text{if } v \overset{\mathcal{H}}{\to} w \text{ then } f(v) \overset{\mathcal{G}}{\to} f(w) & \qquad & d: & d(v) = d(f(v)) \\ \rightsquigarrow : & \text{if } v \overset{\mathcal{H}}{\rightsquigarrow} w \text{ then } f(v) \overset{\mathcal{G}}{\rightsquigarrow} f(w) & & \ell: & \ell(v) = \ell(f(v)) \end{array}$$

A *(modal) skew fibration* is a modal arena homomorphism $f \colon \mathcal{H} \to \mathcal{G}$ which:

- preserves $\curlywedge$: if $v \curlywedge_{\mathcal{H}} w$ then $f(v) \curlywedge_{\mathcal{G}} f(w)$;
- is a *skew lifting*: if $f(v) \curlywedge_{\mathcal{G}} w$, then there exists u with $u \curlywedge_{\mathcal{H}} v$ and $f(u) \not\curlywedge_{\mathcal{G}} w$.
- is a *modal lifting*: if $f(v) \overset{\mathcal{G}}{\rightsquigarrow} f(w)$, then there exists u with $u \overset{\mathcal{H}}{\rightsquigarrow} w$ and $f(u) = f(v)$.

Lemma 6.2 *The composition of two skew fibrations is a skew fibration.*

Proof. By definition, the composition preserves $\to$, $\rightsquigarrow$, ℓ and d. Then the preservation of $\curlywedge$ and the skew lifting condition of the composition are guaranteed as consequence of the preservation of d and $\to$. Similarly, the modal lifting condition of the composition is guaranteed as consequence of the preservation of $\rightsquigarrow$ and the fact that source and target of a skew fibration are modal arenas. □

In order to prove the correspondence between $\{\mathsf{c}^{\bullet}_{\downarrow}, \mathsf{w}^{\bullet}_{\downarrow}\}$ derivations and skew fibrations, we provide the following definition.

[3] In [17] the definition of skew fibration only demands the weaker *root preserving* condition (that is, if $v \in \overrightarrow{R}_{\mathcal{H}}$ then $f(v) \in \overrightarrow{R}_{\mathcal{G}}$) instead of the depth preserving condition $d(v) = d(f(v))$ that we use here. However, in the same paper it is proven that root preservation is equivalent to depth preservation.

Definition 6.3 If $f_1 : \mathcal{H}_1 \to \mathcal{G}_1$ and $f_2 : \mathcal{H}_2 \to \mathcal{G}_2$ are modal arena homomorphisms such that $\mathcal{H}_1$ and $\mathcal{H}_2$ are disjoint modal arenas, we define the following modal arena homomorphisms:

$$\begin{array}{rcllll} f_1 + f_2 & = & f_1 \cup f_2 : \mathcal{H}_1 + \mathcal{H}_2 & \to & \mathcal{G}_1 + \mathcal{G}_2 & \\ f_1 \multimap f_2 & = & f_1 \cup f_2 : \mathcal{H}_1 \multimap \mathcal{H}_2 & \to & \mathcal{G}_1 \multimap \mathcal{G}_2 & (\mathcal{H}_2,\, \mathcal{G}_2 \neq \emptyset) \\ f_1 \rightsquigarrow f_2 & = & f_1 \cup f_2 : \mathcal{H}_1 \rightsquigarrow \mathcal{H}_2 & \to & \mathcal{G}_1 \rightsquigarrow \mathcal{G}_2 & \\ [f_1, f_2] & = & f_1 \cup f_2 : \mathcal{H}_1 + \mathcal{H}_2 & \to & \mathcal{G} & (\mathcal{G}_1 = \mathcal{G}_2 = \mathcal{G}) \end{array} \quad (5)$$

Lemma 6.4 *The operations from Definition 6.3 preserve skew fibration properties.*

Proof. It suffices to check that if f_1 and f_2 are skew fibrations, then also $f_1 + f_2$, $f_1 \multimap f_2$, $f_1 \cup f_2$, $f_1 \rightsquigarrow f_2$ and $[f_1, f_2]$ are. □

Lemma 6.5 *Let H' and H be* P*-formulas. If $H' \overset{\{\mathsf{c}^\bullet_\downarrow, \mathsf{w}^\bullet_\downarrow\}}{\vdash} H$, then there is a skew fibration $f \colon [\![H']\!] \to [\![H]\!]$.*

Proof. After Lemma 6.2, it suffices to prove that if $\frac{H'}{H}\,\rho$ for a $\rho \in \{\mathsf{w}^\bullet_\downarrow, \mathsf{c}^\bullet_\downarrow\}$. then there is a skew fibration $f \colon [\![H']\!] \to [\![H]\!]$. This is immediate after Lemma 6.4 after remarking that any map $\emptyset_{\mathcal{G}} \colon \emptyset \to \mathcal{G}$ is a skew fibration. □

To prove the converse, we need some additional definitions and results.

Definition 6.6 Two distinct vertices v and w in a modal arena they are *disjunct*, denoted $v \curlyvee w$, if their meeting depth is even. An *odd skew fibration* is either a map $\emptyset_{\mathcal{G}} \colon \emptyset \to \mathcal{G}$, or a modal arena homomorphism $f \colon \mathcal{H} \to \mathcal{G}$ which:

- preserves $\curlyvee$: if $v \curlyvee_{\mathcal{H}} w$ then $f(v) \curlyvee_{\mathcal{G}} f(w)$;
- is a *odd skew lifting*: if $f(v) \curlyvee_{\mathcal{G}} w$, then there exists u with $v \curlyvee_{\mathcal{H}} u$ and $f(u) \not\curlyvee_{\mathcal{G}} w$.

Lemma 6.7 *If $f \colon \mathcal{H} \to \mathcal{G}$ be a modal arena homomorphism and $\mathcal{G} = \mathcal{G}_1 + \mathcal{G}_2$, then $f = f_1 + f_2$ with $f_1 \colon \mathcal{H}_1 \to \mathcal{G}_1$ and $f_2 \colon \mathcal{H}_2 \to \mathcal{G}_2$ modal arena homomorphisms for some $\mathcal{H}_1, \mathcal{H}_2$ such that $\mathcal{H} = \mathcal{H}_1 + \mathcal{H}_2$.*

Proof. Since f preserves $\to$, then if $v \to^* w$ for a $w \in \vec{R}_{\mathcal{G}}$ then $f(v) \to^* f(w)$. Thus if $\mathcal{G} = \mathcal{G}_1 + \mathcal{G}_2$, then there is a partition $\vec{R}_{\mathcal{G}} = \vec{R}_{\mathcal{G}_1} \uplus \vec{R}_{\mathcal{G}_2}$. As remarked in the proof of Theorem 4.9, in construction such partition, because of $\rightsquigarrow$-coherence, whenever $v \rightsquigarrow w$ then v and w belong to the same subset. Then we can define $V_{\mathcal{H}_1}$ and $V_{\mathcal{H}_2}$ as the sets of vertices of $\mathcal{H}$ which images by f admit a $\to$-path to a vertex in $\vec{R}_{\mathcal{G}_1}$ and $\vec{R}_{\mathcal{G}_2}$ respectively. The modal arenas $\mathcal{H}_1$ and $\mathcal{H}_2$ are defined from $\mathcal{H}$ by the sets $V_{\mathcal{H}_1}$ and $V_{\mathcal{H}_2}$ respectively. □

Lemma 6.8 *Let $f \colon \mathcal{H} \to \mathcal{G}$ be a skew fibration or an odd skew fibration, with $\mathcal{G} = \mathcal{G}_1 \multimap \mathcal{G}_2$ and $\mathcal{G}_1$ modal arenas. If there exist two modal arenas $\mathcal{H}'$ and $\mathcal{H}''$ such that $\mathcal{H} = \mathcal{H}' \multimap \mathcal{H}''$ and $\mathcal{H}''$ cannot be written as $\multimap$ of two modal arenas, then $f(v) \in V_{\mathcal{G}_2}$ for all $v \in V_{\mathcal{H}''}$.*

Proof. Let $v \in \mathcal{H}''$ such that $f(v) \in \mathcal{G}_1$. Since f preserves d, then $v \notin \vec{R}_{\mathcal{H}}$. Thus $\mathcal{H}''$ cannot be a single-vertex modal arena. If $\mathcal{H}''$ is a + of two modal arenas, then there is $z \in \overrightarrow{R}_{\mathcal{H}''}$ such that $v \not\to^* z$, hence $v \curlywedge z$ in $\mathcal{H}$ but $f(v) \not\curlywedge f(z)$ in $\mathcal{G}$. Therefore f is not a skew fibration. Let $f(z) = w$. Then $f(v) \curlyvee w$ because $f(v) \in \mathcal{G}_1$ and $w \in \overrightarrow{R}_{\mathcal{G}}$. If there is a u with $v \curlyvee u$ in $\mathcal{H}$ then there is $x \in V_{\mathcal{H}}$ such that $u \to x^\circ$ and $v \to x^\circ$. Since $x \to^* w$ we have $f(u) \curlyvee w$, which means that f cannot be an odd skew fibration either. Then $\mathcal{H}''$ has to be of the shape $w \rightsquigarrow\!\triangleright \mathcal{H}_2''$ and $f(w) \in \mathcal{G}_2$ because $v \in \overrightarrow{R}_{\mathcal{H}}$. We can conclude as for the previous case that f is not an even or odd skew fibration. Contradiction. □

Lemma 6.9 *Let $f\colon \mathcal{H} \to \mathcal{G}$ is an odd skew fibration, with $\mathcal{G} = \mathcal{G}_1 \multimap\!\triangleright \mathcal{G}_2$ for a modal arena $\mathcal{G}_1$. If there is a modal arena $\mathcal{H}'$ such that $\mathcal{H} = \mathcal{H}' \multimap\!\triangleright \mathcal{H}''$, then there are $\mathcal{H}_1$ and $\mathcal{H}_2$ such that $\mathcal{H} = \mathcal{H}_1 \multimap\!\triangleright \mathcal{H}_2$ and $f = f_1 \multimap\!\triangleright f_2$ where $f_1\colon \mathcal{H}_1 \to \mathcal{G}_1$ and $f_2\colon \mathcal{H}_2 \to \mathcal{G}_2$ are modal arena homomorphisms.*

Proof. By hypothesis, we can assume that $\mathcal{H}$ is of the form $\mathcal{H} = \mathcal{H}' \multimap\!\triangleright \mathcal{H}''$ where $\mathcal{H}''$ is not a $\multimap\!\triangleright$ of two modal arenas. We conclude by Lemma 6.8 that $f(v) \in V_{\mathcal{G}_2}$ for any $v \in V_{\mathcal{H}''}$. If $V_{\mathcal{G}_2} = f(V_{\mathcal{H}''})$, then we conclude that $\mathcal{H}_1 = \mathcal{H}'$ and $\mathcal{H}_2 = \mathcal{H}''$. Otherwise, let $\mathcal{H}' = \mathcal{H}_1' + \cdots + \mathcal{H}_n'$ such that $\mathcal{H}_i'$ is a $+$ of two modal arenas for no $i \in \{1, \ldots, n\}$. If $v, w \in V_{\mathcal{H}'}$, then there is a $(\leftrightarrow \cup \leftrightsquigarrow)$-path from v to w in $V_{\mathcal{H}'}$ iff there is $i \in \{1, \ldots, n\}$ such that $v, w \in V_{\mathcal{H}_i'}$. Since $\overrightarrow{R}_{\mathcal{G}} \subset f(V_{\mathcal{H}''})$, this implies that if there is $i \in \{1, \ldots, n\}$ such that $v, w \in V_{\mathcal{H}_i'}$, then there is $(\leftrightarrow \cup \leftrightsquigarrow)$-path from $f(v)$ to $f(w)$ in $V_{\mathcal{G}} \setminus \overrightarrow{R}_{\mathcal{G}}$. That is, $f(V_{\mathcal{H}_i'})$ is either a subset of $V_{\mathcal{G}_1}$ or a subset of $V_{\mathcal{G}_2}$ for all $i \in \{1, \ldots, n\}$. Without loss of generality we assume there is j such that that $f(V_{\mathcal{H}_i'}) \subset V_{\mathcal{G}_1}$ for all $i \leq j$. We conclude that $\mathcal{H}_1 = \mathcal{H}_1' + \cdots + \mathcal{H}_j'$ and $\mathcal{H}_2 = (\mathcal{H}_{j+1}' + \cdots + \mathcal{H}_n') \multimap\!\triangleright \mathcal{H}''$. □

Lemma 6.10 *Let $f\colon \mathcal{H} \to \mathcal{G}$ be a modal arena homomorphism and $\mathcal{G} = v \rightsquigarrow\!\triangleright \mathcal{G}'$. If f is a skew fibration then, $\mathcal{H} = w \rightsquigarrow\!\triangleright \mathcal{H}'$ and $f = 1_w \rightsquigarrow\!\triangleright f'$ with $f'\colon \mathcal{H}' \to \mathcal{G}'$ a skew fibration. If f is odd skew fibration, then*

- *either $\mathcal{H} = w \rightsquigarrow\!\triangleright \mathcal{H}_2$ and $f = 1_w \rightsquigarrow\!\triangleright f_2$ with $f_2\colon \mathcal{H}_2 \to \mathcal{G}_2$ an odd skew fibration;*
- *or $\mathcal{H} = (w \rightsquigarrow\!\triangleright \mathcal{H}_1) + \mathcal{H}_2$ and $f = [f_1, f_2]$ with $f_1\colon (w \rightsquigarrow\!\triangleright \mathcal{H}_1) \to (v \rightsquigarrow\!\triangleright \mathcal{G}_2)$ and $f_2\colon \mathcal{H}_2 \to (v \rightsquigarrow\!\triangleright \mathcal{G}_2)$.*

Proof. If f is a skew fibration, then to conclude it suffices to remark there is a unique w such that $f(w) = v$ since $v \in \overrightarrow{R}_{\mathcal{G}}$. If f is an odd skew fibration, let w such that $f(w) = v$. If $V_{\mathcal{H}} \setminus \{w\} = \mathsf{Scope}(w)$, then we can conclude. Otherwise we conclude with $\mathcal{H}_2$ be the modal arena with vertices in $V_{\mathcal{H}} \setminus (\{w\} \cup \mathsf{Scope}(w))$. □

Lemma 6.11 *Every skew fibration is of the form $1_{\mathcal{G}}$, $f^\circ + g^\circ$, $f^\bullet \multimap\!\triangleright g^\circ$ or $1_v \rightsquigarrow\!\triangleright g^\circ$. Every odd skew fibration is of the form $1_{\mathcal{G}}$, $[f^\bullet, g^\bullet]$, $f^\bullet + g^\bullet$, $f^\circ \multimap\!\triangleright g^\bullet$, $1_v \rightsquigarrow\!\triangleright g^\bullet$ or $\emptyset_{\mathcal{G}}$, where f° and g° are skew fibrations, $f^\bullet$ and $g^\bullet$ are odd skew fibrations, $v \in V_{[\![H]\!]}^{\Box\Diamond}$, and $\mathcal{G}$ can be any modal arena.*

Proof. By case analysis, let $f\colon \mathcal{H} \to \mathcal{G}$ be a modal arena homomorphism, remarking that for any modal arena $\mathcal{G}$, the identity map $1_{\mathcal{G}}$ is by definition an even and an odd skew fibration. If $f^\circ\colon \mathcal{H} \to \mathcal{G}$ is a skew fibration, then

- if $\mathcal{G}$ is a single-vertex modal arena, then $\mathcal{H}$ cannot be either of the shape $\mathcal{H}_1 + \mathcal{H}_2$ or $\mathcal{H}_1 \mathrel{\sim\!\!\triangleright} \mathcal{H}_2$ otherwise f would not preserve $\curlywedge$, or of the shape $\mathcal{H}_1 \mathrel{-\!\!\triangleright} \mathcal{H}_2$ otherwise it would not preserve d. Then $f = 1_v$ with v the unique vertex in $V_{\mathcal{H}} = V_{\mathcal{G}}$;
- if $\mathcal{G} = \mathcal{G}_1 + \mathcal{G}_2$, then by Lemma 6.7 we have that $f^\circ = f_1 + f_2$ with f_1 and f_2 arena homomorphisms. Since f° is an even skew fibration, it follows by definition of $+$ that f_1 and f_2 are skew fibrations;
- if $\mathcal{G} = \mathcal{G}_1 \mathrel{-\!\!\triangleright} \mathcal{G}_2$, then we define $V_1 = \{v \in V_{\mathcal{H}} \mid f(v) \in \mathcal{G}_1\}$ and $V_2 = \{v \in V_{\mathcal{H}} \mid f(v) \in \mathcal{G}_2\}$. We have that $V_2 \neq \emptyset$ since f preserve d. If $V_1 = \emptyset$, then $f = \emptyset_{\mathcal{G}_1} \mathrel{-\!\!\triangleright} f_2$ with $f_2 \colon \mathcal{H} \to \mathcal{G}_2$. Otherwise, $V_1 \neq \emptyset$ and $\mathcal{H}$ cannot be a single vertex. Similarly, $\mathcal{H}$ cannot be of the shape $\mathcal{H}_1 + \mathcal{H}_2$ otherwise f would not preserve $\curlywedge$, nor of the shape $v \mathrel{\sim\!\!\triangleright} \mathcal{H}_2$ otherwise f would not be modal. We conclude by Lemma 6.9 that $f = f_1 \mathrel{-\!\!\triangleright} f_2$. Moreover, since f is a skew fibration if follows that f_2 also preserves $\curlywedge$ and satisfies skew lifting while f_1 preserve $\curlyvee$ and satisfies odd skew lifting;
- if $\mathcal{G} = v \mathrel{\sim\!\!\triangleright} \mathcal{G}_2$, we conclude by Lemma 6.10 .

If $f^\bullet \colon \mathcal{H} \to \mathcal{G}$ is an odd skew fibration, then we proceed similarly. If $\mathcal{G}$ is a single-vertex modal arena, then $\mathcal{H}$ cannot be of the shape $\mathcal{H}_1 \mathrel{\sim\!\!\triangleright} \mathcal{H}_2$ otherwise f it would not be modal, or of the shape $\mathcal{H}_1 \mathrel{-\!\!\triangleright} \mathcal{H}_2$ otherwise it would not preserve d. Let $\mathcal{H} = \mathcal{H}_1 + \mathcal{H}_2$ such that $\mathcal{H}_1 \neq \mathcal{H}'_1 + \mathcal{H}''_1$. Since $f^\bullet$ preserve d and $\rightsquigarrow$, then $\mathcal{H}_1$ is a single-vertex modal arenas. Moreover, $f_2 \colon \mathcal{H}_2 \to \mathcal{G}_2$ is an odd skew fibration by definition. Then $f = [1_v, f_2]$ with v the unique vertex in $V_{\mathcal{H}} = V_{\mathcal{G}}$;

If $\mathcal{G} = \mathcal{G}_1 + \mathcal{G}_2$, $\mathcal{G} = \mathcal{G}_1 \mathrel{-\!\!\triangleright} \mathcal{G}_2$ or $\mathcal{G} = v \mathrel{\sim\!\!\triangleright} \mathcal{G}_2$ we apply a similar reasoning in the case of f skew fibration. □

Lemma 6.11 is now enough to complete the proof of Theorem 6.12: Given a skew fibration f, we can decompose f as an expression with the operations in Lemma 6.11, which can then be immediately transformed into a deep inference derivation using only $\mathsf{w}^\bullet_\downarrow$ and $\mathsf{c}^\bullet_\downarrow$. (This is a standard operation in deep inference, see e.g. [15].)

Theorem 6.12 *Let H and H' be* P*-formulas.*

$$H' \overset{\{\mathsf{c}^\bullet_\downarrow, \mathsf{w}^\bullet_\downarrow\}}{\vdash} H \iff \textit{there is a skew fibration } f \colon [\![H']\!] \to [\![H]\!]$$

Proof. To prove the "if" direction, it suffices to prove that if $\dfrac{H'}{H}\,\rho$ for a $\rho \in \{\mathsf{w}^\bullet_\downarrow, \mathsf{c}^\bullet_\downarrow\}$, then there is a skew fibration $f \colon [\![H']\!] \to [\![H]\!]$. Then we conclude by Lemma 6.2.

By Lemma 6.11, we have that any skew fibration can be written as composition from $1_v \colon v \to v$ and $\emptyset_{\mathcal{G}} \colon \emptyset \to \mathcal{G}$ via the operations in (5) above. In particular, each $\emptyset_{\mathcal{G}}$ occurring in the decomposition corresponds to an application of a $\mathsf{w}^\bullet_\downarrow$, while each occurrence of $[-,-]$ corresponds to an application of a $\mathsf{c}^\bullet_\downarrow$. We conclude by reconstructing a derivation in $\{\mathsf{c}^\bullet_\downarrow, \mathsf{w}^\bullet_\downarrow\}$ using this decomposition and the correspondence between P-formulas and modal arenas (Theorem 4.9). □

$$\dfrac{\dfrac{\Gamma, A, B, C \vdash D}{\Gamma, A, B \vdash C \supset D}\,{\supset}^{\mathsf{R}}}{\Gamma, A \wedge B \vdash C \supset D}\,{\wedge}^{\mathsf{L}} \quad \equiv \quad \dfrac{\dfrac{\Gamma, A, B, C \vdash D}{\Gamma, A \wedge B, C \vdash D}\,{\wedge}^{\mathsf{L}}}{\Gamma, A \wedge B \vdash C \supset D}\,{\supset}^{\mathsf{R}} \qquad \dfrac{\dfrac{\Gamma, A, B \vdash C}{\Gamma, A \wedge B \vdash C}\,{\wedge}^{\mathsf{R}} \quad \Delta \vdash D}{\Gamma, \Delta, A \wedge B \vdash C \wedge D}\,{\wedge}^{\mathsf{L}} \quad \equiv \quad \dfrac{\dfrac{\Gamma, A, B \vdash C \quad \Delta \vdash D}{\Gamma, \Delta, A, B \vdash C \wedge D}\,{\wedge}^{\mathsf{L}}}{\Gamma, \Delta, A \wedge B \vdash C \wedge D}\,{\wedge}^{\mathsf{R}}$$

Fig. 9. Examples of independent rule permutations.

7 Combinatorial Proofs

We can now combine the results of the previous sections, to define combinatorial proofs for the logics CK and CD, and to prove their soundness and completeness.

Definition 7.1 A *modal intuitionistic combinatorial proof* is a skew fibration $f : \mathcal{G} \to [\![F]\!]$ from a modal arena net $\mathcal{G}$ to the modal arena of a formula F. We say f is a CK-*intuitionistic combinatorial proof*, or CK-ICP, (resp. CD-*intuitionistic combinatorial proof*, or CD-ICP) if $\mathcal{G}$ is a CK-arena net (resp. CD-arena net).

The *intuitionistic combinatorial proofs* (or ICPs) from [17] are the special cases where no modalities occur, that is, an ICP is a skew fibration $f : \mathcal{G} \to [\![F]\!]$ from an arena net $\mathcal{G}$ to the arena of a modality-free formula F.

Theorem 7.2 (Soundness and Completeness) *If F is a formula and* $\mathsf{X} \in \{\mathsf{CK}, \mathsf{CD}\}$, *then* $\stackrel{\mathsf{LX}}{\vdash} F$ *iff there is an* X-ICP $f : \mathcal{G} \to [\![F]\!]$.

Proof. By Theorem 3.1 there are P-formulas H and H' such that $F = \lfloor H \rfloor$ and H' is clean and $\stackrel{\mathsf{LX}}{\vdash} F \iff \stackrel{\mathsf{LX}^{\bullet}_{\ell}}{\vdash} H' \stackrel{\{\mathsf{c}^{\bullet}_{\downarrow}, \mathsf{w}^{\bullet}_{\downarrow}\}}{\vdash} H$. In Theorem 5.6 we have shown that $\stackrel{\mathsf{LX}^{\bullet}_{\ell}}{\vdash} H'$ iff there is an X-arena net $\mathcal{G}$ with H' the formula associated to $\mathcal{G}$. In Section 6 we have shown that $H' \stackrel{\{\mathsf{c}^{\bullet}_{\downarrow}, \mathsf{w}^{\bullet}_{\downarrow}\}}{\vdash} H$ iff there is a skew fibration $f : [\![\mathcal{H}']\!] \to [\![H]\!]$. This is equivalent to having an X-ICP $f : \mathcal{G} \to [\![F]\!]$, since $[\![H]\!] = [\![F]\!]$. □

Lemma 7.3 *Let* $\mathsf{X} \in \{\mathsf{CK}, \mathsf{CD}\}$. *If $\mathcal{H}$ and $\mathcal{G}$ are* 2-dags *and $f : V_{\mathcal{H}} \to V_{\mathcal{G}}$, then it can be checked in polynomial time (in the size of $\mathcal{H}$ and $\mathcal{G}$) whether f is an* X-ICP.

Proof. All the following checks can be done in polynomial time: that a 2-dag $\mathcal{G}$ is a modal arena; that a modal arena is an X-arena net; and that a map between two modal arenas is a skew fibration. □

Corollary 7.4 *Let* $\mathsf{X} \in \{\mathsf{CK}, \mathsf{CD}\}$. *Then the* X-ICP*s form a sound and complete proof system in the sense of Cook and Reckhow [10].*

8 On Proof Equivalence for Constructive Modal Logics

Let us now compare various notions of proof equivalence in constructive modal logics, building from the previous results in [17], where the authors show that intuitionistic combinatorial proofs capture a finer notion of proof equivalence than the one induced by the *simply typed lambda calculus* or by *winning innocent strategies* from games semantics [1,22,27].

In the following, we use $\equiv$ to denote the proof equivalence over derivations generated by *independent rule permutations*, that is, permutations of infer-

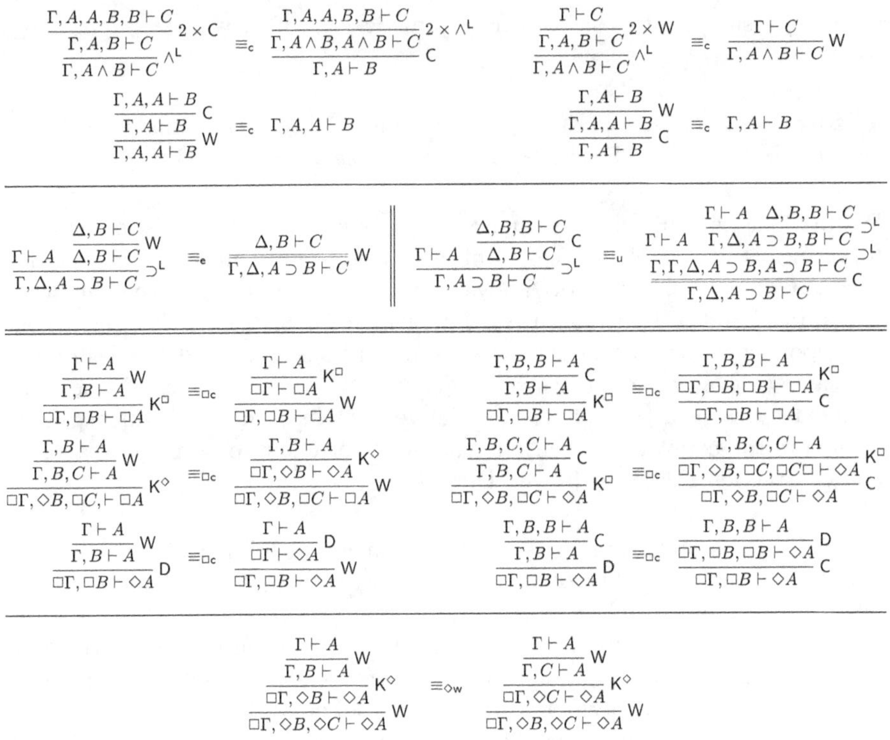

$$\dfrac{\dfrac{\Gamma,A,A,B,B\vdash C}{\Gamma,A,B\vdash C}\,2\times\mathsf{C}}{\Gamma,A\wedge B\vdash C}\,\wedge^{\mathsf{L}} \quad\equiv_{\mathsf{c}}\quad \dfrac{\dfrac{\Gamma,A,A,B,B\vdash C}{\Gamma,A\wedge B,A\wedge B\vdash C}\,2\times\wedge^{\mathsf{L}}}{\Gamma,A\vdash B}\,\mathsf{C} \qquad \dfrac{\dfrac{\Gamma\vdash C}{\Gamma,A,B\vdash C}\,2\times\mathsf{W}}{\Gamma,A\wedge B\vdash C}\,\wedge^{\mathsf{L}} \quad\equiv_{\mathsf{c}}\quad \dfrac{\Gamma\vdash C}{\Gamma,A\wedge B\vdash C}\,\mathsf{W}$$

$$\dfrac{\dfrac{\Gamma,A,A\vdash B}{\Gamma,A\vdash B}\,\mathsf{C}}{\Gamma,A,A\vdash B}\,\mathsf{W} \quad\equiv_{\mathsf{c}}\quad \Gamma,A,A\vdash B \qquad \dfrac{\dfrac{\Gamma,A\vdash B}{\Gamma,A,A\vdash B}\,\mathsf{W}}{\Gamma,A\vdash B}\,\mathsf{C} \quad\equiv_{\mathsf{c}}\quad \Gamma,A\vdash B$$

$$\dfrac{\Gamma\vdash A \quad \dfrac{\Delta,B\vdash C}{\Delta,B\vdash C}\,\mathsf{W}}{\Gamma,\Delta,A\supset B\vdash C}\,\supset^{\mathsf{L}} \quad\equiv_{\mathsf{e}}\quad \dfrac{\Delta,B\vdash C}{\Gamma,\Delta,A\supset B\vdash C}\,\mathsf{W}$$

$$\dfrac{\Gamma\vdash A \quad \dfrac{\Delta,B,B\vdash C}{\Delta,B\vdash C}\,\mathsf{C}}{\Gamma,A\supset B\vdash C}\,\supset^{\mathsf{L}} \quad\equiv_{\mathsf{u}}\quad \dfrac{\dfrac{\Gamma\vdash A \quad \dfrac{\Gamma\vdash A \quad \Delta,B,B\vdash C}{\Gamma,\Delta,A\supset B,B\vdash C}\,\supset^{\mathsf{L}}}{\Gamma,\Gamma,\Delta,A\supset B,A\supset B\vdash C}\,\supset^{\mathsf{L}}}{\Gamma,\Delta,A\supset B\vdash C}\,\mathsf{C}$$

$$\dfrac{\dfrac{\Gamma\vdash A}{\Gamma,B\vdash A}\,\mathsf{W}}{\Box\Gamma,\Box B\vdash\Box A}\,\mathsf{K}^{\Box} \quad\equiv_{\Box\mathsf{c}}\quad \dfrac{\dfrac{\Gamma\vdash A}{\Box\Gamma\vdash\Box A}\,\mathsf{K}^{\Box}}{\Box\Gamma,\Box B\vdash\Box A}\,\mathsf{W} \qquad \dfrac{\dfrac{\Gamma,B,B\vdash A}{\Gamma,B\vdash A}\,\mathsf{C}}{\Box\Gamma,\Box B\vdash\Box A}\,\mathsf{K}^{\Box} \quad\equiv_{\Box\mathsf{c}}\quad \dfrac{\dfrac{\Gamma,B,B\vdash A}{\Box\Gamma,\Box B,\Box B\vdash\Box A}\,\mathsf{K}^{\Box}}{\Box\Gamma,\Box B\vdash\Box A}\,\mathsf{C}$$

$$\dfrac{\dfrac{\Gamma,B\vdash A}{\Gamma,B,C\vdash A}\,\mathsf{W}}{\Box\Gamma,\Diamond B,\Box C,\vdash\Box A}\,\mathsf{K}^{\Diamond} \quad\equiv_{\Box\mathsf{c}}\quad \dfrac{\dfrac{\Gamma,B\vdash A}{\Box\Gamma,\Diamond B\vdash\Diamond A}\,\mathsf{K}^{\Diamond}}{\Box\Gamma,\Diamond B,\Box C\vdash\Box A}\,\mathsf{W} \qquad \dfrac{\dfrac{\Gamma,B,C,C\vdash A}{\Gamma,B,C\vdash A}\,\mathsf{C}}{\Box\Gamma,\Diamond B,\Box C\vdash\Diamond A}\,\mathsf{K}^{\Box} \quad\equiv_{\Box\mathsf{c}}\quad \dfrac{\dfrac{\Gamma,B,C,C\vdash A}{\Box\Gamma,\Diamond B,\Box C,\Box C\Box\vdash\Diamond A}\,\mathsf{K}^{\Box}}{\Box\Gamma,\Diamond B,\Box C\vdash\Diamond A}\,\mathsf{C}$$

$$\dfrac{\dfrac{\Gamma\vdash A}{\Gamma,B\vdash A}\,\mathsf{W}}{\Box\Gamma,\Box B\vdash\Diamond A}\,\mathsf{D} \quad\equiv_{\Box\mathsf{c}}\quad \dfrac{\dfrac{\Gamma\vdash A}{\Box\Gamma\vdash\Diamond A}\,\mathsf{D}}{\Box\Gamma,\Box B\vdash\Diamond A}\,\mathsf{W} \qquad \dfrac{\dfrac{\Gamma,B,B\vdash A}{\Gamma,B\vdash A}\,\mathsf{C}}{\Box\Gamma,\Box B\vdash\Diamond A}\,\mathsf{D} \quad\equiv_{\Box\mathsf{c}}\quad \dfrac{\dfrac{\Gamma,B,B\vdash A}{\Box\Gamma,\Box B,\Box B\vdash\Diamond A}\,\mathsf{D}}{\Box\Gamma,\Box B\vdash\Diamond A}\,\mathsf{C}$$

$$\dfrac{\dfrac{\dfrac{\Gamma\vdash A}{\Gamma,B\vdash A}\,\mathsf{W}}{\Box\Gamma,\Diamond B\vdash\Diamond A}\,\mathsf{K}^{\Diamond}}{\Box\Gamma,\Diamond B,\Diamond C\vdash\Diamond A}\,\mathsf{W} \quad\equiv_{\Diamond\mathsf{w}}\quad \dfrac{\dfrac{\dfrac{\Gamma\vdash A}{\Gamma,C\vdash A}\,\mathsf{W}}{\Box\Gamma,\Diamond C\vdash\Diamond A}\,\mathsf{K}^{\Diamond}}{\Box\Gamma,\Diamond B,\Diamond C\vdash\Diamond A}\,\mathsf{W}$$

Fig. 10. Non independent rule permutations

ence rules whose active formulas and principal formulas are disjoint, as in the examples shown in Figure 9. Then, in Figure 10, we show examples of rule permutations which are non-independent. Based on these, we define the following proof equivalences:

$$\equiv_{\mathsf{ICP}} := (\equiv \cup \equiv_{\mathsf{e}} \cup \equiv_{\mathsf{c}}) \qquad \equiv_{\lambda} := (\equiv_{\mathsf{ICP}} \cup \equiv_{\mathsf{u}}) \qquad \equiv_{\mathsf{WIS}} := (\equiv_{\lambda} \cup \equiv_{\Box\mathsf{c}})$$

Note that $\equiv_{\mathsf{ICP}}$ and $\equiv_{\lambda}$ can be defined in the same way for the non-modal case. Indeed, it has been shown in [17] that $\equiv_{\mathsf{ICP}}$ is the proof equivalence induced by intuitionistic combinatorial proofs. We extend this result to the modal case in Theorem 8.1 below. In the modality-free case, $\equiv_{\lambda}$ corresponds to the proof identifications made by the simply-typed λ-calculus, and we conjecture that in the case with modalities, the proof equivalence $\equiv_{\lambda}$ is the same as the one induced by λ-terms/natural deduction proofs presented in [7] for constructive modal logics. We also conjecture that the proof equivalence $\equiv_{\mathsf{WIS}}$ is the same as the one induced by the winning strategies presented in [3]. However, it is worth remarking that even though $\equiv_{\Diamond\mathsf{w}}$ seems to be in the same spirit as $\equiv_{\Box\mathsf{c}}$, this permutation is even beyond the winning strategies of [3]. For this reason it is listed separately in Figure 10. Note that according to our conjecture, there is no one-to-one correspondence between winning innocent strategies and λ-terms (natural deduction proofs) for constructive modal logics. This is in contrast with the result in propositional intuitionistic logic [22], where $\equiv_\lambda$ and $\equiv_{\mathsf{WIS}}$ coincide.

Theorem 8.1 *Let* $\mathsf{X} \in \{\mathsf{CK}, \mathsf{CD}\}$ *and let* $\mathfrak{D}$ *and* $\mathfrak{D}'$ *be derivations in* LX*. Then we have* $\mathfrak{D} \equiv_{\mathsf{ICP}} \mathfrak{D}'$ *iff* $\mathfrak{D}$ *and* $\mathfrak{D}'$ *are represented by the same* X-ICP.

Proof. This is a direct consequence of the result on intuitionistic combinatorial proofs in [17]. It suffices to observe that if weakening and contraction rules could be permuted below/above K-rules, this would change the number of modalities handled by these rules. However, in CK-ICPs there is a one-to-one correspondence between $\sim$-classes and applications of K- and D-rules and at the same time a one-to-one correspondence between the number of modalities handled by each of these rules and the vertices in such $\sim$-equivalence class. Therefore the equivalence relation $\equiv_{\mathsf{ICP}}$ does not allow to permute weakening and contraction rules over K- and D-rules. □

The rule permutations in $\equiv_{\Box\mathsf{c}}$ are well-known in linear logic, as they correspond to the possibility of moving both weakening and contraction gates outside a !?-box in multiplicative exponential proof nets (see the notion of generalized ?-nodes introduced in [11] allowing to capture both the rule permutations in $\equiv_{\mathsf{c}}$ and $\equiv_{\Box\mathsf{c}}$). It has been observed before (see, e.g., [2]), that including the $\equiv_{\Box\mathsf{c}}$ rule permutations in the proof equivalence of classical linear logic makes proof equivalence **PSPACE**-complete. This immediately follows from the result in [16] about **PSPACE**-completeness of proof equivalence for multiplicative linear logic with units. More precisely, multiplicative linear logic proof nets require each $\bot$-gate to be attached to an axiom by a so-called "jump" in order to guarantee a polynomial proof equivalence.

A similar phenomenon occurs in the constructive modal logics studied in our paper since in combinatorial proof each $\top^\bullet$-rule instance is linked to a K- or a D-rule instance occurring below it. Since rule permutations in $\equiv_{\Box\mathsf{c}}$ or $\equiv_{\Diamond\mathsf{w}}$ trigger a "jump-rewiring" mechanism similar to the one observed in [16], we conjecture that proof equivalence including these permutations is **PSPACE**-hard.

9 Conclusions and Future Works

We have presented the syntax of combinatorial proofs for the disjunction-free fragment of the constructive modal logics CK and CD. We have proved that (1) this syntax is a sound and complete proof system in the sense of Cook and Reckhow [10], and that (2) it enforces a notion of proof equivalence which is finer than the one provided by natural deduction proofs, but still coarser than plain sequent calculus.

In future work we want to further investigate the various notions of proof equivalence for constructive modal logics, and in a next step study combinatorial proofs for other variants of intuitionistic modal logics.

References

[1] Abramsky, S., P. Malacaria and R. Jagadeesan, *Full abstraction for pcf*, in: *International Symposium on Theoretical Aspects of Computer Software*, Springer, 1994, pp. 1–15.

[2] Acclavio, M., *Exponentially handsome proof nets and their normalization*, Electronic Proceedings in Theoretical Computer Science **353** (2021), pp. 1–25.
URL https://doi.org/10.4204%2Feptcs.353.1

[3] Acclavio, M., D. Catta and L. Straßburger, *Game semantics for constructive modal logic*, in: *International Conference on Automated Reasoning with Analytic Tableaux and Related Methods*, Springer, 2021, pp. 428–445.

[4] Acclavio, M. and L. Straßburger, *From syntactic proofs to combinatorial proofs*, in: *International Joint Conference on Automated Reasoning*, Springer, 2018, pp. 481–497.

[5] Acclavio, M. and L. Straßburger, *On combinatorial proofs for logics of relevance and entailment*, in: *International Workshop on Logic, Language, Information, and Computation*, Springer, 2019, pp. 1–16.

[6] Acclavio, M. and L. Straßburger, *On combinatorial proofs for modal logic*, in: S. Cerrito and A. Popescu, editors, *Automated Reasoning with Analytic Tableaux and Related Methods* (2019), pp. 223–240.

[7] Bellin, G., V. De Paiva and E. Ritter, *Extended Curry-Howard correspondence for a basic constructive modal logic*, in: *In Proceedings of Methods for Modalities*, 2001.

[8] Benjamin, R. and L. Straßburger, *Towards a combinatorial proof theory*, in: *Tableaux 2019*, Springer, 2019.

[9] Bierman, G. M. and V. C. de Paiva, *On an intuitionistic modal logic*, Studia Logica **65** (2000), pp. 383–416.

[10] Cook, S. A. and R. A. Reckhow, *The relative efficiency of propositional proof systems*, J. of Symb. Logic **44** (1979), pp. 36–50.

[11] Danos, V. and L. Regnier, *Proof-nets and the Hilbert space*, in: *Proceedings of the Workshop on Advances in Linear Logic* (1995), p. 307–328.

[12] Davies, R. and F. Pfenning, *A modal analysis of staged computation*, Journal of the ACM **48** (2001), pp. 555–604.

[13] Fairtlough, M. and M. Mendler, *Propositional lax logic*, Information and Computation **137** (1997), pp. 1–33.

[14] Fitch, F. B., *Intuitionistic modal logic with quantifiers*, Portugaliae mathematica **7** (1948), pp. 113–118.

[15] Guglielmi, A., T. Gundersen and M. Parigot, *A Proof Calculus Which Reduces Syntactic Bureaucracy*, in: C. Lynch, editor, *Proceedings of the 21st International Conference on Rewriting Techniques and Applications*, Leibniz International Proceedings in Informatics (LIPIcs) **6** (2010), pp. 135–150.
URL http://drops.dagstuhl.de/opus/volltexte/2010/2649

[16] Heijltjes, W. and R. Houston, *No proof nets for MLL with units: proof equivalence in MLL is PSPACE-complete*, in: T. A. Henzinger and D. Miller, editors, *Joint Meeting of the Twenty-Third EACSL Annual Conference on Computer Science Logic (CSL) and the Twenty-Ninth Annual ACM/IEEE Symposium on Logic in Computer Science (LICS), CSL-LICS '14, Vienna, Austria, July 14 - 18, 2014* (2014), pp. 50:1–50:10.

[17] Heijltjes, W., D. Hughes and L. Straßburger, *Intuitionistic proofs without syntax*, in: *LICS 2019 - 34th Annual ACM/IEEE Symposium on Logic in Computer Science* (2019), pp. 1–13.
URL https://hal.inria.fr/hal-02386878

[18] Heilala, S. and B. Pientka, *Bidirectional decision procedures for the intuitionistic propositional modal logic IS4*, in: *International Conference on Automated Deduction*, Springer, 2007, pp. 116–131.

[19] Hughes, D., *Proofs without syntax*, Annals of Math. **164** (2006), pp. 1065–1076.

[20] Hughes, D. J. D., *First-order proofs without syntax* (2019).

[21] Hughes, D. J. D., L. Straßburger and J. Wu, *Combinatorial proofs and decomposition theorems for first-order logic*, in: *36th Annual ACM/IEEE Symposium on Logic in*

Computer Science, LICS 2021, Rome, Italy, June 29 - July 2, 2021 (2021), pp. 1–13.
URL `https://doi.org/10.1109/LICS52264.2021.9470579`

[22] Hyland, J. and C.-H. Ong, *On full abstraction for PCF: I, II, and III*, Information and Computation **163** (2000), pp. 285 – 408.
URL `http://www.sciencedirect.com/science/article/pii/S0890540100929171`

[23] Kojima, K., "Semantical study of intuitionistic modal logics," Ph.D. thesis, Kyoto University (2012).

[24] Kuznets, R., S. Marin and L. Straßburger, *Justification logic for constructive modal logic*, Journal of Applied Logics: IfCoLog Journal of Logics and their Applications **8** (2021), pp. 2313–2332.
URL `https://hal.inria.fr/hal-01614707`

[25] Lamarche, F., *Proof nets for intuitionistic linear logic: Essential nets* (2008).
URL `https://hal.inria.fr/inria-00347336`

[26] Lamarche, F. and C. Retoré, *Proof nets for the Lambek-calculus — an overview*, in: V. M. Abrusci and C. Casadio, editors, *Proceedings of the Third Roma Workshop "Proofs and Linguistic Categories"* (1996), pp. 241–262.

[27] McCusker, G., *Games and full abstraction for FPC*, Information and Computation **160** (2000), pp. 1 – 61.
URL `http://www.sciencedirect.com/science/article/pii/S0890540199928456`

[28] Mendler, M. and S. Scheele, *Cut-free Gentzen calculus for multimodal CK*, Information and Computation **209** (2011), pp. 1465–1490.

[29] Plotkin, G. and C. Stirling, *A framework for intuitionistic modal logics*, in: *Proceedings of the 1st Conference on Theoretical Aspects of Reasoning about Knowledge (TARK)*, 1986, pp. 399–406.

[30] Prawitz, D., "Natural deduction: A proof-theoretical study," Courier Dover Publications, 2006.

[31] Simpson, A. K., "The proof theory and semantics of intuitionistic modal logic," Ph.D. thesis, University of Edinburgh. College of Science and Engineering (1994).

[32] Straßburger, L., *Cut elimination in nested sequents for intuitionistic modal logics*, in: F. Pfenning, editor, *FoSSaCS'13*, LNCS **7794** (2013), pp. 209–224.

[33] Troelstra, A. S. and H. Schwichtenberg, "Basic proof theory," Cambridge University Press, 2000.

Saturation-Based Uniform Interpolation for Multi-Modal Logics

Ruba Alassaf, Renate A. Schmidt, and Uli Sattler

University of Manchester
Oxford Road, Manchester
United Kingdom

Abstract

Uniform interpolation has been the subject of many recent research papers due to its link to Craig interpolation and its potential use in knowledge-based and agent-based systems. In this paper, we present a saturation-based system that computes a *local* uniform interpolant for a formula and a "keep" signature in the multi-modal logic K_n. The system works by exhaustively applying a set of rules to generate a sufficient number of local consequences, which are then filtered to remove those that contain symbols outside the keep signature. We show that the system is guaranteed to terminate and is sound and uniform interpolation complete. We further prove that we can extend the system to compute uniform interpolants for formulas in multi-modal logics of serial and reflexive frames D_n and T_n.

Keywords: Uniform Interpolation, Resolution, Bisimulations

1 Introduction

A formula ϕ' is said to be a *uniform* Σ*-interpolant* of ϕ, if for any ψ over the signature Σ, ψ is implied by ϕ' if and only if ψ is implied by ϕ. Uniform interpolation is closely related to the notion of Craig interpolation. In Craig interpolation, two formulae ϕ and ψ such that ϕ implies ψ are given, and the task is to compute an intermediate formula ϕ' such that ϕ implies ϕ' and ϕ' implies ψ. A logic is said to have the uniform interpolation (respectively, Craig interpolation) property if, for any formula and signature (respectively, two formulas), a uniform interpolant (respectively, Craig interpolant) can be computed.

The notion of uniform interpolation is stronger than Craig interpolation for two reasons. The first is that any logic that has the uniform interpolation property also has the Craig interpolation property. The second is that uniform interpolation can be used to compute Craig interpolants. This can be achieved by "keeping" consequences over the shared signature during interpolation.

Regarding applications, uniform interpolation and Craig interpolation have been investigated for a range of modal logics, and the closely related description logics, which underlie ontology languages such as OWL [2]. Ontology engineers benefit from uniform interpolation in ontology debugging, versioning and summarisation [17,13]. In agent-based applications, uniform interpolation is used to update the knowledge of an agent by making them ignorant of certain propositional formulas [1]. Knowledge-sharing applications may use uniform interpolation to facilitate knowledge exchange among agents with different domain specialisations [20].

Several variants of the uniform interpolation problem have been studied. For classical logic, the problem is reduced to the second-order quantifier elimination problem [7,8], which aims to eliminate given predicate symbols from the formula. Uniform interpolation has been investigated in description logic sometimes under the name of deductive forgetting. Uniform interpolation and forgetting are dual notions as the former aims to compute a formula by keeping a signature Σ, while the aim of latter is to compute a formula that eliminates the complement of Σ. Most studies have focused on TBox forgetting [17,13,14,16] sometimes with an ABox [15], with [19] considering on concept forgetting.

The difference between TBox forgetting and concept forgetting (or forgetting for a local modal K_n formula) is that a TBox is a set of axioms that are universally quantified, whereas a concept or a modal K_n-formula is local, i.e., a modal K_n-formula is instantiated to a particular set of individuals/worlds. Uniform interpolation of a TBox is thus a global interpolation problem whereas uniform interpolation of a concept or a K_n-formula is a local interpolation problem. There are also differences in the complexity of the two problems. It was shown that TBox forgetting is triple exponential in the size of the input [17], whereas concept forgetting is in ExpSpace [19].

We are interested in the local form of uniform interpolation for modal logics. The modal logic K was shown to have the uniform interpolation property via constructive proofs [9,21]. An implementable approach to constructing uniform interpolants was given in [3] for the modal logics K and T. Wolter [22] proved that the modal logic $S5$ has the uniform interpolation property, and that uniform interpolation for any normal mono-modal logic can be generalised to its multi-modal case. Recently, it was shown that $K45_n$ and $KD45_n$ have the uniform interpolation property [5]. It is known that $S4$ and $K4$ do not have the uniform interpolation property [10,3].

The aim of our research is to develop resolution-based systems for computing local uniform interpolants in modal logics that are suitable for implementation. In this paper, we present such a system for the multi-modal logics K_n, D_n, and T_n. We prove that the system is guaranteed to terminate, is sound and uniform interpolation complete. We established the complexity of the system, and discuss the relationship to other works. The main contributions of this paper are:

- The system is the first provably correct resolution-based system for computing uniform interpolants in the modal logics K_n, D_n and T_n. Our work extends and improves the work presented in [11] for modal logic K.
- The idea of our completeness proof is novel for resolution-based uniform interpolation systems. Completeness proofs for previous systems which

use resolution are based on proof-theoretic arguments [11,14,15,16]. Our completeness proof is based on a model-theoretic argument in which bisimulation takes a central role.

2 Getting Started

We assume the reader is familiar with the multi-modal logic K_n [4,12]. We fix $\mathcal{A}$ a set of strings, and $P = \{p, q, r, \dots\}$ a countable possibly infinite set of propositional symbols. A K_n formula ϕ over a given signature is defined inductively as follows: $\phi ::= p \mid \top \mid \bot \mid \neg\phi \mid \phi \vee \phi \mid \phi \wedge \phi \mid \Box_a \phi \mid \Diamond_a \phi$ where $a \in \mathcal{A}$.

We use $\mathcal{F} = (\mathcal{W}, R)$ to denote a Kripke frame where $\mathcal{W}$ is a nonempty set of worlds and R a mapping from elements of $\mathcal{A}$ to binary relations over $\mathcal{W}$. To abbreviate notation, we use R_a instead of $R(a)$, and we say that u is *a-accessible* from w if $R_a(w, u)$. We use $\mathcal{M} = (\mathcal{W}, R, V)$ to denote a Kripke model where $(\mathcal{W}, R)$ is a Kripke frame, and V is a valuation function that assigns each propositional symbol p in P a subset $V(p)$ of $\mathcal{W}$.

A formula ϕ is (locally) *satisfiable in a model* $\mathcal{M}$, if there is a world w in $\mathcal{W}$ at which ϕ is true. We use $\mathcal{M}, w \models \phi$ to denote that ϕ is true at w in $\mathcal{M}$. A formula ϕ is (unconditionally) *satisfiable* if it is true at some world in some model. A formula ϕ is globally satisfied (or true) in a model $\mathcal{M}$, denoted $\mathcal{M} \models \phi$, if it is true at every w in $\mathcal{M}$. A formula ϕ is valid, denoted $\models \phi$, if it is satisfied in all models over any frame $\mathcal{F}$. A set of formulae N is globally satisfied by a model $\mathcal{M}$, denoted $\mathcal{M} \models N$, if for each formula ϕ in N, $\mathcal{M}$ globally satisfies ϕ.

In this paper, we consider two logics which extend K_n, namely D_n and T_n, the multi-modal logics of serial and reflexive frames, respectively. In a D_n model, each accessibility relation R_a is serial, i.e. for each $w \in \mathcal{W}$, and each $a \in \mathcal{A}$, there exists a $w' \in \mathcal{W}$ such that $R_a(w, w')$ is true in the model. In a T_n model, each accessibility relation R_a is reflexive, i.e. for each $w \in \mathcal{W}$, and each $a \in \mathcal{A}$, $R_a(w, w)$ is true in the model. The modal logic D_n is axiomatized by the axioms of K_n and the axiom schema $(D) = \Box_a \phi \to \Diamond_a \phi$. Similarly, T_n is axiomatized by adding the axiom schema $(T) = \Box_a \phi \to \phi$.

We are interested in the problem of computing local uniform interpolant of a formula and a signature.

Definition 2.1 Given a formula ϕ, a *uniform interpolant* of ϕ with respect to a signature Σ of propositional symbols is a formula ϕ' such that:

(i) ϕ' does not contain symbols outside of Σ, and

(ii) for any modal formula ψ over Σ, we have that for all models $\mathcal{M}$, $\mathcal{M} \models \phi \to \psi$ iff for all models $\mathcal{M}$, $\mathcal{M} \models \phi' \to \psi$.

3 Uniform Interpolation Method

We start with a high-level description of our uniform interpolation method for multi-modal logic K_n. Without loss of generality, we assume that the input formula ϕ is given in negation normal form.

Overview. The system is based on resolution with modal logic adaptations. The idea behind our approach is the following: for each propositional symbol x outside the given signature Σ, we generate a sufficient set of clauses for the given formula and subsequently eliminate any formulae that contain x. We repeat the process for all propositional symbols outside Σ.

The system uses special *world symbols*, or *W-symbols* for short, which are propositional symbols that help in two related ways:

(i) They are used to flatten the input formula to surface some parts of it. E.g., $\Box(\psi \vee \Diamond\phi)$ becomes $\Box W_1$, $W_1 \Rightarrow \psi \vee \Diamond W_2$ and $W_2 \Rightarrow \phi$.

(ii) They allow the inferences to be restricted to subformulae labelled with the same W-symbol. E.g., $\Box(x \wedge (\neg x \vee p))$ becomes $\Box W$, $W \Rightarrow x$ and $W \Rightarrow \neg x \vee p$. Later on, we see that one of our rules allows us to apply a resolution step on x.

Initially, we can think of W-symbols as constants representing worlds in a labelled tableau algorithm.

For a formula ϕ, a signature Σ, and an ordering $\succ$ over the symbols outside the input signature Σ, the system is provided a clause set $N_0 = \{W_0 \Rightarrow \phi\}$ as input, and applies its rules exhaustively to the formulae in the set until no rules can be applied, resulting in a clause set of the form $N_n = \{W_0 \Rightarrow \phi_1, ..., W_0 \Rightarrow \phi_m\}$. The formula $\phi' = \phi_1 \wedge \ldots \wedge \phi_m$ is then a uniform Σ-interpolant of ϕ, which is proved later.

The role of W_0 is to represent a specific world that satisfies the given formula ϕ. Any model $\mathcal{M}$ that satisfies ϕ at point w can be extended in a non-vacuous way to one that satisfies W_0 and $W_0 \Rightarrow \phi$ by setting $w \in V(W_0)$. In this extended model, $W_0 \Rightarrow \phi$ is globally witnessed as non-vacuously true.

The process of constructing a uniform interpolant is iterative with respect to the symbols outside Σ, and the ordering $\succ$ fixes the order in which these symbols are eliminated. For some uniform interpolation problems, a good ordering may allow the system to solve a problem in far fewer steps. For simplicity, and since the ordering does not improve any worst-case complexity results, we can assume this ordering is arbitrary. We use x to denote the maximal propositional symbol occurring in the current clause set N_i at the ith step.

The System. The rules of our uniform interpolant system are given in Figures 1, 2 and 3. Each rule has a premise, some conditions and a conclusion. The rules are structured with the premise above a horizontal line and the conclusion below it. The premise (respectively conclusion) can be one or more clauses depending on which rule is being applied. There are three types of rules in the system: preprocessing rules, resolution rules, and elimination rules.

The preprocessing rules and the elimination rules are replacement rules; they replace the premise in the current working clause set with the conclusion. The resolution rules are saturation rules; they keep the premise and extend the clause set with the conclusion. The rules can be applied in any order as long as the conditions for each rule are met.

The clauses obtained and handled by our system are in a normal form. They are all labelled with a W-symbol in the condition of the implication. We can have a formula or another W-symbol in the consequence of the implication. Concretely, for some W-symbols W_i and W_j, and some modal formula ψ, a clause can have the form $W_i \Rightarrow \psi$ or $W_i \Rightarrow W_j$. If ψ is a disjunction of modal formulae, we assume that it is a set, i.e., there is no repetition among the disjuncts. This is essential for the correctness of the method. We use the notation $\Rightarrow$, in contrast to $\rightarrow$, to highlight that an implication is generated by our system to maintain our normal form. Semantically, they are identical.

To describe the different types of W-symbols, we introduce some terminology and the functions *Def* and *Corr* which are used in the conditions of our system, and later on in the proofs.

Definition 3.1 Given a set N of clauses, the set S_w is the set of W-symbols introduced for subformulas appearing under a modal operator via the WI rule. We call these symbols *base W-symbols.*

We use an injective function Def that maps base W-symbols to subformulas of clauses in N, and that is extended each time we introduce a new W-symbol.

The set C_w is the set of W-symbols introduced by the RES $\Box\bigcirc$ rule. We call these symbols *combinatory W-symbols.*

We define a function $Corr$ that maps W-symbols to subsets of S_w as follows:

$$Corr(W_i) = \begin{cases} \{W_i\}, & \text{if } W_i \in S_w \\ Corr(W_n) \cup Corr(W_m), & \text{if } W_i \in C_w \text{ where } W_n \text{ and } W_m \text{ come} \\ & \text{from the premise of the RES } \Box\bigcirc \\ & \text{rule that has introduced } W_i. \end{cases}$$

It is easy to see that the definition of *Corr* is well-founded. Intuitively, a base W-symbol is introduced to represent a subformula, and a combinatory W-symbol can be seen as a unique representative of a subset of the base W-symbols.

We now describe the three groups of rules which together make up our system. We use N to refer to the current working clause set. We assume that x is the current symbol we would like to eliminate, i.e., it is the maximal symbol in the current set with respect to the given ordering $\succ$. The W-symbol W_i is the ith W-symbol introduced during the inference process.

Preprocessing. The purpose of the preprocessing rules is to apply transformations to the members of the working clause set so that they can be handled by the other rules. Generally, the idea is to surface symbols appearing in ϕ that are not in Σ, i.e., to surface x in ϕ. The rules are applied in a lazy manner which means their application can be deferred to whenever they are necessary. The preprocessing rules are provided in Figure 1. The clausification rule distributes disjunction over conjunction. The world introduction rule performs structural transformation that flattens modal formulae.

Resolution. The purpose of the resolution rules is to deduce a sufficient number of clauses/formulas to generate a uniform interpolant. The rules are given in Figure 2.

Clausification:

$$\frac{N, W_i \Rightarrow (\phi_1 \wedge \phi_2) \vee \phi_3}{N, W_i \Rightarrow \phi_1 \vee \phi_3, W_i \Rightarrow \phi_2 \vee \phi_3}$$

provided that either ϕ_1 or ϕ_2 contain x. ϕ_3 may be empty.

World Introduction (WI):

$$\frac{N, W_i \Rightarrow \bigcirc_a \phi_1 \vee \phi_2}{N, W_i \Rightarrow \bigcirc_a W_j \vee \phi_2, W_j \Rightarrow \phi_1}$$

provided that
(i) $\bigcirc \in \{\Box, \Diamond\}$,
(ii) ϕ_1 must contain x,
(iii) if ϕ_2 contains x then x must occur under a modal operator, and
(iv) if there is a W_k such that $Def(W_k) = \phi$, then $W_j = W_k$, otherwise W_j is a fresh W-symbol, $Corr(W_j) = \{W_j\}$ and $Def(W_j) = \{\phi\}$.

ϕ_2 may be empty.

Fig. 1. The preprocessing rules of the UI_{K_n} system for the modal logic K_n.

Literal Resolution (RES):

$$\frac{W_i \Rightarrow \psi_1 \vee x \quad W_i \Rightarrow \psi_2 \vee \neg x}{W_i \Rightarrow \psi_1 \vee \psi_2}$$

ψ_1 and/or ψ_2 may be empty.

World Resolution (RES W):

$$\frac{W_i \Rightarrow \psi \quad W_j \Rightarrow W_i}{W_j \Rightarrow \psi}$$

provided that ψ contains x.

$\Box\bigcirc$ **Resolution** (RES $\Box\bigcirc$):

$$\frac{W_i \Rightarrow \psi_1 \vee \Box_a W_n \quad W_i \Rightarrow \psi_2 \vee \bigcirc_a W_m}{W_i \Rightarrow \psi_1 \vee \psi_2 \vee \bigcirc_a W_j,\ W_j \Rightarrow W_n,\ W_j \Rightarrow W_m}$$

provided that:
(i) $\bigcirc \in \{\Box, \Diamond\}$,
(ii) $Corr(W_n) \cap Corr(W_m)$ is empty,
(iii) if there is a W_k such that $Corr(W_k) = Corr(W_n) \cup Corr(W_m)$ then $W_j = W_k$, otherwise W_j is a fresh W-symbol, and $Corr(W_j) = Corr(W_n) \cup Corr(W_m)$.

ψ_1 and/or ψ_2 may be empty.

Fig. 2. The resolution rules of the UI_{K_n} system for modal logic K_n.

The literal resolution rule is the heart of our system; it computes a formula by resolving on a maximal symbol x if the premise is labelled with the same W-symbol. The world resolution rule is used to propagate formulas labelled by another W-symbol, which is essentially a resolution step between world symbols. The $\Box\bigcirc$ resolution rule is used to capture combinations of successor relations. The second and third conditions are the blocking conditions; they aim to ensure that the rule application is not redundant which is important

Positive Purification ($+$PUR):

$$\frac{N, W_i \Rightarrow \psi \vee x}{N, W_i \Rightarrow \psi \vee \top}$$

provided that no more non-purification rules can be applied to the named clause in the premise. ψ may be empty.

Negative Purification ($-$PUR):

$$\frac{N, W_i \Rightarrow \psi \vee \neg x}{N, W_i \Rightarrow \psi \vee \top}$$

provided that no more non-purification rules can be applied to the named clause in the premise. ψ may be empty.

World Elimination (ELM W):

$$\frac{N, W_i \Rightarrow \psi_1, \ldots, W_i \Rightarrow \psi_n}{N^{W_i}_{(\psi_1 \wedge \cdots \wedge \psi_n)}}$$

provided that $i \neq 0$, $\psi_1, \ldots, \psi_n$ do not contain x or any W-symbol, and N only contains W_i on the right hand side of $\Rightarrow$ clauses. The expression N^{ϕ}_{ψ} denotes the set of clauses that is obtained by replacing each occurrence of ϕ in N by ψ.

Fig. 3. The purification and elimination rules of the UI_{K_n} system for modal logic K_n.

to control the complexity, and that the system does not infinitely introduce W-symbols which is essential for termination.

Elimination. The elimination rules are responsible for eliminating symbols outside of $\Sigma \cup \{W_0\}$. They are applied once we have exhaustively applied the resolution rules to compute conclusions over Σ. The rules are given in Figure 3.

The positive and negative purification rules replace a maximal symbol x, occurring either positively or negatively, with $\top$. The world elimination rule collects modal formulas labelled with the same W-symbol, and replaces right hand side occurrences of the W-symbol with the conjunction of these formulas, effectively eliminating the W-symbol from the set of clauses.

3.1 Examples

In the following examples, we demonstrate how the UI_{K_n} system is used to compute a uniform interpolant with respect to $\Sigma = \{p, q\}$. Starting from $i = 0$, we use N_i to refer to the clause set that is obtained after applying the ith step in the derivation.

Example 3.2 Consider a formula $\phi = (\neg p \vee \Diamond x) \wedge (\neg x \vee \Box q)$.

The input to the system is the set $N_0 = \{W_0 \Rightarrow (\neg p \vee \Diamond x) \wedge (\neg x \vee \Box q)\}$. The only rule applicable to N_0 is the clausification rule which gives

$$N_1 = \{W_0 \Rightarrow \neg p \vee \Diamond x, W_0 \Rightarrow \neg x \vee \Box q\}.$$

Now we apply the world introduction rule to get

$$N_2 = \{W_0 \Rightarrow \neg p \vee \Diamond W_1, W_1 \Rightarrow x, W_0 \Rightarrow \neg x \vee \Box q\}.$$

The only applicable rules are the positive and negative purification rules. We achieve

$$N_3 = \{W_0 \Rightarrow \neg p \vee \Diamond W_1, W_1 \Rightarrow \top, W_0 \Rightarrow \top \vee \Box q\}.$$

Eliminating W_1, we obtain

$$N_4 = \{W_0 \Rightarrow \neg p \vee \Diamond \top, W_0 \Rightarrow \top \vee \Box q\}.$$

The Σ-uniform interpolant is $\phi' = (\neg p \vee \Diamond \top) \wedge (\top \vee \Box q)$.

Notice that this example illustrates the local flavour of the system. We see that the occurrences of x at two different modal levels do not interact via any resolution rule.

Example 3.3 Consider a formula $\phi = (\neg p \vee \Diamond x) \wedge \Box(\neg x \vee \Box q)$. We start with the set $N_0 = \{W_0 \Rightarrow (\neg p \vee \Diamond x) \wedge \Box(\neg x \vee \Box q)\}$. Applying the clausification rule to N_0 we get

$$N_1 = \{W_0 \Rightarrow \neg p \vee \Diamond x, W_0 \Rightarrow \Box(\neg x \vee \Box q)\}.$$

By applying the world introduction rule twice, we have

$$N_3 = \{W_0 \Rightarrow \neg p \vee \Diamond W_1, W_1 \Rightarrow x, W_0 \Rightarrow \Box W_2, W_2 \Rightarrow \neg x \vee \Box q\}.$$

The only applicable rule is the $\Box\Diamond$ rule, and it yields

$$N_4 = N_3 \cup \{W_0 \Rightarrow \neg p \vee \Diamond W_3, W_3 \Rightarrow W_1, W_3 \Rightarrow W_2\}.$$

By applying the world resolution rule twice, we obtain

$$N_6 = N_4 \cup \{W_3 \Rightarrow x, W_3 \Rightarrow \neg x \vee \Box q\}.$$

Now, we can apply the literal resolution rule which yields

$$N_7 = N_6 \cup \{W_3 \Rightarrow \Box q\}.$$

We apply the positive and negative purification rules (4 applications) and achieve

$$\begin{aligned} N_{11} = \{ & W_0 \Rightarrow \neg p \vee \Diamond W_1, && W_1 \Rightarrow \top, && W_0 \Rightarrow \Box W_2, \\ & W_2 \Rightarrow \top \vee \Box q, && W_0 \Rightarrow \neg p \vee \Diamond W_3, && W_3 \Rightarrow W_1, \\ & W_3 \Rightarrow W_2, && W_3 \Rightarrow \top, && W_3 \Rightarrow \top \vee \Box q, \\ & W_3 \Rightarrow \Box q \}. \end{aligned}$$

Now, x does not appear anywhere. We eliminate the world variables W_1, W_2, W_3 via the world elimination rule.

To eliminate W_1, we look for clauses labelled with W_1, in this case we only have $W_1 \Rightarrow \top$. We remove $W_1 \Rightarrow \top$ and replace each occurrence of W_1 on the right hand side of $\Rightarrow$ with $\top$ as follows:

$$\begin{aligned} N_{12} = \{ & W_0 \Rightarrow \neg p \vee \Diamond \top, && W_0 \Rightarrow \Box W_2, && W_2 \Rightarrow \top \vee \Box q, \\ & W_0 \Rightarrow \neg p \vee \Diamond W_3, && W_3 \Rightarrow \top, && W_3 \Rightarrow W_2, \\ & W_3 \Rightarrow \top \vee \Box q, && W_3 \Rightarrow \Box q \}. \end{aligned}$$

Similarly for W_2, we remove $W_2 \Rightarrow \top \vee \Box q$, and replace the other occurrences of W_2 with $\top \vee \Box q$.

$$\begin{aligned} N_{13} = \{ & W_0 \Rightarrow \neg p \vee \Diamond \top, && W_0 \Rightarrow \Box(\top \vee \Box q), && W_0 \Rightarrow \neg p \vee \Diamond W_3, \\ & W_3 \Rightarrow \top, && W_3 \Rightarrow \top \vee \Box q, && W_3 \Rightarrow \Box q \}. \end{aligned}$$

Finally, we eliminate W_3,

$$\begin{aligned} N_{14} = \{ W_0 \Rightarrow \neg p \vee \Diamond \top, \quad & W_0 \Rightarrow \Box(\top \vee \Box q), \\ & W_0 \Rightarrow \neg p \vee \Diamond(\top \wedge (\top \vee \Box q) \wedge \Box q) \}. \end{aligned}$$

The uniform interpolant is

$$\phi' = (\neg p \vee \Diamond \top) \wedge (\Box(\top \vee \Box q)) \wedge (\neg p \vee \Diamond(\top \wedge (\top \vee \Box q) \wedge \Box q)),$$

which is equivalent to $\phi' = (\neg p \vee \Diamond \Box q)$ by standard simplifications.

4 Correctness and Complexity

The first lemmas in this section are relevant to termination. We prove termination by showing that any derivation uses a finite number of symbols, and we argue that because of this, the system will stop generating new clauses.

Lemma 4.1 *The number of W-symbols introduced in a run of the UI_{K_n} system is bounded by $2^n - 1$ for n being the length of the input.*

Proof. Let S_w be the set of base W-symbols. S_w is finite because the number of modal operators in the input formula is finite, and the role of the world introduction rule is to replace each subformula (not containing a W-symbol) appearing under a modal operator with a W-symbol.

Let C_w be the set of combinatory W-symbols introduced by the RES $\Box\bigcirc$ rule. We show inductively that for each W-symbol W_i in C_w, the set $Corr(W_i)$ corresponds to a unique set of W-symbols from S_w.

Since the function $Corr$ has a finite range, which is the powerset of S_w, we can prove that the domain is finite by showing that $Corr$ is injective. Let W_n and W_m be two W-symbols. We show inductively that if $Corr(W_n) = Corr(W_m)$, then $W_n = W_m$. The proof is given in the appendix.

Since S_w is a finite set, and by condition (*ii*) of the $\Box\bigcirc$ resolution rule, the number of possible W-symbols is bounded by the number of unique combinations of symbols from S_w. The upper bound is equal to $2^{|S_w|}$-1, where $|S_w|$ is the size of S_w. □

Lemma 4.2 *The UI_{K_n} system will stop generating new clauses.*

Proof. The input formula ϕ contains a finite number of propositional symbols and modal operators. By Lemma 4.1, the system introduces a finite number of W-symbols. The resolution rules do not produce results with increased modal depth. □

Lemma 4.3 *The UI_{K_n} system will not reintroduce a W-symbol that was eliminated before.*

Proof. This is because a W-symbol is introduced to surface a formula that has symbols not in Σ, and only when a W-symbol no longer labels non-Σ symbols, the world elimination rule is allowed to be applied. □

From Lemmas 4.1, 4.2 and 4.3, we conclude the following theorem.

Theorem 4.4 (Termination) *Given a formula ϕ and a signature Σ, the UI_{K_n} system computes a formula ϕ' in a finite number of steps.*

The following lemma considers the space complexity of our system.

Lemma 4.5 *The space complexity of the UI_{K_n} calculus is double exponentially bounded in the length of the input.*

Proof. The root source of the complexity is that the system generates an exponential number of combinatory symbols. The complexity argument is built upon three claims. Using a linear number of propositional symbols, a linear number of base W-symbols, and an exponential number of combinatory W-symbols:

- C1 the size of each clause, before eliminating any W-symbols, has an exponential upper bound in the size of the input,
- C2 the number of clauses we could generate has a double exponential upper bound in the size of the input, and
- C3 the size of each clause, after eliminating W-symbols, has a double exponential upper bound in the size of the input.

We show the three claims in the appendix. □

The next lemmas show that the signature of ϕ' is Σ.

Lemma 4.6 *For a given formula ϕ and a signature Σ, the UI_{K_n} system will always be able to eliminate every W-symbol that is not W_0, using the world elimination rule.*

Proof. The normal form that is used in our system dictates that only one W-symbol can be present on the left hand side of $\Rightarrow$. Eliminating a W-symbol will replace occurrences of a W in the working clause set with a combination of clauses that do not contain W-symbols. The only situation that may prevent the elimination rule from being applied is if a clause contains the same W-symbol on both sides of the $\Rightarrow$, but it can be shown inductively that this cannot happen. □

Lemma 4.7 *For a given formula ϕ and a signature Σ, the UI_{K_n} system will always be able to eliminate symbols in the signature of ϕ that are not in Σ.*

Proof. Let x be the maximal symbol in the signature of ϕ but outside Σ. The world introduction rule is a replacement rule that aims to surface x. When no resolution rule is applicable, the calculus applies a purification rule that eliminates x. □

Theorem 4.8 (Soundness) *Given a formula ϕ and a signature Σ, the UI_{K_n} system computes a formula ϕ' such that for any formula ψ over Σ, we have that if $\models \phi' \rightarrow \psi$ then $\models \phi \rightarrow \psi$.*

Proof. Consider ϕ, ϕ' as in Theorem 4.8. We use $\mathcal{M}|_\Sigma$ for the reduction of $\mathcal{M}$ to Σ, i.e. the model obtained from $\mathcal{M}$ by ignoring all symbols outside Σ.

We say that two models $\mathcal{M}', \mathcal{M}$ are Σ-*inseparable* if $\mathcal{M}'|_\Sigma = \mathcal{M}|_\Sigma$. To show the theorem, we use the following claims:

C4 For each model $\mathcal{M}$ and world w_0 such that $\mathcal{M}, w_0 \models \phi$, there exists a model $\mathcal{M}'$ such that $\mathcal{M}$ and $\mathcal{M}'$ are Σ-inseparable, $\mathcal{M}' \models W_0 \Rightarrow \phi$, and $V(W_0) = \{w_0\}$.

C5 Each rule in our calculus is Σ-preserving, that is, if a rule is applied to a set of clauses N to produce N' and $\mathcal{M}$ is a model that globally satisfies N, then there exists a model $\mathcal{M}'$ that is Σ-inseparable from $\mathcal{M}$ that satisfies N'.

C6 Let $N = \{W_0 \Rightarrow \psi_i \mid 1 \leq i \leq n\}$ be the set obtained at the end of our derivation. For each model $\mathcal{M}$ that globally satisfies N and for each $w_0 \in V(W_0)$, we have $\mathcal{M}, w_0 \models \phi'$ with $\phi' = \bigwedge_{i \leq |N|} \psi_i$.

The claims are shown in the appendix.

Via C4, C5, and C6, we can show that for any model of the input formula, there exists a Σ-inseparable model for the output formula. And, thus, for any formula ψ over Σ, if $\phi \to \psi$ is true in all models, then $\phi' \to \psi$ is true in all models. □

For our completeness proof, we are interested in understanding models that are invariant up to the satisfaction of Σ-modal formulas. Σ-modal formulas are modal formulas described using a signature of propositional symbols Σ. For this purpose, we use the following notion.

Definition 4.9 Let $(\mathcal{M}, w)$ and $(\mathcal{M}', w')$ be two Kripke models where $\mathcal{M} = (\mathcal{W}, R, V)$ and $\mathcal{M}' = (\mathcal{W}', R', V')$. A Σ-*bisimulation* between $\mathcal{M}$ and $\mathcal{M}'$ is a relation $\rho \subseteq \mathcal{W} \times \mathcal{W}'$ such that $w \rho w'$, and whenever $u \rho u'$, the following holds:
atoms u and u' satisfy the same propositional symbols from Σ;
forth For all a, if uR_at, then there is a t' such that $u'R'_at'$ and $t\rho t'$;
back For all a, if $u'R'_at'$, then there is a t such that uR_at and $t\rho t'$.

The following is our completeness theorem.

Theorem 4.10 (Completeness) *Given a K_n formula ϕ and a signature Σ, the UI_{K_n} system computes a K_n formula ϕ' such that, for any K_n formula ψ over Σ, we have that if $\models \phi \to \psi$ then $\models \phi' \to \psi$.*

Proof. Consider ϕ, ϕ' and ψ as in Theorem 4.10. Assume $\models \phi \to \psi$ but $\not\models \phi' \to \psi$. The assumption implies that there exists a counter model $\mathcal{M}'$ and a world w_0 such that, $\mathcal{M}', w_0 \models \phi'$ and $\mathcal{M}', w_0 \not\models \psi$.

The idea of our proof is to inductively show that a Σ-bisimilar model $\mathcal{M}$ can be defined based on $\mathcal{M}'$ such that, $\mathcal{M}, w_0 \models \phi$, and $\mathcal{M}, w_0 \not\models \psi$. This will contradict our assumption $\models \phi \to \psi$.

Let n be the number of steps used for generating ϕ' from ϕ via our calculus. Let N_k denote the set of formulae obtained after applying k steps to ϕ. For each i, starting from $\mathcal{M}_n = \mathcal{M}'$, we will construct a model $\mathcal{M}_i = (\mathcal{W}_i, R_i, V_i)$ for N_i.

We construct $\mathcal{M}_i$ by inductively extending $\mathcal{M}_{i+1}$ and ensuring that $\mathcal{M}_i$ and $\mathcal{M}_{i+1}$ are Σ-bisimilar. In this view, $N_1 = \{W_0 \Rightarrow \phi\}$, and $N_n = \{W_0 \Rightarrow \phi'_1, \dots, W_0 \Rightarrow \phi'_m\}$ where $\phi' = \phi'_1 \wedge \dots \wedge \phi'_m$. This will lead to a counter-model

$\mathcal{M}_0 = \mathcal{M}$ as described above.

Before we state our claims, we introduce a terminology: given a set N of clauses and W-symbols W_i and W_j, W_i is said to be *directly related* to W_j if $W_i \Rightarrow W_j \in N$ or $W_j \Rightarrow W_i \in N$. Two W-symbols W_i and W_j are *indirectly related* if there is a W-symbol W_k such that $W_k \Rightarrow \gamma_1 \vee \bigcirc_a W_i$ and $W_k \Rightarrow \gamma_2 \vee \Box_a W_j$ where γ_1, γ_2 may be empty. We say that two W-symbols are *related* if they are directly related or indirectly related.

We prove the following claims simultaneously by backward induction: the first claim is at the heart of our completeness, and the other three claims are invariants that will help in our induction.

Claim 1: for all $k \leq n$, there exists a model $\mathcal{M}_k$ such that $\mathcal{M}_k, w_0 \models W_0$ and $\mathcal{M}_k \models N_k$ but $\mathcal{M}_k, w_0 \not\models \psi$.

Claim 2: for all $0 \leq k < n$, $\mathcal{M}_k$ and $\mathcal{M}_{k+1}$ are Σ-bisimilar.

Claim 3: for all $k \leq n$, $\mathcal{M}_k$ has the W-symbol independence property; that is that either two W-symbols are *related* in N_k or their interpretations are disjoint.

We assume w.l.o.g. that the valuation function in $\mathcal{M}'$ maps any symbol not in Σ to the empty set.

Base case ($k = n$): To show Claim 1, we define the model $\mathcal{M}_n$ as an extension of $\mathcal{M}'$, to satisfy N_n by setting the interpretation of W_0 to be true in w_0, i.e., $V_n(W_0) = \{w_0\}$. Claim 2 holds trivially because $k = n$, and Claim 3 holds because there is only one W-symbol and that is W_0.

Step case: Assuming all claims hold for $k + 1$, we will show that there exists a Σ-bisimilar model $\mathcal{M}_k$ such that $\mathcal{M}_k \models N_k$ but $\mathcal{M}_k, w_0 \not\models \psi$, and that this model has the W-symbol independence property.

To this end, we will consider the effect of each rule and show that if it was applied as the k-th step, it is possible to construct a model $\mathcal{M}_k$ from $\mathcal{M}_{k+1}$ that satisfies our claims.

For the preprocessing rules, except for the world introduction rule, the model remains the same. For the world introduction rule, we must reset the interpretation of the W-symbol that was introduced to maintain Claim 3 and 4, i.e., we make $V_k(W_j) = \emptyset$.

Similarly, for the literal resolution rule and the world resolution rule, the model remains the same, and the argument is that N_k is a subset of N_{k+1}, so a model that satisfies N_{k+1} satisfies N_k, but for the $\Box\bigcirc$ resolution rule, we must reset the interpretation of the W-symbol that was introduced to maintain Claim 3.

We give proofs for the world elimination and purification rules.

World Elimination: Let N_{k+1} be the set obtained after eliminating the symbol W_i from N_k.

Assume without loss of generality (w.l.o.g.) that the following formulas are the only formulas that have left hand side (l.h.s.) occurrences of W_i in N_k:

$$W_i \Rightarrow \gamma_1, \ldots, W_i \Rightarrow \gamma_m \tag{1}$$

The formulas where W_i can occur on the right hand side (r.h.s.) are of the forms:

$$W_j \Rightarrow W_i \quad \text{or} \quad W_j \Rightarrow \gamma \vee \bigcirc_a W_i \quad \text{where } \bigcirc \in \{\Diamond, \Box\}. \tag{2}$$

In N_{k+1}, i.e. after the application of the world elimination rule, the formulas in Equation (1) are removed. The formulas of the forms in Equation (2) are replaced with
$W_j \Rightarrow (\gamma_1 \wedge \cdots \wedge \gamma_m)$ and $W_j \Rightarrow \gamma \vee \bigcirc_a(\gamma_1 \wedge \cdots \wedge \gamma_m)$ where $\bigcirc \in \{\Diamond, \Box\}$ respectively. Given $\mathcal{M}_{k+1}$, a model for N_{k+1}, our aim is to expand the model into $\mathcal{M}_k$ for N_k by giving W_i an appropriate assignment.

By assumption, this assignment is empty, and to achieve an appropriate interpretation, we incrementally populate the assignment in the following way.

First, to satisfy a new formula of the form $W_j \Rightarrow W_i$ in N_k, we must expand the valuation mapping of W_i in the model being constructed $\mathcal{M}_k$ to include every world in the mapping of W_j. Explicitly, the mapping of W_i must include the following worlds: $\{w \mid w \in V_{k+1}(W_j)\}$.

Second, we consider the two cases for satisfying a formula of the form $W_j \Rightarrow \gamma \vee \bigcirc_a W_i$ where $\bigcirc \in \{\Diamond, \Box\}$.

Case 1: $\bigcirc$ denotes a $\Diamond$ operator. We consider each $w \in V_{k+1}(W_j)$, if $\mathcal{M}_{k+1}, w \models \Diamond_a(\gamma_1 \wedge \cdots \wedge \gamma_m)$ then there is a world u such that $wR_a u$ is in R_a, and $\mathcal{M}_{k+1}, u \models \gamma_1 \wedge \cdots \wedge \gamma_m$. We extend the domain $\mathcal{W}_k$ with a fresh world u'. We start by making the interpretation of u' in $\mathcal{M}_k$ identical to the interpretation of u for symbols from Σ, i.e., for all p in Σ, if u is in $V_{k+1}(p)$, then we include u' in $V_k(p)$. We extend the frame with successors in the following way: for all a, and all t, if $uR_a t$ is true in $\mathcal{M}_{k+1}$, then $u'R_a t$ is made true in $\mathcal{M}_k$, and if $tR_a u$ is true in $\mathcal{M}_{k+1}$, then $tR_a u'$ is made true in $\mathcal{M}_k$. This is to ensure that the new model $\mathcal{M}_k$ maintains Σ-bisimilarity. We extend the interpretation of $V_k(W_i)$ to include u'. To maintain Claim 3, we extend the interpretation of the V_k for W-symbols related to W_i in the following way: for each W_m, if u is in $V_k(W_m)$, and W_m is related to W_i then we include u' in the valuation of u'.

Case 2: $\bigcirc$ denotes a $\Box$ operator. Similar to the first case, we consider each $w \in V_{k+1}(W_j)$. If $\mathcal{M}_{k+1}, w \models \Box_a(\gamma_1 \wedge \cdots \wedge \gamma_m)$, then for every world u connected to w via an a-successor, we check if we can make W_i true in u while maintaining Claim 3, i.e. we check if the following condition holds: for all W_m, if u is in $V_{k+1}(W_m)$, then W_m and W_i must be related. If this is not possible, we extend the domain $\mathcal{W}_k$ with a fresh world u', and we define V_k as in the case 1, but the accessibility relation is defined as follows:

$$R_{a_k} = \begin{cases} wR_{a_k}u' & \text{if for some } w,\ wR_{a_{k+1}}u \in R_{a_{k+1}} \\ u'R_{a_k}w & \text{if for some } w,\ uR_{a_{k+1}}w \in R_{a_{k+1}} \\ wR_{a_k}w' & \text{if for some } w, w' \text{ s.t. } w \neq u \text{ and } w \neq w',\ wR_{a_{k+1}}w' \in R_{a_{k+1}} \end{cases}$$

The difference between the construction in case 1 and the one here is that in case 1, $\mathcal{M}_k$ is an extension of $\mathcal{M}_{k+1}$, whereas, in this case, this would not help because of the box operator, e.g. consider $\phi = \neg x \wedge \Box(p \wedge x)$, and let $\mathcal{M}_n = (\{w_0\}, \{R(w_0, w_0)\}, V_n)$ where $V_n(p) = \{w_0\}$.

Now, we have completed the construction of the model $\mathcal{M}_k$. It is clearly a model for N_k, and is Σ-bisimilar to $\mathcal{M}_{k+1}$. The world independence property is maintained as, in our construction, we ensure that if there is a W-symbol that is unrelated to W_i in some world u, they, and their related worlds, are separated using a new world u'.

Positive Purification: Let $W_i \Rightarrow \top \vee \gamma_1$ be the clause obtained after applying purification to $W_i \Rightarrow x \vee \gamma_1$ from N_k. We define a model $\mathcal{M}_k = (\mathcal{W}_{k+1}, R_{k+1}, V_k)$ where V_k extends the map of x in the following way

$$V_k(x) = V_{k+1}(x) \cup \{w \mid w \in V_{k+1}(W_i) \textit{ and } \mathcal{M}_{k+1}, w \not\models \gamma_1\}.$$

The deleted clause $W_i \Rightarrow x \vee \gamma_1$ is now true by the construction of $\mathcal{M}_k$. It remains to check that $\mathcal{M}_k \models N_k$. Assume $\mathcal{M}_k \not\models N_k$. Since the only change was an extension to the interpretation of x, there must be a clause which contains x negatively in N_k that was true in $\mathcal{M}_{k+1}$ but is not true in $\mathcal{M}_k$. This clause must be of the form $W_j \Rightarrow \neg x \vee \gamma_2$.

Since making the clause $W_i \Rightarrow x \vee \gamma_1$ true in $\mathcal{M}_k$ made $W_j \Rightarrow \neg x \vee \gamma_2$ false, we can infer that there is a common world w in $V_k(W_i) \cap V_k(W_j)$ such that $\mathcal{M}_k, w \models x \vee \gamma_1$ but $\mathcal{M}_k, w \not\models \neg x \vee \gamma_2$.

There are four cases to consider.

Case 1: $i = j$. By literal resolution, this implies that $W_i \Rightarrow \gamma_1 \vee \gamma_2$ is in N_k and N_{k+1}, and is satisfied by $\mathcal{M}_{k+1}$ but not by $\mathcal{M}_k$. This is impossible because the two models agree up to x. γ_1 and γ_2 may contain x but only under a modal operator which, by how we defined $\mathcal{M}_k$ for the World Elimination rule, means that they cannot be realised by w.

Case 2: W_i and W_j are directly related in N_k. Without loss of generality, let us assume $W_i \Rightarrow W_j$ is in N_k. This implies that (i) $V(W_i) \subseteq V(W_j)$, and (ii) by world resolution, $W_i \Rightarrow \neg x \vee \gamma_2$ is in N_k. The problem is now reduced to what has been discussed in the first case. Therefore, we can similarly conclude that this case cannot happen.

Case 3: W_i and W_j are indirectly related in N_k. Without loss of generality, assume there exists a W-symbol W_k such that $W_k \Rightarrow \gamma_1 \vee \bigcirc_a W_i$ and $W_k \Rightarrow \gamma_2 \vee \Box_a W_j$ are in N_k. By the $\Box\bigcirc$ resolution rule, this implies that $W_k \Rightarrow \gamma_1 \vee \gamma_2 \vee \bigcirc W_{ij}$, $W_{ij} \Rightarrow W_i$ and $W_{ij} \Rightarrow W_j$ is in N_k. This problem is now reduced to what has been discussed in case 2, and hence it cannot occur.

Case 4: $i \neq j$ and W_i and W_j are not related in N_k. By Claim 3, their interpretations are disjoint, i.e., $V_k(W_i) \cap V_k(W_j)$ is empty. Hence, this case cannot occur.

Finally, since we have not changed the structure of the model nor the interpretation of the elements in Σ, $\mathcal{M}_{k+1}$ and $\mathcal{M}_k$ are Σ-bisimilar. The third claim is maintained since the interpretation of the W-symbols has not changed.

As for the case of *negative purification*, we define a model $\mathcal{M}_k$ to be an identical copy of $\mathcal{M}_{k+1}$. What remains is to show that $\mathcal{M}_k \models W_i \Rightarrow \neg x \vee \gamma_1$. The proof is analogous to the proof given for positive purification. □

5 Extensions

We consider extensions of K_n, namely D_n and T_n, and show that our UI_{K_n} system could be used or extended to cover them.

Multi-Modal Logic D_n. We claim that the UI_{K_n} system computes uniform interpolants for the multi-modal logic D_n. To show our claim, we prove the completeness theorem.

Theorem 5.1 (Completeness) *Given a D_n formula ϕ and a signature Σ, the UI_{K_n} system computes a D_n formula ϕ' such that, for any D_n formula ψ over Σ, we have that if $\models \phi \rightarrow \psi$ then $\models \phi' \rightarrow \psi$.*

Proof. The proof uses a similar argument to the one for K_n except, here we assume that the accessibility relations in the initial model are serial.

In addition to the three claims from the proof of Theorem 4.10, we prove the following claim:

Claim 4: For all $k \leq n$, each accessibility relation in $\mathcal{M}_k$ is serial.

Observe that $\mathcal{M}_k$ uses the frame underlying the model $\mathcal{M}_{k+1}$ in the proofs given for each of the rules, except for the world elimination rule. Therefore, for these rules, Claim 4 is established. Consider the case of the world elimination rule. We construct $\mathcal{M}_k$ to be Σ-bisimilar to $\mathcal{M}_{k+1}$. By the induction hypothesis, $\mathcal{M}_{k+1}$ is a serial model, and by the "forth" condition of definition of Σ-bisimilation, $\mathcal{M}_k$ must be serial, too. □

Multi-Modal Logic T_n. In this part, we introduce the UI_{T_n} system, and show that it is terminating, sound and uniform interpolation complete for T_n. The system extends UI_{K_n} by generalising the world resolution rule and adding the reflexivity rule as a new saturation rule. These rules are shown in Figure 4.

World Resolution (RES W):

$$\frac{W_i \Rightarrow \psi_1 \quad W_j \Rightarrow \psi_2 \vee W_i}{W_j \Rightarrow \psi_1 \vee \psi_2}$$

provided that ψ_1 contains x. ψ_2 may be a W-symbol.

Reflexivity (T):

$$\frac{W_i \Rightarrow \psi \vee \Box_a W_m}{W_i \Rightarrow \psi \vee W_m}$$

provided that W_m is a base symbol.

Fig. 4. The reflexivity rule and the world resolution rule of the UI_{T_n} system for modal logic T_n.

The soundness theorem and proof are analogous to Theorem 4.8 and its proof. The termination and complexity arguments are identical to the ones in Theorem 4.4 and Lemma 4.5. It remains to show the completeness theorem.

Theorem 5.2 (Completeness) *Given a formula ϕ and a signature Σ, the UI_{T_n} system computes a formula ϕ' such that, for any formula ψ over Σ, we have that if $\models \phi \rightarrow \psi$ then $\models \phi' \rightarrow \psi$.*

Proof. This proof is an extension of the proof of Theorem 4.10. We assume that each accessibility relation is reflexive in $\mathcal{M}'$. We generalise our definition of *directly related*. Given N a set of clauses and W-symbols W_i and W_j, W_i is said to be *directly related* to W_j if, for some γ, $W_i \Rightarrow \gamma \vee W_j \in N$ or $W_j \Rightarrow \gamma \vee W_i \in N$.

Our aim is to prove the same three claims and the following claim.

Claim 4: For all $k \leq n$, each accessibility relation R_i is reflexive in $\mathcal{M}_k$.

Base case ($k = n$): To show Claim 4, we observe that a change happened to the valuation of W_0, and that this does not change the frame underlying the model. Therefore, the new model $\mathcal{M}_n$ remains reflexive.

Because the set of reflexive models is a subset of the models considered in the proof for Theorem 4.10, the three claims remain true for all the shared rules. We focus on showing the claims for the reflexivity rule and the generalised world resolution rule, and show that Claim 4 is true for all the remaining rules.

Step case: Assuming all claims hold for $k + 1$, we show that there exists a model $\mathcal{M}_k$ that satisfies all four claims.

For all saturation rules, including the reflexivity rule and the world resolution rule, the model $\mathcal{M}_k$ is defined to be identical to $\mathcal{M}_{k+1}$. For the world elimination rule, we repeat the model construction of $\mathcal{M}_k$ as shown in the proof of Theorem 4.10, with the consideration that now we have the general form $W_j \Rightarrow \gamma \vee W_i$, but we extend it to make $R_a(u', u')$ true for each $R_a \in R_k$. Observe that this change still maintains the Σ-bisimulation, but will make $\mathcal{M}_k$ reflexive. Clearly, $\mathcal{M}_k$ is a model for N_k, and the world independence property is maintained.

For the positive and negative purification rules, observe that the frame underlying the model $\mathcal{M}_k$ is identical to that of $\mathcal{M}_{k+1}$, so Claim 4 holds. □

6 Related Work

Fang et al. [6] use what Moss [18] called canonical formulas. They exploit the fact that an arbitrary modal formula is equivalent to a disjunction of a finite set of satisfiable canonical formulas, and prove that a uniform interpolant can be constructed via literal elimination. Although the proof is constructive, as the authors explicitly mention, the method is unpractical because the size of a canonical formula is non-elementary.

Other uniform interpolation methods can be divided into two groups based on whether they use a conjunctive normal form (CNF) (e.g., [11,15]) or a disjunctive normal form (DNF) (e.g., [3,19]). Uniform interpolants can be easily computed for formulae in disjunctive normal form, in fact the method in [19] was shown to have an exponential worst case space complexity, which they prove to be a tight upper bound.

Comparing the two types of approaches, we notice that methods that use the conjunctive normal form, which are often based on resolution, struggle with the following type of problem.

(i) $\Box(x \vee q1) \vee \Box(\neg x \vee q2)$ (ii) $\Box(\neg x \vee r1) \vee \Box(x \vee r2)$

This is due to the $\Box\Box$ resolution rule. We claim that using our method, this problem generates a number of clauses which is double exponentially bounded by the input. Other resolution methods have a rule for combining $\Box$ operators and work in a similar way [11,15]. The work that the other rules in resolution systems do is comparable to those that use transformation to disjunctive normal form. However, even though the complexity of our system is less optimal than [19], we perform better in general cases of the following example:

(i) $\neg x \vee \Box(\neg x \vee q1) \vee \Box(\neg x \vee q2)$ (ii) $\Box(x \vee r1) \vee \Box(x \vee r2)$

Because x, the symbol that we want to eliminate, appears at level zero in the first clause but only under a modal operator in the second clause, no resolution rule is applicable. Hence, $\neg x$ in the first clause can be purified, and by extending the system with standard simplification rules, the clause is replaced with $\top$.

Different to [11], we use additional propositional symbols to flatten our input, and different to [14], we do not use unification for first-order variables, which was used there because the problem is slightly different: they look at global (TBox) uniform interpolation with local (ABox) formulae. More broadly, completeness proofs for resolution-based uniform interpolation systems are traditionally shown via proof-theoretic arguments. Our proofs show that a model-theoretic argument, using Σ-bisimulation, can be made for proving uniform interpolation completeness of resolution-based systems. We notice that the proofs for D_n and T_n required only very modest extensions.

7 Conclusion

We presented a resolution-based method to compute uniform interpolants for the multi-modal logic K_n. We proved that our method terminates, and is sound and uniform interpolation complete. The space complexity was proven to be at most double exponential in the length of the input. We showed that the method can be used for computing uniform interpolants in D_n, and can be extended to compute uniform interpolants in T_n.

For future work, we would like to study logics which are known to have the uniform interpolation property, and show that the presented system can be extended to solve the uniform interpolation problem for more modal logics. Furthermore, it would be interesting to implement the UI_{K_n} system and perform an empirical comparison between this system and a system that transforms the input into disjunctive normal form (e.g., [19,3]).

Appendix

A Proofs

Lemma A.1 *The number of W-symbols introduced by the UI_{K_n} is bounded by 2^n-1 for n being the length of the input.*

Proof. Let S_w be the set of base W-symbols. S_w is finite because the number of modal operators in the input formula is finite, and the role of the world introduction rule is to replace each subformula (not containing a W-symbol) appearing under a modal operator with a W-symbol.

Let C_w be the set of combinatory W-symbols introduced by the RES□○ rule. We show inductively that for each W-symbol W_i in C_w, the set $Corr(W_i)$ corresponds to a unique set of W-symbols from S_w.

Since the function $Corr$ has a finite range, which is the powerset of S_w, we can prove that the domain is finite by showing that $Corr$ is injective. Let W_n and W_m be two W-symbols. We will show inductively that if $Corr(W_n) = Corr(W_m)$, then $W_n = W_m$.

Base case: W_n is in S_w. The assumption is that $Corr(W_n) = Corr(W_m)$. Since W_n was produced as a result of an application of the world introduction rule, we have that $Corr(W_n) = \{W_n\}$. By assumption, this means that $Corr(W_m) = \{W_n\}$. Since $Corr(W_m)$ corresponds to a singleton set, it must be introduced via the world introduction rule as well, and hence, $W_n = W_m$.

Step case: W_n is in C_w. The cardinality of $Corr(W_n)$ must be greater than 1; this is because when W_n is introduced after a □○ rule application, $Corr(W_n)$ is defined as the combination of two sets that share no elements. Assuming that $Corr(W_n) = Corr(W_m)$ entails that W_m must be in C_w as two equal sets have the same cardinality. Since W_n and W_m are in C_w, they must have been introduced after applying the □○ rule. Assume w.l.o.g. that W_m was introduced first. Let $Corr(W_n) = Corr(W_i) \cup Corr(W_j)$ for a W_i and W_j that uniquely correspond to subsets of S_w symbols. By condition (iii) of the □○ rule, there cannot be a W_k such that $Corr(W_k) = Corr(W_i) \cup Corr(W_j)$, therefore W_n must be equal to W_m. Since S_w is a finite set, and by condition (ii) of the □○ resolution rule, the number of possible W-symbols is bounded by the number of unique combinations of symbols from S_w. The upper bound is equal to $2^{|S_w|}$-1, where $|S_w|$ is the size of S_w. □

Lemma A.2 *The space complexity of the UI_{K_n} calculus is double exponentially bounded in the length of the input.*

Proof. The root source of the complexity is that the system generates an exponential number of combinatory symbols. The complexity argument is built upon three claims. Using a linear number of propositional symbols, a linear number of base W-symbols, and an exponential number of combinatory W-symbols:

C1 the size of each clause, before eliminating any W-symbols, has an exponential upper bound in the size of the input,

C2 the number of clauses we could generate has a double exponential upper bound in the size of the input, and

C3 the size of each clause, after eliminating W-symbols, has a double exponential upper bound in the size of the input.

For C1, before eliminating any W-symbols, a clause has propositional variables, a disjunction ψ of subformulas of ϕ, and a maximally exponential number of W-symbols appearing under a box or a diamond operator, or both. The number of propositional variables is linear in the size of the input (since they must appear in the input), the size of ψ is linear too for the same reason, and since the number of W-symbols is in $O(2^n)$ (n is the size of the input), then the size of a clause before eliminating any W-symbol is in $O(2^n)$.

For C2, since for each clause we have an exponential number of (propositional or W-symbol) variables that can either appear positively or negatively, the number of clauses that could be generated before eliminating any W-symbols is at most double exponential in the size of the input. (Note that we assume that the system incorporates simplification rules such as subsumption elimination.)

For C3, since the number of clauses is at most double exponential, and each clause has a size that is of at most an exponential size, eliminating W-symbols will result in a formula that is maximally bounded by $O(2^n * 2^{2^n})$. □

Theorem A.3 (Soundness) *Given a formula ϕ and a signature Σ, the UI_{K_n} system computes a formula ϕ' such that for any formula ψ over Σ, we have that if $\models \phi' \to \psi$ then $\models \phi \to \psi$.*

Proof. Consider ϕ, ϕ' as in Theorem 4.8. We use $\mathcal{M}|_\Sigma$ for the reduction of $\mathcal{M}$ to Σ, i.e. the model obtained from $\mathcal{M}$ by ignoring all symbols outside Σ. We say that two models $\mathcal{M}', \mathcal{M}$ are *Σ-inseparable* if $\mathcal{M}'|_\Sigma = \mathcal{M}|_\Sigma$. To show the theorem, we show the following claims:

C4 For each model $\mathcal{M}$ and world w_0 such that $\mathcal{M}, w_0 \models \phi$, there exists a model $\mathcal{M}'$ such that $\mathcal{M}$ and $\mathcal{M}'$ are Σ-inseparable, $\mathcal{M}' \models W_0 \Rightarrow \phi$, and $V(W_0) = \{w_0\}$.

C5 Each rule in our calculus is Σ-preserving, that is, if a rule is applied to a set of clauses N to produce N' and $\mathcal{M}$ is a model that globally satisfies N, then there exists a model $\mathcal{M}'$ that is Σ-inseparable from $\mathcal{M}$ that satisfies N'.

C6 Let $N = \{W_0 \Rightarrow \psi_i \mid 1 \leq i \leq n\}$ be the set obtained at the end of our derivation. For each model $\mathcal{M}$ that globally satisfies N and for each $w_0 \in V(W_0)$, we have $\mathcal{M}, w_0 \models \phi'$ with $\phi' = \bigwedge_{i \leq |N|} \psi_i$.

For C4, take a model $\mathcal{M}$ and a world w_0 such that $\mathcal{M}, w_0 \models \phi$. Extending $\mathcal{M}$ by setting $V(W_0) = \{w_0\}$ results in a Σ-inseparable model in which $W_0 \Rightarrow \phi$ is globally satisfied.

For C5, we consider two of our rules: the world introduction rule and the □○ resolution rule. Proofs for the remaining rules are standard.

World Introduction: Let N' be the set obtained by replacing $W_i \Rightarrow \phi_1 \vee \bigcirc_a \phi_2$ in N with $W_i \Rightarrow \phi_1 \vee \bigcirc_a W_j$ and $W_j \Rightarrow \phi_2$.

Given $\mathcal{M}$, a model for N, our aim is to expand the model into $\mathcal{M}'$ for N' by giving W_j an appropriate assignment. Consider the following case distinction.

Case 1: ○ denotes a ◇ operator. We consider each $w \in V(W_i)$, if $\mathcal{M}, w \models \Diamond_a \phi_2$ then, there exists a world u, connected to w via an a-successor, such that $\mathcal{M}, u \models \phi_2$. We include u in $V'(W_j)$, i.e. ,

$$V'(W_j) = \{u \mid \mathcal{M}, u \models \phi_2 \text{ and } \exists w \text{ s.t. } w \in V(W_i) \text{ and } wR_a u \text{ is true in } \mathcal{M}\}.$$

By giving W_j the above interpretation, $W_i \Rightarrow \phi_1 \vee \Diamond_a W_j$ and $W_j \Rightarrow \phi_2$ become globally satisfiable in $\mathcal{M}'$.

Case 2: $\bigcirc$ denotes a $\Box$ operator. Similar to the first case, we consider each $w \in V(W_i)$. If $\mathcal{M}, w \models \Box_a \phi_2$, then for every world u connected to w via an a-successor, W_j must be made true in u. We extend the definition of $V'(W_j)$ to include u.

By giving W_j the above interpretation, $W_i \Rightarrow \phi_1 \vee \Box_a W_j$ and $W_j \Rightarrow \phi_2$ become globally satisfiable in $\mathcal{M}'$.

Now, we have completed the construction of the model $\mathcal{M}'$. We show that $\mathcal{M}'$ is a model for N'. The model $\mathcal{M}'$ satisfies all formulas in N' that do not include W_j. Indeed, this is because the interpretation of all symbols apart from W_j has not been changed. From Cases 1 and 2, it globally satisfies all clauses that contain W_j. Therefore, $\mathcal{M}'$ is a model for N'. The two models are Σ-inseparable because they are defined over the same frame, and the valuation function was not changed for propositional symbols in Σ.

$\Box\bigcirc$ **resolution:** Let N' be the set obtained by applying the $\Box\bigcirc$ resolution rule to $W_i \Rightarrow \gamma_1 \vee \Box_a W_n$ and $W_i \Rightarrow \gamma_2 \vee \bigcirc_a W_m$ attaining $W_i \Rightarrow \gamma_1 \vee \gamma_2 \vee \bigcirc_a W_j$ and $W_j \Rightarrow W_n$ and $W_j \Rightarrow W_m$.

Given $\mathcal{M}$, a model for N, our aim is to expand the model into $\mathcal{M}'$ for N' by giving W_j an appropriate assignment. We consider the two cases:

Case 1: $\bigcirc$ denotes a $\Diamond$ operator. We consider each $w \in V(W_i)$, if $\mathcal{M}, w \models \Box_a W_n$ and $\mathcal{M}, w \models \Diamond_a W_m$ then, there exists a world u, connected to w via an a-successor, such that $\mathcal{M}, u \models W_m$, and $\mathcal{M}, u \models W_n$. We include u in $V'(W_j)$.

Case 2: $\bigcirc$ denotes a $\Box$ operator. We consider each $w \in V(W_i)$. If $\mathcal{M}, w \models \Box_a W_n$ and $\mathcal{M}, w \models \Box_a W_m$, then for every world u connected to w via an a-successor, W_j must be made true in u. We include u in $V'(W_j)$.

We show that $\mathcal{M}'$ is a model for N'. Since the only change was to the interpretation of W_j, $\mathcal{M}'$ satisfies all clauses in N' without W_j. From Cases 1 and 2, $\mathcal{M}'$ satisfies clauses that contain W_j. The two models are Σ-inseparable because they are defined over the same frame, and the valuation function was not changed for propositional symbols in Σ.

For C6, the argument is trivial. Consider N as in the claim. Let $\mathcal{M}$ be a model that globally satisfies N, and w_0 be a world in $V(W_0)$. By the definition of global satisfiability, all clauses in N are satisfiable at w_0. By the semantics of implication, we have that $\mathcal{M}, w_0 \models \psi_i$ for $1 \leq i \leq |N|$. By the semantics of conjunction, $\mathcal{M}, w_0 \models \bigwedge_{i \leq |N|} \psi_i$.

We showed via C4, C5, and C6 that for any model of the input formula, there exists a Σ-inseparable model for the output formula. And, thus, for any formula ψ over Σ, if $\phi \rightarrow \psi$ is true in all models, then $\phi' \rightarrow \psi$ is true in all models. $\Box$

References

[1] Baral, C. and Y. Zhang, *Knowledge updates: Semantics and complexity issues*, Artificial Intelligence **164** (2005), pp. 209–243.

[2] Bechhofer, S., F. van Harmelen, J. Hendler, I. Horrocks, D. McGuinness, P. Patel-Schneijder and L. A. Stein, *OWL Web Ontology Language Reference*, Recommendation,

World Wide Web Consortium (W3C) (2004), see `http://www.w3.org/TR/owl-ref/`.

[3] Bílková, M., *Uniform interpolation and propositional quantifiers in modal logics*, Studia Logica (2007), pp. 1–31.

[4] Blackburn, P., M. d. Rijke and Y. Venema, "Modal Logic," Cambridge Tracts in Theoretical Computer Science, Cambridge University Press, 2001.

[5] Fang, L., Y. Liu and H. Van Ditmarsch, *Forgetting in Multi-agent Modal Logics*, in: *Proc. IJCAI 2016* (2016), pp. 1066–1073.

[6] Fang, L., Y. Liu and H. Van Ditmarsch, *Forgetting in multi-agent modal logics*, Artificial Intelligence **266** (2019), pp. 51–80.

[7] Gabbay, D. M. and H. J. Ohlbach, *Quantifier elimination in second-order predicate logic*, in: *Proceedings of the Third International Conference on Principles of Knowledge Representation and Reasoning*, KR'92 (1992), pp. 425–435.

[8] Gabbay, D. M., R. A. Schmidt and A. Szałas, "Second-Order Quantifier Elimination: Foundations, Computational Aspects and Applications," College Publications, 2008.

[9] Ghilardi, S., *An algebraic theory of normal forms*, Annals of Pure and Applied Logic **71** (1995), pp. 189–245.

[10] Ghilardi, S. and M. Zawadowski, *Undefinability of propositional quantifiers in the modal system S4*, Studia Logica **55** (1995), pp. 259–271.

[11] Herzig, A. and J. Mengin, *Uniform interpolation by resolution in modal logic*, in: *European Workshop on Logics in Artificial Intelligence*, Springer, 2008, pp. 219–231.

[12] Horrocks, I., U. Hustadt, U. Sattler and R. Schmidt, *Computational modal logic*, in: *Handbook of Modal Logic*, Elsevier, 2007 pp. 181–245.

[13] Konev, B., D. Walther and F. Wolter, *Forgetting and uniform interpolation in large-scale description logic terminologies.*, in: *IJCAI* (2009), pp. 830–835.

[14] Koopmann, P., "Practical uniform interpolation for expressive description logics," Ph.D. thesis, The University of Manchester (2015).

[15] Koopmann, P. and R. A. Schmidt, *Uniform interpolation and forgetting for $\mathcal{ALC}$ ontologies with ABoxes*, in: *Proceedings of the AAAI Conference on Artificial Intelligence*, 2015, pp. 175–181.

[16] Ludwig, M. and B. Konev, *Practical uniform interpolation and forgetting for $\mathcal{ALC}$ TBoxes with applications to logical difference*, in: *Principles of Knowledge Representation and Reasoning: Proceedings of the Fourteenth International Conference, KR 2014*, 2014, pp. 318–327.

[17] Lutz, C. and F. Wolter, *Foundations for uniform interpolation and forgetting in expressive description logics*, in: *Twenty-Second International Joint Conference on Artificial Intelligence*, 2011, pp. 989–995.

[18] Moss, L. S., *Finite models constructed from canonical formulas*, Journal of Philosophical Logic **36** (2007), pp. 605–640.

[19] Ten Cate, B., W. Conradie, M. Marx, Y. Venema et al., *Definitorially complete description logics.*, KR (2006), pp. 79–89.

[20] Toluhi, D., R. Schmidt and B. Parsia, *Concept description and definition extraction for the anemone system*, in: *International Workshop on Engineering Multi-Agent Systems*, Springer, 2021, pp. 352–372.

[21] Visser, A., *Bisimulations, model descriptions and propositional quantifiers*, Logic Group Preprint Series **161** (1996).

[22] Wolter, F., *Fusions of modal logics revisited*, in: *Advances in Modal Logic* (1998), pp. 361–379.

Provability Logics of Hierarchies

Amirhossein Akbar Tabatabai [1]

University of Groningen, Bernoulliborg,
Nijenborgh 9 9747 AG Groningen, The Netherlands

Abstract

Provability logic is a framework to investigate the provability behavior of the mathematical theories. More precisely, it studies the relationship between a mathematical theory T and a modal logic L via the provability interpretations that read the modality as a provability predicate of T. In this paper, we will extend this relationship from one single theory to a hierarchy of theories capturing the philosophical intuition of the hierarchy of meta-theories one may use to talk about the theories themselves. More precisely, using the modal language with infinitely many modalities, $\{\Box_n\}_{n=0}^{\infty}$, we will first define the hierarchical counterparts of the classical modal logics $\mathsf{K4}$, $\mathsf{KD4}$ and GL. Then, we will show that they are sound and complete with respect to their provability interpretations in the class of all hierarchies, the hierarchies of consistent theories and the constant hierarchies, respectively. We will also show that none of the extensions of the hierarchical counterpart of $\mathsf{KD45}$ has a provability interpretation.

Keywords: provability logic, provability interpretation, Solovay's completeness theorems.

1 Introduction

Provability logic is a framework that identifies the key modal aspects of the provability predicates of the mathematical theories. This modal approach was roughly initiated by Gödel's short note [11] on the interpretation of the Brouwerian constructions as the usual classical proofs. It then gained power when Löb [16] identified the modal properties of a provability predicate required in the usual proof of Gödel's second incompleteness theorem. He also added his own key generalization, the well-known Löb's axiom $\Box(\Box A \to A) \to \Box A$ that is valid under all provability interpretations interpreting $\Box$ as a provability predicate for a strong enough theory T. Finding such non-trivial modal formulas asked for the characterization of all such formulas, the provability logic of the theory T, where the key step was taken by Solovay in his seminal paper [20]. He invented the internalization technique embedding Kripke frames into the formal arithmetic in order to prove that the provability logic of Peano arithmetic is GL, the logic $\mathsf{K4}$ plus the Löb's axiom. Inspired by this seminal work,

[1] Support by the Netherlands Organisation for Scientific Research under grant 639.073.807 and also the FWF project P 33548 are gratefully acknowledged.

a series of deep investigations were initiated to study the different layers of the provability behavior of the theories, from first-order provability logics [14] and interpretability logics [22,14] to the provability logics of intuitionistic theories and the bimodal and polymodal provability logics addressing more than one provability predicates at the same time [7,10,19,15,13]. In this paper, we are taking a similar route as in the latter to employ a polymodal language to reflect the provability behavior of a hierarchy of theories. The most well-known provability logic in this sense is GLP, introduced by Japaridze in [15] alongside its elaborate Solovay-style arithmetical completeness theorem (also see [8]) and studied extensively later, from many different angles, from topological semantics [6] to computational complexity [18]. However, our motivation and hence our setting is somewhat different. To explain the motivation, let us come back to the original Gödel's interpretation of Brouwerian constructions.

To have a formal language for classical informal provability, in [11], Gödel proposed the modal logic S4. The axioms are all valid under the intuitive interpretation of $\Box$ as the informal provability predicate. The axiom $(K) : \Box(A \to B) \to (\Box A \to \Box B)$ states that the provability predicate is closed under modus ponens. The axiom $(4) : \Box A \to \Box\Box A$ states that "the provability of a provable statement is also provable" which seems a reasonable condition to have and finally $(T) : \Box A \to A$ states that the proofs are all sound. However, as Gödel observed himself, S4 is not sound with respect to the formal provability interpretation that reads $\Box$ as Pr_T, for some strong enough theory T. Because, $\mathsf{S4} \vdash \neg\Box\bot \wedge \Box\neg\Box\bot$ and hence the formula should be valid under the provability interpretation while its interpretation $\neg\mathrm{Pr}_T(\bot) \wedge \mathrm{Pr}_T(\neg\mathrm{Pr}_T(\bot))$ contradicts with Gödel's own second incompleteness theorem. Having that observation, one may wonder if there is any formalization for the intuitive provability interpretation.

To find the source of the mismatch between the formal and the informal provability interpretations, one should look into the role of the nested modalities. Nested modalities intuitively capture the nested use of the provability predicates to express the statements such as "the provability of p", "the provability of "the provability of p"" and so on. These different layers of provability predicates naturally refer to different layers of theories, meta-theories, meta-meta-theories and so on. But the usual provability interpretation reads all of them as the provability predicate for a fixed theory. Philosophically speaking, there is no reason to assume that all the layers of our meta-theories are the same. Quite the contrary, in the actual practice of proof theory, sometimes we need to have more powerful meta-theories to investigate the behavior of the theory itself. For instance, in the aforementioned problematic formula $\neg\Box\bot \wedge \Box\neg\Box\bot$, observing that the inner box refers to a theory T while the outer box refers to its meta-theory U, transforms the contradictory interpretation of the formula to $\neg\mathrm{Pr}_T(\bot) \wedge \mathrm{Pr}_U(\neg\mathrm{Pr}_T(\bot))$ which simply states the safe and intuitive claim that T is consistent and its consistency is provable in its meta-theory U.

Having that observation, [1] proposed using a hierarchy of theories to formalize the different layers of meta-theories instead of using just one theory for

all the levels. Following that approach and using some natural classes of the hierarchies of the arithmetical theories, we found some natural interpretations for some modal logics such as $\mathsf{K4}$, $\mathsf{KD4}$ and $\mathsf{S4}$ and hence a formalization for Brouwer-Heyting-Kolmogorov interpretation.
This framework extension suggests a reverse problem of characterizing the provability behavior of a given class of hierarchies of theories rather than providing a provability interpretation for a given modal logic. The present paper is devoted to this problem. We employ the polymodal language $\mathcal{L}_\infty$ with infinitely many modalities $\{\Box_n\}_{n=0}^{\infty}$ to capture the different layers of the meta-theories' hierarchy. However, our polymodal approach deviates from the usual polymodal approach by making the syntactical restrictions to avoid using the lower boxes over themselves or the higher ones. This captures the intuition that a theory can not refer to itself or its meta-theories. Using this restriction, the modal logics naturally avoids GL-style principles to transparently reflect the provability behavior of the hierarchies rather than the somewhat peculiar behavior of the single theories. Employing this restriction, we will introduce the hierarchical counterparts of the logics $\mathsf{K4}$, $\mathsf{KD4}$, $\mathsf{S4}$, $\mathsf{KD45}$, $\mathsf{S5}$ and GL, denoted by $\mathsf{K4}_\infty$, $\mathsf{KD4}_\infty$, $\mathsf{S4}_\infty$, $\mathsf{KD45}_\infty$, $\mathsf{S5}_\infty$ and GL_∞, respectively. Then, we will introduce the provability interpretation for some of these new logics with respect to the hierarchies of theories. We will see that $\mathsf{K4}_\infty$ is sound and complete with respect to the class of all hierarchies, while $\mathsf{KD4}_\infty$ and GL_∞ capture all consistent and constant hierarchies, respectively. We will also show that no extension of $\mathsf{KD45}_\infty$, including $\mathsf{S5}_\infty$ has a provability interpretation. To prove the completeness results, unfortunately, it seems impossible to imitate Solovay's technique directly. However, we will present a reduction method that reduces the required completeness to Solovay's theorem. It is also possible to develop a similar result for $\mathsf{S4}_\infty$ but its technique is beyond what we employ in this paper. The logics and their connection to hierarchies were introduced in the unpublished preprint [2], where we used the results in [1] to provide the required completeness theorems. In this paper, we present a somewhat different presentation of the systems and a self-contained direct completeness proofs independent of the results in [1].

2 Preliminaries

In this section, we will recall some basic preliminary facts about the modal logic GL, its sequent-style proof system and its provability interpretations. Let $\mathcal{L} = \{\wedge, \vee, \bot, \rightarrow, \Box\}$ be the language of modal logics. We use $\neg A$ and $\top$ as abbreviations for $A \rightarrow \bot$ and $\bot \rightarrow \bot$, respectively. The only modal logic we work with in this paper is Gödel-Löb logic GL defined as the smallest set of formulas in $\mathcal{L}$ containing all classical tautologies, the axioms $(K) : \Box(A \rightarrow B) \rightarrow (\Box A \rightarrow \Box B)$, $(4) : \Box A \rightarrow \Box\Box A$ and $(L) : \Box(\Box A \rightarrow A) \rightarrow \Box A$ and closed under the rules $(MP) : A, A \rightarrow B \vdash B$ and $(NC) : A \vdash \Box A$.
By a sequent over $\mathcal{L}$, we mean an expression in the form $S = \Gamma \Rightarrow \Delta$, where Γ and Δ are multisets of formulas in $\mathcal{L}$. Define **GGL** as the system consisting of the rules depicted in Figure 1. **GGL** is equivalent to the system defined in

$$\frac{}{A \Rightarrow A}\;Ax \qquad\qquad \frac{}{\bot \Rightarrow}\;L\bot$$

$$\frac{\Gamma, A, A \Rightarrow \Delta}{\Gamma, A \Rightarrow \Delta}\;Lc \qquad\qquad \frac{\Gamma \Rightarrow A, A, \Delta}{\Gamma \Rightarrow A, \Delta}\;Rc$$

$$\frac{\Gamma \Rightarrow \Delta}{\Gamma, A \Rightarrow \Delta}\;Lw \qquad\qquad \frac{\Gamma \Rightarrow \Delta}{\Gamma \Rightarrow A, \Delta}\;Rw$$

$$i \in \{0,1\}\;\frac{\Gamma, A_i \Rightarrow \Delta}{\Gamma, A_0 \wedge A_1 \Rightarrow \Delta}\;L\wedge \qquad \frac{\Gamma \Rightarrow A, \Delta \qquad \Gamma \Rightarrow B, \Delta}{\Gamma \Rightarrow A \wedge B, \Delta}\;R\wedge$$

$$\frac{\Gamma, A \Rightarrow \Delta \qquad \Gamma, B \Rightarrow \Delta}{\Gamma, A \vee B \Rightarrow \Delta}\;L\vee \qquad i \in \{0,1\}\;\frac{\Gamma \Rightarrow A_i, \Delta}{\Gamma \Rightarrow A_0 \vee A_1, \Delta}\;R\vee$$

$$\frac{\Gamma \Rightarrow A, \Delta \qquad \Gamma, B \Rightarrow \Delta}{\Gamma, A \to B \Rightarrow \Delta}\;L\to \qquad \frac{\Gamma, A \Rightarrow B, \Delta}{\Gamma \Rightarrow A \to B, \Delta}\;R\to$$

$$\frac{\Gamma, \Box\Gamma, \Box A \Rightarrow A}{\Box\Gamma \Rightarrow \Box A}\;GL$$

Fig. 1. The sequent calculus **GGL**.

[12] and hence is complete for GL. Later, for some technical reasons, we will extend the language $\mathcal{L}$ by a sequence of fresh atomic variables $Q = \{q_n\}_{n=0}^{\infty}$. We will denote this language, the logic and the sequent system for it by $\mathcal{L}(Q)$, $\mathsf{GL}(Q)$ and $\mathbf{GGL}(Q)$, respectively.

The second ingredient we need is the arithmetical theories and the provability interpretation they provide for the logic GL. We only recall some important points and for the rest refer the reader to [4]. Let $\mathcal{L}_{\mathrm{PA}} = \{\leq, s, +, \cdot, \exp, 0\}$ be the usual language of Peano arithmetic augmented with the symbol exp with the intended meaning $\exp(n) = 2^n$. The expressions $\forall x \leq t\, \phi(x)$ and $\exists x \leq t\, \phi(x)$ abbreviate $\forall x(x \leq t \to \phi(x))$ and $\exists x(x \leq t \wedge \phi(x))$, respectively. The occurrence of the quantifiers in these formulas are called bounded. By Σ_1, we mean the least class of formulas in $\mathcal{L}_{\mathrm{PA}}$ containing the atomic formulas and their negations and closed under conjunction, disjunction, bounded quantifiers and existential quantifiers. By the abuse of notation, we extend Σ_1 to include any formula logically equivalent to a formula in Σ_1. The formulas in Σ_1 describe recursively enumerable sets and vice verse. By $I\Sigma_1$, we mean a basic quantifier-free theory defining the symbols of the language [4], extended by the induction axiom $\phi(0) \wedge \forall x(\phi(x) \to \phi(s(x))) \to \forall x \phi(x)$, where $\phi(x) \in \Sigma_1$. The theory $I\Sigma_1$ enjoys Σ_1-completeness, meaning that for any sentence $\phi \in \Sigma_1$, if $\mathbb{N} \vDash \phi$ then $I\Sigma_1 \vdash \phi$. A theory T is called Σ_1-sound if $T \vdash \phi$ implies $\mathbb{N} \vDash \phi$, for any sentence $\phi \in \Sigma_1$.

One of the interesting properties of $I\Sigma_1$ is its power to formalize a basic amount of meta-mathematics. Let $\ulcorner \phi \urcorner$ be one of the natural Gödel numberings for the formulas of $\mathcal{L}_{\mathrm{PA}}$ and set $\mathrm{Pr}(x) \in \Sigma_1$ as a predicate satisfying

(i) $I\Sigma_1 \vdash \phi$ iff $\mathbb{N} \vDash \mathrm{Pr}(\ulcorner \phi \urcorner)$,

(ii) $I\Sigma_1 \vdash \mathrm{Pr}(\ulcorner \phi \to \psi \urcorner) \to (\mathrm{Pr}(\ulcorner \phi \urcorner) \to \mathrm{Pr}(\ulcorner \psi \urcorner))$,

(iii) (formalized Σ_1-completeness) $I\Sigma_1 \vdash \phi \to \mathrm{Pr}(\ulcorner\phi\urcorner)$, for any $\phi \in \Sigma_1$.

For such a predicate, see [4]. We fix this predicate throughout the paper as the provability predicate for $I\Sigma_1$. Now, let T be a recursively enumerable theory over $\mathcal{L}_{\mathrm{PA}}$ extending $I\Sigma_1$. By a provability predicate for T, we mean a formula $\mathrm{Pr}_T(x) \in \Sigma_1$ such that:

(i) $T \vdash \phi$ iff $\mathbb{N} \vDash \mathrm{Pr}_T(\ulcorner\phi\urcorner)$,

(ii) $I\Sigma_1 \vdash \mathrm{Pr}_T(\ulcorner\phi \to \psi\urcorner) \to (\mathrm{Pr}_T(\ulcorner\phi\urcorner) \to \mathrm{Pr}_T(\ulcorner\psi\urcorner))$,

(iii) $I\Sigma_1 \vdash \mathrm{Pr}(\ulcorner\phi\urcorner) \to \mathrm{Pr}_T(\ulcorner\phi\urcorner)$.

For simplicity, we usually write $\mathrm{Pr}_T(\phi)$ for $\mathrm{Pr}_T(\ulcorner\phi\urcorner)$. In this paper, we only work with recursively enumerable theories T extending $I\Sigma_1$. Our provability predicates also formally reflect this fact as the part *(iii)* demands. For the future reference, to address *(iii)*, if Pr_T is clear from the context, we say that T is extending $I\Sigma_1$, provably in $I\Sigma_1$. It is easy to see that any provability predicate Pr_T satisfies the following conditions:

(i) $T \vdash \phi$ iff $I\Sigma_1 \vdash \mathrm{Pr}_T(\phi)$,

(ii) $I\Sigma_1 \vdash \mathrm{Pr}_T(\phi \to \psi) \to (\mathrm{Pr}_T(\phi) \to \mathrm{Pr}_T(\psi))$,

(iii) $I\Sigma_1 \vdash \mathrm{Pr}_T(\phi) \to \mathrm{Pr}_T(\mathrm{Pr}_T(\phi))$.

The first is a consequence of Σ_1-completeness of $I\Sigma_1$ and the third is a consequence of formalized Σ_1-completeness of $I\Sigma_1$ together with the fact that T extends $I\Sigma_1$, provably in $I\Sigma_1$. It is routine to see that these conditions imply $I\Sigma_1 \vdash \mathrm{Pr}_T(\mathrm{Pr}_T(\phi) \to \phi) \to \mathrm{Pr}_T(\phi)$, for any sentence $\phi \in \mathcal{L}_{\mathrm{PA}}$ and specifically, the formalized Gödel's second incompleteness theorem, i.e., $I\Sigma_1 \vdash \mathrm{Pr}_T(\neg\mathrm{Pr}_T(\bot)) \to \mathrm{Pr}_T(\bot)$. Denoting $\neg\mathrm{Pr}_T(\bot)$ by Cons_T, we have $I\Sigma_1 \vdash \mathrm{Cons}_T \to \neg\mathrm{Pr}_T(\mathrm{Cons}_T)$ [4].

Definition 2.1 By an arithmetical substitution σ, we mean a function assigning arithmetical sentences to the atomic formulas of $\mathcal{L}$. Let $T \supseteq I\Sigma_1$ be a theory, Pr_T be a provability predicate for T and $A \in \mathcal{L}$ be a modal formula. Then, by $A^{\mathrm{Pr}_T,\sigma}$, we mean an arithmetical sentence resulting by substituting the atoms of A according to σ and interpreting its boxes as Pr_T.

In the following, we will present the uniform version of Solovay's characterization of GL [20] investigated in [17,3,21,9,5]. For a clear exposition and the generality we use here, see [4].

Theorem 2.2 *(Uniform Solovay's Theorem) If* $\mathsf{GL} \vdash A$*, then* $I\Sigma_1 \vdash A^{\mathrm{Pr}_T,\sigma}$*, for any arithmetical theory* $T \supseteq I\Sigma_1$*, any provability predicate* Pr_T *for* T *and any arithmetical substitution* σ*. Conversely, for any* Σ_1*-sound theory* $T \supseteq I\Sigma_1$*, there is a provability predicate* Pr_T *and an arithmetical substitution* $*$ *such that for any modal formula* $A \in \mathcal{L}$*, if* $T \vdash A^{\mathrm{Pr}_T,*}$ *then* $\mathsf{GL} \vdash A$*.*

Notice the uniformity in the completeness part of the theorem that provides one arithmetical substitution for all modal formulas. This property will play a crucial role in our completeness results later in Section 5.

3 Hierarchical Modal Logics

In this section, we first introduce a polymodal language to reflect the provability predicates for a hierarchy of theories rather than just one theory. Then, we will introduce the hierarchical counterparts of some basic modal logics.

Definition 3.1 Let $\mathcal{L}_\infty = \{\wedge, \vee, \bot, \rightarrow\} \cup \{\Box_n\}_{n=0}^{\infty}$ be a modal language with infinitely many modalities. The set of formulas in this language, also denoted by $\mathcal{L}_\infty$, is defined as the least set of expressions containing the atomic formulas and $\bot$ and closed under all propositional operations and the following operation: If $A \in \mathcal{L}_\infty$ and n is *strictly greater than* the index of any box occurring in A, then $\Box_n A \in \mathcal{L}_\infty$. By the rank of $A \in \mathcal{L}_\infty$, denoted by $r(A)$, we mean the greatest index of the boxes occurring in A. If there is none, then set $r(A) = -1$. Finally, for a multiset Γ of formulas in $\mathcal{L}_\infty$, define $r(\Gamma)$ as the maximum of the ranks of its elements.

Notice the difference between the formulas in $\mathcal{L}_\infty$ and the usual polymodal formulas. In the former case, we impose a syntactic restriction that only allows a box in a formula if its index is greater than all the indices of the boxes lying in its scope. For instance, the expression $\Box_1(\Box_0 p \rightarrow p)$ is a formula in $\mathcal{L}_\infty$ with rank one, while the expression $\Box_1\Box_1 p$ is not a formula. From now on, we implicitly assume that any polymodal formula used in this paper belongs to the set $\mathcal{L}_\infty$. For instance, whenever we consider an axiom, we only allow the substitutions that result in a formula in $\mathcal{L}_\infty$. As an example, in the axiom $\Box_1 p \rightarrow \Box_1 p$, the formula p can be substituted by $\Box_0 q \rightarrow r$ but not $\Box_2 q$.

Definition 3.2 Consider the following set of axioms:

(H) $\Box_n A \rightarrow \Box_{n+1} A$,

(K_∞) $\Box_n(A \rightarrow B) \rightarrow (\Box_n A \rightarrow \Box_n B)$,

(4_∞) $\Box_n A \rightarrow \Box_{n+1}\Box_n A$,

(D_∞) $\neg\Box_n \bot$,

(T_∞) $\Box_n A \rightarrow A$,

(5_∞) $\neg\Box_n A \rightarrow \Box_{n+1}\neg\Box_n A$,

(L_∞) $\Box_{n+1}(\Box_n A \rightarrow A) \rightarrow \Box_n A$.

Let $\mathcal{A}$ be a set of these axioms. Define the set of $L(\mathcal{A})$-proofs as the least set of finite sequences of formulas containing the sequences with length one of classical tautologies over the language $\mathcal{L}_\infty$ or instances of the axioms in $\mathcal{A}$ and closed under the following two rules:

(MP) If $\{A_i\}_{i=1}^m$ and $\{B_j\}_{j=1}^l$ are $L(\mathcal{A})$-proofs such that $A_m = D$ and $B_l = D \rightarrow E$, then $\{C_k\}_{k=1}^{m+l+1}$ is an $L(\mathcal{A})$-proof, where $C_k = A_k$, for $1 \leq k \leq m$, $C_k = B_{k-m}$, for $m+1 \leq k \leq m+l$ and $C_{m+l+1} = E$,

(NC_∞) If $\{A_i\}_{i=1}^m$ is an $L(\mathcal{A})$-proof such that $A_m = D$ and $r(A_i) < n$, for any $1 \leq i \leq m$, then $\{B_k\}_{k=1}^{m+1}$ is an $L(\mathcal{A})$-proof, where $B_k = A_k$, for $1 \leq k \leq m$ and $B_{m+1} = \Box_n D$,

By the rank of an $L(\mathcal{A})$-proof, we mean the maximum of the ranks of the formulas it contains. If an $L(\mathcal{A})$-proof ends with the formula A, we call it an $L(\mathcal{A})$-proof for A. If there exists an $L(\mathcal{A})$-proof for A, we write $L(\mathcal{A}) \vdash A$. For any (not necessarily finite) set $\Gamma \cup \{A\} \subseteq \mathcal{L}_\infty$, by $L(\mathcal{A}) \vdash \Gamma \Rightarrow A$, we mean the existence of a finite set $\Delta \subseteq \Gamma$ such that $L(\mathcal{A}) \vdash \bigwedge \Delta \to A$. We denote the following $L(\mathcal{A})$'s by their usual modal terminology: $\mathsf{K4}_\infty = L(H, K_\infty, 4_\infty)$, $\mathsf{KD4}_\infty = L(H, K_\infty, 4_\infty, D_\infty)$, $\mathsf{GL}_\infty = L(H, K_\infty, 4_\infty, L_\infty)$, $\mathsf{KD45}_\infty = L(H, K_\infty, 4_\infty, D_\infty, 5_\infty)$, and $\mathsf{S5}_\infty = L(H, K_\infty, 4_\infty, T_\infty, 5_\infty)$.

The only point to clarify is the deviation of (NC_∞) from the usual necessitation rule. To explain, assume we already provided a proof for a statement A using formulas with maximum rank $n-1$. This argument, as it refers to the meta-theories up to the level $n-1$, must live in a higher meta-theory. Hence, it is reasonable to conclude "the provability of A" in the level n or higher, i.e., $\Box_m A$, for $m \geq n$. Note that even if $\Box_k A \in \mathcal{L}_\infty$, for some $k \leq n-1$, we can not use (NC_∞) to conclude $\Box_k A$ as the whole proof lives in the level n or higher. In this sense, our necessitation is a global operation depending on the whole proof. One may wonder if it is the case that any provable formula A has a proof with rank bounded by $r(A)$. To prove this form of analyticity, we need to design cut-free sequent calculi for our logics which is beyond the scope of this paper, see [2]. However, we use an indirect method to prove it for $\mathbf{K4}_\infty$ in Section 7.

4 Provability Models

The canonical notion of model for the introduced hierarchical modal logics must consist of a classical model to interpret the box-free formulas and a hierarchy of theories to interpret the boxes.

Definition 4.1 A provability model is a tuple $\mathfrak{M} = (M, \{T_n\}_{n=0}^\infty, \{\mathrm{Pr}_n\}_{n=0}^\infty)$, where M is a model of $I\Sigma_1$, $\{T_n\}_{n=0}^\infty$ is a hierarchy of recursively enumerable arithmetical theories, all extending $I\Sigma_1$ and $\{\mathrm{Pr}_n\}_{n=0}^\infty$ is a sequence of provability predicates such that for any $n \geq 0$, Pr_n is a provability predicate for T_n and $T_n \subseteq T_{n+1}$, provably in $I\Sigma_1$, i.e., $I\Sigma_1 \vdash \mathrm{Pr}_n(\phi) \to \mathrm{Pr}_{n+1}(\phi)$, for any arithmetical sentence ϕ. We denote M, T_n and Pr_n, by $|\mathfrak{M}|$, $T_n^{\mathfrak{M}}$ and $\mathrm{Pr}_n^{\mathfrak{M}}$, respectively.

Remark 4.2 Here are some remarks. First, we assume that all theories extend the basic theory $I\Sigma_1$ and $M \vDash I\Sigma_1$ as we want our theories and our model to have the power to implement and understand the basic meta-mathematical theorems, respectively. As long as the base theory is powerful enough, the choice of $I\Sigma_1$ is immaterial. Secondly, from now on, as a theory is uniquely determined by its provability predicate, by dropping $\{T_n\}_{n=0}^\infty$, we only use the pair $\mathfrak{M} = (M, \{\mathrm{Pr}_n\}_{n=0}^\infty)$ to denote a provability model.

Definition 4.3 (i) The class of all provability models is denoted by $\mathbf{PrM}$.

(ii) A provability model $(M, \{\mathrm{Pr}_n\}_{n=0}^\infty)$ is called *consistent*, if for any $n \geq 0$, the model M thinks that T_n is consistent and $T_{n+1} \vdash \mathrm{Cons}(T_n)$, i.e., $M \vDash$

$\text{Cons}(T_n)$ and $M \models \text{Pr}_{n+1}(\text{Cons}(T_n))$. The class of all consistent provability models is denoted by **Cons**.

(*iii*) A provability model $(M, \{\text{Pr}_n\}_{n=0}^{\infty})$ is *constant*, if for any n and m, $(M, \{\text{Pr}_n\}_{n=0}^{\infty})$ thinks that $T_n = T_m$, i.e., $M \models \text{Pr}_m(\phi) \leftrightarrow \text{Pr}_n(\phi)$ and $M \models \text{Pr}_0(\text{Pr}_m(\phi) \leftrightarrow \text{Pr}_n(\phi))$, for any sentence $\phi \in \mathcal{L}_{\text{PA}}$. The class of all constant provability models is denoted by **Cst**.

Definition 4.4 By an arithmetical substitution, we mean a function assigning an arithmetical sentence to any atomic formula of $\mathcal{L}_\infty$. If $\mathfrak{M} = (M, \{\text{Pr}_n\}_{n=0}^{\infty})$ is a provability model, $A \in \mathcal{L}_\infty$ is a formula and σ is an arithmetical substitution, then by $A^{\mathfrak{M},\sigma}$, we mean the arithmetical sentence resulting from substituting the atomic formulas in A according to σ and interpreting $\Box_n$ in A as Pr_n. If Γ is a set of formulas, by $\Gamma^{\mathfrak{M},\sigma}$, we mean the set $\{A^{\mathfrak{M},\sigma} \mid A \in \Gamma\}$.

Definition 4.5 Let $\mathfrak{M}$ be a provability model and $A \in \mathcal{L}_\infty$ be a formula. Then, A is satisfied in the model $\mathfrak{M}$, denoted by $\mathfrak{M} \models A$, if $|\mathfrak{M}| \models A^{\mathfrak{M},\sigma}$, for any arithmetical substitution σ. Moreover, if $\Gamma \cup \{A\}$ is a (not necessarily finite) set of formulas, $\mathcal{C}$ is a class of provability models and σ is an arithmetical substitution, we write $\mathcal{C} \models \Gamma^\sigma \Rightarrow A^\sigma$ when $|\mathfrak{M}| \models \bigwedge \Gamma^{\mathfrak{M},\sigma}$ implies $|\mathfrak{M}| \models A^{\mathfrak{M},\sigma}$, for any $\mathfrak{M} \in \mathcal{C}$ and we write $\mathcal{C} \models \Gamma \Rightarrow A$ if $\mathcal{C} \models \Gamma^\sigma \Rightarrow A^\sigma$, for any arithmetical substitution σ.

Example 4.6 Define $T_0 = I\Sigma$ and $T_{n+1} = T_n + \text{Cons}(T_n)$, for any $n \geq 0$ and set $\text{Pr}_0 = \text{Pr}$ and $\text{Pr}_{n+1}(\phi) = \text{Pr}_n(\text{Cons}_{T_n} \to \phi)$. Then, the pair $(\mathbb{N}, \{\text{Pr}_n\}_{n=0}^{\infty})$ is clearly a consistent provability model. To have an example of satisfaction of a formula in a provability model, note that $(\mathbb{N}, \{T_n\}_{n=0}^{\infty}) \models \Box_{n+1}(\neg\Box_n p \vee \neg\Box_n \neg p)$, as for any arithmetical substitution σ, as $T_{n+1} \vdash \neg\text{Pr}_n(\bot)$, $I\Sigma_1 \subseteq T_{n+1}$ and Pr_n is a provability predicate, we have $T_{n+1} \vdash \neg\text{Pr}_n(p^\sigma) \vee \neg\text{Pr}_n(\neg p^\sigma)$ which implies $\mathbb{N} \models \text{Pr}_{n+1}(\neg\text{Pr}_n(p^\sigma) \vee \neg\text{Pr}_n(\neg p^\sigma))$.

Lemma 4.7 *Let $(L, \mathcal{C})$ be one of the pairs* $(\mathsf{K4}_\infty, \mathbf{PrM})$, $(\mathsf{KD4}_\infty, \mathbf{Cons})$, *or* $(\mathsf{GL}_\infty, \mathbf{Cst})$. *If A has an L-proof with rank n, then $|\mathfrak{M}| \models A^{\mathfrak{M},\sigma}$ and $|\mathfrak{M}| \models \text{Pr}_m(A^{\mathfrak{M},\sigma})$, for any provability model $\mathfrak{M} \in \mathcal{C}$, any $m > n$ and any arithmetical substitution σ.*

Proof. Fix σ and $\mathfrak{M} = (M, \{\text{Pr}_n\}_{n=0}^{\infty}) \in \mathcal{C}$ and denote $D^{\mathfrak{M},\sigma}$ by D^σ, for any $D \in \mathcal{L}_\infty$. Now, use a structural induction on the set of L-proofs to prove the claim. If A is a classical tautology or an instance of the axioms (H), (K_∞), or (4_∞), we first show $I\Sigma_1 \vdash A^\sigma$. The case for the classical tautology, (H) and (K_∞) are easy. For (4_∞), we have $A = \Box_{n-1}B \to \Box_n\Box_{n-1}B$, for some B. Therefore, $A^\sigma = \text{Pr}_{n-1}(B^\sigma) \to \text{Pr}_n(\text{Pr}_{n-1}(B^\sigma))$. As $\text{Pr}_{n-1} \in \Sigma_1$, by the formalized Σ_1-completeness, we have $I\Sigma_1 \vdash \text{Pr}_{n-1}(B^\sigma) \to \text{Pr}(\text{Pr}_{n-1}(B^\sigma))$. Finally, as $I\Sigma \subseteq T_n$ provably in $I\Sigma_1$, we have $I\Sigma_1 \vdash \text{Pr}_{n-1}(B^\sigma) \to \text{Pr}_n(\text{Pr}_{n-1}(B^\sigma))$. Now, as $I\Sigma_1 \vdash A^\sigma$ for a classical tautology or an instance of the axioms (H), (K_∞) or (4_∞), we have $M \models A^\sigma$ as $M \models I\Sigma_1$. On the other hand, as $I\Sigma_1 \subseteq T_m$, we have $T_m \vdash A^\sigma$. By Σ_1-completeness, we reach $I\Sigma_1 \vdash \text{Pr}_m(A^\sigma)$ which implies $M \models \text{Pr}_m(A^\sigma)$, again by $M \models I\Sigma_1$.

For the axiom (D_∞), we have $A = \neg\Box_n\bot$ and $\mathfrak{M} \in \mathbf{Cons}$. Hence, $M \models$

$\neg\mathrm{Pr}_n(\bot)$ and $M \models \mathrm{Pr}_{n+1}(\neg\mathrm{Pr}_n(\bot))$, by definition. As $m \geq n+1$ and the hierarchy is increasing, provably in $I\Sigma_1$, we have $I\Sigma_1 \vdash \mathrm{Pr}_{n+1}(\neg\mathrm{Pr}_n(\bot)) \to \mathrm{Pr}_m(\neg\mathrm{Pr}_n(\bot))$ and hence, we reach $M \models \mathrm{Pr}_m(\neg\mathrm{Pr}_n(\bot))$.
For the axiom (L_∞), we have $A = \Box_n(\Box_{n-1}B \to B) \to \Box_{n-1}B$, for some B and $\mathfrak{M} \in \mathbf{Cst}$. Let $\phi = \mathrm{Pr}_{n-1}(B^\sigma) \to B^\sigma$. Then, $M \models \mathrm{Pr}_n(\phi) \leftrightarrow \mathrm{Pr}_{n-1}(\phi)$ and $M \models \mathrm{Pr}_0(\mathrm{Pr}_n(\phi) \leftrightarrow \mathrm{Pr}_{n-1}(\phi))$ as $\mathfrak{M}$ is constant. Therefore, as $T_0 \subseteq T_m$, provably in $I\Sigma_1$, we have $M \models \mathrm{Pr}_m(\mathrm{Pr}_n(\phi) \leftrightarrow \mathrm{Pr}_{n-1}(\phi))$. Hence, M thinks that A^σ and $\mathrm{Pr}_m(A^\sigma)$ are equivalent to $\mathrm{Pr}_{n-1}(\mathrm{Pr}_{n-1}(B^\sigma) \to B^\sigma) \to \mathrm{Pr}_{n-1}(B^\sigma)$ and $\mathrm{Pr}_m(\mathrm{Pr}_{n-1}(\mathrm{Pr}_{n-1}(B^\sigma) \to B^\sigma) \to \mathrm{Pr}_{n-1}(B^\sigma))$, respectively. However, as Pr_{n-1} is a provability predicate, we have $I\Sigma_1 \vdash \mathrm{Pr}_{n-1}(\mathrm{Pr}_{n-1}(B^\sigma) \to B^\sigma) \to \mathrm{Pr}_{n-1}(B^\sigma)$ and hence $M \models \mathrm{Pr}_{n-1}(\mathrm{Pr}_{n-1}(B^\sigma) \to B^\sigma) \to \mathrm{Pr}_{n-1}(B^\sigma)$. Moreover, we have $I\Sigma_1 \vdash \mathrm{Pr}(\mathrm{Pr}_{n-1}(\mathrm{Pr}_{n-1}(B^\sigma) \to B^\sigma) \to \mathrm{Pr}_{n-1}(B^\sigma))$, by formalized Σ_1-completeness. As $I\Sigma_1 \subseteq T_m$, provably in $I\Sigma_1$, we finally reach $I\Sigma_1 \vdash \mathrm{Pr}_m(\mathrm{Pr}_{n-1}(\mathrm{Pr}_{n-1}(B^\sigma) \to B^\sigma) \to \mathrm{Pr}_{n-1}(B^\sigma))$ which implies $M \models \mathrm{Pr}_m(\mathrm{Pr}_{n-1}(\mathrm{Pr}_{n-1}(B^\sigma) \to B^\sigma) \to \mathrm{Pr}_{n-1}(B^\sigma))$.
For the rules, if A is the result of the modus ponens rule over the L-proofs for B and $B \to A$, by the induction hypothesis, we have $M \models B^\sigma$, $M \models \mathrm{Pr}_m(B^\sigma)$, $M \models B^\sigma \to A^\sigma$, and $M \models \mathrm{Pr}_m(B^\sigma \to A^\sigma)$. Hence, $M \models A^\sigma$ and $M \models \mathrm{Pr}_m(A^\sigma)$, as Pr_m is a provability predicate. If A is a consequence of the rule (NC_∞), then $A = \Box_n B$, for some B and n is greater than the rank of all formulas in the proof prior to B. By the induction hypothesis, we have $M \models \mathrm{Pr}_n(B^\sigma)$. Moreover, by the formalized Σ_1-completeness, we have $I\Sigma_1 \vdash \mathrm{Pr}_n(B^\sigma) \to \mathrm{Pr}(\mathrm{Pr}_n(B^\sigma))$. As $I\Sigma_1 \subseteq T_m$, provably in $I\Sigma_1$, we reach $I\Sigma_1 \vdash \mathrm{Pr}_n(B^\sigma) \to \mathrm{Pr}_m(\mathrm{Pr}_n(B^\sigma))$ which proves $M \models \mathrm{Pr}_m(\mathrm{Pr}_n(B^\sigma))$, as $M \models I\Sigma_1$. □

Theorem 4.8 *(Soundness Theorem)*

(i) If $\mathsf{K4}_\infty \vdash \Gamma \Rightarrow A$ *then* $\mathbf{PrM} \models \Gamma \Rightarrow A$.

(ii) If $\mathsf{KD4}_\infty \vdash \Gamma \Rightarrow A$ *then* $\mathbf{Cons} \models \Gamma \Rightarrow A$.

(iii) If $\mathsf{GL}_\infty \vdash \Gamma \Rightarrow A$ *then* $\mathbf{Cst} \models \Gamma \Rightarrow A$.

Proof. We only prove the case of $\mathsf{K4}_\infty$. The rest are similar. If $\mathsf{K4}_\infty \vdash \Gamma \Rightarrow A$, then, there exists a finite set $\Delta \subseteq \Gamma$ such that $\mathsf{K4}_\infty \vdash \bigwedge \Delta \to A$. Then, by Lemma 4.7, for any $\mathfrak{M} \in \mathbf{PrM}$ and any arithmetical substitution σ, we have $|\mathfrak{M}| \models \bigwedge \Delta^{\mathfrak{M},\sigma} \to A^{\mathfrak{M},\sigma}$. Therefore, $\mathfrak{M} \models \Gamma \Rightarrow A$. □

5 Completeness Results

In this section, we will provide the completeness results for the provability interpretations we have provided before.

5.1 Logics $\mathsf{K4}_\infty$ and $\mathsf{KD4}_\infty$

For the completeness of $\mathsf{K4}_\infty$ and $\mathsf{KD4}_\infty$, our strategy is using a translation between GL and $\mathsf{K4}_\infty$ to reduce the completeness to uniform Solovay's theorem.

Definition 5.1 Let $Q = \{q_n\}_{n=0}^\infty$ be a sequence of fresh atomic formulas occurring nowhere in the formulas of $\mathcal{L}$. Define the translation function

$t : \mathcal{L}_\infty \to \mathcal{L}(Q)$ as follows: $\bot^t = \bot$, $p^t = p$, for any atomic formula p, $(B \circ C)^t = B^t \circ C^t$, for any $\circ \in \{\wedge, \vee, \to\}$ and $(\Box_n B)^t = \Box(\bigwedge_{i=0}^n q_i \to B^t)$.

The translation t is a syntactical way to connect the provability interpretations of $\mathsf{K4}_\infty$ to that of GL by interpreting a (suitable) *hierarchy of theories* as *a base theory* extended by a sequence of formulas. The following lemma provides the connection we are seeking. The proof is the main machinery of the present paper and shall be given in a separate Section 7.

Lemma 5.2 *(Reduction Lemma) If* $\mathsf{GL}(Q) \vdash A^t$ *then* $\mathsf{K4}_\infty \vdash A$.

Theorem 5.3 *(Uniform Strong Completeness) Let* $(L, \mathcal{C})$ *be one of the pairs* $(\mathsf{K4}_\infty, \mathbf{PrM})$ *or* $(\mathsf{KD4}_\infty, \mathbf{Cons})$. *Then, for any* Σ_1*-sound recursively enumerable arithmetical theory* $T \supseteq I\Sigma_1$, *there exist a hierarchy of theories* $\{T_n\}_{n=0}^\infty$, *all extending* T, *a hierarchy of provability predicates* $\{\mathrm{Pr}_n\}_{n=0}^\infty$ *for* $\{T_n\}_{n=0}^\infty$ *and an arithmetical substitution* τ *such that* $T_n \subseteq T_{n+1}$, *provably in* $I\Sigma_1$, *for any* n *and for any set (not necessarily finite)* $\Gamma \cup \{A\}$ *of formulas in* $\mathcal{L}_\infty$, *if* $\{(M, \{\mathrm{Pr}_n\}_{n=0}^\infty) \in \mathcal{C} \mid M \vDash T\} \vDash \Gamma^\tau \Rightarrow A^\tau$, *then* $L \vdash \Gamma \Rightarrow A$.

Proof. We first prove the claim for $L = \mathsf{K4}_\infty$. Let Pr_T and $*$ be the provability predicate and the substitution that the uniform Solovay's theorem, Theorem 2.2, provides. Therefore, $T \vdash B^{\mathrm{Pr}_T, *}$ iff $\mathsf{GL}(Q) \vdash B$, for any formula $B \in \mathcal{L}(Q)$. For any $n \geq 0$, set $T_n = T + \{q_i^*\}_{i=0}^n$ with the provability predicate $\mathrm{Pr}_n(\phi) = \mathrm{Pr}_T(\bigwedge_{i=0}^n q_i^* \to \phi)$. We claim that $\tau = *$ together with the hierarchies $\{T_n\}_{n=0}^\infty$ and $\{Pr_n\}_{n=0}^\infty$ satisfies the properties the theorem claims. First, note that by definition, $T \subseteq T_n$ and $T_n \subseteq T_{n+1}$, provably in $I\Sigma_1$, for any n. Then, let M be an arbitrary model of T and fix $\mathfrak{M}_M = (M, \{\mathrm{Pr}_n\}_{n=0}^\infty)$. Recall that the translation between $\mathsf{K4}_\infty$ and $\mathsf{GL}(Q)$ interprets $\Box_m C$ as $\Box(\bigwedge_{i=0}^m q_i \to C^t)$. Therefore, it is easy to see that $D^{\mathfrak{M}_M, \tau} = (D^t)^{\mathrm{Pr}_T, *}$, for any $D \in \mathcal{L}_\infty$. Now, if for any $M \vDash T$, we have $M \vDash \Gamma^{\mathfrak{M}_M, \tau} \Rightarrow A^{\mathfrak{M}_M, \tau}$, we reach $M \vDash (\Gamma^t)^{\mathrm{Pr}_T, *} \Rightarrow (A^t)^{\mathrm{Pr}_T, *}$ which implies $T \cup (\Gamma^t)^{\mathrm{Pr}_T, *} \vdash (A^t)^{\mathrm{Pr}_T, *}$. Hence, there is a finite $\Delta \subseteq \Gamma$ such that $T \vdash \bigwedge(\Delta^t)^{\mathrm{Pr}_T, *} \to (A^t)^{\mathrm{Pr}_T, *}$. Note that $\Delta^t \cup \{A^t\} \subseteq \mathcal{L}(Q)$. Hence, by uniform Solovay's theorem, $\mathsf{GL}(Q) \vdash \bigwedge \Delta^t \to A^t$. Finally, by using Lemma 5.2, we reach $\mathsf{K4}_\infty \vdash \bigwedge \Delta \to A$ and hence $\mathsf{K4}_\infty \vdash \Gamma \Rightarrow A$.
For $L = \mathsf{KD4}_\infty$, let $\Pi = \{\neg \Box_n \bot, \Box_{n+1} \neg \Box_n \bot\}_{n \in \mathbb{N}}$. Then, it is easy to see that a provability model $\mathfrak{M}$ satisfies all the elements of Π iff it is consistent. Therefore, if $\{(M, \{\mathrm{Pr}_n\}_{n=0}^\infty) \in \mathbf{Cons} \mid M \vDash T\} \vDash \Gamma^\tau \Rightarrow A^\tau$, we can conclude $\{(M, \{\mathrm{Pr}_n\}_{n=0}^\infty) \in \mathbf{PrM} \mid M \vDash T\} \vDash \Pi^\tau, \Gamma^\tau \Rightarrow A^\tau$. Hence, by the first part we have $\mathsf{K4}_\infty \vdash \Gamma \cup \Pi \Rightarrow A$. Since $\mathsf{KD4}_\infty$ proves all formulas in Π, we finally reach $\mathsf{KD4}_\infty \vdash \Gamma \Rightarrow A$. □

Corollary 5.4 *(Strong Completeness)*

(i) If $\mathbf{PrM} \vDash \Gamma \Rightarrow A$ *then* $\mathsf{K4}_\infty \vdash \Gamma \Rightarrow A$.

(ii) If $\mathbf{Cons} \vDash \Gamma \Rightarrow A$ *then* $\mathsf{KD4}_\infty \vdash \Gamma \Rightarrow A$.

5.2 Logic GL_∞

For the logic GL_∞, the canonical strategy is reducing the completeness of GL_∞ directly to Solovay's result. For that purpose, we need the forgetful translation

$f : \mathcal{L}_\infty \to \mathcal{L}$ that keeps the atomic formulas and the propositional connectives intact and maps $\Box_n$ to $\Box$.

Lemma 5.5 *If* $\mathsf{GL} \vdash A^f$ *then* $\mathsf{GL}_\infty \vdash A$.

Proof. We prove the following, where the part (iii) is our main claim. The other two are required to prove (iii).

(i) For any $A \in \mathcal{L}_\infty$ and any natural numbers $m, n > r(A)$, $\mathsf{GL}_\infty \vdash \Box_m A \leftrightarrow \Box_n A$.

(ii) For any $A, B \in \mathcal{L}_\infty$, if $A^f = B^f$, then $\mathsf{GL}_\infty \vdash A \leftrightarrow B$.

(iii) If $\mathsf{GL} \vdash A^f$ then $\mathsf{GL}_\infty \vdash A$.

For (i), it is enough to show that if $n > r(A)$, we have $\mathsf{GL}_\infty \vdash \Box_n A \leftrightarrow \Box_{n+1} A$. The direction $\Box_n A \to \Box_{n+1} A$ is an instance of the axiom (H). For the other direction, by (L_∞), we have $\mathsf{GL}_\infty \vdash \Box_{n+1}(\Box_n A \to A) \to \Box_n A$. Using (K_∞), it is easy to see that $\mathsf{GL}_\infty \vdash \Box_{n+1} A \to \Box_{n+1}(\Box_n A \to A)$. which implies $\mathsf{GL}_\infty \vdash \Box_{n+1} A \to \Box_n A$.
For (ii), use induction on the structure of A. The atomic and propositional cases are straightforward. For the modal case, assume $A = \Box_m C$ which implies $B^f = A^f = \Box C^f$. Hence, there must be a formula $D \in \mathcal{L}_\infty$ such that $B = \Box_n D$ and $C^f = D^f$. By induction hypothesis, $\mathsf{GL}_\infty \vdash C \leftrightarrow D$. Therefore, for a large enough k, we can use (NC_∞) to prove $\mathsf{GL}_\infty \vdash \Box_k(C \leftrightarrow D)$. Hence, by (K_∞), we have $\mathsf{GL}_\infty \vdash \Box_k C \leftrightarrow \Box_k D$. By (i), we have $\mathsf{GL}_\infty \vdash \Box_k C \leftrightarrow \Box_m C$ and $\mathsf{GL}_\infty \vdash \Box_k D \leftrightarrow \Box_n D$. Hence, $\mathsf{GL}_\infty \vdash \Box_m C \leftrightarrow \Box_n D$ which means $\mathsf{GL}_\infty \vdash A \leftrightarrow B$.
For (iii), first consider the translation $g : \mathcal{L} \to \mathcal{L}_\infty$ as follows: $\bot^g = \bot$, $p^g = p$, for any atomic formula p, $(B \circ C)^g = B^g \circ C^g$, for any $\circ \in \{\wedge, \vee, \to\}$ and $(\Box B)^g = \Box_n B^g$, where $n = r(B^g) + 1$. It is clear that $(B^g)^f = B$. Now, we show that if $\mathsf{GL} \vdash B$, then there exists a formula $B' \in \mathcal{L}_\infty$ such that $\mathsf{GL}_\infty \vdash B'$ and $B'^f = B$. For that purpose, use induction on the length of the proof of B in GL. If B is a classical tautology, set $B' = B^g$. It is easy to see that B^g is also a classical tautology. For the axiom (K), if $B = \Box(C \to D) \to (\Box C \to \Box D)$, then set $B' = \Box_n(C^g \to D^g) \to (\Box_n C^g \to \Box_n D^g)$, where $n = max\{r(C^g), r(D^g)\} + 1$. The proof for the other axioms are similar. For the modus ponens, if $\mathsf{GL} \vdash C$ and $\mathsf{GL} \vdash C \to B$, by induction hypothesis, there are formulas C', C'', and B' such that $C'^f = C$, $C''^f = C$, $B'^f = B$, $\mathsf{GL}_\infty \vdash C'$ and $\mathsf{GL}_\infty \vdash C'' \to B'$. By part (ii), we have $\mathsf{GL}_\infty \vdash C' \leftrightarrow C''$. Hence, $\mathsf{GL}_\infty \vdash B'$. For necessitation, we must have $B = \Box C$ and $\mathsf{GL} \vdash C$. By induction hypothesis, there exists C' such that $\mathsf{GL}_\infty \vdash C'$ and $C'^f = C$. Then, for a large enough n, by (NC_∞), we have $\mathsf{GL}_\infty \vdash \Box_n C'$.
Now, it is easy to prove (iii). If $\mathsf{GL} \vdash A^f$, then, there exists a formula B' such that $\mathsf{GL}_\infty \vdash B'$ and $B'^f = A^f$. Hence, by part (ii), we have $\mathsf{GL}_\infty \vdash A \leftrightarrow B'$. Hence, $\mathsf{GL}_\infty \vdash A$. ☐

Theorem 5.6 *(Uniform Strong Completeness) For any Σ_1-sound recursively enumerable arithmetical theory $T \supseteq I\Sigma_1$, there is a provability predicate Pr_T and an arithmetical substitution τ such that for any set (not necessarily finite)*

$\Gamma \cup \{A\}$ *of formulas in* $\mathcal{L}_\infty$, *if* $\{(M, \{\mathrm{Pr}_T\}_{n=0}^{\infty}) \mid M \models T\} \models \Gamma^\tau \Rightarrow A^\tau$, *then* $\mathsf{GL}_\infty \vdash \Gamma \Rightarrow A$.

Proof. First, notice that for any arithmetical substitution σ, any formula $B \in \mathcal{L}_\infty$ and any $M \models T$, if we set $\mathfrak{M}_M = (M, \{\mathrm{Pr}_T\}_{n=0}^{\infty})$, then $B^{\mathfrak{M}_M, \sigma} = (B^f)^{\mathrm{Pr}_T, \sigma}$ simply because all provability predicates are equal to Pr_T. Therefore, $\{(M, \{\mathrm{Pr}_T\}_{n=0}^{\infty}) \mid M \models T\} \models \Gamma^\tau \Rightarrow A^\tau$ implies $M \models (\Gamma^f)^{\mathrm{Pr}_T, \tau} \Rightarrow (A^f)^{\mathrm{Pr}_T, \tau}$, for any $M \models T$. Hence, $T \cup (\Gamma^f)^{\mathrm{Pr}_T, \tau} \vdash (A^f)^{\mathrm{Pr}_T, \tau}$. Therefore, there is a finite set $\Delta \subseteq \Gamma$ such that $T \vdash \bigwedge (\Delta^f)^{\mathrm{Pr}_T, \tau} \to (A^f)^{\mathrm{Pr}_T, \tau}$. By Theorem 2.2, we have $\mathsf{GL} \vdash \bigwedge \Delta^f \to A^f$. Finally, by Lemma 5.5, we have $\mathsf{GL}_\infty \vdash \Delta \Rightarrow A$ and hence $\mathsf{GL}_\infty \vdash \Gamma \Rightarrow A$. □

Corollary 5.7 *(Strong Completeness) If* **Cst** $\models \Gamma \Rightarrow A$ *then* $\mathsf{GL}_\infty \vdash \Gamma \Rightarrow A$.

6 The Extensions of $\mathsf{KD45}_\infty$

The logic $\mathsf{S5}_\infty$ is too strong to have a provability interpretation. The axioms (T_∞), (4_∞), and (5_∞) together imply that $\Box_n A \leftrightarrow \Box_{n+1}\Box_n A$ and $\neg\Box_n A \leftrightarrow \Box_{n+1}\neg\Box_n A$ which informally state that the provability in T_n is decidable in T_{n+1} and as T_{n+1} is recursively enumerable, we reach the decidability of T_n that is impossible. In this section, we will prove a stronger version that generalizes the result to $\mathsf{KD45}_\infty$.

Theorem 6.1 *There is no provability model for any extension of the logic* $\mathsf{KD45}_\infty$. *Specially,* $\mathsf{S5}_\infty$ *has no provability model.*

Proof. Assume $(M, \{\mathrm{Pr}_n\}_{n=0}^{\infty}) \models \mathsf{KD45}_\infty$. Then, for any arithmetical substitution σ, we have $M \models \neg\mathrm{Pr}_n(p^\sigma) \to \mathrm{Pr}_{n+1}(\neg\mathrm{Pr}_n(p^\sigma))$. Pick an arithmetical substitution that maps p to the arithmetical sentence $\mathrm{Pr}_{n+1}(\bot)$. Hence,

$$M \models \neg\mathrm{Pr}_n(\mathrm{Pr}_{n+1}(\bot)) \to \mathrm{Pr}_{n+1}(\neg\mathrm{Pr}_n(\mathrm{Pr}_{n+1}(\bot))). \qquad (*)$$

On the other hand, by the formalized Σ_1-completeness and the fact that $I\Sigma_1 \subseteq T_n$, provably in $I\Sigma_1$, we have $I\Sigma_1 \vdash \mathrm{Pr}_{n+1}(\bot) \to \mathrm{Pr}_n(\mathrm{Pr}_{n+1}(\bot))$ and hence $I\Sigma_1 \vdash \neg\mathrm{Pr}_n(\mathrm{Pr}_{n+1}(\bot)) \to \neg\mathrm{Pr}_{n+1}(\bot)$. Thus, $T_{n+1} \vdash \neg\mathrm{Pr}_n(\mathrm{Pr}_{n+1}(\bot)) \to \neg\mathrm{Pr}_{n+1}(\bot)$. Moreover, by Σ_1-completeness, we have $I\Sigma_1 \vdash \mathrm{Pr}_{n+1}(\neg\mathrm{Pr}_n(\mathrm{Pr}_{n+1}(\bot)) \to \neg\mathrm{Pr}_{n+1}(\bot))$. Therefore, $I\Sigma_1 \vdash \mathrm{Pr}_{n+1}(\neg\mathrm{Pr}_n(\mathrm{Pr}_{n+1}(\bot))) \to \mathrm{Pr}_{n+1}(\neg\mathrm{Pr}_{n+1}(\bot))$. And since $M \models I\Sigma_1$, we have $M \models \mathrm{Pr}_{n+1}(\neg\mathrm{Pr}_n(\mathrm{Pr}_{n+1}(\bot))) \to \mathrm{Pr}_{n+1}(\neg\mathrm{Pr}_{n+1}(\bot))$. Therefore, using $(*)$, we have $M \models \neg\mathrm{Pr}_n(\mathrm{Pr}_{n+1}(\bot)) \to \mathrm{Pr}_{n+1}(\neg\mathrm{Pr}_{n+1}(\bot))$. By the formalized Gödel's second incompleteness theorem, we have $I\Sigma_1 \vdash \neg\mathrm{Pr}_{n+1}(\bot) \to \neg\mathrm{Pr}_{n+1}(\neg\mathrm{Pr}_{n+1}(\bot))$. Therefore, $M \models \neg\mathrm{Pr}_n(\mathrm{Pr}_{n+1}(\bot)) \to \mathrm{Pr}_{n+1}(\bot)$. However, the provability model $(M, \{\mathrm{Pr}_n\}_{n=0}^{\infty})$ is a model of (D_∞). Hence, $M \models \neg\mathrm{Pr}_{n+1}(\bot)$. Hence, $M \models \mathrm{Pr}_n(\mathrm{Pr}_{n+1}(\bot))$. Since $T_n \subseteq T_{n+2}$, provably in $I\Sigma_1$, we reach $M \models \mathrm{Pr}_{n+2}(\mathrm{Pr}_{n+1}(\bot))$. Again, since the provability model is a model for the logic $\mathsf{KD4}_\infty$, it satisfies the formula $\Box_{n+2}\neg\Box_{n+1}\bot$. Hence, $M \models \mathrm{Pr}_{n+2}(\neg\mathrm{Pr}_{n+1}(\bot))$. Therefore, $M \models \mathrm{Pr}_{n+2}(\bot)$, which contradicts with an instance of the axiom (D_∞). □

It is worth mentioning that the main reason behind this lack of provability models for $\mathsf{KD45}_\infty$ is the assumption that all the theories in a provability model are recursively enumerable. In the well-known polymodal provability logic GLP, this condition is relaxed [13] and hence having the axiom (5_∞) does not make any problem. However, we believe that the restriction we use is philosophically justified, as the theories for such interpretations must be human understandable.

7 Proof of the Reduction Lemma

The main strategy to prove the reduction lemma, Lemma 5.2, is using a cut-free proof for A^t in $\mathsf{GL}(Q)$ to construct a $\mathsf{K4}_\infty$-proof for A. For that purpose, we need to prove a stronger version of necessitation in $\mathsf{K4}_\infty$. Let us start with this task. Let $n \geq 0$ be a given natural number and set the n-truncation as the function over $\mathcal{L}_\infty$ defined as follows: $\bot^n = \bot$, $p^n = p$, for any atomic formula p, $(B \circ C)^n = B^n \circ C^n$, for any $\circ \in \{\wedge, \vee, \to\}$ and $(\Box_i B)^n = \Box_i B$, for $i < n$ and $(\Box_i B)^n = \top$, for $i \geq n$. Moreover, for any sequence of formulas $\{A_i\}_{i=1}^m$, define $(\{A_i\}_{i=1}^m)^n = \{A_i^n\}_{i=1}^m$. Observe that for any formula A, if $r(A) < n$, then $A^n = A$ and $r(B^n) < n$, for any formula $B \in \mathcal{L}_\infty$.

Lemma 7.1 *If π is a $\mathsf{K4}_\infty$-proof for A, then π^n is a $\mathsf{K4}_\infty$-proof for A^n. Specially, if $\mathsf{K4}_\infty \vdash A$, then A has a $\mathsf{K4}_\infty$-proof with rank $r(A)$.*

Proof. We use structural induction on the set of the $\mathsf{K4}_\infty$-proofs. If A is a classical tautology, as the translation commutes with the propositional connectives, A^n would also be a classical tautology and hence there is nothing to prove. If A is an instance of the axiom (H), then $A = \Box_k B \to \Box_{k+1} B$. If $k+1 < n$, then $A^n = A$ and hence there is nothing to prove. If $n \leq k+1$, then $A^n = (\Box_k B)^n \to \top$ which is a classical tautology. If A is an instance of the axiom (K_∞), then $A = \Box_k(B \to C) \to (\Box_k B \to \Box_k C)$. If $k < n$, we have $A^n = A$ and hence, there is nothing to prove. If $k \geq n$, we have $A^n = \top \to (\top \to \top)$ which is a classical tautology. If A is an instance of the axiom (4_∞), then $A = \Box_k B \to \Box_{k+1}\Box_k B$. The case $k+1 < n$ is trivial. If $n \leq k+1$, we have $A^n = (\Box_k B)^n \to \top$ which is again a classical tautology. For the rules, if the last rule is modus ponens, there is nothing to prove as the n-truncation commutes with implication. If it is the necessitation rule, then $\pi = \{A_i\}_{i=1}^{m+1}$, $A_{m+1} = \Box_k B$ and $r(A_i) < k$, for any $i \leq m$. Set $\pi' = \{A_i\}_{i=1}^m$. If $k < n$, then $r(\pi) = k < n$. Hence, π remains intact under the n-truncation and hence there is nothing to prove. If $k \geq n$, then using the induction hypothesis, π'^n is a $\mathsf{K4}_\infty$-proof. As $(\Box_k B)^n = \top$, the sequence π^n is just π'^n with one classical tautology $\top$ added to its end. Therefore, π^n is clearly a $\mathsf{K4}_\infty$-proof.
For the second part, if $\mathsf{K4}_\infty \vdash A$, then there is a $\mathsf{K4}_\infty$-proof π for A. Set $n = r(A)+1$. Then, as $r(A) < n$, we have $A^n = A$. By the first part, π^n is a proof for $A^n = A$ and $r(\pi^n) < n = r(A)+1$. $\Box$

Theorem 7.2 *(Strong Necessitation) Let I and J be some finite sets. Then, if $\mathsf{K4}_\infty \vdash \{\Box_{n_i} A_i\}_{i \in I}, \{\Box_n B_j\}_{j \in J} \Rightarrow A$, where $r(A) < n$ and $n_i < n$, for any $i \in I$, then $\mathsf{K4}_\infty \vdash \{\Box_{n_i} A_i\}_{i \in I} \Rightarrow \Box_n A$.*

Proof. Assume π is a $\mathsf{K4}_\infty$-proof for $\bigwedge_{i\in I}\Box_{n_i}A_i \wedge \bigwedge_{j\in J}\Box_n B_j \to A$. Therefore, by Lemma 7.1, π^n is a $\mathsf{K4}_\infty$-proof for $\bigwedge_{i\in I}(\Box_{n_i}A_i)^n \wedge \bigwedge_{j\in J}(\Box_n B_j)^n \to A^n$. As $r(A), n_i < n$, we have $A^n = A$, $(\Box_{n_i}A_i)^n = \Box_{n_i}A_i$ and $(\Box_n B_j)^n = \top$. Hence, π^n is a $\mathsf{K4}_\infty$-proof for $\bigwedge_{i\in I}\Box_{n_i}A_i \wedge \bigwedge_{j\in J}\top \to A$. As $r(\pi^n) < n$, by necessitation, we have $\mathsf{K4}_\infty \vdash \Box_n(\bigwedge_{i\in I}\Box_{n_i}A_i \wedge \bigwedge_{j\in J}\top \to A)$. Hence, by using the axiom (K_∞), we have $\mathsf{K4}_\infty \vdash \bigwedge_{i\in I}\Box_n\Box_{n_i}A_i \wedge \bigwedge_{j\in J}\Box_n\top \to \Box_n A$. Finally, using the axioms (4_∞) and (H) and the fact that $\mathsf{K4}_\infty \vdash \Box_n\top$, we have $\mathsf{K4}_\infty \vdash \bigwedge_{i\in I}\Box_{n_i}A_i \to \Box_n A$. $\Box$

Now, as we have proved the strong necessitation, we are ready to prove the reduction lemma. Define the set $X \subseteq \mathcal{L}(Q)$ as the least set of modal formulas containing $\bot$ and all the atomic formulas (including the atoms in Q) and closed under the propositional connectives and the following rule: If $A \in X$, then:

- $\Box(\bigwedge_{i=0}^n q_i \to A) \in X$, if n is greater than all the indices of q_j's occurring in A.
- $\Box(\bigwedge_{i=0}^m q_i \wedge \bigwedge_{i=m+1}^n \bot \to A) \in X$, if m is greater than or *equal to* all the indices of q_j's occurring in A and $m < n$.

Set X_0 and X_1 as the sets of all formulas in forms $\Box(\bigwedge_{i=0}^n q_i \to A)$ and $\Box(\bigwedge_{i=0}^m q_i \wedge \bigwedge_{i=m+1}^n \bot \to A)$ in X, respectively. Note that these are the only formulas in the form $\Box B$ in X. It is easy to check that X includes all subformulas of formulas D^t, for any $D \in \mathcal{L}_\infty$. The proof is by induction on the structure of D. The only non-trivial case is $D = \Box_n E$. In this case, $D^t = \Box(\bigwedge_{i=0}^n q_i \to E^t)$ in which n is greater than all indices of q_j's occurring in E. Hence, $D^t \in X_0$. It is easy to see that all proper subformulas of D^t also belong to X.

Here are some terminology. An X-proof is a cut-free proof in the system $\mathbf{GGL}(Q)$ consisting only of the formulas in X. By the rank of a formula $A \in X$, denoted by $r(A)$, we mean the greatest number n such that q_n occurs in the formula A. If there is none, set $r(A) = -1$. For any multiset $\Gamma \subseteq X$, by $r(\Gamma)$, we mean the maximum of the ranks of the elements of Γ. An X-proof is called nice, if $r(\Box\Gamma) \leq r(\Box A)$, for any occurrence of the rule

$$\frac{\Gamma, \Box\Gamma, \Box A \Rightarrow A}{\Box\Gamma \Rightarrow \Box A}\ GL$$

in the proof. Note that the equality $r(\Box\Gamma) = r(\Box A)$ is also allowed.

Let n be a natural number and σ_n be the substitution that maps q_i to $\bot$, for $i > n$ and keeps the other atomic formulas intact. It is easy to see that $r(\sigma_n(A)) \leq n$, for any formula $A \in X$. Moreover, for a boxed formula $A \in X$, if $r(A) > n$, we have $r(\sigma_n(A)) = n$. The latter is easy to prove by checking the two forms of the modal formulas in X.

Denoting the result of applying σ_n on a sequence π by $\sigma_n(\pi)$, we have:

Lemma 7.3 *The sets X and X_1 are closed under σ_n and if $r(A) > n$ and $\Box A \in X_0$, then $\sigma_n(\Box A) \in X_1$. Moreover, if π is a nice X-proof, then so is $\sigma_n(\pi)$.*

Proof. For the first part, to show the closure of X under σ_n, use a simple structural induction. We only explain the box case. If $A \in X_0$, we have $A = \Box(\bigwedge_{i=0}^{m} q_i \to B)$. If $n \geq m$, then $\sigma_n(A) = A$, as there is no q_i in A such that $i > n$. Therefore, $\sigma_n(A) \in X$. If $n < m$, then $\sigma_n(A) = \Box(\bigwedge_{i=0}^{n} q_i \wedge \bigwedge_{i=n+1}^{m} \bot \to \sigma_n(B))$. By induction hypothesis, $\sigma_n(B) \in X$. As $n \geq r(\sigma_n(B))$, we have $\sigma_n(A) \in X_1$. If $A \in X_1$, we have $A = \Box(\bigwedge_{i=0}^{m} q_i \wedge \bigwedge_{i=m+1}^{k} \bot \to B)$, where $m < k$. If $n \geq m$, we have $\sigma_n(A) = A$, as there is no q_i in B such that $i > n$. Therefore, $\sigma_n(A) \in X$. If $n < m$, then $\sigma_n(A) = \Box(\bigwedge_{i=0}^{n} q_i \wedge \bigwedge_{i=n+1}^{k} \bot \to \sigma_n(B))$. By induction hypothesis, we have $\sigma_n(B) \in X$ which implies $\sigma_n(A) \in X_1$, as $n \geq r(\sigma_n(B))$. Note that our argument shows that $\sigma_n(A) \in X_1$, for $A \in X_1$ and if $r(A) > n$ and $\Box A \in X_0$, then $\sigma_n(\Box A) \in X_1$. For the second part, note that if π is an X-proof then as proofs are closed under substitutions, by the first part, we know that $\sigma_n(\pi)$ is also an X-proof. For niceness, since π is a nice X-proof, $r(\Box\Gamma) \leq r(\Box A)$, for any occurrence of the rule (GL) in π:

$$\frac{\Gamma, \Box\Gamma, \Box A \Rightarrow A}{\Box\Gamma \Rightarrow \Box A}\, GL$$

If $r(\Box A) \leq n$, we also have $r(\Box\Gamma) \leq n$ and hence $\Box\Gamma$ and $\Box A$ have no q_j with $j > n$. Therefore, $\Box\Gamma$, Γ, $\Box A$ and A remain intact and hence there is nothing to prove. If $r(\Box A) > n$, then as $r(\sigma_n(\Box A)) = n$, we have to show that $r(\sigma_n(\Box\Gamma)) \leq n$ which is trivial. □

Lemma 7.4 *If $\Gamma \Rightarrow \Delta$ has an X-proof, then there exists $\Box\Sigma \subseteq X_1$ such that $\Box\Sigma, \Gamma \Rightarrow \Delta$ has a nice X-proof.*

Proof. We use induction on the length of the X-proof of $\Gamma \Rightarrow \Delta$. If the last rule is an axiom, a structural rule or a propositional rule, then the claim is obvious from the induction hypothesis. For the modal rule (GL), we know that $\Box\Gamma \Rightarrow \Box A$ is proved by $\Gamma, \Box\Gamma, \Box A \Rightarrow A$. By the induction hypothesis, there exists $\Box\Sigma \subseteq X_1$ such that $\Box\Sigma, \Gamma, \Box\Gamma, \Box A \Rightarrow A$ has a nice X-proof. Call it π. Note that since $\Box\Sigma \subseteq X_1$, we also have $\Sigma \subseteq X$. Set $r(A) = n$. Divide Γ into two parts, Γ_0 and Γ_1 in a way that $r(\Gamma_0) \leq n$ and $r(\gamma) > n$, for any $\gamma \in \Gamma_1$. Notice that σ_n does not change Γ_0 and A, as their ranks are bounded by n. By Lemma 7.3, we know that $\sigma_n(\pi)$ is a nice X-proof for $\Box\sigma_n(\Sigma), \Gamma_0, \Box\Gamma_0, \sigma_n(\Gamma_1), \Box\sigma_n(\Gamma_1), \Box A \Rightarrow A$. Set $\Sigma' = \sigma_n(\Sigma) \cup \sigma_n(\Gamma_1)$ and note that $\Sigma' \subseteq X$, as X is closed under σ_n, by Lemma 7.3. Hence, by left weakening to add Σ', we have a nice X-proof for $\Sigma', \Box\Sigma', \Gamma_0, \Box\Gamma_0, \Box A \Rightarrow A$. Now, use (GL) to prove $\Box\Sigma', \Box\Gamma_0 \Rightarrow \Box A$. Again, by the left weakening for $\Box\Gamma_1$, we have $\Box\Sigma', \Box\Gamma_0, \Box\Gamma_1 \Rightarrow \Box A$. Therefore, we have provided a proof for $\Box\Sigma', \Box\Gamma \Rightarrow \Box A$. It is clear that this proof is an X-proof. To show that it is nice, notice that the use of (GL) is allowed, as $r(\Box\Gamma_0) \leq n$, by definition, and $r(\Box\Sigma') \leq n$, as $\Box\Sigma' = \sigma_n(\Box\Sigma \cup \Box\Gamma_1)$. Finally, we must show $\Box\Sigma' \subseteq X_1$. First, note that as $\Box\Sigma \subseteq X_1$ and X_1 is closed under σ_n, we have $\sigma_n(\Box\Sigma) \subseteq X_1$. On the other hand, for any formula $\gamma \in \Gamma_1$, if $\Box\gamma \in X_1$, then $\sigma_n(\Box\gamma) \in X_1$ by the closure of X_1 under σ_n and if $\Box\gamma \in X_0$, then as $r(\gamma) > n$, we have $\sigma_n(\Box\gamma) \in X_1$, by Lemma 7.3. □

Define the translation function $s : X \to \mathcal{L}_\infty$ as follows: $\bot^s = \bot$, $p^s = p$ and $q_i^s = \top$, for any $i \geq 0$, $(B \circ C)^s = B^s \circ C^s$, for any $\circ \in \{\wedge, \vee, \to\}$ and if $A = \Box(\bigwedge_{i=0}^n q_i \to B)$ then $A^s = \Box_n B^s$ and if $A = \Box(\bigwedge_{i=0}^m q_i \wedge \bigwedge_{i=m+1}^n \bot \to B)$ then $A^s = \top$. Here are some basic properties of the translation s. First of all, s is well-defined, meaning that $A^s \in \mathcal{L}_\infty$, for any $A \in X$. To prove, we show the stronger claim that if $A \in X$, then $A^s \in \mathcal{L}_\infty$ and $r(A^s) \leq r(A)$. The proof is by structural induction on X. The only non-trivial case is when $A \in X_0$. If $A = \Box(\bigwedge_{i=0}^n q_i \to B)$, then $n > r(B)$, by definition. By induction hypothesis, $B^s \in \mathcal{L}_\infty$ and $r(B^s) \leq r(B)$. Therefore, $\Box_n B^s \in \mathcal{L}_\infty$ and $r(A^s) = r(\Box_n B^s) = n = r(A)$. Secondly, if $A = \Box B \in X_1$, then B^s is provably equivalent to $\top$ in $\mathsf{K4}_\infty$, because $B = \bigwedge_{i=0}^m q_i \wedge \bigwedge_{i=m+1}^n \bot \to C$ has a $\bot$ in its premises, which means that B^s is equivalent to $\top$. Thirdly, note that s is a left inverse for the translation t, meaning that for any $A \in \mathcal{L}_\infty$, we have $(A^t)^s = A$.

Lemma 7.5 *If* $\Gamma \Rightarrow \Delta$ *has a nice* X*-proof, then* $\mathsf{K4}_\infty \vdash \bigwedge \Gamma^s \to \bigvee \Delta^s$.

Proof. The proof is based on an induction on the length of the nice X-proof. If the last rule is an axiom, a structural rule or a propositional rule, then the claim follows from the induction hypothesis. The reason is that s commutes with the propositional connectives and $\mathsf{K4}_\infty$ proves all propositional tautologies. For the modal rule (GL), if $\Box\Gamma \Rightarrow \Box A$ is proved by $\Box\Gamma, \Gamma, \Box A \Rightarrow A$, by induction hypothesis, we have $\mathsf{K4}_\infty \vdash (\Box\Gamma)^s, \Gamma^s, (\Box A)^s \Rightarrow A^s$. We want to show $\mathsf{K4}_\infty \vdash (\Box\Gamma)^s \Rightarrow (\Box A)^s$. For that purpose, we must investigate the form of $(\Box A)^s$ and the elements of $(\Box\Gamma)^s$. If $\Box A \in X_1$, then by definition $(\Box A)^s = \top$ and hence there is nothing to prove. Therefore, assume $\Box A \in X_0$. Hence, A has the form $A = \bigwedge_{i=0}^m q_i \to B$ and $r(B) < r(A) = m$. As $r(B^s) \leq r(B)$, by Lemma 7.1, the formula $A^s = \bigwedge_{i=0}^m \top \to B^s$ is equivalent to B^s, by a $\mathsf{K4}_\infty$-proof with rank $r(B)$. Therefore, by (NC_∞) and (K_∞) and using the fact that $r(B) < m$, it is easy to see that $(\Box A)^s$ is equivalent to $\Box_m B^s$, provably in $\mathsf{K4}_\infty$. On the other hand, for any $\Box\gamma \in \Box\Gamma \subseteq X$, if $\Box\gamma \in X_1$, then γ^s is equivalent to $\top$, provably in $\mathsf{K4}_\infty$. Moreover, $(\Box\gamma)^s = \top$, by definition. Therefore, we can ignore this kind of boxed formulas in $\Box\Gamma$ and w.l.o.g., assume $\Box\Gamma \subseteq X_0$. Therefore, γ has the form $\gamma = \bigwedge_{i=0}^{n_\gamma} q_i \to \beta_\gamma$ and hence $r(\gamma) = n_\gamma > r(\beta_\gamma)$. Therefore, $\gamma^s = \bigwedge_{i=0}^{n_\gamma} \top \to \beta_\gamma^s$ and as $r(\beta_\gamma^s) \leq r(\beta_\gamma)$, by Lemma 7.1, the formulas γ^s and β_γ^s are equivalent with a $\mathsf{K4}_\infty$-proof with rank $r(\beta_\gamma)$ and hence as before, since $r(\beta_\gamma) < n_\gamma$, we know that $(\Box\gamma)^s$ is equivalent to $\Box_{n_\gamma} \beta_\gamma^s$, provably in $\mathsf{K4}_\infty$. Now, the result of the induction hypothesis, i.e., $\mathsf{K4}_\infty \vdash (\Box\Gamma)^s, \Gamma^s, (\Box A)^s \Rightarrow A^s$ implies $\mathsf{K4}_\infty \vdash \{\Box_{n_\gamma} \beta_\gamma^s, \beta_\gamma^s\}_{\gamma \in \Gamma}, \Box_m B^s \Rightarrow B^s$. As the X-proof is nice, we have $m = r(A) = r(\Box A) \geq n_\gamma = r(\gamma) = r(\Box\gamma)$. Split Γ into Γ_1 consisting of $\gamma \in \Gamma$ such that $n_\gamma = m$ and $\Gamma_2 = \Gamma - \Gamma_1$. Hence, $\mathsf{K4}_\infty \vdash \{\Box_m \beta_\gamma^s\}_{\gamma \in \Gamma_1}, \Box_m B^s, \{\Box_{n_\gamma} \beta_\gamma^s\}_{\gamma \in \Gamma_2} \Rightarrow \bigwedge_{\gamma \in \Gamma} \beta_\gamma^s \to B^s$. As $r(\beta_\gamma) < r(\gamma) \leq m$ and $r(B) < r(A) = m$, we have $r(\beta_\gamma^s) \leq r(\beta_\gamma) < m$ and $r(B^s) \leq r(B) < m$. By strong necessitation, Theorem 7.2, we reach $\mathsf{K4}_\infty \vdash \{\Box_{n_\gamma} \beta_\gamma^s\}_{\gamma \in \Gamma_2} \Rightarrow \Box_m(\bigwedge_{\gamma \in \Gamma} \beta_\gamma^s \to B^s)$. By a simple application of (K_∞), we will have $\mathsf{K4}_\infty \vdash \{\Box_{n_\gamma} \beta_\gamma^s\}_{\gamma \in \Gamma_2}, \{\Box_m \beta_\gamma^s\}_{\gamma \in \Gamma} \Rightarrow \Box_m B^s$. As $n_\gamma \leq m$, for any $\gamma \in \Gamma$, by (H), $\mathsf{K4}_\infty \vdash \{\Box_{n_\gamma} \beta_\gamma^s\}_{\gamma \in \Gamma} \Rightarrow \Box_m B^s$ which completes the proof. □

Proof. [of Lemma 5.2] If $\mathsf{GL}(Q) \vdash A^t$, there is a cut-free proof of $(\Rightarrow A^t)$ in $\mathbf{GGL}(Q)$. Therefore, every formula in the proof is a subformula of A^t and hence it is in X. Hence, $(\Rightarrow A^t)$ has an X-proof. By Lemma 7.4, there is a set $\Box\Sigma \subseteq X_1$ such that $\Box\Sigma \Rightarrow A^t$ has a nice X-proof. Then, by Lemma 7.5, $\mathsf{K4}_\infty \vdash \bigwedge(\Box\Sigma)^s \Rightarrow (A^t)^s$. We know that $(A^t)^s = A$. Since $\Box\Sigma \subseteq X_1$, we have $(\Box\sigma)^s = \top$, for any $\sigma \in \Sigma$. Therefore, $\mathsf{K4}_\infty \vdash A$. $\Box$

References

[1] Akbar Tabatabai, A., *Provability interpretation of propositional and modal logics*, arXiv preprint arXiv:1704.07677 (2017).

[2] Akbar Tabatabai, A., *Provability logics of hierarchies*, arXiv preprint arXiv:1704.07678 (2017).

[3] Artemov, S., "Extensions of arithmetic and modal logics," Ph.D. thesis, PhD thesis, Steklov Mathematical Insitute, Moscow, 1979. In Russian (1979).

[4] Artemov, S. N. and L. D. Beklemishev, *Provability logic*, in: *Handbook of Philosophical Logic, 2nd Edition*, Springer, 2005 pp. 189–360.

[5] Avron, A., *On modal systems having arithmetical interpretations*, J. Symbolic Logic **49** (1984), pp. 935–942.
URL https://doi-org.proxy.library.uu.nl/10.2307/2274147

[6] Beklemishev, L. and D. Gabelaia, *Topological completeness of the provability logic GLP*, Ann. Pure Appl. Logic **164** (2013), pp. 1201–1223.

[7] Beklemishev, L. D., *Bimodal logics for extensions of arithmetical theories*, J. Symbolic Logic **61** (1996), pp. 91–124.

[8] Beklemishev, L. D., *A simplified proof of the arithmetical completeness theorem for the provability logic GLP*, Tr. Mat. Inst. Steklova **274** (2011), pp. 32–40.

[9] Boolos, G., *Extremely undecidable sentences*, J. Symbolic Logic **47** (1982), pp. 191–196.
URL https://doi-org.proxy.library.uu.nl/10.2307/2273393

[10] Carlson, T., *Modal logics with several operators and provability interpretations*, Israel J. Math. **54** (1986), pp. 14–24.
URL https://doi-org.proxy.library.uu.nl/10.1007/BF02764872

[11] Gödel, K., *Eine interpretation des intuitionistis-chen aussagenkalkuls*, Ergebnisse eines mathematisches Kolloquiums **4** (1933), pp. 39–40.

[12] Goré, R. and R. Ramanayake, *Valentini's cut-elimination for provability logic resolved*, Rev. Symb. Log. **5** (2012), pp. 212–238.
URL https://doi.org/10.1017/S1755020311000323

[13] Japaridze, G., *The polymodal provability logic*, in: *Intensional logics and logical structure of theories: material from the Fourth Soviet-Finnish Symposium on Logic*, 1988, pp. 16–48.

[14] Japaridze, G. and D. de Jongh, *The logic of provability*, in: *Handbook of proof theory*, Stud. Logic Found. Math. **137**, North-Holland, Amsterdam, 1998 pp. 475–546.
URL https://doi.org/10.1016/S0049-237X(98)80022-0

[15] Japaridze, G. K., "The modal logical means of investigation of provability," Ph.D. thesis, Thesis in Philosophy, in Russian, Moscow (1986).

[16] Löb, M. H., *Solution of a problem of Leon Henkin*, J. Symbolic Logic **20** (1955), pp. 115–118.
URL https://doi-org.proxy.library.uu.nl/10.2307/2266895

[17] Montagna, F., *On the diagonalizable algebra of Peano arithmetic*, Boll. Un. Mat. Ital. B (5) **16** (1979), pp. 795–812.

[18] Pakhomov, F., *On the complexity of the closed fragment of Japaridze's provability logic*, Arch. Math. Logic **53** (2014), pp. 949–967.

[19] Smoryński, C., "Self-reference and modal logic," Universitext, Springer-Verlag, New York, 1985, xii+333 pp.

[20] Solovay, R. M., *Provability interpretations of modal logic*, Israel J. Math. **25** (1976), pp. 287–304.
[21] Visser, A., *Numerations, λ-calculus & arithmetic*, in: *To H. B. Curry: essays on combinatory logic, lambda calculus and formalism*, Academic Press, London-New York, 1980 pp. 259–284.
[22] Visser, A., *An overview of interpretability logic*, in: *Advances in modal logic, Vol. 1 (Berlin, 1996)*, CSLI Lecture Notes **87**, CSLI Publ., Stanford, CA, 1998 pp. 307–359.

Uniform Lyndon interpolation for intuitionistic monotone modal logic

Amirhossein Akbar Tabatabai Rosalie Iemhoff Raheleh Jalali [1]

University of Groningen
Bernoulliborg, Nijenborgh 9
9747 AG Groningen, the Netherlands

Utrecht University
Janskerkhof 13
3512 BL Utrecht, the Netherlands

Abstract

In this paper we show that the intuitionistic monotone modal logic iM has the uniform Lyndon interpolation property (ULIP). The logic iM is a non-normal modal logic on an intuitionistic basis, and the property ULIP is a strengthening of interpolation in which the interpolant depends only on the premise or the conclusion of an implication, respecting the polarities of the propositional variables. Our method to prove ULIP yields explicit uniform interpolants and makes use of a terminating sequent calculus for iM that we have developed for this purpose. As far as we know, the results that iM has ULIP and a terminating sequent calculus are the first of their kind for an intuitionistic non-normal modal logic. However, rather than proving these particular results, our aim is to show the flexibility of the constructive proof-theoretic method that we use for proving ULIP. It has been developed over the last few years and has been applied to substructural, intermediate, classical (non-)normal modal and intuitionistic normal modal logics. In light of these results, intuitionistic non-normal modal logics seem a natural next class to try to apply the method to, and we take the first step in that direction in this paper.

Keywords: intuitionistic monotone modal logic, uniform interpolation, uniform Lyndon interpolation.

1 Introduction

Over the last years a method to prove uniform (Lyndon) interpolation has been developed by the authors that applies to various (intuitionistic) modal and intermediate logics [9,10,12,1,2,3]. Uniform interpolation is a strengthening of interpolation in which the interpolant depends only on the premise or the conclusion of an implication. It is Lyndon whenever the interpolant in addition

[1] Support by the Netherlands Organisation for Scientific Research under grant 639.073.807 and by the FWF project P 33548 is gratefully acknowledged.

respects the polarities of the propositional variables involved. Our method to prove the property is based on sequent calculi for the given logics. Until now, it has been applied to classical modal logics, normal as well as non-normal, but in the intuitionistic setting only to intermediate logics and to intuitionistic modal logics that are normal.

In this paper, we try to show the general applicability of the method by applying it to a well-known intuitionistic non-normal modal logic namely the intuitionistic monotone modal logic iM, which is axiomatized over intuitionistic propositional logic IPC by the following axiom and rule [5]:

$$\Box(\varphi \wedge \psi) \rightarrow \Box\varphi \wedge \Box\psi \qquad \frac{\varphi \rightarrow \psi \qquad \psi \rightarrow \varphi}{\Box\varphi \rightarrow \Box\psi}\, E$$

The axiom is one direction of the principle $\Box(\varphi \wedge \psi) \leftrightarrow \Box\varphi \wedge \Box\psi$ that holds in every normal modal logic. We show that iM has uniform Lyndon interpolation, which, to our knowledge, is the first result of this kind, meaning the first result stating that an intuitionistic non-normal modal logic has uniform (Lyndon) interpolation. Our method is effective in that it provides explicit (existential and universal) interpolants and it makes use of a terminating sequent calculus for the logic. The calculus is an extension of the calculus **G4ip**, which has been introduced by Dyckhoff as a variant of **G3ip** in which proof search terminates (without extra conditions on the search) [6]. The terminating calculus that we develop here seems to be the first terminating calculus for the logic iM. Our method to prove uniform interpolation is inspired by the first syntactic proof of uniform interpolation, given by Pitts for intuitionistic propositional logic [18].

As can be seen from [5], the semantics for intuitionistic non-normal modal logic that combines the semantics of intuitionistic logic and classical non-normal modal logic is not simple. In this light it is somewhat surprising that the proof-theoretic method developed in this paper is essentially not more complicated than the one for its normal counterpart, that, we have to admit, is already quite complicated in itself.

In the literature there are many syntactic proofs of Craig interpolation, most of them connected in some way or another to the well-known syntactic *Maehara method* [16]. Proofs of uniform interpolation are less common, and syntactic proofs of uniform interpolation even more so. Most of the existing proofs are inspired by Pitts' syntactic proof of uniform interpolation for IPC [18] mentioned earlier. Some proof systems seem to lend themselves better for syntactic proofs of (uniform) interpolation than others. Especially for nested sequents there are several syntactic results. There are nested sequent systems for certain tense logics and bi-intuitionistic logic that allow for a syntactic proof of Craig interpolation [15]. A similar statement holds for various modal and intermediate logics, although in this case the method is no longer purely syntactical but also contains semantic elements [7,13].

Uniform interpolation has applications in computer science, in particular in description logics [14], but our interest in the property stems from a project in universal proof theory where we aim to develop methods to prove that certain (classes of) logics cannot have certain well-behaved proof systems, in our case sequent calculi [3,10]. Here we make use of the fact that uniform interpolation seems to be a rare property among logics. For example, only seven intermediate logics have this property [8,17]. In fact, in this case also only seven intermediate logics have Craig interpolation, but in modal logics these two properties in general do not coincide. And while in this class the property is equivalent to interpolation, this is certainly not the case for modal logics. The logics K4 and S4 are examples of logics that have Craig interpolation but not uniform interpolation [4,8]. The result in this paper is meant as a first step to also consider the class of intuitionistic non-normal modal logics in the project. The reason that we only treat one logic in that class and only prove that it has uniform Lyndon interpolation without taking the further generalization steps needed for the project, is more for reasons of space than anything else. We hope to take these further steps and cover more intuitionistic non-normal modal logics in the future.

This paper is built up as follows. Section 2 contains the preliminaries, in which the interpolation properties, the intuitionistic non-normal modal logic iM, and the sequent calculi **G3iM^w^** and **G4w** are defined. In Section 3, the terminating calculus **G4iM** is introduced and is shown to be equivalent to **G3iM^w^**, which implies that it is a terminating calculus for iM. In Section 4, it is proved that iM has uniform Lyndon interpolation property.

2 Preliminaries

The language we use is $\mathcal{L} = \{\wedge, \vee, \to, \Box, \bot\}$, and $\top$ is an abbreviation for $\bot \to \bot$, as usual. We use small Roman letters $p, q, \ldots$ for atomic formulas, small Greek letters $\varphi, \psi, \ldots$ to denote formulas, and capital Greek letters $\Sigma, \Delta, \ldots$ and also $\bar{\varphi}, \bar{\psi}, \ldots$ to denote finite multisets of formulas. The *weight* of a formula is defined as follows, which is a combination of the definitions given in [4] and [6]: $w(p) = w(\bot) = w(\top) = 1$, for any atomic formula p, $w(\varphi \odot \psi) = w(\varphi) + w(\psi) + 1$, for $\odot \in \{\vee, \to\}$, $w(\varphi \wedge \psi) = w(\varphi) + w(\psi) + 2$, and $w(\Box\varphi) = w(\varphi) + 1$. This weight function induces an ordering on the multisets: $\Gamma \prec \Delta$ if Γ is obtained from Δ by replacing one or more formulas of Δ by zero or more formulas, each of which is of a strictly lower weight. Note that this order is well-founded.

Definition 2.1 The sets of positive and negative variables of a formula $\varphi \in \mathcal{L}$, denoted by $V^+(\varphi)$ and $V^-(\varphi)$, respectively, are defined recursively by:

- $V^+(p) = \{p\}, V^-(p) = V^+(\top) = V^-(\top) = V^+(\bot) = V^-(\bot) = \varnothing$ for atom p,
- $V^+(\varphi \odot \psi) = V^+(\varphi) \cup V^+(\psi)$, $V^-(\varphi \odot \psi) = V^-(\varphi) \cup V^-(\psi)$, for $\odot \in \{\wedge, \vee\}$,
- $V^+(\varphi \to \psi) = V^-(\varphi) \cup V^+(\psi)$ and $V^-(\varphi \to \psi) = V^+(\varphi) \cup V^-(\psi)$,
- $V^+(\Box\varphi) = V^+(\varphi)$ and $V^-(\Box\varphi) = V^-(\varphi)$.

Define $V(\varphi)$ as $V^+(\varphi) \cup V^-(\varphi)$ and set $V^+(\Gamma) = \bigcup_{\gamma \in \Gamma} V^+(\gamma)$ and $V^-(\Gamma) = \bigcup_{\gamma \in \Gamma} V^-(\gamma)$, for a multiset Γ. For an atomic formula p, a formula φ is called *p^+-free* (*p^--free*), if $p \notin V^+(\varphi)$ ($p \notin V^-(\varphi)$). It is called *p-free* if $p \notin V(\varphi)$. Note that a formula is p-free iff p does not occur anywhere in the formula.

We will need the following notations: if we want to refer to both $V^+(\varphi)$ and $V^-(\varphi)$, we use $V^\dagger(\varphi)$, with the condition "for any $\dagger \in \{+,-\}$". When we want to refer to both $V^+(\varphi)$ and $V^-(\varphi)$ but also to their duals, we use the notation $V^\circ(\varphi)$ for the one we intend and $V^\diamond(\varphi)$ [2] for its dual, respectively. Therefore, by the sentence 'if $p \in V^\circ(\varphi)$, then $p \notin V^\diamond(\varphi)$, for any $\circ, \diamond \in \{+,-\}$', we mean 'if $p \in V^+(\varphi)$, then $p \notin V^-(\varphi)$ and if $p \in V^-(\varphi)$, then $p \notin V^+(\varphi)$'.

Definition 2.2 A *logic* L is a set of formulas of $\mathcal{L}$ extending the set of intuitionistic tautologies, IPC, and closed under substitution and modus ponens $\varphi, \varphi \to \psi \vdash \psi$.

Definition 2.3 A logic L has the *Lyndon interpolation property (LIP)* if for any formulas $\varphi, \psi \in \mathcal{L}$ such that $L \vdash \varphi \to \psi$, there is a formula $\theta \in \mathcal{L}$ such that $V^\dagger(\theta) \subseteq V^\dagger(\varphi) \cap V^\dagger(\psi)$, for any $\dagger \in \{+,-\}$ and $L \vdash \varphi \to \theta$ and $L \vdash \theta \to \psi$. A logic has *Craig interpolation (CIP)* if it has the above properties, omitting all the superscripts $\dagger \in \{+,-\}$.

Definition 2.4 A logic L has the *uniform Lyndon interpolation property (ULIP)* if for any formula $\varphi \in \mathcal{L}$, atom p, and $\circ \in \{+,-\}$, there are p°-free formulas, $\exists^\circ p\varphi$ and $\forall^\circ p\varphi$, such that $V^\dagger(\exists^\circ p\varphi) \subseteq V^\dagger(\varphi)$ and $V^\dagger(\forall^\circ p\varphi) \subseteq V^\dagger(\varphi)$, for any $\dagger \in \{+,-\}$ and

(*i*) $L \vdash \varphi \to \exists^\circ p\varphi$, and

(*ii*) for any p°-free formula ψ if $L \vdash \varphi \to \psi$ then $L \vdash \exists^\circ p\varphi \to \psi$,

(*iii*) $L \vdash \forall^\circ p\varphi \to \varphi$,

(*iv*) for any p°-free formula ψ if $L \vdash \psi \to \varphi$ then $L \vdash \psi \to \forall^\circ p\varphi$,

A logic has *uniform interpolation property (UIP)* if it has all the above properties, omitting the superscripts $\circ, \dagger \in \{+,-\}$, everywhere. Note that although the interpolants are indicated by expressions that contain symbols that do not belong to $\mathcal{L}$, they do stand for formulas in the language $\mathcal{L}$.

Theorem 2.5 *If a logic L has ULIP, then it has both LIP and UIP.*

Proof. For UIP, define $\forall p\varphi = \forall^+ p\forall^- p\varphi$ and $\exists p\varphi = \exists^+ p\exists^- p\varphi$. We will show that $\forall p\varphi$ acts as the uniform interpolant for φ. The case for $\exists p\varphi$ is similar. By definition, $V^\dagger(\forall^+ p\forall^- p\varphi) \subseteq V^\dagger(\forall^- p\varphi) \subseteq V^\dagger(\varphi)$, for any $\dagger \in \{+,-\}$. Therefore, $V(\forall p\varphi) \subseteq V(\varphi)$. Moreover, $\forall^+ p\forall^- p\varphi$ is p^+-free, by definition. Suppose $p \in V^-(\forall^+ p\forall^- p\varphi)$. Then, $p \in V^-(\forall^- p\varphi)$, which is a contradiction, as $\forall^- p\varphi$ is p^--free. Hence, $\forall p\varphi$ is p-free.

Condition (*iii*) is easy, as we have $L \vdash \forall^+ p\forall^- p\varphi \to \forall^- p\varphi$ and $L \vdash \forall^- p\varphi \to \varphi$, by Definition 2.4. Therefore, $L \vdash \forall p\varphi \to \varphi$. For condition (*iv*), let ψ be a p-free formula such that $L \vdash \psi \to \varphi$. Then, as ψ is p^--free, we have $L \vdash \psi \to \forall^- p\varphi$ and as ψ is p^+-free, we get $L \vdash \psi \to \forall^+ p\forall^- p\varphi$.

For LIP, assume $L \vdash \varphi \to \psi$. Define $\theta = \exists^+ P^+ \exists^- P^- \varphi$, where $P^\dagger = V^\dagger(\varphi) - [V^\dagger(\varphi) \cap V^\dagger(\psi)]$, for any $\dagger \in \{+,-\}$ and by $\exists^\dagger\{p_1, \ldots, p_n\}^\dagger$ we mean $\exists p_1^\dagger \ldots \exists p_n^\dagger$. Since θ is $p^\dagger$-free for any $p \in P^\dagger$ and any $\dagger \in \{+,-\}$, we have

[2] The superscript $\diamond$ has nothing to do with the usual modality $\Diamond$.

$$\Gamma, p \Rightarrow p \quad Ax \qquad\qquad \Gamma, \bot \Rightarrow \varphi \quad L\bot$$

$$\frac{\Gamma, \varphi, \psi \Rightarrow \theta}{\Gamma, \varphi \wedge \psi \Rightarrow \theta}\ L\wedge \qquad\qquad \frac{\Gamma \Rightarrow \varphi \quad \Gamma \Rightarrow \psi}{\Gamma \Rightarrow \varphi \wedge \psi}\ R\wedge$$

$$\frac{\Gamma, \varphi \Rightarrow \theta \quad \Gamma, \psi \Rightarrow \theta}{\Gamma, \varphi \vee \psi \Rightarrow \theta}\ L\vee \qquad\qquad \frac{\Gamma \Rightarrow \varphi_i}{\Gamma \Rightarrow \varphi_0 \vee \varphi_1}\ R\vee\ (i = 0, 1)$$

$$\frac{\Gamma, \varphi \to \psi \Rightarrow \varphi \quad \Gamma, \psi \Rightarrow \theta}{\Gamma, \varphi \to \psi \Rightarrow \theta}\ L\to \qquad\qquad \frac{\Gamma, \varphi \Rightarrow \psi}{\Gamma \Rightarrow \varphi \to \psi}\ R\to$$

$$\frac{\varphi \Rightarrow \psi}{\Gamma, \Box\varphi \Rightarrow \Box\psi}\ M_{\Box}^{seq}$$

Fig. 1. The sequent calculus $\mathbf{G3iM^w}$. In Ax, p must be an atom.

$V^{\dagger}(\theta) \subseteq V^{\dagger}(\varphi) - P^{\dagger} \subseteq V^{\dagger}(\varphi) \cap V^{\dagger}(\psi)$. For the other condition, it is clear that $L \vdash \varphi \to \theta$ and as ψ is $p^{\dagger}$-free, for any $p \in P^{\dagger}$, we have $L \vdash \theta \to \psi$. □

2.1 Sequent calculi

A *sequent* S is any expression of the form $\Gamma \Rightarrow \Delta$, where Γ and Δ are two multisets of formulas called the *antecedent* and the *succedent* of the sequent, denoted by S^a and S^s, respectively. A sequent is called *single-conclusion* if its succedent has at most one formula. The *multiplication* of the sequents S and T is defined by $S \cdot T = (S^a \cup T^a) \Rightarrow (S^s \cup T^s)$. The set of *positive variables* (V^+) and the set of *negative variables* (V^-) of a sequent S are defined by $V^{\circ}(S) = V^{\circ}(S^a) \cup V^{\circ}(S^s)$, for any $\circ \in \{+, -\}$ and $V(S) = V(S^a) \cup V(S^s)$. In case it is clear from the notation which set we mean we omit the words "positive" and "negative". The ordering $\prec$ can be extended to sequents by $S \prec T := S^a \cup S^s \prec T^a \cup T^s$. If $S \prec T$, we say that S is *lower* than T. Sequents and multisets can also be compared with each other in an expected way. For instance, $\Sigma \prec S$ means $\Sigma \prec (S^a \cup S^s)$. A *rule* is an expression of the form $\dfrac{S_1 \ \dots \ S_n}{S}$ where $S_1, \dots, S_n$, and S are sequents called the *premises* and the *conclusion* of the rule, respectively. If a sequent S is the conclusion of an instance of a rule, we say that the rule is *backward applicable* to S. A *sequent calculus* is a set of rules. In this paper, we consider single-conclusion sequent calculi, where only single-conclusion sequents are allowed.

Let us introduce three sequent calculi that we need throughout the paper. The first is the sequent calculus $\mathbf{G3iM^w}$ [3] presented in Figure 1. The system was introduced in [5] under the name G.□ – IM. The logic of $\mathbf{G3iM^w}$, i.e., the set of all formulas such that ($\Rightarrow \varphi$) is derivable in $\mathbf{G3iM^w}$, is called iM, the intuitionistic monotone modal logic. The second system is **G4ip**, a sequent calculus for IPC presented in Figure 2 and introduced in [6]. If we add the following weakening rules to **G4ip**, we get the third system **G4w**:

$$\frac{\Gamma \Rightarrow \Delta}{\Gamma, \varphi \Rightarrow \Delta}\ Lw \qquad \frac{\Gamma \Rightarrow}{\Gamma \Rightarrow \varphi}\ Rw$$

[3] The use of superscript w becomes clear in the next section.

$$\Gamma, p \Rightarrow p \quad Ax \qquad\qquad \Gamma, \bot \Rightarrow \varphi \quad L\bot$$

$$\frac{\Gamma, \varphi, \psi \Rightarrow \Delta}{\Gamma, \varphi \wedge \psi \Rightarrow \Delta} L\wedge \qquad\qquad \frac{\Gamma \Rightarrow \varphi \qquad \Gamma \Rightarrow \psi}{\Gamma \Rightarrow \varphi \wedge \psi} R\wedge$$

$$\frac{\Gamma, \varphi \Rightarrow \Delta \qquad \Gamma, \psi \Rightarrow \Delta}{\Gamma, \varphi \vee \psi \Rightarrow \Delta} L\vee \qquad\qquad \frac{\Gamma \Rightarrow \varphi_i}{\Gamma \Rightarrow \varphi_0 \vee \varphi_1} R\vee_i \; (i = 0, 1)$$

$$\frac{\Gamma, p, \psi \Rightarrow \Delta}{\Gamma, p, p \to \psi \Rightarrow \Delta} Lp\!\to \qquad\qquad \frac{\Gamma, \varphi_1 \to (\varphi_2 \to \psi) \Rightarrow \Delta}{\Gamma, (\varphi_1 \wedge \varphi_2) \to \psi \Rightarrow \Delta} L\wedge\!\to$$

$$\frac{\Gamma, \varphi_1 \to \psi, \varphi_2 \to \psi \Rightarrow \Delta}{\Gamma, \varphi_1 \vee \varphi_2 \to \psi \Rightarrow \Delta} L\vee\!\to \qquad\qquad \frac{\Gamma, \varphi \Rightarrow \psi}{\Gamma \Rightarrow \varphi \to \psi} R\!\to$$

$$\frac{\Gamma, \varphi_2 \to \psi \Rightarrow \varphi_1 \to \varphi_2 \qquad \Gamma, \psi \Rightarrow \Delta}{\Gamma, (\varphi_1 \to \varphi_2) \to \psi \Rightarrow \Delta} L\!\to\to$$

Fig. 2. The sequent calculus **G4ip** from [6]. In Ax and ($Lp\!\to$), p is an atom.

The weakening rules are admissible in **G4ip** and hence there is no need to include them explicitly. However, as we will work with an extension of the system **G4ip**, we will need the explicit weakening rules later.

As the final part of this section, let us mention some of the properties of the rules in **G4w** that we will need later. First, note that in any of the rules of **G4w**, Γ and Δ are free for any multiset substitution. We call this property the *free-context property*. For later reference, we call any premise of a rule with Δ in its succedent *contextual* and the other premises *non-contextual*. Second, if we denote the set of rules in **G4w** minus the rule ($Lp\!\to$) by **G4w**$^-$, then all the rules of **G4w**$^-$, have one of the following general forms:

$$\frac{\{\Gamma, \bar{\varphi}_i \Rightarrow \bar{\delta}_i\}_{i \in I} \qquad \{\Gamma, \bar{\psi}_j \Rightarrow \Delta\}_{j \in J}}{\Gamma, \varphi \Rightarrow \Delta} \qquad\qquad \frac{\{\Gamma, \bar{\varphi}_i \Rightarrow \bar{\delta}_i\}_{i \in I}}{\Gamma \Rightarrow \varphi}$$

where I and J are some finite (possibly empty) sets, Γ and Δ are free for any multiset substitution and $\bar{\varphi}_i$'s, $\bar{\psi}_i$'s and $\bar{\delta}_i$'s are (possibly empty) multisets of formulas, where $\bar{\delta}_i$'s are either empty or a singleton. The formula φ is called the *main* formula and the formulas in $\bar{\varphi}_i, \bar{\psi}_j$, and $\bar{\delta}_i$ are called the *active* formulas of the rule. If the main formula is in the antecedent (succedent), the rule is called a *left* (*right*) rule. Third, notice that each rule in **G4w**$^-$ enjoys the *local variable preserving property*, i.e., given $\circ \in \{+, -\}$, for the left rule, we have $\bigcup_i \bigcup_{\theta \in \bar{\varphi}_i} V^{\circ}(\theta) \cup \bigcup_j \bigcup_{\theta \in \bar{\psi}_j} V^{\circ}(\theta) \cup \bigcup_i \bigcup_{\theta \in \bar{\delta}_i} V^{\diamond}(\theta) \subseteq V^{\circ}(\varphi)$, and for the right one, $\bigcup_i \bigcup_{\theta \in \bar{\varphi}_i} V^{\circ}(\theta) \cup \bigcup_i \bigcup_{\theta \in \bar{\delta}_i} V^{\circ}(\theta) \subseteq V^{\circ}(\varphi)$. This property ensures the crucial condition $\bigcup_{i=1}^{n} V^{\circ}(S_i) \subseteq V^{\circ}(S)$, for any instance of the rule $\frac{S_1 \cdots S_n}{S}$ in **G4w**$^-$ and any $\circ \in \{+, -\}$. We call this weaker property the *variable preserving property*. Note that the rule ($Lp \to$) also enjoys this property. Finally, noticc that in any rule in **G4w**, each of the premises is lower than the conclusion in the order $\prec$.

3 A Terminating sequent calculus

In this section, we provide a terminating single-conclusion sequent calculus for iM. Define the system **G4iM** as **G4w** extended by the following rules:

$$\frac{\varphi \Rightarrow \psi}{\Box\varphi \Rightarrow \Box\psi}\, M \qquad \frac{\varphi \Rightarrow \psi \qquad \Gamma, \Box\varphi, \theta \Rightarrow \Delta}{\Gamma, \Box\varphi, \Box\psi \to \theta \Rightarrow \Delta}\, LM{\to}$$

Note that each premise of any rule in **G4iM** is lower than the conclusion. Consequently, **G4iM** is terminating, i.e., any proof search terminates.

The remainder of this section is devoted to the proof that $\mathbf{G3iM^w}$ and **G4iM** are equivalent, the main part of which consists of a proof that $\mathbf{G3iM^w}$ and $\mathbf{G4iM^w}$ are equivalent. This proof is an adaptation of a similar result for **G3i** and **G4i** in [6]. We start with some preliminaries.

3.1 Strict proofs in $\mathbf{G3iM^w}$

Lemma 3.1 *All rules in* $\mathbf{G3iM^w}$ *except* $R\vee$ *and* $L\to$ *are invertible, and* $\mathbf{G3iM^w}$ *is closed under weakening, contraction and* implication inversion, *i.e. the following rule is admissible:*

$$\frac{\Gamma, \varphi \to \psi \Rightarrow \Delta}{\Gamma, \psi \Rightarrow \Delta}$$

Proof. Closures under the structural rules and implication inversion are proved with induction to the depth of the derivation. □

A multiset is *irreducible* if it has no element that is a disjunction or a conjunction or falsum and for no atom p does it contain both $p \to \psi$ and p. A sequent S is *irreducible* if S^a is. A proof is *sensible* if its last inference does not have a principal formula on the left of the form $p \to \psi$ for some atom p and formula ψ.[4] A proof in $\mathbf{G3iM^w}$ is *strict* if in the last inference, in case it is an instance of $L\to$ with principal formula $\Box\varphi \to \psi$, the left premise is an axiom or the conclusion of an application of the modal rule.

Lemma 3.2 *Every irreducible sequent that is provable in* $\mathbf{G3iM^w}$ *has a sensible strict proof in* $\mathbf{G3iM^w}$.

Proof. This is proved in the same way as the corresponding lemma (Lemma 1) in [6]. Arguing by contradiction, assume that among all provable irreducible sequents that have no sensible strict proofs, S is such a sequent with the shortest proof, $\mathcal{D}$, where the *length* of a proof is the length of its leftmost branch. Thus the last inference in the proof is an application

$$\frac{\begin{matrix}\mathcal{D}_1 \\ \Gamma, \varphi \to \psi \Rightarrow \varphi\end{matrix} \qquad \begin{matrix}\mathcal{D}_2 \\ \Gamma, \psi \Rightarrow \Delta\end{matrix}}{\Gamma, \varphi \to \psi \Rightarrow \Delta}$$

of L$\to$, where φ is an atom or a modal formula. Since S^a is irreducible, $\bot \notin S^a$ and if φ is an atom, $\varphi \notin S^a$. Therefore the left premise cannot be an instance

[4] In [11] the requirement that the principal formula be on the left was erroneously omitted.

of an axiom and hence is the conclusion of a rule, say $\mathcal{R}$. Since the succedent of the conclusion of $\mathcal{R}$ consists of an atom or a modal formula, $\mathcal{R}$ is a left rule or a right modal rule. The latter case cannot occur, since the proof then would be strict and sensible. Thus $\mathcal{R}$ is a left rule.

We proceed as in [6]. Sequent $(\Gamma, \varphi \to \psi \Rightarrow \varphi)$ is irreducible and has a shorter proof than S. Thus its subproof $\mathcal{D}_1$ is strict and sensible. Since the sequent is irreducible and φ is an atom or a pure modal formula, the last inference of $\mathcal{D}_1$ is L$\to$ with a principal formula $\varphi' \to \psi'$ such that φ' is not an atom. Let $\mathcal{D}'$ be the proof of the left premise $(\Gamma, \varphi \to \psi \Rightarrow \varphi')$. Thus the last part of $\mathcal{D}$ looks as follows, where $\Pi, \varphi' \to \psi' = \Gamma$.

$$\dfrac{\dfrac{\dfrac{\mathcal{D}'}{\Pi, \varphi \to \psi, \varphi' \to \psi' \Rightarrow \varphi'} \quad \dfrac{\mathcal{D}''}{\Pi, \varphi \to \psi, \psi' \Rightarrow \varphi}}{\Pi, \varphi \to \psi, \varphi' \to \psi' \Rightarrow \varphi} \qquad \dfrac{\mathcal{D}_2}{\Pi, \psi, \varphi' \to \psi' \Rightarrow \Delta}}{\Pi, \varphi \to \psi, \varphi' \to \psi' \Rightarrow \Delta}$$

Consider the following proof of S.

$$\dfrac{\dfrac{\mathcal{D}'}{\Pi, \varphi \to \psi, \varphi' \to \psi' \Rightarrow \varphi'} \qquad \dfrac{\dfrac{\mathcal{D}''}{\Pi, \varphi \to \psi, \psi' \Rightarrow \varphi} \quad \dfrac{\mathcal{D}'''}{\Pi, \psi, \psi' \Rightarrow \Delta}}{\Pi, \varphi \to \psi, \psi' \Rightarrow \Delta}}{\Pi, \varphi \to \psi, \varphi' \to \psi' \Rightarrow \Delta}$$

The existence of $\mathcal{D}'''$ follows from Lemma 3.1 and the existence of $\mathcal{D}_2$. The obtained proof is strict and sensible: In case φ' is not a modal formula, this is straightforward. In case φ' is a modal formula, it follows from the fact that was observed above, namely that $\mathcal{D}_1$ is strict and sensible. □

Theorem 3.3 **G3iM**$^{\mathbf{w}}$ *and* **G4iM**$^{\mathbf{w}}$ *are equivalent (derive exactly the same sequents).*

Proof. The proof is an adaptation of the proof of Theorem 3.4 in [11], which again is an adaptation of Theorem 1 in [6]. Under the assumptions in the theorem we have to show that for all sequents S: $\vdash_{\mathbf{G3iM^w}} S$ if and only if $\vdash_{\mathbf{G4iM^w}} S$.

The proof of the direction from right to left is straightforward because **G3iM**$^{\mathbf{w}}$ is closed under the structural rules. For weakening and contraction this is easy to see, and cut-elimination for **G3iM**$^{\mathbf{w}}$ is proved in [5]. For details, see [11]

The other direction, left to right, is proved by induction on the order $\ll$ with respect to which **G4iM**$^{\mathbf{w}}$ is terminating, in a similar manner as in [11]. So suppose **G3iM**$^{\mathbf{w}}$ $\vdash S$. Sequents lowest in the order do not contain connectives or modal operators by definition of the weight function underlying $\ll$. Thus such sequents have to be instances of axioms, and since **G3iM**$^{\mathbf{w}}$ and **G4iM**$^{\mathbf{w}}$ have the same axioms, S is provable in **G4iM**$^{\mathbf{w}}$.

We turn to the case that S is not the lowest in the order. If S^a contains a conjunction, say $S = (\Gamma, \varphi_1 \wedge \varphi_2 \Rightarrow \Delta)$, then $S' = (\Gamma, \varphi_1, \varphi_2 \Rightarrow \Delta)$ is provable in

G3iM$^{\mathbf{w}}$ by Lemma 3.1. As **G4iM**$^{\mathbf{w}}$ contains $L\wedge$ and **G4iM**$^{\mathbf{w}}$ is terminating, $S' \ll S$ follows. Hence S' is provable in **G4iM**$^{\mathbf{w}}$ by the induction hypothesis. Thus so is $(\Gamma, \varphi_1 \wedge \varphi_2 \Rightarrow \Delta)$. A disjunction in S^a as well as the case that both p and $p \to \varphi$ belong to S^a, can be treated in the same way.

Thus only the case that S is irreducible remains, and by Lemmas 3.1 and 3.2 we may assume its proof in **G3iM**$^{\mathbf{w}}$ to be sensible and strict. The irreducibility of S implies that the last inference of the proof is an application of a rule, $\mathcal{R}$, that is either a nonmodal right rule, a modal rule or $L \to$. In the first two cases, $\mathcal{R}$ belongs to both calculi and the fact that **G4iM**$^{\mathbf{w}}$ is terminating implies that the premise(s) of $\mathcal{R}$ is lower in the order $\ll$ than S. Thus the induction hypothesis implies that the premise(s) is derivable in **G4iM**$^{\mathbf{w}}$, and since $\mathcal{R}$ belongs to **G4iM**$^{\mathbf{w}}$, the conclusion S is derivable in **G4iM**$^{\mathbf{w}}$ as well.

We turn to the third case. Suppose that the principal formula of the last inference is $(\gamma \to \psi)$ and $S = (\Gamma, \gamma \to \psi \Rightarrow \Delta)$. Since the proof is sensible, γ is not atomic. We distinguish according to the main connective of γ.

If $\gamma = \bot$, then $\Gamma \Rightarrow \Delta$ is derivable in **G3iM**$^{\mathbf{w}}$ because of the closure under cut: **G3iM**$^{\mathbf{w}}$ derives $(\Rightarrow \bot \to \psi)$, and so the cut

$$\frac{\Rightarrow \bot \to \psi \quad \Gamma, \bot \to \psi \Rightarrow \Delta}{\Gamma \Rightarrow \Delta}$$

shows that $\Gamma \Rightarrow \Delta$ is derivable in **G3iM**$^{\mathbf{w}}$. Since $(\Gamma \Rightarrow \Delta) \ll S$, it follows that $\Gamma \Rightarrow \Delta$ is derivable in **G4iM**$^{\mathbf{w}}$ by the induction hypothesis. As **G4iM**$^{\mathbf{w}}$ is closed under weakening, S is derivable in **G4iM**$^{\mathbf{w}}$ too.

If $\gamma = \varphi_1 \wedge \varphi_2$, then the fact that S is derivable in **G3iM**$^{\mathbf{w}}$ implies the same for $S' = (\Gamma, \varphi_1 \to (\varphi_2 \to \psi) \Rightarrow \Delta)$, as **G3iM**$^{\mathbf{w}}$ is closed under cut. The fact that **G4iM**$^{\mathbf{w}}$ is terminating and contains $L\wedge\!\to$ implies $S' \ll S$. Hence S' is derivable in **G4iM**$^{\mathbf{w}}$ by the induction hypothesis. Thus so is $\Gamma, \varphi_1 \wedge \varphi_2 \to \psi \Rightarrow \Delta$ by an application of $L\wedge\!\to$. The case that $\gamma = \varphi_1 \vee \varphi_2$ is analogous.

If $\gamma = \varphi_1 \to \varphi_2$, then because $\gamma \to \psi$ is the principal formula, both premises $S_1 = (\Gamma, \psi \Rightarrow \Delta)$ and $\Gamma, \gamma \to \psi \Rightarrow \gamma$ are derivable in **G3iM**$^{\mathbf{w}}$. Thus so is $\Gamma, \gamma \to \psi, \varphi_1 \Rightarrow \varphi_2$ by the invertibility of $R\!\to$ (Lemma 3.1). It is not hard to see that $\Gamma, \varphi_2 \to \psi, \varphi_1, \varphi_1 \to \varphi_2 \Rightarrow \psi$ is derivable in **G3iM**$^{\mathbf{w}}$. Hence so is $\Gamma, \varphi_2 \to \psi, \varphi_1 \Rightarrow \gamma \to \psi$. Together with $\Gamma, \gamma \to \psi, \varphi_1 \Rightarrow \varphi_2$ and the fact that **G3iM**$^{\mathbf{w}}$ is closed under cut, this gives the derivability of $\Gamma, \varphi_2 \to \psi, \varphi_1 \Rightarrow \varphi_2$ in **G3iM**$^{\mathbf{w}}$, which implies that $S_2 = \Gamma, \varphi_2 \to \psi \Rightarrow \gamma$ is derivable in **G3iM**$^{\mathbf{w}}$.

Since S_1 and S_1 are the premises of $L\!\to\!\to$ with conclusion S in **G4iM**$^{\mathbf{w}}$, they both are lower in the order $\ll$ than S. Therefore they are derivable in **G4iM**$^{\mathbf{w}}$ by the induction hypothesis. And thus so is S by an application of $L\!\to\!\to$.

If $\gamma = \Box\varphi$, then the fact that the proof is strict and S is irreducible implies that the left premise is the conclusion of the modal rule $\mathcal{R}$ with premises $\chi \Rightarrow \varphi$.

Thus the derivation looks as follows:

$$\dfrac{\dfrac{\begin{array}{c}\mathcal{D}_0\\ \chi \Rightarrow \varphi\end{array}}{\Gamma, \Box\varphi \to \psi, \Box\chi \Rightarrow \Box\varphi} \qquad \begin{array}{c}\mathcal{D}_2\\ \Gamma, \Box\chi, \psi \Rightarrow \Delta\end{array}}{\Gamma, \Box\varphi \to \psi, \Box\chi \Rightarrow \Delta}$$

Observe that **G4iM^w** contains the rule

$$\dfrac{\chi \Rightarrow \varphi \qquad \Gamma, \Box\chi, \psi \Rightarrow \Delta}{\Gamma, \Box\varphi \to \psi, \Box\chi \Rightarrow \Delta}$$

Since **G4iM^w** is terminating, it follows that both premises are below S in the ordering. By the induction hypothesis they are derivable in **G4iM^w**, say with derivations $\mathcal{D}_0'$ and $\mathcal{D}_2'$. Then the following is a proof of S in **G4iM^w**:

$$\dfrac{\begin{array}{c}\mathcal{D}_0'\\ \chi \Rightarrow \varphi\end{array} \qquad \begin{array}{c}\mathcal{D}_2'\\ \Gamma, \Box\chi, \psi \Rightarrow \Delta\end{array}}{\Gamma, \Box\varphi \to \psi, \Box\chi \Rightarrow \Delta}$$

□

Lemma 3.4 **G4iM^w** *is closed under the structural rules, including cut.*

Theorem 3.5 *The systems* **G3iM^w** *and* **G4iM** *are equivalent.*

Proof. To show that **G3iM^w** and **G4iM** are equivalent, it suffices to show that **G4iM^w** and **G4iM** are equivalent. That every sequent derivable in **G4iM^w** is derivable in **G4iM** is clear. For the other direction it suffices to show that weakening is admissible in **G4iM^w** which already has been investigated. □

4 Uniform Lyndon interpolation

In this section, we will prove that the logic iM enjoys ULIP. To this end, we will provide a stronger variant of ULIP for the sequent calculus **G4iM** and prove that the system has that property. From now on, when we say a sequent is derivable, we mean it is derivable in **G4iM**, unless specified otherwise.

Theorem 4.1 **G4iM** *has* ULIP, *i.e., for any sequent S, multiset Σ, atom p, and $\circ \in \{+,-\}$, there exist formulas $\forall^\circ pS$ and $\exists^\circ p\Sigma$ such that:*

(*var*) *$\forall^\circ pS$ and $\exists^\circ p\Sigma$ are p°-free and $V^\dagger(\forall^\circ pS) \subseteq V^\dagger(S)$ and $V^\dagger(\exists^\circ p\Sigma) \subseteq V^\dagger(\Sigma)$, for any $\dagger \in \{+,-\}$,*

(i) *$\Sigma \Rightarrow \exists^\circ p\Sigma$ is derivable,*

(ii) *for any sequent $\bar{C} \Rightarrow \bar{D}$ where $\bar{D}$ has at most one formula and $p \notin V^\circ(\bar{C} \Rightarrow \bar{D})$ if $\Sigma, \bar{C} \Rightarrow \bar{D}$ is derivable, then $(\exists^\circ p\Sigma, \bar{C} \Rightarrow \bar{D})$ is also derivable.*

(iii) *$S \cdot (\forall^\circ pS \Rightarrow)$ is derivable,*

(iv) *for any multiset $\bar{C}$ such that $p \notin V^\circ(\bar{C})$ if $S \cdot (\bar{C} \Rightarrow)$ is derivable, then $(\bar{C}, \exists^\circ pS^a \Rightarrow \forall^\circ pS)$ is also derivable,*

$\forall^\circ pS$ *(resp.,* $\exists^\circ p\Sigma$*) is called a* uniform $\forall^\circ_p$-interpolant of S *(resp.,* uniform $\exists^\circ_p$-interpolant of Σ*).*

Let us first derive the main result of the paper as an immediate corollary.

Corollary 4.2 iM *has ULIP. Hence, it also has UIP and LIP.*

Proof. By Theorem 4.1, **G4iM** has ULIP. Hence, for $\Sigma = \varnothing$, by (i), we know that $(\Rightarrow \exists^\circ p\varnothing)$ is derivable. Therefore, $\exists^\circ p\varnothing$ is provably equivalent to $\top$. Now, set $\forall^\circ pA = \forall^\circ p(\Rightarrow A)$ and $\exists^\circ pA = \exists^\circ p\{A\}$. First, by (iii), we know that $\forall^\circ p(\Rightarrow A) \Rightarrow A$ is derivable. Hence, $\mathsf{iM} \vdash \forall^\circ pA \to A$. Secondly, for any p°-free formula B, if $\mathsf{iM} \vdash B \to A$, then $B \Rightarrow A$ is derivable in **G4iM**. Therefore, by (iv), when $\bar{C} = \{B\}$, we get the derivability of $B, \exists^\diamond p\varnothing \Rightarrow Ap(\Rightarrow A)$. Since $\exists^\diamond p\varnothing$ is provably equivalent to $\top$, we get $B \Rightarrow \forall^\circ p(\Rightarrow A)$ and hence, $\mathsf{iM} \vdash B \to \forall^\circ pA$. Therefore, $\forall^\circ pA$ satisfies the conditions in Definition 2.4. The case for $\exists^\circ pA$ is easier and will be skipped here. The second part of the corollary is a result of Theorem 2.5. □

Now, to prove Theorem 4.1, we need the following lemma.

Lemma 4.3 **G4iM** *enjoys* ULIP with respect to the axioms, *i.e., for any sequent S, multiset Σ, atom p, and $\circ \in \{+, -\}$, there exist formulas $\forall^\circ_{ax}pS$ and $\exists^\circ_{ax}p\Sigma$ such that they satisfy conditions (var),* (i), *and* (iii) *in Theorem 4.1 and*

(ii′) for any sequent $\bar{C} \Rightarrow \bar{D}$ such that $p \notin V^\circ(\bar{C} \Rightarrow \bar{D})$, if $\Sigma, \bar{C} \Rightarrow \bar{D}$ is an axiom *in* **G4iM** *then $(\exists^\circ_{ax}p\Sigma, \bar{C} \Rightarrow \bar{D})$ is derivable,*

(iv′) for any multiset $\bar{C}$ such that $p \notin V^\circ(\bar{C})$, if $S \cdot (\bar{C} \Rightarrow)$ is an axiom *in* **G4iM** *then $(\bar{C} \Rightarrow \forall^\circ_{ax}pS)$ is derivable.*

Proof. Define $\exists^\circ_{ax}p\Sigma$ as the conjunction of all p°-free formulas in Σ and $\forall^\circ_{ax}pS$ as the following: if S is provable, define it as $\top$, otherwise, define $\forall^\circ_{ax}pS$ as the disjunction of all p°-free formulas in S^s. We will show that $\exists^\circ_{ax}p\Sigma$ and $\forall^\circ_{ax}pS$ satisfy the conditions (*var*), (i) and (iii) of Theorem 4.1, and condition (*ii′*) and (*iv′*). Clearly, $\exists^\circ_{ax}p\Sigma$ and $\forall^\circ_{ax}pS$ are p°-free, $V^\dagger(\exists^\circ_{ax}p\Sigma) \subseteq V^\dagger(\Sigma)$ and $V^\dagger(\forall^\circ_{ax}pS) \subseteq V^\dagger(S)$, for any $\dagger \in \{+, -\}$ and $\Sigma \Rightarrow \exists^\circ_{ax}p\Sigma$ and $S \cdot (\forall^\circ_{ax}pS \Rightarrow)$ are derivable.

For (*ii′*), if $\Sigma, \bar{C} \Rightarrow \bar{D}$ is an axiom, it is either of the form $\Gamma, q \Rightarrow q$, where q is an atom, or $\Gamma, \bot \Rightarrow \Delta$. In the first case, as $\bar{D} = \{q\}$, the atom q is p°-free. If $q \in \bar{C}$, then $\bar{C}, \exists^\circ_{ax}p\Sigma \Rightarrow \bar{D}$ is an instance of (Ax) and hence provable. If $q \in \Sigma$, then q appears as a conjunct in $\exists^\circ_{ax}p\Sigma$ and $\bar{C}, \exists^\circ_{ax}p\Sigma \Rightarrow \bar{D}$ is provable. If $\Sigma, \bar{C} \Rightarrow \bar{D}$ is of the form $\Gamma, \bot \Rightarrow \Delta$, then if $\bot \in \bar{C}$, then $\bar{C}, \exists^\circ_{ax}p\Sigma \Rightarrow \bar{D}$ is an instance of $(L\bot)$ and provable. If $\bot \in \Sigma$, then $\bot$ appears as a conjunct in $\exists^\circ_{ax}p\Sigma$ and hence $\bar{C}, \exists^\circ_{ax}p\Sigma \Rightarrow \bar{D}$ is provable.

For (*iv′*), suppose a p°-free multiset $\bar{C}$ is given such that $S \cdot (\bar{C} \Rightarrow)$ is an axiom in **G4iM**. If S is provable, then as $\forall^\circ_{ax}pS = \top$, we have $\bar{C} \Rightarrow \forall^\circ_{ax}pS$. If S is not provable, then there are two cases to consider. First, suppose $S \cdot (\bar{C} \Rightarrow)$ is of the form $\Gamma, q \Rightarrow q$, where q is an atom. If $q \notin \bar{C}$, then $q \in S^a$ which implies that S is provable. Therefore, $q \in \bar{C}$. Hence q is p°-free and appears as a disjunct in $\forall^\circ_{ax}pS$. Therefore, as $q \in \bar{C}$, we get $\bar{C} \Rightarrow \forall^\circ_{ax}pS$. Second, suppose

$S \cdot (\bar{C} \Rightarrow)$ is of the form $\Gamma, \bot \Rightarrow \Delta$. If $\bot \in S^a$, then S is provable. Therefore, $\bot \notin S^a$. Hence, $\bot \in \bar{C}$ which implies $\bar{C} \Rightarrow \forall^{\circ}_{ax} pS$. □

Proof. [of Theorem 4.1] Let us fix some notations. Let S_ψ denote $(\Rightarrow \psi)$. We use $\Sigma_{q,\psi}$(respectively $\Sigma_{\Box\psi,\theta}$) to denote the multiset obtained from Σ by replacing one instance of $q \to \psi$ (respectively $\Box\psi \to \theta$) in Σ by ψ (respectively θ). Accordingly, for $S = (\Sigma \Rightarrow \Delta)$, define $S_{q,\psi} = (\Sigma_{q,\psi} \Rightarrow \Delta)$ and $S_{\Box\psi,\theta} = (\Sigma_{\Box\psi,\theta} \Rightarrow \Delta)$. Define $I^{\circ}_{at}(\Sigma) = \{(q,\psi) \mid q \to \psi \in \Sigma$ and q is an atom and p°-free$\}$ and $I_m(\Sigma) = \{(\varphi,\psi,\theta) \mid \Box\varphi, \Box\psi \to \theta \in \Sigma$ and $\mathbf{G4iM} \vdash \varphi \Rightarrow \psi\}$. Here are some remarks:

(R_1) If $q \to \psi \in \Sigma$ (resp. $\Box\psi \to \theta \in \Sigma$), then $\Sigma_{q,\psi} \prec \Sigma$ (resp. $\Sigma_{\Box\psi,\theta} \prec \Sigma$) and $V^{\dagger}(\Sigma_{q,\psi}) \subseteq V^{\dagger}(\Sigma)$, (resp. $V^{\dagger}(\Sigma_{\Box\psi,\theta}) \subseteq V^{\dagger}(\Sigma)$), for any $\dagger \in \{+,-\}$. Similarly, if $q \to \psi \in S^a$ (resp. $\Box\psi \to \theta \in S^a$), then $S_{q,\psi} \prec S$ (resp. $S_{\Box\psi,\theta} \prec S$) and $V^{\dagger}(S_{q,\psi}) \subseteq V^{\dagger}(S)$, (resp. $V^{\dagger}(S_{\Box\psi,\theta}) \subseteq V^{\dagger}(S)$).

(R_2) If $\Box\psi \to \theta \in \Sigma$, then $S_\psi \prec \Sigma$ and $V^{\dagger}(S_\psi) \subseteq V^{\dagger}(\Sigma)$, for any $\dagger \in \{+,-\}$. Similarly, if $\Box\psi \to \theta \in S^a$, then $S_\psi \prec S$ and $V^{\dagger}(S_\psi) \subseteq V^{\dagger}(S)$.

(R_3) If $(q,\psi) \in I^{\circ}_{at}(\Sigma)$, then q is p°-free.

Define the four formulas $\exists^+ p\Sigma$, $\exists^- p\Sigma$, $\forall^+ pS$ and $\forall^- pS$ simultaneously by recursion on the well-founded order $\prec$ over the set of all multisets and sequents: if $\Sigma = \varnothing$, define $\exists^{\circ} p\Sigma$ as $\top$. Otherwise, define it as:

$$\bigwedge_{R \in \mathcal{LR}} \Big(\bigwedge_i (\exists^{\circ} pS_i^a \to \forall^{\diamond} pS_i) \to \bigvee_j \exists^{\circ} p\Sigma_j\Big) \wedge (\exists^{\circ}_{ax} p\Sigma) \wedge (\exists^{\circ}_{at} p\Sigma) \wedge (\exists^{\circ}_{m} p\Sigma)$$

The first conjunction is over all the left rules R in $\mathbf{G4w}^-$ and the rule $(Lp\to)$, backward applicable to $(\Sigma \Rightarrow)$, where $(\Sigma \Rightarrow)$ is the conclusion, the S_i's are the non-contextual premises and $(\Sigma_j \Rightarrow)$'s are the contextual premises. The second conjunct is provided by Lemma 4.3. The rest are defined as

$$\exists^{\circ}_{at} p\Sigma = \bigwedge_{(q,\psi) \in I^{\circ}_{at}(\Sigma)} q \to \exists^{\circ} p\Sigma_{q,\psi} \quad ,$$

$$\exists^{\circ}_{m} p\Sigma = \bigwedge_{\Box\varphi \in \Sigma} \Box\exists^{\circ} p\varphi \wedge \bigwedge_{\Box\psi \to \theta \in \Sigma} (\Box\forall^{\diamond} pS_\psi \to \exists^{\circ} p\Sigma_{\Box\psi,\theta}) \wedge \bigwedge_{(\varphi,\psi,\theta) \in I_m(\Sigma)} \exists^{\circ} p\Sigma_{\Box\psi,\theta}$$

For $\forall^{\circ} pS$, if S is provable define it as $\top$, otherwise, define $\forall^{\circ} pS$ as:

$$\bigvee_R \Big(\bigwedge_i (\exists^{\diamond} pS_i^a \to \forall^{\circ} pS_i)\Big) \vee (\forall^{\circ}_{ax} pS) \vee (\forall^{\circ}_{at} pS) \vee (\forall^{\circ}_{m} pS)$$

The first disjunction is over all rules R in $\mathbf{G4w}$ backward applicable to S, where S_i's are the premises. The second disjunct is provided by Lemma 4.3. The third is defined as

$$\forall^{\circ}_{at} pS = \bigwedge_{(q,\psi) \in I^{\circ}_{at}(S^a)} q \wedge (\exists^{\diamond} pS^a_{q,\psi} \to \forall^{\circ} pS_{q,\psi}),$$

For $\forall^{\circ}_{m} pS$, if $S = (\Rightarrow \Box\psi)$, define $\forall^{\circ}_{m} pS = \Box\forall^{\circ} pS_\psi$. Otherwise, define

$$\forall^{\circ}_{m} pS = \bigvee_{\Box\psi \to \theta \in S^a} (\forall^{\circ} pS_{\Box\psi,\theta} \wedge \Box\forall^{\circ} pS_\psi) \vee \bigvee_{(\varphi,\psi,\theta) \in I_m(S^a)} \forall^{\circ} pS_{\Box\psi,\theta}$$

We use induction on the well-founded order $\prec$ to prove that $\exists^{\circ} p\Sigma$ and $\forall^{\circ} pS$

have all the properties of Theorem 4.1. In fact, in the induction step, we assume that for a sequent S (resp. multiset Σ), all $\exists^+ p\Sigma'$, $\exists^- p\Sigma'$, $\forall^+ pS'$ and $\forall^- pS'$ exist for any sequent S' and multiset Σ' lower than S (resp. Σ).
To prove that the recursive definition is well-defined, note that in any rule in **G4w**, if S_i's are the premises and S is the conclusion, we have $S_i \prec S$. Therefore, using the remarks (R_1) and (R_2) above, we conclude that both $\forall^\circ pS$ and $\exists^\circ pS$ are well-defined.

To prove (var), using the induction hypothesis and the remark (R_3), it is clear that $\forall^\circ pS$ and $\exists^\circ p\Sigma$ are p°-free. Moreover, as every rule in **G4iM** enjoys the variable preserving property, it is enough to use the induction hypothesis, Lemma 4.3 and the remark (R_1) and (R_2) to prove $V^\dagger(\forall^\circ pS) \subseteq V^\dagger(S)$ and $V^\dagger(\exists^\circ p\Sigma) \subseteq V^\dagger(\Sigma)$, for any $\dagger \in \{+, -\}$.

To prove conditions (i), (ii), (iii) and (iv), as the cases that $\Sigma = \varnothing$ and S is provable are easy, from now on, we assume that $\Sigma \neq \varnothing$ and S is not provable.

To prove (i), it is enough to show that the following formulas

$$\bigwedge_{R \in \mathcal{LR}} (\bigwedge_i (\exists^\circ pS_i^a \to \forall^\diamond pS_i) \to \bigvee_j \exists^\circ p\Sigma_j) \; (1) \; , \; \exists^\circ_{ax} p\Sigma \; (2) \; , \; \exists^\circ_{at} p\Sigma \; (3) \; , \; \exists^\circ_m p\Sigma \; (4)$$

are all derivable from Σ. For (1), assume that the left rule R of **G4w**$^-$ is backward applicable to $(\Sigma \Rightarrow)$ with S_i's as the non-contextual premises and $(\Sigma_j \Rightarrow)$'s as the contextual premises. Since S_i's and Σ_j's are lower than Σ, by the induction hypothesis, we have $S_i \cdot (\forall^\diamond pS_i \Rightarrow)$, $(S_i^a \Rightarrow \exists^\circ pS_i^a)$ and $(\Sigma_j \Rightarrow \exists^\circ p\Sigma_j)$ for all i and j. Therefore, $S_i \cdot (\bigwedge_i (\exists^\circ pS_i^a \to \forall^\diamond pS_i) \Rightarrow)$ and $\Sigma_j \Rightarrow \bigvee_j \exists^\circ p\Sigma_j$ are derivable. By the free-context property of R, we can add $\bigwedge_i (\exists^\circ pS_i^a \to \forall^\diamond pS_i)$ to the antecedents of the premises and conclusion and put $\bigvee_j \exists^\circ p\Sigma_j$ in the succedents of the contextual premises and the conclusion. The rule will become:

$$\frac{\{\bigwedge_i (\exists^\circ pS_i^a \to \forall^\diamond pS_i), S_i^a \Rightarrow S_i^s\}_{i \in I} \qquad \{\Sigma_j \Rightarrow \bigvee_j \exists^\circ p\Sigma_j\}_{j \in J}}{\Sigma, \bigwedge_i (\exists^\circ pS_i^a \to \forall^\diamond pS_i) \Rightarrow \bigvee_j \exists^\circ p\Sigma_j}$$

Hence, we get $\Sigma \Rightarrow (\bigwedge_i (\exists^\circ pS_i^a \to \forall^\diamond pS_i) \to \bigvee_j \exists^\circ p\Sigma_j)$. The case where the last rule is $(Lp\!\to)$ is similar.

For (2), we use Lemma 4.3. For (3), if q is $p^\diamond$-free and $q \to \psi \in \Sigma$, then we have $\Sigma = \Sigma' \cup \{q \to \psi\}$ and $\Sigma_{q,\psi} = \Sigma' \cup \{\psi\}$. As $\Sigma_{q,\psi}$ is lower than Σ, by the induction hypothesis $\Sigma_{q,\psi} \Rightarrow \exists^\circ p\Sigma_{q,\psi}$. Therefore, $\Sigma', \psi \Rightarrow \exists^\circ p\Sigma_{q,\psi}$ which implies $\Sigma', q, q \to \psi \Rightarrow \exists^\circ p\Sigma_{q,\psi}$. Hence, we get $\Sigma \Rightarrow q \to \exists^\circ p\Sigma_{q,\psi}$.

For (4), we have to show that each conjunct in $\exists^\circ_m p\Sigma$ is derivable from Σ. For the first conjunct, suppose $\Box\varphi \in \Sigma$, i.e., $\Sigma = \Pi, \Box\varphi$. As φ is lower than Σ, by the induction hypothesis we have $\varphi \Rightarrow \exists^\circ p\varphi$. Therefore, by the rule (M), we have $\Box\varphi \Rightarrow \Box\exists^\circ p\varphi$. By (Lw), we get $\Pi, \Box\varphi \Rightarrow \Box\exists^\circ p\varphi$, which is $\Sigma \Rightarrow \Box\exists^\circ p\varphi$. For the second conjunct, suppose $\Box\psi \to \theta \in \Sigma$, i.e., $\Sigma = \Pi, \Box\psi \to \theta$. As $\Sigma_{\Box\psi,\theta}$ and S_ψ are lower than Σ, by the induction hypothesis, we have $\Sigma_{\Box\psi,\theta} \Rightarrow \exists^\circ p\Sigma_{\Box\psi,\theta}$ and $S_\psi \cdot (\forall^\diamond pS_\psi \Rightarrow)$, which are $\Pi, \theta \Rightarrow \exists^\circ p\Sigma_{\Box\psi,\theta}$ and $\forall^\diamond pS_\psi \Rightarrow \psi$. By (Lw), we have $\Pi, \Box\forall^\diamond pS_\psi, \theta \Rightarrow \exists^\circ p\Sigma_{\Box\psi,\theta}$. Applying the rule $(LM \to)$, we get $\Pi, \Box\forall^\diamond pS_\psi, \Box\psi \to \theta \Rightarrow \exists^\circ p\Sigma_{\Box\psi,\theta}$, which implies

$\Sigma \Rightarrow \Box\forall^{\circ}pS_{\psi} \to \exists^{\circ}p\Sigma_{\Box\psi,\theta}$. For the last conjunct, as $(\varphi, \psi, \theta) \in I_m(\Sigma)$, we know that $\Box\varphi, \Box\psi \to \theta \in \Sigma$ and the sequent $\varphi \Rightarrow \psi$ is provable. Therefore, $\Sigma = \Pi, \Box\varphi, \Box\psi \to \theta$. As $\Sigma_{\Box\psi,\theta}$ is lower than Σ, by the induction hypothesis, we have $\Sigma_{\Box\psi,\theta} \Rightarrow \exists^{\circ}p\Sigma_{\Box\psi,\theta}$, which is $\Pi, \Box\varphi, \theta \Rightarrow \exists^{\circ}p\Sigma_{\Box\psi,\theta}$. Since $\varphi \Rightarrow \psi$ is also provable, we can apply the rule $(LM{\to})$ to obtain $\Pi, \Box\varphi, \Box\psi \to \theta \Rightarrow \exists^{\circ}p\Sigma_{\Box\psi,\theta}$, which is $\Sigma \Rightarrow \exists^{\circ}p\Sigma_{\Box\psi,\theta}$.

For (ii), we assume $p \notin V^{\diamond}(\bar{C}), p \notin V^{\circ}(\bar{D})$ and $\Sigma, \bar{C} \Rightarrow \bar{D}$ is derivable and we use induction on the length of its proof. If $\Sigma, \bar{C} \Rightarrow \bar{D}$ is an axiom, we have $\exists^{\circ}_{ax}p\Sigma, \bar{C} \Rightarrow \bar{D}$, by Lemma 4.3, and hence $\exists^{\circ}p\Sigma, \bar{C} \Rightarrow \bar{D}$. If the last rule is a left rule in **G4w**$^-$, it is of the form:

$$\frac{\{\Gamma, \bar{\varphi}_i \Rightarrow \bar{\delta}_i\}_{i\in I} \qquad \{\Gamma, \bar{\psi}_j \Rightarrow \bar{D}\}_{j\in J}}{\Gamma, \varphi \Rightarrow \bar{D}}$$

Then, there are two cases to consider, i.e., either $\varphi \in \bar{C}$ or $\varphi \in \Sigma$. If $\varphi \in \bar{C}$, set $\bar{C}' = \bar{C} - \{\varphi\}$. Since $\varphi \in \bar{C}$, it is $p^{\diamond}$-free by the assumption, and by the local variable preserving property $\bar{\varphi}_i$'s and $\bar{\psi}_j$'s are $p^{\diamond}$-free and $\bar{\delta}_i$'s are p°-free. By the induction hypothesis, as $p \notin V^{\circ}(\bar{C}', \bar{\varphi}_i \Rightarrow \bar{\delta}_i)$, $p \notin V^{\circ}(\bar{C}', \bar{\psi}_j \Rightarrow \bar{D})$, and $(\Sigma, \bar{C}', \bar{\varphi}_i \Rightarrow \bar{\delta}_i)$ and $(\Sigma, \bar{C}', \bar{\psi}_j \Rightarrow \bar{D})$ have shorter proofs, we have $(\exists^{\circ}p\Sigma, \bar{C}', \bar{\varphi}_i \Rightarrow \bar{\delta}_i)$ and $(\exists^{\circ}p\Sigma, \bar{C}', \bar{\psi}_j \Rightarrow \bar{D})$ are derivable, for each $i \in I$ and $j \in J$. By using the rule itself, we get $\exists^{\circ}p\Sigma, \bar{C}', \varphi \Rightarrow \bar{D}$, which implies $\exists^{\circ}p\Sigma, \bar{C} \Rightarrow \bar{D}$.

If $\varphi \in \Sigma$, set $\Sigma' = \Sigma - \{\varphi\}$. Hence, the last rule is of the form:

$$\frac{\{\Sigma', \bar{C}, \bar{\varphi}_i \Rightarrow \bar{\delta}_i\}_{i\in I} \qquad \{\Sigma', \bar{C}, \bar{\psi}_j \Rightarrow \bar{D}\}_{j\in J}}{\Sigma', \bar{C}, \varphi \Rightarrow \bar{D}}$$

Note that neither $\bar{C}$ nor $\bar{D}$ contain any active formulas. By the free-context property, if we delete $\bar{C}$ and $\bar{D}$ from the premises and conclusion of the last rule, the rule remains valid and it changes to:

$$\frac{\{\Sigma', \bar{\varphi}_i \Rightarrow \bar{\delta}_i\}_{i\in I} \qquad \{\Sigma', \bar{\psi}_j \Rightarrow\}_{j\in J}}{\Sigma', \varphi \Rightarrow}$$

Hence, the rule is backward applicable to $(\Sigma \Rightarrow)$. Set $S_i = (\Sigma', \bar{\varphi}_i \Rightarrow \bar{\delta}_i)$ and $\Sigma_j = \Sigma', \bar{\psi}_j$. As $S_i \cdot (\bar{C} \Rightarrow)$ and $(\Sigma_j, \bar{C} \Rightarrow \bar{D})$ are provable and S_i's and Σ_j's are lower than $(\Sigma \Rightarrow)$, by the induction hypothesis, we have $(\exists^{\circ}pS_i^a, \bar{C} \Rightarrow \forall^{\diamond}pS_i)$ and $(\exists^{\circ}p\Sigma_j, \bar{C} \Rightarrow \bar{D})$. Hence $\bar{C} \Rightarrow \bigwedge_i(\exists^{\circ}pS_i^a \to \forall^{\diamond}pS_i)$ and $(\bigvee_j \exists^{\circ}p\Sigma_j, \bar{C} \Rightarrow \bar{D})$ are derivable. Therefore, we have $(\bigwedge_i(\exists^{\circ}pS_i^a \to \forall^{\diamond}pS_i) \to \bigvee_j \exists^{\circ}p\Sigma_j, \bar{C} \Rightarrow \bar{D})$. As $\bigwedge_i(\exists^{\circ}pS_i^a \to \forall^{\diamond}pS_i) \to \bigvee_j \exists^{\circ}p\Sigma_j$ is a conjunct in $\exists^{\circ}p\Sigma$, we have $\exists^{\circ}p\Sigma, \bar{C} \Rightarrow \bar{D}$.

If the last rule of the proof is a right rule, then it is of the form:

$$\frac{\{\Sigma, \bar{C}, \bar{\varphi}_i \Rightarrow \bar{\psi}_i\}_{i\in I}}{\Sigma, \bar{C} \Rightarrow \varphi}$$

and $\bar{D} = \{\varphi\}$. Hence, φ is p°-free. By the local variable preserving property, $\bar{\varphi}_i$'s are $p^\diamond$-free and $\bar{\psi}_i$'s are p°-free. By the induction hypothesis, as $(\bar{C}, \bar{\varphi}_i \Rightarrow \bar{\psi}_i)$ is p°-free and $(\Sigma, \bar{C}, \bar{\varphi}_i \Rightarrow \bar{\psi}_i)$ has a shorter proof, we have $(\exists^\circ p\Sigma, \bar{C}, \bar{\varphi}_i \Rightarrow \bar{\psi}_i)$. Using the rule itself, we get $\exists^\circ p\Sigma, \bar{C} \Rightarrow \varphi$ which is $\exists^\circ p\Sigma, \bar{C} \Rightarrow \bar{D}$.

If the last rule is the rule $(Lp\rightarrow)$, then it is of the form

$$\frac{\Gamma, q, \psi \Rightarrow \bar{D}}{\Gamma, q, q \rightarrow \psi \Rightarrow \bar{D}}$$

There are four cases to consider, depending on whether q or $q \rightarrow \psi$ are in $\bar{C}$.

If $q, q \rightarrow \psi \in \bar{C}$, then set $\bar{C}' = \bar{C} - \{q, q \rightarrow \psi\}$. As $\Sigma, \bar{C}', q, \psi \Rightarrow \bar{D}$ has a shorter proof and q and ψ are $p^\diamond$-free, then by the induction hypothesis, $\exists^\circ p\Sigma, \bar{C}', q, \psi \Rightarrow \bar{D}$. Hence, by the rule itself, we have $\exists^\circ p\Sigma, \bar{C}', q, q \rightarrow \psi \Rightarrow \bar{D}$.

If $q, q \rightarrow \psi \notin \bar{C}$, then the premise of the rule is of the form $\Sigma_{q,\psi}, \bar{C} \Rightarrow \bar{D}$, and by the induction hypothesis, we have $\exists^\circ p\Sigma_{q,\psi}, \bar{C} \Rightarrow \bar{D}$. However, by the free-context property, we can delete $\bar{C}$ and $\bar{D}$ in the premise and conclusion and the rule remains valid and has the form $\dfrac{\Sigma_{q,\psi} \Rightarrow}{\Sigma \Rightarrow}$. Therefore, this rule is backward applicable to $(\Sigma \Rightarrow)$ and as this rule has no non-contextual premise, $\top \rightarrow \exists^\circ p\Sigma_{q,\psi}$ appears as a conjunct in $\exists^\circ p\Sigma$. Hence, $\exists^\circ p\Sigma, \bar{C} \Rightarrow \bar{D}$ is derivable.

If $q \rightarrow \psi \notin \bar{C}$ and $q \in \bar{C}$, then q is $p^\diamond$-free. Set $\bar{C}' = \bar{C} - \{q\}$. Then, the premise of the rule is of the form $\Sigma_{q,\psi}, q, \bar{C}' \Rightarrow \bar{D}$. As $q \rightarrow \psi \in \Sigma$, we have $\Sigma_{q,\psi} \prec \Sigma$, by (R_1). As $(\bar{C}', q \Rightarrow \bar{D})$ is p°-free, by the induction hypothesis, $\bar{C}', q, \exists^\circ p\Sigma_{q,\psi} \Rightarrow \bar{D}$. Using the rule $(Lp\rightarrow)$, we get $\bar{C}', q, q \rightarrow \exists^\circ p\Sigma_{q,\psi} \Rightarrow \bar{D}$. As $q \rightarrow \exists^\circ p\Sigma_{q,\psi}$ is a conjunct in $\exists^\circ p\Sigma$, we have $\exists^\circ p\Sigma, \bar{C} \Rightarrow \bar{D}$.

If $q \notin \bar{C}$ and $q \rightarrow \psi \in \bar{C}$, then ψ is $p^\diamond$-free and q is p°-free. Set $\bar{C}' = \bar{C} - \{q \rightarrow \psi\}$. Then, the premise of the rule is of the form $\Sigma, \psi, \bar{C}' \Rightarrow \bar{D}$. By the induction hypothesis, we have $\exists^\circ p\Sigma, \psi, \bar{C}' \Rightarrow \bar{D}$. Moreover, by (Ax), we have $\Sigma \Rightarrow q$. Since q is p°-free, we have $\exists^\circ_{ax} p\Sigma \Rightarrow q$ and hence, $\exists^\circ p\Sigma \Rightarrow q$. Therefore, $\exists^\circ p\Sigma, q \rightarrow \psi, \bar{C}' \Rightarrow \bar{D}$, which is $\exists^\circ p\Sigma, \bar{C} \Rightarrow \bar{D}$.

If the last rule in the proof is the modal rule (M), then it is of the form

$$\frac{\varphi \Rightarrow \psi}{\Box\varphi \Rightarrow \Box\psi}\, M$$

As $\Sigma \neq \varnothing$, we have $\bar{C} = \varnothing$ and $\Sigma = \Box\varphi$. Since $\bar{D} = \Box\psi$, the formula ψ is p°-free. As $(\varphi \Rightarrow \psi)$ is provable and φ is lower than Σ, by the induction hypothesis, we have $\exists^\circ p\varphi \Rightarrow \psi$ and by (M), $\Box\exists^\circ p\varphi \Rightarrow \Box\psi$. As $\Box\exists^\circ p\varphi$ appears as a conjunct in the definition of $\exists^\circ p\Sigma$, we get $\exists^\circ p\Sigma, \bar{C} \Rightarrow \bar{D}$.

If the last rule in the proof is the modal rule $(LM\rightarrow)$, then it is of the form

$$\frac{\varphi \Rightarrow \psi \qquad \Gamma, \Box\varphi, \theta \Rightarrow \bar{D}}{\Gamma, \Box\varphi, \Box\psi \rightarrow \theta \Rightarrow \bar{D}}\, LM\rightarrow$$

There are four cases to consider based on which formulas are in $\bar{C}$.

If $\Box\varphi, \Box\psi \rightarrow \theta \in \bar{C}$, then φ and θ are $p^\diamond$-free and ψ is p°-free. Set $\bar{C}' = \bar{C} - \{\Box\varphi, \Box\psi \rightarrow \theta\}$. The right premise of the rule is of the form $\Sigma, \bar{C}', \Box\varphi, \theta \Rightarrow \bar{D}$.

Hence, by the induction hypothesis $\exists^{\circ}p\Sigma, \bar{C}', \Box\varphi, \theta \Rightarrow \bar{D}$. By $(LM \to)$ on the latter sequent and $\varphi \Rightarrow \psi$, we get $\exists^{\circ}p\Sigma, \bar{C}', \Box\varphi, \Box\psi \to \theta \Rightarrow \bar{D}$ which is $\exists^{\circ}p\Sigma, \bar{C} \Rightarrow \bar{D}$.

If $\Box\varphi, \Box\psi \to \theta \notin \bar{C}$, then set $\Sigma' = \Sigma - \{\Box\varphi, \Box\psi \to \theta\}$. Hence, the right premise of the rule is of the form $\Sigma', \Box\varphi, \theta, \bar{C} \Rightarrow \bar{D}$, or equivalently $\Sigma_{\Box\psi,\theta}, \bar{C} \Rightarrow \bar{D}$. Therefore, by the induction hypothesis $\exists^{\circ}p\Sigma_{\Box\psi,\theta}, \bar{C} \Rightarrow \bar{D}$. However, $\exists^{\circ}p\Sigma_{\Box\psi,\theta}$ appears as a conjunct in the definition of $\exists^{\circ}p\Sigma$, since $\Box\varphi, \Box\psi \to \theta \in \Sigma$ and $\mathbf{G4iM} \vdash \varphi \Rightarrow \psi$. Consequently, $\exists^{\circ}p\Sigma, \bar{C} \Rightarrow \bar{D}$.

If $\Box\psi \to \theta \in \bar{C}$ and $\Box\varphi \notin \bar{C}$, then θ is $p^{\diamond}$-free and ψ is p°-free. Set $\bar{C}' = \bar{C} - \{\Box\psi \to \theta\}$. Then, the premise has the form $\Sigma, \bar{C}', \theta \Rightarrow \bar{D}$. By the induction hypothesis, we have $\exists^{\circ}p\Sigma, \bar{C}', \theta \Rightarrow \bar{D}$ and by (Lw), we have $\exists^{\circ}p\Sigma, \bar{C}', \Box\exists^{\circ}p\varphi, \theta \Rightarrow \bar{D}$. Moreover, as $\varphi \Rightarrow \psi$ is provable, ψ is p°-free, and φ is lower than Σ, by the induction hypothesis, we have $\exists^{\circ}p\varphi \Rightarrow \psi$. Applying $(LM \to)$ on $\exists^{\circ}p\varphi \Rightarrow \psi$ and $\exists^{\circ}p\Sigma, \bar{C}', \Box\exists^{\circ}p\varphi, \theta \Rightarrow \bar{D}$, we get $\exists^{\circ}p\Sigma, \bar{C}', \Box\exists^{\circ}p\varphi, \Box\psi \to \theta \Rightarrow \bar{D}$. Note that as $\Box\varphi \in \Sigma$, by definition $\Box\exists^{\circ}p\varphi$ appears as a conjunct in $\exists^{\circ}p\Sigma$. Therefore, $\exists^{\circ}p\Sigma, \bar{C} \Rightarrow \bar{D}$.

Finally, if $\Box\varphi \in \bar{C}$ and $\Box\psi \to \theta \notin \bar{C}$, then φ is $p^{\diamond}$-free. Set $\bar{C}' = \bar{C} - \{\Box\varphi\}$. Therefore, the right premise is of the form $\bar{C}', \Box\varphi, \Sigma_{\Box\psi,\theta} \Rightarrow \bar{D}$. Since $\bar{C}'$ and φ are $p^{\diamond}$-free and $\bar{D}$ is p°-free, by the induction hypothesis, we have $\bar{C}', \Box\varphi, \exists^{\circ}p\Sigma_{\Box\psi,\theta} \Rightarrow \bar{D}$. Moreover, for the premise $\varphi \Rightarrow \psi$, as S_{ψ} is lower than Σ and φ is $p^{\diamond}$-free, by the induction hypothesis, we get $\varphi, \exists^{\circ}pS^{a}_{\psi} \Rightarrow \forall^{\diamond}pS_{\psi}$. However, as $S^{a}_{\psi} = \varnothing$, by definition we have $\exists^{\circ}pS^{a}_{\psi} = \top$. Hence, $\varphi \Rightarrow \forall^{\diamond}pS_{\psi}$. Applying $(LM \to)$ on $\varphi \Rightarrow \forall^{\diamond}pS_{\psi}$ and $\bar{C}', \Box\varphi, \exists^{\circ}p\Sigma_{\Box\psi,\theta} \Rightarrow \bar{D}$ we get $\bar{C}', \Box\varphi, \Box\forall^{\diamond}pS_{\psi} \to \exists^{\circ}p\Sigma_{\Box\psi,\theta} \Rightarrow \bar{D}$. Therefore, as $\Box\psi \to \theta \in \Sigma$, the formula $\Box\forall^{\diamond}pS_{\psi} \to \exists^{\circ}p\Sigma_{\Box\psi,\theta}$ is a conjunct in $\exists^{\circ}p\Sigma$, and hence $\exists^{\circ}p\Sigma, \bar{C} \Rightarrow \bar{D}$ is derivable.

To prove (iii), it is enough to show that the following are provable:

$$S \cdot (\bigwedge_i (\exists^{\diamond}pS^{a}_i \to \forall^{\circ}pS_i) \Rightarrow) \quad (1), \qquad S \cdot (\forall^{\circ}_{ax}pS \Rightarrow) \quad (2),$$
$$S \cdot (\forall^{\circ}_{at}pS \Rightarrow) \quad (3), \qquad S \cdot (\forall^{\circ}_{m}pS \Rightarrow) \quad (4).$$

For (1), assume that the rule R in **G4w** is backward applicable to S and the premises of R are S_i's. As S_i's are lower than S, by the induction hypothesis $S_i \cdot (\forall^{\circ}pS_i \Rightarrow)$ and $S^{a}_i \Rightarrow \exists^{\diamond}pS^{a}_i$. Therefore, $S_i \cdot (\exists^{\diamond}pS^{a}_i \to \forall^{\circ}pS_i \Rightarrow)$. Hence, by weakening, we have $S_i \cdot (\{\exists^{\diamond}pS^{a}_i \to \forall^{\circ}pS_i\}_i \Rightarrow)$. Since any rule in **G4w** has the free-context property, we can add $\{\exists^{\diamond}pS^{a}_i \to \forall^{\circ}pS_i\}_i$ to the antecedents of the premises and conclusion and by the rule itself, we have $S \cdot (\{\exists^{\diamond}pS^{a}_i \to \forall^{\circ}pS_i\}_i \Rightarrow)$ and hence we get $S \cdot (\bigwedge_i (\exists^{\diamond}pS^{a}_i \to \forall^{\circ}pS_i) \Rightarrow)$.

For (2), see Lemma 4.3. For (3), if $(q, \psi) \in I^{\circ}_{at}(S^{a})$, then $S = (\Gamma, q \to \psi \Rightarrow \Delta)$ and $S_{q,\psi} = (\Gamma, \psi \Rightarrow \Delta)$. As $S_{q,\psi} \prec S$, by the induction hypothesis $\Gamma, \psi, \forall^{\circ}pS_{q,\psi} \Rightarrow \Delta$ and $\Gamma, \psi \Rightarrow \exists^{\diamond}pS^{a}_{q,\psi}$. Hence, $\Gamma, \psi, \exists^{\diamond}pS^{a}_{q,\psi} \to \forall^{\circ}pS_{q,\psi} \Rightarrow \Delta$. Therefore, $\Gamma, q, q \to \psi, \exists^{\diamond}pS^{a}_{q,\psi} \to \forall^{\circ}pS_{q,\psi} \Rightarrow \Delta$ which implies $\Gamma, q \to \psi, q \wedge (\exists^{\diamond}pS^{a}_{q,\psi} \to \forall^{\circ}pS_{q,\psi}) \Rightarrow \Delta$, and we get $S \cdot (q \wedge (\exists^{\diamond}pS^{a}_{q,\psi} \to \forall^{\circ}pS_{q,\psi}) \Rightarrow)$.

For (4), if $S = (\Rightarrow \Box\psi)$, then by definition $\forall^{\circ}_{m}pS = \Box\forall^{\circ}pS_{\psi}$. As $S_{\psi} \prec S$, by the induction hypothesis $S_{\psi} \cdot (\forall^{\circ}pS_{\psi} \Rightarrow) = (\forall^{\circ}pS_{\psi} \Rightarrow \psi)$ is provable. By

(M) we get $\Box\forall^{\circ}pS_{\psi} \Rightarrow \Box\psi$ which is $S \cdot (\forall^{\circ}_{m}pS \Rightarrow)$. If S is not of the form $(\Rightarrow \Box\psi)$, then $\forall^{\circ}_{m}pS$ is defined by a disjunction over two families of formulas. We have to show that adding any such disjunct to the antecedent of S makes it provable.

For the first family of disjuncts, if $\Box\psi \to \theta \in S^{a}$, then S is in form $(\Gamma, \Box\psi \to \theta \Rightarrow \Delta)$. As S_{ψ} and $S_{\Box\psi,\theta}$ are lower than S, by the induction hypothesis we have $(\forall^{\circ}pS_{\psi} \Rightarrow \psi)$ and $(\forall^{\circ}pS_{\Box\psi,\theta}, \Gamma, \theta \Rightarrow \Delta)$. By (Lw), we get $(\Box\forall^{\circ}pS_{\psi}, \forall^{\circ}pS_{\Box\psi,\theta}, \Gamma, \theta \Rightarrow \Delta)$ and by $(LM \to)$, $\Box\forall^{\circ}pS_{\psi}, \forall^{\circ}pS_{\Box\psi,\theta}, \Gamma, \Box\psi \to \theta \Rightarrow \Delta$, which implies $S \cdot (\Box\forall^{\circ}pS_{\psi} \wedge \forall^{\circ}pS_{\Box\psi,\theta} \Rightarrow)$.

For the second family of disjuncts, if $(\varphi, \psi, \theta) \in I_{m}(S^{a})$, then S has the form $S = (\Gamma, \Box\varphi, \Box\psi \to \theta \Rightarrow \Delta)$ and **G4iM** $\vdash \varphi \Rightarrow \psi$. Since $S_{\Box\psi,\theta} = (\Gamma, \Box\varphi, \theta \Rightarrow \Delta)$ is lower than S, by the induction hypothesis we have $\forall^{\circ}pS_{\Box\psi,\theta}, \Gamma, \Box\varphi, \theta \Rightarrow \Delta$. Applying $(LM \to)$ on the latter sequent and $\varphi \Rightarrow \psi$, we get $S \cdot (\forall^{\circ}pS_{\Box\psi,\theta} \Rightarrow)$.

For (iv), we use induction on the length of the proof of $S \cdot (\bar{C} \Rightarrow)$. If $S \cdot (\bar{C} \Rightarrow)$ is an axiom, by Lemma 4.3 we have $\bar{C} \Rightarrow \forall^{\circ}_{ax}pS$, and hence $\exists^{\diamond}pS^{a}, \bar{C} \Rightarrow \forall^{\circ}pS$. If the last rule is a left one in $\mathbf{G4w}^{-}$, it is of the form:

$$\frac{\{\Gamma, \bar{\varphi}_i \Rightarrow \bar{\delta}_i\}_{i \in I} \qquad \{\Gamma, \bar{\psi}_j \Rightarrow \Delta\}_{j \in J}}{\Gamma, \varphi \Rightarrow \Delta}$$

There are two cases to consider, either $\varphi \in \bar{C}$ or $\varphi \in S^{a}$. If $\varphi \in \bar{C}$, then it is p°-free and by the local variable preserving property, $\bar{\varphi}_i$'s and $\bar{\psi}_j$'s are p°-free and $\bar{\delta}_i$'s are $p^{\diamond}$-free. Set $\bar{C}' = \bar{C} - \{\varphi\}$. As the sequent $S \cdot (\bar{C}', \bar{\psi}_j \Rightarrow)$ has a shorter proof, by the induction hypothesis we have $(\exists^{\diamond}pS^{a}, \bar{C}', \bar{\psi}_j \Rightarrow \forall^{\circ}pS)$. Now, we want to use the induction hypothesis to prove $(\exists^{\diamond}pS^{a}, \bar{C}', \bar{\varphi}_i \Rightarrow \bar{\delta}_i)$. Note that it might be the case that $S^{s} = \varnothing$ and hence S^{a} is not lower than S. However, we already saw that for any multiset Σ, having the conditions (i), (ii), (iii) and (iv) for all multisets and sequents below Σ proves part (ii) for Σ. Putting $\Sigma = S^{a}$, as any multiset or sequent below S^{a} is also below S, by the induction hypothesis we have all four conditions for them and hence we have (ii) for S^{a}. Now, as $(S^{a}, \bar{C}', \bar{\varphi}_i \Rightarrow \bar{\delta}_i)$ is provable and $p \notin V^{\diamond}(\bar{C}', \bar{\varphi}_i \Rightarrow \bar{\delta}_i)$, by (ii), we have $(\exists^{\diamond}pS^{a}, \bar{C}', \bar{\varphi}_i \Rightarrow \bar{\delta}_i)$. By the rule itself, we have

$$\frac{\{\exists^{\diamond}pS^{a}, \bar{C}', \bar{\varphi}_i \Rightarrow \bar{\delta}_i\}_{i \in I} \qquad \{\exists^{\diamond}pS^{a}, \bar{C}', \bar{\psi}_j \Rightarrow \forall^{\circ}pS\}_{j \in J}}{\exists^{\diamond}pS^{a}, \bar{C}', \varphi \Rightarrow \forall^{\circ}pS}$$

which is $\exists^{\diamond}pS^{a}, \bar{C} \Rightarrow \forall^{\circ}pS$.
If $\varphi \notin \bar{C}$, then it does not contain any active formulas of the rule. Set $\Gamma' = S^{a} - \{\varphi\}$. The last rule is of the form:

$$\frac{\{\Gamma', \bar{C}, \bar{\varphi}_i \Rightarrow \bar{\delta}_i\}_{i \in I} \qquad \{\Gamma', \bar{C}, \bar{\psi}_j \Rightarrow \Delta\}_{j \in J}}{\Gamma', \bar{C}, \varphi \Rightarrow \Delta}$$

By the free-context property, we can delete $\bar{C}$ from the premises and conclusion of the rule which remains valid and changes to:

$$\frac{\{\Gamma', \bar{\varphi}_i \Rightarrow \bar{\delta}_i\}_{i \in I} \qquad \{\Gamma', \bar{\psi}_j \Rightarrow \Delta\}_{j \in J}}{\Gamma', \varphi \Rightarrow \Delta}$$

Therefore, the rule is backward applicable to $S = (\Gamma', \varphi \Rightarrow \Delta)$. Set $S_i = (\Gamma', \bar{\varphi}_i \Rightarrow \bar{\delta}_i)$ and $T_j = (\Gamma', \bar{\psi}_j \Rightarrow \Delta)$. As S_i's and T_j's are lower than S and $S_i \cdot (\bar{C} \Rightarrow)$ and $T_j \cdot (\bar{C} \Rightarrow)$ are provable, by the induction hypothesis, we have $\exists^\diamond p S_i^a, \bar{C} \Rightarrow \forall^\circ p S_i$ and $\exists^\diamond p T_j^a, \bar{C} \Rightarrow \forall^\circ p T_j$. Hence, $\bar{C} \Rightarrow \bigwedge_i (\exists^\diamond p S_i^a \to \forall^\circ p S_i) \wedge \bigwedge_j (\exists^\diamond p T_j^a \to \forall^\circ p T_j)$ and as $\bigwedge_i (\exists^\diamond p S_i^a \to \forall^\circ p S_i) \wedge \bigwedge_j (\exists^\diamond p T_j^a \to \forall^\circ p T_j)$ appears as a disjunct in $\forall^\circ p S$, we have $\bar{C} \Rightarrow \forall^\circ p S$ and hence $\exists^\diamond p S^a, \bar{C} \Rightarrow \forall^\circ p S$. The case where the last rule is a right one in **G4w**$^-$ is similar. If the last rule is the rule $(Lp\to)$, then it has the form

$$\frac{\Gamma, q, \psi \Rightarrow \Delta}{\Gamma, q, q \to \psi \Rightarrow \Delta}$$

There are four cases to consider, depending on whether q or $q \to \psi$ are in $\bar{C}$.

If $q, q \to \psi \in \bar{C}$, then q and ψ are p°-free. Set $\bar{C'} = \bar{C} - \{q, q \to \psi\}$. As the premise $S \cdot (\bar{C'}, q, \psi \Rightarrow)$ has a shorter proof, by the induction hypothesis $\exists^\diamond p S^a, \bar{C'}, q, \psi \Rightarrow \forall^\circ p S$ and by $(Lp\to)$ we have $\exists^\diamond p S^a, \bar{C'}, q, q \to \psi \Rightarrow \forall^\circ p S$.

If $q, q \to \psi \notin \bar{C}$, then by the free-context property we can delete $\bar{C}$ from the premise and the conclusion and the rule remains valid and changes to

$$\frac{\Gamma - \bar{C}, q, \psi \Rightarrow \Delta}{\Gamma - \bar{C}, q, q \to \psi \Rightarrow \Delta}$$

Note that the conclusion is S. Therefore, the rule is backward applicable to S. Denote the premise by S'. By the induction hypothesis we have $\exists^\diamond p S'^a, \bar{C} \Rightarrow \forall^\circ p S'$. As $\exists^\diamond p S'^a \to \forall^\circ p S'$ appears as a disjunct in $\forall^\circ p S$, we have $\exists^\diamond p S^a, \bar{C} \Rightarrow \forall^\circ p S$.

If $q \to \psi \notin \bar{C}$ and $q \in \bar{C}$, then $q \to \psi \in S^a$, q is p°-free and the premise is of the form $S_{q,\psi} \cdot (\bar{C} \Rightarrow)$. By (R_1), we know $S_{q,\psi} \prec S$. Hence, by the induction hypothesis $\exists^\diamond p S^a_{q,\psi}, \bar{C} \Rightarrow \forall^\circ p S_{q,\psi}$. Hence, as $q \in \bar{C}$, we get $\bar{C} \Rightarrow q \wedge (\exists^\diamond p S^a_{q,\psi} \to \forall^\circ p S_{q,\psi})$. As $q \to \psi \in S^a$ and q is p°-free, we have $(q, \psi) \in I^\circ_{at}(S^a)$. Hence, $q \wedge (\exists^\diamond p S^a_{q,\psi} \to \forall^\circ p S_{q,\psi})$ is a disjunct in $\forall^\circ p S$, and we have $\exists^\diamond p S^a, \bar{C} \Rightarrow \forall^\circ p S$.

If $q \notin \bar{C}$ and $q \to \psi \in \bar{C}$, then $q \in S^a$, q is $p^\diamond$-free and ψ is p°-free. Set $\bar{C'} = \bar{C} - \{q \to \psi\}$. As the premise is of the form $S \cdot (\bar{C'}, \psi \Rightarrow)$ and it has a shorter proof, by the induction hypothesis we have $\exists^\diamond p S^a, \bar{C'}, \psi \Rightarrow \forall^\circ p S$. As $S^a \Rightarrow q$ is an instance of (Ax) and q is $p^\diamond$-free, we reach $\exists^\diamond_{ax} p S^a \Rightarrow q$. Therefore, as $\exists^\diamond_{ax} p S^a$ is a conjunct in $\exists^\diamond p S^a$, we have $\exists^\diamond p S^a, q \to \psi, \bar{C'} \Rightarrow \forall^\circ p S$.

If the last rule in the proof is the modal rule (M), then it is of the form

$$\frac{\varphi \Rightarrow \psi}{\Box\varphi \Rightarrow \Box\psi}\, M$$

If $\bar{C} = \varnothing$, then $S = (\Box\varphi \Rightarrow \Box\psi)$ is provable which contradicts with the assumption that S is not provable. Hence, $\bar{C} = \Box\varphi$. Therefore, φ is p°-free and $S = (\Rightarrow \Box\psi)$. By definition $\forall^\circ_m p S = \Box \forall^\circ p S_\psi$. Since φ is p°-free and $\varphi \Rightarrow \psi$ is provable, by the induction hypothesis $\exists^\diamond p S^a_\psi, \varphi \Rightarrow \forall^\circ p S_\psi$. However, since $S^a_\psi = \varnothing$, we have $\exists^\diamond p S^a_\psi = \top$. Therefore, $\varphi \Rightarrow \forall^\circ p S_\psi$, and by (M) and then (Lw), we get $\exists^\diamond p S^a, \bar{C} \Rightarrow \forall^\circ_m p S$. As $\forall^\circ_m p S$ is one of the disjuncts in the definition of $\forall^\circ p S$, we get $\exists^\diamond p S^a, \bar{C} \Rightarrow \forall^\circ p S$.

If the last rule in the proof is the modal rule $(LM\to)$, then it is of the form

$$\frac{\varphi \Rightarrow \psi \qquad \Gamma, \Box\varphi, \theta \Rightarrow \Delta}{\Gamma, \Box\varphi, \Box\psi \to \theta \Rightarrow \Delta}\ LM{\to}$$

There are four cases to consider based on which formulas are in $\bar{C}$.

If $\Box\varphi, \Box\psi \to \theta \in \bar{C}$, then φ and θ are p°-free. Set $\bar{C}' = \bar{C} - \{\Box\varphi, \Box\psi \to \theta\}$. The right premise is of the form $S \cdot (\bar{C}', \Box\varphi, \theta \Rightarrow)$. Therefore, by the induction hypothesis $\exists^\diamond p S^a, \bar{C}', \Box\varphi, \theta \Rightarrow \forall^\circ p S$. Applying $(LM{\to})$ on the latter sequent and $\varphi \Rightarrow \psi$, we get $(\exists^\diamond p S^a, \bar{C}', \Box\varphi, \Box\psi \to \theta \Rightarrow \forall^\circ p S) = (\exists^\diamond p S^a, \bar{C} \Rightarrow \forall^\circ p S)$.

Note that in the other three cases below, we have $S^a \neq \varnothing$. Hence, S is not of the form $(\Rightarrow \Box\psi)$ and hence $\forall^\circ_m p S$ is in the form of the big disjunction.

Suppose $\Box\varphi, \Box\psi \to \theta \notin \bar{C}$. As the right premise is of the form $S_{\Box\psi,\theta} \cdot (\bar{C} \Rightarrow)$ and by remark (R_1), we have $S_{\Box\psi,\theta} \prec S$, by the induction hypothesis we have $\exists^\diamond p S^a_{\Box\psi,\theta}, \bar{C} \Rightarrow \forall^\circ p S_{\Box\psi,\theta}$. Since both $\Box\varphi$ and $\Box\psi \to \theta$ are in S^a and $\varphi \Rightarrow \psi$ is provable, we get $(\varphi, \psi, \theta) \in I_m(S^a)$, which implies that $\exists^\diamond p S^a_{\Box\psi,\theta}$ is a conjunct in $\exists^\diamond p S^a$ and $\forall^\circ p S_{\Box\psi,\theta}$ a disjunct in $\forall^\circ p S$. Therefore, we get $\exists^\diamond p S^a, \bar{C} \Rightarrow \forall^\circ p S$.

If $\Box\psi \to \theta \in \bar{C}$ and $\Box\varphi \notin \bar{C}$, then ψ is $p^\diamond$-free and θ is p°-free. Set $\bar{C}' = \bar{C} - \{\Box\psi \to \theta\}$. Since $\varphi \Rightarrow \psi$ is provable and ψ is $p^\diamond$-free, by the induction hypothesis condition (ii) we have $\exists^\diamond p \varphi \Rightarrow \psi$. As $S \cdot (\bar{C}', \theta \Rightarrow)$ is a premise of the rule and has a shorter proof, by the induction hypothesis we have $\exists^\diamond p S^a, \bar{C}', \theta \Rightarrow \forall^\circ p S$ and by (Lw), $\Box\exists^\diamond p\varphi, \exists^\diamond p S^a, \bar{C}', \theta \Rightarrow \forall^\circ p S$. Applying $(LM{\to})$ on the latter sequent and on $\exists^\diamond p\varphi \Rightarrow \psi$ we get $\Box\exists^\diamond p\varphi, \exists^\diamond p S^a, \bar{C}', \Box\psi \to \theta \Rightarrow \forall^\circ p S$. As $\Box\varphi \in S^a$, by definition $\Box\exists^\diamond p\varphi$ appears as a conjunct in $\exists^\diamond p S^a$. Hence, $\exists^\diamond p S^a, \bar{C} \Rightarrow \forall^\circ p S$.

If $\Box\varphi \in \bar{C}$ and $\Box\psi \to \theta \notin \bar{C}$, set $\bar{C}' = \bar{C} - \{\Box\varphi\}$. As $S_\psi \cdot (\varphi \Rightarrow) = (\varphi \Rightarrow \psi)$ is provable and φ is p°-free, by the induction hypothesis $\exists^\diamond p S^a_\psi, \varphi \Rightarrow \forall^\circ p S_\psi$, or equivalently $\varphi \Rightarrow \forall^\circ p S_\psi$, as $S^a_\psi = \varnothing$ and hence $\exists^\diamond p S^a_\psi = \top$. Therefore, by (M),

$$\Box\varphi \Rightarrow \Box\forall^\circ p S_\psi \qquad (1)$$

On the other hand, as the right premise is $S_{\Box\psi,\theta} \cdot (\bar{C}', \Box\varphi \Rightarrow)$ and $\bar{C}'$ and $\Box\varphi$ are p°-free, by the induction hypothesis, condition (iv), we get

$$\exists^\diamond p S^a_{\Box\psi,\theta}, \bar{C}', \Box\varphi \Rightarrow \forall^\circ p S_{\Box\psi,\theta}. \qquad (2)$$

Therefore, using (1) and (2) we get

$$\Box\forall^\circ p S_\psi \to \exists^\diamond p S^a_{\Box\psi,\theta}, \bar{C}', \Box\varphi \Rightarrow \forall^\circ p S_{\Box\psi,\theta}.$$

As $\Box\psi \to \theta \in S^a$, the formula $\Box\forall^\circ p S_\psi \to \exists^\diamond p S^a_{\Box\psi,\theta}$ appears as a conjunct in the definition of $\exists^\diamond p S^a$. Hence, $\exists^\diamond p S^a, \bar{C}', \Box\varphi \Rightarrow \forall^\circ p S_{\Box\psi,\theta}$. Together with (1) we get $\exists^\diamond p S^a, \bar{C}', \Box\varphi \Rightarrow \forall^\circ p S_{\Box\psi,\theta} \wedge \Box\forall^\circ p S_\psi$, which implies $\exists^\diamond p S^a, \bar{C} \Rightarrow \forall^\circ p S$, as $\forall^\circ p S_{\Box\psi,\theta} \wedge \Box\forall^\circ p S_\psi$ is a disjunct in $\forall^\circ p S$, again by $\Box\psi \to \theta \in S^a$. □

Acknowledgements

We thank three anonymous referees for valuable comments on an earlier version of this paper, and one in particular for the careful reading of the manuscript and the detailed suggestions for improvement.

References

[1] Akbar Tabatabai, A., R. Iemhoff and R. Jalali, *Uniform lyndon interpolation for basic non-normal modal logics*, in: *International Workshop on Logic, Language, Information, and Computation*, Springer, 2021, pp. 287–301.

[2] Akbar Tabatabai, A. and R. Jalali, *Universal proof theory: Semi-analytic rules and craig interpolation*, arXiv preprint arXiv:1808.06256 (2018).

[3] Akbar Tabatabai, A. and R. Jalali, *Universal proof theory: semi-analytic rules and uniform interpolation*, arXiv preprint arXiv:1808.06258 (2018).

[4] Bílková, M., *Uniform interpolation and propositional quantifiers in modal logics*, Studia Logica (2007), pp. 1–31.

[5] Dalmonte, T., C. Grellois and N. Olivetti, *Intuitionistic non-normal modal logics: A general framework*, Journal of Philosophical Logic **49** (2020), pp. 833–882.

[6] Dyckhoff, R., *Contraction-free sequent calculi for intuitionistic logic.*

[7] Fitting, M. and R. Kuznets, *Modal interpolation via nested sequents*, Annals of Pure and Applied Logic **166** (2015), pp. 274–305.

[8] Ghilardi, S. and M. Zawadowski, *Undefinability of propositional quantifiers in the modal system S4*, Studia Logica **55** (1995), pp. 259–271.

[9] Iemhoff, R., *Uniform interpolation and sequent calculi in modal logic*, Archive for Mathematical Logic **58** (2019), pp. 155–181.

[10] Iemhoff, R., *Uniform interpolation and the existence of sequent calculi*, Annals of Pure and Applied Logic **170** (2019), pp. 1701–1712.

[11] Iemhoff, R., *The G4i analogue of a G3i calculus*, Studia Logica (2022), accepted for publication.

[12] Iemhoff, R., *Proof theory for Lax Logic*, in: *Dick de Jongh on Intuitionistic and Provability Logic*, Outstanding Contributions to Logic **24**, Springer, 2022 To appear.

[13] Kuznets, R. and B. Lellmann, *Interpolation for intermediate logics via hyper- and linear nested sequents*, in: *Advances in Modal Logic*, 2018.

[14] Lutz, C. and F. Wolter, *Foundations for uniform interpolation and forgetting in expressive description logics*, in: *IJCAI'11: Proceedings of the twenty-second international joint conference on Artificial Intelligence* (2011), pp. 989–995.

[15] Lyon, T., A. Tiu, R. Goré and R. Clouston, *Syntactic Interpolation for Tense Logics and Bi-Intuitionistic Logic via Nested Sequents*, in: M. Fernández and A. Muscholl, editors, *28th EACSL Annual Conference on Computer Science Logic (CSL 2020)*, Leibniz International Proceedings in Informatics (LIPIcs) **152** (2020), pp. 28:1–28:16.

[16] Maehara, S., *On the interpolation theorem of craig*, Sûgaku **12** (1960), pp. 235–237.

[17] Maksimova, L., *Craig's theorem in superintuitionistic logics and amalgamated varieties of pseudo-boolean algebras*, Algebra Logika **16** (1977), pp. 643–681.

[18] Pitts, A., *On an interpretation of second order quantification in first order intuitionistic propositional logic*, Journal of Symbolic Logic **57** (1992), pp. 33–52.

Parametrized modal logic I: an introduction

Philippe Balbiani [1]   Saúl Fernández González [2]

[1,2] *Institut de recherche en informatique de Toulouse*
CNRS-INPT-UT3, Toulouse University, Toulouse, France

Abstract

In this paper, within the context of a two-typed parametrized modal language, we define a parametrized modal logic as a couple whose components are sets of formulas containing, in their respective types, all propositional tautologies and the distribution axiom and closed, in their respective types, under modus ponens, uniform substitution and generalization. We axiomatically introduce different two-typed parametrized modal logics and we prove their completeness with respect to appropriate classes of two-typed frames by means of an adaptation of the canonical model construction.

Keywords: Parametrized modal logic. Complete axiomatization. Canonical model construction. Bounded morphism Lemma. Epistemic reasoning.

1 Introduction

In an application domain such as reasoning about knowledge where states and agents have been identified as the primitive entities of interest, one usually considers relational structures of the form $(S, \equiv)$ where S is a nonempty set of states and $\equiv$ is a function associating an equivalence relation $\equiv_a$ on S to every element a of a fixed set A of agents [8,9,17]. In that setting, for all $a \in A$, two states s and t are equivalent modulo $\equiv_a$ exactly when a cannot distinguish between s and t. When one wants to reason about distributed knowledge, it is of interest to assume that $\equiv$ is also a function associating an equivalence relation $\equiv_B$ on S to every $B \in \wp(A)$ in such a way that for all $B \in \wp(A)$, $\equiv_B = \bigcap\{\equiv_a\colon\ a \in B\}$.

The modal language interpreted over relational structures of the form $(S, \equiv)$ traditionally consists of one type of formulas: state-formulas — to be interpreted by sets of states. State-formulas are constructed over the Boolean connectives and the modal connectives $[B]$ — B ranging over $\wp(A)$. The state-formula $[B]\varphi$ is true in a state s of some model if the state-formula φ is true in every state of that model that can be distinguished from s by no B-agents.

[1] Email address: philippe.balbiani@irit.fr.
[2] Email address: saul.fgonzalez@irit.fr.

Although agents are omnipresent in the standard syntax of modal languages used for talking about relational structures of the form $(S, \equiv)$, states are the only entities that these relational structures take as first-class citizens. However, there is no need to force oneself to find examples where, in addition to equivalence relations between states parametrized by sets of agents, one would like to have at hand binary relations between agents parametrized by sets of states.

Indeed, in many situations, one would like to use relational structures of the form $(S, A, \equiv, \triangleright)$ where on top of the above-considered elements S and $\equiv$, one can find a nonempty set A of agents and a function $\triangleright$ associating a binary relation $\triangleright_s$ on A to every element s of S. Which situations? Situations where relationships between agents such as the following ones have to be taken into account: "agent a trusts agent b in state s", "agent a is a friend of agent b in state s", etc[3]. In these situations, for all $s \in S$, two agents a and b are related by $\triangleright_s$ exactly when a trusts b in state s, a is a friend of b in state s, etc. Moreover, on top of the assumption that $\equiv$ is also a function associating an equivalence relation $\equiv_B$ on S to every $B \in \wp(A)$ in such a way that for all $B \in \wp(A)$, $\equiv_B = \bigcap\{\equiv_a\colon\ a \in B\}$, one will naturally assume that $\triangleright$ is also a function associating a binary relation $\triangleright_T$ on A to every $T \in \wp(S)$ in such a way that for all $T \in \wp(S)$, $\triangleright_T = \bigcap\{\triangleright_s :\ s \in T\}$.

The modal language interpreted over relational structures of the form $(S, A, \equiv, \triangleright)$ will naturally consist of two types of formulas: state-formulas — to be interpreted by sets of states — and agent-formulas — to be interpreted by sets of agents. State-formulas will be constructed over the Boolean connectives and the modal connectives $[\alpha]$ — α ranging over the set of all agent-formulas — whereas agent-formulas will be constructed over the Boolean connectives and the modal connectives $[\varphi]$ — φ ranging over the set of all state-formulas. The state-formula $[\alpha]\varphi$ will be true in a state s of some model if the state-formula φ is true in every state of that model that can be distinguished from state s by no α-agents whereas the agent-formula $[\varphi]\alpha$ will be true in an agent a of some model if the agent-formula α is true in every agent of that model that is trusted by agent a at all φ-states.

In this paper, within the context of a two-typed parametrized modal language, we define a parametrized modal logic as a couple whose components are sets of formulas containing, in their respective types, all propositional tautologies and the distribution axiom and closed, in their respective types, under modus ponens, uniform substitution and generalization. We axiomatically introduce different two-typed parametrized modal logics and we prove

[3] See [4,14,25] and [15,24,28] for examples of situations in which one would like to use relationships such as "trusts", "is a friend of", etc.

their completeness with respect to appropriate classes of two-typed frames by means of an adaptation of the canonical model construction[4].

2 Syntax

From now on, when we write "$\underline{1}$" we will mean "2" and when we write "$\underline{2}$" we will mean "1". For all $i \in \{1,2\}$, let $\mathcal{P}_i$ be a countably infinite set (with typical members denoted p_i, q_i, etc). For all $i \in \{1,2\}$, members of $\mathcal{P}_i$ will be called *atomic i-formulas.* An *atomic formula* is either an atomic 1-formula, or an atomic 2-formula. We will always assume that $\mathcal{P}_1$ and $\mathcal{P}_2$ are disjoint. A *tip* is a couple (Σ_1, Σ_2) where for all $i \in \{1,2\}$, Σ_i is a set of finite words over the alphabet $\mathcal{P}_1 \cup \mathcal{P}_2 \cup \{\bot_1, \bot_2, \neg_1, \neg_2, \vee_1, \vee_2, \Box_1, \Box_2, (,)\}$ (with typical members denoted φ, ψ, etc). Let $\sqsubseteq$ be the partial order between tips defined by

- $(\Sigma_1, \Sigma_2) \sqsubseteq (\Delta_1, \Delta_2)$ if and only if for all $i \in \{1,2\}$, $\Sigma_i \subseteq \Delta_i$.

Let $(\mathcal{L}_1, \mathcal{L}_2)$ be the least tip such that for all $i \in \{1,2\}$, $\mathcal{P}_i \subseteq \mathcal{L}_i$ and

- $\bot_i \in \mathcal{L}_i$,
- for all $\varphi \in \mathcal{L}_i$, $\neg_i \varphi \in \mathcal{L}_i$,
- for all $\varphi, \psi \in \mathcal{L}_i$, $(\varphi \vee_i \psi) \in \mathcal{L}_i$,
- for all $\varphi \in \mathcal{L}_{\underline{i}}$ and for all $\psi \in \mathcal{L}_i$, $(\varphi \Box_i \psi) \in \mathcal{L}_i$.

Obviously, $\mathcal{L}_1$ and $\mathcal{L}_2$ are disjoint. For all $i \in \{1,2\}$, members of $\mathcal{L}_i$ will be called *i-formulas.* A *formula* is either a 1-formula, or a 2-formula. Let $\mathcal{L}_{\mathbf{PML}}$ — the *language of parametrized modal logic* — be the set of all formulas. For all $i \in \{1,2\}$, the Boolean connectives $\top_i$, $\wedge_i$, $\rightarrow_i$ and $\leftrightarrow_i$ are defined as usual. For all $i \in \{1,2\}$, the modal connective $\Diamond_i$ is defined by $(\varphi \Diamond_i \psi) ::= \neg_i(\varphi \Box_i \neg_i \psi)$, where φ ranges over $\mathcal{L}_{\underline{i}}$ and ψ ranges over $\mathcal{L}_i$. For all $i \in \{1,2\}$, for all $\varphi \in \mathcal{L}_{\underline{i}}$ and for all $\psi \in \mathcal{L}_i$, we will write "$[\varphi]_i \psi$" instead of "$(\varphi \Box_i \psi)$". For all $i \in \{1,2\}$, for all $\varphi \in \mathcal{L}_{\underline{i}}$ and for all $\psi \in \mathcal{L}_i$, we will write "$\langle \varphi \rangle_i \psi$" instead of "$(\varphi \Diamond_i \psi)$". For all $i \in \{1,2\}$, for all $\varphi \in \mathcal{L}_{\underline{i}}$ and for all sets Σ_i of i-formulas, let $[\varphi]\Sigma_i = \{\psi : [\varphi]_i \psi \in \Sigma_i\}$. A tip (Σ_1, Σ_2) is *readable* if $(\Sigma_1, \Sigma_2) \sqsubseteq (\mathcal{L}_1, \mathcal{L}_2)$. When writing formulas, most of the times, we will not make explicit the types of the connectives constituting them: if we know the type of a formula then we can inductively determine the types of its constituents. Therefore, for all $i \in \{1,2\}$, the result of uniformly replacing the atomic formulas of a given Boolean formula by arbitrary i-formulas can be seen as a i-formula. As a result, for all $i \in \{1,2\}$, we will talk about "the i-formula $(p \vee \neg q)$" instead of talking about "the formula $(p_i \vee_i \neg_i q_i)$", we will talk about "the i-formula $(p \vee (\neg q \Box \bot))$" instead of talking about "the formula $(p_i \vee_i (\neg_{\underline{i}} q_{\underline{i}} \Box_i \bot_i))$", etc. We adopt the standard rules for omission of the parentheses. A *substitution* is a couple (σ_1, σ_2) of functions $\sigma_1 : \mathcal{L}_1 \longrightarrow \mathcal{L}_1$ and $\sigma_2 : \mathcal{L}_2 \longrightarrow \mathcal{L}_2$ such that for all $i \in \{1,2\}$,

- $\sigma_i(\bot) = \bot$,

[4] See Propositions 5.5, 5.6, 6.9, 6.10, 6.20, and 6.21 for the main completeness results of the paper. A sketch of an alternative proof of Proposition 6.11 is included in the Appendix. The proofs of immediate consequences of the definitions are not given.

- $\sigma_i(\neg\varphi) = \neg\sigma_i(\varphi)$,
- $\sigma_i(\varphi \vee \psi) = \sigma_i(\varphi) \vee \sigma_i(\psi)$,
- $\sigma_i([\varphi]\psi) = [\sigma_{\underline{i}}(\varphi)]\sigma_i(\psi)$.

As a result, $\sigma_i(\langle\varphi\rangle\psi) = \langle\sigma_{\underline{i}}(\varphi)\rangle\sigma_i(\psi)$.

3 Relational semantics

A *frame* is a 4-tuple (W_1, W_2, R_1, R_2) where for all $i \in \{1,2\}$, W_i is a nonempty set and $R_i : \wp(W_{\underline{i}}) \longrightarrow \wp(W_i \times W_i)$. A *frame of indiscernibility* is a frame (W_1, W_2, R_1, R_2) such that for all $i \in \{1,2\}$ and for all $A \in \wp(W_{\underline{i}})$, $R_i(A)$ is an equivalence relation on W_i. A frame (W_1, W_2, R_1, R_2) is *conjunctive* if for all $i \in \{1,2\}$ and for all $A \in \wp(W_{\underline{i}})$, $R_i(A) = \bigcap\{R_i(\{s_{\underline{i}}\}) : s_{\underline{i}} \in A\}$.

Example 3.1 *Let* **Ob** *be a nonempty set of* objects, **At** *be a nonempty set of* attributes, **Val** *be a nonempty set of* values *and* $m : \mathbf{Ob} \times \mathbf{At} \longrightarrow \wp(\mathbf{Val})$. *Systems such as the 4-tuple* $(\mathbf{Ob}, \mathbf{At}, \mathbf{Val}, m)$ *have been introduced and developed by Orłowska and Pawlak within the context of analysis of data and representation of nondeterministic information [18,20,21,23]. In* $(\mathbf{Ob}, \mathbf{At}, \mathbf{Val}, m)$*, the objects* o *and* o' *are* equivalent for the attribute a *if* $m(o,a) = m(o',a)$ *whereas the attributes* a *and* a' *are* equivalent for the object o *if* $m(o,a) = m(o,a')$. *Obviously, the frame* (W'_1, W'_2, R'_1, R'_2) *where*

- $W'_1 = \mathbf{Ob}$,
- $W'_2 = \mathbf{At}$,
- *for all* $At \in \wp(\mathbf{At})$, $R'_1(At)$ *is the binary relation on* **Ob** *defined by*
  - $oR'_1(At)o'$ *if and only if for all* $a \in At$, $m(o,a) = m(o',a)$,
- *for all* $Ob \in \wp(\mathbf{Ob})$, $R'_2(Ob)$ *is the binary relation on* **At** *defined by*
  - $aR_2(Ob)a'$ *if and only if for all* $o \in Ob$, $m(o,a) = m(o,a')$.

is conjunctive. In this frame, two objects are related by a set of attributes if and only if these objects are equivalent for all attributes in that set whereas two attributes are related by a set of objects if and only if these attributes are equivalent for all objects in that set[5] .

A frame (W_1, W_2, R_1, R_2) is *unitary* if for all $i \in \{1,2\}$, $R_i(\emptyset) = W_i \times W_i$, i.e. $R_i(\emptyset)$ is the *universal relation on* W_i. Obviously, every conjunctive frame is unitary. A unitary frame (W_1, W_2, R_1, R_2) is *preconjunctive* if for all $i \in \{1,2\}$ and for all $A, B \in \wp(W_{\underline{i}})$, $R_i(A \cup B) = R_i(A) \cap R_i(B)$. A unitary frame (W_1, W_2, R_1, R_2) is *paraconjunctive* if for all $i \in \{1,2\}$ and for all $A, B \in \wp(W_{\underline{i}})$, if $A \subseteq B$ then $R_i(A) \supseteq R_i(B)$.

Proposition 3.2 *Every conjunctive frame is preconjunctive.*

Proposition 3.3 *Every preconjunctive frame is paraconjunctive.*

---

[5] For all sets At of attributes, the above-defined binary relation $R'_1(At)$ between objects is exactly the *strong indiscernibility relation* considered by Demri, Orłowska and Vakarelov [7,32,33]: two objects are in the relation $R'_1(At)$ if and only if they have the same values for all attributes in At.

Example 3.4 *Let* (W_1, W_2, R_1, R_2) *be the frame where* $W_1 = \mathbb{N}$, $W_2 = \mathbb{N}$, *for all* $A \in \wp(\mathbb{N})$, *if* A *is finite then* $R_1(A) = \mathbb{N} \times \mathbb{N}$ *else* $R_1(A) = \emptyset$ *and for all* $A \in \wp(\mathbb{N})$, *if* A *is finite then* $R_2(A) = \mathbb{N} \times \mathbb{N}$ *else* $R_2(A) = \emptyset$. *Obviously, this frame is preconjunctive. However, it is not conjunctive, seeing that for all* $i \in \{1, 2\}$, $R_i(\mathbb{N}) = \emptyset$ *and* $\bigcap\{R_i(\{s_{\underline{i}}\}) : s_{\underline{i}} \in \mathbb{N}\} = \mathbb{N} \times \mathbb{N}$.

Example 3.5 *Let* (W_1, W_2, R_1, R_2) *be the frame where* $W_1 = \mathbb{N}$, $W_2 = \mathbb{N}$, *for all* $A \in \wp(\mathbb{N})$, *if* $\mathtt{Card}(A) < 2$ *then* $R_1(A) = \mathbb{N} \times \mathbb{N}$ *else* $R_1(A) = \emptyset$ *and for all* $A \in \wp(\mathbb{N})$, *if* $\mathtt{Card}(A) < 2$ *then* $R_2(A) = \mathbb{N} \times \mathbb{N}$ *else* $R_2(A) = \emptyset$. *Obviously, this frame is paraconjunctive. However, it is not preconjunctive, seeing that for all* $i \in \{1, 2\}$, $R_i(\{0, 1\}) = \emptyset$ *and* $R_i(\{0\}) \cap R_i(\{1\}) = \mathbb{N} \times \mathbb{N}$.

A *valuation on a frame* (W_1, W_2, R_1, R_2) is a couple (V_1, V_2) of functions $V_1 : \mathcal{L}_1 \longrightarrow \wp(W_1)$ and $V_2 : \mathcal{L}_2 \longrightarrow \wp(W_2)$ such that for all $i \in \{1, 2\}$,

- $V_i(\bot) = \emptyset$,
- $V_i(\neg\varphi) = W_i \setminus V_i(\varphi)$,
- $V_i(\varphi \vee \psi) = V_i(\varphi) \cup V_i(\psi)$,
- $V_i([\varphi]\psi) = \{s_i \in W_i : \forall t_i \in W_i \ (s_i R_i(V_{\underline{i}}(\varphi)) t_i \ \Rightarrow \ t_i \in V_i(\psi))\}$.

As a result, $V_i(\langle\varphi\rangle\psi) = \{s_i \in W_i : \exists t_i \in W_i \ (s_i R_i(V_{\underline{i}}(\varphi)) t_i \ \& \ t_i \in V_i(\psi))\}$. A *model* is a 6-tuple consisting of a frame and a valuation on that frame. A model is *conjunctive (resp., unitary, preconjunctive, paraconjunctive)* if it is based on a conjunctive (resp., unitary, preconjunctive, paraconjunctive) frame.

Lemma 3.6 *For all unitary models* $(W_1, W_2, R_1, R_2, V_1, V_2)$, *for all* $i \in \{1, 2\}$ *and for all i-formulas* φ, *the i-formula* $[\bot]\varphi$ *is such that*

- *if* $V_i(\varphi) = W_i$ *then* $V_i([\bot]\varphi) = W_i$,
- *otherwise,* $V_i([\bot]\varphi) = \emptyset$.

For all $i \in \{1, 2\}$, a i-formula φ is *true in a model* $(W_1, W_2, R_1, R_2, V_1, V_2)$ (in symbols $(W_1, W_2, R_1, R_2, V_1, V_2) \models \varphi$) if $V_i(\varphi) = W_i$.

Lemma 3.7 *For all* $i \in \{1, 2\}$ *and for all i-formulas* φ, *the i-formulas* $[\bot]\varphi \to \varphi$ *and* $\langle\bot\rangle\varphi \to [\bot]\langle\bot\rangle\varphi$ *are true in any unitary model.*

Lemma 3.8 *For all* $i \in \{1, 2\}$, *for all* $\underline{i}$*-formulas* φ, ψ *and for all i-formulas* χ, *the i-formula* $\langle\varphi \vee \psi\rangle\chi \to \langle\varphi\rangle\chi \wedge \langle\psi\rangle\chi$ *is true in any paraconjunctive model.*

A formula φ is *valid on a frame* (W_1, W_2, R_1, R_2) (in symbols $(W_1, W_2, R_1, R_2) \models \varphi$) if for all (W_1, W_2, R_1, R_2)-valuations (V_1, V_2), $(W_1, W_2, R_1, R_2, V_1, V_2) \models \varphi$.

Example 3.9 *In a frame* (W_1, W_2, R_1, R_2), *one may easily prove that for all* $i \in \{1, 2\}$,

- *the i-formula* $[\bot]p \to p$ *is valid if and only if for all* $s_i \in W_i$, $s_i R_i(\emptyset) s_i$,
- *the i-formula* $[\bot]p \to [\bot][\bot]p$ *is valid if and only if for all* $s_i, t_i, u_i \in W_i$, *if* $s_i R_i(\emptyset) t_i$ *and* $t_i R_i(\emptyset) u_i$ *then* $s_i R_i(\emptyset) u_i$,
- *the i-formula* $\langle\bot\rangle p \to [\bot]\langle\bot\rangle p$ *is valid if and only if for all* $s_i, t_i, u_i \in W_i$, *if*

$s_iR_i(\emptyset)t_i$ and $s_iR_i(\emptyset)u_i$ then $t_iR_i(\emptyset)u_i$.

Example 3.10 *In a conjunctive frame (W_1, W_2, R_1, R_2), one may easily prove that for all $i \in \{1, 2\}$,*

- *the i-formula $\langle p\rangle q \to [p]q$ is valid if and only if $\mathtt{Card}(W_i) \leq 1$,*
- *the i-formula $[p]q \to q$ if and only if for all $s_i \in W_i$ and for all $t_{\underline{i}} \in W_{\underline{i}}$, $s_iR_i(\{t_{\underline{i}}\})s_i$,*
- *the i-formula $[p]q \to [p][p]q$ if and only if for all $s_i, t_i, u_i \in W_i$ and for all $v_{\underline{i}} \in W_{\underline{i}}$, if $s_iR_i(\{v_{\underline{i}}\})t_i$ and $t_iR_i(\{v_{\underline{i}}\})u_i$ then $s_iR_i(\{v_{\underline{i}}\})u_i$,*
- *the i-formula $\langle p\rangle q \to [p]\langle p\rangle q$ if and only if for all $s_i, t_i, u_i \in W_i$ and for all $v_{\underline{i}} \in W_{\underline{i}}$, if $s_iR_i(\{v_{\underline{i}}\})t_i$ and $s_iR_i(\{v_{\underline{i}}\})u_i$ then $t_iR_i(\{v_{\underline{i}}\})u_i$.*

A formula φ is *valid on a class $\mathcal{C}$ of frames* (in symbols $\mathcal{C} \models \varphi$) if for all frames (W_1, W_2, R_1, R_2) in $\mathcal{C}$, $(W_1, W_2, R_1, R_2) \models \varphi$.

Lemma 3.11 *For all frames (W_1, W_2, R_1, R_2), (W_1, W_2, R_1, R_2) is a frame of indiscernibility if and only if for all $i \in \{1, 2\}$, for all $\underline{i}$-formulas φ and for all i-formulas ψ, $(W_1, W_2, R_1, R_2) \models [\varphi]\psi \to \psi$ and $(W_1, W_2, R_1, R_2) \models \langle\varphi\rangle\psi \to [\varphi]\langle\varphi\rangle\psi$.*

A *bounded morphism from a model $(W_1, W_2, R_1, R_2, V_1, V_2)$ to a model $(W_1', W_2', R_1', R_2', V_1', V_2')$* is a couple (f_1, f_2) of functions $f_1 : W_1 \longrightarrow W_1'$ and $f_2 : W_2 \longrightarrow W_2'$ such that for all $i \in \{1, 2\}$,

Atomic condition: for all $p \in \mathcal{P}_i$, $f_i^{-1}[V_i'(p)] = V_i(p)$,

Forward condition: for all $s_i, t_i \in W_i$ and for all $\underline{i}$-formulas φ, if $s_iR_i(V_{\underline{i}}(\varphi))t_i$ then $f_i(s_i)R_i'(V_{\underline{i}}'(\varphi))f_i(t_i)$,

Backward condition: for all $s_i \in W_i$, for all $t_i' \in W_i'$ and for all $\underline{i}$-formulas φ, if $f_i(s_i)R_i'(V_{\underline{i}}'(\varphi))t_i'$ then there exists $t_i \in W_i$ such that $f_i(t_i) = t_i'$ and $s_iR_i(V_{\underline{i}}(\varphi))t_i$.

Proposition 3.12 *For all models $(W_1, W_2, R_1, R_2, V_1, V_2)$ and $(W_1', W_2', R_1', R_2', V_1', V_2')$, for all bounded morphisms (f_1, f_2) from $(W_1, W_2, R_1, R_2, V_1, V_2)$ to $(W_1', W_2', R_1', R_2', V_1', V_2')$, for all $i \in \{1, 2\}$ and for all i-formulas φ, $f_i^{-1}[V_i'(\varphi)] = V_i(\varphi)$.*

Proof. Similar to the proof of Bounded Morphism Lemma [5, Proposition 2.14]. □

Proposition 3.13 is an immediate consequence of Proposition 3.12.

Proposition 3.13 *Let (W_1, W_2, R_1, R_2) and (W_1', W_2', R_1', R_2') be frames and (f_1, f_2) be a couple of surjective functions $f_1 : W_1 \longrightarrow W_1'$ and $f_2 : W_2 \longrightarrow W_2'$. If for all (W_1', W_2', R_1', R_2')-valuations (V_1', V_2'), there exists a (W_1, W_2, R_1, R_2)-valuation (V_1, V_2) such that (f_1, f_2) is a bounded morphism from $(W_1, W_2, R_1, R_2, V_1, V_2)$ to $(W_1', W_2', R_1', R_2', V_1', V_2')$ then for all formulas φ, if $(W_1, W_2, R_1, R_2) \models \varphi$ then $(W_1', W_2', R_1', R_2') \models \varphi$.*

4 Parametrized modal logics

A *parametrized modal logic (PML)* is a readable tip $(\mathbf{L}_1, \mathbf{L}_2)$ such that for all $i \in \{1,2\}$, $\mathbf{L}_i$ satisfies the following conditions:

($\mathbf{Taut}_i$) $\mathbf{L}_i$ contains all i-formulas obtained from propositional tautologies after having uniformly replaced their atomic formulas by arbitrary i-formulas,

($\mathbf{Dist}_i$) $\mathbf{L}_i$ contains all i-formulas of the form $[\varphi](\psi \to \chi) \to ([\varphi]\psi \to [\varphi]\chi)$,

($\mathbf{MP}_i$) if $\varphi \in \mathbf{L}_i$ and $\varphi \to \psi \in \mathbf{L}_i$ then $\mathbf{L}_i$ contains the i-formula ψ,

($\mathbf{Gen}_i$) if $\varphi \in \mathbf{L}_i$ then $\mathbf{L}_i$ contains all i-formulas of the form $[\psi]\varphi$,

($\mathbf{RE}_i$) if $\varphi \leftrightarrow \psi \in \mathbf{L}_{\underline{i}}$ then $\mathbf{L}_i$ contains all i-formulas of the form $[\varphi]\chi \leftrightarrow [\psi]\chi$,

($\mathbf{UOI}_i$) if $\bot \in \mathbf{L}_{\underline{i}}$ then $\mathbf{L}_i$ contains the i-formula $\bot$,

($\mathbf{US}_i$) if $\varphi \in \mathbf{L}_i$ and (σ_1, σ_2) is a substitution then $\mathbf{L}_i$ contains the i-formula $\sigma_i(\varphi)$.

There is a greatest PML, namely $(\mathcal{L}_1, \mathcal{L}_2)$. There is also a least PML (denoted $(\mathbf{K}_1, \mathbf{K}_2)$), seeing that for all collections $(\mathbf{L}_1^m, \mathbf{L}_2^m)_{m \in I}$ of PMLs, $(\bigcap\{\mathbf{L}_1^m : m \in I\}, \bigcap\{\mathbf{L}_2^m : m \in I\})$ is a PML. Let $(\mathbf{Equiv}_1^g, \mathbf{Equiv}_2^g)$ be the least PML such that for all $i \in \{1,2\}$, $\mathbf{Equiv}_i^g$ contains all i-formulas of the form $[\varphi]\psi \to \psi$ and $\langle\varphi\rangle\psi \to [\varphi]\langle\varphi\rangle\psi$. A PML $(\mathbf{L}_1, \mathbf{L}_2)$ is *paraconjunctive* if for all $i \in \{1,2\}$, $\mathbf{L}_i$ satisfies the following conditions:

($\mathbf{UNI}_i$) $\mathbf{L}_i$ contains all i-formulas of the form $[\bot]\varphi \to \varphi$ and $\langle\bot\rangle\varphi \to [\bot]\langle\bot\rangle\varphi$,

($\mathbf{RI}_i^-$) if $\bigwedge_{k=1\ldots m}[\bot](\langle\psi_k'\rangle\chi_k' \to \langle\varphi_k'\rangle\chi_k') \to \bigvee_{l=1\ldots n}[\bot](\varphi_l \to \psi_l) \in \mathbf{L}_{\underline{i}}$ then $\mathbf{L}_i$ contains all i-formulas of the form $\bigwedge_{k=\ldots m}[\bot](\varphi_k' \to \psi_k') \to \bigvee_{l=1\ldots n}[\bot](\langle\psi_l\rangle\chi_l \to \langle\varphi_l\rangle\chi_l)$.

Let $(\mathbf{ParCon}_1, \mathbf{ParCon}_2)$ be the least paraconjunctive PML. Let $(\mathbf{Equiv}_1^p, \mathbf{Equiv}_2^p)$ be the least paraconjunctive PML such that for all $i \in \{1,2\}$, $\mathbf{Equiv}_i^p$ contains all i-formulas of the form $[\varphi]\psi \to \psi$ and $\langle\varphi\rangle\psi \to [\varphi]\langle\varphi\rangle\psi$. For all PMLs $(\mathbf{L}_1, \mathbf{L}_2)$ and for all readable tips (Σ_1, Σ_2), let $(\mathbf{L}_1, \mathbf{L}_2) + (\Sigma_1, \Sigma_2)$ be the least PML containing $(\mathbf{L}_1 \cup \Sigma_1, \mathbf{L}_2 \cup \Sigma_2)$.

Problem 4.1 *We do not know if there exists a finite readable tip* (Σ_1, Σ_2) *such that* $(\mathbf{ParCon}_1, \mathbf{ParCon}_2) = (\mathbf{K}_1, \mathbf{K}_2) + (\Sigma_1, \Sigma_2)$.

Problem 4.2 *We do not know if there exists a finite readable tip* (Σ_1, Σ_2) *such that* $(\mathbf{Equiv}_1^p, \mathbf{Equiv}_2^p) = (\mathbf{K}_1, \mathbf{K}_2) + (\Sigma_1, \Sigma_2)$.

A PML $(\mathbf{L}_1, \mathbf{L}_2)$ is *consistent* if for all $i \in \{1,2\}$, $\mathbf{L}_i \neq \mathcal{L}_i$. Obviously, thanks to the conditions $(\mathbf{UOI}_1)$ and $(\mathbf{UOI}_2)$, $(\mathcal{L}_1, \mathcal{L}_2)$ is the one and only inconsistent PML. For all PMLs $(\mathbf{L}_1, \mathbf{L}_2)$ and for all $i \in \{1,2\}$, we will say that a set s_i of i-formulas is $(\mathbf{L}_1, \mathbf{L}_2)$-*consistent,* if for all $n \in \mathbb{N}$ and for all $\varphi_1, \ldots, \varphi_n \in s_i$, $\neg(\varphi_1 \wedge \ldots \wedge \varphi_n) \notin \mathbf{L}_i$.

Lemma 4.3 *For all PMLs* $(\mathbf{L}_1, \mathbf{L}_2)$, *for all* $i \in \{1,2\}$ *and for all* $(\mathbf{L}_1, \mathbf{L}_2)$-*consistent sets* s_i *of* i-*formulas, there exists a maximal* $(\mathbf{L}_1, \mathbf{L}_2)$-*consistent set* t_i *of* i-*formulas such that* $s_i \subseteq t_i$.

Proof. Similar to the proof of Lindenbaum's Lemma [6, Lemma 5.1]. □

Lemma 4.4 *For all PMLs* $(\mathbf{L}_1, \mathbf{L}_2)$*, for all* $i \in \{1,2\}$*, for all maximal* $(\mathbf{L}_1, \mathbf{L}_2)$*-consistent sets* s_i *of* i*-formulas, for all* $\underline{i}$*-formulas* φ *and for all* i*-formulas* ψ*, if* $[\varphi]\psi \notin s_i$ *then* $[\varphi]s_i \cup \{\neg\psi\}$ *is a* $(\mathbf{L}_1, \mathbf{L}_2)$*-consistent set of* i*-formulas.*

Proof. Similar to the proof of Existence Lemma [13, Proposition 2.8.4]. □

Notice that for all consistent PMLs $(\mathbf{L}_1, \mathbf{L}_2)$ and for all $i \in \{1,2\}$, $\mathbf{L}_i$ is a $(\mathbf{L}_1, \mathbf{L}_2)$-consistent set of i-formulas. A PML $(\mathbf{L}_1, \mathbf{L}_2)$ is *sound with respect to a class* $\mathcal{C}$ *of frames* if for all $i \in \{1,2\}$ and for all i-formulas φ, if $\varphi \in \mathbf{L}_i$ then $\mathcal{C} \models \varphi$. A PML $(\mathbf{L}_1, \mathbf{L}_2)$ is *complete with respect to a class* $\mathcal{C}$ *of frames* if for all $i \in \{1,2\}$ and for all i-formulas φ, if $\mathcal{C} \models \varphi$ then $\varphi \in \mathbf{L}_i$. The proofs of the soundness statements expressed in Proposition 4.5 are as expected.

Proposition 4.5 *In Table 1, the PMLs listed in the left column are sound with respect to the corresponding classes of frames listed in the right column.*

PMLs	**Classes of frames**
$(\mathbf{K}_1, \mathbf{K}_2)$	All frames
$(\mathbf{Equiv}_1^g, \mathbf{Equiv}_2^g)$	All frames of indiscernibility
$(\mathbf{ParCon}_1, \mathbf{ParCon}_2)$	All paraconjunctive frames All preconjunctive frames All conjunctive frames
$(\mathbf{Equiv}_1^p, \mathbf{Equiv}_2^p)$	All paraconjunctive frames of indiscernibility All preconjunctive frames of indiscernibility All conjunctive frames of indiscernibility

Table 1

As for the proofs of the corresponding completeness statements, they are not so obvious, especially when the considered PMLs are paraconjunctive[6]. In Sections 5 and 6, we adapt the ordinary canonical model construction to the context of our parametrized relational semantics.

5 Completeness: the general case

From now on in this section, we will assume that $(\mathbf{L}_1, \mathbf{L}_2)$ is a consistent PML. Let $(W_1^g, W_2^g, R_1^g, R_2^g)$ be the 4-tuple where for all $i \in \{1,2\}$,

- W_i^g is the set of all maximal $(\mathbf{L}_1, \mathbf{L}_2)$-consistent sets of i-formulas,
- $R_i^g : \wp(W_{\underline{i}}^g) \longrightarrow \wp(W_i^g \times W_i^g)$ is such that for all $A \in \wp(W_{\underline{i}}^g)$ and for all $s_i, t_i \in W_i^g$, $s_i R_i^g(A) t_i$ if and only if for all $\underline{i}$-formulas φ, if $\widehat{\varphi} = A$ then $[\varphi]s_i \subseteq t_i$,

[6] The problem with paraconjunctive PMLs is that the operation of intersection — which is used in conjunctive frames for the interpretation of the modalities — is not modally definable. See [1] and [22] for investigations about the intersection of modalities in epistemic logics and dynamic logics.

where for all $i \in \{1,2\}$ and for all $\underline{i}$-formulas φ, $\widehat{\varphi} = \{u_{\underline{i}} \in W_{\underline{i}}^g : \varphi \in u_{\underline{i}}\}$. Since for all $i \in \{1,2\}$, $\mathbf{L}_i$ is a $(\mathbf{L}_1, \mathbf{L}_2)$-consistent set of i-formulas, by Lemma 4.3, for all $i \in \{1,2\}$, W_i^g is nonempty.

Lemma 5.1 *The 4-tuple* $(W_1^g, W_2^g, R_1^g, R_2^g)$ *is a frame.*

The 4-tuple $(W_1^g, W_2^g, R_1^g, R_2^g)$ will be called *general canonical frame for* $(\mathbf{L}_1, \mathbf{L}_2)$. Lemma 5.2 is an immediate consequence of the fact that for all $i \in \{1,2\}$, $\mathbf{Equiv}_i^g$ contains all i-formulas of the form $[\varphi]\psi \rightarrow \psi$ and $\langle\varphi\rangle\psi \rightarrow [\varphi]\langle\varphi\rangle\psi$.

Lemma 5.2 *If* $(\mathbf{L}_1, \mathbf{L}_2)$ *contains* $(\mathbf{Equiv}_1^g, \mathbf{Equiv}_2^g)$ *then the general canonical frame for* $(\mathbf{L}_1, \mathbf{L}_2)$ *is a frame of indiscernibility.*

Lemma 5.3 is an immediate consequence of Lemma 4.3.

Lemma 5.3 *For all* $i \in \{1,2\}$ *and for all i-formulas* φ, ψ, $\widehat{\varphi} = \widehat{\psi}$ *if and only if* $\varphi \leftrightarrow \psi \in \mathbf{L}_i$.

For all $i \in \{1,2\}$, let $V_i^g : \mathcal{L}_i \longrightarrow \wp(W_i^g)$ be such that for all $\varphi \in \mathcal{L}_i$, $V_i^g(\varphi) = \widehat{\varphi}$. The 6-tuple $(W_1^g, W_2^g, R_1^g, R_2^g, V_1^g, V_2^g)$ will be called *general canonical model for* $(\mathbf{L}_1, \mathbf{L}_2)$.

Lemma 5.4 (Truth Lemma: the general case) *The general canonical model for* $(\mathbf{L}_1, \mathbf{L}_2)$ *is a model.*

Proof. The proof that for all $i \in \{1,2\}$, V_i^g satisfies the conditions for $\bot$, $\neg$ and $\vee$ is as expected. We only show that for all $i \in \{1,2\}$, V_i^g satisfies the condition for $[\cdot]$. Let $i \in \{1,2\}$. Let φ be a $\underline{i}$-formula and ψ be a i-formula. Let $s_i \in W_i^g$. We only demonstrate $s_i \in V_i^g([\varphi]\psi)$ if for all $t_i \in W_i^g$, if $s_i R_i^g(V_{\underline{i}}^g(\varphi))t_i$ then $t_i \in V_i^g(\psi)$, the "only if" direction being left as an exercise for the reader. Suppose $s_i \notin V_i^g([\varphi]\psi)$. We demonstrate there exists $t_i \in W_i^g$ such that $s_i R_i^g(V_{\underline{i}}^g(\varphi))t_i$ and $t_i \notin V_i^g(\psi)$. Since $s_i \notin V_i^g([\varphi]\psi)$, $[\varphi]\psi \notin s_i$. Let $t_i^0 = [\varphi]s_i \cup \{\neg\psi\}$. Notice that $[\varphi]s_i \subseteq t_i^0$ and $\neg\psi \in t_i^0$. By Lemma 4.4, t_i^0 is a $(\mathbf{L}_1, \mathbf{L}_2)$-consistent set of i-formulas. Hence, by Lemma 4.3, let t_i be a maximal $(\mathbf{L}_1, \mathbf{L}_2)$-consistent set of i-formulas such that $t_i^0 \subseteq t_i$. Since $[\varphi]s_i \subseteq t_i^0$ and $\neg\psi \in t_i^0$, $[\varphi]s_i \subseteq t_i$ and $\neg\psi \in t_i$.

Claim $s_i R_i^g(V_{\underline{i}}^g(\varphi))t_i$.

Proof. We demonstrate for all $\underline{i}$-formulas φ', if $\widehat{\varphi'} = V_{\underline{i}}^g(\varphi)$ then $[\varphi']s_i \subseteq t_i$. Let φ' be a $\underline{i}$-formula. Suppose $\widehat{\varphi'} = V_{\underline{i}}^g(\varphi)$. We demonstrate $[\varphi']s_i \subseteq t_i$. Let ψ' be a i-formula. Suppose $[\varphi']\psi' \in s_i$. We demonstrate $\psi' \in t_i$. Since $\widehat{\varphi'} = V_{\underline{i}}^g(\varphi)$, by Lemma 5.3, $\varphi' \leftrightarrow \varphi \in \mathbf{L}_{\underline{i}}$. Since $\mathbf{L}_i$ satisfies the closure condition $(\mathbf{RE}_i)$, $[\varphi']\psi' \leftrightarrow [\varphi]\psi' \in \mathbf{L}_i$. Since $[\varphi']\psi' \in s_i$, $[\varphi]\psi' \in s_i$. Thus, $\psi' \in [\varphi]s_i$. Since $[\varphi]s_i \subseteq t_i$, $\psi' \in t_i$. □

Finally, the reader may easily verify that $t_i \notin V_i^g(\psi)$. Here finishes the proof of Lemma 5.4.

Proposition 5.5 is an immediate consequence of Lemmas 4.3, 5.1 and 5.4.

Proposition 5.5 $(\mathbf{K}_1, \mathbf{K}_2)$ *is complete with respect to the class of all frames.*

Proposition 5.6 is an immediate consequence of Lemmas 4.3, 5.2 and 5.4.

Proposition 5.6 $(\mathbf{Equiv}_1^g, \mathbf{Equiv}_2^g)$ *is complete with respect to the class of all frames of indiscernibility.*

6 Completeness: the paraconjunctive case

From now on in this section, we will assume that $(\mathbf{L}_1, \mathbf{L}_2)$ is a consistent paraconjunctive PML. Therefore, for all $i \in \{1,2\}$, $\mathbf{L}_i$ contains all i-formulas of the form $[\perp]\varphi \to \varphi$ and $\langle\perp\rangle\varphi \to [\perp]\langle\perp\rangle\varphi$. As a result, for all $i \in \{1,2\}$, $\mathbf{L}_i$ contains all i-formulas of the form $[\perp]\varphi \to [\perp][\perp]\varphi$.

6.1 Preliminaries

Lemmas 6.1 and 6.2 are immediate consequences of the fact that for all $i \in \{1,2\}$, $\mathbf{L}_i$ contains all i-formulas of the form $[\perp]\varphi \to \varphi$ and $\langle\perp\rangle\varphi \to [\perp]\langle\perp\rangle\varphi$.

Lemma 6.1 *For all $i \in \{1,2\}$ and for all maximal $(\mathbf{L}_1, \mathbf{L}_2)$-consistent sets s_i of i-formulas, $[\perp]s_i$ is $(\mathbf{L}_1, \mathbf{L}_2)$-consistent.*

Lemma 6.2 *For all $i \in \{1,2\}$ and for all maximal $(\mathbf{L}_1, \mathbf{L}_2)$-consistent sets s_i, t_i, u_i of i-formulas, if $[\perp]s_i \subseteq t_i$ and $[\perp]s_i \subseteq u_i$ then $[\perp]t_i \subseteq u_i$.*

A readable tip (s_1, s_2) is *paraconjunctive* if for all $i \in \{1,2\}$, s_i is a maximal $(\mathbf{L}_1, \mathbf{L}_2)$-consistent set of i-formulas such that for all $\underline{i}$-formulas φ, ψ, if $[\perp](\varphi \to \psi) \in s_{\underline{i}}$ then for all i-formulas χ, $[\perp](\langle\psi\rangle\chi \to \langle\varphi\rangle\chi) \in s_i$.

Lemma 6.3 *For all $i \in \{1,2\}$ and for all maximal $(\mathbf{L}_1, \mathbf{L}_2)$-consistent sets s_i of i-formulas, there exists a maximal $(\mathbf{L}_1, \mathbf{L}_2)$-consistent set $s_{\underline{i}}$ of $\underline{i}$-formulas such that the readable tip (s_1, s_2) is paraconjunctive.*

Proof. Let $i \in \{1,2\}$ and s_i be a maximal $(\mathbf{L}_1, \mathbf{L}_2)$-consistent set of i-formulas. We demonstrate there exists a maximal $(\mathbf{L}_1, \mathbf{L}_2)$-consistent set $s_{\underline{i}}$ of $\underline{i}$-formulas such that the readable tip (s_1, s_2) is paraconjunctive. Let $s_{\underline{i}}^0$ be the set consisting of the following $\underline{i}$-formulas:

- $\neg[\perp](\varphi \to \psi)$ for each $\varphi, \psi \in \mathcal{L}_{\underline{i}}$ and for each $\chi \in \mathcal{L}_i$ such that $\neg[\perp](\langle\psi\rangle\chi \to \langle\varphi\rangle\chi) \in s_i$,
- $[\perp](\langle\psi'\rangle\chi' \to \langle\varphi'\rangle\chi')$ for each $\varphi', \psi' \in \mathcal{L}_i$ and for each $\chi' \in \mathcal{L}_{\underline{i}}$ such that $[\perp](\varphi' \to \psi') \in s_i$.

Claim $s_{\underline{i}}^0$ is a $(\mathbf{L}_1, \mathbf{L}_2)$-consistent set of $\underline{i}$-formulas.

Proof. For the sake of the contradiction, suppose $s_{\underline{i}}^0$ is not a $(\mathbf{L}_1, \mathbf{L}_2)$-consistent set of $\underline{i}$-formulas. Hence, let m, n in $\mathbb{N}$, $\varphi_1, \ldots, \varphi_m$, $\psi_1, \ldots, \psi_m$, $\chi'_1, \ldots, \chi'_n$ in $\mathcal{L}_{\underline{i}}$ and $\varphi'_1, \ldots, \varphi'_n$, $\psi'_1, \ldots, \psi'_n$, $\chi_1, \ldots, \chi_m$ in $\mathcal{L}_i$ be such that

- for all $k \in \{1, \ldots, m\}$, $\neg[\perp](\langle\psi_k\rangle\chi_k \to \langle\varphi_k\rangle\chi_k) \in s_i$,
- for all $l \in \{1, \ldots, n\}$, $[\perp](\varphi'_l \to \psi'_l) \in s_i$,

- $\neg(\bigwedge_{k=1\ldots m}\neg[\bot](\varphi_k \to \psi_k) \wedge \bigwedge_{l=1\ldots n}[\bot](\langle\psi'_l\rangle\chi'_l \to \langle\varphi'_l\rangle\chi'_l)) \in \mathbf{L}_{\underline{i}}$.

Thus, $\bigwedge_{l=1\ldots n}[\bot](\langle\psi'_l\rangle\chi'_l \to \langle\varphi'_l\rangle\chi'_l) \to \bigvee_{k=1\ldots m}[\bot](\varphi_k \to \psi_k) \in \mathbf{L}_{\underline{i}}$. Since $\mathbf{L}_i$ satisfies the closure condition $(\mathbf{RI}_i^-)$, $\mathbf{L}_i$ contains the i-formula $\bigwedge_{l=1\ldots n}[\bot](\varphi'_l \to \psi'_l) \to \bigvee_{k=1\ldots m}[\bot](\langle\psi_k\rangle\chi_k \to \langle\varphi_k\rangle\chi_k)$. Since for all $l \in \{1,\ldots,n\}$, $[\bot](\varphi'_l \to \psi'_l) \in s_i$, there exists $k \in \{1,\ldots,m\}$ such that $[\bot](\langle\psi_k\rangle\chi_k \to \langle\varphi_k\rangle\chi_k) \in s_i$. Consequently, there exists $k \in \{1,\ldots,m\}$ such that $\neg[\bot](\langle\psi_k\rangle\chi_k \to \langle\varphi_k\rangle\chi_k) \notin s_i$: a contradiction. □

Thus, by Lemma 4.3, let $s_{\underline{i}}$ be a maximal $(\mathbf{L}_1,\mathbf{L}_2)$-consistent set of $\underline{i}$-formulas such that $s^0_{\underline{i}} \subseteq s_{\underline{i}}$.

Claim The readable tip (s_1,s_2) is paraconjunctive.

Proof. For the sake of the contradiction, suppose the readable tip (s_1,s_2) is not paraconjunctive. Since for all $j \in \{1,2\}$, s_j is a maximal $(\mathbf{L}_1,\mathbf{L}_2)$-consistent set of j-formulas, there exists $j \in \{1,2\}$ and there exists $\underline{j}$-formulas φ,ψ such that $[\bot](\varphi \to \psi) \in s_{\underline{j}}$ and there exists a j-formula χ such that $[\bot](\langle\psi\rangle\chi \to \langle\varphi\rangle\chi) \notin s_j$. Consider the following two cases: $j = i$ and $j = \underline{i}$. In the former case, since $[\bot](\langle\psi\rangle\chi \to \langle\varphi\rangle\chi) \notin s_j$, $\neg[\bot](\langle\psi\rangle\chi \to \langle\varphi\rangle\chi) \in s_i$. Hence, $\neg[\bot](\varphi \to \psi) \in s^0_{\underline{i}}$. Since $s^0_{\underline{i}} \subseteq s_{\underline{i}}$, $\neg[\bot](\varphi \to \psi) \in s_{\underline{i}}$. Thus, $[\bot](\varphi \to \psi) \notin s_{\underline{j}}$: a contradiction. In the latter case, since $[\bot](\varphi \to \psi) \in s_{\underline{j}}$, $[\bot](\varphi \to \psi) \in s_i$. Consequently, $[\bot](\langle\psi\rangle\chi \to \langle\varphi\rangle\chi) \in s^0_{\underline{i}}$. Since $s^0_{\underline{i}} \subseteq s_{\underline{i}}$, $[\bot](\langle\psi\rangle\chi \to \langle\varphi\rangle\chi) \in s_{\underline{i}}$. Hence, $[\bot](\langle\psi\rangle\chi \to \langle\varphi\rangle\chi) \in s_j$: a contradiction. □

Here finishes the proof of Lemma 6.3.

6.2 Paraconjunctive case: first set of completeness results

For a while, let us fix a paraconjunctive readable tip (s_1,s_2). Let $(W^c_1,W^c_2,R^c_1,R^c_2)$ be the 4-tuple where for all $i \in \{1,2\}$,

- W^c_i is the set of all maximal $(\mathbf{L}_1,\mathbf{L}_2)$-consistent sets t_i of i-formulas such that $[\bot]s_i \subseteq t_i$,
- $R^c_i : \wp(W^c_{\underline{i}}) \longrightarrow \wp(W^c_i \times W^c_i)$ is such that for all $A \in \wp(W^c_{\underline{i}})$ and for all $t_i,u_i \in W^c_i$, $t_iR^c_i(A)u_i$ if and only if for all $\underline{i}$-formulas φ, if $\widehat{\varphi} \subseteq A$ then $[\varphi]t_i \subseteq u_i$,

where for all $i \in \{1,2\}$ and for all $\underline{i}$-formulas φ, $\widehat{\varphi} = \{v_{\underline{i}} \in W^c_{\underline{i}} : \varphi \in v_{\underline{i}}\}$. Since for all $i \in \{1,2\}$, s_i is a maximal $(\mathbf{L}_1,\mathbf{L}_2)$-consistent set of i-formulas, by Lemmas 4.3 and 6.1, for all $i \in \{1,2\}$, W^c_i is nonempty. Moreover, by Lemma 6.2, for all $i \in \{1,2\}$ and for all $t_i,u_i \in W^c_i$, $[\bot]t_i \subseteq u_i$.

Lemma 6.4 *For all $i \in \{1,2\}$ and for all i-formulas φ, if $\widehat{\varphi} = W^c_i$ then $[\bot]\varphi \in s_i$.*

Proof. Let $i \in \{1,2\}$ and φ be a i-formula. Suppose $\widehat{\varphi} = W^c_i$. We demonstrate $[\bot]\varphi \in s_i$. For the sake of the contradiction, suppose $[\bot]\varphi \notin s_i$. Let $u^0_i = [\bot]s_i \cup \{\neg\varphi\}$. Notice that $[\bot]s_i \subseteq u^0_i$ and $\neg\varphi \in u^0_i$. By Lemma 4.4, u^0_i is a

$(\mathbf{L}_1, \mathbf{L}_2)$-consistent set of i-formulas. Hence, by Lemma 4.3, let u_i be a maximal $(\mathbf{L}_1, \mathbf{L}_2)$-consistent set of i-formulas such that $u_i^0 \subseteq u_i$. Since $[\bot]s_i \subseteq u_i^0$ and $\neg\varphi \in u_i^0$, $[\bot]s_i \subseteq u_i$ and $\neg\varphi \in u_i$. Thus, $u_i \in W_i^c$ and $\varphi \notin u_i$. Since $\widehat{\varphi} = W_i^c$, $\varphi \in u_i$: a contradiction. □

Lemma 6.5 *For all $i \in \{1, 2\}$ and for all $\underline{i}$-formulas φ, if $\widehat{\varphi} = \emptyset$ then for all $t_i, u_i \in W_i^c$, $[\varphi]t_i \subseteq u_i$.*

Proof. Let $i \in \{1, 2\}$ and φ be a $\underline{i}$-formula. Suppose $\widehat{\varphi} = \emptyset$. We demonstrate for all $t_i, u_i \in W_i^c$, $[\varphi]t_i \subseteq u_i$. Let $t_i, u_i \in W_i^c$. For the sake of the contradiction, suppose $[\varphi]t_i \not\subseteq u_i$. Hence, let ψ be a i-formula such that $[\varphi]\psi \in t_i$ and $\psi \notin u_i$. Thus, $\psi \notin [\bot]t_i$. Consequently, $[\bot]\psi \notin t_i$. Since $[\varphi]\psi \in t_i$, $\langle\bot\rangle\neg\psi \rightarrow \langle\varphi\rangle\neg\psi \notin t_i$. Hence, $[\bot](\langle\bot\rangle\neg\psi \rightarrow \langle\varphi\rangle\neg\psi) \notin s_i$. Since (s_1, s_2) is paraconjunctive, $[\bot](\varphi \rightarrow \bot) \notin s_{\underline{i}}$ Let $w_{\underline{i}}^0 = [\bot]s_{\underline{i}} \cup \{\varphi\}$. Notice that $[\bot]s_{\underline{i}} \subseteq w_{\underline{i}}^0$ and $\varphi \in w_{\underline{i}}^0$. By Lemma 4.4, $w_{\underline{i}}^0$ is a $(\mathbf{L}_1, \mathbf{L}_2)$-consistent set of $\underline{i}$-formulas. Hence, by Lemma 4.3, let $w_{\underline{i}}$ be a maximal $(\mathbf{L}_1, \mathbf{L}_2)$-consistent set of $\underline{i}$-formulas such that $w_{\underline{i}}^0 \subseteq w_{\underline{i}}$. Since $[\bot]s_{\underline{i}} \subseteq w_{\underline{i}}^0$ and $\varphi \in w_{\underline{i}}^0$, $[\bot]s_{\underline{i}} \subseteq w_{\underline{i}}$ and $\varphi \in w_{\underline{i}}$. Thus, $w_{\underline{i}} \in W_{\underline{i}}^c$. Since $\widehat{\varphi} = \emptyset$, $\varphi \notin w_{\underline{i}}$: a contradiction. Here finishes the proof of Lemma 6.5. □

Lemma 6.6 *The 4-tuple $(W_1^c, W_2^c, R_1^c, R_2^c)$ is a paraconjunctive frame.*

Proof. By Lemma 6.5, for all $i \in \{1, 2\}$, for all $t_i, u_i \in W_i^c$ and for all $\underline{i}$-formulas φ, if $\widehat{\varphi} = \emptyset$ then $[\varphi]t_i \subseteq u_i$. Hence, for all $i \in \{1, 2\}$ and for all $t_i, u_i \in W_i^c$, $t_i R_i^c(\emptyset) u_i$. Thus, for all $i \in \{1, 2\}$, $R_i^c(\emptyset) = W_i^c \times W_i^c$. Moreover, for all $i \in \{1, 2\}$ and for all $A, B \in \wp(W_{\underline{i}}^c)$, if $A \subseteq B$ then for all $\underline{i}$-formulas φ, if $\widehat{\varphi} \subseteq A$ then $\widehat{\varphi} \subseteq B$. Consequently, for all $i \in \{1, 2\}$ and for all $A, B \in \wp(W_{\underline{i}}^c)$, if $A \subseteq B$ then for all $t_i, u_i \in W_i^c$, if $t_i R_i^c(B) u_i$ then $t_i R_i^c(A) u_i$. Hence, for all $i \in \{1, 2\}$ and for all $A, B \in \wp(W_{\underline{i}}^c)$, if $A \subseteq B$ then $R_i^c(A) \supseteq R_i^c(B)$. □

The 4-tuple $(W_1^c, W_2^c, R_1^c, R_2^c)$ will be called *paraconjunctive canonical frame for $(\mathbf{L}_1, \mathbf{L}_2)$ determined by (s_1, s_2)*. Lemma 6.7 is an immediate consequence of the fact that for all $i \in \{1, 2\}$, $\mathbf{Equiv}_i^p$ contains all i-formulas of the form $[\varphi]\psi \rightarrow \psi$ and $\langle\varphi\rangle\psi \rightarrow [\varphi]\langle\varphi\rangle\psi$.

Lemma 6.7 *If $(\mathbf{L}_1, \mathbf{L}_2)$ contains $(\mathbf{Equiv}_1^p, \mathbf{Equiv}_2^p)$ then the paraconjunctive canonical frame for $(\mathbf{L}_1, \mathbf{L}_2)$ determined by (s_1, s_2) is a paraconjunctive frame of indiscernibility.*

For all $i \in \{1, 2\}$, let $V_i^c : \mathcal{L}_i \longrightarrow \wp(W_i^c)$ be such that for all $\varphi \in \mathcal{L}_i$, $V_i^c(\varphi) = \widehat{\varphi}$. The 6-tuple $(W_1^c, W_2^c, R_1^c, R_2^c, V_1^c, V_2^c)$ will be called *paraconjunctive canonical model for $(\mathbf{L}_1, \mathbf{L}_2)$ determined by (s_1, s_2)*.

Lemma 6.8 (Truth Lemma: the paraconjunctive case) *The paraconjunctive canonical model for $(\mathbf{L}_1, \mathbf{L}_2)$ determined by (s_1, s_2) is a model.*

Proof. The proof that for all $i \in \{1, 2\}$, V_i^c satisfies the conditions for $\bot$, $\neg$ and $\vee$ is as expected. We only show that for all $i \in \{1, 2\}$, V_i^c satisfies the condition for $[\cdot]$. Let $i \in \{1, 2\}$. Let φ be a $\underline{i}$-formula and ψ be a i-formula. Let $t_i \in W_i^c$. We only demonstrate $t_i \in V_i^c([\varphi]\psi)$ if for all $u_i \in W_i^c$, if $t_i R_i^c(V_{\underline{i}}^c(\varphi)) u_i$ then $u_i \in V_i^c(\psi)$, the "only if" direction being left as an exercise for the reader. Suppose $t_i \notin V_i^c([\varphi]\psi)$. We demonstrate there exists $u_i \in W_i^c$ such that $t_i R_i^c(V_{\underline{i}}^c(\varphi))u_i$ and $u_i \notin V_i^c(\psi)$. Since $t_i \notin V_i^c([\varphi]\psi)$, $[\varphi]\psi \notin t_i$. Let $u_i^0 = [\varphi]t_i \cup \{\neg\psi\}$. Notice that $[\varphi]t_i \subseteq u_i^0$ and $\neg\psi \in u_i^0$. By Lemma 4.4, u_i^0 is a $(\mathbf{L}_1, \mathbf{L}_2)$-consistent set of i-formulas. Hence, by Lemma 4.3, let u_i be a maximal $(\mathbf{L}_1, \mathbf{L}_2)$-consistent set of i-formulas such that $u_i^0 \subseteq u_i$. Since $[\varphi]t_i \subseteq u_i^0$ and $\neg\psi \in u_i^0$, $[\varphi]t_i \subseteq u_i$ and $\neg\psi \in u_i$.

Claim $u_i \in W_i^c$.

Proof. For the sake of the contradiction, suppose $u_i \notin W_i^c$. Hence, $[\bot]s_i \not\subseteq u_i$. Thus, let χ be a i-formula such that $[\bot]\chi \in s_i$ and $\chi \notin u_i$. Since $\mathbf{L}_i$ contains all i-formulas of the form $[\bot]\chi' \rightarrow [\bot][\bot]\chi'$, $[\bot][\bot]\chi \in s_i$. Consequently, $[\bot]\chi \in [\bot]s_i$. Since $[\bot]s_i \subseteq t_i$, $[\bot]\chi \in t_i$. Since $\mathbf{L}_{\underline{i}}$ contains the $\underline{i}$-formula $\bot \rightarrow \varphi$ and $\mathbf{L}_{\underline{i}}$ satisfies the closure condition $(\mathbf{Gen}_{\underline{i}})$, $\mathbf{L}_{\underline{i}}$ contains the $\underline{i}$-formula $[\bot](\bot \rightarrow \varphi)$. Since $\mathbf{L}_i$ satisfies the closure condition $(\mathbf{RI}_i^-)$, $\mathbf{L}_i$ contains the i-formula $[\bot]([\bot]\chi \rightarrow [\varphi]\chi)$ [7]. Since $\mathbf{L}_i$ contains all i-formulas of the form $[\bot]\phi \rightarrow \phi$, $\mathbf{L}_i$ contains the i-formula $[\bot]\chi \rightarrow [\varphi]\chi$. Since $[\bot]\chi \in t_i$, $[\varphi]\chi \in t_i$. Hence, $\chi \in [\varphi]t_i$. Since $[\varphi]t_i \subseteq u_i$, $\chi \in u_i$: a contradiction. □

Claim $t_i R_i^c(V_{\underline{i}}^c(\varphi))u_i$.

Proof. For the sake of the contradiction, suppose not $t_i R_i^c(V_{\underline{i}}^c(\varphi))u_i$. Thus, there exists a $\underline{i}$-formula φ' such that $\widehat{\varphi'} \subseteq V_{\underline{i}}^c(\varphi)$ and $[\varphi']t_i \not\subseteq u_i$. Consequently, let χ be a i-formula such that $[\varphi']\chi \in t_i$ and $\chi \notin u_i$. Since $\widehat{\varphi'} \subseteq V_{\underline{i}}^c(\varphi)$, for all $v_{\underline{i}} \in W_{\underline{i}}^c$, if $\varphi' \in v_{\underline{i}}$ then $\varphi \in v_{\underline{i}}$. Hence, for all $v_{\underline{i}} \in W_{\underline{i}}^c$, $\varphi' \rightarrow \varphi \in v_{\underline{i}}$. Thus, by Lemma 6.4, $[\bot](\varphi' \rightarrow \varphi) \in s_{\underline{i}}$. Since (s_1, s_2) is paraconjunctive, $[\bot]([\varphi']\chi \rightarrow [\varphi]\chi) \in s_i$. Since $\mathbf{L}_i$ contains all i-formulas of the form $[\bot]\phi \rightarrow [\bot][\bot]\phi$, $[\bot][\bot]([\varphi']\chi \rightarrow [\varphi]\chi) \in s_i$. Consequently, $[\bot]([\varphi']\chi \rightarrow [\varphi]\chi) \in t_i$. Since $\mathbf{L}_i$ contains all i-formulas of the form $[\bot]\phi \rightarrow \phi$, $[\varphi']\chi \rightarrow [\varphi]\chi \in t_i$. Since $[\varphi']\chi \in t_i$, $[\varphi]\chi \in t_i$. Hence, $\chi \in [\varphi]t_i$. Since $[\varphi]t_i \subseteq u_i$, $\chi \in u_i$: a contradiction. □

Finally, the reader may easily verify that $u_i \notin V_i^c(\psi)$. Here finishes the proof of Lemma 6.8.

Proposition 6.9 is an immediate consequence of Lemmas 4.3, 6.3, 6.6 and 6.8.

Proposition 6.9 $(\mathbf{ParCon}_1, \mathbf{ParCon}_2)$ *is complete with respect to the class of all paraconjunctive frames.*

Proposition 6.10 is an immediate consequence of Lemmas 4.3, 6.3, 6.7 and 6.8.

Proposition 6.10 $(\mathbf{Equiv}_1^p, \mathbf{Equiv}_2^p)$ *is complete with respect to the class of all paraconjunctive frames of indiscernibility.*

[7] Here, we are using the closure condition $(\mathbf{RI}_i^-)$ with $m = 0$ and $n = 1$.

6.3 Paraconjunctive case: second set of completeness results

As for the completeness of ($\mathbf{ParCon}_1, \mathbf{ParCon}_2$) with respect to the class of all conjunctive frames and the completeness of ($\mathbf{Equiv}_1^p, \mathbf{Equiv}_2^p$) with respect to the class of all conjunctive frames of indiscernibility, we will show in Propositions 6.11 and 6.19 that every paraconjunctive frame is the bounded morphic image of a conjunctive frame and every paraconjunctive frame of indiscernibility is the bounded morphic image of a conjunctive frame of indiscernibility [8].

Proposition 6.11 *Let (W_1, W_2, R_1, R_2) be a paraconjunctive frame. There exists a conjunctive frame (W_1', W_2', R_1', R_2') and a couple (f_1, f_2) of surjective functions $f_1 : W_1' \longrightarrow W_1$ and $f_2 : W_2' \longrightarrow W_2$ such that for all (W_1, W_2, R_1, R_2)-valuations (V_1, V_2), there exists a (W_1', W_2', R_1', R_2')-valuation (V_1', V_2') such that (f_1, f_2) is a bounded morphism from $(W_1', W_2', R_1', R_2', V_1', V_2')$ to $(W_1, W_2, R_1, R_2, V_1, V_2)$.*

Proof. For all $i \in \{1, 2\}$, let

- $\det_i : \wp(W_{\underline{i}}) \times W_i \times W_i \longrightarrow \wp(W_i)$ be such that for all $A \in \wp(W_{\underline{i}})$ and for all $t_i, u_i \in W_i$, if $t_i R_i(A) u_i$ then $\det(A, t_i, u_i) = \emptyset$ else $\det(A, t_i, u_i) = W_i$ [9].

For all $i \in \{1, 2\}$, let Λ_i be the set of all $\tau_i : \wp(W_{\underline{i}}) \times W_{\underline{i}} \longrightarrow \wp(W_i)$ such that for all $A \in \wp(W_{\underline{i}})$, $\{t_{\underline{i}} \in W_{\underline{i}} : \tau_i(A, t_{\underline{i}}) \neq \emptyset\}$ is finite [10]. Let (W_1', W_2', R_1', R_2') be the 4-tuple where for all $i \in \{1, 2\}$,

- $W_i' = W_i \times \Lambda_i$,
- $R_i' : \wp(W_{\underline{i}}') \longrightarrow \wp(W_i' \times W_i')$ is such that for all $A' \in \wp(W_{\underline{i}}')$ and for all $(t_i, \tau_i), (u_i, \mu_i) \in W_i'$, $(t_i, \tau_i) R_i'(A')(u_i, \mu_i)$ if and only if for all $A \in \wp(W_{\underline{i}})$ [11],
  · if $A' \cap (A \times \Lambda_{\underline{i}}) \neq \emptyset$ then $\bigoplus_i \{\tau_i(A, v_{\underline{i}}) \oplus_i \mu_i(A, v_{\underline{i}}) : v_{\underline{i}} \in A\} = \det_i(A, t_i, u_i)$,
  · for all $(v_{\underline{i}}, \nu_{\underline{i}}) \in A' \cap (A \times \Lambda_{\underline{i}})$, $\tau_i(A, v_{\underline{i}}) \oplus_i \mu_i(A, v_{\underline{i}}) = \emptyset$.

Claim 6.12 *For all $i \in \{1, 2\}$ and for all $A' \in \wp(W_{\underline{i}}')$, $R_i'(A') = \bigcap\{R_i'(\{(v_{\underline{i}}, \nu_{\underline{i}})\}) : (v_{\underline{i}}, \nu_{\underline{i}}) \in A'\}$.*

Proof. Let $i \in \{1, 2\}$ and $A' \in \wp(W_{\underline{i}}')$. The proof that $R_i'(A') \subseteq \bigcap\{R_i'(\{(v_{\underline{i}}, \nu_{\underline{i}})\}) : (v_{\underline{i}}, \nu_{\underline{i}}) \in A'\}$ being a simple application of the definitions, it is left as an exercise for the reader. We demonstrate $R_i'(A') \supseteq \bigcap\{R_i'(\{(v_{\underline{i}}, \nu_{\underline{i}})\}) : (v_{\underline{i}}, \nu_{\underline{i}}) \in A'\}$. For the sake of the contradiction, suppose $R_i'(A') \not\supseteq \bigcap\{R_i'(\{(v_{\underline{i}}, \nu_{\underline{i}})\}) : (v_{\underline{i}}, \nu_{\underline{i}}) \in A'\}$. Hence, there exists $(t_i, \tau_i), (u_i, \mu_i) \in W_i'$ such that not $(t_i, \tau_i) R_i'(A')(u_i, \mu_i)$ and for all $(v_{\underline{i}}, \nu_{\underline{i}}) \in A'$, $(t_i, \tau_i) R_i'(\{(v_{\underline{i}}, \nu_{\underline{i}})\})(u_i, \mu_i)$. Thus, for all $(v_{\underline{i}}, \nu_{\underline{i}}) \in A'$ and for all $A \in \wp(W_{\underline{i}})$,

---

[8] The sketch of an alternative proof of Proposition 6.11 is presented in the Appendix.

[9] Notice that for all $A \in \wp(W_{\underline{i}})$ and for all $t_i, u_i \in W_i$, $\det(A, t_i, u_i) = \emptyset$ if and only if $t_i R_i(A) u_i$.

[10] Notice that for all $\tau_i, \mu_i \in \Lambda_i$ and for all $A \in \wp(W_{\underline{i}})$, $\{v_{\underline{i}} \in A : \tau_i(A, v_{\underline{i}}) \neq \mu_i(A, v_{\underline{i}})\}$ is finite.

[11] Here, $\oplus_i$ is the operation of symmetric difference in $\wp(W_i)$. Moreover, $(v_{\underline{i}}^1, \ldots, v_{\underline{i}}^N)$ being the list of all $v_{\underline{i}} \in A$ such that $\tau_i(A, v_{\underline{i}}) \neq \mu_i(A, v_{\underline{i}})$, $\bigoplus_i \{\tau_i(A, v_{\underline{i}}) \oplus_i \mu_i(A, v_{\underline{i}}) : v_{\underline{i}} \in A\}$ denotes $\tau_i(A, v_{\underline{i}}^1) \oplus_i \mu_i(A, v_{\underline{i}}^1) \oplus_i \ldots \oplus_i \tau_i(A, v_{\underline{i}}^N) \oplus_i \mu_i(A, v_{\underline{i}}^N)$.

- if $\{(v_{\underline{i}},\nu_{\underline{i}})\} \cap (A \times \Lambda_{\underline{i}}) \neq \emptyset$ then $\bigoplus_i \{\tau_i(A, w_{\underline{i}}) \oplus_i \mu_i(A, w_{\underline{i}}) : w_{\underline{i}} \in A\} = \det_i(A, t_i, u_i)$,
- for all $(w_{\underline{i}}, \omega_{\underline{i}}) \in \{(v_{\underline{i}},\nu_{\underline{i}})\} \cap (A \times \Lambda_{\underline{i}})$, $\tau_i(A, w_{\underline{i}}) \oplus_i \mu_i(A, w_{\underline{i}}) = \emptyset$.

Consequently, for all $A \in \wp(W_{\underline{i}})$,

- if $A' \cap (A \times \Lambda_{\underline{i}}) \neq \emptyset$ then $\bigoplus_i \{\tau_i(A, w_{\underline{i}}) \oplus_i \mu_i(A, w_{\underline{i}}) : w_{\underline{i}} \in A\} = \det_i(A, t_i, u_i)$,
- for all $(w_{\underline{i}}, \omega_{\underline{i}}) \in A' \cap (A \times \Lambda_{\underline{i}})$, $\tau_i(A, w_{\underline{i}}) \oplus_i \mu_i(A, w_{\underline{i}}) = \emptyset$.

Hence, $(t_i, \tau_i) R'_i(A')(u_i, \mu_i)$: a contradiction. □

Claim 6.13 is an immediate consequence of Claim 6.12.

Claim 6.13 *The 4-tuple (W'_1, W'_2, R'_1, R'_2) is a conjunctive frame.*

Let $f_1 : W'_1 \longrightarrow W_1$ be such that for all $(v_1, \nu_1) \in W'_1$, $f_1(v_1, \nu_1) = v_1$ and $f_2 : W'_2 \longrightarrow W_2$ be such that for all $(v_2, \nu_2) \in W'_2$, $f_2(v_2, \nu_2) = v_2$.

Claim 6.14 *f_1 and f_2 are surjective.*

Let (V_1, V_2) be a (W_1, W_2, R_1, R_2)-valuation. For all $i \in \{1, 2\}$, let $V'_i : \mathcal{L}_i \longrightarrow \wp(W'_i)$ be such that for all $\varphi \in \mathcal{L}_i$, $V'_i(\varphi) = f_i^{-1}[V_i(\varphi)]$. Notice that for all $i \in \{1, 2\}$ and for all $\varphi \in \mathcal{L}_i$, $V'_i(\varphi) = V_i(\varphi) \times \Lambda_i$.

Claim 6.15 *For all $(t_i, \tau_i), (u_i, \mu_i) \in W'_i$ and for all $\underline{i}$-formulas φ, if $(t_i, \tau_i) R'_i(V'_{\underline{i}}(\varphi))(u_i, \mu_i)$ then $t_i R_i(V_{\underline{i}}(\varphi)) u_i$.*

Proof. Let $(t_i, \tau_i), (u_i, \mu_i) \in W'_i$ and φ be a $\underline{i}$-formula. Suppose $(t_i, \tau_i) R'_i(V'_{\underline{i}}(\varphi))(u_i, \mu_i)$. For the sake of the contradiction, suppose not $t_i R_i(V_{\underline{i}}(\varphi)) u_i$. Hence, $V_{\underline{i}}(\varphi) \neq \emptyset$. Thus, $V'_{\underline{i}}(\varphi) \cap (V_{\underline{i}}(\varphi) \times \Lambda_i) \neq \emptyset$. Since $(t_i, \tau_i) R'_i(V'_{\underline{i}}(\varphi))(u_i, \mu_i)$, $\bigoplus_i \{\tau_i(V_{\underline{i}}(\varphi), v_{\underline{i}}) \oplus_i \mu_i(V_{\underline{i}}(\varphi), v_{\underline{i}}) : v_{\underline{i}} \in V_{\underline{i}}(\varphi)\} = \det_i(V_{\underline{i}}(\varphi), t_i, u_i)$. Moreover, for all $(v_{\underline{i}}, \nu_{\underline{i}}) \in V'_{\underline{i}}(\varphi) \cap (V_{\underline{i}}(\varphi) \times \Lambda_{\underline{i}})$, $\tau_i(V_{\underline{i}}(\varphi), v_{\underline{i}}) \oplus_i \mu_i(V_{\underline{i}}(\varphi), v_{\underline{i}}) = \emptyset$. Consequently, for all $v_{\underline{i}} \in V_{\underline{i}}(\varphi)$, $\tau_i(V_{\underline{i}}(\varphi), v_{\underline{i}}) \oplus_i \mu_i(V_{\underline{i}}(\varphi), v_{\underline{i}}) = \emptyset$. Hence, $\bigoplus_i \{\tau_i(V_{\underline{i}}(\varphi), v_{\underline{i}}) \oplus_i \mu_i(V_{\underline{i}}(\varphi), v_{\underline{i}}) : v_{\underline{i}} \in V_{\underline{i}}(\varphi)\} = \emptyset$. Since $\bigoplus_i \{\tau_i(V_{\underline{i}}(\varphi), v_{\underline{i}}) \oplus_i \mu_i(V_{\underline{i}}(\varphi), v_{\underline{i}}) : v_{\underline{i}} \in V_{\underline{i}}(\varphi)\} = \det_i(V_{\underline{i}}(\varphi), t_i, u_i)$, $\det_i(V_{\underline{i}}(\varphi), t_i, u_i) = \emptyset$. Thus, $t_i R_i(V_{\underline{i}}(\varphi)) u_i$: a contradiction. □

Claim 6.16 *For all $(t_i, \tau_i) \in W'_i$, for all $u_i \in W_i$ and for all $\underline{i}$-formulas φ, if $t_i R_i(V_{\underline{i}}(\varphi)) u_i$ then there exists $\mu_i \in \Lambda_i$ such that $(t_i, \tau_i) R'_i(V'_{\underline{i}}(\varphi))(u_i, \mu_i)$.*

Proof. Let $(t_i, \tau_i) \in W'_i$, $u_i \in W_i$ and φ be a $\underline{i}$-formula. Suppose $t_i R_i(V_{\underline{i}}(\varphi)) u_i$. We demonstrate there exists $\mu_i \in \Lambda_i$ such that $(t_i, \tau_i) R'_i(V'_{\underline{i}}(\varphi))(u_i, \mu_i)$. Indeed, we are looking for $\mu_i : \wp(W_{\underline{i}}) \times W_{\underline{i}} \longrightarrow \wp(W_i)$ such that for all $A \in \wp(W_{\underline{i}})$,

($\mathbf{C_1}$) $\{w_{\underline{i}} \in W_{\underline{i}} : \mu_i(A, w_{\underline{i}}) \neq \emptyset\}$ is finite,

($\mathbf{C_2}$) if $V'_{\underline{i}}(\varphi) \cap (A \times \Lambda_{\underline{i}}) \neq \emptyset$ then $\bigoplus_i \{\tau_i(A, w_{\underline{i}}) \oplus_i \mu_i(A, w_{\underline{i}}) : w_{\underline{i}} \in A\} = \det_i(A, t_i, u_i)$,

($\mathbf{C_3}$) for all $(w_{\underline{i}}, \omega_{\underline{i}}) \in V'_{\underline{i}}(\varphi) \cap (A \times \Lambda_{\underline{i}})$, $\tau_i(A, w_{\underline{i}}) \oplus_i \mu_i(A, w_{\underline{i}}) = \emptyset$.

For all $A \in \wp(W_{\underline{i}})$, let $\mu_i^A : W_{\underline{i}} \longrightarrow \wp(W_i)$ be defined as follows:

Case "$A \subseteq V_{\underline{i}}(\varphi)$": for all $w_{\underline{i}} \in W_{\underline{i}}$, let $\mu_i^A(w_{\underline{i}}) = \tau_i(A, w_{\underline{i}})$,

Case "$A \not\subseteq V_{\underline{i}}(\varphi)$": let $w_{\underline{i}}^A \in W_{\underline{i}}$ be such that $w_{\underline{i}}^A \in A$ and $w_{\underline{i}}^A \notin V_{\underline{i}}(\varphi)$ and for all $w_{\underline{i}} \in W_{\underline{i}}$,

- if $w_{\underline{i}} \neq w_{\underline{i}}^A$ then let $\mu_i^A(w_{\underline{i}}) = \tau_i(A, w_{\underline{i}})$,
- otherwise, let $\mu_i^A(w_{\underline{i}}) = \tau_i(A, w_{\underline{i}}) \oplus_i \det_i(A, t_i, u_i)$.

Let $\mu_i : \wp(W_{\underline{i}}) \times W_{\underline{i}} \longrightarrow \wp(W_i)$ be such that for all $A \in \wp(W_{\underline{i}})$ and for all $w_{\underline{i}} \in W_{\underline{i}}$, $\mu_i(A, w_{\underline{i}}) = \mu_i^A(w_{\underline{i}})$. Now, we just have to verify that for all $A \in \wp(W_{\underline{i}})$, $(\mathbf{C_1})$, $(\mathbf{C_2})$ and $(\mathbf{C_3})$ hold. Let $A \in \wp(W_{\underline{i}})$. Concerning $(\mathbf{C_1})$, it holds, seeing that $\mu_i^A(w_{\underline{i}}) = \tau_i(A, w_{\underline{i}})$ for every $w_{\underline{i}} \in W_{\underline{i}}$ except when $A \not\subseteq V_{\underline{i}}(\varphi)$ and $w_{\underline{i}} = w_{\underline{i}}^A$. About $(\mathbf{C_2})$, suppose $V_{\underline{i}}'(\varphi) \cap (A \times \Lambda_{\underline{i}}) \neq \emptyset$ and consider the following two cases: $A \subseteq V_{\underline{i}}(\varphi)$ and $A \not\subseteq V_{\underline{i}}(\varphi)$. In the former case, since $t_i R_i(V_{\underline{i}}(\varphi)) u_i$, $t_i R_i(A) u_i$. Hence, $\det_i(A, t_i, u_i) = \emptyset$. Since $A \subseteq V_{\underline{i}}(\varphi)$, for all $w_{\underline{i}} \in W_{\underline{i}}$, $\mu_i^A(w_{\underline{i}}) = \tau_i(A, w_{\underline{i}})$. Thus, for all $w_{\underline{i}} \in W_{\underline{i}}$, $\tau_i(A, w_{\underline{i}}) \oplus_i \mu_i(A, w_{\underline{i}}) = \emptyset$. Consequently, $\bigoplus_i \{\tau_i(A, w_{\underline{i}}) \oplus_i \mu_i(A, w_{\underline{i}}) : w_{\underline{i}} \in A\} = \emptyset$. Since $\det_i(A, t_i, u_i) = \emptyset$, $(\mathbf{C_2})$ holds. In the latter case, $\mu_i^A(w_{\underline{i}}) = \tau_i(A, w_{\underline{i}})$ for every $w_{\underline{i}} \in W_{\underline{i}}$ except when $w_{\underline{i}} = w_{\underline{i}}^A$. Hence, $\bigoplus_i \{\tau_i(A, w_{\underline{i}}) \oplus_i \mu_i(A, w_{\underline{i}}) : w_{\underline{i}} \in A\} = \tau_i(A, w_{\underline{i}}^A) \oplus_i \mu_i(A, w_{\underline{i}}^A)$. Since $\mu_i^A(w_{\underline{i}}^A) = \tau_i(A, w_{\underline{i}}^A) \oplus_i \det_i(A, t_i, u_i)$, $(\mathbf{C_2})$ holds. As for $(\mathbf{C_3})$, it holds, seeing that for all $w_{\underline{i}} \in W_{\underline{i}}$, if $w_{\underline{i}} \in A$ and $w_{\underline{i}} \in V_{\underline{i}}(\varphi)$ then $\mu_i^A(w_{\underline{i}}) = \tau_i(A, w_{\underline{i}})$. □

Claim 6.17 *(V_1', V_2') is a (W_1', W_2', R_1', R_2')-valuation.*

Proof. The proof that for all $i \in \{1, 2\}$, V_i' satisfies the conditions for $\bot$, $\neg$ and $\vee$ is as expected. The proof that for all $i \in \{1, 2\}$, V_i' satisfies the condition for $[\cdot]$ being a simple application of Claims 6.15 and 6.16, it is left as an exercise for the reader. □

Claim 6.18 is an immediate consequence of Claims 6.15 and 6.16.

Claim 6.18 *(f_1, f_2) is a bounded morphism from $(W_1', W_2', R_1', R_2', V_1', V_2')$ to $(W_1, W_2, R_1, R_2, V_1, V_2)$.*

Here finishes the proof of Proposition 6.11. □

Proposition 6.19 *Let (W_1, W_2, R_1, R_2) be a paraconjunctive frame of indiscernibility. There exists a conjunctive frame of indiscernibility (W_1', W_2', R_1', R_2') and a couple (f_1, f_2) of surjective functions $f_1 : W_1' \longrightarrow W_1$ and $f_2 : W_2' \longrightarrow W_2$ such that for all (W_1, W_2, R_1, R_2)-valuations (V_1, V_2), there exists a (W_1', W_2', R_1', R_2')-valuation (V_1', V_2') such that (f_1, f_2) is a bounded morphism from $(W_1', W_2', R_1', R_2', V_1', V_2')$ to $(W_1, W_2, R_1, R_2, V_1, V_2)$.*

Proof. For all $i \in \{1, 2\}$, let

- $\det_i : \wp(W_{\underline{i}}) \times W_i \times W_i \longrightarrow \wp(W_i)$ be such that for all $A \in \wp(W_{\underline{i}})$ and for all $t_i, u_i \in W_i$, $\det(A, t_i, u_i) = [t_i]_{R_i(A)} \oplus_i [u_i]_{R_i(A)}$ where $[t_i]_{R_i(A)}$ and $[u_i]_{R_i(A)}$ are the equivalence classes of t_i and u_i modulo $R_i(A)$ [12].

[12]Notice that for all $A \in \wp(W_{\underline{i}})$ and for all $t_i, u_i \in W_i$, $\det(A, t_i, u_i) = \emptyset$ if and only if $t_i R_i(A) u_i$.

Now, the rest of the proof is similar to the corresponding rest of the proof of Proposition 6.11, the main difference being that one has to verify here that the considered frames are frames of indiscernibility, an exercise that we leave for the reader. □

Proposition 6.20 is an immediate consequence of Propositions 3.2, 3.13, 6.9 and 6.11.

Proposition 6.20 ($\mathbf{ParCon}_1, \mathbf{ParCon}_2$) *is complete with respect to the class of all preconjunctive frames and the class of all conjunctive frames.*

Proposition 6.21 is an immediate consequence of Propositions 3.2, 3.13, 6.10 and 6.19.

Proposition 6.21 ($\mathbf{Equiv}_1^p, \mathbf{Equiv}_2^p$) *is complete with respect to the class of all preconjunctive frames of indiscernibility and the class of all conjunctive frames of indiscernibility.*

7 A short case study: social epistemic logic

Formalizing epistemic reasoning in social networks, Seligman *et al.* [28] introduce relational structures of the form $(S, A, \equiv, \triangleright)$ such as the ones considered in our introduction. They also introduce a modal language $\mathcal{L}_{\mathbf{SEL}}$ — the *language of social epistemic logic* — consisting of one type of formulas: state-agent formulas — to be interpreted by sets of state-agent couples. Such formulas (typically denoted φ, ψ, etc) are constructed over the Boolean connectives and the modal connectives K and F as follows:

- $\varphi ::= p \mid \bot \mid \neg\varphi \mid (\varphi \vee \psi) \mid K\varphi \mid F\varphi$,

p ranging over a countably infinite set of atomic formulas[13]. The formula $K\varphi$ (read "I know that φ holds") is true in a state-agent couple (s, a) of some model if the formula φ is true in every state-agent couple (t, a) such that $s \equiv_a t$ whereas the formula $F\varphi$ ("for all my friends, φ holds") is true in a state-agent couple (s, a) of some model if the formula φ is true in every state-agent couple (s, b) such that $a \triangleright_s b$. See [3,27] and [10, Chapter 5].

In the relational structures of the form $(S, A, \equiv, \triangleright)$ considered in our introduction, assuming that $\equiv$ is also a function associating an equivalence relation $\equiv_B$ on S to every $B \in \wp(A)$ in such a way that for all $B \in \wp(A)$, $\equiv_B = \bigcap\{\equiv_a : a \in B\}$ and $\triangleright$ is also a function associating a binary relation $\triangleright_T$ on A to every $T \in \wp(S)$ in such a way that for all $T \in \wp(S)$, $\triangleright_T = \bigcap\{\triangleright_s : s \in T\}$, the language defined in Section 2 can be interpreted as explained in Section 3. A $\mathcal{L}_{\mathbf{SEL}}$-formula φ is said to be $\mathcal{L}_{\mathbf{PML}}$*-definable in a class* $\mathcal{C}$ *of frames* if there exists $\varphi' \in \mathcal{L}_{\mathbf{PML}}$ such that for all $\mathcal{C}$-frames $(S, A, \equiv, \triangleright)$, $(S, A, \equiv, \triangleright) \models \varphi$ (in the sense of [28]) if and only if $(S, A, \equiv, \triangleright) \models \varphi'$ (in the sense of Section 3). A $\mathcal{L}_{\mathbf{PML}}$-formula φ is said to be $\mathcal{L}_{\mathbf{SEL}}$*-definable in a class* $\mathcal{C}$ *of frames* if there

[13] Indeed, the modal language introduced by Seligman *et al.* contains as well nominals used to give names to agents. In this section, however, we only consider its nominal-free fragment.

exists $\varphi' \in \mathcal{L}_{\mathbf{SEL}}$ such that for all $\mathcal{C}$-frames $(S, A, \equiv, \rhd)$, $(S, A, \equiv, \rhd) \models \varphi$ (in the sense of Section 3) if and only if $(S, A, \equiv, \rhd) \models \varphi'$ (in the sense of [28]).

Let $\mathcal{C}_0$ be the class of all frames $(S, A, \equiv, \rhd)$ such that for all $a \in A$, $\equiv_a$ is an equivalence relation and for all $s \in S$, $\rhd_s$ is irreflexive and symmetric [14]. In a $\mathcal{C}_0$-frame $(S, A, \equiv, \rhd)$, for all $a, b \in A$, we say that a and b are *strong friends* if for all $s \in S$, $a \rhd_s b$. In a $\mathcal{C}_0$-frame $(S, A, \equiv, \rhd)$, for all $a \in A$, we say that *a is aware of her friends* if for all $s, t \in S$ and for all $b \in A$, if $s \equiv_a t$ then $a \rhd_s b$ if and only if $a \rhd_t b$. One may easily verify that for all $\mathcal{C}_0$-frames $(S, A, \equiv, \rhd)$,

- $(S, A, \equiv, \rhd) \models [\top_1]_2 \bot_2$ if and only if no agent has strong friends,
- $(S, A, \equiv, \rhd) \models \langle \top_1 \rangle_2 \top_2$ if and only if every agent has strong friends,
- $(S, A, \equiv, \rhd) \models p_2 \wedge_2 \langle p'_1 \rangle_2 q_2 \rightarrow_2 \langle \langle p_2 \rangle_1 p'_1 \rangle_2 q_2$ if and only if every agent is aware of her friends.

Moreover, the elementary conditions "no agent has strong friends", "every agent has strong friends" and "every agent is aware of her friends" correspond to no $\mathcal{L}_{\mathbf{SEL}}$-formula in $\mathcal{C}_0$ [15]. Hence,

Proposition 7.1 *The $\mathcal{L}_{\mathbf{PML}}$-formulas $[\top_1]_2 \bot_2$, $\langle \top_1 \rangle_2 \top_2$ and $p_2 \wedge_2 \langle p'_1 \rangle_2 q_2 \rightarrow_2 \langle \langle p_2 \rangle_1 p'_1 \rangle_2 q_2$ are not $\mathcal{L}_{\mathbf{SEL}}$-definable in $\mathcal{C}_0$.*

Problem 7.2 *We do not know if there exists $\mathcal{L}_{\mathbf{SEL}}$-formulas which are not $\mathcal{L}_{\mathbf{PML}}$-definable in $\mathcal{C}_0$.*

8 Conclusion

What has been done in this paper? Within the context of a two-typed parametrized modal language, we have defined a PML as a couple whose components are sets of formulas containing, in their respective types, all propositional tautologies and closed, in their respective types, under modus ponens and uniform substitution. Assuming the normality condition, these components also contain the distribution axiom and are closed under the generalization rule. We have axiomatically introduced different two-typed PMLs and we have given the proofs of their completeness with respect to appropriate classes of relational structures by means of an adaptation of the

---

[14] Although the choice of reflexive, symmetric and transitive relations between states for the epistemic modalities is standard, the choice of irreflexive and symmetric relations between agents for the friendship modalities is debatable. In this paper, we simply follow Seligman *et al.* [28] in their assumptions about the friendship modalities.

[15] For instance, in order to show that the elementary conditions "no agent has strong friends" and "every agent has strong friends" correspond to no $\mathcal{L}_{\mathbf{SEL}}$-formula in $\mathcal{C}_0$, it suffices to consider the frames $(S', A', \equiv', \rhd')$ and $(S'', A'', \equiv'', \rhd'')$ where $S' = \{1\}$, $A' = \{a, b, c, d\}$, $\rhd'_1 = \{(a, b), (b, a), (c, d), (d, c)\}$, $S'' = \{2\}$, $A'' = \{a, b, c, d\}$ and $\rhd''_2 = \{(a, c), (b, d), (c, a), (d, b)\}$. Let $(S, A, \equiv, \rhd)$ be their disjoint union. By induction on $\varphi \in \mathcal{L}_{\mathbf{SEL}}$, one may easily verify that $(S, A, \equiv, \rhd) \models \varphi$ if and only if $(S', A', \equiv', \rhd') \models \varphi$ and $(S'', A'', \equiv'', \rhd'') \models \varphi$: a contradiction with the obvious fact that no agent has strong friends in $(S, A, \equiv, \rhd)$ whereas every agent has strong friends both in $(S', A', \equiv', \rhd')$ and in $(S'', A'', \equiv'', \rhd'')$.

ordinary canonical model construction. The operation of intersection — which is used in conjunctive frames for the interpretation of the modalities — being not modally definable, these proofs of completeness are not so obvious when the considered PMLs are paraconjunctive.

What challenges remain? Much of this paper revolves around one goal: the definition of a modal language interpreted over relational structures including different types of entities as well as different kinds of relationships between them. As far as we are aware, the modal languages realizing that goal are scarce, even if the first ones were proposed about 25 years ago within the context of spatial logics and arrow logics [16,26,34]. Therefore, much remains to be done. For instance,

- to investigate the computability of the membership problem in PMLs (filtration method, tableaux-based approach, etc),
- to develop the model theory of PMLs (bisimulations, saturated models, etc),
- to construct the duality theory of PMLs (Boolean algebras with operators, general frames, etc),
- to elaborate the correspondence theory of PMLs (Chagrova's Theorem, Sahlqvist Correspondence Theorem, etc),
- to show how a multi-typed parametrized modal language can be used for solving the formalization problems facing those who have to take into account relationships such as the following ones: "trusts", "is a friend of", etc.

Other avenues of research might consist in considering that frames are 4-tuples of the form $(W_1, W_2, \tau_1, \tau_2)$ where for all $i \in \{1, 2\}$, W_i is a nonempty set and $\tau_i : \wp(W_{\underline{i}}) \longrightarrow \wp(\wp(W_i))$ is such that for all $i \in \{1, 2\}$ and for all $A \in \wp(W_{\underline{i}})$, $\tau_i(A)$ is a topology on W_i. In that case, a *valuation on a frame* $(W_1, W_2, \tau_1, \tau_2)$ will be a couple (V_1, V_2) of functions $V_1 : \mathcal{L}_1 \longrightarrow \wp(W_1)$ and $V_2 : \mathcal{L}_2 \longrightarrow \wp(W_2)$ such that for all $i \in \{1, 2\}$,

- $V_i([\varphi]\psi) = \{s_i \in W_i : \exists \mathcal{O}_i \in \tau_i(V_{\underline{i}}(\varphi)) \ (s_i \in \mathcal{O}_i \ \& \ \mathcal{O}_i \subseteq V_i(\psi))\}$,

among other conditions. As a result, $V_i(\langle\varphi\rangle\psi) = \{s_i \in W_i : \forall \mathcal{O}_i \in \tau_i(V_{\underline{i}}(\varphi)) \ (s_i \in \mathcal{O}_i \Rightarrow \mathcal{O}_i \cap V_i(\psi) \neq \emptyset)\}$. Further investigations are needed for obtaining the PML that will completely axiomatize the validities thus defined.

Acknowledgements

Special acknowledgement is heartily granted to our colleagues of the Toulouse Institute of Computer Science Research (Toulouse, France). We also make a point of strongly thanking the referees for their feedback.

References

[1] Ågotnes, T., Wáng, Y.: *Resolving distributed knowledge.* Artificial Intelligence **252** (2017) 1–21.

[2] Balbiani, P.: *Reasoning about vague concepts in the theory of property systems.* Logic & Analyse **47** (2004) 445–460.

[3] Balbiani, P., Fernández González, S.: *Indexed frames and hybrid logics.* In *Advances in Modal Logic.* College Publications (2020) 53–72.

[4] Ben-naim, J., Longin, D., Lorini, E.: *Formalization of cognitive-agent systems, trust, and emotions.* In *A Guided Tour of Artificial Intelligence Research.* Springer (2020) 629–650.

[5] Blackburn, P., de Rijke, M., Venema, Y.: *Modal Logic.* Cambridge University Press (2001).

[6] Chagrov, A., Zakharyaschev, M.: *Modal Logic.* Oxford University Press (1997).

[7] Demri, S., Orłowska, E.: *Logical analysis of indiscernibility.* In *Incomplete Information: Rough Set Analysis.* Springer (1998) 347–380.

[8] Van Ditmarsch, H., Kooi, B., van der Hoek, W.: *Dynamic Epistemic Logic.* Springer (2008).

[9] Fagin, R., Halpern, J., Moses, Y., Vardi, M.: *Reasoning about Knowledge.* MIT Press (1995).

[10] Fernández González, S.: *Logics for Social Networks. Asynchronous Announcements in Orthogonal Structures.* Doctoral Thesis of Toulouse University (2021).

[11] Gargov, G., Passy, S.: *A note on Boolean modal logic.* In *Mathematical Logic.* Plenum Press (1990) 299–309.

[12] Gargov, G., Passy, S., Tinchev, T.: *Modal environment for Boolean speculations.* In *Mathematical Logic and its Applications.* Plenum Press (1987) 253–263.

[13] Kracht, M.: *Tools and Techniques in Modal Logic.* Elsevier (1999).

[14] Liu, F., Lorini, E.: *Reasoning about belief, evidence and trust in a multi-agent setting.* In *PRIMA 2017: Principles and Practice of Multi-Agent Systems.* Springer (2017) 71–89.

[15] Liu, F., Seligman, J., Girard, P.: *Logical dynamics of belief change in the community.* Synthese **191** (2014) 2403–2431.

[16] Marx, M.: *Dynamic arrow logic.* In *Arrow Logic and Multi-Modal Logic.* Center for the Study of Language and Information (1996) 109–123.

[17] Meyer, J.-J., van der Hoek, W.: *Epistemic Logic for AI and Computer Science.* Cambridge University Press (1995).

[18] Orłowska, E.: *Logic of nondeterministic information.* Studia Logica **44** (1985) 91–100.

[19] Orłowska, E.: *Kripke models with relative accessibility and their applications to inferences from incomplete information.* In *Mathematical Problems in Computation Theory.* PWN – Polish Scientific, Banach Center (1988) 329–339.

[20] Orłowska, E.: *Kripke semantics for knowledge representation logics.* Studia Logica **49** (1990) 255–272.

[21] Orłowska, E., Pawlak, Z.: *Representation of nondeterministic information.* Theoretical Computer Science **29** (1984) 27–39.

[22] Passy, S., Tinchev, T.: *An essay in combinatory dynamic logic.* Information and Computation **93** (1991) 263–332.

[23] Pawlak, Z.: *Information systems theoretical foundations.* Information Systems **6** (1981) 205–218.

[24] Pedersen, M., Smets, S., Ågotnes, T.: *Modal logics and group polarization.* Journal of Logic and Computation **31** (2021) 2240–2269.

[25] Perrotin, E., Galimullin, R., Canu, Q., Alechina, N.: *Public group announcements and trust in doxastic logic.* In *Logic, Rationality, and Interaction.* Springer (2019) 199-213.

[26] de Rijke, M.: *The logic of Peirce algebras.* Journal of Logic, Language and Information **4** (1995) 227–250.

[27] Sano, K.: *Axiomatizing epistemic logic of friendship via tree sequent calculus.* In *Logic, Rationality, and Interaction.* Springer (2017) 224–239.

[28] Seligman, J., Liu, F., Girard, P.: *Logic in the community.* In *Logic and its Applications.* Springer (2011) 178–188.

[29] Vakarelov, D.: *A modal logic for similarity relations in Pawlak knowledge representation systems.* Fundamenta Informaticæ **15** (1991) 61–79.

[30] Vakarelov, D.: *Modal logics for knowledge representation systems.* Theoretical Computer Science **90** (1991) 433–456.

[31] Vakarelov, D.: *A duality between Pawlak's knowledge representation systems and BI-consequence systems.* Studia Logica **55** (1995) 205–228.

[32] Vakarelov, D.: *Information systems, similarity relations and modal logics.* In *Incomplete Information: Rough Set Analysis.* Springer (1998) 492–550.

[33] Vakarelov, D.: *A modal characterization of indiscernibility and similarity relations in Pawlak's information systems.* In *Rough Sets, Fuzzy Sets, Data Mining, and Granular Computing.* Springer (2005) 12–22.

[34] Venema, Y.: *Points, lines and diamonds: a two-sorted modal logic for projective planes.* Journal of Logic and Computation **9** (1999) 601–621.

Appendix

Alternative proof of Proposition 6.11. The proof of Proposition 6.11 included in the body of the paper may seem unnecessarily complicated. Its good point is that it can be easily converted into a proof of Proposition 6.19. Now, we present the sketch of a simpler proof of Proposition 6.11 that, unfortunately, does not seem to be easily convertible into a proof of Proposition 6.19. For all $i \in \{1,2\}$, let Λ_i be the set of all $\tau_i : \wp(W_{\underline{i}}) \times W_{\underline{i}} \longrightarrow \{0,1\}$. Let (W'_1, W'_2, R'_1, R'_2) be the 4-tuple where for all $i \in \{1,2\}$,

- $W'_i = W_i \times \Lambda_i$,
- $R'_i : \wp(W'_{\underline{i}}) \longrightarrow \wp(W'_i \times W'_i)$ is such that for all $A' \in \wp(W'_{\underline{i}})$ and for all $(t_i, \tau_i), (u_i, \mu_i) \in W'_i$, $(t_i, \tau_i)R'_i(A')(u_i, \mu_i)$ if and only if for all $A \in \wp(W_{\underline{i}})$,
  - if $A' \cap (A \times \Lambda_{\underline{i}}) \neq \emptyset$ then $t_i R_i(A) u_i$ if and only if for all $v_{\underline{i}} \in A$, $\tau_i(A, v_{\underline{i}}) = \mu_i(A, v_{\underline{i}})$,
  - for all $(v_{\underline{i}}, \nu_{\underline{i}}) \in A' \cap (A \times \Lambda_{\underline{i}})$, $\tau_i(A, v_{\underline{i}}) = \mu_i(A, v_{\underline{i}})$.

The reader may easily verify that

- for all $i \in \{1,2\}$ and for all $A' \in \wp(W'_{\underline{i}})$, $R'_i(A') = \bigcap\{R'_i(\{(v_{\underline{i}}, \nu_{\underline{i}})\}) : (v_{\underline{i}}, \nu_{\underline{i}}) \in A'\}$,
- the 4-tuple (W'_1, W'_2, R'_1, R'_2) is a conjunctive frame.

Let $f_1 : W'_1 \longrightarrow W_1$ be such that for all $(v_1, \nu_1) \in W'_1$, $f_1(v_1, \nu_1) = v_1$ and $f_2 : W'_2 \longrightarrow W_2$ be such that for all $(v_2, \nu_2) \in W'_2$, $f_2(v_2, \nu_2) = v_2$. The reader may easily verify that f_1 and f_2 are surjective. Let (V_1, V_2) be a (W_1, W_2, R_1, R_2)-valuation. For all $i \in \{1,2\}$, let $V'_i : \mathcal{L}_i \longrightarrow \wp(W'_i)$ be such that for all $\varphi \in \mathcal{L}_i$, $V'_i(\varphi) = f_i^{-1}[V_i(\varphi)]$. Notice that for all $i \in \{1,2\}$ and for all $\varphi \in \mathcal{L}_i$, $V'_i(\varphi) = V_i(\varphi) \times \Lambda_i$. Finally, the reader may easily verify that

- for all $(t_i, \tau_i), (u_i, \mu_i) \in W'_i$ and for all $\underline{i}$-formulas φ, if $(t_i, \tau_i)R'_i(V'_{\underline{i}}(\varphi))(u_i, \mu_i)$ then $t_i R_i(V_{\underline{i}}(\varphi)) u_i$,
- for all $(t_i, \tau_i) \in W'_i$, for all $u_i \in W_i$ and for all $\underline{i}$-formulas φ, if $t_i R_i(V_{\underline{i}}(\varphi)) u_i$ then there exists $\mu_i \in \Lambda_i$ such that $(t_i, \tau_i)R'_i(V'_{\underline{i}}(\varphi))(u_i, \mu_i)$,
- (V'_1, V'_2) is a (W'_1, W'_2, R'_1, R'_2)-valuation,
- (f_1, f_2) is a bounded morphism from $(W'_1, W'_2, R'_1, R'_2, V'_1, V'_2)$ to $(W_1, W_2, R_1, R_2, V_1, V_2)$.

□

Projective unification through duality

Philippe Balbiani [1] Quentin Gougeon [2]

[1,2] *Institut de recherche en informatique de Toulouse*
CNRS-INPT-UT3, Toulouse University, Toulouse, France

Abstract

Unification problems can be formulated and investigated in an algebraic setting, by identifying substitutions to modal algebra homomorphisms. This opens the door to applications of the notorious duality between modal algebras and descriptive frames. Through substantial use of this correspondence, we give a necessary and sufficient condition for modal formulas to be projective. Applying this result to a number of different logics, we then obtain concise and lightweight proofs of their projective – or non-projective – character. In particular, we prove that the projective extensions of **K5** are exactly the extensions of **K45**. This resolves the open question of whether **K5** is projective.

Keywords: Normal modal logics. Elementary unification. Projective formulas. Duality theory.

1 Introduction

In a propositional language, substitutions can be defined as functions mapping variables to formulas. For reasons related to Unification Theory [2, Section 2], it is usually considered that such functions are almost everywhere equal to the identity function. As a result, one can see a substitution as a function $\sigma : \mathcal{L}_P \to \mathcal{L}_Q$ where $\mathcal{L}_P$ (resp. $\mathcal{L}_Q$) is the set of all formulas with variables in a finite set P (resp. Q), and satisfying ($\blacklozenge$) $\sigma(\circ(\varphi_1, \ldots, \varphi_n)) = \circ(\sigma(\varphi_1), \ldots, \sigma(\varphi_n))$ for all n-ary connectives $\circ$ of the language and all formulas $\varphi_1, \ldots, \varphi_n \in \mathcal{L}_P$. According to this point of view, which is the one usually considered within the context of modal logics [3,7,10], two substitutions $\sigma : \mathcal{L}_P \to \mathcal{L}_Q$ and $\tau : \mathcal{L}_P \to \mathcal{L}_{Q'}$ are said to be equivalent with respect to a propositional logic $\mathbf{L}$ (in symbols $\sigma \simeq_{\mathbf{L}} \tau$) if for all $p \in P$, the formulas $\sigma(p)$ and $\tau(p)$ are $\mathbf{L}$-equivalent.

A formula $\varphi \in \mathcal{L}_P$ is $\mathbf{L}$-unifiable if $\mathbf{L}$ contains instances of φ. In that case, any substitution $\sigma : \mathcal{L}_P \to \mathcal{L}_Q$ such that $\sigma(\varphi) \in \mathbf{L}$ counts as a $\mathbf{L}$-unifier of φ. A $\mathbf{L}$-unifiable formula $\varphi \in \mathcal{L}$ is $\mathbf{L}$-projective if it possesses a projective $\mathbf{L}$-unifier, that is to say a $\mathbf{L}$-unifier σ such that $\varphi \vdash_{\mathbf{L}} \sigma(p) \leftrightarrow p$ holds for all $p \in P$. Such unifiers are interesting because they constitute by themselves minimal complete

[1] Email address: `philippe.balbiani@irit.fr`.

[2] Email address: `quentin.gougeon@irit.fr`.

sets of unifiers [3,7,10]. For this reason, it is of the utmost importance to be able to determine if a given formula is **L**-projective.

Now, condition (♦) may evoke homomorphism properties. Following this observation, Unification Theory was also formalized and studied in an algebraic setting [9,20]. Indeed, let us consider the Lindenbaum-Tarski algebra [3] $\mathbf{A}_P$ obtained by taking the quotient of $\mathcal{L}_P$ modulo the relation $\equiv_{\mathbf{L}}$ of **L**-equivalence. One can associate to a substitution $\sigma : \mathcal{L}_P \to \mathcal{L}_Q$ the map $\sigma^{\triangleright} : \mathbf{A}_P \to \mathbf{A}_Q$ by setting $\sigma^{\triangleright}([\varphi]_{\mathbf{L}}) := [\sigma(\varphi)]_{\mathbf{L}}$ for any formula $\varphi \in \mathcal{L}_P$, whose equivalence class modulo $\equiv_{\mathbf{L}}$ is denoted by $[\varphi]_{\mathbf{L}}$. In this perspective, condition (♦) then truly expresses the homomorphic character of $\sigma^{\triangleright}$. Obviously, this association between substitutions and homomorphisms of Lindenbaum algebras is one-to-one modulo $\simeq_{\mathbf{L}}$: substitutions associated to the same homomorphism are equivalent modulo $\simeq_{\mathbf{L}}$. Then various properties of substitutions, such as being a **L**-unifier of a formula, admit an algebraic counterpart too.

In this paper, we combine this correspondence with a more traditional one, provided by *Duality Theory*. For any set P of variables, there is indeed a tight connection between the Lindenbaum algebra $\mathbf{A}_P$ and the canonical frame $\mathfrak{F}_P$ of **L** over P, determined by the set of all ultrafilters [4] on $\mathbf{A}_P$. Homomorphisms between Lindenbaum algebras are then in correspondence with bounded morphisms between canonical frames. See [4, Chapter 5], [5, Chapter 7] and [16, Chapter 4] for a general introduction to this subject. Given a finite set P of variables, we make essential use of this duality to construct a necessary and sufficient condition for $\varphi \in \mathcal{L}_P$ to be **L**-projective: the existence of a bounded morphism $f : \mathfrak{F}_P \to \mathfrak{F}_P$ such that the image of f is contained in $\widehat{\varphi}^{\infty}$ and all elements of $\widehat{\varphi}^{\infty}$ are fixpoints of f, where $\widehat{\varphi}^{\infty}$ denotes the set of all points in $\mathfrak{F}_P$ containing $[\Box^n\varphi]_{\mathbf{L}}$ for all $n \in \mathbb{N}$.

This paper is structured as follows. In Section 2, we introduce some basics of modal logic [5] and Unification Theory [6]. Section 3 introduces modal algebras and in particular Lindenbaum algebras, and explains how to 'algebraize' unification problems. In Section 4, we develop some basics of Duality Theory in modal logics, concentrating on the bijective correspondence between bounded morphisms of canonical frames and homomorphisms of Lindenbaum algebras. We then apply these tools to establish the above-mentioned necessary and sufficient condition for a formula to be projective. In Section 5, we use this characterization to investigate the projective character of the extensions

[3] Or Lindenbaum algebra for short.

[4] The points of $\mathfrak{F}_P$ are usually defined as maximal **L**-consistent sets of formulas instead of ultrafilters, but as explained in Section 5, this makes no difference.

[5] We follow the same conventions as in [4,5,16] for talking about modal logics: **KT** is the least modal logic containing the formula usually denoted (**T**), **S4** is the least modal logic containing the formulas usually denoted (**T**) and (**4**), etc.

[6] We usually distinguish between elementary unification and unification with parameters. In elementary unification, all variables are likely to be replaced by formulas when one applies a substitution [1]. In unification with parameters, some variables — called parameters — remain unchanged [19, Chapter 6]. In this paper, we only interest in elementary unification.

of three selected logics: $\mathbf{K4}_n\mathbf{B}_k$, $\mathbf{K4D1}$, and $\mathbf{K5}$.

2 Background

2.1 Some functional vocabulary

Let $f : X \to Y$ be a function. If $A \subseteq X$, we write $f[A] := \{f(x) \mid x \in A\}$. If $B \subseteq Y$, we write $f^{-1}[B] := \{x \in X \mid f(x) \in B\}$. We denote by $\mathrm{Im}\, f := f[X]$ the *image* of f. We denote by $\mathrm{fp}\, f := \{x \in X \mid f(x) = x\}$ the set of *fixpoints* of f. Given two functions $f : X \to Y$ and $g : Y \to Z$ we denote by $gf : X \to Z$ the *composition* of f and g, defined by $gf : x \mapsto g(f(x))$.

2.2 Modal logics

Let Prop be an infinite countable set of propositional variables. If $P \subseteq \mathsf{Prop}$ we define the modal language $\mathcal{L}_P$ over P by the following grammar:

$$\varphi ::= p \mid \bot \mid \neg\varphi \mid (\varphi \wedge \varphi) \mid \Box\varphi$$

where $p \in P$. We write $\mathcal{L} := \mathcal{L}_{\mathsf{Prop}}$. The abbreviations $\top, \vee, \to, \leftrightarrow, \Diamond$ are defined as usual. Given $\varphi \in \mathcal{L}$ we denote by $\mathsf{var}(\varphi)$ the set of variables occurring in φ. If $n \in \mathbb{N}$ we define inductively $\Box^n\varphi$ and $\Box^{\leq n}\varphi$ by:

- $\Box^0\varphi := \varphi$ and $\Box^{\leq 0}\varphi := \varphi$,
- for all $n \in \mathbb{N}$, $\Box^{n+1}\varphi := \Box\Box^n\varphi$ and $\Box^{\leq n+1}\varphi := \Box\Box^n\varphi \wedge \Box^{\leq n}\varphi$.

We then define $\Diamond^n\varphi := \neg\Box^n\neg\varphi$ and $\Diamond^{\leq n}\varphi := \neg\Box^{\leq n}\neg\varphi$.

Definition 2.1 A *normal modal logic* is a set $\mathbf{L}$ of formulas such that:

- $\mathbf{L}$ is closed under *uniform substitution* (for all formulas φ, ψ, if $\varphi \in \mathbf{L}$ and ψ is obtained from φ by uniformly replacing variables in φ by arbitrary formulas then $\psi \in \mathbf{L}$),
- $\mathbf{L}$ contains all propositional tautologies,
- $\mathbf{L}$ is closed under *modus ponens* (for all formulas φ, ψ, if $\varphi \in \mathbf{L}$ and $\varphi \to \psi \in \mathbf{L}$ then $\psi \in \mathbf{L}$),
- $\mathbf{L}$ contains all formulas of the form $\Box(p \to q) \to (\Box p \to \Box q)$,
- $\mathbf{L}$ is closed under *generalization* (for all formulas φ, if $\varphi \in \mathbf{L}$ then $\Box\varphi \in \mathbf{L}$).

From now on we fix a normal modal logic $\mathbf{L}$. Instead of $\varphi \in \mathbf{L}$ we may also write $\vdash_{\mathbf{L}} \varphi$. Given $\varphi, \psi \in \mathcal{L}$, we write $\varphi \equiv_{\mathbf{L}} \psi$ in case $\vdash_{\mathbf{L}} \varphi \leftrightarrow \psi$. Then $\equiv_{\mathbf{L}}$ is an equivalence relation, and we denote by $[\varphi]_{\mathbf{L}}$ the equivalence class of φ modulo $\equiv_{\mathbf{L}}$. We call $\mathbf{L}$ *locally tabular* if for all finite sets $P \subseteq \mathsf{Prop}$, there are only finitely many equivalence classes modulo $\equiv_{\mathbf{L}}$ [7]. A set Σ of formulas is $\mathbf{L}$-*consistent* if there are no formulas $\varphi_1, \ldots, \varphi_n \in \Sigma$ such that $\vdash_{\mathbf{L}} \neg(\varphi_1 \wedge \ldots \wedge \varphi_n)$.

Given $\varphi, \psi \in \mathcal{L}$, we call a *derivation of ψ from φ in* $\mathbf{L}$ a sequence of formulas $\chi_0, \ldots, \chi_n \in \mathcal{L}$ such that $\chi_n = \psi$ and for all $i \in \{0, \ldots, n\}$, at least one of the following conditions holds:

[7] Locally tabular modal logics possess interesting properties, in particular when it comes to decidability [17].

- $\chi_i \in \mathbf{L}$,
- $\chi_i = \varphi$,
- there exists $j, k \in \{0, \dots, n\}$ such that $i > j, k$ and χ_i is obtained from χ_j and χ_k by modus ponens,
- there exists $j \in \{0, \dots, n\}$ such that $i > j$ and χ_i is obtained from χ_j by generalization.

If there exists a derivation of ψ from φ in $\mathbf{L}$, we shall say that ψ is *deducible from* φ *in* $\mathbf{L}$, and write $\varphi \vdash_{\mathbf{L}} \psi$. A more concise characterization of derivable formulas is given by the following result:

Proposition 2.2 *The following conditions are equivalent:*[8]

(i) $\varphi \vdash_{\mathbf{L}} \psi$,

(ii) *there exists* $k \in \mathbb{N}$ *such that* $\vdash_{\mathbf{L}} \Box^{\leq k}\varphi \to \psi$.

We call an *extension* of $\mathbf{L}$ any normal modal logic $\mathbf{L}'$ such that $\mathbf{L} \subseteq \mathbf{L}'$. If $\mathbf{L}$ is a normal modal logic and $\Sigma \subseteq \mathcal{L}$, we denote by $\mathbf{L} + \Sigma$ the smallest extension of $\mathbf{L}$ containing Σ.

2.3 Unification

A *substitution* is a function $\sigma : \mathcal{L}_P \to \mathcal{L}_Q$ with $P, Q \subseteq \mathsf{Prop}$ finite and such that for all $\varphi, \psi \in \mathcal{L}_P$ we have $\sigma(\bot) = \bot$, $\sigma(\neg\varphi) = \neg\sigma(\varphi)$, $\sigma(\varphi \wedge \psi) = \sigma(\varphi) \wedge \sigma(\psi)$, and $\sigma(\Box\varphi) = \Box\sigma(\varphi)$. Let $\mathcal{S}$ be the set of all substitutions. The equivalence relation $\simeq_{\mathbf{L}}$ on $\mathcal{S}$ is defined by

$$\sigma \simeq_{\mathbf{L}} \tau \text{ if and only if } \sigma(p) \equiv_{\mathbf{L}} \tau(p) \text{ for all } p \in P$$

where $\sigma, \tau : \mathcal{L}_P \to \mathcal{L}_Q$. Then, we define the preorder $\preccurlyeq_{\mathbf{L}}$ on $\mathcal{S}$ by

$$\sigma \preccurlyeq_{\mathbf{L}} \tau \text{ iff there exists a substitution } \mu : \mathcal{L}_Q \to \mathcal{L}_R \text{ such that } \mu\sigma \simeq_{\mathbf{L}} \tau$$

where $\sigma : \mathcal{L}_P \to \mathcal{L}_Q$ and $\tau : \mathcal{L}_P \to \mathcal{L}_R$. Given $\varphi \in \mathcal{L}_P$ we say that $\sigma : \mathcal{L}_P \to \mathcal{L}_Q$ is a $\mathbf{L}$-*unifier* of φ if we have $\vdash_{\mathbf{L}} \sigma(\varphi)$. A $\mathbf{L}$-*unifier* $\sigma : \mathcal{L}_P \to \mathcal{L}_Q$ of φ is called *concise* if $P = \mathsf{var}(\varphi)$. The formula φ is $\mathbf{L}$-*unifiable* if there exists a $\mathbf{L}$-unifier of φ. A set $\mathcal{T}$ of concise $\mathbf{L}$-unifiers of φ is said to be *complete* if for all concise $\mathbf{L}$-unifiers σ of φ, there exists $\tau \in \mathcal{T}$ such that $\tau \preccurlyeq_{\mathbf{L}} \sigma$. In addition, we call $\mathcal{T}$ a *basis* if $\sigma, \tau \in \mathcal{T}$ and $\sigma \preccurlyeq_{\mathbf{L}} \tau$ implies $\sigma = \tau$.

Proposition 2.3 *For all* $\varphi \in \mathcal{L}$, *if* φ *is* $\mathbf{L}$-*unifiable then for all bases* $\mathcal{T}, \mathcal{U}$ *of concise* $\mathbf{L}$-*unifiers of* φ, *we have* $\mathtt{Card}(\mathcal{T}) = \mathtt{Card}(\mathcal{U})$ [9, Section 2].

A central problem in unification theory is whether a $\mathbf{L}$-unifiable formula possesses a basis of concise $\mathbf{L}$-unifiers. When this is the case, Proposition 2.3 raises a more refined question, that is, how large is such a basis? This gives rise to a full classification of logics based on the cardinality of these bases.

---

[8] This is a simplified version of the so-called Deduction Theorem in modal logics. When $\mathbf{L} = \mathbf{K}$, see [5, Theorem 3.51] for a proof of it, which can be easily adapted to the general case. See [11] for an interesting discussion of this theorem.

Definition 2.4 For all **L**-unifiable $\varphi \in \mathcal{L}$:

- φ is **L**-*nullary* if there exists no basis of concise **L**-unifiers of φ,
- φ is **L**-*infinitary* if there exists an infinite basis of concise **L**-unifiers of φ,
- φ is **L**-*finitary* if there exists a basis of concise **L**-unifiers of φ with finite cardinality ≥ 2,
- φ is **L**-*unitary* if there exists a basis of concise **L**-unifiers of φ with cardinality 1.

We shall say that:

- **L** is *nullary* if there exists a **L**-nullary **L**-unifiable formula,
- **L** is *infinitary* if every **L**-unifiable formula is either **L**-infinitary, or **L**-finitary, or **L**-unitary and there exists a **L**-infinitary **L**-unifiable formula,
- **L** is *finitary* if every **L**-unifiable formula is either **L**-finitary, or **L**-unitary and there exists a **L**-finitary **L**-unifiable formula,
- **L** is *unitary* if every **L**-unifiable formula is **L**-unitary.

A special case that deserves our attention is that of projective unifiers. Let $\varphi \in \mathcal{L}_P$ with $P \subseteq \mathsf{Prop}$ finite. We call a *φ-projective substitution* a substitution $\sigma : \mathcal{L}_P \to \mathcal{L}_P$ such that for all $p \in P$ we have $\varphi \vdash_{\mathbf{L}} \sigma(p) \leftrightarrow p$. We then say that φ is **L**-*projective* if there exists a φ-projective **L**-unifier of φ. Finally, we shall say that **L** is projective if every **L**-unifiable formula is **L**-projective. For an introduction to projective unification, see [3,7,10]. The logic **K** has been shown to be nullary by Jeřábek [13] who has proved that the unifiable formula $p \to \Box p$ has no basis of unifiers in **K**. The logic **S5**, however, is known to be unitary since many years [1,6]. The truth is that if $\sigma : \mathcal{L}_P \to \mathcal{L}_P$ is a **S5**-unifier of a formula φ then the substitution $\epsilon : \mathcal{L}_P \to \mathcal{L}_P$ defined by $\epsilon(p) = (\Box\varphi \wedge p) \vee (\Diamond\neg\varphi \wedge \sigma(p))$ is a projective unifier of φ in **S5**. Projective unifiers have been used by Ghilardi [10] within the context of a transitive modal logic **L** like **K4**, **S4**, etc. In fact, Ghilardi has shown that if a formula φ possesses a unifier then it possesses a finite basis of unifers, this basis being the set of projective unifiers of a finite set of projective formulas of modal degree at most equal to the modal degree of φ, having the same propositional variables as φ and implying φ in **L**. See also [12] for a syntactic approach to unification in transitive modal logics. Recently, Dzik and Wojtylak [8] have proved that the projective extensions of **S4** are exactly the extensions of **S4.3**, a result improved by Kost [14] who has demonstrated that the projective extensions of **K4** are exactly the extensions of **K4D1**. Finally, Kostrzycka [15] has considered the modal logics of the form $\mathbf{K4}_n\mathbf{B}_k$ (see Section 5) and has proved that they are projective. For the proofs of Propositions 2.5–2.8 below, see [10, Section 2].

Proposition 2.5 *Let $\varphi \in \mathcal{L}_P$, and $\sigma : \mathcal{L}_P \to \mathcal{L}_P$ be a substitution. Then σ is φ-projective if and only if for all formulas $\psi \in \mathcal{L}_P$, we have $\varphi \vdash_{\mathbf{L}} \sigma(\psi) \leftrightarrow \psi$.*

Proposition 2.6 *For all formulas $\varphi \in \mathcal{L}_P$, for all φ-projective substitutions $\sigma : \mathcal{L}_P \to \mathcal{L}_P$ and for all **L**-unifiers $\tau : \mathcal{L}_P \to \mathcal{L}_Q$ of φ, we have $\sigma \preccurlyeq_{\mathbf{L}} \tau$.*

Proposition 2.7 *For all* **L**-*unifiable* $\varphi \in \mathcal{L}$, *if* φ *is* **L**-*projective then* φ *is* **L**-*unitary.*

Proposition 2.8 *If* **L** *is projective then* **L** *is unitary.*

As a side note, the extension of a projective logic is, up to our knowledge, not necessarily projective.

Remark 2.9 If $\varphi \in \mathcal{L}$, there are obviously infinitely many subsets P of Prop such that $\varphi \in \mathcal{L}_P$. For this reason, many authors require a **L**-unifier of φ to be of the form $\sigma : \mathcal{L}_{\mathsf{var}(\varphi)} \to \mathcal{L}_Q$ (or 'concise' in our terminology). In our setting, we also allow one to talk about the unifying or φ-projective character of any substitution of the form $\sigma : \mathcal{L}_P \to \mathcal{L}_Q$ with $\mathsf{var}(\varphi) \subseteq P$. This offers more flexibility, which will be helpful in Section 5.

As a result, our definition of **L**-unifiable and **L**-projective formulas are non-standard[9], but this is harmless. Indeed, if $\mathsf{var}(\varphi) \subseteq P \subseteq P'$ and $\sigma : \mathcal{L}_P \to \mathcal{L}_Q$ is a substitution, one can define the substitution $\sigma' : \mathcal{L}_{P'} \to \mathcal{L}_Q$ by setting $\sigma'(p) := \sigma(p)$ for all $p \in P$ and $\sigma'(p) := p$ for all $p \in P' \setminus P$, and it is clear that σ is a **L**-unifier of φ (resp. is φ-projective) if and only if σ' is a **L**-unifier of φ (resp. is φ-projective). Likewise, if $\mathsf{var}(\varphi) \subseteq P \subseteq P'$ and $\sigma : \mathcal{L}_{P'} \to \mathcal{L}_Q$ is a substitution, let us denote by $\sigma' := \sigma_{|\mathcal{L}_P}$ the restriction of σ to $\mathcal{L}_P$. Then σ is a **L**-unifier of φ (resp. is φ-projective) if and only if σ' is a **L**-unifier of φ (resp. is φ-projective).

2.4 General frames

A *general frame* is a pair $\mathfrak{F} = (X, R, \mathcal{A})$ with X a set of possible worlds, R a binary relation on X, and $\mathcal{A} \subseteq \mathcal{P}(X)$ such that:

- $\emptyset \in \mathcal{A}$,
- $A \in \mathcal{A}$ implies $X \setminus A \in \mathcal{A}$,
- $A, B \in \mathcal{A}$ implies $A \cap B \in \mathcal{A}$,
- $A \in \mathcal{A}$ implies $\Box_R A \in \mathcal{A}$, with $\Box_R A := \{x \in X \mid \forall y \in X, xRy \Rightarrow y \in A\}$.

Further, we call $\mathfrak{F}$ *differentiated* if for all $x, y \in X$ such that $x \neq y$, there exists $A \in \mathcal{A}$ such that $x \in \mathcal{A}$ and $y \notin \mathcal{A}$. We call $\mathfrak{F}$ *tight* if for all $x, y \in X$ such that *not* xRy, there exists $A \in \mathcal{A}$ such that $x \in \Box_R A$ and $y \notin A$. We call $\mathfrak{F}$ *compact* if for all $\mathcal{B} \subseteq \mathcal{A}$, if $\bigcap \mathcal{B}' \neq \emptyset$ whenever $\mathcal{B}' \subseteq \mathcal{B}$ is finite, then $\bigcap \mathcal{B} \neq \emptyset$. Finally, we say that $\mathfrak{F}$ is a *descriptive frame* if $\mathfrak{F}$ is tight, compact and differentiated.

If $\mathfrak{F} = (X, R, \mathcal{A})$ and $\mathfrak{F}' = (X', R', \mathcal{A}')$ are two general frames, we call a *bounded morphism* from $\mathfrak{F} = (X, R, \mathcal{A})$ to $\mathfrak{F}' = (X', R', \mathcal{A}')$ a map $f : X \to X'$ such that:

- if xRy then $f(x)Rf(y)$,
- if $f(x)R'y'$ then there exists $y \in X$ such that $f(y) = y'$ and xRy,
- if $A' \in \mathcal{A}'$ then $f^{-1}[A'] \in \mathcal{A}$.

---

[9] Note that the unification type *does* remain standard, since it is defined with respect to concise unifiers only.

3 An algebraic perspective

The algebraic aspects of unification have already been investigated in e.g. [9,20]. Here we give a lightweight, self-sufficient account of them. The goal of this section is essentially to state Proposition 3.5, a modest but inspiring starting point.

Definition 3.1 A *modal algebra* is a structure $\mathbf{A} = (A, 0, \neg, \wedge, \Box)$ with $(A, 0, \neg, \wedge)$ a Boolean algebra and $\Box : A \to A$ an operator satisfying $\Box 1 = 1$ and $\Box(a \wedge b) = \Box a \wedge \Box b$ for all $a, b \in \mathbf{A}$. For convenience we will identify $\mathbf{A}$ to its underlying set A.

Definition 3.2 A *homomorphism* from a modal algebra $\mathbf{A}$ to a modal algebra $\mathbf{B}$ is a map $\alpha : \mathbf{A} \to \mathbf{B}$ such that for all $a, b \in \mathbf{A}$ we have $\alpha(0) = 0$, $\alpha(\neg a) = \neg\alpha(a)$, $\alpha(a \wedge b) = \alpha(a) \wedge \alpha(b)$, and $\alpha(\Box a) = \Box\alpha(a)$. We denote by $\text{Ker } \alpha := \{(a, b) \in \mathbf{A}^2 \mid \alpha(a) = \alpha(b)\}$ the *kernel* of α.

Definition 3.3 Let $\mathbf{A}$ be a modal algebra. An equivalence relation $\sim$ on $\mathbf{A}$ is called a *congruence* on $\mathbf{A}$ if for all $a, a', b, b' \in \mathbf{A}$:

- $a \sim a'$ implies $\neg a \sim \neg a'$,
- $a \sim a'$ and $b \sim b'$ implies $(a \wedge b) \sim (a' \wedge b')$,
- $a \sim a'$ implies $\Box a \sim \Box a'$.

Then $\sim$ induces a *quotient algebra* $\mathbf{A}_{/\sim}$ over the set of all equivalence classes of $\sim$. If for all $a \in \mathbf{A}$ we denote by $\pi(a)$ the equivalence class of a modulo $\sim$, we obtain a surjective homomorphism $\pi : \mathbf{A} \to \mathbf{A}_{/\sim}$ [16, Section 1.2].

If $P \subseteq \mathsf{Prop}$, a particularly interesting modal algebra is the *Lindenbaum algebra* of $\mathbf{L}$ over P, defined as $\mathbf{A}_P := (\mathcal{L}_{P/\equiv_{\mathbf{L}}}, 0, \neg, \wedge, \Box)$ with

- $0 := [\bot]_{\mathbf{L}}$,
- $\neg [\varphi]_{\mathbf{L}} := [\neg\varphi]_{\mathbf{L}}$,
- $[\varphi]_{\mathbf{L}} \wedge [\psi]_{\mathbf{L}} := [\varphi \wedge \psi]_{\mathbf{L}}$,
- $[\Box\varphi]_{\mathbf{L}} := \Box [\varphi]_{\mathbf{L}}$.

Notice that if $\mathbf{L}$ is locally tabular and P is finite, then $\mathbf{A}_P$ is finite. A substitution $\sigma : \mathcal{L}_P \to \mathcal{L}_Q$ then naturally induces a homomorphism $\sigma^{\triangleright} : \mathbf{A}_P \to \mathbf{A}_Q$ defined by $\sigma^{\triangleright}([\varphi]_{\mathbf{L}}) := [\sigma(\varphi)]_{\mathbf{L}}$. Conversely, one can recover σ from $\sigma^{\triangleright}$ (up to equivalence modulo $\simeq_{\mathbf{L}}$) since we have $\sigma(\varphi) \equiv_{\mathbf{L}} \psi$ for any $\psi \in \sigma^{\triangleright}([\varphi]_{\mathbf{L}})$. There is thus a one-to-one correspondence between homomorphisms and substitutions (up to equivalence modulo $\simeq_{\mathbf{L}}$). For convenience, we will then identify the two: the symbol σ will indifferently denote the substitution σ and the homomorphism $\sigma^{\triangleright}$.

Properties of substitutions can also be expressed algebraically. Given $\varphi \in \mathcal{L}_P$, let $\equiv_\varphi$ be the least congruence on $\mathbf{A}_P$ such that $[\varphi]_{\mathbf{L}} \equiv_\varphi 1$. We then denote by $\pi_\varphi : \mathbf{A}_P \to \mathbf{A}_{P/\equiv_\varphi}$ the homomorphism associated to $\equiv_\varphi$ (as introduced in definition 3.3). Obviously the kernel of π_φ is $\equiv_\varphi$ itself.

Proposition 3.4 *Given $\psi, \theta \in \mathcal{L}_P$, the following are equivalent:*

(i) $\varphi \vdash_{\mathbf{L}} \psi \leftrightarrow \theta$,

(ii) $[\psi]_{\mathbf{L}} \equiv_{\varphi} [\theta]_{\mathbf{L}}$.

Proof. **(i)** $\Rightarrow$ **(ii)**: Suppose that $\varphi \vdash_{\mathbf{L}} \psi \leftrightarrow \theta$. Then by Proposition 2.2 there exists $n \in \mathbb{N}$ such that $\vdash_{\mathbf{L}} \Box^{\leq n}\varphi \to (\psi \leftrightarrow \theta)$. Thus $[\Box^{\leq n}\varphi]_{\mathbf{L}} \leq [\psi \leftrightarrow \theta]_{\mathbf{L}}$, and it follows that $\pi_{\varphi}([\Box^{\leq n}\varphi]_{\mathbf{L}}) \leq \pi_{\varphi}([\psi \leftrightarrow \theta]_{\mathbf{L}})$. It is easily proved by induction on n that $\pi_{\varphi}([\Box^{\leq n}\varphi]_{\mathbf{L}}) = 1$, and therefore $\pi_{\varphi}([\psi \leftrightarrow \theta]_{\mathbf{L}}) = 1$ too. Hence $\pi_{\varphi}([\psi]_{\mathbf{L}}) = \pi_{\varphi}([\theta]_{\mathbf{L}})$, or equivalently $[\psi]_{\mathbf{L}} \equiv_{\varphi} [\theta]_{\mathbf{L}}$.
(ii) $\Rightarrow$ **(i)**: Let us write $[\psi]_{\mathbf{L}} \sim [\theta]_{\mathbf{L}}$ whenever $\varphi \vdash_{\mathbf{L}} \psi \leftrightarrow \theta$. It is easily verified that $\sim$ is a congruence on $\mathbf{A}_P$, and that $[\varphi]_{\mathbf{L}} \sim 1$. By construction, $\equiv_{\varphi}$ is then included in $\sim$, and this proves the claim. □

We may now connect the projective or unifying character of a substitution to its algebraic properties.

Proposition 3.5 *Let $P \subseteq$ Prop finite and $\varphi \in \mathcal{L}_P$. Then:*

(i) *A substitution $\sigma : \mathcal{L}_P \to \mathcal{L}_Q$ is a $\mathbf{L}$-unifier of φ iff* $\mathrm{Ker}\ \pi_{\varphi} \subseteq \mathrm{Ker}\ \sigma$.

(ii) *A substitution $\sigma : \mathcal{L}_P \to \mathcal{L}_P$ is φ-projective iff $\pi_{\varphi}\sigma = \pi_{\varphi}$.*

Proof.

(i) Suppose that σ is a $\mathbf{L}$-unifier of φ. Then $\vdash_{\mathbf{L}} \sigma(\varphi)$, or equivalently $\sigma([\varphi]_{\mathbf{L}}) = 1$. Thus $\mathrm{Ker}\ \sigma$ is a congruence on $\mathbf{A}_P$ containing $([\varphi]_{\mathbf{L}}, 1)$, so by construction it contains $\equiv_{\varphi}$. Conversely, if $\mathrm{Ker}\ \pi_{\varphi} \subseteq \mathrm{Ker}\ \sigma$ then in particular $([\varphi]_{\mathbf{L}}, 1) \in \mathrm{Ker}\ \sigma$ and therefore $\vdash_{\mathbf{L}} \sigma(\varphi)$.

(ii) We have

$$
\begin{array}{lll}
 & \sigma \text{ is } \varphi\text{-projective} & \\
\text{iff} & \forall \psi \in \mathcal{L}_P,\ \varphi \vdash_{\mathbf{L}} \sigma(\psi) \leftrightarrow \psi & \text{by Proposition 2.5} \\
\text{iff} & \forall \psi \in \mathcal{L}_P,\ [\sigma(\psi)]_{\mathbf{L}} \equiv_{\varphi} [\psi]_{\mathbf{L}} & \text{by Proposition 3.4} \\
\text{iff} & \forall \psi \in \mathcal{L}_P,\ \pi_{\varphi}\sigma([\psi]_{\mathbf{L}}) = \pi_{\varphi}([\psi]_{\mathbf{L}}) & \\
\text{iff} & \pi_{\varphi}\sigma = \pi_{\varphi} &
\end{array}
$$

□

4 Duality

Now it is time to let duality play its role. In this section we introduce some rudiments of Duality Theory, and apply them to our setting. For more details we refer to [4, Chapter 5], [5, Chapter 7] and [16, Chapter 4]. This investigation will ultimately lead to theorem 4.5. First, let $\mathbf{A}$ be a modal algebra. Given a set $F \subseteq \mathbf{A}$, we call F an *ultrafilter* on $\mathbf{A}$ if it satisfies the following conditions:

- $0 \notin F$,
- $a \in F$ and $a \leq b$ implies $b \in F$,
- $a, b \in F$ implies $a \wedge b \in F$,
- for all $a \in \mathbf{A}$, either $a \in F$ or $\neg a \in F$.

Then the *dual* of $\mathbf{A}$ is the general frame $\mathbf{A}^* := (X_{\mathbf{A}}, R_{\mathbf{A}}, \mathcal{A}_{\mathbf{A}})$ with:

- $X_{\mathbf{A}} := \{F \subseteq \mathbf{A} \mid F \text{ is an ultrafilter}\}$,

- $R_{\mathbf{A}} := \{(F, F') \in X^2 \mid \forall a \in \mathbf{A}, \Box a \in F \Rightarrow a \in F'\}$,
- $\mathcal{A}_{\mathbf{A}} := \{\{F \in X \mid a \in F\} \mid a \in \mathbf{A}\}$.

The frame $\mathbf{A}^*$ is, in fact, a descriptive frame (see Section 2.4). Further, if $\alpha : \mathbf{A} \to \mathbf{B}$ is a homomorphism, we define a bounded morphism $\alpha^* : \mathbf{B}^* \to \mathbf{A}^*$ by

$$\alpha^*(F) := \alpha^{-1}[F].$$

If α and β are two appropriate homomorphisms, the identity

$$(\alpha\beta)^* = \beta^*\alpha^*$$

is easily verified. One can prove that for all descriptive frames $\mathfrak{F}$ there exists a unique modal algebra $\mathbf{A}$ (up to isomorphism) such that $\mathfrak{F}$ and $\mathbf{A}^*$ are isomorphic. Likewise, if $f : \mathbf{B}^* \to \mathbf{A}^*$ is a bounded morphism, there exists a unique homomorphism $\alpha : \mathbf{A} \to \mathbf{B}$ such that $\alpha^* = f$ [10]. In what follows we are going to make extensive use of this correspondence.

Naturally, the dual of the Lindenbaum algebra of $\mathbf{L}$ is of central interest to us. So if $\mathbf{A} = \mathbf{A}_P$ for some $P \subseteq \mathsf{Prop}$, we write $(X_{\mathbf{A}}, R_{\mathbf{A}}, \mathcal{A}_{\mathbf{A}}) = (X_P, R_P, \mathcal{A}_P)$ for simplicity. Given $\varphi \in \mathcal{L}_P$, we also write $\widehat{\varphi} := \{F \in X_P \mid [\varphi]_{\mathbf{L}} \in F\}$, and we then see that $\mathcal{A}_P = \{\widehat{\varphi} \mid \varphi \in \mathcal{L}_P\}$. Also note that if $\mathbf{L}$ is locally tabular and P is finite, then X_P is finite and $\mathcal{A}_P = \mathcal{P}(X_P)$.

Remark 4.1 The tight similarity between $\mathcal{L}_P$ and $\mathbf{A}_P$ materializes itself in a one-to-one correspondence between the *maximal consistent subsets* of $\mathcal{L}_P$ [4, Section 4.2] and the ultrafilters of $\mathbf{A}_P$, realized by the mapping

$$\Gamma \mapsto \{[\varphi]_{\mathbf{L}} \mid \varphi \in \Gamma\}$$

where $\Gamma \subseteq \mathcal{L}_P$ is maximal consistent. In fact, this correspondence induces an isomorphism between the frame $\mathfrak{F}_P := (X_P, R_P)$ and the *canonical frame* of $\mathbf{L}$ over P (as pointed out in [4, Section 5.3]).

Now assume that P is finite and let $\varphi \in \mathcal{L}_P$. In order to characterize the unifiable or projective character of φ, it is crucial to understand the behaviour of $\pi_\varphi^* : (\mathbf{A}_{P/\equiv_\varphi})^* \to \mathbf{A}_P^*$. In the algebraic setting, the relevant information was contained in the kernel of π_φ [11]. In the dual setting, this information turns out to be carried by the image of π_φ^*, which we proceed to describe. We prove that this image coincides with the set $\widehat{\varphi}^\infty := \bigcap_{n\in\mathbb{N}} \widehat{\Box^n\varphi}$.

Lemma 4.2 *Let $F \in X_P$. Then the following are equivalent:*

(i) *F is closed under $\equiv_\varphi$;*

(ii) *$F \in \mathrm{Im}\ \pi_\varphi^*$.*

---

[10] In categorical terms, we thus say that $(\cdot)^*$ is a *dual equivalence* between the category of modal algebras with homomorphisms and the category of descriptive frames with bounded morphisms.

[11] In Proposition 3.5, the kernel of π_φ is not mentioned in item (ii), but it still appears implicitly since $\pi_\varphi\sigma = \pi_\varphi$ can also be phrased as $\{(\sigma(a), a) \mid a \in \mathbf{A}_P\} \subseteq \mathrm{Ker}\ \pi_\varphi$.

Proof.
(i) $\Rightarrow$ **(ii)**: Suppose that F is closed under $\equiv_\varphi$. We introduce $G := \pi_\varphi[F]$ and prove that G is an ultrafilter on $\mathbf{A}_{P/\equiv_\varphi}$.

- Suppose that $0 \in G$. Then $0 = \pi_\varphi(a)$ for some $a \in F$. Hence $\pi_\varphi(0) = 0 = \pi_\varphi(a)$, which entails $0 \equiv_\varphi a$, and thus $0 \in F$ by assumption, contradicting the fact that F is an ultrafilter. Therefore $0 \notin G$.
- Suppose that $a' \in G$ and $a' \leq b'$. Then $a' = \pi_\varphi(a)$ for some $a \in F$. In addition, since π_φ is surjective, we have $b' = \pi_\varphi(b)$ for some $b \in \mathbf{A}_P$. From $a' \leq b'$ we obtain $a' = a' \wedge b'$, and thus $\pi_\varphi(a) = \pi_\varphi(a) \wedge \pi_\varphi(b) = \pi_\varphi(a \wedge b)$. Hence $a \equiv_\varphi a \wedge b$, and since $a \in F$ our assumption entails $a \wedge b \in F$. From $a \wedge b \leq b$ we obtain $b \in F$. Therefore $b' \in G$.
- Let $a', b' \in G$. We have $a' = \pi_\varphi(a)$ and $b' = \pi_\varphi(b)$ with $a, b \in F$. Then $a' \wedge b' = \pi_\varphi(a \wedge b)$ with $a \wedge b \in F$. Therefore $a' \wedge b' \in G$.
- Let $a' \in \mathbf{A}_{P/\equiv_\varphi}$. Then $a' = \pi_\varphi(a)$ for some $a \in \mathbf{A}_P$. Since F is an ultrafilter we have either $a \in F$ or $\neg a \in F$. Therefore, we have either $a' \in G$ or $\neg a' \in G$.

To prove that $F \in \mathrm{Im}\ \pi_\varphi^*$ we then show that $\pi_\varphi^*(G) = F$, that is, $\pi_\varphi^{-1}[\pi_\varphi[F]] = F$. The inclusion from right to left is trivial. From left to right, suppose that $a \in \pi_\varphi^{-1}[\pi_\varphi[F]]$. Then $\pi_\varphi(a) \in \pi_\varphi[F]$, that is, $\pi_\varphi(a) = \pi_\varphi(b)$ for some $b \in F$. Since F is closed under $\equiv_\varphi$ we obtain $a \in F$ and we are done.
(ii) $\Rightarrow$ **(i)**: Let $F \in \mathrm{Im}\ \pi_\varphi^*$. Then there exists an ultrafilter $G \in (\mathbf{A}_{P/\equiv_\varphi})^*$ such that $F = \pi_\varphi^*(G) = \pi_\varphi^{-1}[G]$. If $a \in F$ and $a \equiv_\varphi b$ then $\pi_\varphi(b) = \pi_\varphi(a) \in G$, and therefore $b \in F$. This proves that F is closed under $\equiv_\varphi$. □

Proposition 4.3 *We have* $\mathrm{Im}\ \pi_\varphi^* = \widehat{\varphi}^\infty$.

Proof. Let $F \in \mathrm{Im}\ \pi_\varphi^*$. Given $n \in \mathbb{N}$, we have $[\Box^n\varphi]_\mathbf{L} \equiv_\varphi \Box^n 1 \equiv_\varphi 1$ with $1 \in F$, so by Lemma 4.2 we obtain $[\Box^n\varphi]_\mathbf{L} \in F$ and thus $F \in \widehat{\Box^n\varphi}$.

Conversely, let $F \in \widehat{\varphi}^\infty$. By Lemma 4.2, it suffices to prove that F is closed under $\equiv_\varphi$. So assume $[\psi]_\mathbf{L} \in F$ and $[\psi]_\mathbf{L} \equiv_\varphi [\theta]_\mathbf{L}$. By Proposition 3.4 we obtain $\varphi \vdash_\mathbf{L} \psi \leftrightarrow \theta$, and then by Proposition 2.2 there exists $n \in \mathbb{N}$ such that $\vdash_\mathbf{L} \Box^{\leq n}\varphi \to (\psi \leftrightarrow \theta)$. Since $F \in \widehat{\Box^{\leq n}\varphi}$ we obtain $[\psi \leftrightarrow \theta]_\mathbf{L} \in F$, and since $[\psi]_\mathbf{L} \in F$ we conclude that $[\theta]_\mathbf{L} \in F$. □

With this result, we are then ready to transition from the algebraic setting to the dual setting.

Proposition 4.4 *Let* $P \subseteq \mathsf{Prop}$ *finite and* $\varphi \in \mathcal{L}_P$. *Then:*

(i) *for any homomorphism* $\sigma : \mathbf{A}_P \to \mathbf{A}_Q$ *we have* $\mathrm{Ker}\ \pi_\varphi \subseteq \mathrm{Ker}\ \sigma$ *iff* $\mathrm{Im}\ \sigma^* \subseteq \widehat{\varphi}^\infty$ *iff* $\mathrm{Im}\ \sigma^* \subseteq \widehat{\varphi}$;

(ii) *for any homomorphism* $\sigma : \mathbf{A}_P \to \mathbf{A}_P$ *we have* $\pi_\varphi\sigma = \pi_\varphi$ *iff* $\widehat{\varphi}^\infty \subseteq \mathrm{fp}\ \sigma^*$.

Proof. Given $\sigma : \mathbf{A}_P \to \mathbf{A}_Q$, recall that we have $\sigma^* : X_Q \to X_P$.

(i) Suppose that $\mathrm{Ker}\ \pi_\varphi \subseteq \mathrm{Ker}\ \sigma$. Let $F \in \mathrm{Im}\ \sigma^*$. Then $F = \sigma^{-1}[G]$ for some $G \in X_Q$. Let $n \in \mathbb{N}$. We have $\pi_\varphi([\Box^n\varphi]_\mathbf{L}) = 1 = \pi_\varphi(1)$, and thus

$\sigma([\Box^n\varphi]_{\mathbf{L}}) = \sigma(1)$ by assumption. Hence $\sigma([\Box^n\varphi]_{\mathbf{L}}) \in G$, and therefore $[\Box^n\varphi]_{\mathbf{L}} \in F$. This proves that $F \in \widehat{\varphi}^\infty$.

That Im $\sigma^* \subseteq \widehat{\varphi}^\infty$ implies Im $\sigma^* \subseteq \widehat{\varphi}$ is immediate. Now suppose that Im $\sigma^* \subseteq \widehat{\varphi}$. First, we prove that $\sigma([\varphi]_{\mathbf{L}}) = 1$. If not, then $\sigma([\neg\varphi]_{\mathbf{L}}) \neq 0$, so by the Ultrafilter Theorem [4, Proposition 5.38] there exists an ultrafilter $G \in X_Q$ such that $\sigma([\neg\varphi]_{\mathbf{L}}) \in G$. It follows that $[\neg\varphi]_{\mathbf{L}} \in \sigma^{-1}[G]$, whereas our assumption entails $\sigma^*(G) \in \widehat{\varphi}$ and thus $[\varphi]_{\mathbf{L}} \in \sigma^{-1}[G]$, a contradiction. Hence $\sigma([\varphi]_{\mathbf{L}}) = 1 = \sigma(1)$, which means that Ker σ is a congruence containing $([\varphi]_{\mathbf{L}}, 1)$. By construction, we then obtain Ker $\pi_\varphi \subseteq$ Ker σ.

(ii) Suppose that $\pi_\varphi\sigma = \pi_\varphi$. Then $\sigma^*\pi^*_\varphi = \pi^*_\varphi$. Now let $F \in \widehat{\varphi}^\infty$. By Proposition 4.3 we have $F \in$ Im π^*_φ. Then there exists an ultrafilter $G \in (\mathbf{A}_{P/\equiv_\varphi})^*$ such that $F = \pi^*_\varphi(G)$. Consequently, $\sigma^*(F) = \sigma^*\pi^*_\varphi(G) = \pi^*_\varphi(G) = F$, and this proves that $F \in$ fp σ^*.

Conversely, suppose that $\widehat{\varphi}^\infty \subseteq$ fp σ^*. Given $G \in (\mathbf{A}_{P/\equiv_\varphi})^*$ we have $\pi^*_\varphi(G) \in \widehat{\varphi}^\infty$ by Proposition 4.3 and thus $\pi^*_\varphi(G) \in$ fp σ^*. Hence $\sigma^*(\pi^*_\varphi(G)) = \pi^*_\varphi(G)$. This proves that $\sigma^*\pi^*_\varphi = \pi^*_\varphi$, and therefore $\pi_\varphi\sigma = \pi_\varphi$.

□

Finally, by combining Proposition 3.5 and Proposition 4.4, we obtain the following characterization.

Theorem 4.5 *Let $\varphi \in \mathcal{L}_P$ with $P \subseteq$ Prop finite. Then:*

(i) *φ is **L**-unifiable if and only if there exists a bounded morphism $f : X_Q \to X_P$ with* Im $f \subseteq \widehat{\varphi}^\infty$;

(ii) *φ is **L**-projective if and only if there exists a bounded morphism $f : X_P \to X_P$ with* Im $f \subseteq \widehat{\varphi}^\infty \subseteq$ fp f.

Accordingly, we will call a *dual unifier* of φ any bounded morphism $f : X_Q \to X_P$ such that Im $f \subseteq \widehat{\varphi}^\infty$, and a *projective dual unifier* of φ any bounded morphism $f : X_P \to X_P$ such that Im $f \subseteq \widehat{\varphi}^\infty \subseteq$ fp f.

5 Applications

In this section we delve into various applications of theorem 4.5, and turn our attention to the following logics (with $n, k \geq 1$) :

$$\begin{array}{rcl}
\mathbf{K4} & := & \mathbf{K} + (\Diamond\Diamond p \to \Diamond p) \\
\mathbf{K5} & := & \mathbf{K} + (\Diamond p \to \Box\Diamond p) \\
\mathbf{K45} & := & \mathbf{K4} + (\Diamond p \to \Box\Diamond p) \\
\mathbf{K4D1} & := & \mathbf{K4} + \Box(\Box p \to q) \vee \Box(\Box q \to p) \\
\mathbf{K4}_n & := & \mathbf{K} + (\Diamond^{n+1} p \to \Diamond^{\leq n} p) \\
\mathbf{K4}_n\mathbf{D1}_n & := & \mathbf{K4}_n + \Box(\Box^{\leq n} p \to q) \vee \Box(\Box^{\leq n} q \to p) \\
\mathbf{K4}_n\mathbf{B}_k & := & \mathbf{K4}_n + (p \to \Box^{\leq k}\Diamond^{\leq k} p)
\end{array}$$

First, we recall some elementary facts and definitions. Let $P \subseteq$ Prop. From now on we abstract away from the nature of the elements of X_P: we see them as points instead of ultrafilters, and denote them with the letters $x, y, z, \ldots$ Given

$X \subseteq X_P$, we write $R_P X := \{y \in X_P \mid \exists x \in X, xR_P y\}$ and $R_P^{-1} X := \{x \in X_P \mid \exists y \in X, xR_P y\}$. We then call X *upward closed* if $R_P X \subseteq X$, and *downward closed* if $R_P^{-1} X \subseteq X$. By recursion, we also define $R_P^0 := \{(x,x) \mid x \in X_P\}$ and $R_P^{n+1} := \{(x,z) \mid (x,y) \in R_P^n \text{ and } (y,z) \in R_P\}$ for all $n \in \mathbb{N}$. Then for all $n \in \mathbb{N}$ we write $R_P^{\leq n} := \bigcup_{k=0}^{n} R_P^k$. The following properties are either well-known, or stem from standard arguments [4,5]:

- if $\mathbf{K4} \subseteq \mathbf{L}$, then $\mathfrak{F}_P$ is transitive;
- if $\mathbf{K5} \subseteq \mathbf{L}$, then $\mathfrak{F}_P$ is *Euclidean*, that is, if $xR_P y$ and $xR_P z$ then $yR_P z$;
- if $\mathbf{K4D1} \subseteq \mathbf{L}$, then $\mathfrak{F}_P$ is transitive and *strongly connected*, that is, if $xR_P y$ and $xR_P z$, then either $yR_P z$ or $zR_P y$;
- if $\mathbf{K4}_n \subseteq \mathbf{L}$, then $\mathfrak{F}_P$ is *n-transitive*, that is, $xR_P^{n+1} y$ implies $xR_P^{\leq n} y$;
- if $\mathbf{K4}_n\mathbf{D1}_n \subseteq \mathbf{L}$, then $\mathfrak{F}_P$ is n-transitive and *strongly n-connected*, that is, if $xR_P y$ and $xR_P z$, then either $yR_P^{\leq n} z$ or $zR_P^{\leq n} y$;
- if $\mathbf{K4}_n\mathbf{B}_k \subseteq \mathbf{L}$ with $n, k \geq 1$, then $\mathfrak{F}_P$ is n-transitive and *k-symmetric*, that is, $xR_P^{\leq k} y$ implies $yR_P^{\leq k} x$ [15].

We first address the logic $\mathbf{K4}_n\mathbf{B}_k$. We propose a relatively short proof of its projective character, thus recovering Kostrzycka's result [15].

Theorem 5.1 *Let $n, k \geq 1$. Every extension of $\mathbf{K4}_n\mathbf{B}_k$ is projective.*

Proof. Let $\mathbf{L}$ be an extension of $\mathbf{K4}_n\mathbf{B}_k$. Let $P \subseteq \mathsf{Prop}$ finite and let $\varphi \in \mathcal{L}_P$ be $\mathbf{L}$-unifiable. Then there exists a $\mathbf{L}$-unifier $\sigma : \mathcal{L}_P \to \mathcal{L}_Q$ of φ. Obviously we can assume $P \subseteq Q$. Then, as explained in Remark 2.9, we can construct a $\mathbf{L}$-unifier $\sigma' : \mathcal{L}_Q \to \mathcal{L}_Q$ of φ. By Propositions 3.5 and 4.4 we then obtain a dual unifier $f := (\sigma')^*$ of φ.

We argue that $\widehat{\varphi}^\infty$ is both upward and downward closed for R_Q. For suppose $xR_Q y$. If $x \in \widehat{\varphi}^\infty$ then for all $i \in \mathbb{N}$ we have $x \in \widehat{\Box^{i+1}\varphi}$ and thus $y \in \widehat{\Box^i \varphi}$, and it follows that $y \in \widehat{\varphi}^\infty$. Conversely, suppose that $y \in \widehat{\varphi}^\infty$. Then since $\mathbf{K4}_n\mathbf{B}_k \subseteq \mathbf{L}$ and $xR_Q^{\leq k} y$ we have $yR_Q^{\leq k} x$. Since $\widehat{\varphi}^\infty$ is upward closed a straightforward recursion yields $x \in \widehat{\varphi}^\infty$. Now let us define $g : X_Q \to X_Q$ by

$$g(x) := \begin{cases} x & \text{if } x \in \widehat{\varphi}^\infty \\ f(x) & \text{otherwise} \end{cases}$$

for all $x \in X_Q$. We prove that g is a bounded morphism. First, assume that $xR_Q y$. We have seen that $x \in \widehat{\varphi}^\infty$ iff $y \in \widehat{\varphi}^\infty$, and we know that f is a bounded morphism, so $g(x)R_Q g(y)$ is immediate. Now suppose that $x \in X_Q$ and $g(x)R_Q y'$. If $x \in \widehat{\varphi}^\infty$ then $g(x) = x$ and $xR_Q y'$, and thus $y' \in \widehat{\varphi}^\infty$ too, leading to $g(y') = y'$. Otherwise we have $x \notin \widehat{\varphi}^\infty$ and $g(x) = f(x)$ with $f(x)R_Q y'$. Since f is a bounded morphism we obtain the existence of $y \in X_Q$ such that $f(y) = y'$ and $xR_Q y$. Then $y \notin \widehat{\varphi}^\infty$, and thus $g(y) = y'$ as desired.

Now let $\psi \in \mathcal{A}_Q$. We have

$$g^{-1}[\widehat{\psi}] = (\widehat{\psi} \cap \widehat{\varphi}^\infty) \cup (f^{-1}[\widehat{\psi}] \setminus \widehat{\varphi}^\infty).$$

Since $\vdash_{\mathbf{L}} \Box^{\leq n} p \to \Box^{n+1} p$ we have $\widehat{\varphi}^{\infty} = \widehat{\Box^{\leq n}\varphi} \in \mathcal{A}_Q$, and therefore $g^{-1}[\widehat{\psi}] \in \mathcal{A}_Q$ too. Finally it is immediate that $\mathrm{Im}\ g \subseteq \widehat{\varphi}^{\infty} \subseteq \mathrm{fp}\ g$. We conclude that φ is **L**-projective. □

As mentioned in Section 2, Kost [14] showed that an extension of **K4** is projective if and only if it contains **K4D1**. In Theorem 5.2, we reprove a weaker version of the right-to-left implication, limited to locally tabular logics. In Theorem 5.3 we reprove the left-to-right implication.

Theorem 5.2 *Every locally tabular extension of* **K4D1** *is projective.*

Proof. Let **L** be a locally tabular extension of **K4D1**. Let $P \subseteq \mathsf{Prop}$ finite and let $\varphi \in \mathcal{L}_P$ be **L**-unifiable. Reasoning as above, we obtain the existence of a dual unifier $f : X_Q \to X_Q$ of φ with $Q \subseteq \mathsf{Prop}$ finite and $P \subseteq Q$. To define $g : X_Q \to X_Q$, we consider $x \in X_Q$ and proceed as follows:

(i) if $x \in \widehat{\varphi}^{\infty}$ we set $g(x) := x$;

(ii) otherwise, if $x \in R_Q^{-1}\widehat{\varphi}^{\infty}$ then we select $g(x)$ in the set $Y := \{y \in \widehat{\varphi}^{\infty} \mid xR_Qy\}$ so that $g(x)R_Qy$ for all $y \in Y$;

(iii) otherwise, we set $g(x) := f(x)$.

Case (ii) requires some justification. Since $\mathbf{K4D1} \subseteq \mathbf{L}$, the frame $\mathfrak{F}_Q$ is strongly connected, so for all $y, z \in Y$ we have either yR_Qz or zR_Qy. In addition, Y is non-empty by assumption, and finite since **L** is locally tabular. This yields the existence of a 'smallest' element $g(x)$ with respect to R_Q – of course such an element is not necessarily unique. We now prove that g is a bounded morphism. First, suppose that xR_Qy. We examine each case for x.

(i) If $x \in \widehat{\varphi}^{\infty}$, then $y \in \widehat{\varphi}^{\infty}$ too, and thus $g(x)R_Qg(y)$ is immediate.

(ii) Otherwise, suppose that $x \in R_Q^{-1}\widehat{\varphi}^{\infty}$. Then $xR_Qg(x)$ and xR_Qy, and since $\mathfrak{F}_Q$ is strongly connected we obtain either $g(x)R_Qy$ or $yR_Qg(x)$. Since $g(x) \in \widehat{\varphi}^{\infty}$ it follows respectively that $y \in \widehat{\varphi}^{\infty}$ or $y \in R_Q^{-1}\widehat{\varphi}^{\infty}$. In both cases we have $yR_Q^{\leq 1}g(y)$. Since xR_Qy and $\mathfrak{F}_Q$ is transitive we then obtain $xR_Qg(y)$. Since $g(y) \in \widehat{\varphi}^{\infty}$, it follows by construction of $g(x)$ that $g(x)R_Q^{\leq 1}g(y)$. If $g(x)R_Qg(y)$ we are done. Otherwise $g(x) = g(y)$. Since $xR_Qg(x)$ and $\mathfrak{F}_Q$ is strongly connected we also have $g(x)R_Qg(x)$. Therefore $g(x)R_Qg(y)$ holds as well.

(iii) Otherwise, we have $g(x) = f(x)$. If $y \in R_Q^{-1}\widehat{\varphi}^{\infty}$, we have $x \in R_Q^{-1}\widehat{\varphi}^{\infty}$ too by transitivity, a contradiction. Thus $y \notin R_Q^{-1}\widehat{\varphi}^{\infty}$ and $g(y) = f(y)$. Since xR_Qy and f is a bounded morphism we obtain $f(x)R_Qf(y)$, and therefore $g(x)R_Qg(y)$.

Now suppose that $g(x)R_Qy$. If x falls in case (i) or case (ii) then $xR_Q^{\leq 1}g(x)$, and by transitivity we obtain xR_Qy. In addition, $g(x) \in \widehat{\varphi}^{\infty}$ entails $y \in \widehat{\varphi}^{\infty}$. Thus xR_Qy with $g(y) = y$, as desired. Otherwise, x falls in case (iii), and we have $g(x) = f(x)$. Then since f is a bounded morphism there exists $z \in X_Q$ such that $f(z) = y$ and xR_Qz. Since $x \notin R_Q^{-1}\widehat{\varphi}^{\infty}$ we obtain $z \notin R_Q^{-1}\widehat{\varphi}^{\infty}$ too, and therefore $g(z) = f(z) = y$.

Finally, for any $A \in \mathcal{A}_Q$ we have obviously $g^{-1}[A] \in \mathcal{A}_Q$ since $\mathcal{A}_Q = \mathcal{P}(X_Q)$. It is also immediate that $\mathrm{Im}\, g \subseteq \widehat{\varphi}^\infty \subseteq \mathrm{fp}\, g$. We conclude that φ is **L**-projective. □

Theorem 5.3 *Any projective extension of* **K4** *is also an extension of* **K4D1**.

Proof. Suppose that $\mathbf{K4} \subseteq \mathbf{L}$. By contraposition suppose that $\mathbf{K4D1} \not\subseteq \mathbf{L}$. We prove that $\varphi := \Box(\Box p \to q) \vee \Box(\Box q \to p)$ is not projective. Let $P := \{p, q\}$. First we have $\nvdash_{\mathbf{L}} \neg\varphi$ by assumption. By the Ultrafilter Theorem [4, Proposition 5.38], there exists an ultrafilter $x \in X_P$ such that $x \in \widehat{\Diamond(\Box p \wedge \neg q) \wedge \Diamond(\Box q \wedge \neg p)}$. Then, by the Existence Lemma [4, Lemma 4.20] (together with remark 4.1), there exist $y, z \in X_P$ such that xR_Py, xR_Pz, $y \in \widehat{\Box p \wedge \neg q}$ and $z \in \widehat{\Box q \wedge \neg p}$. Suppose toward a contradiction that there exists a projective dual unifier $f : X_P \to X_P$ of φ.

Since $y \in \widehat{\Box p}$ and $\mathbf{K4} \subseteq \mathbf{L}$, we have $y \in \widehat{\Box^{n+1} p}$ and thus $y \in \widehat{\Box^n \Box(\Box q \to p)}$ for all $n \in \mathbb{N}$. Therefore $y \in \widehat{\varphi}^\infty$ and $f(y) = y$. Since f is a bounded morphism and xR_Py, we obtain $f(x)R_Py$. Likewise, we can prove that $f(x)R_Pz$. Then since $f(x) \in \widehat{\varphi}^\infty$ we have in particular $f(x) \in \widehat{\varphi}$, and thus either $f(x) \in \widehat{\Box(\Box p \to q)}$ or $f(x) \in \widehat{\Box(\Box q \to p)}$. In the former case we obtain $y \in \widehat{\Box p \to q}$, and in the latter we obtain $z \in \widehat{\Box q \to p}$. Both outcomes are contradictions, and this concludes the proof. □

Interestingly, the proof of theorem 5.3 can easily be adapted to derive an analogous new result for extensions of $\mathbf{K4}_n$.

Theorem 5.4 *Any projective extension of* $\mathbf{K4}_n$ *is also an extension of* $\mathbf{K4}_n\mathbf{D1}_n$.

Proof. Suppose that $\mathbf{K4}_n \subseteq \mathbf{L}$. By contraposition suppose that $\mathbf{K4}_n\mathbf{D1}_n \not\subseteq \mathbf{L}$. We prove that $\varphi := \Box(\Box^{\leq n} p \to q) \vee \Box(\Box^{\leq n} q \to p)$ is not projective. Let $P := \{p, q\}$. We have $\nvdash_{\mathbf{L}} \neg\varphi$ by assumption, and arguing as above we obtain an ultrafilter $x \in X_P$ such that $x \in \widehat{\Diamond(\Box^{\leq n} p \wedge \neg q) \wedge \Diamond(\Box^{\leq n} q \wedge \neg p)}$. Then there exist $y, z \in X_P$ such that xR_Py, xR_Pz, $y \in \widehat{\Box^{\leq n} p \wedge \neg q}$ and $z \in \widehat{\Box^{\leq n} q \wedge \neg p}$. Suppose that there exists a projective dual unifier $f : X_P \to X_P$ of φ.

Since $y \in \widehat{\Box^{\leq n} p}$ and $\mathbf{K4}_n \subseteq \mathbf{L}$, we have $y \in \widehat{\Box^{k+1} p}$ and thus $y \in \widehat{\Box^k \Box(\Box^{\leq n} q \to p)}$ for all $k \in \mathbb{N}$. Therefore $y \in \widehat{\varphi}^\infty$ and $f(y) = y$. Since f is a bounded morphism and xR_Py, we obtain $f(x)R_Py$. Likewise, we can prove that $f(x)R_Pz$. Then since $f(x) \in \widehat{\varphi}^\infty$ we have either $f(x) \in \widehat{\Box(\Box^{\leq n} p \to q)}$ or $f(x) \in \widehat{\Box(\Box^{\leq n} q \to p)}$. In the former case we obtain $y \in \widehat{\Box^{\leq n} p \to q}$, and in the latter we obtain $z \in \widehat{\Box^{\leq n} q \to p}$. This yields a contradiction. □

Obviously Theorem 5.2 is weaker than Kost's result, but still covers a decent range of logics. In particular, it is enough to conclude that all extensions of **K45** are projective, since **K5** is locally tabular [18, Corollary 5]. We then refine this result by showing that the projective extensions of **K5** are, in fact, *exactly* the extensions of **K45**. We thus obtain a complete description of the landscape of projective logics above **K5**, which was only partially known prior to our work.

Theorem 5.5 *Let* **L** *be an extension of* **K5**. *Then* **L** *is projective if and only if* **K45** $\subseteq$ **L**.

Proof. We already know that if **K45** $\subseteq$ **L** then **L** is projective. Conversely, suppose that **K45** $\not\subseteq$ **L**. We prove that $\varphi := \Diamond\Diamond p \to \Diamond p$ is not projective. Let $P := \{p\}$. First we have $\nvdash_{\mathbf{L}} \Diamond\Diamond p \to \Diamond p$ by assumption. Arguing as before, we obtain an ultrafilter $x \in X_P$ such that $x \in \widehat{\Diamond\Diamond p \wedge \Box\neg p}$. Then there exists $y \in X_P$ such that $xR^2_P y$ and $y \in \widehat{p}$. Suppose toward a contradiction that there exists a projective dual unifier $f : X_P \to X_P$ of φ. Then $f(x)R^2_P f(y)$. Since y has a predecessor, it belongs to a final cluster (see [18] for a comprehensive description of Euclidean frames). Consequently $y \in \widehat{\varphi}^\infty$, and thus $f(y) = y$. Hence $f(x) \in \widehat{\Diamond\Diamond p}$, and since $f(x) \in \widehat{\varphi}^\infty$ we obtain $f(x) \in \widehat{\Diamond p}$. Therefore there exists $z \in X_P$ such that $f(x)R_P z$ and $z \in \widehat{p}$. Since f is a bounded morphism there exists $t \in X_P$ such that $xR_P t$ and $f(t) = z$. Again $t \in \widehat{\varphi}^\infty$ and therefore $f(t) = t$. Hence $xR_P z$, contradicting $x \in \widehat{\Box\neg p}$. This concludes the proof. $\Box$

6 Conclusion

In this paper, through substantial use of the duality between descriptive frames and modal algebras, we have given a necessary and sufficient condition for modal formulas to be projective. Applying it to the extensions of $\mathbf{K4}_n\mathbf{B}_k$ and **K4D1**, we have reproved known results obtained by Kost [14] and Kostrzycka [15]. Applying it to the extensions of **K5**, we have proved the new result saying that the projective extensions of **K5** are exactly the extensions of **K45**. It should be noted that our proofs are fairly lightweight and concise, as opposed to syntactic methods, which often involve all sorts of technical twists. Of course, this is only a first insight of what duality has to offer. Apart from the results about the unification types of modal logics mentioned in Section 5 and the new result about **K5** extensions, very little is known. For example, the unification types of $\mathbf{KD} := \mathbf{K} + \Diamond\top$ and $\mathbf{KT} := \mathbf{K} + \Box p \to p$ are just known to be non-unitary, seeing that the substitutions $\sigma_\top$ and $\sigma_\bot$ on the propositional variable p defined by $\sigma_\top(p) := \top$ and $\sigma_\bot(p) := \bot$ constitute both a basis of concise **KD**-unifiers and a basis of concise **KT**-unifiers of $\Box p \vee \Box\neg p$. This is an immediate consequence of the fact that **KD** and **KT** possess the modal disjunction property saying that for all formulas φ, ψ, if $\Box\varphi \vee \Box\psi$ is in **KD** (resp. **KT**) then either φ or ψ is in **KD** (resp. **KT**). To take another example, the unification types of $\mathbf{DAlt}_1 := \mathbf{KD} + \Diamond p \to \Box p$ and $\mathbf{KB} := \mathbf{K} + p \to \Box\Diamond p$ are not known either. Therefore, much remain to be done and further investigations are needed for obtaining, by means of our duality approach, the unification types of modal logics such as **KD**, **KT**, $\mathbf{DAlt}_1$ and **KB**.

References

[1] BAADER, F., and S. GHILARDI, 'Unification in modal and description logics', *Logic Journal of the IGPL* **19** (2011) 705–730.

[2] BAADER, F., and W. SNYDER, 'Unification theory', In: *Handbook of Automated Reasoning,* Elsevier (2001) 439–526.

[3] BABENYSHEV, S., and V. RYBAKOV, 'Unification in linear temporal logic **LTL**', *Annals of Pure and Applied Logic* **162** (2011) 991–1000.

[4] BLACKBURN, P., M. DE RIJKE, and Y. VENEMA, *Modal Logic,* Cambridge University Press (2001).

[5] CHAGROV, A., and M. ZAKHARYASCHEV, *Modal Logic,* Oxford University Press (1997).

[6] DZIK, W., 'Unitary unification of **S5** modal logics and its extensions', *Bulletin of the Section of Logic* **32** (2003) 19–26.

[7] DZIK, W., *Unification Types in Logic,* Wydawnicto Uniwersytetu Slaskiego (2007).

[8] DZIK, W., and P. WOJTYLAK, 'Projective unification in modal logic', *Logic Journal of the IGPL* **20** (2012) 121–153.

[9] GHILARDI, S., 'Unification through projectivity', *Journal of Logic and Computation* **7** (1997) 733–752.

[10] GHILARDI, S., 'Best solving modal equations', *Annals of Pure and Applied Logic* **102** (2000) 183–198.

[11] HAKLI, R., and S. NEGRI, 'Does the deduction theorem fail for modal logic?', *Synthese* **187** (2012) 849–867.

[12] IEMHOFF, R., 'A syntactic approach to unification in transitive reflexive modal logics', *Notre Dame Journal of Formal Logic* **57** (2016) 233–247.

[13] JEŘÁBEK, E., 'Blending margins: the modal logic **K** has nullary unification type', *Journal of Logic and Computation* **25** (2015) 1231–1240.

[14] KOST, S., 'Projective unification in transitive modal logics', *Logic Journal of the IGPL* **26** (2018) 548–566.

[15] KOSTRZYCKA, Z., 'Projective unification in weakly transitive and weakly symmetric modal logics', *Journal of Logic and Computation* doi.org/10.1093/logcom/exab081doi.

[16] KRACHT, M., *Tools and Techniques in Modal Logic,* Elsevier (1999).

[17] MIYAZAKI, Y., 'Normal modal logics containing **KTB** with some finiteness conditions', In: *Advances in Modal Logic,* College Publications (2004) 171–190.

[18] NAGLE, M. and S. THOMASON, 'The extensions of the modal logic **K5**', *Journal of Symbolic Logic* **50** (1985) 102–109.

[19] RYBAKOV, V., *Admissibility of Logical Inference Rules,* Elsevier (1997).

[20] SŁOMCZYŃSKA, K., 'Unification and projectivity in Fregean Varieties', *Logic Journal of the IGPL* **20** (2012) 73–93.

Local Dependence and Guarding

Johan van Benthem

ILLC University of Amsterdam, Stanford University, Tsinghua University

Balder ten Cate [1]

ILLC University of Amsterdam

Raoul Koudijs

ILLC University of Amsterdam

Abstract

We study LFD, a base logic of functional dependence introduced by Baltag and van Benthem (2021) and its connections with the guarded fragment GF of first-order logic. Like other logics of dependence, the semantics of LFD uses teams: sets of permissible variable assignments. What sets LFD apart is its ability to express local dependence between variables and local dependence of statements on variables.

Known features of LFD include decidability, explicit axiomatization, finite model property, and a bisimulation characterization. Others, including the complexity of satisfiability, remained open so far. More generally, what has been lacking is a good understanding of what makes the LFD approach to dependence computationally well-behaved, and how it relates to other decidable logics. In particular, how do allowing variable dependencies and guarding quantifiers compare as logical devices?

We provide a new compositional translation from GF into LFD, and conversely, we translate LFD into GF in an 'almost compositional' manner. Using these two translations, we transfer known results about GF to LFD in a uniform manner, yielding, e.g., tight complexity bounds for LFD satisfiability, as well as Craig interpolation. Conversely, e.g., the finite model property of LFD transfers to GF. Thus, local dependence and guarding turn out to be intricately entangled notions.

Keywords: Logic, Dependence, Guarded Fragment

1 Introduction: from guarding to dependence

The Guarded Fragment GF of first-order logic, [2], is a well-known formalism for quantifying over tuples of objects that are locally *guarded* by predicates. Guarded existential quantification has the format $\exists\bar{y}(G(\bar{x},\bar{y}) \wedge \varphi(\bar{x},\bar{y}))$, where

[1] Supported by the European Union's Horizon 2020 research and innovation programme under grant MSCA-101031081.

$\bar{x}$, $\bar{y}$ are finite sets or tuples of variables, the atom $G(\bar{x}, \bar{y})$ is the guard, and $\varphi(\bar{x}, \bar{y})$ is a guarded formula having at most the free variables displayed. For further literature on GF and its extensions, see [7], [5].

Guarding quantifiers in truth conditions for existing semantics is a general device for lowering complexity of logical systems, [3]. But there are more such devices. In particular, [2] connects guarding with 'generalized assignment models' for first-order logic [to be called 'teams' henceforth in accordance with modern terminology] where not all maps from variables to objects need to be present. These 'gaps' in the full function space can be seen as modeling *dependence*, a major topic in the current logical literature, [1], [16]. With assignments missing, correlations arise, as changing the value of one variable x may only be feasible in the given space by also changing that of another variable y. In contrast, in standard models for first-order logic all variables take their values independently. The origins of this semantics lie in 'cylindric relativized set algebra' CRS in algebraic logic where standard set algebras are relativized to one fixed relation over the domain, [15]. Unlike first-order logic, CRS logic is decidable.

That guarding can encode dependence was proved in [2]. First-order formulas in CRS semantics translate compositionally into guarded formulas over standard first-order models by coding the available tuples of values for finite tuples of relevant variables $\bar{v}$ as one new guard predicate $G(\bar{v})$. Conversely, [18] gave an effective reduction for the satisfiability problem for GF formulas φ to that for first-order formulas over team models, though the translation employed is not compositional. [2] Thus, dependence models can also encode guarding. The main topic of this paper is a further technical analysis of how far this analogy goes, now based on modern dependence logics rather than CRS.

CRS-style semantics treats dependence implicitly by giving up classical FO laws such as Commutativity $\exists x \exists y\, \varphi(x, y) \leftrightarrow \exists y \exists x\, \varphi(x, y)$ which express independence of the variables x and y. In a next step, [16] introduced explicit syntactic atoms $D_X y$ expressing functional dependence of variable y on the set of variables X. The first generation of dependence logics over this richer language was second-order and non-classical, but many simpler fragments have been studied. A systematic use of GF-style guarding to lower the complexity of dependence logics is found in [8]. However, the present paper is concerned with a simpler base system for dependence logic.

A new base logic LFD for a first-order language with explicit dependence atoms was recently proposed in [4] as a minimal way of reasoning about dependence on a classical base relying on a modal local semantics. LFD is a dualized CRS logic [in a sense to be defined below] plus dependence atoms, which is decidable, axiomatizable, and allows for natural language extensions, e.g., to function symbols and independence modalities. Many of these properties involve GF-style methods such as the use of type models and representation theorems. Thus, the question arises whether the connection between guarding

[2] Related results and further general analysis are found in [13].

and dependence extends to this new setting. But there are obstacles here. The first-order translation for LFD lands in the guarded fragment, except for the dependence atoms, cf. [9]. In this paper, we offer the following new results.

- We give a compositional translation from GF into LFD without dependence atoms, considerably improving the known SAT reduction.
- We give an effective SAT reduction from LFD into GF, answering a question left open in the literature.
- We show how these results allow for transfer of important properties between LFD and GF. For instance, the finite model property for GF [7], can be derived from that for LFD, established in [12]. Conversely, the decidability of LFD follows from that of GF. Moreover, using our translations, we solve the open problem of the computational complexity of LFD.
- Beyond single examples, we also discuss the general transfer of system properties made possible by our translations, and discuss more general consequences for relating decidable fragments of first-order logic.

All missing proofs can be found in the appendix.

2 Preliminaries

2.1 The Logic of Functional Dependence LFD

The Logic of Functional Dependencies (LFD) was introduced in [4], to which we refer for all definitions and results stated in this section.

Definition 2.1 (Syntax of LFD) Fix a finite set of variables V_{LFD}, and a relational signature (i.e., a set of relation symbols with associated arity) $\mathbf{S}$. The formulas of LFD are recursively generated by the following grammar.

$$\psi ::= P(v_1, \ldots, v_n) \mid D_V u \mid \psi \wedge \psi \mid \neg\psi \mid \mathbb{E}_V \psi$$

with $v_1, \ldots, v_n, u \in V_{LFD}$, $V \subseteq V_{LFD}$, and $P \in \mathbf{S}$ an n-ary relation symbol.

We read $D_V u$ as 'u locally depends (only) on V' or 'u is locally determined by V'. For notational convenience, we write $D_V U$ for the conjunction $\bigwedge_{u \in U} D_V u$. We also write $D_v u$ as a shorthand for $D_{\{v\}} u$, and $\mathbb{E}_v \psi$ for $\mathbb{E}_{\{v\}} \psi$.

The semantics of LFD uses *dependence models*, also known as 'generalized assignment models'. These are pairs $\mathbb{M} = (M, A)$ where M is a standard model over the relational signature $\mathbf{S}$, and $A \subseteq dom(M)^{V_{LFD}}$ is a collection of admissible variable assignments, also known as a 'team'.

Definition 2.2 (Semantics of LFD)
Truth of a formula φ in a dependence model $\mathbb{M} = (M, A)$ under an assignment $s \in A$ is defined as follows, where $s =_V t$ means that $s(v) = t(v)$ for all $v \in V$:

$$
\begin{array}{ll}
\mathbb{M}, s \models P(v_1, \ldots, v_n) & \text{iff } s(v_1, \ldots, v_n) \in P^M \\
\mathbb{M}, s \models \varphi \wedge \psi & \text{iff } \mathbb{M}, s \models \varphi \text{ and } \mathbb{M}, s \models \psi \\
\mathbb{M}, s \models \neg\varphi & \text{iff not } \mathbb{M}, s \models \varphi \\
\mathbb{M}, s \models \mathbb{E}_V \varphi & \text{iff for some } t \in A \text{ , } s =_V t \text{ and } \mathbb{M}, t \models \varphi \\
\mathbb{M}, s \models D_V u & \text{iff for all } t \in A, \ s =_V t \text{ implies } s =_u t
\end{array}
$$

LFD formulas come with a notion of *free variable*, being a variable whose value can affect the truth of the formula, in the semantics to follow.

Definition 2.3 (Free Variables)
The free variables of an LFD formula ψ are defined as follows:

$$
\begin{array}{ll}
\text{free}(P(v_1...v_n)) & = \{v_1, ..., v_n\} \\
\text{free}(\neg\varphi) & = \text{free}(\varphi) \\
\text{free}(\varphi \wedge \psi) & = \text{free}(\varphi) \cup \text{free}(\psi) \\
\text{free}(D_V u) = \text{free}(\mathbb{E}_V \psi) & = V
\end{array}
$$

It might seem odd that $free(\mathbb{E}_x\varphi) = \{x\}$ while $free(\exists x\varphi) = free(\varphi) \setminus \{x\}$. The difference, however, is a matter of perspective, since the two notions of free variable coincide on their interpretation as “a variable whose value can affect the truth of the formula”. With the preceding definitions, it is easy to show that LFD, just like FO, satisfies a property called *variable locality*: any two assignments s, t that agree on some set of variables X agree on the truth values of all formulas whose free variables are contained in X (plus all variables that locally depend on X at s, t).

The difference with FO free variables arises because the LFD modality is dual to that of FO, and displays variables whose values are kept fixed, rather than those whose values are allowed to vary. More generally, LFD shares a feature with CRS that makes it distinct from FO in its standard semantics. While in first-order formulas quantified variables are arbitrary placeholders, in LFD formulas variables have an individual character given their possibly different roles in the assignment space. [3]

Standard first-order models induce special ‘full’ dependence models (M, M^V) where *all* assignments are available. On such models there are no non-trivial dependencies (atoms $D_X y$ are always false unless $y \in X$) and the dependence quantifier $\mathbb{E}$ collapses to the standard existential quantifier $\exists$, recapturing FO. Conversely, dependence models can also be encoded as standard models, supporting a faithful translation into an expanded version of FO.

Concretely, there is a translation $tr(\cdot)$ from LFD formulas to FO formulas (over an expanded signature), and a contravariant function T that bijectively maps models of $tr(\varphi)$ to dependence models satisfying φ. In a picture:

[3] Accordingly, renaming bound variables in one model is not generally possible in LFD, though one can shift to new variables when allowing changing the current model [4].

$$\begin{array}{lcl} \text{LFD-formula } \varphi & \xrightarrow{\ tr\ } & \text{first-order formula } tr(\varphi) \\ \text{dependence model } T(M) & \underset{\text{(bijective)}}{\xleftarrow{\ T\ }} & \text{first-order model } M \end{array}$$

Since we will see variants of this diagram of contravariant maps in later sections of this paper, we will spell things out in detail, as it will serve as an example for our later translation schemes.

Definition 2.4 (First-Order Translation)
Let $V_{LFD} = \{v_1, \ldots, v_n\}$. The translation function $tr(\cdot)$ maps LFD formulas over V_{LFD} to FO formulas using the same variables $v_1, \ldots, v_n$ and auxiliary variables $v'_1, \ldots, v'_n$, and using an additional n-ary relation symbol A.

$$\begin{aligned} tr(P(u_1, \ldots, u_n)) &:= P(u_1, \ldots, u_n) \\ tr(\varphi \wedge \psi) &:= tr(\varphi) \wedge tr(\psi) \\ tr(\neg\varphi) &:= \neg tr(\varphi) \\ tr(D_V u) &:= \forall \bar{v}'(A(\bar{v}') \wedge \bigwedge_{v \in V} (v = v' \to u = u')) \\ tr(\mathbb{E}_V \psi) &:= \exists \bar{z}(A(\bar{v}) \wedge tr(\psi)) \end{aligned}$$

where $\bar{v} = v_1, \ldots, v_n$, $\bar{v}' = v'_1, \ldots, v'_n$, and $\bar{z}$ enumerates $V_{LFD} \setminus V$.

In this translation A is a new predicate encoding the admissible assignments on the relevant variables.

Definition 2.5 (Model Transformation)
Let $V_{LFD} = \{v_1, \ldots, v_n\}$. For a standard model M over the extended signature with the A relation, $T(M)$ is the dependence model obtained by extracting a team $A := \{s : V \to M \mid (s(v_1), \ldots, s(v_n)) \in A^M\}$ on M and removing the relation A^M from the model.

Note that the map T is bijective. Its inverse T^{-1} maps a dependence model $\mathbb{M} = (M, A)$ to the standard model over the extended signature obtained by interpreting A as $\{(s(v_1), \ldots, s(v_n)) \mid s \in A\}$.

Theorem 2.6 *For every LFD formula ψ, standard model M, assignment s : $\mathrm{free}(\psi) \to M$ such that $M, s \models A(v_1, \ldots, v_n)$, we have the equivalence*

$$T(M), s \models \psi \quad \textit{iff} \quad M, s \models tr(\psi) \wedge A(v_1, \ldots, v_n)$$ [4]

Since $T(\cdot)$ is bijective, Theorem 2.6 implies, in particular, that $tr(\varphi) \wedge A(v_1, \ldots, v_n)$ and φ are equi-satisfiable. As consequences of this translation, one can also derive an effective reduction of semantic consequence, Compactness, Löwenheim-Skolem, recursive enumerability, all of them inherited by LFD from FO. Other properties require additional work in reproving classical results, much as happens with the standard FO translation for modal logic [6].

[4] The added conjunct $A(v_1, \ldots, v_n)$ makes sure that s is an admissible assignment in M. We could also work this information into the compositional clauses of the translation.

Remark 2.7 (LFD and CRS) *Dependence models are the generalised semantic structures first introduced for the algebraic version CRS of FO logic, [15], with $M, s \models \exists x\, \varphi$ if some available assignment $t =^x s$ in M satisfies φ. Here the relation $=^x$ is dual to the $=_x$ used for LFD, requiring that s, t agree on the values for all variables* except x. *Thus CRS quantifiers, like the first-order ones, display the variables that are allowed to vary, rather than those kept fixed. A polyadic version $\exists \bar{x}\, \varphi$ is defined analogously, which, unlike in standard FO semantics, does not reduce to iterated single existential quantifiers. When the total set V of variables is finite, the existential quantifiers of CRS and the existential modality of LFD are intertranslatable, e.g., $\exists x\, \varphi = \mathbb{E}_{V \setminus \{x\}}\, \varphi$. Thus, the fragment of LFD without dependence atoms amounts to a version of CRS with 'dual quantifiers', more detailed discussion can be found in [4]. Thus, some results in what follows also specialize to new connections between GF and CRS.*

2.2 Dependence Bisimulations, and Type Models

We need to introduce two more technical notions related to LFD that will play an important role in our proofs in subsequent sections.

Given its similarities with modal logic, LFD comes with a notion of *dependence-bisimulations*, which leaves the truth values of formulas invariant [9,12]. For each set of variables V, dependence model $\mathbb{M}$, and assignment s, the *dependence-closure* $D^s_V = \{u \in V_{LFD} \mid \mathbb{M}, s \models D_V u\}$ is the set of variables locally determined by V at s. Clearly, $V \subseteq D^s_V$. We call a set V *dependence-closed* at assignment s if $D^s_V = V$. Also, for any two assignments, $V^{s,t} = \{v \in V_{LFD} \mid s =_v t\}$ denotes the set of variables on which s, t agree.

Definition 2.8 (Dependence Bisimulations) Let $(M, A), (M', A')$ be dependence models. A relation $Z \subseteq A \times A'$ is a *dependence bisimulation* if for every pair $(s, s') \in Z$, the following conditions are satisfied: [5]

(atom) $s(\bar{u}) \in P^M$ iff $s'(\bar{u}) \in P^{M'}$ for all tuples $\bar{u}$ of LFD variables

(forth) for every $t \in A$, there is a $t' \in A'$ such that (i) $V^{s,t} \subseteq V^{s',t'}$, (ii) $(t, t') \in Z$ and (iii) the set $V^{s,t}$ is dependence-closed at s'

(back) for every $t' \in A'$, there is a $t \in A$ such that (i) $V^{s',t'} \subseteq V^{s,t}$, (ii) $(t, t') \in Z$ and (iii) the set $V^{s',t'}$ is dependence-closed at s

LFD formulas do not distinguish between dependence-bisimilar models. Indeed, it was shown in [9] that a first-order formula is invariant for dependence bisimulations iff it is equivalent to the $tr(\cdot)$ translation of an LFD formula. [6]

Dependence bisimulations allow us to show that LFD is 'blind' for precisely how assignments assign values to different variables. This is captured formally by the *distinguished model property of LFD*. A dependence model is *distin-*

[5] Note that dependence bisimulations Z are always *total*, i.e., $dom(Z) = A$ and $cod(Z) = A'$.

[6] The definition of LFD-bisimulations in [9] is stated differently: it does not have the dependence-closed condition (iii) in the forth and back-clauses but instead includes the requirement that $s \models D_X y$ iff $s' \models D_X y$ into the atomic condition. As it turns out, these definitions are equivalent, but the one in [9] also works for type models.

guished if each variable only takes values in its own range different from the ranges of all other variables. Thus for every object in the domain there is a unique variable to which it can be assigned by an admissible assignment.

Proposition 2.9 *There is a transformation* $(\cdot)^d$ *taking any dependence model* M *into a dependence-bisimilar distinguished dependence model* M^d.

Proof. For each admissible assignment $s \in A$, let s^d be the assignment defined by $s^d(v) := (v, s(v))$. Now take the team $A_d := \{s^d \mid s \in A\}$ on the domain $\bigcup_{s \in A} s^d[V_{LFD}]$. We interpret relations on the new value domain by setting

$$R^{M_d} := \{(u_1, s(u_1)), ..., (u_m, s(u_m)) \mid (s(u_1), ..., s(u_m)) \in R^M\},$$

making it into a standard model M_d. Note that each $s^d : V_{LFD} \to M_d$ and hence (M_d, A_d) is a dependence model which is distinguished by construction. It is now straightforward to check that the relation $\{(s, s_d) \mid s \in A\}$ is a dependence bisimulation between (M, A) and (M_d, A_d). □

The distinguished model property has to do with the lack of an equality in the language. LFD extended with equality atoms $v_i = v_j$ is undecidable [9].

Finally, we review an abstract type model semantics, also known as 'quasi-models' [19,18], originally used to prove the decidability of LFD in analogy with type models for GF. One can think of type models as finite certificates for satisfiability, or as information reductions of 'real' models with objects.

Definition 2.10 (Types)
The *closure* of an LFD formula ψ, denoted $Cl(\psi)$, is the smallest set containing ψ that is closed under subformulas and single negations [7] and that contains all atomic formulas of the form $D_V u$. A subset $\Delta \subseteq Cl(\psi)$ is a *type* (for ψ) if it satisfies the following three properties for the logical operations plus two further constraints on dependence atoms:

- **($\neg$-Consistency)** $\neg\chi \in \Delta$ iff not $\chi \in \Delta$
- **($\wedge$-Consistency)** $\chi \wedge \xi \in \Delta$ iff $\chi \in \Delta$ and $\xi \in \Delta$
- **($\mathbb{E}$-Consistency)** $\chi \in \Delta$ implies $\mathbb{E}_V \chi \in \Delta$ whenever $\mathbb{E}_V \chi \in Cl(\psi)$
- **(Projection)** $D_V u \in \Delta$ for all $V \subseteq V_{LFD}$ and $u \in V$
- **(Transitivity)** $D_V U, D_U W \in \Delta$ implies $D_V W \in \Delta$ for all $V, U, W \subseteq V_{LFD}$

For types Δ, Δ' and sets $V \subseteq V_{LFD}$, we will henceforth write $\Delta \sim_V \Delta'$ if $\{\psi \in \Delta \mid \text{free}(\psi) \subseteq V\} = \{\psi \in \Delta' \mid \text{free}(\psi) \subseteq V\}$.

Definition 2.11 (LFD Type Models)
Fix an LFD formula ψ. An LFD type model $\mathfrak{M}$ is a set of types (for ψ) satisfying the following two conditions:

(witness) if $\mathbb{E}_V \varphi \in \Delta$ then $\exists \Delta' \in \mathfrak{M}$ with $\Delta \sim_{D^\Delta_V} \Delta'$ and $\varphi \in \Delta'$

(universal) $\sim_\emptyset$ is the universal relation on $\mathfrak{M}$

---

[7] The single negation $\sim(\psi) = \psi'$ if $\psi = \neg\psi'$ for some ψ', and $\sim(\psi) = \neg\psi$ otherwise.

with $D_V^\Delta = \{u \in V_{LFD} \mid D_V u \in \Delta\}$ the 'dependence-closure' of V w.r.t Δ.[8]

We say $\mathfrak{M}$ is a type model *for* ψ if there is a type $\Delta \in \mathfrak{M}$ with $\psi \in \Delta$. As it turns out, LFD type models always encode the existence of a real dependence model, so they may serve as certificates for the satisfiability of an LFD formula.

Theorem 2.12 *There is an 'unravelling' operation R such that, for each LFD type model $\mathfrak{M}$, $R(\mathfrak{M})$ is a dependence model realizing precisely the types in $\mathfrak{M}$.*

In fact, there is a dependence bisimulation (in the sense of [9]) from $R(\mathfrak{M})$ onto $\mathfrak{M}$. Theorem 2.12 implies that LFD is also complete w.r.t. type models.[9]

2.3 The Guarded Fragment of First-Order Logic

By a "standard model" we mean an ordinary first-order structure, and for FO formulas φ, $\varphi(\bar{x})$ indicates that free$(\varphi) \subseteq \{\bar{x}\}$. We will be concerned with the Guarded Fragment GF [2]. This is the fragment of FO consisting of formulas in which all quantification is *guarded*, i.e., of the form $\exists\bar{y}(G(\bar{x},\bar{y}) \wedge \psi(\bar{x},\bar{y}))$, where $G(\bar{x},\bar{y})$ is an atomic formula and free$(\psi) \subseteq$ free$(G(\bar{x},\bar{y})) = \{\bar{x},\bar{y}\}$. We will write guarded quantifiers $\exists\bar{y}(G(\bar{x},\bar{y}) \wedge \varphi)$ when it is clear from context that φ is guarded. We also use $\forall\bar{y}(G(\bar{x},\bar{y}) \to \psi(\bar{x},\bar{y}))$ as syntactic sugar for $\neg\exists\bar{y}(G(\bar{x},\bar{y}) \wedge \neg\psi(\bar{x},\bar{y}))$. For simplicity, we will be working with the *equality-free* guarded fragment. Our results extend to full GF with equality, cf. Section 6.

For later reference, observe that the translation $tr(\cdot)$ from LFD to FO that we discussed earlier lands in GF except for the dependence atoms: the formula $tr(D_X y)$ uses unguarded quantification.

For the properties of the Guarded Fragment used in this paper, we refer to the earlier-mentioned sources. In particular, the satisfiability problem for GF is decidable, while the language has a characterization as a fragment of FO in terms of invariance for *guarded bisimulations*, [2].

3 A compositional translation from GF into LFD

It was shown in [18] that satisfiability in the Guarded Fragment reduces effectively to satisfiability in the logic CRS, The latter, as we mentioned before, corresponds to the fragment of LFD without dependence atoms. The translation achieving this used conjunctions of (a) a compositional translation for GF formulas φ, (b) a 'set-up component' that records the requirements for a GF type model for φ, both stated as CRS formulas. In contrast, here, we will present a compositional translation from GF to LFD (in fact, to the fragment of the logic LFD without dependence atoms), allowing us to transfer properties such as Craig interpolation. Our surprisingly simple translation also neatly exhibits a conceptual difference between first-order variables and LFD variables.

Remark 3.1 (Two views of variables) *Given the duality between the LFD modality $\mathbb{E}_X$ and the first-order quantifier $\exists X$, it is tempting to translate FO formulas to LFD by replacing $\exists x\varphi$ by $\mathbb{E}_Y\varphi$ with Y the set of all free variables*

---

[8] Note that if $\Delta \sim_V \Delta'$ then $D_V^\Delta = D_V^{\Delta'}$ because free$(D_V u) = V$.

[9] This analysis can be rephrased modal-style in terms of 'general relational models' [4].

of φ except x. But this does not work. Consider the unsatisfiable first-order sentence $\exists x P(x) \wedge \neg \exists y P(y)$. Its translation, following the recipe just described, is $\mathbb{E}_{\emptyset} P(x) \wedge \neg \mathbb{E}_{\emptyset} P(y)$. However, this LFD formula is satisfied in a dependence model $\mathbb{M} = (M, A)$ where M has domain $\{a, b\}$ with just one atomic fact $P(a)$, and every assignment in A maps x to a and y to b. The crux lies in the fact that, in first-order logic, variables are interchangeable placeholders, whereas in LFD, variables have individual behavior, at least within a dependence model.

To deal with this discrepancy, the following translation τ separates the sets of relevant variables V_{LFD} used for LFD and V_{GF} for GF, and then connects these explicitly via a mapping that can be seen as an abstract substitution.

Definition 3.2 Let φ be a FO formula and $\rho : \text{free}(\varphi) \to V_{LFD}$. The translation $\tau_\rho(\varphi)$ is defined as follows, with $\rho =_X \rho'$ for $\rho(x) = \rho'(x)$ for all $x \in X$:

$$\begin{aligned}
\tau_\rho(P(x_1, \ldots, x_n)) &:= P(\rho(x_1), \ldots, \rho(x_n)) \\
\tau_\rho(\varphi \wedge \psi) &:= \tau_\rho(\varphi) \wedge \tau_\rho(\psi) \\
\tau_\rho(\neg\varphi) &:= \neg\tau_\rho(\varphi) \\
\tau_\rho(\exists \bar{y}\ \varphi(\bar{x}, \bar{y})) &:= \bigvee_{\substack{\rho' : \{\bar{x}, \bar{y}\} \to V_{LFD} \\ \rho =_{\bar{x}} \rho'}} \mathbb{E}_{\rho(\bar{x})}\ \tau_{\rho'}(\varphi)
\end{aligned}$$

With this new compositional translation, and assuming that $V_{LFD} = \{v_1, v_2\}$, the above FO formula $\exists x P(x) \wedge \neg \exists y P(y)$ translates to the LFD-formula $(\mathbb{E}_{\emptyset} P(v_1) \vee \mathbb{E}_{\emptyset} P(v_2)) \wedge \neg(\mathbb{E}_{\emptyset} P(v_1) \vee \mathbb{E}_{\emptyset} P(v_2))$, which, indeed, is unsatisfiable. Even so, τ is not an effective satisfiability reduction from FO to LFD, since LFD is decidable. But as we will show, τ preserves satisfiability for formulas in the Guarded Fragment.

Our next step is to define a corresponding operation on models. The overall scheme will look like this:

$$\begin{array}{ccc}
\text{GF-formula } \varphi & \xrightarrow{\ \tau\ } & \text{LFD formula } \tau_\rho(\varphi) \\
\text{standard model } G(\mathbb{M}) & \underset{\text{(surjective)}}{\xleftarrow{\ G\ }} & \text{dependence model } \mathbb{M}
\end{array}$$

The model transformation G, consists of throwing away 'unnamed' facts.

Definition 3.3 Given any dependence model $\mathbb{M} = (M, A)$, the standard model $G(\mathbb{M})$ has for its domain $dom(M)$ and for its predicate interpretations $R^{G(\mathbb{M})} := \{\bar{m} \in R^M \mid \{\bar{m}\} \subseteq s[V_{LFD}] \text{ for some } s \in A\}$, where $s[V_{LFD}]$ is the image of $s : V_{LFD} \to M$, i.e. $s[V_{LFD}] = \{s(v) \mid v \in V_{LFD}\}$.

The transformation G is surjective because each standard model M is the G image of its matching 'full' dependence model $F(M) = (M, M^{V_{LFD}})$. Note that, while $G(F(M)) = M$, in general $F(G(\mathbb{M}))$ can be very different from $\mathbb{M}$, since $F(G(\mathbb{M}))$ is always full. A further complication is that, while the operation G is defined on all dependence models, the crucial equivalence theorem that connects τ with G, which we will present next, applies only for *distinguished* dependence models. We will address this issue separately.

Theorem 3.4 *For every distinguished dependence model* $\mathbb{M} = (M, A)$*, assignment* $s \in A$ *and GF formula* φ *with map* $\rho : \text{free}(\varphi) \to V_{LFD}$ *we have*

$$\mathbb{M}, s \models \tau_\rho(\varphi) \qquad \textit{iff} \qquad G(\mathbb{M}), s \circ \rho \models \varphi$$

Proof. By induction on the complexity of the formula φ. The atomic case says that $(M, A), s \models P\rho(x_1)...\rho(x_n)$ holds iff $G(M, A), s \circ \rho \models Px_1...x_n$. This holds because, by our definition, the fact $P(s(\rho(x_1)), ..., s(\rho(x_n)))$ does not get removed from (M, A) by the transformation G. The Boolean cases are routine, so the only important case to analyze is that of guarded quantification.

From left to right, suppose $(M, A), s \models \tau_\rho(\exists \bar{y}(G(\bar{x}, \bar{y}) \wedge \varphi)$ (where $\rho : \{\bar{x}\} \to V_{LFD}$). This means that there is some $\rho' : \{\bar{x}, \bar{y}\} \to V_{LFD}$ extending ρ such that $(M, A), s \models \mathbb{E}_{\rho[\bar{x}]}(\tau_{\rho'}(G(\bar{x}, \bar{y}) \wedge \varphi))$, since one of the disjuncts in our translation clause is true. By the LFD semantics, there is then some witnessing assignment $t \in A$ with $s =_{\rho[\bar{x}]} t$ and $t \models \tau_{\rho'}(G(\bar{x}, \bar{y}) \wedge \varphi)$. By the inductive hypothesis, it now follows that $G(M, A),\ t \circ \rho' \models G(\bar{x}, \bar{y}) \wedge \varphi$. But since $\rho[\bar{x}] = \rho'[\bar{x}]$ and $s =_{\rho[\bar{x}]} t$ also $s \circ \rho =_{\bar{x}} t \circ \rho'$. The GF semantics yields that $G(M, A),\ s \circ \rho \models \exists \bar{y}(G(\bar{x}, \bar{y}) \wedge \varphi)$ where the objects $(t \circ \rho')[\bar{y}]$ witness the bound variables $\bar{y}$.

From right to left, we need to appeal to the guard in the quantification. Suppose that $G(M, A), s \circ \rho \models \exists \bar{y}(G(\bar{x}, \bar{y}) \wedge \varphi)$. By the standard FO semantics, there are objects $\bar{m} = (m_1, ..., m_n)$ s.t. $G(M, A), (s \circ \rho)\,[\bar{m}/\bar{y}] \models G(\bar{x}, \bar{y}) \wedge \varphi$. By distinguishedness, for each $m \in \bar{m}$ there is a *unique* LFD-variable v_m s.t. there is some admissible assignment $s \in A$ with $s(v_m) = m$. Now consider the true guard fact $G((s \circ \rho)(\bar{x}), \bar{m})$ (which is a fact of $G(M, A)$). Given the uniqueness of variable names in M, this fact must have been witnessed by some available assignment $t \in A$ using some atomic formula $G(\rho(\bar{x}), \bar{v_m})$, where $t \circ \rho(x) = s \circ \rho(x)$ for all $x \in \bar{x}$. It follows that $s =_{\rho(\bar{x})} t$. Now let $\rho' : \{\bar{x}, \bar{y}\} \to V_{LFD}$ extend ρ by mapping $y_i \mapsto v_{m_i}$ for each $1 \leq i \leq n$. It follows that (i) $s =_{\rho(\bar{x})} t$ in the model (M, A), and (ii) $t \circ \rho' =_{\{\bar{x}, \bar{y}\}} s \circ \rho\,[\bar{m}/\bar{y}]$ as assignments in the model $G(M, A)$. In particular then, since the truth of a FO-formula under an assignment depends only on what the assignment maps its free variables to, $G(M, A), t \circ \rho' \models G(\bar{x}, \bar{y}) \wedge \varphi$. Applying the inductive hypothesis, we see that $M, t \models \tau_{\rho'}(G(\bar{x}, \bar{y}) \wedge \varphi)$ and hence $M, s \models \mathbb{E}_{\rho(\bar{x})}(\tau_{\rho'}(G(\bar{x}, \bar{y}) \wedge \varphi)$. This is one of the disjuncts in the translation $\tau_\rho(\exists \bar{y}(G(\bar{x}, \bar{y}) \wedge \varphi))$, and we are done. □

As noted before, the map G from dependence models to standard models is surjective. However, not every standard model is the G-image of a *distinguished* dependence model. This is an issue, because Theorem 3.4 applies only to distinguished dependence models. Fortunately, we have the following:

Theorem 3.5 *Let the number of variables in* V_{LFD} *be at least as great as the maximum arity of relations in the signature. Then, for every standard model* M *and guarded assignment* s*, there is a distinguished dependence model* $\mathbb{M}$ *and an admissible LFD assignment* t *in* $\mathbb{M}$ *such that* $(G(\mathbb{M}), t \circ \rho)$ *is GF-bisimilar to* (M, s)*, for some map* ρ.

Proof. Let $\mathbb{M} = (M', A)$ where M', A are as follows:

- $Dom(M') = Dom(M) \times V_{LFD}$, and
- $((m_1, u_1) \ldots, (m_k, u_k)) \in R^{M'}$ iff $(m_1, \ldots, m_k) \in R^M$ and for all $i, j \leq k$, if $m_i \neq m_j$ then $u_i \neq u_j$.
- $A \subseteq Dom(M')^{V_{LFD}}$ consist of all assignments that map each LFD-variable v to a pair of the form (m, v) for $m \in dom(M)$ some object in the old domain.

It is clear from the construction that $\mathbb{M}$ is distinguished.

Since s is a guarded assignment, there exists a tuple $(m_1, \ldots, m_k) \in R^M$ for some relation R, such that s maps every variable to an element of this tuple. We know that $|V_{LFD}| \geq k$. Let ρ be an arbitrary mapping from FO variables to LFD variables, such that $s(x) \neq s(y)$ iff $\rho(x) \neq \rho(y)$. Finally, let t be an LFD assignment that maps each $\rho(x)$ to the pair $(s(x), \rho(x))$. In this way, we have that, for each FO variable x, $(t \circ \rho)(x) = (s(x), \rho(x))$. Furthermore, by construction, t is an admissible assignment for $\mathbb{M}$.

It remains only to show that $(G(\mathbb{M}, t \circ \rho)$ is GF-bisimilar to (M, s). The GF bisimulation in question consists of all partial functions $f : Dom(M') \to Dom(M)$ such that (i) $dom(f)$ is a guarded subset of $dom(M')$, and (ii) f is the natural projection on its domain, i.e., $f((m, v) = m$. It can be easily verified that this satisfies all requirements of a GF-bisimulation. □

We state one immediate consequence, further implications of our translations will be discussed in Section 5 below.

Corollary 3.6 *Let the number of variables in V_{LFD} be at least as great as the maximum arity of relations in the signature. Then, a GF-formula φ is satisfiable on standard models iff, for some function ρ : free(φ) $\to V_{LFD}$, $\tau_\rho(\varphi)$ is satisfiable on dependence models.*

Proof. From right to left, if the LFD-formula $\tau_\rho(\varphi)$ is satisfiable, it is also satisfiable (by Proposition 2.9) on a pointed distinguished model $(M, A), s$, and so, by Theorem 3.4 we also have $G(M, A), s \circ \rho \models \varphi$.

From left to right, suppose $M, t \models \varphi$. Since every GF-formula is a Boolean combination of self-guarded formulas, we can assume without loss of generality that t is a guarded assignment. It then follows from Theorem 3.5 together with Theorem 3.4 and the invariance of GF-formulas under GF bisimulations, that, for some map ρ, $\tau_\rho(\varphi)$ is satisfied in a (distinguished) dependence model. □

Given the compositional nature of our translations τ_ρ on the Boolean operations, this immediately also implies a faithful reduction of semantic consequence in GF to semantic consequence in LFD.

4 A translation from LFD to GF

We now present a satisfiability-preserving translation from LFD to GF. Note that the translation $tr(\cdot)$ from LFD to FO given in Section 2 yields FO formulas that are, in general, not guarded. For example, if $V_{LFD} = \{x, y, z\}$, then $tr(D_x y)$ is the first-order formula $\varphi(x) = \forall y' z' (A(x, y', z') \to y = y')$, which uses unguarded quantification. Note, though, that the tr-translation of an

LFD formula without dependence atoms is guaranteed to be a GF formula. In this section, we present a satisfiability-preserving translation from full LFD, including the dependence atoms, to GF. This translation works by replacing dependence atoms by relational atoms using fresh relations, and adding guarded conditions that force these relations to behave like dependence atoms. [10]

Definition 4.1 ($\widehat{\mathbf{S}}$) LFD was defined relative to a relational signature $\mathbf{S}$ and a finite set of variables V_{LFD}. We now denote by $\widehat{\mathbf{S}}$ the relational signature that extends $\mathbf{S}$ with an n-ary relation A, for $n = |V_{LFD}|$, and with a fresh k-ary relation symbol $R^{X,Y}$ for each pair of subsets $X, Y \subseteq V_{LFD}$, where $k = |X|$.

The overall scheme of our translation will look like this:

$$\begin{array}{ccc} \text{LFD-formula } \varphi \text{ over } \mathbf{S} & \xrightarrow{\quad\sigma\quad} & \text{GF formula } \sigma(\varphi) \text{ over } \widehat{\mathbf{S}} \\ \text{dependence model } R(\mathfrak{M}_M) & & \text{standard model } M \\ \nwarrow & & \swarrow \\ & \text{LFD type model } \mathfrak{M}_M & \end{array}$$

In particular, our operation on models takes a detour through type models. In this diagram, R is the mapping from type models to dependence models given by Theorem 2.12.

Next we define the matching formula translation σ.

Definition 4.2 Let $V_{LFD} = \{v_1, \ldots, v_n\}$. For an LFD-formula ψ over a signature $\mathbf{S}$, we define $\sigma(\psi)$ as the following GF-formula over the signature $\widehat{\mathbf{S}}$:

$$\sigma(\psi) \quad := \quad tr^{\bullet}(\psi) \wedge setup(\psi) \wedge A(v_1, \ldots, v_n)$$

where

- $tr^{\bullet}(\psi)$ is defined inductively in the same way as $tr(\psi)$, with the only difference that $tr^{\bullet}(D_V U)) = R^{V,U}(\bar{v})$. (Here, for the sake of presentation, we assume a fixed, canonical ordering on variables, so that it is clear which element of the tuple $\bar{v}$ corresponds to which element of the set V.)
- $setup(\psi)$ is the conjunction of all GF-sentences of the form

$$\begin{array}{l} \forall v_1, \ldots, v_n(A(v_1, \ldots, v_n) \to \bigwedge_{u \in V} R^{V,u}(\bar{v})) \\ \forall v_1, \ldots, v_n(A(v_1, \ldots, v_n) \to (R^{V,U}(\bar{v}) \wedge R^{U,W}(\bar{u}) \to R^{V,W}(\bar{v})) \\ \forall v_1, \ldots, v_n(A(v_1, \ldots, v_n) \to [R^{V,U}(\bar{v}) \wedge tr^{\bullet}(\mathbb{E}_V \xi) \to tr^{\bullet}(\mathbb{E}_{V \cup U} \xi)]) \end{array}$$

where $V, U, W \subseteq V_{LFD}$; $\bar{v}, \bar{u}, \bar{w}$ are canonical enumerations of these sets and $\xi \in Cl(\psi)$ is a formula whose main connective is not conjunction or negation.

---

[10] This same trick with additional predicates is used in proving the finite dependence model property for LFD by an appeal to Herwig's theorem [12].

The formulas in the "set up" part of the above translation can be viewed as instances of the Projection, Transitivity and Transfer axioms in the axiomatization of LFD given in [4].

Next, we define our operation H, which takes a standard model and produces an LFD type model.

Definition 4.3 Let $V_{LFD} = \{v_1, \ldots, v_n\}$ and let ψ be an LFD formula. For a standard model $M \models setup(\psi)$, we define $\mathfrak{M}^{\psi}_{M}$ as:

$$\mathfrak{M}^{\psi}_{M} := \{\text{type}^{\psi}(M, s) \mid s : V_{LFD} \to Dom(M) \text{ such that } M, s \models A(v_1, \ldots, v_n)\}$$

with $\text{type}^{\psi}(M, s) = \{\varphi \in Cl(\psi) \mid M, s \models tr^{\bullet}(\varphi)\}$, where $Cl(\psi)$ is the closure of φ, as defined in Section 2.2.

Lemma 4.4 *If $M \models setup(\psi)$, then $\mathfrak{M}^{\psi}_{M}$ is an LFD type model.*

Proof. First, we show that each $\Delta \in \mathfrak{M}^{\psi}_{M}$ is a type. The $\neg$-consistency, $\wedge$-consistency, and $\mathbb{E}$-consistency properties are immediate from the construction of $\mathfrak{M}^{\psi}_{M}$ and the definition of $tr^{\bullet}$. Projection and transitivity hold by virtue of the corresponding conjuncts of $setup(\psi)$.

The *universal* property, i.e., that $\sim_{\emptyset}$ is the universal relation on $\mathfrak{M}$, also follows immediately from the construction of $\mathfrak{M}^{\psi}_{M}$ using the fact that $tr^{\bullet}$-translation of an LFD-sentence is a GF-sentence.

Finally, for the *witness* property, suppose that $\mathbb{E}_V \varphi \in \Delta$. Let s be a witnessing assignment for Δ. Then, in particular, $M, s \models A(v_1, \ldots, v_n) \wedge tr^{\bullet}(\mathbb{E}_V \varphi)$. It follows by the definition of $tr^{\bullet}$ that there is an assignment t with $t =_V s$, such that $M, t \models A(v_1, \ldots, v_n) \wedge tr^{\bullet}(\varphi)$. Let $\Delta' \in \mathfrak{M}^{\psi}_{M}$ be the type obtained from t as in Definition 4.3. Then $\varphi \in \Delta'$. Moreover, since $t =_V s$, and the translation $tr^{\bullet}$ preserves the free variables of the formula, we have that $\Delta \sim_V \Delta'$. In order to satisfy the witness property of type models, we must show something slightly stronger, namely that $\Delta \sim_{D^{\Delta}_{V}} \Delta'$. This can be shown in two steps: (i) it is easy to see that $D^{\Delta'}_{V} = D^{\Delta}_{V}$ by virtue of the fact that $\text{free}(tr^{\bullet}(D_V U)) = V$; (ii) the third conjunct of $setup(\psi)$ now allows us to lift the $\Delta \sim_V \Delta'$ relation to the $\Delta \sim_{D^{\Delta}_{V}} \Delta'$. To see this, first note that, by the $\neg$-consistency and $\wedge$-consistency properties of types, it is enough to show that Δ and Δ' agree on "non-decomposable" formulas with free variables in D^{Δ}_{V}, where we call a formula decomposable if its main connective is a conjunction or a negation. Let $\xi \in \Delta$ with $Y = \text{free}(\xi) \subseteq D^{\Delta}_{V}$, and assume that ξ is non-decomposable. Then $M, s \models tr^{\bullet}(\xi)$ and $M, s \models R^{V,Y}(\bar{v})$. By existential generalisation (cf. the definition of $tr^{\bullet}$) we get that $M, s \models tr^{\bullet}(\mathbb{E}_V \xi)$. Since $s =_V t$, this implies $M, t \models tr^{\bullet}(\mathbb{E}_V \xi)$ and $M, t \models R^{V,Y}(\bar{v})$ as well. Since we know that $M, t \models setup(\psi)$, we can now apply the 'transfer' condition to get that $M, t \models tr^{\bullet}(\mathbb{E}_{V \cup Y} \xi)$ which implies that $M, t \models \xi$ as $\text{free}(\xi) \subseteq V \cup Y$. Hence $\xi \in \Delta'$. The converse direction follows from a symmetric argument. □

We now obtain our dependence model $H(M)$ by applying the unraveling operation R from Section 2.2 to $\mathfrak{M}^{\psi}_{M}$.

Definition 4.5 Fix an LFD formula ψ over signature $\mathbf{S}$, let M be a standard model for $\widehat{\mathbf{S}}$ s.t. $M \models setup(\psi)$, and let $s : V_{LFD} \to dom(M)$ such that $M, s \models A(v_1, \dots, v_n)$.

(i) We define $H(M)$ as $R(\mathfrak{M}^{\psi}_{M})$.

(ii) We define $H(s)$ as an (arbitrarily chosen) admissible LFD assignment on $H(M)$ that realizes $\text{type}^{\psi}(M, s)$ in $H(M)$.

While not reflected in our notation, the model $H(M)$ depends on ψ. Also, note that $\text{type}^{\psi}(M, s) \in \mathfrak{M}^{\psi}_{M}$, and, by Theorem 2.12, $H(M)$ realizes every type in $\mathfrak{M}^{\psi}_{M}$, making $H(s)$ indeed well-defined.

Theorem 4.6 *Fix an LFD formula ψ over signature $\mathbf{S}$, and let M be standard model over the signature $\widehat{\mathbf{S}}$ such that $M \models setup(\psi)$, and let $s : V_{LFD} \to Dom(M)$, such that $M, s \models A(v_1, \dots, v_n)$. Then, for all formulas $\varphi \in Cl(\psi)$,*

$$H(M), H(s) \models \varphi \qquad \textit{iff} \qquad M, s \models tr^{\bullet}(\varphi)$$

Proof. If $H(M), H(s) \models \varphi$, then by the definition of $H(s)$, we have that $\varphi \in type^{\psi}(M, s)$, i.e., $M, s \models tr^{\bullet}(\varphi)$. Conversely, if $M, s \models tr^{\bullet}(\varphi)$, then $\varphi \in type^{\psi}(M, s)$. Hence, again by the definition of $H(s)$, we have that $H(M), H(s) \models \varphi$. □

Corollary 4.7 (i) *An LFD formula ψ is satisfiable iff its GF-translation $\sigma(\psi)$ is satisfiable.*

(ii) *An LFD entailment $\varphi \models \psi$ is valid iff the GF-entailment $setup(\varphi) \wedge tr^{\bullet}(\varphi) \wedge A(v_1, \dots, v_n) \models setup(\psi) \to tr^{\bullet}(\psi)$ is valid.*

Proof. From left to right, if $(M, A), s \models \psi$, then let M' be the $\widehat{\mathbf{S}}$-expansion of M obtained by setting $A^{M'}$ as $\{(s(v_1), \dots, s(v_n)) \mid s \in A\}$, and each $(R^{V,U})^{M'} = \{s(\bar{v}) \mid (M, A), s \models D_V U(\bar{v})\}$. A straightforward formula induction shows that, for all LFD formulas φ, $(M, A), s \models \varphi$ iff $M', s \models A(v_1, \dots, v_n) \wedge tr^{\bullet}(\varphi)$. Furthermore, $M' \models setup(\psi)$ since all auxiliary predicates have their intended interpretation. Therefore, we have that $M', s \models \psi$. The converse direction follows immediately from Theorem 4.6.

(ii) follows from (i), because (as careful inspection shows) $setup(\varphi \wedge \neg\psi)$ is logically equivalent to $setup(\varphi) \wedge setup(\psi)$ (to see this, note that, if $\varphi \wedge \neg\psi$ has a subformula that is non-decomposible, then the latter must be a subformula of φ or of ψ) and the fact that $tr^{\bullet}$ commutes with the Boolean connectives. □

5 Transfer results from our translations

In addition to shedding light on the relationship between LFD and GF, our translations can also be used to transfer results from GF to LFD and vice versa. We give several examples connecting known results, but also a new result, viz. a complete complexity analysis of LFD.

As a warming up, we reprove a known result, namely the finite model property for GF [7], Recall that a logic has the finite model property (FMP) if every satisfiable formula has a finite model. It has been long known that

GF has the FMP [7], and FMP was only recently shown to hold for LFD [12]. Therefore, the following transfer fact does not give us a new result, but it does provide an illustration of the value of our translations.

Theorem 5.1 (FMP for GF) *GF has the finite model property.*

Proof. Let φ be a satisfiable guarded formula. By corollary 3.6, since φ is satisfiable on standard models, it follows that $\tau_\rho(\varphi)$ is satisfiable on dependence models, for some $\rho : \text{free}(\varphi) \to V_{LFD}$ (for V_{LFD} sufficiently large). By the finite model property of LFD [12], there is a finite dependence model $\mathbb{M}$ satisfying $\tau_\rho(\varphi)$ at some $s \in A$. We can take $\mathbb{M}$ to be distinguished by Proposition 2.9 (which preserves finiteness). In fact, inspection of the proof in [12] shows that its construction via Herwig's theorem already ensures distinguishedness. By Theorem 3.4, then, $G(\mathbb{M})$ is a finite model of φ. □

Theorem 5.2 (Complexity of LFD satisfiability) *For a finite set of LFD-variables V_{LFD}, the satisfiability problem for LFD-formulas in V_{LFD} is ExpTime-complete. The same problem is 2ExpTime-complete if V_{LFD} is considered as part of the input.*

Proof. Upper bound: by reduction to GF. The satisfiability problem for GF is 2ExpTime-complete [7]. More precisely, the satisfiability of a GF formula with k FO-variables can be checked in time $2^{O(|\varphi|\cdot k^k)}$ (cf. [14]). A careful analysis of our translation from LFD to GF shows it can be performed in time $O(|\varphi|) \cdot 2^{O(k)}$, where $k = |V_{LFD}|$. Indeed, computing the modified first-order translation $tr^\bullet(\varphi)$ can be done in linear time, so only the set-up part brings all the complexity. The formula $setup(\varphi)$ consist of 2^k many conjuncts encoding a couple of projection axiom instances on the proxy-dependence atoms $R^{X,y}$ as well as 2^{3k} many conjuncts encoding a transitivity axiom instance and $O(|\varphi|) \cdot 2^{2k}$ many conjuncts encoding a transfer axiom instance for some 'indecomposable' formulas in the closure $Cl(\varphi)$. Putting this together, we obtain that the satisfiability problem for LFD is in ExpTime if V_{LFD} is treated as fixed in the complexity analysis, and that it is in 2ExpTime if V_{LFD} is part of the input (due to the fact that, even though the translation from LFD to GF is exponential, it does not increase the number of variables).

Lower bound: by reduction from GF. Observe that our translation from GF to LFD is exponential due to the disjunction in the translation clause for the guarded quantifiers. Exponential in the nesting depth of (polyadic) guarded-quantifications, to be precise. For this reason, it is not immediately clear that complexity upper bounds for GF transfer to LFD as is. However, we can make use of a result from [7], according to which there is a satisfiability-preserving translation from GF-formulas to a "Scott-normal form" (which can be performed in polynomial time and without increasing the number of variables), where the normal form uses only two levels of polyadic guarded quantification. Our translation is polynomial when applied to GF-formulas in this normal form. As we mentioned earlier, the satisfiability problem for GF-formulas with a bounded number of variables is ExpTime-complete, and with an unbounded number of variables it is 2ExpTime-complete. Since our translation from GF to LFD does not require more LFD variables than the number of variables in the input formula, this establishes our lower bounds. □

In fact, we can improve this result a little. By the *monadic fragment* of LFD we mean LFD-formulas that only use unary relation symbols.

Theorem 5.3 (Complexity of the monadic fragment of LFD) *Fix a finite set V_{LFD} of LFD-variables with $|V_{LFD}| \geq 2$ The satisfiability problem for the monadic fragment of LFD with V_{LFD} is ExpTime-complete.*

Proof. [sketch] The upper bound follows from Theorem 5.2. For the lower bound, we reduce from the satisfiability problem of the basic modal logic **K** extended with the global modality (which we will denote as $\mathbf{K}+U$). This logic is known to be ExpTime-complete (cf. [6]). By a simple encoding trick due to Halpern and Vardi [10], the satisfiability problem for $\mathbf{K}+U$ reduces to the satisfiability problem for the multi-modal logic $\mathbf{S5}_2+U$ (where $\mathbf{S5}_2$ is the bi-modal "fusion" logic that has two S5 modality without interaction axioms). The coding trick in question consists of replacing every occurrence of $\Diamond$ by $\Diamond_1\Diamond_2$ where $\Diamond_1$ and $\Diamond_2$ are the two **S5**-modalities (cf. [10] for the proof that this preserves satisfiability). The satisfiability problem for $\mathbf{S5}_2+U$, finally, embeds straightforwardly into the monadic fragment of LFD (even without using dependence atoms): each proposition letter becomes a unary predicate, $\Diamond_1$ becomes $\mathbb{E}_x$ and $\Diamond_2$ becomes $\mathbb{E}_y$ for $x, y \in V_{LFD}$ two distinct LFD-variables, whereas the global modality (in its existential form) becomes $\mathbb{E}_\emptyset$. □

Remark 5.4 Our translation from GF to (the dependence-atom-free fragment of) LFD can also be composed with the *tr* translation from LFD to GF to obtain a SAT-reduction from GF to its "universal guard" fragment (i.e., GF-formulas with a single guard predicate that occurs only in guard position). [11]

As our final example, we prove that LFD has Craig interpolation, offering a model-theoretic alternative to the sequent calculus-based proof in [4].

Theorem 5.5 (Craig interpolation for LFD) *For every valid LFD implication $\models \varphi \to \psi$, there is an LFD formula ϑ such that*

(i) $\models \varphi \to \vartheta$,

(ii) $\models \vartheta \to \psi$, *and*

(iii) *all relation symbols occurring in ϑ occur both in φ and in ψ.*

Proof. [sketch] Let $\varphi \to \psi$ be valid in LFD. By Corollary 4.7, the implication $(setup(\varphi) \wedge tr^\bullet(\varphi) \wedge A(v_1,\dots,v_n)) \to ((setup(\psi) \wedge A(v_1,\dots,v_n)) \to tr^\bullet(\psi))$ is valid in GF. Note that, in this GF formula, all quantification is guarded by A. Furthermore, the antecedent and the negation of the consequent are self-guarded (meaning that the free variables are guarded by an atomic conjunct), and the free variables of the consequent form a subset of the free variables of the antecedent (viz. $\{v_1,\dots,v_n\}$). Hence, by the weak Craig interpolation theorem [11, Thm. 4.5], there is a GF-formula χ such that

[11] In fact, to its fragment where all quantifiers are guarded by the same atom $A(v_1,\dots,v_n)$.

(i) $\models (setup(\varphi) \wedge tr^\bullet(\varphi) \wedge A(v_1, \ldots, v_n)) \to \chi$,

(ii) $\models \chi \to ((setup(\psi) \wedge A(v_1, \ldots, v_n)) \to tr^\bullet(\psi))$, and

(iii) All relation symbols occurring in χ except possibly for A, occur both in $(setup(\varphi) \wedge tr^\bullet(\varphi))$ and in $((setup(\psi) \wedge A(v_1, \ldots, v_n)) \to tr^\bullet(\psi))$.

(iv) The free variables of χ belong to V_{LFD}.

It suffices only to translate χ back to LFD. This is not entirely trivial, since χ is now a GF-formula over the expanded signature $\widehat{\mathbf{S}}$. That is, it will in general contain the A and $R^{V,U}$ relations. Fortunately, it turns out that the translation τ from GF to LFD that we gave in Section 3 can be extended in a straightforward way to the full signature $\widehat{\mathbf{S}}$, as follows:

$$\tau_\rho(A(\bar{x})) \quad = \top \text{ if } \rho(\bar{x}) \text{ is precisely the sequence } v_1, \ldots, v_n;\ \bot \text{ otherwise}$$

$$\tau_\rho(R^{U,V}(\bar{u})) = D_{\rho[U]}\rho[V]$$

For any dependence model $\mathbb{M} = (M, A)$, let $\widehat{\mathbb{M}}$ be the standard model that is the $\widehat{\mathbf{S}}$-expansion of M obtained in the natural way (as we did in the proof of Corollary 4.7). Then, it can be shown that, with the above modification to τ,

(a) For all distinguished $\mathbb{M}$ and admissible s, and for all GF-formulas ξ over $\widehat{\mathbf{S}}$, we have $\widehat{\mathbb{M}}, s \circ \rho \models \xi$ iff $\mathbb{M}, s \models \tau_\rho(\xi)$.

(b) For every LFD formula ξ, $\tau_\rho(tr^\bullet(\xi))$ is equivalent to ξ over distinguished dependence models, and hence over all dependence models, if we take ρ to be the identity map on V_{LFD}.

Putting this all together, it follows that $\vartheta := \tau_\rho(\chi)$ (under the above modified translation function τ, and with ρ the identity map on V_{LFD}) is a Craig interpolant for our original implication $\varphi \to \psi$ on distinguished dependence models, and hence, by Proposition 2.9, on all dependence models. □

LFD also has Craig interpolants that only use shared variables, [4]. To obtain this further information, the preceding analysis should be refined.

6 Further Directions

Language extensions One natural question is if our results extend to richer languages than the ones considered here. Here is an example.

The translation τ_ρ from GF into LFD presented in Section 3 can be extended to deal with *identity atoms* $x = y$ between GF variables x, y. It suffices to add two clauses: (a) $\tau_\rho(x = y) = \top$ if $\rho(x) = \rho(y)$ [the mapping to LFD variables enforces identity throughout], (b) $\tau_\rho(x = y) = \bot$ if $\rho(x) \neq \rho(y)$ [the values will always be distinct in distinguished dependence models]. Thus GF(=), too, translates compositionally into the decidable logic LFD without identity.

Another natural extension from an LFD perspective is adding *local independence atoms* $I_X y$ saying that fixing the local values of the variables X at assignment s in the current dependence model puts no constraint on y: which can still take any value in its range on the X-restricted subset of assignments.

Adding atoms $I_X y$ to LFD results in an undecidable logic [4]. Still, even from the original motivation for CRS, analyzing true FO quantifiers in terms of their independence behavior versus the dependence behavior of CRS quantifiers makes sense, suggesting a study of richer fragments of FO.[12]

Translation patterns Our three translations show some general patterns.

Translating LFD into the FO language via $tr(.)$ was compositional, while there was an inverse model transformation T supporting the usual contravariant translation equivalence $M, s \models tr(\varphi)$ iff $T(M), s \models \varphi$. Thus there is an adjunction between the maps $tr(.), T$ whose transfer of properties between models has been determined in an abstract setting in [17]. In fact, T was onto and bijective, which allows for additional transfer of properties between LFD and FO to the extent that LFD can be identified with a fragment of first-order logic.

The compositional translation τ_ρ from GF into LFD also came with an inverse model transformation G supporting a contravariant equivalence, but this time, it only worked for the special classes of distinguished LFD models and correlated distinguished GF models. However, all models for LFD were LFD-bisimilar to these distinguished models, and an analogous result held for GF. This relaxed notion of translation up to bisimulation for the relevant two languages seems an interesting generalization of the standard case which still facilitates a good deal of transfer. We saw this for logical consequence, it also works straightforwardly for decidability, and we even saw how, putting together the translation with its matching model transformation allowed us to transfer the finite model property.

Finally, our translation from LFD into GF was not entirely compositional, as we also needed to carry special conditions for LFD type models for the formula being translated. This amounts to translating from LFD on its ordinary models into GF on a special class of models satisfying a theory consisting of special conditions for type models. This theory allowed for a transformation of its GF models into LFD type models, which can then, in a modular fashion, be represented as ordinary LFD dependence models.

Several general questions arise here. One is about the range of the third type of translation technique. We believe that it can deal quite generally with modal-style logics that have an effective 'finite quasimodel property' plus a representation theorem.[13] We leave this matter for further exploration.[14]

[12]Other natural language extensions to explore would introduce *fixed-point operators*, [7], a device whose power has not yet been studied for the dependence logic LFD.

[13]As an illustration, consider the modal logic K4 which is complete for transitive models. We cannot translate this directly into GF since transitivity is not guarded. This can be overcome by suitably translating into the guarded fixed-point logic μ(GF), but a simpler solution is this. Translate the requirements on finite filtrations for K4 models, where in particular, if $\Box\varphi$ belongs to a type, then both $\Box\varphi, \varphi$ belong to accessible types. The result is a guarded description of filtration models.

[14]A further interesting issue would be an abstract general transfer analysis of the generalized types of translation that we have provided in this paper.

7 Conclusion

The guarded fragment GF models restricted quantification, the modal dependence logic LFD models local dependence between variables. We have presented two new translation results. One is a faithful compositional translation from GF into the dual CRS fragment of LFD, the other is an effective reduction from full LFD with dependence atoms to GF. We then demonstrated a number of consequences for transfer of known properties between GF and LFD, and also derived some new results, such as a determination of the computational complexity of satisfiability in LFD. In summary, local dependence and guarding are much closer as semantic notions than what may have been thought previously. Moreover, the techniques that we introduce to prove our results may be of wider interest, and we provided some pointers in our final discussion.

References

[1] Abiteboul, S., R. Hull and V. Vianu, "Foundations of Databases," Addison-Wesley, 1995. URL `http://webdam.inria.fr/Alice/`

[2] Andréka, H., I. Németi and J. van Benthem, *Modal languages and bounded fragments of predicate logic*, Journal of Philosophical Logic **27** (1998).

[3] Andréka, H., J. van Benthem, N. Bezhanishvili and Í. Németi, "Changing a Semantics: Opportunism or Courage?" Birkhäuser Verlag, Basel, 2007 pp. 307–337.

[4] Baltag, A. and J. van Benthem, *A simple logic of functional dependence*, J. Philos. Log. **50** (2021), pp. 939–1005.
URL `https://doi.org/10.1007/s10992-020-09588-z`

[5] Barany, V., B. ten Cate and L. Segoufin, *Guarded negation*, Journal of the ACM (JACM), Association for Computing Machinery (2015), pp. 22.1–22:26.

[6] Blackburn, P., M. de Rijke and Y. Venema, "Modal Logic," Cambridge University Press, Cambridge UK, 2001.

[7] Grädel, E., *On the restraining power of guards*, Journal of Symbolic Logic **64** (1998), pp. 1719–1742.

[8] Grädel, E. and M. Otto, "Guarded Teams: The Horizontally Guarded Case," Leibniz-Zentrum für Informatik, Dagstuhl, Germany, 2020 pp. 22:1–22:17.

[9] Grädel, E. and P. Pützstück, *Logics of dependence and independence: The local variants*, Journal of Logic and Computation **31** (2021), pp. 1690–1715.

[10] Halpern, J. Y. and M. Y. Vardi, *The complexity of reasoning about knowledge and time. i. lower bounds*, Journal of Computer and System Sciences **38** (1989), pp. 195–237.
URL `https://www.sciencedirect.com/science/article/pii/0022000089900391`

[11] Hoogland, E. and M. Marx, *Interpolation and definability in guarded fragments*, Studia Logica **70** (2002), pp. 373–409.

[12] Koudijs, R., *Finite model property and bisimulation for lfd*, Electronic Proceedings in Theoretical Computer Science **346** (2021), pp. 166–178.
URL `http://dx.doi.org/10.4204/EPTCS.346.11`

[13] Marx, M., *Taming first order logic: Relating the semantic and the syntactic approach.*
URL `https://festschriften.illc.uva.nl/j50/contribs/marx/index.html`

[14] Marx, M. and Y. Venema, "Local Variations on a Loose Theme: Modal Logic and Decidability," Springer Berlin Heidelberg, Berlin, Heidelberg, 2007 pp. 371–429.

[15] Németi, Í., "Free algebras and decidability in algebraic logic," Doctoral Dissertation for the Hungarian Academy of Sciences, 1986, translation of the relevant chapter as: Decidability of weakened versions of first-order logic. In Logic Colloquium 92, Stanford, CSLI Pubications, 1995. pp.177–241.

[16] Väänänen, J., "Dependence Logic," Cambridge University Press, Cambridge UK, 2009.

[17] van Benthem, J., *Information transfer across chu spaces*, Logic Journal of the IGPL **6** (2000), pp. 719–731.

[18] van Benthem, J., *Guards, bounds, and generalized semantics*, J. Log. Lang. Inf. **14** (2005), pp. 263–279.
URL `https://doi.org/10.1007/s10849-005-5786-y`

[19] Van Benthem, J., "Crs and Guarded Logics," Bolyai Society Mathematical Studies 22, Mathematical Institute, Hungarian Academy of Sciences, Budapest, 2012 pp. 273 – 301.

A multi-modal logic for Galois connections

Alfredo Burrieza [1]

Departamento de Filosofía
Universidad de Málaga, Spain

Inmaculada Pérez de Guzmán [2]

Departamento de Matemática Aplicada
Universidad Málaga, Spain

Antonio Yuste-Ginel [3]

Departamento de Filosofía
Universidad de Málaga, Spain

Abstract

This paper provides an axiomatic characterization of Galois connections by means of a sound and complete system for a multi-modal language. Semantically, our frames consist of families of partially ordered sets (posets) which are (possibly) related by Galois connections. Syntactically, we use a language containing two types of modalities: One to move around inside each poset and the other to jump functionally from one poset to another. The completeness proof follows a step-by-step argument with some interesting particularities.

Keywords: Galois connections, modal logic, temporal×modal logic, multi-modal logic, step-by-step, completeness

---

1 Introduction

Galois connections are a very well-known concept within diverse areas of mathematics such as algebra, geometry and topology (see [8] for a monograph). Roughly speaking, a Galois connection is a couple of order-preserving or order-inverting maps between two ordered sets. This tool has also been extensively used in theoretical computer science, as it plays a fundamental role in the development of diverse theories with immediate applications (see [18]), Formal Concept Analysis (FCA) [11] being one of the most notable.

---

[1] burrieza@uma.es

[2] pguzman@uma.es

[3] antonioyusteginel@gmail.com

Concurrently, multi-modal logics are an excellent tool for studying properties of mathematical theories (with [12,16,17] as pioneering works in this field). Hence, it is unsurprising that, during the last few years, there has been a growing interest in the analysis of the links between (multi-)modal logics and Galois connections (e.g., [14,15,19,20,22]).

This paper is inserted in the previous tradition. However, and in contrast to the quoted approaches, our main objective is not to introduce operators which, due to its behaviour, form a Galois connection, nor to extend logics with new operators and to structure their semantics using Galois connections. Rather, we aim at characterising axiomatically what a Galois connection is. This objective is pursued through the development of a multi-modal logic, which we denote as $\mathcal{L}_{GP}$.[4] In other words, our main goal is to define a multi-modal logic provided with a Kripke semantics in which frames are partially ordered sets which are in turn related by means of Galois connections, and to give a sound and complete axiomatic system for such a logic.

As we will see, our representation naturally calls for the use of two types of modal operators: One of them is used to move forward and back inside each poset, while the other one connects these posets functionally. In this sense, our framework builds on the spirit and style of [4,3,5,6], where a very similar setting is used to axiomatise different properties of functions between linear orders. Moreover, and differently to what is usually done, (*i*) we use partial functions instead of total functions; and (*ii*) we study Galois connections between arbitrary families of posets, instead of restricting our attention to a finite number of them (usually one or two).

The rest of this paper is organised as follows. Section 2 presents the needed mathematical background. In particular, we recall the notion of Galois connections for partial functions. In Section 3, we introduce the language and semantics of our logic $\mathcal{L}_{GP}$. We present and comment on our axiomatisation in Section 4. The main technical contributions of the paper are contained in Section 5, where we provide a completeness proof for the mentioned axiom system following an elaborated step-by-step construction. In Section 6, we briefly discuss some closely related work. Finally, we close the paper in Section 7 by depicting open paths for future work. Some of the proofs are to be found in the Appendix.

2 Mathematical Preliminaries

In this section, we provide the necessary ingredients for a proper understanding of our semantics. Essentially, we introduce (alternative characterisations of) the notion of a Galois connection for partial functions. Although this notion makes sense for weaker order-theoretic structures, such as preorders (see e.g., [8, Chapter 1]), we restrict our attention to the most common case of partially ordered sets. So, let us first of all recall the definition.

Definition 2.1 Given a non-empty set A and a binary relation R on A, we

---

[4] As an acronym of "$\mathcal{L}$"**ogic of "G"alois connections between "P"osets**.

say that (A, R) is a **partially ordered set** (*poset*, for short) if R is reflexive, antisymmetric and transitive.

Notation: In what follows we shall use the following notation: If $(A, \leq_A)$ is a poset and $a \in A$:

$$(i)\ a{\uparrow} = \{a' \in A \mid a \leq_A a'\}; \qquad (ii)\ a{\downarrow} = \{a' \in A \mid a' \leq_A a\}.$$

We adopt the following definition of a Galois connection, which generalises the usual one from total functions to partial ones: [5]

Definition 2.2 Let $(A, \leq_A)$ and $(B, \leq_B)$ be two posets and $f\colon (A, \leq_A) \longrightarrow (B, \leq_B)$, $g : (B, \leq_B) \longrightarrow (A, \leq_A)$ be partial functions. We say that the pair (f, g) is a **Galois connection** (between $(A, \leq_A)$ and $(B, \leq_B)$) iff:

1.1 *$Im(f) \subseteq Dom(g)$ and $Im(g) \subseteq Dom(f)$; and*

1.2 For all $a \in Dom(f)$ and for all $b \in Dom(g)$, we have that:

$$a \leq_A g(b) \quad \textit{if and only if} \quad f(a) \leq_B b.$$

Given a Galois connection (f, g), we say that f is a **residuated function** (sometimes denoted by $f^{\rightarrow}$) and g is called its **residual function** (sometimes denoted by $f^{\leftarrow}$).

Our definition is well-behaved since, once item **1.1** is assumed, then a typical, alternative characterisation of Galois connections is equivalent to **1.2**. Let us introduce a couple of preliminary, needed notions.

Definition 2.3 Let $(A, \leq_A)$ and $(B, \leq_B)$ be posets and $f : A \to B$ a partial function. We say that f is **monotone** if, for all $a_1, a_2 \in Dom(f)$, we have that, if $a_1 \leq_A a_2$, then $f(a_1) \leq_B f(a_2)$.

Definition 2.4 Let $(A, \leq)$ be a poset and $f : A \to A$ a partial function. We say that f is **inflationary** if for all $a \in Dom(f)$ we have that $a \leq f(a)$. We say that f is **deflationary** if for all $a \in Dom(f)$, we have that $f(a) \leq a$

Proposition 2.5 *Let $(A, \leq_A)$ and $(B, \leq_B)$ be two posets and $f\colon (A, \leq_A) \longrightarrow (B, \leq_B)$ a partial function. Then, the following conditions are equivalent:*

1. *There exists a partial function $g\colon (B, \leq_B) \longrightarrow (A, \leq_A)$ s.t. (f, g) is a Galois connection.*
2. *$f : (A, \leq_A) \longrightarrow (B, \leq_B)$ is monotone and there exists monotone $g : (B, \leq_B) \longrightarrow (A, \leq_A)$ s.t.*

   2.1 $Im(f) \subseteq Dom(g)$ and $Im(g) \subseteq Dom(f)$.

   2.2 $g \circ f$ is inflationary and $f \circ g$ is deflationary.

Remark 2.6 [Lack of uniqueness of residual partial functions] In Galois connections for total functions, each residuated function uniquely determines its residual (see e.g., [10]). Interestingly, this property is lost when our definition for partial function is adopted. Figure 1 provides an example of two different Galois connections between posets $\mathbb{P}_0 = (A_0, \leq_0)$ and $\mathbb{P}_1 = (A_1, \leq_1)$ which,

[5] See [21] for further generalizations of the notion.

however, have the same resituated function. Each element of $\leq_0$ and $\leq_1$ is represented as a simple, directed arrow, except for reflexive arrows, which have been omitted for the sake of clarity. Residuated functions are depicted as double arrows, while residual functions are depicted as dashed, double arrows.

Fig. 1. Counterexample to uniqueness of residual partial functions.

Example 1 (Concept forming operators in FCA) In *Formal Concept Analysis* (FCA) [11], a *formal context* is a tuple $(\mathcal{O}, \mathcal{A}, \mathcal{R})$ where $\mathcal{O}$ is a set of objects, $\mathcal{A}$ is a set of attributes and $\mathcal{R} \subseteq \mathcal{O} \times \mathcal{A}$ is a relation among objects and attributes. The concept forming operator $\uparrow_R$: $2^{\mathcal{O}} \longrightarrow 2^{\mathcal{A}}$ is defined as $\uparrow_R (O) = \{a \in \mathcal{A} \mid \forall o \in O, (o, a) \in \mathcal{R}\}$ for any $O \subseteq \mathcal{O}$ (and something analogous is done for $\downarrow_R$: $2^{\mathcal{A}} \longrightarrow 2^{\mathcal{O}}$). These operators are used to set up the notion of a *formal concept*, i.e., a pair (O, A) s.t. $\uparrow_R (O) = A$ and $\downarrow (A) = O$. As it has been thoroughly exploited within FCA, we have that $(\uparrow_R, \downarrow_R)$ is a Galois connection between $(2^{\mathcal{O}}, \subseteq)$ and $(2^{\mathcal{A}}, \supseteq)$.

3 The logic $\mathcal{L}_{GP}$

In this section we introduce the multi-modal logic $\mathcal{L}_{GP}$. We start by introducing the language of this logic, and then move to define an adequate class of models for it.

3.1 Syntax

We assume a denumerable set of atoms $\mathcal{V}$ as fixed from now on. The language L_{GP} is the one generated by the following grammar:

$$A ::= \perp \mid p \mid \neg A \mid (A \wedge A) \mid FA \mid PA \mid \langle \overset{ij}{\rightarrow} \rangle A \mid \langle \overset{ij}{\leftarrow} \rangle A$$

where p ranges over $\mathcal{V}$, and both i and j range over $\mathbb{N}$.

FA reads "A is true at a state of the current poset which is $\leq$-accessible from the current state", while PA reads "A is true at a state of the current poset from which the current state is $\leq$-accessible". Moreover, $\langle \overset{ij}{\rightarrow} \rangle A$ reads "we are at the state s_i of poset $\mathbb{P}_i$ and A is true at the image of s_i in poset $\mathbb{P}_j$ (by a residuated function)", while the meaning of $\langle \overset{ij}{\leftarrow} \rangle A$ is "we are at the state s_j of poset $\mathbb{P}_j$ and A is true at the image of s_j in poset $\mathbb{P}_i$ (by a residual function)". The rest of Boolean connectives are defined and read as usual. The duals of $F, P, \langle \overset{ij}{\rightarrow} \rangle, \langle \overset{ij}{\leftarrow} \rangle$ are denoted $G, H, [\overset{ij}{\rightarrow}], [\overset{ij}{\leftarrow}]$ and defined $\neg F \neg, \neg P \neg, \neg \langle \overset{ij}{\rightarrow} \rangle \neg, \neg \langle \overset{ij}{\leftarrow} \rangle \neg$ respectively. The need of taking into account both

$\langle \overset{ij}{\rightarrow} \rangle$ and $\langle \overset{ij}{\leftarrow} \rangle$ as primitive operators will be clear when we define the semantics of L_{GP} in Section 3.2.

We also introduce the notion of the **Galois mirror image** of a formula. If A is a formula, its Galois mirror image is a formula A' obtained from A by replacing in A each occurrence of $P, F, H, G, \langle \overset{ij}{\rightarrow} \rangle, \langle \overset{ij}{\leftarrow} \rangle, [\overset{ij}{\rightarrow}]$ and $[\overset{ij}{\leftarrow}]$ by $F, P, G, H, \langle \overset{ij}{\leftarrow} \rangle, \langle \overset{ij}{\rightarrow} \rangle, [\overset{ij}{\leftarrow}]$ and $[\overset{ij}{\rightarrow}]$ respectively.

Example 2 The Galois mirror image of $[\overset{01}{\rightarrow}]\, H\, p \vee \langle \overset{21}{\leftarrow} \rangle\, q$ is $[\overset{01}{\leftarrow}]\, G\, p \vee \langle \overset{21}{\rightarrow} \rangle\, q$.

3.2 Semantics of L_{GP}

As announced, the skeleton of our models consists of a family of posets connected by residuated and residual functions. A formal definition follows below:

Definition 3.1 A **Galois frame** for L_{GP} is a tuple $\Sigma = (\Lambda, \mathcal{P}_{oset}, \mathcal{F})$ s.t.:

(i) $\varnothing \neq \Lambda \subseteq \mathbb{N}$, whose elements are called **labels**.

(ii) $\mathcal{P}_{oset} = \{(\mathbb{P}_i, \leq_i) \mid i \in \Lambda\}$ is a non-empty set of pairwise disjoint posets s.t. $\mathbb{P}_i \neq \varnothing$ for every label $i \in \Lambda$.
The elements of the disjoint union $\mathbb{S}_\Lambda = \bigsqcup_{i \in \Lambda} \mathbb{P}_i$, denoted by $s, s', etc.$, are called **states**. If we want to specify that a state s belongs to a poset $\mathbb{P}_i$ we denote it by s_i.

(iii) $\mathcal{F} \subseteq \{f : \mathbb{P}_i \longrightarrow \mathbb{P}_j \mid i, j \in \Lambda\}$ is a set of partial functions s.t.:
  (a) for each $f \in \mathcal{F}$, we have that $Dom(f) \neq \varnothing$.
  (b) for an arbitrary pair $i, j \in \Lambda$, it holds that:
  - if $i \neq j$, then there is at most one function $f \in \mathcal{F}$ s.t. $f\colon \mathbb{P}_i \longrightarrow \mathbb{P}_j$.
  - if $i = j$, then there are at most two functions $f, f' \in \mathcal{F}$ s.t. $f\colon \mathbb{P}_i \longrightarrow \mathbb{P}_j$ and $f'\colon \mathbb{P}_i \longrightarrow \mathbb{P}_j$.

  (c) for every $f \in \mathcal{F}$, f is either a residuated function or its residual. If f is a residuated function from $\mathbb{P}_i$ to $\mathbb{P}_j$, then we denote it as $f_{\overrightarrow{ij}}$ and its residual as $f_{\overleftarrow{ij}}$. Moreover, for every $i, j \in \Lambda$, we have that $f_{\overrightarrow{ij}} \in \mathcal{F}$ if and only if $f_{\overleftarrow{ij}} \in \mathcal{F}$. In the special case where $f, f' \in \mathcal{F}$, $f : \mathbb{P}_i \longrightarrow \mathbb{P}_j$ and $f' : \mathbb{P}_j \longrightarrow \mathbb{P}_i$ and both (f, f') and (f', f) are Galois connections (e.g., when f is an isomorphism and $f' = f^{-1}$), then we have to explicitly indicate which of the two functions is considered residuated and which one is considered residual.

Definition 3.2 A **Galois model** for L_{GP} is a tuple $\mathcal{M} = (\Sigma, h)$, where $\Sigma = (\Lambda, \mathcal{P}_{oset}, \mathcal{F})$ is a Galois frame and h is a function, called an **interpretation**, assigning to each atom $p \in \mathcal{V}$ a subset of $\mathbb{S}_\Lambda$. An interpretation h is recursively extended to a function (still denoted by h) defined for every formula of L_{GP}, by interpreting Boolean constants and connectives in a standard way and satisfying the following conditions:

- $h(FA) = \{s \in \mathbb{S}_\Lambda \mid s{\uparrow} \cap h(A) \neq \varnothing\}$
- $h(PA) = \{s \in \mathbb{S}_\Lambda \mid s{\downarrow} \cap h(A) \neq \varnothing\}$

- $h(\langle \overset{ij}{\rightarrow} \rangle A) = \{s_i \in \mathbb{P}_i \mid f_{ij}^{\rightarrow} \in \mathcal{F},\ s_i \in Dom(f_{ij}^{\rightarrow}) \textit{ and } f_{ij}^{\rightarrow}(s_i) \in h(A)\}$
- $h(\langle \overset{ij}{\leftarrow} \rangle A) = \{s_j \in \mathbb{P}_j \mid f_{ij}^{\leftarrow} \in \mathcal{F}, s_j \in Dom(f_{ij}^{\leftarrow}) \textit{ and } f_{ij}^{\leftarrow}(s_j) \in h(A)\}$

Hence, we can deduce the semantics of non-primitive connectives:

- $h(GA) = \{s \in \mathbb{S}_\Lambda \mid s\!\uparrow \subseteq h(A)\}$
- $h(HA) = \{s \in \mathbb{S}_\Lambda \mid s\!\downarrow \subseteq h(A)\}$
- $h([\overset{ij}{\rightarrow}]A) = \{s_i \in \mathbb{P}_i \mid \text{If } f_{ij}^{\rightarrow} \in \mathcal{F} \text{ and } s_i \in Dom(f_{ij}^{\rightarrow}), \text{then } f_{ij}^{\rightarrow}(s_i) \in h(A)\} \cup \{s_k \in \mathbb{P}_k \mid k \neq i\}$
- $h([\overset{ij}{\leftarrow}]A) = \{s_j \in \mathbb{P}_j \mid \text{If } f_{ij}^{\leftarrow} \in \mathcal{F} \text{ and } s_j \in Dom(f_{ij}^{\leftarrow}), \text{ then } f_{ij}^{\leftarrow}(s_j) \in h(A)\} \cup \{s_k \in \mathbb{P}_k \mid k \neq j\}$

The class of all Galois-models is denoted by $\mathcal{M}_{GP}$. The semantic notions of *satisfiability*, *validity* and related ones are defined as usual (see e.g., [1]).

Example 3 Figure 2 depicts a Galois frame with three posets, $(\mathbb{P}_0, \leq_0)$, $(\mathbb{P}_1, \leq_1)$ and $(\mathbb{P}_2, \leq_2)$. Each element of $\leq_i$ is represented as a simple, directed arrow, except for reflexive and transitive arrows, which have been omitted for the sake of clarity. Residuated functions are depicted as double arrows, while residual functions are depicted as dashed, double arrows.

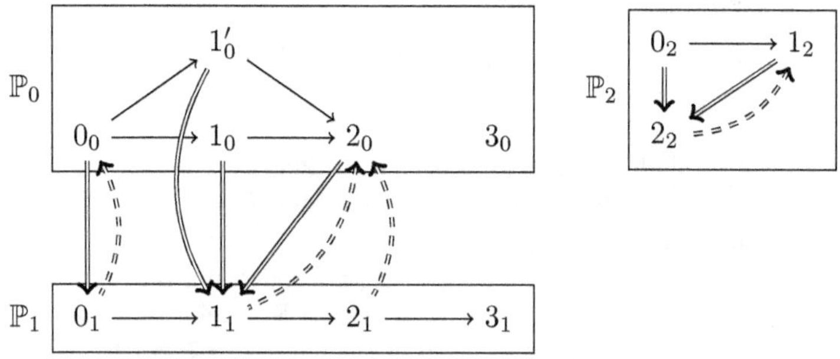

Fig. 2. Galois frame of Example 3.

Let us now define a model based on the previous Galois frame by adding the interpretation h defined as follows $h(p) = \{1_0, 1_1, 2_1, 3_1, 1_2\}$ and $h(q) = \mathbb{P}_0 \cup \mathbb{P}_1 \cup \mathbb{P}_2$ (for all $q \neq p$). Let us evaluate some formulae in this model:

Formula	**True at**	**Formula**	**True at**
$\langle \overset{01}{\rightarrow} \rangle p$	$1_0, 1_0', 2_0$	$\langle \overset{12}{\leftarrow} \rangle p$	*none*
$[\overset{01}{\rightarrow}] p$	*every state except* 0_0	$\langle \overset{22}{\leftarrow} \rangle p$	2_2
$\langle \overset{01}{\leftarrow} \rangle \neg p$	$0_1, 1_1, 2_1$	$PF\langle \overset{01}{\leftarrow} \rangle \langle \overset{01}{\rightarrow} \rangle p$	$0_1, 1_1, 2_1, 3_1$
$[\overset{12}{\rightarrow}] p$	*every state*	$F\langle \overset{01}{\rightarrow} \rangle H[\overset{01}{\leftarrow}] \neg p$	$0_0, 1_0, 1_0', 2_0$

4 An axiom system for L_{GP}

We now introduce an axiomatic calculus intended to produce as theorems all formulae that are valid in $\mathcal{M}_{GP}$. We denote this axiom system by S_{GP}. In what follows, we will use the symbol λ to denote a finite sequence (included the empty one) of the operators F and P.

Axiom schemata

1. A, where A is a truth-functional tautology
2. Axiom schemata for non-indexed connectives:
   - 2.1 $G(A \rightarrow B) \rightarrow (GA \rightarrow GB)$
   - 2.2 $A \rightarrow GPA$
   - 2.3 $GA \rightarrow GGA$
   - 2.4 $GA \rightarrow A$
   - 2.5 The Galois mirror images of 2.1–2.4
3. Axiom schemata for indexed connectives: For each $i, j \in \mathbb{N}$,
   - 3.1 $[\overset{ij}{\rightarrow}](A \rightarrow B) \rightarrow ([\overset{ij}{\rightarrow}]A \rightarrow [\overset{ij}{\rightarrow}]B)$
   - 3.2 $\langle\overset{ij}{\rightarrow}\rangle A \rightarrow [\overset{ij}{\rightarrow}]A$
   - 3.3 For all $i, j, k, l \in \mathbb{N}$:

$$\begin{cases} 3.3.1 & \langle\overset{ij}{\rightarrow}\rangle\lambda\langle\overset{kl}{\leftarrow}\rangle\top \rightarrow \bot, & \text{if } j \neq l \\ 3.3.2 & \langle\overset{ij}{\rightarrow}\rangle\lambda\langle\overset{kl}{\rightarrow}\rangle\top \rightarrow \bot & \text{if } j \neq k \\ 3.3.3 & \langle\overset{ij}{\leftarrow}\rangle\lambda\langle\overset{kl}{\leftarrow}\rangle\top \rightarrow \bot & \text{if } i \neq l \\ 3.3.4 & \langle\overset{ij}{\leftarrow}\rangle\lambda\langle\overset{kl}{\rightarrow}\rangle\top \rightarrow \bot & \text{if } i \neq k \end{cases}$$

$$\begin{cases} 3.3.5 & (\langle\overset{ij}{\rightarrow}\rangle\top \wedge \lambda\langle\overset{kl}{\leftarrow}\rangle\top) \rightarrow \bot & \text{if } i \neq l \\ 3.3.6 & (\langle\overset{ij}{\rightarrow}\rangle\top \wedge \lambda\langle\overset{kl}{\rightarrow}\rangle\top) \rightarrow \bot & \text{if } i \neq k \\ 3.3.7 & (\langle\overset{ij}{\leftarrow}\rangle\top \wedge \lambda\langle\overset{kl}{\leftarrow}\rangle\top) \rightarrow \bot & \text{if } j \neq l \end{cases}$$

   - 3.4 For all $i, j \in \mathbb{N}$: $\langle\overset{ij}{\rightarrow}\rangle\top \rightarrow \langle\overset{ij}{\rightarrow}\rangle\langle\overset{ij}{\leftarrow}\rangle\top$
   - 3.5 For all $i, j \in \mathbb{N}$: $\langle\overset{ij}{\rightarrow}\rangle F \langle\overset{ij}{\leftarrow}\rangle A \rightarrow FA$
   - 3.6 The Galois mirror images of 3.1, 3.2, 3.4 and 3.5.
   - 3.7 For all $i, j \in \mathbb{N}$: $(\langle\overset{ij}{\rightarrow}\rangle\top \wedge \lambda\langle\overset{ji}{\leftarrow}\rangle\top) \rightarrow \bot$ if $i \neq j$
4. Duality axioms: $\blacklozenge A \leftrightarrow \neg\blacksquare\neg A$ where $\blacklozenge \in \{F, P, \langle\overset{ij}{\rightarrow}\rangle, \langle\overset{ij}{\leftarrow}\rangle\}$ and $\blacksquare$ denotes the dual of $\blacklozenge$ (e.g., if $\blacklozenge = \langle\overset{ij}{\rightarrow}\rangle$, then $\blacksquare = [\overset{ij}{\rightarrow}]$)

Inference rules

$$(MP): \quad \frac{A,\ A \rightarrow B}{B}$$

$$(NG): \frac{A}{GA} \qquad (NH): \frac{A}{HA}$$

$$(N[\overset{ij}{\rightarrow}]): \frac{A}{[\overset{ij}{\rightarrow}]A} \qquad (N[\overset{ij}{\leftarrow}]): \frac{A}{[\overset{ij}{\leftarrow}]A} \qquad \text{(for all } i, j \in \mathbb{N})$$

Remark 4.1 Let us comment a bit on some of the schemata.

Schemata 2.1–2.5 conform the standard axiomatisation of posets in a temporal modal language (see e.g., [2]).

Schema 3.2 expresses the fact that each $f_{ij}^{\rightarrow}$ is a partial function. The same happens with $f_{ij}^{\leftarrow}$ and the Galois mirror image of 3.2.

Schemata 3.3.1–3.3.7 captures the fact that labels are unique for each poset, so that the same poset is not named twice.

Schema 3.4. and its Galois mirror image express the required conditions in item 1.1 of Definition 2.2. Specifically, they express $Im(f_{ij}^{\rightarrow}) \subseteq Dom(f_{ij}^{\leftarrow})$ and $Im(f_{ij}^{\leftarrow}) \subseteq Dom(f_{ij}^{\rightarrow})$, respectively.

Schema 3.5 and its Galois mirror image express the two implications contained in item 1.2 of Definition 2.2. More in detail, they capture, respectively, (a) if $f_{ij}^{\rightarrow}(s_i) \leq_j s_j$, then $s_i \leq_i f_{ij}^{\leftarrow}(s_j)$, and (b) if $s_i \leq_i f_{ij}^{\leftarrow}(s_j)$, then $f_{ij}^{\rightarrow}(s_i) \leq_j s_j$.

Finally, schema 3.7 tells us that, given two different indices i, j, we cannot have at the same time a resituated function and a residual from i to j.

5 Soundness and completeness of S_{GP}

This section contains the main technical results of the paper. The notions of *proof* in S_{GP} and *theorem* of S_{GP} (noted $\vdash \varphi$) are standard (see, e.g., [1, Chapter 1]). Let us first formally state the soundness of our axiom system:

Theorem 5.1 (Soundness) *S_{GP} is sound w.r.t. $\mathcal{M}_{GP}$, that is, every formula A of L_{GP} that is provable in S_{GP} is valid in $\mathcal{M}_{GP}$.*

More interestingly, our system is complete w.r.t. $\mathcal{M}_{GP}$.

Theorem 5.2 (Completeness) *S_{GP} is complete w.r.t. $\mathcal{M}_{GP}$, that is, every formula A of L_{GP} that is valid in $\mathcal{M}_{GP}$ is provable in S_{GP}.*

The rest of this section is devoted to the proof of this theorem, which is based on the step-by-step method (see e.g., [2] or [1, Chapter 4.6] for an introduction to this kind of constructions). In short, we will build, through a sequence of steps, a model satisfying each consistent formula A. At each step we will have a finite frame that does not necessarily satisfy all the properties of a Galois frame. However, this process approaches to a limit satisfying all the desired requirements. Moreover, at each step we have a frame which is "good enough". The step-by-step method is useful for dealing with frame properties that are not definable in the entertained modal language. In our case, the need of the method is ultimately triggered by the antisymmetry of each poset within a Galois frame, which is clearly not definable with the operators we take into account. Moreover, the idiosyncrasy of our frames makes the main argument a bit more elaborated since we have to take care that indexed modalities are well-behaved through the construction.

The proof is structured as follows. We first present a couple of theorems of S_{GP} that will be useful for the rest of the work (Proposition 5.3). After that, we will state and prove some properties of maximally consistent sets of formulae which are specific to our system. Finally, we will proceed to build our step-by-step model.

Proposition 5.3 *The following formulae are theorems of S_{GP}:*

T1. $(\langle \overset{ij}{\rightarrow} \rangle A \wedge \langle \overset{ij}{\rightarrow} \rangle B) \rightarrow \langle \overset{ij}{\rightarrow} \rangle (A \wedge B)$

T2. $(\langle \overset{ij}{\leftarrow} \rangle A \wedge \langle \overset{ij}{\leftarrow} \rangle B) \rightarrow \langle \overset{ij}{\leftarrow} \rangle (A \wedge B)$

Results about maximal consistent sets

The syntactical notions of *consistency* of a set of formulae Γ (denoted $\Gamma \nvdash \bot$) and *maximal consistency* of Γ in S_{GP} are defined in the usual way. Familiarity with basic properties of maximally consistent sets (*mc*-sets, for short) is assumed (see [1, Chapter 4]). We denote by $\mathcal{MC}$ the class of all *mc*-sets in S_{GP}.

Definition 5.4 Let $\Gamma_1, \Gamma_2 \in \mathcal{MC}$, and $i, j \in \mathbb{N}$. Then we define:

(*a*) $\Gamma_1 \preceq_{\mathbb{P}} \Gamma_2$ iff $\{A \mid GA \in \Gamma_1\} \subseteq \Gamma_2$.

(*b*) $\Gamma_1 \prec_{ij}^{\rightarrow} \Gamma_2$ iff $\varnothing \neq \{A \mid \langle \overset{ij}{\rightarrow} \rangle A \in \Gamma_1\} \subseteq \Gamma_2$.

(*c*) $\Gamma_1 \prec_{ij}^{\leftarrow} \Gamma_2$ iff $\varnothing \neq \{A \mid \langle \overset{ij}{\leftarrow} \rangle A \in \Gamma_1\} \subseteq \Gamma_2$.

As a consequence of this definition we have:

Proposition 5.5 *Let $\Gamma_1, \Gamma_2 \in \mathcal{MC}$, and $i, j \in \mathbb{N}$. Then:*

(i) $\Gamma_1 \preceq_{\mathbb{P}} \Gamma_2$ *iff* $\{FA \mid A \in \Gamma_2\} \subseteq \Gamma_1$ *iff* $\{A \mid HA \in \Gamma_2\} \subseteq \Gamma_1$ *iff* $\{PA \mid A \in \Gamma_1\} \subseteq \Gamma_2$.

(ii) $\Gamma_1 \prec_{ij}^{\rightarrow} \Gamma_2$ *iff* $\{A \mid [\overset{ij}{\rightarrow}]A \in \Gamma_1\} \subseteq \Gamma_2$ *iff* $\{\langle \overset{ij}{\rightarrow} \rangle A \mid A \in \Gamma_2\} \subseteq \Gamma_1$.

(iii) $\Gamma_1 \prec_{ij}^{\leftarrow} \Gamma_2$ *iff* $\{A \mid [\overset{ij}{\leftarrow}]A \in \Gamma_1\} \subseteq \Gamma_2$ *iff* $\{\langle \overset{ij}{\leftarrow} \rangle A \mid A \in \Gamma_2\} \subseteq \Gamma_1$.

We move to state some typical basic results. Namely, the *Lindenbaum's Lemma*, the so-called *Existence Lemma* for each of our diamond modalities, and the definability of reflexivity and transitivity in the basic temporal language.

Proposition 5.6 *The following properties are satisfied:*

1. *(Lindenbaum's Lemma) Any consistent set of formulae in S_{GP} can be extended to a mc-set in S_{GP}.*
2. *(Existence Lemmas) Let $\Gamma_1 \in \mathcal{MC}$ and $i, j \in \mathbb{N}$, then we have:*
   (a) If $FA \in \Gamma_1$, then there exists $\Gamma_2 \in \mathcal{MC}$ s.t. $\Gamma_1 \preceq_{\mathbb{P}} \Gamma_2$ and $A \in \Gamma_2$.
   (b) If $PA \in \Gamma_1$, then there exists $\Gamma_2 \in \mathcal{MC}$ s.t. $\Gamma_2 \preceq_{\mathbb{P}} \Gamma_1$ and $A \in \Gamma_2$.
   (c) If $\langle \overset{ij}{\rightarrow} \rangle A \in \Gamma_1$, then there exists $\Gamma_2 \in \mathcal{MC}$ s.t. $\Gamma_1 \prec_{ij}^{\rightarrow} \Gamma_2$ and $A \in \Gamma_2$.
   (d) If $\langle \overset{ij}{\leftarrow} \rangle A \in \Gamma_1$, then there exists $\Gamma_2 \in \mathcal{MC}$ s.t. $\Gamma_1 \prec_{ij}^{\leftarrow} \Gamma_2$ and $A \in \Gamma_2$.
3. *($\preceq_{\mathbb{P}}$ is a preorder) Let $\Gamma_1, \Gamma_2, \Gamma_3 \in \mathcal{MC}$. Then we have:*

(*a*) $\Gamma_1 \preceq_{\mathbb{P}} \Gamma_1$.

(*b*) *If* $\Gamma_1 \preceq_{\mathbb{P}} \Gamma_2$ *and* $\Gamma_2 \preceq_{\mathbb{P}} \Gamma_3$, *then* $\Gamma_1 \preceq_{\mathbb{P}} \Gamma_3$.

The following proposition states that the relations between mc-sets, $\prec_{ij}^{\rightarrow}$ and $\prec_{ij}^{\leftarrow}$, satisfy condition 1.1 of Definition 2.2.

Proposition 5.7 *Let* $\Gamma_1, \Gamma_2 \in \mathcal{MC}$ *and* $i, j \in \mathbb{N}$. *Then we have:*

(i) *If* $\Gamma_1 \prec_{ij}^{\rightarrow} \Gamma_2$, *then there exists* $\Gamma_3 \in \mathcal{MC}$ *s.t.* $\Gamma_2 \prec_{ij}^{\leftarrow} \Gamma_3$.

(ii) *If* $\Gamma_1 \prec_{ij}^{\leftarrow} \Gamma_2$, *then there exists* $\Gamma_3 \in \mathcal{MC}$ *s.t.* $\Gamma_2 \prec_{ij}^{\rightarrow} \Gamma_3$.

The following proposition states that the relations between mc-sets, $\prec_{ij}^{\rightarrow}$ and $\prec_{ij}^{\leftarrow}$, satisfy property 1.2 of Definition 2.2.

Proposition 5.8 *Let* $\Gamma_1, \Gamma_2, \Gamma_3, \Gamma_4 \in \mathcal{MC}$ *and* $i, j \in \mathbb{N}$. *Then:*

(i) *If* $\Gamma_1 \prec_{ij}^{\rightarrow} \Gamma_2, \Gamma_2 \preceq_{\mathbb{P}} \Gamma_3$ *and* $\Gamma_3 \prec_{ij}^{\leftarrow} \Gamma_4$, *then* $\Gamma_1 \preceq_{\mathbb{P}} \Gamma_4$.

(ii) *If* $\Gamma_1 \prec_{ij}^{\leftarrow} \Gamma_2, \Gamma_3 \preceq_{\mathbb{P}} \Gamma_2$ *and* $\Gamma_3 \prec_{ij}^{\rightarrow} \Gamma_4$, *then* $\Gamma_4 \preceq_{\mathbb{P}} \Gamma_1$.

Step-by-step method

It is now time to make formally precise what does it mean that, at each step of our construction, although our frame might not be a Galois frame, it is "good enough":

Definition 5.9 Let $(A, \leq_A)$ and $(B, \leq_B)$ be two posets and $f : (A, \leq_A) \longrightarrow (B, \leq_B)$, $g : (B, \leq_B) \longrightarrow (A, \leq_A)$ a pair of partial functions. We say that the pair (f, g) is a **quasi-Galois connection** if for all $a \in Dom(f)$ and for all $b \in Dom(g)$, we have that $a \leq_A g(b)$ iff $f(a) \leq_B b$. If (f, g) is a quasi-Galois connection, we call f **quasi-residuated**, and denote it by $f^{q,\rightarrow}$, and the function g is called the **quasi-residual of** f, and denoted by $f^{q,\leftarrow}$.

Remark 5.10 Since the previous definition does not require that $Im(f) \subseteq Dom(g)$ and $Im(g) \subseteq Dom(f)$, then by Definition 2.2 we have that a quasi-Galois connection, (f, g), is not always a Galois connection. Moreover, according to the proof of Proposition 2.5, neither have we assured that f and g are monotone; nor, likewise, that $g \circ f$ is inflationary and $f \circ g$ deflationary.

Definition 5.11 A **quasi-Galois frame** for L_{GP} is a tuple $\Sigma = (\Lambda, \mathcal{P}_{oset}, \mathcal{F})$ where every component is just as in a Galois frame (Definition 3.1) except for $\mathcal{F}$, where the condition (iii)(c) of Definition 3.1 is replaced by:

(c') for every $f \in \mathcal{F}$, it is either a quasi-residuated function or its quasi-residual. If f is a quasi-residuated function from $\mathbb{P}_i$ to $\mathbb{P}_j$, then we denote it by $f_{ij}^{q,\rightarrow}$, and we use $f_{ij}^{q,\leftarrow}$ to denote its quasi-residual.

Note that, unlike what we did in (iii)(c) (Definition 3.1), we do not require $f^{q,\rightarrow} \in \mathcal{F}$ iff $f^{q,\leftarrow} \in \mathcal{F}$ for quasi-Galois frames.

In our construction, the quasi-Galois frame entertained at each step will be an *extension* of the previous one. Let us make this notion precise:

Definition 5.12 Let $\Sigma_1 = (\Lambda_1, \mathcal{P}_{oset1}, \mathcal{F}_1)$ and $\Sigma_2 = (\Lambda_2, \mathcal{P}_{oset2}, \mathcal{F}_2)$ be a pair of quasi-Galois frames. We say that Σ_2 is an **extension** of Σ_1 if the following conditions hold:

(i) $\Lambda_1 \subseteq \Lambda_2$.

(ii) for any $(\mathbb{P}_i, \leq_i) \in \mathcal{P}_{oset1}$, the poset $(\mathbb{P}'_i, \leq'_i) \in \mathcal{P}_{oset2}$ satisfies:

- $\mathbb{P}_i \subseteq \mathbb{P}'_i$;
- $\leq_i = \leq'_i \cap (\mathbb{P}_i \times \mathbb{P}_i)$.

We also define a special type of quasi-Galois frame that will be useful in the construction.

Definition 5.13 Given $k \in \mathbb{N}$, a **simple quasi-Galois frame at** k is a quasi-Galois frame $\Sigma^k = (\Lambda, \mathcal{P}_{oset}, \mathcal{F})$ where $\mathcal{F} = \varnothing$, $\Lambda = \{k\}$, and $\mathcal{P}_{oset} = \{(\mathcal{P}_k, \leq_k)\}$. Given $j \in \mathbb{N}$, we say that a simple quasi-Galois frame is a j**-renamed frame of** Σ^k, denoted $\Sigma^{j/k}$, if it is obtained by replacing in Σ^k every occurrence of k by j.

As usual in the *step-by-step* completeness method, we introduce a function (called *trace*) that associates elements of $\mathcal{MC}$ to states of a quasi-Galois frame.

Definition 5.14 Let $\Sigma = (\Lambda, \mathcal{P}_{oset}, \mathcal{F})$ be a quasi-Galois frame for L_{GP}. A **trace** of Σ is a function $\Phi_\Sigma : \mathbb{S}_\Lambda \to \mathcal{MC}$. Moreover, if Σ^k is a simple quasi-Galois frame and Φ_{Σ^k} is a trace of it, the **renamed trace** $\Phi_{\Sigma^{j/k}}$ is obtained by replacing all occurrences of k in the domain of Φ_{Σ^k} by j.

We now introduce the desired properties of a trace:

Definition 5.15 Let Φ_Σ be a trace of a quasi-Galois frame $\Sigma = (\Lambda, \mathcal{P}_{oset}, \mathcal{F})$. Then Φ_Σ is called:

- **nominally-coherent**, if for all $s \in \mathbb{S}_\Lambda$ and $i, j \in \mathbb{N}$ we have that:
  (nc_1) : if $\lambda\langle \overset{ij}{\rightarrow} \rangle A \in \Phi_\Sigma(s)$, then $s = s_i$.
  (nc_2) : if $\lambda\langle \overset{ij}{\leftarrow} \rangle A \in \Phi_\Sigma(s)$, then $s = s_j$.
- **poset-coherent**, if for all $s, s' \in \mathbb{S}_\Lambda$ we have that:
  if $s' \in s\!\uparrow$, then $\Phi_\Sigma(s) \preceq_{\mathbb{P}} \Phi_\Sigma(s')$.
- **functionally-coherent**, if for all $s_i, s_j \in \mathbb{S}_\Lambda$ with $i, j \in \Lambda$ we have that:
  (fc_1) : *if* $s_j = f^{\rightarrow}_{ij}(s_i)$, *then* $\Phi_\Sigma(s_i) \prec^{\rightarrow}_{ij} \Phi_\Sigma(s_j)$.
  (fc_2) : *if* $s_i = f^{\leftarrow}_{ij}(s_j)$, *then* $\Phi_\Sigma(s_j) \prec^{\leftarrow}_{ij} \Phi_\Sigma(s_i)$.
- **$\uparrow$-projectable** if for all $A \in L_{GP}$ and $s \in \mathbb{S}_\Lambda$ we have that:
  if $FA \in \Phi_\Sigma(s)$, there exists $s' \in s\!\uparrow$ s.t. $A \in \Phi_\Sigma(s')$.
- **$\downarrow$-projectable** if for all $A \in L_{GP}$ and $s \in \mathbb{S}_\Lambda$ we have that:
  if $PA \in \Phi_\Sigma(s)$, there exists $s' \in s\!\downarrow$ s.t. $A \in \Phi_\Sigma(s')$.
- **$\langle\rightarrow\rangle$-projectable** if for all $A \in L_{GP}$, $i, j \in \mathbb{N}$ and $s_k \in \mathbb{S}_\Lambda$ we have that:
  if $\langle \overset{ij}{\rightarrow} \rangle A \in \Phi_\Sigma(s_k)$, then there exists $s_j = f^{\rightarrow}_{kj}(s_k)$ s.t. $A \in \Phi_\Sigma(s_j)$.
- **$\langle\leftarrow\rangle$-projectable** if for all $A \in L_{GP}$, $i, j \in \mathbb{N}$ and $s_k \in \mathbb{S}_\Lambda$ we have that:
  if $\langle \overset{ij}{\leftarrow} \rangle A \in \Phi_\Sigma(s_k)$, then there exists $s_i = f^{\leftarrow}_{ik}(s_k)$ s.t. $A \in \Phi_\Sigma(s_i)$.
- **quasi-coherent** if it is poset-coherent, functionally-coherent (but not necessarily nominally-coherent).

- **coherent** if it is nominally-coherent, poset-coherent and functionally-coherent.
- **full** if it is coherent, $\uparrow$-projectable, $\downarrow$-projectable, $\langle\rightarrow\rangle$-projectable, and $\langle\leftarrow\rangle$-projectable.

Remark 5.16 It is worth noting the following consideration in the definition of the nominally-coherent condition: If $\lambda\langle\overset{ij}{\rightarrow}\rangle A \in \Phi_\Sigma(s)$, then any formula in $\Phi_\Sigma(s)$ other than $\lambda\langle\overset{ij}{\rightarrow}\rangle A$ of the form $\lambda'\langle\overset{kl}{\rightarrow}\rangle B$ implies $i = k$ and of the form $\lambda'\langle\overset{kl}{\leftarrow}\rangle B$ implies $i = l$ (by axioms 3.3.5 and 3.3.6 and the fact that $\Phi_\Sigma(s)$ is an mc-set). Similar considerations can be done if $\lambda\langle\overset{ij}{\leftarrow}\rangle A \in \Phi_\Sigma(s)$.

Remark 5.17 We will refer to the conditionals introduced in the previous definition (e.g., in the definition of $\uparrow$-projectable trace) just as we refer to a trace Φ_Σ that satisfies them. In general, we will also use the expression "conditional for Φ_Σ" to mean that it is a $\uparrow$-projectable ($\downarrow$-projectable, $\langle\rightarrow\rangle$-projectable or $\langle\leftarrow\rangle$-projectable) conditional for Φ_Σ. Moreover, given a conditional (α) for Φ_Σ, if we replace the index Σ' by Σ, where Σ' is an extension of Σ, then (α) is a conditional for Φ'_Σ, but we can say that we refer to the same conditional in both cases.

Definition 5.18 Let Φ_Σ be a trace of a quasi-Galois frame.

- Consider a $\uparrow$-projectable conditional:

  "*If* $FA \in \Phi_\Sigma(s)$, *then there exists* $s' \in s\uparrow$ *s.t.* $A \in \Phi_\Sigma(s')$"

  We say that it is **active** for Φ_Σ if $FA \in \Phi_\Sigma(s)$, but there is no $s' \in s\uparrow$ s.t. $A \in \Phi_\Sigma(s')$. On the other hand, we say that it is **exhausted** for Φ_Σ if there exists a state s' s.t. $s' \in s\uparrow$ and $A \in \Phi_\Sigma(s')$.

  The notions of activeness and exhaustedness are defined in a similar way for $\downarrow$-projectable conditionals.

- Consider a $\langle\rightarrow\rangle$-projectable conditional:

  "*If* $\langle\overset{ij}{\rightarrow}\rangle A \in \Phi_\Sigma(s_k)$, *then there exists* $s_j = f^{\rightarrow}_{kj}(s_k)$ *s.t.* $A \in \Phi_\Sigma(s_j)$."

  We say that it is **active** for Φ_Σ if $\langle\overset{ij}{\rightarrow}\rangle A \in \Phi_\Sigma(s_k)$, but there is no $s_j = f^{\rightarrow}_{kj}(s_k)$ s.t. $A \in \Phi_\Sigma(s_j)$. On the other hand, the conditional is **exhausted** for Φ_Σ if there exists $s_j = f^{\rightarrow}_{kj}(s_k)$ s.t. $A \in \Phi_\Sigma(s_j)$.

  The notions of activeness and exhaustedness are defined in a similar way for $\langle\leftarrow\rangle$-projectable conditionals.

Lemma 5.19 (Truth) *Let* Φ_Σ *be a full trace of a Galois frame* Σ. *Let* h *be an interpretation assigning to each propositional variable* p *the set* $h(p) = \{s \in \mathbb{S}_\Sigma \mid p \in \Phi_\Sigma(s)\}$. *Then, for any formula* A, *we have* $h(A) = \{s \in \mathbb{S}_\Sigma \mid A \in \Phi_\Sigma(s)\}$.

As we will see, at every stage of the step-by-step construction, the trace of the constructed quasi-Galois frame is quasi-coherent or coherent. However, this is not necessarily so for the projectable properties, which are only satisfied at the end of the process. Let us then state the lemmas that allow exhausting active conditionals at each step.

Lemma 5.20 (Basic exhausting lemma) *Let Φ_{Σ^k} be a quasi-coherent trace of a simple quasi-Galois frame Σ^k and let (α) be an active conditional for Φ_{Σ^k}, then:*

(i) *If (α) is a $\langle\rightarrow\rangle$-($\langle\leftarrow\rangle$-)conditional, then there is $i \in \mathbb{N}$ s.t. $\Phi_{\Sigma^{i/k}}$ is coherent.*

(ii) *If (α) is a $\uparrow$-($\downarrow$-)conditional, then there is an extension of Σ^k, Σ^k_1, and a quasi-coherent trace $\Phi_{\Sigma^k_1}$ of Σ^k_1 s.t. $\Phi_{\Sigma^k_1} \subseteq \Phi_{\Sigma^k}$ and (α) is exhausted for $\Phi_{\Sigma^k_1}$. Moreover, if Φ_{Σ^k} was coherent, then $\Phi_{\Sigma^k_1}$ is coherent.*

Lemma 5.21 (General exhausting lemma) *Let Φ_{Σ_1} be a coherent trace of a finite quasi-Galois frame Σ_1 and let (α) be an active conditional for Φ_{Σ_1}. Then there is a finite quasi-Galois frame Σ_2, which is an extension of Σ_1, and a coherent trace Φ_{Σ_2} of Σ_2 s.t. $\Phi_{\Sigma_1} \subseteq \Phi_{\Sigma_2}$ and (α) is exhausted for Φ_{Σ_2}.*

We are finally able to give a **proof of Theorem 5.2**:

Proof. [Sketch] By a standard argument, it suffices to show that given a consistent formula A, this formula is satisfiable. In order to do so, we will build a model $\mathcal{M} = (\Sigma, h)$ where $\Sigma = (\Lambda, \mathcal{P}_{oset}, \mathcal{F})$ is a Galois frame. Given that A is consistent, there exists a *mc*-set Γ_0 containing A (*Lindenbaum's Lemma*). We start our construction with the finite simple quasi-Galois frame $\Sigma_0 = (\{0\}, \{\{\{s_0\}, \{(s_0, s_0)\}\}\}, \varnothing)$. The corresponding trace Φ_{Σ_0} is defined as $\Phi_{\Sigma_0}(s_0) = \Gamma_0$. It is straightforward to show that Φ_{Σ_0} is quasi-coherent. Now, we aim at constructing:

- a denumerable sequence, $\Sigma_0, \Sigma_1, \ldots, \Sigma_n \ldots$, of finite quasi-Galois frames whose union will be a Galois frame, and
- a denumerable sequence of the corresponding quasi-coherent or coherent traces, $\Phi_{\Sigma_0}, \Phi_{\Sigma_1}, \ldots, \Phi_{\Sigma_n}, \ldots$.

In order to do so, let $A_0, A_1, \ldots$ be an enumeration of all the existential formulae of the language [6] in which every formula occurs infinitely many times. Assume that $\Sigma_n = (\Lambda_n, \mathcal{P}_{oset_n}, \mathcal{F}_n)$ and Φ_{Σ_n} (with $n \geq 0$) are given and take the existential formula A_n of the above enumeration. Consider the finite set $\mathbb{S}_{\Sigma_n}$ and $S \subseteq \mathbb{S}_{\Sigma_n}$ the set of all states s.t. for every $s \in S$, "$A_n \in \Phi_{\Sigma_n}(s)$" is the antecedent of an active conditional. Let us now define Σ_{n+1} inductively:

(A) If $S = \varnothing$, then we establish $\Sigma_{n+1} = \Sigma_n$ and $\Phi_{\Sigma_{n+1}} = \Phi_{\Sigma_n}$, and continue the process considering the existential formula A_{n+1}. Clearly, if Φ_{Σ_n} is (quasi-)coherent, then so it is $\Phi_{\Sigma_{n+1}}$.

(B) If $S \neq \varnothing$, $\Lambda_n = \{0\}$ and A_n is of the form $\langle\overset{i\,j}{\rightarrow}\rangle B$ or $\langle\overset{j\,i}{\leftarrow}\rangle B$ with $i \neq 0$, then we apply Lemma 5.20.(i) by setting $\Sigma_{n+1} = \Sigma^{i/0}$, and we reconsider A_n, so that we reconfigure the enumeration of existential formulae by setting $A_{n+1} \mapsto A_n$, $A_{n+2} \mapsto A_{n+1}$,... It is easy to see that Σ_n is quasi-coherent but

[6] These are formulae with the prefixes F, P, $\langle\overset{i\,j}{\rightarrow}\rangle$ or $\langle\overset{i\,j}{\leftarrow}\rangle$ (for $i, j \in \mathbb{N}$).

not coherent, and Σ_{n+1} is coherent.

(C) If $S \neq \varnothing$, $\Lambda_n = \{0\}$ and A_n is of the form FB or PB, then, by several applications of Lemma 5.20.(ii), we can obtain a sequence, $\Sigma_{n_1}, \ldots, \Sigma_{n_m}$, of finite, simple quasi-Galois frames s.t. each of them is an extension of the previous one, and a corresponding sequence of (quasi-)coherent traces $\Phi_{\Sigma_{n_1}}, \ldots, \Phi_{\Sigma_{n_m}}$ s.t. $\Phi_{\Sigma_n} = \Phi_{\Sigma_{n_1}} \subseteq \ldots \subseteq \Phi_{\Sigma_{n_m}}$, so that each active conditional $(\alpha)_i$ (for $1 \leq i < m$) with antecedent $A_n \in \Phi_{\Sigma_{n_i}}(s)$ is exhausted for $\Phi_{\Sigma_{n_{i+1}}}$. Moreover, we set $\Sigma_{n+1} = \Sigma_{n_m}$ and $\Phi_{\Sigma_{n+1}} = \Phi_{\Sigma_{n_m}}$.

(D) If $S \neq \varnothing$ and either (i) $\Lambda_n \neq \{0\}$ or (ii) $\Lambda_n = \{0\}$ and A_n is of the form $\langle \overset{0\,j}{\rightarrow} \rangle B$ or $\langle \overset{j\,0}{\leftarrow} \rangle B$, then we can guarantee that Φ_{Σ_n} is coherent, so that we can do the same as in the previous case, but applying Lemma 5.21. Moreover, the coherence of $\Phi_{\Sigma_{n+1}}$ is also ensured.

Finally, we define $\Sigma = (\Lambda, \mathcal{P}_{oset}, \mathcal{F})$ with $\Lambda = \bigcup_{n\in\omega} \Lambda_n$, $\mathcal{P}_{oset} = \bigcup_{n\in\omega} \mathcal{P}_{oset_n}$, where $\bigcup_{n\in\omega} \mathcal{P}_{oset_n}$ means the pointwise union of all posets with the same index, and $\mathcal{F} = \bigcup_{n\in\omega} \mathcal{F}_n$. Let us show that Σ is a Galois frame. If $\mathcal{F} = \varnothing$, then this is trivial. Otherwise:

- Conditions (i), (ii) and (iii)(a) of Definition 3.1 are straightforwardly guaranteed by construction. As for condition (iii)(b), details are left to the reader, but note that for each state $s \in \Lambda$ and each $i, j \in \mathbb{N}$ with $i \neq j$ we found at most one active $\langle \rightarrow \rangle$-($\langle \leftarrow \rangle$-)conditional during the construction process –axiom 3.7. is needed to show this– (and at most two when $i = j$).
- Regarding condition (iii)(c), we have to show that Σ satisfies both requirements of Definition 2.2 (1.1 and 1.2) and that a residuated function is defined in $\mathcal{F}$ if and only if its residual is also in $\mathcal{F}$ (in symbols, $f_{ij}^{\rightarrow} \in \mathcal{F}$ iff $f_{ij}^{\leftarrow} \in \mathcal{F}$). In effect, assume that, at a given step of the construction, we have a quasi-residuated function $f_{ij}^{q,\rightarrow}$ with $f_{ij}^{q,\rightarrow}(s_i) = s_j$, but $f_{ij}^{q,\leftarrow}(s_j)$ does not exist, so that $Im(f_{ij}^{q,\rightarrow}) \not\subseteq Dom(f_{ij}^{q,\leftarrow})$. Then at a later step we will create a new point $f_{ij}^{q,\leftarrow}(s_j)$ where $f_{ij}^{q,\leftarrow}(s_j)$ will be associated with a new mc-set. It occurs similarly if we consider a quasi-residual function. Both cases are justified by Proposition 5.7 and the construction. This ensures property 1.1. Moreover, this argument also guarantees that $f_{ij}^{\rightarrow} \in \mathcal{F}$ iff $f_{ij}^{\leftarrow} \in \mathcal{F}$. Finally, Σ satisfies property 1.2 of Definition 2.2, because all members of the sequence of quasi-Galois frames satisfy that property.

Let us now show that Φ_Σ is coherent. If Φ_{Σ_0} is coherent, then it is clear that coherence is preserved through the construction. If Φ_{Σ_0} is not coherent, then it is not nominally coherent which means that there exists some formula in $\Phi_{\Sigma_0}(s_0)$ of the form $\lambda\langle \overset{i\,j}{\rightarrow} \rangle A$ or $\langle \overset{j\,i}{\leftarrow} \rangle A$ (with $i \neq 0$). Hence, (B) is reached at some point of the construction, so that we obtain a coherent trace (by Lemma 5.20.(i)), and this is preserved through the rest of the construction. Note that (B) is reached at most once in the whole process. Furthermore, by lemmas 5.20.(ii) and 5.21, each active conditional for a given Φ_{Σ_n} is exhausted sooner or later for a given Φ_{Σ_m}. Thus Φ_Σ is full.

Finally, we can define a Galois model (Σ, h) where $h(p) = \{s \in \mathbb{S}_\Sigma \mid p \in \Phi_\Sigma(s)\}$ for every propositional variable p. So, by the Truth Lemma (5.19), A is satisfiable. □

6 Related work

The current paper can be understood as an integration of two different lines of research: the modal study of Galois connections (e.g., [13,15,19,22]) and the investigation of temporal×modal functional frames (e.g., [3,4,5,6]). Let us comment briefly on both, focusing on how they compare to our approach.

In the first place, there are several works in the literature looking at some interactions between modal operators and Galois connections (among others, [7,19,20,14,22,15,13]). In general, these works are based on the identification of a certain Galois connection within a standard Kripke frame (i.e., a pair (W, R) where $W \neq \emptyset$ and R is a relation on W), usually followed by a deep algebraic analysis (e.g., in [22,19,13]). Such identifications have been used with diverse particular purposes. For instance, [15] presents the *Information Logic of Galois Connections* (ILGC) as a means for reasoning about approximate information (represented mathematically as *rough sets*). ILGC is essentially the basic tense logic K_t taking F and H as primitive operators. The axiomatisation of ILGC is based on the fact that, when interpreted on Kripke frames, F and H form a Galois connection. More precisely, given a Kripke frame (W, R), we can look at F and H as semantic operators $2^W \longrightarrow 2^W$ by setting $F(X) = \{w \in W \mid \exists u.(w,u) \in R \quad \text{and} \quad u \in X\}$ and $H(X) = \{w \in W \mid \forall u, (u,w) \in R \quad \text{implies} \quad u \in X\}$; and it is clear that (F, H) is a Galois connection in $(2^W, \subseteq)$. In [13], the author does something similar, but using two-sorted Kripke frames (frames where the domain and range of the relation R are disjoint and form a partition of W). As a third example, [19] uses these sorted frames, which are nothing but formal contexts, in order to provide a logical tool for Formal Concept Analysis (see Example 1).

The second line of research is the one initiated in [4], where functional temporal×modal frames are introduced, and later developed in a series of papers ([3,5,6] among others). Our language is essentially the one used in [6] for the axiomatisation of surjective functions. The changes in the semantics of both works can be identified at first sight, the most important being the addition of condition (iii) in our definition of frame (Definition 3.1), and in the semantic clause for $\langle \overset{ij}{\leftarrow} \rangle$.

By integrating both branches of research, we generalize existing frameworks for the modal study of Galois connections in at least three well-differentiated aspects. First, we consider Galois connections for partial functions, instead of the more usual and restricted definition that looks only at total functions. Second, our frames take into account connections among an arbitrary number of posets, instead of considering a single connection among two (possibly different) posets. Third, these frames can accommodate the representation of *any* poset $(\mathbb{P}_i, \leq_i)$, while in the quoted works these are always the complete lattice generated by a power set of worlds and the set inclusion relation among them $(2^W, \subseteq)$. Moreover, the pre-orders $\leq_i$ are first-class citizen in our semantics, described in turn with the temporal modalities F and P, while the subset inclusion among sets of worlds is left implicit in the logical treatment of the quoted papers.

7 Future work

Let us just mention three open lines for future work. First, computational aspects of reasoning tasks associated to S_{GP}, e.g., decidability and complexity of the provability problem, have been left out of this work. We believe, however, that the logic is indeed decidable. This would open the door to using our results as a first step towards an automated modal prover for the study of Galois connections. Second, as mentioned elsewhere in this paper, the notion of a Galois connection makes also sense among pre-ordered sets (sets endowed with a reflexive and transitive relation), hence a natural extension of our work consists in relaxing our set of assumptions on $\leq_i$ and studying the resulting logic. Finally, the link between our approach and Gaggle theory [9], a natural generalisation of Galois connections that has been shown very fruitful for logic, remains to be studied too.

Acknowledgements

We are very thankful for the AiML2022 reviewers' work, whose comments helped us to spot some mistakes in the initial version of this paper, as well as to substantially improve its final outlook. The research activity of Antonio Yuste-Ginel was partially funded by a grant from "Plan Propio de Investigación de la Universidad de Málaga". The authors also gratefully acknowledge funding from the projects PID2020-117871GB-I00 (Spanish Ministry of Science and Innovation) and PY20_01140 (Regional Government of Andalusia).

Appendix

In this section, we shall use the following conventions: by PC (Propositional Calculus), we denote proofs in the classical logic; and by ML we denote proofs in the basic multi-modal logic.

[**Proposition 2.5**]

Proof. For $1 \Rightarrow 2$: Let g be the function whose existence is affirmed in item 1.

- Let us see that $g \circ f$ is inflationary. In effect, given that for all $a \in Dom(f)$ we have that $f(a) \leq_B f(a)$ and, by hypothesis, we have that $Im(f) \subseteq Dom(g)$, item 1 ensures that $a \leq_A g(f(a))$.
- Let us see that $f \circ g$ is deflationary. In effect, given that for all $b \in Dom(g)$ we have that $g(b) \leq_A g(b)$ and, by hypothesis, we have that $Im(g) \subseteq Dom(f)$, item 1 guarantees that $f(g(b)) \leq_B b$.
- Let us see that f is monotone. Let $a_1, a_2 \in Dom(f)$ be s.t. $a_1 \leq_A a_2$,

since $g \circ f$ is inflationary we obtain $a_2 \leq_A g(f(a_2))$ and, as a consequence, $a_1 \leq_A g(f(a_2))$. Now, the hypothesis ensures that $f(a_1) \leq_B f(a_2)$.

- Let us see that g is monotone. Let $b_1, b_2 \in Dom(g)$ be s.t. $b_1 \leq_B b_2$, since $f \circ g$ is deflationary, we have that $f(g(b_1)) \leq_B b_1$ and, as a consequence, $f(g(b_1)) \leq_B b_2$. Now, the hypothesis ensures that $g(b_1) \leq_A g(b_2)$.

As for $2 \Rightarrow 1$: Let us see that for all $b \in Dom(g)$ we have that $f^{-1}(b\downarrow) = g(b)\downarrow \cap Dom(f)$ where g is the function whose existence is affirmed in item 2.

Assume $a \in f^{-1}(b\downarrow)$. Then $f(a) \leq_B b$. Since g is monotone, $(g \circ f)(a)$ is defined (given that $Im(f) \subseteq Dom(g)$) and $b \in Dom(g)$, we have that $g(f(a)) \leq_A g(b)$. Now, since $g \circ f$ is inflationary, we obtain $a \leq_A g(f(a)) \leq_A g(b)$, that is, $a \in g(b)\downarrow$.

Reciprocally, assume that $a \in g(b)\downarrow \cap Dom(f)$, that is, $a \leq_A g(b)$. Since f is monotone, $a \in Dom(f)$ and $Im(g) \subseteq Dom(f)$, we have that $f(a) \leq_B f(g(b))$ and given that $f \circ g$ is deflationary we obtain $f(a) \leq_B f(g(b)) \leq_B b$. Therefore, $f(a) \in b\downarrow$ and $a \in f^{-1}(b\downarrow)$.

Let us see that $a \leq_A g(b)$ if and only if $f(a) \leq_B b$.

- Assume $a \leq_A g(b)$, that is, $a \in g(b)\downarrow \cap Dom(f) = f^{-1}(b\downarrow)$ and, as a consequence, $f(a) \leq_B b$.
- Assume $f(a) \leq_B b$, that is, $f(a) \in b\downarrow$. Now, given that the function $f^{-1} : (2^B, \subseteq) \longrightarrow (2^A, \subseteq)$ defined by $f^{-1}(Y) = \{a \in A \mid f(a) \in Y\}$ is monotone we obtain $a \in f^{-1}(f(a)) \subseteq f^{-1}(b\downarrow) = g(b)\downarrow \cap Dom(f)$ and, as a consequence, $a \leq_A g(b)$.

□

[**Theorem 5.1**]

Proof. Soundness follows S_{GP} from a standard inductive argument on the length of derivations. Let us just show, as an illustration, the validity of schema 3.3.1: $\langle \overset{ij}{\rightarrow} \rangle \lambda \langle \overset{kl}{\leftarrow} \rangle \top \rightarrow \bot$, where $j \neq l$.

Let $(\Lambda, \mathcal{P}_{oset}, \mathcal{F}, h)$ be a Galois model, $s \in \mathbb{S}_\Lambda$ and $j \neq l$. Suppose, for the sake of contradiction, that $s \in h(\langle \overset{ij}{\rightarrow} \rangle \lambda \langle \overset{kl}{\leftarrow} \rangle \top)$, which implies that there is an $f_{ij}^{\rightarrow} \in \mathcal{F}$ s.t. $f_{ij}^{\rightarrow}(s) \in h(\lambda \langle \overset{kl}{\leftarrow} \rangle \top)$. Since $f_{ij}^{\rightarrow}(s) \in h(\lambda \langle \overset{kl}{\leftarrow} \rangle \top)$, there is a $s_j \in \mathbb{P}_j$ s.t. $s_j \in h(\langle \overset{kl}{\leftarrow} \rangle \top)$ (this can be shown by induction on the length of λ). But then, by the semantic clause for $\langle \overset{kl}{\leftarrow} \rangle$, there is an $f_{kl}^{\leftarrow} \in \mathcal{F}$ s.t. $s_j \in Dom(f_{kl}^{\leftarrow})$, which in turn implies that $s_j \in \mathbb{P}_l$ (with $j \neq l$), and this is absurd. □

[**Proposition 5.3**]

Proof. We only prove **T1** (**T2** is its Galois mirror image):

1.	$\langle\overset{ij}{\rightarrow}\rangle A \rightarrow [\overset{ij}{\rightarrow}]A$	Axiom 3.2
2.	$\langle\overset{ij}{\rightarrow}\rangle B \rightarrow [\overset{ij}{\rightarrow}]B$	Axiom 3.2
3.	$(\langle\overset{ij}{\rightarrow}\rangle A \wedge \langle\overset{ij}{\rightarrow}\rangle B) \rightarrow ([\overset{ij}{\rightarrow}]A \wedge [\overset{ij}{\rightarrow}]B)$	from 1, 2 by *PC*
4.	$([\overset{ij}{\rightarrow}]A \wedge [\overset{ij}{\rightarrow}]B) \rightarrow [\overset{ij}{\rightarrow}](A \wedge B)$	*ML*
5.	$(\langle\overset{ij}{\rightarrow}\rangle A \wedge \langle\overset{ij}{\rightarrow}\rangle B) \rightarrow [\overset{ij}{\rightarrow}](A \wedge B)$	from 3, 4 by *PC*
6.	$(\langle\overset{ij}{\rightarrow}\rangle A \wedge \langle\overset{ij}{\rightarrow}\rangle B \wedge [\overset{ij}{\rightarrow}](A \wedge B)) \rightarrow \langle\overset{ij}{\rightarrow}\rangle(A \wedge B)$	*ML*
7.	$(\langle\overset{ij}{\rightarrow}\rangle A \wedge \langle\overset{ij}{\rightarrow}\rangle B) \rightarrow \langle\overset{ij}{\rightarrow}\rangle(A \wedge B)$	from 5, 6 by *PC*

□

[Proposition 5.5]

Proof. The proof of (i) is standard in modal logic. With respect to (ii) we first show that $\Gamma_1 \prec^{\rightarrow}_{ij} \Gamma_2$ iff $\{A \mid [\overset{ij}{\rightarrow}]A \in \Gamma_1\} \subseteq \Gamma_2$.

The left-to-right direction is proved as follows: Assume $\Gamma_1 \prec^{\rightarrow}_{ij} \Gamma_2$ and also $[\overset{ij}{\rightarrow}]A \in \Gamma_1$. We have to show that $A \in \Gamma_2$. Now suppose the contrary, i.e., $A \notin \Gamma_2$. Hence $\langle\overset{ij}{\rightarrow}\rangle A \notin \Gamma_1$, because $\Gamma_1 \prec^{\rightarrow}_{ij} \Gamma_2$ iff $\varnothing \neq \{A \mid \langle\overset{ij}{\rightarrow}\rangle A \in \Gamma_1\} \subseteq \Gamma_2$ by Definition 5.4(b), and so $[\overset{ij}{\rightarrow}]\neg A \in \Gamma_1$ by *ML*, hence $[\overset{ij}{\rightarrow}]\bot \in \Gamma_1$ by *ML* again (since $[\overset{ij}{\rightarrow}]A \in \Gamma_1$). Since the set $\{A \mid \langle\overset{ij}{\rightarrow}\rangle A \in \Gamma_1\}$ is non-empty, it should be clear that $\langle\overset{ij}{\rightarrow}\rangle\top \in \Gamma_1$ which, by *ML*, leads to a contradiction.

For the right-to-left direction, suppose $\{A \mid [\overset{ij}{\rightarrow}]A \in \Gamma_1\} \subseteq \Gamma_2$ (†). We first show that $\{A \mid \langle\overset{ij}{\rightarrow}\rangle A \in \Gamma_1\} \neq \varnothing$. Assume the contrary, then we have that $\langle\overset{ij}{\rightarrow}\rangle\top \notin \Gamma_1$, hence $[\overset{ij}{\rightarrow}]\bot \in \Gamma_1$ by *ML*, so $\bot \in \Gamma_2$ by (†), and Γ_2 would be inconsistent, which is impossible. Now we shall show that $\{A \mid \langle\overset{ij}{\rightarrow}\rangle A \in \Gamma_1\} \subseteq \Gamma_2$. Consider $\langle\overset{ij}{\rightarrow}\rangle A \in \Gamma_1$, given axiom 3.2 we obtain $[\overset{ij}{\rightarrow}]A \in \Gamma_1$, so by (†) we get $A \in \Gamma_2$. This completes the proof of that direction. Moreover, the proof of $\{A \mid [\overset{ij}{\rightarrow}]A \in \Gamma_1\} \subseteq \Gamma_2$ iff $\{\langle\overset{ij}{\rightarrow}\rangle A \mid A \in \Gamma_2\} \subseteq \Gamma_1$ is standard in modal logic.

The proof of item (iii) is similar to that of (ii). □

[Proposition 5.7]

Proof. For (i), assume $\Gamma_1 \prec^{\rightarrow}_{ij} \Gamma_2$. All we have to prove is that the set $\{A \mid [\overset{ij}{\leftarrow}]A \in \Gamma_2\}$ is consistent. If not, then there are formulae $A_1, \dots, A_n$ such that $[\overset{ij}{\leftarrow}]A_1, \dots [\overset{ij}{\leftarrow}]A_n \in \Gamma_2$ and $\vdash \neg(A_1 \wedge \dots \wedge A_n)$. Then, by *ML*, $\vdash ([\overset{ij}{\leftarrow}]A_1 \wedge \dots \wedge [\overset{ij}{\leftarrow}]A_{n-1}) \rightarrow [\overset{ij}{\leftarrow}]\neg A_n$. Therefore $[\overset{ij}{\leftarrow}]\neg A_n \in \Gamma_2$, hence $[\overset{ij}{\leftarrow}]\bot \in \Gamma_2$ using *ML* again (since $[\overset{ij}{\leftarrow}]A_n \in \Gamma_2$). Now, given $\Gamma_1 \prec^{\rightarrow}_{ij} \Gamma_2$, we obtain $\langle\overset{ij}{\rightarrow}\rangle[\overset{ij}{\leftarrow}]\bot \in \Gamma_1$ (using $[\overset{ij}{\leftarrow}]\bot \in \Gamma_2$ and Proposition 5.5(ii)). Moreover, by the assumption and Definition 5.4(b), we also obtain $\langle\overset{ij}{\rightarrow}\rangle\top \in \Gamma_1$ (because $\top \in \Gamma_2$). So, by axiom 3.4, we get $\langle\overset{ij}{\rightarrow}\rangle\langle\overset{ij}{\leftarrow}\rangle\top \in \Gamma_1$. Thus, by **T1.** of Proposition 5.3 and the fact that $\langle\overset{ij}{\rightarrow}\rangle[\overset{ij}{\leftarrow}]\bot \in \Gamma_1$, we get $\langle\overset{ij}{\rightarrow}\rangle(\langle\overset{ij}{\leftarrow}\rangle\top \wedge [\overset{ij}{\leftarrow}]\bot) \in \Gamma_1$ and so $\langle\overset{ij}{\rightarrow}\rangle\langle\overset{ij}{\leftarrow}\rangle\bot \in \Gamma_1$, by

ML, which leads to a contradiction using items 2(c) and 2(d) of Proposition 5.6, since there would then exist an mc-set Γ such that $\bot \in \Gamma$, which is impossible. The proof of item (ii) is similar. □

[**Proposition 5.8**]

Proof. We shall prove item (i). Assume $\Gamma_1 \prec_{ij}^{\rightarrow} \Gamma_2$, $\Gamma_2 \preceq_{\mathbb{P}} \Gamma_3$ and $\Gamma_3 \prec_{ij}^{\leftarrow} \Gamma_4$ and $GA \in \Gamma_1$. We will show that $A \in \Gamma_4$. Now, the axiom 3.5 establish that $\langle \overset{ij}{\rightarrow} \rangle F \langle \overset{ij}{\leftarrow} \rangle X \to FX$, so, from $GA \in \Gamma_1$ we obtain that $[\overset{ij}{\rightarrow}]G[\overset{ij}{\leftarrow}]A \in \Gamma_1$ (by ML). Since $\Gamma_1 \prec_{ij}^{\rightarrow} \Gamma_2$, from Proposition 5.5(ii), we get $G[\overset{ij}{\leftarrow}]A \in \Gamma_2$, and from $\Gamma_2 \preceq_{\mathbb{P}} \Gamma_3$, by Definition 5.4(a), we obtain $[\overset{ij}{\leftarrow}]A \in \Gamma_3$. Finally, since $\Gamma_3 \prec_{ij}^{\leftarrow} \Gamma_4$, by Proposition 5.5(iii), we get $A \in \Gamma_4$ as required. Thus $\Gamma_1 \preceq_{\mathbb{P}} \Gamma_4$.

The proof of item (ii) is similar. □

[**Lemma 5.20**]

Proof. For item (i), let $\lambda\langle \overset{ij}{\rightarrow} \rangle B \in \Phi_{\Sigma^k}(s)$ (resp. $\lambda\langle \overset{ji}{\leftarrow} \rangle \in \Phi_{\Sigma^k}(s)$) the antecedent of (α), then the renaming we are looking for is just $\Sigma^{i/k}$. Item (ii) is proved using the standard construction for temporal logic [2]. □

[**Lemma 5.21**]

Proof. Let Φ_{Σ_1} be a coherent trace of a finite quasi-Galois frame $\Sigma_1 = (\Lambda_1, \mathcal{P}_{oset_1}, \mathcal{F}_1)$, and let (α) be an active conditional for Φ_{Σ_1}. We want to construct an extension of Σ_1, call it Σ_2, together with a coherent trace Φ_{Σ_2} s.t. $\Phi_{\Sigma_1} \subseteq \Phi_{\Sigma_2}$ and (α) is exhausted for Φ_{Σ_2}.

In order to do so, if (α) is either a $\uparrow$-projectable or a $\downarrow$-projectable conditional, then such a construction is carried out following the standard way in temporal logic (see e.g., [2] or [1, Chapter 4.6]). So, let us consider only the case in which (α) is a $\langle \rightarrow \rangle$-projectable conditional.

Hence, assume $i, j \in \mathbb{N}$ and let (α) be the following active $\langle \rightarrow \rangle$-projectable conditional for Φ_Σ:

"If $\langle \overset{ij}{\rightarrow} \rangle A \in \Phi_{\Sigma_1}(s_i)$, then there exists $s_j = f_{ij}^{\rightarrow}(s_i)$ such that $A \in \Phi_{\Sigma_1}(s_j)$" [7]

Thus, we have that $\langle \overset{ij}{\rightarrow} \rangle A \in \Phi_{\Sigma_1}(s_i)$, but there is no $s_j = f_{ij}^{\rightarrow}(s_i)$ s.t. $A \in \Phi_{\Sigma_1}(s_j)$. Moreover, by item 2(c) of Proposition 5.6, there exists an mc-set, Γ, s.t. $\Phi_{\Sigma_1}(s_i) \prec_{ij}^{\rightarrow} \Gamma$ and $A \in \Gamma$. Furthermore, as $\langle \overset{ij}{\rightarrow} \rangle A \in \Phi_{\Sigma_1}(s_i)$, we have $i \in \Lambda_1$, so we continue by cases:

Case 1: $j \notin \Lambda_1$.

Then, we need a new poset labelled with j, namely $\mathbb{P}_j$, which requires extending Λ_1 and which contains one element, called s_j, associated with Γ. We also need to introduce a new function, $f_{ij}^{q,\rightarrow}$, extending $\mathcal{F}_1$ so that s_j is the image of s_i in $\mathbb{P}_j$. That is, we define $\Sigma_2 = (\Lambda_2, \mathcal{P}_{oset_2}, \mathcal{F}_2)$, an extension of Σ_1, and Φ_{Σ_2}, an extension of Φ_{Σ_1}, as follows:

---

[7] The match between the indices in the formula $\langle \overset{ij}{\rightarrow} \rangle A$ and the state s_i are guaranteed by the assumptions that (α) is active and Φ_{Σ_1} is coherent.

- $\Lambda_2 = \Lambda_1 \cup \{j\}$;
- $\mathcal{P}_{oset_2} = \mathcal{P}_{oset_1} \cup \{(\mathbb{P}_j, \leq_j)\}$, where $\mathbb{P}_j = \{s_j\}$ and $\leq_j = \{(s_j, s_j)\}$;
- $\mathcal{F}_2 = \mathcal{F}_1 \cup \{f_{ij}^{q,\rightarrow}\}$, where $f_{ij}^{q,\rightarrow} = \{(s_i, s_j)\}$;
- $\Phi_{\Sigma_2} = \Phi_{\Sigma_1} \cup \{(s_j, \Gamma)\}$.

It should be clear that Σ_2, as defined, is a quasi-Galois-frame, given that Σ_1 is a quasi-Galois frame. Let us show that Φ_{Σ_2} preserves coherency.

- Φ_{Σ_2} is poset-coherent by Proposition 5.6(3).
- Moreover, taking into account the definition of Φ_{Σ_2}, it is easy to see that it is functionally-coherent.
- As for the preservation of nominal coherence, the only new element in the frame Σ_2 is s_j, therefore we will focus our attention exclusively on it. The fact that $\langle \overset{ij}{\rightarrow} \rangle A \in \Phi_{\Sigma_1}(s_i)$ together with axioms 3.3.1, 3.3.2, 3.3.5 and 3.3.6 prevent that a formula of the form $\lambda\langle \overset{kl}{\rightarrow} \rangle \top$ or of the form $\lambda\langle \overset{lk}{\leftarrow} \rangle \top$ (being $k \neq j$) appears in $\Phi_{\Sigma_1}(s_j)(= \Gamma)$, hence Φ_{Σ_2} is nominally coherent too.

  The fact that $\langle \overset{ij}{\rightarrow} \rangle A \in \Phi_{\Sigma_1}(s_i)$ and the comments of Remark 5.16 ensure us that Φ_{Σ_2} is nominally coherent too.

Case 2: $j \in \Lambda_1$.

We distinguish two relevant subcases:

Case 2.1: $f_{ij}^{q,\leftarrow} \notin \mathcal{F}_1$.

We define $\Sigma_2 = (\Lambda_2, \mathcal{P}_{oset_2}, \mathcal{F}_2)$ analogously to what we did in **Case 1**. However, we do not need to create a new poset $\mathbb{P}_j$, because it already exists, but just to introduce a new point s_j that will be the $f_{ij}^{q,\rightarrow}$-image of s_i. So, we define:

- $\Lambda_2 = \Lambda_1$;
- $\mathcal{P}_{oset_2} = (\mathcal{P}_{oset_1} \setminus \{(\mathbb{P}_j, \leq_j)\}) \cup \{(\mathbb{P}'_j, \leq'_j)\}$ where $\mathbb{P}'_j = \mathbb{P}_j \cup \{s_j\}$ and $\leq'_j = \leq_j \cup \{(s_j, s_j)\}$;
- $$\mathcal{F}_2 = \begin{cases} (\mathcal{F}_1 \setminus \{f_{ij}^{q,\rightarrow}\}) \cup \{f_{ij}^{\prime q,\rightarrow}\}, \text{ where } f_{ij}^{\prime q,\rightarrow} = f_{ij}^{q,\rightarrow} \cup \{(s_i, s_j)\} & \text{if } f_{ij}^{q,\rightarrow} \in \mathcal{F}_1; \\ \mathcal{F}_1 \cup \{f_{ij}^{q,\rightarrow'}\}, \text{ where } f_{ij}^{q,\rightarrow'} = \{(s_i, s_j)\} & \text{otherwise.} \end{cases}$$
- $\Phi_{\Sigma_2} = \Phi_{\Sigma_1} \cup \{(s_j, \Gamma)\}$.

It is easy to check that Σ_2 is a quasi-Galois frame. Note that we don't have to check that $(f_{ij}^{\prime q,\rightarrow}, f_{ij}^{\prime q,\leftarrow'})$ form a quasi-Galois connection, because $f_{ij}^{\prime q,\leftarrow'}$ does not exist by hypothesis. Moreover, it is also easy to check that Φ_{Σ_2} is coherent.

Case 2.2: $f_{ij}^{q,\leftarrow} \in \mathcal{F}_1$.

Let us define the set:

$$X = \{s \in Dom(f_{ij}^{q,\leftarrow}) \mid f_{ij}^{q,\leftarrow}(s) \in s_i \uparrow\}.$$

Now, we define $\Sigma_2 = (\Lambda_2, \mathcal{P}_{oset_2}, \mathcal{F}_2)$ and Φ_{Σ_2}, where Λ_2, $\mathcal{F}_2$, and Φ_{Σ_2} are just as in **Case 2.1**, and $\mathcal{P}_{oset_2} = (\mathcal{P}_{oset_1} - \{(P_j, \leq_j)\}) \cup \{(P'_j, \leq'_j)\}$, where $\mathbb{P}'_j = \mathbb{P}_j \cup \{s_j\}$ and $\leq'_j$ is the transitive closure of the relation $\leq_j \cup \{(s_j, s_j)\} \cup \{(s_j, s) \mid s \in X\}$.

The frame so defined is a quasi-Galois frame. It suffices to show that $(f'^{q,\rightarrow}_{ij}, f'^{q,\leftarrow}_{ij})$ form a quasi-Galois connection. So let $x_i \in Dom(f'^{q,\rightarrow}_{ij}), x_j \in Dom(f'^{q,\leftarrow}_{ij})$, we need to show that

$$x_i \leq'_i f'^{q,\leftarrow}_{ij}(x_j) \text{ iff } f'^{q,\rightarrow}_{ij}(x_i) \leq'_j x_j.$$

For the left-to-right direction, suppose $x_i \leq'_i f'^{q,\leftarrow}_{ij}(x_j)$. We analyse two cases. First, if $x_i \neq s_i$, then we have $x_i \leq_i f^{q,\leftarrow}_{ij}(x_j)$ (because $f'^{q,\leftarrow}_{ij} = f^{q,\leftarrow}_{ij}$), and then $f^{q,\rightarrow}_{ij}(x_i) \leq_j x_j$ (because $(f^{q,\rightarrow}_{ij}, f^{q,\leftarrow}_{ij})$ is a quasi-Galois connection, by hypothesis), which implies $f'^{q,\rightarrow}_{ij}(x_i) \leq'_j x_j$ (because $f^{q,\rightarrow}_{ij} \subseteq f'^{q,\rightarrow}_{ij}$ and $\leq_j \subseteq \leq'_j$ by construction). If $x_i = s_i$, then $f'^{q,\rightarrow}_{ij}(x_i) \leq'_j x_j$ follows immediately by definition of $\leq'_j$ (because $x_j \in X$). The right-to-left direction is analogous.

As for the treatment of $\langle\leftarrow\rangle$-projectable conditionals, it is similar to the previous case. However, it is worth noticing that when we arrive to the case that is analogous to **Case 2.2** above (i.e., $f^{q,\rightarrow}_{ij} \in \mathcal{F}_1$, with $\langle \overset{ij}{\leftarrow} \rangle A \in \Phi_\Sigma(s_j)$ and $s_i = f'^{q,\leftarrow}_{ij}(s_j)$) we have to consider the set $Y = \{s \in Dom(f^{q,\rightarrow}_{ij}) \mid f^{q,\rightarrow}_{ij}(s) \in s_j \downarrow\}$ instead of X. Moreover, $\leq'_i$ is the transitive closure of the relation $\leq_i \cup \{(s_i, s_i)\} \cup \{(s, s_i) \mid s \in Y\}$. □

References

[1] Blackburn, P., M. De Rijke and Y. Venema, "Modal Logic," Cambridge University Press, 2002.

[2] Burgess, J. P., *Basic tense logic*, in: D. M. Gabbay and F. Guenthner, editors, *Handbook of Philosophical Logic* (2002), pp. 1–42.

[3] Burrieza, A. and I. P. de Guzmán, *A functional approach for temporalXmodal logics*, Acta informatica **39** (2003), pp. 71–96.

[4] Burrieza, A., I. P. de Guzman and E. Muñoz, *Indexed flows in temporalXmodal logic with functional semantics*, in: M. Fisher and A. Artale, editors, *Proceedings Ninth International Symposium on Temporal Representation and Reasoning*, IEEE, 2002, pp. 146–153.

[5] Burrieza, A., I. P. de Guzmán and E. Muñoz-Velasco, *Analyzing completeness of axiomatic functional systems for temporal×modal logics*, Mathematical Logic Quarterly **56** (2010), pp. 89–102.

[6] Burrieza, A., I. Fortes and I. P. de Guzmán, *Completeness of a functional system for surjective functions*, Mathematical Logic Quarterly **63** (2017), pp. 574–597.

[7] Demri, S. P. and E. Orlowska, "Incomplete information: Structure, inference, complexity," Springer, 2013.

[8] Denecke, K., M. Erné and S. L. Wismath, "Galois connections and applications," Springer, 2004.

[9] Dunn, J. M., *Gaggle theory: An abstraction of galois connections and residuation, with applications to negation, implication, and various logical operators*, in: J. van Eijck, editor, *Logics in AI* (1991), pp. 31–51.

[10] Erné, M., J. Koslowski, A. Melton and G. E. Strecker, *A primer on galois connections*, Annals of the New York Academy of Sciences **704** (1993), pp. 103–125.

[11] Ganter, B. and R. Wille, "Formal concept analysis: mathematical foundations," Springer, 2012.

[12] Gödel, K., *Eine interpretation des intuitionistischen aussagenkalküls*, in: E. Dierker and K. Sigmund, editors, *Ergebnisse eines mathematischen Kolloquiums 4 (1933)*, pp. 39–40, reprinted and translated in: S. Feferman et al.(eds.), Kurt Gödel. Collected Works. Vol. 1, 1986.

[13] Hartonas, C., *Lattice logic as a fragment of (2-sorted) residuated modal logic*, Journal of Applied Non-Classical Logics **29** (2019), pp. 152–170.

[14] Horn, A., *Dynamic epistemic algebra with post-conditions to reason about robot navigation*, in: L. D. Beklemishev and R. J. G. B. de Queiroz, editors, *Proceedings of WoLLIC 2011*, Lecture Notes in Computer Science **6642** (2011), pp. 161–175.

[15] Järvinen, J., M. Kondo and J. Kortelainen, *Logics from galois connections*, Int. J. Approx. Reason. **49** (2008), pp. 595–606.
URL `https://doi.org/10.1016/j.ijar.2008.06.003`

[16] McKinsey, J. C. C., *A solution of the decision problem for the Lewis systems S2 and s4, with an application to topology*, J. Symb. Log. **6** (1941), pp. 117–134.

[17] McKinsey, J. C. C. and A. Tarski, *The algebra of topology*, Annals of mathematics (1944), pp. 141–191.

[18] Melton, A., D. A. Schmidt and G. E. Strecker, *Galois connections and computer science applications*, in: D. Pitt, S. Abramsky, A. Poigné and D. Rydeheard, editors, *Category Theory and Computer Programming* (1986), pp. 299–312.

[19] Orlowska, E. and I. Rewitzky, *Context algebras, context frames, and their discrete duality*, Trans. Rough Sets **9** (2008), pp. 212–229.

[20] Orlowska, E. and I. Rewitzky, *Algebras for Galois-style connections and their discrete duality*, Fuzzy Sets Syst. **161** (2010), pp. 1325–1342.

[21] Pöschel, R., *Galois connections for operations and relations*, in: K. Denecke, M. Erné and S. L. Wismath, editors, *Galois Connections and Applications*, Springer, 2004 pp. 231–258.

[22] von Karger, B., *Temporal algebra*, Math. Struct. Comput. Sci. **8** (1998), pp. 277–320.

Completeness for an Intuitionistic Modal Logic of Vagueness

Ahmee Christensen [1]

University of California, Berkeley

Abstract

Wright has long advocated for an intuitionistic solution to the Sorites paradox. Recently, Bobzien and Rumfitt have suggested an extension to this solution by introducing the modality 'it is borderline whether', in part intended to provide the intuitionist with alternatives to assenting, dissenting, and remaining silent when asked questions about vague predicates (e.g., 'Is this tube red?'). Their proposal includes a collection of formulas and inference rules that they argue an intuitionistic modal logic of vagueness ought to prove. This paper proposes a logic meeting Bobzien and Rumfitt's desiderata, establishes a semantics for which the logic is sound and complete, and then uses completeness to prove a metatheorem asserting the equivalence of three notions of when the logic settles the matter of some formula. We then consider the addition of an axiom ruling out clear borderline cases, which is endorsed by proponents of columnar vagueness like Bobzien. Leaning heavily on a topological analogy, we show that the semantics can be adapted to accommodate this extension of the logic (and the corresponding view on higher-order vagueness) without losing completeness.

Keywords: Intutionism, intuitionistic modal logic, vagueness, completeness

1 Introduction

Bobzien and Rumfitt [3] defend Wright's [8] proposal to use the intuitionistic propositional calculus when reasoning about vague statements. If we were to present the intuitionist with an array of one hundred tubes whose colors imperceptibly shift, from the first tube to the last, from red to orange, she would not be obligated to accept $\mathrm{R}a_n \vee \neg \mathrm{R}a_n$ for each n, where R is a predicate for redness and a_n refers to the nth tube. Hence, even though the first tube is clearly red and the last is clearly not, the intuitionist does not find herself in the classicist's predicament of being forced to hold that, while each pair of consecutive tubes is indiscriminable, there exists a consecutive pair of tubes where the first of the pair is red and the second is not. As Bobzien and Rumfitt explain, "An intuitionist like Wright is unwilling to assert certain instances of the Law [of Excluded Middle], such as $\mathrm{R}a_{50} \vee \neg \mathrm{R}a_{50}$ with a_{50} supposed to be a borderline case of R"[3, p. 237].

[1] achris@berkeley.edu

This picture misses something, however. When a tube of questionable redness is presented to the intuitionist, her silence is not an indication of her having nothing to say on the matter. Bobzien and Rumfitt write that in cases where "[the intuitionist] does not assert '$Ra_{50} \vee \neg Ra_{50}$'...she may invoke borderlineness" [3, p. 240]. Thus, there is a pull to extend the language so that it can express as much. Bobzien and Rumfitt argue for a number of principles that should govern a borderlineness modality ∇. Adding ∇ to the language of propositional logic, they take ∇A to mean that it is *borderline whether* A. The modalities $\Box$ and $\Diamond$ are then defined as $\Box A \equiv A \wedge \neg\nabla A$ and $\Diamond A \equiv A \vee \nabla A$. $\Box A$ can be taken to mean *it is clear that* A, while $\Diamond A$ can be taken to mean *it cannot be ruled out that* A [3, p. 242].

A number of axioms are suggested by Bobzien and Rumfitt, though they leave open whether this list is complete. In this paper, we will propose an intuitionistic modal logic of vagueness that strengthens one of their axioms, provide a formal semantics for their language, and prove the corresponding soundness and completeness theorems. These results will then be used to prove a metatheorem for our logic that establishes an equivalence among three candidate notions of a logic settling the matter of a formula φ.

For convenience, in what follows we will take $\Box$ and $\Diamond$ to be the primitive modalities, though this is of no material difference, since the axioms proposed by Bobzien and Rumfitt are strong enough to define ∇ in terms of $\Box$ and $\Diamond$. In particular, any system that they endorse will have $\vdash \nabla\varphi \leftrightarrow (\Diamond\varphi \wedge \neg\Box\varphi)$. After fixing a countably infinite set $\mathbf{P} = \{p, q, \ldots\}$ of propositional variables, we define the language $\mathcal{L}$ recursively, as follows:

$$\varphi ::= \bot \,|\, p \,|\, (\varphi \wedge \varphi) \,|\, (\varphi \vee \varphi) \,|\, (\varphi \to \varphi) \,|\, \Box\varphi \,|\, \Diamond\varphi$$

where $p \in \mathbf{P}$. $\mathcal{L}_0$ will denote the subset of modal-free formulas—those formulas with no occurrences of $\Box$ or $\Diamond$. We take $\neg\varphi$, $\top$, and $\varphi \leftrightarrow \psi$ to be shorthands for $\varphi \to \bot$, $\bot \to \bot$, and $(\varphi \to \psi) \wedge (\psi \to \varphi)$, respectively. The axioms amassed by Bobzien and Rumfitt are the axioms in Figure 1 appearing above the second dashed line, as well as the *stable nabla* axiom $\mathbf{S}\nabla$ $(\neg\neg\nabla p \to \nabla p)$. [2]

For the logic of vagueness proposed in this paper, we first argue that $\mathbf{S}\nabla$ ought to be strengthened to $\neg\neg\Diamond p \to \Diamond p$ ($\mathbf{S}\Diamond$). Because we have an equivalence between ∇p and $\Diamond p \wedge \neg\Box p$, the antecedent of $\mathbf{S}\nabla$ can be written as $\neg\neg(\Diamond p \wedge \neg\Box p)$, which is in turn equivalent to $\neg\neg\Diamond p \wedge \neg\Box p$ by an intuitionistically acceptable argument. The consequent of $\mathbf{S}\nabla$ is similarly equivalent to $\Diamond p \wedge \neg\Box p$, so we can equivalently express the axiom as $(\neg\neg\Diamond p \wedge \neg\Box p) \to \Diamond p$. But now it seems that we may as well strengthen the axiom by dropping $\neg\Box p$ from the antecedent. After all, if we are trying to capture conditions sufficient for concluding *it cannot be ruled out that* p, why would knowing that p is not clearly true help our case?

---

[2] It should be noted that the first Fischer Servi axiom **FS1** is mentioned but not defended in [3]. We will leave the matter unsettled as they did.

This brings us to the system IVL (intuitionistic vagueness logic), which comprises the aforementioned axioms along with three uncontroversial deduction rules. For the rest of this paper, we will write $\vdash$ for $\vdash_{\mathsf{IVL}}$. We present IVL in Figure 1, where the Fischer Servi logic FS can be obtained by restricting to the axioms above the first dashed line (but keeping all of the inference rules).

The System IVL	
I	any theorem of IPC
K□a	$\Box(p \wedge q) \leftrightarrow (\Box p \wedge \Box q)$
K□b	$\Box\top$
K◊a	$\Diamond(p \vee q) \leftrightarrow (\Diamond p \vee \Diamond q)$
K◊b	$\neg\Diamond\bot$
FS1	$(\Diamond p \rightarrow \Box q) \rightarrow \Box(p \rightarrow q)$
FS2	$\Diamond(p \rightarrow q) \rightarrow (\Box p \rightarrow \Diamond q)$
T□	$\Box p \rightarrow p$
T◊	$p \rightarrow \Diamond p$
4□	$\Box p \rightarrow \Box\Box p$
4◊	$\Diamond\Diamond p \rightarrow \Diamond p$
S◊	$\neg\neg\Diamond p \rightarrow \Diamond p$
MP	from $\varphi \rightarrow \psi$ and φ infer ψ
US	from φ infer $\varphi[\psi/p]$
Reg	from $\varphi \rightarrow \psi$ infer $\bigcirc\varphi \rightarrow \bigcirc\psi$

Fig. 1. φ and ψ are any formulas in $\mathcal{L}$ and $\bigcirc \in \{\Box, \Diamond\}$.

When we present deductions, we will make free use of intuitionistic reasoning without spelling out every line. Such moves are, of course, just a number of instances of **I** and applications of MP and US. Where possible, we will give deductions in the weaker system FS to highlight which theorems do not depend on any S4-like properties. We will now offer deductions of a couple well-known theorems of FS, both of which will be of use to us in later sections.

Lemma 1.1 $\vdash_{\mathsf{FS}} (\Box p \wedge \Diamond q) \rightarrow \Diamond(p \wedge q)$.

Proof.

$$\vdash_{\mathsf{FS}} q \rightarrow (p \rightarrow p \wedge q) \qquad (1)$$

$$\vdash_{\mathsf{FS}} \Diamond q \rightarrow \Diamond(p \rightarrow p \wedge q) \qquad \text{Reg}, (1) \qquad (2)$$

$$\vdash_{\mathsf{FS}} \Diamond q \rightarrow (\Box p \rightarrow \Diamond(p \wedge q)) \qquad \mathbf{FS2}, (2) \qquad (3)$$

$$\vdash_{\mathsf{FS}} (\Box p \wedge \Diamond q) \rightarrow \Diamond(p \wedge q) \qquad (3) \qquad (4)$$

□

Proposition 1.2 $\vdash_{\mathsf{FS}} \neg\Diamond p \leftrightarrow \Box\neg p$.

Proof.

$\vdash_{\mathsf{FS}} \neg\Diamond p \to (\Diamond p \to \Box\bot)$		(1)
$\vdash_{\mathsf{FS}} \neg\Diamond p \to \Box(p \to \bot)$	**FS1**, (1)	(2)
$\vdash_{\mathsf{FS}} \neg\Diamond p \to \Box\neg p$	(2)	(3)
$\vdash_{\mathsf{FS}} (\Box\neg p \wedge \Diamond p) \to \Diamond(p \wedge \neg p)$	Lemma 1.1	(4)
$\vdash_{\mathsf{FS}} (\Box\neg p \wedge \Diamond p) \to \Diamond\bot$	(4)	(5)
$\vdash_{\mathsf{FS}} \neg\Diamond\bot$	**K$\Diamond$b**	(6)
$\vdash_{\mathsf{FS}} \Box\neg p \to \neg\Diamond p$	(5), (6)	(7)
$\vdash_{\mathsf{FS}} \neg\Diamond p \leftrightarrow \Box\neg p$	(3), (7)	(8)

□

2 Semantics

2.1 Relational Preliminaries

As the formal semantics we develop will be relational, we will first establish some conventions for relations and operations on them. A relation R on a set A is a subset of A^2, and we write $a\ R\ b$ if $(a, b) \in R$. If R is a relation on A, we define $R^{-1} = \{(a, b) \in A^2 : b\ R\ a\}$. We also have a notion of relation composition. For two relations R_1 and R_2 on A, we set

$$R_1 \circ R_2 = \{(a, b) \in A^2 : \text{there exists } x \text{ such that } a\ R_2\ x \text{ and } x\ R_1\ b\}.$$

If $a \in A$ and R is a relation on A, we set $R\ (a) = \{b \in A : a\ R\ b\}$. In a few places, we will need to appeal to the transitive closure R^* of a relation R, which is the smallest transitive relation extending R.

Turning to partial orders specifically, if $(A, \preccurlyeq)$ is a poset, we denote by $\mathrm{Up}(A, \preccurlyeq)$ the collection of all upwardly closed subsets (up-sets) of A. For any $B \subseteq A$, we denote the upward closure of B as $\uparrow(B)$. The *principal up-set generated by* a is just $\uparrow(\{a\})$, which we will shorten to $\uparrow(a)$ when it is clear that a is an element of the underlying set of the partial order in question. Finally, for $C \subseteq B \subseteq A$, we say that C is *cofinal* in B if for every $b \in B$, there exists $c \in C$ with $b \preccurlyeq c$.

2.2 Frames

The semantics proposed is the same as for the Fischer Servi logic FS, restricting to a smaller class of frames to accommodate the additional axioms.

Definition 2.1 *A Fischer Servi frame is a triple* $(W, \preccurlyeq, R)$ *where* $\preccurlyeq$ *is a partial order and* R *is a binary relation that satisfy the following conditions:*

$$\begin{array}{ll} (FC1) & (\preccurlyeq \circ R) \subseteq (R \circ \preccurlyeq); \\ (FC2) & (\preccurlyeq \circ R^{-1}) \subseteq (R^{-1} \circ \preccurlyeq). \end{array}$$

A Fischer Servi model is a Fischer Servi frame equipped with a function

$v : \mathbf{P} \to \mathrm{Up}(W, \preccurlyeq)$. For such a model $\mathcal{M} = (W, \preccurlyeq, R, v)$, the semantics follows:

$$
\begin{array}{lll}
\mathcal{M}, w \nVdash \bot & & \\
\mathcal{M}, w \Vdash p & \text{iff} & w \in v(p) \\
\mathcal{M}, w \Vdash (\varphi \wedge \psi) & \text{iff} & \mathcal{M}, w \Vdash \varphi \text{ and } \mathcal{M}, w \Vdash \psi \\
\mathcal{M}, w \Vdash (\varphi \vee \psi) & \text{iff} & \mathcal{M}, w \Vdash \varphi \text{ or } \mathcal{M}, w \Vdash \psi \\
\mathcal{M}, w \Vdash (\varphi \to \psi) & \text{iff} & \text{for all } x \succcurlyeq w,\ \mathcal{M}, x \nVdash \varphi \text{ or } \mathcal{M}, x \Vdash \psi \\
\mathcal{M}, w \Vdash \Box\varphi & \text{iff} & \text{for all } x \in (R \circ \preccurlyeq)(w),\ \mathcal{M}, x \Vdash \varphi \\
\mathcal{M}, w \Vdash \Diamond\varphi & \text{iff} & \text{there exists } x \in R\,(w) \text{ such that } \mathcal{M}, x \Vdash \varphi.
\end{array}
$$

Fischer Servi studied extensions of the logic FS using the above semantics in, for example, [4] and [5], with one of her foundational results in the study of such logics being the completeness of FS with respect to the class of Fischer Servi frames. Looking at the forcing clauses, the relation $\preccurlyeq$ plays the roll of the partial order that appears in Kripke models for IPC and the relation R acts similarly to the relation that appears in Kripke models for classical normal modal logics. Accordingly, we might refer to $\preccurlyeq$ as the *intuitionistic relation* and R as the *modal relation.* As it stands, our class of models is too large to obtain a soundness result, so we will have to make further demands on frames.

Definition 2.2 *An* S4 *Fischer Servi frame is a Fischer Servi frame* $(W, \preccurlyeq, R)$ *where* R *is moreover a quasi-order.*

We call this frame class $\mathcal{S}$. It is already known that the logic FSS4 = FS $\oplus$ $\{\mathbf{T}\Diamond, \mathbf{T}\Box, \mathbf{4}\Box, \mathbf{4}\Diamond\}$ is complete with respect to this class [1].

At this point, it will be useful to introduce diagrams, which may provide better intuition than the relation composition notation used thus far. For this paper, we will use the convention that single-line arrows indicate the intuitionistic relation and double-line arrows indicate the modal relation. (Where possible, the intuitionistic arrows will point up and the modal arrows will point to the side.) From this point on, we are concerned with only birelational structures where both relations are quasi-orders, so we can and will unambiguously omit all self-loops as well as arrows whose existences are implied by transitivity. Finally, dotted arrows and hollow points are used to mark existential quantifiers and instances of relations appearing in the consequent. Figure 2 illustrates conditions (FC1) and (FC2).

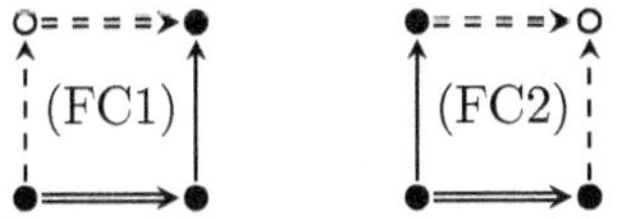

Fig. 2. The Fischer Servi frame conditions from Definition 2.1

To obtain IVL from FSS4 we need only extend by the axiom $\mathbf{S}\Diamond$. Analogously, we need just one additional imposition on our frame class to obtain a soundness result. Consider the following first-order frame condition:

(FC3-weak) For every $w \in W$, there exists $w' \succcurlyeq w$ such that $(R \circ \preccurlyeq)(w') \subseteq (\succcurlyeq \circ R)(w)$.

Perhaps easier to understand is the diagram in Figure 3.

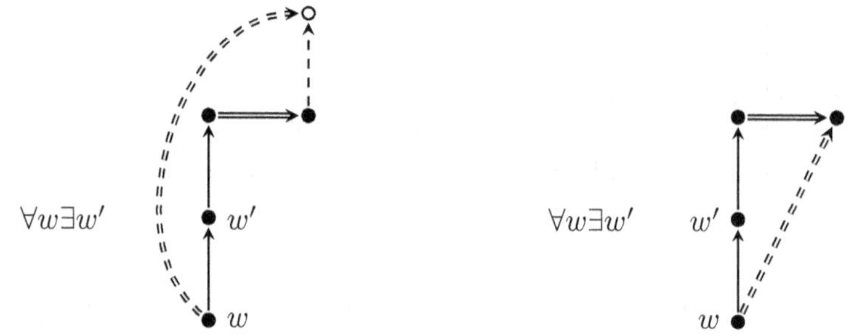

Fig. 3. The diagrams for conditions (FC3-weak) (left) and (FC3) (right)

Proposition 2.3 (Correspondence) *A frame in $\mathcal{S}$ validates* $\mathbf{S}\Diamond$ *if and only if it satisfies* (FC3-weak).

Proof. It is straightforward to check that any S4 Fischer Servi frame satisfying (FC3-weak) will force $\neg\neg\Diamond p \rightarrow \Diamond p$ at every point.

We will now check the other direction. For convenience, let $\Phi(w, x, y)$ abbreviate the first-order property ($x\ R \circ \preccurlyeq\ y$) and $\neg(\exists y')(y' \succcurlyeq y$ and $w\ R\ y')$. Suppose that some S4 Fischer Servi frame $\mathcal{F} = (W, \preccurlyeq, R)$ fails to satisfy (FC3-weak). Then there is a point $w \in W$ such that set $A = \{x \in W : (\exists y)\Phi(w, x, y)\}$ is cofinal in the principal up-set of w. For each element $x \in A \cap \uparrow(w)$ choose a $y(x)$ witnessing $(\exists y)\Phi(w, x, y)$, and fix a valuation with $v(p) = \uparrow(\{y(x) : x \in A \cap \uparrow(w)\})$. This will yield $w \Vdash \neg\neg\Diamond p$, as the points above w that have a modal successor forcing p are cofinal in $\uparrow(w)$. On the other hand, $w \nVdash \Diamond p$, as forcing $\Diamond p$ is equivalent to having a modal successor that is an intuitionistic successor of some $y(x)$, and each $y(x)$ was picked so that this is impossible. □

As it turns out, we can actually get away with a slightly stronger—and easier to work with—condition than (FC3-weak) without losing completeness, so we will end up taking our class of frames to be slightly smaller than the class of all S4 Fischer Servi frames validating $\mathbf{S}\Diamond$.

(FC3) For every $w \in W$, there exists $w' \succcurlyeq w$ such that $(R \circ \preccurlyeq)(w') \subseteq R\,(w)$.

We will sometimes refer to the point w' in condition (FC3) as a *diamond-reflection point for* w. (FC3) can then be thought of as simply asserting that every point w has a diamond-reflection point.

Definition 2.4 *An* intuitionistic vagueness *(IV) frame is an* S4 *Fischer Servi frame satisfying the condition (FC3).*

We call the class of IV frames $\mathcal{V}$.

2.3 Intuitionistic Vagueness Models

An IV model is an IV frame equipped with a function $v : \mathbf{P} \rightarrow \mathrm{Up}(W, \preccurlyeq)$.

Proposition 2.5 *For any IV model $\mathcal{M} = (W, \preccurlyeq, R, v)$, $\varphi \in \mathcal{L}$, and $w \preccurlyeq w'$, if $w \Vdash \varphi$ then $w' \Vdash \varphi$.*

Proof. The proof is by induction on formula complexity. The argument for each non-modal case is the standard one given for intuitionistic logic. The $\Box$ case is built into the semantics. For the $\Diamond$ case, suppose that $w \Vdash \Diamond\psi$. Then there is some $x \in W$ with $w\ R\ x$ and $x \Vdash \psi$. By frame condition (FC2), there is an $x' \succcurlyeq x$ with $w'\ R\ x'$. By the induction hypothesis, $x' \Vdash \psi$, so $w' \Vdash \Diamond\psi$. □

Theorem 2.6 (Soundness) IVL *is sound with respect to $\mathcal{V}$.*

Proof. We will just check that **S**$\Diamond$ holds at every point in every IV model, as all of the other axioms and the rules are not novel to this paper. Fix a model $\mathcal{M} = (W, \preccurlyeq, R, v)$ and suppose that $w \Vdash \neg\neg\Diamond\varphi$ for some $w \in W$ and $\varphi \in \mathcal{L}$. By condition (FC3), we can find a $w' \succcurlyeq w$ such that whenever there is a point x with $w' \preccurlyeq w''\ R\ x$, it is the case that $w\ R\ x$. There must be a point $w''' \succcurlyeq w'$ with $w''' \vDash \Diamond\varphi$. w''' sees a point x forcing φ, but by condition (FC3), w also sees x. We conclude $w \Vdash \Diamond\varphi$. □

The argument above sheds light on the term *diamond-reflection.* In any model where w' is a diamond-reflection point of w, $w \Vdash \Diamond\varphi$ if and only if $w' \Vdash \Diamond\varphi$, for every formula φ. Further, since any successor of w' is also a diamond-reflection point of w, we have $w \nVdash \Diamond\varphi$ if and only if $w' \Vdash \neg\Diamond\varphi$.

The soundness theorem allows us to quickly verify that IVL is not, in a few senses, too strong. *A priori* it seems possible that the underlying propositional calculus of our system is stronger than IPC via some deduction making use of the modalities. Another concern is the status of $\Box p \vee \neg\Box p$. Bobzien and Rumfitt begin their exploration by extending the basic propositional language by $\Box$. Now, the classicist can deny $\Box p \vee \Box\neg p$; even classically, some propositions are neither clearly the case nor clearly not the case. This does not solve the problem of vagueness for them, however, as this move alone still demands commitment to $\Box p \vee \neg\Box p$.

We can give a very simple argument to address the first concern.

Corollary 2.7 IVL *is a conservative extension of* IPC.

Proof. Let $\varphi \in \mathcal{L}_0$ be a non-theorem of IPC. By completeness of IPC [6], we can find some model $\mathcal{M} = (W, \preccurlyeq, v)$ with $\mathcal{M}, x \nVdash \varphi$ for some $x \in W$. Extend this to an intuitionistic vagueness model by setting $\mathcal{M}^* = (W, \preccurlyeq, W^2, v)$. Clearly, we have $\mathcal{M}^*, x \nVdash \varphi$. Hence by Proposition 2.6, φ is not a theorem of IVL. □

Turning to the second concern, we introduce the technique of drawing models, which is an efficient method of producing corollaries of soundness. Our conventions for these drawings will be largely the same as for frame condition diagrams. For arguments involving a certain subset of a model, dotted arrows and hollow points will be reserved for applications of the frame conditions (FC1) and (FC2), but all quantification should be made clear by the accompanying text. We will now employ this technique to dispel the second concern.

Corollary 2.8 IVL *does not prove* $\Box p \vee \neg\Box p$.

Proof. We present a countermodel.

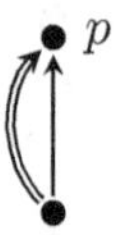

The lower point forces neither $\Box p$ nor $\neg\Box p$. □

A surprising feature of IVL is that it does not have the disjunction property: there are formulas φ and ψ such that $\vdash \varphi \vee \psi$, but $\nvdash \varphi$ and $\nvdash \psi$.

Proposition 2.9 *For any* φ, $\vdash \Diamond\varphi \vee \Diamond\neg\varphi$.

Proof.

$$
\begin{array}{lrr}
\vdash \neg\neg(\varphi \vee \neg\varphi) & & (1) \\
\vdash \Diamond\neg\neg(\varphi \vee \neg\varphi) & \mathbf{T}\Diamond, (1) & (2) \\
\vdash \neg\neg\Diamond\neg\neg(\varphi \vee \neg\varphi) & (2) & (3) \\
\vdash \neg\Box\neg\neg\neg(\varphi \vee \neg\varphi) & \text{Proposition 1.2}, (3) & (4) \\
\vdash \neg\Box\neg(\varphi \vee \neg\varphi) & (4) & (5) \\
\vdash \neg\neg\Diamond(\varphi \vee \neg\varphi) & \text{Proposition 1.2}, (5) & (6) \\
\vdash \Diamond(\varphi \vee \neg\varphi) & \mathbf{S}\Diamond, (6) & (7) \\
\vdash \Diamond\varphi \vee \Diamond\neg\varphi & \mathbf{K}\Diamond\mathbf{a}, (7) & (8)
\end{array}
$$

□

Corollary 2.10 IVL *does not have the disjunction property.*

Proof. By Proposition 2.9, $\vdash \Diamond p \vee \neg\Diamond p$. By the soundness theorem for IVL, we have both $\nvdash \Diamond p$ and $\nvdash \Diamond\neg p$ as there is an obvious one-point countermodel in each case. □

2.4 Model Constructions

IV models are decidedly less flexible than Fischer Servi models. For instance, given two Fischer Servi models, one can take their disjoint union and add a new universal intuitionistic predecessor that does not stand in the modal relation with any other points. It is straightforward to check that this is still a Fischer Servi model, which, alongside soundness and completeness, furnishes a neat argument for the disjunction property of the logic FS. In the case of FS extended by the S4 principles, the same argument works, with the small caveat that one lets the new point stand in the modal relation with itself. IVL does not have the disjunction property, however, so we know that when we consider the disjoint union of two IV models with a new universal predecessor, there can be no general procedure to choose the modal relation such that both the new structure is an IV model and the forcing relation is preserved at the old points. Two typical constructions are still available to us, however.

Definition 2.11 *For each i in some index set I, let $\mathcal{M}_i = (W_i, \preccurlyeq_i, R_i, v_i)$ be an IV model. We define their disjoint union as*

$$\bigsqcup_{i\in I} \mathcal{M}_i = \left(\bigsqcup_{i\in I} W_i, \bigsqcup_{i\in I} \preccurlyeq_i, \bigsqcup_{i\in I} R_i, \bigsqcup_{i\in I} v_i \right).$$

The following proposition is straightforward to verify:

Proposition 2.12 *Let $\{\mathcal{M}_i\}_{i\in I}$ be a collection of IV models. Then $\bigsqcup_{i\in I} \mathcal{M}_i$ is an IV model and for each $j \in I$, $w \in W_i$, and $\varphi \in \mathcal{L}$, $\mathcal{M}_j, w \Vdash \varphi$ if and only if $\bigsqcup_{i\in I} \mathcal{M}_i, w \Vdash \varphi$.*

Definition 2.13 *Let $\mathcal{M} = (W, \preccurlyeq, R, v)$ be an IV model with $w \in W$. Then the submodel generated by w is $\mathcal{M}(w)$ where we have replaced W with the image of w under $(\preccurlyeq \cup R)^*$ and restricted the relations and valuation to this set.*

Also easily proved is the natural analog of Proposition 2.12 for generated submodels, which we now state.

Proposition 2.14 *Let $\mathcal{M}$ be an IV model with $w \in W$. Then $\mathcal{M}(w)$ is an IV model, and for any formula φ and x in the universe of $\mathcal{M}(w)$, we have $\mathcal{M}, x \Vdash \varphi$ if and only $\mathcal{M}(w), x \Vdash \varphi$.*

Because of the Fischer Servi frame condition (FC1) and the transitivity of both the modal and intuitionistic relations in an IV model, we can also obtain a simple characterization of the universe of a generated submodel.

Proposition 2.15 *Let $\mathcal{M} = (W, \preccurlyeq, R, v)$ be an IV model with $w \in W$. Then $(\preccurlyeq \cup R)^*(w) = (R \circ \preccurlyeq)(w)$.*

Proof. It is clear that any point in $(R \circ \preccurlyeq)(w)$ is in the transitive closure of the union of $\preccurlyeq$ and R. In the other direction, suppose that $x \in (\preccurlyeq \cup R)^*(w)$. Then there is a sequence of points and relations $w \sim_1 x_1 \sim_2 x_2 \cdots x_{n-1} \sim_n x$, where each $\sim_i$ is either R or $\preccurlyeq$. By frame condition (FC1), whenever $\sim_i$ is R and $\sim_{i+1}$ is $\preccurlyeq$, we can replace x_i with some new point and swap the relations. Iterate this process until each instance of $\preccurlyeq$ occurs before each instance of R in the sequence. That is, we get some sequence $w \preccurlyeq y_1 \cdots y_{k-1} \preccurlyeq y_k \; R \; y_{k+1} \cdots y_{n-1} \; R \; x$. Note that w and x are left intact, as we only swap out points that occur between two relations in the sequence. By the transitivity of each relation, we can now contract our sequence, so that we simply have $w \preccurlyeq y_k \; R \; x$. In other words, $x \in (R \circ \preccurlyeq)(w)$. □

3 Completeness

3.1 The Completeness Theorem

We will prove completeness of IVL with respect to the semantics in §2.2 using a canonical model argument. First we will derive the traditional modal inference rule of necessitation in our system.

Lemma 3.1 *The rule* `Nec` *(from φ infer $\Box\varphi$) is derivable in* IVL.

Proof.

$\vdash \varphi$	assumption	(1)
$\vdash \top \to \varphi$	(1)	(2)
$\vdash \Box\top \to \Box\varphi$	Reg, (2)	(3)
$\vdash \Box\top$	$\mathbf{K\Box b}$	(4)
$\vdash \Box\varphi$	MP, (3), (4)	(5)

□

Next we define the theories over $\mathcal{L}$ that will serve as the points in our canonical model. This definition, and the shape of the completeness proof in general, is essentially just a modalized version of the standard proof for the completeness of IPC, as found, e.g., in [6].

Definition 3.2 *$\Gamma \subseteq \mathcal{L}$ is a prime theory if it is deductively closed: if $\Gamma \vdash \varphi$, then $\varphi \in \Gamma$; consistent: $\Gamma \nvdash \bot$; and disjunctive: if $\varphi \vee \psi \in \Gamma$, then $\varphi \in \Gamma$ or $\psi \in \Gamma$.*

Definition 3.3 *For $\Gamma, \Delta \subseteq \mathcal{L}$, (Γ, Δ) is a consistent pair if for every $\{\varphi_i\}_{i=1}^n \subseteq \Gamma$ and $\{\psi_j\}_{j=1}^m \subseteq \Delta$, we have $\nvdash \wedge_{i=1}^n \varphi_i \to \vee_{j=1}^m \psi_i$.*

Lemma 3.4 *If (Γ, Δ) is a consistent pair, then there is a prime theory $\Gamma' \supseteq \Gamma$ such that $\Gamma' \cap \Delta = \varnothing$.*

Proof. A standard recursive argument suffices. □

If Γ' is a prime theory and $(\Gamma', \mathcal{L} \setminus \Gamma')$ extends (Γ, Δ), we might say simply that the prime theory Γ' extends the pair (Γ, Δ) without making reference to $\mathcal{L} \setminus \Gamma'$.

For convenience, we fix a few pieces of notation for some set Γ of formulas: $\Gamma^\Box = \{\varphi \in \mathcal{L} : \Box\varphi \in \Gamma\}$; $\Gamma^\Diamond = \{\varphi \in \mathcal{L} : \Diamond\varphi \in \Gamma\}$; $B(\Gamma) = \{\Box\psi \in \mathcal{L} : \psi \in \Gamma\}$; $D(\Gamma) = \{\Diamond\psi \in \mathcal{L} : \psi \in \Gamma\}$; and $N(\Gamma) = \{\nabla\psi \in \mathcal{L} : \psi \in \Gamma\}$. These will significantly streamline the notation in the definition of the canonical model, as well as in the proofs of the completeness theorem and some of its corollaries.

Definition 3.5 *The canonical model for IVL is $\mathcal{M}^\mathsf{C} = (W^\mathsf{C}, \preccurlyeq^\mathsf{C}, R^\mathsf{C}, v^\mathsf{C})$ where:*

(a) $\mathcal{M}^\mathsf{C}$ is the set of all prime theories over IVL;

(b) $\Gamma \preccurlyeq^\mathsf{C} \Gamma'$ if and only $\Gamma \subseteq \Gamma'$;

(c) $\Gamma\, R^\mathsf{C}\, \Delta$ if and only if $\Gamma^\Box \subseteq \Delta \subseteq \Gamma^\Diamond$;

(d) $v^\mathsf{C}(p) = \{\Gamma \in W^\mathsf{C} : p \in \Gamma\}$.

Proposition 3.6 *The canonical model is an intuitionistic vagueness model.*

Proof. It is clear that $v^\mathsf{C}(p)$ is an upset for each p. It is also immediate that $\preccurlyeq^\mathsf{C}$ is a partial order, as it is just set containment.

First we check that R^C is a quasi-order. It is reflexive since $\Gamma^\Box \subseteq \Gamma \subseteq \Gamma^\Diamond$ by the deductive closure of Γ and the axioms $\mathbf{T\Diamond}$ and $\mathbf{T\Box}$. For transitivity, suppose that $\Gamma\, R^\mathsf{C}\, \Delta\, R^\mathsf{C}\, \Theta$. If $\Box\varphi \in \Gamma$ then $\Box\Box\varphi \in \Gamma$ by $\mathbf{4\Box}$, so $\Box\varphi \in \Delta$, so

$\varphi \in \Theta$. On the other hand, if $\varphi \in \Theta$, then $\Diamond\varphi \in \Delta$, so $\Diamond\Diamond\varphi \in \Gamma$, so $\Diamond\varphi \in \Gamma$ by **4**$\Diamond$.

Now we check the frame conditions:

(FC1) Suppose that $\Gamma \; R^{\mathsf{C}} \; \Delta \; \preccurlyeq^{\mathsf{C}} \; \Delta'$. We first argue that $(\Gamma \cup D(\Delta'), B(\mathcal{L} \setminus \Delta'))$ is a consistent pair. If it were not, then we would have $\Gamma \Vdash \bigwedge_{i=1}^{n} \Diamond\varphi_i \to \bigvee_{j=1}^{m} \Box\psi_j$, where $\varphi_i \in \Delta'$ and $\psi_j \notin \Delta'$. Since $\Vdash \Diamond \bigwedge_{i=1}^{n} \varphi_i \to \bigwedge_{i=1}^{n} \Diamond\varphi_i$ and $\Vdash \bigvee_{j=1}^{m} \Box\psi_j \to \Box \bigvee_{j=1}^{m} \psi_j$, we actually have the more usable

$$\Gamma \Vdash \Diamond \bigwedge_{i=1}^{n} \varphi_i \to \Box \bigvee_{j=1}^{m} \psi_j.$$

The conjunction in the antecedent is in Δ' by deductive closure, and the disjunction in the consequent is not in Δ' by disjunctivity, so we can rewrite this as $\Gamma \Vdash \Diamond\varphi \to \Box\psi$ with $\varphi \in \Delta$ and $\psi \notin \Delta$. By **FS1**, we then obtain $\Gamma \Vdash \Box(\varphi \to \psi)$, so $\varphi \to \psi \in \Delta$ and subsequently $\varphi \to \psi \in \Delta'$. By deductive closure, $\psi \in \Delta'$, which is a contradiction. Therefore, $(\Gamma \cup D(\Delta'), B(\mathcal{L} \setminus \Delta'))$ is a consistent pair. Using Propositions 3.4, we can find a prime theory Γ' extending this pair. By construction, $\Gamma \subseteq \Gamma'$, and $\Gamma'^{\Box} \subseteq \Delta' \subseteq \Gamma'^{\Diamond}$. We see that $\Gamma \preccurlyeq^{\mathsf{C}} \Gamma' \; R^{\mathsf{C}} \; \Delta'$, as desired.

(FC2) Now suppose that $\Gamma \preccurlyeq^{\mathsf{C}} \Gamma'$ and $\Gamma \; R^{\mathsf{C}} \; \Delta$. We want to find Δ' such that $\Delta \preccurlyeq^{\mathsf{C}} \Delta'$ and $\Gamma' \; R^{\mathsf{C}} \; \Delta'$. In this case, we want to check that $(\Delta \cup \Gamma'^{\Box}, \mathcal{L} \setminus \Gamma'^{\Diamond})$ is a consistent pair. If it were inconsistent, we would have $\Delta \Vdash \varphi \to \psi$ where $\varphi \in \Gamma'^{\Box}$ and $\psi \notin \Gamma'^{\Diamond}$. By the definition of R^{C}, $\Gamma \Vdash \Diamond(\varphi \to \psi)$. By **FS2**, $\Gamma \Vdash \Box\varphi \to \Diamond\psi$. Γ' inherits this, and since $\Box\varphi \in \Gamma'$, deductive closure ensures $\Gamma' \Vdash \Diamond\psi$, which is a contradiction. Any prime theory Δ' extending the pair in question will then witness this instance of condition (FC2).

(FC3) Let $\Gamma \in W^{\mathsf{C}}$. We first want to show that $\Gamma \cup \{\neg\Diamond\varphi \in \mathcal{L} : \Diamond\varphi \notin \Gamma\}$ is consistent, since any prime theory extending it will be a diamond-reflection point for Γ. If it is not consistent, then for some $\{\neg\Diamond\varphi_i\}_{i=1}^{n} \subseteq \{\neg\Diamond\varphi \in \mathcal{L} : \Diamond\varphi \notin \Gamma\}$, we have $\Gamma \vdash \neg \bigwedge_{i=1}^{n} \neg\Diamond\varphi_i$. We argue as follows:

$\Gamma \vdash \neg \bigwedge_{i=1}^{n} \neg\Diamond\varphi_i$	assumption	(1)
$\Gamma \vdash \neg\neg \bigvee_{i=1}^{n} \Diamond\varphi_i$	(1)	(2)
$\Gamma \Vdash \neg\neg\Diamond \bigvee_{i=1}^{n} \varphi_i$	**K**$\Diamond$**a**, (2)	(3)
$\Gamma \Vdash \Diamond \bigvee_{i=1}^{n} \varphi_i$	**S**$\Diamond$, (3)	(4)
$\Gamma \Vdash \bigvee_{i=1}^{n} \Diamond\varphi_i$	**K**$\Diamond$**a**, (4)	(5)
$\bigvee_{i=1}^{n} \Diamond\varphi_i \in \Gamma$	deductive closure, (5)	(6)
$\Diamond\varphi_k \in \Gamma$ for some $1 \le k \le n$	disjunctivity, (6)	(7)

This is contradiction, so we conclude that $\Gamma \cup \{\neg\Diamond\varphi \in \mathcal{L} : \Diamond\varphi \notin \Gamma\}$ is consistent and choose Γ' to be an element of W^{C} containing $\Gamma \cup \{\neg\Diamond\varphi \in \mathcal{L} : \Diamond\varphi \notin \Gamma\}$. We check $(\mathcal{M}^{\mathsf{C}} \circ \preccurlyeq^{\mathsf{C}})(\Gamma') \subseteq \mathcal{M}^{\mathsf{C}}(\Gamma)$. For any successor $\Gamma' \preccurlyeq^{\mathsf{C}} \Gamma''$ with $\Gamma'' \; R^{\mathsf{C}} \; \Delta$, we have $\Gamma^{\Box} \subseteq \Gamma''^{\Box} \subseteq \Delta$. On the other hand, if $\varphi \in \Delta$, we must have $\Diamond\varphi \in \Gamma$, since if not, $\neg\Diamond\varphi \in \Gamma''$, which would mean $\Delta \not\subseteq \Gamma''^{\Diamond}$.

□

Lemma 3.7 (Truth lemma) $\varphi \in \Gamma$ *if and only if* $\mathcal{M}^{\mathsf{C}}, \Gamma \Vdash \varphi$.

Proof. We proceed by induction on formula complexity. The atomic case, as well as the inductive steps for $\wedge$, $\vee$, and $\rightarrow$ are the same as in the completeness proof for IPC.

- Suppose $\varphi = \Box\psi$. Then, if $\varphi \in \Gamma$, by the definition of R^{C} and the forcing rule for $\Box$, $\mathcal{M}^{\mathsf{C}}, \Gamma \Vdash \Box\psi$.

  For the other direction, suppose $\varphi \notin \Gamma$. We want to find $\Gamma \preccurlyeq^{\mathsf{C}} \Gamma' \; R^{\mathsf{C}} \; \Delta$ such that $\psi \notin \Delta$. We construct Δ first. Note that $(\Gamma^{\Box}, \{\psi\})$ is a consistent pair, since otherwise $\Box\psi \in \Gamma$ by a standard argument. Take Δ to be a prime theory extending this pair. Next, we construct Γ'. Consider the pair

$$(\Gamma \cup \{\Diamond\chi : \chi \in \Delta\}, \{\Box\theta : \theta \notin \Delta\}).$$

  This pair is consistent by the same argument that we used to check condition (FC1) in Proposition 3.6. Take Γ' to be a prime theory extending this pair. We have $\Gamma \subseteq \Gamma'$, so $\Gamma \preccurlyeq^{\mathsf{C}} \Gamma'$. Additionally, by construction $\Gamma'^{\Box} \subseteq \Delta \subseteq \Gamma'^{\Diamond}$, so $\Gamma' \; R^{\mathsf{C}} \; \Delta$.

- Suppose $\varphi = \Diamond\psi$. If $\varphi \notin \Gamma$, by the definition of R^{C} and the forcing rule for $\Diamond$, $\mathcal{M}^{\mathsf{C}}, \Gamma \nVdash \Diamond\psi$.

  If $\varphi \in \Gamma$, then we want to check that $(\Gamma^{\Box} \cup \{\psi\}, \mathcal{L} \setminus \Gamma^{\Diamond})$ is a consistent pair. If not, then for some $\theta \in \Gamma^{\Box}$ and $\chi \in \mathcal{L} \setminus \Gamma^{\Diamond}$ we would have $\vdash \theta \wedge \psi \rightarrow \chi$.[3] Reg then affords us $\vdash \Diamond(\theta \wedge \psi) \rightarrow \Diamond\chi$. As $\Box\theta \in \Gamma$, Lemma 1.1 implies $\Diamond(\theta \wedge \psi) \in \Gamma$. By deductive closure, $\Diamond\chi \in \Gamma$, which is a contradiction.

□

As usual, verifying that our canonical model satisfies the truth lemma immediately grants us completeness.

Theorem 3.8 IVL *is strongly complete with respect to* $\mathcal{V}$.

3.2 An Application of Completeness

Using soundness, one can quickly establish that for any φ, we have $\nvdash \nabla\varphi$; simply note that any one-point model can never force a borderline statement, as the borderlineness of φ requires at least two points, one forcing φ and one not. This raises the opposite question of when the system *settles the matter* of φ. There seem to be three natural candidates for how this should be formalized: $\vdash \varphi$ or $\vdash \neg\varphi$; $\vdash \neg\nabla\varphi$; and $\vdash \varphi \vee \neg\varphi$. In fact, these are all equivalent.

In order to prove this equivalence, we first establish two procedures for producing new models. The first is the construction of the *omnispective expansion*.

[3] We can take single formulas here since $\Gamma^{\Box}$ is closed under conjunction and $\mathcal{L} \setminus \Gamma^{\Diamond}$ is closed under disjunction.

Definition 3.9 *Let $\mathcal{M} = (W, \preccurlyeq, R, v)$ be an IV model and let o be some point not in W. Then the* omnispective expansion *of $\mathcal{M}$ by o is defined as $\mathcal{M}^o = (W \cup \{o\}, \preccurlyeq \cup \{(o,o)\}, R \cup (\{o\} \times W), v)$.*

The omnispective expansion of a model is then just the result of appending a new point that can access all of the pre-existing points via the modal relation but has no intuitionistic interaction with them.

Lemma 3.10 *Let $\mathcal{M} = (W, \preccurlyeq, R, v)$ be a vagueness model. Then $\mathcal{M}^o$ is a vagueness model. Moreover, if there are points $w_1, w_2 \in W$ with $\mathcal{M}, w_1 \Vdash \varphi$ and $\mathcal{M}, w_2 \nVdash \neg\varphi$, then $o \Vdash \nabla\varphi$.*

Corollary 3.11 *If there exists a model $\mathcal{M}$ with a point forcing φ and a point not forcing φ, then $\nvdash \neg\nabla\varphi$.*

Proof. $\mathcal{M}^o, o \Vdash \nabla\varphi$, so we are done by soundness. □

Corollary 3.12 *Let $\varphi \in \mathcal{L}$. Then at least one of φ and $\neg\varphi$ must be consistent with every set of the form $D(\Psi)$ where Ψ is a set of formulas whose negations are not theorems.*

Proof. Toward a contradiction, assume that φ is inconsistent with $D(\Psi_1)$ and $\neg\varphi$ is inconsistent with $D(\Psi_2)$ for some $\Psi_1, \Psi_2 \subseteq \mathcal{L}$ whose individual formulas are not refuted. By completeness, take a collection of models $\mathcal{M}_{i \in I}$ such that for each $\psi \in \Psi_1 \cup \Psi_2$, there is some i such that some point in $\mathcal{M}_i$ forces ψ. Now consider the omnispective expansion $(\bigsqcup_{i \in I} \mathcal{M}_i)^o$. The point o must force either φ or $\neg\varphi$ and also forces every formula in $D(\Psi_1)$ and $D(\Psi_2)$, which is a contradiction. □

We are now situated for the promised application of the completeness theorem.

Theorem 3.13 *The following are equivalent: (1) $\vdash \varphi$ or $\vdash \neg\varphi$; (2) $\vdash \neg\nabla\varphi$; and (3) $\vdash \varphi \vee \neg\varphi$.*

Proof. First we check (1) $\Longrightarrow$ (2). Suppose $\vdash \varphi$ or $\vdash \neg\varphi$. Then we have either $\vdash \Box\varphi$ or $\vdash \Box\neg\varphi$. In the second case, we additionally get $\vdash \neg\Diamond\varphi$. Both cases are then plainly inconsistent with $\nabla\varphi$, which is just shorthand for $\Diamond\varphi \wedge \neg\Box\varphi$.

For (2) $\Longrightarrow$ (3), we proceed by contraposition. If $\nvdash \varphi \vee \neg\varphi$, then there is a model $\mathcal{M}$ with a point x not forcing $\varphi \vee \neg\varphi$. Since $\neg\varphi$ is not forced, there is also some $x' \succcurlyeq x$ with $x' \Vdash \varphi$. Then we are done by Corollary 3.11.

For (3) $\Longrightarrow$ (1), we again use contraposition. By Corollary 3.12, only one of φ and $\neg\varphi$ can be inconsistent with a set of formulas of the form $D(\Psi)$. Assume that $\neg\varphi$ is consistent with all such sets. In the case that we have to choose φ for this role, the argument will be identical.

Take some maximally consistent set Γ containing φ. $D(\Gamma)$ is then consistent with $\neg\varphi$, so by strong completeness we get a model $\mathcal{N}'$ with a point y forcing $D(\Gamma) \cup \{\neg\varphi\}$. We also get a model $\mathcal{M}'$ forcing Γ at some point x. Note that we can take x and y to be maximal with respect to the intuitionistic relation since every point is underneath a maximal point in the canonical model. For ease of notation set $\mathcal{M} = \mathcal{M}'(x)$ and $\mathcal{N} = \mathcal{N}'(y)$. Then, writing $\mathcal{M} = (W_{\mathcal{M}}, \preccurlyeq_{\mathcal{M}}, R_{\mathcal{M}}, v_{\mathcal{M}})$ and $\mathcal{N} = (W_{\mathcal{N}}, \preccurlyeq_{\mathcal{N}}, R_{\mathcal{N}}, v_{\mathcal{N}})$, we define a new model $\mathcal{O} = (W, \preccurlyeq, R, v)$ as follows: $W = W_{\mathcal{N}} \sqcup W_{\mathcal{N}} \sqcup \{w\}$; $w \preccurlyeq w$, $w \preccurlyeq x$, and $w \preccurlyeq y$; for any $a \in W_{\mathcal{M}}$, $w\ R\ a$; for any $a \in W_{\mathcal{M}}$ and $z\ R_{\mathcal{N}}\ y$, $z\ R\ a$; the restriction of $\preccurlyeq$ to $\mathcal{M}$ is $\preccurlyeq_{\mathcal{M}}$ and the restriction of R to $\mathcal{M}$ is $R_{\mathcal{M}}$; the restriction of $\preccurlyeq$ to $\mathcal{N}$ is $\preccurlyeq_{\mathcal{N}}$, and the restriction of R to $\mathcal{N}$ is $R_{\mathcal{N}}$; no other instances of relations occur; and $v(p) = v_{\mathcal{M}}(p) \cup v_{\mathcal{N}}(p)$ for all $p \in \mathbf{P}$.

This construction is perhaps easiest understood pictorially, as presented in Figure 4. As an intuition pump, we want to mimic the standard argument of the disjunction property for IPC, so we add a new intuitionistic predecessor for x and y. Condition (FC3) forces us to allow w to be able to take modal steps to all points in $W_{\mathcal{M}}$. Then, (FC2) forces us to allow y to take modal steps to all points in $W_{\mathcal{M}}$. Finally, transitivity forces us to allow all z with $z\ R_{\mathcal{N}}\ y$ to be able to take modal steps to all points in $W_{\mathcal{M}}$.

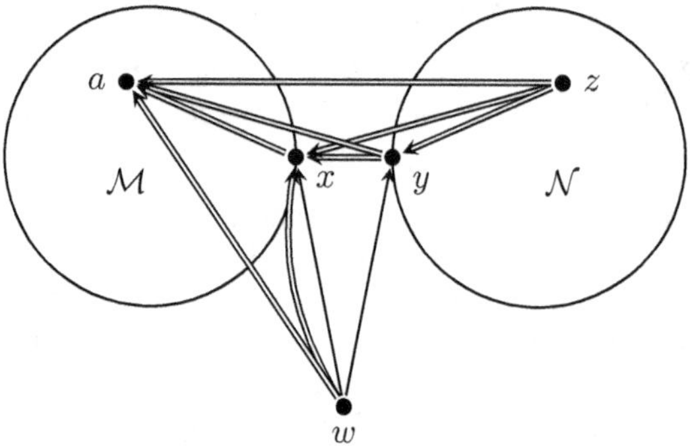

Fig. 4. The model $\mathcal{O}$ where some arbitrary $a \in W_{\mathcal{M}}$ and arbitrary z with $z\ R_{\mathcal{N}}\ y$ are shown

First we need to check that $\mathcal{O}$ is an IV model. This is almost entirely routine, with the exception that we are worried about satisfying (FC2) in the case where we have $z' \succcurlyeq z\ R\ a$. If this is the case, we must also have $z' \succcurlyeq_{\mathcal{N}} z\ R_{\mathcal{N}}\ y$. Then, by (FC2) for $\mathcal{N}$, there is some $y' \succcurlyeq_{\mathcal{N}} y$ such that $z'\ R_{\mathcal{N}}\ y'$. Since y is maximal, $z'\ R_{\mathcal{N}}\ y$. This means that $z'\ R\ a$, so a itself is a witness for this instance of (FC2).

It is clear that the forcing relation on $W_{\mathcal{M}(x)}$ is preserved, as no outgoing arrows were added for those points. We will prove by induction on the complexity of φ that for all $z \in W_{\mathcal{N}(y)}$, $\mathcal{N}, z \Vdash \varphi$ if and only if $\mathcal{O}, z \Vdash \varphi$.

- The atomic case is trivial by the construction of $\mathcal{O}$.
- The disjunction, conjunction, and conditional cases are immediate from the induction hypothesis, noting for the conditional case that there are no new outgoing instances of intuitionistic relations for points in $W_{\mathcal{N}}$.
- Suppose $\mathcal{N}, z \Vdash \Diamond\psi$. Then there is some $b \in \mathcal{N}$ with $\mathcal{N}, b \Vdash \psi$ and $z\ R\ b$. Such a b will still force ψ in $\mathcal{O}$ by the induction hypothesis, so $\mathcal{O}, z \Vdash \Diamond\psi$. In the other direction, suppose $\mathcal{O}, z \Vdash \Diamond\psi$. If this is witnessed in $W_{\mathcal{N}}$, we are done by the induction hypothesis, so suppose that $z\ R\ a$ for some $a \in \mathcal{M}$

with $\mathcal{O}, a \Vdash \psi$. We have $x\ R\ a$, so $\mathcal{M}, x \Vdash \Diamond\psi$, and $\Diamond\psi \in \Gamma$. Therefore, $\Diamond\Diamond\psi \in D(\Gamma)$, so $\mathcal{N}, y \Vdash \Diamond\psi$ by **4**$\Diamond$ and soundness. Now, since $z\ R\ a$, we can conclude that $z\ R_{\mathcal{N}}\ y$, as well, so $\mathcal{N}, z \vDash \Diamond\Diamond\psi$, implying our desired conclusion $\mathcal{N}, z \vDash \Diamond\psi$.

- Suppose $\mathcal{O}, z \Vdash \Box\psi$. We immediately get that $\mathcal{N}, z \Vdash \Box\psi$, by the induction hypothesis. In the other direction, we assume $\mathcal{N}, z \Vdash \Box\psi$. We need to make sure that there is no point $z' \succcurlyeq z$ with $z\ R\ a$ for some $a \in W_{\mathcal{M}}$ with $\mathcal{O}, a \nVdash \psi$. Suppose that this does happen. Then, by the maximality of x, $\mathcal{O}, x \Vdash \neg\Box\psi$. As we observed in the previous case of the inductive argument, $z'\ R\ a$ as well, so $\mathcal{O}, z' \Vdash \Diamond\neg\Box\psi$. This is inconsistent with $\mathcal{O}, z' \Vdash \Box\Box\psi$, which we have by **4**$\Box$ and soundness, so we arrive at a contradiction.

□

That the three most natural notions of φ being settled coincide in IVL seems to attest to the naturalness of the logic itself.

4 Higher-Order Vagueness

4.1 Stable Columnarity

Mormann [7] observes, in the classical setting, that when a logic at least as strong as S4 is assumed, all propositions are *stably columnar* [7]. That is, for any formula φ, $\nabla\nabla\varphi$ is provably equivalent to $\nabla\nabla\nabla\varphi$. By topological completeness for S4, this claim is just a redressing of the well-known fact that $\partial\partial A = \partial\partial\partial A$ where A is any subset of a topological space X and ∂ is the boundary (in the sense of closure minus interior) operator on X. This feature is sufficient for side-stepping paradoxes of higher-order vagueness, and it has the added thrust of not requiring the denial of clearly borderline cases. We can now verify that IVL proves that all formulas are stably columnar.

Lemma 4.1 $\vdash \Diamond\nabla p \to \nabla p$.

Proof. We can verify this quickly by drawing a picture and appealing to completeness.

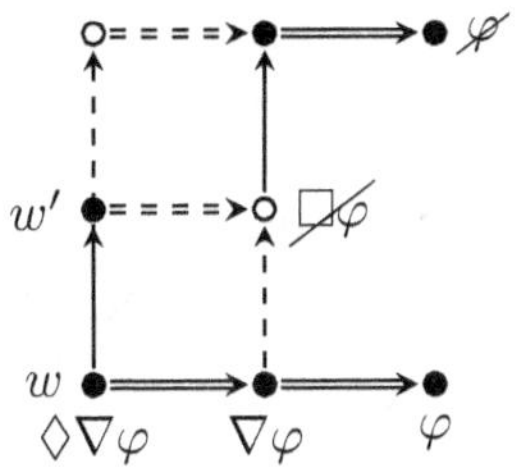

As illustrated above, for any w forcing $\Diamond\nabla\varphi$, it must force $\Diamond\varphi$ by transitivity. Also, by (FC1), (FC2), and transitivity, any $w' \succcurlyeq w$ cannot force $\Box\varphi$. Therefore, $\vdash \Diamond\nabla\varphi \to \nabla\varphi$. □

The above lemma then allows us derive a more useful theorem of the system.

Proposition 4.2 $\vdash \neg\Box\nabla\nabla p$.

Proof.

$$\begin{array}{lrr}
\vdash \Diamond\nabla p \to \nabla p & \text{Lemma 4.1} & (1)\\
\vdash \Box\Diamond\nabla p \to \Box\nabla p & \texttt{Reg}, (1) & (2)\\
\vdash \Box\nabla p \to \Diamond\Box\nabla p & \mathbf{T}\Diamond & (3)\\
\vdash \Box\Diamond\nabla p \to \Diamond\Box\nabla p & (2), (3) & (4)\\
\vdash \Box\nabla\nabla p \to \Box\Diamond\nabla p & \mathbf{K}\Box\mathbf{a} & (5)\\
\vdash \Box\nabla\nabla p \to \Diamond\Box\nabla p & (4), (5) & (6)\\
\vdash \Box\nabla\nabla p \to \Box\neg\Box\nabla p & \mathbf{K}\Box\mathbf{a} & (7)\\
\vdash \Box\neg\Box\nabla p \to \neg\Diamond\Box\nabla p & \text{Proposition 1.2} & (8)\\
\vdash \Box\nabla\nabla p \to \neg\Diamond\Box\nabla p & (7), (8) & (9)\\
\vdash \neg\Box\nabla\nabla p & (6), (9) & (10)
\end{array}$$

□

At this point we need only put together some established facts to prove the stable columnarity theorem for IVL.

Theorem 4.3 $\vdash \nabla\nabla p \leftrightarrow \nabla\nabla\nabla p$.

Proof. The theorem is of course equivalent to the pair of claims $\vdash \nabla\nabla\nabla p \to \nabla\nabla p$ and $\vdash \nabla\nabla p \to \nabla\nabla\nabla p$. To verify the first claim, we simply note that $\nabla\nabla\nabla p \to \Diamond\nabla\nabla p$ and $\Diamond\nabla\nabla p \to \nabla\nabla p$ are both theorems, the first by conjunction elimination and the second by Lemma 4.1. For the second claim, since $\nabla\nabla\nabla p$ is shorthand for $\Diamond\nabla\nabla p \wedge \neg\Box\nabla\nabla p$, we just need to verify that $\nabla\nabla p \to \Diamond\nabla\nabla p$ and $\nabla\nabla p \to \neg\Box\nabla\nabla p$ are theorems. As $\nabla\nabla p \to \Diamond\nabla\nabla p$ is just an instance of $\mathbf{T}\Diamond$ and $\vdash \nabla\nabla p \to \neg\Box\nabla\nabla p$ holds trivially in light of Proposition 4.2, we are done. □

4.2 Axiom M

Bobzien and Rumfitt are also concerned with paradoxes of higher-order vagueness. Partially in an effort to block such problems, they consider and defend the axiom $\neg\Box\nabla p$ (**M**). (Note that this is classically equivalent to the McKinsey axiom $\Box\Diamond p \to \Diamond\Box p$.) Part of their charge for an adequate intuitionistic modal logic of vagueness is that it validates the equivalence of $\nabla p \leftrightarrow \nabla\nabla p$ (the $\nabla\nabla$ principle) and **M**. Therefore, by endorsing **M**, they deny that there is a "real hierarchy" of higher-order vagueness, a move that is in keeping with Wright's position [3, p. 244][9]. Since Bobzien and Wright prove that their axioms and rules are strong enough to derive this equivalence, we can be certain that IVL is sufficiently strong, as well, as it is a strengthening. We will show that our methods can accommodate the intuitionist who assents to **M**.

We define the logic $\mathsf{IVLM} = \mathsf{IVL} \oplus \{\mathbf{M}\}$. For the sake cleaner notation, $\vdash_{\mathsf{M}}$ will denote provability in IVLM. Semantics for IVLM are easy to come by, since we can simply keep the forcing clauses from our semantics for IVL and restrict to a smaller class of frames. First we define a new kind of point that can occur in S4 Fischer Servi frames.

Definition 4.4 *Let $(W, \preccurlyeq, R)$ be an* S4 *Fischer Servi frame. We say that $x \in W$ is quasi-isolated if for all y with $x\ R\ y$ there exists an $x'' \succcurlyeq x$ such that for all $x''' \succcurlyeq x''$ and for all z with $x'''\ R\ z$, we have $y \preccurlyeq z$.*

As the name suggests, quasi-isolated points are meant to be a birelational stand-in for isolated points in a topological space or maximal points in a unirelational structure. In the classical setting, such points are the ones that validate $\Diamond\varphi \to \Box\varphi$. This is classically equivalent to $\neg\nabla\varphi$, so the hope is that in our setting, quasi-isolated points similarly can resist making borderline statements true. The following proposition illustrates this.

Proposition 4.5 *Let $\mathcal{M} = (W, \preccurlyeq, R, v)$ be an* S4 *Fischer-Servi model. Then, if x is quasi-isolated, $x \nVdash \nabla\varphi$ for any formula φ.*

Proof. We may assume that $x \Vdash \Diamond\varphi$. If this is the case, there is y with $x\ R\ y$ such that $y \Vdash \varphi$. Because x is quasi-isolated, we can find a point $x'' \succcurlyeq x$ such that every modal successor of an intuitionistic successor of x'' is additionally an intuitionistic successor of y. By persistence, we must have that $x'' \Vdash \Box\varphi$. This means that $x \nVdash \neg\Box\varphi$, so $x \nVdash \nabla\varphi$. □

Another useful fact is that over the class $\mathcal{V}$, there is a simpler equivalent condition to quasi-isolatedness.

Proposition 4.6 *Let $(W, \preccurlyeq, R)$ be an IV frame. Then, a point x is quasi-isolated if and only if for every y with $x\ R\ y$, there is an $x' \succcurlyeq x$ such that for all z with $x'\ R\ z$, we have $y \preccurlyeq z$.*

Proof. The left-to-right direction is trivial, as the new condition is evidently weaker. In the other direction, suppose that x satisfies the weaker condition and that $x\ R\ y$. We get an $x' \succcurlyeq x$ such that all of its modal successors are intuitionistic successors of y. Because we are working in an IV frame, x' has a diamond-reflection point x''. Now, for any x''' and z with $x'' \preccurlyeq x'''\ R\ z$, we have $x'\ R\ z$. But by assumption, this means that $y \preccurlyeq z$, so x is quasi-isolated. □

We are now in a position to write down our new frame condition:

(FC4-weak) For every $w \in W$ there exist w' and x with $w \preccurlyeq w'\ R\ x$ such that x is quasi-isolated.

We leave open whether this frame condition actually corresponds to **M**. Analogous to the situation with (FC3-weak) and (FC3), we can actually strengthen this condition without losing completeness.

(FC4) For every $w \in W$ there exists x with $w\ R\ x$ such that x is quasi-isolated.

This condition is illustrated in Figure 5. An IV frame satisfying (FC4) is called a *weakly quasi-scattered intuitionistic vagueness* (WQSIV) frame. We call the class of such frames $\mathcal{W}$.

Theorem 4.7 IVLM *is sound with respect to $\mathcal{W}$.*

Proof. This follows from Theorem 2.6 and Propositions 4.5 and 4.6. □

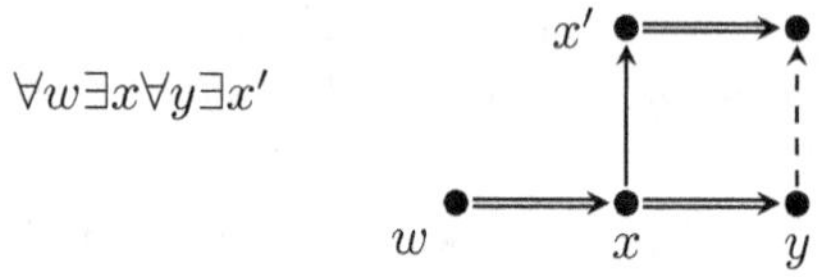

Fig. 5. The diagram for condition (FC4)

As with IVL, we use soundness to prove that IVLM is conservative over IPC.

Corollary 4.8 IVLM *is a conservative extension of* IPC.

Proof. Let $\varphi \in \mathcal{L}_0$ by a non-theorem of IPC. Then there is a some model $\mathcal{M} = (W, \preccurlyeq, v)$ with $\mathcal{M}, w \nVdash \varphi$ for some $w \in W$. For some $c \notin W$, we define $\mathcal{M}^\dagger = (W \cup \{c\}, \preccurlyeq \cup \{c, c\}, W^2 \cup (W \times \{c\}), v)$. One can check that $\mathcal{M}^\dagger$ is a WQSIV model, and since the intuitionistic relationship was left alone, we have $\mathcal{M}^\dagger, w \nVdash \varphi$. By soundness, φ is not a theorem of IVLM. □

4.3 Ersatz Topology

Before we can set out to prove a completeness theorem for IVLM and $\mathcal{W}$, we will need to prove a handful of topologically-inspired facts. As we will see, the system IVL is strong enough to force our modalities $\Box$, $\Diamond$, and ∇ to behave sufficiently similarly to the topological operators interior, closure, and boundary, respectively. In this subsection, as well as in §4.3, we will use completeness of IVL for $\mathcal{V}$ to avoid particularly arduous derivations. Of course, none of these applications are strictly speaking necessary. In all of the following results, $\mathsf{L} = \mathsf{IVL} \oplus \Gamma$, where Γ is some set of formulas.

Definition 4.9 *A formula* φ *is* L-nowhere dense *if* $\vdash_{\mathsf{L}} \neg\Box\Diamond\varphi$.

Lemma 4.10 *If* φ *is* L*-nowhere dense, then* $\vdash_{\mathsf{L}} \varphi \to \nabla\varphi$.

Proof. By $\mathbf{T}\Diamond$, $\vdash_{\mathsf{L}} \varphi \to \Diamond\varphi$. Then by Reg, we see $\vdash_{\mathsf{L}} \Box\varphi \to \Box\Diamond\varphi$. Contraposition yields $\vdash_{\mathsf{L}} \neg\Box\Diamond\varphi \to \neg\Box\varphi$, so by MP, $\vdash \neg\Box\varphi$. We can conclude $\vdash_{\mathsf{L}} \varphi \to \nabla\varphi$. □

Lemma 4.11 L*-nowhere denseness is preserved by disjunction.*

Proof. Let φ and ψ be L-nowhere dense formulas. By necessitation, we have both $\vdash_{\mathsf{L}} \Box\neg\Box\Diamond\varphi$ and $\vdash_{\mathsf{L}} \Box\neg\Box\Diamond\psi$. We will just need to prove $\vdash ((\Box\neg\Box\Diamond\varphi) \wedge (\Box\neg\Box\Diamond\psi)) \to \neg\Box\Diamond(\varphi \vee \psi)$. Note the use of $\vdash$. Regardless of which logic L is, we will prove this implication in the weaker system IVL so that we can avail ourselves of the completeness theorem that we already proved. We proceed by contradiction. Then $\Box\neg\Box\Diamond\varphi$, $\Box\neg\Box\Diamond\psi$, and $\Box\Diamond(\varphi\vee\psi)$ are mutually consistent, so we can find a model $\mathcal{M}$ with a point w forcing all three. Since w forces $\neg\Box\Diamond\varphi$, we can find $w' \succcurlyeq w$ and x with $w'\ R\ x$ such that $x \nVdash \Diamond\varphi$. Now, let x' be a diamond-reflection point of x. We must have $x' \Vdash \neg\Diamond\varphi$. As $x' \Vdash \neg\Box\Diamond\psi$, it has an intuitionistic successor x'' which in turn has a modal successor y, with $y \nVdash \Diamond\psi$. Again, we take a diamond-reflection point y' of y. Then $y' \Vdash \neg\Diamond\varphi \wedge \neg\Diamond\psi$, but this is equivalent to $\neg\Diamond(\varphi \vee \psi)$, contradicting that $w \Vdash \Box\Diamond(\varphi \vee \psi)$. Figure

6 provides a helpful visual reference for this argument. □

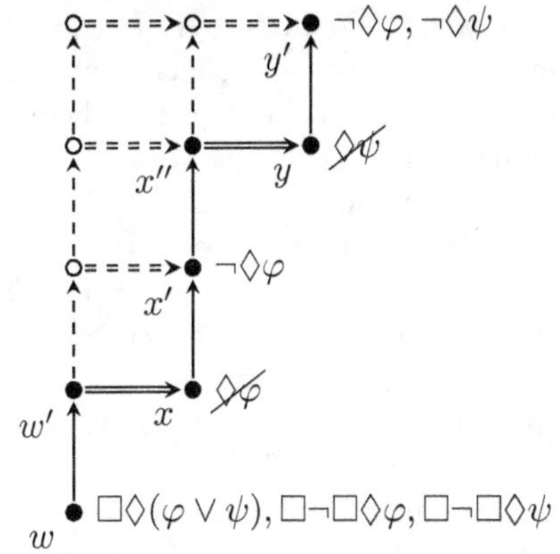

Fig. 6. For the proof of Lemma 4.11, a partial view of $\mathcal{M}$, where x' and y' are diamond-reflection points for x and y, respectively.

There is also a natural analog of topological closedness for a system L, which will furnish elegant proofs of facts about ∇.

Definition 4.12 *A formula φ is L-closed if $\vdash_{\mathsf{L}} \Diamond\varphi \to \varphi$.*

We can promptly make a couple of observations about L-closed formualas. First, by $\mathbf{T}\Diamond$, an L-closed formula φ is always provably equivalent to $\Diamond\varphi$. Second, any formula of the form $\Diamond\varphi$ is L-closed by $\mathbf{4}\Diamond$. Finally, by Lemma 4.1, all formulas of the form $\nabla\varphi$ are L-closed.

We can also prove a proposition about such formulas mirroring the topological fact that the boundary of a closed set A is always a subset of A.

Proposition 4.13 *If φ is L-closed, then $\vdash_{\mathsf{L}} \nabla\varphi \to \varphi$.*

Proof. Clearly we have $\vdash \nabla\varphi \to \Diamond\varphi$, so $\vdash \nabla\varphi \to \varphi$, as φ is L-closed. □

Corollary 4.14 $\vdash_{\mathsf{L}} \nabla\nabla p \to \nabla p$.

Proof. This follows immediately from Lemma 4.1 and Proposition 4.13. □

As a corollary, we obtain Bobzien and Rumfitt's $\nabla\nabla$ principle.

Corollary 4.15 $\vdash_{\mathsf{M}} \nabla p \leftrightarrow \nabla\nabla p$.

Proof. Since ∇p is IVLM-closed, $\vdash_{\mathsf{M}} \neg\Box\nabla p \to \neg\Box\Diamond\nabla p$. Therefore, $\vdash_{\mathsf{M}} \neg\Box\Diamond\nabla p$, so ∇p is IVLM-nowhere dense. The result then follows from Lemma 4.10 and Corollary 4.14. □

4.4 Another Completeness Theorem

We have now positioned ourselves to prove a completeness theorem for IVLM.

Theorem 4.16 IVLM *is strongly complete with respect to $\mathcal{W}$.*

Proof. We define the canonical model $\mathcal{M}^{\mathsf{CM}} = (W^{\mathsf{CM}}, \preccurlyeq^{\mathsf{CM}}, R^{\mathsf{CM}}, v^{\mathsf{CM}})$ for IVLM almost exactly as we defined $\mathcal{M}^{\mathsf{C}}$, with the only difference being that our prime theories are now defined in terms of $\vdash_{\mathsf{M}}$ instead of $\vdash$. All of the previous results still hold, but we need to check that the canonical model now also satisfies condition (FC4). Let Γ be a prime theory. First we claim that we need only find a prime theory Δ such that $\Gamma^{\Box} \subseteq \Delta \subseteq \Gamma^{\Diamond}$ and Δ has the property $\Diamond\varphi \in \Delta \Longrightarrow \neg\Box\varphi \notin \Delta$ for all $\varphi \in \mathcal{L}$.

Suppose that such a Δ exists, and suppose further that $\Delta\ R^{\mathsf{CM}}\ \Theta$. We need to check that there is some extension Δ' of Δ such that $\Delta'^{\Box} \supseteq \Theta$. This is exactly the condition guaranteeing that any modal successor of Δ' is an intuitionistic successor of Θ. By Lemma 3.4, this is equivalent to the consistency of $\Delta \cup B(\Theta)$. Toward a contradiction, suppose that $\Delta \cup B(\Theta)$ is inconsistent. Then $\Delta \vdash_{\mathsf{M}} \neg(\wedge_{i=1}^{k}\Box\varphi_i)$ for some set of formulas $\{\varphi_i\}_{i=1}^{k} \subseteq \Theta$. Using $\mathbf{K}\Box\mathbf{a}$ and noting that Θ is closed under conjunction, we can rewrite this as $\Delta \vdash \neg\Box\varphi$ for some $\varphi \in \Theta$. Since $\Delta\ R^{\mathsf{CM}}\ \Theta$, however, $\Diamond\varphi \in \Delta$. Therefore, by the assumption on Δ, $\neg\Box\varphi \notin \Delta$, which is our desired contradiction.

Now we obtain the desired set Δ. The condition that we want Δ to satisfy is equivalent to Δ not containing any formulas of the form $\nabla\varphi$, so we consider the pair $(\Gamma^{\Box}, (\mathcal{L} \setminus \Gamma^{\Diamond}) \cup N(\mathcal{L}))$. If we can prove the consistency of this pair, we are again done by Lemma 3.4. Suppose that the pair is not consistent. Then there are formulas $\varphi \in \Gamma^{\Box}$ and $\psi \notin \Gamma^{\Diamond}$ and a set of formulas $\{\theta_i\}_{i=1}^{k} \subseteq N(\Theta)$ such that $\vdash \varphi \to \psi \vee (\vee_{i=1}^{k}\theta_i)$. Intuitionistic reasoning grants us $\vdash (\varphi \wedge \neg\psi) \to \vee_{i=1}^{k}\theta_i$. Applying `Reg` twice, $\vdash \Diamond\Box(\varphi \wedge \neg\psi) \to \Diamond\Box(\vee_{i=1}^{k}\nabla\theta_i)$, which is equivalent to $\vdash \Diamond(\Box\varphi \wedge \Box\neg\psi) \to \Diamond\Box(\vee_{i=1}^{k}\nabla\theta_i)$. We now check that $\Gamma \vdash \Diamond\Box\neg\psi$. If not, then by Lemma 2.9 and the disjunctivity of Γ, we would have $\Gamma \vdash \Diamond\neg\Box\neg\psi$, which is equivalent by Lemma 1.2 to $\Gamma \vdash \Diamond\neg\neg\Diamond\psi$. By $\mathbf{S}\Diamond$, this is equivalent to $\Gamma \vdash \Diamond\Diamond\psi$, which, using $\mathbf{4}\Diamond$, implies $\Gamma \vdash \Diamond\psi$, which is a contradiction. Therefore, $\Gamma \vdash \Diamond\Box\neg\psi$. By $\mathbf{4}\Box$, $\Gamma \vdash \Box\Box\varphi$, so by Lemma 1.1, $\Gamma \vdash \Diamond(\Box\varphi \wedge \Box\neg\psi)$. Therefore, $\Gamma \vdash \Diamond\Box(\vee_{i=1}^{k}\nabla\theta_i)$. Each $\nabla\theta_i$ is IVLM-nowhere dense, so by Proposition 4.11 the disjunction is as well. By Proposition 4.10, $\vdash_{\mathsf{M}} (\vee_{i=1}^{k}\nabla\theta_i) \to \nabla(\vee_{i=1}^{k}\nabla\theta_i)$. Again, applying `Reg` twice, we have $\vdash_{\mathsf{M}} \Diamond\Box(\vee_{i=1}^{k}\nabla\theta_i) \to \Diamond\Box\nabla(\vee_{i=1}^{k}\nabla\theta_i)$. Hence, $\Gamma \vdash_{\mathsf{M}} \Diamond\Box\nabla\theta$ for some θ. But $\mathbf{M}$ affords us $\vdash \Box\nabla\theta \leftrightarrow \bot$, so we have $\Gamma \vdash \Diamond\bot$, which contradicts $\mathbf{K}\Diamond\mathbf{b}$. Γ is then inconsistent, which is a contradiction. □

Thus, we have shown that the semantics developed in Section 2 can be easily adapted for use by proponents of both columnar vagueness and an intuitionistic logic of vagueness without losing the mathematical power of completeness. In particular, this gives a strong formal grounding for the modal extension of Wright's view.

5 Conclusion

In this paper, we have furthered the work of Bobzien and Rumfitt to formalize and modalize Wright's intuitionsitic position on the Sorites paradox. Like they did, we propose a deductive system, but we also establish a semantics for which our system is sound and complete, thereby allowing us to establish results of philosophical import, like verifying that the underlying sentential calculus really is intuitionistic.

We have left open a few mathematical questions. Establishing the finite model property for IVL or IVLM and settling whether there are first-order correspondents for the axiom $\mathbf{M}$ over the class $\mathcal{S}$ both seem to be natural next steps. Additionally, while we do not think $\mathbf{S}\nabla$ is philosophically compelling in the absence of $\mathbf{S}\Diamond$, establishing whether there is a first-order definable class of frames for which we have a completeness theorem could be of technical interest.

In [2], Bobzien works with semantics for a predicate extension of S4.1 and manages to give an account of the frames in terms of viewpoints. Although our semantics have proven useful, we have not given any sort of intuitive gloss on them, so a natural extension of Bobzien's recent work would be to argue for the philosophical meaningfulness of the semantics. Additionally, performing a parallel analysis of a predicate extension of IVL or IVLM would bolster the intuitionist's formal foothold.

References

[1] Amati, G. and A. Pirri, *A uniform tableau method for intuitionistic logics i**, in: *Studia Logica 53*, 1994.

[2] Bobzien, S., *A generic solution to the sorites paradox based on an extension of the normal modal logic s4m (s4.1)*, in: *On the Sorites Paradox: New Essays* (2021).

[3] Bobzien, S. and I. Rumfitt, *Intuitionism and the modal logic of vagueness*, in: *J Philos Logic 49*, 2020, pp. 221–248.

[4] Fischer Servi, G., *On modal logic with an intuitionistic base*, in: *Studia Logica 36*, 1977, pp. 141–149.

[5] Fischer Servi, G., *Semantics for a class of intuitionistic modal calculi*, in: *Italian Studies in the Philosophy of Science*, 1980, pp. 59–72.

[6] Kuznetsov, S., *Propositional intuitionistic logic (lecture notes)* (2017).

[7] Mormann, T., *Topological models of columnar vagueness*, in: *Erkenntnis 87*, 2022, pp. 693–716.

[8] Wright, C., *On being in a quandary: Relativism vagueness logical revisionism*, in: *Mind 437*, 2001, pp. 45–98.

[9] Wright, C., *The illusion of higher-order vagueness*, in: R. Dietz and S. Moruzzi, editors, *Cuts and Clouds* (2010).

Taming Bounded Depth with Nested Sequents

Agata Ciabattoni

Vienna University of Technology

Lutz Straßburger

Inria Saclay & LIX, Ecole Polytechnique

Matteo Tesi

Scuola Normale Superiore di Pisa

Abstract

Bounded depth refers to a property of Kripke frames that serve as semantics for intuitionistic logic. We introduce nested sequent calculi for the intermediate logics of bounded depth. Our calculi are obtained in a modular way by adding suitable structural rules to a variant of Fitting's calculus for intuitionistic propositional logic, for which we present the first syntactic cut elimination proof. This proof modularly extends to the new nested sequent calculi introduced in this paper.

Keywords: Nested Sequents, Cut Elimination, Intermediate Logics, Bounded Depth Kripke Models.

1 Introduction

Nested sequents are a natural extension of ordinary Gentzen sequents, (re)discovered several times in different contexts [15,27,2]. Whereas Gentzen sequents are lists (or multisets) of formulas, nested sequents are trees of multisets of formulas. The tree structure makes nested sequents well-suited to handle logics having Kripke-style semantics. Indeed, nested sequents have been employed to provide internal analytic calculi for modal logics for which this was not possible before, for example, for the modal logic KB in [2]. Here, *internal* means that there are no "external" semantic constructs (like labels) in the syntax, and *analytic* means that all derivations in these calculi have the property that they only contain formulas which are subformulas of the conclusion (subformula property). This feature of analytic calculi renders them a natural starting point for proving meta-logical properties as decidability, complexity, and interpolation, and for developing automated reasoning methods.

Nested sequents have been effective for classical modal logics in the S5-cube [2], as well as for intuitionistic modal logics in the IS5-cube [28,24,17], and more generally, for all extensions of IK with Horn-Scott-Lemmon axioms [22]. Furthermore, they have been used to provide focused proof systems for classical and intuitionistic modal logics [3,4], to construct interpolation proofs [21], to implement proof search [11,26], and to "tame" modal logics with path-axioms [12], where "taming" means to have proof systems suited for proof search. In all these calculi, the tree structure of the nested sequents corresponds to the accessibilty relation between the worlds in the Kripke frame.

An analytic nested calculus has also been proposed for intuitionistic propositional logic (IPL) by Fitting [8]; there the tree structure of the nested sequents corresponds to the future relation between the worlds in the Kripke frame. His calculus, that has been obtained as a notational variant of prefixed tableaux, is not equipped with a direct cut elimination proof, which seemed hard to define.

This leads to the first contribution of this paper. We import insights from the nested sequent calculi for modal logics to introduce NIPL — a variant of Fitting's calculus for IPL that allows for a direct cut elimination proof. In the spirit of Belnap's conditions for cut elimination in display calculi [1], our proof relies on the abstract conditions (N1)–(N5), which will be presented in Section 3. The rules of NIPL are all invertible, and therefore NIPL offers a purely formula-driven approach to proof search.

The next question we address is: Can we extend NIPL by additional structural rules to capture intermediate logics (i.e., logics between intuitionistic and classical logic), in a similar way as the nested systems for the modal logics K and IK are extended with additional rules corresponding to additional axioms [24,13]? Moreover, can this be done in such a way that the cut elimination property is preserved, and can this be proved in a modular way, i.e., can we reuse the existing proof and only add the cases concerning the new rules? [1]

For hypersequents—which are disjunctions of ordinary sequents—we have such results, provided the additional axioms for intermediate logics follow a certain shape [5]. This is not the case for the family of intermediate logics characterized by Kripke models of bounded depth, usually denoted by $\mathsf{BD_n}$. The logic $\mathsf{BD_1}$ is just classical logic and the logic $\mathsf{BD_2}$ is one of the seven interpolable intermediate logics [23]. Analytic hypersequent calculi for $\mathsf{BD_n}$ are indeed provably not obtainable [18] by using the methodology in [5] which extends the base hypersequent calculus for IPL with suitable structural rules corresponding to the additional axioms.

Modular analytic calculi for $\mathsf{BD_n}$ have been defined using frameworks more powerful than hypersequents (and nested sequents); however, the objects that the resulting calculi manipulate (labelled or display sequents [25,6]) cannot be translated into formulas of the logic, and hence there is no real subformula property even when the calculus is analytic; moreover the rules of these calculi are not invertible.

This brings us to our second contribution: modular nested sequent calculi

[1] Girard has argued in [10] that the lack in modularity is one of the main technical limitations in structural proof theory.

for the logics $\mathsf{BD_n}$ (for all $n \geq 1$). The calculi are obtained by adding suitable structural rules to NIPL. We show that the cut elimination proof for NIPL scales to the new systems, as our conditions (N1)–(N5) are preserved. It is interesting to note that the additional rules for $\mathsf{BD_n}$, with $n \geq 2$, have more than one premise. To our knowledge, this is a novelty. So far, there are no nested sequent systems with multi-premise structural rules.

2 Preliminaries

The *formulas* of intuitionistic propositional logic (IPL), denoted by $A, B, C, \ldots$, are generated from a countable set of atoms $\{p, q, \ldots\}$ via the grammar

$$A ::= p \mid \bot \mid A \vee A \mid A \wedge A \mid A \supset A$$

We define the *degree* of a formula to be the number of connectives in it. A *nested sequent* is a finite tree of multisets of formulas. In ordinary sequents for intuitionistic logic we distinguish between the left and the right hand side of the turnstile. To make this distinction in nested sequents, we use polarities on formulas. There are two polarities, input (intuitively as if on the left of the turnstile in the conventional sequent calculus), denoted by a $\bullet$ superscript and output (intuitively as if on the right of the turnstile), denoted by a $\circ$ superscript. Now, a nested sequent can be written as:

$$\Gamma = A_1^\bullet, ..., A_m^\bullet, B_1^\circ, ..., B_n^\circ, [\Gamma_1], ..., [\Gamma_k] \tag{1}$$

where $A_1^\bullet, ..., A_m^\bullet, B_1^\circ, ..., B_n^\circ$ is the multiset of formulas at the root of the sequent tree of Γ, and where $\Gamma_1, \ldots, \Gamma_k$ are its immediate subtrees. We use $\varnothing$ to denote the *empty sequent*, i.e., where $m = n = k = 0$ in (1) above. We use capital Greek letters Γ, Δ, Σ, ..., to denote nested sequents, and we assume that the associativity and commutativity of the comma is implicit in our systems, and that $\varnothing$ acts as its unit. We write $\Gamma^\bullet$ for $A_1^\bullet, ..., A_m^\bullet$ and Γ° for $B_1^\circ, ..., B_n^\circ, [\Gamma_1], ..., [\Gamma_k]$ if Γ is as in (1) above. In other words, for every nested sequent Γ we have that $\Gamma = \Gamma^\bullet, \Gamma^\circ$. More generally, we will write $\Gamma^\bullet$, $\Delta^\bullet$, $\Sigma^\bullet$, ..., for multisets of input formulas (i.e., all formulas have $\bullet$-polarity, and there are no nestings), and we will write Γ°, Δ°, Σ°, ..., for sequents that have only $\circ$-formulas at their root nodes (i.e., there are no $\bullet$-formulas at the root, but there can be nestings with $\bullet$-formulas inside).

The *corresponding formula* of the sequent in (1) above is defined as

$$fm(\Gamma) = \bigwedge_{i=1}^{m} A_i \supset \left(\bigvee_{j=1}^{n} B_j \vee \bigvee_{l=1}^{k} fm(\Gamma_l) \right) \tag{2}$$

A *(sequent) context* is a nested sequent with a hole $\{\ \}$, taking the place of a formula. Contexts are denoted by $\Gamma\{\ \}$, and $\Gamma\{\Delta\}$ is the sequent obtained from $\Gamma\{\ \}$ by replacing the occurrence of $\{\ \}$ with Δ. We write $\Gamma\{\varnothing\}$ for the sequent obtained from $\Gamma\{\ \}$ by removing the $\{\ \}$ (i.e., the hole is filled with nothing). The depth of a context $\Gamma\{\ \}$, denoted by $depth(\Gamma\{\ \})$, is the length

Initial Sequents

$$\frac{}{\Gamma\{p^\bullet, \Delta\{p^\circ\}\}}\;\mathsf{ax} \qquad\qquad \frac{}{\Gamma\{\bot^\bullet\}}\;\bot^\bullet$$

Logical Rules

$$\frac{\Gamma\{A^\bullet, B^\bullet\}}{\Gamma\{A \wedge B^\bullet\}}\;\wedge^\bullet \qquad\qquad \frac{\Gamma\{A^\circ\} \qquad \Gamma\{B^\circ\}}{\Gamma\{A \wedge B^\circ\}}\;\wedge^\circ$$

$$\frac{\Gamma\{A^\bullet\} \qquad \Gamma\{B^\bullet\}}{\Gamma\{A \vee B^\bullet\}}\;\vee^\bullet \qquad\qquad \frac{\Gamma\{A^\circ, B^\circ\}}{\Gamma\{A \vee B^\circ\}}\;\vee^\circ$$

$$\frac{\Gamma\{A \supset B^\bullet, \Delta\{\Sigma, A^\circ\}\} \qquad \Gamma\{A \supset B^\bullet, \Delta\{\Sigma, B^\bullet\}\}}{\Gamma\{A \supset B^\bullet, \Delta\{\Sigma\}\}}\;\supset^\bullet \qquad\qquad \frac{\Gamma\{[A^\bullet, B^\circ]\}}{\Gamma\{A \supset B^\circ\}}\;\supset^\circ$$

Figure 1. The calculus NIPL

$$\frac{\Gamma\{A^\circ\} \qquad \Gamma\{A^\bullet\}}{\Gamma\{\varnothing\}}\;\mathsf{cut}$$

Figure 2. The cut rule

$$\frac{\Gamma\{\varnothing\}}{\Gamma\{\Delta\}}\;\mathsf{w} \qquad \frac{\Gamma\{\Delta, \Delta\}}{\Gamma\{\Delta\}}\;\mathsf{c} \qquad \frac{\Gamma\{[\Delta]\}}{\Gamma\{\Delta\}}\;\mathsf{t} \qquad \frac{\Gamma\{\Sigma^\circ\}}{\Gamma\{[\Sigma^\circ]\}}\;\mathsf{4} \qquad \frac{\Gamma\{[\Sigma^\bullet, \Delta]\}}{\Gamma\{\Sigma^\bullet, [\Delta]\}}\;\mathsf{lift} \qquad \frac{\Gamma\{\Sigma^\circ, [\Delta]\}}{\Gamma\{[\Sigma^\circ, \Delta]\}}\;\mathsf{lower}$$

Figure 3. Admissible structural rules

of the path in the sequent tree from the root to the hole $\{\ \}$. It is defined inductively as follows: $depth(\{\ \}) = 0$ and $depth(\Gamma', \Gamma\{\ \}) = depth(\Gamma\{\ \})$ and $depth([\Gamma\{\ \}]) = depth(\Gamma\{\ \}) + 1$.

We will also use the notation $\Gamma\{\!\!\{\Delta\}\!\!\}$ as abbreviation for $\Gamma\{[\Delta]\}$.

Example 2.1 Let $\Gamma\{\ \} = A^\bullet, B^\circ, [\{\ \}, [D^\bullet, C^\circ]]$. We have that

$$\Gamma\{B^\circ\} = A^\bullet, B^\circ, [B^\circ, [D^\bullet, C^\circ]]$$

and

$$\Gamma\{\varnothing\} = A^\bullet, B^\circ, [[D^\bullet, C^\circ]].$$

Let $\Delta = F^\bullet, [G^\circ]$, then

$$\Gamma\{\Delta\} = A^\bullet, B^\circ, [F^\bullet, [G^\circ], [D^\bullet, C^\circ]]$$

and

$$\Gamma\{\!\!\{\Delta\}\!\!\} = A^\bullet, B^\circ, [[F^\bullet, [G^\circ]], [D^\bullet, C^\circ]].$$

3 Nested Sequent Calculus for IPL

An elegant nested sequent calculus for Intuitionistic Propositional Logic IPL was introduced by Fitting [8], as a notational variant of prefixed tableaux. The lack of a direct cut elimination proof in his calculus has prevented its extension to cover intermediate logics. Indeed, to the best of our knowledge there are no analytic nested calculi for any intermediate logic (other than classical and intuitionistic logic). Even methods to extract such calculi from more powerful frameworks (like the structural refinement method in [22,20] that used labelled calculi as starting point), do not seem to work for intermediate logics.[2] In order to define analytic nested calculi for intermediate logics, the nested sequent formalism has been extended in various ways, giving rise to, e.g., linear nested calculi [19], and injective nested calculi [16].

In general, proving syntactic cut elimination for nested calculi is harder than for other proof theoretic formalisms, e.g., (hyper)sequent or display calculus. Often this result is obtained by translating the nested calculus at hand to other formalisms, as e.g. in [13,8]. The few existing cut elimination proofs for nested calculi are indeed tailored to specific systems [27,2,24,28], and their proofs do not seem to be generalizable (in particular, to deal with multi-premise structural rules).

In this section we present NIPL, a variant of Fitting's calculus for IPL designed to have all invertible rules, and to admit a direct cut elimination proof. The system NIPL, whose rules are shown in Figure 1, is obtained from Fitting's calculus by using multisets instead of sets and by absorbing the rule lift into the initial sequents and the rule $\supset^\bullet$. Observe indeed that ax and $\supset^\bullet$ can be simulated in Fitting's calculus by repeated applications of lift. As an immediate consequence we obtain the soundness of NIPL with respect to IPL.

Terminology: As in standard sequent calculi, we call *context* the part left unchanged from premises to conclusions, we call *principal* the introduced formula in a logical rule, and the rest *active* part/formulas (active formulas in the initial sequents are $p^\bullet$, p°, and $\bot^\bullet$).

As we will show in the next section, NIPL satisfies the following properties that guarantee a relatively simple proof of the elimination of the cut rule depicted in Figure 2.

(N1) *All rules are height-preserving invertible.*

(N2) *Dedicated structural rules are height-preserving admissible.* These rules, displayed in Figure 3, are the usual weakening (w) and contraction (c), the lift-rule from [8], variations of the rules for the modal axioms t and 4, from [24], and the new lower-rule which can be seen as the inverse of lift.

(N3) *A cut over formulas that are not principal can be shifted upwards over its premises.* This condition is implied by Belnap's sufficient conditions (C2)–(C7) for cut elimination in display calculi [1].

(N4) *All logical rules are reductive.* This means that they allow the replacement of a cut whose cut formula is principal in the left and right premise of the cut rule by cuts on smaller formulas (possibly using the dedicated structural rules from (N2)). This property is the nested sequent formulation of Belnap's (C8) condition [1].

[2] See also [14] for the correspondence between labeled systems and nesed sequents.

(N5) *Cuts having an initial sequent as one of their premises can be removed.*

Let us mention two useful features of NIPL. The first is standard in well-designed sequent-style calculi: the general form of the ax-rule is derivable.

Lemma 3.1 (Axiom expansion) *The sequent* $\Gamma\{A^\bullet, \Pi\{A^\circ, \Delta\}\}$ *is derivable in* NIPL *for every context* Γ, Π, Δ *and every formula* A.

Proof By induction on the degree of the formula A. We detail the case in which A is of the shape $B \supset C$, the other cases being similar.

$$\dfrac{\dfrac{\Gamma\{B \supset C^\bullet, \Pi\{[B^\bullet, B^\circ, C^\circ], \Delta\}\} \qquad \Gamma\{B \supset C^\bullet, \Pi\{[B^\bullet, C^\bullet, C^\circ], \Delta\}\}}{\Gamma\{B \supset C^\bullet, \Pi\{[B^\bullet, C^\circ], \Delta\}\}}\supset^\bullet}{\Gamma\{B \supset C^\bullet, \Pi\{B \supset C^\circ, \Delta\}\}}\supset^\circ$$

The premises are derivable by induction hypothesis. □

The second feature concerns the admissibility of the necessitation rule

$$\dfrac{\Gamma}{[\Gamma]}\,\mathsf{nec}$$

which will be used in the heuristic for generating the structural rules for $\mathsf{BD_n}$ in Section 5. Note that unlike all other rules, nec is shallow, as it cannot be applied inside a context.

Proposition 3.2 *If a sequent* Γ *is derivable, then so is* $[\Gamma]$.

Proof The proof is by induction on the height n of the derivation of Γ. If Γ is an initial sequent, so is $[\Gamma]$. If $n > 0$, then apply the induction hypothesis to the premise(s) of the rule and then rule again. □

4 Cut elimination for NIPL

We are going to show that NIPL satisfies conditions (N1)–(N5) and how these conditions entail the cut elimination theorem.

The preservation of the height of a derivation is crucial for all our arguments. Formally, the *height* of a derivation is the length of the longest path in the tree from its root to one of its leaves. A inference rule with premises $\Gamma_1, \ldots, \Gamma_n$ and conclusion Γ is *height-preserving invertible*, if for every derivation of Γ, there are derivations of each of $\Gamma_1, \ldots, \Gamma_n$ with at most the same height. The rule is *height-preserving admissible* if, whenever the premises are derivable, the conclusion has a derivation whose height is not bigger than any derivation of a premise.

Lemma 4.1 *The weakening rule* w *is height-preserving admissible in* NIPL.

Proof By induction on the height n of the derivation of $\Gamma\{\varnothing\}$. If $n = 0$, then $\Gamma\{\varnothing\}$ is an initial sequent and so is $\Gamma\{\Delta\}$. If $n > 0$, we apply the induction hypothesis to the premise(s) of the last rule applied and then the rule again. □

Lemma 4.2 *Every rule in* NIPL *is height-preserving invertible.*

Proof By induction on the height n of the derivation of the conclusion of each rule. The proofs for conjunction and disjunction are standard. The rule $\supset^\bullet$ is height-preserving invertible by the height-preserving admissibility of the rule of weakening. We discuss the rule $\supset^\circ$. If $\Gamma\{A \supset B^\circ\}$ is an initial sequent, then $\Gamma\{[A^\bullet, B^\circ]\}$ is an initial sequent too. If $n > 0$, then we apply the induction hypothesis to each of the premise(s) and then we apply the rule again. □

Lemma 4.3 *The contraction rule* c *is height-preserving admissible in* NIPL.

Proof By induction on the height n of the derivation. If $\Gamma\{\Delta, \Delta\}$ is an initial sequent, the conclusion easily follows. If $n > 0$ and the principal formula is not in Δ, we apply the induction hypothesis to each of the premises and then the rule again. If $n > 0$ and the principal formula is in Δ we exploit the height-preserving invertibility of the logical rules as shown below, where ρ stands for an arbitrary rule instance and $\bar{\rho}$ stands for its inversion (which does not count for the overall height, as ρ is heigt-preserving invertible):

$$\dfrac{\dfrac{\Gamma\{\Delta', \Delta\}}{\Gamma\{\Delta, \Delta\}}\,\rho}{\Gamma\{\Delta\}}\,\mathsf{c} \quad\rightsquigarrow\quad \dfrac{\dfrac{\dfrac{\Gamma\{\Delta', \Delta\}}{\Gamma\{\Delta', \Delta'\}}\,\bar{\rho}}{\Gamma\{\Delta'\}}\,\mathsf{c}}{\Gamma\{\Delta\}}\,\rho$$

The application of c is removed invoking the induction hypothesis. The case where ρ is a binary rule is analogous and we omit the details. □

The way we formulated the rules in NIPL allows us to establish the admissibility of the lift-rule. A variant of this rule was instead explicitly present in Fitting's system. Its absence (in combination with w and c) permits the use of the additive version of cut, which simplifies the cut elimination argument.

Lemma 4.4 *The* lift*-rule is height-preserving admissible in* NIPL.

Proof Proceed by induction on the height n of the derivation of the premise $\Gamma\{[\Sigma^\bullet, \Delta]\}$ of the rule. If $n = 0$ and no formula in $\Sigma^\bullet$ is active, then we can remove it. Otherwise, $\Gamma\{\Sigma^\bullet, [\Delta]\}$ is again an instance of ax. If $n > 0$ and no formula in Σ is principal, we apply the induction hypothesis to the premise(s) of the rule and then the rule again.
If a formula $A^\bullet$ in $\Sigma^\bullet$ is principal in $\wedge^\bullet$ or $\vee^\bullet$, we apply the induction hypothesis (possibly twice). E.g.,

$$\dfrac{\Gamma\{[\Sigma'^\bullet, A^\bullet, B^\bullet, \Delta]\}}{\Gamma\{[\Sigma'^\bullet, A \wedge B^\bullet, \Delta]\}}\,\wedge^\bullet \quad\rightsquigarrow\quad \dfrac{\dfrac{\Gamma\{[\Sigma'^\bullet, A^\bullet, B^\bullet, \Delta]\}}{\Gamma\{\Sigma'^\bullet, A^\bullet, B^\bullet, [\Delta]\}}\,\mathsf{lift}}{\Gamma\{\Sigma'^\bullet, A \wedge B^\bullet, [\Delta]\}}\,\wedge^\bullet$$

If a formula $A^\bullet$ in $\Sigma^\bullet$ is principal in $\supset^\bullet$ as in

$$\dfrac{\Gamma\{[\Sigma'^\bullet, A \supset B^\bullet, \Delta\{\Pi, A^\circ\}]\} \qquad \Gamma\{[\Sigma'^\bullet, A \supset B^\bullet, \Delta\{\Pi, B^\bullet\}]\}}{\Gamma\{[\Sigma'^\bullet, A \supset B^\bullet, \Delta\{\Pi\}]\}}\,\supset^\bullet$$

we apply the induction hypothesis and the rule $\supset^\bullet$, as in

$$\dfrac{\dfrac{\Gamma\{[\Sigma'^\bullet, A \supset B^\bullet, \Delta\{\Pi, A^\circ\}]\}}{\Gamma\{\Sigma'^\bullet, A \supset B^\bullet, [\Delta\{\Pi, A^\circ\}]\}}\,\mathsf{lift} \qquad \dfrac{\Gamma\{[\Sigma'^\bullet, A \supset B^\bullet, \Delta\{\Pi, B^\bullet\}]\}}{\Gamma\{\Sigma'^\bullet, A \supset B^\bullet, [\Delta\{\Pi, B^\bullet\}]\}}\,\mathsf{lift}}{\Gamma\{\Sigma'^\bullet, A \supset B^\bullet, [\Delta\{\Pi\}]\}}\,\supset^\bullet$$

□

Note that with the admissibility of the lift-rule we immediately obtain completeness of NIPL with respect to IPL via Fitting's system [8].

Lemma 4.5 *The* 4-*rule is height-preserving admissible in* NIPL.

Proof By induction on the height n of the derivation of the rule premise. If $\Gamma\{\Sigma^\circ\}$ is an initial sequent, then so is $\Gamma\{[\Sigma^\circ]\}$. If $n > 0$ we assume that a formula in Σ° is principal, otherwise the proof is trivial. We apply the induction hypothesis to the premise(s) of the rule and then the rule again. For example, if the last rule applied is $\supset^\circ$, we have:

$$\dfrac{\Gamma\{\Delta, [A^\bullet, B^\circ]\}}{\Gamma\{\Delta, A \supset B^\circ\}}\,\supset^\circ \quad \rightsquigarrow \quad \dfrac{\dfrac{\Gamma\{\Delta, [A^\bullet, B^\circ]\}}{\Gamma\{[\Delta, [A^\bullet, B^\circ]]\}}\,4}{\Gamma\{[\Delta, A \supset B^\circ]\}}\,\supset^\circ$$

□

Lemma 4.6 *The* lower-*rule is height-preserving admissible in* NIPL.

Proof The lower-rule is derivable with the following height-preserving steps:

$$\dfrac{\dfrac{\dfrac{\Gamma\{\Sigma^\circ, [\Delta]\}}{\Gamma\{[\Sigma^\circ], [\Delta]\}}\,4}{\Gamma\{[\Sigma^\circ, \Delta], [\Sigma^\circ, \Delta]\}}\,\mathsf{w}}{\Gamma\{[\Sigma^\circ, \Delta]\}}\,\mathsf{c}$$

□

Lemma 4.7 *The* t-*rule is height-preserving admissible in* NIPL.

Proof By induction on the height n of the premise $\Gamma\{[\Delta]\}$. If $n = 0$, then $\Gamma\{[\Delta]\}$ is an initial sequent and so is $\Gamma\{\Delta\}$. If $n > 0$, we apply the induction hypothesis to the premise(s) and then the rule again. As an example, consider the case in which the last rule applied is $\supset^\bullet$ and formulas are introduced (bottom-up) in $[\Delta]$. We have:

$$\dfrac{\Gamma\{A \supset B^\bullet, [\Delta, A^\circ]\} \qquad \Gamma\{A \supset B^\bullet, [\Delta, B^\bullet]\}}{\Gamma\{A \supset B^\bullet, [\Delta]\}}\,\supset^\bullet$$

We construct the following derivation:

$$\dfrac{\dfrac{\Gamma\{A \supset B^\bullet, [\Delta, A^\circ]\}}{\Gamma\{A \supset B^\bullet, \Delta, A^\circ\}}\,\mathsf{t} \qquad \dfrac{\Gamma\{A \supset B^\bullet, [\Delta, B^\bullet]\}}{\Gamma\{A \supset B^\bullet, \Delta, B^\bullet\}}\,\mathsf{t}}{\Gamma\{A \supset B^\bullet, \Delta\}}\,\supset^\bullet$$

where the applications of t are removed by induction hypothesis. □

This completes the proof of the properties (N1) and (N2). To eliminate cut, we also need (N3)–(N5), which will be shown below.

Theorem 4.8 (Cut elimination) *The* cut-*rule is admissible for* NIPL.

Proof We consider a uppermost cut and proceed by induction on the lexicographically ordered pair (c, n) where c is the degree of its cut formula and n is the height of the derivation of $\Gamma\{A^\bullet\}$.[3]

(N5) If $n = 0$, then $\Gamma\{A^\bullet\}$ is an initial sequent. If $A^\bullet$ is not active, $\Gamma\{\varnothing\}$ is an initial sequent too. If $A^\bullet$ is active in ax, we have $A^\bullet = p^\bullet$ for some p, and

$$\dfrac{\Gamma\{p^\circ, \Delta\{p^\circ\}\} \qquad \dfrac{}{\Gamma\{p^\bullet, \Delta\{p^\circ\}\}}\,\mathsf{ax}}{\Gamma\{\Delta\{p^\circ\}\}}\,\mathsf{cut}$$

The cut is eliminated as follows:

$$\dfrac{\dfrac{\Gamma\{p^\circ, \Delta\{p^\circ\}\}}{\Gamma\{\Delta\{p^\circ, p^\circ\}\}}\,\mathsf{lower}}{\Gamma\{\Delta\{p^\circ\}\}}\,\mathsf{c}$$

The case of the axiom $\bot^\bullet$ is handled similarly, noticing that from the derivability in NIPL of $\Gamma\{\bot^\circ\}$ follows the derivability of $\Gamma\{\varnothing\}$.

(N3) If $n > 0$ and $A^\bullet$ is not principal, we apply the invertibility of the corresponding rule to $\Gamma\{A^\circ\}$, permute the cut upwards, and remove it by secondary induction hypothesis. For example, consider the following derivation, where ρ is some binary rule from NIPL:

$$\dfrac{\Gamma\{A^\circ\} \qquad \dfrac{\Gamma'\{A^\bullet\} \qquad \Gamma''\{A^\bullet\}}{\Gamma\{A^\bullet\}}\,\rho}{\Gamma\{\varnothing\}}\,\mathsf{cut}$$

We construct the following derivation, where we again use the height-preserving invertibility of ρ:

$$\dfrac{\dfrac{\dfrac{\Gamma\{A^\circ\}}{\Gamma'\{A^\circ\}}\,\bar{\rho} \qquad \Gamma'\{A^\bullet\}}{\Gamma'\{\varnothing\}}\,\mathsf{cut} \qquad \dfrac{\dfrac{\Gamma\{A^\circ\}}{\Gamma''\{A^\circ\}}\,\bar{\rho} \qquad \Gamma''\{A^\bullet\}}{\Gamma''\{\varnothing\}}\,\mathsf{cut}}{\Gamma\{\varnothing\}}\,\rho$$

(N4) If $A^\bullet$ is principal in $\wedge$ or $\vee$, the case is handled in the usual way using the invertibility of the rules. For example

$$\dfrac{\Gamma\{B \vee C^\circ\} \qquad \dfrac{\Gamma\{B^\bullet\} \qquad \Gamma\{C^\bullet\}}{\Gamma\{B \vee C^\bullet\}}\,\vee^\bullet}{\Gamma\{\varnothing\}}\,\mathsf{cut}$$

is eliminated as follows (where $\bar{\vee}^\circ$ is the inversion of $\vee^\circ$ and each cut is on a formula of lesser degree):

$$\dfrac{\dfrac{\dfrac{\Gamma\{B \vee C^\circ\}}{\Gamma\{B^\circ, C^\circ\}}\,\bar{\vee}^\circ \qquad \dfrac{\Gamma\{B^\bullet\}}{\Gamma\{B^\bullet, C^\circ\}}\,\mathsf{w}}{\Gamma\{C^\circ\}}\,\mathsf{cut} \qquad \Gamma\{C^\bullet\}}{\Gamma\{\varnothing\}}\,\mathsf{cut}$$

---

[3] It is enough to consider only the height of the left premise as every right rule is invertible.

The case below in which $A^\bullet$ is principal in $\supset^\bullet$:

$$\dfrac{\Gamma\{B \supset C^\circ, \Pi\{\Sigma\}\} \qquad \dfrac{\Gamma\{B \supset C^\bullet, \Pi\{B^\circ, \Sigma\}\} \qquad \Gamma\{B \supset C^\bullet, \Pi\{C^\bullet, \Sigma\}\}}{\Gamma\{B \supset C^\bullet, \Pi\{\Sigma\}\}}\,\supset^\bullet}{\Gamma\{\Pi\{\Sigma\}\}}\,\mathsf{cut}$$

is handled using some of the structural rules from (N2). We first construct a derivation of $\Gamma\{\Pi\{B^\circ, \Sigma\}\}$:

$$\dfrac{\dfrac{\Gamma\{B \supset C^\circ, \Pi\{\Sigma\}\}}{\Gamma\{B \supset C^\circ, \Pi\{B^\circ, \Sigma\}\}}\,\mathsf{w} \qquad \Gamma\{B \supset C^\bullet, \Pi\{B^\circ, \Sigma\}\}}{\Gamma\{\Pi\{B^\circ, \Sigma\}\}}\,\mathsf{cut}$$

The cut is removed by secondary induction hypothesis. A symmetrical derivation yields $\Gamma\{\Pi\{C^\bullet, \Sigma\}\}$, and the reduction is completed as follows:

$$\dfrac{\dfrac{\dfrac{\Gamma\{\Pi\{B^\circ, \Sigma\}\}}{\Gamma\{\Pi\{B^\circ, C^\circ, \Sigma\}\}}\,\mathsf{w} \qquad \dfrac{\dfrac{\dfrac{\Gamma\{B \supset C^\circ, \Pi\{\Sigma\}\}}{\Gamma\{\Pi\{B \supset C^\circ, \Sigma\}\}}\,\mathsf{lower}}{\Gamma\{\Pi\{[B^\bullet, C^\circ], \Sigma\}\}}\,\supset^\circ}{\Gamma\{\Pi\{B^\bullet, C^\circ, \Sigma\}\}}\,\mathsf{t}}{\Gamma\{\Pi\{C^\circ, \Sigma\}\}}\,\mathsf{cut} \qquad \Gamma\{\Pi\{C^\bullet, \Sigma\}\}}{\Gamma\{\Pi\{\Sigma\}\}}\,\mathsf{cut}$$

□

We can now show completeness independently from Fitting's calculus:

Corollary 4.9 NIPL *is complete with respect to* IPL.

Proof It is easy to check that every axiom of IPL can be proved in NIPL and modus ponens can be simulated by cut. The claim follows by Theorem 4.8. □

5 Intermediate Logics of Bounded Depth

We introduce nested calculi for the propositional intermediate logics $\mathsf{BD_n}$ (Bounded Depth n) which are semantically characterized by intuitionistic Kripke frames in which every chain is of length less or equal to n. Our calculi are defined in a modular way by extending NIPL with suitable structural rules, which preserve conditions (N1)–(N5) and therefore cut elimination.

Definition 5.1 A *bounded depth n Kripke frame* is a pair $\langle P, \leq \rangle$, where P is a non empty set of *worlds*, denoted by $x, y, z, \ldots$, and $\leq$ is a partial order on P with

$$\forall x_0 \ldots x_n . \left(\bigwedge_{0 \leq i \leq n-1} x_i \leq x_{i+1} \supset \bigvee_{0 \leq i \leq n-1} x_{i+1} \leq x_i \right). \tag{3}$$

A *bounded depth Kripke model* is a triple $\langle P, \leq, v \rangle$, where $\langle P, \leq \rangle$ is a bounded depth Kripke frame and v a function which maps propositional atoms to subsets of P, such that for all worlds x and y, if $x \leq y$ and $x \in v(p)$ then

$y \in v(p)$ (monotonicity condition). Truth conditions for a formula in a world are defined as usual, see, e.g. [9]. A formula is *true in a model* if it is true in every world of the model. A formula is *valid* if it is true in every bounded depth Kripke model. A sequent is true in a world (resp. true in a model, resp. valid) if the corresponding formula is.

The logics $\mathsf{BD_n}$ are axiomatized by the schemata

$$\mathbf{bd_1} := p_1 \vee (p_1 \to \bot) \qquad \text{and} \qquad \mathbf{bd_{n+1}} := p_{n+1} \vee (p_{n+1} \to \mathbf{bd}_n)$$

where with an abuse of notation here and below we use $p_{(i)}, q$ as metavariables, to be substituted by arbitrary formulas. For each n, the logic $\mathsf{BD_n}$ is sound and complete with respect to the class of bounded depth n Kripke frames [9]. $\mathsf{BD_1}$ turns out to be classical logic, and $\mathsf{BD_2}$ one of the seven interpolable intermediate logics [23].

Before presenting the peculiar structural rule capturing $\mathsf{BD_n}$, we sketch its genesis, using $\mathsf{BD_2}$ as case study. The starting point for the rule's definition is (an adaptation to nested sequents of) the algorithm for transforming Hilbert axioms into structural hypersequent rules given in [5], and into display rules in [6,7]. Indeed, we start with the axiom schema $\mathbf{bd_2} := p \vee (p \supset (q \vee \neg q))$ characterizing $\mathsf{BD_2}$. By using the invertible rules of NIPL, the axiom is decomposed into the equivalent nested sequent in which all connectives are removed:

$$p^\circ, [p^\bullet, q^\circ, [q^\bullet]] \tag{4}$$

This sequent can be transformed into the equivalent (interderivable) rule

$$\frac{\Delta_0, [\Delta_0, \Delta_1, \Sigma, [\Delta_2]] \qquad \Delta_0, [\Delta_1, \Sigma, [\Delta_2, \Sigma]]}{\Delta_0, [\Delta_1, \Sigma, [\Delta_2]]}\ bd_2'$$

following the steps below:

- First, move p° and q° to the premise, as shown below left, and then $p^\bullet$ and $q^\bullet$, as in the rule below right (Δ_i, Σ are *fresh* metavariables for sequents):

$$\frac{\Sigma, q^\bullet \quad \Delta_0, p^\bullet}{\Delta_0, [p^\bullet, \Sigma, [q^\bullet]]} \qquad \frac{\Sigma, q^\bullet \quad \Delta_0, p^\bullet \quad \Delta_0, [\Delta_1, p^\circ, \Sigma, [\Delta_2]] \quad \Delta_0, [\Delta_1, \Sigma, [\Delta_2, q^\circ]]}{\Delta_0, [\Delta_1, \Sigma, [\Delta_2]]}\ r_2$$

  It is easy to see that each of these rules derives the sequent (4) in NIPL (instantiate $\Delta_0 := p^\circ$ and $\Delta_1 := p^\bullet$ and $\Sigma := q^\circ$ and $\Delta_2 := q^\bullet$). The converse direction follows by nec (Proposition 3.2), w, c and cut with the sequent in (4).

- The rule (bd_2') is obtained from the (r_2) rule above by cutting the premises of the latter in all possible ways (using first nec and w). This ensures that (bd_2') derives (r_2) in NIPL. For the converse direction we show that from any instance of the premisses of (bd_2') we can derive an instance of the premises of (r_2) for suitable p and q. Take $p := fm(\Delta_0^\circ)$, and $q := fm(\Sigma^\circ)$.

As shown below, the (bd_2') rule allows the derivation of the axiom $\mathbf{bd_2}$, and therefore its addition to NIPL result in a complete calculus for the $\mathsf{BD_2}$ logic.

$$\dfrac{\dfrac{\dfrac{\dfrac{\dfrac{A^\circ,[A^\circ,A^\bullet,B^\circ,[B^\bullet]] \qquad A^\circ,[A^\bullet,B^\circ,[B^\bullet,B^\circ]]}{A^\circ,[A^\bullet,B^\circ,[B^\bullet]]}\,bd'_2}{A^\circ,[A^\bullet,B^\circ,\neg B^\circ]}\supset^\circ}{A^\circ,[A^\bullet,B\vee\neg B^\circ]}\vee^\circ}{A^\circ,A\supset(B\vee\neg B)^\circ}\supset^\circ}{A\vee(A\supset(B\vee\neg B))^\circ}\vee^\circ$$

However, (bd'_2) does not preserve cut elimination. Prior identifying Δ_1 and Σ, we modify (bd'_2) from shallow to *deep* (using the terminology in [13]), to permit its application on structures in any arbitrary node in the tree. This leads to:

$$\dfrac{\Gamma\{\Delta_0\{\Delta_0\{\varnothing\},\Delta_1\{\Delta_2\}\}\} \qquad \Gamma\{\Delta_0\{\Delta_1\{\Delta_1\{\varnothing\},\Delta_2\}\}\}}{\Gamma\{\Delta_0\{\Delta_1\{\Delta_2\}\}\}}\,\mathsf{bd_2}$$

The argument applies also to $\mathsf{BD_n}$ $(n \geq 1)$ resulting in a rule having n premises.

Definition 5.2 $\mathsf{NIPLBD_n}$ is the calculus NIPL extended with the rule $\mathsf{bd_n}$:

$$\dfrac{\Gamma\{\Delta_0\{\ldots\{\Delta_{i-1}\{\varnothing\},\Delta_i\{\ldots\{\Delta_n\}\ldots\}\}\ldots\}\} \quad \Big| \quad 1\leqslant i\leqslant n}{\Gamma\{\Delta_0\{\Delta_1\{\ldots\Delta_{n-1}\{\Delta_n\}\ldots\}\}\}}\,\mathsf{bd_n}$$

Proposition 5.3 *For every n, the rule $\mathsf{bd_n}$ is sound with respect to intuitionistic Kripke frames with depth bounded by n.*

Proof By contradiction. We assume that $\Gamma\{\Delta_0\{\ldots\{\Delta_n\}\ldots\}\}$ is not valid. Hence there are worlds $x_0,\ldots,x_n$ such that:

- $x_i \leqslant x_{i+1}$, with $0 \leqslant i \leqslant n-1$.
- $x_i \Vdash \bigwedge \Delta_i^\bullet$ and $x_i \nVdash \bigvee \Delta_i^\circ$.

where $\Delta_i^\bullet$,(Δ_i°) are the input (output) formulas (formulas and boxed sequents) in Δ_i. By (3) we have $x_{i+1} \leqslant x_i$ for some $i \in \{0,...,n-1\}$. In each case, by monotonicity (w.r.t. compound formulas) we get $x_{i+1} \Vdash \bigwedge \Delta_i^\bullet$ and thus $x_{i+1} \Vdash \bigvee \Delta_i^\circ$, which (again by monotonicity) yields $x_i \Vdash \bigvee \Delta_i^\circ$, which is a contradiction. □

6 Cut Elimination for $\mathsf{NIPLBD_n}$

We show that cut elimination holds for $\mathsf{NIPLBD_n}$. The proof extends that for NIPL in a modular way: only the cases concerning the new rules need to be considered. We start by showing properties (N1) and (N2): invertibility of all rules, and height-preserving admissibility of the dedicated rules in Fig. 3.

Lemma 6.1 *The weakening rule is height-preserving admissible in* $\mathsf{NIPLBD_n}$.

Proof Proceeds as the proof of Lemma 4.1. □

Lemma 6.2 *Every rule is height-preserving invertible in* $\mathsf{NIPLBD_n}$.

Proof By induction on the height of the derivation for every rule of the system. The structural rule $\mathsf{bd_n}$ is height-preserving invertible by using Lemma 6.1. The addition of the rule $\mathsf{bd_n}$ preserves the invertibility of the other rules. The strategy consists in applying (possibly twice, due to the repetition of the contexts) the induction hypothesis to each premise of the rule. □

Lemma 6.3 *The contraction rule is height-preserving admissible in* $\mathsf{NIPLBD_n}$.

Proof By induction on the height of the derivation. The only additional case concerns the rule $\mathsf{bd_n}$. Since the principal formulas are repeated in each premise of $\mathsf{bd_n}$ we just apply the induction hypothesis and then the rule again. We give a concrete example of this qualitative analysis; to improve the readability we consider the particular case of the $\mathsf{bd_2}$ rule.

$$\frac{\Gamma\{\Delta_0\{[\Delta_0\{\varnothing\},\Delta_1\{\!\{\Delta_2\}\!\}],[\Delta_1\{\!\{\Delta_2\}\!\}]\}\} \qquad \Gamma\{\Delta_0\{[\Delta_1\{\!\{\Delta_1\{\varnothing\},\Delta_2\}\!\}],[\Delta_1\{\!\{\Delta_2\}\!\}]\}\}}{\Gamma\{\Delta_0\{[\Delta_1\{\!\{\Delta_2\}\!\}],[\Delta_1\{\!\{\Delta_2\}\!\}]\}\}}\ \mathsf{bd_2}$$

We construct the following derivation:

$$\dfrac{\dfrac{\dfrac{\Gamma\{\Delta_0\{[\Delta_0\{\varnothing\},\Delta_1\{\!\{\Delta_2\}\!\}],[\Delta_1\{\!\{\Delta_2\}\!\}]\}\}}{\Gamma\{\Delta_0\{[\Delta_0\{\varnothing\},\Delta_1\{\!\{\Delta_2\}\!\}],[\Delta_0\{\varnothing\},\Delta_1\{\!\{\Delta_2\}\!\}]\}\}}\ \mathsf{w}}{\Gamma\{\Delta_0\{[\Delta_0\{\varnothing\},\Delta_1\{\!\{\Delta_2\}\!\}]\}\}}\ \mathsf{c} \qquad \dfrac{\dfrac{\Gamma\{\Delta_0\{[\Delta_1\{\!\{\Delta_1\{\varnothing\},\Delta_2\}\!\}],[\Delta_1\{\!\{\Delta_2\}\!\}]\}\}}{\Gamma\{\Delta_0\{[\Delta_1\{\!\{\Delta_1\{\varnothing\},\Delta_2\}\!\}],[\Delta_1\{\!\{\Delta_1\{\varnothing\},\Delta_2\}\!\}]\}\}}\ \mathsf{w}}{\Gamma\{\Delta_0\{[\Delta_1\{\!\{\Delta_1\{\varnothing\},\Delta_2\}\!\}]\}\}}\ \mathsf{c}}{\Gamma\{\Delta_0\{[\Delta_1\{\!\{\Delta_2\}\!\}]\}\}}\ \mathsf{bd_2}$$

□

Lemma 6.4 *The rule* lift *is height-preserving admissible in* $\mathsf{NIPLBD_n}$.

Proof By induction on the height of the derivation. To simplify the notation we consider only $\mathsf{bd_2}$; the generalization to $\mathsf{bd_n}$ is immediate.

Let $\Gamma\{[\Sigma^\bullet,\Delta]\}$ be the conclusion of $\mathsf{bd_2}$, we need to consider two subcases. Either $[\Sigma^\bullet,\Delta]$ is moved by the rule or not. In the latter case we simply apply the induction hypothesis to the premises of the rule and then the rule again. In the former case we need to distinguish two further subcases. Either $[\Sigma^\bullet,\Delta]$ is Δ_i for $i \in \{0,1,2\}$ or not. In the latter case we apply the induction hypothesis to the premises and then the rule again. In the former case assume $[\Delta_1\{\ \}] = [\Sigma^\bullet,\Delta\{\ \}]$. We have:

$$\frac{\Gamma\{\Delta_0\{\!\{\Delta_0\{\varnothing\},\Sigma^\bullet,\Delta\{\!\{\Delta_2\}\!\}\}\!\}\} \qquad \Gamma\{\Delta_0\{\!\{\Sigma^\bullet,\Delta\{\!\{\Sigma^\bullet,\Delta\{\varnothing\},\Delta_2\}\!\}\}\!\}\}}{.\qquad\qquad \Gamma\{\Delta_0\{\!\{\Sigma^\bullet,\Delta\{\!\{\Delta_2\}\!\}\}\!\}\}}\ \mathsf{bd_2}$$

We construct the following derivation:

$$\dfrac{\dfrac{\dfrac{\Gamma\{\Delta_0\{\!\{\Delta_0\{\varnothing\},\Sigma^\bullet,\Delta\{\!\{\Delta_2\}\!\}\}\!\}\}}{\Gamma\{\Delta_0\{\Sigma^\bullet,[\Delta_0\{\varnothing\},\Delta\{\!\{\Delta_2\}\!\}]\}\}}\ \mathsf{lift}}{\Gamma\{\Delta_0\{\Sigma^\bullet,[\Delta_0\{\Sigma^\bullet\},\Delta\{\!\{\Delta_2\}\!\}]\}\}}\ \mathsf{w} \qquad \dfrac{\dfrac{\Gamma\{\Delta_0\{\!\{\Sigma^\bullet,\Delta\{\!\{\Sigma^\bullet,\Delta\{\varnothing\},\Delta_2\}\!\}\}\!\}\}}{\Gamma\{\Delta_0\{\Sigma^\bullet,\Sigma^\bullet,[\Delta\{\!\{\Delta\{\varnothing\},\Delta_2\}\!\}]\}\}}\ \text{several lift}}{\Gamma\{\Delta_0\{\Sigma^\bullet,[\Delta\{\!\{\Delta\{\varnothing\},\Delta_2\}\!\}]\}\}}\ \mathsf{c}}{\Gamma\{\Delta_0\{\Sigma^\bullet,[\Delta\{\!\{\Delta_2\}\!\}]\}\}}\ \mathsf{bd_2}$$

□

Lemma 6.5 *The rule 4 is height-preserving admissible in* $\mathsf{NIPLBD_n}$.

Proof By induction on the height of the derivation. Assume the last applied rule is $\mathsf{bd_n}$. We observe that the general form of Σ° is $A_1^\circ,\ldots,A_n^\circ,[\Sigma_1],\ldots,[\Sigma_m]$. We distinguish three cases: either Σ° does not move, or it entirely moves to another boxed sequent, or different components of Σ° move to different sequents. In the first and second case we simply apply the induction hypothesis to the premises and then the rule again. In the latter case we have that Σ° is of shape $\Sigma^\circ = \Sigma'\{\!\{\Sigma''\{\Sigma'''\}\}\!\}$ and we have

$$\dfrac{\cdots \qquad \Gamma\{\Delta_0\{\!\{\ldots\{\!\{\Delta'_i, \Sigma'\{\!\{\Delta'_i, \Sigma'\{\varnothing\}, \Sigma''\{\ldots\{\!\{\Delta_n\}\!\}\ldots\}\}\!\}\ldots\}\!\}\}\!\}\} \qquad \cdots}{\dfrac{\Gamma\{\Delta_0\{\!\{\ldots\{\!\{\Delta'_i, \Sigma'\{\!\{\Sigma''\{\ldots\{\!\{\Delta_n\}\!\}\ldots\}\}\!\}\ldots\}\!\}\}\!\}\}}{\Gamma\{\Delta_0\{\!\{\ldots\{\!\{\Delta'_i, [\Sigma'\{\!\{\Sigma''\{\ldots\{\!\{\Delta_n\}\!\}\ldots\}\}\!\}]\ldots\}\!\}\}\!\}\}}\,\mathsf{4}}\,\mathsf{bd_n}$$

and we can proceed as follows:

$$\dfrac{\cdots \qquad \dfrac{\dfrac{\Gamma\{\Delta_0\{\!\{\ldots\{\!\{\Delta'_i, \Sigma'\{\!\{\Delta'_i, \Sigma'\{\varnothing\}, \Sigma''\{\ldots\{\!\{\Delta_n\}\!\}\ldots\}\}\!\}\ldots\}\!\}\}\!\}\}}{\Gamma\{\Delta_0\{\!\{\ldots\{\!\{\Delta'_i, [\Sigma'\{\!\{\Delta'_i, \Sigma'\{\varnothing\}, \Sigma''\{\ldots\{\!\{\Delta_n\}\!\}\ldots\}\}\!\}]\ldots\}\!\}\}\!\}\}}\,\mathsf{4}}{\Gamma\{\Delta_0\{\!\{\ldots\{\!\{\Delta'_i, [\Sigma'\{\!\{\Delta'_i, [\Sigma'\{\varnothing\}], \Sigma''\{\ldots\{\!\{\Delta_n\}\!\}\ldots\}\}\!\}]\ldots\}\!\}\}\!\}\}}\,\mathsf{4} \qquad \cdots}{\Gamma\{\Delta_0\{\!\{\ldots\{\!\{\Delta'_i, [\Sigma'\{\!\{\Sigma''\{\ldots\{\!\{\Delta_n\}\!\}\ldots\}\}\!\}]\ldots\}\!\}\}\!\}\}}\,\mathsf{bd_n}$$

□

Lemma 6.6 *The rule* lower *is height-preserving admissible in* $\mathsf{NIPLBD_n}$.

Proof This proof is literally the same as for NIPL, applying 4, w and c. □

We now need to prove the admissibility of the t rule, which removes boxes. To simplify the proof in the presence of $\mathsf{bd_n}$, we consider the auxiliary rule lift* below.

Lemma 6.7 *The following rule is derivable with* $\{\mathsf{t}, \mathsf{c}, \mathsf{4}, \mathsf{lift}, \mathsf{w}\}$.

$$\dfrac{\Gamma\{\Sigma\{\!\{\Delta, \Sigma\{\Delta\}\}\!\}\}}{\Gamma\{\Sigma\{\!\{\Delta, \Sigma\{\varnothing\}\}\!\}\}}\,\mathsf{lift}^*$$

Proof For better readability we only show the case where the depth of $\Sigma\{\ \}$ is 1, i.e., $\Sigma\{\ \} = \Sigma_1, [\Sigma_2^\bullet, \Sigma_2^\circ, \{\ \}]$:

$$\begin{array}{cl}
\Gamma\{\Sigma_1, [\Sigma_2^\bullet, \Sigma_2^\circ, [\Delta, \Sigma_1, [\Sigma_2^\bullet, \Sigma_2^\circ, \Delta]]]\} & \\ \hline
\Gamma\{\Sigma_1, [{\Sigma_2}^\bullet, {\Sigma_2}^\circ, [\Delta, \Sigma_1, {\Sigma_2}^\bullet, {\Sigma_2}^\circ, \Delta]]\} & \mathsf{t} \\ \hline
\Gamma\{\Sigma_1, [{\Sigma_2}^\bullet, {\Sigma_2}^\circ, [\Delta, \Sigma_1, {\Sigma_2}^\bullet, {\Sigma_2}^\circ]]\} & \mathsf{c} \\ \hline
\Gamma\{\Sigma_1, [{\Sigma_2}^\bullet, {\Sigma_2}^\bullet, {\Sigma_2}^\circ, [\Delta, \Sigma_1, {\Sigma_2}^\circ]]\} & \mathsf{lift} \\ \hline
\Gamma\{\Sigma_1, [{\Sigma_2}^\bullet, {\Sigma_2}^\circ, [\Delta, \Sigma_1, {\Sigma_2}^\circ]]\} & \mathsf{c} \\ \hline
\Gamma\{\Sigma_1, [{\Sigma_2}^\bullet, {\Sigma_2}^\circ, [\Delta, \Sigma_1, [{\Sigma_2}^\circ]]]\} & \mathsf{4} \\ \hline
\Gamma\{\Sigma_1, [{\Sigma_2}^\bullet, {\Sigma_2}^\circ, [\Delta, \Sigma_1, [{\Sigma_2}^\bullet, {\Sigma_2}^\circ]]]\} & \mathsf{w}
\end{array}$$

If the depth of $\Sigma\{\ \}$ is n, all steps in this derivation have to repeated n times.□

Lemma 6.8 *The rule* t *is height-preserving admissible in* $\mathsf{NIPLBD_n}$.

Proof The only new case to consider is the $\mathsf{bd_n}$-rule. Again, for the sake of clarity, we discuss $\mathsf{bd_2}$. We distinguish the following subcases: either we apply t to $[\Delta_i]$ for some i$\in \{1, 2\}$, or to some other boxed sequent. In the latter case, we apply the induction hypothesis (possibly twice) to the premises and then the rule again. In the former case we assume w.l.o.g. that we apply t to Δ_1:

$$\dfrac{\dfrac{\Gamma\{\Delta_0\{\!\!\{\Delta_0\{\varnothing\}, \Delta_1\{\!\!\{\Delta_2\}\!\!\}\}\!\!\}\} \qquad \Gamma\{\Delta_0\{\!\!\{\Delta_1\{\!\!\{\Delta_1\{\varnothing\}, \Delta_2\}\!\!\}\}\!\!\}\}}{\Gamma\{\Delta_0\{\!\!\{\Delta_1\{\!\!\{\Delta_2\}\!\!\}\}\!\!\}\}}\,\mathsf{bd_2}}{\Gamma\{\Delta_0\{\Delta_1\{\!\!\{\Delta_2\}\!\!\}\}\}}\,\mathsf{t}$$

We have two subcases to distinguish here:

- $[\Delta_1]$ is an immediate child of Δ_0 in the nested sequent tree. We have:

$$\dfrac{\dfrac{\Gamma\{\Delta_0, [\Delta_0, \Delta_1\{\!\!\{\Delta_2\}\!\!\}]\}}{\Gamma\{\Delta_0, \Delta_0, \Delta_1\{\!\!\{\Delta_2\}\!\!\}\}}\,\mathsf{t}}{\Gamma\{\Delta_0, \Delta_1\{\!\!\{\Delta_2\}\!\!\}\}}\,\mathsf{c}$$

  by applying the induction hypothesis to the left subproof, discarding the right one.

- $[\Delta_1]$ is not an immediate child of Δ_0 in the nested sequent tree, so we can assume that Δ_0 contains at least one nesting, thus we have:

$$\Delta_0\{\ \} = \Delta_0'\{\!\!\{\Delta_0''\{\ \}\}\!\!\}$$

  for some contexts $\Delta_0'\{\ \}$ and $\Delta_0''\{\ \}$. We construct the derivation

$$\dfrac{\dfrac{\dfrac{\dfrac{\Gamma\{\Delta_0'\{\!\!\{\Delta_0'', [\Delta_0'\{\!\!\{\Delta_0''\{\varnothing\}\}\!\!\}, \Delta_1\{\!\!\{\Delta_2\}\!\!\}]\}\!\!\}\}}{\Gamma\{\Delta_0'\{\!\!\{\Delta_0'', \Delta_0'\{\!\!\{\Delta_0''\{\varnothing\}\}\!\!\}, \Delta_1\{\!\!\{\Delta_2\}\!\!\}\}\!\!\}\}}\,\mathsf{t}}{\Gamma\{\Delta_0'\{\!\!\{\Delta_0'', \Delta_0'\{\Delta_0''\{\varnothing\}\}, \Delta_1\{\!\!\{\Delta_2\}\!\!\}\}\!\!\}\}}\,\mathsf{t}}{\Gamma\{\Delta_0'\{\!\!\{\Delta_0'', \Delta_0'\{\varnothing\}, \Delta_1\{\!\!\{\Delta_2\}\!\!\}\}\!\!\}\}}\,\mathsf{lift^*} \qquad \dfrac{\dfrac{\Gamma\{\Delta_0'\{\!\!\{\Delta_0'', [\Delta_1\{\!\!\{\Delta_1\{\varnothing\}, \Delta_2\}\!\!\}]\}\!\!\}\}}{\Gamma\{\Delta_0'\{\!\!\{\Delta_0'', \Delta_1\{\!\!\{\Delta_1\{\varnothing\}, \Delta_2\}\!\!\}\}\!\!\}\}}\,\mathsf{t}}{\Gamma\{\Delta_0'\{\!\!\{\Delta_0'', \Delta_1\{\!\!\{\Delta_0'', \Delta_1\{\varnothing\}, \Delta_2\}\!\!\}\}\!\!\}\}}\,\mathsf{w}}{\Gamma\{\Delta_0'\{\!\!\{\Delta_0''\{\Delta_1\{\!\!\{\Delta_2\}\!\!\}\}\}\!\!\}\}}\,\mathsf{bd_2}$$

  which yields the desired conclusion.

□

This completes the proofs of conditions (N1) and (N2).

Theorem 6.9 (Cut elimination) *The cut rule is eliminable in* $\mathsf{NIPLBD_n}$.

Proof The proof proceeds as for NIPL. We only need to check that the cut rule can be shifted up over $\mathsf{bd_n}$. That means we only have to verify (N3), as (N4) and (N5) are not affected by adding new structural rules to the system (and (N1) and (N2) still hold).

If the cut formula $A^\bullet$ is not active in $\mathsf{bd_n}$, we can permute the cut upwards and remove it by induction hypothesis. If $A^\bullet$ is active, assume w.l.o.g. that the cut formula is in $[\Delta_i]$. We distinguish two cases according to its position.

- If $A^\bullet$ is in the branch from Δ_0 to Δ_n, we have:

$$\dfrac{\Gamma\{\Delta_0\{\!\!\{A^\circ, \Delta_1\{\!\!\{\Delta_2 \ldots \{\!\!\{\Delta_n\}\!\!\}\}\!\!\}\}\!\!\}\} \qquad \dfrac{\ldots \qquad \Gamma\{\Delta_0\{\!\!\{A^\bullet, \Delta_1\{\!\!\{A^\bullet, \Delta_1\{\varnothing\}, \Delta_2 \ldots \{\!\!\{\Delta_n\}\!\!\}\}\!\!\}\}\!\!\}\} \qquad \ldots}{\Gamma\{\Delta_0\{\!\!\{A^\bullet, \Delta_1\{\!\!\{\Delta_2 \ldots \{\!\!\{\Delta_n\}\!\!\}\}\!\!\}\}\!\!\}\}}\,\mathsf{bd_n}}{\Gamma\{\Delta_0\{\!\!\{\Delta_1\{\!\!\{\Delta_2 \ldots \{\!\!\{\Delta_n\}\!\!\}\}\!\!\}\}\!\!\}\}}\,\mathsf{cut}$$

  (we display the premise in which the position of $A^\bullet$ changes). For each premise different from $\Gamma\{\Delta_0\{\!\!\{A^\bullet, \Delta_1\{\!\!\{A^\bullet, \Delta_1\{\varnothing\}, \Delta_2 \ldots \{\!\!\{\Delta_n\}\!\!\}\}\!\!\}\}\!\!\}\}$, we proceed by height-preserving admissibility of w and cross-cuts which are removed by induction hypothesis. Then we construct the following derivation:

$$\dfrac{\dfrac{\Gamma\{\Delta_0\{\!\{A^\circ, \Delta_1\{\!\{\Delta_2 \ldots \{\!\{\Delta_n\}\!\}\}\!\}\}\!\}\}}{\Gamma\{\Delta_0\{\!\{A^\circ, \Delta_1\{\!\{\Delta_1\{\varnothing\}, \Delta_2 \ldots \{\!\{\Delta_n\}\!\}\}\!\}\}\!\}\}}\,\mathsf{w} \qquad \dfrac{\Gamma\{\Delta_0\{\!\{A^\bullet, \Delta_1\{\!\{A^\bullet, \Delta_1\{\varnothing\}, \Delta_2 \ldots \{\!\{\Delta_n\}\!\}\}\!\}\}\!\}\}}{\Gamma\{\Delta_0\{\!\{A^\bullet, \Delta_1\{\!\{\Delta_1\{\varnothing\}, \Delta_2 \ldots \{\!\{\Delta_n\}\!\}\}\!\}\}\!\}\}}\,\mathsf{lift}}{\Gamma\{\Delta_0\{\!\{\Delta_1\{\!\{\Delta_1\{\varnothing\}, \Delta_2 \ldots \{\!\{\Delta_n\}\!\}\}\!\}\}\!\}\}}\,\mathsf{cut}$$

to which we can now apply the rule $\mathsf{bd_n}$ and obtain the desired conclusion.

- If $A^\bullet$ is not in the branch from Δ_0 to Δ_n, then $A^\bullet$ must be inside a bracket in this branch, as in:

$$\dfrac{\Gamma\{\Delta_0\{\!\{[\Delta\{A^\circ\}], \Delta_1\{\!\{\Delta_2 \ldots \{\!\{\Delta_n\}\!\}\}\!\}\}\!\}\} \qquad \dfrac{\ldots \quad \Gamma\{\Delta_0\{\!\{[\Delta\{A^\bullet\}], \Delta_1\{\!\{[\Delta\{A^\bullet\}], \Delta_1\{\varnothing\}, \Delta_2 \ldots \{\!\{\Delta_n\}\!\}\}\!\}\}\!\}\} \quad \ldots}{\Gamma\{\Delta_0\{\!\{[\Delta\{A^\bullet\}], \Delta_1\{\!\{\Delta_2 \ldots \{\!\{\Delta_n\}\!\}\}\!\}\}\!\}\}}\,\mathsf{bd_n}}{\Gamma\{\Delta_0\{\!\{[\Delta\{\varnothing\}], \Delta_1\{\!\{\Delta_2 \ldots \{\!\{\Delta_n\}\!\}\}\!\}\}\!\}\}}\,\mathsf{cut}$$

We construct the following derivation:

$$\dfrac{\dfrac{\dfrac{\Gamma\{\Delta_0\{\!\{[\Delta\{A^\circ\}], \Delta_1\{\!\{\Delta_2 \ldots \{\!\{\Delta_n\}\!\}\}\!\}\}\!\}\}}{\Gamma\{\Delta_0\{\!\{\Delta_1\{\!\{[\Delta\{A^\circ\}], \Delta_2 \ldots \{\!\{\Delta_n\}\!\}\}\!\}\}\!\}\}}\,\mathsf{lower}}{\Gamma\{\Delta_0\{\!\{\Delta_1\{\!\{[\Delta\{A^\circ\}], \Delta_1\{\varnothing\}, \Delta_2 \ldots \{\!\{\Delta_n\}\!\}\}\!\}\}\!\}\}}\,\mathsf{w} \qquad \dfrac{\Gamma\{\Delta_0\{\!\{[\Delta\{A^\bullet\}], \Delta_1\{\!\{[\Delta\{A^\bullet\}], \Delta_1\{\varnothing\}, \Delta_2 \ldots \{\!\{\Delta_n\}\!\}\}\!\}\}\!\}\}}{\Gamma\{\Delta_0\{\!\{\Delta_1\{\!\{[\Delta\{A^\bullet\}], \Delta_1\{\varnothing\}, \Delta_2 \ldots \{\!\{\Delta_n\}\!\}\}\!\}\}\!\}\}}\,\mathsf{lower, c}}{\Gamma\{\Delta_0\{\!\{\Delta_1\{\!\{[\Delta\{\varnothing\}], \Delta_1\{\varnothing\}, \Delta_2 \ldots \{\!\{\Delta_n\}\!\}\}\!\}\}\!\}\}}\,\mathsf{cut}$$

The cut is removed invoking the secondary induction hypothesis. With respect to the other premises we permute the cut upwards and we remove it applying the secondary induction hypothesis. The desired sequent then follows by an application of $\mathsf{bd_n}$. □

Corollary 6.10 $\mathsf{NIPLBD_n}$ *is complete with respect to* $\mathsf{BD_n}$.

Proof Follows from Corollary 4.9, the derivability of the axiom $\mathbf{bd_n}$, and the cut elimination theorem. □

Concluding Remark

The cut elimination proof contained in this paper makes use of the auxiliary rules t and $\mathsf{4}$, originally introduced in the modal logics context [24]. Our results can be seen as "transfer of knowledge" from (the proof theory of) modal logics to intermediate logics. We consider it an important aspect or our future research to also initiate a transfer back to modal logics. In fact, notice that our cut elimination proof holds for NIPL extended by any structural rule that preserves properties (N1)–(N5). This paves the way for the definition of an algorithm, along the line of that in [5,6], to introduce analytic nested calculi for a large class of intermediate logics starting from their axiomatizations. Many such intermediate logics have indeed modal counterparts that have not yet been investigated from a proof theoretical point of view. We plan to do so using our work on nested sequents.

References

[1] N. D. Belnap, Jr. Display logic. *J. Philos. Logic*, 11(4):375–417, 1982.

[2] K. Brünnler. Deep sequent systems for modal logic. *Archive for Mathematical Logic*, 48(6):551–577, 2009.

[3] K. Chaudhuri, S. Marin, and L. Straßburger. Focused and synthetic nested sequents. In B. Jacobs and C. Löding, editors, *Proceedings of FoSSaCS*, 2016.

[4] K. Chaudhuri, S. Marin, and L. Straßburger. Modular focused proof systems for intuitionistic modal logics. In D. Kesner and B. Pientka, editors, *Proceedings of FSCD*, volume 52 of *LIPIcs*, pages 16:1–16:18. Schloss Dagstuhl - Leibniz-Zentrum fuer Informatik, 2016.

[5] A. Ciabattoni, N. Galatos, and K. Terui. From axioms to analytic rules in nonclassical logics. In *Proceedings of LICS'08*, pages 229–20. IEEE, 2008.

[6] A. Ciabattoni and R. Ramanayake. Power and limits of structural display rules. *ACM Trans. Comput. Log.*, 17(3):17:1–17:39, 2016.

[7] W. Conradie and A. Palmigiano. Algorithmic correspondence and canonicity for distributive modal logic. *Annals of Pure and Applied Logic*, 163(3):338–376, 2012.

[8] M. Fitting. Nested sequents for intuitionistic logics. *Notre Dame Journal of Formal Logic*, 55(1):41–61, 2014.

[9] D. Gabbay. *Semantical Investigations in Heyting's Intuitionistic Logic.* Reidel, 1983.

[10] J.-Y. Girard. *Proof Theory and Logical Complexity, Volume I*, volume 1 of *Studies in Proof Theory.* Bibliopolis, edizioni di filosofia e scienze, 1987.

[11] M. Girlando and L. Straßburger. MOIN: A nested sequent theorem prover for intuitionistic modal logics (system description). In N. Peltier and V. Sofronie-Stokkermans, editors, *Proceedings of IJCAR 2020*, volume 12167 of *LNCS*, pages 398–407. Springer, 2020.

[12] R. Goré, L. Postniece, and A. Tiu. Taming displayed tense logics using nested sequents with deep inference. In M. Giese and A. Waaler, editors, *Automated Reasoning with Analytic Tableaux and Related Methods*, volume 5607 of *LNCS*, pages 189–204. Springer, 2009.

[13] R. Goré, L. Postniece, and A. Tiu. On the correspondence between display postulates and deep inference in nested sequent calculi for tense logics. *Log. Methods Comput. Sci.*, 7(2):2:8, 38, 2011.

[14] R. Goré and R. Ramanayake. Labelled tree sequents, tree hypersequents and nested (deep) sequents. In T. Bolander, T. Braüner, S. Ghilardi, and L. S. Moss, editors, *Advances in Modal Logic 9, papers from the ninth conference on "Advances in Modal Logic," held in Copenhagen, Denmark, 22-25 August 2012*, pages 279–299. College Publications, 2012.

[15] R. Kashima. Cut-free sequent calculi for some tense logics. *Studia Logica*, 53(1):119–136, 1994.

[16] R. Kuznets and B. Lellmann. Interpolation for intermediate logics via injective nested sequents. *J. Log. Comput.*, 31(3):797–831, 2021.

[17] R. Kuznets and L. Straßburger. Maehara-style modal nested calculi. *Arch. Math. Log.*, 58(3-4):359–385, 2019.

[18] F. Lauridsen. Intermediate logics admitting a structural hypersequent calculus. *Studia Logica*, 107:247–282, 2019.

[19] B. Lellmann. Linear nested sequents, 2-sequents and hypersequents. In H. de Nivelle, editor, *Automated Reasoning with Analytic Tableaux and Related Methods TABLEAUX*, volume 9323 of *LNCS*, pages 135–150. Springer, 2015.

[20] T. Lyon. On the correspondence between nested calculi and semantic systems for intuitionistic logics. *J. Log. Comput.*, 31(1):213–265, 2021.

[21] T. Lyon, A. Tiu, R. Goré, and R. Clouston. Syntactic interpolation for tense logics and bi-intuitionistic logic via nested sequents. In M. Fernández and A. Muscholl, editors, *Computer Science Logic, CSL 2020*, volume 152 of *LIPIcs*, pages 28:1–28:16. Leibniz-Zentrum für Informatik, 2020.

[22] T. S. Lyon. Nested sequents for intuitionistic modal logics via structural refinement. In A. Das and S. Negri, editors, *Automated Reasoning with Analytic Tableaux and Related Methods*, volume 12842 of *LNCS*, pages 409–427. Springer, 2021.

[23] L. Maksimova. Craig's theorem in superintuitionistic logics and amalgamable varieties of pseudo-boolean algebras. *Algebra and Logic*, 16:427–455, 1977.

[24] S. Marin and L. Straßburger. Label-free modular systems for classical and intuitionistic modal logics. In *Advances in Modal Logic 10*, pages 387–406. College Publications, 2014.

[25] S. Negri. Proof analysis in modal logics. *Journal of Philosophical Logic*, 34(5-6):507–544, 2005.

[26] N. Olivetti and G. L. Pozzato. Nescond: An implementation of nested sequent calculi for conditional logics. In S. Demri, D. Kapur, and C. Weidenbach, editors, *Proceedings of IJCAR 2014*, pages 511–518. Springer, 2014.

[27] F. Poggiolesi. The method of tree-hypersequents for modal propositional logic. In *Towards mathematical philosophy*, volume 28 of *Trends Log. Stud. Log. Libr.*, pages 31–51. Springer, Dordrecht, 2009.

[28] L. Straßburger. Cut elimination in nested sequents for intuitionistic modal logics. In F. Pfenning, editor, *FoSSaCS'13*, volume 7794 of *LNCS*, pages 209–224. Springer, 2013.

Describing neighborhoods in inquisitive modal logic

Ivano Ciardelli [1]

Munich Center for Mathematical Philosophy
LMU Munich

Abstract

We introduce and investigate an inquisitive modal logic interpreted on neighborhood models. The logic extends propositional inquisitive logic with a binary modality $\Rrightarrow$, where $(\varphi \Rrightarrow \psi)$ is a statement true at a world iff every neighborhood that supports φ also supports ψ. This logic provides a natural language to describe properties of neighborhood structures, has an interpretation as a logic of ability, and generalizes previous versions of inquisitive modal logic. We give an appropriate notion of bisimulation, establish a Hennessy-Milner theorem, and relate a fragment of the logic to the instantial neighborhood logic of [6] by truth-preserving translations. We also prove a completeness theorem, showing that our modal conditional is axiomatized by four simple principles familiar from the study of strict conditionals.

Keywords: Inquisitive logic, neighborhood structures, bisimulation, axiomatization, strict conditional, logic of ability, team semantics, instantial neighborhood logic.

1 Introduction

This paper introduces and investigates an inquisitive modal logic InqCM which extends propositional inquisitive logic with a conditional modality denoted $\Rrightarrow$. This logic can be motivated from three different directions.

One line of motivation starts from considering neighborhood structures. A standard version of neighborhood semantics for modal logic [2,7,8,21,24] interprets a formula $\Box\varphi$ as true at a world w if φ is true throughout some neighborhood s of w (in this case, we say that s *supports* φ). [2] However, the standard modal language with this interpretation does not allow us to express much about the configurations that arise in the neighborhoods of a given point. Consider three worlds w_1, w_2, w_3 associated with sets of neighborhoods $\Sigma(w_i)$ as depicted in Figure 1. Here are some respects in which the situation at these

[1] Gefördert durch die Deutsche Forschungsgemeinschaft (DFG) - Projektnummer 446711878. Funded by the German Research Foundation (DFG) - Project number 446711878.

[2] This is different from the Scott-Montague neighborhood semantics [23,27], which interprets $\Box\varphi$ as true at w if the set of worlds satisfying φ is a neighborhood of w, though the two are equivalent if the set of neighborhoods is closed under supersets. See [24] for discussion.

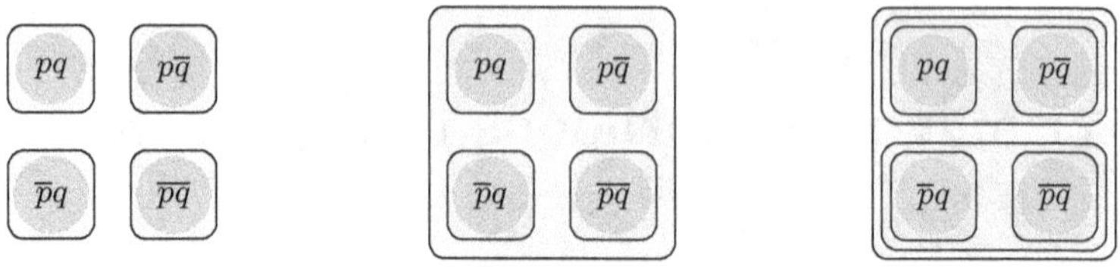

(a) $\Sigma(w_1)$ (b) $\Sigma(w_2)$ (c) $\Sigma(w_3)$

Fig. 1. Sets of neighborhoods associated with three worlds.

worlds is different: every neighborhood of w_1 settles whether p (i.e., the truth value of p is constant within each neighborhood) while this is not the case for w_2 and w_3; moreover, every neighborhood of w_2 that settles whether p also settles whether q, whereas this is not the case for w_3. These facts cannot be expressed in the standard modal language with the above semantics.

More generally, in a neighborhood structure it is natural to regard two worlds as equivalent only if any neighborhood of the one is 'matched' by some neighborhood of the other; in turn, it is natural to regard two neighborhoods as matching if every world in one is equivalent to some world in the other. This leads to a natural notion of bisimilarity for neighborhood structures. It seems worthwhile to pursue a modal language that, at least in restriction to finite models, can distinguish points that are not bisimilar in this sense. The *instantial neighborhood logic* recently developed in [6] is such a language. Modal formulas of this logic have the form $\Box(\psi_1, \ldots, \psi_n; \varphi)$ for $n \geq 0$ and express the existence of a neighborhood s that supports φ and is compatible with each ψ_i.

We will explore a different way to set up a language that can express the relevant properties. Our approach uses a modal conditional $\Rrightarrow$, where a modal formula $\varphi \Rrightarrow \psi$ is true at a world if all neighborhoods supporting φ also support ψ. Crucially, however, φ and ψ are allowed to be questions, and for this reason, the relevant modal conditional is added to an underlying inquisitive logic. In our logic, the notion of support at a set of worlds is taken as primitive, while still boiling down to global truth for a large class of formulas. Thus, e.g., our language will include a formula $?p$ which is supported by a set of worlds s just in case s settles whether p, i.e., if the truth value of p is constant in s. Then the fact that every neighborhood settles whether p can be expressed as $\top \Rrightarrow ?p$ (true at w_1 but not at w_2, w_3), whereas the fact that every neighborhood that settles whether p also settles whether q can be expressed as $?p \Rrightarrow ?q$ (true at w_1, w_2 but not at w_3). One attraction of this approach is that the modal implication $\Rrightarrow$ is very natural both conceptually and formally. Its logic is completely axiomatized by the following four properties, familiar from the study of strict conditionals: (i) if φ entails ψ, then $\varphi \Rrightarrow \psi$ is a logical truth; (ii) $\Rrightarrow$ is transitive; (iii) if φ implies two formulas, it implies their conjunction; (iv) if ψ is implied by two formulas, it is implied by their (inquisitive) disjunction.

A related but more concrete take on our logic arises from thinking about action and ability. Consider three scenarios in which an agent a is deliberating whether to make p and q true or false. In the first scenario, a must make a choice about each of p and q. In the second scenario, a can make both choices, or she can delegate both choices to another agent. In the third scenario, a can make both choices, or make the choice about p and delegate the one about q, or she can delegate both choices. The three scenarios correspond to the pictures in Fig. 1, where each neighborhood corresponds to an action available to a, and the worlds in a neighborhood represent the possible outcomes of that action.

Standard analyses of strategic ability [1,3,4,7,25] focus on the powers of agents to force certain outcomes. From the perspective of these analyses, these situations are the same: in all of them, a is in effect a dictator who can force any of the outcomes, while other agents cannot prevent any outcome. Yet, there is a clear sense in which these situations are very different. In the first scenario, a not only *can*, but also *must* decide on p and q, while in the other two scenarios, she can delegate (which may well be the optimal action—perhaps the other agent is in a better position to make the relevant decisions). Moreover, in the second scenario, a must decide on q if she wants to decide on p, whereas in the third scenario she can can choose on p while delegating the choice on q. Thus, while these situations are the same in terms of what the agent *can* force, they differ in terms of what she *must* force, or what she must force if she wants to force something else. The logic that we develop in this paper will allow us to express these significant facts in a natural way. So, for instance, the fact that the agent is forced to choose on p is expressed by $\top \Rrightarrow ?p$, while the fact that she is forced to choose on q if she wants to choose on p is expressed by $?p \Rrightarrow ?q$.

A third way to view the logic is as an extension of previous work on inquisitive modal logic [9,10,15,17,22,26]. In that work, formulas are interpreted over *downward-closed* neighborhood models, i.e., models where the set of neighborhoods is closed under subsets; the key modality, denoted $\boxplus$, is a universal quantifier over neighborhoods. The language with this modality was shown to be expressively adequate for the appropriate notion of bisimulation in [13]. But while the downward closure constraint is well-motivated under some interpretations of the model (as in inquisitive epistemic logic, [15]), it may not be under other interpretations. This paper thus generalizes inquisitive modal logic by dropping the downward closure requirement. As we show, in this more general setting the universal modality $\boxplus$ is no longer sufficient to have an expressively adequate language. Instead, we obtain an adequate language by taking the modal conditional $\Rrightarrow$ as primitive (one can then define $\boxplus\varphi$ as $(\top \Rrightarrow \varphi)$).

The paper is structured as follows: in Section 2 we introduce and illustrate the logic InqCM; in Section 3 we introduce the notion of bisimulation for which our language is invariant and show a Hennessy-Milner theorem; in Section 4 we describe a Hilbert-style proof system for InqCM; in Section 5 we adapt the standard notion of resolutions and state some useful technical lemmas; in Section 6 we relate a fragment of InqCM to instantial neighborhood logic by defining truth-preserving translations in both directions; in Section 7 we show the completeness of our proof system; finally, in Section 8 we discuss a number of directions in which the present work can be extended.

2 Inquisitive neighborhood logic

In this section we introduce our logic InqCM. Given a set $\mathcal{P}$ of atomic sentences, the language $\mathcal{L}$ of the logic is given by the following BNF definition:

$$\mathcal{L} \qquad \varphi ::= p \mid \bot \mid (\varphi \wedge \varphi) \mid (\varphi \to \varphi) \mid (\varphi \vee\!\!\!\vee \varphi) \mid (\varphi \Rrightarrow \varphi) \qquad\qquad p \in \mathcal{P}$$

As standard in inquisitive logic, we also use the following defined connectives: $\neg\varphi := (\varphi \to \bot)$, $\top := \neg\bot$, $\varphi \vee \psi := \neg(\neg\varphi \wedge \neg\psi)$, $?\varphi := \varphi \vee\!\!\!\vee \neg\varphi$.

The connective $\vee\!\!\!\vee$, called *inquisitive disjunction*, is thought of as a question-forming disjunction. Thus, e.g., whereas $p \vee \neg p$ stands for the tautological statement that p or $\neg p$ is the case, $p \vee\!\!\!\vee \neg p$ (denoted $?p$) stands for the polar question whether p or $\neg p$ is the case, i.e., whether p is true or false.

The propositional fragment of $\mathcal{L}$, consisting of formulas without the binary modality $\Rrightarrow$, is just inquisitive propositional logic [10,14]. Propositional formulas without $\vee\!\!\!\vee$ can be simply identified with standard propositional formulas.

As anticipated in the introduction, we interpret this language over neighborhood models, with the only constraint that neighborhoods be nonempty.

Definition 2.1 An *inhabited neighborhood model* (*in-model* for short) is a tuple $M = \langle W, \Sigma, V\rangle$ where W is a nonempty set, $\Sigma : W \to \wp\wp(W)$ is a map such that $\emptyset \notin \Sigma(w)$ for all $w \in W$, and $V : \mathcal{P} \to \wp(W)$ is a valuation function. We refer to elements $w \in W$ as *worlds* or *points*, to subsets $s \subseteq W$ as *information states* or simply *states*, and to the elements $s \in \Sigma(w)$ as the *neighborhoods* of w.

As mentioned in the introduction, one possible interpretation of these models is as follows. Worlds $w \in W$ are stages of a dynamic process unfolding over time. At each stage w, the relevant agent has at her disposal a set of actions, each of which may produce a range of different outcomes depending on factors such as randomness and the choices of other agents. Each neighborhood $s \in \Sigma(w)$ collects all the outcomes that may result from a given action. The constraint $\emptyset \notin \Sigma(w)$ captures the idea that an action must allow at least one outcome.

To interpret statements like p and questions like $?p$ uniformly, the semantics is not given by a recursive definition of truth at a world, but by a recursive definition of *support* at an information state, as customary in inquisitive logic.

Definition 2.2 [Support] Given an in-model M and a state $s \subseteq W$ we define:

- $M, s \models p \iff s \subseteq V(p)$
- $M, s \models \bot \iff s = \emptyset$
- $M, s \models \varphi \wedge \psi \iff M, s \models \varphi$ and $M, s \models \psi$
- $M, s \models \varphi \vee\!\!\!\vee \psi \iff M, s \models \varphi$ or $M, s \models \psi$
- $M, s \models \varphi \to \psi \iff \forall t \subseteq s : M, t \models \varphi$ implies $M, t \models \psi$
- $M, s \models \varphi \Rrightarrow \psi \iff \forall w \in s\, \forall t \in \Sigma(w) : M, t \models \varphi$ implies $M, t \models \psi$

When no confusion arises, we suppress reference to M and write simply $s \models \varphi$. We assume that the propositional clauses are familiar from previous work in inquisitive logic (see, e.g., [10,12]). The novelty is the presence of the modality

$\Rrightarrow$. Before discussing it in detail, it will be useful to recall some standard facts and notions. First, the following standard properties of inquisitive logics hold:

- Persistency: if $M, s \models \varphi$ and $t \subseteq s$ then $M, t \models \varphi$;
- Empty state property: $M, \emptyset \models \varphi$ for all $\varphi \in \mathcal{L}$.

Logical entailment and equivalence are defined in terms of support, as follows.

Definition 2.3 Let $\Phi \cup \{\psi\} \subseteq \mathcal{L}$. We say that Φ *entails* ψ, denoted $\Phi \models \psi$, if for all in-models M and states s, if $M, s \models \varphi$ for all $\varphi \in \Phi$, then $M, s \models \psi$. We say that ψ is *valid* if $\emptyset \models \psi$. We say that $\varphi, \psi \in \mathcal{L}$ are *logically equivalent*, denoted $\varphi \equiv \psi$, if φ and ψ are supported by the same states in every in-model.

Although the semantics is defined in terms of support, a notion of truth at a world is recovered by defining $M, w \models \varphi$ as $M, \{w\} \models \varphi$. One can then check that all standard connectives have their classical truth-functional behavior, for instance $M, w \models \neg\varphi \iff M, w \not\models \varphi$. In particular, this means that all standard (i.e., $\vee\!\!\!\vee$-free) propositional formulas have their usual truth conditions. The set of worlds where φ is true is denoted $|\varphi|_M$, or simply $|\varphi|$ if M is clear.

For many formulas in the language, support at a state s simply boils down to truth at each world in s. We call such formulas *truth-conditional.*

Definition 2.4 A formula $\varphi \in \mathcal{L}$ is truth-conditional if for all in-models M and states $s \subseteq W$ we have $M, s \models \varphi \iff \forall w \in s : M, w \models \varphi$.

We now define a set $\mathcal{L}_!$ of declarative formulas (or *declaratives*) as follows:

$$\mathcal{L}_! \qquad \alpha ::= p \mid \bot \mid (\alpha \wedge \alpha) \mid (\alpha \to \alpha) \mid (\varphi \Rrightarrow \varphi) \qquad\qquad p \in \mathcal{P}, \varphi \in \mathcal{L}$$

In words, a formula α is a declarative if all occurrences of $\vee\!\!\!\vee$ in α are within the scope of the modality $\Rrightarrow$ (either in the 'antecedent' or in the 'consequent'). In the following, we will use $\alpha, \beta, \gamma, \delta$ as meta-variables ranging over $\mathcal{L}_!$, while we use φ, ψ, χ for arbitrary formulas in $\mathcal{L}$.

Up to logical equivalence, $\mathcal{L}_!$ is exactly the truth-conditional fragment of $\mathcal{L}$.

Proposition 2.5 *Every $\alpha \in \mathcal{L}_!$ is truth-conditional. Conversely, if $\varphi \in \mathcal{L}$ is truth-conditional, $\varphi \equiv \alpha$ for some $\alpha \in \mathcal{L}_!$.*

Proof. The first part of the statement is proved by a plain induction on $\alpha \in \mathcal{L}_!$. For the second part, given any $\varphi \in \mathcal{L}$, let φ^c be the declarative obtained by replacing any occurrence of $\vee\!\!\!\vee$ which is not in the scope of $\Rrightarrow$ by $\vee$. A simple induction shows that φ and φ^c are true at the same worlds in any model. If φ is truth-conditional, then since φ and φ^c are both truth-conditional with the same truth conditions, they are equivalent. □

Note that all standard (i.e., $\vee\!\!\!\vee$-free) propositional formulas are declaratives and thus truth-conditional. For these formulas, our support semantics is equivalent to the standard truth-conditional semantics. For an example of a formula which is *not* truth-conditional, take $?p$. This is supported if the truth value of p is the same is all worlds in s, that is, we have $M, s \models ?p \iff s \subseteq |p|$ or $s \cap |p| = \emptyset$.

With these preliminaries in place, let us now turn to the modality $\Rrightarrow$. This modality can be applied to $\varphi, \psi \in \mathcal{L}$ to produce the modal formula $\varphi \Rrightarrow \psi$. This formula is a declarative and so truth-conditional by Proposition 2.5. Hence, to understand its semantics it suffices to focus on its truth conditions, which are:

$$M, w \models \varphi \Rrightarrow \psi \iff \forall t \in \Sigma(w) : M, t \models \varphi \text{ implies } M, t \models \psi$$

In words, $\varphi \Rrightarrow \psi$ is true at a world w if every neighborhood of w that supports φ also supports ψ. To appreciate what can be expressed by modal formulas of this form, it is helpful to consider a number of special cases.

First, suppose α, β are declaratives. By Proposition 2.5, the formula $\alpha \Rrightarrow \beta$ expresses a kind of global consequence from the perspective of the world w: for any neighborhood s, if α is true everywhere in s, then β is true everywhere in s. More generally, if $\alpha, \beta_1, \ldots, \beta_n$ are declaratives, $\alpha \Rrightarrow (\beta_1 \mathbin{\rotatebox[origin=c]{-90}{$\geqslant$}} \cdots)$ expresses a kind of multiple-conclusion global consequence: for any neighborhood s of w, if α is true everywhere in s then some β_i is true everywhere in s:

$$w \models (\alpha \Rrightarrow \bigvee\!\!\!\!\bigvee_{i \leq n} \beta_i) \iff \forall s \in \Sigma(w) : \text{if } s \subseteq |\alpha| \text{ then } (s \subseteq |\beta_i| \text{ for some } i)$$

Note that, as a consequence, the negation $\neg(\alpha \Rrightarrow (\neg\beta_1 \mathbin{\mathbb{V}} \cdots \mathbin{\mathbb{V}} \neg\beta_n))$ expresses the existence of a neighborhood s such that α is true everywhere in s and for each $i \leq n$, β_i is true somewhere in s. This is exactly the meaning of a modal formula of the form $\Box(\beta_1, \ldots, \beta_n; \alpha)$ in instantial neighborhood semantics [6]; we will come back to the connection in Section 6. As another example, consider the formula $(?p \Rrightarrow ?q)$. This expresses the fact that any neighborhood that settles whether p also settles whether q, in symbols:

$$w \models (?p \Rrightarrow ?q) \iff \forall s \in \Sigma(w) : \text{if } (s \subseteq |p| \text{ or } s \subseteq \overline{|p|}) \text{ then } (s \subseteq |q| \text{ or } s \subseteq \overline{|q|})$$

Looking at Figure 1, this formula is true at worlds w_1 and w_2, but not at w_3.

An interesting feature of the language is the interplay of the two conditionals $\Rrightarrow$ and $\to$, that can be used to specify global and local restriction respectively: while $\Rrightarrow$ allows us to restrict the class of neighborhoods under consideration, $\to$ allows us to restrict the worlds in each neighborhood. Consider, e.g., the formula $(p \Rrightarrow (q \to ?r))$. This expresses the following fact: if we restrict to those neighborhoods that support p and then we restrict each of these neighborhoods to the q-worlds, all the resulting states settle whether r:

$$w \models (p \Rrightarrow (q \to ?r)) \iff \forall s \in \Sigma(w) : s \subseteq |p| \text{ implies } s \cap |q| \models ?r$$

Similarly, the formula $(?p \Rrightarrow (?q \to ?r))$ expresses that if we restrict to neighborhoods that settle whether p, and then look at the parts of such neighborhoods where the truth value of q is settled, each of these parts settles whether r.

$$w \models (?p \Rrightarrow (?q \to ?r)) \iff \forall s \in \Sigma(w) : (s \subseteq |p| \text{ or } s \subseteq \overline{|p|}) \text{ implies } (s \cap |q| \models ?r \text{ and } s \cap \overline{|q|} \models ?r)$$

Another interesting observation is that from the binary modality $\Rrightarrow$, we can define two unary modalities as follows:

$$\boxplus\varphi := (\top \Rrightarrow \varphi) \qquad \Diamond\!\!\!+\varphi := \neg(\varphi \Rrightarrow \bot)$$

Since all neighborhoods of a point support $\top$ and none support $\bot$ (note: here we use the requirement that neighborhoods are nonempty!) we find that $\boxplus$ and $\diamondplus$ are, respectively, a universal and an existential modality over neighborhoods:

$$w \models \boxplus\varphi \iff \forall s \in \Sigma(w) : s \models \varphi \qquad w \models \diamondplus\varphi \iff \exists s \in \Sigma(w) : s \models \varphi$$

In particular, if $\alpha \in \mathcal{L}_!$ then $\diamondplus\alpha$ is true at a world w if there is some neighborhood s of w such that α is true everywhere in s; as discussed in the introduction, this is the sort of fact expressed by modal formulas in a standard version of neighborhood semantics. When $\alpha \in \mathcal{L}_!$, the formula $\boxplus\alpha$ is not especially interesting: it says that α is true throughout every neighborhood of the given point w, which simply means that α is true everywhere in the set $R_\Sigma(w) = \bigcup\Sigma(w)$. However, the modality $\boxplus$ becomes very interesting when applied to a question. For instance, consider the formula $\boxplus ?p$: this is true if the truth value of p is settled, one way or the other, in every neighborhood of the given world. Looking at Figure 1, this formula is true at world w_1, but false at w_2 and w_3. Previous work on inquisitive modal logic [9,10,13,15,17,22,26] focused on a language where $\boxplus$ is the main primitive modality.[3] However, the semantics was based on models that are downward closed in the following sense.[4]

Definition 2.6 Let $M = \langle W, \Sigma, V\rangle$ be an in-model. Its downward-closure is $M^\downarrow = \langle W, \Sigma^\downarrow, V\rangle$, where $\Sigma^\downarrow(w) = \{t \subseteq W \mid t \neq \emptyset \text{ and } t \subseteq s \text{ for some } s \in \Sigma(w)\}$. We say that M is *downward closed* if $M = M^\downarrow$.

Now consider the language $\mathcal{L}_\boxplus$ given by the following BNF definition:

$$\mathcal{L}_\boxplus \qquad \eta ::= p \mid \bot \mid (\eta \wedge \eta) \mid (\eta \to \eta) \mid (\eta \mathbin{\vee\!\!\!\vee} \eta) \mid \boxplus\eta \qquad\qquad p \in \mathcal{P}$$

In the context of downward-closed models, $\mathcal{L}_\boxplus$ is equi-expressive with $\mathcal{L}$. This is a consequence of the following proposition (the easy proof is left as an exercise).

Proposition 2.7 *If M is a downward-closed in-model then for any $s \subseteq W$, $M, s \models (\varphi \Rrightarrow \psi) \iff M, s \models \boxplus(\varphi \to \psi)$.*

In general, however, $\mathcal{L}_\boxplus$ is strictly less expressive than $\mathcal{L}$. Indeed, formulas of $\mathcal{L}_\boxplus$ cannot distinguish a model from its downward closure (cf. Prop. 2.6 in [22]).

Proposition 2.8 *Let $\eta \in \mathcal{L}_\boxplus$. Then for any M, s: $M, s \models \eta \iff M^\downarrow, s \models \eta$.*

Proof. By induction on η. The key case is the one for $\eta = \boxplus\theta$. Suppose $M, s \models \boxplus\theta$. Take any $w \in s$ and any state $t \in \Sigma^\downarrow(w)$. Then $t \subseteq t'$ for some $t' \in \Sigma(w)$. Since $M, s \models \boxplus\theta$ and $w \in s$ we have $M, t' \models \theta$ and so by persistency $M, t \models \theta$, which by induction hypothesis gives $M^\downarrow, t \models \theta$. This shows that $M^\downarrow, s \models \boxplus\theta$. The converse implication is immediate since $\Sigma(w) \subseteq \Sigma^\downarrow(w)$. □

---

[3] Normally, a second primitive modality $\Box$ is considered as well, but at least in propositional modal logic, it can be removed from the language without loss of expressive power (see [10]).

[4] A subtlety: in previous work $\Sigma(w)$ is required to always contain $\emptyset$, whereas here we require $\Sigma(w)$ to never contain $\emptyset$. However, the presence or absence of $\emptyset$ in $\Sigma(w)$ is immaterial to the semantics of $\mathcal{L}$ and $\mathcal{L}_\boxplus$ by the empty state property. Thus, the models considered in previous work can be safely identified with their counterparts where $\emptyset$ is removed from all $\Sigma(w)$.

It is now easy to see that many $\varphi \in \mathcal{L}$ are not equivalent to any $\eta \in \mathcal{L}_{\boxplus}$. For instance, consider a model M with $W = \{v, v'\}$, $\Sigma(v) = \{W\}$ and $V(p) = \{v\}$. Then $M, v \not\models \oplus p$ but $M^{\downarrow}, v \models \oplus p$. Thus, $\oplus p$ is not equivalent to any $\eta \in \mathcal{L}_{\boxplus}$.[5]

3 Bisimulation and expressive power

In a neighborhood model, it is natural to regard two worlds as equivalent if they agree on atomic sentences and every neighborhood of the one is equivalent to some neighborhood of the other. In turn, it is natural to regard two neighborhoods as equivalent if every world in the one is equivalent to some world in the other. This leads naturally to the following notion of bisimulation.

Definition 3.1 [(cf. [6,13])] Given models $M = \langle W, \Sigma, V \rangle, M' = \langle W', \Sigma', V' \rangle$, a relation $Z \subseteq W \times W'$ is a *bisimulation* if whenever wZw' holds we have:

Atoms: $w \in V(p) \iff w' \in V'(p)$ for all $p \in \mathcal{P}$;

Forth: $\forall s \in \Sigma(w)\, \exists s' \in \Sigma'(w')$ with $s\overline{Z}s'$;

Back: $\forall s' \in \Sigma'(w')\, \exists s \in \Sigma(w)$ with $s\overline{Z}s'$.

Here, $\overline{Z}$ is the Egli-Milner lifting of Z, that is, $s\overline{Z}s'$ holds just in case we have:

$$(\forall w \in s\, \exists w' \in s' : wZw') \text{ and } (\forall w' \in s'\, \exists w \in s : wZw')$$

Two worlds w, w' are *bisimilar* ($w \sim w'$) if there is a bisimulation Z with wZw'; two states s, s' are bisimilar ($s \sim s'$) if there is a bisimulazion Z with $s\overline{Z}s'$.

Bisimilarity can be characterized in terms of a game with two players, S(poiler) and D(uplicator). The game alternates between world-positions $\langle w, w' \rangle \in W \times W'$ and state-positions $\langle s, s' \rangle \in \wp(W) \times \wp(W')$. Playing from a world position, S picks a neighborhood of either world and D responds with a neighborhood of the other world, leading to a state position. Playing from a state position, S picks a world in either state and D responds with a world in the other state. If D is unable to make a move, or if a world-position is reached where the worlds disagree on some atomic sentence, S wins; in all other cases, D wins. Then worlds w, w' (respectively, states s, s') are bisimilar iff D has a winning strategy in the game starting from position $\langle w, w' \rangle$ (respectively, $\langle s, s' \rangle$).

This notion of bisimulation is especially useful since it allows for standard model-theoretic constructions like disjoint unions and (partial) tree unfoldings (see [6] for the details), while preserving the satisfaction of InqCM formulas.

Definition 3.2 [Modal equivalence] Given models M, M', we say that two states $s \subseteq W, s' \subseteq W'$ are *modally equivalent* ($s \leftrightsquigarrow s'$) if they support the same formulas of InqCM. Similarly, two worlds $w \in W$ and $w' \in W'$ are modally equivalent ($w \leftrightsquigarrow w'$) if they make the same formulas true.

---

[5] There is also a tight relation between our operator $\Rrightarrow$ and the inquisitive strict conditional $\Rightarrow$ studied in [11]. With every Kripke model $M = \langle W, R, V \rangle$ we can associate a neighborhood model $M^n = \langle W, \Sigma_R, V \rangle$ where $\Sigma_R(w) = \{s \subseteq R[w] \mid s \neq \emptyset\}$. Then the semantics of a formula $\varphi \Rightarrow \psi$ in the setting of M coincides with the semantic of $\varphi \Rrightarrow \psi$ in M^n, and the logic $\mathsf{inq}^{\Rightarrow}$ of [11] can be viewed as the special case of InqCM over a restricted class of models.

Proposition 3.3 (Invariance under bisimulation) *For any worlds w, w', $w \sim w'$ implies $w \leftrightsquigarrow w'$; for any states s, s', $s \sim s'$ implies $s \leftrightsquigarrow s'$.*

The proof is straightforward; we include it in Appendix A for completeness.

Conversely, we can show a Hennessy-Milner theorem. For an in-model M, we say M is *image-finite* if the set $\bigcup \Sigma(w)$ is finite for every $w \in W$. In such models, modal equivalence implies bisimilarity for worlds and for finite states.

Theorem 3.4 *If M, M' are image-finite, then for any worlds w, w', $w \leftrightsquigarrow w'$ implies $w \sim w'$, and for all finite states s, s', $s \leftrightsquigarrow s'$ implies $s \sim s'$.*

Proof. We will show that the modal equivalence relation $\leftrightsquigarrow$ on worlds is a bisimulation. Suppose $w \leftrightsquigarrow w'$. Clearly, w and w' satisfy the same atoms. We show that the Forth condition is satisfied (the proof for Back is analogous). Take a state $s \in \Sigma(w)$. By image-finiteness we can write $s = \{w_1, \ldots, w_n\}$ and $\bigcup \Sigma'(w') = \{v_1, \ldots, v_m\}$. Note that $n \geq 1$ since neighborhoods are required to be nonempty, and $m \geq 1$ since if $m = 0$ we would have $\Sigma'(w') = \emptyset$, in which case w and w' are distinguished by the formula $\top \Rrightarrow \bot$. Now for each i, j, if $w_i \leftrightsquigarrow v_j$ we define $\delta_{ij} = \top$, while if $w_i \not\leftrightsquigarrow v_j$ we let δ_{ij} be a declarative such that $M, w_i \models \delta_{ij}$ and $M, v_j \models \neg\delta_{ij}$. Note that such a declarative exists: if $w_i \not\leftrightsquigarrow v_j$, then w_i and v_j disagree about the truth of some formula $\xi \in \mathcal{L}$. Now ξ has the same truth conditions as the declarative ξ^c defined as in the proof of Proposition 2.5, so w_i and v_i disagree on the truth of ξ^c. We can choose $\delta = \xi^c$ if $w_i \models \xi^c$ and $\delta = \neg\xi^c$ otherwise.

Let $\gamma_i := \bigwedge_{j \leq m} \delta_{ij}$. For all $j \leq m$ we have $M', v_j \models \gamma_i \iff w_i \leftrightsquigarrow v_j$. Now consider the formula:

$$\varphi := (\bigvee_{i=1}^{n} \gamma_i) \Rrightarrow (\mathbb{V}_{i=1}^{n} \neg\gamma_i)$$

It is easy to check that s supports the antecedent of φ but not the consequent, so $M, w \not\models \varphi$. Since $w \leftrightsquigarrow w'$, we have that $M', w' \not\models \varphi$. So there is a $s' \in \Sigma'(w')$ that supports the antecedent of φ but not the consequent. We are now going to show that every world in s is modally equivalent to a world in s' and vice versa, thus showing that the Forth condition holds for $\leftrightsquigarrow$. Since $s' \models \bigvee_{i \leq n} \gamma_i$, by persistency every world $v \in s'$ satisfies some formula γ_i and thus is modally equivalent to some world in s, namely w_i. For the converse, take a world $w_i \in s$. Since s' does not support the consequent of φ we have $s' \not\models \neg\gamma_i$. Since $\neg\gamma_i$ is a declarative and thus truth-conditional, there is a world $v \in s'$ such that $v \not\models \neg\gamma_i$ and so $v \models \gamma_i$, which means that v is modally equivalent to w_i.

This proves the claim for worlds. For the claim about states, let s, s' be finite states with $s \leftrightsquigarrow s'$. Since $\leftrightsquigarrow$ on worlds is a bisimulation, it suffices to show that every world in s is modally equivalent to a world in s' and vice versa. So, take $w \in s$. Proceeding as above, we can define a declarative γ_w such that $w \models \gamma_w$ and for all $v \in s'$ we have $v \models \gamma_w \iff w \leftrightsquigarrow v$. Now since $w \in s$, by persistency we have $s \not\models \neg\gamma_w$ and since $s \leftrightsquigarrow s'$ also $s' \not\models \neg\gamma_w$. Since $\neg\gamma_w$ is a declarative and thus truth-conditional, there is some $w' \in s'$ with $w' \not\models \neg\gamma_w$ and so $w' \models \gamma_w$, which implies $w \leftrightsquigarrow w'$. The converse is proved analogously. □

It is not hard to construct examples showing that the result fails if M, M' are not image-finite, or if s, s' are infinite states (even within image-finite models).

We can also introduce notions of n-step bisimilarity and show that, if $\mathcal{P}$ is finite, two worlds/states are n-step bisimilar iff they cannot be distinguished by a formula of modal depth up to n. We omit the details due to space constraints.

4 Axiomatization

In this section we describe a Hilbert-style proof system for InqCM. The propositional basis for the system consists of all instances of the axioms for intuitionistic propositional logic, with $\mathbin{\rotatebox[origin=c]{90}{$\geqslant$}}$ in the role of intuitionistic disjunction, and all instances of the following two axioms, where $\alpha \in \mathcal{L}_!$ and $\varphi, \psi \in \mathcal{L}$:

(DDN) $\quad \neg\neg\alpha \to \alpha$

(Split) $\quad (\alpha \to \varphi \mathbin{\rotatebox[origin=c]{90}{$\geqslant$}} \psi) \to (\alpha \to \varphi) \mathbin{\rotatebox[origin=c]{90}{$\geqslant$}} (\alpha \to \psi)$

An explicit list of the propositional axiom schemata is included in Appendix B. We then have three axiom schemata for $\Rrightarrow$, namely:

(Tran) $\quad (\varphi \Rrightarrow \psi) \wedge (\psi \Rrightarrow \chi) \to (\varphi \Rrightarrow \chi)$

(Conj) $\quad (\varphi \Rrightarrow \psi) \wedge (\varphi \Rrightarrow \chi) \to (\varphi \Rrightarrow (\psi \wedge \chi))$

(Disj) $\quad (\varphi \Rrightarrow \chi) \wedge (\psi \Rrightarrow \chi) \to ((\varphi \mathbin{\rotatebox[origin=c]{90}{$\geqslant$}} \psi) \Rrightarrow \chi)$

The inference rules are modus ponens and a conditional version of necessitation:

$$\frac{\varphi \quad \varphi \to \psi}{\psi}\ (\mathtt{MP}) \qquad\qquad \frac{\varphi \to \psi}{\varphi \Rrightarrow \psi}\ (\mathtt{CN})$$

If $\Phi, \Psi \subseteq \mathcal{L}$, we write $\Phi \vdash \Psi$ if there are $\varphi_1, \ldots, \varphi_n \in \Phi$ and $\psi_1, \ldots, \psi_m \in \Psi$ such that the formula $\varphi_1 \wedge \cdots \wedge \varphi_n \to \psi_1 \mathbin{\rotatebox[origin=c]{90}{$\geqslant$}} \cdots \mathbin{\rotatebox[origin=c]{90}{$\geqslant$}} \psi_m$ is derivable in the system (we allow for $n = 0$, in which case the relevant antecedent is $\top$, and for $m = 0$, in which case the consequent is $\bot$). We write $\varphi_1, \ldots, \varphi_n \vdash \psi_1, \ldots, \psi_m$ instead of $\{\varphi_1, \ldots, \varphi_n\} \vdash \{\psi_1, \ldots, \psi_m\}$, and we write $\varphi \dashv\vdash \psi$ in case $\varphi \vdash \psi$ and $\psi \vdash \varphi$. As usual, soundness is proved by checking that the axioms are valid and the inference rules preserve validity. We omit the straightforward proof.

Proposition 4.1 (Soundness) *For all $\Phi \cup \{\psi\} \subseteq \mathcal{L}$, $\Phi \vdash \psi$ implies $\Phi \models \psi$.*

The completeness of the system will be shown in Section 7. For that, we need to introduce one more notion which is standard in inquisitive logic: resolutions.

5 Resolutions

Following a standard recipe, we will associate to each $\varphi \in \mathcal{L}$ a finite nonempty set of declaratives $\mathcal{R}(\varphi) \subseteq \mathcal{L}_!$ such that φ is equivalent to $\mathbin{\rotatebox[origin=c]{90}{$\geqslant$}}\mathcal{R}(\varphi)$.

Definition 5.1 [Resolutions (cf. Definition 7.1.24 in [10])]

- $\mathcal{R}(\alpha) = \{\alpha\}$ if α is an atom, $\bot$, or a modal formula $(\varphi \Rrightarrow \psi)$
- $\mathcal{R}(\varphi \wedge \psi) = \{\alpha \wedge \beta \mid \alpha \in \mathcal{R}(\varphi), \beta \in \mathcal{R}(\psi)\}$
- $\mathcal{R}(\varphi \mathbin{\rotatebox[origin=c]{90}{$\geqslant$}} \psi) = \mathcal{R}(\varphi) \cup \mathcal{R}(\psi)$

- $\mathcal{R}(\varphi \to \psi) = \{\bigwedge_{\alpha \in \mathcal{R}(\varphi)}(\alpha \to f(\alpha)) \mid f : \mathcal{R}(\varphi) \to \mathcal{R}(\psi)\}$

It is immediate to verify that for every declarative $\alpha \in \mathcal{L}_!$ we have $\mathcal{R}(\alpha) = \{\alpha\}$.

The following provable normal form result uses only the propositional component of the proof system, and can thus be proved in exactly the same way as for inquisitive propositional logic (see Lemma 3.3.4 in [10] for the details).

Lemma 5.2 *For all $\varphi \in \mathcal{L}$, $\varphi \dashv\vdash \bigvee\!\!\!\!\bigvee \mathcal{R}(\varphi)$.*

As an immediate corollary we have the following fact.

Lemma 5.3 *For all $\varphi \in \mathcal{L}$ and all $\alpha \in \mathcal{R}(\varphi)$, $\alpha \vdash \varphi$.*

The next lemma says that if a set of declaratives derives a formula, it derives a particular resolution of it (note that the converse holds by the previous lemma). The proof is again standard (see, e.g., Lemma 5.6 in [11]), but we include it in Appendix C for the sake of completeness.

Lemma 5.4 *Let $\Gamma \subseteq \mathcal{L}_!$ and $\varphi \in \mathcal{L}$. If $\Gamma \vdash \varphi$, then $\Gamma \vdash \alpha$ for some $\alpha \in \mathcal{R}(\varphi)$.*

We will also need a notion of resolutions for sets of formulas. A resolution of a set Φ is a set obtained by replacing each element $\varphi \in \Phi$ by a resolution of it.

Definition 5.5 Given $\Phi \subseteq \mathcal{L}$, a resolution function for Φ is a function f that associates to each $\varphi \in \Phi$ some $f(\varphi) \in \mathcal{R}(\varphi)$. A resolution of Φ is the image of Φ under a resolution function: $\mathcal{R}(\Phi) = \{f[\Phi] \mid f \text{ a resolution function for } \Phi\}$.

Note that if Γ is a set of declaratives then $\mathcal{R}(\Gamma) = \{\Gamma\}$.

The next lemma says that if Φ fails to derive Ψ, then Φ can be strengthened to a resolution $\Gamma \in \mathcal{R}(\Phi)$ that still fails to derive Ψ. The proof is again standard. Since it uses only Lemma 5.2 and the propositional axioms for $\bigvee\!\!\!\!\bigvee$, we omit it and refer, e.g., to pp. 87-88 in [10], where the argument is given in detail.

Lemma 5.6 *For all $\Phi, \Psi \subseteq \mathcal{L}$, if $\Phi \nvdash \Psi$ then there is $\Delta \in \mathcal{R}(\Phi)$ s.t. $\Delta \nvdash \Psi$.*

With these standard results at hand, we are now ready to prove the completeness of our system. Before turning to that, however, we will make use of resolutions to relate InqCM more precisely to instantial neighborhood logic.

6 Translating to and from instantial neighborhood logic

As discussed in the introduction, instantial neighborhood logic (INL, [6]) is a modal language interpreted on neighborhood structures which is invariant under the notion of bisimulation that we discussed in Section 3. The language $\mathcal{L}_{\mathsf{INL}}$ of INL is a modal language with primitive connectives $\neg$ and $\wedge$, and where modal formulas have the form $\Box(\rho_1, \ldots, \rho_n; \sigma)$ with $n \geq 0$. The semantics is given by a standard definition of truth at a world, where the modal clause is:

$$M, w \models \Box(\rho_1, \ldots, \rho_n; \sigma) \iff \exists s \in \Sigma(w) : (\forall v \in s : M, v \models \sigma) \text{ and } (\forall i \leq n \exists v \in s : M, v \models \rho_i)$$

We will show that over inhabited neighborhood models, INL has the same expressive power as the declarative fragment of InqCM. We prove this by defining two translations that preserve truth conditions. We define a translation

$(\cdot)^* : \mathcal{L}_{\mathsf{INL}} \to \mathcal{L}_!$ by letting $p^* = p$, $(\neg\sigma)^* = \neg\sigma^*$, $(\rho \wedge \sigma)^* = \rho^* \wedge \sigma^*$ and, crucially,

$$\Box(\rho_1, \dots, \rho_n; \sigma)^* = \neg(\sigma^* \Rrightarrow (\neg\rho_1^* \vee\!\!\!\vee \cdots \vee\!\!\!\vee \neg\rho_n^*))$$

As a straightforward induction shows, this map preserves truth conditions.

Proposition 6.1 *Let $\sigma \in \mathcal{L}_{\mathsf{INL}}$. For every in-model M and every world w, $M, w \models \sigma \iff M, w \models \sigma^*$.*

Translating declaratives of InqCM to INL is more tricky. Take a modal formula $(\varphi \Rrightarrow \psi)$: in general, φ and ψ are not declaratives, so the translation will not be defined on them. Instead, we first compute the resolutions of φ and ψ, and then assemble a translation from the translations of these. To make this precise, we need a non-standard notion of complexity. Given $\alpha, \beta \in \mathcal{L}_!$, we let $\alpha \prec \beta$ in case either α has lower modal depth than β, or α and β have the same modal depth and α is a subformula of β. Clearly, $\prec$ is well-founded and thus suitable for induction. Now we define $(\cdot)^\star : \mathcal{L}_! \to \mathcal{L}_{\mathsf{INL}}$ recursively on $\prec$ as follows: $p^\star = p$; $\bot^\star = (p \wedge \neg p)$ for an arbitrary $p \in \mathcal{P}$; $(\alpha \wedge \beta)^\star = \alpha^\star \wedge \beta^\star$; $(\alpha \to \beta)^\star = \neg(\alpha^\star \wedge \neg\beta^\star)$; and finally:

$$(\varphi \Rrightarrow \psi)^\star = \bigwedge_{i=1}^{n} \neg\Box(\neg\beta_1^\star, \dots, \neg\beta_m^\star; \alpha_i^\star)$$

where $\{\alpha_1, \dots, \alpha_n\} = \mathcal{R}(\varphi)$ and $\{\beta_1, \dots, \beta_m\} = \mathcal{R}(\psi)$. Note that $\alpha_i^\star$ is defined since $\alpha_i \prec (\varphi \Rrightarrow \psi)$: this is because α_i has the same modal depth as φ, which is lower that the modal depth of $(\varphi \Rrightarrow \psi)$. A similar argument goes for β_j.

Proposition 6.2 *Let $\alpha \in \mathcal{L}_!$ be any declarative in InqCM. For any in-model M and world w we have $M, w \models \alpha \iff M, w \models \alpha^\star$.*

Proof. By induction on $\prec$. We focus on the induction step for $\alpha = (\varphi \Rrightarrow \psi)$. Let $\mathcal{R}(\varphi) = \{\alpha_1, \dots, \alpha_n\}$ and $\mathcal{R}(\psi) = \{\beta_1, \dots, \beta_m\}$. Take any in-model M and any world w in M. We will show that $M, w \not\models (\varphi \Rrightarrow \psi)$ iff $M, w \not\models (\varphi \Rrightarrow \psi)^\star$. The second step uses Lemma 5.2, while the fourth step uses Proposition 2.5.

$$\begin{aligned}
& w \not\models (\varphi \Rrightarrow \psi) \\
\iff & \exists s \in \Sigma(w) : s \models \varphi \text{ and } s \not\models \psi \\
\iff & \exists s \in \Sigma(w) : s \models \textstyle\bigvee\!\!\!\bigvee_{i \le n} \alpha_i \text{ and } s \not\models \bigvee\!\!\!\bigvee_{j \le m} \beta_j \\
\iff & \exists s \in \Sigma(w) : \exists i \le n (s \models \alpha_i) \text{ and } \forall j \le m (s \not\models \beta_j) \\
\iff & \exists s \in \Sigma(w) : \exists i \le n (\forall v \in s : v \models \alpha_i) \text{ and } \forall j \le m (\exists v \in s : v \not\models \beta_j) \\
\iff & \exists i \le n \exists s \in \Sigma(w) : (\forall v \in s : v \models \alpha_i) \text{ and } \forall j \le m (\exists v \in s : v \models \neg\beta_j) \\
\iff & \exists i \le n \exists s \in \Sigma(w) : (\forall v \in s : v \models \alpha_i^\star) \text{ and } \forall j \le m (\exists v \in s : v \models \neg\beta_j^\star) \\
\iff & \exists i \le n : w \models \Box(\neg\beta_1^\star, \dots, \neg\beta_m^\star; \alpha_i^\star) \\
\iff & w \not\models \textstyle\bigwedge_{i=1}^{n} \neg\Box(\neg\beta_1^\star, \dots, \neg\beta_m^\star; \alpha_i^\star) \\
\iff & w \not\models (\varphi \Rrightarrow \psi)^\star
\end{aligned}$$

Again, note that we can use the induction hypothesis on α_i since $\alpha_i \prec (\varphi \Rrightarrow \psi)$: indeed, α_i has the same modal depth as φ, which is lower than the one of $(\varphi \Rrightarrow \psi)$. A similar argument goes for β_j. □

A couple of remarks on this translation. First, note that given a formula $\sigma \in \mathcal{L}_{\mathsf{INL}}$, the size of σ^* grows linearly on the size of σ. By contrast, since the number of resolutions of a formula $\varphi \in \mathcal{L}$ grows exponentially in the length of φ due to the clause for implication, the size of the translation of a formula $\alpha \in \mathcal{L}_!$ may grow exponentially in the size of α. It seems natural to conjecture that this is inevitable for such a translation, and thus that InqCM is exponentially more succinct than INL, but we will not try to prove this here.

It is also worth noting that this strategy, that relies crucially on resolutions, would not be viable in the setting of inquisitive predicate logic, where no analogue of resolutions is available. It is natural to conjecture that a first-order version of InqCM would be strictly more expressive than a first-order version of INL (for readers familiar with inquisitive logic, a challenge would be to translate formulas like $\boxplus(\forall x?Px \to \forall x?Qx)$, which say that in every neighborhood s of w the extension of Q is functionally determined by the extension of P).

7 Completeness

We will now prove that our axiomatization is complete by constructing a canonical model. Call a set $\Gamma \subseteq \mathcal{L}_!$ of declaratives a *complete theory of declaratives* (CTD for short) if (i) Γ is deductively closed w.r.t. declaratives: if $\Gamma \vdash \alpha$ and $\alpha \in \mathcal{L}_!$ then $\alpha \in \Gamma$; (ii) Γ is complete: for all $\alpha \in \mathcal{L}_!$, exactly one of α and $\neg\alpha$ is in Γ (thus, $\bot \notin \Gamma$, since $\top \in \Gamma$ by (i)). Now if S is a set of CTDs, we let $\bigcap S = \{\alpha \in \mathcal{L}_! \mid \alpha \in \Gamma \text{ for all } \Gamma \in S\}$ (thus, in particular, $\bigcap \emptyset = \mathcal{L}_!$).

Remark 7.1 For any set S of CTDs, $\bigcap S \vdash \alpha$ and $\alpha \in \mathcal{L}_!$ implies $\alpha \in \bigcap S$.

With any set $\Delta \subseteq \mathcal{L}_!$ of declaratives we can associate a set of CTDs, namely, the set of its complete extensions: $S_\Delta = \{\Gamma \mid \Gamma \text{ is a CTD and } \Delta \subseteq \Gamma\}$. The following Lindenbaum-type lemma is proved by the usual saturation argument.

Lemma 7.2 *If $\Delta \subseteq \mathcal{L}_!$ and $\Delta \not\vdash \bot$, then $S_\Delta \neq \emptyset$.*

For any $\Delta \subseteq \mathcal{L}_!$, the sets Δ and $\bigcap S_\Delta$ prove the same formulas.

Lemma 7.3 *For any $\Delta \subseteq \mathcal{L}_!$ and $\varphi \in \mathcal{L}$: $\Delta \vdash \varphi \iff \bigcap S_\Delta \vdash \varphi$.*

Proof. The direction $\Rightarrow$ is clear as $\Delta \subseteq \bigcap S_\Delta$. For the converse, suppose for a contradiction that for some φ we had $\bigcap S_\Delta \vdash \varphi$ but $\Delta \not\vdash \varphi$. Since $\bigcap S_\Delta$ is a set of declaratives, by Lemma 5.4 we have $\bigcap S_\Delta \vdash \alpha$ for some $\alpha \in \mathcal{R}(\varphi)$. Since $\Delta \not\vdash \varphi$, it follows by Lemma 5.3 that $\Delta \not\vdash \alpha$. By the axiom $\neg\neg\alpha \to \alpha$, this implies $\Delta \not\vdash \neg\neg\alpha$, and therefore $\Delta, \neg\alpha \not\vdash \bot$. So by Lemma 7.2 there is a CTD Γ' such that $\Delta \cup \{\neg\alpha\} \subseteq \Gamma'$. Since $\Gamma' \in S_\Delta$ and $\alpha \notin \Gamma'$ we have $\alpha \notin \bigcap S_\Delta$, so by Remark 7.1 we have $\bigcap S_\Delta \not\vdash \alpha$, which is a contradiction. □

We now define a canonical model based on complete theories of declaratives.

Definition 7.4 The canonical model for InqCM is $M^c = \langle W^c, \Sigma^c, V^c \rangle$ where:

- W^c is the set of complete theories of declaratives;
- $\Sigma^c(\Gamma) = \{S \neq \emptyset \mid \forall \varphi, \psi \in \mathcal{L} : (\varphi \Rrightarrow \psi) \in \Gamma \text{ and } \bigcap S \vdash \varphi \text{ implies } \bigcap S \vdash \psi\}$
- $V^c(p) = \{\Gamma \in W^c \mid p \in \Gamma\}$

The following analogue of the standard existence lemma is the key to the completeness proof.

Lemma 7.5 (Existence Lemma) *If* Γ *is a CTD and* $(\varphi \Rrightarrow \psi) \notin \Gamma$, *there exists a state* $S \in \Sigma^c(\Gamma)$ *such that* $\bigcap S \vdash \varphi$ *and* $\bigcap S \nvdash \psi$.

Proof. Here we provide the outline of the argument, referring to Appendix D for some technical details. Given two sets $\Phi, \Psi \subseteq \mathcal{L}$, we write $\Phi \Rrightarrow_\Gamma \Psi$ if there are $\varphi_1, \ldots, \varphi_n \in \Phi$ and $\psi_1, \ldots, \psi_m \in \Psi$ with $((\bigwedge_{i \leq n} \varphi_i) \Rrightarrow (\bigvee\!\!\!\bigvee_{i \leq m} \psi_i)) \in \Gamma$.

Step 1. We show that $\Phi \cup \{\chi\} \Rrightarrow_\Gamma \Psi$ and $\Phi \Rrightarrow_\Gamma \Psi \cup \{\chi\}$ implies $\Phi \Rrightarrow_\Gamma \Psi$.

This is where the modal part of the proof system is used in a crucial way. We refer to to Appendix D for the details of this step.

Step 2. We partition $\mathcal{L}$ into two sets L, R with $\varphi \in \mathsf{L}$, $\psi \in \mathsf{R}$, and $\mathsf{L} \not\Rrightarrow_\Gamma \mathsf{R}$.

For this, we use Step 1 within a saturation procedure familiar from intuitionistic and inquisitive logic (cf. §3.3 of [16] and [20]). See Appendix D for the details.

Step 3. We construct the required state S.

Since $\mathsf{L} \not\Rrightarrow_\Gamma \mathsf{R}$, the rule (CN) guarantees that $\mathsf{L} \nvdash \mathsf{R}$. By Lemma 5.6 we can find a set $\Delta \in \mathcal{R}(\mathsf{L})$ with $\Delta \nvdash \mathsf{R}$. We can now take $S = S_\Delta = \{\Gamma' \in W^c \mid \Delta \subseteq \Gamma'\}$. We need to verify that (i) $\bigcap S \vdash \varphi$, (ii) $\bigcap S \nvdash \psi$ and (iii) $S \in \Sigma^c(\Gamma)$.

- For (i), recall $\varphi \in \mathsf{L}$. Since $\Delta \in \mathcal{R}(\mathsf{L})$, for some $\alpha \in \mathcal{R}(\varphi)$ we have $\alpha \in \Delta$. By Lemma 5.3, $\Delta \vdash \varphi$, and thus by Lemma 7.3 $\bigcap S \vdash \varphi$.
- For (ii), recall $\psi \in \mathsf{R}$. Since $\Delta \nvdash \mathsf{R}$ we have $\Delta \nvdash \psi$. By Lemma 7.3, $\bigcap S \nvdash \psi$.
- For (iii), first note that since $\Delta \nvdash \mathsf{R}$, we have $\Delta \nvdash \bot$, so by Lemma 7.2 $S \neq \emptyset$. Next, suppose $(\chi \Rrightarrow \xi) \in \Gamma$ and $\bigcap S \vdash \chi$. By Lemma 7.3, $\Delta \vdash \chi$. Since by construction $\Delta \nvdash \mathsf{R}$, it follows that $\chi \notin \mathsf{R}$, so $\chi \in \mathsf{L}$. Now we must have $\xi \in \mathsf{L}$ as well, for if we had $\xi \in \mathsf{R}$ it would follow from $(\chi \Rrightarrow \xi) \in \Gamma$ that $\mathsf{L} \Rrightarrow_\Gamma \mathsf{R}$, contrary to what we saw. Since $\xi \in \mathsf{L}$ and $\Delta \in \mathcal{R}(\mathsf{L})$, for some $\alpha \in \mathcal{R}(\xi)$ we have $\alpha \in \Delta$, so by Lemma 5.3 $\Delta \vdash \xi$. Finally, Lemma 7.3 gives $\bigcap S \vdash \xi$.

□

The bridge between derivability in our proof system and semantics in M^c is given by the following support lemma, which generalizes the usual truth lemma.

Lemma 7.6 *For all states* $S \subseteq W^c$ *and all* $\varphi \in \mathcal{L}$: $M^c, S \models \varphi \iff \bigcap S \vdash \varphi$.

Proof. By induction on φ. The cases for atoms and connectives are standard (cf. pp. 90-91 in [10]). We spell out the inductive step for $\varphi = (\psi \Rrightarrow \chi)$.

Suppose $\bigcap S \vdash (\psi \Rrightarrow \chi)$. Take a world $\Gamma \in S$ and a state $T \in \Sigma^c(\Gamma)$ with $M^c, T \models \psi$. By induction hypothesis we have $\bigcap T \vdash \psi$. Since $\Gamma \in S$ we have $\bigcap S \subseteq \Gamma$, so $\Gamma \vdash (\psi \Rrightarrow \chi)$. Since $(\psi \Rrightarrow \chi) \in \mathcal{L}_!$, it follows that $(\psi \Rrightarrow \chi) \in \Gamma$. By definition of Σ^c, from $(\psi \Rrightarrow \chi) \in \Gamma$ and $\bigcap T \vdash \psi$ we can conclude $\bigcap T \vdash \chi$, which by induction hypothesis gives $M^c, T \models \chi$. Hence, $M^c, S \models (\psi \Rrightarrow \chi)$.

For the converse, suppose $\bigcap S \nvdash (\psi \Rrightarrow \chi)$. Then there is some $\Gamma \in S$ such that $(\psi \Rrightarrow \chi) \notin \Gamma$. By the Existence Lemma (Lemma 7.5) there is a state

$T \in \Sigma^c(\Gamma)$ such that $\bigcap T \vdash \psi$ and $\bigcap T \nvdash \chi$, which by induction hypothesis means that $M^c, T \models \psi$ and $M^c, T \not\models \chi$. Hence, $M^c, S \not\models (\psi \Rrightarrow \chi)$. □

Finally, we use this lemma to establish the strong completeness of our system.

Theorem 7.7 (Completeness) *For all $\Phi \cup \{\psi\} \subseteq \mathcal{L}$, $\Phi \models \psi$ implies $\Phi \vdash \psi$.*

Proof. Suppose $\Phi \nvdash \psi$. By Lemma 5.6 we can find a $\Delta \in \mathcal{R}(\Phi)$ with $\Delta \nvdash \psi$. Note that since $\Delta \in \mathcal{R}(\Phi)$, for all $\varphi \in \Phi$ there is some $\alpha \in \mathcal{R}(\varphi)$ with $\alpha \in \Delta$, which by Lemma 5.3 gives $\Delta \vdash \varphi$. Now take $S_\Delta = \{\Gamma' \in W^c \mid \Delta \subseteq \Gamma'\}$. By Lemma 7.3, $\bigcap S_\Delta \vdash \varphi$ for all $\varphi \in \Phi$, while $\bigcap S_\Delta \nvdash \psi$. By the support lemma, in the model M^c the state S_Δ supports all formulas in Φ but not ψ, so $\Phi \not\models \psi$.□

8 Further work

We close by outlining a number of directions for future work. First, it would be natural to relate the expressive power of InqCM to that of first-order logic over neighborhood models, viewed as two-sorted structures. One could define a standard translation and aim for a van Benthem-style characterization of InqCM as the bisimulation-invariant fragment of first-order logic. For inquisitive modal logic over downward closed models, such a result was proved in [13].

Second, it would be interesting to develop modal correspondence theory for InqCM, relating the validity of InqCM-schemata over a neighborhood frame $\langle W, \Sigma \rangle$ to corresponding frame properties. For instance, given an appropriate notion of validity, one can show that the schema $(\varphi \Rrightarrow \psi) \to \boxplus(\varphi \to \psi)$ is valid on a finite neighborhood frame $\langle W, \Sigma \rangle$ if and only if Σ is downward closed, in the sense that $\Sigma = \Sigma^{\downarrow}$ (cf. Definition 2.6). Relatedly, it would be interesting to investigate the logic of frame classes arising from natural constraints on Σ.

One should also explore what happens when we allow empty neighborhoods. Presumably, this makes little difference, provided the clause for $\Rrightarrow$ is explicitly re-cast as "for all *non-empty* $s \in \Sigma(w)$...". Our language $\mathcal{L}$ would, of course, be unable to tell whether or not $\emptyset \in \Sigma(w)$, but this could be fixed by adding a modal atom empty, which is truth-conditional with truth conditions given by $w \models \mathsf{empty} \iff \emptyset \in \Sigma(w)$. We expect that all the results in this paper can be replicated straightforwardly for this extended language.

Finally, it would be interesting to look at concrete interpretations of InqCM. One salient interpretation that we mentioned is in terms of action and ability. This interpretation naturally calls for an extension to the multi-agent setting; in fact, it would be natural to consider not only the abilities of single agents, but also those of groups, leading to an inquisitive extension of coalition logic [25] capable of expressing facts such as "by acting, the coalition of a and b is bound to settle whether p one way or the other" or "the coalition is bound to force p if they want to force q" (see [5,18] for parallel multi-agent extensions of instantial neighborhood logic). Such a logic could, in turn, be further extended with the resources of temporal logic to talk about long-term strategic abilities, leading for instance to an inquisitive extension of ATL [3,19].

Appendix

A Proof of Proposition 3.3 (bisimulation invariance).

In this section we give the proof of Proposition 3.3. It suffices to prove the claim for information states, since the claim about worlds is obtained as a special case (note that $(w \sim w') \iff (\{w\} \sim \{w'\})$ and $(w \leftrightsquigarrow w') \iff (\{w\} \leftrightsquigarrow \{w'\})$).

So, take two in-models M, M'. We will prove that for every $\varphi \in \mathcal{L}$ we have:

$$\forall s \subseteq W \forall s' \subseteq W' : (s \sim s') \text{ implies } (M, s \models \varphi \iff M', s' \models \varphi)$$

We show this by induction on φ. The inductive cases for $\wedge$ and $\mathbin{\rotatebox[origin=c]{90}{$\geqslant$}}$ are immediate; we spell out the remaining cases.

- φ is an atom $p \in \mathcal{P}$. Suppose $s \sim s'$ and let Z be a bisimulation with $s\overline{Z}s'$. Suppose $M, s \models p$. Then $s \subseteq V(p)$. Now take any $w' \in s'$. Since $s\overline{Z}s'$ there is $w \in s$ with wZw', and since Z is a bisimulation, $w \in V(p) \iff w' \in V'(p)$. Since $w \in V(p)$, we have $w' \in V'(p)$. Since w' was arbitrary in s', it follows that $s' \subseteq V(p)$ and so $M', s' \models p$. The converse direction is analogous.
- $\varphi = \bot$. Simply observe that if $s \sim s'$, either s, s' are both empty or neither is.
- $\varphi = (\psi \to \chi)$. Suppose $s \sim s'$. We first show that every $t \subseteq s$ is bisimilar to some $t' \subseteq s'$ and vice versa. Let Z be a bisimulation with $s\overline{Z}s'$. Given $t \subseteq s$, define $t' = \{w' \in s' \mid \exists w \in t : wZw'\}$. We claim that $t\overline{Z}t'$. To see that this is the case, take $w \in t$. Since $s\overline{Z}s'$ we have wZw' for some $w' \in s'$, and then $w' \in t'$ by definition of t'. So for all $w \in t$ there is a $w' \in t'$ with wZw'. The converse is obvious from the definition of t'. So $t \sim t'$. The claim that every subset of s' is bisimilar to some subset of s is proved analogously.

  Now suppose $M, s \not\models \psi \to \chi$. Then there is a state $t \subseteq s$ with $M, t \models \psi$ and $M, t \not\models \chi$. By the previous argument there is $t' \subseteq s'$ with $t \sim t'$. By induction hypothesis we have $M', t' \models \psi$, and $M', t' \not\models \chi$, so $M', s' \not\models \psi \to \chi$. The converse is proved analogously.
- $\varphi = (\psi \Rrightarrow \chi)$. Suppose $s \sim s'$ and let Z be a bisimulation with $s\overline{Z}s'$. Suppose $M, s \not\models (\psi \Rrightarrow \chi)$. So there are a world $w \in s$ and a state $t \in \Sigma(w)$ such that $M, t \models \psi$ but $M, t \not\models \chi$. Since $s\overline{Z}s'$, there is a world $w' \in s'$ with wZw', and then by the Forth condition there is a $t' \in \Sigma'(w')$ with $t\overline{Z}t'$, and thus with $t \sim t'$. By induction hypothesis we have $M', t' \models \psi$ but $M', t' \not\models \chi$, which shows that $M', s' \not\models (\varphi \Rrightarrow \psi)$. The converse is proved analogously. □

B Propositional axioms for **InqCM**

The propositional axioms of InqCM are all instances of the following schemata, where $\varphi, \psi, \chi \in \mathcal{L}$ and $\alpha \in \mathcal{L}_!$:

- $\varphi \to (\psi \to \varphi)$
- $(\varphi \to (\psi \to \chi)) \to (\varphi \to \psi) \to (\varphi \to \chi)$
- $\varphi \to (\psi \to \varphi \wedge \psi)$
- $\varphi \wedge \psi \to \varphi, \quad \varphi \wedge \psi \to \psi$

- $\varphi \to \varphi \vee \psi, \quad \psi \to \varphi \vee \psi$
- $(\varphi \to \chi) \to ((\psi \to \chi) \to (\varphi \vee \psi \to \chi))$
- $\bot \to \varphi$
- $\neg\neg\alpha \to \alpha$
- $(\alpha \to \varphi \vee \psi) \to (\alpha \to \varphi) \vee (\alpha \to \psi)$

The first seven schemata are simply schemata for intuitionistic logic with $\vee$ identified with intuitionistic disjunction, while the last two schemata are specific to inquisitive logic (see [10] for discussion of their significance).

C Proof of Lemma 5.4

We first show the following claim:

$$\text{If } \vdash \varphi, \text{ then } \vdash \alpha \text{ for some } \alpha \in \mathcal{R}(\varphi) \tag{C.1}$$

We proceed by induction on the length of the shortest proof of φ. If φ is provable with a proof of length 1, φ is an axiom. Suppose φ is a propositional axiom: then we can check case-by-case that some resolution of φ is a classical tautology. By way of example, suppose φ is an instance of $\psi \to (\chi \to \psi)$; then the following resolution of φ is a classical tautology:

$$\bigwedge_{\alpha\in\mathcal{R}(\psi)} (\alpha \to \bigwedge_{\beta\in\mathcal{R}(\chi)} (\beta \to \alpha))$$

But in restriction to declaratives, our system contains a complete set of axioms for classical propositional logic, and so it proves all classical tautologies.

Next, suppose φ is an instance of one of the modal schemata (`Tran`), (`Conj`), or (`Disj`). Then φ is a declarative, and so $\mathcal{R}(\varphi) = \{\varphi\}$ (recall that $\mathcal{R}(\alpha) = \{\alpha\}$ for every declarative α). Thus, (C.1) holds trivially.

Now consider the inductive case. We have only two cases to consider:

(i) $\varphi = (\psi \Rrightarrow \chi)$ can be obtained by (`CN`). In this case, φ is a declarative, so $\mathcal{R}(\varphi) = \{\varphi\}$ and (C.1) holds trivially.

(ii) φ can be obtained by (`MP`) from formulas $\psi \to \varphi$ and ψ which are provable with shorter proofs. By induction hypothesis, the system proves a resolution γ of $\psi \to \varphi$ and a resolution β_0 of ψ. By definition of resolutions of an implication, γ has the form $\bigwedge_{\beta\in\mathcal{R}(\psi)}(\beta \to f(\beta))$ for some $f : \mathcal{R}(\psi) \to \mathcal{R}(\varphi)$. But then, clearly, $f(\beta_0)$ is a provable resolution of φ.

This completes the inductive proof of (C.1). To prove Lemma 5.4, suppose $\Gamma \subseteq \mathcal{L}_!$ and $\Gamma \vdash \varphi$. This means that $\vdash \gamma_1 \wedge \cdots \wedge \gamma_n \to \varphi$ for some $\gamma_1, \ldots, \gamma_n \in \Gamma$. Let $\gamma := \gamma_1 \wedge \cdots \wedge \gamma_n$: since γ is a declarative we have $\mathcal{R}(\gamma) = \{\gamma\}$. Thus, by definition of resolutions of an implication, we have:

$$\mathcal{R}(\gamma \to \varphi) = \{\gamma \to f(\gamma) \mid f : \{\gamma\} \to \mathcal{R}(\varphi)\} = \{\gamma \to \alpha \mid \alpha \in \mathcal{R}(\varphi)\}$$

Since $\vdash \gamma \to \varphi$, it follows from C.1 that $\vdash \gamma \to \alpha$ for some resolution $\alpha \in \mathcal{R}(\varphi)$, which shows that $\Gamma \vdash \alpha$. □

D Proof of Lemma 7.5 (existence lemma).

In this appendix we provide the missing details from the proof of Lemma 7.5.

Step 1. Show that if $\Phi \cup \{\chi\} \Rrightarrow_\Gamma \Psi$ and $\Phi \Rrightarrow_\Gamma \Psi \cup \{\chi\}$, then $\Phi \Rrightarrow_\Gamma \Psi$.

Suppose $\Phi, \chi \Rrightarrow_\Gamma \Psi$ and $\Phi \Rrightarrow_\Gamma \Psi, \chi$. So there are $\varphi_1, \ldots, \varphi_n, \varphi_{n+1}, \ldots, \varphi_{n+m} \in \Phi$ and $\psi_1, \ldots, \psi_h, \psi_{h+1}, \ldots, \psi_{h+k} \in \Psi$ such that:

$$((\chi \wedge \bigwedge_{1=i}^{n} \varphi_i) \Rrightarrow \bigvee\!\!\!\bigvee_{1=i}^{h} \psi_i) \in \Gamma \qquad ((\bigwedge_{i=1}^{m} \varphi_{n+i}) \Rrightarrow (\chi \mathbin{\vee\!\!\!\vee} \bigvee\!\!\!\bigvee_{i=1}^{k} \psi_{h+i})) \in \Gamma$$

We show that Γ contains $((\bigwedge_{i\leq n+m} \varphi_i) \Rrightarrow (\bigvee\!\!\!\bigvee_{i\leq h+k} \psi_i))$, witnessing $\Phi \Rrightarrow_\Gamma \Psi$. To ease notation, we spell out the details for the case $n = m = h = k = 1$, but the general case is completely analogous. It suffices to show that:

$$(\varphi_1 \wedge \chi \Rrightarrow \psi_1), (\varphi_2 \Rrightarrow \psi_2 \mathbin{\vee\!\!\!\vee} \chi) \vdash (\varphi_1 \wedge \varphi_2 \Rrightarrow \psi_1 \mathbin{\vee\!\!\!\vee} \psi_2)$$

Since the formulas on the left-hand-side are in Γ and Γ is closed under deduction of declaratives, so is the conclusion.

First note that the following formula is provable in the propositional component of the proof system using the standard axioms for $\wedge$ and $\mathbin{\vee\!\!\!\vee}$:

$$\varphi_1 \wedge (\psi_2 \mathbin{\vee\!\!\!\vee} \chi) \rightarrow \psi_2 \mathbin{\vee\!\!\!\vee} (\varphi_1 \wedge \chi) \tag{D.1}$$

In the following derivation, we indicate explicitly only the modal axioms and rules involved in the reasoning, omitting reference to propositional axioms and rules. For simplicity, we use the formulas $(\varphi_1 \wedge \chi \Rrightarrow \psi_1)$ and $(\varphi_2 \Rrightarrow \psi_2 \mathbin{\vee\!\!\!\vee} \chi)$ as premises; this is legitimate since we will not use conditional necessitation on these formulas or anything inferred from them. Rewriting the argument with the relevant formulas used as antecedents is tedious but straightforward.

1.	$\varphi_1 \wedge \chi \Rrightarrow \psi_1$	(premise)
2.	$\varphi_2 \Rrightarrow \psi_2 \mathbin{\vee\!\!\!\vee} \chi$	(premise)
3.	$\varphi_1 \wedge \varphi_2 \Rrightarrow \varphi_1$	(CN) from axiom $\varphi_1 \wedge \varphi_2 \rightarrow \varphi_1$
4.	$\varphi_1 \wedge \varphi_2 \Rrightarrow \varphi_2$	(CN) from axiom $\varphi_1 \wedge \varphi_2 \rightarrow \varphi_2$
5.	$\varphi_1 \wedge \varphi_2 \Rrightarrow \psi_2 \mathbin{\vee\!\!\!\vee} \chi$	(Tran), 4, 2
6.	$\varphi_1 \wedge \varphi_2 \Rrightarrow \varphi_1 \wedge (\psi_2 \mathbin{\vee\!\!\!\vee} \chi)$	(Conj), 3, 5
7.	$\varphi_1 \wedge (\psi_2 \mathbin{\vee\!\!\!\vee} \chi) \Rrightarrow \psi_2 \mathbin{\vee\!\!\!\vee} (\varphi_1 \wedge \chi)$	(CN) from (D.1)
8.	$\varphi_1 \wedge \varphi_2 \Rrightarrow \psi_2 \mathbin{\vee\!\!\!\vee} (\varphi_1 \wedge \chi)$	(Tran), 6, 7
9.	$\psi_1 \Rrightarrow \psi_1 \mathbin{\vee\!\!\!\vee} \psi_2$	(CN) from axiom $\psi_1 \rightarrow \psi_1 \mathbin{\vee\!\!\!\vee} \psi_2$
10.	$\psi_2 \Rrightarrow \psi_1 \mathbin{\vee\!\!\!\vee} \psi_2$	(CN) from axiom $\psi_2 \rightarrow \psi_1 \mathbin{\vee\!\!\!\vee} \psi_2$
11.	$\varphi_1 \wedge \chi \Rrightarrow \psi_1 \mathbin{\vee\!\!\!\vee} \psi_2$	(Tran), 1, 9
12.	$\psi_2 \mathbin{\vee\!\!\!\vee} (\varphi_1 \wedge \chi) \Rrightarrow \psi_1 \mathbin{\vee\!\!\!\vee} \psi_2$	(Disj), 10, 11
13.	$\varphi_1 \wedge \varphi_2 \Rrightarrow \psi_1 \mathbin{\vee\!\!\!\vee} \psi_2$	(Tran), 8, 12

Step 2. We partition $\mathcal{L}$ into two sets L, R with $\varphi \in \mathsf{L}$, $\psi \in \mathsf{R}$, and $\mathsf{L} \not\Rrightarrow_\Gamma \mathsf{R}$.

Fix an enumeration $(\chi_n)_{n\in\mathbb{N}}$ of all formulas in $\mathcal{L}$ (we assume $\mathcal{L}$ is countable, though this is not strictly necessary). We define a sequence of sets $(\mathsf{L}_n)_{n\in\mathbb{N}}$ and $(\mathsf{R}_n)_{n\in\mathbb{N}}$ as follows:

- $\mathsf{L}_0 = \{\varphi\}, \mathsf{R}_0 = \{\psi\}$
- if $\mathsf{L}_n \cup \{\chi_n\} \not\Rrightarrow_\Gamma \mathsf{R}_n$ we let $\mathsf{L}_{n+1} := \mathsf{L}_n \cup \{\chi_n\}$ and $\mathsf{R}_{n+1} = \mathsf{R}_n$
- if $\mathsf{L}_n \cup \{\chi_n\} \Rrightarrow_\Gamma \mathsf{R}_n$ we let $\mathsf{L}_{n+1} := \mathsf{L}_n$ and $\mathsf{R}_{n+1} = \mathsf{R}_n \cup \{\chi_n\}$

We will show by induction on n that $\mathsf{L}_n \not\Rrightarrow_\Gamma \mathsf{R}_n$. For $n = 0$ this amounts to $(\varphi \Rrightarrow \psi) \notin \Gamma$, which is true by assumption. Now suppose this is true for n and consider $n+1$. If $\mathsf{L}_n \cup \{\chi_n\} \not\Rrightarrow_\Gamma \mathsf{R}_n$, the claim is obvious from the definition. So, suppose $\mathsf{L}_n \cup \{\chi_n\} \Rrightarrow_\Gamma \mathsf{R}_n$. Since by induction hypothesis $\mathsf{L}_n \not\Rrightarrow_\Gamma \mathsf{R}_n$, Step 1 above implies $\mathsf{L}_n \not\Rrightarrow_\Gamma \mathsf{R}_n \cup \{\chi_n\}$, which by definition amounts to $\mathsf{L}_{n+1} \not\Rrightarrow_\Gamma \mathsf{R}_{n+1}$.

Now let $\mathsf{L} = \bigcup_{n\in\mathbb{N}} \mathsf{L}_n$ and $\mathsf{R} = \bigcup_{n\in\mathbb{N}} \mathsf{R}_n$. We have $\mathsf{L} \not\Rrightarrow_\Gamma \mathsf{R}$, since otherwise there would be an n such that $\mathsf{L}_n \Rrightarrow_\Gamma \mathsf{R}_n$. Moreover, L and R form a partition of $\mathcal{L}$: by construction, every formula occurs in either set, and no formula can occur in both (if $\chi \in \mathsf{L} \cap \mathsf{R}$, then since $(\chi \Rrightarrow \chi) \in \Gamma$ we would have $\mathsf{L} \Rrightarrow_\Gamma \mathsf{R}$).

References

[1] Abdou, J. and H. Keiding, "Effectivity Functions in Social Choice," Kluwer, 1991.

[2] Aiello, M., J. Van Benthem and G. Bezhanishvili, *Reasoning about space: the modal way*, Journal of Logic and Computation **13** (2003), pp. 889–920.

[3] Alur, R., T. Henzinger and O. Kupferman, *Alternating-time temporal logic*, Journal of the ACM **49** (2002), pp. 672–713.

[4] van Benthem, J., "Logic in games," MIT press, 2014.

[5] van Benthem, J., N. Bezhanishvili and S. Enqvist, *A new game equivalence, its logic and algebra*, Journal of Philosophical Logic **48** (2019), pp. 649–684.

[6] van Benthem, J., N. Bezhanishvili, S. Enqvist and J. Yu, *Instantial neighbourhood logic*, The Review of Symbolic Logic **10** (2017), pp. 116–144.

[7] Brown, M. A., *On the logic of ability*, Journal of Philosophical Logic **17** (1988), pp. 1–26.

[8] Chellas, B. F., "Modal logic: an introduction," Cambridge university press, 1980.

[9] Ciardelli, I., *Modalities in the realm of questions: axiomatizing inquisitive epistemic logic*, in: R. Goré, B. Kooi and A. Kurucz, editors, *Advances in Modal Logic (AIML)* (2014), pp. 94–113.

[10] Ciardelli, I., "Questions in logic," Ph.D. thesis, Institute for Logic, Language and Computation, University of Amsterdam (2016).

[11] Ciardelli, I., *Dependence statements are strict conditionals*, in: G. Bezhanishvili, G. D'Agostino, G. Metcalfe and T. Studer, editors, *Advances in Modal Logic (AIML)* (2018), pp. 123–142.

[12] Ciardelli, I., J. Groenendijk and F. Roelofsen, "Inquisitive Semantics," Oxford University Press, 2018.

[13] Ciardelli, I. and M. Otto, *Inquisitive bisimulation*, The Journal of Symbolic Logic **86** (2021), pp. 77–109.

[14] Ciardelli, I. and F. Roelofsen, *Inquisitive logic*, Journal of Philosophical Logic **40** (2011), pp. 55–94.

[15] Ciardelli, I. and F. Roelofsen, *Inquisitive dynamic epistemic logic*, Synthese **192** (2015), pp. 1643–1687.

[16] Gabbay, D., "Semantical investigations in Heyting's intuitionistic logic," Springer, 1981.

[17] van Gessel, T., *Action models in inquisitive logic*, Synthese **197** (2020), pp. 3905–3945.

[18] Goranko, V. and S. Enqvist, *Socially friendly and group protecting coalition logics*, in: *17th International Conference on Autonomous Agents and Multiagent Systems (AAMAS 2018)*, The International Foundation for Autonomous Agents and Multiagent Systems, 2018, pp. 372–380.

[19] Goranko, V. and G. van Drimmelen, *Complete axiomatization and decidability of alternating-time temporal logic*, Theoretical Computer Science **353** (2006), pp. 93–117.

[20] Grilletti, G., *Completeness for the classical antecedent fragment of inquisitive first-order logic*, Journal of Logic, Language, and Information **30** (2021), pp. 725–751.

[21] Hansen, H. H., "Monotonic modal logics," Ph.D. thesis, Universiteit van Amsterdam (2003).

[22] Meißner, S. and M. Otto, *A first-order framework for inquisitive modal logic*, The Review of Symbolic Logic (2021), pp. 1–23.

[23] Montague, R., *Universal grammar*, Theoria **36** (1970), pp. 373–398.

[24] Pacuit, E., "Neighborhood Semantics for Modal Logic," Springer, 2017.

[25] Pauly, M., *A modal logic for coalitional power in games*, Journal of logic and computation **12** (2002), pp. 149–166.

[26] Punčochář, V. and I. Sedlár, *Epistemic extensions of substructural inquisitive logics*, Journal of Logic and Computation **31** (2021), pp. 1820–1844.

[27] Scott, D., *Advice on modal logic*, in: K. Lambert, editor, *Philosophical Problems in Logic: Some Recent Developments*, Springer Netherlands, 1970 pp. 142–173.

Algebraic Semantics for One-Variable Lattice-Valued Logics

Petr Cintula [1]

Institute of Computer Science of the Czech Academy of Sciences
Prague, Czech Republic
cintula@cs.cas.cz

George Metcalfe [2] Naomi Tokuda

Mathematical Institute, University of Bern, Switzerland
{george.metcalfe,naomi.tokuda}@unibe.ch

Abstract

The one-variable fragment of any first-order logic may be considered as a modal logic, where the universal and existential quantifiers are replaced by a box and diamond modality, respectively. In several cases, axiomatizations of algebraic semantics for these logics have been obtained: most notably, for the modal counterparts S5 and MIPC of the one-variable fragments of first-order classical logic and intuitionistic logic, respectively. Outside the setting of first-order intermediate logics, however, a general approach is lacking. This paper provides the basis for such an approach in the setting of first-order lattice-valued logics, where formulas are interpreted in algebraic structures with a lattice reduct. In particular, axiomatizations are obtained for modal counterparts of one-variable fragments of a broad family of these logics by generalizing a functional representation theorem of Bezhanishvili and Harding for monadic Heyting algebras. An alternative proof-theoretic proof is also provided for one-variable fragments of first-order substructural logics that have a cut-free sequent calculus and admit a certain bounded interpolation property.

Keywords: Modal Logic, Substructural Logics, Lattice-Valued Logics, One-Variable Fragment, Superamalgamation, Sequent Calculus, Interpolation.

1 Introduction

The *one-variable fragment* of any first-order logic — the valid formulas built using one variable x, unary relation symbols, propositional connectives, and quantifiers $(\forall x)$ and $(\exists x)$ — may be studied as an "S5-like" modal logic. Just replace each occurrence of an atom $P(x)$ with a propositional variable p, and

[1] Supported by Czech Science Foundation project GA22-01137S and by RVO 67985807.

[2] Supported by Swiss National Science Foundation grant 200021_184693.

$(\forall x)$ and $(\exists x)$ with $\Box$ and $\Diamond$, respectively. The first-order semantics typically induces a relational semantics for this modal logic, but an axiomatization of its algebraic semantics may be rather elusive; in particular, an axiomatization of the first-order logic does not directly yield a Hilbert-style axiomatization of (the modal counterpart of) its one-variable fragment.

Such axiomatizations have been obtained in several notable cases. Monadic Boolean algebras [17] and monadic Heyting algebras [4, 20] correspond to the modal counterparts S5 and MIPC of the one-variable fragments of first-order classical logic and intuitionistic logic, respectively. More generally, varieties of monadic Heyting algebras corresponding to modal counterparts of one-variable fragments of first-order intermediate logics (based on frames with and without constant domains) have been investigated in [1, 3, 5–7, 22, 24, 25]. One-variable fragments of certain first-order many-valued logics have also been studied in some depth. In particular, modal counterparts of the one-variable fragments of first-order Łukasiewicz logic and Abelian logic correspond to monadic MV-algebras [8, 13, 23] and monadic Abelian ℓ-groups [19], respectively.

In this paper, we take first steps towards a general approach to addressing this axiomatization problem. As a starting point, we introduce in Section 2 (first-order) one-variable lattice-valued logics, where formulas are interpreted in structures defined over complete $\mathcal{L}$-lattices: algebraic structures for a given algebraic signature $\mathcal{L}$ that have a lattice reduct. In Section 3, we then define an m-$\mathcal{L}$-lattice to be an $\mathcal{L}$-lattice expanded with modalities $\Box$ and $\Diamond$ satisfying certain natural equations, and for a class $\mathcal{K}$ of $\mathcal{L}$-lattices, let $m\mathcal{K}$ denote the class of m-$\mathcal{L}$-lattices with an $\mathcal{L}$-lattice reduct in $\mathcal{K}$. (In particular, if $\mathcal{K}$ is the variety of Boolean algebras or Heyting algebras, $m\mathcal{K}$ is the variety of monadic Boolean algebras or monadic Heyting algebras, respectively.) Generalizing previous results in the literature (see, e.g., [1, 27]), we obtain a one-to-one correspondence between m-$\mathcal{L}$-lattices and $\mathcal{L}$-lattices equipped with a subalgebra that satisfies a certain relative completeness condition.

Given a variety $\mathcal{V}$ of $\mathcal{L}$-lattices, equational consequence in the variety $m\mathcal{V}$ always implies consequence in the one-variable lattice-valued logic based on the complete members of $\mathcal{V}$. In Section 4, we show that the converse also holds if $\mathcal{V}$ admits regular completions and has the superamalgamation property. The key tool in this proof is a generalization of Bezhanishvili and Harding's functional representation theorem for monadic Heyting algebras [3]. In Section 5, we show that this theorem applies to certain varieties of $\mathrm{FL_e}$-algebras, yielding axiomatizations of the modal counterparts of one-variable fragments of certain first-order substructural logics. In Section 6, we provide an alternative proof of these results for substructural logics by establishing a bounded interpolation property for cut-free sequent calculi for the one-variable fragments.

Finally, in Section 7 we sketch a broader perspective for one-variable lattice-valued logics based on an arbitrary class $\mathcal{K}$ of complete $\mathcal{L}$-lattices, observing that when $\mathcal{K}$ consists of the complete members of a variety $\mathcal{V}$, the corresponding class of m-$\mathcal{L}$-lattices need not in general be $m\mathcal{V}$ or even a variety.

2 One-Variable Lattice-Valued Logics

Let $\mathcal{L}_n$ denote the set of operation symbols of an algebraic signature $\mathcal{L}$ of arity $n \in \mathbb{N}$, and call $\mathcal{L}$ *lattice-oriented* if $\mathcal{L}_2$ contains distinct symbols $\wedge$ and $\vee$. We will assume throughout this paper that $\mathcal{L}$ is a fixed lattice-oriented signature.

An $\mathcal{L}$-*lattice* is an algebraic structure $\mathbf{A} = \langle A, \{\star^{\mathbf{A}} \mid n \in \mathbb{N}, \star \in \mathcal{L}_n\}\rangle$ such that $\langle A, \wedge^{\mathbf{A}}, \vee^{\mathbf{A}}\rangle$ is a lattice with order $x \leq^{\mathbf{A}} y :\Longleftrightarrow x \wedge^{\mathbf{A}} y = x$ and $\star^{\mathbf{A}}$ is an n-ary operation on A for each $\star \in \mathcal{L}_n$ ($n \in \mathbb{N}$). As usual, we omit superscripts when these are clear from the context.

We call $\mathbf{A}$ *complete* if its lattice reduct $\langle A, \wedge, \vee\rangle$ is complete, i.e., $\bigwedge X$ and $\bigvee X$ exist in A for all $X \subseteq A$. Given a class $\mathcal{K}$ of $\mathcal{L}$-lattices, we denote by $\overline{\mathcal{K}}$ the class of its complete members and say that $\mathcal{K}$ *admits regular completions* if for any $\mathbf{A} \in \mathcal{K}$, there exist a $\mathbf{B} \in \overline{\mathcal{K}}$ and an embedding $f\colon \mathbf{A} \to \mathbf{B}$ that preserves all existing meets and joins of $\mathbf{A}$.

Example 2.1 The varieties $\mathcal{BA}$ and $\mathcal{HA}$ of Boolean algebras and Heyting algebras, respectively, are closed under MacNeille completions and hence admit regular completions. Although these are the only two non-trivial varieties of Heyting algebras closed under MacNeille completions [2], a broad family of varieties that provide semantics for substructural logics (see Section 5) also have this property [10]. Moreover, for a still broader family of varieties, including the variety $\mathcal{GA}$ of Gödel algebras (Heyting algebras that satisfy the prelinearity axiom $(x \to y) \vee (y \to x) \approx 1$), the class of their subdirectly irreducible members is closed under MacNeille completions [9]. Note, however, that neither the variety $\mathcal{MV}$ of MV-algebras nor the class of its subdirectly irreducible members admit regular completions [15].

First-order formulas with propositional connectives in $\mathcal{L}$ can be defined for an arbitrary predicate language as usual (see, e.g., [11, Section 7.1]). We restrict our attention here, however, to the set $\mathrm{Fm}^1_\forall(\mathcal{L})$ of *one-variable* $\mathcal{L}$-*formulas* $\varphi, \psi, \chi, \ldots$, built from a countably infinite set of unary predicates $\{P_i\}_{i\in\mathbb{N}}$, a variable x, connectives in $\mathcal{L}$, and quantifiers $\forall, \exists$. Given $\varphi, \psi \in \mathrm{Fm}^1_\forall(\mathcal{L})$, we refer to $\varphi \approx \psi$ as an $\mathrm{Fm}^1_\forall(\mathcal{L})$-*equation* and let $\varphi \leq \psi$ denote $\varphi \wedge \psi \approx \varphi$.[3]

Let $\mathbf{A}$ be a complete $\mathcal{L}$-lattice. An $\mathbf{A}$-*structure* is an ordered pair $\mathfrak{S} = \langle S, \mathcal{I}\rangle$ such that S is a non-empty set and $\mathcal{I}(P_i)$ is a map from S to A for every $i \in \mathbb{N}$. For each $u \in S$, we define a map $\|\cdot\|^{\mathfrak{S}}_u : \mathrm{Fm}^1_\forall(\mathcal{L}) \to A$ inductively as follows:

$$\begin{aligned}
\|P_i(x)\|^{\mathfrak{S}}_u &= \mathcal{I}(P_i)(u) && i \in \mathbb{N}\\
\|\star(\varphi_1, \ldots, \varphi_n)\|^{\mathfrak{S}}_u &= \star^{\mathbf{A}}(\|\varphi_1\|^{\mathfrak{S}}_u, \ldots, \|\varphi_n\|^{\mathfrak{S}}_u) && n \in \mathbb{N}, \star \in \mathcal{L}_n\\
\|(\forall x)\varphi\|^{\mathfrak{S}}_u &= \bigwedge\{\|\varphi\|^{\mathfrak{S}}_v \mid v \in S\}\\
\|(\exists x)\varphi\|^{\mathfrak{S}}_u &= \bigvee\{\|\varphi\|^{\mathfrak{S}}_v \mid v \in S\}.
\end{aligned}$$

[3] To avoid confusion, let us emphasize that $\varphi \approx \psi$ is a primitive syntactic object that relates two formulas and not terms. In some settings (e.g., first-order classical or intuitonistic logic), the validity of $\varphi \approx \psi$ can be expressed using the validity of a formula such as $\varphi \leftrightarrow \psi$ and we may define semantical consequence between formulas, but this is not always the case.

We say that an $\mathrm{Fm}^1_\forall(\mathcal{L})$-equation $\varphi \approx \psi$ is *valid* in $\mathfrak{S}$, denoted by $\mathfrak{S} \models \varphi \approx \psi$, if $\|\varphi\|^{\mathfrak{S}}_u = \|\psi\|^{\mathfrak{S}}_u$ for all $u \in S$. We also say that an $\mathrm{Fm}^1_\forall(\mathcal{L})$-equation $\varphi \approx \psi$ is a *(sentential) semantical consequence* of a set of $\mathrm{Fm}^1_\forall(\mathcal{L})$-equations T with respect to a class $\mathcal{K}$ of complete $\mathcal{L}$-lattices, denoted by $T \models^\forall_\mathcal{K} \varphi \approx \psi$, if $\mathfrak{S} \models \varphi \approx \psi$ for any $\mathbf{A} \in \mathcal{K}$ and $\mathbf{A}$-structure $\mathfrak{S}$ satisfying $\mathfrak{S} \models \varphi' \approx \psi'$ for all $\varphi' \approx \psi' \in T$.[4]

Now let $\mathrm{Fm}_\Box(\mathcal{L})$ be the set of propositional formulas $\alpha, \beta, \ldots$ constructed using propositional variables $\{p_i\}_{i\in\mathbb{N}}$, connectives in $\mathcal{L}$, and unary connectives $\Box, \Diamond$, and call $\alpha \approx \beta$ an $\mathrm{Fm}_\Box(\mathcal{L})$-*equation* for any $\alpha, \beta \in \mathrm{Fm}_\Box(\mathcal{L})$. The standard translation functions $(-)^*$ and $(-)^\circ$ between $\mathrm{Fm}^1_\forall(\mathcal{L})$ and $\mathrm{Fm}_\Box(\mathcal{L})$ are defined inductively as follows:

$$
\begin{aligned}
(P_i(x))^* &= p_i & p_i^\circ &= P_i(x) & \\
(\star(\varphi_1,\ldots,\varphi_n))^* &= \star(\varphi_1^*,\ldots,\varphi_n^*) & (\star(\alpha_1,\ldots,\alpha_n))^\circ &= \star(\alpha_1^\circ,\ldots,\alpha_n^\circ) & \star \in \mathcal{L}_n\\
((\forall x)\varphi)^* &= \Box\varphi^* & (\Box\alpha)^\circ &= (\forall x)\alpha^\circ & \\
((\exists x)\varphi)^* &= \Diamond\varphi^* & (\Diamond\alpha)^\circ &= (\exists x)\alpha^\circ. &
\end{aligned}
$$

These translations extend in the obvious way also to (sets of) $\mathrm{Fm}^1_\forall(\mathcal{L})$-equations and $\mathrm{Fm}_\Box(\mathcal{L})$-equations.

Clearly $(\varphi^*)^\circ = \varphi$ for any $\varphi \in \mathrm{Fm}^1_\forall(\mathcal{L})$ and $(\alpha^\circ)^* = \alpha$ for any $\alpha \in \mathrm{Fm}_\Box(\mathcal{L})$; hence we can alternate between first-order and modal notations as convenient. In particular, given any class $\mathcal{K}$ of complete $\mathcal{L}$-lattices, we obtain an equational consequence relation on $\mathrm{Fm}_\Box(\mathcal{L})$ corresponding to $\models^\forall_\mathcal{K}$. Therefore, to find an algebraic semantics for a one-variable lattice-valued logic, we seek a (natural) axiomatization of a variety $\mathcal{V}$ of algebras in the signature of $\mathrm{Fm}_\Box(\mathcal{L})$ such that $\models^\forall_\mathcal{K}$ corresponds, via the above translations, to equational consequence in $\mathcal{V}$. More precisely, let us call a homomorphism from the formula algebra with universe $\mathrm{Fm}_\Box(\mathcal{L})$ to $\mathbf{A} \in \mathcal{V}$ an $\mathbf{A}$-*evaluation*, and define for any set $\Sigma \cup \{\alpha \approx \beta\}$ of $\mathrm{Fm}_\Box(\mathcal{L})$-equations,

$$
\begin{aligned}
\Sigma \models_\mathcal{V} \alpha \approx \beta \;:\Longleftrightarrow\; & f(\alpha) = f(\beta) \text{ for every } \mathbf{A} \in \mathcal{V} \text{ and } \mathbf{A}\text{-evaluation } f\\
& \text{such that } f(\alpha') = f(\beta') \text{ for all } \alpha' \approx \beta' \in \Sigma.
\end{aligned}
$$

Then $\mathcal{V}$ should satisfy for any set of $\mathrm{Fm}^1_\forall(\mathcal{L})$-equations $T \cup \{\varphi \approx \psi\}$,

$$T \models^\forall_\mathcal{K} \varphi \approx \psi \iff T^* \models_\mathcal{V} \varphi^* \approx \psi^*.$$

In Section 4, we solve this problem for the case where $\mathcal{K}$ consists of the complete members of a variety of $\mathcal{L}$-lattices that admits regular completions and has the superamalgamation property (Corollary 4.2).

[4] These notions can be extended to arbitrary (classes of) $\mathcal{L}$-lattices by saying that $\|(\forall x)\varphi\|^{\mathfrak{S}}_u$ or $\|(\exists x)\varphi\|^{\mathfrak{S}}_u$ is *undefined* if the corresponding infimum or supremum fails to exist and that $\mathfrak{S}$ is *safe* if $\|\varphi\|^{\mathfrak{S}}_u$ is defined for all $\varphi \in \mathrm{Fm}^1_\forall(\mathcal{L})$ and $u \in S$. Clearly, if $\mathcal{K}$ is a class of complete $\mathcal{L}$-lattices, then the two notions of semantical consequence coincide, and if $\mathcal{K}$ admits regular completions, then $\models^\forall_\mathcal{K} = \models^\forall_{\overline{\mathcal{K}}}$.

Example 2.2 If $\mathcal{K}$ is $\overline{\mathcal{BA}}$ or $\overline{\mathcal{HA}}$, then $\models^{\forall}_{\mathcal{K}}$ is semantical consequence in the one-variable fragment of first-order classical logic or intuitionistic logic, and corresponds to equational consequence in monadic Boolean algebras [17] or monadic Heyting algebras [4, 20], respectively. Similarly, if $\mathcal{K}$ is $\overline{\mathcal{GA}}$, then $\models^{\forall}_{\mathcal{K}}$ is semantical consequence in the one-variable fragment of the first-order logic of linear frames [12], which corresponds to equational consequence in prelinear monadic Heyting algebras [6]. On the other hand, if $\mathcal{K}$ is the class of totally ordered members of $\overline{\mathcal{GA}}$, then $\models^{\forall}_{\mathcal{K}}$ is semantical consequence in the one-variable fragment of first-order Gödel logic, the first-order logic of linear frames with constant domains, which corresponds to equational consequence in monadic Gödel algebras, i.e., prelinear monadic Heyting algebras satisfying the constant domain axiom $\Box(\Box x \vee y) \approx \Box x \vee \Box y$ [7]. Finally, semantical consequence in the one-variable fragment of first-order Łukasiewicz logic is obtained by taking $\mathcal{K}$ to be the class of totally ordered members of $\overline{\mathcal{MV}}$ and corresponds to equational consequence in monadic MV-algebras [8, 13, 23].

3 An Algebraic Semantics

An *m-lattice* is an algebraic structure $\langle L, \wedge, \vee, \Box, \Diamond \rangle$ such that $\langle L, \wedge, \vee \rangle$ is a lattice and the following equations are satisfied:

$$
\begin{array}{llll}
(\mathrm{L1}_\Box) & \Box x \wedge x \approx \Box x & (\mathrm{L1}_\Diamond) & \Diamond x \vee x \approx \Diamond x \\
(\mathrm{L2}_\Box) & \Box(x \wedge y) \approx \Box x \wedge \Box y & (\mathrm{L2}_\Diamond) & \Diamond(x \vee y) \approx \Diamond x \vee \Diamond y \\
(\mathrm{L3}_\Box) & \Box\Diamond x \approx \Diamond x & (\mathrm{L3}_\Diamond) & \Diamond\Box x \approx \Box x.
\end{array}
$$

Recalling that $\alpha \leq \beta$ stands for $\alpha \wedge \beta \approx \alpha$ and implies $\alpha \vee \beta \approx \beta$ in any lattice, it is easy to check that every m-lattice satisfies the following (quasi-)equations:

$$
\begin{array}{llll}
(\mathrm{L4}_\Box) & \Box\Box x \approx \Box x & (\mathrm{L4}_\Diamond) & \Diamond\Diamond x \approx \Diamond x \\
(\mathrm{L5}_\Box) & x \leq y \implies \Box x \leq \Box y & (\mathrm{L5}_\Diamond) & x \leq y \implies \Diamond x \leq \Diamond y.
\end{array}
$$

Let $\mathcal{L}$ again be a fixed lattice-oriented signature. An *m-$\mathcal{L}$-lattice* is an algebraic structure $\langle \mathbf{A}, \Box, \Diamond \rangle$ such that $\mathbf{A}$ is an $\mathcal{L}$-lattice, $\langle A, \wedge, \vee, \Box, \Diamond \rangle$ is an m-lattice, and the following equation is satisfied for each $\star \in \mathcal{L}_n$ ($n \in \mathbb{N}$):

$$(\star_\Box) \qquad \Box(\star(\Box x_1, \ldots, \Box x_n)) \approx \star(\Box x_1, \ldots, \Box x_n).$$

It follows from $(\star_\Box)$, $(\mathrm{L3}_\Box)$, and $(\mathrm{L3}_\Diamond)$ that $\langle \mathbf{A}, \Box, \Diamond \rangle$ also satisfies the following equation for each $\star \in \mathcal{L}_n$ ($n \in \mathbb{N}$):

$$(\star_\Diamond) \qquad \Diamond(\star(\Diamond x_1, \ldots, \Diamond x_n)) \approx \star(\Diamond x_1, \ldots, \Diamond x_n).$$

Given any class $\mathcal{K}$ of $\mathcal{L}$-lattices, we let $m\mathcal{K}$ denote the class of all m-$\mathcal{L}$-lattices $\langle \mathbf{A}, \Box, \Diamond \rangle$ such that $\mathbf{A} \in \mathcal{K}$. Clearly, if $\mathcal{K}$ is a variety, then also $m\mathcal{K}$ is a variety.

Example 3.1 It is easily checked that $m\mathcal{BA}$ and $m\mathcal{HA}$ are the varieties of monadic Boolean algebras [17] and monadic Heyting algebras [20], respectively. Similarly, $m\mathcal{GA}$ is the variety of prelinear monadic Heyting algebras [6], while the subvariety of $m\mathcal{GA}$ satisfying the constant domain equation is the variety of monadic Gödel algebras [7]. However, $m\mathcal{MV}$ is not the variety of monadic MV-algebras considered in [8,13,23], which satisfy the equation $\Diamond x \cdot \Diamond x \approx \Diamond(x \cdot x)$. To see this, consider the MV-algebra $\text{Ł}_3 = \langle \{0, \frac{1}{2}, 1\}, \wedge, \vee, \cdot, \to, 0, 1 \rangle$ (in the language of FL_e-algebras) with the usual order, where $x \cdot y := \max(0, x+y-1)$, $x \to y := \min(1, 1-x+y)$. Let $\Box 0 = \Box \frac{1}{2} = \Diamond 0 = 0$ and $\Box 1 = \Diamond \frac{1}{2} = \Diamond 1 = 1$. Then $\langle \text{Ł}_3, \Box, \Diamond \rangle \in m\mathcal{MV}$, but $\Diamond \frac{1}{2} \cdot \Diamond \frac{1}{2} = 1 \cdot 1 = 1 \neq 0 = \Diamond 0 = \Diamond(\frac{1}{2} \cdot \frac{1}{2})$.

We now establish a useful representation theorem for m-$\mathcal{L}$-lattices that will be crucial in the proof of the functional representation theorem for certain varieties in the next section.

Lemma 3.2 *Let $\langle \mathbf{A}, \Box, \Diamond \rangle$ be any m-$\mathcal{L}$-lattice. Then $\Box A := \{\Box a \mid a \in A\}$ is a subuniverse of $\mathbf{A}$ satisfying $\Box A = \Diamond A := \{\Diamond a \mid a \in A\}$ and for any $a \in A$,*

$$\Box a = \max\{b \in \Box A \mid b \leq a\} \quad \text{and} \quad \Diamond a = \min\{b \in \Box A \mid a \leq b\}.$$

We let $\Box \mathbf{A}$ denote the subalgebra of $\mathbf{A}$ with universe $\Box A$.

Proof. The fact that $\Box A$ is a subuniverse of $\mathbf{A}$ is a direct consequence of $(\star_\Box)$, and the fact that $\Box A = \Diamond A$ follows from $(\text{L3}_\Box)$ and $(\text{L3}_\Diamond)$. Moreover, if $b \in \Box A$ satisfies $b \leq a$, then $b = \Box b \leq \Box a$, by $(\text{L4}_\Box)$ and $(\text{L5}_\Box)$. But also $\Box a \leq a$, by $(\text{L1}_\Box)$, so $\Box a = \max\{b \in \Box A \mid b \leq a\}$. Similarly, $\Diamond a = \min\{b \in \Box A \mid a \leq b\}$. □

A sublattice $\mathbf{L}_0$ of a lattice $\mathbf{L}$ is said to be *relatively complete* if for any $a \in L$, the set $\{b \in L_0 \mid b \leq a\}$ contains a maximum and the set $\{b \in L_0 \mid a \leq b\}$ contains a minimum. Equivalently, $\mathbf{L}_0$ is relatively complete if the inclusion map f_0 from L_0 to L has left and right adjoints: that is, there exist order-preserving maps $\Box \colon L \to L_0$, $\Diamond \colon L \to L_0$ satisfying for $a \in L$, $b \in L_0$,

$$f_0(b) \leq a \iff b \leq \Box a \quad \text{and} \quad a \leq f_0(b) \iff \Diamond a \leq b.$$

For convenience, we also say that a subalgebra $\mathbf{A}_0$ of an $\mathcal{L}$-lattice $\mathbf{A}$ is relatively complete if this is the case for the lattice reducts. By Lemma 3.2, the subalgebra $\Box \mathbf{A}$ of $\mathbf{A}$ is relatively complete for any m-$\mathcal{L}$-lattice $\langle \mathbf{A}, \Box, \Diamond \rangle$. The following result establishes a converse.

Lemma 3.3 *Let $\mathbf{A}_0$ be a relatively complete subalgebra of an $\mathcal{L}$-lattice $\mathbf{A}$, and define $\Box_0 a := \max\{b \in A_0 \mid b \leq a\}$ and $\Diamond_0 a := \min\{b \in A_0 \mid a \leq b\}$ for each $a \in A$. Then $\langle \mathbf{A}, \Box_0, \Diamond_0 \rangle$ is an m-$\mathcal{L}$-lattice and $\Box_0 A = \Diamond_0 A = A_0$.*

Proof. It is straightforward to check that $\langle A, \wedge, \vee, \Box_0, \Diamond_0 \rangle$ is an m-lattice; for example, it satisfies $(\text{L2}_\Box)$, since for any $a_1, a_2 \in A$,

$$\begin{aligned}
\Box_0(a_1 \wedge a_2) &= \max\{b \in A_0 \mid b \leq a_1 \wedge a_2\} \\
&= \max\{b \in A_0 \mid b \leq a_1 \text{ and } b \leq a_2\} \\
&= \max\{b \in A_0 \mid b \leq a_1\} \wedge \max\{b \in A_0 \mid b \leq a_2\} \\
&= \Box_0 a_1 \wedge \Box_0 a_2.
\end{aligned}$$

Since $\mathbf{A}_0$ is a subalgebra of $\mathbf{A}$, clearly $\langle \mathbf{A}, \Box_0, \Diamond_0 \rangle$ also satisfies $(\star_\Box)$. Hence $\langle \mathbf{A}, \Box_0, \Diamond_0 \rangle$ is an m-$\mathcal{L}$-lattice and $\Box_0 A = \Diamond_0 A = A_0$. □

Combining Lemmas 3.2 and 3.3, we obtain the following representation theorem for m-$\mathcal{L}$-lattices.

Theorem 3.4 *Let $\mathcal{K}$ be any class of $\mathcal{L}$-lattices. Then there exists a one-to-one correspondence between the members of $m\mathcal{K}$ and ordered pairs $\langle \mathbf{A}, \mathbf{A}_0 \rangle$, where $\mathbf{A} \in \mathcal{K}$ and $\mathbf{A}_0$ is a relatively complete subalgebra of $\mathbf{A}$, implemented by the maps $\langle \mathbf{A}, \Box, \Diamond \rangle \mapsto \langle \mathbf{A}, \Box\mathbf{A} \rangle$ and $\langle \mathbf{A}, \mathbf{A}_0 \rangle \mapsto \langle \mathbf{A}, \Box_0, \Diamond_0 \rangle$.*

We now show that m-$\mathcal{L}$-lattices encompass the algebraic semantics of the one-variable lattice-valued logics defined in Section 2.

Proposition 3.5 *Let $\mathbf{A}$ be any complete $\mathcal{L}$-lattice and let W be any set. Then $\langle \mathbf{A}^W, \Box, \Diamond \rangle$ is an m-$\mathcal{L}$-lattice, where the operations of $\mathbf{A}^W$ are defined pointwise and for each $f \in A^W$ and $u \in W$,*

$$\Box f(u) = \bigwedge_{v \in W} f(v) \quad \textit{and} \quad \Diamond f(u) = \bigvee_{v \in W} f(v).$$

Moreover, if $\mathbf{A} \in \mathcal{V}$ for some variety $\mathcal{V}$ of $\mathcal{L}$-lattices, then $\langle \mathbf{A}^W, \Box, \Diamond \rangle \in m\mathcal{V}$.

Proof. Since $\mathbf{A}$ is an $\mathcal{L}$-lattice, $\mathbf{A}^W$, with operations defined pointwise, is also an $\mathcal{L}$-lattice. It is also easy to check that $\langle A^W, \wedge, \vee, \Box, \Diamond \rangle$ is an m-lattice; for example, $\langle A^W, \wedge, \vee, \Box, \Diamond \rangle$ satisfies (L2$_\Box$), since for any $f, g \in A^W$ and $u \in W$,

$$\Box(f \wedge g)(u) = \bigwedge_{v \in W} (f \wedge g)(v) = (\bigwedge_{v \in W} f(v)) \wedge (\bigwedge_{v \in W} g(v)) = (\Box f \wedge \Box g)(u).$$

Moreover, for any $\star \in \mathcal{L}_n$ ($n \in \mathbb{N}$), $f_1, \ldots, f_n \in A^W$, and $u \in W$,

$$\begin{aligned}
\Box(\star(\Box f_1, \ldots, \Box f_n))(u) &= \bigwedge_{v \in W} \star(\Box f_1, \ldots, \Box f_n)(v) \\
&= \bigwedge_{v \in W} \star(\Box f_1(v), \ldots, \Box f_n(v)) \\
&= \star(\Box f_1(u), \ldots, \Box f_n(u)) \\
&= \star(\Box f_1, \ldots, \Box f_n)(u),
\end{aligned}$$

noting that in the last-but-one equality we have used the fact that $\Box f_i(v) = \Box f_i(u)$ for each $v \in W$ and $i \in \{1, \ldots, n\}$. Hence $\langle \mathbf{A}^W, \Box, \Diamond \rangle$ satisfies ($\star_\Box$). Finally, if $\mathbf{A} \in \mathcal{V}$ for some variety $\mathcal{V}$ of $\mathcal{L}$-lattices, then, since varieties are closed under taking direct products, also $\mathbf{A}^W \in \mathcal{V}$, and so $\langle \mathbf{A}^W, \Box, \Diamond \rangle \in m\mathcal{V}$. □

Let us call an m-$\mathcal{L}$-lattice $\langle \mathbf{A}^W, \Box, \Diamond \rangle$ defined as described in Proposition 3.5 *full functional*, and say that an m-$\mathcal{L}$-lattice is *functional* if it is isomorphic to a subalgebra of a full functional m-$\mathcal{L}$-lattice. The following identification of the semantics of one-variable lattice-valued logics with evaluations into full functional m-$\mathcal{L}$-lattices is a direct consequence of Proposition 3.5.

Corollary 3.6 *Let* $\mathbf{A}$ *be any complete* $\mathcal{L}$*-lattice.*

(a) *Given any* $\mathbf{A}$*-structure* $\mathfrak{S} = \langle S, \mathcal{I}\rangle$*, the evaluation* f *for the full functional* m*-*$\mathcal{L}$*-lattice* $\langle \mathbf{A}^S, \Box, \Diamond\rangle$ *defined by* $f(p_i) := \mathcal{I}(P_i)$ *for each* $i \in \mathbb{N}$ *satisfies for all* $\varphi, \psi \in \mathrm{Fm}^1_\forall(\mathcal{L})$ *and* $u \in S$*,*

$$f(\varphi^*)(u) = \|\varphi\|^{\mathfrak{S}}_u \quad \text{and} \quad \mathfrak{S} \models \varphi \approx \psi \iff f(\varphi^*) = f(\psi^*).$$

(b) *Given any evaluation* g *for a full functional* m*-*$\mathcal{L}$*-lattice* $\langle \mathbf{A}^W, \Box, \Diamond\rangle$*, the* $\mathbf{A}$*-structure* $\mathfrak{W} = \langle W, \mathcal{J}\rangle$*, where* $\mathcal{J}(P_i) := g(p_i)$ *for each* $i \in \mathbb{N}$*, satisfies for all* $\varphi, \psi \in \mathrm{Fm}^1_\forall(\mathcal{L})$ *and* $u \in W$*,*

$$g(\varphi^*)(u) = \|\varphi\|^{\mathfrak{W}}_u \quad \text{and} \quad \mathfrak{W} \models \varphi \approx \psi \iff g(\varphi^*) = g(\psi^*).$$

Corollary 3.7 *For any variety* $\mathcal{V}$ *of* $\mathcal{L}$*-lattices and set of* $\mathrm{Fm}^1_\forall(\mathcal{L})$*-equations* $T \cup \{\varphi \approx \psi\}$*,*

$$T^* \models_{m\mathcal{V}} \varphi^* \approx \psi^* \quad \Longrightarrow \quad T \models^{\forall}_{\mathcal{V}} \varphi \approx \psi.$$

Moreover, the converse also holds if every member of $m\mathcal{V}$ *is functional.*

4 A Functional Representation Theorem

In this section, we establish a functional representation theorem for $m\mathcal{K}$ when $\mathcal{K}$ is a class of $\mathcal{L}$-lattices satisfying certain conditions, following very closely a proof of the same result for monadic Heyting algebras [3, Theorem 3.6].

Let $\mathcal{K}$ be any class of $\mathcal{L}$-lattices. A *V-formation* in $\mathcal{K}$ is a 5-tuple $\langle \mathbf{A}, \mathbf{B}_1, \mathbf{B}_2, f_1, f_2\rangle$ consisting of $\mathbf{A}, \mathbf{B}_1, \mathbf{B}_2 \in \mathcal{K}$ and embeddings $f_1 \colon \mathbf{A} \to \mathbf{B}_1$, $f_2 \colon \mathbf{A} \to \mathbf{B}_2$. An *amalgam* in $\mathcal{K}$ of this V-formation is a triple $\langle \mathbf{C}, g_1, g_2\rangle$ consisting of $\mathbf{C} \in \mathcal{K}$ and embeddings $g_1 \colon \mathbf{B}_1 \to \mathbf{C}$, $g_2 \colon \mathbf{B}_2 \to \mathbf{C}$ such that $g_1 \circ f_1 = g_2 \circ f_2$. It is called a *superamalgam* if for any $b_1 \in B_1$, $b_2 \in B_2$ and distinct $i, j \in \{1, 2\}$ such that $g_i(b_i) \leq g_j(b_j)$, there exists an $a \in A$ such that $g_i(b_i) \leq g_i \circ f_i(a) = g_j \circ f_j(a) \leq g_j(b_j)$. The class $\mathcal{K}$ is said to have the *superamalgamation property* if any V-formation in $\mathcal{K}$ has a superamalgam in $\mathcal{K}$.

Theorem 4.1 *Let* $\mathcal{K}$ *be a class of* $\mathcal{L}$*-lattices that is closed under subalgebras and direct limits, admits regular completions, and has the superamalgamation property. Then any member of* $m\mathcal{K}$ *is functional.*

Proof. Consider any $\langle \mathbf{A}, \Box, \Diamond\rangle \in m\mathcal{K}$. Then $\mathbf{A} \in \mathcal{K}$ and, since $\mathcal{K}$ is closed under subalgebras, also $\Box\mathbf{A} \in \mathcal{K}$. We let W denote the set of positive natural numbers and define inductively a sequence of $\mathcal{L}$-lattices $\langle \mathbf{A}_i\rangle_{i\in W}$ in $\mathcal{K}$ and sequences of embeddings $\langle f_i \colon \Box\mathbf{A} \to \mathbf{A}_i\rangle_{i\in W}$, $\langle g_i \colon \mathbf{A} \to \mathbf{A}_i\rangle_{i\in W}$, and $\langle s_i \colon \mathbf{A}_{i-1} \to \mathbf{A}_i\rangle_{i\in W}$. Let $\mathbf{A}_0 := \mathbf{A}$. By assumption, there exists for each $i \in W$, a superamalgam $\langle \mathbf{A}_i, s_i, g_i\rangle$ of the V-formation $\langle \Box\mathbf{A}, \mathbf{A}_{i-1}, \mathbf{A}, f_{i-1}, f_0\rangle$, where $f_0 \colon \Box A \to A$ is the inclusion map and $f_i := s_i \circ f_{i-1} = g_i \circ f_0 = g_i|_{\Box A}$.

Now let $\mathbf{L}$ be the direct limit of the system $\langle\langle \mathbf{A}_i, s_{i+1}\rangle\rangle_{i\in W}$ with associated embeddings $\langle l_i \colon \mathbf{A}_i \to \mathbf{L}\rangle_{i\in W}$. Then, by assumption, $\mathbf{L} \in \mathcal{K}$, and there exist a complete $\mathcal{L}$-lattice $\overline{\mathbf{L}} \in \mathcal{K}$ and an embedding $h \colon \mathbf{L} \to \overline{\mathbf{L}}$ that preserves all existing meets and joins of $\mathbf{L}$. We depict the first two amalgamation steps of this construction in the following diagram:

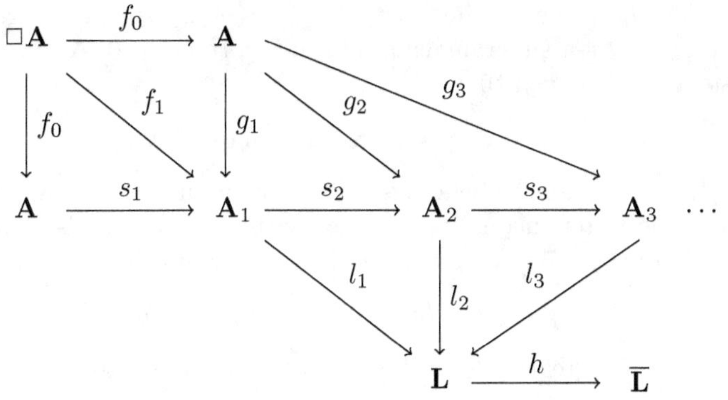

To show that $\langle \mathbf{A}, \Box, \Diamond \rangle$ is functional, it suffices to prove that the following map into the full functional m-$\mathcal{L}$-lattice $\langle \overline{\mathbf{L}}^W, \Box, \Diamond \rangle$ is an embedding:

$$f\colon \langle \mathbf{A}, \Box, \Diamond \rangle \to \langle \overline{\mathbf{L}}^W, \Box, \Diamond \rangle; \quad a \mapsto \langle h \circ l_i \circ g_i(a) \rangle_{i \in W}.$$

It is easy to see that f is an embedding of $\mathcal{L}$-lattices, so it remains to show that $f(\Box a) = \Box f(a)$ and $f(\Diamond a) = \Diamond f(a)$ for all $a \in A$. Fix some $a \in A$. Considering just the case of $\Box$ (since the case of $\Diamond$ is analogous), we obtain

$$\begin{aligned} f(\Box a) &= \langle h \circ l_i \circ g_i(\Box a) \rangle_{i \in W} \\ &\overset{(1)}{=} \langle h(\bigwedge_{j \in W} l_j \circ g_j(a)) \rangle_{i \in W} \\ &\overset{(2)}{=} \langle \bigwedge_{j \in W} h \circ l_j \circ g_j(a) \rangle_{i \in W} \\ &= \Box \langle h \circ l_i \circ g_i(a) \rangle_{i \in W} \\ &= \Box f(a). \end{aligned}$$

To justify (1), it suffices to fix an $i \in W$ and show that $l_i \circ g_i(\Box a)$ is the greatest lower bound of the set $S := \{l_j \circ g_j(a) \mid j \in W\}$. This implies that S has an infimum and so also (2) follows by the definition of regular completions.

We start by showing that $l_i \circ g_i(\Box a) = l_j \circ g_j(\Box a)$ for all $j \in W$. Clearly, this follows from the fact that for any $k \in W$,

$$l_k \circ g_k(\Box a) = l_k \circ f_k(\Box a) = l_{k+1} \circ s_{k+1} \circ f_k(\Box a) = l_{k+1} \circ g_{k+1}(\Box a),$$

where the first and the last equality are due to the definition of f_k and the second is due to the definition of direct limits. Using this fact, we can easily show that $l_i \circ g_i(\Box a)$ is a lower bound of S: just observe that since $\Box a \leq a$, for each $j \in W$,

$$l_i \circ g_i(\Box a) = l_j \circ g_j(\Box a) \leq l_j \circ g_j(a).$$

Finally, suppose that $c \in L$ is any lower bound of S. By the definition of direct limits, there exist a $k \in W$ and a $d \in A_k$ such that

$$l_{k+1} \circ s_{k+1}(d) = l_k(d) = c \leq l_{k+1} \circ g_{k+1}(a).$$

Moreover, since l_{k+1} is an embedding, $s_{k+1}(d) \leq g_{k+1}(a)$. Therefore, because $\langle \mathbf{A}_{k+1}, s_{k+1}, g_{k+1} \rangle$ is a superamalgam of the V-formation $\langle \Box\mathbf{A}, \mathbf{A}_k, \mathbf{A}, f_k, f_0 \rangle$, there exists a $b \in \Box A$ such that

$$s_{k+1}(d) \leq s_{k+1} \circ f_k(b) = g_{k+1} \circ f_0(b) \leq g_{k+1}(a).$$

Since s_{k+1} and g_{k+1} are embeddings and f_0 is the inclusion map, $d \leq f_k(b)$ and $b \leq a$. The latter inequality together with $b \in \Box A$ entails that $b \leq \Box a$. Hence also $f_k(b) \leq f_k(\Box a) = g_k(\Box a)$ and, using the first inequality,

$$c = l_k(d) \leq l_k \circ f_k(b) \leq l_k \circ g_k(\Box a) = l_i \circ g_i(\Box a).$$

So $l_i \circ g_i(\Box a)$ is the greatest lower bound of the set S as required. □

Combining Theorem 4.1 with Corollary 3.7 yields the following result.

Corollary 4.2 *Let $\mathcal{V}$ be a variety of $\mathcal{L}$-lattices that admits regular completions and has the superamalgamation property. Then for any set $T \cup \{\varphi \approx \psi\}$ of* $\mathrm{Fm}^1_\forall(\mathcal{L})$*-equations,*

$$T \models^\forall_\mathcal{V} \varphi \approx \psi \quad \iff \quad T^* \models_{m\mathcal{V}} \varphi^* \approx \psi^*.$$

Example 4.3 The variety of lattices admits regular completions and has the superamalgamation property [16]. Hence, by Theorem 4.1, every m-lattice is functional, and, by Corollary 4.2, consequence in the one-variable fragment of first-order lattice logic corresponds to equational consequence in m-lattices.

5 One-Variable Substructural Logics

In this section, we turn our attention to one-variable fragments of first-order substructural logics. Let $\mathcal{L}_s$ be the lattice-ordered signature consisting of binary connectives $\vee$, $\wedge$, $\cdot$, and $\to$, and constant symbols f and e. An $\mathrm{FL_e}$*-algebra* is an $\mathcal{L}_s$-lattice $\mathbf{A} = \langle A, \vee, \wedge, \cdot, \to, \mathrm{f}, \mathrm{e} \rangle$ such that $\langle A, \cdot, \mathrm{e} \rangle$ is a commutative monoid and $\to$ is the residuum of $\cdot$, i.e., $a \cdot b \leq c \iff a \leq b \to c$ for all $a, b, c \in A$.

The class of $\mathrm{FL_e}$-algebras forms a variety $\mathcal{FL}_e$ that serves as an algebraic semantics for the Full Lambek Calculus with exchange (see, e.g., [14]), and subvarieties of $\mathcal{FL}_e$ provide algebraic semantics for other substructural logics. In particular, the varieties $\mathcal{FL}_{ew}$ and $\mathcal{FL}_{ec}$ consist of $\mathrm{FL_e}$-algebras satisfying $\mathrm{f} \leq x \leq \mathrm{e}$ and $x \leq x \cdot x$, respectively. The variety $\mathcal{FL}_{ew} \cap \mathcal{FL}_{ec}$ is term-equivalent to $\mathcal{HA}$ (just identify $\cdot$ and $\wedge$), and $\mathcal{BA}$ and $\mathcal{GA}$ are term-equivalent to the subvarieties of $\mathcal{FL}_{ew} \cap \mathcal{FL}_{ec}$ axiomatized by $(x \to \mathrm{f}) \to \mathrm{f} \approx x$ and $(x \to y) \vee (y \to x) \approx \mathrm{e}$, respectively. $\mathcal{MV}$ is term-equivalent to the subvariety of $\mathcal{FL}_{ew}$ satisfying $x \vee y \approx (x \to y) \to y$.

In combination with residuation, the defining equations of m-$\mathcal{L}_s$-lattices yield further relationships between the propositional and modal connectives.

Proposition 5.1 *Let* $\mathbf{A}$ *be an* $\mathrm{FL_e}$*-algebra. Then any m-$\mathcal{L}_s$-lattice $\langle \mathbf{A}, \Box, \Diamond \rangle$ satisfies the equations*

$$(\mathrm{L6}_\Box) \quad \Box(x \to \Box y) \approx \Diamond x \to \Box y \qquad\qquad (\mathrm{L6}_\Diamond) \quad \Box(\Box x \to y) \approx \Box x \to \Box y.$$

That is, $\langle \mathbf{A}, \Box, \Diamond \rangle$ is a monadic $\mathrm{FL_e}$*-algebra in the sense of* [27, Definition 2.1].

Proof. We just prove ($L6_\Box$), as the proof for ($L6_\Diamond$) is very similar. Consider any $a, b \in A$. Since $a \le \Diamond a$, by ($L1_\Diamond$), also $\Diamond a \to \Box b \le a \to \Box b$. Hence, using ($L5_\Box$), ($L3_\Box$), and ($\star_\Box$),

$$\Diamond a \to \Box b = \Box\Diamond a \to \Box b = \Box(\Box\Diamond a \to \Box b) = \Box(\Diamond a \to \Box b) \le \Box(a \to \Box b).$$

Conversely, since $\Box(a \to \Box b) \le a \to \Box b$, by ($L1_\Box$), it follows by residuation that $a \le \Box(a \to \Box b) \to \Box b$ and hence, using ($L5_\Diamond$), ($L3_\Diamond$), and ($\star_\Diamond$),

$$\Diamond a \le \Diamond(\Box(a \to \Box b) \to \Box b) = \Box(a \to \Box b) \to \Box b.$$

By residuation again, $\Box(a \to \Box b) \le \Diamond a \to \Box b$. □

The varieties $\mathcal{FL}_e$, $\mathcal{FL}_{ew}$, and $\mathcal{FL}_{ec}$ are closed under MacNeille completions and have the superamalgamation property (see, e.g., [14]). Hence Theorem 4.1 and Corollary 4.2 yield the following result.

Theorem 5.2 *Let $\mathcal{V} \in \{\mathcal{FL}_e, \mathcal{FL}_{ew}, \mathcal{FL}_{ec}\}$.*

(a) *Any member of $m\mathcal{V}$ is functional.*

(b) *For any set $T \cup \{\varphi \approx \psi\}$ of $\mathrm{Fm}^1_\forall(\mathcal{L}_s)$-equations,*

$$T \models^\forall_\mathcal{V} \varphi \approx \psi \quad \iff \quad T \models_{m\mathcal{V}} \varphi^* \approx \psi^*.$$

In [10] it was proved that a variety of $\mathrm{FL_e}$-algebras axiomatized relative to $\mathcal{FL}_e$ by equations of a certain simple syntactic form (called "$\mathcal{N}_2$-equations") is closed under MacNeille completions if and only if it has an analytic sequent calculus of a certain form. It has also been proved that many varieties of $\mathrm{FL_e}$-algebras have the superamalgamation property (equivalently, the Craig interpolation property) (see, e.g., [14, 26]), but a precise characterization is not known.

6 A Proof-Theoretic Approach

In this section, we apply proof-theoretic methods to obtain an alternative proof of Theorem 5.2(b). Although no new results are proved, the approach described here may be used to deal with varieties of $\mathcal{L}$-lattices that either do not admit regular completions or lack the superamalgamation property, and may also be used to tackle decidability and complexity issues for one-variable lattice logics.

Let $\mathrm{Fm}^{1+}_\forall(\mathcal{L}_s)$ be the set of first-order $\mathcal{L}_s$-formulas $\varphi, \psi, \chi, \ldots$ constructed inductively from unary predicates $\{P_i\}_{i\in\mathbb{N}}$, variables $\{x\} \cup \{x_i\}_{i\in\mathbb{N}}$, connectives in $\mathcal{L}_s$, and quantifiers $(\forall x)$ and $(\exists x)$ such that no variable x_i is in the scope of a quantifier. Clearly, $\mathrm{Fm}^1_\forall(\mathcal{L}_s) \subseteq \mathrm{Fm}^{1+}_\forall(\mathcal{L}_s)$. We also write $\varphi(\bar{y})$ to denote that the free variables of $\varphi \in \mathrm{Fm}^{1+}_\forall(\mathcal{L}_s)$ belong to the set $\bar{y}$, recalling that a sentence is a formula with no free variables.

For the purposes of this paper, a *sequent* is an ordered pair of finite multisets of first-order $\mathcal{L}_s$-formulas, denoted by $\Gamma \Rightarrow \Delta$, such that Δ contains a most one $\mathcal{L}_s$-formula. We also define for $n \in \mathbb{N}^{>0}$ and $\varphi_1, \ldots, \varphi_n, \psi \in \mathrm{Fm}^{1+}_\forall(\mathcal{L}_s)$,

$$\prod(\varphi_1, \ldots, \varphi_n) := \varphi_1 \cdots \varphi_n, \quad \prod() := \mathrm{e}, \quad \sum(\psi) := \psi, \quad \sum() := \mathrm{f}.$$

The (cut-free) sequent calculus $\forall^+_1\mathsf{FL_e}$ is presented in Figure 1, subject to the following side-conditions:

Axioms

$$\frac{}{\varphi \Rightarrow \varphi}\ (\text{ID}) \qquad \frac{}{\mathsf{f} \Rightarrow}\ (\mathsf{f}\Rightarrow) \qquad \frac{}{\Rightarrow \mathsf{e}}\ (\Rightarrow\mathsf{e})$$

Operation Rules

$$\frac{\Gamma \Rightarrow \Delta}{\Gamma, \mathsf{e} \Rightarrow \Delta}\ (\mathsf{e}\Rightarrow) \qquad \frac{\Gamma \Rightarrow}{\Gamma \Rightarrow \mathsf{f}}\ (\Rightarrow\mathsf{f})$$

$$\frac{\Gamma_1 \Rightarrow \varphi \quad \Gamma_2, \psi \Rightarrow \Delta}{\Gamma_1, \Gamma_2, \varphi \to \psi \Rightarrow \Delta}\ (\to\Rightarrow) \qquad \frac{\Gamma, \varphi \Rightarrow \psi}{\Gamma \Rightarrow \varphi \to \psi}\ (\Rightarrow\to)$$

$$\frac{\Gamma, \varphi, \psi \Rightarrow \Delta}{\Gamma, \varphi \cdot \psi \Rightarrow \Delta}\ (\cdot\Rightarrow) \qquad \frac{\Gamma_1 \Rightarrow \varphi \quad \Gamma_2 \Rightarrow \psi}{\Gamma_1, \Gamma_2 \Rightarrow \varphi \cdot \psi}\ (\Rightarrow\cdot)$$

$$\frac{\Gamma, \varphi \Rightarrow \Delta}{\Gamma, \varphi \wedge \psi \Rightarrow \Delta}\ (\wedge\Rightarrow)_1 \qquad \frac{\Gamma \Rightarrow \varphi}{\Gamma \Rightarrow \varphi \vee \psi}\ (\Rightarrow\vee)_1$$

$$\frac{\Gamma, \psi \Rightarrow \Delta}{\Gamma, \varphi \wedge \psi \Rightarrow \Delta}\ (\wedge\Rightarrow)_2 \qquad \frac{\Gamma \Rightarrow \psi}{\Gamma \Rightarrow \varphi \vee \psi}\ (\Rightarrow\vee)_2$$

$$\frac{\Gamma, \varphi \Rightarrow \Delta \quad \Gamma, \psi \Rightarrow \Delta}{\Gamma, \varphi \vee \psi \Rightarrow \Delta}\ (\vee\Rightarrow) \qquad \frac{\Gamma \Rightarrow \varphi \quad \Gamma \Rightarrow \psi}{\Gamma \Rightarrow \varphi \wedge \psi}\ (\Rightarrow\wedge)$$

$$\frac{\Gamma, \varphi(t) \Rightarrow \Delta}{\Gamma, (\forall x)\varphi(x) \Rightarrow \Delta}\ (\forall\Rightarrow) \qquad \frac{\Gamma \Rightarrow \psi(y)}{\Gamma \Rightarrow (\forall x)\psi(x)}\ (\Rightarrow\forall)$$

$$\frac{\Gamma, \varphi(y) \Rightarrow \Delta}{\Gamma, (\exists x)\varphi(x) \Rightarrow \Delta}\ (\exists\Rightarrow) \qquad \frac{\Gamma \Rightarrow \psi(t)}{\Gamma \Rightarrow (\exists x)\psi(x)}\ (\Rightarrow\exists)$$

Fig. 1. The Sequent Calculus $\forall_1^+\mathsf{FL_e}$

(i) the term t occurring in $(\forall\Rightarrow)$ and $(\Rightarrow\exists)$ is a variable occurring in the conclusion of the rule;

(ii) the variable y occurring in the premise of $(\Rightarrow\forall)$ and $(\exists\Rightarrow)$ does not occur freely in the conclusion of the rule.

$\forall_1^+\mathsf{FL_{ew}}$ and $\forall_1^+\mathsf{FL_{ec}}$ are defined as the extensions of $\forall_1^+\mathsf{FL_e}$ with, respectively,

$$\frac{\Gamma_1 \Rightarrow \Delta_1}{\Gamma_1, \Gamma_2 \Rightarrow \Delta_1, \Delta_2}\ (\text{w}) \qquad \text{and} \qquad \frac{\Gamma_1, \Gamma_2, \Gamma_2 \Rightarrow \Delta}{\Gamma_1, \Gamma_2 \Rightarrow \Delta}\ (\text{c})$$

If there exists a derivation d of a sequent $\Gamma \Rightarrow \Delta$ in a sequent calculus C, we write either $d \vdash_{\mathsf{C}} \Gamma \Rightarrow \Delta$ or $\vdash_{\mathsf{C}} \Gamma \Rightarrow \Delta$, and let $\text{md}(d)$ denote the maximum number of applications of $(\forall\Rightarrow)$, $(\Rightarrow\forall)$, $(\exists\Rightarrow)$, $(\Rightarrow\exists)$ on a branch of d.

The next result follows directly from well-known completeness and cut-elimination results for sequent calculi for (first-order) substructural logics.

Proposition 6.1 ([18,21]) *Let $\mathcal{V}$ be $\mathcal{FL}_e$, $\mathcal{FL}_{ew}$, or $\mathcal{FL}_{ec}$, and let* C *be* $\forall_1^+\mathsf{FL_e}$, $\forall_1^+\mathsf{FL_{ew}}$, *or* $\forall_1^+\mathsf{FL_{ec}}$, *respectively. For any sequent* $\Gamma \Rightarrow \Delta$ *consisting only of formulas from* $\mathrm{Fm}^1_\forall(\mathcal{L}_s)$,

$$\vdash_{\mathsf{C}} \Gamma \Rightarrow \Delta \quad \Longleftrightarrow \quad \models^{\forall}_{\mathcal{V}} \prod \Gamma \leq \sum \Delta.$$

We now prove that each of $\forall_1^+\mathsf{FL_e}$, $\forall_1^+\mathsf{FL_{ew}}$, and $\forall_1^+\mathsf{FL_{ec}}$ satisfies a certain bounded interpolation property.

Theorem 6.2 *Let* $\mathsf{C} \in \{\forall_1^+\mathsf{FL_e}, \forall_1^+\mathsf{FL_{ew}}, \forall_1^+\mathsf{FL_{ec}}\}$. *If* $d \vdash_{\mathsf{C}} \Gamma(\bar{y}), \Pi(\bar{z}) \Rightarrow \Delta(\bar{z})$, *where* $\bar{y} \cap \bar{z} = \emptyset$, *then there exist a sentence* χ *and derivations* d_1, d_2 *such that* $\mathrm{md}(d_1), \mathrm{md}(d_2) \leq \mathrm{md}(d)$, $d_1 \vdash_{\mathsf{C}} \Gamma(\bar{y}) \Rightarrow \chi$, *and* $d_2 \vdash_{\mathsf{C}} \Pi(\bar{z}), \chi \Rightarrow \Delta(\bar{z})$.

Proof. We proceed by induction on the height of d and consider the last rule applied. Note that if $\bar{y}$ or $\bar{z}$ are empty — in particular, if $\Gamma(\bar{y}), \Pi(\bar{z}) \Rightarrow \Delta(\bar{z})$ is an instance of an axiom — we can take $\chi := \prod \Gamma$ or $\chi := \prod \Pi$, respectively, so we may assume that this is not the case. Note also that the additional cases of (W) for $\forall_1^+\mathsf{FL_{ew}}$ and (C) for $\forall_1^+\mathsf{FL_{ec}}$ follow directly from an application of the induction hypothesis. We just consider here the quantifier rules ($\forall\Rightarrow$) and ($\Rightarrow\forall$), dealing with the remaining rules in the appendix.

- ($\forall\Rightarrow$): Suppose first that $\Gamma(\bar{y})$ is $\Gamma'(\bar{y}), (\forall x)\varphi(x)$ and

$$d' \vdash_{\mathsf{C}} \Gamma'(\bar{y}), \varphi(u), \Pi(\bar{z}) \Rightarrow \Delta(\bar{z}),$$

where $\mathrm{md}(d') = \mathrm{md}(d) - 1$. For subcase (i), suppose that $u \in \bar{y}$. By the induction hypothesis, there exist a sentence χ and derivations d_1', d_2 such that $\mathrm{md}(d_1'), \mathrm{md}(d_2) \leq \mathrm{md}(d')$ and

$$d_1' \vdash_{\mathsf{C}} \Gamma'(\bar{y}), \varphi(u) \Rightarrow \chi, \qquad d_2 \vdash_{\mathsf{C}} \Pi(\bar{z}), \chi \Rightarrow \Delta(\bar{z}).$$

Using ($\forall\Rightarrow$), there exists also a derivation d_1 such that $\mathrm{md}(d_1) = \mathrm{md}(d_1') + 1 \leq \mathrm{md}(d') + 1 = \mathrm{md}(d)$ and

$$d_1 \vdash_{\mathsf{C}} \Gamma'(\bar{y}), (\forall x)\varphi(x) \Rightarrow \chi.$$

For subcase (ii), suppose that $u \in \bar{z}$. By the induction hypothesis, there exist a sentence χ' and derivations d_1', d_2' such that $\mathrm{md}(d_1'), \mathrm{md}(d_2') \leq \mathrm{md}(d')$ and

$$d_1' \vdash_{\mathsf{C}} \Gamma'(\bar{y}) \Rightarrow \chi', \qquad d_2' \vdash_{\mathsf{C}} \Pi(\bar{z}), \varphi(u), \chi' \Rightarrow \Delta(\bar{z}).$$

We define $\chi := \chi' \cdot (\forall x)\varphi(x)$ and obtain derivations d_1, d_2 satisfying $\mathrm{md}(d_1) = \mathrm{md}(d_1') \leq \mathrm{md}(d') < \mathrm{md}(d)$, $\mathrm{md}(d_2) = \mathrm{md}(d_2') + 1 \leq \mathrm{md}(d') + 1 = \mathrm{md}(d)$, and

$$d_1 \vdash_{\mathsf{C}} \Gamma'(\bar{y}), (\forall x)\varphi(x) \Rightarrow \chi' \cdot (\forall x)\varphi(x), \qquad d_2 \vdash_{\mathsf{C}} \Pi(\bar{z}), \chi' \cdot (\forall x)\varphi(x) \Rightarrow \Delta(\bar{z}).$$

Suppose next that $\Pi(\bar{z})$ is $\Pi'(\bar{z}), (\forall x)\varphi(x)$ and

$$d' \vdash_{\mathsf{C}} \Gamma(\bar{y}), \Pi'(\bar{z}), \varphi(u) \Rightarrow \Delta(\bar{z}).$$

The case of $u \in \bar{z}$ is very similar to subcase (i) above, so suppose that $u \in \bar{y}$. By the induction hypothesis, there exist a sentence χ' and derivations d_1', d_2' such that $\mathrm{md}(d_1'), \mathrm{md}(d_2') \leq \mathrm{md}(d')$ and

$$d_1' \vdash_{\mathsf{C}} \Gamma'(\bar{y}), \varphi(u) \Rightarrow \chi', \qquad d_2' \vdash_{\mathsf{C}} \Pi'(\bar{z}), \chi' \Rightarrow \Delta(\bar{z}).$$

We let $\chi := (\forall x)\varphi(x) \to \chi'$ and obtain derivations d_1, d_2 satisfying $\mathrm{md}(d_1) = \mathrm{md}(d_1') + 1 \leq \mathrm{md}(d') + 1 = \mathrm{md}(d)$, $\mathrm{md}(d_2) = \mathrm{md}(d_2') < \mathrm{md}(d)$, and

$$d_1 \vdash_{\mathsf{C}} \Gamma(\bar{y}) \Rightarrow (\forall x)\varphi(x) \to \chi', \quad d_2 \vdash_{\mathsf{C}} \Pi'(\bar{z}), (\forall x)\varphi(x), (\forall x)\varphi(x) \to \chi' \Rightarrow \Delta(\bar{z}).$$

• $(\Rightarrow\forall)$: Suppose that $\Delta(\bar{z})$ is $(\forall x)\varphi(x)$ and for some variable u that does not occur freely in $\Gamma(\bar{y}), \Pi(\bar{z}) \Rightarrow (\forall x)\varphi(x)$,

$$d' \vdash_{\mathsf{C}} \Gamma(\bar{y}), \Pi(\bar{z}) \Rightarrow \varphi(u),$$

where $\mathrm{md}(d') = \mathrm{md}(d) - 1$. By the induction hypothesis, there exist a sentence χ and derivations d_1, d_2' such that $\mathrm{md}(d_1), \mathrm{md}(d_2') \leq \mathrm{md}(d')$ and

$$d_1 \vdash_{\mathsf{C}} \Gamma(\bar{y}) \Rightarrow \chi, \qquad d_2' \vdash_{\mathsf{C}} \Pi(\bar{z}), \chi \Rightarrow \varphi(u).$$

An application of $(\Rightarrow\forall)$ yields a derivation d_2 satisfying $\mathrm{md}(d_2) = \mathrm{md}(d_2') + 1 \leq \mathrm{md}(d') + 1 = \mathrm{md}(d)$ and $d_2 \vdash_{\mathsf{C}} \Pi(\bar{z}), \chi \Rightarrow (\forall x)\varphi(x)$. □

Alternative proof of Theorem 5.2(b). The right-to-left direction follows from Corollary 3.7. For the converse, let $\mathcal{V}$ be $\mathcal{FL}_e$, $\mathcal{FL}_{ew}$, or $\mathcal{FL}_{ec}$, and let C be $\forall_1^+\mathsf{FL_e}$, $\forall_1^+\mathsf{FL_{ew}}$, or $\forall_1^+\mathsf{FL_{ec}}$, respectively. We note first that due to compactness and the local deduction theorem for $\models_{\mathcal{V}}^{\forall}$ (see [11, Sections 4.6, 4.8]), we can restrict to the case where $T = \emptyset$. Hence, by Proposition 6.1, it suffices to prove that for any sequent $\Gamma \Rightarrow \Delta$ consisting only of formulas from $\mathrm{Fm}_{\forall}^1(\mathcal{L}_s)$,

$$d \vdash_{\mathsf{C}} \Gamma \Rightarrow \Delta \quad \Longrightarrow \quad \models_{m\mathcal{V}} (\textstyle\prod \Gamma)^* \leq (\sum \Delta)^*.$$

We proceed by induction on the lexicographically ordered pair $\langle \mathrm{md}(d), \mathrm{ht}(d) \rangle$, where $\mathrm{ht}(d)$ is the height of the derivation d. The base cases are clear and all the cases for the last application of a rule in d except $(\Rightarrow\forall)$ and $(\exists\Rightarrow)$ follow by applying the induction hypothesis and the equations defining $m\mathcal{V}$. Just note that for each such rule, the premises contain only formulas from $\mathrm{Fm}_{\forall}^1(\mathcal{L}_s)$ with at least one fewer symbol. In particular, for $(\forall\Rightarrow)$ and $(\Rightarrow\exists)$, the term t occurring in the premise must be x and the result follows using $(\mathrm{L1}_\Box)$ or $(\mathrm{L1}_\Diamond)$.

Consider now a last application of $(\Rightarrow\forall)$ in d, where Δ is $(\forall x)\psi(x)$. Then $d' \vdash_{\mathsf{C}} \Gamma \Rightarrow \psi(y)$ for some variable y that does not occur freely in $\Gamma \Rightarrow (\forall x)\psi(x)$, where $\mathrm{md}(d') = \mathrm{md}(d) - 1$. If $y = x$, then x does not occur freely in Γ and the result follows by an application of the induction hypothesis and equations defining $m\mathcal{V}$. Suppose that $y \neq x$. By Theorem 6.2, there exist a sentence χ and derivations d_1, d_2 such that $d_1 \vdash_{\mathsf{C}} \Gamma \Rightarrow \chi$ and $d_2 \vdash_{\mathsf{C}} \chi \Rightarrow \psi(y)$ with $\mathrm{md}(d_1), \mathrm{md}(d_2) \leq \mathrm{md}(d')$. Since χ is a sentence and $y \neq x$, we can substitute in d_2 every free occurrence of x by some new variable z, and then every occurrence of y by x, to obtain a derivation d_2' of $\chi \Rightarrow \psi(x)$ with $\mathrm{md}(d_2') = \mathrm{md}(d_2)$. By the induction hypothesis, $\models_{m\mathcal{V}} (\prod\Gamma)^* \leq \chi^*$ and $\models_{m\mathcal{V}} \chi^* \leq \psi(x)^*$. Since χ is a sentence, the equations defining $m\mathcal{V}$ yield also $\models_{m\mathcal{V}} \chi^* \leq ((\forall x)\psi(x))^*$. So $\models_{m\mathcal{V}} \Gamma^* \leq ((\forall x)\psi(x))^*$.

Consider finally a last application of $(\exists\Rightarrow)$ in d, where Γ is $\Gamma', (\exists x)\psi(x)$. Then $d' \vdash_{\mathsf{c}} \Gamma', \psi(y) \Rightarrow \Delta$ for some variable y that does not occur freely in $\Gamma', (\exists x)\psi(x) \Rightarrow \Delta$, where $\mathrm{md}(d') < \mathrm{md}(d)$. If $y = x$, then x does not occur freely in Γ' or Δ and the result follows by applying the induction hypothesis and equations defining $m\mathcal{V}$. Suppose that $y \neq x$. By Theorem 6.2, there exist a sentence χ and derivations d_1, d_2 such that $d_1 \vdash_{\mathsf{c}} \psi(y) \Rightarrow \chi$ and $d_2 \vdash_{\mathsf{c}} \Gamma', \chi \Rightarrow \Delta$ with $\mathrm{md}(d_1), \mathrm{md}(d_2) \leq \mathrm{md}(d')$. Since χ is a sentence and $y \neq x$, we can substitute in d_1 every free occurrence of x by some new variable z, and then every occurrence of y by x, to obtain a derivation d_1' of $\psi(x) \Rightarrow \chi$ with $\mathrm{md}(d_1') = \mathrm{md}(d')$. By the induction hypothesis, $\models_{m\mathcal{V}} \psi(x)^* \leq \chi^*$ and $\models_{m\mathcal{V}} (\prod(\Gamma', \chi))^* \leq (\sum\Delta)^*$. Since χ is a sentence, the equations defining $m\mathcal{V}$ yield also $\models_{m\mathcal{V}} ((\exists x)\psi(x))^* \leq \chi^*$. So $\models_{m\mathcal{V}} (\prod(\Gamma', (\exists x)\psi(x)))^* \leq (\sum\Delta)^*$. □

7 Concluding Remarks

Let us conclude this paper by sketching a broader perspective. Given some class $\mathcal{K}$ of complete $\mathcal{L}$-lattices, the challenge is to find a (natural) axiomatization of the generalized quasivariety generated by the class of full functional m-$\mathcal{L}$-lattices of the form $\langle \mathbf{A}^W, \Box, \Diamond\rangle$ for some $\mathbf{A} \in \mathcal{K}$ and set W. Corollary 4.2 shows that in the case where $\mathcal{K}$ is the class of complete members of a variety $\mathcal{V}$ that admits regular completions and has the superamalgamation property, this generalized quasivariety is in fact the variety $m\mathcal{V}$. In general, however, we may need to add further axioms to obtain a proper subvariety or even proper subquasivariety or sub-generalized quasivariety of $m\mathcal{V}$.

For example, let $\mathcal{V}$ be a variety of *semilinear* $\mathrm{FL_e}$-algebras, that is, algebras that are isomorphic to a subdirect product of totally ordered $\mathrm{FL_e}$-algebras. In general, such varieties may not admit regular completions (e.g., as shown in [15] for $\mathcal{V} = \mathcal{MV}$), so Corollary 4.2 may not apply. Moreover, using the fact that $\mathcal{V}$ is generated by totally ordered $\mathrm{FL_e}$-algebras, $\models^{\forall}_{\mathcal{V}} (\exists x)\varphi \cdot (\exists x)\varphi \approx (\exists x)(\varphi \cdot \varphi)$, while, as proved in Example 3.1, if $\mathbf{Ł}_3 \in \mathcal{V}$ (e.g., if $\mathcal{V}$ is $\mathcal{MV}$ or the variety of all semilinear $\mathrm{FL_e}$-algebras), then $\not\models_{m\mathcal{V}} \Diamond x \cdot \Diamond x \approx \Diamond(x \cdot x)$.

In fact, for a variety $\mathcal{V}$ of semilinear $\mathrm{FL_e}$-algebras, first-order semantical consequence is typically defined with respect to the class of its complete totally ordered members. In this case, the corresponding generalized quasivariety will satisfy also the constant domain axiom $\Box(\Box x \vee y) \approx \Box x \vee \Box y$. Indeed, if $\mathcal{V}$ is $\mathcal{MV}$, this generalized quasivariety is the variety of monadic MV-algebras axiomatized as the subvariety of $m\mathcal{MV}$ satisfying $\Diamond x \cdot \Diamond x \approx \Diamond(x \cdot x)$ and the constant domain axiom [23]. Interestingly, a proof of this latter result is given in [8] using the fact that the class of totally ordered MV-algebras has the amalgamation property (see also [19, 27] for related results), suggesting that the approach developed in this paper can be adapted to a broader class of one-variable lattice logics.

References

[1] G. Bezhanishvili, *Varieties of monadic Heyting algebras - part I*, Studia Logica **61** (1998), no. 3, 367–402.

[2] G. Bezhanishvili and J. Harding, *MacNeille completions of Heyting algebras*, Houston J. Math. **30** (2004), 937–952.

[3] ———, *Functional monadic Heyting algebras*, Algebra Universalis **48** (2002), 1–10.

[4] R.A. Bull, *MIPC as formalisation of an intuitionist concept of modality*, J. Symb. Log. **31** (1966), 609–616.

[5] X. Caicedo, G. Metcalfe, R. Rodríguez, and J. Rogger, *Decidability in order-based modal logics*, J. Comput. System Sci. **88** (2017), 53–74.

[6] X. Caicedo, G. Metcalfe, R. Rodríguez, and O. Tuyt, *One-variable fragments of intermediate logics over linear frames*, 2022. Inform. and Comput., to appear.

[7] X. Caicedo and R. Rodríguez, *Bi-modal Gödel logic over* $[0,1]$*-valued Kripke frames*, J. Logic Comput. **25** (2015), no. 1, 37–55.

[8] D. Castaño, C. Cimadamore, J.P.D. Varela, and L. Rueda, *Completeness for monadic fuzzy logics via functional algebras*, Fuzzy Sets and Systems **407** (2021), 161–174.

[9] A. Ciabattoni, N. Galatos, and K. Terui, *MacNeille completions of FL-algebras*, Algebra Universalis **66** (2011), no. 4, 405–420.

[10] ———, *Algebraic proof theory for substructural logics: Cut-elimination and completions*, Ann. Pure Appl. Logic **163** (2012), no. 3, 266–290.

[11] P. Cintula and C. Noguera, *Logic and implication*, Springer, 2021.

[12] G. Corsi, *Completeness theorem for Dummett's* LC *quantified*, Studia Logica **51** (1992), 317–335.

[13] A. di Nola and R. Grigolia, *On monadic* MV*-algebras*, Ann. Pure Appl. Logic **128** (2004), no. 1-3, 125–139.

[14] N. Galatos, P. Jipsen, T. Kowalski, and H. Ono, *Residuated lattices: An algebraic glimpse at substructural logics*, Elsevier, 2007.

[15] M. Gehrke and H.A. Priestley, *Non-canonicity of* MV*-algebras*, Houston J. Math. **28** (2002), no. 3, 449–456.

[16] G. Grätzer, *General lattice theory*, 2nd ed., Birkhäuser, 1998.

[17] P.R. Halmos, *Algebraic logic, I. Monadic Boolean algebras*, Compos. Math. **12** (1955), 217–249.

[18] Y. Komori, *Predicate logics without the structural rules*, Studia Logica **45** (1986), no. 4, 393–104.

[19] G. Metcalfe and O. Tuyt, *A monadic logic of ordered abelian groups*, Proc. AiML 2020, 2020, pp. 441–457.

[20] A. Monteiro and O. Varsavsky, *Algebras de Heyting monádicas*, Actas de las X Jornadas de la Unión Matemática Argentina, Bahía Blanca (1957), 52–62.

[21] H. Ono and Y. Komori, *Logic without the contraction rule*, J. Symb. Log. **50** (1985), 169–201.

[22] H. Ono and N.-Y. Suzuki, *Relations between intuitionistic modal logics and intermediate predicate logics*, Rep. Math. Logic **22** (1988), 65–87.

[23] J.D. Rutledge, *A preliminary investigation of the infinitely many-valued predicate calculus*, Ph.D. Thesis, 1959.

[24] N.-Y. Suzuki, *An algebraic approach to intuitionistic modal logics in connection with intermediate predicate logics,* Studia Logica (1989), 141–155.

[25] ———, *Kripke bundles for intermediate predicate logics and Kripke frames for intuitionistic modal logics*, Studia Logica **49** (1990), no. 3, 289–306.

[26] A.A. Tabatabai and R. Jalali, *Universal proof theory: Semi-analytic rules and interpolation*, 2022. arXiv:1808.06258.

[27] O. Tuyt, *One-variable fragments of first-order many-valued logics*, Ph.D. Thesis, 2021.

A Missing cases for the proof of Theorem 6.2

• $(\exists\Rightarrow)$: Suppose first that $\Gamma(\bar{y})$ is $\Gamma'(\bar{y}), (\exists x)\varphi(x)$ and for some variable u that does not occur freely in $\Gamma'(\bar{y}), (\exists x)\varphi(x), \Pi(\bar{z}) \Rightarrow \Delta(\bar{z})$,

$$d' \vdash_{\mathsf{c}} \Gamma'(\bar{y}), \varphi(u), \Pi(\bar{z}) \Rightarrow \Delta(\bar{z}),$$

where $\mathrm{md}(d') = \mathrm{md}(d) - 1$. Let $\bar{y'} := \bar{y} \cup \{u\}$ and $\hat{\Gamma}(\bar{y'}) := \Gamma'(\bar{y}) \cup \{\varphi(u)\}$. By the induction hypothesis, we obtain a sentence χ and derivations d_1', d_2 such that $\mathrm{md}(d_1'), \mathrm{md}(d_2) \leq \mathrm{md}(d') = \mathrm{md}(d) - 1$ and

$$d_1' \vdash_{\mathsf{c}} \hat{\Gamma}(\bar{y'}) \Rightarrow \chi, \qquad d_2 \vdash_{\mathsf{c}} \Pi(\bar{z}), \chi \Rightarrow \Delta(\bar{z}).$$

The derivation d_1' together with an application of $(\exists\Rightarrow)$ yields a derivation d_1 such that $\mathrm{md}(d_1) = \mathrm{md}(d_1') + 1 \leq \mathrm{md}(d') + 1 = \mathrm{md}(d)$ and

$$d_1 \vdash_{\mathsf{c}} \Gamma'(\bar{y}), (\exists x)\varphi(x) \Rightarrow \chi.$$

Now suppose that $\Pi(\bar{z})$ is $\Pi'(\bar{z}), (\exists x)\varphi(x)$ and for some variable u that does not occur freely in $\Gamma(\bar{y}), \Pi'(\bar{z}), (\exists x)\varphi(x) \Rightarrow \Delta(\bar{z})$,

$$d' \vdash_{\mathsf{c}} \Gamma(\bar{y}), \Pi'(\bar{z}), \varphi(u) \Rightarrow \Delta(\bar{z}),$$

where $\mathrm{md}(d') = \mathrm{md}(d) - 1$. We let $\bar{z'} := \bar{z} \cup \{u\}$ and $\hat{\Pi}(\bar{z'}) := \Pi'(\bar{z}) \cup \{\varphi(u)\}$. Note that $\Delta(\bar{z'}) = \Delta(\bar{z})$. By the induction hypothesis, we obtain a sentence χ and derivations d_1, d_2' such that $\mathrm{md}(d_1), \mathrm{md}(d_2') \leq \mathrm{md}(d') = \mathrm{md}(d) - 1$ and

$$d_1 \vdash_{\mathsf{c}} \Gamma(\bar{y}) \Rightarrow \chi, \qquad d_2' \vdash_{\mathsf{c}} \hat{\Pi}(\bar{z'}), \chi \Rightarrow \Delta(\bar{z'}).$$

The derivation d_2' together with an application of $(\exists\Rightarrow)$ yields a derivation d_2 such that $\mathrm{md}(d_2) = \mathrm{md}(d_2') + 1 \leq \mathrm{md}(d') + 1 = \mathrm{md}(d)$ and

$$d_2 \vdash_{\mathsf{c}} \Pi'(\bar{z}), (\exists x)\varphi(x), \chi \Rightarrow \Delta(\bar{z}).$$

• $(\Rightarrow\exists)$: Suppose that $\Delta(\bar{z})$ is $(\exists x)\varphi(x)$ and there is a derivation d' such that $\mathrm{md}(d') = \mathrm{md}(d) - 1$ and

$$d' \vdash_{\mathsf{c}} \Gamma(\bar{y}), \Pi(\bar{z}) \Rightarrow \varphi(u).$$

There are two subcases. For subcase (i), suppose that $u \in \bar{y}$. By the induction hypothesis, there exist a sentence χ' and derivations d_1', d_2' such that $\mathrm{md}(d_1'), \mathrm{md}(d_2') \leq \mathrm{md}(d') = \mathrm{md}(d) - 1$ and

$$d_1' \vdash_{\mathsf{c}} \Pi(\bar{z}) \Rightarrow \chi', \qquad d_2' \vdash_{\mathsf{c}} \Gamma(\bar{y}), \chi' \Rightarrow \varphi(u).$$

Let $\chi := \chi' \to (\exists x)\varphi(x)$. Then d_1', together with the derivation $\hat{d} \vdash_{\mathsf{c}} (\exists x)\varphi(x) \Rightarrow (\exists x)\varphi(x)$ and an application of $(\to\Rightarrow)$, yields a derivation d_2, and d_2', together with applications of $(\Rightarrow \exists)$ and $(\Rightarrow\to)$, yields a derivation d_1, satisfying $\mathrm{md}(d_2) = \mathrm{md}(d_1')$, $\mathrm{md}(d_1) = \mathrm{md}(d_2') + 1 \leq \mathrm{md}(d') + 1 = \mathrm{md}(d)$ and

$$d_2 \vdash_{\mathsf{c}} \Pi(\bar{z}), \chi' \to (\exists x)\varphi(x) \Rightarrow (\exists x)\varphi(x), \qquad d_1 \vdash_{\mathsf{c}} \Gamma(\bar{y}) \Rightarrow \chi' \to (\exists x)\varphi(x).$$

For subcase (ii), suppose that $u \in \bar{z}$. By the induction hypothesis, there exist a sentence χ and derivations d_1, d_2' such that $\mathrm{md}(d_1), \mathrm{md}(d_2') \leq \mathrm{md}(d') = \mathrm{md}(d) - 1$ and

$$d_1 \vdash_{\mathsf{c}} \Gamma(\bar{y}) \Rightarrow \chi, \qquad d_2' \vdash_{\mathsf{c}} \Pi(\bar{z}), \chi \Rightarrow \varphi(u).$$

The derivation d_2' with an application of $(\Rightarrow\exists)$ yields a derivation d_2 such that $\mathrm{md}(d_2) = \mathrm{md}(d_2') + 1 \leq \mathrm{md}(d') + 1 = \mathrm{md}(d)$ and

$$d_2 \vdash_{\mathsf{c}} \Pi(\bar{z}), \chi \Rightarrow (\exists x)\varphi(x).$$

- $(\to\Rightarrow)$: Suppose first that $\Gamma(\bar{y})$ is $\Gamma_1(\bar{y}), \Gamma_2(\bar{y}), \varphi(\bar{y}) \to \psi(\bar{y})$ and $\Pi(\bar{z})$ is $\Pi_1(\bar{z}), \Pi_2(\bar{z})$, and

$$d_1' \vdash_{\mathsf{c}} \Gamma_1(\bar{y}), \Pi_1(\bar{z}) \Rightarrow \varphi(\bar{y}), \qquad d_2' \vdash_{\mathsf{c}} \Gamma_2(\bar{y}), \psi(\bar{y}), \Pi_2(\bar{z}) \Rightarrow \Delta(\bar{z}).$$

By the induction hypothesis, there exist sentences χ_1, χ_2 and derivations $d_{11}', d_{12}', d_{21}', d_{22}'$ such that

$$\begin{array}{ll} d_{11}' \vdash_{\mathsf{c}} \Gamma_1(\bar{y}), \chi_1 \Rightarrow \varphi(\bar{y}), & d_{12}' \vdash_{\mathsf{c}} \Pi_1(\bar{z}) \Rightarrow \chi_1, \\ d_{21}' \vdash_{\mathsf{c}} \Gamma_2(\bar{y}), \psi(\bar{y}) \Rightarrow \chi_2, & d_{22}' \vdash_{\mathsf{c}} \Pi_2(\bar{z}), \chi_2 \Rightarrow \Delta(\bar{z}). \end{array}$$

Let $\chi := \chi_1 \to \chi_2$. Then the derivations d_{11}', d_{21}', together with applications of $(\to\Rightarrow)$ and $(\Rightarrow\to)$, yield a derivation d_1, and the derivations d_{12}', d_{22}', together with an application of $(\to\Rightarrow)$ yield a derivation d_2 satisfying

$$\begin{array}{l} d_1 \vdash_{\mathsf{c}} \Gamma_1(\bar{y}), \Gamma_2(\bar{y}), \varphi(\bar{y}) \to \psi(\bar{y}) \Rightarrow \chi_1 \to \chi_2, \\ d_2 \vdash_{\mathsf{c}} \Pi_1(\bar{z}), \Pi_2(\bar{z}), \chi_1 \to \chi_2 \Rightarrow \Delta(\bar{z}). \end{array}$$

Clearly, the constraints on $\mathrm{md}(d_1)$ and $\mathrm{md}(d_2)$ are satisfied.

Now suppose that $\Gamma(\bar{y})$ is $\Gamma_1(\bar{y}), \Gamma_2(\bar{y})$ and $\Pi(\bar{z})$ is $\Pi_1(\bar{z}), \Pi_2(\bar{z}), \varphi(\bar{z}) \to \psi(\bar{z})$, and

$$d_1' \vdash_{\mathsf{c}} \Gamma_1(\bar{y}), \Pi_1(\bar{z}) \Rightarrow \varphi(\bar{z}), \qquad d_2' \vdash_{\mathsf{c}} \Gamma_2(\bar{y}), \psi(\bar{z}), \Pi_2(\bar{z}) \Rightarrow \Delta(\bar{z}).$$

By the induction hypothesis, there exist sentences χ_1, χ_2 and derivations $d_{11}', d_{12}', d_{21}', d_{22}'$ such that

$$\begin{array}{ll} d_{11}' \vdash_{\mathsf{c}} \Gamma_1(\bar{y}) \Rightarrow \chi_1, & d_{12}' \vdash_{\mathsf{c}} \Pi_1(\bar{z}), \chi_1 \Rightarrow \varphi(\bar{z}), \\ d_{21}' \vdash_{\mathsf{c}} \Gamma_2(\bar{y}) \Rightarrow \chi_2, & d_{22}' \vdash_{\mathsf{c}} \Pi_2(\bar{z}), \psi(\bar{z}), \chi_2 \Rightarrow \Delta(\bar{z}). \end{array}$$

Let $\chi := \chi_1 \cdot \chi_2$. Then the derivations d_{11}', d_{21}', together with an application of $(\Rightarrow\cdot)$, and the derivations d_{12}', d_{22}', together with applications of $(\to\Rightarrow)$ and $(\cdot\Rightarrow)$, yield derivations d_1 and d_2, respectively, such that

$$\begin{array}{l} d_1 \vdash_{\mathsf{c}} \Gamma_1(\bar{y}), \Gamma_2(\bar{y}) \Rightarrow \chi_1 \cdot \chi_2, \\ d_2 \vdash_{\mathsf{c}} \Pi_1(\bar{z}), \Pi_2(\bar{z}), \varphi(\bar{z}) \to \psi(\bar{z}), \chi_1 \cdot \chi_2 \Rightarrow \Delta(\bar{z}). \end{array}$$

Again, the constraints on $\mathrm{md}(d_1)$ and $\mathrm{md}(d_2)$ are clearly satisfied in this case.

• $(\Rightarrow\to)$: Suppose that $\Delta(\bar{z})$ is $\varphi(\bar{z}) \to \psi(\bar{z})$ and

$$d' \vdash_{\mathsf{c}} \Gamma(\bar{y}), \Pi(\bar{z}), \varphi(\bar{z}) \Rightarrow \psi(\bar{z}).$$

By the induction hypothesis, there exist a sentence χ and derivations d_1, d_2' such that

$$d_1 \vdash_{\mathsf{c}} \Gamma(\bar{y}) \Rightarrow \chi, \qquad d_2' \vdash_{\mathsf{c}} \Pi(\bar{z}), \varphi(\bar{z}), \chi \Rightarrow \psi(\bar{z}).$$

The derivation d_2' with an application of $(\Rightarrow\to)$ yields a derivation d_2 such that

$$d_2 \vdash_{\mathsf{c}} \Pi(\bar{z}), \chi \Rightarrow \varphi(\bar{z}) \to \psi(\bar{z}).$$

The constraints on $\mathrm{md}(d_1)$ and $\mathrm{md}(d_2)$ clearly hold.

• $(\vee\Rightarrow)$: Suppose first that $\Gamma(\bar{y})$ is $\Gamma'(\bar{y}), \varphi_1(\bar{y}) \vee \varphi_2(\bar{y})$ and

$$d_1' \vdash_{\mathsf{c}} \Gamma'(\bar{y}), \varphi_1(\bar{y}), \Pi(\bar{z}) \Rightarrow \Delta(\bar{z}), \qquad d_2' \vdash_{\mathsf{c}} \Gamma'(\bar{y}), \varphi_2(\bar{y}), \Pi(\bar{z}) \Rightarrow \Delta(\bar{z}).$$

By the induction hypothesis, there exist sentences χ_1, χ_2 and derivations $d_{11}', d_{12}', d_{21}', d_{22}'$ such that

$$\begin{array}{ll} d_{11}' \vdash_{\mathsf{c}} \Gamma'(\bar{y}), \varphi_1(\bar{y}) \Rightarrow \chi_1, & d_{12}' \vdash_{\mathsf{c}} \Pi(\bar{z}), \chi_1 \Rightarrow \Delta(\bar{z}), \\ d_{21}' \vdash_{\mathsf{c}} \Gamma'(\bar{y}), \varphi_2(\bar{y}) \Rightarrow \chi_2, & d_{22}' \vdash_{\mathsf{c}} \Pi(\bar{z}), \chi_2 \Rightarrow \Delta(\bar{z}). \end{array}$$

Define $\chi := \chi_1 \vee \chi_2$. The derivations d_{11}', d_{21}', together with applications of $(\Rightarrow\vee)_1$, $(\Rightarrow\vee)_2$, and $(\vee\Rightarrow)$, yield a derivation d_1, and the derivations d_{12}', d_{22}', together with an application of $(\vee\Rightarrow)$, yield a derivation d_2, satisfying

$$d_1 \vdash_{\mathsf{c}} \Gamma'(\bar{y}), \varphi_1(\bar{y}) \vee \varphi_2(\bar{y}) \Rightarrow \chi_1 \vee \chi_2, \qquad d_2 \vdash_{\mathsf{c}} \Pi(\bar{z}), \chi_1 \vee \chi_2 \Rightarrow \Delta(\bar{z}).$$

The constraints on $\mathrm{md}(d_1)$ and $\mathrm{md}(d_2)$ clearly hold.

Suppose now that $\Pi(\bar{z})$ is $\Pi'(\bar{z}), \varphi_1(\bar{z}) \vee \varphi_2(\bar{z})$ and

$$d_1' \vdash_{\mathsf{c}} \Gamma(\bar{y}), \Pi'(\bar{z}), \varphi_1(\bar{z}) \Rightarrow \Delta(\bar{z}), \qquad d_2' \vdash_{\mathsf{c}} \Gamma(\bar{y}), \Pi'(\bar{z}), \varphi_2(\bar{z}) \Rightarrow \Delta(\bar{z}).$$

By the induction hypothesis, there exist sentences χ_1, χ_2 and derivations $d_{11}', d_{12}', d_{21}', d_{22}'$ such that

$$\begin{array}{ll} d_{11}' \vdash_{\mathsf{c}} \Gamma(\bar{y}) \Rightarrow \chi_1, & d_{12}' \vdash_{\mathsf{c}} \Pi'(\bar{z}), \varphi_1(\bar{z}), \chi_1 \Rightarrow \Delta(\bar{z}), \\ d_{21}' \vdash_{\mathsf{c}} \Gamma(\bar{y}) \Rightarrow \chi_2, & d_{22}' \vdash_{\mathsf{c}} \Pi'(\bar{z}), \varphi_2(\bar{z}), \chi_2 \Rightarrow \Delta(\bar{z}). \end{array}$$

Let $\chi := \chi_1 \wedge \chi_2$. Then the derivations d_{11}', d_{21}', together with an application of $(\Rightarrow\wedge)$, and the derivations d_{12}', d_{22}', together with applications of $(\wedge\Rightarrow)_1$, $(\wedge\Rightarrow)_2$, and $(\vee\Rightarrow)$, yield derivations d_1 and d_2, respectively, such that

$$d_1 \vdash_{\mathsf{c}} \Gamma(\bar{y}) \Rightarrow \chi_1 \wedge \chi_2, \qquad d_2 \vdash_{\mathsf{c}} \Pi'(\bar{z}), \varphi_1(\bar{z}) \vee \varphi_2(\bar{z}), \chi_1 \wedge \chi_2 \Rightarrow \Delta(\bar{z}).$$

The constraints on $\mathrm{md}(d_1)$ and $\mathrm{md}(d_2)$ again clearly hold.

- $(\Rightarrow\vee)_i$ $(i \in \{1,2\})$: Suppose that $\Delta(\bar{z})$ is $\varphi_1(\bar{z}) \vee \varphi_2(\bar{z})$ and

$$d' \vdash_{\mathsf{c}} \Gamma(\bar{y}), \Pi(\bar{z}) \Rightarrow \varphi_i(\bar{z}).$$

By the induction hypothesis, there exist a sentence χ and derivations d_1, d_2' such that

$$d_1 \vdash_{\mathsf{c}} \Gamma(\bar{y}) \Rightarrow \chi, \qquad d_2' \vdash_{\mathsf{c}} \Pi(\bar{z}), \chi \Rightarrow \varphi_i(\bar{z}).$$

The derivation d_2' together with an application of $(\Rightarrow\vee)_i$ yields a derivation d_2 such that

$$d_2 \vdash_{\mathsf{c}} \Pi(\bar{z}), \chi \Rightarrow \varphi_1(\bar{z}) \vee \varphi_2(\bar{z}).$$

The constraints on $\mathrm{md}(d_1)$ and $\mathrm{md}(d_2)$ clearly hold.

- $(\Rightarrow\wedge)$: Suppose that $\Delta(\bar{z})$ is $\psi_1(\bar{z}) \wedge \psi_2(\bar{z})$ and

$$d_1' \vdash_{\mathsf{c}} \Gamma(\bar{y}), \Pi(\bar{z}) \Rightarrow \psi_1(\bar{z}), \qquad d_2' \vdash_{\mathsf{c}} \Gamma(\bar{y}), \Pi(\bar{z}) \Rightarrow \psi_2(\bar{z}).$$

By the induction hypothesis, there exist sentences χ_1, χ_2 and derivations $d_{11}', d_{12}', d_{21}', d_{22}'$ such that

$$\begin{array}{ll} d_{11}' \vdash_{\mathsf{c}} \Gamma(\bar{y}) \Rightarrow \chi_1, & d_{12}' \vdash_{\mathsf{c}} \Pi(\bar{z}), \chi_1 \Rightarrow \psi_1(\bar{z}), \\ d_{21}' \vdash_{\mathsf{c}} \Gamma(\bar{y}) \Rightarrow \chi_2, & d_{22}' \vdash_{\mathsf{c}} \Pi(\bar{z}), \chi_2 \Rightarrow \psi_2(\bar{z}). \end{array}$$

Let $\chi := \chi_1 \wedge \chi_2$. Then the derivations d_{11}', d_{21}', together with an application of $(\Rightarrow\wedge)$, and the derivations d_{12}', d_{22}', together with applications of $(\wedge\Rightarrow)_1$, $(\wedge\Rightarrow)_2$, and $(\Rightarrow\wedge)$, yield derivations d_1 and d_2, respectively, such that

$$d_1 \vdash_{\mathsf{c}} \Gamma(\bar{y}) \Rightarrow \chi_1 \wedge \chi_2, \qquad d_2 \vdash_{\mathsf{c}} \Pi(\bar{z}), \chi_1 \wedge \chi_2 \Rightarrow \psi_1(\bar{z}) \wedge \psi_2(\bar{z}).$$

Clearly, the constraints on $\mathrm{md}(d_1)$ and $\mathrm{md}(d_2)$ are satisfied in this case.

- $(\wedge\Rightarrow)_i$ $(i \in \{1,2\})$: Suppose first that $\Gamma(\bar{y})$ is $\Gamma'(\bar{y}), \varphi_1(\bar{y}) \wedge \varphi_2(\bar{y})$ and

$$d' \vdash_{\mathsf{c}} \Gamma'(\bar{y}), \varphi_i(\bar{y}), \Pi(\bar{z}) \Rightarrow \Delta(\bar{z}).$$

By the induction hypothesis, there exist a sentence χ and derivations d_1', d_2 such that

$$d_1' \vdash_{\mathsf{c}} \Gamma'(\bar{y}), \varphi_i(\bar{y}) \Rightarrow \chi, \qquad d_2 \vdash_{\mathsf{c}} \Pi(\bar{z}), \chi \Rightarrow \Delta(\bar{z}).$$

The derivation d_1' and an application of $(\wedge\Rightarrow)_i$ yield a derivation d_1 satisfying

$$d_1 \vdash_{\mathsf{c}} \Gamma'(\bar{y}), \varphi_1(\bar{y}) \wedge \varphi_2(\bar{y}) \Rightarrow \chi.$$

The constraints on $\mathrm{md}(d_1)$ and $\mathrm{md}(d_2)$ clearly hold.

Now suppose that $\Pi(\bar{z})$ is $\Pi'(\bar{z}), \varphi_1(\bar{z}) \wedge \varphi_2(\bar{z})$ and

$$d' \vdash_{\mathsf{c}} \Gamma(\bar{y}), \Pi'(\bar{z}), \varphi_i(\bar{z}) \Rightarrow \Delta(\bar{z}).$$

By the induction hypothesis, there exist a sentence χ and derivations d_1, d_2' such that

$$d_1 \vdash_{\mathsf{c}} \Gamma(\bar{y}) \Rightarrow \chi, \qquad d_2' \vdash_{\mathsf{c}} \Pi'(\bar{z}), \varphi_i(\bar{z}), \chi \Rightarrow \Delta(\bar{z}).$$

The derivation d_2' together with an application of $(\wedge\Rightarrow)_i$ yields a derivation d_2 such that

$$d_2 \vdash_{\mathsf{c}} \Pi'(\bar{z}), \varphi_1(\bar{z}) \wedge \varphi_2(\bar{z}), \chi \Rightarrow \Delta(\bar{z}).$$

Again, the constraints on $\mathrm{md}(d_1)$ and $\mathrm{md}(d_2)$ hold.

• $(\Rightarrow \cdot)$: Suppose that $\Delta(\bar{z})$ is $\varphi(\bar{z}) \cdot \psi(\bar{z})$, $\Gamma(\bar{y})$ is $\Gamma_1(\bar{y}), \Gamma_2(\bar{z})$, and $\Pi(\bar{z})$ is $\Pi_1(\bar{z}), \Pi_2(\bar{z})$, and

$$d_1' \vdash_{\mathsf{c}} \Gamma_1(\bar{y}), \Pi_1(\bar{z}) \Rightarrow \varphi(\bar{z}), \qquad d_2' \vdash_{\mathsf{c}} \Gamma_2(\bar{y}), \Pi_2(\bar{z}) \Rightarrow \psi(\bar{z}).$$

By the induction hypothesis, there exist sentences χ_1, χ_2 and derivations $d_{11}', d_{12}', d_{21}', d_{22}'$ such that

$$\begin{array}{ll} d_{11}' \vdash_{\mathsf{c}} \Gamma_1(\bar{y}) \Rightarrow \chi_1, & d_{12}' \vdash_{\mathsf{c}} \Pi_1(\bar{z}), \chi_1 \Rightarrow \varphi(\bar{z}), \\ d_{21}' \vdash_{\mathsf{c}} \Gamma_2(\bar{y}) \Rightarrow \chi_2, & d_{22}' \vdash_{\mathsf{c}} \Pi_2(\bar{z}), \chi_2 \Rightarrow \psi(\bar{z}). \end{array}$$

Let $\chi := \chi_1 \cdot \chi_2$. Then the derivations d_{11}', d_{21}', together with an application of $(\Rightarrow \cdot)$, and the derivations d_{12}', d_{22}', together with applications of $(\Rightarrow \cdot)$ and $(\cdot \Rightarrow)$, yield derivations d_1 and d_2, respectively, such that

$$d_1 \vdash_{\mathsf{c}} \Gamma_1(\bar{y}), \Gamma_2(\bar{y}) \Rightarrow \chi_1 \cdot \chi_2, \qquad d_2 \vdash_{\mathsf{c}} \Pi_1(\bar{z}), \Pi_2(\bar{z}), \chi_1 \cdot \chi_2 \Rightarrow \varphi(\bar{z}) \cdot \psi(\bar{z}).$$

The constraints on $\mathrm{md}(d_1)$ and $\mathrm{md}(d_2)$ clearly hold.

• $(\cdot \Rightarrow)$: Suppose first that $\Gamma(\bar{y})$ is $\Gamma'(\bar{y}), \varphi(\bar{y}) \cdot \psi(\bar{y})$ and

$$d' \vdash_{\mathsf{c}} \Gamma'(\bar{y}), \varphi(\bar{y}), \psi(\bar{y}), \Pi(\bar{z}) \Rightarrow \Delta(\bar{z}).$$

By the induction hypothesis, there exist a sentence χ and derivations d_1', d_2 such that

$$d_1' \vdash_{\mathsf{c}} \Gamma'(\bar{y}), \varphi(\bar{y}), \psi(\bar{y}) \Rightarrow \chi, \qquad d_2 \vdash_{\mathsf{c}} \Pi(\bar{z}), \chi \Rightarrow \Delta(\bar{z}).$$

Then d_1' and an application of $(\cdot \Rightarrow)$ yield a derivation d_1 such that

$$d_1 \vdash_{\mathsf{c}} \Gamma'(\bar{y}), \varphi(\bar{y}) \cdot \psi(\bar{y}) \Rightarrow \chi.$$

The constraints on $\mathrm{md}(d_1)$ and $\mathrm{md}(d_2)$ clearly hold.

Now suppose that $\Pi(\bar{z})$ is $\Pi'(\bar{z}), \varphi(\bar{z}) \cdot \psi(\bar{z})$ and

$$d' \vdash_{\mathsf{c}} \Gamma(\bar{y}), \Pi'(\bar{z}), \varphi(\bar{z}), \psi(\bar{z}) \Rightarrow \Delta(\bar{z}).$$

By the induction hypothesis, there exist a sentence χ and derivations d_1, d_2' such that

$$d_1 \vdash_{\mathsf{c}} \Gamma(\bar{y}) \Rightarrow \chi \quad \text{and} \quad d_2' \vdash_{\mathsf{c}} \Pi'(\bar{z}), \varphi(\bar{z}), \psi(\bar{z}), \chi \Rightarrow \Delta(\bar{z}).$$

Taking d_2' and applying $(\cdot \Rightarrow)$ then yields a derivation d_2 such that

$$d_2 \vdash_{\mathsf{c}} \Pi'(\bar{z}), \varphi(\bar{z}) \cdot \psi(\bar{z}), \chi \Rightarrow \Delta(\bar{z}).$$

Again, $\mathrm{md}(d_1)$ and $\mathrm{md}(d_2)$ satisfy the constraints.

Modal inverse correspondence via ALBA

Willem Conradie

School of Mathematics, University of the Witwatersrand, Johannesburg

Mattia Panettiere

Vrije Universiteit, Amsterdam

Abstract

We reformulate Kracht's theory of internal descriptions in the algebraic language of the correspondence algorithm ALBA and, within this language, we characterize (modulo standard translation) the class of first-order correspondents of modal inductive formulas as a suitable subclass of Kracht formulas for tense logic. Our result provides an alternative strategy to Kikot's generalization to the inductive (or 'generalized Sahlqvist') modal formulas of Kracht's inverse correspondence theorem for Sahlqvist formulas. This highlights and makes explicit the order-theoretic mechanisms underlying Kracht's algorithm and thereby paves the way to a generalization of inverse correspondence to modal logics on non-classical base including polyadic intuitionistic, distributive and non-distributive modal logics.

Keywords: Inverse correspondence, Unified correspondence, Kracht's Theorem, ALBA, Classical modal logic.

1 Introduction

Sahlqvist correspondence theory effectively connects a large, syntactically defined class of modal formulas with the first-order conditions their validity impose on Kripke frames. This is an immensely useful and powerful results when one's starting point is a logic axiomatized by modal axioms. However, when seeking an axiomatization for a first-order definable class of frames, one needs a result that goes in the other way, namely one that identifies a large class of first-order conditions which are modally definable and effectively associates them with their modal definitions. This is precisely what Kracht's 'inverse correspondence' theorem [17], based on his calculus of internal descriptions [16,18], provides. The MSQIN second-order quantifier introduction algorithm [9] constitutes and alternative, top-down approach to obtaining finding the modal Sahlqvist equivalents to Kracht formulas. In [12] Goranko and Vakarelov introduce the class of inductive formulas which essentially extends the class of Sahlqvist formulas, and in [14] Kikot establishes the corresponding generalisation of Kracht's theorem. The research programme of 'unified correspondence' (see e.g. [4,5,6]) has greatly generalised Goranko and Vakarelov's result and has established a definition of inductive formulas which can be applied to arbitrary logics algebraically captured by classes of lattice expansions (LE logics) and any relational semantics linked to these classes via an appropriate duality. This definition is based purely on the order-theoretic properties of the algebraic operations interpreting the connectives. A core tool in this research programme is the algorithm ALBA, which applies a set of equivalence preserving rewrite rules to transform modal formulas into 'pure' ones (in an extended language with the adjoints an residuals of all connectives) by eliminating the propositional variables in in favour of special variables, called nominals and co-nominals, which are constrained to range over the join and meet irreducible elements of the algebras which correspond (via duality) to first-order definable subsets of the relational semantics. This brings about a modularization of the correspondence theory, where correspondents are computed in the pure extended algebraic language, independent of any particular choice of relational semantics, and can thence be translated into first-order formulas via standard translations appropriate to particular choices of dual relational semantics.

In this paper, we initiate a line of research aimed at reformulating and extending inverse correspondence from classical modal logic to general LE logics. Key to this extension is a reformulation of the main engine of Kracht's result in the environment of unified correspondence which gives us access to conceptual and algorithmic tools developed there which, as mentioned above, apply across signatures and relational semantics. Specifically, in this paper we focus on the original setting of classical modal logic, where we formulate and prove an inverse correspondence result which characterizes the class of pure formulas in the extended modal language which can effectively be shown to correspond to inductive formulas. The utility of this is two-fold. Firstly, this reformulation helps to distil the order-theoretic information underlying the Kracht-Kikot model-theoretic results, thus paving the way to the required generalization to inverse correspondence for general LE logics.

Secondly, our strategy further modularizes the characterization of pure modal (and thence first-order) correspondents of the inductive formulas: we first characterize the pure correspondents of the Sahlqvist formulas in tense logic and then rely on the fact that every inductive formula in the language of classical modal logic is semantically equivalent to some scattered *very simple Sahlqvist* formula in the language of *tense* logic (cf. Lemma 3.6). We characterize the syntactic shape of such tense formulas, and suitably *restrict* the class of Kracht formulas that target tense Sahlqvist formulas. Thus, our proposed definition also features backward-looking restricted quantifiers.

Structure of the paper. In Section 2 we collect some brief preliminaries on the languages used in the paper, Sahlqvist and inductive formulas, Kracht formulas and the ALBA algorithm. Section 3 presents a characterization of the very simple Sahlqvist formulas in the language of tense logic which are equivalent to inductive formulas in the basic modal language. Our main results are presented in Section 4 where we define Kracht ML^K formulas in an extended pure hybrid modal language and show that they correspond on frames exactly to the very simple scattered Sahlqvist formulas in the language of tense logic, and then to inductive modal formulas.

2 Preliminaries

2.1 Modal languages

The *basic classical modal language* ML is defined using a set of propositional variables AtProp; its well formed formulas ϕ are given by the rule

$$\phi ::= p \mid \neg p \mid \phi \wedge \phi \mid \phi \vee \phi \mid \phi \to \phi \mid \phi \succ\!\!- \phi \mid \Diamond\phi \mid \Box\phi,$$

where p ranges over AtProp, and $\succ\!\!-$ is co-implication. It will be convenient to consider all the connectives as primitives. This is interpreted with the standard Kripke semantics, and the standard algebraic semantics is based on Boolean algebras with operators; furthermore it can be naturally expanded into the language ML* with the two additional unary connectives ◆ and ■, which should be interpreted as the adjoints of □ and ◇ respectively.

The language ML^+ expands ML* with two sorts of variables: nominals (usually denoted by $\mathbf{h}, \mathbf{i}, \mathbf{j}, \mathbf{k}$) and conominals (usually denoted by $\mathbf{l}, \mathbf{m}, \mathbf{n}, \mathbf{o}$). In perfect BAOs, (co)nominals are interpreted as (co)atoms. In what follows, we will denote ML^+-terms with the lower case letters s and t.

We will often consider ML^+-inequalities $s \leq t$; the language of such inequalities is $\mathrm{ML}_{\leq}$. Finally, the language for correspondence ML^K is built upon $\mathrm{ML}_{\leq}$ through the following rules:

$$\xi ::= s \leq t \mid \xi \;\&\; \xi \mid \xi \parr \xi \mid \sim\xi \mid \xi \Rightarrow \xi \mid \forall \mathbf{j}\, \xi \mid \forall \mathbf{m}\, \xi \mid \exists \mathbf{j}\, \xi \mid \exists \mathbf{m}\, \xi,$$

where & denotes conjunction, ⅋ disjunction, ⇒ implication, and ∼ negation.

2.2 Kracht's inverse correspondence

In what follows, FO denotes the frame correspondence language of classical modal logic. An FO-formula is *clean* (cf. [1, Chapter 3]) if no variable occurs both free and bound, and no two distinct (occurrences of) quantifiers bind the same variable. The definition of Kracht FO-formulas relies on the concept of *restricted quantifier*, i.e., quantifiers of the form $(\forall x \rhd y)\beta \equiv \forall x(yRx \to \beta)$ and $(\exists x \rhd y)\beta \equiv \exists x(yRx \wedge \beta)$. When we wish to suppress the restrictor, we will write $\forall^R x\; \beta$ and $\exists^R x\; \beta$.

Definition 2.1 [Kracht formulas] A *Kracht formula*[1] is a clean FO-formula in prenex normal form with a single free variable x_0 and shape:

$$\forall^R x_1 \cdots \forall^R x_n Q_1^R y_1 \cdots Q_m^R y_m\; \beta(x_0, x_1, \ldots, x_n, y_1, \ldots, y_m),$$

where $Q_i \in \{\forall, \exists\}$ (for $1 \leq i \leq m$), variables in $X = \{x_1, \ldots, x_n\}$ and $Y = \{y_1, \ldots, y_m\}$ are called *inherently universal* and *non-inherently universal* respectively; β is an unquantified formula in DNF whose atoms are of the

[1] The definition we present is commonly referred to as *type 1 Kracht formula*. As is well known (cf. [1]), type 1 Kracht formulas are Kracht formulas in prenex normal form where the matrix is rewritten in DNF.

form: $\top$, $\bot$, uRx, xRu, $x = u$ where $x \in X \cup \{x_0\}$ and $u \in X \cup Y$.

Theorem 2.2 ([1]) *Any Kracht formula can be effectively shown to be the first order correspondent of some Sahlqvist formula.*

2.3 Inductive and very simple Sahlqvist inequalities

Inductive formulas are introduced by Goranko and Vakarelov in [10,11,12],and are referred to as generalized Sahlqvist formulas by Kikot [14]. We present an alternative definition that will be convenient for the results in this paper.

Definition 2.3 [**Signed Generation Tree**] The *positive* (resp. *negative*) *generation tree* of any ML-formula s is defined by labelling the root node of the generation tree of s with the sign $+$ (resp. $-$), and then propagating the labelling on each remaining node as follows: for any

- node labelled with $\vee$, $\wedge$, $\Diamond$ or $\Box$ assign the same sign to its children nodes,
- $\neg$-node, assign the opposite sign to its child,
- $\rightarrow$-node, assign the opposite (resp. same) sign to the left (resp. right) child,
- $>\!\!-$-node, assign the opposite (resp. same) sign to the right (resp. left) child.

Nodes in signed generation trees are *positive* (resp. *negative*) if they are signed $+$ (resp. $-$).

Signed generation trees will be used in the context of formula inequalities $s \leq t$. In this context we will typically consider the positive generation tree $+s$ for the left-hand side and the negative one $-t$ for the right-hand side. In this case we will speak of *signed generation trees of inequalities.* An *order type* over $p_1, \ldots, p_n$ is a map $\varepsilon : \{p_1, \ldots, p_n\} \to \{1, \partial\}$. A term-inequality $s \leq t$ is *uniform* in a given variable p if all occurrences of p in both $+s$ and $-t$ have the same sign, and $s \leq t$ is ε-*uniform* in a (sub)array $\overline{p}$ of its variables if $s \leq t$ is uniform in p, occurring with the sign indicated by ε, for every p in $\overline{p}$. Given $\rho \in \{1, \partial\}$, and terms s and t, the notation $s \leq_\rho t$ indicates the inequality $s \leq t$ when $\rho = 1$, and $t \leq s$ otherwise.

For any term $s(p_1, \ldots p_n)$, any order type ε over n, and any $1 \leq i \leq n$, an ε-*critical node* in a signed generation tree of s is a leaf node $+p_i$ with $\varepsilon_i = 1$ or $-p_i$ with $\varepsilon_i = \partial$. An ε-*critical branch* in the tree is a branch ending in an ε-critical node.

We will write $\phi(!x)$ (resp. $\phi(!\overline{x})$) to indicate that the variable x (resp. each variable x in $\overline{x}$) occurs exactly once in ϕ. Accordingly, we will write $\phi(\gamma/!x)$ (resp. $\phi(\overline{\gamma}/!\overline{x})$) to indicate the formula obtained from ϕ by substituting γ (resp. each variable γ in $\overline{\gamma}$) for the unique occurrence of (its corresponding variable) x in ϕ.

Definition 2.4 [Inductive inequality] For any order type ε, and any strict order $<_\Omega$ on the variables (called *dependency order*), a formula is (Ω, ε)-inductive if:

- every ε-critical branch is a concatenation of two (possibly empty) paths P_1 and P_2 from leaf to root, such that, excluding the leaf, P_1 consists of *PIA* nodes, i.e. nodes in $\{-\wedge, +\vee, -\Diamond, +\Box, +\rightarrow\}$; and P_2 consists of *skeleton* nodes, i.e. nodes in $\{+\wedge, -\vee, +\Diamond, -\Box, -\rightarrow\}$[2].

- each subtree rooted in a $+\rightarrow$, $-\wedge$, or $+\vee$ node contains at most one ε-critical variable p and all the other variables q in the subtree are such that $q <_\Omega p$.

An *inductive* inequality is (Ω, ε)-inductive for some ε and $<_\Omega$. In what follows, we will refer to a formula χ such that $+\chi$ (resp. $-\chi$) consists only of skeleton nodes as a *positive* (resp. *negative*) *skeleton*; and we dub formulas ζ as positive (resp. negative) PIA if there is a path from a leaf to the root of $+\zeta$ (resp. $-\zeta$) consisting only of PIA nodes. For every positive (definite) PIA formula[3] $\varphi = \varphi(!x, \overline{z})$ and negative PIA formula $\psi = \psi(!x, \overline{z})$ where x is a leaf of a PIA-path to the root, we define the formulas $\mathsf{LA}(\varphi)(u, \overline{z})$ and $\mathsf{RA}(\psi)(u, \overline{z})$ (with u a new fresh variable) by simultaneous recursion:

$$
\begin{array}{cc}
\mathsf{LA}(x) = u & \mathsf{RA}(x) = u \\
\mathsf{LA}(\Box\varphi(x,\overline{z})) = \mathsf{LA}(\varphi)(\blacklozenge u, \overline{z}) & \mathsf{RA}(\Diamond\psi, \overline{z}) = \mathsf{RA}(\psi)(\blacksquare u, \overline{z}) \\
\mathsf{LA}(\psi(\overline{z}) \rightarrow \varphi(x,\overline{z})) = \mathsf{LA}(\varphi)(\psi(\overline{z}) \wedge u, \overline{z}) & \mathsf{RA}(\psi(x,\overline{z}) \succ\!\!- \varphi(\overline{z})) = \mathsf{RA}(\psi)(\varphi(\overline{z}) \vee u, \overline{z}) \\
\mathsf{LA}(\psi(x,\overline{z}) \rightarrow \varphi(\overline{z})) = \mathsf{RA}(\psi)(u \rightarrow \varphi(\overline{z}), \overline{z}) & \mathsf{RA}(\psi(\overline{z}) \succ\!\!- \varphi(x,\overline{z})) = \mathsf{LA}(\varphi)(\psi(\overline{z}) \succ\!\!- u, \overline{z}) \\
\mathsf{LA}(\psi_1(x,\overline{z}) \vee \psi_2(\overline{z})) = \mathsf{LA}(\psi_1)(u \succ\!\!- \psi_2(\overline{z})) & \mathsf{RA}(\varphi_1(x,\overline{z}) \wedge \varphi_2(\overline{z})) = \mathsf{RA}(\varphi_1)(\varphi_2(\overline{z}) \rightarrow u)
\end{array}
$$

The definition of inductive inequality is expanded to ML* by adding $-\blacklozenge$ and $+\blacksquare$ as PIA nodes and adding $+\blacklozenge$ and $-\blacksquare$ as Skeleton nodes.

Example 2.5 The formula $p \wedge \Box(\Diamond p \rightarrow \Box q) \leq \Diamond\Box\Box q$ is inductive for $\varepsilon_p = \varepsilon_q = 1$ and $p <_\Omega q$. Its signed generation tree is the following

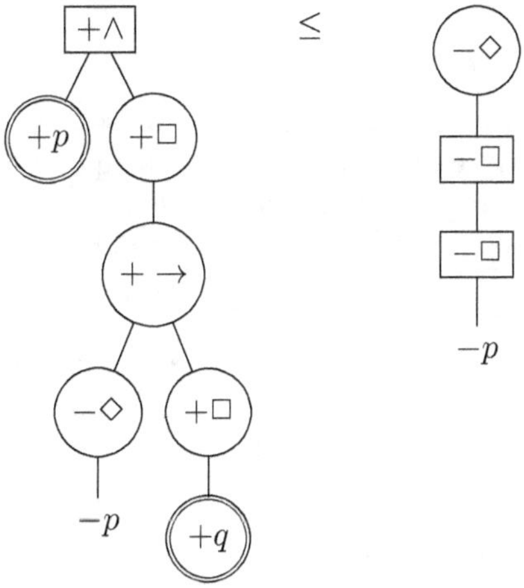

where nodes inside a box are skeleton, nodes inside circles are PIAs, and nodes inside double-circles are the critical occurrences of the variables. The left adjoint of the maximal PIA formula $\varphi(q,p) \equiv \Box(\Diamond p \rightarrow q)$ is $\mathsf{LA}(\varphi(q,p)) =$

[2] This definition of inductive inequality is not the most general one and, in fact, refers to *definite* inductive inequalities (cf. [13]). However, modulo exhaustively distributing $\vee$ and $\wedge$ over the other connectives, every (general) inductive inequality is equivalent to a conjunction of such inequalities.

[3] Here definite refers to the fact that $\vee$ and $\wedge$ have been exhaustively distributed over the other connectives; therefore no $-\vee$, nor $+\wedge$ node occurs in the PIA formula.

$\blacklozenge(\Diamond p \wedge \blacklozenge u)$.

The notion of a *Sahlqvist* inequality is obtained by restricting the nodes that are allowed in the PIA parts of critical branches, while that of a *very simple Sahlqvist* formula eliminates the PIA parts altogether:

Definition 2.6 [(Very simple) Sahlqvist inequalities - version 1] A *Sahlqvist* inequality (in ML or ML*) is any inductive inequality where the P_1 parts of all ε-critical branches consists only of nodes in $\{+\wedge, -\vee, -\Diamond, +\Box\}$. A *very simple Sahlqvist* inequality is any inductive inequality where all ε-critical branches consists only of skeleton nodes (i.e. the P_1 parts are empty).

Following [2], we will often represent (Ω, ε)-inductive inequalities as follows:

$$(\varphi \leq \psi)[\overline{\alpha}/!\overline{x}, \overline{\beta}/!\overline{y}, \overline{\gamma}/!\overline{z}, \overline{\delta}/!\overline{w}],$$

where $(\varphi \leq \psi)[!\overline{x}, !\overline{y}, !\overline{z}, !\overline{w}]$ contains only skeleton nodes, is positive (resp. negative) in $!\overline{x}$ and $!\overline{z}$ (resp. $!\overline{y}$ and $!\overline{w}$), and it is *scattered*, i.e. each variable occurs only once; each α in $\overline{\alpha}$ (resp. β in $\overline{\beta}$) is a positive (resp. negative) PIA.

Definition 2.7 [(Very simple) Sahlqvist inequality] An inductive inequality $(\varphi \leq \psi)[\overline{\alpha}/!\overline{x}, \overline{\beta}/!\overline{y}, \overline{\gamma}/!\overline{z}, \overline{\delta}/!\overline{w}]$, is *Sahlqvist*[4] if every α in $\overline{\alpha}$ and β in $\overline{\beta}$ contains only unary connectives. It is *very simple Sahlqvist* if every α and β is a propositional variable.

2.4 ALBA

ALBA is a calculus for correspondence that is based on the Ackermann lemma and which successfully reduces all inductive inequalities [4,5,6]. (Please see Appendix A for some more details on ALBA.) One can prove (cf. [7]) the output of ALBA on an inductive inequality $(\varphi \leq \psi)[\overline{\alpha}/!\overline{x}, \overline{\beta}/!\overline{y}, \overline{\gamma}/!\overline{z}, \overline{\delta}/!\overline{w}]$ is

$$\forall \overline{\mathbf{j}} \forall \overline{\mathbf{m}} \left((\varphi \leq \psi)[!\overline{\mathbf{j}}/!\overline{x}, !\overline{\mathbf{m}}/!\overline{y}, \overline{\gamma}\left[\overline{\bigvee \mathsf{Mv}(p)}/\overline{p}, \overline{\bigwedge \mathsf{Mv}(q)}/\overline{q}\right]/!\overline{z}, \overline{\delta}\left[\overline{\bigvee \mathsf{Mv}(p)}/\overline{p}, \overline{\bigwedge \mathsf{Mv}(q)}/\overline{q}\right]/!\overline{w}] \right),$$

where $\overline{p}$ (resp. $\overline{q}$) are the variables occuring in positive (resp. negative) position, $\mathsf{Mv}(p)$ and $\mathsf{Mv}(q)$ are defined by recursion on the dependency order as follows: for $<_\Omega$-minimal variables p and q,

- $\mathsf{Mv}(p) := \{\mathsf{LA}(\alpha_p)[\mathbf{j}_k/u], \mathsf{RA}(\beta_p)[\mathbf{m}_h/u] \mid 1 \leq k \leq n_{i_1}, 1 \leq h \leq n_{i_2}\}$
- $\mathsf{Mv}(q) := \{\mathsf{LA}(\alpha_q)[\mathbf{j}_h/u], \mathsf{RA}(\beta_q)[\mathbf{m}_k/u \mid 1 \leq h \leq m_{j_1}, 1 \leq k \leq m_{j_2}\}$

where, n_{i_1} (resp. n_{i_2}) is the number of occurrences of p in αs (resp. in βs) for every $p \in \overline{p}$, and m_{j_1} (resp. m_{j_2}) is the number of occurrences of q in αs (resp. in βs) for every $q \in \overline{q}$; the subscript p in α_p denotes the only critical occurrence under α. For non $<_\Omega$-minimal variables p and q,

[4] Analogously, this definition of Sahlqvist inequality is not the most general one and it refers to *definite* Sahlqvist inequalities (cf. [13]). Again, modulo exhaustively distributing $\vee$ and $\wedge$ over the other connectives, every (general) Sahlqvist inequality is equivalent to a conjunction of such inequalities.

$$\begin{aligned}\mathsf{Mv}(p) := \{&\mathsf{RA}(\alpha_p)[\mathbf{j}_k/u, \overline{\mathsf{mv}(p)}/\overline{p}, \overline{\mathsf{mv}(q)}/\overline{q}], \mathsf{RA}(\beta_p)[\mathbf{m}_h/u, \overline{\mathsf{mv}(p)}/\overline{p}, \overline{\mathsf{mv}(q)}/\overline{q}] \\ &\mid 1 \leq k \leq n_{i_1}, 1 \leq h \leq n_{i_2}, \overline{\mathsf{mv}(p)} \in \overline{\mathsf{Mv}(p)}, \overline{\mathsf{mv}(q)} \in \overline{\mathsf{Mv}(q)}\} \\ \mathsf{Mv}(q) := \{&\mathsf{LA}(\alpha_q)[\mathbf{j}_h/u, \overline{\mathsf{mv}(p)}/\overline{p}, \overline{\mathsf{mv}(q)}/\overline{q}], \mathsf{RA}(\beta_q)[\mathbf{m}_k/u, \overline{\mathsf{mv}(p)}/\overline{p}, \overline{\mathsf{mv}(q)}/\overline{q}) \\ &\mid 1 \leq h \leq m_{j_1}, 1 \leq k \leq m_{j_2}, \overline{\mathsf{mv}(p)} \in \overline{\mathsf{Mv}(p)}, \overline{\mathsf{mv}(q)} \in \overline{\mathsf{Mv}(q)}\},\end{aligned}$$

where, n_{i_1} (resp. n_{i_2}) is the number of occurrences of p in αs (resp. in βs) for every $p \in \overline{p}$, and m_{j_1} (resp. m_{j_2}) is the number of occurrences of q in αs (resp. in βs) for every $q \in \overline{q}$.

3 Inductive formulas in ML as very-simple Sahlqvist with residuals

Definition 3.1 A branch in a signed generation tree $\pm s$ is called *splittable* if it is the concatenation of two paths Q_1 and Q_2, one of which may possibly be of length 0, such that Q_1 is a path from the leaf consisting (apart from variable nodes) only of nodes in $\{+\blacksquare, -\blacklozenge, +\vee, + \rightarrow, -\wedge, - >\!-\}$ and Q_2 consists of (any) ML-nodes.

Definition 3.2 Given an order type ε, a strict partial order Ω on propositional variables, a signed generation tree $\pm\phi$ is called (Ω, ε)-*unpackable* if $\varepsilon^\partial(\pm\phi)$ and

(i) ϕ is a propositional variable or constant, or

(ii) If p_0 is maximal in $var(\phi)$ with respect to Ω, then

(a) the path Q in $\pm\phi$ ending in p_0 is splittable, and

(b) wherever Q passes through a node in $\{+\vee, + \rightarrow, -\wedge, - >\!-\}$, the subtree $\pm\gamma$ corresponding to the argument through which Q does *not* pass is (Ω, ε)-unpackable.

Definition 3.3 An ML*-inequality $\phi \leq \psi$ is called a *crypto $\mathcal{L}$-inductive* if it is a very simple ε-Sahlqvist inequality in ML* and in the signed generation trees $+\phi$ and $-\psi$:

(i) All ε-critical branches contain only signed connectives from ML,

(ii) There exists a strict partial order Ω on the propositional variables occurring in $\phi \leq \psi$, such that for every ε-non-critical branch the signed subtree rooted at the topmost (closest to the root) node on the branch properly belonging to ML* is (Ω, ε)-unpackable.

Proposition 3.4 *Every crypto* ML**-inductive inequality is frame-equivalent to an inductive inequality in* ML.

Proof. Suppose $\phi \leq \psi$ crypto ML*-inductive and let ε and Ω be an order type and a strict partial order satisfying Definition 3.3. Suppose that $var(\phi \leq \psi) = \{p_1, \ldots p_n\}$. We may assume w.l.o.g. that $p_i <_\Omega p_j$ implies $i < j$. Starting from propositional variables p_i minimal with respect to Ω, apply the inverse Ackermann rules to extract the subformulas corresponding to the subtree rooted at the topmost (closest to the root) node on the branch properly belonging to ML*. This transforms $\phi \leq \psi$ into a quasi-inequality of the form

$$q_1 \leq \alpha_1, \ldots, q_m \leq \alpha_m, \beta_{m+1} \leq r_{m+1}, \ldots, \beta_\ell \leq r_\ell \Rightarrow (\phi' \leq \psi')[\overline{q}/!\overline{x}, \overline{r}/!\overline{y}]$$

where $\phi' \leq \psi'$ contains only connectives from ML, the q_i and r_i are new variables, each α_i contains exactly one variable among $p_1, \ldots p_n$ which was Ω-maximal in the extracted subtree from which α originates. Applying adjunction and residuation rules this can be transformed into

$$\mathsf{LA}(\alpha_1)(q_1) \leq_{\varepsilon(p_{i_1})} p_{i_1}, \ldots, \mathsf{LA}(\alpha_m)(q_m) \leq_{\varepsilon(p_{i_m})} p_{i_m},$$
$$p_{i_{m+1}} \leq_{\varepsilon(p_{i_{m+1}})} \mathsf{RA}(\beta_1)(r_{m+1}), \ldots, p_{i_\ell} \leq_{\varepsilon(p_{i_\ell})} \mathsf{RA}(\beta_\ell)(r_\ell) \Rightarrow (\phi' \leq \psi')[\overline{q}/!\overline{x}, \overline{r}/!\overline{y}]$$

Note that each $\mathsf{LA}(\alpha_i)(q_i)$ and $\mathsf{RA}(\beta_i)(r_i)$ is a ML-formula.

This is now in Ackermann-shape w.r.t. the variables $p_1, \ldots p_n$. Applying the Ackermann rules produces $(\phi' \leq \psi')[\overline{q}/!\overline{x}, \overline{r}/!\overline{y}, \overline{\xi}/\overline{p}]$, which is an (Ω', ε) inductive inequality in ML where $q_\ell <_{\Omega'} q_j$ iff $p_{i_\ell} <_\Omega p_{i_j}$, for each ℓ and j.

□

Example 3.5 The ML*-inequality $p_1 \wedge p_2 \leq \Diamond\Box\Box\blacklozenge(\blacklozenge p_2 \wedge \Diamond p_1)$ is crypto-inductive for $\varepsilon(p_1, p_2) = (1, 1)$ and $p_1 <_\Omega p_2$, and is equivalent to $p_1 \leq q_1, \blacklozenge(\blacklozenge p_2 \wedge \Diamond q_1) \leq q_2 \implies p_1 \wedge p_2 \leq \Diamond\Box\Box q_2$, which is equivalent to $p_1 \leq q_1, p_2 \leq \Box(\Diamond q_1 \to \Box q_2) \implies p_1 \wedge p_2 \leq \Diamond\Box\Box q_2$, which becomes $q_1 \wedge \Box(\Diamond q_1 \to \Box q_2) \leq \Diamond\Box\Box q_2$.

Lemma 3.6 *Every inductive inequality* $(\varphi \leq \psi)[\overline{\alpha}/!\overline{x}, \overline{\beta}/!\overline{y}, \overline{\gamma}/!\overline{z}, \overline{\delta}/!\overline{w}]$ *is equivalent to some crypto-*ML**-inductive inequality.*

Proof. Given a definite inductive formula $(\varphi \leq \psi)[\overline{\alpha}/\overline{x}, \overline{\beta}/\overline{y}, \overline{\gamma}/\overline{z}, \overline{\delta}/\overline{w}]$, an ALBA run on it yields

$$\forall \overline{\mathbf{j}} \forall \overline{\mathbf{m}} \forall \overline{\mathbf{i}} \forall \overline{\mathbf{n}} \Big(\overline{\mathbf{i}} \leq \overbrace{\overline{\gamma}\left[\overline{\bigvee \mathsf{Mv}(p)}/\overline{p}, \overline{\bigwedge \mathsf{Mv}(q)}/\overline{q}\right]}^{\overline{\gamma}^{mv}} \;\&\; \overbrace{\overline{\delta}\left[\overline{\bigvee \mathsf{Mv}(p)}/\overline{p}, \overline{\bigwedge \mathsf{Mv}(q)}/\overline{q}\right]}^{\overline{\delta}^{mv}} \leq \overline{\mathbf{n}} \Rightarrow$$
$$(\varphi \leq \psi)[!\overline{\mathbf{j}}/!\overline{x}, !\overline{\mathbf{m}}/!\overline{y}, !\overline{\mathbf{i}}/!\overline{z}, !\overline{\mathbf{n}}/!\overline{w}] \Big),$$

Consider now the inequality

$$\Big((\varphi \leq \psi)[!\overline{\mathbf{j}}/!\overline{x}, !\overline{\mathbf{m}}/!\overline{y}, !\overline{\gamma}^{mv}/!\overline{z}, !\overline{\delta}^{mv}/!\overline{w}] \Big) [\overline{p_j}/\overline{\mathbf{j}}, \overline{p_m}/\overline{\mathbf{m}}], \qquad (1)$$

with $\overline{p_j}$ (resp. $\overline{q_m}$) fresh variables, one for each nominal in $\overline{\mathbf{j}}$ (resp. conominal in $\overline{\mathbf{m}}$). Clearly, the inequality is very simple Sahlqvist in ML* for ε such that each for each p_j (resp. q_m), $\varepsilon(p_j) = 1$ (resp. $\varepsilon(q_m) = \partial$) and some inductive order type $<_\Omega$. More precisely, in the ALBA run each PIA in $\overline{\alpha}$ (resp. $\overline{\beta}$) is approximated by some nominal in $\overline{\mathbf{j}}$ (resp. $\overline{\mathbf{m}}$), let τ be the map that given a variable in $\overline{\mathbf{j}}$ and $\overline{\mathbf{m}}$, yields the critical variable in the corresponding PIA formula. Let $<_{\Omega'}$ the inductive order used in the ALBA run. The inequality (1) is very simple Sahlqvist for the inductive order $<_\Omega$ such that for every $r, t \in \{p, q\}$ and $\mathbf{u}, \mathbf{v}$ in $\overline{\mathbf{j}}$ or $\overline{\mathbf{m}}$, $r_u \leq t_v$ iff $\tau(u) \leq \tau(v)$. It is clear how the ε-critical branches contain only connectives in ML, as the only connectives found there are the ones found in the skeleton of the original inductive inequality. Hence, it remains to show condition (2) of Definition 3.3. We show that every signed subtree rooted at the topmost connective properly in ML* is (ε, Ω)-unpackable for the ε and Ω defined above. Since the operators properly belonging to ML* can only occur in the minimal valuations, any of such nodes has to occur in some formula in $\overline{\gamma}^{mv}$ or $\overline{\delta}^{mv}$ inside some $\mathsf{Mv}(p)$ (resp. $\mathsf{Mv}(q)$) for some variable p (resp. q); hence the paths passing through these nodes ending in $<_\Omega$-maximal variables are splittable. Suppose that one of such paths passes through the jth coordinate of some m-ary SRR node in a non ε-critical branch $\circledast(\gamma_1,\ldots,\gamma_{j-1},\beta,\gamma_{j+1}\ldots,\gamma_m)$, and let i be any index in $\{1,\ldots,m\}\setminus\{j\}$. The subformula γ_i is (ε,Ω)-unpackable as it is of course splittable, and, inductively, every topmost node not properly in ML* is part of some $\mathsf{Mv}(p')$ (resp. $\mathsf{Mv}(q')$) for some p' (resp. q') preceding p (resp. q). □

4 Inverse correspondence in ALBA

This section presents the main results of the paper. We define the Kracht ML^K formulas and show that they correspond on frames exactly to the very simple scattered Sahlqvist formulas in the language of tense logic and, via the results of the previous section, to inductive modal formulas. Any proofs not given in this section can be found in the appendix.

Through the remainder of this section we will treat literals $\neg\mathbf{m}$ (resp. $\neg\mathbf{j}$) as nominals (resp. conominals). Let NL (resp. CNL) be the collection of nominals (resp. conominals) and negated conominals (resp. nominals).

4.1 Original Kracht formulas in ALBA's language

Definition 4.1 [Flat and restricting inequalities] *Flat inequalities* are ML^+-inequalities of the following form:

$$\mathbf{i}\leq\Diamond\mathbf{v},\quad \mathbf{i}\leq\blacklozenge\mathbf{v},\quad \Box\mathbf{v}\leq\mathbf{m},\quad \blacksquare\mathbf{v}\leq\mathbf{m},\quad \mathbf{u}\leq\mathbf{v},\quad \mathbf{u}\to\mathbf{v}\leq\mathbf{n},\quad \mathbf{i}\leq\mathbf{u}>\!\!-\mathbf{v}$$

where $\mathbf{u},\mathbf{v}\in\mathrm{NL}\cup\mathrm{CNL}$, and the two variables in the inequality are different. *Restricting inequalities* are flat inequalities of the form

$$\mathbf{j}\leq\Diamond\mathbf{i},\quad \mathbf{j}\leq\blacklozenge\mathbf{i},\quad \Box\mathbf{n}\leq\mathbf{m},\quad \blacksquare\mathbf{n}\leq\mathbf{m},\quad \mathbf{i}\leq\mathbf{j},\quad \mathbf{n}\leq\mathbf{m},$$
$$\mathbf{j}\leq\mathbf{i}>\!\!-\mathbf{n},\qquad \mathbf{i}\to\mathbf{n}\leq\mathbf{m}$$

The nominals $\mathbf{j}$ and conominals $\mathbf{m}$ are the *restricting* pure variables, while $\mathbf{i}$ and $\mathbf{n}$ are the *restricted* pure variables.

Restricting inequalities encode the type of atoms that can appear in the matrix of a Kracht formula. Indeed, nominals can be thought of as worlds of the frame $x,y,z,\ldots$, and conominals as their complements $x^c,y^c,z^c,\ldots$. Then, with some abuse of notation, including omitting curly braces when writing singletons and their complements, and reading the order $\leq$ as set-theoretic inclusion, the following equivalences hold in (complex algebras of) Kripke frames:

xRy	$x\leq\Diamond y$ iff $\Box y^c\leq x^c$ iff $y\leq\blacklozenge x$ iff $\blacksquare x^c\leq y^c$
$x=y$	$x\leq y$ iff $y^c\leq x^c$ iff $x\not\leq y^c$
$x=y=z$	$x\to y^c\leq z^c$ iff $x\leq y>\!\!-z^c$

Example 4.2 The inequality $p\wedge\Box(\Diamond p\to\Box q)\leq\Diamond\Box\Box q$ (cf. [12]) is not Sahlqvist for any order type, but it is inductive w.r.t. the order-type $\varepsilon(p,q)=(1,1)$ and $p<_\Omega q$. Running ALBA on it yields

$$\begin{array}{ll} & \forall p\forall q(p\wedge\Box(\Diamond p\to\Box q)\leq\Diamond\Box\Box q)\\ \text{iff} & \forall\mathbf{j}\forall\mathbf{m}[\Diamond\Box\Box\blacklozenge(\Diamond\mathbf{j}\wedge\blacklozenge\mathbf{j})\leq\mathbf{m}\Rightarrow\mathbf{j}\leq\mathbf{m}]\\ \text{iff} & \forall\mathbf{j}[\mathbf{j}\leq\Diamond\Box\Box\blacklozenge(\Diamond\mathbf{j}\wedge\blacklozenge\mathbf{j})]\end{array}$$

As the nominal $\mathbf{j}$ represents a world x of the Kripke frame, it is equivalent to:

$$
\begin{array}{ll}
 & \forall x(x \in [\![\Diamond\Box\Box\blacklozenge(\Diamond\mathbf{j}\wedge\blacklozenge\mathbf{j})]\!][\mathbf{j}:=x]) \\
\text{iff} & \forall x\exists y(xRy \ \&\ y \in [\![\Box\Box\blacklozenge(\Diamond\mathbf{j}\wedge\blacklozenge\mathbf{j})]\!][\mathbf{j}:=x]) \\
\text{iff} & \forall x\exists y(xRy \ \&\ \forall z(yR^2z \Rightarrow z \in [\![\blacklozenge(\Diamond\mathbf{j}\wedge\blacklozenge\mathbf{j})]\!][\mathbf{j}:=x])) \\
\text{iff} & \forall x\exists y(xRy \ \&\ \forall z(yR^2z \Rightarrow \exists w(wRz \ \&\ wRx \ \&\ xRw))).
\end{array}
$$

This last condition can equivalently be rewritten in three ways:

$$
\begin{array}{ll}
 & \forall x(\exists y \rhd x)(\forall z_1 \rhd y)(\forall z \rhd z_1)(\exists w \blacktriangleright z)(wRx \ \&\ xRw) \\
\text{iff} & \forall x(\exists y \rhd x)(\forall z_1 \rhd y)(\forall z \rhd z_1)(\exists w \rhd x)(wRz \ \&\ wRx) \\
\text{iff} & \forall x(\exists y \rhd x)(\forall z_1 \rhd y)(\forall z \rhd z_1)(\exists w \blacktriangleright x)(wRz \ \&\ xRw),
\end{array}
$$

where $(\exists w \blacktriangleright z)$ quantifies w over the *predecessors* of z. The second and third ones are not Kracht formulas, as the atom wRz has no inherently universal variables in it (the only inherently universal is x). The first one is a *tense* Kracht formula. This consideration suggests that in order to express first-order conditions in Kracht shape for an inductive formula, we need to admit the presence of operators in the fully residuated language ML*, thus allowing for backwards looking restricted quantifiers.

The example above hints at the need to expand the notation for restricted quantifiers to include the residuals $\blacksquare$ and $\blacklozenge$. *Restricted quantifiers* in ML* are defined in the following way (β being any ML$^+$-formula):

$$
\begin{array}{rcl@{\qquad}rcl}
(\forall\mathbf{i} \rhd \mathbf{j})\beta & \equiv & \forall\mathbf{i}(\mathbf{j} \leq \Diamond\mathbf{i} \Rightarrow \beta) & (\forall\mathbf{i} \blacktriangleright \mathbf{j})\beta & \equiv & \forall\mathbf{i}(\mathbf{j} \leq \blacklozenge\mathbf{i} \Rightarrow \beta) \\
(\exists\mathbf{i} \rhd \mathbf{j})\beta & \equiv & \exists\mathbf{i}(\mathbf{j} \leq \Diamond\mathbf{i} \ \&\ \beta) & (\exists\mathbf{i} \blacktriangleright \mathbf{j})\beta & \equiv & \exists\mathbf{i}(\mathbf{j} \leq \blacklozenge\mathbf{i} \ \&\ \beta) \\
(\forall\mathbf{n} \rhd \mathbf{m})\beta & \equiv & \forall\mathbf{n}(\Box\mathbf{n} \leq \mathbf{m} \Rightarrow \beta) & (\forall\mathbf{n} \blacktriangleright \mathbf{m})\beta & \equiv & \forall\mathbf{n}(\blacksquare\mathbf{n} \leq \mathbf{m} \Rightarrow \beta) \\
(\exists\mathbf{n} \rhd \mathbf{m})\beta & \equiv & \exists\mathbf{n}(\Box\mathbf{n} \leq \mathbf{m} \ \&\ \beta) & (\exists\mathbf{n} \blacktriangleright \mathbf{m})\beta & \equiv & \exists\mathbf{n}(\blacksquare\mathbf{n} \leq \mathbf{m} \ \&\ \beta)
\end{array}
$$

We also consider binary restricted quantifiers whose corresponding restricting inequalities (cf. Definition 4.1) contain a (co)implication.

$$
\begin{array}{c}
(\forall\mathbf{i},\mathbf{n} \rhd \mathbf{m})\beta \equiv \forall\mathbf{i}\forall\mathbf{n}(\mathbf{i} \to \mathbf{n} \leq \mathbf{m} \ \Rightarrow\ \beta) \\
(\exists\mathbf{i},\mathbf{n} \rhd \mathbf{m})\beta \equiv \exists\mathbf{i}\exists\mathbf{n}(\mathbf{i} \to \mathbf{n} \leq \mathbf{m} \ \&\ \beta) \\
(\forall\mathbf{i},\mathbf{n} \rhd \mathbf{j})\beta \equiv \forall\mathbf{i}\forall\mathbf{n}(\mathbf{j} \leq \mathbf{i} \succ\!\!- \mathbf{n} \ \Rightarrow\ \beta) \quad (\exists\mathbf{i},\mathbf{n} \rhd \mathbf{j})\beta \equiv \exists\mathbf{i}\exists\mathbf{n}(\mathbf{j} \leq \mathbf{i} \succ\!\!- \mathbf{n} \ \&\ \beta)
\end{array}
$$

4.2 Kracht MLK-formulas

Definition 4.3 [Kracht disjunct] A *Kracht disjunct* is a formula $\theta(\mathbf{w})$ in ML^K defined inductively together with its *main pure variable* $\mathbf{w} \in$ NL $\cup$ CNL. It either is:

- a flat inequality (cf. Definition 4.1) $s \leq \mathbf{w}$ or $\mathbf{w} \leq t$;
- $(\exists\ \mathbf{u} \rhd \mathbf{w})\theta(\mathbf{u})$, $(\forall\ \mathbf{u} \rhd \neg\mathbf{w})\theta(\neg\mathbf{u})$, $(\exists\ \mathbf{u} \blacktriangleright \mathbf{w})\theta(\mathbf{u})$, or $(\forall\ \mathbf{u} \blacktriangleright \neg\mathbf{w})\theta(\neg\mathbf{u})$, where $\theta(\mathbf{u})$ is a Kracht disjunct where $\mathbf{w}$ does not occur [5];
- $\theta(\mathbf{m}) := (\exists\ \mathbf{i},\mathbf{n} \rhd \mathbf{m})(\theta_1(\mathbf{i}) \ \&\ \theta_2(\mathbf{n}))$ or $\theta(\neg\mathbf{m}) := (\forall\ \mathbf{i},\mathbf{n} \rhd \mathbf{m})(\theta_1(\neg\mathbf{i}) \ ⅋\ \theta_2(\neg\mathbf{n}))$, where $\theta_1(\mathbf{i})$ and $\theta_2(\mathbf{n})$ are Kracht disjuncts such that $\mathbf{m}$ does not occur in them;

[5] Note that the types of $\mathbf{u}$ and $\mathbf{w}$ (nominal or conominal) in these clauses are governed by the typing conventions in the definitions of the restricted quantifiers.

- $\theta(\mathbf{m}) := (\exists\, \mathbf{i}, \mathbf{n} \rhd \mathbf{j})(\theta_1(\mathbf{i})\ \&\ \theta_2(\mathbf{n}))$ or $\theta(\neg\mathbf{j}) := (\forall\, \mathbf{i}, \mathbf{n} \rhd \mathbf{j})(\theta_1(\neg\mathbf{i}) \parr \theta_2(\neg\mathbf{n}))$, where $\theta_1(\mathbf{i})$ and $\theta_2(\mathbf{n})$ are Kracht disjuncts where $\mathbf{m}$ does not occur;
- $\theta_1(\mathbf{w})\ \&\ \theta_2(\mathbf{w})\ \& \cdots \&\theta_n(\mathbf{w})$ where all the θ_i (with $1 \leq i \leq n$) are Kracht disjuncts
- $\theta_1(\mathbf{w}) \parr \theta_2(\mathbf{w}) \parr \cdots \parr \theta_n(\mathbf{w})$ where all the θ_i (with $1 \leq i \leq n$) are Kracht disjuncts.

Furthermore, in the generation trees of all the flat inequalities of $\theta(\mathbf{w})$, each nominal (resp. conominal) different from $\mathbf{w}$ occurs in negative (resp. positive) polarity if it is under the scope of an even number of universal quantifiers, the opposite otherwise.

Definition 4.4 A *Kracht antecedent* in ML^K is a an ML^K-formula $\eta(\mathbf{j}, \mathbf{m})$ which is a conjunction of inequalities of the form $\mathbf{i} \leq \mathbf{h}$ and $\mathbf{o} \leq \mathbf{n}$, plus a single negated inequality $\mathbf{j} \nleq \mathbf{m}$ called a *pivotal inequality*; the variables $\mathbf{j}$ and $\mathbf{m}$ are the *pivotal pure variables* of the antecedent.

Next we introduce the notion of a Kracht ML^K-formula. The reader might find it useful to refer to Example 4.9 while reading this definition.

Definition 4.5 [Kracht ML^K-formula] A closed ML^K-formula is *Kracht* if it is of the following shape:

$$\forall \mathbf{j} \forall \mathbf{m} \forall \overline{\mathbf{h}} \forall \overline{\mathbf{o}} \forall^R \overline{\mathbf{i}}, \overline{\mathbf{n}} (\eta(\mathbf{j}, \mathbf{m}) \Rightarrow \theta_1(\mathbf{w}_1)\ \parr\ \cdots\ \parr\ \theta_n(\mathbf{w}_n)), \tag{2}$$

where $\eta(\mathbf{j}, \mathbf{m})$ is a Kracht antecedent, each θ_i is a Kracht disjunct, and $\forall^R \overline{\mathbf{i}}, \overline{\mathbf{n}}$ denotes a sequence of restricted universal quantifiers introducing the (co)nominals in $\overline{\mathbf{i}}$ and $\overline{\mathbf{n}}$. The variables quantified in the prefix are *inherently universal* variables. The formula also has to satisfy the following conditions:

(i) Each nominal in $\overline{\mathbf{h}}$ (or, resp., conominal in $\overline{\mathbf{o}}$) must appear on the right (resp. left) hand side of exactly one non-pivotal inequality in $\eta(\mathbf{j}, \mathbf{m})$ (and nowhere else in $\eta(\mathbf{j}, \mathbf{m})$).

(ii) the non-main variables (cf. Definition 4.3) in each atom in the consequent are all inherently universal.

(iii) Quantifiers in $\forall^R \overline{\mathbf{i}}, \overline{\mathbf{n}}$ must be of either of the following types: *type 1* quantifiers bind variables occurring in the consequent, but not in the antecedent or as restrictors in the prefix; *type 2* quantifiers bind variables that occur either in the antecedent or as restrictors (exactly once) in the prefix, but not in the consequent.

Remark 4.6 Note that the pivotal inequality in the antecedent $\mathbf{j} \nleq \mathbf{m}$ is not technically necessary as it just translates to $\mathbf{j} = \neg\mathbf{m}$; hence it is possible to apply the same arguments to formulas with just a single pivotal nominal $\mathbf{j}$ and substituting every occurrence of the related pivotal conominal with $\neg\mathbf{j}$. Nevertheless, we shall keep both the pivotal variables in the definition to simplify some of the proofs in the remainder. It follows from Definitions 4.4 and 4.5(i) that the variables in $\overline{\mathbf{h}}$ and $\overline{\mathbf{o}}$ provide alternative names for either pivotal variables or restricted variables in the prefix. This is why we will sometimes refer to them as *aliases*.

Henceforth, we will refer to formulas defined in Definition 4.5 as *Kracht formulas.*

Lemma 4.7 *Every Kracht formula is equivalent to some Kracht formula where the pivotal variables do not occur in the consequent.*

Proof. Any Kracht formula has the following form

$$\forall \mathbf{j} \forall \mathbf{m} \forall \overline{\mathbf{h}} \forall \overline{\mathbf{o}} \forall^{R} \overline{\mathbf{i}}, \overline{\mathbf{n}} (\eta' \ \& \ \mathbf{j} \not\leq \mathbf{m} \Rightarrow \theta_1(\mathbf{w}_1) \ ⅋ \ \cdots \ ⅋ \ \theta_n(\mathbf{w}_n)),$$

and hence it can be equivalently rewritten as the following Kracht formula

$$\forall \mathbf{j}' \forall \mathbf{m}' \forall \mathbf{j} \forall \mathbf{m} \forall \overline{\mathbf{h}} \forall \overline{\mathbf{o}} \forall^{R} \overline{\mathbf{i}}, \overline{\mathbf{n}} (\eta' \ \& \ \mathbf{j}' \leq \mathbf{j} \ \& \ \mathbf{m} \leq \mathbf{m}' \ \& \ \mathbf{j}' \not\leq \mathbf{m}' \Rightarrow \theta_1(\mathbf{w}_1) \ ⅋ \ \cdots \ ⅋ \ \theta_n(\mathbf{w}_n)),$$

where $\mathbf{j}'$ and $\mathbf{m}'$ are fresh variables, and therefore they do not occur in the consequent. The variables $\mathbf{j}$ and $\mathbf{m}$ become part of the $\overline{\mathbf{h}}$ and $\overline{\mathbf{o}}$ respectively of the new formula. □

Lemma 4.8 *Any Kracht formula is equivalent to some Kracht formula such that each alias variable occurs in the consequent.*

Proof. Suppose that an alias nominal $\mathbf{h}$ (resp. conominal $\mathbf{o}$) does not occur in the consequent. By definition of Kracht formulas, it occurs exactly once in the antecedent in an inequality of shape $\mathbf{k_h} \leq \mathbf{h}$ (resp. $\mathbf{o} \leq \mathbf{l_o}$). As it does not occur in the consequent, the universal quantifier that introduces it can be rewritten as an existential quantifier in the antecedent. Now the formula $\exists \mathbf{h}(\mathbf{k_h} \leq \mathbf{h})$ (resp. $\exists \mathbf{o}(\mathbf{o} \leq \mathbf{l_o})$) is equivalent to $\top$, and, therefore it can be eliminated from the antecedent. □

Thanks to Lemmas 4.7 and 4.8, we will henceforth consider only Kracht formulas where the pivotal variables do not occur in the consequent and whose unrestricted non-pivotal variables occur in the consequent. We will also assume that the variables introduced by *type 1* restricted quantifier occur in the consequent, since, otherwise, the formula would be equivalent to the same formula without those quantifiers. We refer to such formulas as *refined Kracht formulas.*

Example 4.9 The following formula

$$\forall \mathbf{j} \forall \mathbf{m} \forall \mathbf{h}_1 \forall \mathbf{h}_2 [\mathbf{j} \leq \mathbf{h}_1 \ \& \ \mathbf{j} \leq \mathbf{h}_2 \ \& \ \mathbf{j} \not\leq \mathbf{m} \Rightarrow (\exists \mathbf{i}_1 \rhd \neg \mathbf{m})(\forall \mathbf{n}_1 \rhd \neg \mathbf{i}_1)(\forall \mathbf{n}_2 \rhd \mathbf{n}_1)(\exists \mathbf{i}_2 \blacktriangleright \neg \mathbf{n}_2)(\mathbf{i}_2 \leq \Diamond \mathbf{h}_1 \ \& \ \mathbf{i}_2 \leq \blacklozenge \mathbf{h}_2)]$$

is Kracht with pivotal variables $\mathbf{j}$ and $\mathbf{m}$, aliases $\mathbf{h}_1$ and $\mathbf{h}_2$, and a single Kracht disjunct. Indeed, in $\mathbf{i}_2 \leq \Diamond \mathbf{h}_1$ and $\mathbf{i}_2 \leq \blacklozenge \mathbf{h}_2$, $\mathbf{h}_1$ and $\mathbf{h}_2$ are inherently universal and they occur in negative polarity while being under the scope of an even number of universal quantifiers. By Lemma 4.7 and by renaming $\mathbf{m}$ to $\mathbf{o}_1$, it is equivalent to the following refined Kracht formula:

$$\forall \mathbf{j} \forall \mathbf{m} \forall \mathbf{h}_1 \forall \mathbf{h}_2 \forall \mathbf{o}_1 [\mathbf{j} \leq \mathbf{h}_1 \ \& \ \mathbf{j} \leq \mathbf{h}_2 \ \& \ \mathbf{o}_1 \leq \mathbf{m} \ \& \ \mathbf{j} \not\leq \mathbf{m} \Rightarrow (\exists \mathbf{i}_1 \rhd \neg \mathbf{o}_1)(\forall \mathbf{n}_1 \rhd \neg \mathbf{i}_1)(\forall \mathbf{n}_2 \rhd \mathbf{n}_1)(\exists \mathbf{i}_2 \blacktriangleright \neg \mathbf{n}_2)(\mathbf{i}_2 \leq \Diamond \mathbf{h}_1 \ \& \ \mathbf{i}_2 \leq \blacklozenge \mathbf{h}_2)]$$

The Kracht formula

$$\forall \mathbf{j} \forall \mathbf{m} \forall \mathbf{h}_1 \forall \mathbf{h}_2 (\forall \mathbf{i}_1 \rhd \mathbf{j})(\forall \mathbf{n}_1 \rhd \mathbf{m})[\mathbf{i}_1 \leq \mathbf{h}_1 \ \& \ \mathbf{i}_1 \leq \mathbf{h}_2 \ \& \ \mathbf{j} \not\leq \mathbf{m} \Rightarrow \neg \mathbf{m} \leq \mathbf{h}_2 \ ⅋ \ (\exists \mathbf{i}_2 \rhd \neg \mathbf{n}_1)(\forall \mathbf{n}_2 \rhd \neg \mathbf{i}_2)(\neg \mathbf{n}_2 \leq \Diamond \mathbf{h}_1)]$$

is equivalent to the following refined Kracht formula (introducing an alias for $\mathbf{m}$)

$$\forall \mathbf{j} \forall \mathbf{m} \forall \mathbf{h}_1 \forall \mathbf{h}_2 \forall \mathbf{o}_1 (\forall \mathbf{i}_1 \rhd \mathbf{j})(\forall \mathbf{n}_1 \rhd \mathbf{m})[\mathbf{i}_1 \leq \mathbf{h}_1 \ \& \ \mathbf{i}_1 \leq \mathbf{h}_2 \ \& \ \mathbf{o}_1 \leq \mathbf{m} \ \& \ \mathbf{j} \nleq \mathbf{m} \Rightarrow \neg \mathbf{o}_1 \leq \mathbf{h}_2 \ \parr \ (\exists \mathbf{i}_2 \rhd \neg \mathbf{n}_1)(\forall \mathbf{n}_2 \rhd \neg \mathbf{i}_2)(\neg \mathbf{n}_2 \leq \Diamond \mathbf{h}_1)]$$

4.3 From Kracht to very simple Sahlqvist with residuals

In this section, we introduce an algorithm that takes refined Kracht formulas as input, and, using ALBA rules, computes very simple Sahlqvist ML*-formulas of which they are first order correspondents, and which are equivalent to inductive ML-inequalities, as discussed in Section 3.

Compaction of the non-inherently universals. By exhaustively applying Ackermann eliminations and inverse splitting, a Kracht disjunct $\theta(\mathbf{w})$ is shown to be equivalent to some inequality that has $\mathbf{w}$ on display.

Algorithm 1 Compaction of a Kracht disjunct $\theta(\mathbf{w})$.

1: **procedure** DisjunctCompaction(θ)
2: **if** θ is a flat inequality **then return** θ
3: **else**
4: Let $\{\theta_1, \ldots, \theta_n\}$ be the set of all the direct sub-disjuncts of θ
5: Let $I = [I_1, \ldots, I_n]$ a list of inequalities
6: **for all** the direct sub-disjuncts θ_i in θ **do**
7: $I_i \leftarrow$ DisjunctCompaction(θ_i)
8: **end for**
9: **if** θ is a (dis/con)junction of disjuncts $\theta_i(\mathbf{j})$ **then**
10: Let $s_1, \ldots, s_n$ be formulas such that I_i is $\mathbf{j} \leq s_i$ for $i = 1, \ldots, n$
11: **return** $\mathbf{j} \leq s_1 \wedge \cdots \wedge s_n$ if conjunction, $\mathbf{j} \leq s_1 \vee \cdots \vee s_n$ otherwise
12: **else if** θ is a (dis/con)junction of disjuncts $\theta_i(\mathbf{m})$ **then**
13: Let $s_1, \ldots, s_n$ be formulas such that I_i is $s_i \leq \mathbf{m}$ for $i = 1, \ldots, n$
14: **return** $s_1 \vee \cdots \vee s_n \leq \mathbf{m}$ if conjunction, $s_1 \wedge \cdots \wedge s_n \leq \mathbf{m}$ otherwise
15: **else if** θ has form $(Q\ \mathbf{u}\ r\ \mathbf{v})(\theta_1)$ with $Q \in \{\forall, \exists\}$ and $r \in \{\rhd, \blacktriangleright\}$ **then**
16: **return** Eliminate $\mathbf{u}$ via Inverse Approximation Rules
17: **else if** θ has form $(\exists \mathbf{i}, \mathbf{n} \rhd \mathbf{m})(\theta_1(\mathbf{i}) \ \& \ \theta_2(\mathbf{n}))$ or $(\forall \mathbf{i}, \mathbf{n} \rhd \mathbf{m})(\theta_1(\neg \mathbf{i}) \ \parr \ \theta_2(\neg \mathbf{n}))$ **then**
18: **return** Eliminate $\mathbf{i}$ and $\mathbf{n}$ via Inverse Approximation Rules
19: **end if**
20: **end if**
21: **end procedure**

Lemma 4.10 *When applied to a Kracht disjunct θ, Algorithm 1 outputs an inequality of shape $\mathbf{k} \leq s$ (resp. $s \leq \mathbf{l}$), where $\mathbf{k}$ (resp. $\mathbf{l}$) is the main pure variable of θ, and it does not occur in s.*

Proof. We proceed by induction on the structure of $\theta(\mathbf{w})$. If $\theta(\mathbf{w})$ is a flat inequality, then the statement holds by definition of Kracht disjunct. If $\theta(\mathbf{w}) := \theta_1(\mathbf{w}) \ \& \ \cdots \ \& \ \theta_n(\mathbf{w})$ (resp. $\theta(\mathbf{w}) := \theta_1(\mathbf{w}) \ \parr \ \cdots \ \parr \ \theta_n(\mathbf{w})$), then the algorithm applies inverse splittingin line 11 if $\mathbf{w}$ is a nominal, line 14 if it is a conominal. As, by inductive hypothesis on each θ_i, $\mathbf{w}$ does not occur in s_i, it does not occur in $\bigwedge_i s_i$ (resp. $\bigvee_i s_i$). Assume that $\theta(\mathbf{w}) := (\exists \mathbf{u}\ r\ \mathbf{w})(\theta_1(\mathbf{u}))$ (resp. $\theta(\mathbf{w}) := (\forall \mathbf{u}\ r\ \neg \mathbf{w})(\theta_1(\neg \mathbf{u}))$), with $r \in \{\rhd, \blacktriangleright\}$, i.e. $\theta(\mathbf{w})$ is of the form $\exists \mathbf{u}(\mathbf{w} \leq f(\mathbf{u}) \ \& \ \theta_1(\mathbf{u}))$ (resp. $\forall \mathbf{u}(\neg \mathbf{w} \leq f(\mathbf{u}) \Rightarrow \theta_1(\neg \mathbf{u}))$) if $\mathbf{u}$ is a nominal, or as $\exists\mathbf{u}(g(\mathbf{u}) \leq \mathbf{w} \ \& \ \theta_1(\mathbf{u}))$ (resp. $\forall\mathbf{u}(g(\mathbf{u}) \leq \neg\mathbf{w} \Rightarrow \theta_1(\neg\mathbf{u})))$ if it is a conominal, where $f \in \{\Diamond, \blacklozenge\}$ (resp. $g \in \{\Box, \blacksquare\}$). In each such case, the induction hypothesis on θ_1 ensures that $\theta(\mathbf{w})$ is in Ackermann shape w.r.t. $\mathbf{u}$, which can then be eliminated by applying the Ackermann rule. The cases $\theta(\mathbf{w}) := (\exists\mathbf{i}, \mathbf{n} \rhd \mathbf{w})(\theta_1(\mathbf{i}) \ \& \ \theta_2(\mathbf{n}))$ and $\theta(\mathbf{w}) := (\forall\mathbf{i}, \mathbf{n} \rhd \neg\mathbf{w})(\theta_1(\neg\mathbf{i}) \parr \theta_2(\neg\mathbf{n}))$ are treated analogously by eliminating $\mathbf{i}$ and $\mathbf{n}$. □

Lemma 4.11 *When applied to some Kracht disjunct $\theta(\mathbf{w})$, Algorithm 1 produces an inequality where the nominals (resp. conominals) different from $\mathbf{w}$ occur in negative (resp. positive) polarity.*

Proof. By definition of Kracht disjunct, the polarity of its non-main variables depends on the number of universal quantifiers under which they are nested. Indeed, polarities are preserved by applications of existential inverse approximation and inverse splitting rules, and are reversed by applications of the universal inverse approximation rule. Thus, in the end nominals (resp. conominals) must occur in negative (resp. positive) polarity. □

Example 4.12 Let us apply Algorithm 1 to the consequent of the formulas in Example 4.9. The first one is $(\exists\mathbf{i}_1 \rhd \neg\mathbf{o}_1)(\forall\mathbf{n}_1 \rhd \neg\mathbf{i}_1)(\forall\mathbf{n}_2 \rhd \mathbf{n}_1)(\exists\mathbf{i}_2 \blacktriangleright \mathbf{n}_2)(\mathbf{i}_2 \leq \Diamond\mathbf{h}_1 \ \& \ \mathbf{i}_2 \leq \blacklozenge\mathbf{h}_2)$. The innermost Kracht disjunct is a conjunction sharing the same main variable, thus we can compact it into one inequality. We then proceed to eliminate the restricted quantifiers: working from the inside out, we first expand them according to their definitions (for the sake of clarity) and then apply appropriate inverse approximation rules:

$$
\begin{array}{lll}
 & (\exists\mathbf{i}_1 \rhd \neg\mathbf{o}_1)(\forall\mathbf{n}_1 \rhd \neg\mathbf{i}_1)(\forall\mathbf{n}_2 \rhd \mathbf{n}_1)(\exists\mathbf{i}_2 \blacktriangleright \neg\mathbf{n}_2)(\mathbf{i}_2 \leq \Diamond\mathbf{h}_1 \wedge \blacklozenge\mathbf{h}_2) & \\
\text{iff} & (\exists\mathbf{i}_1 \rhd \neg\mathbf{o}_1)(\forall\mathbf{n}_1 \rhd \neg\mathbf{i}_1)(\forall\mathbf{n}_2 \rhd \mathbf{n}_1)\exists\mathbf{i}_2(\neg\mathbf{n}_2 \leq \blacklozenge\mathbf{i}_2 \ \& \ \mathbf{i}_2 \leq \Diamond\mathbf{h}_1 \wedge \blacklozenge\mathbf{h}_2) & \text{Inv. Split.} \\
\text{iff} & (\exists\mathbf{i}_1 \rhd \neg\mathbf{o}_1)(\forall\mathbf{n}_1 \rhd \neg\mathbf{i}_1)(\forall\mathbf{n}_2 \rhd \mathbf{n}_1)(\neg\mathbf{n}_2 \leq \blacklozenge(\Diamond\mathbf{h}_1 \wedge \blacklozenge\mathbf{h}_2)) & \text{Inv. Appr.} \\
\text{iff} & (\exists\mathbf{i}_1 \rhd \neg\mathbf{o}_1)(\forall\mathbf{n}_1 \rhd \neg\mathbf{i}_1)\forall\mathbf{n}_2(\Box\mathbf{n}_2 \leq \mathbf{n}_1 \implies \neg\mathbf{n}_2 \leq \blacklozenge(\Diamond\mathbf{h}_1 \wedge \blacklozenge\mathbf{h}_2)) & \\
\text{iff} & (\exists\mathbf{i}_1 \rhd \neg\mathbf{o}_1)(\forall\mathbf{n}_1 \rhd \neg\mathbf{i}_1)\forall\mathbf{n}_2(\blacklozenge(\Diamond\mathbf{h}_1 \wedge \blacklozenge\mathbf{h}_2) \leq \mathbf{n}_2 \implies \neg\mathbf{n}_1 \leq \Box\mathbf{n}_2) & \text{Contrap.} \\
\text{iff} & (\exists\mathbf{i}_1 \rhd \neg\mathbf{o}_1)(\forall\mathbf{n}_1 \rhd \neg\mathbf{i}_1)(\neg\mathbf{n}_1 \leq \Box\blacklozenge(\Diamond\mathbf{h}_1 \wedge \blacklozenge\mathbf{h}_2)) & \text{Inv. Appr.} \\
\text{iff} & (\exists\mathbf{i}_1 \rhd \neg\mathbf{o}_1)\forall\mathbf{n}_1(\Box\mathbf{n}_1 \leq \neg\mathbf{i}_1 \implies \neg\mathbf{n}_1 \leq \Box\blacklozenge(\Diamond\mathbf{h}_1 \wedge \blacklozenge\mathbf{h}_2)) & \\
\text{iff} & (\exists\mathbf{i}_1 \rhd \neg\mathbf{o}_1)\forall\mathbf{n}_1(\Box\blacklozenge(\Diamond\mathbf{h}_1 \wedge \blacklozenge\mathbf{h}_2) \leq \mathbf{n}_1 \implies \mathbf{i}_1 \leq \Box\mathbf{n}_1) & \text{Contrap.} \\
\text{iff} & (\exists\mathbf{i}_1 \rhd \neg\mathbf{o}_1)(\mathbf{i}_1 \leq \Box\Box\blacklozenge(\Diamond\mathbf{h}_1 \wedge \blacklozenge\mathbf{h}_2)) & \text{Inv. Appr} \\
\text{iff} & \exists\mathbf{i}_1(\neg\mathbf{o}_1 \leq \Diamond\mathbf{i}_1 \ \& \ \mathbf{i}_1 \leq \Box\Box\blacklozenge(\Diamond\mathbf{h}_1 \wedge \blacklozenge\mathbf{h}_2)) & \\
\text{iff} & \neg\mathbf{o}_1 \leq \Diamond\Box\Box\blacklozenge(\Diamond\mathbf{h}_1 \wedge \blacklozenge\mathbf{h}_2) & \text{Inv. Appr.}
\end{array}
$$

Now for the consequent of the second formula in Example 4.9, which is

$$\neg\mathbf{o}_1 \leq \mathbf{j}_2 \parr (\exists\mathbf{i}_2 \rhd \neg\mathbf{n}_1)(\forall\mathbf{n}_2 \rhd \neg\mathbf{i}_2)(\neg\mathbf{n}_2 \leq \Diamond\mathbf{j}_1).$$

For the sake of brevity, we will not write out the expansion of bounded quantifiers and contrapositive steps for this example. The first Kracht disjunct is already flat, in the second Kracht disjunct the algorithm yields

$$(\exists\mathbf{i}_2 \rhd \neg\mathbf{n}_1)(\forall\mathbf{n}_2 \rhd \neg\mathbf{i}_2)(\neg\mathbf{n}_2 \leq \Diamond\mathbf{j}_1) \quad \text{iff} \quad (\exists\mathbf{i}_2 \rhd \neg\mathbf{n}_1)(\mathbf{i}_2 \leq \Box\Diamond\mathbf{j}_1) \quad \text{iff} \quad \neg\mathbf{n}_1 \leq \Diamond\Box\Diamond\mathbf{j}_1 .$$

Compaction of the antecedent. After compacting the consequent, the (refined) input formula (2) has the following shape:

$$\forall\mathbf{j}\forall\mathbf{m}\forall\overline{\mathbf{h}}\forall\overline{\mathbf{o}}\forall^R\overline{\mathbf{i}},\overline{\mathbf{n}}\left(\eta(\mathbf{j},\mathbf{m}) \Rightarrow \bigparr(\overline{\mathbf{k} \leq \delta} \parr \overline{\gamma \leq \mathbf{l}})\right), \tag{3}$$

where each $\mathbf{k}$ (resp. $\mathbf{l}$) is either an alias variable, or is bound by some *type 1* quantifier. Furthermore, as the input formula is refined, each alias and each *type 1* variable occurs at least once in the consequent. Let us abbreviate SUCC := $\bigparr(\overline{\mathbf{k} \leq \delta} \parr \overline{\gamma \leq \mathbf{l}})$.

Each restricted quantifier binds one nominal $\mathbf{i}$, one conominal $\mathbf{n}$, or one nominal and one conominal; in each case, the quantifier comes equipped with a restricting inequality in the antecedent, which has shape $\mathbf{k_i} \leq \Diamond\mathbf{i}$ in the first case, $\Box\mathbf{n} \leq \mathbf{l_n}$ in the second case, and $\mathbf{i} \rightarrow \mathbf{n} \leq \mathbf{l_n}$ or $\mathbf{k_i} \leq \mathbf{i} \succ\!\!-\mathbf{n}$ in the third case, for some nominal $\mathbf{k_i}$ and conominal $\mathbf{l_n}$. By currying, for any formula σ,

$$(\forall \mathbf{i} \rhd \mathbf{k_i})(\sigma \Rightarrow \text{SUCC}) \quad \text{i.e.} \quad \forall \mathbf{i}(\mathbf{k_i} \leq \Diamond\mathbf{i} \Rightarrow (\sigma \Rightarrow \text{SUCC})) \quad \text{iff}$$
$$\forall \mathbf{i}((\mathbf{k_i} \leq \Diamond\mathbf{i}\ \&\ \sigma) \Rightarrow \text{SUCC}),$$

and similarly for the other two cases. Let us apply this procedure exhaustively, so to rewrite the antecedent of (3) by conjoining it with all the restricting inequalities of *type 2* quantifiers and of *type 1* quantifiers restricted by variables bound by *type 2* quantifiers. The next lemma shows that all *type 2* quantifiers can be eliminated by proceeding from the rightmost to the leftmost via an inverse approximation rule. We suggest that the reader glance at Example 4.16 while reading the following two lemmas.

Lemma 4.13 *After exhaustively currying, the antecedent of* (3) *is in the right shape for the elimination of the rightmost quantifier via inverse approximation, and, after the elimination, it is again in the right shape for the elimination of the successive quantifiers.*

Proof. After currying, the variables $\mathbf{i}$ and/or $\mathbf{n}$ bound by a a *type 2* quantifier can either occur in the antecedent in inequalities $\mathbf{i} \leq \mathbf{h}$ (resp. $\mathbf{o} \leq \mathbf{n}$) for some alias variable $\mathbf{h}$ (resp. $\mathbf{o}$), or in restricting inequalities. Notice that $\mathbf{i}$ (resp. $\mathbf{n}$) occurs negatively (resp. positively) only in the restricting inequality of the quantifier that binds it (let us call it I_1), and occurs in the opposite polarity in any other restricting inequality where it is a restrictor. Therefore, we can merge via inverse splitting all the inequalities involving aliases and the restricting inequalities where $\mathbf{i}$ (resp. $\mathbf{n}$) occur as restrictor, thus obtaining an inequality I_2. When eliminating the rightmost quantifier, we only have to consider two inequalities in the antecedent, I_1 and I_2, which are in Ackermann shape for the elimination of $\mathbf{i}$ and/or $\mathbf{n}$ (considering that they cannot occur in the consequent). The variable $\mathbf{u}$ on display in the resulting inequality is the restrictor of I_1, and, if it is a nominal (resp. conominal) it occurs on the left (resp. right) hand side of the inequality; furthermore it occurs only once in the inequality. The variable $\mathbf{u}$ can either be a pivotal variable or a variable bound by another *type 2* quantifiers. In the latter case, $\mathbf{u}$ will be eliminated in a later stage by repeating the same procedure. At that stage, this inequality will be merged via inverse splitting with the ones where $\mathbf{u}$ is on display on the left (resp. right) hand side if it is a nominal (resp. conominal). □

After eliminating all the variables bound by *type 2* quantifiers, the shape of the antecedent reduces to the inequality $\mathbf{j} \not\leq \mathbf{m}$ in conjunction with inequalities of the form $\mathbf{j} \leq \theta$ or $\eta \leq \mathbf{m}$, and, moreover, the remaining *type 1* quantifiers can only be restricted by $\mathbf{j}$ and $\mathbf{m}$. Hence, by expanding these remaining quantifiers, exhaustively currying, and applying inverse splitting, the antecedent equivalently reduces to the following conjunction of inequalities $\mathbf{j} \leq \theta_1 \wedge \cdots \wedge \theta_n$ & $\eta_1 \vee \cdots \vee \eta_m \leq \mathbf{m}$ & $\mathbf{j} \not\leq \mathbf{m}$.

Lemma 4.14 *After the elimination of* type 2 *restricted quantifiers and the expansion of* type 1 *quantifiers, the antecedent has form*

$$\mathbf{j} \leq \bigwedge_{i=1}^{n} \theta_i \ \ \& \ \ \bigvee_{i=1}^{m} \eta_i \leq \mathbf{m} \ \& \ \ \mathbf{j} \not\leq \mathbf{m}, \tag{4}$$

where $+\bigwedge_{i=1}^{n} \theta_i$ *and* $-\bigwedge_{i=1}^{m} \eta_i$ *are pure scattered Skeleton formulas where* $\mathbf{j}$ *and* $\mathbf{m}$ *do not occur, and any nominal (resp. conominal) occurs in positive (resp. negative) polarity.*

Proof. It is sufficient to show that every $+\theta_i$ and $-\eta_i$ is made of Skeleton nodes and that each variable occurs only once, since $\mathbf{j}$ and $\mathbf{m}$ clearly cannot occur there as they are not in $\overline{\mathbf{h}}$ or $\overline{\mathbf{o}}$ and they are not even restricted variables. The conjuncts that come from the *type 1* restricted quantifiers clearly satisfy the statement, hence it remains to show that the algorithm for the elimination of *type 2* quantifiers produces conjuncts with the same property. We proceed by induction. Let us consider the case in which we eliminate a quantifier of the kind $(\forall \mathbf{i}, \mathbf{n} \rhd \mathbf{l})$. Before the inverse approximation of an iteration is performed, we have a restricting inequality $\mathbf{i} \to \mathbf{n} \leq \mathbf{l}$ and the two inequalities $\mathbf{i} \leq \varphi$ and $\psi \leq \mathbf{n}$. In the base case, the last two are either restricting inequalities or inequalities without connectives nor operators; in both the cases $+\varphi$ and $-\psi$ are skeleton nodes and each variable occurs only once, as the pure variables in $\overline{\mathbf{h}}$ and $\overline{\mathbf{o}}$ occur only once in the antecedent and the variables in a restricting inequalities of a restricted quantifier are all different. Applying inverse approximation we have $\varphi \to \psi \leq \mathbf{l}$, and clearly $-(\varphi \to \psi)$ is a scattered Skeleton formula where the constraints on the polarity of the variables are met. In the inductive case, we proceed in the same way noting that, by applying inductive hypothesis, $\mathbf{i} \leq \varphi$ or $\psi \leq \mathbf{n}$ are scattered Skeleton formulas that meet the constraints on the polarities; hence, as in the base case, also the result of inverse approximation meets the same requirements. The remaining cases are proved similarly. □

Remark 4.15 The variables occurring $+\bigwedge_{i=1}^{n} \theta_i$ and $-\bigwedge_{i=1}^{m} \eta_i$ are exactly all the ones in $\overline{\mathbf{h}}$, $\overline{\mathbf{o}}$ and the ones bound by *type 1* quantifiers. The former variables are indeed captured because each one of them occurs at least (exactly) once in the antecedent, whilst the latter variables are clearly captured by writing the expansion of the quantifier.

Example 4.16 When treating the antecedent of the first formula in Example 4.9 as further processed in Example 4.12, we do not have inherently universal restricted quantifiers; hence this step consists in a straightforward application of the inverse splitting rule

$$\begin{array}{rl} & \forall \mathbf{j} \forall \mathbf{m} \forall \mathbf{h}_1 \forall \mathbf{h}_2 \forall \mathbf{o}_1 [\mathbf{j} \leq \mathbf{h}_1 \ \& \ \mathbf{j} \leq \mathbf{h}_2 \ \& \ \mathbf{o}_1 \leq \mathbf{m} \ \& \ \mathbf{j} \not\leq \mathbf{m} \Rightarrow \neg \mathbf{o}_1 \leq \Box\Box\blacklozenge(\Diamond \mathbf{h}_1 \wedge \blacklozenge \mathbf{h}_2)] \\ \text{iff} & \forall \mathbf{j} \forall \mathbf{m} \forall \mathbf{h}_1 \forall \mathbf{h}_2 \forall \mathbf{o}_1 [\mathbf{j} \leq \mathbf{h}_1 \wedge \mathbf{h}_2 \ \& \ \mathbf{o}_1 \leq \mathbf{m} \ \& \ \mathbf{j} \not\leq \mathbf{m} \Rightarrow \neg \mathbf{o}_1 \leq \Box\Box\blacklozenge(\Diamond \mathbf{h}_1 \wedge \blacklozenge \mathbf{h}_2)] \end{array}$$

As for the second formula from Examples 4.9 and 4.12, namely

$$\forall \mathbf{j}\forall \mathbf{m}\forall \mathbf{h}_1\forall \mathbf{h}_2\forall \mathbf{o}_1(\forall \mathbf{i}_1 \rhd \mathbf{j})(\forall \mathbf{n}_1 \rhd \mathbf{m})[\mathbf{i}_1 \leq \mathbf{h}_1 \ \&\ \mathbf{i}_1 \leq \mathbf{h}_2 \ \&\ \mathbf{o}_1 \leq \mathbf{m} \ \&\ \mathbf{j} \not\leq \mathbf{m} \Rightarrow \neg \mathbf{o}_1 \leq \mathbf{h}_2 \ \parr\ \neg \mathbf{n}_1 \leq \Diamond\Box\Diamond \mathbf{h}_1],$$

the quantifier $(\forall \mathbf{n}_1 \rhd \mathbf{m})$ is of *type 1*, while $(\forall \mathbf{i}_1 \rhd \mathbf{j})$ is of *type 2*. We start by eliminating the latter and then we merge the inequalities of the antecedent with the one of the restricted quantifier of *type 1*.

$$\forall \mathbf{j}\forall \mathbf{m}\forall \mathbf{h}_1\forall \mathbf{h}_2\forall \mathbf{o}_1\forall \mathbf{i}_1\forall \mathbf{n}_1[\mathbf{i}_1 \leq \mathbf{h}_1 \& \mathbf{i}_1 \leq \mathbf{h}_2 \& \mathbf{j} \leq \Diamond \mathbf{i}_1 \& \mathbf{o}_1 \leq \mathbf{m} \& \Box \mathbf{n}_1 \leq \mathbf{m} \& \mathbf{j} \not\leq \mathbf{m} \Rightarrow \neg \mathbf{o}_1 \leq \mathbf{h}_2 \ \parr\ \neg \mathbf{n}_1 \leq \Diamond\Box\Diamond \mathbf{h}_1]$$

$$\text{iff } \forall \mathbf{j}\forall \mathbf{m}\forall \mathbf{h}_1\forall \mathbf{h}_2\forall \mathbf{o}_1\forall \mathbf{i}_1\forall \mathbf{n}_1[\mathbf{i}_1 \leq \mathbf{h}_1 \wedge \mathbf{h}_2 \& \mathbf{j} \leq \Diamond \mathbf{i}_1 \& \mathbf{o}_1 \leq \mathbf{m} \& \Box \mathbf{n}_1 \leq \mathbf{m} \& \mathbf{j} \not\leq \mathbf{m} \Rightarrow \neg \mathbf{o}_1 \leq \mathbf{h}_2 \ \parr\ \neg \mathbf{n}_1 \leq \Diamond\Box\Diamond \mathbf{h}_1]$$

$$\text{iff } \forall \mathbf{j}\forall \mathbf{m}\forall \mathbf{h}_1\forall \mathbf{h}_2\forall \mathbf{o}_1\forall \mathbf{n}_1[\mathbf{j} \leq \Diamond(\mathbf{h}_1 \wedge \mathbf{h}_2) \& \mathbf{o}_1 \leq \mathbf{m} \& \Box \mathbf{n}_1 \leq \mathbf{m} \& \mathbf{j} \not\leq \mathbf{m} \Rightarrow \neg \mathbf{o}_1 \leq \mathbf{h}_2 \ \parr\ \neg \mathbf{n}_1 \leq \Diamond\Box\Diamond \mathbf{h}_1]$$

$$\text{iff } \forall \mathbf{j}\forall \mathbf{m}\forall \mathbf{h}_1\forall \mathbf{h}_2\forall \mathbf{o}_1\forall \mathbf{n}_1[\mathbf{j} \leq \Diamond(\mathbf{h}_1 \wedge \mathbf{h}_2) \& \mathbf{o}_1 \vee \Box \mathbf{n}_1 \leq \mathbf{m} \& \mathbf{j} \not\leq \mathbf{m} \Rightarrow \neg \mathbf{o}_1 \leq \mathbf{h}_2 \ \parr\ \neg \mathbf{n}_1 \leq \Diamond\Box\Diamond \mathbf{h}_1]$$

Elimination of pivotal variables. After the elimination of *type 2* quantifiers, the contrapositive of the formula is

$$\forall \mathbf{j}\forall \mathbf{m}\forall \overline{\mathbf{h}}\forall \overline{\mathbf{o}}\forall \overline{\mathbf{i}}'\forall \overline{\mathbf{n}}' \left(\overline{\delta \leq \neg \mathbf{k}} \ \&\ \overline{\neg \mathbf{l} \leq \gamma} \Rightarrow \left(\mathbf{j} \leq \bigwedge_{i=1}^{n} \theta_i \ \ \&\ \ \bigvee_{i=1}^{m} \eta_i \leq \mathbf{m} \Rightarrow \ \mathbf{j} \leq \mathbf{m}\right)\right),$$

where $\overline{\mathbf{i}}'$ and $\overline{\mathbf{n}}'$ are the variables originally introduced by *type 1* restricted quantifiers. By applying inverse approximation to eliminate $\mathbf{j}$ and $\mathbf{m}$, and by putting $\varphi := \bigwedge_{i=1}^{n} \theta_i$ and $\psi := \bigwedge_{i=1}^{m} \eta_i$, the formula above is equivalent to

$$\forall \overline{\mathbf{h}}\forall \overline{\mathbf{o}}\forall \overline{\mathbf{i}}'\forall \overline{\mathbf{n}}' \left(\overline{\delta \leq \neg \mathbf{k}} \ \&\ \overline{\neg \mathbf{l} \leq \gamma} \Rightarrow \varphi \leq \psi\right).$$

If one of the literals in $\overline{\mathbf{k}}$ (resp. in $\overline{\mathbf{l}}$) is already negated, namely it is of the form $\neg \mathbf{i}$ (resp. $\neg \mathbf{n}$) for some nominal $\mathbf{i}$ (resp. conominal $\mathbf{n}$), we can just apply self adjunction of negation to obtain a formula $\mathbf{i} \leq \theta$. Hence, we apply this procedure obtaining a formula with shape

$$\forall \overline{\mathbf{h}}\forall \overline{\mathbf{o}}\forall \overline{\mathbf{i}'}\forall \overline{\mathbf{n}'} \left(\overline{\mathbf{k}' \leq \gamma'} \ \&\ \overline{\delta' \leq \mathbf{l}'} \Rightarrow \varphi \leq \psi\right). \tag{5}$$

We can assume that each variable in $\overline{\mathbf{k}'}$ (resp. $\overline{\mathbf{l}'}$) is different, since if it occurs in more than one inequality, these two inequalities can be merged via inverse splitting.

Very simple Sahlqvist in ML^***.** To simplify notation, we drop the apostrophe in $\overline{\mathbf{k}'}$ and $\overline{\mathbf{l}'}$, and we let $\overline{\mathbf{i}}$ (resp. $\overline{\mathbf{n}}$) denote all the other nominals (resp. conominals), i.e. the ones occurring in $\overline{\gamma}$ and $\overline{\delta}$. The formula (5) is thus equivalent to:

$$\forall \overline{\mathbf{k}}\forall \overline{\mathbf{l}}\forall \overline{\mathbf{i}}\forall \overline{\mathbf{n}} \left(\overline{\mathbf{k} \leq \gamma} \ \&\ \overline{\delta \leq \mathbf{l}} \Rightarrow \varphi \leq \psi\right). \tag{6}$$

By Lemma 4.14, we know that the inequality $\varphi \leq \psi$ is a scattered Skeleton inequality containing every variable quantified in the prefix. Furthermore, each nominal in $\overline{\mathbf{i}}$ and $\overline{\mathbf{k}}$ occurs in positive polarity in it, and each conominal in $\overline{\mathbf{n}}$ and $\overline{\mathbf{l}}$ occurs in negative polarity; hence we may eliminate of each $\mathbf{k}$ and $\mathbf{l}$ through inverse approximation (see [5, Section 4.2]). Therefore (6) is equivalent to $\forall \overline{\mathbf{i}}\forall \overline{\mathbf{n}} \left(\varphi[\overline{\gamma}/\overline{\mathbf{k}}, \overline{\delta}/\overline{\mathbf{l}}] \leq \psi[\overline{\gamma}/\overline{\mathbf{k}}, \overline{\delta}/\overline{\mathbf{l}}]\right)$. For each (co)nominal in $\mathbf{i}$ (resp. in $\mathbf{n}$) we introduce a new variable $p_{\mathbf{i}}$ (resp. $q_{\mathbf{n}}$). Let

$$\varphi' := \left(\varphi\left[\overline{\gamma}/\overline{\mathbf{k}}, \overline{\delta}/\overline{\mathbf{l}}\right]\right)\left[\overline{p_{\mathbf{i}}}/\overline{\mathbf{i}}, \overline{q_{\mathbf{n}}}/\overline{\mathbf{n}}\right] \qquad \psi' := \left(\psi\left[\overline{\gamma}/\overline{\mathbf{k}}, \overline{\delta}/\overline{\mathbf{l}}\right]\right)\left[\overline{p_{\mathbf{i}}}/\overline{\mathbf{i}}, \overline{q_{\mathbf{n}}}/\overline{\mathbf{n}}\right].$$

By Lemma 4.11, nominals (resp. conominals) in each $+\gamma$ in $\overline{\gamma}$ and $-\delta$ in $\overline{\delta}$ occur in negative (resp. positive) polarity; hence every $+\gamma$ and $-\delta$ is an ε^∂-uniform subtree in $+\varphi'$ and $-\psi'$, where ε is the order type on $\overline{p}_\mathbf{i}$ and $\overline{q}_\mathbf{n}$ such that $\varepsilon(p_\mathbf{i}) = 1$ and $\varepsilon(q_\mathbf{n}) = \partial$.

Hence, $\varphi' \leq \psi'$ is a scattered very simple ε-Sahlqvist inequality in ML*, and, moreover, ALBA reduces it to (6), as shown below $\forall \overline{p}_\mathbf{i} \forall \overline{q}_\mathbf{n} (\varphi' \leq \psi')$ is equivalent to

$$\forall \overline{p}_\mathbf{i} \forall q_\mathbf{n} \forall \bar{\mathbf{j}} \forall \overline{\mathbf{m}} \forall \bar{\mathbf{j}}' \forall \overline{\mathbf{m}}' \Big(\bar{\mathbf{j}} \leq \overline{\gamma}[\overline{p}_\mathbf{i}/\bar{\mathbf{i}}, \overline{q}_\mathbf{n}/\overline{\mathbf{n}}] \ \&\ \overline{\delta}[\overline{p}_\mathbf{i}/\bar{\mathbf{i}}, \overline{q}_\mathbf{n}/\overline{\mathbf{n}}] \leq \overline{\mathbf{m}} \ \&\ \bar{\mathbf{j}}' \leq \overline{p}_\mathbf{i} \ \&\ \overline{p}_\mathbf{n} \leq \mathbf{m}' \\ \Rightarrow \varphi[\bar{\mathbf{j}}/\overline{\mathbf{k}}, \overline{\mathbf{m}}/\bar{\mathbf{l}}] \leq \psi[\bar{\mathbf{j}}/\overline{\mathbf{k}}, \overline{\mathbf{m}}/\bar{\mathbf{l}}]\Big),$$

which is equivalent to

$$\forall \bar{\mathbf{j}} \forall \overline{\mathbf{m}} \forall \bar{\mathbf{j}}' \forall \overline{\mathbf{m}}' \Big(\bar{\mathbf{j}} \leq \overline{\gamma}[\bar{\mathbf{j}}'/\bar{\mathbf{i}}, \overline{\mathbf{m}}'/\overline{\mathbf{n}}] \ \&\ \overline{\delta}[\bar{\mathbf{j}}'/\bar{\mathbf{i}}, \overline{\mathbf{m}}'/\overline{\mathbf{n}}] \leq \overline{\mathbf{m}} \Rightarrow \varphi[\bar{\mathbf{j}}/\overline{\mathbf{k}}, \overline{\mathbf{m}}/\bar{\mathbf{l}}] \leq \psi[\bar{\mathbf{j}}/\overline{\mathbf{k}}, \overline{\mathbf{m}}/\bar{\mathbf{l}}]\Big)$$

From the above discussion, the main result follows.

Theorem 4.17 *Every (refined) Kracht* ML^K *formula can be effectively associated with a scattered very simple Sahlqvist inequality in* ML* *to which it is equivalent on Kripke frames/complete and atomic BAOs.*

Example 4.18 In Example 4.16 we had

$\forall \mathbf{j} \forall \mathbf{m} \forall \mathbf{h}_1 \forall \mathbf{h}_2 \forall \mathbf{o}_1 [\mathbf{j} \leq \mathbf{h}_1 \wedge \mathbf{h}_2 \ \&\ \mathbf{o}_1 \leq \mathbf{m} \ \&\ \mathbf{j} \not\leq \mathbf{m}_1 \Rightarrow \neg \mathbf{o}_1 \leq \Box\Box \blacklozenge(\Diamond \mathbf{h}_1 \wedge \blacklozenge \mathbf{h}_2)]$.

After the contrapositive step it becomes $\forall \mathbf{h}_1 \forall \mathbf{h}_2 \forall \mathbf{o}_1 [\Box\Box \blacklozenge(\Diamond \mathbf{h}_1 \wedge \blacklozenge \mathbf{h}_2) \leq \mathbf{o}_1 \Rightarrow \mathbf{h}_1 \wedge \mathbf{h}_2 \leq \mathbf{o}_1]$, which, by the previous discussion, is equivalent to the very simple Sahlqvist formula $\forall p_{h_1} \forall p_{h_2} [p_{h_1} \wedge p_{h_2} \leq \Box\Box \blacklozenge(\Diamond p_{h_1} \wedge \blacklozenge p_{h_2})]$.

Taking the second formula in Example 4.16, i.e.

$$\forall \mathbf{j} \forall \mathbf{m} \forall \mathbf{h}_1 \forall \mathbf{h}_2 \forall \mathbf{o}_1 \forall \mathbf{n}_1 [\mathbf{j} \leq \Diamond(\mathbf{h}_1 \wedge \mathbf{h}_2) \ \&\ \mathbf{o}_1 \vee \Box \mathbf{n}_1 \leq \mathbf{m} \ \&\ \mathbf{j} \not\leq \mathbf{m} \Rightarrow \\ \neg \mathbf{o}_1 \leq \mathbf{h}_2 \parr \neg \mathbf{n}_1 \leq \Diamond\Box\Diamond \mathbf{h}_1],$$

after the contrapositive step we obtain $\forall \mathbf{h}_1 \forall \mathbf{h}_2 \forall \mathbf{o}_1 \forall \mathbf{n}_1 [\mathbf{h}_2 \leq \mathbf{o}_1 \ \&\ \Diamond\Box\Diamond \mathbf{h}_1 \leq \mathbf{n}_1 \Rightarrow \Diamond(\mathbf{h}_1 \wedge \mathbf{h}_2) \leq \mathbf{o}_1 \vee \Box \mathbf{n}_1]$, which in turn is equivalent to the very simple Sahlqvist $\forall p_{h_1} \forall q_{o_1} [\Diamond(p_{h_1} \wedge q_{o_1}) \leq q_{o_1} \vee \Box\Diamond\Box\Diamond p_{h_1}]$.

From inductive inequalities to Kracht ML^K **formulas.** By applying an ALBA inductive formula, taking the contrapositive of the resulting pure quasi-inequality, and then applying approximation rules to obtain flat inequalities in the whole formula, the following result can be proved.

Theorem 4.19 *Every inductive inequality is equivalent to some Kracht* ML^K *formula.*

From Kracht ML^K **to inductive.** As, by Lemma 3.6 and Proposition 3.4, the class of inductive formulas is equivalent to the one of crypto ML*-inductive, it is sufficient to restrict ourselves to the class of Kracht ML^K-formulas which correspond to crypto-inductive formulas. To do so, it is sufficient to note that the only condition to enforce is that the Kracht ML-disjuncts starting with an operator in ML* are reduced (by Algorithm 1) to inequalities whose non-main side is an (ε, Ω)-unpackable formula for some ε and Ω. This is easily achieved by imposing the same restrictions as in Definition 3.2 to the operators in the restricted quantifiers of the branch.

5 Conclusion

We have established an inverse correspondence result between the Sahlqvist formulas in tense logic and the inductive formulas in modal logic on the one hand, and a class of quantified pure hybrid tense formulas on the other. The order-theoretic perspective we have introduced in the present paper lays the groundwork for a generalization of Kracht's and Kikot's inverse correspondence theory to general logics algebraically captured by classes of (distributive) lattice expansions or (D)LE-logics (see [3]). One of the main advantages of the latter is that it is a general, modular result which is largely independent of any particular choice of relational semantics for a (D)LE logic but links to such particular choices (and thence to *first-order* inverse correspondence) via duality and the accompanying standard translation.

In closing, we will mention only one of the various further directions that remain to be developed in this line of research: We have focused on correspondence between first-order formulas in one free variable and modal formula, but one may generalize this to *n-correspondence* between first-order formulas in n free variables and n-tuples of modal formulas [18]. Inverse n-correspondence is studied in [15], and a characterization is obtained of the first-order formulas built from relational atoms using conjunction and existential quantification for which modal n-correspondents exist. This problem is also considered in [8] from the perspective of description logics and for, what amounts to, a more general class of first-order formulas. In future work we will generalize and apply the methods of the present paper to the problem of n-correspondence for general DLE-languages.

Appendix

A The rules of ALBA

The algorithm ALBA applies a set of invertable re-write rules to transform (while maintaining equivalence on algebras and frames) quasi-inequalities (of formulas in LE-languages, like ML^+) into sets of pure quasi-inequalities, i.e. ones in which all propositional variables have been eliminated in favour of nominals and co-nominals. We refer the reader to [5] and [6] for the full specification of ALBA including its rules. Because they are of particular importance to the present paper, we here recall only the inverses of the splitting and approximation rules:

Inverse splitting rules.

$$\frac{\alpha \leq \beta \wedge \gamma}{\alpha \leq \beta \quad \alpha \leq \gamma} \qquad \frac{\alpha \vee \beta \leq \gamma}{\alpha \leq \gamma \quad \beta \leq \gamma}$$

Inverse approximation rules. The following are special cases of the general approximation rules given in [5] and [6]. Let Γ be an arbitrary conjunction of ML^+ inequalities and $\phi, \psi \in \mathrm{ML}^+$ such that, in each of the following rules, the quantified nominal or co-nominal does not in occur in the conclusion, and $\diamondsuit \in \{\Diamond, \blacklozenge\}$ and $\boxdot \in \{\Box, \blacksquare\}$:

$$\frac{\exists \mathbf{i}(\mathbf{j} \leq \diamondsuit\, \mathbf{i} \;\&\; \mathbf{i} \leq \phi \;\&\; \Gamma)}{\mathbf{j} \leq \diamondsuit\, \phi \;\&\; \Gamma} \qquad \frac{\exists \mathbf{n}(\boxdot \mathbf{n} \leq \mathbf{m} \;\&\; \phi \leq \mathbf{n} \;\&\; \Gamma)}{\boxdot \phi \leq \mathbf{m} \;\&\; \Gamma}$$

$$\frac{\forall \mathbf{i}(\mathbf{i} \leq \phi \;\&\; \Gamma \Rightarrow \diamondsuit\, \mathbf{i} \leq \psi)}{\Gamma \Rightarrow \diamondsuit\, \phi \leq \psi} \qquad \frac{\forall \mathbf{n}(\phi \leq \mathbf{n} \;\&\; \Gamma \Rightarrow \psi \leq \boxdot \mathbf{n})}{\Gamma \Rightarrow \psi \leq \boxdot \phi}$$

$$\frac{\forall \mathbf{j} \forall \mathbf{m}(\mathbf{j} \leq \phi \;\&\; \psi \leq \mathbf{m} \;\&\; \Gamma \Rightarrow \mathbf{j} \leq \mathbf{m})}{\Gamma \Rightarrow \phi \leq \psi}$$

B From inductive in ML to Kracht

It is well known (cf. [7]) that an ALBA run on a definite inductive formula $(\varphi \leq \psi)[\overline{\alpha}, \overline{x}, \overline{\beta}/\overline{y}, \overline{\gamma}/\overline{z}, \overline{\delta}/\overline{w}]$ yields

$$\forall \overline{\mathbf{j}} \forall \overline{\mathbf{m}} \forall \overline{\mathbf{i}} \forall \overline{\mathbf{n}} \Big(\overline{\mathbf{i}} \leq \overbrace{\overline{\gamma}\left[\overline{\bigvee \mathsf{Mv}(p)}/\overline{p}, \overline{\bigwedge \mathsf{Mv}(q)}/\overline{q}\right]}^{\overline{\gamma}^{mv}} \;\&\; \overbrace{\overline{\delta}\left[\overline{\bigvee \mathsf{Mv}(p)}/\overline{p}, \overline{\bigwedge \mathsf{Mv}(q)}/\overline{q}\right]}^{\overline{\delta}^{mv}} \leq \overline{\mathbf{n}} \Rightarrow (\varphi \leq \psi)[!\overline{\mathbf{j}}/!\overline{x}, !\overline{\mathbf{m}}/!\overline{y}, !\overline{\mathbf{i}}/!\overline{z}, !\overline{\mathbf{n}}/!\overline{w}] \Big),$$

which is equivalent to its contrapositive
$\forall \overline{\mathbf{j}} \forall \overline{\mathbf{m}} \forall \overline{\mathbf{i}} \forall \overline{\mathbf{n}} \left((\varphi \not\leq \psi)[!\overline{\mathbf{j}}/!\overline{x}, !\overline{\mathbf{m}}/!\overline{y}, !\overline{\mathbf{i}}/!\overline{z}, !\overline{\mathbf{n}}/!\overline{w}] \Rightarrow \bigparr_{i=1}^{n} \gamma_i^{mv} \leq \neg \mathbf{i}_i \;\;\parr\;\; \bigparr_{i=1}^{m} \neg \mathbf{n}_i \leq \delta_i^{mv} \right).$

We put $(\varphi' \leq \psi') \equiv (\varphi \leq \psi)[!\overline{\mathbf{j}}/!\overline{x}, !\overline{\mathbf{m}}/!\overline{y}, !\overline{\mathbf{i}}/!\overline{z}, !\overline{\mathbf{n}}/!\overline{w}]$, by approximating the antecedent we have:

$$\forall \mathbf{j}' \forall \mathbf{m}' \forall \overline{\mathbf{j}} \forall \overline{\mathbf{m}} \forall \overline{\mathbf{i}} \forall \overline{\mathbf{n}} \left(\mathbf{j}' \leq \varphi' \;\&\; \psi' \leq \mathbf{m}' \;\&\; \mathbf{j}' \not\leq \mathbf{m}' \Rightarrow \bigparr_{i=1}^{n} \gamma_i^{mv} \leq \neg \mathbf{i}_i \;\;\parr\;\; \bigparr_{i=1}^{m} \neg \mathbf{n}_i \leq \delta_i^{mv} \right). \tag{B.1}$$

The variables $\mathbf{j}'$ and $\mathbf{m}'$ will be the pivotal pure variables of the inductive Kracht formula that we will compute.

Lemma B.1 *Each $\gamma_i^{mv} \leq \neg \mathbf{i}_i$ and $\neg \mathbf{n}_i \leq \delta_i^{mv}$ in* (B.1) *is equivalent to some Kracht disjunct.*

Proof. By induction on the structure of γ_i^{mv} (resp. δ_i^{mv}). When $\gamma_i^{mv} \leq \neg \mathbf{i}_i$ (resp. $\neg \mathbf{n}_i \leq \delta_i^{mv}$) is flat inequalities, there is nothing to do. When it is not flat, $\gamma_i^{mv} \leq \neg \mathbf{i}_i$ (resp. $\neg \mathbf{n}_i \leq \delta_i^{mv}$) can be rewritten in one of the following ways:

- $\Box \theta \leq \neg \mathbf{i}_i$ (resp. $\neg \mathbf{n}_i \leq \Diamond \theta$) for some formula θ. Ackermann lemma yields $(\exists \mathbf{l} \rhd \neg \mathbf{i}_i)\theta \leq \mathbf{l}$ (resp. $(\exists \mathbf{k} \rhd \neg \mathbf{n}_i)\mathbf{k} \leq \theta$) with $\mathbf{l}$ (resp. $\mathbf{k}$) fresh.
- $\blacksquare \theta \leq \neg \mathbf{i}_i$ (resp. $\neg \mathbf{n}_i \leq \blacklozenge \theta$) for some formula θ. Ackermann lemma yields $(\exists \mathbf{l} \blacktriangleright \neg \mathbf{i}_i)\theta \leq \mathbf{l}$ (resp. $(\exists \mathbf{k} \blacktriangleright \neg \mathbf{n}_i)\mathbf{k} \leq \theta$) with $\mathbf{l}$ (resp. $\mathbf{k}$) fresh.
- $\theta \to \eta \leq \neg \mathbf{i}_i$ (resp. $\neg \mathbf{n}_i \leq \theta - \eta$) for some formulas θ and η. Ackermann lemma yields $(\exists \mathbf{k}, \mathbf{l} \rhd \neg \mathbf{i}_i)(\mathbf{k} \leq \theta \;\&\; \eta \leq \mathbf{l})$ (resp. $(\exists \mathbf{k}, \mathbf{l} \rhd \neg \mathbf{n}_i)(\mathbf{k} \leq \theta \;\&\; \eta \leq \mathbf{l})$) with $\mathbf{k}$ and $\mathbf{l}$ fresh.

- $\Diamond\theta \leq \neg\mathbf{i}_i$ (resp. $\neg\mathbf{n}_i \leq \Box\theta$) for some formula θ. Ackermann lemma yields $(\forall\mathbf{k} \rhd \mathbf{i}_i)\theta \leq \neg\mathbf{k}$ (resp. $(\forall\mathbf{l} \rhd \mathbf{n}_i)\neg\mathbf{l} \leq \theta$) with $\mathbf{k}$ (resp. $\mathbf{l}$) fresh.
- $\blacklozenge\theta \leq \neg\mathbf{i}_i$ (resp. $\neg\mathbf{n}_i \leq \blacksquare\theta$) for some formula θ. Ackermann lemma yields $(\forall\mathbf{k} \blacktriangleright \mathbf{i}_i)\theta \leq \neg\mathbf{k}$ (resp. $(\forall\mathbf{l} \blacktriangleright \mathbf{n}_i)\neg\mathbf{l} \leq \theta$) with $\mathbf{k}$ (resp. $\mathbf{l}$) fresh.
- $\theta - \eta \leq \neg\mathbf{i}_i$ (resp. $\neg\mathbf{n}_i \leq \theta \to \eta$) for some formulas θ and η. Ackermann lemma yields $(\forall\mathbf{l}, \mathbf{k} \rhd \mathbf{i}_i)(\neg\mathbf{l} \leq \theta \parr \eta \leq \neg\mathbf{k})$ (resp. $(\forall\mathbf{l}, \mathbf{k} \rhd \mathbf{n}_i)(\neg\mathbf{l} \leq \theta \parr \eta \leq \neg\mathbf{k})$) with $\mathbf{k}$ and $\mathbf{l}$ fresh.
- $\theta \wedge \eta \leq \neg\mathbf{i}_i$ (resp. $\neg\mathbf{n}_i \leq \theta \vee \eta$) for some formulas θ and η. The inverse splitting rule yields $\theta \leq \neg\mathbf{i}_i \parr \eta \leq \neg\mathbf{i}_i$ (resp. $\neg\mathbf{n}_i \leq \theta \parr \neg\mathbf{n}_i \leq \eta$).
- $\theta \vee \eta \leq \neg\mathbf{i}_i$ (resp. $\neg\mathbf{n}_i \leq \theta \wedge \eta$) for some formulas θ and η. The inverse splitting rule yields $\theta \leq \neg\mathbf{i}_i \;\&\; \eta \leq \neg\mathbf{i}_i$ (resp. $\neg\mathbf{n}_i \leq \theta \;\&\; \neg\mathbf{n}_i \leq \eta$).

As the inductive hypothesis holds on the subformulae generated either by inverse splitting or Ackermann lemma, the statement holds. □

Lemma B.2 *The algorithm in Lemma B.1 applied to* $\mathbf{j}' \leq \varphi'$ *and* $\psi' \leq \mathbf{m}'$ *yields only existential quantifiers and (conjunctions) of restricting inequalities.*

Proof. By induction on the structure of φ' (resp. ψ'). If either φ' (resp. ψ') is a pure variable or has one single operator/connective, the statement is true as, being positive Skeleton, the operator would either $\Diamond$, $\blacklozenge$, or $\vee$ (resp. $\Box$, $\blacksquare$, $\to$, $\wedge$). The same thing applies to the uppermost connective in more complex formula, which, when simplified via Ackermann lemma, yield smaller skeleton formulas while outputting a single existential quantifier (it is sufficient to check that the possible cases output existential quantifier). Note that the cases that produce (meta) disjunctions can never occur, as they would require either a $\wedge$ on the left or a $\vee$ on the right hand side of the inequality. □

Consider the formula obtained after the procedures described in the two lemmas above are executed. The existential quantifiers in the antecedent can be rewritten as universal ones in the prefix: as they have been added during the procedure in Lemma B.2, they cannot occur in the consequent. Indeed, they are exactly the *type 2* quantifiers of the formula. The remaining restricting inequalities in the antecedent can be written directly in their corresponding quantifiers: they make up for the *type 1* quantifiers of the formula as they occur in the consequent, but not in the antecedent (since the skeleton in scattered, the restricting inequality originating the quantifier can be the only one containing it). The remaining inequalities in the antecedent do not contain operators/connectives, and the non-pivotal variables occur exactly once (always because the skeleton is scattered). The requirements on polarity do also hold as in the skeleton formula $\varphi' \leq \psi'$ nominals occur in positive position and conominals in negative position due to how the first approximation step of ALBA works. The requirement on the polarities in the consequent is respected as universal quantifiers, as can be verified by looking at the cases in Lemma B.1, flip the polarities of their maximal subformulae (the θs and the ηs in the lemma). Therefore, the claim that every inductive inequality is equivalent to some inductive Kracht formula readily follows from the above discussion.

References

[1] Blackburn, P., M. d. Rijke and Y. Venema, "Modal Logic," Cambridge Tracts in Theoretical Computer Science, Cambridge University Press, 2001.

[2] Chen, J., G. Greco, A. Palmigiano and A. Tzimoulis, *Syntactic completeness of proper display calculi*, arXiv preprint arXiv:2102.11641 (2021).

[3] Conradie, W., A. De Domenico, G. Greco, A. Palmigiano, M. Panettiere and A. Tzimoulis, *Unified inverse correspondence for DLE-logics*, arXiv preprint arXiv:2203.09199 (2022).

[4] Conradie, W., S. Ghilardi and A. Palmigiano, *Unified Correspondence*, in: A. Baltag and S. Smets, editors, *Johan van Benthem on Logic and Information Dynamics*, Outstanding Contributions to Logic **5**, Springer International Publishing, 2014 pp. 933–975.

[5] Conradie, W. and A. Palmigiano, *Algorithmic correspondence and canonicity for distributive modal logic*, Ann. Pure Appl. Log. **163** (2012), pp. 338–376.

[6] Conradie, W. and A. Palmigiano, *Algorithmic correspondence and canonicity for non-distributive logics*, Annals of Pure and Applied Logic **170** (2019), pp. 923–974.

[7] De Rudder, L. and A. Palmigiano, *Slanted canonicity of analytic inductive inequalities*, ACM Trans. Comput. Logic **22** (2021).

[8] Feier, C., C. Lutz and F. Wolter, *From conjunctive queries to instance queries in ontology-mediated querying*, in: J. Lang, editor, *Proceedings of the Twenty-Seventh International Joint Conference on Artificial Intelligence, IJCAI 2018, July 13-19, 2018, Stockholm, Sweden* (2018), pp. 1810–1816.
URL `https://doi.org/10.24963/ijcai.2018/250`

[9] Gabbay, D. M., R. A. Schmidt and A. Szalas, "Second-Order Quantifier Elimination. Foundations, Computational Aspects and Applications," College Publications, 2008.

[10] Goranko, V. and D. Vakarelov, *Sahlqvist formulas unleashed in polyadic modal languages*, in: *Conference: Advances in Modal Logic 3*, 2000, pp. 221–240.

[11] Goranko, V. and D. Vakarelov, *Sahlqvist formulas in hybrid polyadic modal logics*, Journal of Logic and Computation **11** (2001).

[12] Goranko, V. and D. Vakarelov, *Elementary canonical formulae: extending Sahlqvist's theorem*, Annals of Pure and Applied Logic **141** (2006), pp. 180–217.

[13] Greco, G., M. Ma, A. Palmigiano, A. Tzimoulis and Z. Zhao, *Unified correspondence as a proof-theoretic tool*, Journal of Logic and Computation **28** (2018), pp. 1367–1442.

[14] Kikot, S., *An extension of Kracht's theorem to generalized Sahlqvist formulas*, Journal of Applied Non-Classical Logics **19** (2010).

[15] Kikot, S. and E. Zolin, *Modal definability of first-order formulas with free variables and query answering*, Journal of Applied Logic **11** (2013), pp. 190–216.

[16] Kracht, M., "Internal definability and completeness in modal logic," Ph.D. thesis, Freien Universität Berlin (1990).

[17] Kracht, M., *How completeness and correspondence theory got married*, in: M. de Rijke, editor, *Diamonds and Defaults*, Springer, 1993 pp. 175–214.

[18] Kracht, M., "Tools and techniques in modal logic," Studies in Logic and the Foundations of Mathematics **142**, Elsevier Amsterdam, 1999.

Wijesekera-style constructive modal logics

Tiziano Dalmonte

Free University of Bozen-Bolzano, Bolzano, Italy

Abstract

We define a family of propositional constructive modal logics corresponding each to a different classical modal system. The logics are defined in the style of Wijesekera's constructive modal logic [38], and are both proof-theoretically and semantically motivated. On the one hand, they correspond to the single-succedent restriction of standard sequent calculi for classical modal logics. On the other hand, they are obtained by incorporating the hereditariness of intuitionistic Kripke models into the classical satisfaction clauses for modal formulas. We show that, for the considered classical logics, the proof-theoretical and the semantical approach return the same constructive systems.

Keywords: Constructive modal logic, intuitionistic modal logic, sequent calculus, neighbourhood semantics.

1 Introduction

Constructive or intuitionistic modal logics are extensions of intuitionistic logic with modalities $\Box$ and $\Diamond$. The motivations for the study of modalities with an intuitionistic basis are manifold, but they can be schematically classified into two kinds. On the one hand, from a theoretical perspective, it comes natural to combine intuitionistic and modal logic [35], considering in particular that both of them can be semantically arranged in terms of possible world models. In addition, the rejection of classical equivalences can allow for a finer analysis of the modalities. On the other hand, intuitionistic or constructive modal logics can be motivated by specific applications in computer science, such as type-theoretic interpretations, verification, and knowledge representation.

A peculiar feature of intuitionistic modal logics is that, similarly to the intuitionistic connectives, $\Box$ and $\Diamond$ are not interdefinable. This allows for the definition of systems in which $\Box$ and $\Diamond$ satisfy distinct principles. At the same time, it makes possible to define different intuitionistic or constructive counterparts of the same classical logic, as it is testified by the several intuitionistic versions of classical K which have been proposed in the literature (see [35] for a survey).

Intuitionistic modal logics have been formulated as monomodal (with only $\Box$ or only $\Diamond$) or bimodal (with both $\Box$ and $\Diamond$) systems. Considering logics including both modalities, two intuitionistic versions of K have been mostly considered: so-called Intuitionistic K (IK) and Constructive K (CK). The first system was introduced by Fischer Servi [13], Ploktin and Stirling [33], and Ewald [12], and can be defined as the set of formulas whose standard translation is derivable in first-order intuitionistic logic [35,40]. The second system, which is weaker that IK, was introduced by Bellin, de Paiva and Ritter [6], and was motivated by type-theoretic interpretations of the modalities and categorical semantics, but also by contextual reasoning [26,28]. Both the semantics [13,33,12,35,5,26,2,9,29,1] and the proof theory [3,36,14,19,25,24,23,6,4,27,10] of IK and CK have been extensively investigated, in particular significant consideration has been devoted to their extensions with standard modal axioms D, T, B, 4, and 5, so that entire families of intuitionistic and constructive modal logics are now available in the literature.

An additional bimodal constructive version of K was proposed by Wijesekera [38]. Wijesekera's logic aimed at representing reasoning with partial information about the states of concurrent transition systems, and was introduced as a modal extension of first-order intuitionistic logic. If we restrict our attention to its propositional fragment, Wijesekera's logic is intermediate between CK and IK. In particular, Wijesekera's logic (we call it WK) can be defined by extending (any axiomatisation of) intuitionistic propositional logic (IPL) with the following modal axioms and rules: [1]

$$nec\,\frac{A}{\Box A} \qquad \begin{array}{ll} K_\Box & \Box(A \supset B) \supset (\Box A \supset \Box B) \\ K_\Diamond & \Box(A \supset B) \supset (\Diamond A \supset \Diamond B) \end{array} \qquad N_\Diamond \quad \neg\Diamond\bot$$

Then CK can be obtained by dropping $\neg\Diamond\bot$, whereas IK is obtained by extending WK with the axioms $\Diamond(A \vee B) \supset \Diamond A \vee \Diamond B$ and $(\Diamond A \supset \Box B) \supset \Box(A \supset B)$. The interest of WK is not limited to its intended interpretation: this logic also exhibits an elegant relation with classical K, both from a semantical and from a proof-theoretical perspective. We now illustrate this relation.

Semantics for intuitionistic modal logics are typically defined by combining intuitionistic Kripke models and possible-world models for modal logics. A crucial requirement is that the resulting models must preserve the hereditary property of intuitionistic models, meaning that if a formula is true in a world w, then it is true also in all worlds reachable from w through the intuitionistic order $\leq$. Such a requirement can be fulfilled essentially in two ways. First, one can establish suitable combinations between $\leq$ and the modal relation $\mathcal{R}$, as it is done for instance in the semantics for IK. Alternatively, one can build the hereditariness into the satisfaction clauses for modal formulas by requiring that the standard clauses hold for all $\leq$-successors. This is the strategy adopted by Wijesekera [38] who presents models with two relations $\leq$ and $\mathcal{R}$ without any specific combination between them, where the modalities are interpreted in the following way: [2]

---

[1] Wijesekera [38] includes also the axiom $\Box A \wedge \Diamond(A \supset B) \supset \Diamond B$ which is derivable from the others.

[2] The semantics of CK is similar but it also requires 'fallible' worlds satisfying $\bot$ (cf. [26]).

$\mathcal{M}, w \Vdash \Box A$ iff for all $v \geq w$, for all u, if $v\mathcal{R}u$, then $\mathcal{M}, u \Vdash A$.
$\mathcal{M}, w \Vdash \Diamond A$ iff for all $v \geq w$, there is u such that $v\mathcal{R}u$ and $\mathcal{M}, u \Vdash A$.

Wijesekera [38] also provides a sequent calculus for WK, which is defined by extending a suitable calculus for IPL with the following modal rules (where $|\Gamma| \geq 0$ and $0 \leq |\Delta| \leq 1$):

$$\mathsf{K}^{\mathsf{i}}_{\Box}\ \frac{\Gamma \Rightarrow A}{\Box\Gamma \Rightarrow \Box A} \qquad \mathsf{K}^{\mathsf{i}}_{\Diamond}\ \frac{\Gamma, A \Rightarrow \Delta}{\Box\Gamma, \Diamond A \Rightarrow \Diamond\Delta}$$

Gentzen [15] showed that, given a suitable sequent calculus for classical logic, its restriction to single-succedent sequents (i.e., sequents with at most one formula in the consequent) provides a sequent calculus for intuitionistic logic. Interestingly, Wijesekera's logic can be seen as the system obtained by restricting to single-succedent sequents a standard sequent calculus for classical K (formulated with explicit $\Box$ and $\Diamond$), [3] so that this correspondence is preserved at the modal level. We then observe that WK displays a clear and elegant relation with classical K, both semantically and proof-theoretically:

- semantically, WK is obtained simply by incorporating hereditariness into the modal satisfaction clauses of K;
- proof-theoretically, it is obtained by restricting a standard sequent calculus for K to single-succedent sequents.

Despite its interest, Wijesekera's logic has received significantly less consideration than CK and IK. In particular, while alternative semantics and proof systems for WK have been studied [39,18,9,10], no systematic investigation of Wijesekera-style systems has been carried out so far.

Filling this gap is precisely the aim of this paper: we define a family of Wijesekera-style logics corresponding each to a different classical modal logic (for lack of a better name we call them *W-logics*), adopting as a guideline for the definition of these systems the semantical and proof-theoretical relation between WK and K just described. In particular, in Sec. 2 we present standard sequent calculi and semantics for a family of classical modal logics. Then we define constructive counterparts of these logics by (i) restricting the calculi to single-succedent sequents (Sec. 3), and (ii) expressing the classical satisfaction clauses for modal formulas over intuitionistic Kripke models, building hereditariness into these conditions (Sec. 4). The main contribution of this paper consists in showing that, despite being mutually independent, for a wide family of classical modal logics the semantical and the proof-theoretical approach return exactly the same constructive systems.

2 Preliminaries on classical modal logics

Let $\mathcal{L}$ be a propositional modal language based on a set *Atm* of countably many propositional variables $p_1, p_2, p_3, \ldots$; the *well-formed formulas* of $\mathcal{L}$ are

[3] Wijesekera [38] considers a multi-succedent calculus for IPL, however an equivalent calculus can be given by adding Wijesekera's modal rules to a single-succedent calculus (cf. [9] and Sec. 3 in this paper).

$nec\ \dfrac{A}{\Box A}$	$K_\Box$	$\Box(A \supset B) \supset (\Box A \supset \Box B)$	$T_\Box$	$\Box A \supset A$	
	$K_\Diamond$	$\Box(A \supset B) \supset (\Diamond A \supset \Diamond B)$	$T_\Diamond$	$A \supset \Diamond A$	
$mon_\Box\ \dfrac{A \supset B}{\Box A \supset \Box B}$	$C_\Box$	$\Box A \wedge \Box B \supset \Box(A \wedge B)$	D	$\Box A \supset \Diamond A$	
	$C_\Diamond$	$\Diamond(A \vee B) \supset \Diamond A \vee \Diamond B$	$P_\Box$	$\neg\Box\bot$	
$mon_\Diamond\ \dfrac{A \supset B}{\Diamond A \supset \Diamond B}$	$N_\Box$	$\Box\top$	$P_\Diamond$	$\Diamond\top$	
	$N_\Diamond$	$\neg\Diamond\bot$			
$dual\ \ \Box A \supset\subset \neg\Diamond\neg A$	$dual_\wedge$	$\neg(\Box A \wedge \Diamond\neg A)$	$dual_\vee$	$\Box A \vee \Diamond\neg A$	

Fig. 1. Modal axioms and rules.

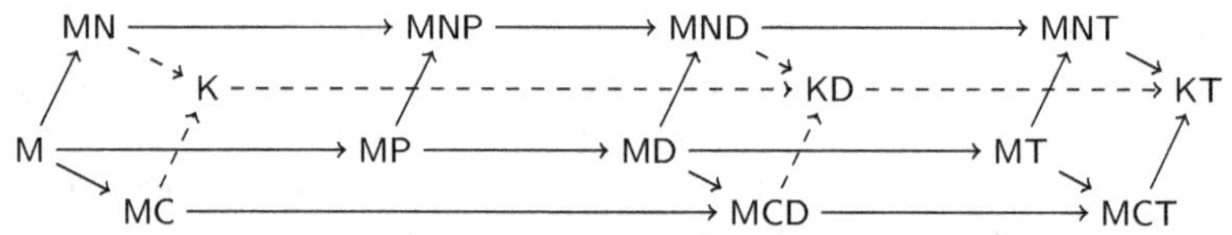

Fig. 2. Dyagram of classical modal logics.

generated by the following grammar, where p_i is any element of Atm:

$$A ::= p_i \mid \bot \mid A \wedge A \mid A \vee A \mid A \supset A \mid \Box A \mid \Diamond A.$$

We also define $\top := \bot \supset \bot$, $\neg A := A \supset \bot$, and $A \supset\subset B := (A \supset B) \wedge (B \supset A)$.

We aim at enriching the family of Wijesekera-style propositional modal logics by defining constructive counterparts of well-known classical modal logics. We consider the following classical systems, which are defined in the language $\mathcal{L}$ extending (any axiomatisation of) classical propositional logic (CPL) with the following modal axioms and rules from Fig. 1:

M := $dual + mon_\Box$	MNP := MN + $P_\Box$	MT := M + $T_\Box$
MN := M + $N_\Box$	MD := M + D	MNT := MN + $T_\Box$
MC := M + $C_\Box$	MND := MN + D	MCT := MC + $T_\Box$
K := M + $N_\Box$ + $C_\Box$	MCD := MC + D	KT := K + $T_\Box$
MP := M + $P_\Box$	KD := K + D	

The considered axiomatisation of K is equivalent to the more standard one with nec and $K_\Box$ (cf. e.g. [8]). The above list contains logics stronger than K as well as weaker (i.e., non-normal) systems. Note that given the duality between $\Box$ and $\Diamond$, the above systems can be equivalently defined by replacing $mon_\Box$, $N_\Box$, $C_\Box$, $T_\Box$, and $P_\Box$, with their $\Diamond$-versions $mon_\Diamond$, $N_\Diamond$, $C_\Diamond$, $T_\Diamond$, $P_\Diamond$ (Fig. 1). The relations among the classical systems are displayed in Fig. 2 (MCP and KP are not considered in the list as they coincide with MCD and KD).

We will define constructive counterparts of classical modal logics by restricting suitable sequent calculi for the classical systems. We consider to this purpose the calculi for classical modal logics defined by the rules in Fig. 3. As usual, we call *sequent* any pair $\Gamma \Rightarrow \Delta$, where Γ and Δ are finite, possibly empty multisets of formulas of $\mathcal{L}$. A sequent $\Gamma \Rightarrow \Delta$ is interpreted as a formula of $\mathcal{L}$ as $\bigwedge\Gamma \supset \bigvee\Delta$ if Γ is non-empty, and it is interpreted as $\bigvee\Delta$ if Γ is empty, where $\bigvee\emptyset$ is interpreted as $\bot$. For every multiset $\Gamma = A_1, ..., A_n$, we denote

Propositional rules $\quad \mathsf{init}\ \Gamma, p \Rightarrow p, \Delta \quad\quad \bot_\mathsf{L}\ \Gamma, \bot \Rightarrow \Delta$

$$\supset_\mathsf{L} \frac{\Gamma \Rightarrow A, \Delta \qquad \Gamma, B \Rightarrow \Delta}{\Gamma, A \supset B \Rightarrow \Delta} \quad \supset_\mathsf{R} \frac{\Gamma, A \Rightarrow B, \Delta}{\Gamma \Rightarrow A \supset B, \Delta} \quad \wedge_\mathsf{R} \frac{\Gamma \Rightarrow A, \Delta \qquad \Gamma \Rightarrow B, \Delta}{\Gamma \Rightarrow A \wedge B, \Delta}$$

$$\wedge_\mathsf{L} \frac{\Gamma, A, B \Rightarrow \Delta}{\Gamma, A \wedge B \Rightarrow \Delta} \quad \vee_\mathsf{R} \frac{\Gamma \Rightarrow A, B, \Delta}{\Gamma \Rightarrow A \vee B, \Delta} \quad \vee_\mathsf{L} \frac{\Gamma, A \Rightarrow \Delta \qquad \Gamma, B \Rightarrow \Delta}{\Gamma, A \vee B \Rightarrow \Delta}$$

Modal rules

$$\mathsf{M}_\Box \frac{A \Rightarrow B}{\Gamma, \Box A \Rightarrow \Box B, \Delta} \quad \mathsf{M}_\Diamond \frac{A \Rightarrow B}{\Gamma, \Diamond A \Rightarrow \Diamond B, \Delta} \quad \mathsf{C}_\Box \frac{\Gamma, A \Rightarrow B, \Delta}{\Gamma', \Box\Gamma, \Box A \Rightarrow \Box B, \Diamond\Delta, \Delta'}$$

$$\wedge\text{-}\mathsf{dual}_\mathsf{M} \frac{A, B \Rightarrow}{\Gamma, \Box A, \Diamond B \Rightarrow \Delta} \quad \mathsf{C}_\Diamond \frac{\Gamma, A \Rightarrow B, \Delta}{\Gamma', \Box\Gamma, \Diamond A \Rightarrow \Diamond B, \Diamond\Delta, \Delta'} \quad \mathsf{D} \frac{A \Rightarrow B}{\Gamma, \Box A \Rightarrow \Diamond B, \Delta}$$

$$\vee\text{-}\mathsf{dual}_\mathsf{M} \frac{\Rightarrow A, B}{\Gamma \Rightarrow \Box A, \Diamond B, \Delta} \quad \wedge\text{-}\mathsf{dual}_\mathsf{C} \frac{\Gamma, A, B \Rightarrow}{\Gamma', \Box\Gamma, \Box A, \Diamond B \Rightarrow \Delta} \quad \mathsf{D}_\Box \frac{A, B \Rightarrow}{\Gamma, \Box A, \Box B \Rightarrow \Delta}$$

$$\vee\text{-}\mathsf{dual}_\mathsf{C} \frac{\Rightarrow A, B, \Delta}{\Gamma \Rightarrow \Box A, \Diamond B, \Diamond\Delta, \Delta'} \quad \mathsf{K}_\Box \frac{\Gamma \Rightarrow A, \Delta}{\Gamma', \Box\Gamma \Rightarrow \Box A, \Diamond\Delta, \Delta'} \quad \mathsf{D}_\Diamond \frac{\Rightarrow A, B}{\Gamma \Rightarrow \Diamond A, \Diamond B, \Delta}$$

$$\mathsf{K}_\Diamond \frac{\Gamma, A \Rightarrow \Delta}{\Gamma', \Box\Gamma, \Diamond A \Rightarrow \Diamond\Delta, \Delta'} \quad \mathsf{N}_\Box \frac{\Rightarrow A}{\Gamma \Rightarrow \Box A, \Delta} \quad \mathsf{N}_\Diamond \frac{A \Rightarrow}{\Gamma, \Diamond A \Rightarrow \Delta} \quad \mathsf{P}_\Box \frac{A \Rightarrow}{\Gamma, \Box A \Rightarrow \Delta}$$

$$\mathsf{P}_\Diamond \frac{\Rightarrow A}{\Gamma \Rightarrow \Diamond A, \Delta} \quad \mathsf{T}_\Box \frac{\Gamma, \Box A, A \Rightarrow \Delta}{\Gamma, \Box A \Rightarrow \Delta} \quad \mathsf{T}_\Diamond \frac{\Gamma \Rightarrow A, \Diamond A, \Delta}{\Gamma \Rightarrow \Diamond A, \Delta} \quad \mathsf{CD} \frac{\Gamma \Rightarrow \Delta}{\Gamma', \Box\Gamma \Rightarrow \Diamond\Delta, \Delta'}$$

Fig. 3. Sequent rules for classical modal logics (where $|\Gamma|, |\Gamma'|, |\Delta|, |\Delta'| \geq 0$).

with $\Box\Gamma$ and $\Diamond\Gamma$ the multisets $\Box A_1, ..., \Box A_n$ and $\Diamond A_1, ..., \Diamond A_n$, respectively. We consider G3-style calculi with all structural rules admissible (cf. [37, Ch. 3]). Moreover, we consider a formulation of the calculi in which both $\Box$ and $\Diamond$ occur explicitly, this formulation will be needed to handle the constructive systems, where the modalities are not interdefinable (for a sequent calculus with explicit $\Box$ and $\Diamond$ see e.g. [37, Ch. 9], for sequent calculi for non-normal modal logics see [20,21]). For each logic L, the corresponding calculus S.L contains the propositional rules and the following modal rules:

$\mathsf{S.M} := \mathsf{M}_\Box + \mathsf{M}_\Diamond + \wedge\text{-}\mathsf{dual}_\mathsf{M} + \vee\text{-}\mathsf{dual}_\mathsf{M}$ $\quad$ $\mathsf{S.MP} := \mathsf{S.M} + \mathsf{P}_\Box + \mathsf{P}_\Diamond$
$\mathsf{S.MN} := \mathsf{S.M} + \mathsf{N}_\Box + \mathsf{N}_\Diamond$ $\quad$ $\mathsf{S.MNP} := \mathsf{S.MN} + \mathsf{P}_\Box + \mathsf{P}_\Diamond$
$\mathsf{S.MC} := \mathsf{C}_\Box + \mathsf{C}_\Diamond + \wedge\text{-}\mathsf{dual}_\mathsf{C} + \vee\text{-}\mathsf{dual}_\mathsf{C}$
$\mathsf{S.K} := \mathsf{K}_\Box + \mathsf{K}_\Diamond$

$\mathsf{S.MD} := \mathsf{S.M} + \mathsf{D} + \mathsf{D}_\Box + \mathsf{D}_\Diamond + \mathsf{P}_\Box + \mathsf{P}_\Diamond$ $\quad$ $\mathsf{S.MT} := \mathsf{S.M} + \mathsf{T}_\Box + \mathsf{T}_\Diamond$
$\mathsf{S.MND} := \mathsf{S.MN} + \mathsf{D} + \mathsf{D}_\Box + \mathsf{D}_\Diamond + \mathsf{P}_\Box + \mathsf{P}_\Diamond$ $\quad$ $\mathsf{S.MNT} := \mathsf{S.MN} + \mathsf{T}_\Box + \mathsf{T}_\Diamond$
$\mathsf{S.MCD} := \mathsf{S.MC} + \mathsf{CD}$ $\quad$ $\mathsf{S.MCT} := \mathsf{S.MC} + \mathsf{T}_\Box + \mathsf{T}_\Diamond$
$\mathsf{S.KD} := \mathsf{S.K} + \mathsf{CD}$ $\quad$ $\mathsf{S.KT} := \mathsf{S.K} + \mathsf{T}_\Box + \mathsf{T}_\Diamond$

Each calculus contains two duality rules $\wedge$-dual and $\vee$-dual (in S.K and its extensions they are obtained as the particular cases of $\mathsf{K}_\Diamond$ and $\mathsf{K}_\Box$ with $|\Delta| = \emptyset$, respectively $|\Gamma| = \emptyset$). The duality rules allow one to derive the Hilbert-style rules

$$Rdual_\wedge \frac{\neg(A \wedge B)}{\neg(\Box A \wedge \Diamond B)} \qquad Rdual_\vee \frac{A \vee B}{\Box A \vee \Diamond B}$$

which are classically equivalent to $dual_\wedge$ and $dual_\vee$ (Fig. 1), and taken together are equivalent to *dual*. The rules $\mathsf{C}_\Box$ and $\mathsf{C}_\Diamond$ can be seen as the generalisation of $\mathsf{M}_\Box$ and $\mathsf{M}_\Diamond$ to n-principal formulas in the antecedent, respectively in the consequent. Differently from $\mathsf{M}_\Box$, the rule $\mathsf{C}_\Box$ involves also $\Diamond$-formulas, similarly $\mathsf{C}_\Diamond$ involves also $\Box$-formulas, this is needed in order to preserve the admissibility of cut in the calculus. Note also that $\mathsf{C}_\Box$ and $\mathsf{C}_\Diamond$ are distinct from $\mathsf{K}_\Box$ and $\mathsf{K}_\Diamond$, since $\mathsf{C}_\Box$ and $\mathsf{C}_\Diamond$ are applicable only to sequents with non-empty antecedent, respectively non-empty consequent, while this is not required for $\mathsf{K}_\Box$ and $\mathsf{K}_\Diamond$. Finally, the calculi S.MD and S.MND contain also the rules $\mathsf{P}_\Box$ and $\mathsf{P}_\Diamond$, this is needed to ensure admissibility of contraction [11,31], and is consistent with the fact that the axioms $P_\Box$ and $P_\Diamond$ are derivable in MD. Each calculus S.L is a calculus for the corresponding logic L in the following sense:

Theorem 2.1 *For every considered classical modal logic* L, $\mathsf{S.L} \vdash \Gamma \Rightarrow \Delta$ *if and only if* $\mathsf{L} \vdash \bigwedge \Gamma \supset \bigvee \Delta$.

We now move to the semantics. Since non-normal logics do not have a (simple) relational semantics,[4] we consider a neighbourhood semantics that uniformly covers all considered systems.

Definition 2.2 A *neighbourhood model* is a tuple $\mathcal{M} = \langle \mathcal{W}, \mathcal{N}, \mathcal{V} \rangle$, where $\mathcal{W}$ is a non-empty set of worlds, $\mathcal{N}$ is a function $\mathcal{P}(\mathcal{W}) \longrightarrow \mathcal{P}(\mathcal{P}(\mathcal{W}))$, called neighbourhood function, and $\mathcal{V}$ is a valuation function $Atm \longrightarrow \mathcal{P}(\mathcal{W})$. The forcing relation $\mathcal{M}, w \Vdash A$ is inductively defined as follows:

$$
\begin{array}{lll}
\mathcal{M}, w \Vdash p & \text{iff} & w \in \mathcal{V}(p). \\
\mathcal{M}, w \nVdash \bot. & & \\
\mathcal{M}, w \Vdash B \wedge C & \text{iff} & \mathcal{M}, w \Vdash B \text{ and } \mathcal{M}, w \Vdash C. \\
\mathcal{M}, w \Vdash B \vee C & \text{iff} & \mathcal{M}, w \Vdash B \text{ or } \mathcal{M}, w \Vdash C. \\
\mathcal{M}, w \Vdash B \supset C & \text{iff} & \mathcal{M}, w \nVdash B \text{ or } \mathcal{M}, w \Vdash C. \\
\mathcal{M}, w \Vdash \Box B & \text{iff} & \text{there is } \alpha \in \mathcal{N}(w) \text{ s.t. for all } v \in \alpha, \mathcal{M}, v \Vdash B. \\
\mathcal{M}, w \Vdash \Diamond B & \text{iff} & \text{for all } \alpha \in \mathcal{N}(w), \text{ there is } v \in \alpha \text{ s.t. } \mathcal{M}, v \Vdash B.
\end{array}
$$

We consider the following properties on neighbourhood models:

(C) If $\alpha, \beta \in \mathcal{N}(w)$, then $\alpha \cap \beta \in \mathcal{N}(w)$.
(N) $\mathcal{N}(w) \neq \emptyset$.
(D) If $\alpha, \beta \in \mathcal{N}(w)$, then $\alpha \cap \beta \neq \emptyset$.
(P) $\emptyset \notin \mathcal{N}(w)$.
(T) If $\alpha \in \mathcal{N}(w)$, then $w \in \alpha$.

We say that $\mathcal{M}$ is a model for a logic L if it satisfies the condition (X) for every modal axiom $X_\Box$ of L (among $C_\Box$, $N_\Box$, $T_\Box$, D, $P_\Box$). As usual, we say that a formula A is valid in a model $\mathcal{M}$, written $\mathcal{M} \models A$, if $\mathcal{M}, w \Vdash A$ for every world w of $\mathcal{M}$.

In the following we simple write $w \Vdash A$ when $\mathcal{M}$ is clear from the context. We also use the following abbreviations:

$\alpha \Vdash^\forall A :=$ for all $w \in \alpha$, $w \Vdash A$; $\qquad \alpha \Vdash^\exists A :=$ there is $w \in \alpha$ s.t. $w \Vdash A$.

---

[4] Cf. [7] for multi-relational semantics for non-normal modal logics, and [34,11] for relational semantics with "non-normal" worlds for the logics containing $C_\Box$ but not $N_\Box$.

Propositional rules $\quad \mathsf{init}^{\mathsf{i}}\ \Gamma, p \Rightarrow p \quad \bot^{\mathsf{i}}_{\mathsf{L}}\ \Gamma, \bot \Rightarrow \Delta$

$$\supset^{\mathsf{i}}_{\mathsf{L}}\ \frac{\Gamma, A \supset B \Rightarrow A \qquad \Gamma, B \Rightarrow \Delta}{\Gamma, A \supset B \Rightarrow \Delta} \quad \supset^{\mathsf{i}}_{\mathsf{R}}\ \frac{\Gamma, A \Rightarrow B}{\Gamma \Rightarrow A \supset B} \quad \wedge^{\mathsf{i}}_{\mathsf{R}}\ \frac{\Gamma \Rightarrow A \qquad \Gamma \Rightarrow B}{\Gamma \Rightarrow A \wedge B}$$

$$\wedge^{\mathsf{i}}_{\mathsf{L}}\ \frac{\Gamma, A, B \Rightarrow \Delta}{\Gamma, A \wedge B \Rightarrow \Delta} \quad \vee^{\mathsf{i}}_{\mathsf{R}}\ \frac{\Gamma \Rightarrow A_i}{\Gamma \Rightarrow A_1 \vee A_2}\ (i \in \{1,2\}) \quad \vee^{\mathsf{i}}_{\mathsf{L}}\ \frac{\Gamma, A \Rightarrow \Delta \qquad \Gamma, B \Rightarrow \Delta}{\Gamma, A \vee B \Rightarrow \Delta}$$

Modal rules

$$\mathsf{M}^{\mathsf{i}}_{\Box}\ \frac{\Gamma, A \Rightarrow B}{\Gamma, \Box A \Rightarrow \Box B} \quad \mathsf{M}^{\mathsf{i}}_{\Diamond}\ \frac{\Gamma, A \Rightarrow B}{\Gamma, \Diamond A \Rightarrow \Diamond B} \quad \wedge\text{-}\mathsf{dual}^{\mathsf{i}}_{\mathsf{M}}\ \frac{A, B \Rightarrow}{\Gamma, \Box A, \Diamond B \Rightarrow \Delta} \quad \mathsf{N}^{\mathsf{i}}_{\Box}\ \frac{\Rightarrow A}{\Gamma \Rightarrow \Box A}$$

$$\mathsf{C}^{\mathsf{i}}_{\Box}\ \frac{\Gamma, A \Rightarrow B}{\Gamma', \Box\Gamma, \Box A \Rightarrow \Box B} \quad \mathsf{C}^{\mathsf{i}}_{\Diamond}\ \frac{\Gamma, A \Rightarrow B}{\Gamma', \Box\Gamma, \Diamond A \Rightarrow \Diamond B} \quad \wedge\text{-}\mathsf{dual}^{\mathsf{i}}_{\mathsf{C}}\ \frac{\Gamma, A, B \Rightarrow}{\Gamma', \Box\Gamma, \Box A, \Diamond B \Rightarrow \Delta}$$

$$\mathsf{N}^{\mathsf{i}}_{\Diamond}\ \frac{A \Rightarrow}{\Gamma, \Diamond A \Rightarrow \Delta} \quad \mathsf{K}^{\mathsf{i}}_{\Box}\ \frac{\Gamma \Rightarrow A}{\Gamma', \Box\Gamma \Rightarrow \Box A} \quad \mathsf{K}^{\mathsf{i}}_{\Diamond}\ \frac{\Gamma, A \Rightarrow B}{\Gamma', \Box\Gamma, \Diamond A \Rightarrow \Diamond B} \quad \mathsf{T}^{\mathsf{i}}_{\Diamond}\ \frac{\Gamma \Rightarrow A}{\Gamma \Rightarrow \Diamond A}$$

$$\wedge\text{-}\mathsf{dual}^{\mathsf{i}}_{\mathsf{K}}\ \frac{\Gamma, A \Rightarrow}{\Gamma', \Box\Gamma, \Diamond A \Rightarrow \Delta} \quad \mathsf{T}^{\mathsf{i}}_{\Box}\ \frac{\Gamma, \Box A, A \Rightarrow \Delta}{\Gamma, \Box A \Rightarrow \Delta} \quad \mathsf{P}^{\mathsf{i}}_{\Box}\ \frac{A \Rightarrow}{\Gamma, \Box A \Rightarrow \Delta} \quad \mathsf{P}^{\mathsf{i}}_{\Diamond}\ \frac{\Rightarrow A}{\Gamma \Rightarrow \Diamond A}$$

$$\mathsf{D}^{\mathsf{i}}\ \frac{A \Rightarrow B}{\Gamma, \Box A \Rightarrow \Diamond B} \quad \mathsf{D}^{\mathsf{i}}_{\Box}\ \frac{A, B \Rightarrow}{\Gamma, \Box A, \Box B \Rightarrow \Delta} \quad \mathsf{CD}^{\mathsf{i}}\ \frac{\Gamma \Rightarrow A}{\Gamma', \Box\Gamma \Rightarrow \Diamond A} \quad \mathsf{CD}^{\mathsf{i}}_{\Box}\ \frac{\Gamma \Rightarrow}{\Gamma', \Box\Gamma \Rightarrow \Delta}$$

Fig. 4. Sequent rules for W-logics (where $|\Gamma|, |\Gamma'| \geq 0$, and $0 \leq |\Delta| \leq 1$).

Using these abbreviations, the satisfaction clauses for modal formulas can be equivalently written as

$$w \Vdash \Box B \quad \text{iff} \quad \text{there is } \alpha \in \mathcal{N}(w) \text{ such that } \alpha \Vdash^{\forall} B.$$
$$w \Vdash \Diamond B \quad \text{iff} \quad \text{for all } \alpha \in \mathcal{N}(w),\ \alpha \Vdash^{\exists} B.$$

The following holds (cf. e.g. [8,32]).

Theorem 2.3 *For every considered classical modal logic* L, $\mathsf{L} \vdash A$ *if and only if* $\mathcal{M} \models A$ *for all neighbourhood models* $\mathcal{M}$ *for* L.

3 Single-succedent calculi and W-logics

We now define a family of Wijesekera-style constructive modal logics corresponding to the classical logics considered in Sec. 2. In particular, we firstly define constructive modal calculi by restricting the classical calculi from Sec. 2 to single-succedent sequents, and study their structural properties. Then we define equivalent axiomatic systems, and prove some fundamental properties of them.

The rules obtained by restricting sequents to at most one formula in the consequent are displayed in Fig. 4. This restriction modifies the classical modal rules in two ways: first, the rules $\vee$-$\mathsf{dual}_{\mathsf{M}}$, $\vee$-$\mathsf{dual}_{\mathsf{R}}$, and $\mathsf{D}_{\Diamond}$ are dropped because they require at least two formulas in the consequent of sequents. Second, the right context is deleted from all rules with a principal formula in the consequent of sequents, namely $\mathsf{T}_{\Diamond}$, $\mathsf{C}_{\Box}$, $\mathsf{C}_{\Diamond}$, and $\mathsf{K}_{\Box}$. Note in particular that $\Diamond\Delta$ is removed from $\mathsf{C}_{\Box}$, $\mathsf{C}_{\Diamond}$, and $\mathsf{K}_{\Box}$, moreover $\Diamond A$ is removed from the premiss of $\mathsf{T}^{\mathsf{i}}_{\Diamond}$ (the copy of $\Diamond A$ into the premiss of $\mathsf{T}_{\Diamond}$ is needed in the classical calculus in order to ensure admissibility of right contraction, which is not expressible in the intuitionistic calculus). Note also that $\mathsf{K}_\diamond$ and CD are split into two rules, respectively $\mathsf{K}^{\mathsf{i}}_\diamond$ and $\wedge$-$\mathsf{dual}^{\mathsf{i}}_{\mathsf{K}}$, and CD^{i} and $\mathsf{CD}^{\mathsf{i}}_\Box$, which correspond to the cases in which the consequent of the premiss of $\mathsf{K}_\diamond$ or CD is or is not empty. Finally, note that $\mathsf{C}^{\mathsf{i}}_\Box$ and $\mathsf{K}^{\mathsf{i}}_\Box$ become equivalent. Concerning the propositional rules, $\supset_{\mathsf{L}}$ is modified as usual by copying the principal implication into the left premiss in order to ensure admissibility of contraction [37], and $\vee_{\mathsf{R}}$ is replaced by its single-succedent version. All other rules remain unchanged. The resulting calculi $\mathsf{S.WL}$ are defined by extending the set of intuitionistic propositional rules with the following modal rules:

$\mathsf{S.WM} := \mathsf{M}^{\mathsf{i}}_\Box + \mathsf{M}^{\mathsf{i}}_\diamond + \wedge\text{-}\mathsf{dual}^{\mathsf{i}}_{\mathsf{M}}$ $\mathsf{S.WMP} := \mathsf{S.WM} + \mathsf{P}^{\mathsf{i}}_\Box + \mathsf{P}^{\mathsf{i}}_\diamond$
$\mathsf{S.WMN} := \mathsf{S.WM} + \mathsf{N}^{\mathsf{i}}_\Box + \mathsf{N}^{\mathsf{i}}_\diamond$ $\mathsf{S.WMNP} := \mathsf{S.WMN} + \mathsf{P}^{\mathsf{i}}_\Box + \mathsf{P}^{\mathsf{i}}_\diamond$
$\mathsf{S.WMC} := \mathsf{C}^{\mathsf{i}}_\Box + \mathsf{C}^{\mathsf{i}}_\diamond + \wedge\text{-}\mathsf{dual}^{\mathsf{i}}_{\mathsf{C}}$
$\mathsf{S.WK} := \mathsf{K}^{\mathsf{i}}_\Box + \mathsf{K}^{\mathsf{i}}_\diamond + \wedge\text{-}\mathsf{dual}^{\mathsf{i}}_{\mathsf{K}}$

$\mathsf{S.WMD} := \mathsf{S.WM} + \mathsf{D}^{\mathsf{i}} + \mathsf{D}^{\mathsf{i}}_\Box + \mathsf{P}^{\mathsf{i}}_\Box + \mathsf{P}^{\mathsf{i}}_\diamond$ $\mathsf{S.WMT} := \mathsf{S.WM} + \mathsf{T}^{\mathsf{i}}_\Box + \mathsf{T}^{\mathsf{i}}_\diamond$
$\mathsf{S.WMND} := \mathsf{S.WMN} + \mathsf{D}^{\mathsf{i}} + \mathsf{D}^{\mathsf{i}}_\Box + \mathsf{P}^{\mathsf{i}}_\Box + \mathsf{P}^{\mathsf{i}}_\diamond$ $\mathsf{S.WMNT} := \mathsf{S.WMN} + \mathsf{T}^{\mathsf{i}}_\Box + \mathsf{T}^{\mathsf{i}}_\diamond$
$\mathsf{S.WMCD} := \mathsf{S.WMC} + \mathsf{CD}^{\mathsf{i}} + \mathsf{CD}^{\mathsf{i}}_\Box$ $\mathsf{S.WMCT} := \mathsf{S.WMC} + \mathsf{T}^{\mathsf{i}}_\Box + \mathsf{T}^{\mathsf{i}}_\diamond$
$\mathsf{S.WKD} := \mathsf{S.WK} + \mathsf{CD}^{\mathsf{i}} + \mathsf{CD}^{\mathsf{i}}_\Box$ $\mathsf{S.WKT} := \mathsf{S.WK} + \mathsf{T}^{\mathsf{i}}_\Box + \mathsf{T}^{\mathsf{i}}_\diamond$

Note that the modal rules of $\mathsf{S.WK}$ coincide with those of Wijesekera [38] (except that they have side context in the conclusion in order to embed weakening in their application). $\mathsf{S.WK}$ coincides with the calculus $\mathsf{G.CCDL^p}$ for WK proposed in [9].

From the point of view of the derivable principles, we observe two main consequences of the restriction of the calculi to single-succedent sequents. First, the rule $Rdual_\vee$ is no longer derivable in the calculi. This is due to the absence of $\vee$-$\mathsf{dual}_{\mathsf{M}}$ and $\vee$-$\mathsf{dual}_{\mathsf{C}}$, and the elimination of $\diamond$-formulas from the conclusion of $\mathsf{K}^{\mathsf{i}}_\Box$. Second, $C_\diamond$ is not derivable in $\mathsf{S.WMC}$, $\mathsf{S.WK}$ and their extensions, this is due to the restriction of $\mathsf{C}^{\mathsf{i}}_\diamond$ and $\mathsf{K}^{\mathsf{i}}_\diamond$ to only one $\diamond$-formula in the right-hand side of the conclusion. By contrast, all other modal principles from Fig. 1 are still derivable in the corresponding calculi (cf. derivations in Fig. 5).

In the following, we denote with $\mathsf{S.W}^*$ any constructive calculus defined above. As usual, we say that a rule is *admissible* in $\mathsf{S.W}^*$ if whenever the premisses are derivable, the conclusion is also derivable, and that a single-premiss rule is *height-preserving admissible* if whenever the premiss is derivable, the conclusion is derivable with a derivation of at most the same height. We now prove that the calculi $\mathsf{S.W}^*$ enjoy admissibility of structural rules and cut, then we present equivalent axiomatic systems.

Proposition 3.1 (Admissibility of structural rules) *The following rules are height-preserving admissible in* $\mathsf{S.W}^*$:

$$\mathsf{Lwk}\ \frac{\Gamma \Rightarrow \Delta}{\Gamma, A \Rightarrow \Delta} \qquad \mathsf{Rwk}\ \frac{\Gamma \Rightarrow}{\Gamma \Rightarrow A} \qquad \mathsf{ctr}\ \frac{\Gamma, A, A \Rightarrow \Delta}{\Gamma, A \Rightarrow \Delta}.$$

Proof. Height-preserving admissibility of Lwk, Rwk, and ctr is proved by induction on the height of the derivation of their premiss, taking into account the last rule applied in the derivation. For Lwk and Rwk the proof is straight-

$$\dfrac{\dfrac{\dfrac{\dfrac{A, A \supset \bot \Rightarrow A \qquad A, \bot \Rightarrow}{A, A \supset \bot \Rightarrow}\,\supset^{\mathsf{i}}_{\mathsf{L}}}{\Box A, \Diamond(A \supset \bot) \Rightarrow \bot}\,\wedge\text{-dual}^{\mathsf{i}}_{\mathsf{M}}}{\Box A \wedge \Diamond(A \supset \bot) \Rightarrow \bot}\,\wedge^{\mathsf{i}}_{\mathsf{L}}}{\Rightarrow \Box A \wedge \Diamond(A \supset \bot) \supset \bot}\,\supset^{\mathsf{i}}_{\mathsf{R}} \qquad \dfrac{\dfrac{\dfrac{\dfrac{A, B \Rightarrow A \qquad A, B \Rightarrow B}{A, B \Rightarrow A \wedge B}\,\wedge^{\mathsf{i}}_{\mathsf{R}}}{\Box A, \Box B \Rightarrow \Box(A \wedge B)}\,\mathsf{C}^{\mathsf{i}}_{\Box}}{\Box A \wedge \Box B \Rightarrow \Box(A \wedge B)}\,\wedge^{\mathsf{i}}_{\mathsf{L}}}{\Rightarrow \Box A \wedge \Box B \supset \Box(A \wedge B)}\,\supset^{\mathsf{i}}_{\mathsf{R}}$$

$$\dfrac{\dfrac{\dfrac{\dfrac{A \supset B, A \Rightarrow A \qquad B, A \Rightarrow B}{A \supset B, A \Rightarrow B}\,\supset^{\mathsf{i}}_{\mathsf{L}}}{\Box(A \supset B), \Box A \Rightarrow \Box B}\,\mathsf{C}^{\mathsf{i}}_{\Box}}{\Box(A \supset B) \Rightarrow \Box A \supset \Box B}\,\supset^{\mathsf{i}}_{\mathsf{R}}}{\Rightarrow \Box(A \supset B) \supset (\Box A \supset \Box B)}\,\supset^{\mathsf{i}}_{\mathsf{R}} \qquad \dfrac{\dfrac{\dfrac{\dfrac{A \supset B, A \Rightarrow A \qquad B, A \Rightarrow B}{A \supset B, A \Rightarrow B}\,\supset^{\mathsf{i}}_{\mathsf{L}}}{\Box(A \supset B), \Diamond A \Rightarrow \Diamond B}\,\mathsf{C}^{\mathsf{i}}_{\Diamond}}{\Box(A \supset B) \Rightarrow \Diamond A \supset \Diamond B}\,\supset^{\mathsf{i}}_{\mathsf{R}}}{\Rightarrow \Box(A \supset B) \supset (\Diamond A \supset \Diamond B)}\,\supset^{\mathsf{i}}_{\mathsf{R}}$$

$$\dfrac{\dfrac{\bot \Rightarrow \bot}{\Rightarrow \bot \supset \bot}\,\supset^{\mathsf{i}}_{\mathsf{R}}}{\Rightarrow \Box(\bot \supset \bot)}\,\mathsf{N}^{\mathsf{i}}_{\Box} \qquad \dfrac{\dfrac{\bot \Rightarrow}{\Diamond\bot \Rightarrow \bot}\,\mathsf{N}^{\mathsf{i}}_{\Diamond}}{\Rightarrow \Diamond\bot \supset \bot}\,\supset^{\mathsf{i}}_{\mathsf{R}} \qquad \dfrac{\dfrac{\Box A, A \Rightarrow A}{\Box A \Rightarrow A}\,\mathsf{T}^{\mathsf{i}}_{\Box}}{\Rightarrow \Box A \supset A}\,\supset^{\mathsf{i}}_{\mathsf{R}} \qquad \dfrac{\dfrac{A \Rightarrow A}{A \Rightarrow \Diamond A}\,\mathsf{T}^{\mathsf{i}}_{\Diamond}}{\Rightarrow A \supset \Diamond A}\,\supset^{\mathsf{i}}_{\mathsf{R}}$$

$$\dfrac{\dfrac{A \Rightarrow A}{\Box A \Rightarrow \Diamond A}\,\mathsf{D}^{\mathsf{i}}}{\Rightarrow \Box A \supset \Diamond A}\,\supset^{\mathsf{i}}_{\mathsf{R}} \qquad \dfrac{\dfrac{\bot \Rightarrow}{\Box\bot \Rightarrow \bot}\,\mathsf{P}^{\mathsf{i}}_{\Box}}{\Rightarrow \Box\bot \supset \bot}\,\supset^{\mathsf{i}}_{\mathsf{R}} \qquad \dfrac{\dfrac{\bot \Rightarrow \bot}{\Rightarrow \bot \supset \bot}\,\supset^{\mathsf{i}}_{\mathsf{R}}}{\Rightarrow \Diamond(\bot \supset \bot)}\,\mathsf{P}^{\mathsf{i}}_{\Diamond}$$

Fig. 5. Derivations of the modal axioms.

forward, we consider some examples for ctr involving the modal rules. The derivations on the left are converted into the derivations on the right, which include applications of ctr which are height-preserving admissible by i.h.:

$$\mathsf{D}^{\mathsf{i}}_{\Box}\,\dfrac{A, A \Rightarrow}{\Gamma, \Box A, \Box A \Rightarrow \Delta} \quad \rightsquigarrow \quad \dfrac{\dfrac{A, A \Rightarrow}{A \Rightarrow}\,\mathsf{ctr}}{\Gamma, \Box A \Rightarrow \Delta}\,\mathsf{P}^{\mathsf{i}}_{\Box}$$

$$\mathsf{T}^{\mathsf{i}}_{\Box}\,\dfrac{\Gamma, \Box A, \Box A, A \Rightarrow \Delta}{\Gamma, \Box A, \Box A \Rightarrow \Delta} \quad \rightsquigarrow \quad \dfrac{\dfrac{\Gamma, \Box A, \Box A, A \Rightarrow \Delta}{\Gamma, \Box A, A \Rightarrow \Delta}\,\mathsf{ctr}}{\Gamma, \Box A \Rightarrow \Delta}\,\mathsf{T}^{\mathsf{i}}_{\Box}$$

$$\mathsf{C}^{\mathsf{i}}_{\Diamond}\,\dfrac{\Gamma, A, A, B \Rightarrow C}{\Gamma', \Box\Gamma, \Box A, \Box A, \Diamond B \Rightarrow \Diamond C} \quad \rightsquigarrow \quad \dfrac{\dfrac{\Gamma, A, A, B \Rightarrow C}{\Gamma, A, B \Rightarrow C}\,\mathsf{ctr}}{\Gamma', \Box\Gamma, \Box A, \Diamond B \Rightarrow \Diamond C}\,\mathsf{C}^{\mathsf{i}}_{\Diamond}$$

$$\mathsf{C}^{\mathsf{i}}_{\Diamond}\,\dfrac{\Gamma, A, B \Rightarrow C}{\Gamma', \Box A, \Box\Gamma, \Box A, \Diamond B \Rightarrow \Diamond C} \quad \rightsquigarrow \quad \dfrac{\Gamma, A, B \Rightarrow C}{\Gamma', \Box\Gamma, \Box A, \Diamond B \Rightarrow \Diamond C}\,\mathsf{C}^{\mathsf{i}}_{\Diamond}$$

In the last example, one occurrence of $\Box A$ is introduced by the first application of $\mathsf{C}^{\mathsf{i}}_{\Diamond}$ as part of the side context, while this is not the case in the second application of $\mathsf{C}^{\mathsf{i}}_{\Diamond}$. □

Theorem 3.2 (Cut admissibility) *The following cut rule is admissible in* S.W**:*

$$\mathsf{cut}\,\dfrac{\Gamma \Rightarrow A \qquad \Gamma', A \Rightarrow \Delta}{\Gamma, \Gamma' \Rightarrow \Delta}\,.$$

Proof. By induction on lexicographically ordered pairs (c, h), where c is the complexity of the cut formula (i.e., the number of binary connectives or modalities occurring in it), and $h = h_1 + h_2$, called cut height, is the sum of the heights of the derivations of the premisses of cut. As usual, we distinguish some cases according to whether the cut formula is or not principal in the last rules applied in the derivation of the premisses of cut. For the cases where the last rules applied in the derivation of the premisses of cut are propositional we refer to [37, Ch. 4]. Here we only show a few most relevant cases involving modal rules, the other cases are similar.

(i) The cut formula is not principal in the last rule application in the derivation of the left premiss of cut. We consider the following two examples, where the derivation on the left is converted into the derivation on the right:

$$\dfrac{\dfrac{A, B \Rightarrow}{\Gamma, \Box A, \Diamond B \Rightarrow C}\,\wedge\text{-dual}^{i}_{M} \qquad \Gamma', C \Rightarrow \Delta}{\Gamma, \Gamma', \Box A, \Diamond B \Rightarrow \Delta}\,\text{cut} \quad \rightsquigarrow \quad \dfrac{A, B \Rightarrow}{\Gamma, \Gamma', \Box A, \Diamond B \Rightarrow \Delta}\,\wedge\text{-dual}^{i}_{M}$$

$$\dfrac{\dfrac{\Gamma, \Box A, A \Rightarrow B}{\Gamma, \Box A \Rightarrow B}\,\mathsf{T}^{i}_{\Box} \qquad \Gamma', B \Rightarrow \Delta}{\Gamma, \Gamma', \Box A \Rightarrow \Delta}\,\text{cut} \quad \rightsquigarrow \quad \dfrac{\dfrac{\Gamma, \Box A, A \Rightarrow B \qquad \Gamma', B \Rightarrow \Delta}{\Gamma, \Gamma', \Box A, A \Rightarrow \Delta}\,\text{cut}}{\Gamma, \Gamma', \Box A \Rightarrow \Delta}\,\mathsf{T}^{i}_{\Box}$$

(ii) The cut formula is not principal in the last rule application in the derivation of the right premiss of cut. We consider the following example:

$$\dfrac{\Gamma \Rightarrow A \qquad \dfrac{\Gamma'', B \Rightarrow C}{\Gamma', A, \Box\Gamma'', \Diamond B \Rightarrow \Diamond C}\,\mathsf{C}^{i}_{\Diamond}}{\Gamma, \Gamma', \Box\Gamma'', \Diamond B \Rightarrow \Diamond C}\,\text{cut} \quad \rightsquigarrow \quad \dfrac{\Gamma'', B \Rightarrow C}{\Gamma, \Gamma', \Box\Gamma'', \Diamond B \Rightarrow \Diamond C}\,\mathsf{C}^{i}_{\Diamond}$$

(iii) The cut formula is principal in the last rule application in the derivations of both premisses of cut. We consider the following three examples, where R^* denotes multiple applications of the rule R:

($\mathsf{C}^{i}_{\Diamond}$; $\wedge\text{-dual}^{i}_{C}$)

$$\dfrac{\dfrac{\Gamma, A \Rightarrow B}{\Gamma', \Box\Gamma, \Diamond A \Rightarrow \Diamond B}\,\mathsf{C}^{i}_{\Diamond} \qquad \dfrac{\Gamma'', C, B \Rightarrow}{\Gamma''', \Box\Gamma'', \Box C, \Diamond B \Rightarrow \Delta}\,\wedge\text{-dual}^{i}_{C}}{\Gamma', \Gamma''', \Box\Gamma, \Box\Gamma'', \Box C, \Diamond A \Rightarrow \Delta}\,\text{cut}$$

$$\wr$$

$$\dfrac{\dfrac{\Gamma, A \Rightarrow B \qquad \Gamma'', C, B \Rightarrow}{\Gamma, \Gamma'', C, A \Rightarrow}\,\text{cut}}{\Gamma', \Gamma''', \Box\Gamma, \Box\Gamma'', \Box C, \Diamond A \Rightarrow \Delta}\,\wedge\text{-dual}^{i}_{C}$$

($\mathsf{T}^{i}_{\Diamond}$; $\mathsf{C}^{i}_{\Diamond}$)

$$\dfrac{\dfrac{\Gamma \Rightarrow A}{\Gamma \Rightarrow \Diamond A}\,\mathsf{T}^{i}_{\Diamond} \qquad \dfrac{\Gamma', A \Rightarrow B}{\Gamma'', \Box\Gamma', \Diamond A \Rightarrow \Diamond B}\,\mathsf{C}^{i}_{\Diamond}}{\Gamma, \Gamma'', \Box\Gamma' \Rightarrow \Diamond B}\,\text{cut} \quad \rightsquigarrow \quad \dfrac{\dfrac{\dfrac{\dfrac{\Gamma \Rightarrow A \qquad \Gamma', A \Rightarrow B}{\Gamma, \Gamma' \Rightarrow B}\,\text{cut}}{\Gamma, \Gamma' \Rightarrow \Diamond B}\,\mathsf{T}^{i}_{\Diamond}}{\Gamma, \Box\Gamma' \Rightarrow \Diamond B}\,\mathsf{T}^{i\,*}_{\Box}}{\Gamma, \Gamma'', \Box\Gamma' \Rightarrow \Diamond B}\,\text{Lwk}^{*}$$

($\mathsf{N}^{i}_{\Box}$; D^{i})

$$\dfrac{\dfrac{\Rightarrow A}{\Gamma \Rightarrow \Box A}\,\mathsf{N}^{i}_{\Box} \qquad \dfrac{A \Rightarrow B}{\Gamma', \Box A \Rightarrow \Diamond B}\,\mathsf{D}^{i}}{\Gamma, \Gamma' \Rightarrow \Diamond B}\,\text{cut} \quad \rightsquigarrow \quad \dfrac{\dfrac{\Rightarrow A \qquad A \Rightarrow B}{\Rightarrow B}\,\text{cut}}{\Gamma, \Gamma' \Rightarrow \Diamond B}\,\mathsf{P}^{i}_{\Diamond}$$

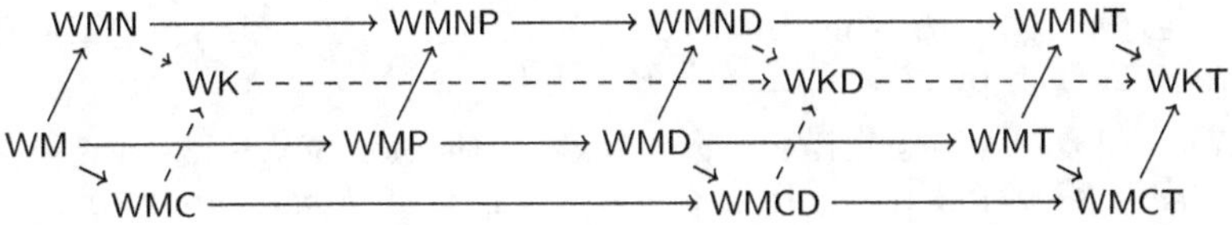

Fig. 6. Dyagram of constructive modal logics.

□

3.1 Axiom systems

For each constructive calculus S.WL, we now define an equivalent axiomatic system. The logics WL are defined in the language $\mathcal{L}$ extending (any axiomatisation of) IPL with the following modal axioms and rules from Fig. 1:

WM := $dual_\wedge + mon_\Box + mon_\Diamond$	WMND := WMN + D
WMN := WM + $N_\Box$	WMCD := WMC + D + $P_\Diamond$
WMC := WM + $C_\Box$ + $K_\Diamond$	WKD := WK + D
WK := WMC + $N_\Box$	WMT := WM + $T_\Box$ + $T_\Diamond$
WMP := WM + $P_\Diamond$	WMNT := WMN + $T_\Box$ + $T_\Diamond$
WMNP := WMN + $P_\Diamond$	WMCT := WMC + $T_\Box$ + $T_\Diamond$
WMD := WM + D + $P_\Diamond$	WKT := WK + $T_\Box$ + $T_\Diamond$

In the following, we will refer to these systems as *W-logics*. Moreover, we denote W* any W-logic, and we denote WC*, resp. WD*, resp. WT* any W-logic with axioms $C_\Box$ and $K_\Diamond$, resp. with axiom D, resp. with axioms $T_\Box$ and $T_\Diamond$. As usual, we say that A is a theorem of W*, written W* $\vdash A$, if there is a finite sequence of formulas ending with A in which every formula is an axiom of W*, or it is obtained from previous formulas by the application of a rule of W*. Moreover, we say that A is deducible in W* from a set of formulas Σ, written $\Sigma \vdash_{\mathsf{W}^*} A$, if there is a finite set $\{B_1, ..., B_n\} \subseteq \Sigma$ such that $\vdash_{\mathsf{W}^*} B_1 \wedge ... \wedge B_n \supset A$. Furthermore, given two axiomatic systems L_1 and L_2, we say that L_1 is included in L_2 if $\mathsf{L}_1 \vdash A$ entails $\mathsf{L}_2 \vdash A$ for all $A \in \mathcal{L}$, and that L_1 and L_2 are equivalent if they derive exactly the same theorems.

Concerning W-logics specifically, note that $C_\Diamond$ is not an axiom of WMC because it is not derivable in S.WMC, $C_\Diamond$ must be replaced by $K_\Diamond$ which is instead derivable in the calculus. Note also that $P_\Diamond$ must be included in the axiomatisation of WMD and WMCD as it is not derivable from D in the intuitionistic systems. The relations among the W-logics are displayed in Fig. 6.

We prove some basic results about W-logics.

Proposition 3.3 *(i)* WM $\vdash Rdual_\wedge$. *(ii)* WMN $\vdash nec$. *(iii)* WMN $\vdash N_\Diamond$. *(iv)* WMC $\vdash K_\Box$. *(v)* WMP $\vdash P_\Box$. *(vi)* WMND $\vdash P_\Diamond$.

Proof. (i) From $\neg(A \wedge B)$, by IPL we obtain $B \supset \neg A$, then by $mon_\Diamond$, $\Diamond B \supset \Diamond\neg A$, thus $\neg\Diamond\neg A \supset \neg\Diamond B$. Moreover from $dual_\wedge$ we have $\Box A \supset \neg\Diamond\neg A$, thus $\Box A \supset \neg\Diamond B$, therefore $\neg(\Box A \wedge \Diamond B)$. (ii) From A, by IPL we have $\top \supset A$, then by $mon_\Box$, $\Box\top \supset \Box A$, then by $N_\Box$, $\Box A$. (iii) By $Rdual_\wedge$, $\Box\top \supset \neg\Diamond\bot$, then by $N_\Box$, $\neg\Diamond\bot$. (iv) By $C_\Box$, $\Box(A \supset B) \wedge \Box A \supset \Box((A \supset B) \wedge A)$, and by $mon_\Box$,

$\Box((A \supset B) \wedge A) \supset \Box B$, thus by IPL, $\Box(A \supset B) \supset (\Box A \supset \Box B)$. (v) By $dual_\wedge$, $\Diamond\top \supset \neg\Box\bot$, then by $P_\Diamond$, $\neg\Box\bot$. (vi) By D, $\Box\top \supset \Diamond\top$, then by $N_\Box$, $\Diamond\top$. □

We recall that Wijesekera's original axiomatisation of WK [38] was given by IPL $+ nec + K_\Box + K_\Diamond + N_\Diamond$. It is easy to verify that $dual_\wedge$, $mon_\Box$, $mon_\Diamond$, $C_\Box$, and $N_\Box$ are all derivable in Wijesekera's original system. Then from Proposition 3.3 it follows in particular that the axiomatisation of WK considered here is equivalent to Wijesekera's one.

We now prove that the systems W* are equivalent to the corresponding calculi.

Theorem 3.4 S.W* $\vdash \Gamma \Rightarrow \Delta$ *if and only if* W* $\vdash \bigwedge\Gamma \supset \bigvee\Delta$.

Proof. From right to left, it is easy to see that all modal axioms are derivable in the corresponding calculi (cf. derivations in Fig. 5), observing that initial sequents init can be generalised as usual to arbitrary formulas A. For $mon_\Box$, from $\Rightarrow A \supset B$ and $A \supset B, A \Rightarrow B$, by cut (which has been proved admissible) we obtain $A \Rightarrow B$, then by $\mathsf{M}^{\mathsf{i}}_\Box$, $\Box A \Rightarrow \Box B$, and by $\supset^{\mathsf{i}}_{\mathsf{R}}$, $\Rightarrow \Box A \supset \Box B$. $mon_\Diamond$ is derived similarly. The derivations of the intuitionistic axioms are standard, moreover modus ponens is simulated by cut in the usual way.

For the other direction, we can consider standard derivations of the propositional rules. Here we show that for every modal sequent rule of S.W* with premiss $\Gamma \Rightarrow \Delta$ and conclusion $\Gamma' \Rightarrow \Delta'$, the Hilbert-style rule $\bigwedge\Gamma \supset \bigvee\Delta$ / $\bigwedge\Gamma' \supset \bigvee\Delta'$ is derivable in the corresponding system W*. We only consider some relevant examples, the other derivations are similar.

($\mathsf{C}^{\mathsf{i}}_\Box$) From $\bigwedge\Gamma \wedge A \supset B$, by $mon_\Box$ we get $\Box(\bigwedge\Gamma \wedge A) \supset \Box B$, moreover by $C_\Box$, $\bigwedge\Box\Gamma \wedge \Box A \supset \Box(\bigwedge\Gamma \wedge A)$, then $\bigwedge\Gamma' \wedge \bigwedge\Box\Gamma \wedge \Box A \supset \Box B$.

($\mathsf{C}^{\mathsf{i}}_\Diamond$) From $\bigwedge\Gamma \wedge A \supset B$, we get $\bigwedge\Gamma \supset (A \supset B)$, then by $mon_\Box$, $\Box\bigwedge\Gamma \supset \Box(A \supset B)$. By $C_\Box$ we have $\bigwedge\Box\Gamma \supset \Box(A \supset B)$, then by $K_\Diamond$, $\bigwedge\Box\Gamma \supset (\Diamond A \supset \Diamond B)$, thus $\bigwedge\Gamma' \wedge \bigwedge\Box\Gamma \wedge \Diamond A \supset \Diamond B$.

($\wedge$-$\mathsf{dual}^{\mathsf{i}}_{\mathsf{C}}$) From $\bigwedge\Gamma \wedge A \wedge B \supset \bot$, we get $\bigwedge\Gamma \wedge A \supset \neg B$, then by $mon_\Box$, $\Box(\bigwedge\Gamma \wedge A) \supset \Box\neg B$, and by $C_\Box$, $\bigwedge\Box\Gamma \wedge \Box A \supset \Box\neg B$. Moreover by $dual_\wedge$, $\Box\neg B \supset \neg\Diamond B$, thus $\bigwedge\Gamma' \wedge \bigwedge\Box\Gamma \wedge \Box A \wedge \Diamond B \supset C$ for any C.

(CD^{i}) If $\Gamma \neq \emptyset$, then from $\bigwedge\Gamma \supset B$, by $mon_\Box$ we get $\Box\bigwedge\Gamma \supset \Box B$, then by $C_\Box$, $\bigwedge\Box\Gamma \supset \Box B$, and by D, $\bigwedge\Box\Gamma \supset \Diamond B$, thus $\bigwedge\Gamma' \wedge \bigwedge\Box\Gamma \supset \Diamond B$. If $\Gamma = \emptyset$, then from B we get $\top \supset B$, then by $mon_\Diamond$, $\Diamond\top \supset \Diamond B$, and by $P_\Diamond$, $\Diamond B$, thus $\bigwedge\Gamma' \supset \Diamond B$.

($\mathsf{CD}^{\mathsf{i}}_\Box$) From $\bigwedge\Gamma \supset \bot$ we get $\bigwedge\Gamma'' \supset \neg B$ for some $B \in \Gamma$ and $\Gamma'' = \Gamma \setminus B$. Then by $mon_\Diamond$, $\Diamond\bigwedge\Gamma'' \supset \Diamond\neg B$, thus by D, $\Box\bigwedge\Gamma'' \supset \Diamond\neg B$, by $C_\Box$, $\bigwedge\Box\Gamma'' \supset \Diamond\neg B$, and by $dual_\wedge$, $\bigwedge\Box\Gamma'' \supset \neg\Box B$, thus $\bigwedge\Gamma' \wedge \bigwedge\Box\Gamma \supset C$ for any C.

□

3.2 Some properties of W-logics

Cut-free sequent calculi are a very powerful tool for the analysis of logical systems. In this subsection, we present some fundamental properties of W-logics that easily follow from those of their sequent calculi.

We start considering the following result, which establishes how classical and Wijesekera-style modal logics are related from the point of view of the axiomatic systems.

Theorem 3.5 *Let* L *be any classical modal logic from Sec. 2, and* WL *be the corresponding W-logic. Then* L *is equivalent to* $\mathsf{WL} + A \vee \neg A + \Box A \vee \Diamond\neg A$.

Proof. From left to right, it is easy to verify that all axioms of WL, as well as $A \vee \neg A$ and $\Box A \vee \Diamond\neg A$, are derivable in L. For the opposite direction, adding $A \vee \neg A$ one derives as usual all theorems of CPL, while adding $\Box A \vee \Diamond\neg A$ one derives *dual*. □

Note that $\mathsf{WL} + A \vee \neg A + \Box A \vee \Diamond\neg A$ is a proper extension of WL, as it is stated by the following proposition.

Proposition 3.6 *For every W-logic* W^*, $\mathsf{W}^* \nvdash p \vee \neg p$ *and* $\mathsf{W}^* \nvdash \Box p \vee \Diamond\neg p$.

Proof. If $\mathsf{W}^* \vdash p \vee \neg p$, then by Theorem 3.4, $\mathsf{S.W}^* \vdash\ \Rightarrow p \vee \neg p$. The only rule of $\mathsf{S.W}^*$ with a consequence of the form $\Rightarrow p \vee \neg p$ is $\vee^{\mathsf{i}}_{\mathsf{R}}$, thus $\vee^{\mathsf{i}}_{\mathsf{R}}$ must be the last rule applied in the derivation, with premiss either $\Rightarrow p$ or $\Rightarrow \neg p$. However, by inspecting the rules of $\mathsf{S.W}^*$ it is easy to verify that $\mathsf{S.W}^* \nvdash\ \Rightarrow p$ and $\mathsf{S.W}^* \nvdash\ \Rightarrow \neg p$, therefore $\mathsf{W}^* \nvdash p \vee \neg p$. $\mathsf{W}^* \nvdash \Box p \vee \Diamond\neg p$ is proved similarly, observing that $\mathsf{S.W}^* \nvdash\ \Rightarrow \Box p$ and $\mathsf{S.W}^* \nvdash\ \Rightarrow \Diamond\neg p$. □

In a similar way we can prove that W-logics satisfy the disjunction property.

Proposition 3.7 (Disjunction property) *For all W-logics* W^* *and all formulas* A, B *of* $\mathcal{L}$, *if* $\mathsf{W}^* \vdash A \vee B$, *then* $\mathsf{W}^* \vdash A$ *or* $\mathsf{W}^* \vdash B$.

Proof. If $\mathsf{W}^* \vdash A \vee B$, then by Theorem 3.4, $\mathsf{S.W}^* \vdash\ \Rightarrow A \vee B$. The only rule of $\mathsf{S.W}^*$ with a consequence of the form $\Rightarrow A \vee B$ is $\vee^{\mathsf{i}}_{\mathsf{R}}$, thus $\vee^{\mathsf{i}}_{\mathsf{R}}$ must be the last rule applied in the derivation, with premiss either $\Rightarrow A$ or $\Rightarrow B$. Then $\mathsf{S.W}^* \vdash\ \Rightarrow A$ or $\mathsf{S.W}^* \vdash\ \Rightarrow B$, therefore $\mathsf{W}^* \vdash A$ or $\mathsf{W}^* \vdash B$. □

Furthermore, we can prove that derivability in W-logics is decidable. To this aim, observe that for every rule R of $\mathsf{S.W}^*$, the premisses of R have a smaller complexity than its conclusion, with the only exceptions of $\supset_{\mathsf{L}}$ and $\mathsf{T}^{\mathsf{i}}_{\Box}$ which copy the principal formula into one premiss. It follows that bottom-up proof search in $\mathsf{S.W}^*$ is not strictly terminating, however, similarly to [37, Ch. 4], termination can be gained by controlling the applications of $\supset_{\mathsf{L}}$ and $\mathsf{T}^{\mathsf{i}}_{\Box}$ with a simple loop-checking in order to avoid redundant applications of these rules, preserving at the same time the completeness of the calculi. Adopting this restriction, it turns out that every proof tree for a root sequent $\Gamma \Rightarrow \Delta$ is finite, moreover there are only finitely many distinct proof trees for it. Then, given the equivalence between $\mathsf{S.W}^*$ and W^*, it follows that derivability of A in W^* is decidable for any A: the decision procedure trivially consists in checking all possible derivations of $\Rightarrow A$ in $\mathsf{S.W}^*$.

Theorem 3.8 (Decidability) *Given a W-logic* W^* *and a formula* A *of* $\mathcal{L}$, *it is decidable whether* A *is derivable in* W^*.

Finally, we can prove that all W-logics enjoy Craig interpolation. For every formula A of $\mathcal{L}$ and every multiset $\Gamma = B_1, ..., B_n$, we define $var(A) = \{\bot\} \cup \{p \in Atm \mid p \text{ occurs in } A\}$, and $var(\Gamma) = var(B_1) \cup ... \cup var(B_n)$. Then Craig interpolation amounts to the following property.

Definition 3.9 A logic W^* enjoys *Craig interpolation* if for all $A, B \in \mathcal{L}$, if $\mathsf{W}^* \vdash A \supset B$, then there is $C \in \mathcal{L}$ such that $\mathsf{W}^* \vdash A \supset C$, $\mathsf{W}^* \vdash C \supset B$, and $var(C) \subseteq var(A) \cap var(B)$.

The proof of Craig interpolation is based on the following lemma.

Lemma 3.10 *For every calculus* $\mathsf{S.W}^*$, *if* $\mathsf{S.W}^* \vdash \Gamma_1, \Gamma_2 \Rightarrow \Delta$, *then there is* $C \in \mathcal{L}$ *such that* $\mathsf{S.W}^* \vdash \Gamma_1 \Rightarrow C$, $\mathsf{S.W}^* \vdash C, \Gamma_2 \Rightarrow \Delta$, *and* $var(C) \subseteq var(\Gamma_1) \cap var(\Gamma_2, \Delta)$.

Proof. By induction on the height h of the derivation of $\Gamma_1, \Gamma_2 \Rightarrow \Delta$, taking into account the last rule applied in the derivation. If $h = 0$ or the last rule applied is propositional we refer to [30]. Here we consider just one significant case involving a modal rule, for the other rules the proof is analogous.

Let $\mathsf{C}^{\mathsf{i}}_{\Diamond}$ be the last rule applied in the derivation. Then $\Gamma_1, \Gamma_2 \Rightarrow \Delta$ has the form $\Gamma'_1, \Box\Gamma, \Diamond A, \Gamma'_2 \Rightarrow \Diamond B$ and it is obtained from the premiss $\Gamma, A \Rightarrow B$. There are four possible partitions of $\Gamma'_1, \Box\Gamma, \Diamond A, \Gamma'_2$ into Γ_1, Γ_2.

(i) $\Gamma_1 = \Gamma'_1$ and $\Gamma_2 = \Box\Gamma, \Diamond A, \Gamma'_2$. Then $\top$ is an interpolant: $\Gamma'_1 \Rightarrow \top$ is derivable, and from $\Gamma, A \Rightarrow B$, by $\mathsf{C}^{\mathsf{i}}_{\Diamond}$ we obtain $\top, \Box\Gamma, \Diamond A, \Gamma'_2 \Rightarrow \Diamond B$.

(ii) $\Gamma_1 = \Gamma'_1, \Box\Gamma, \Diamond A$ and $\Gamma_2 = \Gamma'_2$. By i.h., there is C such that $\Gamma, A \Rightarrow C$ and $C \Rightarrow B$ are derivable, and $var(C) \subseteq var(\Gamma, A) \cap var(B)$. Then by $\mathsf{C}^{\mathsf{i}}_{\Diamond}$ we obtain $\Gamma'_1, \Box\Gamma, \Diamond A \Rightarrow \Diamond C$ and $\Diamond C, \Gamma'_2 \Rightarrow \Diamond B$. Since $var(\Diamond C) = var(C)$, $\Diamond C$ is an interpolant.

The following two partitions are possible if $\Gamma = D_1, ..., D_n$ and $n \geq 2$.

(iii) $\Gamma_1 = \Gamma'_1, \Box D_1, ..., \Box D_k$ and $\Gamma_2 = \Box D_{k+1}, ..., \Box D_n, \Diamond A, \Gamma'_2$ (for $1 \leq k < n$). By i.h., there is C such that $D_1, ..., D_k \Rightarrow C$ and $C, D_{k+1}, ..., D_n, A \Rightarrow B$ are derivable, and $var(C) \subseteq var(D_1, ..., D_k) \cap var(D_{k+1}, ..., D_n, A, B)$. Then by $\mathsf{C}^{\mathsf{i}}_{\Box}$, $\Gamma'_1, \Box D_1, ..., \Box D_k \Rightarrow \Box C$ is derivable, and by $\mathsf{C}^{\mathsf{i}}_{\Diamond}$, $\Box C, \Box D_{k+1}, ..., \Box D_n, \Diamond A, \Gamma'_2 \Rightarrow \Diamond B$ is derivable. Then $\Box C$ is an interpolant.

(iv) $\Gamma_1 = \Gamma'_1, \Box D_1, ..., \Box D_k, \Diamond A$ and $\Gamma_2 = \Box D_{k+1}, ..., \Box D_n, \Gamma'_2$ (for $1 \leq k < n$). By i.h., there is C such that $D_1, ..., D_k, A \Rightarrow C$ and $C, D_{k+1}, ..., D_n \Rightarrow B$ are derivable, and $var(C) \subseteq var(D_1, ..., D_k, A) \cap var(D_{k+1}, ..., D_n, B)$. Then by $\mathsf{C}^{\mathsf{i}}_{\Diamond}$, $\Gamma'_1, \Box D_1, ..., \Box D_k, \Diamond A \Rightarrow \Diamond C$ and $\Diamond C, \Box D_{k+1}, ..., \Box D_n, \Gamma'_2 \Rightarrow \Diamond B$ are derivable. Then $\Diamond C$ is an interpolant.

□

Theorem 3.11 *Every W-logic* W^* *enjoys Craig interpolation.*

Proof. Suppose that $\mathsf{W}^* \vdash A \supset B$. Then $\mathsf{S.W}^* \vdash A \Rightarrow B$. By Lemma 3.10, there is $C \in \mathcal{L}$ such that $var(C) \subseteq var(A) \cap var(B)$, $\mathsf{S.W}^* \vdash A \Rightarrow C$, and $\mathsf{S.W}^* \vdash C \Rightarrow B$, thus $\mathsf{W}^* \vdash A \supset C$ and $\mathsf{W}^* \vdash C \supset B$. □

4 Semantics

We now define constructive neighbourhood models (CNMs) that characterise the constructive modal logics defined in Sec. 3. CNMs are defined analogously to Wijesekera's relational models [38]: we enrich intuitionistic Kripke models with a neighbourhood function (rather than a binary relation as in Wijesekera's models), moreover we generalise the classical satisfaction clauses for modal formulas to all $\leq$-successors, so that hereditariness is built into the clauses: in order that w satisfies $\Box A$, we require that for all successors of w there is a neighbourhood α such that $\alpha \Vdash^{\forall} A$, and similarly for $\Diamond A$. We show that, for every classical logic characterised by neighbourhood models satisfying some conditions from Def. 2.2, the corresponding W-logic is characterised by the CNMs satisfying exactly the same conditions. CNMs are defined as follows.

Definition 4.1 A *constructive neighbourhood model* (CNM) is a tuple $\mathcal{M} = \langle \mathcal{W}, \leq, \mathcal{N}, \mathcal{V} \rangle$, where $\mathcal{W}$ is a non-empty set of worlds, $\leq$ is a preorder on $\mathcal{W}$, $\mathcal{N} : \mathcal{P}(\mathcal{W}) \longrightarrow \mathcal{P}(\mathcal{P}(\mathcal{W}))$ is a neighbourhood function, and $\mathcal{V} : Atm \longrightarrow \mathcal{P}(\mathcal{W})$ is a hereditary valuation function (i.e., if $w \in \mathcal{V}(p)$ and $w \leq v$, then $v \in \mathcal{V}(p)$). The forcing relation $\mathcal{M}, w \Vdash A$ is defined as in Def. 2.2 for $A = p, \bot, B \wedge C, B \vee C$, otherwise it is as follows:

$\mathcal{M}, w \Vdash B \supset C$ iff for all $v \geq w$, $\mathcal{M}, v \Vdash B$ implies $\mathcal{M}, v \Vdash C$.
$\mathcal{M}, w \Vdash \Box B$ iff for all $v \geq w$, there is $\alpha \in \mathcal{N}(v)$ such that $\alpha \Vdash^{\forall} B$.
$\mathcal{M}, w \Vdash \Diamond B$ iff for all $v \geq w$, for all $\alpha \in \mathcal{N}(v)$, $\alpha \Vdash^{\exists} B$.

We consider the following properties on CNMs:

(C) If $\alpha, \beta \in \mathcal{N}(w)$, then $\alpha \cap \beta \in \mathcal{N}(w)$.
(D) If $\alpha, \beta \in \mathcal{N}(w)$, then $\alpha \cap \beta \neq \emptyset$.
(T) If $\alpha \in \mathcal{N}(w)$, then $w \in \alpha$.
(N) $\mathcal{N}(w) \neq \emptyset$.
(P) $\emptyset \notin \mathcal{N}(w)$.

We say that $\mathcal{M}$ is a model for a logic W^*, or is a W^*-model, if for every modal axiom X of W^* (among $C_\Box$, $N_\Box$, $T_\Box$, D, $P_\Diamond$), $\mathcal{M}$ satisfies the corresponding condition (X).

CNMs represent the simplest way of combining intuitionistic Kripke models and neighbourhood models. From the definition of $\mathcal{V}$ and of the satisfaction clauses, it immediately follows that CNMs enjoy the hereditary property.

Proposition 4.2 (Hereditary property) *For all $A \in \mathcal{L}$ and all CNMs $\mathcal{M}$, if $\mathcal{M}, w \Vdash A$ and $w \leq v$, then $\mathcal{M}, v \Vdash A$.*

Proof. By induction on the construction of A. We only consider the inductive cases $A = \Box B, \Diamond B$ as the other cases are standard. ($A = \Box B$) If $w \Vdash \Box B$, then for all $u \geq w$, there is $\alpha \in \mathcal{N}(u)$ such that $\alpha \Vdash^{\forall} B$, then for all $u \geq v$, there is $\alpha \in \mathcal{N}(u)$ such that $\alpha \Vdash^{\forall} B$, thus $v \Vdash \Box B$. ($A = \Diamond B$) If $w \Vdash \Diamond B$, then for all $u \geq w$, for all $\alpha \in \mathcal{N}(u)$, $\alpha \Vdash^{\exists} B$, then for all $u \geq v$, for all $\alpha \in \mathcal{N}(u)$, $\alpha \Vdash^{\exists} B$, thus $v \Vdash \Diamond B$. □

We show that, for any classical modal logic characterised by neighbourhood models satisfying some conditions among (C), (N), (T), (D), (P), the corresponding W-logic is characterised by the CNMs satisfying the same conditions. We first show that W-logics are sound with respect to their classes of CNMs, then prove their completeness by a canonical model construction.

Theorem 4.3 (Soundness) *For every W-logic* W^*, *if* $\mathsf{W}^* \vdash A$, *then* $\mathcal{M} \models A$ *for all* W^**-models* $\mathcal{M}$.

Proof. As usual, we need to show that the axioms of W^* are valid in all W^*-models, and that the rules of W^* preserve the validity in W^*-models.

($\mathsf{M}_\Box$) Assume that $\mathcal{M} \models A \supset B$ and $w \Vdash \Box A$. Then for all $v \geq w$, there is $\alpha \in \mathcal{N}(v)$ such that $\alpha \Vdash^\forall A$. If follows that $\alpha \Vdash^\forall B$. Therefore $w \Vdash \Box B$.

($\mathsf{M}_\Diamond$) Assume that $\mathcal{M} \models A \supset B$ and $w \Vdash \Diamond A$. Then for all $v \geq w$, for all $\alpha \in \mathcal{N}(v)$, $\alpha \Vdash^\exists A$. If follows that $\alpha \Vdash^\exists B$. Therefore $w \Vdash \Diamond B$.

($dual_\wedge$) Assume that $w \Vdash \Box A \wedge \Diamond\neg A$. Then for all $v \geq w$, there is $\alpha \in \mathcal{N}(v)$ such that $\alpha \Vdash^\forall A$, and for all $\beta \in \mathcal{N}(v)$, $\beta \Vdash^\exists \neg A$. Thus there is $\gamma \in \mathcal{N}(w)$ such that $\gamma \Vdash^\exists A \wedge \neg A$, which is impossible. Therefore $\mathcal{M} \models \neg(\Box A \wedge \Diamond\neg A)$.

($C_\Box$) Assume that $\mathcal{M}$ satisfies condition (C) and $w \Vdash \Box A \wedge \Box B$. Then for all $v \geq w$, there is $\alpha \in \mathcal{N}(v)$ such that $\alpha \Vdash^\forall A$, and there is $\beta \in \mathcal{N}(v)$ such that $\beta \Vdash^\forall B$. By (C), $\alpha \cap \beta \in \mathcal{N}(v)$, moreover $\alpha \cap \beta \Vdash^\forall A \wedge B$. Thus $w \Vdash \Box(A \wedge B)$.

($K_\Diamond$) Assume by contradiction that $\mathcal{M}$ satisfies (C), $w \Vdash \Box(A \supset B)$, $w \Vdash \Diamond A$ and $w \nVdash \Diamond B$. By $w \nVdash \Diamond B$, there are $v \geq w$ and $\alpha \in \mathcal{N}(v)$ such that $\alpha \nVdash^\exists B$. Then by $w \Vdash \Box(A \supset B)$, there is $\beta \in \mathcal{N}(v)$ such that $\beta \Vdash^\forall A \supset B$. Thus by (C), $\alpha \cap \beta \in \mathcal{N}(v)$. It follows $\alpha \cap \beta \nVdash^\exists B$ and $\alpha \cap \beta \Vdash^\forall A \supset B$. However by $w \Vdash \Diamond A$, $\alpha \cap \beta \Vdash^\exists A$, thus $\alpha \cap \beta \Vdash^\exists B$, which gives a contradiction.

($N_\Box$) If $\mathcal{M}$ satisfies the condition (N), then for all w there is $\alpha \in \mathcal{N}(w)$. Moreover, $\alpha \Vdash^\forall \top$, therefore $\mathcal{M} \models \Box\top$.

($P_\Diamond$) If $\mathcal{M}$ satisfies the condition (P), then for all w and all $\alpha \in \mathcal{N}(v)$, $\alpha \neq \emptyset$, thus $\alpha \Vdash^\exists \top$. Therefore $\mathcal{M} \models \Diamond\top$.

(D) Assume by contradiction that $\mathcal{M}$ satisfies the condition (D), $w \Vdash \Box A$, and $w \nVdash \Diamond A$. Then there are $v \geq w$ and $\alpha \in \mathcal{N}(v)$ such that $\alpha \nVdash^\exists A$. Moreover, there is $\beta \in \mathcal{N}(v)$ such that $\beta \Vdash^\forall A$. By (D), there is $u \in \alpha\cap\beta$. Thus $u \nVdash A$ and $u \Vdash A$. Therefore $w \Vdash \Diamond A$.

($T_\Box$, $T_\Diamond$) Suppose that $\mathcal{M}$ satisfies the condition (T). Then if $w \Vdash \Box A$, there is $\alpha \in \mathcal{N}(w)$ such that $\alpha \Vdash^\forall A$. By (T), $w \in \alpha$, thus $w \Vdash A$. Moreover, if $w \Vdash A$, then by Prop. 4.2, $v \Vdash A$ for all $v \geq w$. By (T), $v \in \alpha$ for all $\alpha \in \mathcal{N}(v)$, thus $\alpha \Vdash^\exists A$. Therefore $w \Vdash \Diamond A$.

□

4.1 Completeness

We now prove that W-logics are complete with respect to the corresponding classes of CNMs. As usual, for every logic W^*, we call W^*-*prime* any set Σ of formulas of $\mathcal{L}$ such that $\Sigma \nvdash_{\mathsf{W}^*} \bot$ (consistency), if $\Sigma \vdash_{\mathsf{W}^*} A$, then $A \in \Sigma$

(closure under derivation), and if $A \vee B \in \Sigma$, then $A \in \Sigma$ or $B \in \Sigma$ (disjunction property). Moreover, for every set of formulas Σ, we denote $\Box^{-}\Sigma$ the set $\{A \mid \Box A \in \Sigma\}$. One can prove in a standard way the following lemma.

Lemma 4.4 (Lindenbaum) *If $\Sigma \nvdash_{\mathsf{W}^*} A$, then there is a W^*-prime set Π such that $\Sigma \subseteq \Pi$ and $A \notin \Pi$.*

We also consider the following notion of segment (we adopt the terminology of [38]), and prove the subsequent lemma that will be needed in the following.

Definition 4.5 For every logic W^*, a W^*-*segment* is a pair $(\Sigma, \mathscr{C})$, where Σ is a W^*-prime set, and $\mathscr{C}$ is a class of sets of W^*-prime sets such that:

- if $\Box A \in \Sigma$, then there is $\mathscr{U} \in \mathscr{C}$ such that for all $\Pi \in \mathscr{U}$, $A \in \Pi$; and
- if $\Diamond A \in \Sigma$, then for all $\mathscr{U} \in \mathscr{C}$, there is $\Pi \in \mathscr{U}$ such that $A \in \Pi$.

WC^*-, WD^*- and WT^*-segments must satisfy also the following conditions: (WC^*) If $\mathscr{U}, \mathscr{U}' \in \mathscr{C}$, then $\mathscr{U} \cap \mathscr{U}' \in \mathscr{C}$. ($\mathsf{WD}^*$) If $\mathscr{U}, \mathscr{U}' \in \mathscr{C}$, then $\mathscr{U} \cap \mathscr{U}' \neq \emptyset$. ($\mathsf{WT}^*$) For all $\mathscr{U} \in \mathscr{C}$, $\Sigma \in \mathscr{U}$.

Lemma 4.6 *For every W^*-prime set Σ, there exists a W^*-segment $(\Sigma, \mathscr{C})$.*

Proof. Given a W^*-prime set Σ, we construct a W^*-segment $(\Sigma, \mathscr{C})$ as follows. If there is no $\Box A \in \Sigma$, we put $\mathscr{C} = \emptyset$. If there is no $\Diamond A \in \Sigma$, we put $\mathscr{C} = \{\emptyset\}$. Otherwise we distinguish two cases.

(i) W^* does not contain $C_\Box$, $K_\Diamond$. Let $\Box A, \Diamond B \in \Sigma$. Then $A, B \nvdash_{\mathsf{W}^*} \bot$ (otherwise by $Rdual_\wedge$, $\Box A, \Diamond B \vdash_{\mathsf{W}^*} \bot$, against the consistency of Σ). Then by Lemma 4.4, there is Σ'_{AB} W^*-prime such that $A, B \in \Sigma'_{AB}$. For all $\Box A \in \Sigma$, we define $\mathscr{U}_A = \{\Sigma'_{AB} \mid \Diamond B \in \Sigma\}$ if W^* does not contain $T_\Box$, and $\mathscr{U}_A = \{\Sigma'_{AB} \mid \Diamond B \in \Sigma\} \cup \{\Sigma\}$ if it contains $T_\Box$. Moreover we define $\mathscr{C} = \{\mathscr{U}_A \mid \Box A \in \Sigma\}$. Then $(\Sigma, \mathscr{C})$ is a W^*-segment: if $\Box A \in \Sigma$, then $\mathscr{U}_A \in \mathscr{C}$ and for all $\Sigma' \in \mathscr{U}_A$, $A \in \Sigma'$. If $\Diamond B \in \Sigma$, then for all $\mathscr{U}_A \in \mathscr{C}$, there is $\Sigma'_{AB} \in \mathscr{U}_A$ such that $B \in \Sigma'_{AB}$. Moreover for WD^*, if $\mathscr{U}_A, \mathscr{U}_B \in \mathscr{C}$, $\mathscr{U}_A \neq \mathscr{U}_B$, then $\Box A, \Box B \in \Sigma$, then by axiom D, $\Diamond A, \Diamond B \in \Sigma$, thus $\Sigma'_{AB} \in \mathscr{U}_A \cap \mathscr{U}_B$.

(ii) W^* contains $C_\Box$, $K_\Diamond$. Let $\Diamond B \in \Sigma$. Then $\Box^{-}\Sigma \cup \{B\} \nvdash_{\mathsf{W}^*} \bot$ (otherwise by $Rdual_\wedge$ and $C_\Box$, $\Sigma \vdash_{\mathsf{W}^*} \bot$). Then by Lemma 4.4, there is Σ'_B W^*-prime such that $\Box^{-}\Sigma \subseteq \Sigma'_B$ and $B \in \Sigma'_B$. We define $\mathscr{U} = \{\Sigma'_B \mid \Diamond B \in \Sigma\}$ if W^* does not contain $T_\Box$, and $\mathscr{U} = \{\Sigma'_B \mid \Diamond B \in \Sigma\} \cup \{\Sigma\}$ if it contains $T_\Box$. Moreover we define $\mathscr{C} = \{\mathscr{U}\}$. It is easy to verify that $(\Sigma, \mathscr{C})$ is a WC^*-segment.

□

We consider the following definition of canonical model.

Definition 4.7 For every logic W^*, the *canonical model* for W^* is the tuple $\mathcal{M} = \langle \mathcal{W}, \leq, \mathcal{N}, \mathcal{V} \rangle$, where:

- $\mathcal{W}$ is the class of all W^*-segments;
- $(\Sigma, \mathscr{C}) \leq (\Sigma', \mathscr{C}')$ if and only if $\Sigma \subseteq \Sigma'$;

- for every set $\mathscr{U}$ of W*-prime sets, $\alpha_{\mathscr{U}} = \{(\Sigma, \mathscr{C}) \mid \Sigma \in \mathscr{U}\}$;
- $\alpha_{\mathscr{U}} \in \mathcal{N}((\Sigma, \mathscr{C}))$ if and only if $\mathscr{U} \in \mathscr{C}$;
- $(\Sigma, \mathscr{C}) \in \mathcal{V}(p)$ if and only if $p \in \Sigma$.

We prove the following two lemmas which entail completeness of W*-logics.

Lemma 4.8 *The canonical model for* W* *is a CNM for* W*.

Proof. We show that the canonical model for W* satisfies the conditions of CNMs for W*. (C), (D), (T) and hereditariness of $\mathcal{V}$ are immediate by Defs. 4.5 and 4.7. (N) For all WN*-prime sets Σ, $\Box\top \in \Sigma$, then for all WN*-segments $(\Sigma, \mathscr{C})$, $\mathscr{C} \neq \emptyset$, thus $\mathcal{N}((\Sigma, \mathscr{C})) \neq \emptyset$. (P) For all WP*-prime sets Σ, $\Diamond\top \in \Sigma$, then for all WP*-segments $(\Sigma, \mathscr{C})$, for all $\mathscr{U} \in \mathscr{C}$, $\mathscr{U} \neq \emptyset$, thus $\emptyset \notin \mathcal{N}((\Sigma, \mathscr{C}))$.□

Lemma 4.9 *Let* W* *be a W-logic, and* $\mathcal{M} = \langle \mathcal{W}, \leq, \mathcal{N}, \mathcal{V} \rangle$ *be the canonical model for* W*. *Then for all* $(\Sigma, \mathscr{C}) \in \mathcal{W}$, $(\Sigma, \mathscr{C}) \Vdash A$ *if and only if* $A \in \Sigma$.

Proof. By induction on the construction of A. If $A = p$ or $A = \bot$, the proof is immediate by definition of $\mathcal{V}$ or by consistency of Σ, moreover for $A = B \wedge C, B \vee C$ the proof is immediate by applying the inductive hypothesis. We consider the remaining cases.

• $A = B \supset C$: If $B \supset C \in \Sigma$, then suppose $(\Sigma, \mathscr{C}) \leq (\Sigma', \mathscr{C}')$ and $(\Sigma', \mathscr{C}') \Vdash B$. Then $\Sigma \subseteq \Sigma'$, thus $B \supset C \in \Sigma'$. Moreover by i.h., $B \in \Sigma'$, then $C \in \Sigma'$, thus by i.h., $(\Sigma', \mathscr{C}') \Vdash C$. Therefore $(\Sigma, \mathscr{C}) \Vdash B \supset C$. If instead $B \supset C \notin \Sigma$, then $\Sigma \not\vdash B \supset C$, thus $\Sigma \cup \{B\} \not\vdash C$. By Lemma 4.4, there is Σ' W*-prime such that $\Sigma \cup \{B\} \subseteq \Sigma'$ and $C \notin \Sigma'$. Then by Lemma 4.6 and Def. 4.7, there is a W*-segment $(\Sigma', \mathscr{C}') \in \mathcal{W}$. By definition, $(\Sigma, \mathscr{C}) \leq (\Sigma', \mathscr{C}')$, and by i.h., $(\Sigma', \mathscr{C}') \Vdash B$ and $(\Sigma', \mathscr{C}') \nVdash C$. Therefore $(\Sigma, \mathscr{C}) \nVdash B \supset C$.

• $A = \Box B$: If $\Box B \in \Sigma$, then for all $(\Sigma', \mathscr{C}') \geq (\Sigma, \mathscr{C})$, $\Box B \in \Sigma'$. By definition of segment, there is $\mathscr{U}' \in \mathscr{C}'$ such that for all $\Sigma'' \in \mathscr{U}'$, $B \in \Sigma''$. Then $\alpha_{\mathscr{U}'} \in \mathcal{N}((\Sigma', \mathscr{C}'))$, moreover by i.h., $(\Sigma'', \mathscr{C}'') \Vdash B$ for all $(\Sigma'', \mathscr{C}'') \in \alpha_{\mathscr{U}'}$. Therefore $(\Sigma, \mathscr{C}) \Vdash \Box B$. Now suppose that $\Box B \notin \Sigma$. If there is no $\Box C \in \Sigma$, then $(\Sigma, \emptyset)$ is a W*-segment, moreover $(\Sigma, \mathscr{C}) \leq (\Sigma, \emptyset)$ and $\mathcal{N}((\Sigma, \emptyset)) = \emptyset$, thus $(\Sigma, \mathscr{C}) \nVdash \Box B$. If instead there is $\Box C \in \Sigma$, we distinguish two cases:

(i) W* does not contain $C_{\Box}$, $K_{\Diamond}$. Then for all $\Box D \in \Sigma$, $D \not\vdash B$ (otherwise by $mon_{\Box}$, $\Box D \vdash \Box B$, whence $\Box B \in \Sigma$). Then there is Σ'_D W*-prime such that $D \in \Sigma'_D$ and $B \notin \Sigma'_D$. Moreover, for all $\Diamond C \in \Sigma$, $C, D \not\vdash \bot$ (otherwise by $Rdual_{\wedge}$, $\Diamond C, \Box D \vdash \bot$, whence $\bot \in \Sigma$). Then there is Σ'_{CD} W*-prime such that $C, D \in \Sigma'_{CD}$. For all $\Box D \in \Sigma$, we define $\mathscr{U}'_D = \{\Sigma'_D\} \cup \{\Sigma'_{CD} \mid \Diamond C \in \Sigma\}$ if W* does not contain $T_{\Box}$, and $\mathscr{U}'_D = \{\Sigma'_D\} \cup \{\Sigma'_{CD} \mid \Diamond C \in \Sigma\} \cup \{\Sigma\}$ if it contains $T_{\Box}$. Moreover, we define $\mathscr{C}' = \{\mathscr{U}'_D \mid \Box D \in \Sigma\}$. It is easy to verify that $(\Sigma, \mathscr{C}')$ is a W*-segment. Moreover, for all $\mathscr{U}'_D \in \mathscr{C}'$, $\Sigma'_D \in \mathscr{U}'_D$ and $B \notin \Sigma'_D$, thus by i.h., $(\Sigma'_D, \mathscr{C}'') \nVdash B$ for any $(\Sigma'_D, \mathscr{C}'') \in \mathcal{W}$. It follows that for all $\alpha_{\mathscr{U}} \in \mathcal{N}((\Sigma, \mathscr{C}'))$, $\alpha_{\mathscr{U}} \nVdash^{\forall} B$. Thus $(\Sigma, \mathscr{C}') \nVdash \Box B$, and since $(\Sigma, \mathscr{C}) \leq (\Sigma, \mathscr{C}')$, $(\Sigma, \mathscr{C}) \nVdash \Box B$.

(ii) W* contains $C_{\Box}$, $K_{\Diamond}$. Then $\Box^{-}\Sigma \not\vdash B$ (otherwise by $mon_{\Box}$ and $C_{\Box}$, $\Sigma \vdash \Box B$), then there is Σ' WC*-prime such that $\Box^{-}\Sigma \subseteq \Sigma'$ and $B \notin \Sigma'$.

Moreover, for all $\Diamond C \in \Sigma$, $\Box^{-}\Sigma \cup \{C\} \nvdash \bot$, then there is Σ'_C WC*-prime such that $\Box^{-}\Sigma \subseteq \Sigma'_C$ and $C \in \Sigma'_C$. We define $\mathscr{U}' = \{\Sigma'\} \cup \{\Sigma'_C \mid \Diamond C \in \Sigma\}$ if WC* does not contain $T_\Box$, and $\mathscr{U}' = \{\Sigma'\} \cup \{\Sigma'_C \mid \Diamond C \in \Sigma\} \cup \{\Sigma\}$ if it contains $T_\Box$. Moreover, we define $\mathscr{C}' = \{\mathscr{U}'\}$. It is easy to verify that $(\Sigma, \mathscr{C}')$ is a WC*-segment. Moreover, since $B \notin \Sigma'$, by i.h., $(\Sigma', \mathscr{C}'') \nVdash B$ for any $(\Sigma', \mathscr{C}'') \in \mathcal{W}$, it follows that for all $\alpha_{\mathscr{U}} \in \mathcal{N}((\Sigma, \mathscr{C}'))$, $\alpha_{\mathscr{U}} \nVdash^{\forall} B$. Thus $(\Sigma, \mathscr{C}') \nVdash \Box B$, and since $(\Sigma, \mathscr{C}) \leq (\Sigma, \mathscr{C}')$, $(\Sigma, \mathscr{C}) \nVdash \Box B$.

• $A = \Diamond B$: If $\Diamond B \in \Sigma$, then for all $(\Sigma', \mathscr{C}') \geq (\Sigma, \mathscr{C})$, $\Diamond B \in \Sigma'$. By definition of segment, for all $\mathscr{U}' \in \mathscr{C}'$, there is $\Sigma'' \in \mathscr{U}'$ such that $B \in \Sigma''$. Then for all $\alpha_{\mathscr{U}'} \in \mathcal{N}((\Sigma', \mathscr{C}'))$, there is $(\Sigma'', \mathscr{C}'') \in \alpha_{\mathscr{U}'}$ such that $B \in \Sigma''$, thus by i.h., $(\Sigma'', \mathscr{C}'') \Vdash B$. It follows that $(\Sigma, \mathscr{C}) \Vdash \Diamond B$. Now suppose that $\Diamond B \notin \Sigma$. If there is no $\Diamond C \in \Sigma$, then $(\Sigma, \{\emptyset\})$ is a W*-segment, moreover $(\Sigma, \mathscr{C}) \leq (\Sigma, \{\emptyset\})$ and $\mathcal{N}((\Sigma, \{\emptyset\})) = \{\emptyset\}$, thus $(\Sigma, \mathscr{C}) \nVdash \Diamond B$. If instead there is $\Diamond C \in \Sigma$, we distinguish two cases:

(i) W* does not contain $C_\Box$, $K_\Diamond$. Then for all $\Diamond C \in \Sigma$, $C \nvdash B$ (otherwise by $mon_\Diamond$, $\Diamond C \vdash \Diamond B$, whence $\Diamond B \in \Sigma$). Then there is Σ'_C W*-prime such that $C \in \Sigma'_C$ and $B \notin \Sigma'_C$. Moreover, for all $\Box D \in \Sigma$, $C, D \nvdash \bot$ (otherwise by $Rdual_\wedge$, $\Diamond C, \Box D \vdash \bot$, whence $\bot \in \Sigma$). Then there is Σ'_{CD} W*-prime such that $C, D \in \Sigma'_{CD}$. If in addition W* does not contain $T_\Box$, we define $\mathscr{U}' = \{\Sigma'_C \mid \Diamond C \in \Sigma\}$, and for all $\Box D \in \Sigma$, $\mathscr{U}'_D = \{\Sigma'_{CD} \mid \Diamond C \in \Sigma\}$; otherwise we define $\mathscr{U}' = \{\Sigma'_C \mid \Diamond C \in \Sigma\} \cup \{\Sigma\}$, and for all $\Box D \in \Sigma$, $\mathscr{U}'_D = \{\Sigma'_{CD} \mid \Diamond C \in \Sigma\} \cup \{\Sigma\}$. Moreover, we define $\mathscr{C}' = \{\mathscr{U}'\} \cup \{\mathscr{U}'_D \mid \Box D \in \Sigma\}$. It is easy to verify that $(\Sigma, \mathscr{C}')$ is a W*-segment: for instance for WD*, if $\mathscr{U}'_D, \mathscr{U}'_E \in \mathscr{C}'$, then $\mathscr{U}'_D \cap \mathscr{U}'_E \neq \emptyset$ (cf. proof of Lemma 4.6), moreover by $P_\Diamond$, $\Diamond\top \in \Sigma$, thus for every $\Box D \in \Sigma$, $\Sigma'_{\top D} \in \mathscr{U}'_D \cap \mathscr{U}'$, then $\mathscr{U}'_D \cap \mathscr{U}' \neq \emptyset$. In addition, by definition we have $\alpha_{\mathscr{U}'} \in \mathcal{N}((\Sigma, \mathscr{C}'))$, and for all $\Sigma' \in \mathscr{U}'$, $B \notin \Sigma'$ (in particular, by $T_\Diamond$, $B \notin \Sigma$). Thus by i.h., for all $\Sigma' \in \mathscr{U}'$ and all $(\Sigma', \mathscr{C}'') \in \mathcal{W}$, $(\Sigma', \mathscr{C}'') \nVdash B$, then $\alpha_{\mathscr{U}'} \nVdash^{\exists} B$. Therefore $(\Sigma, \mathscr{C}') \nVdash \Diamond B$, and since $(\Sigma, \mathscr{C}) \leq (\Sigma, \mathscr{C}')$, $(\Sigma, \mathscr{C}) \nVdash \Diamond B$.

(ii) W* contains $C_\Box$, $K_\Diamond$. Then for all $\Diamond C \in \Sigma$, $\Box^{-}\Sigma \cup \{C\} \nvdash B$ (otherwise by $mon_\Box$, $C_\Box$ and $K_\Diamond$, $\Sigma \vdash \Diamond B$). Then there is Σ'_C WC*-prime such that $\Box^{-}\Sigma \cup \{C\} \subseteq \Sigma'_C$ and $B \notin \Sigma'_C$. We define $\mathscr{U}' = \{\Sigma'_C \mid \Diamond C \in \Sigma\}$ if WC* does not contain $T_\Box$, and $\mathscr{U}' = \{\Sigma'_C \mid \Diamond C \in \Sigma\} \cup \{\Sigma\}$ if it contains $T_\Box$. Moreover we define $\mathscr{C}' = \{\mathscr{U}'\}$. It is easy to verify that $(\Sigma, \mathscr{C}')$ is a WC*-segment. Moreover, for all $\Sigma' \in \mathscr{U}'$, $B \notin \Sigma'$, then by i.h., for all $(\Sigma', \mathscr{C}'') \in \mathcal{W}$, $(\Sigma', \mathscr{C}'') \nVdash B$, thus $\alpha_{\mathscr{U}'} \nVdash^{\exists} B$. It follows that $(\Sigma, \mathscr{C}') \nVdash \Diamond B$, and since $(\Sigma, \mathscr{C}) \leq (\Sigma, \mathscr{C}')$, $(\Sigma, \mathscr{C}) \nVdash \Diamond B$.

□

Theorem 4.10 (Completeness) *For every W-logic* W*, *if* $\mathcal{M} \models A$ *for all* W*-*models* $\mathcal{M}$, *then* W* $\vdash A$.

Proof. Suppose that W* $\nvdash A$. Then by Lemma 4.4, there is a W*-prime set Σ such that $A \notin \Sigma$, thus by Lemma 4.6, there exists a W*-segment $(\Sigma, \mathscr{C})$. By Def. 4.7, $(\Sigma, \mathscr{C})$ belongs to the canonical model $\mathcal{M}$ for W*. Then by Lemma 4.9,

$(\Sigma, \mathscr{C}) \not\Vdash A$, and by Lemma 4.8, $\mathcal{M}$ is a W^*-model. Therefore A is not valid in all W^*-models. □

5 Discussion and future work

In this paper, we have defined a family of 14 constructive modal logics both proof-theoretically and semantically motivated, corresponding each to a different classical modal logic. On the one hand, the logics correspond to the single-succedent restriction of standard sequent calculi for classical modal logics. On the other hand, the same logics are obtained by considering over intuitionistic Kripke models a natural generalisation of the classical satisfaction clauses for modal formulas in the neighbourhood semantics. The main result of this paper is that, despite being mutually independent, for the considered logics the two approaches return exactly the same systems.

In addition, we have provided some preliminary analysis of W-logics. First, we have shown how W-logics are related to the corresponding classical modal logics from the point of view of the axiomatic systems: each classical modal logic considered in this paper can be obtained by extending the corresponding W-logic with both excluded middle $A \vee \neg A$ and disjunctive duality $\Box A \vee \Diamond \neg A$. Moreover, basing on their sequent calculi we have proved some fundamental properties of W-logics, such as the disjunction property, decidability and Craig interpolation.

Simpson [35, Ch. 3] listed some requirements that one expects to be satisfied by any intuitionistic modal logic: they must be conservative over IPL; they must contain all axioms of IPL (over the whole language) and be closed under modus ponens; they must satisfy the disjunction property; the modalities must be independent; the addition of the axiom $A \vee \neg A$ must yield a standard classical modal logic. Basing on the results presented in this paper, it is easy to verify that all W-logics satisfy the first four requirements, by contrast they do not satisfy the last one.[5] However, it comes natural to ask whether there could be some modal principle, additional to excluded middle, that distinguishes between constructive and classical modalities. As a matter of fact, it is easy to identify such a principle for W-logics: as we observed above this principle is precisely $\Box A \vee \Diamond \neg A$.

This relation between classical and W-logics is not entirely trivial. For instance, the same relation does not hold between CK and K, in particular CK must be extended also with $\neg(\Box A \wedge \Diamond \neg A)$ (or equivalently with $\neg \Diamond \bot$) in order to obtain classical K. Moreover, we believe that failure of $\Box A \vee \Diamond \neg A$ is justifiable from a constructive perspective, as it can be seen as a modalised form of excluded middle.

Concerning the semantics of W-logics, the choice of considering neighbourhood models is motivated by the possibility to uniformly cover all considered logics, which include both normal and non-normal systems. However, WK,

---

[5] This requirement has been sometimes criticised as being too strong, see [22] for an argument against this requirement based on negative translations.

WKD and WKT have an equivalent characterisation in terms of constructive bi-relational models [38], and we conjecture that an analogous characterisation can be given for WMC and its extensions in terms of relational models with non-normal worlds (cf. e.g. [34]). As a byproduct of this work, we have provided a new semantics for WK alternative to its original relational semantics [38] and to the neighbourhood semantics in [18,9].

The possibility to define constructive counterparts of both normal and non-normal classical logics can be seen as providing additional justification for the present approach. To make a comparison, it is not obvious how to extend the family of intuitionistic modal logics (IK and extensions) to non-normal systems, given that their definition ultimately relies on the standard translation of modal formulas into first-order sentences, which in turn is based on the relational semantics. Interestingly, the constructive counterparts of non-normal logics that we have obtained are not entirely new. In particular, WM and WMN coincide with the logics IM and $\mathsf{IMN}_\Box$ introduced in [9], where they are given an alternative semantics with distinct neighbourhood functions for $\Box$ and $\Diamond$. By contrast, WMC is not equivalent to IMC in [9], since WMC contains $K_\Diamond$ which is not a theorem of IMC.

The results presented in this paper can be extended in several directions. In future work we plan to study the complexity of W-logics, possibly extending some optimal calculi for IPL (G3-style calculi are not adequate to establish good complexity bounds for constructive logics). Moreover, we would like to study whether Iemhoff's proof-theoretical method for proving uniform interpolation [16] can be adapted to W-logics. We would also like to define calculi for W-logics that allow for a direct extraction of countermodels from failed proofs, along the lines of [17,10,11].

Furthermore, one can extend the present analysis to further classical modal logics in order to enrich the family of W-logics, but also to inspect the limits of our approach. An obvious limit concerns the logics for which no standard cut-free Gentzen calculi exist, such as S5. For these logics one can study whether a similar analysis could be based on alternative kinds of calculi, like hyper- or nested sequent calculi. At the same time, it is known that incorporating hereditariness into the satisfaction clauses is not sufficient to provide a semantics for some constructive systems, this is the case for instance of the logics with axiom 4 [2]. Concerning instead weaker systems, it seems that for non-normal logic E [8] this approach returns a very weak form of duality analogous to the one of IE_1 in [9], but this requires further study.

Acknowledgements. This work was supported by the SPGAS and CompRAS projects at the Free University of Bozen-Bolzano, and by the EU H2020 project INODE (grant agreement No 863410).

References

[1] Acclavio, M., D. Catta and L. Straßburger, *Game semantics for constructive modal logic*, in: *Proceedings of TABLEAUX 2021* (2021), pp. 428–445.

[2] Alechina, N., M. Mendler, V. de Paiva and E. Ritter, *Categorical and kripke semantics for constructive S4 modal logic*, in: *Proceedings of CSL 2001* (2001), pp. 292–307.

[3] Amati, G. and F. Pirri, *A uniform tableau method for intuitionistic modal logics I*, Studia Logica **53** (1994), pp. 29–60.

[4] Arisaka, R., A. Das and L. Straßburger, *On nested sequents for constructive modal logics*, Log. Methods Comput. Sci. **11** (2015).

[5] Balbiani, P., M. Diéguez and D. Fernández-Duque, *Some constructive variants of S4 with the finite model property*, in: *Proceedings of LICS 2021* (2021), pp. 1–13.

[6] Bellin, G., V. de Paiva and E. Ritter, *Extended Curry-Howard correspondence for a basic constructive modal logic*, in: *Proceedings of M4M-2*, 2001.

[7] Calardo, E. and A. Rotolo, *Variants of multi-relational semantics for propositional non-normal modal logics*, J. Appl. Non Class. Logics **24** (2014), pp. 293–320.

[8] Chellas, B. F., "Modal Logic - An Introduction," Cambridge University Press, 1980.

[9] Dalmonte, T., C. Grellois and N. Olivetti, *Intuitionistic non-normal modal logics: A general framework*, J. Philos. Log. **49** (2020), pp. 833–882.

[10] Dalmonte, T., C. Grellois and N. Olivetti, *Terminating calculi and countermodels for constructive modal logics*, in: *Proceedings of TABLEAUX 2021* (2021), pp. 391–408.

[11] Dalmonte, T., B. Lellmann, N. Olivetti and E. Pimentel, *Hypersequent calculi for non-normal modal and deontic logics: countermodels and optimal complexity*, J. Log. Comput. **31** (2021), pp. 67–111.

[12] Ewald, W. B., *Intuitionistic tense and modal logic*, J. Symb. Log. **51** (1986), pp. 166–179.

[13] Fischer Servi, G., *Semantics for a class of intuitionistic modal calculi*, in: *Italian Studies in the Philosophy of Science* (1980), pp. 57–72.

[14] Galmiche, D. and Y. Salhi, *Tree-sequent calculi and decision procedures for intuitionistic modal logics*, J. Log. Comput. **28** (2018), pp. 967–989.

[15] Gentzen, G., *Untersuchungen über das logische Schliessen I*, Matematische Zeitschrift **39** (1935), pp. 176–210.

[16] Iemhoff, R., *Uniform interpolation and sequent calculi in modal logic*, Archive for Mathematical Logic **58** (2019), pp. 155–181.

[17] Iemhoff, R., *The G4i analogue of a G3i calculus*, arXiv preprint arXiv:2011.11847 (2020).

[18] Kojima, K., *Relational and neighborhood semantics for intuitionistic modal logic*, Reports Math. Log. **47** (2012), pp. 87–113.

[19] Kuznets, R. and L. Straßburger, *Maehara-style modal nested calculi*, Arch. Math. Log. **58** (2019), pp. 359–385.

[20] Lavendhomme, R. and T. Lucas, *Sequent calculi and decision procedures for weak modal systems*, Stud Logica **66** (2000), pp. 121–145.

[21] Lellmann, B. and E. Pimentel, *Modularisation of sequent calculi for normal and non-normal modalities*, ACM Trans. Comput. Log. **20** (2019), pp. 7:1–7:46.

[22] Litak, T., M. Polzer and U. Rabenstein, *Negative translations and normal modality*, in: *Proceedings of FSCD 2017*, Schloss Dagstuhl-Leibniz-Zentrum für Informatik, 2017.

[23] Lyon, T. S., *Nested sequents for intuitionistic modal logics via structural refinement*, in: *Proceedings of TABLEAUX 2021* (2021), pp. 409–427.

[24] Marin, S., M. Morales and L. Straßburger, *A fully labelled proof system for intuitionistic modal logics*, J. Log. Comput. **31** (2021), pp. 998–1022.

[25] Marin, S. and L. Straßburger, *Proof theory for indexed nested sequents*, in: *Proceedings of TABLEAUX 2017* (2017), pp. 81–97.

[26] Mendler, M. and V. de Paiva, *Constructive CK for contexts*, in: *Proceedings of CRR'05* (2005).

[27] Mendler, M. and S. Scheele, *Cut-free gentzen calculus for multimodal CK*, Inf. Comput. **209** (2011), pp. 1465–1490.

[28] Mendler, M. and S. Scheele, *On the computational interpretation of CK_n for contextual information processing*, Fundam. Informaticae **130** (2014), pp. 125–162.

[29] Mendler, M., S. Scheele and L. Burke, *The Došen square under construction: A tale of four modalities*, in: *Proceedings of TABLEAUX 2021* (2021), pp. 446–465.
[30] Ono, H., *Proof-theoretic methods in nonclassical logic–an introduction*, in: *Theories of types and proofs* (1998), pp. 207–254.
[31] Orlandelli, E., *Sequent calculi and interpolation for non-normal modal and deontic logics*, Logic and Logical Philosophy **30** (2020), pp. 139–183.
[32] Pacuit, E., "Neighborhood Semantics for Modal Logic," Springer, 2017.
[33] Plotkin, G. D. and C. Stirling, *A framework for intuitionistic modal logics*, in: *Proceedings of TARK'86* (1986), pp. 399–406.
[34] Priest, G., "An Introduction to Non-Cassical Logic: From If to Is," Cambridge University Press, 2008.
[35] Simpson, A. K., "The proof theory and semantics of intuitionistic modal logic," Ph.D. thesis, University of Edinburgh, UK (1994).
[36] Straßburger, L., *Cut elimination in nested sequents for intuitionistic modal logics*, in: F. Pfenning, editor, *Proceedings of FOSSACS 2013* (2013), pp. 209–224.
[37] Troelstra, A. S. and H. Schwichtenberg, "Basic proof theory, Second Edition," Cambridge tracts in theoretical computer science **43**, Cambridge University Press, 2000.
[38] Wijesekera, D., *Constructive modal logics I*, Ann. Pure Appl. Log. **50** (1990), pp. 271–301.
[39] Wijesekera, D. and A. Nerode, *Tableaux for constructive concurrent dynamic logic*, Ann. Pure Appl. Log. **135** (2005), pp. 1–72.
[40] Wolter, F. and M. Zakharyaschev, *Intuitionistic modal logic*, in: *Logic and Foundations of Mathematics* (1999), pp. 227–238.

Comparative plausibility in neighbourhood models: axiom systems and sequent calculi

Tiziano Dalmonte[a] Marianna Girlando[b] [1]

[a] *Free University of Bozen-Bolzano, Bolzano, Italy*
[b] *University of Birmingham, Birmingham, UK*

Abstract

We introduce a family of comparative plausibility logics over neighbourhood models, generalising Lewis' comparative plausibility operator over sphere models. We provide axiom systems for the logics, and prove their soundness and completeness with respect to the semantics. Then, we introduce two kinds of analytic proof systems for several logics in the family: a multi-premisses sequent calculus in the style of Lellmann and Pattinson, for which we prove cut admissibility, and a hypersequent calculus based on structured calculi for conditional logics by Girlando et al., tailored for counter-model construction over failed proof search. Our results constitute the first steps in the definition of a unified proof theoretical framework for logics equipped with a comparative plausibility operator.

Keywords: Comparative plausibility, neighbourhood semantics, sequent calculus, hypersequent calculus, countermodel construction.

1 Introduction

In the seminal work *Counterfactuals* [16], besides the well-known analysis of counterfactual sentences, David Lewis defined a notion of comparative plausibility which has then become a standard.[2] Specifically, Lewis introduced a *comparative plausibility operator* $A \preccurlyeq B$, read "A is at least as plausible as B", which is evaluated on thc plausibility ordering of worlds of a model.

Lewis' notion of comparative plausibility is defined over *sphere models*. These are possible-world models in which every world x is endowed with a system of spheres $S(x)$, that is, a set of sets of worlds such that for every two sets in the class, one of the two is included in the other (if $\alpha, \beta \in S(x)$, then $\alpha \subseteq \beta$ or $\beta \subseteq \alpha$). This property, known as *nesting*, determines a total ordering

[1] This work was supported by the UKRI Future Leaders Fellowship 'Structure vs Invariants in Proofs' MR/S035540/1, by the SPGAS and CompRAS projects at the Free University of Bozen-Bolzano, and by the EU H2020 project INODE (grant agreement No 863410).

[2] Lewis [16] refers to $\preccurlyeq$ as the operator for *comparative possibility*. In the literature, the same or similar operators also go under the names of *entrenchment* [15], *comparative similarity* [25] or *relative likelihood* [12]. Here we adopt the terminology of, e.g., [22].

over the set of worlds belonging to a system of spheres, where worlds in the inner spheres are taken to be more plausible than worlds in the outer spheres. Then, $A \preccurlyeq B$ is true at a world x if the innermost sphere in $S(x)$ containing a world which forces B also contains a world that forces A. The operator $\preccurlyeq$ is interdefinable with Lewis' *conditional operator* $A > B$ expressing counterfactual sentences (formally, $A > B$ is equivalent to $(\bot \preccurlyeq A) \lor \neg((A \land \neg B) \preccurlyeq (A \land B))$).

Other than in Lewis' work, several operators expressing forms of similarity or closeness between states of affairs or concepts have been studied in the literature, and find applications in many areas of computer science and philosophy. In knowledge representation, Sheremet et al. developed in [25,26] the *logic of comparative concept similarity*, evaluated over distance models, which implements a description logic-like formalism for reasoning about similarity of concepts in ontologies. Refer to [1] for a Lewis-style semantics for this logic. Moreover, similarity operators can be used in deontic reasoning to express degrees of urgency of obligations [2] or, more recently, to express the preferred scenario an agent would choose in an ethical decision-making process [18]. In philosophical logic, a logic equipped with an operator to express *ceteris paribus* preference between states of affairs was introduced by Von Wright in [28], and formalised in [27]. Moreover, a logic expressing *ceteris paribus* preferences in a deontic setting was recently defined in [17].

A natural semantics to express generalized forms of Lewis' comparative plausibility is *preferential semantics.* Preferential models consist of a set of worlds equipped with an explicit preorder relation $\leq_x$ for every world x, encoding similarity or preference among worlds. These models represent a generalisation of sphere models, where totality of the ordering is not assumed, and have been studied as a semantics for a family of conditional logic weaker than Lewis' counterfactual logic, called *Preferential Conditional Logics* [3,6], strongly related to non-monotonic logic P from [13]. In [12], Halpern proposes partially ordered preferential structures as a general framework to represent forms of preference or similarity.

We here propose a setting even more general than preferential semantics, by interpreting the comparative plausibility operator over *neighbourhood models* (Sec. 2). These possible-worlds models are endowed with a neighbourhood function which assigns to every world x a set of sets of worlds, $N(x)$, where nesting is not assumed. In this weaker setting, the truth condition for $\preccurlyeq$ can be taken to express forms of similarity of closeness between states or concepts, which are not assumed to be totally ordered. Neighbourhood models were introduced to define a semantics for non-normal modal logics [24,20] and, among other applications, have been employed as a semantics for conditional logics [19,21,11].

We introduce axiom systems for the family of *logics of Comparative Plausibility in Neighbourhood models* (CPN logics), and prove their adequacy with respect to some relevant classes of neighbourhood models (Sec. 3). We then study the proof theory of CPN logics, by defining two kinds of proof systems for them. Our calculi are inspired from analytic proof systems for Lewis' conditional logics introduced in the literature.

We first present a *multi-premisses* sequent calculus in the style of Lellmann and Pattinson [15,14] (Sec. 4). The rules of these calculi display a number of premisses which depends on the number of comparative plausibility formulas occurring in the conclusion. The calculi for CPN logics represent simpler fragments of the calculi for Lewis' logics presented in [14]. We prove cut-admissibility for the multi-premisses calculi. While these calculi have strong proof-theoretical properties, they are not best suited for root-first proof search: due to the fact that the comparative plausibility rules are not invertible, a heavy use of backtracking is needed to construct derivations.

This motivates the introduction of a second family of proof systems, based on *hypersequents* (Sec. 5). The calculi are inspired from the structured calculi for Lewis' logics introduced in [22,8], which introduce an additional structural connective to Gentzen-style sequents representing $\preccurlyeq$-formulas. Following a strategy adopted e.g. in [4] in the context of non-normal modal logics, we further enrich the structure of sequents from [8] by introducing hypersequent-style calculi, and show that they simulate the multi-premisses calculi. Thanks to this richer structure we obtain invertibility of *all* the rules in the calculus, which we would not have using the sequent structure from [8], and a more direct construction of countermodels from branches of failed proof search trees. We conclude by discussing related works and further research directions (Sec. 6).

2 Neighbourhood semantics

For $Atm = \{p_0, p_1, p_2, ...\}$ denumerable set of propositional variables, we consider the formulas of $\mathscr{L}$ be defined by the BNF grammar $A ::= p \mid \bot \mid A \to A \mid A \preccurlyeq A$, where p is any element of Atm, and $\preccurlyeq$ is the operator for comparative plausibility. We assume $\top, \neg, \wedge, \vee$ to be defined as usual in terms of $\bot, \to$.

Definition 2.1 A *neighbourhood model* is a tuple $\mathcal{M} = \langle W, N, V \rangle$, where W is a non-empty set of worlds, V is a valuation function $Atm \longrightarrow \mathcal{P}(W)$, and N is a function $W \longrightarrow \mathcal{P}(\mathcal{P}(W))$, called *neighbourhood function*, satisfying the non-emptiness condition: for all $w \in W$: $\emptyset \notin N(w)$.[3] For all $w \in W$ and $A \in \mathscr{L}$, the forcing relation $\mathcal{M}, w \Vdash A$ is defined inductively as follows:

$$
\begin{array}{lll}
\mathcal{M}, w \Vdash p & \text{iff} & w \in V(p). \\
\mathcal{M}, w \nVdash \bot. & & \\
\mathcal{M}, w \Vdash B \to C & \text{iff} & \text{if } \mathcal{M}, w \Vdash B, \text{ then } \mathcal{M}, w \Vdash C. \\
\mathcal{M}, w \Vdash B \preccurlyeq C & \text{iff} & \text{for all } \alpha \in N(w), \text{ if there is } v \in \alpha \text{ s.t. } \mathcal{M}, v \Vdash C, \\
 & & \text{then there is } u \in \alpha \text{ s.t. } \mathcal{M}, u \Vdash B.
\end{array}
$$

We say that A is *valid in a model* $\mathcal{M}$, written $\mathcal{M} \models A$, if $\mathcal{M}, w \Vdash A$ for all worlds w of $\mathcal{M}$, and it is *valid on a class of models* $\mathcal{C}$ if $\mathcal{M} \models A$ for all $\mathcal{M} \in \mathcal{C}$.

In the following we simply write $w \Vdash A$ when $\mathcal{M}$ is clear from the context.

---

[3] Non-emptiness could be dropped as it has no impact on the satisfiability of $\preccurlyeq$-formulas [16]. We assume it as it allows for a clean formulation of the conditions for the extensions, and for uniformity with the neighbourhood semantics of conditional logics from [21,11].

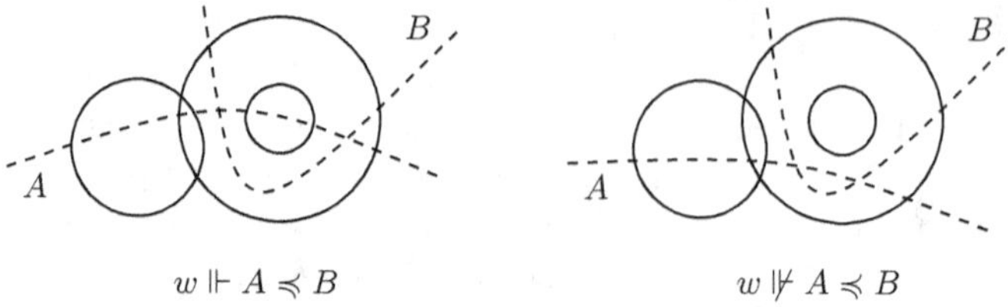

Fig. 1. Representation of the forcing clause for $\preccurlyeq$-formulas. Dashed lines represent the extensions of A and B.

We shall also use $\alpha \Vdash^{\exists} A$ as an abbreviation for 'there is $w \in \alpha$ such that $w \Vdash A$'. Thus, we can rewrite the forcing clause of $\preccurlyeq$-formulas, graphically represented in Fig. 1, as follows:

$$w \Vdash B \preccurlyeq C \quad \text{iff} \quad \text{for all } \alpha \in N(w), \text{ if } \alpha \Vdash^{\exists} C, \text{ then } \alpha \Vdash^{\exists} B.$$

We observe that unary modalities can be defined on the basis of $\preccurlyeq$, namely $\Box_K A := \bot \preccurlyeq \neg A$, and $\Box_M A := \neg(\neg A \preccurlyeq \top)$, where $\Box_K$ and $\Box_M$ are the Box modalities of respectively logic K and non-normal logic M (cf. e.g [23]). Note also that sphere models can be recovered by adding the condition of *nesting* to the neighbourhood function: for all $\alpha, \beta \in N(w)$, either $\alpha \subseteq \beta$ or $\beta \subseteq \alpha$. Lewis considered in [16] several additional properties, which turn out to be of interest when formulated on the neighbourhood function. We consider here classes of neighbourhood models satisfying combinations of the following properties:

N	$N(w) \neq \emptyset$.	(*Normality*)
T	There is $\alpha \in N(w)$ such that $w \in \alpha$.	(*Total reflexivity*)
W	$N(w) \neq \emptyset$ and for all $\alpha \in N(w)$, $w \in \alpha$.	(*Weak centering*)
C	$\{w\} \in N(w)$ and for all $\alpha \in N(w)$, $w \in \alpha$.	(*Centering*)
U	If $v \in \alpha$ and $\alpha \in N(w)$, then $\bigcup N(v) = \bigcup N(w)$.	(*Uniformity*)
A	If $v \in \alpha$ and $\alpha \in N(w)$, then $N(v) = N(w)$.	(*Absoluteness*)

The condition of absoluteness can also be formulated as follows:

$\mathsf{A+}$	For all $v, w \in W$, $N(v) = N(w)$.	(*Strong Absoluteness*)

Equivalence of A and $\mathsf{A+}$ over formulas validity can be easily established using the same strategy described by Lewis [16, p. 122]. We will use $\mathsf{A+}$ in Sec. 5.

Neighbourhood semantics can be used to express a variety of situations. By means of example, let $\alpha, \beta, \gamma, ...$ in $N(w)$ represent sources of information available at w which are not arranged in any priority or reliability order. In this setting, $A \preccurlyeq B$ expresses that A is at least as plausible as B in that whenever w receives information B, it also receives information A. Then, model conditions represent natural assumptions about the information sources: by *normality*, every w has a source of information available, while according to *reflexivity* or *weak centering*, w belongs to some or all of the sources available to itself (e.g. online forums w must be registered at). Moreover, *uniformity* and *absoluteness* express kinds of information bubbles, since if v belongs to a source available to w, then w and v have access to the same sources of information.

cpr	$\dfrac{A \to B}{B \preccurlyeq A}$	tr	$(A \preccurlyeq B) \wedge (B \preccurlyeq C) \to (A \preccurlyeq C)$
		or	$(A \preccurlyeq B) \wedge (A \preccurlyeq C) \to (A \preccurlyeq B \vee C)$
n	$\neg(\bot \preccurlyeq \top)$	u-	$\neg(\bot \preccurlyeq A) \to (\bot \preccurlyeq (\bot \preccurlyeq A))$
t	$(\bot \preccurlyeq A) \to \neg A$	u	$(\bot \preccurlyeq A) \to (\bot \preccurlyeq \neg(\bot \preccurlyeq A))$
w	$A \to (A \preccurlyeq \top)$	a-	$(A \preccurlyeq B) \to (\bot \preccurlyeq \neg(A \preccurlyeq B))$
c	$(A \preccurlyeq \top) \to A$	a	$\neg(A \preccurlyeq B) \to (\bot \preccurlyeq (A \preccurlyeq B))$

Fig. 2. Axioms and rules for CPN logics.

3 Axiom systems for CPN logics

In this section we present the logics of Comparative Plausibility in Neighbourhood models (CPN logics in the following) corresponding to the classes of neighbourhood models introduced in Sec. 2. We propose axiom systems for CPN logics and show their soundness and completeness. Then, we compare CPN logics with Lewis' logics of comparative plausibility in sphere models.

Definition 3.1 CPN logics are defined by extending classical propositional logic (CPL) formulated in $\mathscr{L}$ with the rules and axioms for $\preccurlyeq$ from Fig. 2:

$$\mathsf{N}_{\preccurlyeq} := \mathsf{CPL} \cup \{\mathsf{tr}, \mathsf{or}, \mathsf{cpr}\} \quad \mathsf{NN}_{\preccurlyeq} := \mathsf{N}_{\preccurlyeq} \cup \{\mathsf{n}\} \quad \mathsf{NT}_{\preccurlyeq} := \mathsf{N}_{\preccurlyeq} \cup \{\mathsf{t}\}$$
$$\mathsf{NW}_{\preccurlyeq} := \mathsf{NT}_{\preccurlyeq} \cup \{\mathsf{w}\} \quad \mathsf{NC}_{\preccurlyeq} := \mathsf{NW}_{\preccurlyeq} \cup \{\mathsf{c}\}$$

Moreover, for $\mathsf{L} \in \{\mathsf{N}_{\preccurlyeq}, \mathsf{NN}_{\preccurlyeq}, \mathsf{NT}_{\preccurlyeq}, \mathsf{NW}_{\preccurlyeq}, \mathsf{NC}_{\preccurlyeq}\}$, we define $\mathsf{LU}_{\preccurlyeq} := \mathsf{L} \cup \{\mathsf{u\text{-}}, \mathsf{u}\}$ and $\mathsf{LA}_{\preccurlyeq} := \mathsf{L} \cup \{\mathsf{a\text{-}}, \mathsf{a}\}$.

The logics generated by this definition are displayed in the lower layer of the lattice of systems in Fig. 3. The axioms of $\mathsf{N}_{\preccurlyeq}$ are those defined by Lewis in [16, Ch.6], while axioms for extensions of $\mathsf{N}_{\preccurlyeq}$ are reformulations of Lewis' axioms in terms of $\preccurlyeq$ [15,8]. In the following, for every logic L from Def. 3.1, we denote L^* any extension of L. As usual, we say that a formula A is *derivable* in $\mathsf{N}^*_{\preccurlyeq}$, written $\vdash_{\mathsf{N}^*_{\preccurlyeq}} A$, if there is a finite sequence of formulas ending with A where every formula is an axiom of $\mathsf{N}^*_{\preccurlyeq}$, or is obtained from previous formulas by *modus ponens* or cpr. Moreover, we say that A is *deducible* in $\mathsf{N}^*_{\preccurlyeq}$ from a set of formulas Φ if there is a finite set $\{B_1, ..., B_n\} \subseteq \Phi$ such that $\vdash_{\mathsf{N}^*_{\preccurlyeq}} B_1 \wedge ... \wedge B_n \to A$.

For each logic $\mathsf{N}^*_{\preccurlyeq}$, we call $\mathsf{N}^*_{\preccurlyeq}$*-model* any neighbourhood model satisfying the conditions corresponding to the letters appearing beside N in the name of the logic. Thus, $\mathsf{N}_{\preccurlyeq}$-models denotes the class of all neighbourhood models, $\mathsf{NN}_{\preccurlyeq}$-models the class of all models satisfying normality, and so on.

We show that each logic $\mathsf{N}^*_{\preccurlyeq}$ is characterised by the class of all $\mathsf{N}^*_{\preccurlyeq}$-models. We first prove that the logics are sound with respect to the corresponding classes of models.

Theorem 3.2 (Soundness) *For every formula A, if A is derivable in $\mathsf{N}^*_{\preccurlyeq}$, then A is valid in all $\mathsf{N}^*_{\preccurlyeq}$-models.*

Proof. We show that the modal axioms and rules of $\mathsf{N}^*_{\preccurlyeq}$ are valid (resp. sound) in the corresponding models, considering some relevant examples. (cpr) Assume $\mathcal{M} \models A \to B$, and $\alpha \in N(w)$, $\alpha \Vdash^{\exists} A$. Then $\alpha \Vdash^{\exists} B$, therefore $w \Vdash B \preccurlyeq A$. (tr) If $w \Vdash (A \preccurlyeq B) \wedge (B \preccurlyeq C)$, then for every $\alpha \in N(w)$, $\alpha \Vdash^{\exists} B$ implies

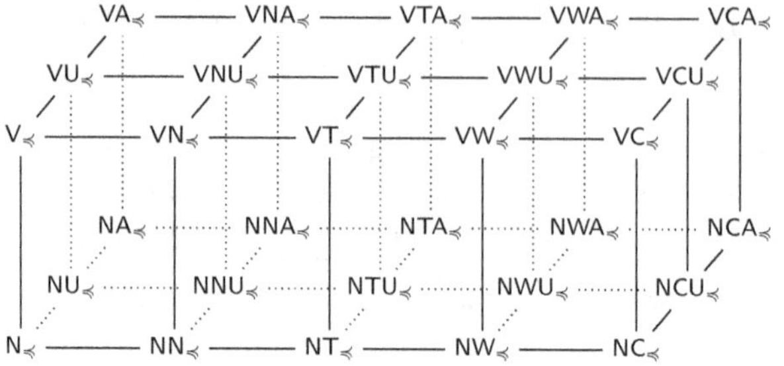

Fig. 3. The family of CPN logics. The system in the upper layer are Lewis' logics.

$\alpha \Vdash^{\exists} A$, and $\alpha \Vdash^{\exists} C$ implies $\alpha \Vdash^{\exists} B$, then $\alpha \Vdash^{\exists} C$ implies $\alpha \Vdash^{\exists} A$, therefore $w \Vdash A \preccurlyeq C$. (or) If $w \Vdash (A \preccurlyeq B) \wedge (A \preccurlyeq C)$, then for every $\alpha \in N(w)$, $\alpha \Vdash^{\exists} B$ implies $\alpha \Vdash^{\exists} A$, and $\alpha \Vdash^{\exists} C$ implies $\alpha \Vdash^{\exists} A$, then $\alpha \Vdash^{\exists} B \vee C$ implies $\alpha \Vdash^{\exists} A$, therefore $w \Vdash A \preccurlyeq B \vee C$. (n) By normality, for all $w \in W$ there is $\alpha \in N(w)$. Moreover, $\alpha \nVdash^{\exists} \bot$ and since $\alpha \neq \emptyset$, $\alpha \Vdash^{\exists} \top$, then $w \Vdash \neg(\bot \preccurlyeq \top)$. (t) Assume $w \Vdash \bot \preccurlyeq A$. Then for all $\alpha \in N(w)$, $\alpha \nVdash^{\exists} A$. Moreover by total reflexivity, there is $\beta \in N(w)$ such that $w \in \beta$. Then $w \nVdash A$, thus $w \Vdash \neg A$. (w) Assume $w \Vdash A$. By weak centering, $N(w) \neq \emptyset$, and $w \in \alpha$ for all $\alpha \in N(w)$. Then for all $\alpha \in N(w)$, $\alpha \Vdash^{\exists} A$, thus $w \Vdash \top \preccurlyeq A$. (c) Assume $w \Vdash A \preccurlyeq \top$. Then for all $\alpha \in N(w)$, $\alpha \Vdash^{\exists} A$. Moreover by centering, $\{w\} \in N(w)$, therefore $w \Vdash A$. □

Using a canonical model construction inspired from [16], we shall now prove that $\mathsf{N}^*_{\preccurlyeq}$ is complete with respect to the class of all $\mathsf{N}^*_{\preccurlyeq}$-models. As usual, for any logic L and set of formulas Φ, we say that Φ is L-*consistent* if $\Phi \nvdash_{\mathsf{L}} \bot$, and that it is L-*maximal consistent* (maxcons) if it is consistent and for every $B \notin \Phi$, $\Phi \cup \{B\} \vdash_{\mathsf{L}} \bot$. The proof of the following Lemma is standard.

Lemma 3.3 *(a) If Φ is a L-consistent set of formulas, then there is a L-maximal consistent set Ψ such that $\Phi \subseteq \Psi$. (b) If Φ is a L-maximal consistent set, then for all $A, B \in \mathscr{L}$, (i) if $\Phi \vdash_{\mathsf{L}} A$, then $A \in \Phi$; (ii) $A \in \Phi$ if and only if $\neg A \notin \Phi$; (iii) if $A \vee B \in \Phi$, then $A \in \Phi$ or $B \in \Phi$.*

We consider the following notion of *cut around*,[4] and prove the subsequent lemma that will be needed in the following.

Definition 3.4 Let Φ be a maximal consistent set of formulas, and Σ be a set of formulas. We say that Σ is a *cut around* Φ if for all finite sets $\{B_1, ..., B_n\} \subseteq \Sigma$ and all $A \notin \Sigma$, $(B_1 \vee ... \vee B_n) \preccurlyeq A \notin \Phi$. Moreover, let $co\Sigma = \{\Psi \text{ maxcons} \mid \Psi \cap \Sigma = \emptyset\}$.

Lemma 3.5 *If Σ is a cut around Φ for some maximal consistent set Φ, then for every formula A, $A \in \Sigma$ if and only if for all $\Psi \in co\Sigma$, $A \notin \Psi$.*

Proof. If $A \in \Sigma$ and $\Psi \in co\Sigma$, then $\Psi \cap \Sigma = \emptyset$, thus $A \notin \Psi$. If instead $A \notin \Sigma$, then suppose by contradiction that $\{\neg B \mid B \in \Sigma\} \cup \{A\} \vdash_{\mathsf{L}} \bot$. Then there

[4] This terminology comes from Lewis [16], but our definition is different from Lewis' one.

are formulas $B_1, ..., B_n \in \Sigma$ such that $\vdash_{\mathsf{L}} \neg B_1 \wedge ... \wedge \neg B_n \to \neg A$, thus $\vdash_{\mathsf{L}} A \to B_1 \vee ... \vee B_n$, therefore $\vdash_{\mathsf{L}} (B_1 \vee ... \vee B_n) \preccurlyeq A$. By closure under derivation, $(B_1 \vee ... \vee B_n) \preccurlyeq A \in \Phi$, but by definition of cut around, $(B_1 \vee ... \vee B_n) \preccurlyeq A \notin \Phi$. We conclude that $\{\neg B \mid B \in \Sigma\} \cup \{A\} \nvdash_{\mathsf{L}} \bot$. Then by Lemma 3.3, there is $\Psi \in W_{\mathsf{L}}$ such that $\{\neg B \mid B \in \Sigma\} \cup \{A\} \subseteq \Psi$, therefore $\Psi \in co\Sigma$ and $A \in \Psi$. □

From Lemma 3.5 it immediately follows that $\bot \in \Sigma$ for all Σ cut around Φ. We now define the canonical model.

Definition 3.6 For every CPN logic L, the *canonical model* for L is the tuple $M_{\mathsf{L}} = \langle W_{\mathsf{L}}, N_{\mathsf{L}}, V_{\mathsf{L}} \rangle$, where:

- W_{L} is the class of all L-maximal consistent sets;
- for all $\Phi \in W_{\mathsf{L}}$, $N_{\mathsf{L}}(\Phi) = \{co\Sigma \mid \Sigma \text{ cut around } \Phi \text{ and } co\Sigma \neq \emptyset\}$;
- $V_{\mathsf{L}}(p) = \{\Phi \in W_{\mathsf{L}} \mid p \in \Phi\}$.

Lemma 3.7 (Truth lemma) *If* L *is a CPN logic and* M_{L} *is the canonical model for* L*, then for all* $A \in \mathscr{L}$ *and all* $\Phi \in W_{\mathsf{L}}$, $\Phi \Vdash A$ *if and only if* $A \in \Phi$.

Proof. By induction on the construction of A. For atomic and propositional formulas the proof is standard. We consider the case $A = B \preccurlyeq C$.

($\Rightarrow$) Suppose $\Phi \Vdash B \preccurlyeq C$. Then for all $\alpha \in N_{\mathsf{L}}(\Phi)$, $\alpha \Vdash^{\exists} C$ implies $\alpha \Vdash^{\exists} B$. By definition, this means that for every Σ cut around Φ, $co\Sigma \Vdash^{\exists} C$ implies $co\Sigma \Vdash^{\exists} B$. Let $\Phi_{B\preccurlyeq} = \{D \mid B \preccurlyeq D \in \Phi\}$. Then since $B \preccurlyeq B \in \Phi$, $B \in \Phi_{B\preccurlyeq}$. Moreover, $\Phi_{B\preccurlyeq}$ is a cut around Φ: if $E_1, ..., E_n \in \Phi_{B\preccurlyeq}$ and $F \notin \Phi_{B\preccurlyeq}$, then $B \preccurlyeq E_1, ..., B \preccurlyeq E_n \in \Phi$, and $B \preccurlyeq F \notin \Phi$. Thus by axiom or and closure under derivation, $B \preccurlyeq (E_1 \vee ... \vee E_n) \in \Phi$, whence by tr, $(E_1 \vee ... \vee E_n) \preccurlyeq F \notin \Phi$. Now suppose by contradiction that $\{\neg D \mid D \in \Phi_{B\preccurlyeq}\} \cup \{C\} \nvdash_{\mathsf{L}} \bot$. Then by Lemma 3.3 there is $\Psi \in W_{\mathsf{L}}$ such that $\{\neg D \mid D \in \Phi_{B\preccurlyeq}\} \cup \{C\} \subseteq \Psi$. We then have $C \in \Psi$ and $\Phi_{B\preccurlyeq} \cap \Psi = \emptyset$, which implies $\Psi \in co\Phi_{B\preccurlyeq}$. By i.h., $\Psi \Vdash C$, thus $co\Phi_{B\preccurlyeq} \Vdash^{\exists} C$, which implies $co\Phi_{B\preccurlyeq} \Vdash^{\exists} B$. This means that there is $\Omega \in co\Phi_{B\preccurlyeq}$ such that $\Omega \Vdash B$, therefore by i.h., $B \in \Omega$. Furthermore, by definition we have $\Omega \cap \Phi_{B\preccurlyeq} = \emptyset$, then $B \notin \Phi_{B\preccurlyeq}$, which contradicts $B \in \Phi_{B\preccurlyeq}$. Therefore $\{\neg D \mid D \in \Phi_{B\preccurlyeq}\} \cup \{C\} \vdash_{\mathsf{L}} \bot$. Then there are $D_1, ..., D_n \in \Phi_{B\preccurlyeq}$ such that $\vdash_{\mathsf{L}} \neg D_1 \wedge ... \wedge \neg D_n \to \neg C$, that is $\vdash_{\mathsf{L}} C \to D_1 \vee ... \vee D_n$, whence by cpr, $(D_1 \vee ... \vee D_n) \preccurlyeq C \in \Phi$. Moreover by definition of $\Phi_{B\preccurlyeq}$, $B \preccurlyeq D_1, ..., B \preccurlyeq D_n \in \Phi$. Then by or, $B \preccurlyeq (D_1 \vee ... \vee D_n) \in \Phi$, finally by tr, $B \preccurlyeq C \in \Phi$.

($\Leftarrow$) Suppose $\Phi \nVdash B \preccurlyeq C$. Then there is $\alpha \in N_{\mathsf{L}}(\Phi)$ such that $\alpha \Vdash^{\exists} C$ and $\alpha \nVdash^{\exists} B$, i.e., there is a Σ cut around Φ with $co\Sigma \Vdash^{\exists} C$ and $co\Sigma \nVdash^{\exists} B$. By i.h. there is $\Psi \in co\Sigma$ such that $C \in \Psi$, and for all $\Omega \in co\Sigma$, $B \notin \Omega$. Then $C \notin \Sigma$, and from Lemma 3.5 it follows that $B \in \Sigma$. Then by definition $B \preccurlyeq C \notin \Phi$. □

Lemma 3.8 (Model lemma) *The canonical model for* $\mathsf{N}^*_{\preccurlyeq}$ *is a* $\mathsf{N}^*_{\preccurlyeq}$*-model.*

Proof. Non-emptiness is immediate. We consider the other conditions.

($\mathsf{NN}^*_{\preccurlyeq}$) For every $\Phi \in W_{\mathsf{NN}_{\preccurlyeq}}$, $\neg(\bot \preccurlyeq \top) \in \Phi$, then by Lemma 3.7, $\Phi \Vdash \neg(\bot \preccurlyeq \top)$, thus there is $\alpha \in N_{\mathsf{NN}_{\preccurlyeq}}(\Phi)$ such that $\alpha \Vdash^{\exists} \top$ and $\alpha \nVdash^{\exists} \bot$.

($\mathsf{NT}^*_{\preccurlyeq}$) For any $\Phi \in W_{\mathsf{NT}_{\preccurlyeq}}$, let $\Sigma = \{A \mid \bot \preccurlyeq A \in \Phi\}$. Σ is a cut around Φ, since for all $B_1, ..., B_n \in \Sigma$ and $C \notin \Sigma$, $\bot \preccurlyeq B_1, ..., \bot \preccurlyeq B_n \in \Phi$ and

$\bot \preccurlyeq C \notin \Phi$, then by or, $\bot \preccurlyeq B_1 \vee ... \vee B_n \in \Phi$, and by tr, $B_1 \vee ... \vee B_n \preccurlyeq C \notin \Phi$. Moreover, for any $A \in \mathscr{L}$, if $A \in \Sigma$, then $\bot \preccurlyeq A \in \Phi$, thus by t, $\neg A \in \Phi$, whence $A \notin \Phi$. Thus $\Sigma \cap \Phi = \emptyset$, which implies $\Phi \in co\Sigma$, and since $co\Sigma \neq \emptyset$, $co\Sigma \in N_{\mathsf{NT}_\preccurlyeq}(\Phi)$.

($\mathsf{NW}^*_\preccurlyeq$) Since $\mathsf{NW}_\preccurlyeq \vdash$ n, by item ($\mathsf{NN}_\preccurlyeq$), $N_{\mathsf{NW}_\preccurlyeq}(\Phi) \neq \emptyset$ for all $\Phi \in W_{\mathsf{NW}_\preccurlyeq}$. Moreover let $co\Sigma \in N_{\mathsf{NW}_\preccurlyeq}(\Phi)$ for a Σ cut around Φ. Then there is $\Psi \in co\Sigma$. Since $\top \in \Psi$, by Lemma 3.5, $\top \notin \Sigma$, then for all $A \in \Sigma$, $A \preccurlyeq \top \notin \Phi$, thus by axiom w, $A \notin \Phi$. This means $\Sigma \cap \Phi = \emptyset$, therefore $\Phi \in co\Sigma$.

($\mathsf{NC}^*_\preccurlyeq$) Since axiom w belongs to $\mathsf{NC}_\preccurlyeq$, by item ($\mathsf{NW}_\preccurlyeq$), for all $\Phi \in W_{\mathsf{NC}_\preccurlyeq}$ and all $\alpha \in N_{\mathsf{NC}_\preccurlyeq}(\Phi)$, $\Phi \in \alpha$. Moreover, let $\Sigma = \{A \mid A \preccurlyeq \top \notin \Phi\}$. Then Σ is a cut around Φ: if $B_1, ..., B_n \in \Sigma$ and $C \notin \Sigma$, then $B_1 \preccurlyeq \top, ..., B_n \preccurlyeq \top \notin \Phi$ and $C \preccurlyeq \top \in \Phi$. By axiom w, $B_1 \notin \Phi$, ..., $B_n \notin \Phi$, thus $B_1 \vee ... \vee B_n \notin \Phi$, then by axiom c, $(B_1 \vee ... \vee B_n) \preccurlyeq \top \notin \Phi$, therefore by tr, $(B_1 \vee ... \vee B_n) \preccurlyeq C \notin \Phi$. Moreover, since $\top \preccurlyeq \top \in \Phi$, $\top \notin \Sigma$, by Lemma 3.5 there is $\Psi \in co\Sigma$, thus $co\Sigma \in N_{\mathsf{NC}_\preccurlyeq}(\Phi)$. Suppose $\Psi \neq \Phi$. Then there is $A \in \mathscr{L}$ such that $A \in \Psi$ and $A \notin \Phi$. Since $\Psi \in co\Sigma$, $A \notin \Sigma$, then $A \preccurlyeq \top \in \Phi$, thus by c, $A \in \Phi$, it follows $\Psi = \Phi$, therefore $co\Sigma = \{\Phi\}$.

($\mathsf{NU}^*_\preccurlyeq$) Suppose $\Psi \in \bigcup N_{\mathsf{NU}^*_\preccurlyeq}(\Phi)$. Then $\Psi \in co\Sigma$ for some Σ cut around Φ. We show that for all $A \in \mathscr{L}$, $\bot \preccurlyeq A \in \Phi$ iff $\bot \preccurlyeq A \in \Psi$. If $\bot \preccurlyeq A \in \Phi$, then by axiom u, $\bot \preccurlyeq \neg(\bot \preccurlyeq A) \in \Phi$, then by Def. 3.4, $\bot \notin \Sigma$ or $\neg(\bot \preccurlyeq A) \in \Sigma$. Since $\bot \in \Sigma$, we have $\neg(\bot \preccurlyeq A) \in \Sigma$, thus $\neg(\bot \preccurlyeq A) \notin \Psi$, then $\bot \preccurlyeq A \in \Psi$. If $\bot \preccurlyeq A \in \Psi$, then $\bot \preccurlyeq A \notin \Sigma$, thus $\bot \preccurlyeq (\bot \preccurlyeq A) \notin \Phi$, then by axiom u-, $\neg(\bot \preccurlyeq A) \notin \Phi$, therefore $\bot \preccurlyeq A \in \Phi$. Let $\Pi = \{A \mid \bot \preccurlyeq A \in \Phi\} = \{A \mid \bot \preccurlyeq A \in \Psi\}$. Then Π is a cut around Φ and Ψ: If $B_1, ..., B_n \in \Pi$ and $C \notin \Pi$, then $\bot \preccurlyeq B_1, ..., \bot \preccurlyeq B_n \in \Phi$ and $\bot \preccurlyeq C \notin \Phi$, thus by or, $\bot \preccurlyeq B_1 \vee ... \vee B_n \in \Phi$, then by tr, $B_1 \vee ... \vee B_n \preccurlyeq C \notin \Phi$. Moreover for all Ω cut around Φ or Ψ, $\Pi \subseteq \Omega$, therefore $co\Omega \subseteq co\Pi$. Then in particular $co\Sigma \subseteq co\Pi$, which implies $\Psi \in co\Pi$, thus $co\Pi \neq \emptyset$. It follows $co\Pi \in N_{\mathsf{NU}^*_\preccurlyeq}(\Phi)$ and $co\Pi \in N_{\mathsf{NU}^*_\preccurlyeq}(\Psi)$, therefore $co\Pi = \bigcup N_{\mathsf{NU}^*_\preccurlyeq}(\Phi) = \bigcup N_{\mathsf{NU}^*_\preccurlyeq}(\Psi)$.

($\mathsf{NA}^*_\preccurlyeq$) Suppose $\Psi \in \bigcup N_{\mathsf{NA}^*_\preccurlyeq}(\Phi)$. Then $\Psi \in co\Sigma$ for some Σ cut around Φ. We show that for all $A, B \in \mathscr{L}$, $A \preccurlyeq B \in \Phi$ iff $A \preccurlyeq B \in \Psi$. If $A \preccurlyeq B \in \Phi$, then by axiom a-, $\bot \preccurlyeq \neg(A \preccurlyeq B) \in \Phi$, then by Def. 3.4, $\bot \notin \Sigma$ or $\neg(A \preccurlyeq B) \in \Sigma$. Since $\bot \in \Sigma$, we have $\neg(A \preccurlyeq B) \in \Sigma$, thus $\neg(A \preccurlyeq B) \notin \Psi$, then $A \preccurlyeq B \in \Psi$. If $A \preccurlyeq B \in \Psi$, then $A \preccurlyeq B \notin \Sigma$, thus $\bot \preccurlyeq (A \preccurlyeq B) \notin \Phi$, then by axiom a, $\neg(A \preccurlyeq B) \notin \Phi$, therefore $A \preccurlyeq B \in \Phi$. It follows that for every Π, Π is a cut around Φ iff Π is a cut around Ψ, thus $co\Pi \in N_{\mathsf{NA}^*_\preccurlyeq}(\Phi)$ iff $co\Pi \in N_{\mathsf{NA}^*_\preccurlyeq}(\Psi)$, therefore $N_{\mathsf{NA}^*_\preccurlyeq}(\Phi) = N_{\mathsf{NA}^*_\preccurlyeq}(\Psi)$. □

As a consequence of the previous lemmas we obtain the following result.

Theorem 3.9 (Completeness) *For every formula A, if A is valid in all $\mathsf{N}^*_\preccurlyeq$-models, then A is derivable in $\mathsf{N}^*_\preccurlyeq$.*

Proof. Suppose $\not\vdash_{\mathsf{N}^*_\preccurlyeq} A$. Then $\{\neg A\}$ is $\mathsf{N}^*_\preccurlyeq$-consistent, thus by Lemma 3.3 there is a $\mathsf{N}^*_\preccurlyeq$-maxcons set Φ such that $\neg A \in \Phi$. By definition, $\Phi \in W_{\mathsf{N}^*_\preccurlyeq}$, and by Lemma 3.7, $\mathcal{M}_{\mathsf{N}^*_\preccurlyeq}, \Phi \not\Vdash A$, moreover by Lemma 3.8, $\mathcal{M}_{\mathsf{N}^*_\preccurlyeq}$ is a $\mathsf{N}^*_\preccurlyeq$-model. □

Let us now turn to the relationship between $\mathsf{N}^*_{\preccurlyeq}$ and Lewis' logics of comparative plausibility over sphere models. Lewis [16] provides two equivalent axiomatisations of the minimal logic $\mathsf{V}_{\preccurlyeq}$, one of the two being $\mathsf{CPL} \cup \{\mathsf{cpr}, \mathsf{cpa}, \mathsf{tr}, \mathsf{co}\}$, where co is the *connection axiom* $(A \preccurlyeq B) \vee (B \preccurlyeq A)$, and cpa is $(A \preccurlyeq A \vee B) \vee (B \preccurlyeq A \vee B)$. We show that a further equivalent axiomatisation of $\mathsf{V}_{\preccurlyeq}$ can be given by extending our minimal logic $\mathsf{N}_{\preccurlyeq}$ with the axiom co:

Proposition 3.10 *For all* $A \in \mathscr{L}$, $\vdash_{\mathsf{V}_{\preccurlyeq}} A$ *if and only if* $\vdash_{\mathsf{N}_{\preccurlyeq} \cup \{\mathsf{co}\}} A$.

Proof. Since $\mathsf{N}_{\preccurlyeq} \cup \{\mathsf{co}\}$ and $\mathsf{V}_{\preccurlyeq} = \mathsf{CPL} \cup \{\mathsf{cpr}, \mathsf{cpa}, \mathsf{tr}, \mathsf{co}\}$ differ only with respect to cpa and or, it suffices to show that (i) $\vdash_{\mathsf{N}_{\preccurlyeq} \cup \{\mathsf{co}\}}$ cpa and (ii) $\vdash_{\mathsf{V}_{\preccurlyeq}}$ or. (i) From $(A \preccurlyeq B) \vee (B \preccurlyeq A)$, by cpr we have $((A \preccurlyeq A) \wedge (A \preccurlyeq B)) \vee ((B \preccurlyeq A) \wedge (B \preccurlyeq B))$, then by or, $(A \preccurlyeq A \vee B) \vee (B \preccurlyeq A \vee B)$. (ii) From $(B \preccurlyeq B \vee C) \vee (C \preccurlyeq B \vee C)$, by tr we have $(A \preccurlyeq B) \wedge (A \preccurlyeq C) \rightarrow (A \preccurlyeq B \vee C) \vee (A \preccurlyeq B \vee C)$, thus $(A \preccurlyeq B) \wedge (A \preccurlyeq C) \rightarrow (A \preccurlyeq B \vee C)$. □

Note also that the extensions of $\mathsf{N}_{\preccurlyeq}$ are defined by the same axioms characterising the extensions of $\mathsf{V}_{\preccurlyeq}$. It follows that each Lewis' logic can be obtained from the corresponding CPN logic by adding the connection axiom co. The relations among these systems are displayed in Fig. 3.

4 Multi-premisses sequent calculi for CPN logics

In this section we present Gentzen-style sequent calculi for the CPN logics $\mathsf{N}_{\preccurlyeq}, \mathsf{NN}_{\preccurlyeq}, \mathsf{NT}_{\preccurlyeq}, \mathsf{NW}_{\preccurlyeq}, \mathsf{NC}_{\preccurlyeq}, \mathsf{NA}_{\preccurlyeq}$, and $\mathsf{NNA}_{\preccurlyeq}$. From now on, let $\mathsf{N}^*_{\preccurlyeq}$ denote any of these systems. For each logic we introduce a calculus $\mathsf{G.N}^*_{\preccurlyeq}$ defined on the basis of the sequent systems for Lewis' logics by Lellmann and Pattinson [14,15]. In these calculi, the rules have up to $m + n$ premisses, where m (resp. n) is the number of $\preccurlyeq$-formulas occurring in the antecedent (resp. consequent) of the conclusion. Calculi for $\mathsf{N}^*_{\preccurlyeq}$ can be provided by restricting the calculi in [14,15] to rules with at most one $\preccurlyeq$-formula in the consequent ($n = 1$), thus obtaining simpler calculi, where each rule introduces at most $m + 1$ premisses.

As usual, we call *sequent* any pair $\Gamma \Rightarrow \Delta$, where Γ and Δ are finite, possibly empty multisets of formulas of $\mathscr{L}$. $\Gamma \Rightarrow \Delta$ is interpreted as the formula $\bigwedge \Gamma \rightarrow \bigvee \Delta$.

The rules of the calculi $\mathsf{G.N}^*_{\preccurlyeq}$ can be found in Fig. 4. Each modal rule simultaneously analyses a number (at least one) of $\preccurlyeq$-formulas appearing in a sequent. The *principal formulas* of each rule R_n are the n or $n+1$ $\preccurlyeq$-formulas in the conclusion which get analysed in the premiss. In some rules the principal $\preccurlyeq$-formulas are copied into the premisses in order to ensure admissibility of contraction. We denote derivability in $\mathsf{G.N}^*_{\preccurlyeq}$ as $\mathsf{G.N}^*_{\preccurlyeq} \vdash \Gamma \Rightarrow \Delta$.

Theorem 4.1 (Soundness) *If* $\mathsf{G.N}^*_{\preccurlyeq} \vdash \Gamma \Rightarrow \Delta$ *then* $\mathsf{N}^*_{\preccurlyeq} \vdash \bigwedge \Gamma \rightarrow \bigvee \Delta$.

Proof. We show that for every rule R of $\mathsf{G.N}^*_{\preccurlyeq}$ with premisses $\Gamma_1 \Rightarrow \Delta_1$, ..., $\Gamma_n \Rightarrow \Delta_n$ and conclusion $\Gamma \Rightarrow \Delta$, the corresponding Hilbert-style rule with premisses $\bigwedge \Gamma_1 \rightarrow \bigvee \Delta_1$, ..., $\bigwedge \Gamma_n \rightarrow \bigvee \Delta_n$ and conclusion $\bigwedge \Gamma \rightarrow \bigvee \Delta$ is derivable in $\mathsf{N}^*_{\preccurlyeq}$. The propositional cases are standard. We use $\mathbf{D}_1^n$ as a shorthand for $D_1, \ldots, D_n$.

$$\mathsf{init}\ \frac{}{\Gamma, p \Rightarrow p, \Delta} \qquad \bot_{\mathsf{L}}\ \frac{}{\Gamma, \bot \Rightarrow \Delta} \qquad \to_{\mathsf{L}}\ \frac{\Gamma \Rightarrow A, \Delta \quad \Gamma, B \Rightarrow \Delta}{\Gamma, A \to B \Rightarrow \Delta} \qquad \to_{\mathsf{R}}\ \frac{\Gamma, A \Rightarrow B, \Delta}{\Gamma \Rightarrow A \to B, \Delta}$$

$$\mathsf{CP}_n\ \frac{\{C_k \Rightarrow A, D_1, \ldots, D_{k-1}\}_{1 \leq k \leq n} \quad B \Rightarrow A, D_1, \ldots, D_n}{\Gamma, C_1 \preccurlyeq D_1, \ldots, C_n \preccurlyeq D_n \Rightarrow A \preccurlyeq B, \Delta}$$

$$\mathsf{N}_n\ \frac{\{C_k \Rightarrow D_1, \ldots, D_{k-1}\}_{1 \leq k \leq n} \quad \Rightarrow D_1, \ldots, D_n}{\Gamma, C_1 \preccurlyeq D_1, \ldots, C_n \preccurlyeq D_n \Rightarrow \Delta}$$

$$\mathsf{T}_n\ \frac{\{C_k \Rightarrow D_1, \ldots, D_{k-1}\}_{1 \leq k \leq n} \quad \Gamma, C_1 \preccurlyeq D_1, \ldots, C_n \preccurlyeq D_n \Rightarrow D_1, \ldots, D_n, \Delta}{\Gamma, C_1 \preccurlyeq D_1, \ldots, C_n \preccurlyeq D_n \Rightarrow \Delta}$$

$$\mathsf{W}_n\ \frac{\{C_k \Rightarrow A, D_1, \ldots, D_{k-1}\}_{1 \leq k \leq n} \quad \Gamma, C_1 \preccurlyeq D_1, \ldots, C_n \preccurlyeq D_n \Rightarrow A, A \preccurlyeq B, D_1, \ldots, D_n, \Delta}{\Gamma, C_1 \preccurlyeq D_1, \ldots, C_n \preccurlyeq D_n \Rightarrow A \preccurlyeq B, \Delta}$$

$$\mathsf{W}_0\ \frac{\Gamma \Rightarrow A \preccurlyeq B, A, \Delta}{\Gamma \Rightarrow A \preccurlyeq B, \Delta} \qquad \mathsf{C}_0\ \frac{\Gamma, A \preccurlyeq B, A \Rightarrow \Delta \quad \Gamma, A \preccurlyeq B \Rightarrow B, \Delta}{\Gamma, A \preccurlyeq B \Rightarrow \Delta}$$

$$\mathsf{A}_n\ \frac{\{\Gamma^{\preccurlyeq}, \Sigma^{\preccurlyeq}, C_k \Rightarrow A \preccurlyeq B, A, D_1, \ldots, D_{k-1}, \Delta^{\preccurlyeq}\}_{1 \leq k \leq n} \quad \Gamma^{\preccurlyeq}, \Sigma^{\preccurlyeq}, B \Rightarrow A \preccurlyeq B, A, D_1, \ldots, D_n, \Delta^{\preccurlyeq}}{\Gamma, C_1 \preccurlyeq D_1, \ldots, C_n \preccurlyeq D_n \Rightarrow A \preccurlyeq B, \Delta}$$

$$\mathsf{N^A}_n\ \frac{\{\Gamma^{\preccurlyeq}, \Sigma^{\preccurlyeq}, C_k \Rightarrow D_1, \ldots, D_{k-1}, \Delta^{\preccurlyeq}\}_{1 \leq k \leq n} \quad \Gamma^{\preccurlyeq}, \Sigma^{\preccurlyeq} \Rightarrow D_1, \ldots, D_n, \Delta^{\preccurlyeq}}{\Gamma, C_1 \preccurlyeq D_1, \ldots, C_n \preccurlyeq D_n \Rightarrow \Delta}$$

where $\Sigma^{\preccurlyeq} = C_1 \preccurlyeq D_1, ..., C_n \preccurlyeq D_n$, $\Gamma^{\preccurlyeq} = \{C' \preccurlyeq D' \mid C' \preccurlyeq D' \in \Gamma\}$, $\Delta^{\preccurlyeq} = \{C' \preccurlyeq D' \mid C' \preccurlyeq D' \in \Delta\}$

$\mathsf{G.N}_{\preccurlyeq} = \{\mathsf{init}, \bot_{\mathsf{L}}, \to_{\mathsf{L}}, \to_{\mathsf{R}}\} \cup \{\mathsf{CP}_n \mid n \geq 0\}$

$\mathsf{G.NN}_{\preccurlyeq} = \mathsf{G.N}_{\preccurlyeq} \cup \{\mathsf{N}_n \mid n \geq 1\}$ $\quad \mathsf{G.NT}_{\preccurlyeq} = \mathsf{G.N}_{\preccurlyeq} \cup \{\mathsf{T}_n \mid n \geq 1\}$

$\mathsf{G.NW}_{\preccurlyeq} = \mathsf{G.N}_{\preccurlyeq} \cup \{\mathsf{W}_n \mid n \geq 0\} \cup \{\mathsf{T}_n \mid n \geq 1\}$ $\quad \mathsf{G.NC}_{\preccurlyeq} = \mathsf{G.N}_{\preccurlyeq} \cup \{\mathsf{W}_0, \mathsf{C}_0\}$

$\mathsf{G.NA}_{\preccurlyeq} = \{\mathsf{init}, \bot_{\mathsf{L}}, \to_{\mathsf{L}}, \to_{\mathsf{R}}\} \cup \{\mathsf{A}_n \mid n \geq 0\}$ $\quad \mathsf{G.NNA}_{\preccurlyeq} = \mathsf{G.NA}_{\preccurlyeq} \cup \{\mathsf{N^A}_n \mid n \geq 1\}$

Fig. 4. Gentzen-style calculi for CPN logics.

(CP_n) Suppose $\vdash C_1 \to A$, $\vdash C_2 \to A \vee D_1$, ..., $\vdash C_n \to A \vee \bigvee \mathbf{D}_1^{n-1}$ and $\vdash B \to A \vee \bigvee \mathbf{D}_1^n$. Then by cpr, we have $\vdash A \preccurlyeq C_1$, $\vdash A \vee D_1 \preccurlyeq C_2$, ..., $\vdash A \vee \bigvee \mathbf{D}_1^{n-1} \preccurlyeq C_n$ and $\vdash A \vee \bigvee \mathbf{D}_1^n \preccurlyeq B$. If $n = 0$, the conclusion immediately follows. From $\vdash C_1 \preccurlyeq D_1$ it follows by tr that $A \preccurlyeq D_1$. Thus, $\vdash (C_1 \preccurlyeq D_1) \to (A \preccurlyeq D_1)$. Since $\vdash A \preccurlyeq A$, by or, we have that $\vdash (C_1 \preccurlyeq D_1) \to (A \preccurlyeq A \vee D_1)$. By tr, $\vdash (C_1 \preccurlyeq D_1) \to (A \preccurlyeq C_2)$. From $\vdash C_2 \preccurlyeq D_2$ it follows by tr that $A \preccurlyeq C_2$. Thus, $\vdash (C_1 \preccurlyeq D_1) \wedge (C_2 \preccurlyeq D_2) \to (A \preccurlyeq D_2)$. By or applied to $\vdash A \preccurlyeq A$ and to $A \preccurlyeq D_1$, we have $\vdash (C_1 \preccurlyeq D_1) \wedge (C_2 \preccurlyeq D_2) \to (A \preccurlyeq A \vee D_1 \vee D_2)$, to which we apply tr twice and conclude $\vdash (C_1 \preccurlyeq D_1) \wedge (C_2 \preccurlyeq D_2) \wedge (C_3 \preccurlyeq D_3) \to (A \preccurlyeq D_3)$. We iterate the steps above until we obtain $\vdash \bigwedge_{i \leq n}(C_i \preccurlyeq D_i) \to (A \preccurlyeq D_n)$. Then, by applications of or to $A \preccurlyeq A$ and to $A \preccurlyeq D_1, \ldots, A \preccurlyeq D_{n-1}$, we obtain $\vdash \bigwedge_{i \leq n}(C_i \preccurlyeq D_i) \to (A \preccurlyeq A \vee \bigvee \mathbf{D}_1^n)$. A final application of tr yields $\vdash \bigwedge_{i \leq n}(C_i \preccurlyeq D_i) \to (A \preccurlyeq B)$. Therefore $\vdash \bigwedge \Gamma \wedge \bigwedge_{i \leq n}(C_i \preccurlyeq D_i) \to (A \preccurlyeq B) \vee \bigvee \Delta$ for every Γ, Δ.

(N_n) Suppose $\vdash C_1 \to \bot$, $\vdash C_2 \to D_1$, ..., $\vdash C_n \to \bigvee \mathbf{D}_1^{n-1}$ and $\vdash \top \to \bigvee \mathbf{D}_1^n$. Then by cpr, $\vdash \bot \preccurlyeq C_1$, $\vdash D_1 \preccurlyeq C_2$, ..., $\vdash \bigvee \mathbf{D}_1^{n-1} \preccurlyeq C_n$ and $\vdash \bigvee \mathbf{D}_1^n \preccurlyeq \top$. Reasoning as in the case of CP_n, we conclude that $\vdash \bigwedge_{i \leq n}(C_i \preccurlyeq D_i) \to (\bot \preccurlyeq \top)$. By n, $\vdash \bigwedge_{i \leq n}(C_i \preccurlyeq D_i) \to \bot$, then $\vdash \bigwedge \Gamma \wedge \bigwedge_{i \leq n}(C_i \preccurlyeq D_i) \to \bigvee \Delta$ for all Γ, Δ.

(T_n) Suppose $\vdash C_1 \to \bot$, $\vdash C_2 \to D_1$, ..., $\vdash C_n \to \bigvee \mathbf{D}_1^{n-1}$ and $\vdash \bigwedge \Gamma \wedge$

$\bigwedge_{i\leq n}(C_i \preccurlyeq D_i) \rightarrow \bigvee \mathbf{D}_1^n \vee \bigvee \Delta$. Then by cpr, $\vdash \bot \preccurlyeq C_1$, $\vdash D_1 \preccurlyeq C_2$, ...$\vdash \bigvee \mathbf{D}_1^{n-1} \preccurlyeq C_n$. By applications of tr and or, we have that $\vdash \bigwedge_{i\leq n}(C_i \preccurlyeq D_i) \rightarrow (\bot \preccurlyeq D_1) \wedge \cdots \wedge (\bot \preccurlyeq D_n)$. By t, $\vdash \bigwedge_{i\leq n}(C_i \preccurlyeq D_i) \rightarrow \neg D_1 \wedge \cdots \wedge \neg D_n$. Then we have $\vdash \bigwedge \Gamma \wedge \bigwedge_{i\leq n}(C_i \preccurlyeq D_i) \rightarrow (\bigvee \mathbf{D}_1^n \bigvee \Delta) \wedge \neg D_1 \wedge \cdots \wedge \neg D_n$, from which we conclude that $\vdash \bigwedge \Gamma \wedge \bigwedge_{i\leq n}(C_i \preccurlyeq D_i) \rightarrow \bigvee \Delta$.

(W_n) Suppose $\vdash C_1 \rightarrow A$, $\vdash C_2 \rightarrow A \vee D_1$, ..., $\vdash C_n \rightarrow A \vee \bigvee \mathbf{D}_1^{n-1}$ and $\vdash \bigwedge \Gamma \wedge \bigwedge_{i\leq n}(C_i \preccurlyeq D_i) \rightarrow A \vee \bigvee \mathbf{D}_1^n \vee A \preccurlyeq B \vee \bigvee \Delta$. Then by cpr, $\vdash A \preccurlyeq C_1$, $\vdash A \vee D_1 \preccurlyeq C_2$, ..., $\vdash A, \bigvee \mathbf{D}_1^{n-1}$. Reasoning as in the case of CP_n, we obtain proofs of the following: $\vdash (C_1 \preccurlyeq D_1) \rightarrow (A \preccurlyeq D_1)$, ..., $\vdash \bigwedge_{i\leq n}(C_i \preccurlyeq D_i) \rightarrow (A \preccurlyeq D_n)$. Moreover by w, $\vdash \bigwedge \Gamma \wedge \bigwedge_{i\leq n}(C_i \preccurlyeq D_i) \rightarrow (A \preccurlyeq \top) \vee \bigvee_{i\leq n}(D_i \preccurlyeq \top) \vee A \preccurlyeq B \vee \bigvee \Delta$. Thus, applying tr to $A \preccurlyeq D_1$, ..., $A \preccurlyeq D_n$, we obtain $\vdash \bigwedge \Gamma \wedge \bigwedge_{i\leq n}(C_i \preccurlyeq D_i) \rightarrow (A \preccurlyeq \top) \vee A \preccurlyeq B \vee \bigvee \Delta$. Since by cpr, $\vdash \top \preccurlyeq B$ for every B, by tr we obtain $\vdash \bigwedge \Gamma \wedge \bigwedge_{i\leq n}(C_i \preccurlyeq D_i) \rightarrow (A \preccurlyeq B) \vee A \preccurlyeq B \vee \bigvee \Delta$.

(W_0) If $\vdash \bigwedge \Gamma \rightarrow A \vee \bigvee \Delta$, then by w, $\vdash \bigwedge \Gamma \rightarrow (A \preccurlyeq \top) \vee \bigvee \Delta$, thus since by cpr, $\vdash \top \preccurlyeq B$ for every B, we have $\vdash \bigwedge \Gamma \rightarrow (A \preccurlyeq B) \vee \bigvee \Delta$.

(C_0) Suppose $\vdash \bigwedge \Gamma \wedge A \rightarrow \bigvee \Delta$ and $\vdash \bigwedge \Gamma \rightarrow B \vee \bigvee \Delta$. Then by w, $\vdash \bigwedge \Gamma \rightarrow (B \preccurlyeq \top) \vee \bigvee \Delta$, thus by tr, $\vdash \bigwedge \Gamma \wedge (A \preccurlyeq B) \rightarrow (A \preccurlyeq \top) \vee \bigvee \Delta$. Then by c, $\vdash \bigwedge \Gamma \wedge (A \preccurlyeq B) \rightarrow A \vee \bigvee \Delta$, therefore $\vdash \bigwedge \Gamma \wedge (A \preccurlyeq B) \rightarrow \bigvee \Delta$.

(A_n) Suppose $\vdash \bigwedge \Gamma^{\preccurlyeq} \wedge \bigwedge_{i\leq n}(C_i \preccurlyeq D_i) \wedge C_1 \rightarrow A \vee (A \preccurlyeq B) \vee \bigvee \Delta^{\preccurlyeq}$, $\vdash \bigwedge \Gamma^{\preccurlyeq} \wedge \bigwedge_{i\leq n}(C_i \preccurlyeq D_i) \wedge C_2 \rightarrow A \vee (A \preccurlyeq B) \vee D_1 \vee \bigvee \Delta^{\preccurlyeq}$, ..., $\vdash \bigwedge \Gamma^{\preccurlyeq} \wedge \bigwedge_{i\leq n}(C_i \preccurlyeq D_i) \wedge C_n \rightarrow A \vee \bigvee \mathbf{D}_1^{n-1} \vee \bigvee \Delta^{\preccurlyeq}$ and $\vdash \bigwedge \Gamma^{\preccurlyeq} \wedge \bigwedge_{i\leq n}(C_i \preccurlyeq D_i) \wedge B \rightarrow A \vee \bigvee \mathbf{D}_1^n \vee \bigvee \Delta^{\preccurlyeq}$. Using the same strategy as in CP_n, we prove that $\vdash \bigwedge_{i\leq n}(C_i \preccurlyeq D_i) \rightarrow (A \preccurlyeq B)$ follows from the simpler set of assumptions where we remove $\bigwedge \Gamma^{\preccurlyeq}$, $\bigwedge_{i\leq n}(C_i \preccurlyeq D_i)$, $\bigvee \Delta^{\preccurlyeq}$ and $A \preccurlyeq B$. From this, we conclude $\vdash \bigwedge \Gamma \wedge \bigwedge_{i\leq n}(C_i \preccurlyeq D_i) \bigwedge_{i\leq n}(C_i \preccurlyeq D_i) \rightarrow (A \preccurlyeq B) \vee \bigvee \Delta$ for any Γ, Δ. ($\mathsf{N^A}_n$) is similar. □

We now show that $\mathsf{G.N}^*_{\preccurlyeq}$ enjoy cut admissibility, where a rule is said to be *(height-preserving) admissible* if, whenever the premisses are derivable, also the conclusion is derivable (with a derivation of at most the same height). We start by considering the following auxiliary result.

Proposition 4.2 *The rules below are height-preserving admissible in* $\mathsf{G.N}^*_{\preccurlyeq}$*:*

$$\mathsf{wk_L}\,\frac{\Gamma \Rightarrow \Delta}{\Gamma, A \Rightarrow \Delta} \qquad \mathsf{wk_R}\,\frac{\Gamma \Rightarrow \Delta}{\Gamma \Rightarrow A, \Delta} \qquad \mathsf{ctr_L}\,\frac{\Gamma, A, A \Rightarrow \Delta}{\Gamma, A \Rightarrow \Delta} \qquad \mathsf{ctr_R}\,\frac{\Gamma \Rightarrow A, A, \Delta}{\Gamma \Rightarrow A, \Delta}$$

Proof. By induction on the height h of the derivation of the premiss of the rules. The cases of $\mathsf{wk_L}$, $\mathsf{wk_R}$ and $\mathsf{ctr_R}$ are immediate. We show admissibility of $\mathsf{ctr_L}$, for $h > 0$ and the last rule applied in the derivation being CP_{n+1}. Let $C_i = C_{i+1}$ and $D_i = D_{i+1}$.

$$\mathsf{ctr_L}\,\dfrac{\mathsf{CP}_n\,\dfrac{\{C_k \Rightarrow A, D_1, \ldots, D_{k-1}\}_{k\leq n} \quad B \Rightarrow A, D_1, \ldots, D_i, D_{i+1}, \ldots, D_n}{G, C_1 \preccurlyeq D_1, \ldots, C_i \preccurlyeq D_i, C_{i+1} \preccurlyeq D_{i+1}, \ldots, C_n \preccurlyeq D_n \Rightarrow A \preccurlyeq B, \Delta}}{G, C_1 \preccurlyeq D_1, \ldots, C_i \preccurlyeq D_i, C_{i+2} \preccurlyeq D_{i+2}, \ldots, C_n \preccurlyeq D_n \Rightarrow A \preccurlyeq B, \Delta}$$

For each $j \geq i+2$ apply contraction to the following sequent of smaller height:

$$\mathsf{ctr_L}\ \frac{C_j \Rightarrow A, D_1, \ldots, D_i, D_{i+1}, \ldots, D_{j-1}}{C_j \Rightarrow A, D_1, \ldots, D_i, D_{i+2}, \ldots, D_{j-1}}$$

A final application of $\mathsf{ctr_R}$ on smaller height and CP_{n-1} yields the desired result:

$$\mathsf{CP}_{n-1}\ \frac{\begin{array}{c}\{C_k \Rightarrow A, D_1, \ldots, D_{k-1}\}_{k \leq i} \\ \{C_j \Rightarrow A, D_1, \ldots, D_i, D_{i+2} \ldots, D_{j-1}\}_{i+2 \leq j \leq n}\end{array} \quad \mathsf{ctr_L}\ \dfrac{B \Rightarrow A, D_1, \ldots, D_i, D_{i+1}, \ldots, D_n}{B \Rightarrow A, D_1, \ldots, D_i, \ldots, D_n}}{G, C_1 \preccurlyeq D_1, \ldots, C_i \preccurlyeq D_i, C_{i+2} \preccurlyeq D_{i+2}, \ldots, C_n \preccurlyeq D_n \Rightarrow A \preccurlyeq B, \Delta}$$

□

Theorem 4.3 *The cut rule is admissible in* $\mathsf{G.N}^*_{\preccurlyeq}$, *where A is the* cut formula*:*

$$\mathsf{cut}\ \frac{\Gamma \Rightarrow A, \Delta \quad \Gamma', A \Rightarrow \Delta'}{\Gamma, \Gamma' \Rightarrow \Delta, \Delta'}$$

Proof. By induction on lexicographically ordered pairs (c, h), where c is the complexity of the cut formula (i.e., the number of binary connectives or modalities occurring in it), and h is the sum of the heights of the derivations of the premisses of cut. We distinguish cases according to whether the cut formula is principal in the last rules applied in the derivation of the premisses of cut.

If the cut formula is not principal in the last rule application of the derivation of one of the two premisses of cut, then the conclusion of cut is standardly obtained by i.h. on h. Suppose the cut formula is principal in the last rule application of the derivations of both premisses of cut.

- Both premisses of cut are derived by CP_n:

$$\mathsf{cut}\ \frac{\mathsf{CP}_n\ \dfrac{\{C_k \Rightarrow A_i, D_1, \ldots, D_{k-1}\}_{k \leq n} \quad B_i \Rightarrow A_i, D_1, \ldots, D_n}{\Gamma, C_1 \preccurlyeq D_1, \ldots, C_n \preccurlyeq D_n \Rightarrow A_i \preccurlyeq B_i, \Delta} \quad \mathsf{CP}_n\ \dfrac{\{A_\ell \Rightarrow E, B_1, \ldots, B_{\ell-1}\}_{\ell \leq m} \quad F \Rightarrow E, B_1, \ldots, B_i, \ldots, B_m}{\Gamma', A_1 \preccurlyeq B_1, \ldots, A_i \preccurlyeq B_i, \ldots, A_m \preccurlyeq B_m \Rightarrow E \preccurlyeq F, \Delta'}}{\Gamma, \Gamma', A_1 \preccurlyeq B_1, \ldots, A_{i-1} \preccurlyeq B_{i-1}, C_1 \preccurlyeq D_1, \ldots, C_n \preccurlyeq D_n, A_{i+1} \preccurlyeq B_{i+1}, \ldots, A_m \preccurlyeq B_m \Rightarrow E \preccurlyeq F, \Delta, \Delta'}$$

The derivation is converted as follows: First, for every $k \leq n$ we obtain the following derivation, by induction on c:

$$\mathsf{cut}\ \frac{C_k \Rightarrow A_i, D_1, \ldots, D_{k-1} \quad A_i \Rightarrow E, B_1, \ldots, B_{i-1}}{C_k \Rightarrow E, B_1, \ldots, B_{i-1}, D_1, \ldots, D_{k-1}}$$

Always by induction on c, for every $1 \leq \ell \leq n - i$ we obtain the following, where the double line denotes several applications of $\mathsf{ctr_R}$:

$$\mathsf{ctr_R}\ \frac{\mathsf{cut}\ \dfrac{A_{i+l} \Rightarrow E, B_1, \ldots, B_i, \ldots, B_{i+l-1} \quad \mathsf{cut}\ \dfrac{B_i \Rightarrow A_i, D_1, \ldots, D_n \quad A_i \Rightarrow E, B_1, \ldots, B_{i-1}}{B_i \Rightarrow E, D_1, \ldots, D_n, B_1, \ldots, B_{i-1}}}{A_{i+l} \Rightarrow E, E, B_1, \ldots, B_{i-1}, B_1, \ldots, B_{i-1}, D_1, \ldots, D_n, B_{i+1}, \ldots, B_{i+l-1}}}{A_{i+l} \Rightarrow E, B_1, \ldots, B_{i-1}, D_1, \ldots, D_n, B_{i+1}, \ldots, B_{i+l-1}}$$

By induction on c we construct the following derivation $\mathcal{S}$:

$$\mathsf{ctr_R}\ \frac{\mathsf{cut}\ \dfrac{F \Rightarrow E, B_1, \ldots, B_i, \ldots, B_m \quad \mathsf{cut}\ \dfrac{B_i \Rightarrow A_i, D_1, \ldots, D_n \quad A_i \Rightarrow E, B_1, \ldots, B_{i-1}}{B_i \Rightarrow E, B_1, \ldots, B_{i-1}, D_1, \ldots, D_n}}{F \Rightarrow E, E, B_1, \ldots, B_{i-1}, B_1, \ldots, B_{i-1}, D_1, \ldots, D_n, B_{i+1}, \ldots, B_m}}{F \Rightarrow E, B_1, \ldots, B_{i-1}, D_1, \ldots, D_n, B_{i+1}, \ldots, B_m}$$

A final application of CP_{n-1} yields a derivation of the conclusion of cut:

$$\mathsf{CP}_{n-1}\ \dfrac{\begin{array}{c}\{A_\ell \Rightarrow E, B_1, ..., B_{\ell-1}\}_{\ell<i} \\ \{C_k \Rightarrow E, B_1, ..., B_{i-1}, D_1, ..., D_{k-1}\}_{k\leq n} \\ \{A_{i+l} \Rightarrow E, B_1, ..., B_{i-1}, D_1, ..., D_n, B_{i+1}, ..., B_{i+l-1}\}_{1\leq \ell \leq n-i}\end{array} \qquad \mathcal{S}}{\Gamma, \Gamma', A_1 \preccurlyeq B_1, ..., A_{i-1} \preccurlyeq B_{i-1}, C_1 \preccurlyeq D_1, ..., C_n \preccurlyeq D_n, A_{i+1} \preccurlyeq B_{i+1}, ..., A_m \preccurlyeq B_m \Rightarrow E \preccurlyeq F, \Delta, \Delta'}$$

- The cut formula is principal in W_n and CP_m:

$$\mathsf{cut}\ \dfrac{\mathsf{W}_m\ \dfrac{\begin{array}{c}\{C_k \Rightarrow A_i, D_1, \dots D_{k-1}\}_{1<k\leq n} \\ \Gamma, C_1 \preccurlyeq D_1, \dots, C_n \preccurlyeq D_n \Rightarrow \Delta, A_i \preccurlyeq B_i, A_i, D_1, \dots, D_n\end{array}}{\Gamma, C_1 \preccurlyeq D_1, \dots, C_n \preccurlyeq D_n \Rightarrow \Delta, A_i \preccurlyeq B_i} \quad \mathsf{CP}_m\ \dfrac{\{A_\ell \Rightarrow E, B_1, \dots, B_{\ell-1}\}_{\ell\leq m} \quad F \Rightarrow E, B_1, \dots, B_i, \dots, B_m}{\Gamma', A_1 \preccurlyeq B_1, ..., A_i \preccurlyeq B_i, ..., A_m \preccurlyeq B_m \Rightarrow \Delta', E \preccurlyeq F}}{\Gamma, \Gamma', A_1 \preccurlyeq B_1, \dots, A_{i-1} \preccurlyeq B_{i-1}, C_1 \preccurlyeq D_1, \dots, C_n \preccurlyeq D_n, A_{i+1} \preccurlyeq B_{i+1}, \dots, A_m \preccurlyeq B_m \Rightarrow \Delta, \Delta', E \preccurlyeq F}$$

We first perform a cut on smaller h between the premiss of W_m and the rightmost premiss of cut, obtaining sequent $\Sigma = \Gamma, \Gamma', A_1 \preccurlyeq B_1, \dots, A_{i-1} \preccurlyeq B_{i-1}, A_i, C_1 \preccurlyeq D_1, \dots, C_n \preccurlyeq D_n, A_{i+1} \preccurlyeq B_{i+1}, \dots, A_m \preccurlyeq B_m \Rightarrow \Delta, \Delta', E \preccurlyeq F, D_1, \dots, D_n$.

Next, we perform a cut by induction on c on Σ and on $A_i \Rightarrow E, B_1, \dots, B_{i-1}$. This yields a derivation of the sequent $\Sigma' = \Gamma, \Gamma', A_1 \preccurlyeq B_1, \dots, A_{i-1} \preccurlyeq B_{i-1}, C_1 \preccurlyeq D_1, \dots, C_n \preccurlyeq D_n, A_{i+1} \preccurlyeq B_{i+1}, \dots, A_m \preccurlyeq B_m \Rightarrow \Delta, \Delta', E \preccurlyeq F, D_1, \dots, D_n, E, B_1, \dots, B_{i-1}$ Then, for $1 < k \leq n$, we generate the following derivation, by induction on c:

$$\mathsf{cut}\ \dfrac{C_k \Rightarrow A_i, D_1, \dots D_{k-1} \quad A_i \Rightarrow E, B_1, \dots, B_{i-1}}{C_k \Rightarrow E, B_1, \dots, B_{i-1}, D_1, \dots, D_{k-1}}$$

A final application of W_j, where $j = n + (i-1)$, yields the desired conclusion:

$$\mathsf{W}_j\ \dfrac{\begin{array}{c}\{A_\ell \Rightarrow E, B_1, \dots, B_{\ell-1}\}_{\ell\leq i} \\ \{C_k \Rightarrow E, B_1, \dots, B_{i-1}, D_1, \dots, D_{k-1}\}_{k\leq n}\end{array} \qquad \Sigma'}{\Gamma, \Gamma', A_1 \preccurlyeq B_1, \dots, A_{i-1} \preccurlyeq B_{i-1}, C_1 \preccurlyeq D_1, \dots, C_n \preccurlyeq D_n, A_{i+1} \preccurlyeq B_{i+1}, \dots, A_m \preccurlyeq B_m \Rightarrow \Delta, \Delta', E \preccurlyeq F}$$

- The cut formula is principal in W_0 and CP_m:

$$\mathsf{cut}\ \dfrac{\mathsf{W}_0\ \dfrac{\Gamma \Rightarrow \Delta, A_i \preccurlyeq B_i, A_i}{\Gamma \Rightarrow \Delta, A_i \preccurlyeq B_i} \quad \mathsf{CP}_m\ \dfrac{\{A_\ell \Rightarrow E, B_1, \dots, B_{\ell-1}\}_{\ell\leq m} \quad F \Rightarrow E, B_1, \dots, B_i, \dots, B_m}{\Gamma', A_1 \preccurlyeq B_1, \dots, A_i \preccurlyeq B_i, \dots A_m \preccurlyeq B_m \Rightarrow \Delta', E \preccurlyeq F}}{\Gamma, \Gamma', A_1 \preccurlyeq B_1, \dots, A_{i-1} \preccurlyeq B_{i-1}, A_{i+1} \preccurlyeq B_{i+1}, \dots, A_m \preccurlyeq B_m \Rightarrow \Delta, \Delta', E \preccurlyeq F}$$

We first perform a cut by induction on h on the premiss of W_0 and on the rightmost premiss of cut, obtaining sequent $\Sigma = \Gamma, \Gamma', A_1 \preccurlyeq B_1, \dots, A_{i-1} \preccurlyeq B_{i-1}, A_{i+1} \preccurlyeq B_{i+1}, \dots A_m \preccurlyeq B_m \Rightarrow \Delta, \Delta', E \preccurlyeq F, A_i$. Then, applying cut to Σ and $A_i \Rightarrow E, B_1, \dots, B_{i-1}$ we obtain $\Sigma' = \Gamma, \Gamma', A_1 \preccurlyeq B_1, \dots, A_{i-1} \preccurlyeq B_{i-1}, A_{i+1} \preccurlyeq B_{i+1}, \dots A_m \preccurlyeq B_m \Rightarrow \Delta, \Delta', E \preccurlyeq F, E, B_1, \dots, B_{i-1}$. Let Γ^* denote the premiss of Σ' and $\Delta^* = \Delta, \Delta'$. We now construct the following derivation, containing $i-1$ applications of C_0:

$$\mathsf{W}_0\ \dfrac{\mathsf{C}_0\ \dfrac{\mathsf{wk}\ \dfrac{A_1 \Rightarrow E}{A_1, \Gamma^* \Rightarrow \Delta^*, E \preccurlyeq F, E} \qquad \begin{array}{c} \mathsf{C}_0\ \dfrac{\mathsf{wk}\ \dfrac{A_{i-2} \Rightarrow E, B_1, \dots, B_{i-2}}{A_{i-2}, \Gamma^* \Rightarrow \Delta^*, E \preccurlyeq F, E, B_1, \dots, B_{i-3}} \quad \mathsf{C}_0\ \dfrac{\mathsf{wk}\ \dfrac{A_{i-1} \Rightarrow E, B_1, \dots, B_{i-2}}{A_{i-1}, \Gamma^* \Rightarrow \Delta^*, E \preccurlyeq F, E, B_1, \dots, B_{i-2}} \quad \Sigma'}{\Gamma^* \Rightarrow \Delta^*, E, B_1, \dots, B_{i-2}}}{\Gamma^* \Rightarrow \Delta^*, E \preccurlyeq F, E, B_1, \dots, B_{i-3}} \\ \vdots \\ \Gamma^* \Rightarrow \Delta^*, E \preccurlyeq F, E, B_1 \end{array}}{\Gamma^* \Rightarrow \Delta^*, E \preccurlyeq F, E}}{\Gamma^* \Rightarrow \Delta^*, E \preccurlyeq F}$$

$$
\dfrac{\dfrac{\dfrac{A \Rightarrow A \quad A \Rightarrow A, B \quad \dfrac{B \Rightarrow A, B, C \quad C \Rightarrow A, B, C}{B \vee C \Rightarrow A, B, C}\,\vee_{\mathsf{L}}}{A \preccurlyeq B, A \preccurlyeq C \Rightarrow A \preccurlyeq B \vee C}\,\mathsf{CP}_2}{(A \preccurlyeq B) \wedge (A \preccurlyeq C) \Rightarrow A \preccurlyeq B \vee C}\,\wedge_{\mathsf{L}}}{\Rightarrow (A \preccurlyeq B) \wedge (A \preccurlyeq C) \rightarrow (A \preccurlyeq B \vee C)}\,\rightarrow_{\mathsf{R}}
$$

$$
\dfrac{\dfrac{\dfrac{A \Rightarrow A \quad B, A \Rightarrow B \quad C \Rightarrow A, B, C}{A \preccurlyeq B, B \preccurlyeq C \Rightarrow A \preccurlyeq C}\,\mathsf{CP}_2}{(A \preccurlyeq B) \wedge (B \preccurlyeq C) \Rightarrow A \preccurlyeq C}\,\wedge_{\mathsf{L}}}{\Rightarrow (A \preccurlyeq B) \wedge (B \preccurlyeq C) \rightarrow (A \preccurlyeq C)}\,\rightarrow_{\mathsf{R}}
\qquad
\dfrac{\dfrac{\Rightarrow A \rightarrow B \quad \dfrac{A \Rightarrow A, B \quad A, B \Rightarrow B}{A, A \rightarrow B \Rightarrow B}\,\rightarrow_{\mathsf{L}}}{A \Rightarrow B}\,\mathsf{cut}}{\Rightarrow B \preccurlyeq A}\,\mathsf{CP}_0
$$

Fig. 5. Derivations of the axioms and rules of $\mathsf{N}_{\preccurlyeq}$ in $\mathsf{G.N}_{\preccurlyeq}$.

The remaining cases are: $\mathsf{CP}_n + \mathsf{W}_m$, which is proved similarly as $\mathsf{W}_m + \mathsf{CP}_n$; $\mathsf{C}_0 + \mathsf{CP}_n$, similar to $\mathsf{CP}_n + \mathsf{W}_0$; $\mathsf{C}_0 + \mathsf{W}_0$, which is immediate, $\mathsf{CP}_n + \mathsf{N}_n$, which is proven in the same way as $\mathsf{CP}_n + \mathsf{CP}_m$, $\mathsf{CP}_n + \mathsf{T}_m$, which is proven as $\mathsf{CP}_n + \mathsf{W}_m$, and the cases for absoluteness, which are proven as their counterpart without absoluteness. □

Thanks to cut-admissibility, we obtain cut-free completeness of the calculi, by deriving the axioms and inference rules of $\mathsf{N}^*_{\preccurlyeq}$ in $\mathsf{G.N}^*_{\preccurlyeq}$.

Corollary 4.4 (Completeness) *If* $\mathsf{N}^*_{\preccurlyeq} \vdash \bigwedge \Gamma \rightarrow \bigvee \Delta$ *then* $\mathsf{G.N}^*_{\preccurlyeq} \vdash \Gamma \Rightarrow \Delta$.

Proof. Derivations of the $\mathsf{N}_{\preccurlyeq}$ axioms in $\mathsf{G.N}_{\preccurlyeq}$ are displayed in Fig. 5. The derivations employ standard propositional rules for $\wedge$ and $\vee$, which can be defined in $\mathsf{G.N}^*_{\preccurlyeq}$. The derivations of the axioms for extensions are straightforward. *Modus ponens* is simulated using cut in the usual way. □

Termination of root-first proof search in $\mathsf{G.N}^*_{\preccurlyeq}$ can be easily proved by observing that non redundant rule applications strictly decrease the complexity of formulas. However, $\mathsf{G.N}^*_{\preccurlyeq}$ are not suited for root-first proof search: the comparative plausibility rules are *not invertible*, meaning that derivability of the conclusion does not imply derivability of the premiss(es) of the rule. As a consequence, backtrack points are generated when constructing root-first a derivation. Next section introduces proof systems having only invertible rules.

5 Hypersequent calculi for CPN logics

In this section we present hypersequent calculi $\mathsf{H.N}^*_{\preccurlyeq}$ for the same family of CPN logics treated in Sec. 4, namely $\mathsf{N}_{\preccurlyeq}, \mathsf{NN}_{\preccurlyeq}, \mathsf{NT}_{\preccurlyeq}, \mathsf{NW}_{\preccurlyeq}, \mathsf{NC}_{\preccurlyeq}, \mathsf{NA}_{\preccurlyeq}$ and $\mathsf{NNA}_{\preccurlyeq}$, always denoted by $\mathsf{N}^*_{\preccurlyeq}$. Disregarding the hypersequent structure, the calculi $\mathsf{H.N}^*_{\preccurlyeq}$ are fragments of the sequent calculi for Lewis' logics by Olivetti and Pozzato [22] and Girlando et al. [8], the difference being that we do not assume the communication rule com. The basic components of the calculi $\mathsf{H.N}^*_{\preccurlyeq}$ are Gentzen-style sequents to which is added the following *block structure* from [22], representing $\preccurlyeq$-formulas in the right-hand side of sequents.

Definition 5.1 A *block* is a structure $[\Sigma \lhd A]$, where Σ is a multiset of formulas and A is a formula. A *sequent with blocks* is a pair $\Gamma \Rightarrow \Delta$, where Γ is a

$$\mathsf{init}\ \frac{}{\mathcal{G} \mid \Gamma, p \Rightarrow p, \Delta} \qquad \bot_{\mathsf{L}}\ \frac{}{\mathcal{G} \mid \Gamma, \bot \Rightarrow \Delta} \qquad \to_{\mathsf{L}}\ \frac{\mathcal{G} \mid \Gamma, A \to B, B \Rightarrow \Delta \quad \mathcal{G} \mid \Gamma, A \to B \Rightarrow \Delta, A}{\mathcal{G} \mid \Gamma, A \to B \Rightarrow \Delta}$$

$$\preccurlyeq_{\mathsf{L}}\ \frac{\mathcal{G} \mid \Gamma, A \preccurlyeq B \Rightarrow \Delta, [B, \Sigma \lhd C] \quad \mathcal{G} \mid \Gamma, A \preccurlyeq B \Rightarrow \Delta, [\Sigma \lhd C], [\Sigma \lhd A]}{\mathcal{G} \mid \Gamma, A \preccurlyeq B \Rightarrow \Delta, [\Sigma \lhd C]}$$

$$\to_{\mathsf{R}}\ \frac{\mathcal{G} \mid \Gamma, A \to B, A \Rightarrow \Delta, B}{\mathcal{G} \mid \Gamma \Rightarrow \Delta, A \to B} \qquad \preccurlyeq_{\mathsf{R}}\ \frac{\mathcal{G} \mid \Gamma \Rightarrow \Delta, A \preccurlyeq B, [A \lhd B]}{\mathcal{G} \mid \Gamma \Rightarrow \Delta, A \preccurlyeq B} \qquad \mathsf{jp}\ \frac{\mathcal{G} \mid \Gamma \Rightarrow \Delta, [\Sigma \lhd A] \mid A \Rightarrow \Sigma}{\mathcal{G} \mid \Gamma \Rightarrow \Delta, [\Sigma \lhd A]}$$

$$\mathsf{N}\ \frac{\mathcal{G} \mid \Gamma \Rightarrow \Delta, [\bot \lhd \top]}{\mathcal{G} \mid \Gamma \Rightarrow \Delta} \qquad \mathsf{T}\ \frac{\mathcal{G} \mid \Gamma, A \preccurlyeq B \Rightarrow \Delta, B \quad \mathcal{G} \mid \Gamma, A \preccurlyeq B \Rightarrow \Delta, [\bot \lhd A]}{\mathcal{G} \mid \Gamma, A \preccurlyeq B \Rightarrow \Delta}$$

$$\mathsf{W}\ \frac{\mathcal{G} \mid \Gamma \Rightarrow \Delta, [\Sigma \lhd A], \Sigma}{\mathcal{G} \mid \Gamma \Rightarrow \Delta, [\Sigma \lhd A]} \qquad \mathsf{C}\ \frac{\mathcal{G} \mid \Gamma, A \preccurlyeq B \Rightarrow \Delta, B \quad \mathcal{G} \mid \Gamma, A \preccurlyeq B, A \Rightarrow \Delta}{\mathcal{G} \mid \Gamma, A \preccurlyeq B \Rightarrow \Delta}$$

$$\mathsf{A_L}\ \frac{\mathcal{G} \mid \Gamma, A \preccurlyeq B \Rightarrow \Delta \mid \Omega, A \preccurlyeq B \Rightarrow \Theta}{\mathcal{G} \mid \Gamma, A \preccurlyeq B \Rightarrow \Delta \mid \Omega \Rightarrow \Theta} \qquad \mathsf{A_R}\ \frac{\mathcal{G} \mid \Gamma \Rightarrow \Delta, A \preccurlyeq B \mid \Omega \Rightarrow \Theta, A \preccurlyeq B}{\mathcal{G} \mid \Gamma \Rightarrow \Delta, A \preccurlyeq B \mid \Omega \Rightarrow \Theta}$$

$\mathsf{H.N}_{\preccurlyeq} = \{\mathsf{init}, \bot_{\mathsf{L}}, \to_{\mathsf{L}}, \to_{\mathsf{R}}\} \cup \{\preccurlyeq_{\mathsf{L}}, \preccurlyeq_{\mathsf{R}}, \mathsf{jp}\}$ $\quad \mathsf{H.NW}_{\preccurlyeq} = \mathsf{H.N}_{\preccurlyeq} \cup \{\mathsf{T}, \mathsf{W}\}$

$\mathsf{H.NN}_{\preccurlyeq} = \mathsf{H.N}_{\preccurlyeq} \cup \{\mathsf{N}\}$ $\quad \mathsf{H.NC}_{\preccurlyeq} = \mathsf{H.N}_{\preccurlyeq} \cup \{\mathsf{W}, \mathsf{C}\}$

$\mathsf{H.NT}_{\preccurlyeq} = \mathsf{H.N}_{\preccurlyeq} \cup \{\mathsf{T}\}$ $\quad \mathsf{H.NA}^*_{\preccurlyeq} = \mathsf{H.N}^*_{\preccurlyeq} \cup \{\mathsf{A_L}, \mathsf{A_R}\}$

Fig. 6. Rules of hypersequent calculi $\mathsf{H.N}^*_{\preccurlyeq}$.

multiset of formulas, and Δ is a multiset of formulas and blocks. Sequents are interpreted in $\mathscr{L}$ as follows (where Δ' does not contain blocks):

$$i(\Gamma \Rightarrow \Delta', [\Sigma_1 \lhd C_1], \dots, [\Sigma_k \lhd C_k]) =$$
$$\bigwedge \Gamma \to \bigvee \Delta' \vee (\bigvee \Sigma_1 \preccurlyeq C_1) \vee \dots \vee (\bigvee \Sigma_k \preccurlyeq C_k).$$

A *hypersequent* $\mathcal{H}$ is a finite multiset of sequents with blocks $\Gamma_1 \Rightarrow \Delta_1 \mid \dots \mid \Gamma_n \Rightarrow \Delta_n$, where $\Gamma_1 \Rightarrow \Delta_1, \dots, \Gamma_n \Rightarrow \Delta_n$ are called the *components* of $\mathcal{H}$. We say that a hypersequent is *valid in a model* $\mathcal{M}$ if it has a component $\Gamma_k \Rightarrow \Delta_k$ such that $\mathcal{M} \models i(\Gamma_k \Rightarrow \Delta_k)$.

While hypersequents do not have a formula interpretation, sequents with blocks are interpreted as formulas of $\mathscr{L}$, in a way different from [22,8]. Specifically, for us $[B_1, \dots, B_n \lhd A]$ is interpreted as $(B_1 \vee \dots \vee B_n) \preccurlyeq A$, while in [22,8] it corresponds to $(B_1 \preccurlyeq A) \vee \dots \vee (B_n \preccurlyeq A)$. These two interpretations are equivalent in $\mathsf{V}_{\preccurlyeq}$ but are not equivalent in $\mathsf{N}_{\preccurlyeq}$.

The calculi $\mathsf{H.N}^*_{\preccurlyeq}$ are defined in Fig. 6. The rules are *cumulative*, meaning that each rule has the principal formula copied in the premisses. Differently from the calculi $\mathsf{G.N}^*_{\preccurlyeq}$ in the previous section, $\mathsf{H.N}^*_{\preccurlyeq}$ have separate left and right rules for $\preccurlyeq$, and all rules have a fixed number of premisses.

We point out that the hypersequent structure is not necessary to define sequent calculi with blocks for CPN logics. Moreover, it can be checked that a hypersequent is derivable if and only if one of its components is derivable. Following the strategy from [4], we chose to employ a hypersequential structure to obtain invertibility of *all* the rules of the calculi, there including the jp rule, which was not invertible in [8]. Together with their cumulative formulation, invertibility of the rules allows to directly construct countermodels from failed proof search, without the need of backtracking or inserting any additional computation. Moreover, differently from [8], the countermodel construction modularly extends to logics with *absoluteness*. The rules for absoluteness are inspired from [9], and correspond to condition A+ from Sec. 2. Soundness of the rules is proved as follows. Let $\mathsf{H.N}^*_{\preccurlyeq} \vdash A$ denote derivability of A in $\mathsf{H.N}^*_{\preccurlyeq}$.

Theorem 5.2 (Soundness) *For every formula A, if A is derivable in $\mathsf{H.N}^*_{\preccurlyeq}$, then A is valid in all $\mathsf{N}^*_{\preccurlyeq}$-models.*

Proof. For every rule R of $\mathsf{H.N}^*_{\preccurlyeq}$, we show that if the premisses of R are valid in a $\mathsf{N}^*_{\preccurlyeq}$-model $\mathcal{M}$, then the conclusion is also valid in $\mathcal{M}$. We only consider some relevant examples of modal rules. ($\preccurlyeq_{\mathsf{L}}$) Suppose $\mathcal{M} \models \mathcal{G} \mid \Gamma, A \preccurlyeq B \Rightarrow \Delta, [B, \Sigma \lhd C]$ and $\mathcal{M} \models \mathcal{G} \mid \Gamma, A \preccurlyeq B \Rightarrow \Delta, [\Sigma \lhd C], [\Sigma \lhd A]$. If $\mathcal{M} \models \mathcal{G}$ we are done. Otherwise $\mathcal{M} \models \bigwedge\Gamma \wedge (A \preccurlyeq B) \to (B \vee \bigvee\Sigma \preccurlyeq C) \vee \bigvee\Delta$ and $\mathcal{M} \models \bigwedge\Gamma \wedge (A \preccurlyeq B) \to (\bigvee\Sigma \preccurlyeq C) \vee (\bigvee\Sigma \preccurlyeq A) \vee \bigvee\Delta$. Then by tr, $\mathcal{M} \models \bigwedge\Gamma \wedge (A \preccurlyeq B) \to (\bigvee\Sigma \preccurlyeq C) \vee (\bigvee\Sigma \preccurlyeq B) \vee \bigvee\Delta$, and by cpr and or, $\mathcal{M} \models \bigwedge\Gamma \wedge (A \preccurlyeq B) \to (\bigvee\Sigma \preccurlyeq C) \vee (\bigvee\Sigma \preccurlyeq B \vee \bigvee\Sigma) \vee \bigvee\Delta$, therefore by tr, $\mathcal{M} \models \bigwedge\Gamma \wedge (A \preccurlyeq B) \to (\bigvee\Sigma \preccurlyeq C) \vee (\bigvee\Sigma \preccurlyeq C) \vee \bigvee\Delta$, thus $\mathcal{M} \models \bigwedge\Gamma \wedge (A \preccurlyeq B) \to (\bigvee\Sigma \preccurlyeq C) \vee \bigvee\Delta$. It follows $\mathcal{M} \models \mathcal{G} \mid \Gamma, A \preccurlyeq B \Rightarrow \Delta, [\Sigma \lhd C]$. ($\mathsf{jp}$) Suppose that $\mathcal{M} \models \mathcal{G} \mid \Gamma \Rightarrow \Delta, [\Sigma \lhd A] \mid A \Rightarrow \Sigma$. If $\mathcal{M} \models \mathcal{G} \mid \Gamma \Rightarrow \Delta, [\Sigma \lhd A]$ we are done, otherwise $\mathcal{M} \models A \to \bigvee\Sigma$. Then by cpr, $\mathcal{M} \models \bigvee\Sigma \preccurlyeq A$, therefore $\mathcal{M} \models \Gamma \Rightarrow \Delta, [\Sigma \lhd A]$. ($\mathsf{T}$) Suppose that $\mathcal{M} \models \mathcal{G} \mid \Gamma, A \preccurlyeq B \Rightarrow \Delta, B$ and $\mathcal{M} \models \mathcal{G} \mid \Gamma, A \preccurlyeq B \Rightarrow \Delta, [\bot \lhd A]$. If $\mathcal{M} \models \mathcal{G}$ we are done, otherwise $\mathcal{M} \models \bigwedge\Gamma \wedge (A \preccurlyeq B) \to \bigvee\Delta \vee (B \wedge (\bot \preccurlyeq A))$. Then by tr, $\mathcal{M} \models \bigwedge\Gamma \wedge (A \preccurlyeq B) \to \bigvee\Delta \vee (B \wedge (\bot \preccurlyeq B))$, and by t, $\mathcal{M} \models \bigwedge\Gamma \wedge (A \preccurlyeq B) \to \bigvee\Delta \vee (B \wedge \neg B)$, therefore $\mathcal{M} \models \bigwedge\Gamma \wedge (A \preccurlyeq B) \to \bigvee\Delta$. □

The calculi $\mathsf{H.N}^*_{\preccurlyeq}$ enjoy admissibility of the following structural properties:

Lemma 5.3 *It holds that all the rules of $\mathsf{H.N}^*_{\preccurlyeq}$ are height-preserving invertible, and that the following rules of weakening and contraction are height-preserving admissible in $\mathsf{H.N}^*_{\preccurlyeq}$, where A in $\mathsf{wk_R}$ or $\mathsf{ctr_R}$ can be a formula or a block.*

$$\mathsf{wk_L}\,\frac{\mathcal{G} \mid \Gamma \Rightarrow \Delta}{\mathcal{G} \mid A, \Gamma \Rightarrow \Delta} \qquad \mathsf{wk_R}\,\frac{\mathcal{G} \mid \Gamma \Rightarrow \Delta}{\mathcal{G} \mid \Gamma \Rightarrow \Delta, A} \qquad \mathsf{wk_C}\,\frac{\mathcal{G}}{\mathcal{G} \mid \mathcal{H}} \qquad \mathsf{wk_B}\,\frac{\mathcal{G} \mid \Gamma \Rightarrow \Delta, [\Sigma \lhd A]}{\mathcal{G} \mid \Gamma \Rightarrow \Delta, [\Sigma, B \lhd A]}$$

$$\mathsf{ctr_L}\,\frac{\mathcal{G} \mid A, A, \Gamma \Rightarrow \Delta}{\mathcal{G} \mid A, \Gamma \Rightarrow \Delta} \qquad \mathsf{ctr_R}\,\frac{\mathcal{G} \mid \Gamma \Rightarrow \Delta, A, A}{\mathcal{G} \mid \Gamma \Rightarrow \Delta, A} \qquad \mathsf{ctr_C}\,\frac{\mathcal{G} \mid \mathcal{H} \mid \mathcal{H}}{\mathcal{G} \mid \mathcal{H}} \qquad \mathsf{ctr_B}\,\frac{\mathcal{G} \mid \Gamma \Rightarrow \Delta, [\Sigma, B, B \lhd A]}{\mathcal{G} \mid \Gamma \Rightarrow \Delta, [\Sigma, B \lhd A]}$$

Proof. Height-preserving admissibility of weakening can be standardly proved by induction on the height of the derivation. Invertibility of all the rules of $\mathsf{H.N}^*_{\preccurlyeq}$ immediately follows. For instance, the premiss of the jp rule can be derived from the conclusion of jp using $\mathsf{wk_C}$. Admissibility of contraction also follows by standard induction on the height of derivations. □

Concerning completeness, a proof can be given by showing that the derivations in the calculi $\mathsf{G.N}^*_{\preccurlyeq}$ can be simulated in $\mathsf{H.N}^*_{\preccurlyeq}$.

Theorem 5.4 (Simulation) *For A $\mathscr{L}$ formula, if $\mathsf{G.N}^*_{\preccurlyeq} \vdash A$ then $\mathsf{H.N}^*_{\preccurlyeq} \vdash A$.*

Proof. We show that the rules of $\mathsf{G.N}^*_{\preccurlyeq}$ can be stepwise simulated by the rules of $\mathsf{H.N}^*_{\preccurlyeq}$. Then the proof of the claim is similar to the one given in [8]. Let $\Sigma_n = C_1 \preccurlyeq D_1, ..., C_n \preccurlyeq D_n$ and $\mathbf{D}_1^n = D_1, \ldots, D_n$. Here follows the translation of CP_n.

$$\mathsf{\preccurlyeq_L}\dfrac{\mathsf{jp}\dfrac{\mathsf{wk_C}\dfrac{B \Rightarrow A, \mathbf{D}_1^n}{\Gamma, \Sigma_n \Rightarrow [A, \mathbf{D}_1^n \lhd B], \Delta \mid B \Rightarrow A, \mathbf{D}_1^n}}{\Gamma, \Sigma_n \Rightarrow [A, \mathbf{D}_1^n \lhd B], \Delta} \quad \mathsf{jp}\dfrac{\mathsf{wk_C}\dfrac{C_n \Rightarrow A, \mathbf{D}_1^{n-1}}{\Gamma, \Sigma_n \Rightarrow [A, \mathbf{D}_1^{n-1} \lhd C_n], \Delta \mid C_n \Rightarrow A, \mathbf{D}_1^{n-1}}}{\Gamma, \Sigma_n \Rightarrow [A, \mathbf{D}_1^{n-1} \lhd C_n], \Delta}}{\Gamma, \Sigma_n \Rightarrow [A, \mathbf{D}^{\lhd}_{n-1} B], \Delta}$$

$$\vdots$$

$$\mathsf{\preccurlyeq_R}\dfrac{\mathsf{\preccurlyeq_L}\dfrac{\mathsf{\preccurlyeq_L}\dfrac{\Gamma, \Sigma_n \Rightarrow [A, D_1, D_2 \lhd B], \Delta}{\Gamma, \Sigma_n \Rightarrow [A, D_1 \lhd B], \Delta} \quad \mathsf{jp}\dfrac{\mathsf{wk_C}\dfrac{C_1 \Rightarrow A}{\Gamma, \Sigma_n \Rightarrow [A \lhd C_1], \Delta \mid C_1 \Rightarrow A}}{\Gamma, \Sigma_n \Rightarrow [A \lhd C_1], \Delta}}{\Gamma, \Sigma_n \Rightarrow [A \lhd B], \Delta}}{\Gamma, \Sigma_n \Rightarrow A \preccurlyeq B, \Delta}$$

Rule N_n is derived in a similar way, by replacing $\mathsf{\preccurlyeq_R}$ with N and removing occurrences of A and B in the derivation above. For A_n and $\mathsf{N^A}_n$ each jp is followed by applications of $\mathsf{A_L}$ and $\mathsf{A_R}$. To derive rule W_n, replace the upper leftmost occurrence of jp with rule W. Rule $\mathsf{W_0}$ is immediately derivable using jp and W, and rule $\mathsf{C_0}$ using C. The case of rule T_n is more complex. We start with the following derivation.

$$\mathsf{T}\dfrac{\Gamma, \Sigma_n \Rightarrow \Delta, \mathbf{D}_2^n, D_1 \quad \mathsf{jp}\dfrac{\mathsf{wk_B,wk}\dfrac{C_1 \Rightarrow}{\Gamma, \Sigma_n \Rightarrow \Delta, \mathbf{D}_2^n, [\bot \lhd C_1] \mid C_1 \Rightarrow \bot}}{\Gamma, \Sigma_n \Rightarrow \Delta, \mathbf{D}_2^n, [\bot \lhd C_1]}}{\Gamma, \Sigma_n \Rightarrow \Delta, \mathbf{D}_2^n}$$

$$\vdots$$

$$\mathsf{T}\dfrac{\mathsf{T}\dfrac{\Gamma, \Sigma_n \Rightarrow \Delta, D_n, D_{n-1}}{\Gamma, \Sigma_n \Rightarrow \Delta, D_n} \quad \Gamma, \Sigma_n \Rightarrow \Delta, [\bot \lhd C_n]}{\Gamma, \Sigma_n \Rightarrow \Delta}$$

The leftmost sequent is premiss $\Gamma, \Sigma_n \Rightarrow \Delta, \mathbf{D}_1^n$ of T_n. We now construct from the remaining premisses of T_n, that is, $\{C_k \Rightarrow D_1, \ldots, D_{k-1}\}$, for $k \leq n$, derivations of sequents $\Gamma, \Sigma_n \Rightarrow \Delta, \mathbf{D}, [\bot \lhd C_k]$, for $1 < k \leq n$, where $\mathbf{D} = D_n, \ldots, D_{k+1}$ if $k < n$, and is empty otherwise. In applications of the jp rule, we omit specifying the leftmost component of the hypersequents.

$$\mathsf{\preccurlyeq_L}\dfrac{\mathsf{jp}\dfrac{\mathsf{wk_B,wk}\dfrac{C_k \Rightarrow \mathbf{D}_1^{k-1}}{\ldots \mid C_k \Rightarrow \mathbf{D}_1^{k-1}, \bot}}{\Gamma, \Sigma_n \Rightarrow \Delta, \mathbf{D}, [\mathbf{D}_1^{k-1}, \bot \lhd C_k]} \quad \mathsf{jp}\dfrac{\mathsf{wk_C,wk}\dfrac{C_{k-1} \Rightarrow \mathbf{D}_1^{k-2}}{\ldots \mid C_{k-1} \Rightarrow \mathbf{D}_1^{k-2}, \bot}}{\Gamma, \Sigma_n \Rightarrow \Delta, \mathbf{D}, [\mathbf{D}_1^{k-2}, \bot \lhd C_k], [\mathbf{D}_1^{k-2}, \bot \lhd C_{k-1}]}}{\Gamma, \Sigma_n \Rightarrow \Delta, \mathbf{D}, [\mathbf{D}_1^{k-2}, \bot \lhd C_k]}$$

$$\vdots$$

$$\mathsf{\preccurlyeq_L}\dfrac{\mathsf{\preccurlyeq_L}\dfrac{\mathsf{\preccurlyeq_L}\dfrac{\Gamma, \Sigma_n \Rightarrow \Delta, \mathbf{D}, [D_1, D_2, D_3, \bot \lhd C_k]}{\Gamma, \Sigma_n \Rightarrow \Delta, \mathbf{D}, [D_1, D_2, \bot \lhd C_k]} \quad \mathsf{jp}\dfrac{\mathsf{wk_C,wk}\dfrac{C_2 \Rightarrow D_1}{\ldots \mid C_2 \Rightarrow D_1, \bot}}{\Gamma, \Sigma_n \Rightarrow \Delta, \mathbf{D}, [D_1, \bot \lhd C_k], [D_1, \bot \lhd C_2]}}{\Gamma, \Sigma_n \Rightarrow \Delta, \mathbf{D}, [D_1, \bot \lhd C_k]}}{\Gamma, \Sigma_n \Rightarrow \Delta, \mathbf{D}, [\bot \lhd C_k]}$$

The rightmost premiss of the lower occurrence of $\mathsf{\preccurlyeq_L}$, not shown, is sequent $\Gamma, \Sigma_n \Rightarrow \Delta, \mathbf{D}, [\bot \lhd C_k], [\bot \lhd C_1]$, which is derivable by jp from premiss $C_1 \Rightarrow$.□

Since $\mathsf{G.N}^*_{\preccurlyeq}$ are complete with respect to $\mathsf{N}^*_{\preccurlyeq}$, this simulation entails that $\mathsf{H.N}^*_{\preccurlyeq}$ are also complete. Here we present in more detail an alternative completeness proof based on the semantics. In particular, we define a terminating bottom-up proof-search strategy in $\mathsf{H.N}^*_{\preccurlyeq}$, and show that whenever the strategy fails, one can directly extract a countermodel of the root formula/hypersequent. The strategy is based on the following notion of saturation.

Definition 5.5 Let $\mathcal{H} = \Gamma_1 \Rightarrow \Delta_1 \mid ... \mid \Gamma_n \Rightarrow \Delta_n$ be a hypersequent occurring in proof for $\mathcal{H}'$ in $\mathsf{H.N}^*_{\preccurlyeq}$. The *saturation conditions* associated to each application of a rule of $\mathsf{H.N}^*_{\preccurlyeq}$ are as follows: (init) $\Gamma_k \cap \Delta_k = \emptyset$. ($\bot_\mathsf{L}$) $\bot \notin \Gamma_k$. ($\rightarrow_\mathsf{L}$) If $A \rightarrow B \in \Gamma_k$, then $A \in \Delta_k$ or $B \in \Gamma_k$. ($\rightarrow_\mathsf{R}$) If $A \rightarrow B \in \Delta_k$, then $A \in \Gamma_k$ and $B \in \Delta_k$. ($\preccurlyeq_\mathsf{L}$) If $A \preccurlyeq B \in \Gamma_k$ and $[\Sigma \lhd C] \in \Delta_k$, then $B \in \Sigma$ or there is $[\Pi \lhd A] \in \Delta_k$ such that $set(\Sigma) \subseteq set(\Pi)$. ($\preccurlyeq_\mathsf{R}$) If $A \preccurlyeq B \in \Delta_k$, then there is $[\Sigma \lhd B] \in \Delta_k$ such that $A \in \Sigma$. (jp) If $[\Sigma \lhd A] \in \Delta_k$, then there is $\Gamma_j \Rightarrow \Delta_j \in \mathcal{H}$ such that $A \in \Gamma_j$ and $set(\Sigma) \subseteq \Delta_j$. ($\mathsf{N}$) There is $[\Sigma \lhd \top] \in \Delta_k$ such that $\bot \in \Sigma$. (T) If $A \preccurlyeq B \in \Gamma_k$, then $B \in \Delta_k$ or there is $[\Sigma, \bot \lhd A] \in \Delta_k$. ($\mathsf{W}$) If $[\Sigma \lhd A] \in \Delta_k$, then $set(\Sigma) \subseteq \Delta_k$. ($\mathsf{C}$) If $A \preccurlyeq B \in \Gamma_k$, then $B \in \Delta_k$ or $A \in \Gamma_k$. ($\mathsf{A_L}$) If $A \preccurlyeq B \in \Gamma_k$, then for all $\Gamma_j \Rightarrow \Delta_j \in \mathcal{H}$, $A \preccurlyeq B \in \Gamma_j$. ($\mathsf{A_R}$) If $A \preccurlyeq B \in \Delta_k$, then for all $\Gamma_j \Rightarrow \Delta_j \in \mathcal{H}$, $A \preccurlyeq B \in \Delta_j$. We say that $\mathcal{H}$ is *saturated with respect to an application of a rule R* if it satisfies the saturation condition (R) for that particular rule application, and it is *saturated with respect to* $\mathsf{H.N}^*_{\preccurlyeq}$ if it is saturated with respect to all possible applications of any rule of $\mathsf{H.N}^*_{\preccurlyeq}$.

The strategy consists simply in applying the rules backward until no additional rule application is possible respecting the following two conditions: (i) no rule can be applied to an initial hypersequent; (ii) the application of a rule is not allowed if the hypersequent is already saturated with respect to that specific rule application. The conditions (i) and (ii) ensure that proof-search terminates for every hypersequent $\mathcal{H}$.

Proposition 5.6 *Proof-search for $\mathcal{H}$ in $\mathsf{H.N}^*_{\preccurlyeq}$ in accordance with the strategy always terminates after a finite number of steps.*

Proof. Let $\mathscr{P}$ be a proof of $\mathcal{H}$ constructed according to the strategy. Then all formulas occurring in $\mathscr{P}$ (both inside and outside blocks) are subformulas of formulas of $\mathcal{H}$ or they are $\bot$ or $\top$, so they are finitely many. Moreover, the saturation conditions prevent duplications of the same formulas (both inside and outside blocks) and the same blocks. It follows that all hypersequents occurring in $\mathscr{P}$ have a finite length, moreover every branch of $\mathscr{P}$ contains only finitely many hypersequents. □

If the strategy succeeds, then it constructs a derivation of the root hypersequent $\mathcal{H}$. Otherwise, a saturated hypersequent will occur in the leaf of a branch. We now prove that the proof-search strategy is complete, showing that whenever the strategy fails, from every saturated hypersequent one can directly construct a countermodel for $\mathcal{H}$.

Proposition 5.7 (Countermodel construction) *Let $\mathcal{H} = \Gamma_1 \Rightarrow \Delta_1 \mid \ldots \mid \Gamma_k \Rightarrow \Delta_k$ be a saturated hypersequent occurring in a proof search tree for $\mathcal{H}_0$*

in $\mathsf{H.N}^*_{\preccurlyeq}$ *built in accordance with the strategy. For* Σ *multiset of formulas, let* $\Sigma^\Delta = \{n \mid \Gamma_n \Rightarrow \Delta_n \in \mathcal{H}$ and $set(\Sigma) \subseteq \Delta_n\}$. *We define* $\mathcal{M} = \langle W, N, V \rangle$*:*

- $W = \{n \mid \Gamma_n \Rightarrow \Delta_n \in \mathcal{H}\}$.
- *For every* $n \in W$, $N(n) = \{\Sigma^\Delta \mid$ there is A such that $[\Sigma \lhd A] \in \Delta_n\}$.
- *For every* $p \in Atm$, $V(p) = \{n \in W \mid p \in \Gamma_n\}$.

Then for all $n \in W$, *(i) if* $A \in \Gamma_n$, *then* $n \Vdash A$*; (ii) if* $A \in \Delta_n$, *then* $n \nVdash A$*; and (iii) if* $[\Sigma \lhd A] \in \Delta_n$, *then* $n \nVdash \bigvee \Sigma \preccurlyeq A$. *Moreover* $\mathcal{M}$ *is a* $\mathsf{N}^*_{\preccurlyeq}$*-model.*

Proof. The claims (i), (ii) and (iii) are proved simultaneously by induction on the following notion of complexity of formulas and blocks: $c(p) = c(\bot) = 1$, $c(A \to B) = c(A \preccurlyeq B) = c(A) + c(B) + 1$, $c([B_1, ..., B_n \lhd A]) = c(B_1) + ... + c(B_n) + c(A)$. For $A = p, \bot, B \to C$ the proof is routine. We consider the case $A = B \preccurlyeq C$. ($B \preccurlyeq C \in \Gamma_n$) Suppose $\alpha \in N(n)$. By definition, $\alpha = \Sigma^\Delta$ for some Σ such that there is $[\Sigma \lhd D] \in \Delta_n$. Then by saturation of $\preccurlyeq_\mathsf{L}$, $C \in \Sigma$ or there is Π such that $set(\Sigma) \subseteq set(\Pi)$ and $[\Pi \lhd B] \in \Delta_n$. In the first case, for every $m \in \Sigma^\Delta$, $C \in \Delta_m$, then by i.h., $m \nVdash C$. Therefore $\Sigma^\Delta \nVdash^\exists C$. In the second case, by saturation of jp there is $m \in W$ such that $B \in \Gamma_m$ and $set(\Pi) \subseteq \Delta_m$, thus $set(\Sigma) \subseteq \Delta_m$. Then by i.h., $m \Vdash B$, and by definition $m \in \Sigma^\Delta$. Therefore $\Sigma^\Delta \Vdash^\exists B$. It follows $n \Vdash B \preccurlyeq C$. ($B \preccurlyeq C \in \Delta_n$) By saturation of $\preccurlyeq_\mathsf{R}$, there is $[\Sigma \lhd C] \in \Delta_n$ such that $B \in \Sigma$. Then by definition, $\Sigma^\Delta \in N(n)$, and by i.h., $m \nVdash B$ for every $m \in \Sigma^\Delta$, that is $\Sigma^\Delta \nVdash^\exists B$. Moreover, by saturation of jp there is $m \in W$ such that $set(\Sigma) \subseteq \Delta_m$ and $C \in \Gamma_m$. Then by i.h., $m \Vdash C$, and by definition $m \in \Sigma^\Delta$, thus $\Sigma^\Delta \Vdash^\exists C$. Therefore $n \nVdash B \preccurlyeq C$. ($[\Sigma \lhd A] \in \Delta_n$) Analogous to the previous item, considering that by i.h. $m \nVdash B$ for all $m \in \Sigma^\Delta$ and $B \in \Sigma$, that is $\Sigma^\Delta \nVdash^\exists \bigvee \Sigma$.

We now show that $\mathcal{M}$ satisfies the conditions of $\mathsf{N}^*_{\preccurlyeq}$-models. (Non-emptyness) If $\alpha \in N(n)$, then $\alpha = \Sigma^\Delta$ for some Σ such that there is $[\Sigma \lhd A] \in \Delta_n$. Then by saturation of jp, there is $m \in W$ such that $A \in \Gamma_m$ and $set(\Sigma) \subseteq \Delta_m$, thus $m \in \Sigma^\Delta$. (Normality) By saturation of N, there is $[\Sigma, \bot \lhd \top] \in \Delta_n$, thus $(\Sigma, \bot)^\Delta \in N(n)$, that is $N(n) \neq \emptyset$. (Total reflexivity) We modify the definition of the neighbourhood function as follows. For all $n \in W$, let $\mathcal{O}(n) = \bigcup N(n) \cup \{n\}$. Then, define $N^\mathsf{T}(n) = N(n) \cup \mathcal{O}(n)$. We show that the claim (i) above still holds ((ii) and (iii) are proved as before): Suppose $B \preccurlyeq C \in \Delta_n$. As before we can prove that $\Sigma^\Delta \nVdash^\exists C$ or $\Sigma^\Delta \Vdash^\exists B$ for all $[\Sigma \lhd A] \in \Delta_n$. Here we show that the same holds for $\mathcal{O}(n)$. If there is $\alpha \in N(n)$ such that $\alpha \Vdash^\exists B$, then $\mathcal{O}(n) \Vdash^\exists B$. If instead there is no $\alpha \in N(n)$ such that $\alpha \Vdash^\exists B$, then $\alpha \nVdash^\exists C$ for all $\alpha \in N(n)$, that is $\bigcup N(n) \nVdash^\exists C$. Assume by contradiction that $\mathcal{O}(n) \Vdash^\exists C$. Then $n \Vdash C$. Moreover by saturation of T, $C \in \Delta_n$ or there is $[\Pi \lhd B] \in \Delta_n$ such that $\bot \in \Pi$. If $C \in \Delta_m$, then by i.h., $n \nVdash C$, contradicting $n \Vdash C$. If $[\Pi \lhd B] \in \Delta_n$, then by saturation of jp there is $m \in W$ such that $B \in \Gamma_m$ and $set(\Pi) \subseteq \Delta_m$. Then by i.h., $m \Vdash B$, moreover $\Pi^\Delta \in N(n)$ and $m \in \Pi^\Delta$, thus $\Pi^\Delta \Vdash^\exists B$, against the hypothesis. Therefore $\mathcal{O}(n) \nVdash^\exists C$. (Weak centering) If $\alpha \in N(n)$, then $\alpha = \Sigma^\Delta$ for some Σ such that there is $[\Sigma \lhd A] \in \Delta_n$. Then by saturation of W, $set(\Sigma) \subseteq \Delta_n$, thus

$n \in \Sigma^{\Delta}$. (Centering) We modify the definition of the neighbourhood function as $N^{\mathsf{C}}(n) = N(n) \cup \{\{n\}\}$. We show that (i) still holds ((ii) and (iii) are as before): Suppose $B \preccurlyeq C \in \Delta_n$. As before we can prove that $\Sigma^{\Delta} \not\Vdash^{\exists} C$ or $\Sigma^{\Delta} \Vdash^{\exists} B$ for all $[\Sigma \lhd A] \in \Delta_n$. Here we show that the same holds for $\{n\}$. By saturation of C, $C \in \Delta_n$ or $B \in \Gamma_n$, thus by i.h., $m \Vdash B$ or $m \not\Vdash C$, therefore $\{n\}^{\Delta} \Vdash^{\exists} B$ or $\{n\} \not\Vdash^{\exists} C$. (Strong absoluteness) We modify the definition of N as $N^{\mathsf{A}}(n) = \{\Sigma^{\Delta} \mid$ there are $m \in W$ and A such that $[\Sigma \lhd A] \in \Delta_m\}$. We show that (i) still holds ((ii) and (iii) are as before). Suppose $B \preccurlyeq C \in \Gamma_n$ and $\alpha \in N(n)$. Then $\alpha = \Sigma^{\Delta}$ for some $[\Sigma \lhd D] \in \Delta_m$ for some $m \in W$. By saturation of $\mathsf{A_L}$, $B \preccurlyeq C \in \Gamma_m$, then by saturation of $\preccurlyeq_{\mathsf{L}}$, $C \in \Sigma$ or there is Π such that $set(\Sigma) \subseteq set(\Pi)$ and $[\Pi \lhd B] \in \Delta_m$. In the first case, $\Sigma^{\Delta} \not\Vdash^{\exists} C$. In the second case, by saturation of jp there is $k \in W$ such that $B \in \Gamma_k$ and $set(\Pi) \subseteq \Delta_k$, thus $set(\Sigma) \subseteq \Delta_m$, therefore $\Sigma^{\Delta} \Vdash^{\exists} B$. □

Note that, since all rules are cumulative, the claims (i) and (ii) of Prop. 5.7 also hold for the root hypersequent $\mathcal{H}_0$, thus $\mathcal{M}$ is a countermodel of $\mathcal{H}_0$. Moreover, since every proof built in accordance with the strategy either provides a derivation of the root hypersequent, or contains a saturated hypersequent, this result entails a constructive proof of the completeness of $\mathsf{H.N}^*_{\preccurlyeq}$.

Theorem 5.8 (Semantic completeness) *For every hypersequent $\mathcal{H}$, if $\mathcal{H}$ is valid in all $\mathsf{N}^*_{\preccurlyeq}$-models, then $\mathcal{H}$ is derivable in $\mathsf{H.N}^*_{\preccurlyeq}$.*

Here follows an example of the countermodel construction.

Example 5.9 We show that axiom co is not derivable in $\mathsf{N}_{\preccurlyeq}$. Here follows a failed proof of $\Rightarrow (p \preccurlyeq q) \vee (q \preccurlyeq p)$ in $\mathsf{H.N}_{\preccurlyeq}$, where $\mathcal{H}$ is saturated, and $\vee_R$ is admissible from the rules of $\mathsf{N}_{\preccurlyeq}$:

$$
\dfrac{\dfrac{\dfrac{\mathcal{H}:\ \Rightarrow [q \lhd p], [p \lhd q], p \preccurlyeq q, q \preccurlyeq p, (p \preccurlyeq q) \vee (q \preccurlyeq p) \mid q \Rightarrow p \mid p \Rightarrow q}{\Rightarrow [q \lhd p], [p \lhd q], p \preccurlyeq q, q \preccurlyeq p, (p \preccurlyeq q) \vee (q \preccurlyeq p)}\ \mathsf{jp}(\times 2)}{\Rightarrow p \preccurlyeq q, q \preccurlyeq p, (p \preccurlyeq q) \vee (q \preccurlyeq p)}\ \preccurlyeq_{\mathsf{R}}(\times 2)}{\Rightarrow (p \preccurlyeq q) \vee (q \preccurlyeq p)}\ \vee_{\mathsf{R}}
$$

We consider the following enumeration of the components of the saturated hypersequent $\mathcal{H}$: 1: $\Rightarrow [q \lhd p], [p \lhd q], p \preccurlyeq q, q \preccurlyeq p, (p \preccurlyeq q) \vee (q \preccurlyeq p)$; 2: $q \Rightarrow p$; and 3: $p \Rightarrow q$. Then, following the construction of Prop. 5.7 we obtain the following countermodel $\mathcal{M} = \langle W, N, V\rangle$: $W = \{1, 2, 3\}$. $N(1) = \{p^{\Delta}, q^{\Delta}\} = \{\{2\}, \{3\}\}$, and $N(2) = N(3) = \emptyset$. $V(p) = \{3\}$ and $V(q) = \{2\}$. Then we have $p^{\Delta} \Vdash^{\exists} q$ and $p^{\Delta} \not\Vdash^{\exists} p$, thus $1 \not\Vdash p \preccurlyeq q$, moreover $q^{\Delta} \Vdash^{\exists} p$ and $q^{\Delta} \not\Vdash^{\exists} q$, thus $1 \not\Vdash q \preccurlyeq p$. Therefore $1 \not\Vdash (p \preccurlyeq q) \vee (q \preccurlyeq p)$.

6 Conclusions

We introduced CPN logics, which are a generalisation of Lewis' logics of comparative plausibility defined over neighbourhood rather than sphere models. As a difference with sphere models, neighbourhoods need not to be nested, allowing to express more general notions of comparative plausibility. From a proof-theoretic viewpoint, CPN logics are captured by suitable restrictions of sequent calculi for Lewis' logics: they coincide to restrictions of calculi from [14,15] to a single principal $\preccurlyeq$-formula in the right-hand side of sequents, and the single-component formulation of their hypersequent calculi corresponds to the structured calculi from [22,8] without the communication rule.

Overall, CPN logics represent a general theory of comparative plausibility with well-understood proof theory and semantics. Differently from stronger logics expressing comparative plausibility, CPN logics allow to model preference or similarity in situations where no priority order is assumed between states of affairs or concepts. Moreover, CPN logics are an expressive framework, encompassing Lewis' logics [16], which are obtained by adding nesting to CPN logics. In future work we plan to investigate the relations between CPN logics and other well-known comparative plausibility logics introduced in the literature, most notably Halpern's comparative plausibility logics defined over preferential structures [12]. We conjecture that Halpern's logics could be obtained by adding the property of *closure under non-empty intersections* to neighbourhood models, which is required to prove equivalence between neighbourhood and preferential structures. Moreover, we wish to relate our systems with the logic of comparative obligation introduced by Brown [2]. Brown's operator is defined on a kind of neighbourhood models containing a function $\mathcal{R} : W \longrightarrow \mathcal{P}(\mathcal{P}(\mathcal{P}(W)))$, representing a degree of urgency of obligation.

Furthermore, CPN logics parallel the preferential conditional logics studied in [3]. These logics generalise Lewis' counterfactual logics, and admit a neighbourhood semantics, introduced in [11]. Interestingly, while comparative plausibility and conditional entailment are interdefinable in sphere models, the two operators are not interdefinable in neighbourhood semantics, giving rise to two independent theories. While in [11] a proof-theoretical analysis of the conditional operator in neighbourhood semantics is proposed, this work explores the behaviour of the comparative plausibility operator in neighbourhood structures. Moreover, having lost the interdefinability between $\preccurlyeq$ and $>$, we wish to study whether alternative and meaningful notions of conditional entailment can be defined in terms of comparative plausibility. We also intend to study applications of CPN logics, possibly related to the analysis of information sources.

Concerning the proof theory for CPN logics, we wish to analyse the complexity of the logics based on the decision procedure induced by the multi-premisses and the hypersequent calculi. Moreover, we plan to automate the proof search and countermodel construction of the hypersequent calculi within a theorem prover, along the lines of what done in [5,10,7]. We will also investigate extensions of the hypersequent calculi to CPN logics with uniformity, possibly adapting the approach proposed in [9] for Lewis' logics to our setting, as well as with other semantic conditions, aiming at developing a uniform proof-theoretic account of CPN logics.

Acknowledgements. We wish to thank Björn Lellmann for his suggestions and contributions to the analysis of the comparative plausibility operator.

References

[1] Alenda, R., N. Olivetti and C. Schwind, *Comparative concept similarity over minspaces: Axiomatisation and tableaux calculus*, in: *International Conference on Automated Reasoning with Analytic Tableaux and Related Methods*, Springer, 2009, pp. 17–31.

[2] Brown, M. A., *A logic of comparative obligation*, Studia Logica **57** (1996), pp. 117–137.

[3] Burgess, J. P., *Quick completeness proofs for some logics of conditionals.*, Notre Dame Journal of Formal Logic **22** (1981), pp. 76–84.

[4] Dalmonte, T., B. Lellmann, N. Olivetti and E. Pimentel, *Hypersequent calculi for non-normal modal and deontic logics: countermodels and optimal complexity*, Journal of Logic and Computation **31** (2020), pp. 67–111.
URL `https://doi.org/10.1093/logcom/exaa072`

[5] Dalmonte, T., N. Olivetti and G. L. Pozzato, *Hypno: theorem proving with hypersequent calculi for non-normal modal logics (system description)*, in: *International Joint Conference on Automated Reasoning*, Springer, 2020, pp. 378–387.

[6] Friedman, N. and J. Y. Halpern, *On the complexity of conditional logics*, in: J. Doyle, E. Sandewall and P. Torasso, editors, *Principles of Knowledge Representation and Reasoning: Proceedings of the Fourth International Conference (KR'94)*, Morgan Kaufmann Pub, 1994, pp. 202–213.

[7] Girlando, M., B. Lellmann, N. Olivetti, S. Pesce and G. L. Pozzato, *Calculi, countermodel generation and theorem prover for strong logics of counterfactual reasoning*, Journal of Logic and Computation (2022).

[8] Girlando, M., B. Lellmann, N. Olivetti and G. L. Pozzato, *Standard sequent calculi for Lewis' logics of counterfactuals*, in: L. Michael and A. C. Kaks, editors, *European Conference on Logics in Artificial Intelligence*, Springer, 2016, pp. 272–287.

[9] Girlando, M., B. Lellmann, N. Olivetti and G. L. Pozzato, *Hypersequent calculi for lewis' conditional logics with uniformity and reflexivity*, in: *International Conference on Automated Reasoning with Analytic Tableaux and Related Methods*, Springer, 2017, pp. 131–148.

[10] Girlando, M., B. Lellmann, N. Olivetti, G. L. Pozzato and Q. Vitalis, *Vinte: an implementation of internal calculi for lewis' logics of counterfactual reasoning*, in: *International Conference on Automated Reasoning with Analytic Tableaux and Related Methods*, Springer, 2017, pp. 149–159.

[11] Girlando, M., S. Negri and N. Olivetti, *Uniform labelled calculi for preferential conditional logics based on neighbourhood semantics*, Journal of Logic and Computation **31** (2021), pp. 947–997.

[12] Halpern, J. Y., *Defining relative likelihood in partially-ordered preferential structures*, Journal of Artificial Intelligence Research **7** (1997), pp. 1–24.

[13] Kraus, S., D. Lehmann and M. Magidor, *Nonmonotonic reasoning, preferential models and cumulative logics*, Artificial intelligence **44** (1990), pp. 167–207.

[14] Lellmann, B., "Sequent Calculi with Context Restrictions and Applications to Conditional Logic," Ph.D. thesis, Imperial College London (2013).
URL `http://hdl.handle.net/10044/1/18059`

[15] Lellmann, B. and D. Pattinson, *Sequent systems for Lewis' conditional logics*, in: L. F. del Cerro, A. Herzig and J. Mengin, editors, *JELIA 2012*, LNAI **7519**, Springer-Verlag Berlin Heidelberg, 2012 pp. 320–332.

[16] Lewis, D., "Counterfactuals," Blackwell, 1973.

[17] Loreggia, A., E. Lorini and G. Sartor, *Modelling ceteris paribus preferences with deontic logic*, Journal of Logic and Computation **32** (2022).

[18] Lorini, E., *A logic of evaluation*, in: *Proceedings of the 20th International Conference on Autonomous Agents and MultiAgent Systems*, 2021, pp. 827–835.

[19] Marti, J. and R. Pinosio, *Topological semantics for conditionals*, The Logica Yearbook (2013).

[20] Montague, R., *Pragmatics and intensional logic*, Synthese **22** (1970), pp. 68–94.
[21] Negri, S. and N. Olivetti, *A sequent calculus for preferential conditional logic based on neighbourhood semantics*, in: *International Conference on Automated Reasoning with Analytic Tableaux and Related Methods*, Springer, 2015, pp. 115–134.
[22] Olivetti, N. and G. L. Pozzato, *A standard internal calculus for Lewis' counterfactual logics*, in: H. de Nivelle, editor, *Proceedings of the 22nd Conference on Automated Reasoning with Analytic Tableaux and Related Methods (Tableaux 2015)*, Lecture Notes in Artificial Intelligence LNAI **9323**, Springer, 2015, pp. 270–286.
[23] Pacuit, E., "Neighborhood semantics for modal logic," Short Textbooks in Logic, Springer, 2017.
[24] Scott, D., *Advice on modal logic*, in: *Philosophical problems in logic*, Springer, 1970 pp. 143–173.
[25] Sheremet, M., D. Tishkovsky, F. Wolter and M. Zakharyaschev, *Comparative similarity, tree automata, and diophantine equations*, in: *International Conference on Logic for Programming Artificial Intelligence and Reasoning*, Springer, 2005, pp. 651–665.
[26] Sheremet, M., D. Tishkovsky, F. Wolter and M. Zakharyaschev, *A logic for concepts and similarity*, Journal of Logic and Computation **17** (2007), pp. 415–452.
[27] Van Benthem, J., P. Girard and O. Roy, *Everything else being equal: A modal logic for ceteris paribus preferences*, Journal of philosophical logic **38** (2009), pp. 83–125.
[28] Von Wright, G. H., *The logic of preference reconsidered*, Theory and Decision **3** (1972), pp. 140–169.

Modal logic and the polynomial hierarchy: from QBFs to K and back

Anupam Das [1]

University of Birmingham
United Kingdom

Sonia Marin

University of Birmingham
United Kingdom

Abstract

In this work we classify formulas of the basic normal modal logic K into fragments complete for each level of the polynomial time hierarchy, with respect to validity. In particular, we identify a pair of encodings, from true Quantified Boolean Formulas (QBFs) to modal logic and vice-versa, whose composition preserves the number of quantifier alternations. This yields a formal analogue of 'quantifier complexity' within modal logic.
Our translation from QBFs to modal formulas is an optimised version of common translations employed in modal logic solving. In the other direction, we encode proof search itself, for a cut-free sequent calculus, as an alternating time predicate. The aforementioned tight bounds are obtained by carefully calibrating both the optimisation (QBFs to modal logic) and the measurement of proof search complexity (modal logic to QBFs). This approach is inspired by recent work achieving a similar result for the exponential-free fragment of Linear Logic (MALL).

Keywords: Modal Logic, Polynomial Hierarchy. Quantified Boolean Formulas, Proof Search, Sequent Calculus

1 Introduction

Ladner's seminal work [13] showed that a large number of modal logics between K and $S4$ are **PSPACE**-complete. Adding further axioms, such as 5, can simplify the underlying complexity of the validity problem, with $S5$ being *co***NP**-complete. Indeed, the 'gap' between *co***NP**-complete and **PSPACE**-complete normal modal logics has formed the subject of several works in recent years [19,12]. [2]

[1] Supported by a UKRI Future Leaders Fellowship, *Structure vs. Invariants in Proofs.*

[2] Ladner's result was rather **NP**-completeness of the *satisfiability* problem for $S5$. In this work we only consider *validity*, which duly exhibits dual complexity bounds to satisfiability.

That said, as far as we know, attempts to characterise fragments of modal logics corresponding to levels of the polynomial hierarchy (**PH**) have not appeared in the literature. **PH** essentially delineates **PSPACE** according to 'bounded quantifier alternation', e.g. by identifying **PSPACE** with the set of true quantified Boolean formulas (QBFs), another well-known **PSPACE**-complete problem. On the other hand, translations from QBFs to modal logic now comprise a fundamental benchmark in modal satisfiability solving [16,17].

There are many known translations from QBFs to the basic normal modal logic K; some of those (whose variants are) employed for benchmarking modal satisfiability solvers include Ladner's original one [13], a more optimal one due to Schmidt-Schauss and Scholka [21], [3] and that of Pan and Vardi designed to reduce modal solving to QBF solving [20]. These translations, their utility for benchmarking, and the approach to modal solving by QBF-encoding, are now well surveyed, e.g. [14,22,16,17].

However, despite the considerable literature relating QBFs and modal logic, their commonly employed complexity measures do not match up. In modal solving the key measure is that of *modal depth*, the maximal number of modalities in a path through the formula tree, cf. [22,14]. For QBFs the key measure is quantifier complexity, i.e. the number of alternations of $\exists$ and $\forall$ in a (prenex) QBF. While it is well-known that the alternation of quantifiers in QBFs corresponds precisely with the levels of the polynomial hierarchy [24,4], Halpern has showed in [10] that the validity problem for K (with any number of agents) for formulas with modal depth bounded by some constant $d \geq 2$ is in fact only *co***NP**-complete (see also [19]). [4]

It is this shortcoming of 'modal depth' that forms our principal motivation: can we identify a measure for modal formulas that coincides with quantifier complexity for QBFs, formally? In other words, can we find fragments of the modal logic K complete for each level of the polynomial hierarchy?

We answer this question positively in the present work by designing an *inverse* translation from true QBFs back into K. Our key idea is to encode modal provability itself as an alternating predicate, and analyse the alternation between 'invertible' and 'non-invertible' rules during proof search so as to delineate theorems according to the polynomial hierarchy. For this to work, we must first carefully devise a particular translation from QBFs to modal logics that is compatible with alternation in the aforementioned proof search predicate. In particular, composing our two translations yields an automorphism on (true) QBFs that *preserves* quantifier complexity.

---

[3] This translation was originally given for the Description Logic ALC, a notational variant of multi-agent modal logic K_m.

[4] Again, both Halpern and Nguyen study the satisfiability problem and state **NP**-completeness, whereas we study the dual problem of validity, inheriting *co***NP**-completeness from their results.

Related work and methodology

The idea of encoding proof search to obtain upper bounds for modal validity or satisfiability is not new, e.g. this is the approach taken in Halpern and Moses' 'guide' [11], only for a tableau system that is related to the sequent system we consider in this work. However none of the aforementioned works on complexity of modal logics give refinements of **PSPACE**-completeness to levels of the polynomial hierarchy, regardless of the method employed. [5]

This paper builds on recent work achieving similar delineations for multiplicative additive linear/affine logic [7,8] and fragments of intuitionistic logic [6], also well-known **PSPACE**-complete logics. Those works leveraged (alternative presentations of) *focussed* systems from structural proof theory (see, e.g., [1,15]), which elegantly control the alternation of invertible and non-invertible rules during proof search, the principal contributor to quantifier alternation in a proof search predicate.

(Normal) modal logics such as K (and, indeed, the entire 'modal $S5$ cube') have also recently received focussed treatments in the setting of *labelled* sequents [18] and *nested* sequents [5]. However nested and labelled systems, while admitting an elegant proof theory, do not enjoy terminating proof search per se, and thus are not adequate for obtaining alternating time bounds (see further discussion in Conclusions, Sec. 7). Instead we work with a standard cut-free sequent calculus for K and give a bespoke analysis of the proof search space according to invertible and non-invertible rules that suffices for our purposes.

2 Preliminaries on modal logic and (true) QBFs

Both of the logics we consider in this work are extensions of usual *classical propositional logic* (CPC). So we shall start by presenting CPC before duly extending it. Our exposition will be brief throughout this preliminary section, but we refer the reader to standard texts on modal logic [3] and proof theory [25] for further details.

We assume a countable set Var of *propositional variables*, written x, y, z etc. *(Propositional) formulas*, written A, B, C etc., are always in De Morgan normal form (i.e. with negation reduced to the variables) and are generated by the following grammar:

$$A, B \quad ::= \quad x \quad | \quad \bar{x} \quad | \quad (A \vee B) \quad | \quad (A \wedge B)$$

We assume usual bracketing conventions, i.e. omitting internal brackets of large disjunctions or conjunctions under associativity and external brackets too. Note that we may recover negation by extending the notation $\bar{\cdot}$ to all formulas by setting:

$$\bar{\bar{x}} := x \qquad \overline{A \vee B} := \bar{A} \wedge \bar{B} \qquad \overline{A \wedge B} := \bar{A} \vee \bar{B}$$

[5] Let us point out too that such delineations do not seem to be known for desription logics either, according to the online *Description Logic Complexity Navigator* [26].

We may employ standard logical abbreviations, e.g. writing $A \supset B$ for $\bar{A} \vee B$ and $A \equiv B$ for $(A \supset B) \wedge (B \supset A)$.

Semantics

A *(Boolean) assignment* is a map $\alpha : \mathsf{Var} \to \{0, 1\}$. We write Ass for the set of all Boolean assignments. We may *evaluate* a Boolean formula with respect to an assignment by setting:

- $\alpha \vDash x$ if $\alpha(x) = 1$.
- $\alpha \vDash \bar{x}$ if $\alpha(x) = 0$.
- $\alpha \vDash A \vee B$ if $\alpha \vDash A$ or $\alpha \vDash B$.
- $\alpha \vDash A \wedge B$ if $\alpha \vDash A$ and $\alpha \vDash B$.

If $\alpha \vDash A$ then we say that α *satisfies* A. Note that, to evaluate a formula A with free variables among vector $\boldsymbol{x}$, we need only consider finite 'partial' assignments with domain containing $\boldsymbol{x}$.

If every $\alpha \in \mathsf{Ass}$ satisfies A, we say that A is *valid* and write simply $\vDash A$. The logic CPC is the set of valid propositional formulas.

Proof theory

CPC admits well-known Hilbert axiomatisations via axioms and rules, but we shall here rather present a standard (one-sided) sequent system for later use. The definitions and results of this subsection are based on analogous ones in [25].

A *sequent*, written Γ, Δ etc., is a multiset of formulas. We usually omit braces when writing such multisets and use the comma for multiset union, e.g. writing simply $A_1, \ldots, A_n$ for the multiset $\{A_1, \ldots, A_n\}$. We may interpret sequents as the disjunction of their formulas, in particular saying that a sequent Γ is valid or satisfied by an assignment just if $\bigvee \Gamma$ is.

Definition 2.1 (System for CPC) The system (one-sided, propositional) G3c is given by the following three rules:

$$\mathsf{id}\,\frac{}{\Gamma, x, \bar{x}} \qquad \vee\,\frac{\Gamma, A, B}{\Gamma, A \vee B} \qquad \wedge\,\frac{\Gamma, A \quad \Gamma, B}{\Gamma, A \wedge B}$$

Derivations and proofs are defined as usual for inference systems. We write $\vdash \Gamma$ if there is a G3c-proof of the sequent Γ.

Remark 2.2 (Judgements) Note that we have indexed neither our satisfaction judgement $\vDash$ nor our provability judgement $\vdash$ by the logic or system in question. This is intentional since later logics and systems we consider are bona fide extensions of these. At the level of proofs, cut-freeness of our systems will guarantee that our notion of provability is unambiguous: any proof of Γ (in any of our systems) will contain only subformulas of Γ.

Proposition 2.3 (Soundness and completeness, CPC) *Let A be a propositional formula. $\vDash A$ if and only if $\vdash A$.*

2.1 Extension to modal logic

Modal formulas are generated by extending the grammar of propositional formulas by:

$$A, B \quad ::= \quad \dots \quad | \quad \Diamond A \quad | \quad \Box A$$

We extend the notation $\bar{A}$ to all modal formulas by setting,

$$\overline{\Diamond A} := \Box \bar{A} \qquad \overline{\Box A} := \Diamond \bar{A}$$

and admit logical abbreviations, e.g. $A \supset B$, as in propositional logic earlier.

We consider usual *relational semantics* for modal formulas, as found in, e.g., [3]. Note that we are only considering the basic normal modal logic K in this work.

Definition 2.4 (Relational semantics) A *(relational) structure* is a binary relation $R \subseteq W \times W$, where W is a set whose elements we call *worlds*. A *(relational) model* $\mathcal{M} = (W, R, \nu)$ is a relational structure $R \subseteq W \times W$ equipped with a *valuation* $\nu : W \to \mathsf{Ass}$. We sometimes write $|\mathcal{M}|$ for W (the *domain*).

Given a model $\mathcal{M} = (W, R, \nu)$ and some $w \in W$, we associate with the pair $(\mathcal{M}, w)$ the assignment $\nu(w)$. In this way, we define the judgement $\mathcal{M}, w \vDash A$ just like for CPC earlier, with the following additional clauses:

- $\mathcal{M}, w \vDash \Diamond A$ if there is $w' \in W$ s.t. wRw' and $\mathcal{M}, w' \vDash A$.
- $\mathcal{M}, w \vDash \Box A$ if, whenever wRw' for $w' \in W$, we have $\mathcal{M}, w' \vDash A$.

Similarly to propositional logic, we say that w *satisfies* A in $\mathcal{M}$ if $\mathcal{M}, w \vDash A$, and that A is *valid* (written $\vDash A$) if A is satisfied by every world in every model. The logic K is the set of valid modal formulas.

Let us write a, b, etc. for formulas of the form x or $\bar{x}$ (called *literals*).

Definition 2.5 (System for K) The calculus G3k is the extension of G3c by the rule:

$$\mathsf{k}\,\frac{\Gamma, A_i}{a, \Diamond\Gamma, \Box A_0, \dots, \Box A_{n-1}}\; i < n$$

We again write $\vdash A$ if there is a G3k-proof of A.

As expected, Prop. 2.3 extends to modal formulas:

Proposition 2.6 (Soundness and completeness, K) *Let A be a modal formula. $\vDash A$ if and only if $\vdash A$.*

2.2 Extension to second-order propositional logic

Quantified Boolean Formulas (QBFs) are generated by extending the grammar of propositional formulas by:

$$A, B \quad ::= \quad \dots \quad | \quad \exists x A \quad | \quad \forall x A$$

We refer to 'free' and 'bound' variables for QBFs in the usual way and write FV(A) for the set of free variables in the QBF A. A QBF A is *closed* if

$\mathrm{FV}(A) = \varnothing$, i.e. A has no free variables. We extend the notation $\bar{A}$ to all QBFs A by setting,

$$\overline{\exists x A} := \forall x \bar{A} \qquad \overline{\forall x A} := \exists x \bar{A}$$

and admit logical abbreviations, e.g. $A \supset B$, as in propositional logic earlier.

We shall often further assume that each quantifier of a QBF binds a distinct variable. Finally, we shall typically only work with QBFs in *prenex normal form*, i.e. of the form $Q_1 x_1 \dots Q_n x_n A$, for some $n \geq 0$, quantifiers $Q_1, \dots, Q_n$, and A quantifier-free (so a propositional formula). It is well known that each QBF is equivalent to one in prenex normal form, in terms of the following semantics:

Definition 2.7 (QBF semantics) Given an assignment α and a QBF A, we define the judgement $\alpha \models A$ just like for CPC, with the following additional clauses,

- $\alpha \models \exists x A$ if $\alpha[x \mapsto 0] \models A$ or $\alpha[x \mapsto 1] \models A$.
- $\alpha \models \forall x A$ if $\alpha[x \mapsto 0] \models A$ and $\alpha[x \mapsto 1] \models A$.

where we write $\alpha[x \mapsto i]$ for the assignment defined just like α but mapping x to $i \in \{0, 1\}$. We duly extend the terminology 'satisfies' and 'valid' from propositional formulas to arbitrary QBFs. A closed QBF is called simply *true* if it is valid (equivalently, satisfiable). The logic CPC2 is the set of valid QBFs.

We will not actually work with proofs for CPC2 in this work, but we include a system for it below for completeness.

Definition 2.8 (System for CPC2) The calculus G3c2 is the extension of G3c by the following rules,

$$\exists \frac{\Gamma, \exists x A, A[B/x]}{\Gamma, \exists x A} \qquad \forall \frac{\Gamma, A[y/x]}{\Gamma, \forall x A} \; y \text{ not free in } \Gamma$$

where we write $A[B/x]$ for the QBF resulting from substituting each free occurrence of x in A by the QBF B.

Again, Prop. 2.3 extends to QBFs as expected:

Proposition 2.9 (Soundness and completeness, CPC2) *Let A be a QBF. $\models A$ if and only if $\vdash A$.*

2.3 Some examples and comments on proof search

Let us take the time to consider some examples of proofs in our systems, in particular to highlight some of the proof search dynamics that will later come into play. We intentionally choose rather simple validities/proofs.

First, consider the modal formula:

$$\Box x \vee \Box y \vee (\Diamond(x \supset y) \wedge \Diamond(y \supset x)) \tag{1}$$

It is not hard to see that this is valid in K, in particular thanks to the following proof in G3k under Prop. 2.6:

$$
2\vee\,\dfrac{\wedge\,\dfrac{\mathsf{k}\,\dfrac{\vee\,\dfrac{\mathsf{id}\,\dfrac{}{x,\bar{x},y}}{x, x\supset y}}{\Box x,\Box y,\Diamond(x\supset y)} \qquad \mathsf{k}\,\dfrac{\vee\,\dfrac{\mathsf{id}\,\dfrac{}{y,\bar{y},x}}{y, y\supset x}}{\Box x,\Box y,\Diamond(y\supset x)}}{\Box x,\Box y,\Diamond(x\supset y)\wedge\Diamond(y\supset x)}}{\Box x\vee\Box y\vee(\Diamond(x\supset y)\wedge\Diamond(y\supset x))} \tag{2}
$$

Viewing proof search here as an alternating predicate, we can see the $\wedge$ rule as *universal branching*: every premiss must be valid. The k rule, on the other hand, is an example of *existential branching*: some choice of premiss must be valid. Finally, the $\vee$ and id rules may be viewed as 'deterministic': they have no computational cost. We will start making these classifications formal in Sec. 5.

We may 'derive' from (1) a valid QBF by mimicking its formula structure,

$$
\forall x A(x)\vee\forall y B(y)\vee(\exists z(A(z)\supset B(z))\wedge\exists z'(B(z')\supset A(z')))
$$

for some arbitrary formulas $A(x)$ and $B(y)$ (possibly with further free variables). This may be put into the following prenex normal form:

$$
\forall x\forall y\exists z\exists z'(A(x)\vee B(y)\vee((A(z)\supset B(z))\wedge(B(z')\supset A(z'))) \tag{3}
$$

In this case, notice that the quantifier prefix above matches the aforementioned universal-existential-branching exhibited during proof search for (1). This is coincidental for this particular case, since we have not yet properly fixed an appropriate translation from modal formulas to QBFs, but arriving at a formal such correspondence constitutes the principal aim of this work.

3 Alternating complexity and some decision problems

We shall assume basic familiarity with (non)deterministic Turing machines (with oracles) and their time and space complexity, for which there are several basic references available, e.g. [2]. Let us also point out that this preliminary section is similar to the analogous one in [8], where there are further details.

For the sake of formality, all languages we consider will be subsets of $\{0,1\}^*$. Throughout we may consider larger (but finite) alphabets than $\{0,1\}$, but these should always be assumed to be adequately coded in binary.

3.1 Some complexity classes

Let us fix **P** (and **PSPACE**) as the class of languages accepted by a deterministic Turing machine in polynomial time (resp., space).

$\mathbf{NP}(L)$ is the class of languages accepted by a nondeterministic Turing Machine with access to an oracle for the language L in polynomial time. Intuitively, such a machine acts just like a usual nondeterministic machine but

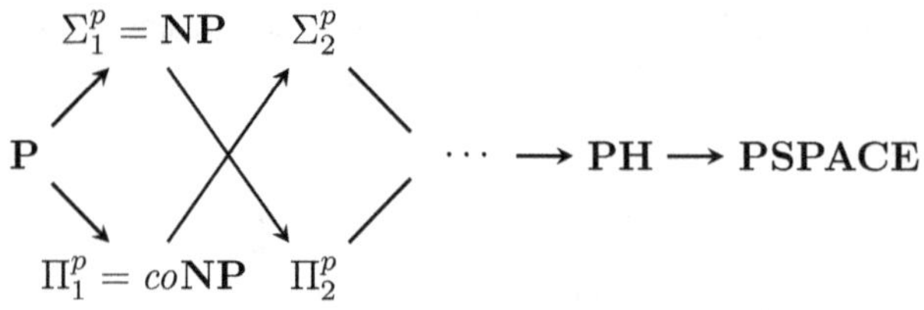

Fig. 1. Relationships between complexity classes. An arrow $\mathcal{C} \to \mathcal{D}$ means the complexity class $\mathcal{C}$ is contained in the complexity class $\mathcal{D}$.

may at any point query (in constant time) whether a string is in L or not. Given a class $\mathcal{C}$ of languages, $\mathbf{NP}(\mathcal{C}) := \bigcup_{L \in \mathcal{C}} \mathbf{NP}(L)$. Finally, given a class $\mathcal{C}$ of languages, $co\mathcal{C}$ is the set of complements of $\mathcal{C}$, i.e. $co\mathcal{C} := \{\{0,1\}^* \setminus L : L \in \mathcal{C}\}$.

Definition 3.1 (Polynomial hierarchy) We define the classes Σ_n^p and Π_n^p, for $n \geq 1$, as follows:

- $\Sigma_1^p := \mathbf{NP}$
- $\Sigma_{n+1}^p := \mathbf{NP}(\Sigma_n^p)$
- $\Pi_n^p := co\Sigma_n^p$

We write $\mathbf{PH} := \bigcup_{n \geq 1} \Sigma_n^p$ (equivalently, $\bigcup_{n \geq 1} \Pi_n^p$).

The following relationships are (almost) immediate from definitions:

(i) $\mathbf{P} \subseteq \mathbf{NP} \cap co\mathbf{NP}$.

(ii) $\Sigma_n^p \subseteq \Pi_{n+1}^p$ and $\Pi_n^p \subseteq \Sigma_{n+1}^p$.

(iii) $\mathbf{PH} \subseteq \mathbf{PSPACE}$.

Formally speaking, the final point is not quite immediate from definitions but follows since $\mathbf{NP}(\mathbf{PSPACE}) \subseteq \mathbf{NPSPACE} \subseteq \mathbf{PSPACE}$ by Savitch's theorem and since **PSPACE**, being a deterministic class, is closed under complementation. These inclusions are visualised in Fig. 1.

3.2 Complexity of QBFs and modal logic

As we have already mentioned, the complexity of checking validity for both modal formulas and QBFs are well-known. The results we state in this subsection are taken from (or implied by) [13] and [4].

Proposition 3.2 K *and* CPC2 *are* **PSPACE**-*complete.*

Note that the same complexity bound holds for the corresponding satisfiability problems, thanks to closure of **PSPACE** under complements, by a usual reduction: A is satisfiable (or valid) if and only if $\bar{A}$ is not valid (resp., satisfiable). On the other hand, the complexity of model checking for the two logics is (presumably) rather different:

Proposition 3.3 (Complexity of satisfaction) *We have the following:*

(i) *Checking* $\mathcal{M}, w \models A$ *is in* **P**, *for a finite model* $\mathcal{M}$, *some* $w \in |\mathcal{M}|$ *and* A *a modal formula.*

(ii) *Checking* $\alpha \models A$ *is* **PSPACE**-*complete, for a finite partial assignment* α *with domain* $\boldsymbol{x}$ *and* A *a QBF with free variables among* $\boldsymbol{x}$.

In the case of QBFs we can give a refinement that will also refine **PSPACE**-completeness of CPC2, and that we shall exploit later in Sec. 5 to reduce proof search to QBF satisfaction. First, we recall another hierarchy:

Definition 3.4 (QBF hierarchy) We define the following classes of QBFs (in prenex normal form):

- $\Sigma_0^q = \Pi_0^q$ is the class of quantifier-free QBFs (i.e. propositional formulas).
- $\Sigma_{n+1}^q := \{\exists \boldsymbol{x} A : A \in \Pi_n^q\}$
- $\Pi_{n+1}^q := \{\forall \boldsymbol{x} A : A \in \Sigma_n^q\}$

It is well-known that the levels of the QBF hierarchy above match up precisely with the levels of the polynomial hierarchy from Dfn. 3.1:

Proposition 3.5 *We have the following, for* $n \geq 1$:

(i) $\{(\alpha, A) : A \in \Sigma_n^q, \alpha : \mathrm{FV}(A) \to \{0,1\}\}$ *is* Σ_n^p-*complete.*

(ii) $\{(\alpha, A) : A \in \Pi_n^q, \alpha : \mathrm{FV}(A) \to \{0,1\}\}$ *is* Π_n^p-*complete.*

An immediate consequence of this, by way of Boolean simplification under an assignment, is the following well-known delineation of CPC2 according to levels of **PH**:

Corollary 3.6 *We have the following, for* $n \geq 1$:

(i) *The set of true* Σ_n^q *sentences is* Σ_n^p-*complete.*

(ii) *The set of true* Π_n^q *sentences is* Π_n^p-*complete.*

One of the points of this work is to establish a similar such delineation for modal logic K.

4 From QBFs to K

In this section we present our first translation, from QBFs to modal formulas, inspired somewhat by Statman's translation from QBFs into intuitionistic propositional logic [23], only avoiding the need for 'extension' variables. The translation is polynomial-time computable, and induces an encoding from CPC2 to K, i.e. it is a polynomial-time reduction from CPC2 to K.

Let us write $e(A)$ for the number of existential quantifiers in a QBF A (typically in prenex form).

Definition 4.1 (The $\cdot^\bullet$-translation) For a prenex QBF A we define a modal formula $A^\bullet$ as follows:

- $A^\bullet := A$ if A is quantifier-free.
- $(\exists x A)^\bullet := (\Diamond \Box^{e(A)} x \wedge \Diamond \Box^{e(A)} \bar{x}) \supset \Diamond A^\bullet$

- $(\forall x A)^\bullet := (\Box^{e(A)} x \vee \Box^{e(A)} \bar{x}) \supset A^\bullet$

Let us point out that it is clear that $\cdot^\bullet$ is polynomial-time computable.

Example 4.2 Consider a (not necessarily valid) QBF $\exists x \forall y \exists z A$, for some arbitrary quantifier-free formula A, possibly with free variables. We have:

$$
\begin{aligned}
A^\bullet &= A \\
(\exists z A)^\bullet &= (\Diamond z \wedge \Diamond \bar{z}) \supset \Diamond A \\
(\forall y \exists z A)^\bullet &= (\Box y \vee \Box \bar{y}) \supset (\Diamond z \wedge \Diamond \bar{z}) \supset \Diamond A \\
(\exists x \forall y \exists z A)^\bullet &= (\Diamond \Box x \vee \Diamond \Box \bar{x}) \supset \Diamond((\Box y \vee \Box \bar{y}) \supset (\Diamond z \wedge \Diamond \bar{z}) \supset \Diamond A)
\end{aligned}
$$

Theorem 4.3 *Let A be a closed QBF. $\vDash A$ if and only if $\vDash A^\bullet$.*

To prove this we need an intermediate result. First we set up some notation. Let us write $x^0 := \bar{x}$ and $x^1 := x$ and, for an assignment α, simply $x^\alpha := x^{\alpha(x)}$. Given variables $\boldsymbol{x} = x_1, \dots, x_k$ we shall write $\Box^n \boldsymbol{x}^\alpha := \Box^n x_1^\alpha, \dots, \Box^n x_k^\alpha$ and similarly $\Diamond^n \boldsymbol{x}^\alpha := \Diamond^n x_1^\alpha, \dots, \Diamond^n x_k^\alpha$. Finally, as a temporary abuse of notation to facilitate readability, we shall write $A_1, \dots, A_n \vdash B$ if $\vdash \bar{A}_1, \dots, \bar{A}_n, B$.[6] Instead of provability, here '$\vdash$' should be (temporarily) considered as the 'sequent arrow' in a two-sided calculus, which is a common notation.

Lemma 4.4 *Let A be a prenex QBF with free variables among $\boldsymbol{x}$ and α an assignment. We have that $\alpha \vDash A$ if and only if $\Box^{e(A)} \boldsymbol{x}^\alpha \vdash A^\bullet$.*

Let us point out that this lemma holds crucially due to the 'balanced' structure of the $\cdot^\bullet$-translation, in terms of modal depth. Before giving the proof, let us revisit the earlier example.

Example 4.5 Consider again a QBF of the form $\exists x \forall y \exists z A$ like in Ex. 4.2, now setting $A = x \wedge (y \equiv z)$ so that we have a validity. Notice that we have both $x, y, z \vdash A$ and $x, \bar{y}, \bar{z} \vdash A$ by some propositional rules. Now several examples of Lem. 4.4 are found within the following proof with grey background corresponding, respectively, to the $\cdot^\bullet$ instances from Ex. 4.2:

$$
{\scriptstyle \vee}\dfrac{{\scriptstyle \mathsf{k}}\dfrac{{\scriptstyle \vee}\dfrac{{\scriptstyle \wedge}\dfrac{{\scriptstyle 2\vee}\dfrac{{\scriptstyle \mathsf{k}}\dfrac{\overline{\overline{x, y, z \vdash A}}}{\Box x, \Box y, \Diamond z, \Diamond \bar{z} \vdash \Diamond A}}{\Box x, \Box y \vdash (\Diamond z \wedge \Diamond \bar{z}) \supset \Diamond A} \qquad {\scriptstyle 2\vee}\dfrac{{\scriptstyle \mathsf{k}}\dfrac{\overline{\overline{x, \bar{y}, \bar{z} \vdash A}}}{\Box x, \Box \bar{y}, \Diamond z, \Diamond \bar{z} \vdash \Diamond A}}{\Box x, \Box \bar{y} \vdash (\Diamond z \wedge \Diamond \bar{z}) \supset \Diamond A}}{\Box x, \Box y \vee \Box \bar{y} \vdash (\Diamond z \wedge \Diamond \bar{z}) \supset \Diamond A}}{\Box x \vdash (\Box y \vee \Box \bar{y}) \supset (\Diamond z \wedge \Diamond \bar{z}) \supset \Diamond A}}{\Diamond \Box x \vdash \Diamond((\Box y \vee \Box \bar{y}) \supset (\Diamond z \wedge \Diamond \bar{z}) \supset \Diamond A)}}{\vdash \Diamond \Box x \supset \Diamond((\Box y \vee \Box \bar{y}) \supset (\Diamond z \wedge \Diamond \bar{z}) \supset \Diamond A)}
$$

Proof of Lem. 4.4. We proceed by induction on the number of quantifiers in A. The base case, when A is quantifier-free, is trivial.

---

[6] Note that this really is an abuse of notation as the deduction theorem fails for K.

In the universal case we have:

$$\begin{array}{lll} & \alpha \vDash \forall x A & \\ \iff & \alpha[x \mapsto 0] \vDash A \quad \text{and } \alpha[x \mapsto 1] \vDash A & \text{by definition of } \vDash \\ \iff & \Box^{e(A)}\boldsymbol{x}^{\alpha}, \Box^{e(A)}\bar{x} \vdash A^{\bullet} \text{ and } \Box^{e(A)}\boldsymbol{x}^{\alpha}, \Box^{e(A)}x \vdash A^{\bullet} & \text{by inductive hypothesis} \\ \iff & \Box^{e(A)}\boldsymbol{x}^{\alpha}, \Box^{e(A)}\bar{x} \vee \Box^{e(A)}x \vdash A^{\bullet} & \text{by } \wedge \text{ rule} \\ \iff & \Box^{e(A)}\boldsymbol{x}^{\alpha} \vdash (\Box^{e(A)}\bar{x} \vee \Box^{e(A)}x) \supset A^{\bullet} & \text{by } \vee \text{ rule} \\ \iff & \Box^{e(\forall x A)}\boldsymbol{x}^{\alpha} \vdash (\forall x A)^{\bullet} & \text{by definition of } \cdot^{\bullet} \end{array}$$

The existential case is a little more subtle, so we treat the two directions separately.

$$\begin{array}{lll} \alpha \vDash \exists x A & \Longrightarrow \alpha[x \mapsto i] \vDash A & \text{for some } i \in \{0,1\} \\ & \Longrightarrow \Box^{e(A)}\boldsymbol{x}^{\alpha}, \Box^{e(A)}x^{i} \vdash A^{\bullet} & \text{by inductive hypothesis} \\ & \Longrightarrow \Box^{e(A)+1}\boldsymbol{x}^{\alpha}, \Diamond\Box^{e(A)}x, \Diamond\Box^{e(A)}\bar{x} \vdash \Diamond A^{\bullet} & \text{by } \mathsf{k} \\ & \Longrightarrow \Box^{e(\exists x A)}\boldsymbol{x}^{\alpha} \vdash (\Diamond\Box^{e(A)}x \wedge \Diamond\Box^{e(A)}\bar{x}) \supset \Diamond A^{\bullet} & \text{by } \vee \text{ rule} \\ & \Longrightarrow \Box^{e(\exists x A)}\boldsymbol{x}^{\alpha} \vdash (\exists x A)^{\bullet} & \text{by definition of } \cdot^{\bullet} \end{array}$$

For the other direction the steps are quite similar, but the justifications are different:

$$\begin{array}{lll} & \Box^{e(\exists x A)}\boldsymbol{x}^{\alpha} \vdash (\exists x A)^{\bullet} & \\ \Longrightarrow & \Box^{e(A)+1}\boldsymbol{x}^{\alpha} \vdash (\Diamond\Box^{e(A)}x \wedge \Diamond\Box^{e(A)}\bar{x}) \supset \Diamond A^{\bullet} & \text{by definition of } \cdot^{\bullet} \\ \Longrightarrow & \Box\Box^{e(A)}\boldsymbol{x}^{\alpha}, \Diamond\Box^{e(A)}x, \Diamond\Box^{e(A)}\bar{x} \vdash \Diamond A^{\bullet} & \text{proof must end with } \vee \text{ steps} \\ \Longrightarrow & \Box^{e(A)}\boldsymbol{x}^{\alpha}, \Box^{e(A)}x^{i} \vdash A^{\bullet} & \text{proof must end with } \mathsf{k} \\ \Longrightarrow & \alpha[x \mapsto i] \vDash A & \text{by inductive hypothesis} \\ \Longrightarrow & \alpha \vDash \exists x A & \end{array}$$

Note that we have alluded several times to proof search in system G3k above, crucially taking advantage of its cut-free nature. □

From here it is easy to deduce the main result of this section:

Proof of Thm. 4.3. Follows directly as a special case of Lem. 4.4, since A has no free variables, under soundness and completeness, Prop. 2.6. □

5 Proof search as an alternating time predicate

In this section we build up some of the theory of alternation complexity of proof search in G3k that we will need to ultimately define our 'inverse' translation to $\cdot^{\bullet}$. In particular, we encode proof search in G3k as a family of predicates, parametrised by their corresponding level in the polynomial hierarchy, and thus inducing polynomial-size families of QBFs computing proof search in G3k.

Some of the notions and results here are similar to ones appearing in [7,8], but are formulated necessarily bespoke to the calculus G3k.

5.1 The proof search hierarchy

First, in what follows, we shall classify the rules of G3k as follows:

- **Deterministic.** The id and $\vee$ rules.

- **Nondeterministic.** The k rule.
- **Co-nondeterministic.** The $\wedge$ rule.

The nomenclature above is suggestive, indicating the alternation cost of each rule application during proof search, as we previously hinted at in Subsec. 2.3. From here we can delineate the proof search space according to the number of alternations between nondeterministic and co-nondeterministic phases of rules.

Definition 5.1 (Proof search hierarchy) We define the classes Σ^s_n and Π^s_n of (provable) sequents as follows:

- $\Sigma^s_0 = \Pi^s_0$ is the class of sequents provable using only deterministic rules.
- Σ^s_{n+1} is the class of sequents derivable from sequents in Π^s_n using only deterministic and nondeterministic rules.
- Π^s_{n+1} is the class of sequents derivable from sequents in Σ^s_n using only deterministic and co-nondeterministic rules.

Example 5.2 Revisiting the examples from Subsec. 2.3, we have that the modal formula (1) is in Π^s_2. To see this, let us inspect the proof of (1) in (2). $\Box x, \Box y, \Diamond(x \supset y)$ and $\Box x, \Box y, \Diamond(y \supset x)$ are in Σ^s_1, since their subproofs consist of only $\mathsf{id}, \vee$ rules (deterministic) and k rules (nondeterministic). From these two sequents (1) is derived using only $\vee$ rules (deterministic) and $\wedge$ rules (co-nondeterministic), and so indeed (1) $\in \Pi^s_2$.

Using similar methods to those in [8,7], we may prove the following:

Theorem 5.3 *For $n \geq 0$ we have:*

(i) Σ^s_n *is decidable in* Σ^p_n*; and,*

(ii) Π^s_n *is decidable in* Π^p_n*,*

Proof. We proceed by induction on $n \geq 0$.

Let $\Gamma \in \Sigma^s_0$, then there exists a sequent Γ' and a variable x such that Γ can be derived from the axiom $\Gamma', x, \bar{x}$ using only disjuction rules.

$$\mathsf{id}\,\frac{}{\Gamma', x, \bar{x}} \atop {\vee \| \atop \Gamma}$$

This means that to check whether a given sequent is in Σ^s_0, given that the disjuction rule is invertible, we can apply it maximally (linearly many steps in the size of Γ) and check whether we end up on a sequent that contains $x, \bar{x}$ for some variable x. Therefore, Σ^s_0 (and Π^s_0) provability is decidable in **P**.

$\Gamma \in \Sigma^s_{n+1}$ just if there exists a sequent $\Gamma_m, A_m \in \Pi^s_n$ such that:

$$
\begin{array}{c}
k\,\dfrac{\Gamma_m, A_m}{\boldsymbol{a}_m, \Diamond\Gamma_m, \Box\Delta_m}\, A_m \in \Delta_m \\
\| \\
\vdots \\
\| \\
k\,\dfrac{\Gamma_1, A_1}{\boldsymbol{a}_1, \Diamond\Gamma_1, \Box\Delta_1}\, A_1 \in \Delta_1 \\
\vee \| \\
\Gamma
\end{array}
$$

This means that to check whether a given sequent Γ is in Σ^s_{n+1} we may guess the (polynomial-size) configuration above and then check that $\Gamma_m, A_m \in \Pi^s_n$. By the inductive hypothesis the latter is possible in Π^p_n and so the entire procedure is in $\mathbf{NP}(\Pi^p_n) = \Sigma^p_{n+1}$ as required.

$\Gamma \in \Pi^s_{n+1}$ just if there exist some sequents $\Gamma_i \in \Sigma^s_n$ such that

$$
\begin{array}{c}
\Gamma_1 \quad \dots \quad \Gamma_m \\
\\
\vee + \wedge \\
\\
\Gamma
\end{array}
$$

Note that we may assume that $\vee$ and $\wedge$ rules are applied maximally above as both rules are invertible and simplify sequents, bottom-up. Now, Γ is *not* Π^s_{n+1} provable just if there exists a (polynomial-size) branch in the finite derivation above from Γ to some Γ_i that is *not* Σ^s_n-provable. By induction hypothesis, non-Σ^s_n-provability is checkable in Π^p_n, and so we conclude that non-Π^s_{n+1}-provability is in $\mathbf{NP}(\Pi^p_n) = \Sigma^p_{n+1}$. Finally, by definition of complements, this means that Π^s_{n+1}-provability is in Π^p_{n+1}, as required. □

An immediate consequence of the Thm. 5.3 above, under Prop. 3.5, is:

Corollary 5.4 *For $k \geq 1$, there are polynomial-size Σ^q_k-QBFs Σ^s_k-*Prov_n*, and Π^q_k-QBFs Π^s_k-*Prov_n*, computing Σ^s_k-provability, and Π^s_k-provability respectively, on formulas A with $|A| \leq n$.*

5.2 (Co-)nondeterministic complexity of proof search

We now give a complexity measure for (provable) sequents that coincides with our proof search hierarchy earlier, but that later admits a feasible 'approximation' that will facilitate our inverse translation from modal formulas to QBFs.

Definition 5.5 Let $\mathcal{D}$ be a proof in G3k. The *nondeterministic complexity* $\sigma(\mathcal{D})$ (resp. *co-nondeterministic complexity* $\pi(\mathcal{D})$) is the maximum number of alternations between k and $\wedge$ steps **bottom-up** along any branch in $\mathcal{D}$ **starting with the k steps** (resp. **starting with the $\wedge$ steps**); in particular,

- $\sigma(\mathcal{D}) = \pi(\mathcal{D}) = 0$, if $\mathcal{D}$ contains just $\vee$ steps,

- $\sigma(\mathcal{D}) = 1$, but $\pi(\mathcal{D}) = 2$, if $\mathcal{D}$ contains just k and $\vee$ steps,
- $\pi(\mathcal{D}) = 1$, but $\sigma(\mathcal{D}) = 2$, if $\mathcal{D}$ contains just $\wedge$ and $\vee$ steps.

For a modal sequent Γ, $\sigma(\Gamma)$ (resp. $\pi(\Gamma)$) is the least $n \in \mathbb{N}$ such that there is a proof of Γ in G3k with $\sigma(\Gamma) = n$ (resp. $\pi(\Gamma) = n$).

It is not hard to see the following:

Proposition 5.6 *Let Γ be a modal sequent. For $n \in \mathbb{N}$, we have that:*

(i) *Γ is in Σ^s_n iff $\sigma(\Gamma) \leq n$*

(ii) *Γ is in Π^s_n iff $\pi(\Gamma) \leq n$*

Proof sketch Both directions are proved by mutual induction on $n \geq 1$. We present only the right-to-left direction. If $\sigma(\Gamma) \leq n$, there exists a proof $\mathcal{D}$ of Γ such that $\sigma(\mathcal{D}) \leq n$, that is a proof of the form,

$$\begin{array}{c} \wedge\,\dfrac{\begin{array}{c}\mathcal{D}_1\\ \Gamma_1\end{array} \qquad \begin{array}{c}\mathcal{D}_2\\ \Gamma_2\end{array}}{\Gamma_0} \\ \vee{+}k \Big\Vert \\ \Gamma \end{array}$$

such that the maximum number of $\wedge/k$-alternations along any branch is at most n. Hence, the proof $\mathcal{D}_0$ of Γ_0 is such that $\pi(\mathcal{D}_0) \leq n - 1$. By (IH), $\Gamma_0 \in \Pi^s_{n-1}$, so by Def. 5.1, $\Gamma \in \Sigma^s_n$.

If $\pi(\Gamma) \leq n$, there exists a proof $\mathcal{D}$ of Γ such that $\pi(\mathcal{D}) \leq n$, that is a proof of the form,

$$\begin{array}{c} \begin{array}{c}\mathcal{D}_1\\ k\,\dfrac{\Gamma_1, A_1}{\Gamma'_1}\end{array} \quad \cdots \quad \begin{array}{c}\mathcal{D}_n\\ k\,\dfrac{\Gamma_n, A_n}{\Gamma'_n}\end{array} \\ \vee{+}\wedge \\ \Gamma \end{array}$$

with $\Gamma'_i = \boldsymbol{a_i}, \Diamond\Gamma_i, \Box\Delta_i$ and $A_i \in \Delta_i$, such that the maximum number of $\wedge/k$-alternations along any branch is at most n. Hence, the proofs $\mathcal{D}'_i$ of Γ'_i are such that $\sigma(\mathcal{D}'_i) \leq n - 1$. By (IH), $\Gamma'_i \in \Sigma^s_{n-1}$, so by Def. 5.1, $\Gamma \in \Pi^s_n$. □

5.3 Feasibly approximating (co-)nondeterministic complexity

The complexity measures σ and π defined in the previous section are evidently costly to compute, since they require us to consider all possible proofs of a sequent. Thus, despite the characterisation of Prop. 5.6, we have not yet arrived at a bona fide polynomial-time encoding from K to CPC2. Instead, we are able to define feasible 'over-approximations' for them directly on the structure of the sequent, without having to look at a particular proof.

Definition 5.7 (Approximation of (co-)nondeterministic complexity) We define the functions $\lceil\sigma\rceil$ and $\lceil\pi\rceil$ on modal sequents as follows:

$$\begin{aligned}
\lceil\sigma\rceil(A) &:= 1 \quad \text{if } A \text{ is propositional} \\
\lceil\sigma\rceil(\Gamma, \boldsymbol{a}) &:= \lceil\sigma\rceil(\Gamma) \\
\lceil\sigma\rceil(\Gamma, A \vee B) &:= \lceil\sigma\rceil(\Gamma, A, B) \\
\lceil\sigma\rceil(\Gamma, A \wedge B) &:= 1 + \lceil\pi\rceil(\Gamma, A \wedge B) \\
\lceil\sigma\rceil(\Diamond\Gamma, \Box\Delta) &:= \lceil\sigma\rceil(\Gamma, A) \quad \text{for } A \in \Delta \text{ with maximal } \lceil\sigma\rceil(A)
\end{aligned}$$

$$\begin{aligned}
\lceil\pi\rceil(A) &:= 1 \quad \text{if } A \text{ is propositional} \\
\lceil\pi\rceil(\Gamma, \boldsymbol{a}) &:= \lceil\pi\rceil(\Gamma) \\
\lceil\pi\rceil(\Gamma, A \vee B) &:= \lceil\pi\rceil(\Gamma, A, B) \\
\lceil\pi\rceil(\Gamma, A \wedge B) &:= \begin{cases} \lceil\pi\rceil(\Gamma, A) & \lceil\pi\rceil(A) \geq \lceil\pi\rceil(B) \\ \lceil\pi\rceil(\Gamma, B) & \text{otherwise} \end{cases} \\
\lceil\pi\rceil(\Diamond\Gamma, \Box\Delta) &:= 1 + \lceil\sigma\rceil(\Diamond\Gamma, \Box\Delta)
\end{aligned}$$

In the above definition, note that some choices are arbitrary, in particular when choosing a disjunction $A \vee B$ or conjunction $A \wedge B$ to evaluate $\lceil\sigma\rceil$ or $\lceil\pi\rceil$. For well-definedness, we could impose that the 'smallest' such formula is chosen, according to some global well-order on formulas. As long as this well-order is polynomial-time computable, so are the functions $\lceil\sigma\rceil$ and $\lceil\pi\rceil$.

However, it turns out that the definitions of $\lceil\sigma\rceil$ and $\lceil\pi\rceil$ are *independent* of such a choice, as was proved for similar measures in [8]. We can exploit this to show that $\lceil\sigma\rceil$ and $\lceil\pi\rceil$ indeed define over-approximations of σ and π:

Proposition 5.8 *For a modal sequent* Γ, $\lceil\sigma\rceil(\Gamma) \geq \sigma(\Gamma)$ *and* $\lceil\pi\rceil(\Gamma) \geq \pi(\Gamma)$.

Proof sketch Suppose $\sigma(\Gamma) = k$ and let $\mathcal{D}$ be a proof of Γ with $\sigma(\mathcal{D}) = k$. We may evaluate $\lceil\sigma\rceil(\Gamma)$ by following the choices of principal formula in $\mathcal{D}$. □

The alternating behaviour of $\lceil\sigma\rceil$ on conjunctions and of $\lceil\pi\rceil$ on modalities allows us to recover the following expected relations, again taking advantage of the fact that $\lceil\sigma\rceil$ and $\lceil\pi\rceil$ are independent of the underlying order on formulas:

Lemma 5.9 *Let A be a QBF and Γ be any modal sequent.*

(i) *If $A \in \Pi^q_k$ for $k \geq 1$, then $\lceil\sigma\rceil(\Gamma, A^\bullet) = 1 + \lceil\pi\rceil(\Gamma, A^\bullet)$.*

(ii) *If $A \in \Sigma^q_k$ for $k \geq 1$, then $\lceil\pi\rceil(\Gamma, A^\bullet) = 1 + \lceil\sigma\rceil(\Gamma, A^\bullet)$.*

Proof. If $A = \forall x B$, then $A^\bullet = (\Box^{e(B)}\bar{x} \vee \Box^{e(B)}x) \supset B^\bullet$ and hence:

$$\begin{aligned}
\lceil\sigma\rceil(\Gamma, A^\bullet) &= \lceil\sigma\rceil(\Gamma, \Diamond^{e(B)}x \wedge \Diamond^{e(B)}\bar{x}, B^\bullet) \\
&= 1 + \lceil\pi\rceil(\Gamma, \Diamond^{e(B)}x \wedge \Diamond^{e(B)}\bar{x}, B^\bullet) \\
&= 1 + \lceil\pi\rceil(\Gamma, A^\bullet)
\end{aligned}$$

If $A = \exists x B$, then $A^\bullet = (\Diamond\Box^{e(B^\bullet)}\bar{x} \wedge \Diamond\Box^{e(B)}x) \supset \Diamond B^\bullet$ and hence:

$$\begin{aligned}
\lceil\pi\rceil(\Gamma, A^\bullet) &= \lceil\pi\rceil(\Gamma, \Box\Diamond^{e(B)}x, \Box\Diamond^{e(B)}\bar{x}, \Diamond B^\bullet) \\
&= 1 + \lceil\sigma\rceil(\Gamma, \Box\Diamond^{e(B)}x, \Box\Diamond^{e(B)}\bar{x}, \Diamond B^\bullet) \\
&= 1 + \lceil\sigma\rceil(\Gamma, A^\bullet) \qquad \Box
\end{aligned}$$

Using this Lemma, we may conduct an inductive argument to show that our approximations accurately follow the QBF hierarchy, with respect to $\cdot^\bullet$.

Lemma 5.10 *Let A be a QBF, for any vector of literals $\boldsymbol{a}$:*

- *If $A \in \Pi_k^q$, then $\lceil\pi\rceil(\Diamond^{e(A)}\boldsymbol{a}, A^\bullet) = k$*
- *If $A \in \Sigma_k^q$, then $\lceil\sigma\rceil(\Diamond^{e(A)}\boldsymbol{a}, A^\bullet) = k$*

Proof. By induction on $k \geq 1$.

Let $A = \forall x_1 \ldots \forall x_n B$ with $B \in \Sigma_{k-1}^q$ and $A^\bullet = (\Box^{e(B)}\bar{x}_1 \vee \Box^{e(B)}x_1) \supset (\forall x_2 \ldots \forall x_n B)^\bullet$, hence

$$\begin{aligned}&\lceil\pi\rceil(\Diamond^{e(A)}\boldsymbol{a}, A^\bullet)\\ &= \lceil\pi\rceil(\Diamond^{e(A)}\boldsymbol{a}, \Diamond^{e(B)}x_1 \wedge \Diamond^{e(B)}\bar{x}_1, (\forall x_2 \ldots \forall x_n B)^\bullet)\\ &= \lceil\pi\rceil(\Diamond^{e(A)}\boldsymbol{a}, \Diamond^{e(B)}x_1, (\forall x_2 \ldots \forall x_n B)^\bullet)\\ &= \ldots = \lceil\pi\rceil(\Diamond^{e(A)}\boldsymbol{a}, \Diamond^{e(B)}x_1, \ldots, \Diamond^{e(B)}x_n, B^\bullet)\\ &= \lceil\pi\rceil(\Diamond^{e(B)}\boldsymbol{a}, \Diamond^{e(B)}\boldsymbol{x}, B^\bullet) \text{ as } e(A) = e(B)\end{aligned}$$

as we always choose x_i in the $\wedge$ case of $\lceil\pi\rceil$ in Def. 5.7 without loss of generality.

This gives us, in the inductive case when $k > 1$ that :

$$\begin{aligned}\lceil\pi\rceil(\Diamond^{e(A)}\boldsymbol{a}, A^\bullet) &= 1 + \lceil\sigma\rceil(\Diamond^{e(B)}\boldsymbol{a}, \Diamond^{e(B)}\boldsymbol{x}, B^\bullet) \text{ by Lemma 5.9}\\ &= 1 + (k-1) = k \text{ by (IH) applied to } B \in \Sigma_{k-1}^q\end{aligned}$$

and in the base case when $k = 1$ that:

$$\lceil\pi\rceil(\Diamond^{e(A)}\boldsymbol{a}, A^\bullet) = \lceil\pi\rceil(\boldsymbol{a}, \boldsymbol{x}, B) = 1 \text{ as } B \text{ is quantifier-free}$$

Let $A = \exists x_1 \ldots \exists x_n B$ with $B \in \Pi_{k-1}^q$ and $A^\bullet = (\Diamond\Box^{e(B)+n-1}\bar{x}_1 \wedge \Diamond\Box^{e(B)+n-1}x_1) \supset \Diamond(\exists x_2 \ldots \exists x_n B)^\bullet$, hence

$$\begin{aligned}&\lceil\sigma\rceil(\Diamond^{e(A)}\boldsymbol{a}, A^\bullet)\\ &= \lceil\sigma\rceil(\Diamond^{e(A)}\boldsymbol{a}, \Box\Diamond^{e(B)+n-1}x_1, \Box\Diamond^{e(B)+n-1}\bar{x}_1, \Diamond(\exists x_2 \ldots \exists x_n B)^\bullet)\\ &= \lceil\sigma\rceil(\Diamond^{e(A)-1}\boldsymbol{a}, \Diamond^{e(B)+n-1}x_1, (\exists x_2 \ldots \exists x_n B)^\bullet)\\ &= \ldots = \lceil\sigma\rceil(\Diamond^{e(A)-n}\boldsymbol{a}, \Diamond^{e(B)}x_1, \ldots, \Diamond^{e(B)}x_n, B^\bullet)\\ &= \lceil\sigma\rceil(\Diamond^{e(B)}\boldsymbol{a}, \Diamond^{e(B)}\boldsymbol{x}, B^\bullet) \text{ as } e(A) = e(B) + n\end{aligned}$$

where we always choose x_i for the $\Box$ case of $\lceil\sigma\rceil$ in Def. 5.7 without loss of generality.

This gives us, in the inductive case when $k > 1$ that:

$$\begin{aligned}\lceil\sigma\rceil(\Diamond^{e(A)}\boldsymbol{a}, A^\bullet) &= 1 + \lceil\pi\rceil(\Diamond^{e(B)}\boldsymbol{a}, \Diamond^{e(B)}\boldsymbol{x}, B^\bullet) \text{ by Lemma 5.9}\\ &= 1 + (k-1) = k \text{ by (IH) applied to } B \in \Pi_{k-1}^q\end{aligned}$$

and in the base case when $k = 1$ that:

$\lceil\sigma\rceil(\Diamond^{e(A)}\boldsymbol{a}, A^\bullet) = \lceil\sigma\rceil(\boldsymbol{a}, \boldsymbol{x}, B) = 1$ as B is quantifier-free $\Box$

Finally, we may conclude that the over-approximation we defined are in fact tight with respect to the $\cdot^\bullet$ encoding of QBFs.

Proposition 5.11 (Tightness) *Let A be a QBF.*

(i) *If $A \in \Pi^q_k$, for $k \geq 1$, then $\lceil\pi\rceil(A^\bullet) = \pi(A^\bullet) = k$.*

(ii) *If $A \in \Sigma^q_k$, for $k \geq 1$, then $\lceil\sigma\rceil(A^\bullet) = \sigma(A^\bullet) = k$.*

6 From K to QBFs via proof search

We are finally ready to present our 'inverse' encoding to $\cdot^\bullet$. By Cor. 5.4, let us henceforth fix polynomial-size Σ^q_k-formulas $\Sigma^s_k\text{-}\mathsf{Prov}_n$, and Π^s_k-formulas $\Pi^s_k\text{-}\mathsf{Prov}_n$, computing Σ^s_k-provability, and Π^s_k-provability respectively, on formulas of size $\leq n$.

Definition 6.1 Let A be a modal formula of size n. We define:

$$A^\circ := \begin{cases} \Sigma^s_k\text{-}\mathsf{Prov}_n(A) & k = \lceil\sigma\rceil(A) \leq \lceil\pi\rceil(A) \\ \Pi^s_k\text{-}\mathsf{Prov}_n(A) & k = \lceil\pi\rceil(A) \leq \lceil\sigma\rceil(A) \end{cases}$$

The main result of this work is:

Theorem 6.2 *We have the following:*

(i) $\cdot^\bullet$ *is a polynomial-time encoding from* CPC2 *to* K.

(ii) $\cdot^\circ$ *is a polynomial-time encoding from* K *to* CPC2.

(iii) *The composition $\cdot^{\bullet\circ}$ preserves quantifier complexity, i.e. for $k \geq 1$, if $A \in \Sigma^q_k$ (or $A \in \Pi^q_k$) then also $A^{\bullet\circ} \in \Sigma^q_k$ (or, respectively, $A^{\bullet\circ} \in \Pi^q_k$).*

Proof sketch. (i) already follows from Thm. 4.3. (ii) follows from the fact that $\lceil\sigma\rceil$ and $\lceil\pi\rceil$ are polynomial-time computable, and by Prop. 5.11 and Prop. 5.6. Finally (iii) follows by Prop. 5.6 and the definition of the QBFs $\Sigma^s_k\text{-}\mathsf{Prov}_n$ and $\Pi^s_k\text{-}\mathsf{Prov}_n$, under Thm. 5.3 Cor. 5.4. □

In particular, we now inherit the following delineation of K:

Corollary 6.3 *We have the following, for $k \geq 1$:*

(i) *$\{A$ modal s.t. $\lceil\sigma\rceil(A) \leq k : \models A\}$ is Σ^p_k-complete.*

(ii) *$\{A$ modal s.t. $\lceil\pi\rceil(A) \leq k : \models A\}$ is Π^p_k-complete.*

Let us emphasise here that checking $\lceil\sigma\rceil(A) \leq k$ or $\lceil\pi\rceil(A) \leq k$ is polynomial-time, and so the above corollary indeed induces a *feasible* delineation of modal formulas into classes where validity is complete for the corresponding level of the polynomial-hierarchy.

7 Conclusions

In this work we classified fragments of modal logic K complete for each level of the polynomial hierarchy. In particular, we defined polynomial-time encodings from CPC2 to K and from K to CPC2 whose composition preserves the quantifier complexity of QBFs. Our translation from K to CPC2 employs a decomposition of the proof search space for K's sequent calculus G3k according to invertible and non-invertible rules in order to control alternation in its proof search predicate.

The fact that G3k admits terminating proof search, bottom-up, is crucial to our argument. We suspect that similar results can be found for extensions of K by axioms t or d, and in a multi-agent setting, all of which also give rise to terminating sequent calculi.

On the other hand, axioms such as 4 do not easily admit such terminating calculi (though there has been some recent progress in this regard [9]). Arguments for their **PSPACE**-membership via proof search rely on a form of loop checking, due to the necessary presence of *contraction* in the system. It would be interesting to try employ alternative formalisms, such as labelled or nested sequents equipped with a form of focussing (cf. [18,5]), to find similar treatments of such logics, building on previous works for fragments of intuitionistic logic [6] and linear logic [7,8].

References

[1] Andreoli, J., *Logic programming with focusing proofs in linear logic*, J. Log. Comput. **2** (1992), pp. 297–347.
URL `http://dx.doi.org/10.1093/logcom/2.3.297`

[2] Arora, S. and B. Barak, "Computational Complexity - A Modern Approach," Cambridge University Press, 2009.
URL `http://www.cambridge.org/catalogue/catalogue.asp?isbn=9780521424264`

[3] Blackburn, P., J. F. A. K. van Benthem and F. Wolter, editors, "Handbook of Modal Logic," Studies in logic and practical reasoning **3**, North-Holland, 2007.
URL `https://www.sciencedirect.com/bookseries/studies-in-logic-and-practical-reasoning/vol/3/suppl/C`

[4] Chandra, A. K., D. C. Kozen and L. J. Stockmeyer, *Alternation*, J. ACM **28** (1981), pp. 114–133.
URL `http://doi.acm.org/10.1145/322234.322243`

[5] Chaudhuri, K., S. Marin and L. Straßburger, *Focused and synthetic nested sequents*, in: B. Jacobs and C. Löding, editors, *Foundations of Software Science and Computation Structures - 19th International Conference, FOSSACS 2016, Held as Part of the European Joint Conferences on Theory and Practice of Software, ETAPS 2016, Eindhoven, The Netherlands, April 2-8, 2016, Proceedings*, Lecture Notes in Computer Science **9634** (2016), pp. 390–407.
URL `https://doi.org/10.1007/978-3-662-49630-5_23`

[6] Das, A., *Alternating time bounds from variants of focussed proof systems* (2017), preprint. `http://anupamdas.com/alt-time-bnds-var-foc-sys.pdf`.

[7] Das, A., *Focussing, MALL and the polynomial hierarchy*, in: D. Galmiche, S. Schulz and R. Sebastiani, editors, *Automated Reasoning - 9th International Joint Conference, IJCAR 2018, Held as Part of the Federated Logic Conference, FloC 2018, Oxford, UK, July 14-17, 2018, Proceedings*, Lecture Notes in Computer Science **10900** (2018), pp. 689–705.
URL `https://doi.org/10.1007/978-3-319-94205-6_45`

[8] Das, A., *From qbfs to MALL and back via focussing*, J. Autom. Reason. **64** (2020), pp. 1221–1245.
URL `https://doi.org/10.1007/s10817-020-09564-x`

[9] Fiorentini, C., *Terminating sequent calculi for proving and refuting formulas in S4*, J. Log. Comput. **25** (2015), pp. 179–205.
URL `https://doi.org/10.1093/logcom/exs053`

[10] Halpern, J. Y., *The effect of bounding the number of primitive propositions and the depth of nesting on the complexity of modal logic*, Artif. Intell. **75** (1995), pp. 361–372.
URL `https://doi.org/10.1016/0004-3702(95)00018-A`

[11] Halpern, J. Y. and Y. Moses, *A guide to completeness and complexity for modal logics of knowledge and belief*, Artif. Intell. **54** (1992), pp. 319–379.
URL `https://doi.org/10.1016/0004-3702(92)90049-4`

[12] Halpern, J. Y. and L. C. Rêgo, *Characterizing the NP-PSPACE gap in the satisfiability problem for modal logic*, J. Log. Comput. **17** (2007), pp. 795–806.
URL `https://doi.org/10.1093/logcom/exm029`

[13] Ladner, R. E., *The computational complexity of provability in systems of modal propositional logic.*, SIAM J. Comput. **6** (1977), pp. 467–480.
URL `http://dblp.uni-trier.de/db/journals/siamcomp/siamcomp6.html#Ladner77`

[14] Lagniez, J., D. L. Berre, T. de Lima and V. Montmirail, *On checking kripke models for modal logic K*, in: P. Fontaine, S. Schulz and J. Urban, editors, *Proceedings of the 5th Workshop on Practical Aspects of Automated Reasoning co-located with International Joint Conference on Automated Reasoning (IJCAR 2016), Coimbra, Portugal, July 2nd, 2016*, CEUR Workshop Proceedings **1635** (2016), pp. 69–81.
URL `http://ceur-ws.org/Vol-1635/paper-07.pdf`

[15] Liang, C. and D. Miller, *Focusing and polarization in linear, intuitionistic, and classical logics*, Theoretical Computer Science **410** (2009), pp. 4747–4768.

[16] Massacci, F., *Design and results of the tableaux-99 non-classical (modal) systems comparison*, in: N. V. Murray, editor, *Automated Reasoning with Analytic Tableaux and Related Methods, International Conference, TABLEAUX '99, Saratoga Springs, NY, USA, June 7-11, 1999, Proceedings*, Lecture Notes in Computer Science **1617** (1999), pp. 14–18.
URL `https://doi.org/10.1007/3-540-48754-9_2`

[17] Massacci, F. and F. M. Donini, *Design and results of TANCS-2000 non-classical (modal) systems comparison*, in: R. Dyckhoff, editor, *Automated Reasoning with Analytic Tableaux and Related Methods, International Conference, TABLEAUX 2000, St Andrews, Scotland, UK, July 3-7, 2000, Proceedings*, Lecture Notes in Computer Science **1847** (2000), pp. 52–56.
URL `https://doi.org/10.1007/10722086_4`

[18] Miller, D. and M. Volpe, *Focused labeled proof systems for modal logic*, in: M. Davis, A. Fehnker, A. McIver and A. Voronkov, editors, *Logic for Programming, Artificial Intelligence, and Reasoning - 20th International Conference, LPAR-20 2015, Suva, Fiji, November 24-28, 2015, Proceedings*, Lecture Notes in Computer Science **9450** (2015), pp. 266–280.
URL `https://doi.org/10.1007/978-3-662-48899-7_19`

[19] Nguyen, L. A., *On the complexity of fragments of modal logics*, in: R. A. Schmidt, I. Pratt-Hartmann, M. Reynolds and H. Wansing, editors, *Advances in Modal Logic 5, papers from the fifth conference on "Advances in Modal logic," held in Manchester, UK, 9-11 September 2004* (2004), pp. 249–268.
URL `http://www.aiml.net/volumes/volume5/Nguyen.ps`

[20] Pan, G. and M. Y. Vardi, *Optimizing a bdd-based modal solver*, in: F. Baader, editor, *Automated Deduction - CADE-19, 19th International Conference on Automated Deduction Miami Beach, FL, USA, July 28 - August 2, 2003, Proceedings*, Lecture Notes in Computer Science **2741** (2003), pp. 75–89.
URL `https://doi.org/10.1007/978-3-540-45085-6_7`

[21] Schmidt-Schauß, M. and G. Smolka, *Attributive concept descriptions with complements*, Artif. Intell. **48** (1991), pp. 1–26.
URL `https://doi.org/10.1016/0004-3702(91)90078-X`

[22] Sebastiani, R. and M. Vescovi, *Automated reasoning in modal and description logics via SAT encoding: the case study of k(m)/alc-satisfiability*, CoRR **abs/1401.3463** (2014).
URL `http://arxiv.org/abs/1401.3463`

[23] Statman, R., *Intuitionistic propositional logic is polynomial-space complete*, Theor. Comput. Sci. **9** (1979), pp. 67–72.
URL `http://dx.doi.org/10.1016/0304-3975(79)90006-9`

[24] Stockmeyer, L. J., *The polynomial-time hierarchy*, Theor. Comput. Sci. **3** (1976), pp. 1–22.
URL `https://doi.org/10.1016/0304-3975(76)90061-X`

[25] Troelstra, A. S. and H. Schwichtenberg, "Basic proof theory, Second Edition," Cambridge tracts in theoretical computer science **43**, Cambridge University Press, 2000.
[26] Zolin, E., *Description logic complexity navigator*, webpage, accessed June 14th 2022. URL `https://www.cs.man.ac.uk/~ezolin/dl/`

A New Hope

Hans van Ditmarsch

Open University of the Netherlands

Krisztina Fruzsa [1] and Roman Kuznets [2]

TU Wien, Austria

Abstract

Knowledge has long been identified as an inherent component of agents' decision-making in distributed systems. However, for agents in fault-tolerant distributed systems with fully byzantine agents, achieving knowledge is, in most cases, unrealistic. If agents can both lie and themselves be mistaken, then a message received is generally not sufficient to create knowledge. This problem is adequately addressed by an epistemic modality named *hope*, which has already been axiomatized. In this paper, we propose an alternative complete axiomatization for the hope modality by removing the reliance on designated atoms denoting correctness of individual agents and show that hope can be described as a $\mathsf{KB4}_n$ system. This additionally brings a more streamlined presentation of the common hope modality (the hope analog of common knowledge). We also combine $\mathsf{KB4}_n$ hope modalities with $\mathsf{S5}_n$ knowledge modalities traditionally used in the epistemic analysis of fault-free distributed systems and present a logic enriched with both common knowledge and common hope. In these logics we formalize as frame-characterizable axioms some of the main properties of fully byzantine distributed systems: bounds on the number of faulty agents and the epistemic limitations due to agents' inability to rule out brain-in-a-vat scenarios.

Keywords: Epistemic logic, distributed systems, byzantine agents.

1 Introduction

Over at least three decades, epistemic analysis has been used as a potent tool [7, 14] for studying distributed systems. It is often based on the *runs and systems framework* that views global states of a distributed system as possible worlds in a Kripke model. The importance of this methodology is underscored by the broadly applicable *Knowledge of Preconditions Principle* [21], recently formulated by Moses, which states that in all models of distributed systems, if φ is a necessary condition for agent i to perform an action, then agent i knowing φ to hold, written $K_i\varphi$, is also a necessary condition for this agent

[1] PhD student in the Austrian Science Fund (FWF) doctoral program LogiCS (W1255).

[2] Funded by the FWF ByzDEL project (P33600).

to perform this action. The agent's complete reliance on its local state as the source of information about the system naturally induces an equivalence relation on the global states, resulting in agents' knowledge being described using multimodal epistemic logic $\mathsf{S5}_n$.

This epistemic analysis via the runs and systems framework was recently [15–17] extended to *fault-tolerant systems* with so-called *byzantine* agents [18]. (Fully) byzantine agents are the worst-case faulty agents to participate in a distributed system: not only can they arbitrarily deviate from their respective protocols, but their perception of their own actions and the events they observe can be corrupted, possibly unbeknownst to them, resulting in false memories. Whether byzantine agents are actually present in a system or not, the very possibility of their presence has drastic and debilitating effects on the epistemic state of all agents, due to their inability to rule out so-called *brain-in-a-vat* scenarios [23]. In a distributed system, a brain-in-a-vat agent is a faulty agent with completely corrupted perceptions that provide no reliable information about the system [16]. It has been shown that agents' inability to rule out being a brain in a vat precludes them from knowing many basic facts, including their own correctness/faultiness, in both asynchronous [16] and synchronous [25] distributed systems.

The extended runs and systems framework was used in [11] to analyze the *Firing Rebels with Relay* (FRR) problem, a simplified version of the *consistent broadcasting* primitive [26], which has been used as a pivotal building block in distributed algorithms, e.g., for byzantine fault-tolerant clock synchronization [5, 12, 24, 26, 28], synchronous consensus [27], and a general reduction of distributed task solvability in byzantine systems to solvability in systems with crash failures [19]. Instead of knowledge (unattainable due to brain-in-a-vat scenarios), the analysis of FRR hinges on a weaker epistemic notion called *hope*, which, in the presence of knowledge modalities, was initially defined as $H_i\varphi := \mathit{correct}_i \to K_i(\mathit{correct}_i \to \varphi)$. Without knowledge, hope was axiomatized in [10] with the help of designated atoms $\mathit{correct}_i$ representing agent i's correctness. The special nature of these atoms precluded the logic from being a normal modal logic.

Contributions and paper organization: In this paper, we provide an alternative axiomatization of hope that deals away with these atoms treating them as abbreviations $\mathit{correct}_i := \neg H_i\bot$ instead. Not only does this make the logic of hope a normal modal logic, but it turns out to coincide with multimodal $\mathsf{KB4}_n$ (Sect. 2). We explore the language with both hope and knowledge modalities by formulating a combined logic of hope and knowledge including their interaction and showing the Kripke completeness (Sect. 3). We also demonstrate the utility of this logic by providing frame-characterizable axioms to represent various properties of fully byzantine agents, including brain-in-a-vat scenarios, and system specifications, e.g., the upper bound on the number of faulty agents (Sect. 4). Working towards the epistemic analysis of group actions, we axiomatize the logic of common hope and common knowledge and provide the completeness theorem (Sect. 5). In Sect. 6, we take an in-depth look

	$P:$	all propositional tautologies			
K^H:	$H_i(\varphi \to \psi) \land H_i\varphi \to H_i\psi$		T'^H	:	$correct_i \to (H_i\varphi \to \varphi)$
4^H :	$H_i\varphi \to H_iH_i\varphi$		F	:	$faulty_i \to H_i\varphi$
5^H :	$\neg H_i\varphi \to H_i\neg H_i\varphi$		H	:	$H_i correct_i$
MP:	$\dfrac{\varphi \quad \varphi \to \psi}{\psi}$		Nec^H:		$\dfrac{\varphi}{H_i\varphi}$

Fig. 1. Axiom system $\mathscr{H}_{\mathrm{co}}$ in language $\mathcal{L}_H^{\mathrm{co}} = \mathcal{L}(\mathsf{Prop} \sqcup \mathsf{Co}, H_1, \dots, H_n)$ from [10]

at the related work. Finally, in Sect. 7, we provide conclusions and directions for future work.

2 Logic of Individual Hope

We fix a finite set $\mathcal{A} = \{1, \dots, n\}$ of *agents*, a countably infinite set Prop of *atomic propositions* (*atoms*), and a finite set $\mathsf{Co} := \{correct_i \mid i \in \mathcal{A}\}$ of designated *correctness atoms* such that $\mathsf{Prop} \cap \mathsf{Co} = \varnothing$. We will consider a number of multimodal languages varying in modalities and atoms. Hence, it pays off to give a general definition.

Definition 2.1 A *multimodal language* $\mathcal{L}(P, \heartsuit_1, \dots, \heartsuit_m)$ with a set P of atoms and with modalities $\heartsuit_j$ is defined according to the following grammar:

$$\varphi ::= p \mid \neg\varphi \mid (\varphi \land \varphi) \mid \heartsuit_j\varphi$$

where $p \in P$ and $j = 1, \dots, m$. We take $\top$ to be an abbreviation for some fixed propositional tautology, $\bot := \neg\top$ and use standard abbreviations for the remaining boolean connectives.

We first consider language $\mathcal{L}_H^{\mathrm{co}} := \mathcal{L}(\mathsf{Prop} \sqcup \mathsf{Co}, H_1, \dots, H_n)$ for *hope modalities* H_i with n designated atoms $correct_i$, one for each $i \in \mathcal{A}$, intended to mean that agent i has not deviated from its normative behavior in the system. For instance, in Kripke models generated based on runs of a fault-tolerant distributed system with perfect recall, $correct_i$ should signify that, by the time of evaluation, agent i has not violated its protocol and has correctly recorded its performed actions and witnessed events. We abbreviate $faulty_i := \neg correct_i$.

Axiom system $\mathscr{H}_{\mathrm{co}}$ is presented in Fig. 1. Here axioms P, K^H, 4^H, and 5^H, along with rules MP and Nec^H represent standard multimodal logic $\mathsf{K45}_n$, in particular, postulating *positive* and *negative introspection* for the hope modality. Axiom T'^H is *factivity* restricted to correct agents. Axioms H and F represent further properties of $correct_i$: namely, that agents always hope to be correct and that the hopes of faulty agents are unrestricted and all encompassing, in particular, alongside tautologies they also hope for contradictions, making their hopes inconsistent. Intuitively, these properties mean that an agent can rely on its perceptions iff the agent is correct. We elaborate more on the origins of this particular axiomatization in the next section, where hope is related to the knowledge modality.

Definition 2.2 A *Kripke frame* for a language $\mathcal{L} = \mathcal{L}(P, \heartsuit_1, \dots, \heartsuit_m)$ is a tuple $F = (W, R_1, \dots, R_m)$ where $W \neq \varnothing$ is the set of *worlds* (or *states*)

$$\begin{array}{llll} & P: \quad \text{all propositional tautologies} & & \\ K^H: & H_i(\varphi \to \psi) \wedge H_i\varphi \to H_i\psi & B^H: & \varphi \to H_i\neg H_i\neg\varphi \\ 4^H: & H_i\varphi \to H_iH_i\varphi & & \\ MP: & \dfrac{\varphi \quad \varphi \to \psi}{\psi} & Nec^H: & \dfrac{\varphi}{H_i\varphi} \end{array}$$

Fig. 2. Axiom system $\mathscr{H}$ in language $\mathcal{L}_H = \mathcal{L}(\mathsf{Prop}, H_1, \dots, H_n)$

and $R_j \subseteq W \times W$ is an *accessibility relation* for the modality $\heartsuit_j$. A *Kripke model* for $\mathcal{L}$ is $M = (F, \pi)$ where F is a Kripke frame and $\pi\colon P \to \mathcal{P}(W)$ is a *valuation function.* Truth for formulas $\varphi \in \mathcal{L}$ in model M is defined as follows: $M, w \models p$ iff $w \in \pi(p)$ for all $p \in P$, negation and conjunction behave classically within each world, and $M, w \models \heartsuit_j\varphi$ iff $M, v \models \varphi$ for all $v \in R_j(w)$ where $R_j(w) := \{w' \mid wR_jw'\}$. *Validity in model* M, denoted $M \models \varphi$, is defined as truth in all worlds of W. *Validity in frame* F, denoted $F \models \varphi$, is defined as validity in all models (F, π). *Validity in a class* $\mathcal{C}$ *of Kripke frames (models)* is defined as validity in all frames (models) of $\mathcal{C}$.

A binary relation R_j is called

- *transitive* if wR_jv whenever wR_ju and uR_jv,
- *symmetric* if wR_jv whenever vR_jw,
- *euclidean* if wR_jv whenever uR_jw and uR_jv, and
- *shift serial* if $R_j(v) \neq \varnothing$ for any $v \in R_j(w)$.

Class $\mathcal{K}45_m$ ($\mathcal{KB}4_m$; $\mathcal{S}5_m$) consists of all models with m transitive and euclidean (transitive and symmetric; equivalence) accessibility relations. A *partial equivalence relation* is any transitive and symmetric binary relation (see [20]).

From now till the end of this section, we set $m = n = |\mathcal{A}|$ and use $R_i = \mathcal{H}_i$ as the accessibility relation for modality $\heartsuit_i = H_i$.

Definition 2.3 *Class* $\mathcal{K}45_n^{\mathrm{co}}$ consists of all Kripke models $M = ((W, \mathcal{H}_1, \dots, \mathcal{H}_n), \pi) \in \mathcal{K}45_n$ such that for every $i \in \mathcal{A}$ and $w, w' \in W$: (i) if $w \in \pi(correct_i)$, then $w\mathcal{H}_iw$; (ii) if $w \notin \pi(correct_i)$, then $\mathcal{H}_i(w) = \varnothing$; and (iii) if $w\mathcal{H}_iw'$, then $w' \in \pi(correct_i)$.

Theorem 2.4 ([10]) *$\mathscr{H}_{\mathrm{co}}$ is sound and complete with respect to $\mathcal{K}45_n^{\mathrm{co}}$.*

Note that class $\mathcal{K}45_n^{\mathrm{co}}$ of models is not based on any class of frames. Our first result in this paper is an alternative axiomatization for hope that deals away with designated atoms $correct_i$ and, hence, enables us to avoid the dependency of accessibility relations $\mathcal{H}_i$ on valuation function π. This is achieved by adopting the definition

$$correct_i := \neg H_i\bot \tag{1}$$

in language $\mathcal{L}_H := \mathcal{L}(\mathsf{Prop}, H_1, \dots, H_n)$ based on the view that $faulty_i$ can be equated to the inconsistent hopes $H_i\bot$. It turns out that the logic of hope in this language is the logic of class $\mathcal{KB}4_n$ of all transitive and symmetric frames and is axiomatized by axiom system $\mathscr{H} = \mathsf{KB4}_n$ presented in Fig. 2. It is well known that

Theorem 2.5 (Completeness for the logic of hope)
$\mathscr{H}$ is sound and complete with respect to $\mathcal{KB}4_n$.

Remark 2.6 The new axiomatization makes it easier to see how hope is different from the notion of belief. Indeed, belief is quite often assumed to be consistent, i.e., satisfying axiom $\neg B\bot$, which fails for hope due to inconsistent hopes of faulty agents. On the other hand, axiom B^H is typically invalid for belief because, together with 4^H, it would preclude agents from having consistent but false beliefs.

We now show that $\mathscr{H}$ is equivalent to $\mathscr{H}_{\text{co}}$ modulo abbreviation (1):

Lemma 2.7
- *$\mathscr{H} \vdash \varphi$ implies $\mathscr{H}_{\text{co}} \vdash \varphi$ for any $\varphi \in \mathcal{L}_H$.*
- *$\mathscr{H}_{\text{co}} \vdash \varphi$ implies $\mathscr{H} \vdash \varphi^\dagger$, where $\varphi^\dagger \in \mathcal{L}_H$ is the result of replacing each $correct_i$ in $\varphi \in \mathcal{L}_H^{\text{co}}$ with $\neg H_i\bot$, according to (1).*

Proof.
- It is sufficient to show $\mathscr{H}_{\text{co}} \vdash B^H$. Using the instance $faulty_i \to H_i\neg H_i\neg\varphi$ of F, by prop. reasoning, $\mathscr{H}_{\text{co}} \vdash faulty_i \to (\varphi \to H_i\neg H_i\neg\varphi)$. On the other hand, from the instance $correct_i \to (H_i\neg\varphi \to \neg\varphi)$ of T'^H, by prop. reasoning $\mathscr{H}_{\text{co}} \vdash correct_i \to (\varphi \to \neg H_i\neg\varphi)$. Coupling this with the instance $\neg H_i\neg\varphi \to H_i\neg H_i\neg\varphi$ of 5^H, we get $\mathscr{H}_{\text{co}} \vdash correct_i \to (\varphi \to H_i\neg H_i\neg\varphi)$. Since $faulty_i \vee correct_i$ is a propositional tautology, $\mathscr{H}_{\text{co}} \vdash \varphi \to H_i\neg H_i\neg\varphi$ by prop. reasoning.
- It is sufficient to show that axiom 5^H, as well as the $\dagger$-translations of axioms T'^H, F, and H are derivable in $\mathscr{H}$. That 5^H can be derived from 4^H and B^H is a well-known fact (any transitive and symmetric relation is euclidean). Thus, we only discuss the axioms involving $correct_i$.
  - The $\dagger$-translation of T'^H is $\neg H_i\bot \to (H_i\psi \to \psi)$ for $\psi = \varphi^\dagger$. It is sufficient to show the derivability $\mathscr{H} \vdash \neg(H_i\psi \to \psi) \to H_i\bot$ for the contrapositive. Firstly, $\mathscr{H} \vdash \neg(H_i\psi \to \psi) \to H_i\psi \wedge \neg\psi$ is a propositional tautology. Further, $\mathscr{H} \vdash H_i\psi \to H_iH_i\psi$ by 4^H and also $\mathscr{H} \vdash \neg\psi \to H_i\neg H_i\psi$ by B^H. Thus, combining these together, $\mathscr{H} \vdash \neg(H_i\psi \to \psi) \to H_iH_i\psi \wedge H_i\neg H_i\psi$. It remains to use the normality of H_i and prop. reasoning to replace $H_iH_i\psi \wedge H_i\neg H_i\psi$ first with $H_i(H_i\psi \wedge \neg H_i\psi)$ and finally with $H_i\bot$.
  - The $\dagger$-translation of F is (modulo a double negation) $H_i\bot \to H_i\psi$, which follows by the normality of H_i from $\bot \to \psi$.
  - The $\dagger$-translation of axiom H is $H_i\neg H_i\bot$, which is easy to obtain from the instance $\neg\bot \to H_i\neg H_i\neg\neg\bot$ of B^H by prop. reasoning. □

Theorem 2.8 (Equivalence of the two logics of individual hope)
Systems $\mathscr{H}$ and $\mathscr{H}_{\text{co}}$ are equivalent representations of the logic of hope.

We demonstrate the utility of this reformulation of the logic of hope by encoding a standard limitation on the number of faulty agents in a fault-tolerant distributed system as a *frame-characterizable* property in logic $\mathscr{H}$. It is typical to design distributed protocols under the assumption that no more than f of the n agents can become faulty ($0 \leq f < n$). This is a natural restriction given that clearly no outcome of agents' protocols can be guaranteed if, e.g., all agents can ignore these protocols. Moreover, byzantine consensus [9, 18] and byzantine clock synchronization [6,9], among others, are unsolvable for $n \leq 3f$, while, e.g., consensus in asynchronous systems with the weakest failure detector Omega is unsolvable already for $n \leq 2f$ [4]. We can encode such requirements in $\mathcal{L}_H$ by an additional axiom

$$Byz_f \quad := \quad \bigvee_{\substack{G \subseteq \mathcal{A} \\ |G|=n-f}} \bigwedge_{i \in G} \neg H_i \bot.$$

Remark 2.9 $Byz_0 = \bigwedge_{i \in \mathcal{A}} \neg H_i \bot$ simply states that all n agents are correct.

Proposition 2.10 (Frame characterization of $\leq f$ faulty agents) *Axiom Byz_f is characterized by the property of frames $F = (W, \mathcal{H}_1, \dots, \mathcal{H}_n)$*

$$(\forall w \in W)(\exists G \subseteq \mathcal{A})\Big(|G| = n - f \wedge (\forall i \in G)\mathcal{H}_i(w) \neq \varnothing\Big),$$

which we call all-but-f-seriality. *In other words, each world must have outgoing arrows for all but f agents.*

Proof. Take an arbitrary frame $F = (W, \mathcal{H}_1, \dots, \mathcal{H}_n)$ for language $\mathcal{L}_H$. We need to show that

$$F \models Byz_f \qquad \Longleftrightarrow \qquad F \text{ is all-but-}f\text{-serial.}$$

We prove the ($\Longrightarrow$) direction by contrapositive. If F is not all-but-f-serial, there is some $w \in W$ such that any group $G \subseteq \mathcal{A}$ of $n-f$ agents has some agent $i_G \in G$ such that $\mathcal{H}_{i_G}(w) = \varnothing$. Since for all these agents, $(F, \pi), w \not\models \neg H_{i_G}\bot$ for any π, we have $(F, \pi), w \not\models Byz_f$ and, hence, $F \not\models Byz_f$.

For the ($\Longleftarrow$) direction, let F be all-but-f-serial. Take an arbitrary $w \in W$. It now follows that there is a group $G \subseteq \mathcal{A}$ of $n-f$ agents such that $\mathcal{H}_i(w) \neq \varnothing$ for all $i \in G$. Therefore, $(F, \pi), w \models \bigwedge_{i \in G} \neg H_i \bot$ and $(F, \pi), w \models Byz_f$ for any π. The validity in F follows because w and π were chosen arbitrarily. □

Definition 2.11 *Class $\mathcal{KB}4_n^{n-f}$* consists of all models from $\mathcal{KB}4_n$ with all-but-f-serial frames.

Corollary 2.12 *$\mathscr{H} + Byz_f$ is sound and complete with respect to $\mathcal{KB}4_n^{n-f}$.*

3 Logic of Individual Hope and Individual Knowledge

In this section, we consider language $\mathcal{L}_{KH} := \mathcal{L}(\mathsf{Prop}, K_1, \dots, K_n, H_1, \dots, H_n)$. This language with both hope and knowledge modalities for each agent is expressive enough to describe most of the epistemic attitudes relevant to distributed systems and explore relationships among them. Accordingly, the semantics used until the end of the paper has $m = 2n = 2|\mathcal{A}|$ accessibility relations where, in addition to $\mathcal{H}_i$ for hope modalities H_i we now use accessibility relations $\mathcal{K}_i$ for knowledge modalities K_i.

In particular, we now recall how hope was initially defined via knowledge and the correctness atoms. The hope modality appeared in the analysis of the Firing Rebels Problem [11], as well as in earlier works [10, 16], in the form of derived modality $H_i\varphi := \mathit{correct}_i \to K_i(\mathit{correct}_i \to \varphi)$, which, translated into language $\mathcal{L}_{KH}$ using (1), becomes axiom

$$KH \quad := \quad H_i\varphi \quad \leftrightarrow \quad (\neg H_i\bot \to K_i(\neg H_i\bot \to \varphi)). \tag{2}$$

Our new language with hope enables us to (almost) characterize this connection axiom by two frame properties for two directions of equivalence (2).

Proposition 3.1 (Characterizing knowledge-to-hope connection) *On the class of frames $F = (W, \mathcal{K}_1, \dots, \mathcal{K}_n, \mathcal{H}_1, \dots, \mathcal{H}_n)$ with shift serial $\mathcal{H}_i$,*

$$KH^{\leftarrow} \quad := \quad (\neg H_i\bot \to K_i(\neg H_i\bot \to \varphi)) \to H_i\varphi \tag{3}$$

is characterized by frame property

$$\mathcal{H}\mathrm{in}\mathcal{K}: \qquad \mathcal{H}_i \subseteq \mathcal{K}_i.$$

Proof. First assume that a frame F with shift serial $\mathcal{H}_i$ satisfies $\mathcal{H}\mathrm{in}\mathcal{K}$ and let $M = (F, \pi)$ for an arbitrary π. Let the antecedent of (3) hold at an arbitrary world $w \in W$. To show that $M, w \models H_i\varphi$, it is sufficient to show that $M, v \models \varphi$ for all $v \in \mathcal{H}_i(w)$. It is vacuously true if $\mathcal{H}_i(w) = \varnothing$. Otherwise, take any such world v. $M, w \models \neg H_i\bot$ because $\mathcal{H}_i(w) \neq \varnothing$, thus, $M, w \models K_i(\neg H_i\bot \to \varphi)$. Since $\mathcal{H}_i(w) \subseteq \mathcal{K}_i(w)$ due to $\mathcal{H}\mathrm{in}\mathcal{K}$, we get $M, v \models \neg H_i\bot \to \varphi$. But $\mathcal{H}_i(v) \neq \varnothing$ due to the shift seriality of $\mathcal{H}_i$. This is sufficient to conclude that $M, v \models \varphi$, completing the proof that $KH^{\leftarrow}$ is valid in F.

For the opposite direction, assume that F violates $\mathcal{H}\mathrm{in}\mathcal{K}$, i.e., that there are worlds $w, v \in W$ with $w\mathcal{H}_i v$ but not $w\mathcal{K}_i v$. Consider any model $M = (F, \pi)$ with a valuation π such that $\pi(p) = W \setminus \{v\}$ for some atom p. We have $M, w \models K_i(\neg H_i\bot \to p)$ because $\mathcal{K}_i(w) \subseteq W \setminus \{v\} = \pi(p)$. Therefore, $M, w \models \neg H_i\bot \to K_i(\neg H_i\bot \to p)$. However, clearly $M, w \not\models H_i p$ because of v. Thus, we have shown that $M, w \not\models KH^{\leftarrow}$ for $\varphi = p$. Note that this direction does not rely on the shift seriality of $\mathcal{H}_i$. □

Proposition 3.2 (Characterizing hope-to-knowledge connection)

$$KH^{\rightarrow} \quad := \quad H_i\varphi \to (\neg H_i\bot \to K_i(\neg H_i\bot \to \varphi)) \tag{4}$$

for frames $F = (W, \mathcal{K}_1, \dots, \mathcal{K}_n, \mathcal{H}_1, \dots, \mathcal{H}_n)$ is characterized by property

$$\mathrm{one}\mathcal{H}: \qquad (\forall w, v \in W)(\mathcal{H}_i(w) \neq \varnothing \land \mathcal{H}_i(v) \neq \varnothing \land w\mathcal{K}_i v \Longrightarrow w\mathcal{H}_i v).$$

Proof. First assume that F satisfies $\mathrm{one}\mathcal{H}$ and let $M = (F, \pi)$ for an arbitrary π. Let $M, w \models H_i\varphi$. The case of $\mathcal{H}_i(w) = \varnothing$ is trivial since $M, w \models H_i\bot$ makes the succedent of (4) true at w. Otherwise, $\mathcal{H}_i(w) \neq \varnothing$. Similarly, for any $v \in \mathcal{K}_i(w)$ with $\mathcal{H}_i(v) = \varnothing$, we have $M, v \models \neg H_i\bot \to \varphi$. Finally, for any $v \in \mathcal{K}_i(w)$ with $\mathcal{H}_i(v) \neq \varnothing$, we have $v \in \mathcal{H}_i(w)$ by $\mathrm{one}\mathcal{H}$. Hence, $M, v \models \varphi$

$$
\begin{array}{lll}
 & P: & \text{all propositional tautologies} \\
H^\dagger: \ H_i\neg H_i\bot & K^K: & K_i(\varphi \to \psi) \land K_i\varphi \to K_i\psi \\
 & 4^K: & K_i\varphi \to K_iK_i\varphi \\
 & 5^K: & \neg K_i\varphi \to K_i\neg K_i\varphi \\
 & T^K: & K_i\varphi \to \varphi \\
MP: \ \dfrac{\varphi \quad \varphi \to \psi}{\psi} & Nec^K: & \dfrac{\varphi}{K_i\varphi} \\
\end{array}
$$

$$KH: \quad H_i\varphi \leftrightarrow (\neg H_i\bot \to K_i(\neg H_i\bot \to \varphi))$$

Fig. 3. Axiom system $\mathscr{KH}$ in language $\mathcal{L}_{KH} = \mathcal{L}(\mathsf{Prop}, K_1, \ldots, K_n, H_1, \ldots, H_n)$

and $M, v \models \neg H_i\bot \to \varphi$. We have shown that $\neg H_i\bot \to \varphi$ is true in all worlds from $\mathcal{K}_i(w)$ and can again conclude that the succedent of (4) is true at w. This completes the proof that $KH^{\to}$ is valid in F.

For the opposite direction, assume that F violates one$\mathcal{H}$, i.e., there are worlds $w, v \in W$ with $\mathcal{H}_i(w) \neq \varnothing$, $\mathcal{H}_i(v) \neq \varnothing$, $w\mathcal{K}_iv$, but not $w\mathcal{H}_iv$. Consider any model $M = (F, \pi)$ with a valuation π such that $\pi(p) = \mathcal{H}_i(w)$ for some atom p. Clearly, $M, w \models H_ip$ and $M, w \models \neg H_i\bot$. However, $M, v \models \neg H_i\bot$ and $M, v \not\models p$. Hence, $M, v \not\models \neg H_i\bot \to p$ and, given $w\mathcal{K}_iv$, we also have $M, w \not\models K_i(\neg H_i\bot \to p)$. Thus, $M, w \not\models KH^{\to}$ for $\varphi = p$. □

Definition 3.3 *Class $\mathcal{KH}$ of* models for knowledge and hope consists of all Kripke models $M = ((W, \mathcal{K}_1, \ldots, \mathcal{K}_n, \mathcal{H}_1, \ldots, \mathcal{H}_n), \pi)$ where (i) every $\mathcal{K}_i$ is an equivalence relation, (ii) every $\mathcal{H}_i$ is shift serial, and (iii) properties $\mathcal{H}$in$\mathcal{K}$ and one$\mathcal{H}$ are satisfied.

Proposition 3.4 *For all $M = ((W, \mathcal{K}_1, \ldots, \mathcal{K}_n, \mathcal{H}_1, \ldots, \mathcal{H}_n), \pi) \in \mathcal{KH}$, each accessibility relation $\mathcal{H}_i$ is symmetric and transitive.*

Proof. To prove transitivity, let $w\mathcal{H}_iv$ and $v\mathcal{H}_iu$. Then $w\mathcal{K}_iv$ and $v\mathcal{K}_iu$ by $\mathcal{H}$in$\mathcal{K}$. Therefore, $w\mathcal{K}_iu$ by the transitivity of $\mathcal{K}_i$. $\mathcal{H}_i(w) \ni v$ is not empty. $\mathcal{H}_i(u) \neq \varnothing$ by the shift seriality of $\mathcal{H}_i$ because $v\mathcal{H}_iu$. Hence, $w\mathcal{H}_iu$ by one$\mathcal{H}$.

To prove symmetry, let $w\mathcal{H}_iv$. Then $w\mathcal{K}_iv$ by $\mathcal{H}$in$\mathcal{K}$. Therefore, $v\mathcal{K}_iw$ by the symmetry of $\mathcal{K}_i$. $\mathcal{H}_i(w) \ni v$ is not empty. $\mathcal{H}_i(v) \neq \varnothing$ by the shift seriality of $\mathcal{H}_i$ because $w\mathcal{H}_iv$. Hence, $v\mathcal{H}_iw$ by one$\mathcal{H}$. □

Remark 3.5 Hence, $\mathcal{H}_i$ are partial equivalence relations, so that property one$\mathcal{H}$ can be described as "no $\mathcal{K}_i$-equivalence class contains more than one $\mathcal{H}_i$-partial-equivalence class."

A natural way of obtaining the combined logic $\mathscr{KH}$ of hope and knowledge would be to combine the axioms and rules for hope from Fig. 2, standard $\mathsf{S5}$ axioms and rules for knowledge, and connection axiom KH. Prop. 3.4, however, indicates that this would create redundancies. As we now show, in the presence of axiom KH, $\mathsf{KB4}$ properties of hope originate from $\mathsf{S5}$ properties of knowledge, albeit with the help of the translation of axiom $H = H_i\mathit{correct}_i$ from $\mathscr{H}_{\text{co}}$ into language $\mathcal{L}_{KH}$. This translation $H^\dagger = H_i\neg H_i\bot$ can be called *necessary consistency* for hope and is known to be characterized by shift seriality. The resulting simplified axiom system is presented in Fig. 3.

Lemma 3.6 *For all* $i \in \mathcal{A}$,
(i) $\mathscr{KH} \vdash K_i\varphi \to H_i\varphi$;
(ii) $\mathscr{KH} \vdash H_i(\varphi \to \psi) \land H_i\varphi \to H_i\psi$;
(iii) *if* $\mathscr{KH} \vdash \varphi$, *then* $\mathscr{KH} \vdash H_i\varphi$.

Proof.

(i) $K_i\varphi \to K_i(\neg H_i\bot \to \varphi)$ is by the normality of K_i. From prop. tautology $K_i(\neg H_i\bot \to \varphi) \to \big(\neg H_i\bot \to K_i(\neg H_i\bot \to \varphi)\big)$ and one direction $\big(\neg H_i\bot \to K_i(\neg H_i\bot \to \varphi)\big) \to H_i\varphi$ of KH, by syllogism, $K_i\varphi \to H_i\varphi$.

(ii) 1. $K_i\big(\neg H_i\bot \to (\varphi \to \psi)\big) \to \big(K_i(\neg H_i\bot \to \varphi) \to K_i(\neg H_i\bot \to \psi)\big)$ normality of K_i

2. $H_i(\varphi \to \psi) \to \Big(\neg H_i\bot \to K_i\big(\neg H_i\bot \to (\varphi \to \psi)\big)\Big)$ axiom KH

3. $H_i\varphi \to \big(\neg H_i\bot \to K_i(\neg H_i\bot \to \varphi)\big)$ axiom KH

4. $H_i(\varphi \to \psi) \to \Big(H_i\varphi \to \big(\neg H_i\bot \to K_i(\neg H_i\bot \to \psi)\big)\Big)$ from 1.–3.

5. $\big(\neg H_i\bot \to K_i(\neg H_i\bot \to \psi)\big) \to H_i\psi$ axiom KH

6. $H_i(\varphi \to \psi) \to (H_i\varphi \to H_i\psi)$ from 4.–5.

(iii) Easily follows from (i). □

Remark 3.7 Given that $K_i\varphi \to H_i\varphi$ is also known to characterize frame property $\mathcal{H}\text{in}\mathcal{K}$, one might ask whether $KH^{\leftarrow}$ is equivalent to $K_i\varphi \to H_i\varphi$. The answer is negative because $KH^{\leftarrow}$ only characterizes $\mathcal{H}\text{in}\mathcal{K}$ under the additional assumption of $\mathcal{H}_i$ being shift serial. For instance, consider a model M with $W = \{w, v\}$, $\mathcal{K}_j = W \times W$ for all $j \in \mathcal{A}$, $\mathcal{H}_j = W \times W$ for all $j \neq i$, non-shift-serial $\mathcal{H}_i = \{(w, v)\}$, $\pi(p) = \{w\}$ for some atom p, and $\pi(q) = W$ for all $q \in \mathsf{Prop} \setminus \{p\}$. Then $M, w \not\models \big(\neg H_i\bot \to K_i(\neg H_i\bot \to p)\big) \to H_ip$ but $M, w \models K_ip \to H_ip$.

Theorem 3.8 (Completeness of the logic of hope and knowledge) $\mathscr{KH}$ *is sound and complete with respect to* $\mathcal{KH}$.

Proof sketch. The soundness of KH follows because all axioms except for $KH^{\leftarrow}$ are frame characterizable and class $\mathcal{KH}$ consists of frames with corresponding properties, one of which is shift seriality, which takes care of the additional restriction for $KH^{\leftarrow}$. The normality of H_i is derived in Lemma 3.6. This enables us to use the standard canonical model $M^C = \big((W^C, \mathcal{K}_1^C, \dots, \mathcal{K}_n^C, \mathcal{H}_1^C, \dots, \mathcal{H}_n^C), \pi^C\big)$ construction for the logic and prove the Truth Lemma, i.e., that $M^C, \Gamma \models \varphi$ iff $\varphi \in \Gamma$ for each maximal $\mathscr{KH}$-consistent set $\Gamma \in W^C$. It remains to show that $M^C \in \mathcal{KH}$. The argument for $\mathcal{K}_i^C$ being equivalence relations and for $\mathcal{H}_i^C$ being shift serial is standard. Property $\mathcal{H}\text{in}\mathcal{K}$, i.e., $\mathcal{H}_i^C \subseteq \mathcal{K}_i^C$, easily follows from $\mathscr{KH} \vdash K_i\varphi \to H_i\varphi$ proved in Lemma 3.6. Thus, we only show property one$\mathcal{H}$, i.e., that $\Gamma\mathcal{K}_i^C\Delta$ implies $\Gamma\mathcal{H}_i^C\Delta$ whenever $\mathcal{H}_i^C(\Gamma) \neq \varnothing$ and $\mathcal{H}_i^C(\Delta) \neq \varnothing$ for any maximal consistent sets $\Gamma, \Delta \in W^C$. Assume $\Gamma\mathcal{K}_i^C\Delta$, $\mathcal{H}_i^C(\Gamma) \neq \varnothing$, and $\mathcal{H}_i^C(\Delta) \neq \varnothing$. Note that $\mathcal{H}_i^C(\Xi) \neq \varnothing$ implies $\neg H_i\bot \in \Xi$ due to the maximal consistency. We need to prove that $H_i\varphi \in \Gamma$ implies $\varphi \in \Delta$. If $H_i\varphi \in \Gamma$, then $\Gamma \vdash_{\mathscr{KH}} K_i(\neg H_i\bot \to \varphi)$ by axiom KH and

$\neg H_i\bot \in \Gamma$. Hence, $K_i(\neg H_i\bot \to \varphi) \in \Gamma$ and $\neg H_i\bot \to \varphi \in \Delta$ by the definition of $\mathcal{K}_i^C$. Since $\neg H_i\bot \in \Delta$, we conclude that $\varphi \in \Delta$ as required. □

Corollary 3.9 *$\mathscr{KH} \vdash H_i\varphi \to H_iH_i\varphi$ and $\mathscr{KH} \vdash \varphi \to H_i\neg H_i\neg\varphi$ for $i \in \mathcal{A}$.*

Proof. It immediately follows from Theorem 3.8 and Prop. 3.4. □

Definition 3.10 *Class $\mathcal{KH}^{n-f}$ consists of all Kripke models from $\mathcal{KH}$ that have all-but-f-serial frames with respect to $\mathcal{H}_i$ relations.*

Corollary 3.11 *$\mathscr{KH} + Byz_f$ is sound and complete with respect to $\mathcal{KH}^{n-f}$.*

Logics $\mathscr{KH}$ and $\mathscr{KH} + Byz_f$ formalize both reliable (knowledge) and unreliable (hope) information agents possess in fault-tolerant distributed systems, the latter with at most f byzantine agents. The following two propositions outline the epistemic attitudes of agents who know that they are faulty and agents who know that they are correct.

Proposition 3.12 *$\mathscr{KH} \vdash K_iH_i\bot \to H_i\varphi$ for all $i \in \mathcal{A}$.*

Proof. We have $H_i\bot \to H_i\varphi$ is by the normality of H_i. From K_i-factivity $K_iH_i\bot \to H_i\bot$ we get $K_iH_i\bot \to H_i\varphi$ by syllogism. □

Proposition 3.13 *$\mathscr{KH} \vdash K_i\neg H_i\bot \to (H_i\varphi \leftrightarrow K_i\varphi)$ for all $i \in \mathcal{A}$.*

Proof. $\mathscr{KH} \vdash K_i\neg H_i\bot \to (K_i\varphi \to H_i\varphi)$ is an easy corollary of Prop. 3.6. We provide a derivation of $K_i\neg H_i\bot \to (H_i\varphi \to K_i\varphi)$. Firstly, we can obtain $K_i\neg H_i\bot \to \big(H_i\varphi \to K_i(\neg H_i\bot \to \varphi)\big)$ from K_i-factivity $K_i\neg H_i\bot \to \neg H_i\bot$ and direction $H_i\varphi \to \big(\neg H_i\bot \to K_i(\neg H_i\bot \to \varphi)\big)$ of axiom KH by propositional reasoning. It remains to use K^K axiom $K_i(\neg H_i\bot \to \varphi) \to (K_i\neg H_i\bot \to K_i\varphi)$ to get $K_i\neg H_i\bot \to (H_i\varphi \to K_i\varphi)$ by propositional reasoning. □

We now turn to properties relevant for analyzing distributed systems. For instance, due to our earlier discussion of unsolvability of many distributed problems if too many agents become faulty and in view of the Knowledge of Preconditions Principle, it is typically necessary for agents to know that there are at least $n - f$ correct agents overall:

Proposition 3.14
- *$\mathscr{KH} + Byz_f \vdash K_iByz_f$ for all $i \in \mathcal{A}$.*
- *$\mathcal{KH}^{n-f} \models K_iByz_f$ for all $i \in \mathcal{A}$.*

Corollary 3.15 (In fault-free systems, hope is knowledge) *Recall that axiom Byz_0 rules out the presence of faulty agents. For any $i \in \mathcal{A}$,*

$$\mathscr{KH} + Byz_0 \vdash H_i\varphi \leftrightarrow K_i\varphi.$$

Proof. Follows from Remark 2.9 and Props. 3.13 and 3.14. □

Prop. 3.12 can be strengthened because a faulty agent hopes for anything even without knowing that it is faulty, i.e., $\mathscr{KH} \vdash H_i\bot \to H_i\varphi$. By contrast, for Prop. 3.13, the knowledge modality cannot be dropped: for a correct agent, hope does not yet mean knowledge, i.e., $\mathscr{KH} \nvdash \neg H_i\bot \to (H_i\varphi \leftrightarrow K_i\varphi)$. Instead, for a correct agent, hope turns out to be equivalent to another epistemic attitude called *belief*, also used in distributed computing [22]. We introduce the following abbreviations for belief B_i, as well as for mutual knowledge E^K_G, mutual belief E^B_G, and mutual hope E^H_G among a group $G \subseteq \mathcal{A}$ of agents:

$$B_i\varphi := K_i(\neg H_i\bot \to \varphi) \qquad E^K_G\varphi := \bigwedge_{i\in G} K_i\varphi$$

$$E^B_G\varphi := \bigwedge_{i\in G} B_i\varphi \qquad E^H_G\varphi := \bigwedge_{i\in G} H_i\varphi$$

Note that this belief B_i is of type K45 rather than KD45.

Remark 3.16 It immediately follows that

$$\mathscr{KH} \vdash \neg H_i\bot \to (H_i\varphi \leftrightarrow B_i\varphi). \tag{5}$$

Thus, it might seem that, for correct agents, the use of hope is superficial and can be replaced by the better studied belief. The subtlety lies in the fact that belief is actionable in the sense of the Knowledge of Preconditions Principle, because the agent always knows its beliefs due to the positive introspection of knowledge. The same is not true regarding hope. And while hope of a correct agent i is equivalent to its belief, agents might not be aware of this equivalence if they are uncertain whether i is, in fact, correct.

We now show, by purely modal means, that in distributed systems with bounded number of faulty agents, mutual belief among a sufficiently large group of agents can be extracted from mutual hope among all agents. This ability to lift information received about hopes of all agents into actionable beliefs of a critical mass of correct agents is at the core of many distributed algorithms, including Firing Rebels [11] (albeit in a more complex temporal setting).

Proposition 3.17 $\mathscr{KH} + Byz_f \quad \vdash \quad E^H_{\mathcal{A}}\varphi \to \bigvee_{\substack{G\subseteq\mathcal{A}\\ |G|=n-f}} E^B_G\varphi.$

Proof. Follows from (5) and axiom Byz_f. □

Another indication of the independence of hope as an epistemic attitude is the fact that hope generally creates neither knowledge of hope nor hope of knowledge.

Proposition 3.18 (Knowledge and hope do not mix) *For any* $i \in \mathcal{A}$,

- *it is not the case that* $\mathcal{KH} \models H_i\varphi \to H_iK_i\varphi$ *for all* $\varphi \in \mathcal{L}_{KH}$,
- *it is not the case that* $\mathcal{KH} \models H_i\varphi \to K_iH_i\varphi$ *for all* $\varphi \in \mathcal{L}_{KH}$.

Proof. We use the same model to refute both statements but refute them for different formulas φ. Let $M = \big((W, \mathcal{K}_1, \dots, \mathcal{K}_n, \mathcal{H}_1, \dots, \mathcal{H}_n), \pi\big)$ such that $W = \{\mathsf{G}, \mathsf{B}\}$, $\mathcal{K}_j = W \times W$ for all $j \in \mathcal{A}$, $\mathcal{H}_i = \{(\mathsf{G}, \mathsf{G})\}$, $\mathcal{H}_j = W \times W$ for all $j \neq i$, and π be arbitrary. Now, all $\mathcal{K}_j$ are equivalence relations, all $\mathcal{H}_j$ are shift serial, $\mathcal{H}_j \subseteq \mathcal{K}_j$ for all $j \in \mathcal{A}$, and one$\mathcal{H}$ holds. In other words, $M \in \mathcal{KH}$. Clearly, agent i is correct in world G, i.e., $\mathcal{H}_i(\mathsf{G}) \neq \varnothing$, and faulty in world B, i.e., $\mathcal{H}_i(\mathsf{B}) = \varnothing$. We now have that, $M, \mathsf{G} \not\models H_i\neg H_i\bot \to H_iK_i\neg H_i\bot$ and $M, \mathsf{B} \not\models H_i\bot \to K_iH_i\bot$. □

Corollary 3.19 *For any $i \in \mathcal{A}$ and any $f > 0$,*
- *it is not the case that $\mathcal{KH}^{n-f} \models H_i\varphi \to H_iK_i\varphi$ for all $\varphi \in \mathcal{L}_{KH}$,*
- *it is not the case that $\mathcal{KH}^{n-f} \models H_i\varphi \to K_iH_i\varphi$ for all $\varphi \in \mathcal{L}_{KH}$.*

We finish this section by showing that designated atoms *correct*$_i$, which can be defined away via the hope modality as $\neg H_i\bot$, are not definable in language $\mathcal{L}_K := \mathcal{L}(\mathsf{Prop}, K_1, \dots, K_n)$ with knowledge modalities only.

Definition 3.20 For a language $\mathcal{L} = \mathcal{L}(P, \heartsuit_1, \dots, \heartsuit_m)$, let Kripke models $M = ((W, R_1, \dots, R_m), \pi)$ and $M' = ((W', R'_1, \dots, R'_m), \pi')$ be given. A non-empty relation $Z \subseteq W \times W'$ is a *bisimulation* between M and M', notation $Z\colon M \underline{\leftrightarrow} M'$, if for all wZw' and $j \in \{1, \dots, m\}$:

atoms $w \in \pi(p)$ iff $w' \in \pi'(p)$ for all $p \in P$;

forth if wR_jv, then there is a $v' \in W'$ such that $w'R'_jv'$ and vZv';

back if $w'R'_jv'$, then there is a $v \in W$ such that wR_jv and vZv'.

We write $M \underline{\leftrightarrow} M'$ if there is a bisimulation $Z\colon M \underline{\leftrightarrow} M'$. We write $(M, w) \underline{\leftrightarrow} (M', w')$ if there is a bisimulation $Z\colon M \underline{\leftrightarrow} M'$ such that wZw'.

A *restricted (to Q) bisimulation* Z^Q is a bisimulation that satisfies **atoms** for all atoms $Q \subseteq P$, notation $Z^Q\colon M \underline{\leftrightarrow}^Q M'$. And, similarly, $(M, w) \underline{\leftrightarrow}^Q (M', w')$ means that wZ^Qw' for some such Z^Q.

Given $Q \subseteq P$, let us write $\mathcal{L}|Q := \mathcal{L}(Q, \heartsuit_1, \dots, \heartsuit_m)$ for the language restricted to atoms from Q only. Further, let us write
- $(M, w) \equiv (M', w')$ to mean 'for all $\varphi \in \mathcal{L}$, $M, w \models \varphi$ iff $M', w' \models \varphi$' and
- $(M, w) \equiv^Q (M', w')$ to mean 'for all $\varphi \in \mathcal{L}|Q$, $M, w \models \varphi$ iff $M', w' \models \varphi$.'

Theorem 3.21 ([2])
- *$(M, w) \underline{\leftrightarrow} (M', w')$ implies $(M, w) \equiv (M', w')$.*
- *$(M, w) \underline{\leftrightarrow}^Q (M', w')$ implies $(M, w) \equiv^Q (M', w')$.*

In language $\mathcal{L}_{KH}$ with knowledge and hope, *correct*$_i$ is definable as $\neg H_i\bot$, making *faulty*$_i$ = $\neg$*correct*$_i$ equivalent to $H_i\bot$. More precisely, every formula in language $\mathcal{L}(\mathsf{Prop} \sqcup \mathsf{Co}, K_1, ..., K_n, H_1, ..., H_n)$ is equivalent to a formula in language $\mathcal{L}_{KH}$ via the †-translation (cf. Lemma 2.7).

We now show that atoms *correct*$_i$ are not definable in the language with modalities for knowledge only.

Proposition 3.22 *Correctness of agents is not definable from knowledge.*

Proof. We now consider languages $\mathcal{L}_K^{\mathsf{co}} := \mathcal{L}(\mathsf{Prop} \sqcup \mathsf{Co}, K_1, ..., K_n)$ and $\mathcal{L}_K := \mathcal{L}_K^{\mathsf{co}}|\mathsf{Prop} = \mathcal{L}(\mathsf{Prop}, K_1, ..., K_n)$. Assume towards a contradiction that there is a $\varphi_i \in \mathcal{L}_K$ such that φ_i is equivalent to *correct*$_i$. Now consider (M, w) and (M', w') (for the set $\mathsf{Prop} \sqcup \mathsf{Co}$ of all propositions) such that $(M, w) \underline{\leftrightarrow}^{\mathsf{Prop}} (M', w')$ but for some agent $i \in \mathcal{A}$ we have $w \in \pi(\textit{correct}_i)$ while, at the same time, $w' \notin \pi'(\textit{correct}_i)$. From Theorem 3.21 we obtain that $(M, w) \equiv^{\mathsf{Prop}} (M', w')$. Therefore, in particular $M, w \models \varphi_i$ iff $M', w' \models \varphi_i$. This contradicts that $M, w \models \textit{correct}_i$ but $M', w' \not\models \textit{correct}_i$. Thus, *correct*$_i$ is not definable in $\mathcal{L}_K$. □

Note that, were the models M and M' above to also have hope relations, then $w \in \pi(correct_i)$ and $w' \notin \pi'(correct_i)$ would imply that $\mathcal{H}_i(w) \neq \varnothing$ whereas $\mathcal{H}'_i(w') = \varnothing$, thus, precluding their bisimilarity restricted to Prop.

4 Modal Representation of Byzantine Behaviors, including Brain-in-a-Vat

The goal of this section is to show the utility of logic $\mathscr{KH}$ of hope and knowledge by providing axiomatic and semantic descriptions of several properties that originate from earlier analyses of fully byzantine distributed systems in [16]. Roughly speaking, an agent is called *fully byzantine* if neither its behavior nor its perceptions are restricted in any way. We already discussed axiom Byz_f restricting the number of such agents to at most f to ensure solvability of distributed problems. Another important feature of distributed systems with fully byzantine agents is that agents can never exclude the possibility of their perceptions being completely fabricated in a so-called *brain-in-a-vat scenario*. For details of the modeling via global runs in fault-tolerant distributed message-passing systems, which also include a temporal component, we refer the reader to [16]. In particular, the following properties were demonstrated:

- Agents cannot reliably establish their own correctness, as formalized by the axiom, for all $i \in \mathcal{A}$,

$$iByz \qquad := \qquad \neg K_i \neg H_i \bot.$$

- A faulty agent lacks any reliable information about other agents.[3] In particular, we generally assume that a faulty agent has no reliable information to decide whether any other agent is correct or faulty, as formalized by the axiom, for all $i \neq j$,

$$BiV \qquad := \qquad H_i \bot \to \neg K_i H_j \bot \wedge \neg K_i \neg H_j \bot.$$

From these two principles, by purely modal means, we can derive that no agent knows whether other agents are correct or faulty, as proved in [16] by complex manipulations of distributed system runs.

Proposition 4.1 $\mathscr{KH} + iByz + BiV \vdash anyByz_{ij} \wedge anyCor_{ij}$ *for all* $i \neq j$ *where*

$$anyByz_{ij}: \quad \neg K_i \neg H_j \bot; \qquad\qquad anyCor_{ij}: \quad \neg K_i H_j \bot.$$

Proof. $\neg K_i \neg H_i \bot \to \neg K_i \neg \neg K_i H_j \bot$ for any $i \neq j$ follows from BiV by propositional reasoning and the normality of K_i (applied to its dual $\neg K_i \neg$). Hence, $\neg K_i \neg H_i \bot \to \neg K_i H_j \bot$ by T^K. Thus, by invoking $iByz$ and using MP we get $\neg K_i H_j \bot = anyCor_{ij}$. The argument for $anyByz_{ij}$ is analogous. □

Proposition 4.2 (Frame characterization of agents' fallibility) *Axiom $iByz$ for $i \in \mathcal{A}$ is characterized by the* i-may-aseriality *property*

[3] An important exception here must be made for *a priori* knowledge of the system, e.g., logical laws, physical laws, or global specifications of the distributed system.

of frames $F = (W, \mathcal{K}_1, \dots, \mathcal{K}_n, \mathcal{H}_1, \dots, \mathcal{H}_n)$ *requiring that each world have a* $\mathcal{K}_i$*-indistinguishable world with no* $\mathcal{H}_i$*-outgoing arrows:*

$$(\forall w \in W)(\exists w' \in \mathcal{K}_i(w)) \quad \mathcal{H}_i(w') = \varnothing.$$

Proof. Take an arbitrary frame $F = (W, \mathcal{K}_1, \dots, \mathcal{K}_n, \mathcal{H}_1, \dots, \mathcal{H}_n)$. We need to show that for $i \in \mathcal{A}$,

$$F \models iByz \qquad \Longleftrightarrow \qquad F \text{ is } i\text{-may-aserial.}$$

($\Longrightarrow$) We prove the contrapositive. If F is not i-may-aserial, there is some world $w \in W$ such that $\mathcal{H}_i(w') \neq \varnothing$ for all $w' \in \mathcal{K}_i(w)$. Independent of a valuation π, for $M = (F, \pi)$ we have $M, w' \models \neg H_i \bot$ for all $w' \in \mathcal{K}_i(w)$. Hence, we get $M, w \models K_i \neg H_i \bot$ and, hence, $F \not\models iByz$ for this i.
($\Longleftarrow$) Let F be i-may-aserial. Take an arbitrary $w \in W$. It now follows that there is $w' \in \mathcal{K}_i(w)$ such that $\mathcal{H}_i(w') = \varnothing$. Therefore, for $M = (F, \pi)$ with any π, we have $M, w' \models H_i \bot$ and $M, w \models \neg K_i \neg H_i \bot$. The validity of $iByz$ in F for this i follows because w and π were chosen arbitrarily. □

Proposition 4.3 (Frame characterization of brain-in-a-vat)
Axiom BiV *for* $i \neq j$ *is characterized by the* BiValence *property of frames* $F = (W, \mathcal{K}_1, \dots, \mathcal{K}_n, \mathcal{H}_1, \dots, \mathcal{H}_n)$

$$(\forall w \in W)\Big(\mathcal{H}_i(w) = \varnothing \Longrightarrow (\exists w', w'' \in \mathcal{K}_i(w))\big(\mathcal{H}_j(w') \neq \varnothing \wedge \mathcal{H}_j(w'') = \varnothing\big)\Big).$$

Proof. Take any frame $F = (W, \mathcal{K}_1, \dots, \mathcal{K}_n, \mathcal{H}_1, \dots, \mathcal{H}_n)$. We need to show that for $i \neq j$

$$F \models BiV \qquad \Longleftrightarrow \qquad F \text{ is BiValent.}$$

($\Longrightarrow$) We prove the contrapositive. If F is not BiValent, there is some $w \in W$ such that $\mathcal{H}_i(w) = \varnothing$ but either $\mathcal{H}_j(w') = \varnothing$ for all $w' \in \mathcal{K}_i(w)$ or $\mathcal{H}_j(w'') \neq \varnothing$ for all $w'' \in \mathcal{K}_i(w)$. Independent of a valuation π, for $M = (F, \pi)$ we then have $M, w \models K_i \neg H_j \bot \vee K_i H_j \bot$ despite $M, w \models H_i \bot$, and, hence, $F \not\models BiV$ for these $i \neq j$.
($\Longleftarrow$) Let F be BiValent. Take an arbitrary $w \in W$ such that $\mathcal{H}_i(w) = \varnothing$. It now follows that there are $w' \in \mathcal{K}_i(w)$ such that $\mathcal{H}_j(w') \neq \varnothing$ and $w'' \in \mathcal{K}_i(w)$ such that $\mathcal{H}_j(w'') = \varnothing$. Therefore, for $M = (F, \pi)$ with any π, we can conclude $M, w \models \neg K_i H_j \bot \wedge \neg K_i \neg H_j \bot$ whenever $M, w \models H_i \bot$. The validity of BiV in F for these $i \neq j$ follows because w and π were chosen arbitrarily. □

We can also easily derive by purely modal means that brain-in-a-vat scenarios are not compatible with fault-free systems:

Proposition 4.4 $\mathscr{KH} + Byz_0 \vdash \neg iByz$ *for each* $i \in \mathcal{A}$.

Proof. By Prop. 3.14 and the normality of K_i, given that $Byz_0 \to \neg H_i \bot$ is a propositional tautology, we have $\mathscr{KH} + Byz_0 \vdash K_i \neg H_i \bot$ for each $i \in \mathcal{A}$. Deriving $\neg iByz$ is now a matter of propositional reasoning. □

Another interesting special case is $f = 1$ (with $n > 1$). On the one hand, half of BiV becomes derivable and, hence, redundant:

Proposition 4.5 *If any agent but no more than one can be faulty, then no agent can establish the faultiness of other agents: for all $i \neq j \in \mathcal{A}$,*

$$\mathscr{KH} + Byz_1 + iByz \quad \vdash \quad \neg K_i H_j \bot.$$

Proof. We have $H_i\bot \to \neg H_j\bot$ by Byz_1 for any $j \neq i$. Thus, we can conclude $\neg K_i \neg\, H_i\bot \to \neg K_i \neg\, \neg H_j\bot$ by the normality of K_i, i.e., $\neg K_i \neg H_i\bot \to \neg K_i H_j\bot$. Since $\neg K_i \neg H_i\bot$ is axiom $iByz$, we conclude $\neg K_i H_j\bot$ by MP. □

On the other hand, the other half of BiV leads to undesirable consequences:

Proposition 4.6 *For $f = 1$, the inability of a faulty agent to establish correctness of somebody else would lead to its inability to establish its own faultiness: for all $i \neq j \in \mathcal{A}$,*

$$\mathscr{KH} + Byz_1 + (H_i\bot \to \neg K_i \neg H_j \bot) \quad \vdash \quad \neg K_i H_i \bot.$$

Proof. A correct agent i considers its own correctness possible by T^K, i.e., $\neg H_i\bot \to \neg K_i H_i\bot$. Formula $H_i\bot \to \neg K_i\neg H_j\bot$ for at least one $j \neq i$ is an assumption. At the same time, $H_j\bot \to \neg H_i\bot$ by Byz_1. As before, $\neg K_i\neg H_j\bot \to \neg K_i H_i\bot$ follows by the normality of K_i, yielding implication $H_i\bot \to \neg K_i H_i\bot$ by syllogism. Since we have derived $\neg K_i H_i\bot$ from both assumptions $\neg H_i\bot$ and $H_i\bot$, we get $\neg K_i H_i\bot$ without any assumptions by propositional reasoning. □

Remark 4.7 Intuitively, if an agent establishes its own faultiness, which does not run afoul of $iByz$ and can be used, e.g., for self-repairing agents, then it will thereby establish the correctness of all other agents. It seems wrong to prohibit this by adopting the respective half of BiV, whereas the other half is derivable anyway. We, therefore, propose using $\mathscr{KH} + Byz_f + BiV + iByz$ for $f \geq 2$ but $\mathscr{KH} + Byz_1 + iByz$ for $f = 1$. (The case of $f = 0$, which can be axiomatized by $\mathscr{KH} + Byz_0$, is more efficiently dealt with in the standard epistemic language.)

5 Common Hope and Common Knowledge

In this section, we introduce the common hope modality by analogy with the common knowledge modality and explore their relationship. We start by extending language $\mathcal{L}_{KH}$ with unary modal operator C^H_G for common hope and unary modal operator C^K_G for common knowledge, where $\varnothing \neq G \subseteq \mathcal{A}$ is an arbitrary group of agents. We will denote this extended language by $\mathcal{L}^C_{KH}$. Similar to common knowledge among a group G, by common hope of φ we intuitively mean mutual hope that φ and mutual hope of mutual hope that φ, etc.:

$$C^H_G \quad \leadsto \quad E^H_G\varphi \wedge E^H_G E^H_G \varphi \wedge E^H_G E^H_G E^H_G \varphi \wedge \dots$$

Axiom system $\mathscr{KHC}$ for common knowledge and common hope consists of all the axioms of $\mathscr{KH}$ (formulated for $\mathcal{L}^C_{KH}$ formulas) plus the following axioms and inference rules for all $\varnothing \neq G \subseteq \mathcal{A}$ and all formulas $\varphi, \psi \in \mathcal{L}_{KH}^{C}$:

$$Mix^H: \quad C_G^H\varphi \to E_G^H(\varphi \wedge C_G^H\varphi); \quad Mix^K: \quad C_G^K\varphi \to E_G^K(\varphi \wedge C_G^K\varphi);$$

$$Ind^H: \quad \frac{\psi \to E_G^H(\varphi \wedge \psi)}{\psi \to C_G^H\varphi}; \quad Ind^K: \quad \frac{\psi \to E_G^K(\varphi \wedge \psi)}{\psi \to C_G^K\varphi}.$$

That common knowledge has the properties of individual knowledge is well-known. Still, it may be surprising that common hope has the properties of individual hope. (Recall that common belief does not have the properties of individual KD45 belief as it lacks negative introspection.) Proofs are standard and omitted.

Proposition 5.1 *For any formulas $\varphi, \psi \in \mathcal{L}_{KH}^{C}$ and any $\varnothing \neq G \subseteq \mathcal{A}$:*

$$\mathscr{KHC} \vdash C_G^H(\varphi \to \psi) \wedge C_G^H\varphi \to C_G^H\psi \quad \mathscr{KHC} \vdash C_G^K(\varphi \to \psi) \wedge C_G^K\varphi \to C_G^K\psi$$

$$\mathscr{KHC} \vdash C_G^H\varphi \to C_G^H C_G^H\varphi \quad \mathscr{KHC} \vdash C_G^K\varphi \to C_G^K C_G^K\varphi$$

$$\mathscr{KHC} \vdash \neg C_G^K\varphi \to C_G^K \neg C_G^K\varphi$$

$$\mathscr{KHC} \vdash \varphi \Longrightarrow \mathscr{KHC} \vdash C_G^H\varphi \quad \mathscr{KHC} \vdash \varphi \Longrightarrow \mathscr{KHC} \vdash C_G^K\varphi$$

$$\mathscr{KHC} \vdash \varphi \to C_G^H \neg C_G^H \neg\varphi \quad \mathscr{KHC} \vdash C_G^K\varphi \to \varphi$$

Proposition 5.2 $\mathscr{KHC} \vdash C_G^K\varphi \to C_G^H\varphi$.

Proof.

1. $C_G^K\varphi \to E_G^K(\varphi \wedge C_G^K\varphi)$ — axiom Mix^K
2. $E_G^K(\varphi \wedge C_G^K\varphi) \to E_G^H(\varphi \wedge C_G^K\varphi)$ — follows from Prop. 3.6
3. $C_G^K\varphi \to E_G^H(\varphi \wedge C_G^K\varphi)$ — by syllogism from 1. and 2.
4. $C_G^K\varphi \to C_G^H\varphi$ — by Ind^H from 3.

□

Formulas of $\mathcal{L}_{KH}^{C}$ are also evaluated on models from $\mathcal{KH}$, with the new clauses for the common knowledge and common hope added as follows:

Definition 5.3 For a model $\big((W, \mathcal{K}_1, \ldots, \mathcal{K}_n, \mathcal{H}_1, \ldots, \mathcal{H}_n), \pi\big) \in \mathcal{KH}$, we define

$$\mathcal{K}_G^C := \left(\bigcup_{i \in G} \mathcal{K}_i\right)^+, \qquad \mathcal{H}_G^C := \left(\bigcup_{i \in G} \mathcal{H}_i\right)^+,$$

where R^+ is the transitive (but not reflexive) closure of a relation R. Then we define $M, w \models C_G^K\varphi$ iff $M, v \models \varphi$ for all $v \in \mathcal{K}_G^C(w)$ and $M, w \models C_G^H\varphi$ iff $M, v \models \varphi$ for all $v \in \mathcal{H}_G^C(w)$.

Theorem 5.4 (Completeness for common hope and knowledge) *$\mathscr{KHC}$ is sound and complete with respect to $\mathcal{KH}$.*

Proof sketch. The proof uses a finite version of the canonical model construction with maximal consistent sets restricted to subsets of an appropriately chosen *finite* set $cl(\varphi)$ of "extended" subformulas of a given formula φ

(cf. Fischer–Ladner closure [8]). Apart from the choice of this closure set, the proof is rather standard, if lengthy and technical. One shows that finite canonical model M^C_φ belongs to class $\mathcal{KH}$. Then the Truth Lemma is established. Finally, if $\mathscr{KHC} \nvdash \varphi$, then $\{\neg\varphi\}$ is consistent and can be extended to a world in M^C_φ, where φ is false. The main difficulty in this proof is finding the closure set $cl(\varphi)$, hence, we provide it below.

The closure set $cl(\varphi)$ for the finite canonical model is defined in several stages. We use $\mathcal{L}^b_i \subset \mathcal{L}^C_{KH}$ to denote all formulas *not* of the form $\neg H_i\bot \to \psi$.

- $cl_0(\varphi)$ is the smallest set that (a) contains φ and $H_i\neg H_i\bot$ for all $i \in \mathcal{A}$, (b) is closed under subformulas, and, for all $\psi \in \mathcal{L}^C_{KH}$ and $\varnothing \neq G \subseteq \mathcal{A}$, (c) contains $E^H_G(\psi \wedge C^H_G\psi)$ whenever $C^H_G\psi \in cl_0(\varphi)$ and (d) contains $E^K_G(\psi \wedge C^K_G\psi)$ whenever $C^K_G\psi \in cl_0(\varphi)$.
- $cl_1(\varphi) := cl_0(\varphi) \cup \{\neg\psi \mid \psi \in cl_0(\varphi)\}$.
- $cl_2(\varphi) := cl_1(\varphi) \cup \{H_i(\neg H_i\bot \to \psi) \mid K_i(\neg H_i\bot \to \psi) \in cl_1(\varphi)\} \cup \{K_i(\neg H_i\bot \to \psi) \mid H_i(\neg H_i\bot \to \psi) \in cl_1(\varphi)\} \cup \{K_i(\neg H_i\bot \to \psi), H_i(\neg H_i\bot \to \psi), \neg H_i\bot \to \psi \mid H_i\psi \in cl_1(\varphi), \psi \in \mathcal{L}^b_i\} \cup \{H_i\psi, K_i(\neg H_i\bot \to \psi), H_i(\neg H_i\bot \to \psi), \neg H_i\bot \to \psi \mid K_i\psi \in cl_1(\varphi), \psi \in \mathcal{L}^b_i\}$
- $cl_3(\varphi) := cl_2(\varphi) \cup \{\neg\psi \mid \psi \in cl_2(\varphi)\}$.
- $cl_4(\varphi) := cl_3(\varphi) \cup \{K_iK_i\psi, H_iK_i\psi \mid K_i\psi \in cl_3(\varphi)\} \cup \{K_i\neg K_i\psi, H_i\neg K_i\psi \mid \neg K_i\psi \in cl_3(\varphi)\}$
- $cl(\varphi) := cl_4(\varphi) \cup \{\neg\psi \mid \psi \in cl_4(\varphi)\}$.

□

Corollary 5.5 (Decidability) *$\mathscr{KHC}$ is conservative over $\mathscr{KH}$. Both have the finite model property (FMP) and, hence, are decidable.*

Proof. If $\mathscr{KHC} \nvdash \varphi$, the finite canonical model M^C_φ from the proof of Theorem 5.4 serves as a finite countermodel. Thus, $\mathscr{KHC}$ has the FMP. If $\mathscr{KH} \nvdash \varphi$ for $\varphi \in \mathcal{L}_{KH}$, then $\mathcal{KH} \not\models \varphi$ by the completeness of $\mathscr{KH}$ and $\mathscr{KHC} \nvdash \varphi$ by the soundness of $\mathscr{KHC}$. This proves the conservativity, which implies the FMP for $\mathscr{KH}$. □

So far, by and large, the relationship between common knowledge and common hope exhibited the same traits as between their individual variants. But the naive generalization of connection axiom KH is invalid for the common modalities (for at least two agents). We recall that $KH^{\to}$ corresponds to property one$\mathcal{H}$ that each knowledge equivalence class contains at most one hope partial equivalence class (Prop. 4). It is easy to see that when lifted to the common modalities, each common knowledge equivalence class may contain more than one common hope (partial) equivalence class, thus, invalidating the generalization. The proof of the proposition below provides a simple four-world countermodel to that effect:

Proposition 5.6 $\mathcal{KH} \not\models C^H_G p \leftrightarrow \big(\neg C^H_G\bot \to C^K_G(\neg C^H_G\bot \to p)\big)$ *if* $|G| \geq 2$.

Proof. To show this, we construct a countermodel from $\mathcal{KH}$. Let $i \neq j \in G$. Consider a Kripke model $M = \big((W, \mathcal{K}_1, \dots, \mathcal{K}_n, \mathcal{H}_1, \dots, \mathcal{H}_n), \pi\big) \in \mathcal{KH}$ such that

- $W = \{\mathsf{G}', \mathsf{G}'', \mathsf{B}', \mathsf{B}''\}$;
- $\mathcal{K}_i$ splits W into equivalence classes $\{\mathsf{G}', \mathsf{B}'\}$ and $\{\mathsf{G}'', \mathsf{B}''\}$;
- $\mathcal{K}_j$ splits W into equivalence classes $\{\mathsf{G}', \mathsf{B}''\}$ and $\{\mathsf{G}'', \mathsf{B}'\}$;
- $\mathcal{K}_l = \mathcal{K}_i$ for all $l \in G \setminus \{i, j\}$;
- all agents from G are faulty in bad worlds B' and B'' and correct in good worlds G' and G'', i.e., given conditions $\mathcal{H}in\mathcal{K}$ and one$\mathcal{H}$, partial equivalence relations $\mathcal{H}_l = \{(\mathsf{G}', \mathsf{G}'), (\mathsf{G}'', \mathsf{G}'')\}$ for all $l \in G$;
- $\pi(p) = \{\mathsf{G}'\}$;
- other elements are arbitrary.

We now have the following: on the one hand, $M, \mathsf{G}' \models C^H_G p$ because, for any $l \in G$, the only world $\mathcal{H}_l$-accessible from G' is G' itself.

On the other hand, $M, w \models C^H_G \bot$ iff $w \in \{\mathsf{B}', \mathsf{B}''\}$. In particular, we have $M, \mathsf{G}' \models \neg C^H_G \bot$ and $M, \mathsf{G}'' \models \neg C^H_G \bot$. Thus, $M, \mathsf{G}'' \not\models \neg C^H_G \bot \to p$ and, consequently, $M, \mathsf{G}' \not\models C^K_G(\neg C^H_G \bot \to p)$. Overall, we can conclude that $M, \mathsf{G}' \not\models \neg C^H_G \bot \to C^K_G(\neg C^H_G \bot \to p)$.

Thus, $\mathcal{KH} \not\models C^H_G p \leftrightarrow \big(\neg C^H_G \bot \to C^K_G(\neg C^H_G \bot \to p)\big)$. □

6 Related Works

It is interesting to observe that many of the "usual suspects" for an epistemic logic do not fit the properties of hope observed in the runs and systems modeling of Firing Rebels with Relay [11] and derived from the properties of knowledge in this paper. For instance, all extensions of $\mathsf{S4}_n$ are ruled out because hope (of faulty agents) is not factive, i.e., $\not\models H_i\varphi \to \varphi$. Similarly, $\mathsf{KD45}_n$ cannot be used because we take the inconsistency statement $H_i\bot$ to be the definition of agent i's faultiness rather than summarily ruling it out by axiom D.

It is notable that, independently, based on algebraic topological modeling, Goubault et al. [13] proposed $\mathsf{KB4}_n$ as an epistemic attitude for synchronous systems where agent malfunctions are restricted to *crash failures*. They call their $\mathsf{KB4}$ modalities 'knowledge,' use K_i for them, and define a dead agent as $K_i\bot$. We call our $\mathsf{KB4}$ modalities 'hope,' use H_i for them, and define an incorrect agent as $H_i\bot$, whereas 'knowledge' for us is a separate modality of type $\mathsf{S5}$. Our derivation of $\mathsf{KB4}$ properties of hope from the standard $\mathsf{S5}$ properties of knowledge helps explain the similarities between their findings for synchronous agents with at most crash failures and the system for fully byzantine asynchronous agents from [10]. This suggests $\mathsf{KB4}_n$ to be a good epistemic basis for studying a wide range of fault-tolerant systems.

Moses and Shoham [22] introduce three binary modal operators describing a single agent's beliefs as a form of knowledge relativized to an assumption (without committing to any type of knowledge or to any particular assumption). The most relevant of the three for us is the first one $B^\alpha_1\varphi := K(\alpha \to \varphi)$, where α is any formula.[4] Thus, dropping the agent subscript for a single agent, our notion of belief $B\varphi = K(\neg H\bot \to \varphi)$, see also [11, 15–17], coincides with their $B^{\neg H\bot}_1\varphi$.

[4] Here subscript 1 means "first operator out of three" rather than agent 1.

Bolander et al. [3] consider a version of public announcement logic (PAL), called *attention-based announcement logic*, where agents need not pay attention to a public announcement. Not being attentive could be viewed as a special type of fault, which is modeled in [3] by designated atoms h_i for each agent i. Thus, much like the knowledge of our byzantine agents depends on whether they are correct, i.e., whether $\neg H_i\bot$ is true, the knowledge of their agents after a public announcement depends on whether h_i is true. Another common concern is agents' introspective properties regarding their faults, with [3] considering systems for both non-fault-introspective and fault-introspective agents, the latter stipulating the *attention introspection property*: an attentive agent believes to be attentive, $h_i \to B_i h_i$, and an inattentive agent believes to be inattentive, $\neg h_i \to B_i \neg h_i$. This results in logic K_n for non-fault-introspective agents and an extension of logic $\mathsf{K45}_n$ for fault-introspective ones. The distinction between [3] and our byzantine agents is their lack of axiom B corresponding to frame symmetry. Note that, by the very nature of their work, [3] deals with dynamic epistemic notions. The authors also introduce an adaptation of relativized common belief [1] called *attention-based relativized common belief* defined as the fixpoint of the equation $x = E^{\chi}_{\mathcal{A}}(\varphi \wedge x)$, where $E^{\chi}_{\mathcal{A}} := \bigwedge_{i\in\mathcal{A}}\bigl(h_i \to B_i(\chi \to \varphi)\bigr)$ and where χ is the relativization formula. This closely resembles our notion of mutual hope $E^{H}_{\mathcal{A}} = \bigwedge_{i\in\mathcal{A}}\bigl(\neg H_i\bot \to K_i(\neg H_i\bot \to \varphi)\bigr)$.

7 Conclusion and Future Work

We provided a description of epistemic views of agents in fault-tolerant distributed systems with fully byzantine agents by means of a multimodal logic with two types of modalities — hope and knowledge — and showed how system specifications and properties of such agents can be represented by frame-characterizable properties. This analysis yielded new insights, for instance, into the distinctions between the case of fault-tolerant systems with at most one vs. several byzantine agents. This distinction was already observed in [16] but the newly provided axiomatic representation explains which of the general properties of byzantine agents are violated when all but one agents are correct.

The extension of our completeness result to the case of common hope and common knowledge is paving the way for the complete analysis of the Firing Rebels with Relay problem, which involves a temporal dimension and relies on a temporal generalization of mutual hope, called *eventual mutual hope* [11]. As for the case of inattentive agents in [3], we also plan to introduce a dynamic component to our logic of byzantine agents, in order to formalize how communication in distributed systems affects agents' epistemic state depending on agents' correctness. We would also like to describe common hope as relativized common knowledge along the lines of [3], but the difficulty is that in our case formula χ would have to depend on i.

All these developments and extensions will be guided by the need to represent specific types of faults commonly considered in distributed systems.

Acknowledgments. We are grateful to the anonymous reviewers for their valuable suggestions on how to improve the paper. Following one suggestion, we have included decidability results. We are also grateful for multiple fruitful discussions with and/or enthusiastic support of Giorgio Cignarale, Rojo Randrianomentsoa, Hugo Rincón Galeana, Thomas Schlögl, and Ulrich Schmid.

References

[1] van Benthem, J., J. van Eijck and B. Kooi, *Logics of communication and change*, Information and Computation **204** (2006), pp. 1620–1662.
URL `https://doi.org/10.1016/j.ic.2006.04.006`

[2] Blackburn, P., M. de Rijke and Y. Venema, "Modal Logic," Cambridge Tracts in Theoretical Computer Science **53**, Cambridge University Press, 2001.
URL `https://doi.org/10.1017/CBO9781107050884`

[3] Bolander, T., H. van Ditmarsch, A. Herzig, E. Lorini, P. Pardo and F. Schwarzentruber, *Announcements to attentive agents*, Journal of Logic, Language and Information **25** (2016), pp. 1–35.
URL `http://dx.doi.org/10.1007/s10849-015-9234-3`

[4] Chandra, T. D., V. Hadzilacos and S. Toueg, *The weakest failure detector for solving consensus*, Journal of the ACM **43** (1996), pp. 685–722.
URL `https://doi.org/10.1145/234533.234549`

[5] Dolev, D., M. Függer, M. Posch, U. Schmid, A. Steininger and C. Lenzen, *Rigorously modeling self-stabilizing fault-tolerant circuits: An ultra-robust clocking scheme for systems-on-chip*, Journal of Computer and System Sciences **80** (2014), pp. 860–900.
URL `https://doi.org/10.1016/j.jcss.2014.01.001`

[6] Dolev, D., J. Y. Halpern and H. R. Strong, *On the possibility and impossibility of achieving clock synchronization*, Journal of Computer and System Sciences **32** (1986), pp. 230–250.
URL `https://doi.org/10.1016/0022-0000(86)90028-0`

[7] Fagin, R., J. Y. Halpern, Y. Moses and M. Y. Vardi, "Reasoning About Knowledge," MIT Press, 1995.

[8] Fischer, M. J. and R. E. Ladner, *Propositional dynamic logic of regular programs*, Journal of Computer and System Sciences **18** (1979), pp. 194–211.
URL `https://doi.org/10.1016/0022-0000(79)90046-1`

[9] Fischer, M. J., N. A. Lynch and M. Merritt, *Easy impossibility proofs for distributed consensus problems*, Distributed Computing **1** (1986), pp. 26–39.
URL `https://doi.org/10.1007/BF01843568`

[10] Fruzsa, K., *Hope for epistemic reasoning with faulty agents!*, in: *ESSLLI 2019 Student Session* (2019).
URL `http://esslli2019.folli.info/wp-content/uploads/2019/08/tentative_proceedings.pdf`

[11] Fruzsa, K., R. Kuznets and U. Schmid, *Fire!*, in: J. Halpern and A. Perea, editors, *Proceedings Eighteenth Conference on Theoretical Aspects of Rationality and Knowledge*, Electronic Proceedings in Theoretical Computer Science **335** (2021), pp. 139–153.
URL `http://dx.doi.org/10.4204/EPTCS.335.13`

[12] Függer, M. and U. Schmid, *Reconciling fault-tolerant distributed computing and systems-on-chip*, Distributed Computing **24** (2012), pp. 323–355.
URL `http://dx.doi.org/10.1007/s00446-011-0151-7`

[13] Goubault, É., J. Ledent and S. Rajsbaum, *A simplicial model for* $\mathbf{KB4_n}$*: Epistemic logic with agents that may die*, in: P. Berenbrink and B. Monmege, editors, *39th International Symposium on Theoretical Aspects of Computer Science (STACS 2022)*, Leibniz International Proceedings in Informatics (LIPIcs) **219** (2022), pp. 33:1–33:20.
URL `https://doi.org/10.4230/LIPIcs.STACS.2022.33`

[14] Halpern, J. Y. and Y. Moses, *Knowledge and common knowledge in a distributed environment*, Journal of the ACM **37** (1990), pp. 549–587.
URL `https://doi.org/10.1145/79147.79161`

[15] Kuznets, R., L. Prosperi, U. Schmid and K. Fruzsa, *Causality and epistemic reasoning in byzantine multi-agent systems*, in: L. S. Moss, editor, *Proceedings Seventeenth Conference on Theoretical Aspects of Rationality and Knowledge*, Electronic Proceedings in Theoretical Computer Science **297** (2019), pp. 293–312.
URL `http://dx.doi.org/10.4204/EPTCS.297.19`

[16] Kuznets, R., L. Prosperi, U. Schmid and K. Fruzsa, *Epistemic reasoning with byzantine-faulty agents*, in: A. Herzig and A. Popescu, editors, *Frontiers of Combining Systems, 12th International Symposium, FroCoS 2019, London, UK, September 4–6, 2019, Proceedings*, Lecture Notes in Artificial Intelligence **11715** (2019), pp. 259–276.
URL `http://dx.doi.org/10.1007/978-3-030-29007-8_15`

[17] Kuznets, R., L. Prosperi, U. Schmid, K. Fruzsa and L. Gréaux, *Knowledge in Byzantine message-passing systems I: Framework and the causal cone*, Technical Report TUW-260549, TU Wien (2019).
URL `https://publik.tuwien.ac.at/files/publik_260549.pdf`

[18] Lamport, L., R. Shostak and M. Pease, *The Byzantine Generals Problem*, ACM Transactions on Programming Languages and Systems **4** (1982), pp. 382–401.
URL `http://dx.doi.org/10.1145/357172.357176`

[19] Mendes, H., C. Tasson and M. Herlihy, *Distributed computability in Byzantine asynchronous systems*, in: *STOC 2014: 46th Annual Symposium on the Theory of Computing* (2014), pp. 704–713.
URL `http://dx.doi.org/10.1145/2591796.2591853`

[20] Mitchell, J. C. and E. Moggi, *Kripke-style models for types lambda calculus*, Annals of Pure and Applied Logic **51** (1991), pp. 99–124.
URL `http://dx.doi.org/10.1016/0168-0072(91)90067-V`

[21] Moses, Y., *Relating knowledge and coordinated action: The Knowledge of Preconditions principle*, in: R. Ramanujam, editor, *Proceedings Fifteenth Conference on Theoretical Aspects of Rationality and Knowledge*, Electronic Proceedings in Theoretical Computer Science **215** (2016), pp. 231–245.
URL `http://dx.doi.org/10.4204/EPTCS.215.17`

[22] Moses, Y. and Y. Shoham, *Belief as defeasible knowledge*, Artificial Intelligence **64** (1993), pp. 299–321.
URL `https://doi.org/10.1016/0004-3702(93)90107-M`

[23] Pessin, A. and S. Goldberg, editors, "The Twin Earth Chronicles: Twenty Years of Reflection on Hilary Putnam's the "Meaning of 'Meaning'"," Routledge, 1996.
URL `https://doi.org/10.4324/9781315284811`

[24] Robinson, P. and U. Schmid, *The Asynchronous Bounded-Cycle model*, Theoretical Computer Science **412** (2011), pp. 5580–5601.
URL `https://doi.org/10.1016/j.tcs.2010.08.001`

[25] Schlögl, T., U. Schmid and R. Kuznets, *The persistence of false memory: Brain in a vat despite perfect clocks*, in: T. Uchiya, Q. Bai and I. Marsá Maestre, editors, *PRIMA 2020: Principles and Practice of Multi-Agent Systems: 23rd International Conference, Nagoya, Japan, November 18–20, 2020, Proceedings*, Lecture Notes in Artificial Intelligence **12568**, Springer, 2021 pp. 403–411.
URL `https://doi.org/10.1007/978-3-030-69322-0_30`

[26] Srikanth, T. K. and S. Toueg, *Optimal clock synchronization*, Journal of the ACM **34** (1987), pp. 626–645.
URL `http://dx.doi.org/10.1145/28869.28876`

[27] Srikanth, T. K. and S. Toueg, *Simulating authenticated broadcasts to derive simple fault-tolerant algorithms*, Distributed Computing **2** (1987), pp. 80–94.
URL `http://dx.doi.org/10.1007/BF01667080`

[28] Widder, J. and U. Schmid, *The Theta-Model: achieving synchrony without clocks*, Distributed Computing **22** (2009), pp. 29–47.
URL `http://dx.doi.org/10.1007/s00446-009-0080-x`

Algorithmic correspondence and analytic rules

Andrea De Domenico

Vrije Universiteit, The Netherlands

Giuseppe Greco [1]

Vrije Universiteit, The Netherlands

Abstract

We introduce the algorithm MASSA which takes classical modal formulas in input, and, when successful, effectively generates: (a) (analytic) geometric rules of the labelled calculus G3K, and (b) cut-free derivations (of a certain 'canonical' shape) of each given input formula in the geometric labelled calculus obtained by adding the rule in output to G3K. We show that MASSA successfully terminates whenever its input formula is a (definite) analytic inductive formula, in which case, the geometric axiom corresponding to the output rule is, modulo logical equivalence, the first-order correspondent of the input formula.

Keywords: Structural proof theory of modal logic, labelled calculi, analytic extensions of labelled calculi, automatic rule-generation, algorithmic correspondence theory.

1 Introduction

The labelled calculus G3K was presented by Sara Negri in [17] as a basic G3-style sequent calculus for the normal modal logic K (see [19, Chapter 3] and [20, Chapter 11] for the genesis of this calculus). The calculus G3K shares many of the characteristic properties of Gentzen's original sequent calculus G3 for classical logic; for instance, all its rules are invertible, and the basic structural rules (weakening, contraction and cut) are admissible. Moreover, in [17], Negri introduces a general method for extending G3K so as to capture a large class of axiomatic extensions of K; namely, all those axiomatic extensions of K which define *elementary* (i.e. first-order definable) classes of Kripke frames, and such that their defining first-order conditions are, modulo logical equivalence, geometric implications. The rules generated by Negri's method for capturing these axiomatic extensions of K are defined on the basis of their corresponding geometric implications, and are referred to as *geometric rules*. Negri uniformly

[1] This research is supported by the NWO grant KIVI.2019.001.

shows that the structural rules (and cut in particular) are admissible in the calculi obtained by extending G3K with geometric rules.

One important subclass of geometric implications is given, modulo logical equivalence, by the first-order correspondents of the class of *analytic inductive formulas* in classical modal logic. General (i.e. not necessarily analytic) inductive formulas have been introduced by Goranko and Vakarelov in [11], and have been shown to have (local) first-order correspondents, which can be effectively computed via an algorithmic correspondence procedure introduced in [5].

In the present paper, we refine Negri's method for extending G3K, and introduce the algorithm MASSA for generating analytic labelled rules uniformly and equivalently capturing the *analytic inductive* axiomatic extensions of K. An important difference between the algorithmic rule-generation method introduced in this paper and Negri's method is that the present method takes *modal formulas* in input, and, if the input formula is analytic inductive (cf. Section 2.2), it computes its equivalent analytic rule *directly* from the input formula, via a computation which incorporates the effective generation of its first-order correspondent, whereas Negri's method starts from geometric implications in the first-order frame correspondence language, and generates rules which are equivalent to those modal formulas which are assumed to have a first-order correspondent which is (logically equivalent to) a geometric implication.

This paper is structured as follows. In Section 2, we collect basic definitions and results on G3K and analytic inductive formulas in classical modal logic; in Section 3, we introduce the algorithm MASSA and provide intuitive motivation for some of its key steps. In Section 4, we illustrate how MASSA works, by running it on some well known modal axioms; in Section 5, we discuss how the present results embed in a wider research context in structural proof theory, which provides motivations for further research directions.

2 Preliminaries

2.1 The labelled calculus G3K

In what follows, we adopt the usual conventions: $p, q, \ldots$ denote proposition variables, $x, y, z, \ldots$ are labels (corresponding to world-variables in the intended interpretation on Kripke frames), given a label x and a modal formula A, well-formed formulas are of the type $x : A$, while $\varphi, \psi, \ldots$ are meta-variables for well-formed formulas. $\Gamma, \Delta, \ldots$ are meta-variables for sets of wffs, and a sequent is an expression of the form $\Gamma \vdash \Delta$. Given a sequent $S = \Gamma \vdash \Delta$, if the formula $\varphi \in \Gamma$ (resp. $\varphi \in \Delta$), we say that φ occurs in *precedent* (resp. *succedent*) position in S.

Below, we list the rules of the labelled relational sequent calculus G3K for the basic normal modal logic K, where cut, weakening, contraction, and necessitation are admissible rules (see for instance [17]). In the list below, we explicitly mention the cut rule and we do not include the rules for negation. The propositional and modal rules are all invertible.

Initial rules and cut rule

$$\bot_L \frac{}{\Gamma, x : \bot \vdash \Delta} \qquad \frac{}{\Gamma, x : p \vdash x : p, \Delta} \mathrm{Id}_{x:p}$$

$$\frac{\Gamma \vdash x : p, \Delta \qquad \Gamma', x : p \vdash \Delta'}{\Gamma, \Gamma' \vdash \Delta, \Delta'} \mathrm{Cut}$$

Invertible propositional rules

$$\wedge_L \frac{\Gamma, x : A, x : B \vdash \Delta}{\Gamma, x : A \wedge B \vdash \Delta} \qquad \frac{\Gamma \vdash x : A, \Delta \qquad \Gamma \vdash x : B, \Delta}{\Gamma \vdash x : A \wedge B, \Delta} \wedge_R$$

$$\vee_L \frac{\Gamma, x : A \vdash \Delta \qquad \Gamma, x : B \vdash \Delta}{\Gamma, x : A \vee B \vdash \Delta} \qquad \frac{\Gamma \vdash x : A, x : B, \Delta}{\Gamma \vdash x : A \vee B, \Delta} \vee_R$$

$$\rightarrow_L \frac{\Gamma \vdash x : A, \Delta \qquad \Gamma, x : B \vdash \Delta}{\Gamma, x : A \rightarrow B \vdash \Delta} \qquad \frac{\Gamma, x : A \vdash x : B, \Delta}{\Gamma \vdash x : A \rightarrow B, \Delta} \rightarrow_R$$

Invertible modal rules*

$$\Box_L \frac{xRy, \Gamma, x : \Box A, y : A \vdash \Delta}{xRy, \Gamma, x : \Box A \vdash \Delta} \qquad \frac{xRy, \Gamma \vdash y : A, \Delta}{\Gamma \vdash x : \Box A, \Delta} \Box_R$$

$$\Diamond_L \frac{xRy, \Gamma, y : A \vdash \Delta}{\Gamma, x : \Diamond A \vdash \Delta} \qquad \frac{xRy, \Gamma \vdash y : A, x : \Diamond A, \Delta}{xRy, \Gamma \vdash x : \Diamond A, \Delta} \Diamond_R$$

Equality rules

$$\text{Eq-Ref} \frac{x = x, \Gamma \vdash \Delta}{\Gamma \vdash \Delta} \qquad \text{Eq-Trans} \frac{y = z, x = y, x = z, \Gamma \vdash \Delta}{x = y, x = z, \Gamma \vdash \Delta}$$

$$\mathrm{Repl}_{R1} \frac{yRz, x = y, xRz, \Gamma \vdash \Delta}{x = y, xRz, \Gamma \vdash \Delta} \qquad \mathrm{Repl}_{R2} \frac{xRz, y = z, xRy, \Gamma \vdash \Delta}{y = z, xRy, \Gamma \vdash \Delta}$$

$$\mathrm{Repl} \frac{x = y, y : A, x : A, \Gamma \vdash \Delta}{x = y, x : A, \Gamma \vdash \Delta}$$

*Side condition: the label y must not occur in the conclusion of $\Box_R$ and $\Diamond_L$.

Remark 2.1 The logical rules above (namely Propositional and Modal rules) reflect the semantic clauses of each connective in the intended Kripke semantics. Logical rules can be grouped together as *tonicity rules* ($\wedge_R, \vee_L, \rightarrow_L, \Box_L, \Diamond_R$) versus *translation rules* ($\wedge_L, \vee_R, \rightarrow_R, \Box_R, \Diamond_L$). Tonicity rules specify the arity of a connective (i.e. a connective of arity n is introduced by a tonicity rule with n premises) and its tonicity (i.e. if the connective is positive or negative in each coordinate). The translation rules convert a proxy occurring in the premise (either the comma or a relational atom) into a logical connective (namely, the main connective of the principal formula occurring in the conclusion).

Below we list the non-invertible versions of the tonicity logical rules. We sometimes refer to them as multiplicative rules.

Non-invertible tonicity propositional rules

$$\vee_L \frac{\Gamma, x : A \vdash \Delta \qquad \Gamma', x : B \vdash \Delta'}{\Gamma, \Gamma', x : A \vee B \vdash \Delta, \Delta'} \qquad \frac{\Gamma \vdash x : A, \Delta \qquad \Gamma' \vdash x : B, \Delta'}{\Gamma, \Gamma', \vdash x : A \wedge B, \Delta, \Delta'} \wedge_R$$

$$\rightarrow_L \frac{\Gamma \vdash x : A, \Delta \qquad \Gamma', x : B \vdash \Delta'}{\Gamma, \Gamma', x : A \rightarrow B \vdash \Delta, \Delta'}$$

Non-invertible tonicity modal rules*

$$\Box_L \frac{xRy, \Gamma, y : A \vdash \Delta}{xRy, \Gamma, x : \Box A \vdash \Delta} \qquad \frac{xRy, \Gamma \vdash y : A, \Delta}{xRy, \Gamma \vdash x : \Diamond A, \Delta} \Diamond_R$$

Lemma 2.2 *For any modal formula A, the sequent $\Gamma, x : A \vdash x : A, \Delta$ is derivable in G3K.*

Proof. By induction on A. The cases of $A := \bot$ and $A := p \in \mathsf{Prop}$ are immediate. If $A := *(\overline{A'})$ where $* \in \{\Box, \rightarrow, \vee\}$, then the required proof is obtained by applying, from bottom to top, $*_R$ to the occurrence of A in succedent position, followed by a bottom-up application of $*_L$ to the occurrence of A in precedent position, and then using the induction hypothesis on each A' in $\overline{A'}$. Similarly, the required proof if $A := *(\overline{A'})$ where $* \in \{\Diamond, \wedge\}$, is obtained by applying, from bottom to top, $*_L$ followed by $*_R$. □

Notice that the derivation generated in the proof of the lemma above introduces *every* subformula of each occurrence of φ via a logical rule, and, modulo renaming variables, we can assume w.l.o.g. that every new label introduced proceeding bottom-up be fresh in the entire derivation (and not just in every branch, as already required by the side conditions of the rule $\Box_R$ and $\Diamond_L$). Below we recall the definition of a geometric implication.

Definition 2.3 (cf. [16, Section 3]) A *geometric implication* is a first-order sentence of the form

$$\forall \overline{x}(s \rightarrow t),$$

where both s and t are *geometric formulas*, i.e. first-order formulas not containing $\rightarrow$ or $\forall$. Geometric implications can be equivalently rewritten as *geometric axioms*, namely, sentences of the type

$$\forall \overline{x}(P_1 \wedge ... \wedge P_m \rightarrow \overline{\exists y_1} M_1 \vee ... \vee \overline{\exists y_n} M_n)$$

where each P_i is an atomic formula with no free occurrences of any variable y in $\overline{y}$, and M_j is a conjunction of atomic formulas $Q_{j_1} \wedge ... \wedge Q_{j_{k_j}}$. The rule scheme corresponding to geometric axioms takes the form

$$\frac{\overline{Q_1}[\overline{y_1}/\overline{z_1}], \overline{P}, \Gamma \vdash \Delta \qquad ... \qquad \overline{Q_n}[\overline{y_n}/\overline{z_n}], \overline{P}, \Gamma \vdash \Delta}{\overline{P}, \Gamma \vdash \Delta} GR$$

where $\overline{Q_i}[\overline{y_i}/\overline{z_i}]$ denotes the simultaneous replacement of each z in $\overline{z_i}$ with the corresponding y in $\overline{y_i}$, in every Q in $\overline{Q_i}$. In this scheme, the eigenvariables in $\overline{y_i}$ are not free in $\overline{P}, \Delta, \Gamma$. Rules corresponding to geometric axioms are referred to as *geometric (labelled) rules.*

A *geometric labelled calculus* is any extension of G3K with geometric labelled rules.

Theorem 2.4 *(cf. [17, Theorem 4.13]) Any geometric labelled calculus preserves cut admissibility.*

2.2 Analytic inductive formulas

In this section, we specialize and adapt the definition of analytic inductive inequality (cf. [12, Definition 55], [9, Definition 2.14], [1, Section 2.3]) to the language and properties of classical modal logic.

The language of the basic normal modal logic K is recursively defined from a set Prop of proposition variables as follows:

$$\varphi ::= p \mid \bot \mid \neg\varphi \mid \varphi \wedge \varphi \mid \varphi \vee \varphi \mid \varphi \to \varphi \mid \Diamond\varphi \mid \Box\varphi,$$

where p ranges over Prop. In what follows, we will need to keep track of the multiplicity of occurrences of proposition variables in formulas, as well as the order-theoretic properties of the various coordinates of the term-functions associated with formulas. Therefore, we will write e.g. $\psi(!\overline{x})$ to signify that each variable in the vector $\overline{x}$ of placeholder variables occurs exactly once in ψ. Moreover, we will write e.g. $\psi(!\overline{x}, !\overline{y})$ to mean that ψ (resp. the term-function $\psi^{\mathbb{A}}$ in a modal algebra $\mathbb{A}$) is positive (resp. monotone) in each x-coordinate and negative (resp. antitone) in each y-coordinate. In other contexts, we will sometimes need to group coordinates according to different criteria. In each context in which this is the case, we will specifically indicate these criteria. Negative (resp. positive) *Skeleton* formulas $\psi(!\overline{x}, !\overline{y})$ (resp. $\varphi(!\overline{x}, !\overline{y})$) are defined by simultaneous recursion as follows:

$$\begin{array}{rcl} \psi(!\overline{x}, !\overline{y}) & ::= & x \mid \neg\varphi \mid \psi \wedge \psi \mid \psi \vee \psi \mid \varphi \to \psi \mid \Box\psi, \\ \varphi(!\overline{x}, !\overline{y}) & ::= & x \mid \neg\psi \mid \varphi \wedge \varphi \mid \varphi \vee \varphi \mid \Diamond\varphi. \end{array}$$

Positive Skeleton formulas will sometimes be referred to as *negative PIA* formulas. *Definite* negative Skeleton (resp. PIA) formulas are defined by simultaneous recursion as follows:

$$\begin{array}{rcl} \psi(!\overline{x}, !\overline{y}) & ::= & x \mid \neg\varphi \mid \psi \vee \psi \mid \varphi \to \psi \mid \Box\psi, \\ \varphi(!\overline{x}, !\overline{y}) & ::= & x \mid \neg\psi \mid \varphi \wedge \varphi \mid \Diamond\varphi. \end{array}$$

Modulo exhaustively distributing all the other connectives over $\vee$ and $\wedge$, any negative Skeleton (resp. PIA) formula can be equivalently rewritten as a conjunction (resp. disjunction) of definite negative Skeleton (resp. PIA) formulas (cf. [1, Lemma 2.9]).

Definition 2.5 A modal formula $\psi'(\overline{p})$ is (negative) *analytic inductive* if its negative normal form (NNF) is $\psi(\overline{\beta}/!\overline{x}, \overline{\delta}/!\overline{y})$ such that:

(i) $\psi(!\overline{x}, !\overline{y})$ (which we refer to as the *Skeleton* of ψ') is a negative Skeleton formula, and is monotone *both* in its x-coordinates and in its y-coordinates;

(ii) each β in $\overline{\beta}$ and δ in $\overline{\delta}$ is a negative PIA formula;

(iii) the term-function $\delta^{\mathbb{A}}(!\overline{x})$ associated with each $\delta(\overline{p}/!\overline{x})$ in $\overline{\delta}$ is monotone in each coordinate;

(iv) the term-function $\beta^{\mathbb{A}}(!\overline{x}, !\overline{y})$ associated with each β in $\overline{\beta}$ is monotone in each x-coordinate and antitone in each y-coordinate;

(v) the transitive closure $<_\Omega$ of the relation Ω (defined below) is a well-founded strict order on $\overline{p}$, where for all p, p' in $\overline{p}$, $(p, p') \in \Omega$ iff some $\beta \in \overline{\beta}$ exists s.t. $\beta = \beta(\overline{p_1}/!\overline{x}, \overline{p_2}/!\overline{y})$ and p' occurs in $\overline{p_1}$ and p occurs in $\overline{p_2}$, and the lowest common node in the branches ending in p' and p in the generation tree of β is a $\wedge$-node.

In an analytic inductive formula ψ' as above, the variable occurrences in the y-coordinates of each β in $\overline{\beta}$ are referred to as the *critical* occurrences[2] in ψ'. All the other variable occurrences are *non-critical*. An analytic inductive formula is *Sahlqvist* if the relation Ω is empty, and is *definite* if its Skeleton is definite.

As discussed above, for any analytic inductive formula $\psi' := \psi(\overline{\beta}/!\overline{x}, \overline{\delta}/!\overline{y})$, any negative PIA subformula β and δ of ψ' can be equivalently rewritten as a disjunction of definite negative PIA formulas (cf. [1, Lemma 2.9]). Hence, once these $\vee$-nodes have reached the root of β by distributing all the other connectives over them, they can all be considered part of the Skeleton of ψ'. Hence, when representing an analytic inductive formula ψ' as $\psi(\overline{\beta}/!\overline{x}, \overline{\delta}/!\overline{y})$, we can assume w.l.o.g. that each β and δ is a *definite* negative PIA formula, and that there is *exactly one* critical occurrence of a proposition variable in each β in $\overline{\beta}$. To emphasise this, we sometimes write β as β_p.

Example 2.6 (i) The formula $\psi'(p) := \Diamond p \to \Box p$ can be rewritten in NNF as $\psi(\beta/x, \delta/y)$ where $\psi(x, y) := \Box x \vee \Box y$, and $\beta(p) := \neg p$, and $\delta(p) := p$, and is hence (negative) analytic Sahlqvist.

(ii) The formula $\psi'(p) := \Box p \to \Diamond p$ can be rewritten in NNF as $\psi(\beta/x, \delta/y)$ where $\psi(x, y) := x \vee y$, and $\beta(p) := \Diamond\neg p$ and $\delta(p) := \Diamond p$, and is hence (negative) analytic Sahlqvist.

(iii) The formula $\psi'(p) := \Diamond\Box p \to \Box\Diamond p$ can be rewritten in NNF as $\psi(\beta/x, \delta/y)$ where $\psi(x, y) := \Box x \vee \Box y$, and $\beta(p) := \Diamond\neg p$ and $\delta(p) := \Diamond p$, and is hence (negative) analytic Sahlqvist.

(iv) The formula $\psi'(p_1, p_2) := \Box(p_1 \to p_2) \to (\Box p_1 \to \Box p_2)$ can be rewritten in NNF as $\psi(\beta_1/x_1, \beta_2/x_2, \delta/y)$ where $\psi(x_1, x_2, y) := x_1 \vee (x_2 \vee \Box_t y)$, and

[2] In the more general setting of (D)LE-logics (see e.g. [7][6]), inductive and Sahlqvist formulas/inequalities are defined parametrically in every order-type ε on the proposition variables occurring in the given formula/inequality. However, in the Boolean setting this is not needed, and the definition given here corresponds to the general definition relative to the order-type $\varepsilon(p) = 1$ for each $p \in \mathsf{Prop}$.

$\beta_1(p_1, p_2) := \Diamond_y(p_1 \wedge \neg p_2)$ and $\beta_2(p_1) := \Diamond\neg p_1$ and $\delta(p_2) := p_2$, and is hence (negative) analytic inductive with $p_1 <_\Omega p_2$.

(v) The formula $\psi'(p_1, p_2) := \Box(\Box p_1 \to p_2) \vee \Box(\Box p_2 \to p_1)$ can be rewritten in NNF as $\psi(\beta_1/x_1, \beta_2/x_2, \delta_1/y_1, \delta_2/y_2)$ where $\psi(x_1, x_2, y_1, y_2) := \Box(x_1 \vee y_2) \vee \Box(x_2 \vee y_1)$, and $\beta_1(p_1) := \Diamond\neg p_1$ and $\beta_2(p_2) := \Diamond\neg p_2$, and $\delta_1(p_1) := p_1$ and $\delta_(p_2) := p_2$, and is hence (negative) analytic Sahlqvist.

Theorem 2.7 *(cf. [11, Theorem 37]) Every (analytic) inductive formula has a first-order correspondent.*

3 The algorithm MASSA

In this section, we describe the algorithm MASSA. The steps (i)-(iv) generate the analytic labelled rule r associated with the input modal formula φ. Step (v) describes how to read off the geometric implication from the rule r.

(i) **Logical rules.** For any modal formula φ, consider the identity end-sequent $x : \varphi^r \vdash x : \varphi^b$ where the formula in precedent position is coloured red and the formula in succedent position is coloured blue. Let π'_φ be a derivation of $x : \varphi^r \vdash x : \varphi^b$ obtained by applying the procedure described in the proof of Lemma 2.2, where, as discussed early on, each subformula of φ^r and φ^b has been introduced in the proof via a logical rule, and every new label introduced proceeding bottom-up must be fresh in the entire proof (and not just in every branch).[3] At each rule application in π'_φ, propagate the colour of the principal formula to the auxiliary formulas. Prune the proof-tree π'_φ, thereby generating a new proof-tree π_φ with the same structure as π'_φ, but such that each tonicity rule is applied in multiplicative form (cf. Section 2.1).

(ii) **Atomic cuts + PIA parts.** Consider the leaves of π_φ and perform all possible cuts on atomic red-coloured formulas $x : p^r$ occurring in π_φ. These cuts generate new axioms of the form $\Gamma, y = z, y : p^b \vdash z : p^b, \Delta$ in which the new relational atom $y = z$ appears in the conclusion of each cut with cut formulas $y : p^r$ and $z : p^r$. If a proposition variable $x : p^r$ occurs only positively or only negatively in φ, then cut either with an atomic initial rule of the form $x : \bot \vdash x : p^r$ or with $x : p^r \vdash \top$. Collect all the conclusions of these cut-applications, and use them as leaves in a (cut-free) forward-chaining proof-search with goal $\vdash x : \varphi^b$, where only tonicity rules are used.[4] Collect all the attempts π^i_φ generated in this way.

(iii) **Skeleton parts.** Perform a backward-chaining proof search on $\vdash x : \varphi^b$

[3] The latter requirement guarantees that all the relevant information contained in the end-sequent is maintained (and exploited in rule form) in π'_φ.

[4] Notice that we are compositionally constructing all the *maximal PIA subformulas*, that here coincide with those subformulas that can be constructed using only tonicity rule. Notice that whenever an atomic subformula of φ^b is uniform, it is substituted with $\bot$ (resp. $\top$) if in succedent (resp. precedent) position, so it is not strictly speaking a subformula of φ^b.

in which only translation rules are used.[5]

(iv) **Skeleton-PIA merging.** A *merging point* is a tuple of sequents $(S_1, ..., S_n, S)$, of which $S_1, ..., S_n$ are the *premises* and S the *conclusion*, and such that the set of labelled formulae of S is the union of the sets of labelled formulae of $S_1, ..., S_n$. Verify whether $(S_1, ..., S_n, S)$ is a merging point, where the S_i are the endsequents of all the proof-trees π^i_φ generated in item (ii), and S is the uppermost sequent of the proof-section generated in item (iii). If $(S_1, ..., S_n, S)$ is a merging point, then it is an application of the rule r in output, which provides the missing step in proof of $\vdash x : \varphi^b$.

Let R_i and R be the relational parts of S_i and S respectively. The rule r associated with the merging point is:

$$r\,\frac{R_1, \Gamma_1 \vdash \Delta_1 \qquad \ldots \qquad R_n, \Gamma_n \vdash \Delta_n}{R, \Gamma_1, \ldots, \Gamma_n \vdash \Delta_1, \ldots, \Delta_n}$$

(v) **Reading off the geometric axiom from the rule.** Let F_i be defined as the conjunction of the relational atoms in R_i in case in S_i there are no occurrences of $y : \bot$ in precedent position or of $y : \top$ in succedent position (an empty conjunction will be regarded as $\top$). Otherwise, let F_i be $\bot$. If in S there are formulas $y : \bot$ (resp. $y : \top$) in precedent (resp. succedent) position, then the required geometric formula is $\top$. Otherwise, the geometric axiom which we can read off from the rule r is:

$$\forall \overline{x}[\bigwedge R(\overline{x}) \rightarrow \bigvee_i \exists \overline{y_i} F_i(\overline{x}, \overline{y_i}))].$$

Steps (i) and (ii) can be intuitively justified as follows. Whenever φ is a theorem of K, the calculus G3K derives $\vdash x : \varphi$ without any additional rule. Otherwise, we need to identify some assumptions Γ which allow us to derive $\Gamma \vdash x : \varphi$. Clearly, the *minimal* set of assumptions Γ under which φ is derivable is $\Gamma = \{x : \varphi^r\}$. Then, at step (i), we equivalently transform the additional assumption $x : \varphi^r$ into pure relational information and also information stored in the atomic propositions of the form $x : p^r$. The cuts performed in step (ii) extract additional pure relational information from these atomic propositions.

Theorem 3.1 *The algorithm MASSA successfully terminates whenever it receives a definite analytic inductive formula of classical modal logic in input, in which case, the geometric axiom read off from the output rule is, modulo logical equivalence, the first-order correspondent of the input formula.*

Proof. see Appendix A. □

4 Examples

In the present section, we illustrate the algorithm MASSA by running it on some definite analytic inductive formulas. Let us start with the Church-Rosser axiom, sometimes also called the directedness axiom (cf. Example 2.6 (iii)).

[5] Here we are compositionally destroying all the Skeleton connectives namely $\Diamond$ and $\wedge$ if occurring in precedent position, and $\Box, \vee$ and $\rightarrow$ if occurring in succedent position.

Step (i). We build the proof π'_φ using (invertible) additive rules. Below we split the derivation tree into three proof sections: the numbers assigned to each sequent allow to uniquely reconstruct the original tree.

$$\dfrac{\dfrac{\dfrac{\dfrac{\dfrac{}{(7.1)\ xRy, yRt, y : \Box A_1^b, t : A_1^b \vdash t : A_3^r, x : \Diamond\Box A_3^r, x : \Box\Diamond A_2^b}\ \mathrm{Id}_{t:A}}{(6.1)\ xRy, yRt, y : \Box A_1^b \vdash t : A_3^r, x : \Diamond\Box A_3^r, x : \Box\Diamond A_2^b}\ \Box_L}{(5.1)\ xRy, y : \Box A_1^b \vdash y : \Box A_3^r, x : \Diamond\Box A_3^r, x : \Box\Diamond A_2^b}\ \Box_R}{(4.1)\ xRy, y : \Box A_1^b \vdash x : \Diamond\Box A_3^r, x : \Box\Diamond A_2^b}\ \Diamond_R}{x : \Diamond\Box A_1^b \vdash x : \Diamond\Box A_3^r, x : \Box\Diamond A_2^b}\ \Diamond_L$$

(3.1)

$$\dfrac{\dfrac{\dfrac{\dfrac{\dfrac{}{(7.2)\ xRz, zRw, x : \Diamond\Box A_1^b, x : \Box\Diamond A_4^r, w : A_4^r \vdash w : A_2^b, z : \Diamond}\ \mathrm{Id}_{w:A}}{(6.2)\ xRz, zRw, x : \Diamond\Box A_1^b, x : \Box\Diamond A_4^r, w : A_4^r \vdash z : \Diamond A_2^b}\ \Diamond_R}{(5.2)\ xRz, x : \Diamond\Box A_1^b, x : \Box\Diamond A_4^r, z : \Diamond A_4^r \vdash z : \Diamond A_2^b}\ \Diamond_L}{(4.2)\ xRz, x : \Diamond\Box A_1^b, x : \Box\Diamond A_4^r \vdash z : \Diamond A_2^b}\ \Box_L}{x : \Diamond\Box A_1^b, x : \Box\Diamond A_4^r \vdash x : \Box\Diamond A_2^b}\ \Box_R$$

(3.2)

$$\dfrac{\dfrac{\begin{matrix}(3.1) & & (3.2)\\ x : \Diamond\Box A_1^b \vdash x : \Diamond\Box A_3^r, x : \Box\Diamond A_2^b & & x : \Diamond\Box A_1^b, x : \Box\Diamond A_4^r \vdash x : \Box\Diamond A_2^b\end{matrix}}{(2)\ x : \Diamond\Box A_1^b, x : \Diamond\Box A_3^r \to \Box\Diamond A_4^r \vdash x : \Box\Diamond A_2^b}\ \to_L}{(1)\ x : \Diamond\Box A_3^r \to \Box\Diamond A_4^r \vdash x : \Diamond\Box A_1^b \to \Box\Diamond A_2^b}\ \to_R$$

We prune the proof tree π'_φ obtaining the "multiplicative" proof tree π_φ.

$$\dfrac{\dfrac{\dfrac{\dfrac{\dfrac{\dfrac{}{(7.1)\ xRy, yRt, t : A_1^b \vdash t : A_3^r}\ \mathrm{Id}_{t:A}}{(6.1)\ xRy, yRt, y : \Box A_1^b \vdash t : A_3^r}\ \Box_L}{(5.1)\ xRy, y : \Box A_1^b \vdash y : \Box A_3^r}\ \Box_R}{(4.1)\ xRy, y : \Box A_1^b \vdash x : \Diamond\Box A_3^r}\ \Diamond_R}{(3.1)\ x : \Diamond\Box A_1^b \vdash x : \Diamond\Box A_3^r}\ \Diamond_L \quad \dfrac{\dfrac{\dfrac{\dfrac{\dfrac{}{(7.2)\ xRz, zRw, w : A_4^r \vdash w : A_2^b}\ \mathrm{Id}_{w:A}}{(6.2)\ xRz, zRw, w : A_4^r \vdash z : \Diamond A_2^b}\ \Diamond_R}{(5.2)\ xRz, z : \Diamond A_4^r \vdash z : \Diamond A_2^b}\ \Diamond_L}{(4.2)\ xRz, x : \Box\Diamond A_4^r \vdash z : \Diamond A_2^b}\ \Box_L}{(3.2)\ x : \Box\Diamond A_4^r \vdash x : \Box\Diamond A_2^b}\ \Box_R}{\dfrac{(2)\ x : \Diamond\Box A_1^b, x : \Diamond\Box A_3^r \to \Box\Diamond A_4^r \vdash x : \Box\Diamond A_2^b}{(1)\ x : \Diamond\Box A_3^r \to \Box\Diamond A_4^r \vdash x : \Diamond\Box A_1^b \to \Box\Diamond A_2^b}\ \to_R}\ \to_L$$

Step (ii). We consider the leaves (7.1) and (7.2) and perform all the atomic cuts on red coloured formulas.

$$\dfrac{\dfrac{}{(7.1)\ xRy, yRt, t : A_1^b \vdash t : A_3^r}\ \mathrm{Id}_{t:A} \quad \dfrac{}{(7.2)\ xRz, zRw, w : A_4^r \vdash w : A_2^b}\ \mathrm{Id}_{w:A}}{xRy, yRt, xRz, zRw, t = w; t : A_1^b \vdash w : A_2^b}\ \mathrm{Cut}(A_3^r, A_4^r)$$

We now construct the upper portion of the proof π_φ^1.[6] In this step, we build up the PIA sub-formulas of φ.

[6] Notice that we could also construct a proof with a different order of rule applications (e.g. in this case, proceeding top down, first we apply $\Box_L$ and then $\Diamond_R$). Such trivial permutations of rules generate, strictly speaking, different syntactic proofs but do not change the merging point. So, it is enough to pick one of those proofs.

$$\Box_L\,\dfrac{\dfrac{\dfrac{\pi^1_\varphi}{xRy, yRt, xRz, zRw, t = w, t : A^b \vdash w : A^b}}{xRy, yRt, xRz, zRw, t = w, t : A^b \vdash z : \Diamond A^b}\,\Diamond_R}{xRy, yRt, xRz, zRw, t = w, y : \Box A^b \vdash z : \Diamond A^b}$$

Step (iii). In this step, we work on the Skeleton of φ.

$$\dfrac{\dfrac{\Diamond_L\,\dfrac{xRz, xRy\,, y : \Box A^b \vdash z : \Diamond A^b}{xRz\,, x : \Diamond\Box A^b \vdash z : \Diamond A^b}}{x : \Diamond\Box A^b \vdash x : \Box\Diamond A^b}\,\Box_R}{\vdash x : \Diamond\Box A^b \to \Box\Diamond A^b}\,\to_R$$

Step (iv). We now reach a merging point, and hence generate the rule Dir:

$$\dfrac{\dfrac{\Diamond_L\,\dfrac{\text{Dir}\,\dfrac{\Box_L\,\dfrac{\dfrac{}{xRy, yRt, xRz, zRw, t = w, t : A^b \vdash w : A^b}}{xRy, yRt, xRz, zRw, t = w, t : A^b \vdash z : \Diamond A^b}\,\Diamond_R}{xRy, yRt, xRz, zRw, t = w, y : \Box A^b \vdash z : \Diamond A^b}}{xRy, xRz\,, y : \Box A^b \vdash z : \Diamond A^b}}{xRz\,, x : \Diamond\Box A^b \vdash z : \Diamond A^b}}{x : \Diamond\Box A^b \vdash x : \Box\Diamond A^b}\,\Box_R \quad \to_R$$

$$\vdash x : \Diamond\Box A^b \to \Box\Diamond A^b$$

Step (v). Finally, the FO-correspondent reads

$$\forall x \forall y \forall z[xRy \wedge xRz \to \exists t \exists w(yRt \wedge zRw \wedge t = w)],$$

which is equivalent to directedness.

Let us execute MASSA on the 'functionality' axiom (cf. Example 2.6 (i)). The pruned proof-tree generated in the first step is the following:

$$\dfrac{\dfrac{\Diamond_L\,\dfrac{\Diamond_R\,\dfrac{\text{Id}_{y:A}\,\dfrac{}{xRy, y : A^b \vdash y : A^r}}{xRy, y : A^b \vdash x : \Diamond A^r}}{x : \Diamond A^b \vdash x : \Diamond A^r} \qquad \dfrac{\dfrac{\dfrac{}{xRz, z : A^r \vdash z : A^b}\,\text{Id}_{z:A}}{xRz, x : \Box A^r \vdash z : A^b}\,\Box_L}{x : \Box A^r \vdash x : \Box A^b}\,\Box_R}{x : \Diamond A^r \to \Box A^r, x : \Diamond A^b \vdash x : \Box A^b}\,\to_L}{x : \Diamond A^r \to \Box A^r \vdash x : \Diamond A^b \to \Box A^b}\,\to_R$$

The leaves on which we perform the only possible cut are written below:

$$xRy, y : A^b \vdash y : A^r \quad xRz, z : A^r \vdash z : A^b.$$

After performing step (ii) and (iii), the merging point is reached, which generates the following derivation and rule (step (iv)):

$$\dfrac{\dfrac{\Diamond_L\,\dfrac{\text{Fun}\,\dfrac{\dfrac{}{y = z, xRy, xRz, y : A^b \vdash z : A^b}}{xRy, xRz, y : A^b \vdash z : A^b}}{xRz, x : \Diamond A^b \vdash z : A^b}}{x : \Diamond A^b \vdash x : \Box A^b}\,\Box_R}{\vdash x : \Diamond A^b \to \Box A^b}\,\to_R$$

from which the first-order correspondent (step (v)) below can be read off:
$$\forall x \forall y \forall z (xRy \land xRz \to y = z).$$

Merging points do not need to be unary. To see this, let us consider the formula $\Box(\Box A \to B) \lor \Box(\Box B \to A)$ (cf. Example 2.6 (v)). After performing step (i), the leaves of π are as follows:

$$xRy, yRz, z : A^b \vdash z : A^r \qquad xRt, t : A^r \vdash t : A^b$$
$$xRt, tRw, w : B^b \vdash w : B^r \qquad xRy, y : B^r \vdash y : B^b$$

After performing steps (ii) and (iii), we generate a binary merging point and we provide the following derivation (step (iv)):

$$\dfrac{\dfrac{xRy, yRz, xRt, z = t, z : A \vdash t : A}{xRy, yRz, xRt, z = t, y : \Box A \vdash t : A} \qquad \dfrac{xRt, tRw, xRy, y = w, w : B \vdash y : B}{xRt, tRw, xRy, y = w, t : \Box B \vdash y : B}}{\dfrac{xRy, xRt, y : \Box A, t : \Box B \vdash y : B, t : A}{\dfrac{xRy, xRt, y : \Box A \vdash y : B, t : \Box B \to A}{\dfrac{xRy, xRt \vdash y : \Box A \to B, t : \Box B \to A}{\dfrac{xRy \vdash y : \Box A \to B, x : \Box(\Box B \to A)}{\dfrac{\vdash x : \Box(\Box A \to B), x : \Box(\Box B \to A)}{\vdash x : \Box(\Box A \to B) \lor \Box(\Box B \to A)}}}}}}$$

The first order correspondent (step (v)) reads
$$\forall x \forall y \forall t (xRy \land xRt \to \exists z(yRz \land z = t) \lor \exists w(tRw \land y = w))$$
which is equivalent to
$$\forall x \forall y \forall t (xRy \land xRt \to yRt \lor tRy).$$

The examples discussed so far are all Sahlqvist. However, MASSA is successful on (definite analytic) formulas which are *properly inductive*, such as the axiom $K := \Box(A \to B) \to (\Box A \to \Box B)$. After performing step (i), the leaves of π_K are as follows:

$$xRz, z : A^b \vdash z : A^r \qquad xRy, y : A^r \vdash y : A^b$$
$$xRy, y : B^b \vdash y : B^r \qquad xRt, t : B^r \vdash t : B^b$$

After performing steps (ii) and (iii), we reach a merging point and hence the rule deriving K as follows (step (iv)):

$$K \ \dfrac{\dfrac{\dfrac{xRz, zRy, y = z, z : A \vdash y : A}{xRz, zRy, y = z, x : \Box A \vdash y : A} \qquad xRy, xRt, t = y, y : B \vdash t : B}{\dfrac{xRt, xRy, xRz, t = y, t = z, y : A \to B, x : \Box A \vdash t : B}{\dfrac{xRt, xRy, xRz, t = y, t = z, x : \Box(A \to B), x : \Box A \vdash t : B}{\dfrac{xRt, x : \Box(A \to B), x : \Box A \vdash t : B}{\dfrac{x : \Box(A \to B), x : \Box A \vdash x : \Box B}{\dfrac{x : \Box(A \to B) \vdash x : \Box A \to \Box B}{\vdash x : \Box(A \to B) \to (\Box A \to \Box B)}}}}}}}$$

The first-order correspondent (step (v)) reads

$$\forall x \forall t(xRt \to \exists y \exists z(xRy \wedge xRz \wedge t = y \wedge t = z))$$

which is equivalent to $\top$ as expected, since the input formula K is derivable in G3K, i.e. is valid in every Kripke frame.

Let us finish this section by discussing a couple of unsuccessful MASSA runs; let us try and run MASSA on the (non inductive and famously non elementary, see [22]) McKinsey formula $\Box\Diamond A \to \Diamond\Box A$. The first step produces the leaves

$$xRy, yRz, z : A^b \vdash z : A^r \qquad xRw, wRt; t : A^r \vdash t : A^b,$$

but after performing the cut, at step (ii) and (iii) we get stuck:

$$\dfrac{\dfrac{\dfrac{xRy, yRz, xRw, wRt, z = t, z : A \vdash t : A}{???}}{x : \Box\Diamond A \vdash x : \Diamond\Box A}}{\vdash x : \Box\Diamond A \to \Diamond\Box A}$$

We cannot proceed bottom-up since we do not have the necessary relational information, and we cannot proceed top-down without violating the side conditions of G3K.

When we take as input the (Sahlqvist but not analytic) formula $A \to \Diamond\Box A$, the first step produces the leaves

$$x : A^b \vdash x : A^r \qquad xRw, wRt; t : A^r \vdash t : A^b.$$

Again, after performing the cut, we cannot proceed further:

$$\dfrac{\dfrac{\dfrac{xRw, wRt, x = t, x : A \vdash t : A}{???}}{x : A \vdash x : \Diamond\Box A}}{\vdash x : A \to \Diamond\Box A}$$

5 Conclusions

Related work. The results in the present paper pertain to a larger line of research in structural proof theory focusing on the uniform generation of analytic rules for classes of axiomatic extensions in different (nonclassical) logics, which includes e.g., [21,23,18,16,17] in the context of sequent and labelled calculi, [2,14,15] in the context of sequent and hypersequent calculi, and [13,3,12] in the context of (proper) display calculi. We refer to [1] for an overview of this literature.

Range of applicability. We conjecture that the present approach extends to the analytic inductive axiomatic extensions of the basic normal and regular LE-logics (cf. [8]) and also to a large class of (substructural) non-normal modal logics. In future work, we plan to explore this direction. Moreover, we plan to define (invertible) translations between proofs of different calculi modulo intermediate translations into a suitable calculus in the language of ALBA (as we have done in Appendix A).

References

[1] Jinsheng Chen, Giuseppe Greco, Alessandra Palmigiano, and Apostolos Tzimoulis. Syntactic completeness of proper display calculi. *Submitted*, arXiv:2102.11641, 2021.

[2] Agata Ciabattoni, Nikolaos Galatos, and Kazushige Terui. From axioms to analytic rules in nonclassical logics. In *Logic in Computer Science*, volume 8, pages 229–240, 2008.

[3] Agata Ciabattoni and Revantha Ramanayake. Power and limits of structural display rules. *ACM Transactions on Computational Logic*, 17(3):17:1–17:39, February 2016.

[4] Willem Conradie, Silvio Ghilardi, and Alessandra Palmigiano. Unified Correspondence. In Alexandru Baltag and Sonja Smets, editors, *Johan van Benthem on Logic and Information Dynamics*, volume 5 of *Outstanding Contributions to Logic*, pages 933–975. Springer International Publishing, 2014.

[5] Willem Conradie, Valentin Goranko, and Dimiter Vakarelov. Algorithmic correspondence and completeness in modal logic. I. The core algorithm SQEMA. *Logical Methods in Computer Science*, 2:1–26, 2006.

[6] Willem Conradie and Alessandra Palmigiano. Algorithmic correspondence and canonicity for distributive modal logic. *Annals of Pure and Applied Logic*, 163(3):338 – 376, 2012.

[7] Willem Conradie and Alessandra Palmigiano. Algorithmic correspondence and canonicity for non-distributive logics. *Annals of Pure and Applied Logic*, 170(9):923–974, 2019.

[8] Willem Conradie and Alessandra Palmigiano. Constructive canonicity of inductive inequalities. *Logical Methods in Computer Science*, 16(3):1–39, 2020.

[9] Laurent De Rudder and Alessandra Palmigiano. Slanted canonicity of analytic inductive inequalities. *ACM Transactions on Computational Logic (TOCL)*, 22(3):1–41, 2021.

[10] Roy Dyckhoff and Sara Negri. Proof analysis in intermediate logics. *Archive for Mathematical Logic*, 51(1):71–92, 2012.

[11] V. Goranko and D. Vakarelov. Elementary canonical formulae: Extending Sahlqvist theorem. *Annals of Pure and Applied Logic*, 141(1-2):180–217, 2006.

[12] Giuseppe Greco, Minghui Ma, Alessandra Palmigiano, Apostolos Tzimoulis, and Zhiguang Zhao. Unified correspondence as a proof-theoretic tool. *Journal of Logic and Computation*, 28(7):1367–1442, 2016.

[13] Marcus Kracht. Power and weakness of the modal display calculus. In *Proof theory of modal logic*, volume 2 of *Applied Logic Series*, pages 93–121. Kluwer, 1996.

[14] Ori Lahav. From frame properties to hypersequent rules in modal logics. In *Proceedings of the 2013 28th Annual ACM/IEEE Symposium on Logic in Computer Science*, pages 408–417. IEEE Computer Society, 2013.

[15] Björn Lellmann. Axioms vs hypersequent rules with context restrictions: theory and applications. In *Automated Reasoning*, volume 8562 of *Lecture Notes in Computer Science*, pages 307–321. Springer, 2014.

[16] Sara Negri. Contraction-free sequent calculi for geometric theories, with an application to Barr's theorem. *Archive for Mathematical Logic*, 42:389–401, 2003.

[17] Sara Negri. Proof analysis in modal logic. *Journal of Philosophical Logic*, 34(5-6):507–544, 2005.

[18] Sara Negri and Jan Von Plato. Cut elimination in the presence of axioms. *The Bullettin of Symbolic Logic*, 4(4):418–435, 1998.

[19] Sara Negri and Jan von Plato. *Structural Proof Theory*. Cambridge Universty Press, Cambridge, 2001.

[20] Sara Negri and Jan Von Plato. *Proof analysis: a contribution to Hilbert's last problem*. Cambridge University Press, 2011.

[21] A. Simpson. *The Proof Theory and Semantics of Intuitionistic Modal Logic*. PhD dissertation, University of Edinburgh, 1994.

[22] J. van Benthem. *Modal correspondence theory*. PhD thesis, Department of Mathematics, University of Amsterdam, Amsterdam, The Netherlands, 1978.

[23] Luca Viganò. *Labelled non-classical logics*. Springer US, 2000.

A Proof of Theorem 3.1

Main goal. In the present section, we show that, if the algorithm MASSA receives a definite analytic inductive formula ψ' in input, it successfully reaches a merging point in step (iv), and hence it outputs a geometric rule r which derives ψ' when added to G3K , and from which the first-order correspondent of ψ' (which exists, cf. [11]) can be read off. In fact, we will prove even more; namely, that a cut-free derivation of ψ' can be effectively generated in G3K+r, and this derivation has a specific shape.

Our proof will make use of the fact that the first-order correspondent of a generic analytic inductive formula can be represented in the language of the algorithm ALBA [4]. To prove the required statement, it is enough to show that the merging point is reached, and the geometric axiom that we read off from r is *effectively recognizable* as the first-order correspondent of ψ'. To guarantee this effective recognizability, we also translate G3K in a format set in the language of ALBA (cf. [9, Section 2.5]).

Following the conventions and notation of Section 2.2, we represent ψ' as $\psi(\overline{\beta}, \overline{\delta})$. The algorithm ALBA is guaranteed to succeed in computing the first-order correspondent of ψ' (cf. [7, Theorem 8.8]), e.g. via the following run, which, for the sake of simplicity, we execute under the assumption that $\psi'(\overline{p})$ is Sahlqvist, and each $p \in \overline{p}$ occurs both positively and negatively. In what follows, (vectors of) variables $\mathbf{i}$, $\mathbf{j}$, $\mathbf{h}$, and $\mathbf{k}$, referred to as *nominal* variables, are interpreted in Kripke frames as possible worlds, or equivalently as atoms (i.e. completely join-irreducible elements) of the complex algebra of any Kripke frame, while (vectors of) variables $\mathbf{l}$, $\mathbf{m}$, $\mathbf{n}$ and $\mathbf{o}$, referred to as *conominal* variables, are interpreted in Kripke frames as complements of possible worlds, or equivalently as co-atoms (i.e. completely meet-irreducible elements) of the complex algebra of any Kripke frame. Moreover, for every definite negative PIA formula β_p (where the subscript indicates the single critical occurrence of a proposition variable p), the term $RA(\beta_p)(!u)$ is a formula in the language of ALBA the associated term function of which on perfect algebras $\mathbb{A}$ (i.e. on complex algebras of Kripke frames) is characterized by the following equivalence: $\beta^{\mathbb{A}}(b/!p) \leq a$ iff $(RA(\beta))^{\mathbb{A}}(a) \leq b$ for every $a, b \in \mathbb{A}$ (cf. [1, Definition 2.15]).

$$
\begin{array}{ll}
 & \forall \overline{p}[\top \leq \psi(\overline{\beta}, \overline{\delta})] \\
\text{iff} & \forall \overline{p} \forall \overline{\mathbf{n}} \forall \overline{\mathbf{o}}[(\overline{\beta_p \leq \mathbf{n}} \;\&\; \overline{\delta(\overline{p}) \leq \mathbf{o}}) \Rightarrow \top \leq \psi(\overline{\mathbf{n}}, \overline{\mathbf{o}})] \\
\text{iff} & \forall \overline{p} \forall \overline{\mathbf{n}} \forall \overline{\mathbf{o}}[(\overline{RA(\beta_p)(\mathbf{n}) \leq p} \;\&\; \overline{\delta(\overline{p}) \leq \mathbf{o}}) \Rightarrow \top \leq \psi(\overline{\mathbf{n}}, \overline{\mathbf{o}})] \\
\text{iff} & \forall \overline{\mathbf{n}} \forall \overline{\mathbf{o}}[\overline{\delta(\overline{\bigvee RA(\beta_p)(\mathbf{n})}/\overline{p}) \leq \mathbf{o}} \Rightarrow \top \leq \psi(\overline{\mathbf{n}}, \overline{\mathbf{o}})] \\
\text{iff} & \forall \overline{\mathbf{n}} \forall \overline{\mathbf{o}}[\overline{\bigvee \delta(\overline{RA(\beta_p)(\mathbf{n})}/\overline{p}) \leq \mathbf{o}} \Rightarrow \top \leq \psi(\overline{\mathbf{n}}, \overline{\mathbf{o}})] \\
\text{iff} & \forall \overline{\mathbf{n}} \forall \overline{\mathbf{o}}[\&(\overline{\delta(\overline{RA(\beta_p)(\mathbf{n})}/\overline{p}) \leq \mathbf{o}}) \Rightarrow \top \leq \psi(\overline{\mathbf{n}}, \overline{\mathbf{o}})] \\
\text{iff} & \forall \mathbf{j} \forall \overline{\mathbf{n}} \forall \overline{\mathbf{o}}[\&(\delta(\overline{RA(\beta_p)(\mathbf{n})}/\overline{p}) \leq \mathbf{o}) \Rightarrow \mathbf{j} \leq \psi(\overline{\mathbf{n}}, \overline{\mathbf{o}})] \\
\text{iff} & \forall \mathbf{j} \forall \overline{\mathbf{n}} \forall \overline{\mathbf{o}}[\psi(\overline{\mathbf{n}}, \overline{\mathbf{o}}) \leq \neg \mathbf{j} \Rightarrow \parr(\neg \mathbf{o} \leq \delta(\overline{RA(\beta_p)(\mathbf{n})}/\overline{p}))]
\end{array}
$$

Once the proposition variables have been eliminated, any of the conditions above can be translated in the first-order frame correspondence language of Kripke frames (see [6] for details). A moment's reflection will convince the reader that the ensuing implication is geometric. Our strategy will hinge on representing the run generating r so as to read off the last line in the computation above.

Labelled calculus in the language of ALBA. In the present section, we introduce rules for a labelled calculus in which the relational information is captured via *pure inequalities* (i.e. inequalities in which the only variables occurring in formulas are nominal and conominals) in the language of ALBA. In order to match the level of generality used to describe ALBA runs on generic definite analytic inductive (Salhqvist) formulas, we find it convenient to define this calculus via left- and right-introduction rules for definite Skeleton and PIA formulas.

In what follows, $\psi(!\overline{x}, !\overline{y})$ (resp. $\varphi(!\overline{x}, !\overline{y})$) is a definite negative (resp. positive) Skeleton formula which is monotone in its x-coordinates and antitone in its y-coordinates. To make notation lighter, we will write e.g. $\psi(\overline{\mathbf{n}}, \overline{\mathbf{h}})$ for $\psi(\overline{\mathbf{n}}/!\overline{x}, \overline{\mathbf{h}}/!\overline{y})$.

$$\frac{\Gamma, \overline{\mathbf{h} \leq A}, \overline{B \leq \mathbf{n}} \vdash \mathbf{j} \leq \psi(\overline{\mathbf{n}}, \overline{\mathbf{h}}), \Delta}{\Gamma \vdash \mathbf{j} \leq \psi(\overline{B}, \overline{A}), \Delta}\ \psi_R \qquad \frac{\Gamma, \overline{\mathbf{h} \leq A}, \overline{B \leq \mathbf{n}} \vdash \varphi(\overline{\mathbf{h}}, \overline{\mathbf{n}}) \leq \mathbf{m}, \Delta}{\Gamma \vdash \varphi(\overline{A}, \overline{B}) \leq \mathbf{m}, \Delta}\ \varphi_L$$

$$\frac{\left(\Gamma, \mathbf{j}{\leq}\psi(\overline{B}, \overline{A}) \vdash \mathbf{h}_j{\leq}A_j, \mathbf{j}{\leq}\psi(\overline{\mathbf{n}}, \overline{\mathbf{h}}), \Delta\right)_j \qquad \left(\Gamma, \mathbf{j}{\leq}\psi(\overline{B}, \overline{A}) \vdash B_i{\leq}\mathbf{n}_i, \mathbf{j}{\leq}\psi(\overline{\mathbf{n}}, \overline{\mathbf{h}}), \Delta\right)_i}{\Gamma, \mathbf{j} \leq \psi(\overline{B}, \overline{A}) \vdash \mathbf{j} \leq \psi(\overline{\mathbf{n}}, \overline{\mathbf{h}}), \Delta}\ \psi_L$$

$$\frac{\left(\Gamma, \varphi(\overline{A}, \overline{B}){\leq}\mathbf{m} \vdash \mathbf{h}_i{\leq}A_i, \varphi(\overline{\mathbf{h}}, \overline{\mathbf{n}}){\leq}\mathbf{m}, \Delta\right)_i \qquad \left(\Gamma, \varphi(\overline{A}, \overline{B}){\leq}\mathbf{m} \vdash B_j{\leq}\mathbf{n}_j, \varphi(\overline{\mathbf{h}}, \overline{\mathbf{n}}){\leq}\mathbf{m}, \Delta\right)_j}{\Gamma, \varphi(\overline{A}, \overline{B}){\leq}\mathbf{m} \vdash \varphi(\overline{\mathbf{h}}, \overline{\mathbf{n}}){\leq}\mathbf{m}, \Delta}\ \varphi_R$$

where the index i (resp. j) ranges over the length of the vector $\overline{x}$ (resp. $\overline{y}$), and no variable in $\overline{\mathbf{n}}$ or in $\overline{\mathbf{h}}$ occurs in the conclusion of ψ_R or φ_l. The soundness of ψ_R hinges on the fact that, on perfect (complex) modal algebras $\mathbb{A}$, the term-function $\psi^{\mathbb{A}}(!\overline{x}, !\overline{y})$ associated with $\psi(!\overline{x}, !\overline{y})$ is completely meet-preserving (resp. join-reversing), hence monotone (resp. antitone), in each x-coordinate (resp. y-coordinate), and moreover, perfect algebras are completely join-generated (resp. meet-generated) by their completely join-irreducible (resp. meet-irreducible) elements. Thus,

$$\psi^{\mathbb{A}}(\overline{B^{\mathbb{A}}}, \overline{A^{\mathbb{A}}}) = \bigwedge\{\psi^{\mathbb{A}}(\overline{n}, \overline{h}) \mid h \in J^{\infty}(\mathbb{A}), n \in M^{\infty}(\mathbb{A}), \overline{h \leq A^{\mathbb{A}}}, \overline{B^{\mathbb{A}} \leq n}\},$$

and hence, for any $j \in J^{\infty}(\mathbb{A})$, the inequality $j \leq \psi^{\mathbb{A}}(\overline{B^{\mathbb{A}}}, \overline{A^{\mathbb{A}}})$ holds iff $j \leq \psi^{\mathbb{A}}(\overline{n}, \overline{h})$ for all $h \in J^{\infty}(\mathbb{A})$ and $n \in M^{\infty}(\mathbb{A})$ such that $h \leq A^{\mathbb{A}}$ and $B^{\mathbb{A}} \leq n$. Similarly, the soundness of φ_L hinges on the fact that the term-function $\varphi^{\mathbb{A}}(!\overline{x}, !\overline{y})$ associated with $\varphi(!\overline{x}, !\overline{y})$ is completely join-preserving (resp. meet-reversing), hence monotone (resp. antitone), in each x-coordinate (resp. y-coordinate). The soundness of ψ_L (resp. φ_R) follows from the coordinate-wise monotonicity/antitonicity of the term-functions $\psi^{\mathbb{A}}$ and $\varphi^{\mathbb{A}}$.

The rule ψ_R can be regarded as a right-introduction rule, which, in particular, can be instantiated to the counterparts of the rules $\Box_R$, $\vee_R$, $\rightarrow_R$ of G3K when $\psi(\overline{B}, \overline{A}) := \Box B$, $\psi(\overline{B}, \overline{A}) := B_1 \vee B_2$, $\psi(\overline{B}, \overline{A}) := A \rightarrow B$, respectively. In this context, the pure inequality $\mathbf{j} \leq \psi^{\mathbb{A}}(\overline{\mathbf{n}}, \overline{\mathbf{h}})$ captures the relational information. For instance, if $\psi(\overline{B}, \overline{A}) := \Box B$, then $\mathbf{j} \leq \psi^{\mathbb{A}}(\overline{\mathbf{n}}, \overline{\mathbf{h}})$ is $\mathbf{j} \leq \Box\mathbf{n}$, which translates on Kripke frames as $\{x\} \subseteq (R^{-1}[\{y\}^{cc}])^c$, i.e. $x \notin R^{-1}[y]$, i.e. $\neg(xRy)$. If $\psi(\overline{B}, \overline{A}) := A \to B$, then $\mathbf{j} \leq \psi^{\mathbb{A}}(\overline{\mathbf{n}}, \overline{\mathbf{h}})$ is $\mathbf{j} \leq \mathbf{h} \to \mathbf{n}$, which is equivalent to $\mathbf{j} \wedge \mathbf{h} \leq \mathbf{n}$, which translates on Kripke frames as $\{x\} \cap \{y\} \subseteq \{z\}^c$, i.e. $x \neq y$ or $x \neq z$. Similarly, φ_L can be regarded as a left-introduction rule and can be instantiated to the counterparts of the rules $\Diamond_L$, $\wedge_L$ of G3K.

Step (i) + cuts. In the present section, we execute the first phase of MASSA as indicated in Section 3, and derive the axiom $\mathbf{j} \leq \psi' \vdash \mathbf{j} \leq \psi'$ for an arbitrary definite negative analytic Sahlqvist formula. For simplicity, we assume that $\psi' := \psi(\overline{\beta}, \overline{\delta})$ is in NNF with $\psi(!\overline{x}, !\overline{y})$ positive in each x in $\overline{x}$ and each y in $\overline{y}$, and for any $q, q_1 \in \mathsf{Prop}$, we let $p := \neg q$, $p_1 := \neg q_1$, etc.

$$\dfrac{\left(\overline{\beta \leq \mathbf{n}}, \overline{\delta \leq \mathbf{o}}, \mathbf{j} \leq \psi(\overline{\beta}, \overline{\delta}) \vdash \mathbf{j} \leq \psi(\overline{\mathbf{n}}, \overline{\mathbf{o}}), \beta_n \leq \mathbf{n}_n\right)_n}{\pi_1}$$

$$\dfrac{\left(\overline{\beta \leq \mathbf{n}}, \overline{\delta \leq \mathbf{o}}, \mathbf{j} \leq \psi(\overline{\beta}, \overline{\delta}) \vdash \mathbf{j} \leq \psi(\overline{\mathbf{n}}, \overline{\mathbf{o}}), \delta_o \leq \mathbf{o}_o\right)_o}{\pi_2}$$

$$\dfrac{\dfrac{\pi_1 \qquad \pi_2}{\overline{\beta \leq \mathbf{n}}, \overline{\delta \leq \mathbf{o}}, \mathbf{j} \leq \psi(\overline{\beta}, \overline{\delta}) \vdash \mathbf{j} \leq \psi(\overline{\mathbf{n}}, \overline{\mathbf{o}})}\,\psi_L}{\mathbf{j} \leq \psi(\overline{\beta}, \overline{\delta}) \vdash \mathbf{j} \leq \psi(\overline{\beta}, \overline{\delta})}\,\psi_R$$

where n (resp. o) ranges over the length of $\overline{x}$ (resp. $\overline{y}$). Before proceeding on each branch, we prune the proof-tree as follows:

$$\dfrac{\dfrac{\left(\beta_n \leq \mathbf{n}_n \vdash \mathbf{j} \leq \psi(\overline{\mathbf{n}}, \overline{\mathbf{o}}), \beta_n \leq \mathbf{n}_n\right)_n \qquad \left(\delta_o \leq \mathbf{o}_o \vdash \mathbf{j} \leq \psi(\overline{\mathbf{n}}, \overline{\mathbf{o}}), \delta_o \leq \mathbf{o}_o\right)_o}{\overline{\beta \leq \mathbf{n}}, \overline{\delta \leq \mathbf{o}}, \mathbf{j} \leq \psi(\overline{\beta}, \overline{\delta}) \vdash \mathbf{j} \leq \psi(\overline{\mathbf{n}}, \overline{\mathbf{o}})}\,\psi_L}{\mathbf{j} \leq \psi(\overline{\beta}, \overline{\delta}) \vdash \mathbf{j} \leq \psi(\overline{\beta}, \overline{\delta})}\,\psi_R$$

Now we proceed on each branch separately. The assumption that the input formula ψ' is Sahlqvist entails that there are no non-critical occurrences of proposition variables in each β in $\overline{\beta}$ (which, as discussed in Section 2.2, can be assumed w.l.o.g. to be a definite negative PIA formula containing exactly one critical occurrence of a proposition variable). Likewise, each δ in $\overline{\delta}$ can be assumed w.l.o.g. to be a definite negative PIA formula which only contains non-critical occurrences. Thus, the branches of the proof-tree continue as follows for each β_n and δ_o up to the leaves:

$$\dfrac{\beta_{nR}\,\dfrac{p \leq \mathbf{l} \vdash p \leq \mathbf{l}, \mathbf{j} \leq \psi(\overline{\mathbf{n}}, \overline{\mathbf{o}}), \beta_n(\varnothing, \mathbf{l}) \leq \mathbf{n}_n}{\beta_n(\varnothing, p) \leq \mathbf{n}_n, p \leq \mathbf{l} \vdash \beta_n(\varnothing, \mathbf{l}) \leq \mathbf{n}_n, \mathbf{j} \leq \psi(\overline{\mathbf{n}}, \overline{\mathbf{o}})}}{\beta_n(\varnothing, p) \leq \mathbf{n}_n \vdash \beta_n(\varnothing, p) \leq \mathbf{n}_n, \mathbf{j} \leq \psi(\overline{\mathbf{n}}, \overline{\mathbf{o}})}\,\beta_{nL}$$

$$\dfrac{\delta_{oR}\,\dfrac{\left(\mathbf{k}_k \leq q_k \vdash \mathbf{k}_k \leq q_k, \mathbf{j} \leq \psi(\overline{\mathbf{n}}, \overline{\mathbf{o}}), \delta_o(\overline{\mathbf{k}}, \varnothing) \leq \mathbf{o}_o\right)_k}{\delta_o(\overline{q}, \varnothing) \leq \mathbf{o}_o, \overline{\mathbf{k} \leq q} \vdash \delta_o(\overline{\mathbf{k}}, \varnothing) \leq \mathbf{o}_o, \mathbf{j} \leq \psi(\overline{\mathbf{n}}, \overline{\mathbf{o}})}}{\delta_o(\overline{q}, \varnothing) \leq \mathbf{o}_o \vdash \delta_o(\overline{q}, \varnothing) \leq \mathbf{o}_o, \mathbf{j} \leq \psi(\overline{\mathbf{n}}, \overline{\mathbf{o}})}\,\delta_{oL}$$

Next, we perform all the possible cuts between the leaves of the proof-tree described above. The cut formulas involved in each of these cuts will necessarily be one critical and one non-critical occurrence of the same proposition variable. Thus, these cuts can be executed as follows:

$$\dfrac{\dfrac{p \leq \mathbf{l} \vdash p \leq \mathbf{l}, \mathbf{j} \leq \psi(\overline{\mathbf{n}}, \overline{\mathbf{o}}), \beta_n(\varnothing, \mathbf{l}) \leq \mathbf{n}_n}{\vdash \neg \mathbf{l} \leq p, p \leq \mathbf{l}, \mathbf{j} \leq \psi(\overline{\mathbf{n}}, \overline{\mathbf{o}}), \beta_n(\varnothing, \mathbf{l}) \leq \mathbf{n}_n}}{\pi_1}$$

$$\dfrac{\dfrac{\mathbf{k}_k \leq q_k \vdash \mathbf{k}_k \leq q_k, \mathbf{j} \leq \psi(\overline{\mathbf{n}}, \overline{\mathbf{o}}), \delta_o(\overline{\mathbf{k}}, \varnothing) \leq \mathbf{o}_o}{\vdash q_k \leq \neg \mathbf{k}_k, \mathbf{k}_k \leq q_k, \mathbf{j} \leq \psi(\overline{\mathbf{n}}, \overline{\mathbf{o}}), \delta_o(\overline{\mathbf{k}}, \varnothing) \leq \mathbf{o}_o}}{\pi_2}$$

$$\dfrac{\dfrac{\dfrac{\pi_1 \qquad \pi_2}{\vdash \neg \mathbf{l} \leq \neg \mathbf{k}_k, \mathbf{k}_k \leq q, p \leq \mathbf{l}, \beta_n(\varnothing, \mathbf{l}) \leq \mathbf{n}_n, \mathbf{j} \leq \psi(\overline{\mathbf{n}}, \overline{\mathbf{o}}), \delta_o(\overline{\mathbf{k}}, \varnothing) \leq \mathbf{o}_o}}{\neg(\neg \mathbf{l} \leq \neg \mathbf{k}_k) \vdash \mathbf{k}_k \leq q, p \leq \mathbf{l}, \beta_n(\varnothing, \mathbf{l}) \leq \mathbf{n}_n, \mathbf{j} \leq \psi(\overline{\mathbf{n}}, \overline{\mathbf{o}}), \delta_o(\overline{\mathbf{k}}, \varnothing) \leq \mathbf{o}_o}}{\mathbf{k}_k \leq \neg \mathbf{l} \vdash \mathbf{k}_k \leq q, p \leq \mathbf{l}, \beta_n(\varnothing, \mathbf{l}) \leq \mathbf{n}_n, \mathbf{j} \leq \psi(\overline{\mathbf{n}}, \overline{\mathbf{o}}), \delta_o(\overline{\mathbf{k}}, \varnothing) \leq \mathbf{o}_o}$$

for any n, o and k, where, for each proposition variable, the index k ranges over the multiplicity of that variable in δ_o, and q_k and p are a negative and a positive occurrence of that variable, while each $\mathbf{k}_k$ is a different nominal variable; however, $\mathbf{k}_k \leq \neg \mathbf{l}$ translates into $\{x\} \subseteq \{y\}^{cc}$, i.e. $x = y$, that is, the cuts above give rise to new axioms.

Steps (ii) - (iv) Next, we start generating the rule corresponding to $\psi(\overline{\beta}, \overline{\delta})$ by applying the right-introduction rule bottom-up to its Skeleton $\psi(!\overline{x}, !\overline{y})$:

$$\dfrac{\overline{\beta \leq \mathbf{n}}, \overline{\delta \leq \mathbf{o}} \vdash \mathbf{j} \leq \psi(\overline{\mathbf{n}}, \overline{\mathbf{o}})}{\vdash \mathbf{j} \leq \psi(\overline{\beta}, \overline{\delta})} \; \psi_R$$

Writing it contrapositively, the relational information generated by the bottom-up application of ψ_R is exactly the antecedent of the last line in the ALBA run executed in Section A, which we report here for the reader's convenience.

$$\forall \mathbf{j} \forall \overline{\mathbf{n}} \forall \overline{\mathbf{o}} [\psi(\overline{\mathbf{n}}, \overline{\mathbf{o}}) \leq \neg \mathbf{j} \Rightarrow \invamp_{n,o} \overline{(\neg \mathbf{o} \leq \delta_o(\overline{RA(\beta_n)(\mathbf{n})}/\overline{p})}].$$

We claim that every inequality $\neg \mathbf{o} \leq \delta_o(\overline{RA(\beta_n)(\mathbf{n})}/\overline{p})$ in the disjunction of the succedent of the implication above provides the relational information of a premise in the rule generated by the algorithm. Indeed, the starting point for proving this claim is the observation that each such disjunct corresponds to a certain subset of the axioms generated by the cuts executed at the end of the first phase. Accordingly, in what follows, we proceed on one such disjunct $\neg \mathbf{o} \leq \delta_o(\overline{RA(\beta_n)(\mathbf{n})})$, by identifying the corresponding axioms, and using them as the leaves of a derivation of the corresponding premise, which we will generate by successive applications of right-introduction rules on the βs and δs. Below, the index k ranges over the length of $\overline{x}$ in $\delta_o(!\overline{x})$.

$$\delta_{oR} \; \dfrac{\dfrac{\dfrac{\left(\mathbf{k}_k \leq \neg \mathbf{l} \vdash \mathbf{k}_k \leq q_k, p_k \leq \mathbf{l}_k, \beta_k(\mathbf{l}_k) \leq \mathbf{n}_k, \delta_o(\overline{\mathbf{k}}) \leq \mathbf{o}, \mathbf{j} \leq \psi(\overline{\mathbf{n}}, \overline{\mathbf{o}})\right)_k}{\delta_o(\overline{q}) \leq \mathbf{o}, \overline{\mathbf{k} \leq \neg \mathbf{l}} \vdash \overline{q \leq \mathbf{l}}, \overline{\beta(\mathbf{l}) \leq \mathbf{n}}, \delta_o(\overline{\mathbf{k}}) \leq \mathbf{o}, \mathbf{j} \leq \psi(\overline{\mathbf{n}}, \overline{\mathbf{o}})}}{\vdots} \; \beta_R}{\overline{\beta(p) \leq \mathbf{n}}, \delta_o(\overline{q}) \leq \mathbf{o}, \overline{\mathbf{k} \leq \neg \mathbf{l}} \vdash \overline{\beta(\mathbf{l}) \leq \mathbf{n}}, \delta_o(\overline{\mathbf{k}}) \leq \mathbf{o}, \mathbf{j} \leq \psi(\overline{\mathbf{n}}, \overline{\mathbf{o}})} \; \beta_R$$

To see that we have reached a merging point (cf. Section 3), observe that, by construction, the set of the non-pure inequalities of $S :=$

$\overline{\beta \leq \mathbf{n}}, \overline{\delta \leq \mathbf{o}} \vdash \mathbf{j} \leq \psi(\overline{\mathbf{n}}, \overline{\mathbf{o}})$ is the union of the non-pure inequalities of the $S_{(o,n)} := \overline{\beta(p) \leq \mathbf{n}}, \delta_o(\overline{q}) \leq \mathbf{o}, \overline{\mathbf{k} \leq \neg \mathbf{l}} \vdash \overline{\beta(\mathbf{l}) \leq \mathbf{n}}, \delta_o(\overline{\mathbf{k}}) \leq \mathbf{o}, \mathbf{j} \leq \psi(\overline{\mathbf{n}}, \overline{\mathbf{o}})$ associated with every disjunct, and the derivations $\pi^{(o,n)}_{\psi'}$ also corresponding to all these disjuncts. Hence, the rule r generated by the algorithm corresponding to $\psi(\overline{\beta}, \overline{\delta})$ is

$$\frac{\left(\Gamma, \overline{\mathbf{k} \leq \neg \mathbf{l}} \vdash \overline{\beta_n(\mathbf{l}) \leq \mathbf{n}}, \delta_o(\overline{\mathbf{k}}) \leq \mathbf{o}, \mathbf{j} \leq \psi(\overline{\mathbf{n}}, \overline{\mathbf{o}}), \Delta\right)_{o,n}}{\Gamma \vdash \mathbf{j} \leq \psi(\overline{\mathbf{n}}, \overline{\mathbf{o}}), \Delta}$$

where o as before ranges over the number of δs, while the index n ranges over all possible combinations of βs whose critical proposition variables occurs in δ_o. The last line in the derivation above is the premise of the rule r corresponding to the disjunct $\neg \mathbf{o} \leq \delta_o(\overline{RA(\beta_n)})$. To complete the proof, let us show that the relational information in this premise is equivalent to $\neg \mathbf{o} \leq \delta_o(\overline{RA(\beta_n)})$:

Lemma A.1 *The following are equivalent for every perfect distributive modal algebra $\mathbb{A}$, and all formulas $\delta(!\overline{x})$ and $\overline{\beta(!y)}$ such that δ is monotone in each x-coordinate and each β in $\overline{\beta}$ is antitone in y:*

(i) $\mathbb{A} \models \forall \overline{\mathbf{k}} \forall \overline{\mathbf{l}} \forall \overline{\mathbf{n}} \forall \mathbf{o} \big(\overline{\mathbf{k} \leq \neg \mathbf{l}} \vdash \overline{\beta(\mathbf{l}) \leq \mathbf{n}}, \delta(\overline{\mathbf{k}}) \leq \mathbf{o}\big)$;

(ii) $\mathbb{A} \models \forall \overline{\mathbf{n}} \forall \mathbf{o} \big(\neg \mathbf{o} \leq \delta(\overline{RA(\beta)(\mathbf{n})}) \vdash \ \big)$.

Proof. From (ii) to (i), it is enough to show that if $\overline{k} \in J^\infty(\mathbb{A})^k$ and $\overline{l} \in M^\infty(\mathbb{A})^l$ and $\overline{n} \in M^\infty(\mathbb{A})^n$ and $o \in M^\infty(\mathbb{A})$, such that $\overline{k \leq \neg l}$ and $\overline{\neg n \leq \beta^{\mathbb{A}}(l)}$ and $\neg o \leq \delta^{\mathbb{A}}(\overline{k})$, then $\neg o \leq \delta^{\mathbb{A}}(\overline{RA(\beta^{\mathbb{A}})(n)})$.

The assumption $\neg o \leq \delta^{\mathbb{A}}(\overline{k})$ and the monotonicity of $\delta^{\mathbb{A}}$ imply that it is enough to show $k \leq RA(\beta^{\mathbb{A}})(n)$ for each coordinate of $\delta^{\mathbb{A}}$. The assumption $\neg n \leq \beta^{\mathbb{A}}(l)$ is equivalent to $\beta^{\mathbb{A}}(l) \nleq n$ which by adjunction is equivalent to $(RA(\beta))^{\mathbb{A}}(n) \nleq l$, i.e. $\neg l \leq (RA(\beta))^{\mathbb{A}}(n)$. The required inequality then follows by transitivity, combining the latter inequality with the assumption $k \leq \neg l$.

Conversely, let $\overline{n} \in M^\infty(\mathbb{A})^n$ and $o \in M^\infty(\mathbb{A})$ such that $\neg o \leq \delta^{\mathbb{A}}(\overline{(RA(\beta))^{\mathbb{A}}(n)})$, and let us find $\overline{k} \in J^\infty(\mathbb{A})^k$ and $\overline{l} \in M^\infty(\mathbb{A})^l$ such that $\overline{k \leq \neg l}$ and $\overline{\neg n \leq \beta^{\mathbb{A}}(l)}$ and $\neg o \leq \delta^{\mathbb{A}}(\overline{k})$.

Since $\delta(!\overline{x})$ is a definite negative PIA formula which is positive in each coordinate, $\delta^{\mathbb{A}}(!\overline{x})$ is completely join-preserving in each coordinate; thus, $\neg o \leq \delta^{\mathbb{A}}(\overline{(RA(\beta))^{\mathbb{A}}(n)})$ can be equivalently rewritten as $\neg o \leq \bigvee\{\delta^{\mathbb{A}}(\overline{k}) \mid \overline{k} \in J^\infty(\mathbb{A}), \overline{k \leq (RA(\beta))^{\mathbb{A}}(n)}\}$. Since $\neg o \in J^\infty(\mathbb{A})$ and is hence completely join-prime, the latter inequality is equivalent to $\neg o \leq \delta^{\mathbb{A}}(\overline{k})$ for some $\overline{k} \in J^\infty(\mathbb{A})^l$ such that $\overline{k \leq (RA(\beta))^{\mathbb{A}}(n)}$. Let $l := \neg k \in M^\infty(\mathbb{A})$ for each k in $\overline{k}$. Then $k \leq (RA(\beta))^{\mathbb{A}}(n)$ iff $(RA(\beta))^{\mathbb{A}}(n) \nleq l$ iff $\beta^{\mathbb{A}}(l) \nleq n$ iff $\neg n \leq \beta^{\mathbb{A}}(l)$, as required. □

The proof of correctness when $\psi(\overline{\beta}, \overline{\delta})$ is properly inductive w.r.t. some strict order Ω is similar. Consider for instance a formula $\psi'(p_1, p_2) := \psi(\beta_1(\varnothing, p_1), \beta_2(q_1, p_2), \delta(q_2, \varnothing))$, with $p_1 <_\Omega p_2$. Running ALBA on $\psi'(p_1, p_2)$ yields

$$\forall \mathbf{n}_1 \forall \mathbf{n}_2 \forall \mathbf{o} \forall \mathbf{j} [\psi(\mathbf{n}_1, \mathbf{n}_2, \mathbf{o}) \leq \neg \mathbf{j} \Rightarrow \neg \mathbf{o} \leq \delta(RA(\beta_2)(RA(\beta_1)(\mathbf{n}_1), \mathbf{n}_2))]$$

The first phase proceeds as described in section 3, and produces the following cuts:

$$\begin{array}{c} p_1 \leq \mathbf{l}_1 \vdash p_1 \leq \mathbf{l}_1, \mathbf{j} \leq \psi(\mathbf{n}_1, \mathbf{n}_2, \mathbf{o}), \beta_1(\varnothing, \mathbf{l}_1) \leq \mathbf{n}_1 \\ \hline \vdash \neg \mathbf{l}_1 \leq p_1, p_1 \leq \mathbf{l}_1, \mathbf{j} \leq \psi(\mathbf{n}_1, \mathbf{n}_2, \mathbf{o}), \beta_1(\varnothing, \mathbf{l}_1) \leq \mathbf{n}_1 \\ \hline \rho_1 \end{array}$$

$$\begin{array}{c} \mathbf{i} \leq q_1 \vdash \mathbf{i} \leq q_1, \mathbf{j} \leq \psi(\mathbf{n}_1, \mathbf{n}_2, \mathbf{o}), \beta_2(\mathbf{i}, \mathbf{l}_2) \leq \mathbf{n}_2 \\ \hline \vdash q_1 \leq \neg \mathbf{i}, \mathbf{i} \leq q_1, \mathbf{j} \leq \psi(\mathbf{n}_1, \mathbf{n}_2, \mathbf{o}), \beta_2(\mathbf{i}, \mathbf{l}_2) \leq \mathbf{n}_2 \\ \hline \rho_2 \end{array}$$

$$\begin{array}{c} \rho_1 \qquad \rho_1 \\ \hline \vdash \neg \mathbf{l}_1 \leq \neg \mathbf{i}, \mathbf{i} \leq q_1, p_1 \leq \mathbf{l}_1, \beta_1(\varnothing, \mathbf{l}_1) \leq \mathbf{n}_1, \mathbf{j} \leq \psi(\mathbf{n}_1, \mathbf{n}_2, \mathbf{o}), \beta_2(\mathbf{i}, \mathbf{l}_2) \leq \mathbf{n}_2 \\ \hline \mathbf{i} \leq \neg \mathbf{l}_1 \vdash \mathbf{i} \leq q_1, p_1 \leq \mathbf{l}_1, \beta_1(\varnothing, \mathbf{l}_1) \leq \mathbf{n}_1, \mathbf{j} \leq \psi(\mathbf{n}_1, \mathbf{n}_2, \mathbf{o}), \beta_2(\mathbf{i}, \mathbf{l}_2) \leq \mathbf{n}_2 \end{array}$$

$$\begin{array}{c} \mathbf{k} \leq q_2 \vdash \mathbf{k} \leq q_2, \mathbf{j} \leq \psi(\mathbf{n}_1, \mathbf{n}_2, \mathbf{o}), \delta(\mathbf{k}, \varnothing) \leq \mathbf{o} \\ \hline \vdash q_2 \leq \neg \mathbf{k}, \mathbf{k} \leq q_2, \mathbf{j} \leq \psi(\mathbf{n}_1, \mathbf{n}_2, \mathbf{o}), \delta(\mathbf{k}, \varnothing) \leq \mathbf{o} \\ \hline \pi_1 \end{array}$$

$$\begin{array}{c} p_2 \leq \mathbf{l}_2 \vdash p_2 \leq \mathbf{l}_2, \mathbf{j} \leq \psi(\mathbf{n}_1, \mathbf{n}_2, \mathbf{o}), \beta_2(\mathbf{i}, \mathbf{l}_2) \leq \mathbf{n}_2 \\ \hline \vdash \neg \mathbf{l}_2 \leq p_2, p_2 \leq \mathbf{l}_2, \mathbf{j} \leq \psi(\mathbf{n}_1, \mathbf{n}_2, \mathbf{o}), \beta_2(\mathbf{i}, \mathbf{l}_2) \leq \mathbf{n}_2 \\ \hline \pi_2 \end{array}$$

$$\begin{array}{c} \pi_1 \qquad \pi_1 \\ \hline \vdash \neg \mathbf{l}_2 \leq \neg \mathbf{k}, \mathbf{k} \leq q_2, p_2 \leq \mathbf{l}_2, \delta(\mathbf{k}, \varnothing) \leq \mathbf{o}, \mathbf{j} \leq \psi(\mathbf{n}_1, \mathbf{n}_2, \mathbf{o}), \beta_2(\mathbf{i}, \mathbf{l}_2) \leq \mathbf{n}_2 \\ \hline \mathbf{k} \leq \neg \mathbf{l}_2 \vdash \mathbf{k} \leq q_2, p_2 \leq \mathbf{l}_2, \delta(\mathbf{k}, \varnothing) \leq \mathbf{o}, \mathbf{j} \leq \psi(\mathbf{n}_1, \mathbf{n}_2, \mathbf{o}), \beta_2(\mathbf{i}, \mathbf{l}_2) \leq \mathbf{n}_2 \end{array}$$

Using the new axioms, let us complete the second phase as follows:

$$\begin{array}{c} \mathbf{i} \leq \neg \mathbf{l}_1 \vdash \mathbf{i} \leq q_1, p_1 \leq \mathbf{l}_1, \beta_1(\varnothing, \mathbf{l}_1) \leq \mathbf{n}_1, \mathbf{j} \leq \psi(\mathbf{n}_1, \mathbf{n}_2, \mathbf{o}), \beta_2(\mathbf{i}, \mathbf{l}_2) \leq \mathbf{n}_2 \\ \hline \mathbf{i} \leq \neg \mathbf{l}_1, \beta_1(\varnothing, p_1) \leq \mathbf{n}_1 \vdash \mathbf{i} \leq q_1, \beta_1(\varnothing, \mathbf{l}_1) \leq \mathbf{n}_1, \mathbf{j} \leq \psi(\mathbf{n}_1, \mathbf{n}_2, \mathbf{o}), \beta_2(\mathbf{i}, \mathbf{l}_2) \leq \mathbf{n}_2 \\ \hline \pi_1 \end{array}$$

$$\begin{array}{c} \mathbf{k} \leq \neg \mathbf{l}_2 \vdash \mathbf{k} \leq q_2, p_2 \leq \mathbf{l}_2, \delta(\mathbf{k}, \varnothing) \leq \mathbf{o}, \mathbf{j} \leq \psi(\mathbf{n}_1, \mathbf{n}_2, \mathbf{o}), \beta_2(\mathbf{i}, \mathbf{l}_2) \leq \mathbf{n}_2 \\ \hline \mathbf{k} \leq \neg \mathbf{l}_2, \delta(q_2, \varnothing) \leq \mathbf{o} \vdash p_2 \leq \mathbf{l}_2, \delta(\mathbf{k}, \varnothing) \leq \mathbf{o}, \mathbf{j} \leq \psi(\mathbf{n}_1, \mathbf{n}_2, \mathbf{o}), \beta_2(\mathbf{i}, \mathbf{l}_2) \leq \mathbf{n}_2 \\ \hline \pi_2 \end{array}$$

$$\begin{array}{c} \pi_1 \qquad \pi_2 \\ \hline \Gamma^* \vdash \beta_1(\varnothing, \mathbf{l}_1) \leq \mathbf{n}_1, \mathbf{j} \leq \psi(\mathbf{n}_1, \mathbf{n}_2, \mathbf{o}), \beta_2(\mathbf{i}, \mathbf{l}_2) \leq \mathbf{n}_2, \delta(\mathbf{k}, \varnothing) \leq \mathbf{o} \\ \hdashline \beta_1(\varnothing, p_1) \leq \mathbf{n}_1, \delta(q_2, \varnothing) \leq \mathbf{o}, \beta_2(q_1, p_2) \leq \mathbf{n}_2 \vdash \mathbf{j} \leq \psi(\mathbf{n}_1, \mathbf{n}_2, \mathbf{o}) \\ \hline \vdash \mathbf{j} \leq \psi(\beta_1(\varnothing, p_1), \beta_2(q_1, p_2), \delta(q_2, \varnothing)) \end{array}$$

Where $\Gamma^* := \mathbf{i} \leq \neg \mathbf{l}_1, \mathbf{k} \leq \neg \mathbf{l}_2, \beta_1(\varnothing, p_1) \leq \mathbf{n}_1, \delta(q_2, \varnothing) \leq \mathbf{o}, \beta_2(q_1, p_2) \leq \mathbf{n}_2$.

Notice that the dashed line above is a successful merging point where there is only one premise. Hence, the output rule is:

$$\frac{\Gamma, \mathbf{i} \leq \neg \mathbf{l}_1, \mathbf{k} \leq \neg \mathbf{l}_2 \vdash \beta_1(\varnothing, \mathbf{l}_1) \leq \mathbf{n}_1, \mathbf{j} \leq \psi(\mathbf{n}_1, \mathbf{n}_2, \mathbf{o}), \beta_2(\mathbf{i}, \mathbf{l}_2) \leq \mathbf{n}_2, \Delta(\mathbf{k}, \varnothing) \leq \mathbf{o}, \Gamma}{\Gamma \vdash \mathbf{j} \leq \psi(\mathbf{n}_1, \mathbf{n}_2, \mathbf{o}), \Delta}$$

With an argument similar to the one used to prove the equivalence in Lemma A.1, it can be shown that the relational information of the premise of the rule above is equivalent to $\neg \mathbf{o} \leq \delta(RA(\beta_2)(RA(\beta_1)(\mathbf{n}_1), \mathbf{n}_2))$.

Submodel Enumeration of Kripke Structures in Modal Logic

Nicolas Fröhlich Arne Meier

Institut für Theoretische Informatik, Leibniz Universität Hannover
Appelstrasse 9A, 30167 Hannover, Germany
{nicolas.froehlich, meier}@thi.uni-hannover.de

Abstract

Enumeration complexity (Johnson et al. 1988) is about finding algorithms that produce all solutions to a given problem. Moreover, one strives for a stream of solutions that should be as uniform as possible without much waiting time in between two output solutions and avoiding duplicates. In this paper, we study the problem of model checking in modal logic from this point of view. We consider a particular submodel satisfaction relation that keeps the reference to the transition relation of the original model. Then, we distinguish between enumerating subtrees and subgraphs of Kripke structures that satisfy a modal logic formula. We devise enumeration algorithms for both problems that sort them into the class DelayP.

Keywords: Enumeration Complexity, DelayP, Kripke Structures, Modal Logic.

1 Introduction

In software verification, one strives to know if a written program obeys the underlying specification. It is well-known that describing software systems via a Kripke [21] structure K, i.e., a labelled and state-based transition system, is a profound way in practice [13,30]. For the specification, one constructs an apropriate formula φ in some logic, e.g., temporal [29] or modal logic [3]. Afterwards, in the algorithmic task of model checking [7], one verifies whether K satisfies φ or not. Initially, during the development of a system, the model does not fit the specification yet. On the way to reaching the goal of a system that satisfies the specification formula, considering some kind of restrictions of the program can help in understanding the current issues of the program. Restrictions of the software in turn give rise to submodels of the considered structure K. In this context, not only a particular submodel of K is of much help, as it might be too restrictive; for instance, an empty model could be a strong satisfaction candidate for φ, although not desired. However, a (complete) list of satisfying submodels assists the software developer in adapting the system into the right direction. Obtaining such a list in a systematic way then is a crucial algorithmic task. Moreover, having a uniform stream of printed solutions produced by an algorithm is a key property of a good enumeration process.

Beyond the initial example from above, the described task of model enumeration is, even on the propositional level, very central in many areas, e.g., bounded model checking [2], image computation [15], system engineering [34], and predicate abstraction [22] to name only a few. For a more elaborative view on further applications, we confer the reader to the article of Biere, Möhle and Sebastiani [26].

More formally, in enumeration complexity [19,33], one studies not the overall runtime of an algorithm alone (which often is of exponential duration) but also its *delay*, i.e., an upper bound for all time intervals between two consecutively output solutions (as well as the time before the first and after the last solution has been printed). Here, one strives for algorithms that solve problems within bounds of the class DelayP obeying a delay polynomial in the input length. This class is seen to contain efficient enumeration problems. For modal logic, to the best of the authors' knowledge, a systematic study of enumeration problems in this area has not been undertaken, yet. We want to solve this gap in research, now, and initiate a study of enumeration complexity for model checking in modal logic. We will see that, though having a decision complexity of modal logic model checking in P [35,20], this is not a free ticket for immediately obtaining efficient enumeration algorithms. That is why we start with considering restrictions of the problems in the beginning, namely in the scope of graph restrictions on the obtained submodels.

Contributions. We introduce a family of problems $\mathcal{E}$-ML-SubTree$_N$ that asks for all subtrees of bounded depth $N \in \mathbb{N}$ of a given Kripke structure that satisfy a given formula. We show that for each fixed depth $N \in \mathbb{N}$, the problem can be solved with a delay of $O(|W|^{2N-2} \cdot |\mathrm{SF}(\varphi)|)$, where $|W|$ is the number of states of the given Kripke structure and $|\mathrm{SF}(\varphi)|$ the number of subformulas of the given formula, sorting the problem into the class DelayP. We then show how to improve this result via a recursive approach to reach a delay of $O(|W|^N \cdot |\mathrm{SF}(\varphi)|)$. Consecutively, we tackle the more general version of this problem where all satisfying subgraphs shall be printed. Here, we devise, again, a recursive algorithm that is having a delay of $O(|R|^2 \cdot |W|^2 \cdot |\mathrm{SF}(\varphi)|)$, with $|R|$ the number of transitions in the Kripke structure. We make a rather harsh restriction on what we consider satisfiable subtrees and subgraphs by requiring the $\Box$ operator to be satisfied if and only if all transitions from the original model are present in the subtree or subgraph.

Related work. Krebs et al. [20] investigated the complexity of CTL model checking on the level of operator fragments. This contains a classification of the modal logic variant. There exists a line of research of enumeration complexity in the area of so-called team logics [25,16]. These logics are not built on Kripke but team semantics. Accordingly, their results do not directly transfer to our setting. Capelli and Strozecki [5] study enumeration problems obeying *incremental delay* which could be interesting for extended versions of our studied problems. Also the technique of *geometric amortization* by Capelli and Strozecki [6] might be helpful in this context. Furthermore, a more fine grained enumeration complexity analysis is possible via the framework of parameterised enumeration [24,10,11].

Organisation. First, we will introduce the required foundations of enumeration complexity and briefly present the formalities around modal logic. Then, we will start with the task of subtree enumeration and continue with the generalisation to subgraphs. Finally, we will conclude and present some open research questions.

2 Preliminaries

We assume basic familiarity with computational complexity [28,27].

Modal Logic. We follow the notation of Blackburn et al. [3]. Let PROP be an infinite, countable set of propositions. The set of well-formed formulas $\mathcal{ML}$ is then defined via the following EBNF

$$\varphi ::= p \mid \bot \mid \top \mid \neg\varphi \mid \varphi \wedge \varphi \mid \varphi \vee \varphi \mid \Box\varphi \mid \Diamond\varphi,$$

with $p \in$ PROP. Here, $\top$ is the constant true whereas $\bot$ symbolises the constant false. Let $\varphi \in \mathcal{ML}$ be a formula, then its *length* $|\varphi|$ is defined as the number of its symbols. Let us denote with $\mathrm{SF}(\varphi)$ the set of subformulas of a given formula $\varphi \in \mathcal{ML}$, containing φ as well. Observe that for every $\varphi \in \mathcal{ML}$ we have that $|\mathrm{SF}(\varphi)| \leq |\varphi|$.

Now we turn to Kripke semantics. That is, let $\mathcal{F}$ be a pair (W, R), where W is a non-empty set of *worlds* (or *states*) and $R \subseteq W \times W$ is a binary transition relation on W. The tuple $\mathcal{F}$ is also called a *Kripke structure*. We define a *model* $\mathcal{M}$ to be a pair $(\mathcal{F}, \eta)$, where $\mathcal{F}$ is a Kripke structure, and $\eta\colon \mathrm{PROP} \to \mathcal{P}(W)$ is a map which assigns to each proposition p a set $\eta(p)$ of states. The satisfaction relation is then defined as follows.

Definition 2.1 Let $\varphi, \psi \in \mathcal{ML}$ be two modal formulas. Let $\mathcal{M} = (W, R, \eta)$ be a model and $w \in W$ be a world. We inductively define the *satisfaction* of a formula in the model $\mathcal{M}$ in the world w:

$\mathcal{M}, w \models \top$ always
$\mathcal{M}, w \models \bot$ never
$\mathcal{M}, w \models p$ iff $w \in \eta(p)$ with $p \in$ PROP
$\mathcal{M}, w \models \neg\varphi$ iff $\mathcal{M}, w \not\models \varphi$
$\mathcal{M}, w \models \varphi \wedge \psi$ iff $\mathcal{M}, w \models \varphi$ and $\mathcal{M}, w \models \psi$
$\mathcal{M}, w \models \varphi \vee \psi$ iff $\mathcal{M}, w \models \varphi$ or $\mathcal{M}, w \models \psi$
$\mathcal{M}, w \models \Box\varphi$ iff for all $v \in W$ with $(w, v) \in R$, we have that $\mathcal{M}, v \models \varphi$
$\mathcal{M}, w \models \Diamond\varphi$ iff there exists a $v \in W$ with $(w, v) \in R$ such that $\mathcal{M}, v \models \varphi$

Definition 2.2 Given a modal formula $\varphi \in \mathcal{ML}$, we define its *modal depth* $\mathrm{md}(\varphi)$ in the obvious way:

$$\begin{aligned}
\mathrm{md}(\bot) &:= \mathrm{md}(\top) := \mathrm{md}(p) := 0\\
\mathrm{md}(\neg\varphi) &:= \mathrm{md}(\varphi)\\
\mathrm{md}(\varphi \wedge \psi) &:= \mathrm{md}(\varphi \vee \psi) := \max\{\mathrm{md}(\varphi), \mathrm{md}(\psi)\}\\
\mathrm{md}(\Box\varphi) &:= \mathrm{md}(\Diamond\varphi) := 1 + \mathrm{md}(\varphi)
\end{aligned}$$

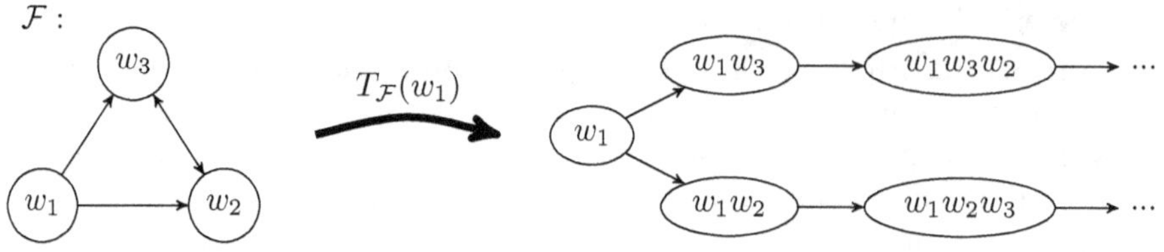

Fig. 1. Example of a computation tree.

Computation Trees. In the following, we define the concept of computation trees that is known from the work of Schnoebelen [31]. Here, a considered Kripke model is unfolded into a computation tree.

First, we need some basic notions. Let $\mathcal{F} = (W, R)$ be a Kripke frame. Furthermore, let a *path (in the Kripke frame $\mathcal{F}$)* $\pi := w_0, w_1, \ldots$ be a sequence of worlds with $(w_i, w_{i+1}) \in R$ for $i = 0, 1, \ldots$. Here, $\pi_i := w_i$ denotes the i-th element on π, and $\Pi_{\mathcal{F}}(w) := \{\, \pi \mid \pi_0 = w \,\}$ is the set of all paths in $\mathcal{F}$ that start in world w. The following definition is comparable to tree unravelings [4, p.15].

Definition 2.3 Let $\mathcal{F} = (W, R)$ be a Kripke frame and $w \in W$ be a world. The *computation tree (from w in $\mathcal{F}$)* $T_{\mathcal{F}}(w) := (V, E)$ is defined as

$$V := \{\, \pi_0\pi_1 \cdots \pi_{|\pi|} \mid \pi \in \Pi_{\mathcal{F}}(w) \,\} \text{ and}$$
$$E := \{\, (\pi, \pi a) \mid \pi, \pi a \in V, a \in W \,\}.$$

By abuse of notation, if $\mathcal{M} = (\mathcal{F}, \eta)$, we interchangeably use also the notion $T_{\mathcal{M}}(w)$ instead of $T_{\mathcal{F}}(w)$.

Figure 1 depicts an example of a Kripke frame that is unfolded into a computation tree for w_1. Note that often computation trees are objects of infinite size, but also can be finitary. Now, we turn towards the subtree notion.

Definition 2.4 Let $\mathcal{F} = (W, R)$ be a Kripke frame, $w \in W$ a world and $T_{\mathcal{F}}(w) = (V, E)$ be the computation tree from w in $\mathcal{F}$. Then, a *subtree* $T^n_{\mathcal{F}}(w') := (V', E')$ is defined as follows

- $V' \subseteq V$ and $E' \subseteq E$,
- $T^n_{\mathcal{F}}(w')$ is a tree with $w' \in T_{\mathcal{F}}(w)$ as root, and
- $|V'| = n \in \mathbb{N}$ is the number of vertices in the subtree which we also call its *size (of the subtree)*.

By abuse of notation, we will also write $T^n_{\mathcal{F}}(w') \subseteq T_{\mathcal{F}}(w)$ for any subtrees of $T_{\mathcal{F}}(w)$. Furthermore, we write $T^n_{\mathcal{F}}(w') \sqsubseteq T_{\mathcal{F}}(w)$, whenever we talk about $T^n_{\mathcal{F}}(w')$ of maximum size.

Notice that in the definition from before, we have $T^m_{\mathcal{F}}(w') \subseteq T^n_{\mathcal{F}}(w)$ for all $m \leq n$, that is, for subtrees with fewer vertices. Also notice that $T^n_{\mathcal{F}}(w') \sqsubseteq T_{\mathcal{F}}(w)$ uniquely determines $T^n_{\mathcal{F}}(w')$.

The *depth* $\mathrm{td}(T^n_{\mathcal{F}})$ of a subtree $T^n_{\mathcal{F}} = (V', E')$ (of a computation tree) then is defined as $\mathrm{td}(T^n_{\mathcal{F}}) := \max_{\pi \in V'}(|\pi|)$, which is the length of its longest path.

The above defined subtrees are a central aspect of this work. They will describe the expected output of the considered enumeration algorithms. We will divide these trees into "satisfiable" and "unsatisfiable" ones. The following satisfaction definition of subtrees will contain a reference to the transition relation of the original structure for the modal operators. We give more insights (see Example 3.7 for this) on this after the presentation of our enumeration algorithm that is used to prove Theorem 3.5.

Definition 2.5 Let $\mathcal{M} = (W, R, \eta)$ be a model, $\varphi \in \mathcal{ML}$ be a formula and $T^n_{\mathcal{M}}(w) = (V', E')$ be a subtree of the computation tree in $w \in W$. Then let $T^n_{\mathcal{M}}(w) \models \varphi$ be inductively defined as

$$
\begin{array}{lll}
T^n_{\mathcal{M}}(w) \models \varphi & \text{iff} & \mathcal{M}, w \models \varphi \text{ for } \varphi \in \{\top, \bot\} \cup \mathrm{PROP}, \\
T^n_{\mathcal{M}}(w) \models \neg\varphi & \text{iff} & T^n_{\mathcal{M}}(w) \not\models \varphi, \\
T^n_{\mathcal{M}}(w) \models \varphi \wedge \psi & \text{iff} & T^n_{\mathcal{M}}(w) \models \varphi \text{ and } T^n_{\mathcal{M}}(w) \models \psi, \\
T^n_{\mathcal{M}}(w) \models \varphi \vee \psi & \text{iff} & T^n_{\mathcal{M}}(w) \models \varphi \text{ or } T^n_{\mathcal{M}}(w) \models \psi, \\
T^n_{\mathcal{M}}(w) \models \Diamond\varphi & \text{iff} & \exists w_0 \text{ with } (w, w_0) \in R : \text{for } T^m_{\mathcal{M}}(w_0) \sqsubseteq T^n_{\mathcal{M}}(w) \\
 & & \text{we have that } (w, ww_0) \in E' \text{ and } T^m_{\mathcal{M}}(w_0) \models \varphi, \\
T^n_{\mathcal{M}}(w) \models \Box\varphi & \text{iff} & \forall w_0 \text{ with } (w, w_0) \in R : \text{for } T^m_{\mathcal{M}}(w_0) \sqsubseteq T^n_{\mathcal{M}}(w) \\
 & & \text{we have that } (w, ww_0) \in E' \text{ and } T^m_{\mathcal{M}}(w_0) \models \varphi.
\end{array}
$$

Notice that for the modal operators $\Box/\Diamond$, we require that for every/one original edge in (W, R) there exists a satisfying subtree. Because of this, it is not possible that $\mathcal{M}, w \not\models \Box\psi$ but a subtree satisfies $\Box\psi$. Clearly, trying to "hide" $\Box$ operators with $\neg\Diamond\neg$ does not work. The reason for that is that for both operators the satisfaction relation for subtrees is defined with respect to the original model. Currently, we need this restriction to limit the number of possibilities in the constructed enumeration algorithms below. We will leave it as a question for future research, whether there exists still a polynomial delay algorithm solving the enumeration problem with an unrestricted satisfaction relation for subtrees.

Enumeration Complexity. In contrast to decision problems, which ask for the existence of solutions to a given instance, for enumeration problems one is concerned with the output of *all* solutions to an instance. Since the number of solutions is usually of exponential size, the running time between the output of two solutions is of particular interest. This elapsed time interval is called the *delay* of an enumeration algorithm and will be defined shortly.

In this context, random access machines are often chosen as the computational model. In this paper, we will use the common one [33,9]. Note that, in this model, one can access particular parts of exponentially large priority queues in polynomial time [19]. The definitions of enumeration problems will follow also recent standard terminology [33,9,24].

Definition 2.6 An *enumeration problem (EP)* is a tuple $\mathcal{E} = (I, \mathrm{Sol})$, where

- I is the set of *instances*, and

- Sol is a function such that for all $x \in I$ the set $\mathrm{Sol}(x)$ is the finite set of *solutions (of x)*.

In this paper, the studied enumeration problems all have the property that we always have a fixed polynomial p such that for every instance $x \in I$ and every solution $y \in \mathrm{Sol}(x)$, we have that $|y| \leq p(|x|)$. Such problems are often classified by a class that is called EnumP [33]. In some sense, this class can be seen as a natural counterpart to NP in the classical setting. Now, we are ready to define enumeration algorithms.

Definition 2.7 Let $\mathcal{E} = (I, \mathrm{Sol})$ be an EP. An algorithm $\mathcal{A}$ is said to be an *enumeration algorithm (EA)* for $\mathcal{E}$, if for every $x \in I$ the algorithm $\mathcal{A}$ obeys the following two properties, where $\mathcal{A}(x)$ denotes the *computation of $\mathcal{A}$ on input x*:

- $\mathcal{A}(x)$ terminates after a finite sequence of steps.
- $\mathcal{A}(x)$ prints exactly $\mathrm{Sol}(x)$ without duplicates.

In the next result, we formally define the notion of a delay of an enumeration algorithm.

Definition 2.8 Let $\mathcal{E} = (I, \mathrm{Sol})$ be an EP, $\mathcal{A}$ be an EA for $\mathcal{E}$, and $x \in I$ be an instance. Then we define

- the *ith delay* of $\mathcal{A}(x)$ as the elapsed time between the output of the ith and $(i+1)$st solution of $\mathrm{Sol}(x)$ by $\mathcal{A}$ on input x,
- the *0th delay* as the *precomputation time*, i.e, the elapsed time before the first output of $\mathcal{A}(x)$, and
- the *mth delay* as the *postcomputation time*, i.e., the elapsed time after the last output of $\mathcal{A}(x)$ until it terminates.

We say that $\mathcal{A}$ has *delay $t(n)$*, for some function $t\colon \mathbb{N} \to \mathbb{N}$, if for all $x \in I$ and all $0 \leq i \leq m$ the ith delay of $\mathcal{A}(x)$ is in $O(t(|x|))$.

After having defined the formalisms around the machine model of enumeration complexity, we will define the complexity class that is relevant in this paper.

Definition 2.9 Let $\mathcal{E} = (I, \mathrm{Sol})$ be an EP and $\mathcal{A}$ be an EA for $\mathcal{E}$. If there exists a polynomial p such that $\mathcal{A}$ has delay $p(n)$, then E belongs to the complexity class DelayP.

3 Enumeration of Subtrees

Before we consider enumeration in modal logic, we first want to take a look at a result by Vardi [35], resp., Clarke and Allen Emerson [8], which shows that model checking for modal formulas can be done in polynomial time.

Proposition 3.1 ([35, Prop. 2.1],[8]) *Let $\mathcal{M} = (W, R, \eta)$ be a model, $w \in W$ a world and φ be a modal formula. Then model checking, i.e., checking whether $(M, w) \models \varphi$ is true, can be verified in time $O(|W|^2 \cdot |\mathrm{SF}(\varphi)|)$.*

Proof. Let $\varphi_1, \ldots, \varphi_m \in \mathrm{SF}(\varphi)$ be the subformulas of φ, for some $m \in \mathbb{N}$, listed in increasing order of length, with ties broken arbitrarily. As a result, we have that $\varphi_m = \varphi$ and if φ_i is a proper subformula of φ_j, then $i < j$. There are at most $|\varphi|$-many subformulas of φ, so we must have that $m \leq |\varphi|$. An induction over k shows that we can label each world $w \in W$ with φ_j or $\neg\varphi_j$, for $j = 1, \ldots, k$, depending on whether or not φ_j is true at w, in time $O(k \cdot |W|)$. The only nontrivial cases are if φ_{k+1} is of the form $\Box\varphi_j$ or $\Diamond\varphi_j$, where $j < k+1$. We label a world w with $\Box\varphi_j$ if and only if each world t such that $(w,t) \in R$ is labelled with φ_j, and with $\Diamond\varphi_j$ if and only if there is a world t such that $(w,t) \in R$ is labelled with φ_j. Assuming inductively that each state has already been labelled with φ_j or $\neg\varphi_j$, this step can clearly be carried out in time $O(|W|^2)$. The total time required is accordingly bound by $O(|W|^2 \cdot |\mathrm{SF}(\varphi)|)$ as desired. □

It is easy to see that this can be used to determine the satisfiability of subtrees $T^n_{\mathcal{M}}(w)$, as they can be seen as acyclic Kripke models themselves. We will now show that the labelling introduced above can easily be updated when removing a leaf from a subtree.

Lemma 3.2 *Let $\mathcal{M} = (W, R, \eta)$ be a model, $w \in W$ be a world, and φ be a modal formula. Given a labelling of satisfied subformulas for all nodes of a subtree $T^n_{\mathcal{M}}(w)$ and subformulas $\mathrm{SF}(\varphi)$, we can correctly update the labelling after removing one leaf in time $O(\mathrm{td}(T^n_{\mathcal{M}}(w)) \cdot |W| \cdot |\mathrm{SF}(\varphi)|)$.*

Proof. Let φ_i or $\neg\varphi_i$ be the labelling for each node in $T^n_{\mathcal{M}}(w)$ and subformulas $\varphi_i \in \mathrm{SF}(\varphi)$. Also let $\ell = \pi_0\pi_1 \ldots \pi_j$ be the leaf removed from $T^n_{\mathcal{M}}(w)$. It should be clear that only nodes $\pi_0\pi_1 \ldots \pi_{j-k}$, with $1 \leq k \leq j$, on the path from root w to leaf ℓ can be effected by the removal. Each of these nodes will have to update $|\mathrm{SF}(\varphi)|$ labels and must check at most $|W|$ nodes for labels $\Box\varphi_i$ and $\Diamond\varphi_i$. The maximum length of a path in $T^n_{\mathcal{M}}(w)$ cannot exceed $\mathrm{td}(T^n_{\mathcal{M}}(w))$. Together the update requires a time of $O(\mathrm{td}(T^n_{\mathcal{M}}(w)) \cdot |W| \cdot |\mathrm{SF}(\varphi)|)$. □

Let us start with a problem that immediately lifts model checking to the enumeration setting.

Problem:	$\mathcal{E}$-ML-SubTree
Input:	$(\mathcal{M}, w, \varphi)$, model $\mathcal{M} = (W, R, \eta)$, world $w \in W$, formula φ
Output:	All subtrees of $T_{\mathcal{M}}(w)$ that satisfy φ

Unfortunately, the next theorem shows that this problem is not an EP as defined in Def. 2.6 as it violates the second property.

Theorem 3.3 *There exists an input $x := (\mathcal{M}, w, \varphi)$ to* $\mathcal{E}$-ML-SubTree *such that $|\mathrm{Sol}(x)| = \infty$.*

Proof. Consider the rather simple input

$$x = ((\{w\}, \{(w,w)\}, \eta), w, \top).$$

The formula $\varphi = \top$ is trivially satisfied in all possible subtrees of $T_{\mathcal{M}}(w)$. As a result, we can construct an infinite number of subtrees $T^n_{\mathcal{M}}(w) \subseteq T_{\mathcal{M}}(w)$, from which we deduce that $|\mathrm{Sol}(x)|$ is infinite. □

Now, we consider a restriction of the previous problem motivated by the last result. Notice that the depth bound $N \in \mathbb{N}$ is part of the problem definition and not part of the instance.

Problem:	$\mathcal{E}$-ML-SubTree$_N$, where $N \in \mathbb{N}$.
Input:	$(\mathcal{M}, w, \varphi)$, model $\mathcal{M} = (W, R, \eta)$, world $w \in W$, formula φ
Output:	All subtrees of $T_{\mathcal{M}}(w)$ with $\mathrm{td}\,(T^n_{\mathcal{M}}(w)) \leq N$ that satisfy φ

We want to mention that on the level of debugging, as explained in the introduction, one usually has that φ is not satisfied by $\mathcal{M}$ in w. This would be a different enumeration problem which asks for falsifying subtrees (to satisfy the formula again which was initially not satisfied). Asking for falsified submodels intuitively makes the problem much harder as we have no real limit on the search space. That is why we currently do not consider this problem and leave this question for future research.

In the following, we show how utilising a priority queue will lead to an enumeration algorithm obeying a polynomial delay. Note that the technique of priority queues goes back to Johnson, Papadimitriou and Yannakakis [19]. As this approach has recurred, the question arose whether the downside of a possibly exponential space can be circumvented [5,6]. The technique of geometric amortization suggested by Capelli and Strozecki [6] can find here an application as well.

Theorem 3.4 *For all fixed $N \in \mathbb{N}$, we have that $\mathcal{E}$-ML-SubTree$_N$ $\in$ DelayP. More precisely, there exists an enumeration algorithm for $\mathcal{E}$-ML-SubTree$_N$ having a delay of $O(|W|^{2N-2} \cdot |\mathrm{SF}(\varphi)|)$, where (W, R, η) is the input model and φ is the given formula.*

Proof. Algorithm 1 uses a priority queue to systematically process the largest subtrees first and avoid printing duplicates. Note, that the largest subtree is uniquely determined.

Let us first turn towards the correctness of the algorithm. Clearly, the algorithm only outputs subtrees which satisfy φ, since only such subtrees are added into the priority queue. In case that the largest subtree $T^n_{\mathcal{M}}(w)$ with depth N does not satisfy φ, the algorithm already does not enter the **if**-condition in line 3 and outputs nothing. Since the size of subtrees added to Q decreases monotonically the algorithm will eventually terminate when there are no more leaves to cut. In line 9, we also only insert subtrees into Q which we have not seen already preventing outputting duplicates. Lastly, the algorithm will reach all eligible subtrees, because it only stops to consider subtrees, that do not satisfy φ. Notice that subtrees of already unsatisfiable subtrees cannot become satisfiable by removing additional nodes (confer Def. 2.5). To summarise, the algorithm will terminate, output all satisfiable subtrees, and avoid printing duplicates.

Now, we classify the delay of the algorithm. Initialising the priority queues S and Q requires constant time. The number of nodes in the largest subtree $T^n_{\mathcal{M}}(w)$ with depth N is bound by $\sum_{i=0}^{N-1}|W|^i$. This bound is reached only for fully connected Kripke structures. In that case, each world has $|W|$ outgoing relations. As a result, the computation tree consists of a root with $|W|$ children, each having $|W|$ children and so forth. The total sum of nodes in such a subtree, with depth N is, accordingly,

$$|W|^0 + |W|^1 + \cdots + |W|^{N-1} = \sum_{i=0}^{N-1}|W|^i = \frac{|W|^N - 1}{|W| - 1} \in O(|W|^{N-1}).$$

Since worlds in not fully connected Kripke structures have fewer outgoing transitions, their computation trees and subtrees will consist of even less nodes. In the worst case, the time required to create the largest subtree $T^n_{\mathcal{M}}(w)$ with depth N in line 2 is thereby bounded by $O(|W|^{N-1})$. The model checking task requires polynomial time (see Prop. 3.1) and line 3 can be done in time $O(|W|^{N-1} \cdot |W| \cdot |\mathrm{SF}(\varphi)|)$, with $|\mathrm{SF}(\varphi)|$ the number of labels per node, $|W|^{N-1}$ the total number of nodes, and $|W|$ the maximum number of worlds considered for subformulas $\Box\varphi_i$ or $\Diamond\varphi_i$. The delay between two consecutively output solutions is now determined by the time of each **while**-iteration. Extracting, outputting, and adding a subtree to S can all be done in time $O(|W|^{N-1})$. The number of leaves with a maximum depth N is equal to or less than $|W|^{N-1}$, therefore line 7 loops for $\leq |W|^{N-1}$ times. In each iteration, we have to model check the new subtree and Lemma 3.2 shows that the time needed in $O(N \cdot |W| \cdot |\mathrm{SF}(\varphi)|)$. Note that N is not part of the input and can therefore be omitted in the following. Checking $T^{n-1}_{\mathcal{M}} \notin Q \cup S$ and potentially adding to Q will still require a time of $O(|W|^{N-1})$. Together, we have an upper bound of

$$O(|W| \cdot |\mathrm{SF}(\varphi)| + |W|^{N-1}) \subseteq O(|W|^{N-1} \cdot |\mathrm{SF}(\varphi)|)$$

for one iteration of the **for**-loop.

Overall the delay is in

$$O\big(|W|^{N-1} \cdot |W|^{N-1} \cdot |\mathrm{SF}(\varphi)|\big) = O\big(|W|^{2N-2} \cdot |\mathrm{SF}(\varphi)|\big)\,,$$

which shows the desired DelayP result. □

In the next step, we will explain how to improve the algorithm from before yielding an EA with a faster delay.

Theorem 3.5 *For all $N \in \mathbb{N}$, there exists an enumeration algorithm for $\mathcal{E}$-ML-SubTree$_N$ having a delay of $O(|W|^N \cdot |\mathrm{SF}(\varphi)|)$, where (W, R, η) is the input model and φ is the given formula.*

Proof. In the following, we will explain how Alg. 2 achieves the desired improvement. Let $T^n_{\mathcal{M}}(w)$ be the largest subtree given a model $\mathcal{M}$ and world w. Furthermore, let $T^m_{\mathcal{M}}(w)$ be any subtree of $T^n_{\mathcal{M}}(w)$ that satisfies $T^m_{\mathcal{M}}(w) \models \varphi$ for a given modal formula φ. Now, we define the set M as follows:

$$M := \{\, k \mid k \text{ is a node in } T^n_{\mathcal{M}}(w), \text{ but not in } T^m_{\mathcal{M}}(w) \,\}.$$

Algorithm 1: Enumeration algorithm for $\mathcal{E}$-ML-SubTree$_N$.

Input: Model $\mathcal{M} = (W, R, \eta)$, world $w \in W$ and formula φ

initialise priority queues $S \leftarrow \emptyset$ and $Q \leftarrow \emptyset$
add largest subtree $T^n_{\mathcal{M}}(w) \sqsubseteq T_{\mathcal{M}}(w)$ with td $(T^n_{\mathcal{M}}(w)) \leq N$ to Q
if $T^n_{\mathcal{M}}(w) \models \varphi$ **then**
  while $Q \neq \emptyset$ **do**
    extract largest subtree $T^n_{\mathcal{M}}(w)$ from Q
    output $T^n_{\mathcal{M}}(w)$ and add it to S
    for all $T^{n-1}_{\mathcal{M}} \subseteq T^n_{\mathcal{M}}(w)$ **do**
      if $T^{n-1}_{\mathcal{M}}(w) \models \varphi$ **and** $T^{n-1}_{\mathcal{M}}(w) \notin Q \cup S$ **then**
        insert $T^{n-1}_{\mathcal{M}}(w)$ into Q
terminate

Algorithm 2: Direct enumeration for $\mathcal{E}$-ML-SubTree$_N$.

Input: Model $\mathcal{M} = (W, R, \eta)$, world $w \in W$ and formula φ

1 determine the largest subtree $T^n_{\mathcal{M}}(w) \sqsubseteq T_{\mathcal{M}}(w)$ with td$(T^n_{\mathcal{M}}(w)) \leq N$
2 label nodes in $T^n_{\mathcal{M}}(w)$ in BFS order
3 **if** $T^n_{\mathcal{M}}(w) \models \varphi$ **then**
4   Output $T^n_{\mathcal{M}}(w)$
5   **for** *all leaves* ℓ *in* $T^n_{\mathcal{M}}(w)$ **do**
6   `EnumSubtreeRec`$(T^n_{\mathcal{M}}(w), \varphi, \ell)$
7 **Function** `EnumSubtreeRec`$(T(w), \varphi, k)$:
8   remove leaf k from $T(w)$
9   **if** $T(w) \models \varphi$ **then**
10   Output $T(w)$
11   **for** *all leaves* ℓ *in* $T(w)$ *with* $\ell <_{\text{BFS}} k$ *in descending order* **do**
12   `EnumSubtreeRec`$(T(w), \varphi, \ell)$

The elements in M are the leaves removed by the recursive calls of ENUMSUBTREEREC. Because leaves are only removed in descending order there is a unique path from $T^n_{\mathcal{M}}(w)$ to $T^m_{\mathcal{M}}(w)$. Also, if we have that $T^m_{\mathcal{M}}(w) \models \varphi$, then the same is true for all subtrees between $T^n_{\mathcal{M}}(w)$ and $T^m_{\mathcal{M}}(w)$, as adding nodes cannot make a subtree falsifying.

Now, we measure the delay of Alg. 2. First, the algorithm constructs the largest subtree $T^n_{\mathcal{M}}(w)$ with depth N, in line 1, which we have seen in the proof of Theorem 3.4 to be bounded by $O(|W|^{N-1})$. Likewise, the labelling of nodes (line 2) and outputting (line 4) can be done in time $O(|W|^{N-1})$, while checking satisfiability (line 3) requires a time of $O(|W|^N \cdot |\text{SF}(\varphi)|)$. Because we immediately output a subtree, if it is satisfiable, the worst delay occurs when iterating over unsatisfiable subtrees. The maximum number of unsatisfiable

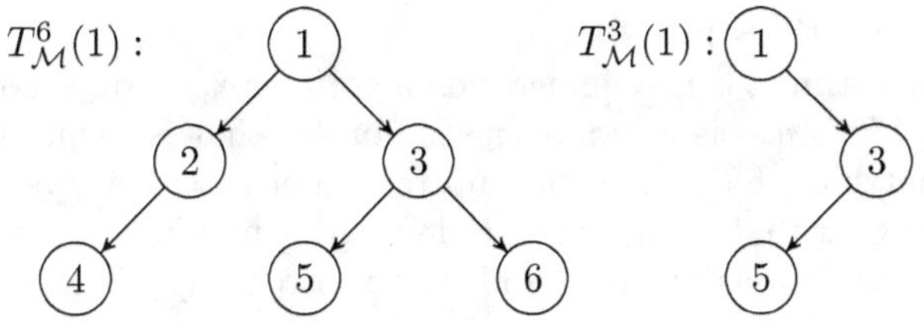

Fig. 2. $M = \{6, 4, 2\}$

subtrees will be in the initial FOR-loop in line 5, with no more than $|W|^{N-1}$ iterations. For each examined subtree we need to check, if it is still satisfiable after removing a leaf in line 8. We have shown in Lemma 3.2 that this can be achieved for each subtree in time $O(N \cdot |W| \cdot |\mathrm{SF}(\varphi)|)$, and again we can omit the N since it is not part of the input. It is also not necessary to consider the depth of recursion here, because if we would find satisfiable subtrees $T^m_{\mathcal{M}}(w)$, all subtrees $T^n_{\mathcal{M}}(w)$ with $T^m_{\mathcal{M}}(w) \subseteq T^n_{\mathcal{M}}(w)$ would be guarantied to be satisfiable and thus reduce the delay. Together the delay of Algorithm 2 is

$$O(|W|^{N-1} \cdot |W| \cdot |\mathrm{SF}(\varphi)|) = O(|W|^N \cdot |\mathrm{SF}(\varphi)|),$$

which shows the desired result. □

The next example illustrates Alg. 2 and the proof of Theorem 3.5.

Example 3.6 In Figure 2, we can see a simplified versions of the subtree $T^6_{\mathcal{M}}(1)$ and of its subtree $T^3_{\mathcal{M}}(1)$. The nodes $6, 4, 2$ are missing from $T^3_{\mathcal{M}}(w)$ and are collected in the set M. The elements from M in descending order show the corresponding calls to ENUMSUBTREEREC(6), ENUMSUBTREEREC(4) and ENUMSUBTREEREC(2) to go from $T^6_{\mathcal{M}}(1)$ to $T^3_{\mathcal{M}}(w)$.

The following example gives more justification for our decision on incorporating the reference to the original transition relation in Definition 2.5. Intuitively, allowing to cut subtrees opens the (problematic) possibility that too many subtrees have to be considered (for instance, in case a clique appears here) that all are not satisfying until the dead-end for the $\Box$ is reached.

Example 3.7 Let $\mathcal{M} = (W, R, \eta)$ be a model with

$$\begin{aligned} W &:= \{w, w_0\} \cup \{w_i \mid 1 \leq i \leq k\}, \\ R &:= \{(w, w_0), (w_0, w_1)\} \cup \{(w_i, w_j) \mid 1 \leq i, j \leq k\}, \\ \eta &:= \{x \mapsto \{w\}\}. \end{aligned}$$

Let $\varphi := \Box x$, and replace in Def. 2.5 the semantics of subtree satisfaction for the modal operators with the following

$$\begin{aligned} T^n_{\mathcal{M}}(w) \models \Diamond\varphi \quad &\text{iff} \quad \exists w_0 \text{ with } (w, ww_0) \in E' : \text{for } T^m_{\mathcal{M}}(w_0) \sqsubseteq T^n_{\mathcal{M}}(w) \\ &\qquad \text{we have that } T^m_{\mathcal{M}}(w_0) \models \varphi, \\ T^n_{\mathcal{M}}(w) \models \Box\varphi \quad &\text{iff} \quad \forall w_0 \text{ with } (w, ww_0) \in E' : \text{for } T^m_{\mathcal{M}}(w_0) \sqsubseteq T^n_{\mathcal{M}}(w) \\ &\qquad \text{we have that } T^m_{\mathcal{M}}(w_0) \models \varphi, \end{aligned}$$

where $T^n_{\mathcal{M}}(w) = (V', E')$.

Then, Algorithm 2 is no enumeration algorithm, as it does not consider the subtree where $T^m_{\mathcal{M}}(w_0)$ is cut off. The reason for that is found in line 3 and 9 as without cutting off $T^m_{\mathcal{M}}(w_0)$, the subtree is not satisfying, yet.

Further notice that analysing the delay of a modification of Algorithm 2 without the restrictions in line 3 and 9 and instead applying the requirement of satisfiability only onto line 4 and 10 directly to prevent printing unsatisfiable subtrees, also does not perform as desired. To see this, first notice that $T^n_{\mathcal{M}}(w) \not\models \varphi$ unless $T^m_{\mathcal{M}}(w_0) \sqsubseteq T^n_{\mathcal{M}}(w)$ is cut from it, because $w_0 \notin \eta(x)$. Also consider that our algorithm starts with the largest possible subtree and successively removes leaves. Since w_0 is the root of $T^m_{\mathcal{M}}(w_0)$ it will naturally be removed last. We therefore have to look at the number of unsatisfiable subtrees the algorithm considers before w_0 becomes a leaf. Observe that $\{\, w_i \mid 1 \leq i \leq k \,\}$ and $\{\, (w_i, w_j) \mid 1 \leq i, j \leq k \,\}$ result in a fully connected subgraphs with k worlds. This will create a full k-ary subtree, which the algorithm as to iterate through. The number of full k-ary trees is recursively given by $a(n+1) = a(n)^k + 1$, with $a(0) = 0$ (this bound follows by a straightforward inductive proof on n and k). It follows that for $a(3) = 2^k + 1$ the number of subtrees is already exponentially in the input size, namely the number of worlds $|W|$ of $\mathcal{M}$ which results in a delay between the largest subtree and the first satisfiable subtree, yielding an exponential delay.

4 Enumeration of subgraphs

Finally, we consider generalising the problem $\mathcal{E}$-ML-SubTree$_N$ to arbitrary subgraphs that do not necessarily have to be trees and also remove the depth bound. By this, we also leave the notion of computation trees that talk about unfolding of a Kripke structure as for the subtree notion. Given a Kripke frame $\mathcal{F} = (W, R)$, a *subgraph* $S_{\mathcal{F}}(R')$ of (W, R) is simply considered with respect to subsets of the edge set $R' \subseteq R$. Furthermore, similar as in Def. 2.5, we define satisfaction of subgraphs. Note that, here again, we need the property regarding modality-prefixed formulas that refer back to the original transition relation R.

Problem:	$\mathcal{E}$-ML-SubGraph
Input:	$(\mathcal{M}, w, \varphi)$, $\mathcal{M} = (W, R, \eta)$ model, φ formula
Output:	All subgraphs of $\mathcal{M}$ that satisfy φ

Theorem 4.1 $\mathcal{E}$-ML-SubGraph $\in$ DelayP.

Proof. Algorithm 3 enumerates all solutions of a model by recursively removing transitions until a resulting subgraphs no longer satisfies the given formula φ. Labelling the transitions uniquely allows us to guarantee no duplicate outputs. As a result, the algorithm is assured to visit each subgraph of $\mathcal{M}$ at most once, analogically to the proof of Theorem 3.5. Subgraphs are not visited, only if a subgraph before already does not satisfy φ, which means that unvisited subgraphs cannot satisfy φ, too. This results in an output of all subgraphs of

Algorithm 3: Subgraph enumeration.

Input: Model $\mathcal{M} = (W, R, \eta)$ and formula φ

label transitions $r \in R$ uniquely
if $S_{\mathcal{M}}(R) \models \varphi$ **then**
  Output $S_{\mathcal{M}}(R)$
  for *all transitions* $r \in R$ **do**
    `EnumSubgraphRec`$(\mathcal{M}, \varphi, \{r\})$

Function `EnumSubgraphRec`$(\mathcal{M}, \varphi, R')$:
  if $S_{\mathcal{M}}(R \setminus R') \models \varphi$ **then**
    Output $\mathcal{M}$
    for *all transitions* $r_0 \in R \setminus R'$ *with* $r_0 < r$ **do**
      `EnumSubgraphRec`$(\mathcal{M}, \varphi, R' \cup \{r_0\})$

$\mathcal{M}$ that satisfy φ without duplicates.

We have already established, that model checking can be done in polynomial time (see Prop. 3.1). Outputting and modifying subgraphs can all be done in linear time, with respect to the size of the initial model. (For the following worst case estimation, also see Example 4.2 below.) The worst case happens if firstly a sequence of ENUMSUBGRAPHREC calls all result in subgraphs, that satisfy φ, which can be at most $|R|$, where $\mathcal{M} = (W, R, \eta)$ is the input instance. Secondly, all remaining calls of ENUMSUBGRAPHREC in all prior recursion steps result in subgraphs not satisfying φ, which is bounded by the maximum recursion depth $|R|$ and the number of possible subgraphs at each step which is $|R| - d$ for d the current depth of the recursion. Together, we have that the number of unsatisfying subgraphs to be checked before we can terminate has to be less than

$$\sum_{i=0}^{|R|} |R| - i = \frac{|R|^2 + |R|}{2} \in O(|R|^2)$$

Each subgraph has to be model checked, which is done in time $O(|W|^2 \cdot |\mathrm{SF}(\varphi)|)$ each, resulting in a total delay of

$$O(|R|^2 \cdot |W|^2 \cdot |\mathrm{SF}(\varphi)|).$$

With this we have shown that the delay of Alg. 3 is requiring polynomial time and, accordingly, $\mathcal{E}$-ML-SubGraph $\in$ DelayP is true. □

Example 4.2 Let $\mathcal{M}$ be a Kripke structure with $R = \{r_0, r_1, r_2, r_3\}$ the set of transitions labelled by their index. Each edge in Fig. 3 represents a call to ENUMSUBGRAPHREC, and nodes are labelled with the transitions of the current subset. The algorithm firstly finds a sequence of satisfiable models going deeper in its recursion until it reaches only unsatisfiable models. It then has to model check all remaining candidates which totals in

$$7 < 10 = \frac{|R|^2 + |R|}{2}.$$

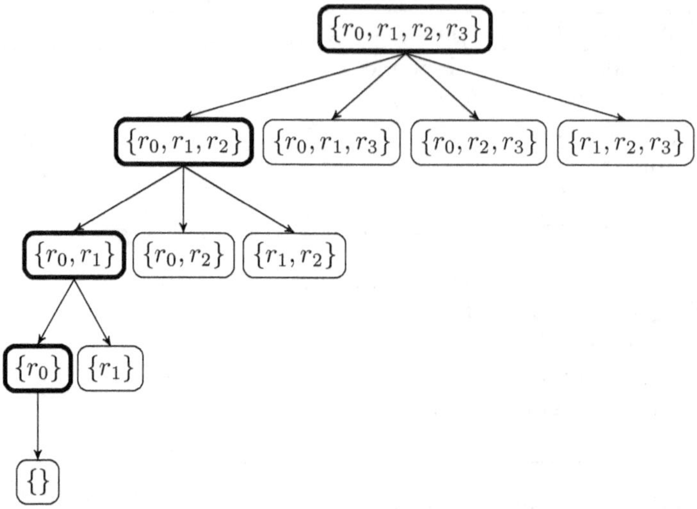

Fig. 3. Recursive calls to EnumSubgraphRec and their resulting transition sets. Nodes marked by a thicker border are satisfiable, while thin borders denote unsatisfiable ones.

This is obviously less than our given upper bound in the proof of Theorem 4.1.

5 Conclusion

In this paper, we classified the enumeration complexity of two problems located in modal logic. First, we studied the family of problems $\mathcal{E}$-ML-SubTree$_N$ that asks, for each $N \in \mathbb{N}$, to print all satisfying subtree models restricted to a depth N of a given Kripke model and showed that this can be efficiently done with a polynomial delay, placing the enumeration problem in the class DelayP. In particular, we showed two algorithmic ways to obtain this result and eventually presented a more efficient version which has a delay of $O(|W|^N \cdot |\mathrm{SF}(\varphi)|)$, where W is the set of worlds of the given Kripke model and $\mathrm{SF}(\varphi)$ are the subformulas of the given modal formula. In this context, it might be worth studying the framework of parameterised complexity [11,10,24] to find more efficient algorithms under possibly the tree-width [12, Cha. 7] parameterisation. Particularly, space-efficient algorithms could be of interest here and placing such problems into classes with very efficient space notions would be very tempting [17]. Another possibility would be to utilise the framework of hard enumeration [9] to show negative answers to this question. Also, a completely different angle on the problems studied here is on the level of parameterised counting [23,14].

Second, we considered a generalisation of this problem, removing the fixed depth bound and asking for satisfying subgraphs (that do not necessarily need to be trees). We showed that an enumeration algorithm exists with a delay of $O(|R|^2 \cdot |W|^2 \cdot |\mathrm{SF}(\varphi)|)$.

An obvious next step would be to investigate ways to weaken our restriction on the satisfiability of the modal operators in subtrees and subgraphs and then still obtain an enumeration algorithm without exponential delay. Also taking a look at other problems, e.g., submodels on the level of subsets of the labelling function η as well as practical implementations and integrations into recent model checkers [18,32,1] might yield interesting insights and results.

Acknowledgements

The authors thank the anonymous reviewers for their valuable comments.

References

[1] Bengtsson, J., K. G. Larsen, F. Larsson, P. Pettersson and W. Yi, *UPPAAL - a tool suite for automatic verification of real-time systems*, in: *Hybrid Systems*, Lecture Notes in Computer Science **1066** (1995), pp. 232–243.

[2] Biere, A., A. Cimatti, E. M. Clarke, O. Strichman and Y. Zhu, *Bounded model checking*, Adv. Comput. **58** (2003), pp. 117–148.

[3] Blackburn, P., M. de Rijke and Y. Venema, "Modal Logic," Cambridge Tracts in Theoretical Computer Science **53**, Cambridge University Press, 2001.

[4] Blackburn, P. and J. van Benthem, *Modal logic: a semantic perspective*, in: P. Blackburn, J. F. A. K. van Benthem and F. Wolter, editors, *Handbook of Modal Logic*, Studies in logic and practical reasoning **3**, North-Holland, 2007 pp. 1–84.
URL `https://doi.org/10.1016/s1570-2464(07)80004-8`

[5] Capelli, F. and Y. Strozecki, *Incremental delay enumeration: Space and time*, Discret. Appl. Math. **268** (2019), pp. 179–190.

[6] Capelli, F. and Y. Strozecki, *Geometric amortization of enumeration algorithms*, CoRR **abs/2108.10208** (2021).

[7] Clarke, E. M., *The birth of model checking*, in: *25 Years of Model Checking*, Lecture Notes in Computer Science **5000** (2008), pp. 1–26.

[8] Clarke, E. M. and E. A. Emerson, *Design and synthesis of synchronization skeletons using branching-time temporal logic*, in: *Logic of Programs*, Lecture Notes in Computer Science **131** (1981), pp. 52–71.

[9] Creignou, N., M. Kröll, R. Pichler, S. Skritek and H. Vollmer, *A complexity theory for hard enumeration problems*, Discret. Appl. Math. **268** (2019), pp. 191–209.

[10] Creignou, N., R. Ktari, A. Meier, J. Müller, F. Olive and H. Vollmer, *Parameterised enumeration for modification problems*, Algorithms **12** (2019), p. 189.

[11] Creignou, N., A. Meier, J. Müller, J. Schmidt and H. Vollmer, *Paradigms for parameterized enumeration*, Theory Comput. Syst. **60** (2017), pp. 737–758.

[12] Cygan, M., F. V. Fomin, L. Kowalik, D. Lokshtanov, D. Marx, M. Pilipczuk, M. Pilipczuk and S. Saurabh, "Parameterized Algorithms," Springer, 2015.

[13] Fix, L., *Fifteen years of formal property verification in Intel*, in: *25 Years of Model Checking*, Lecture Notes in Computer Science **5000** (2008), pp. 139–144.

[14] Flum, J. and M. Grohe, *The parameterized complexity of counting problems*, in: *FOCS* (2002), p. 538.

[15] Gupta, A., Z. Yang, P. Ashar and A. Gupta, *Sat-based image computation with application in reachability analysis*, in: *FMCAD*, Lecture Notes in Computer Science **1954** (2000), pp. 354–371.

[16] Haak, A., A. Meier, F. Müller and H. Vollmer, *Enumerating teams in first-order team logics*, CoRR **abs/2006.06953** (2020).

[17] Haak, A., A. Meier, O. Prakash and B. V. R. Rao, *Parameterised counting in logspace*, in: *STACS*, LIPIcs **187** (2021), pp. 40:1–40:17.

[18] Holzmann, G. J., "The SPIN Model Checker - primer and reference manual," Addison-Wesley, 2004.

[19] Johnson, D. S., C. H. Papadimitriou and M. Yannakakis, *On generating all maximal independent sets*, Inf. Process. Lett. **27** (1988), pp. 119–123.
[20] Krebs, A., A. Meier and M. Mundhenk, *The model checking fingerprints of CTL operators*, Acta Informatica **56** (2019), pp. 487–519.
[21] Kripke, S., *Semantical considerations on modal logic*, Acta Philosophica Fennica **16** (1963), pp. 83–94.
[22] Lahiri, S. K., R. Nieuwenhuis and A. Oliveras, *SMT techniques for fast predicate abstraction*, in: *CAV*, Lecture Notes in Computer Science **4144** (2006), pp. 424–437.
[23] McCartin, C., *Parameterized counting problems*, in: *MFCS*, Lecture Notes in Computer Science **2420** (2002), pp. 556–567.
[24] Meier, A., "Parametrised enumeration," 2020, habilitation thesis, Leibniz Universität Hannover, `https://doi.org/10.15488/9427`.
[25] Meier, A. and C. Reinbold, *Enumeration complexity of poor man's propositional dependence logic*, in: *FoIKS*, Lecture Notes in Computer Science **10833** (2018), pp. 303–321.
[26] Möhle, S., R. Sebastiani and A. Biere, *On enumerating short projected models*, CoRR **abs/2110.12924** (2021).
[27] Papadimitriou, C. H., "Computational complexity," Academic Internet Publ., 2007.
[28] Pippenger, N., "Theories of computability," Cambridge University Press, 1997.
[29] Pnueli, A., *The temporal logic of programs*, in: *FOCS* (1977), pp. 46–57.
[30] Pnueli, A. and A. Zaks, *On the merits of temporal testers*, in: *25 Years of Model Checking*, Lecture Notes in Computer Science **5000** (2008), pp. 172–195.
[31] Schnoebelen, P., *The complexity of temporal logic model checking*, in: *Advances in Modal Logic* (2002), pp. 393–436.
[32] Schwarick, M., M. Heiner and C. Rohr, *MARCIE - model checking and reachability analysis done efficiently*, in: *QEST* (2011), pp. 91–100.
[33] Strozecki, Y., *Enumeration complexity*, Bull. EATCS **129** (2019).
[34] Sullivan, A., D. Marinov and S. Khurshid, *Solution enumeration abstraction: A modeling idiom to enhance a lightweight formal method*, in: *ICFEM*, Lecture Notes in Computer Science **11852** (2019), pp. 336–352.
[35] Vardi, M., *Why is modal logic so robustly decidable?*, Descriptive Complexity and Finite Models DIMACS Series in Discrete Mathematics and Theoretical Computer Science (1997), p. 149–183.

Robinson consistency in many-sorted hybrid first-order logics

Daniel Găină [1]

Kyushu University

Guillermo Badia

University of Queensland

Tomasz Kowalski

Jagiellonian University

Abstract

In this paper we prove a Robinson consistency theorem for a class of many-sorted hybrid logics as a consequence of an Omitting Types Theorem. An important corollary of this result is an interpolation theorem.

Keywords: Institution, Hybrid logic, Robinson consistency, Interpolation.

1 Introduction

Robinson's Joint Consistency Theorem [25] gives a sufficient condition in the context of first-order logic for two theories to have a common model. This result was originally proved by A. Robinson with the aim of providing a new purely model-theoretic proof of the Beth definability property. Robinson's theorem was a historical forerunner of Craig's celebrated interpolation theorem, to which it is famously classically equivalent. In the context of classical first-order logic, it is known since Lindström's work [22] that in the presence of compactness, the Robinson Consistency Property is a consequence of the Omitting Types Theorem. Following in Lindström's footsteps, we use an Omitting Types Theorem for many-sorted hybrid-dynamic first-order logics established in [18] to obtain a Robinson Consistency Theorem. However, our results rely on compactness, so they apply to any star-free fragment of this logic.

In [2], Areces et al. solved the interpolation problem positively for hybrid propositional logic, and in [3], they establish a similar result for hybrid predicate logic with constant domains (also called here rigid domains). Our results,

[1] The work presented in this paper has been partially supported by Japan Society for the Promotion of Science, grant number 20K03718.

although similar, do not follow from theirs. For one thing, the framework of [3] is limited to constant domain quantification, whereas we allow variable domains. Moreover, as usual in the area of algebraic specification, we work in a many-sorted setting and, importantly, we consider arbitrary pushouts of signatures (see, e.g., [26]); not only inclusions. This seemingly small change splits one-sorted interpolation and many-sorted interpolation apart. One can have the former but not the latter, as Example 5.13 shows. The same holds for Robinson consistency.

Our approach is based on institution theory, an abstract framework introduced in [12] for reasoning about properties of logical systems from a meta-perspective. Institutional setting achieves generality appropriate for the development of abstract model theory, yet it is geared towards applications, particularly applications to specification and verification of systems. However, for the sake of simplicity, we work in a concrete example of hybrid logic, applying the modularization principles advanced by institution theory, which have at the core the notion of signature morphism and the satisfaction condition (the truth is invariant w.r.t. change of notation). This brings about certain peculiarities, such as regarding variables as special constants, and omnipresence of signature morphisms, which are simply maps between signatures. For more on institution theory, we refer the reader to [8].

The article is structured in the usual way. Section 2 introduces Hybrid First-Order Logic (HFOL) with rigid symbols in an institutional setting, giving the expected definitions of Kripke structures, reducts, and (local and global) satisfaction relation. Sections 3 and 4 give some preliminary results, most importantly the Lifting Lemma 3.7 and Proposition 4.5. The bulk of the work is in Section 5 where, unsurprisingly, we prove Robinson Joint Consistency Theorem for HFOL. Our proof is modelled after Lindström [22], and uses our earlier result from [18] establishing Omitting Types Theorem for HFOL.

2 Hybrid First-Order Logic with rigid symbols (HFOL)

In this section, we present hybrid first-order logic with rigid symbols.

Signatures The signatures are of the form $\Delta = (\Sigma^{\mathtt{n}}, \Sigma^{\mathtt{r}}, \Sigma)$:

- $\Sigma = (S, F, P)$ is a many-sorted first-order signature such that (a) S is a set of sorts, (b) F is a set of function symbols of the form $\sigma : \mathtt{ar} \to s$, where $\mathtt{ar} \in S^*$ is called the *arity* of σ and $s \in S$ is called the *sort* of σ, and (c) P is a set of relation symbols of the form $\pi : \mathtt{ar}$, where $\mathtt{ar} \in S^*$ is called the arity of π. [2]
- $\Sigma^{\mathtt{r}} = (S^{\mathtt{r}}, F^{\mathtt{r}}, P^{\mathtt{r}})$ is a many-sorted first-order signature of *rigid* symbols such that $\Sigma^{\mathtt{r}} \subseteq \Sigma$.
- $\Sigma^{\mathtt{n}} = (S^{\mathtt{n}}, F^{\mathtt{n}}, P^{\mathtt{n}})$ is a single-sorted first-order signature such that $S^{\mathtt{n}} = \{\mathtt{n}\}$, $F^{\mathtt{n}}$ is a set of constants called *nominals*, and $P^{\mathtt{n}}$ is a set of unary or binary relation symbols called *modalities*.

[2] S^* denotes the set of all strings with elements from S.

We usually write $\Delta = (\Sigma^{\mathtt{n}}, \Sigma^{\mathtt{r}} \subseteq \Sigma)$ rather than $\Delta = (\Sigma^{\mathtt{n}}, \Sigma^{\mathtt{r}}, \Sigma)$. Throughout this paper, we let Δ and Δ^i range over signatures of the form $(\Sigma^{\mathtt{n}}, \Sigma^{\mathtt{r}} \subseteq \Sigma)$ and $(\Sigma_i^{\mathtt{n}}, \Sigma_i^{\mathtt{r}} \subseteq \Sigma_i)$, respectively. A *signature morphism* $\chi\colon \Delta \to \Delta^1$ consists of a pair of first-order signature morphisms $\chi^{\mathtt{n}}\colon \Sigma^{\mathtt{n}} \to \Sigma_1^{\mathtt{n}}$ and $\chi\colon \Sigma \to \Sigma_1$ such that $\chi(\Sigma^{\mathtt{r}}) \subseteq \Sigma_1^{\mathtt{r}}$. [3]

Fact 2.1 HFOL *signature morphisms form a category* $\mathtt{Sig}^{\mathsf{HFOL}}$ *under the component-wise composition as first-order signature morphisms.*

Kripke structures For every signature Δ, the class of Kripke structures over Δ consists of pairs (W, M), where

- W is a first-order structure over $\Sigma^{\mathtt{n}}$, called a *frame*, with the universe $|W|$ consisting of a non-empty set of possible worlds, and
- $M\colon |W| \to |\mathtt{Mod}^{\mathsf{FOL}}(\Sigma)|$ is a mapping from the universe of W to the class of first-order Σ-structures such that the rigid symbols are interpreted in the same way across worlds: $M_{w_1}\!\upharpoonright_{\Sigma^{\mathtt{r}}} = M_{w_2}\!\upharpoonright_{\Sigma^{\mathtt{r}}}$ for all $w_1, w_2 \in |W|$, where M_{w_i} denotes the first-order Σ-structure corresponding to w_i, and $M_{w_i}\!\upharpoonright_{\Sigma^{\mathtt{r}}}$ is the reduct of M_{w_i} to the signature $\Sigma^{\mathtt{r}}$.

A *homomorphism* $h\colon (W, M) \to (V, N)$ over a signature Δ is also a pair

$$(W \xrightarrow{h} V, \{M_w \xrightarrow{h_w} N_{h(w)}\}_{w\in|W|})$$

consisting of first-order homomorphisms such that the mappings corresponding to rigid sorts are shared across the worlds, that is, $h_{w_1,s} = h_{w_2,s}$ for all possible worlds $w_1, w_2 \in |W|$ and all rigid sorts $s \in S^{\mathtt{r}}$.

Fact 2.2 *For any signature* Δ*, the* Δ*-homomorphisms form a category* $\mathtt{Mod}^{\mathsf{HFOL}}(\Delta)$ *under the component-wise composition.*

Reducts Every signature morphism $\chi\colon \Delta \to \Delta^1$ induces appropriate *reductions of models*: every Δ^1-model (V, N) is reduced to a Δ-model $(V, N)\!\upharpoonright_\chi$ that interprets every symbol x in Δ as $(V, N)_{\chi(x)}$. When χ is an inclusion, we usually denote $(V, N)\!\upharpoonright_\chi$ by $(V, N)\!\upharpoonright_\Delta$ – in this case, the model reduct simply forgets the interpretation of those symbols in Δ^1 that do not belong to Δ.

Fact 2.3 *For each signature morphism* $\chi\colon \Delta \to \Delta'$ *and each Kripke structure* (W, M) *over* Δ'*, the map* $\mathtt{Mod}^{\mathsf{HFOL}}$ *from* $\mathtt{Sig}^{\mathsf{HFOL}}$ *to* $\mathbb{C}at^{op}$*, defined by* $\mathtt{Mod}^{\mathsf{HFOL}}(\chi)(W, M) = (W, M)\!\upharpoonright_\chi$*, is a functor.*

Hybrid terms For any signature Δ, we make the following notational conventions: (a) $S^{\mathtt{e}} := S^{\mathtt{r}} \cup \{\mathtt{n}\}$ the extended set of rigid sorts, where $\mathtt{n}$ is the sort of nominals, (b) $S^{\mathtt{f}} := S \setminus S^{\mathtt{r}}$ the subset of flexible sorts, (c) $F^{\mathtt{f}} := F \setminus F^{\mathtt{r}}$ the subset of flexible function symbols, (d) $P^{\mathtt{f}} := P \setminus P^{\mathtt{r}}$ the subset of flexible relation symbols. The *rigidification* of Σ with respect to $F^{\mathtt{n}}$ is the signature $@\Sigma = (@S, @F, @P)$, where (a) $@S := \{@_k\, s \mid k \in F^{\mathtt{n}} \text{ and } s \in S\}$,

[3] A first-order signature morphism $\chi : (S, F, P) \to (S_1, F_1, P_1)$, is a triple $(\chi : S \to S_1, \chi : F \to F_1, \chi : P \to P_1)$ which maps each function symbol $\sigma : s_1 \ldots s_n \to s \in F$ to $\chi(\sigma) : \chi(s_1) \ldots \chi(s_n) \to \chi(s) \in F_1$ and each relation symbol $\pi : \mathtt{ar} \in P$ to $\chi(\pi) : \chi(\mathtt{ar}) \in P_1$.

(b) $@F := \{@_k \sigma \colon @_k \mathtt{ar} \to @_k s \mid k \in F^{\mathtt{n}}$ and $(\sigma \colon \mathtt{ar} \to s) \in F\}$, and (c) $@P := \{@_k \pi \colon @_k \mathtt{ar} \mid k \in F^{\mathtt{n}}$ and $(\pi \colon \mathtt{ar}) \in P\}$. [4] Since rigid symbols have the same interpretation across worlds, we let $@_k x = x$ for all nominals $k \in F^{\mathtt{n}}$ and all rigid symbols x in $\Sigma^{\mathtt{r}}$. The set of *rigid* Δ*-terms* is $T_{@\Sigma}$, while the set of *open* Δ*-terms* is T_Σ. The set of *hybrid* Δ*-terms* is $T_{\overline{\Sigma}}$, where $\overline{\Sigma} = (\overline{S}, \overline{F}, \overline{P})$, $\overline{S} = S \cup @S^{\mathtt{f}}$, $\overline{F} = F \cup @F^{\mathtt{f}}$, and $\overline{P} = P \cup @P^{\mathtt{f}}$.

The interpretation of the hybrid terms in Kripke structures is uniquely defined as follows: for any Δ-model (W, M), and any possible world $w \in |W|$,

1) $M_{w,\sigma(t)} = M_{w,\sigma}(M_{w,t})$, where $(\sigma \colon \mathtt{ar} \to s) \in F$, and $t \in T_{\overline{\Sigma},\mathtt{ar}}$, [5]

2) $M_{w,(@_k \sigma)(t)} = M_{w',\sigma}(M_{w,t})$, where $(@_k \sigma \colon @_k \mathtt{ar} \to @_k s) \in @F^{\mathtt{f}}$, $t \in T_{\overline{\Sigma},@_k \mathtt{ar}}$ and $w' = W_k$.

Sentences The simplest sentences defined over a signature Δ, usually referred to as atomic, are given by

$$\rho ::= k \mid \varrho \mid t_1 = t_2 \mid \varpi(t)$$

where (a) $k \in F^{\mathtt{n}}$ is a nominal, (b) $(\varrho : \mathtt{n}) \in P^{\mathtt{n}}$ is a unary modality, (c) $t_i \in T_{\overline{\Sigma},s}$ are hybrid terms, $s \in \overline{S}$ is a hybrid sort, (d) $\varpi : \mathtt{ar} \in \overline{P}$ and $t \in T_{\overline{\Sigma},\mathtt{ar}}$. We call *hybrid equations* sentences of the form $t_1 = t_2$, and *hybrid relations* sentences of the form $\varpi(t)$. The set $\mathsf{Sen}^{\mathsf{HFOL}}(\Delta)$ of *full sentences* over Δ are given by the following grammar:

$$\varphi ::= \rho \mid @_k \varphi \mid \neg\varphi \mid \vee\Phi \mid {\downarrow} z \cdot \varphi' \mid \exists X \cdot \varphi'' \mid \langle\lambda\rangle\varphi$$

where (a) ρ is an atomic sentence, (b) $k \in F^{\mathtt{n}}$ is a nominal, (c) Φ is a finite set of sentences over Δ, (d) z is a nominal variable for Δ and φ' is a sentence over the signature $\Delta(z)$ obtained from Δ by adding z as a new constant to $F^{\mathtt{n}}$, (e) X is a set of variables for Δ of sorts from the extended set of rigid sorts $S^{\mathtt{e}}$ and φ'' is a a sentence over the signature $\Delta(X)$ obtained from Δ by adding the variables in X as new constants to $F^{\mathtt{n}}$ and $F^{\mathtt{r}}$, and (f) $(\lambda : \mathtt{n}\,\mathtt{n}) \in P^{\mathtt{n}}$ is a binary modality. Other than the first kind of sentences (*atoms*), we refer to the sentence-building operators, as *retrieve*, *negation*, *disjunction*, *store*, *existential quantification* and *possibility*, respectively. Other Boolean connectives and the universal quantification can be defined as abbreviations of the above sentence building operators.

Each signature morphism $\chi \colon \Delta \to \Delta^1$ induces *sentence translations*: any Δ-sentence φ is translated to a Δ^1-sentence $\chi(\varphi)$ by replacing, in an inductive manner, the symbols in Δ with symbols from Δ^1 according to χ.

Fact 2.4 $\mathsf{Sen}^{\mathsf{HFOL}}$ *is a functor* $\mathsf{Sig}^{\mathsf{HFOL}} \to \mathbb{S}\mathrm{et}$ *which maps each signature* Δ *to the set of sentences over* Δ.

Local satisfaction relation Given a Δ-model (W, M) and a world $w \in |W|$, we define the *satisfaction of* Δ*-sentences at* w by structural induction as follows: *For atomic sentences*:

[4] $@_k(s_1 \ldots s_n) := @_k s_1 \ldots @_k s_n$ for all arities $s_1 \ldots s_n$.

[5] $M_{w,(t_1,\ldots,t_2)} := M_{w,t_1}, \ldots, M_{w,t_n}$ for all tuples of hybrid terms $(t_1, \ldots, t_n)$.

- $(W, M) \models^w k$ iff $W_k = w$;
- $(W, M) \models^w \varrho$ iff $w \in W_\varrho$;
- $(W, M) \models^w t_1 = t_2$ iff $M_{w,t_1} = M_{w,t_2}$;
- $(W, M) \models^w \varpi(t)$ iff $M_{w,t} \in M_{w,\varpi}$.

For full sentences:

- $(W, M) \models^w @_k \varphi$ iff $(W, M) \models^{w'} \varphi$, where $w' = W_k$;
- $(W, M) \models^w \neg\varphi$ iff $(W, M) \not\models^w \varphi$;
- $(W, M) \models^w \vee\Phi$ iff $(W, M) \models^w \varphi$ for some $\varphi \in \Phi$;
- $(W, M) \models^w \downarrow z \cdot \varphi$ iff $(W^{z \leftarrow w}, M) \models^w \varphi$, where $(W^{z \leftarrow w}, M)$ is the unique $\Delta(z)$-expansion of (W, M) that interprets the variable z as w; [6]
- $(W, M) \models^w \exists X \cdot \varphi$ iff $(W', M') \models^w \varphi$ for some expansion (W', M') of (W, M) to the signature $\Delta(X)$; [6]
- $(W, M) \models^w \langle\lambda\rangle\varphi$ iff $(W, M) \models^{w'} \varphi$ for some $w' \in |W|$ s.t. $(w, w') \in W_\lambda$.

The following *satisfaction condition* can be proved by induction on the structure of Δ-sentences. The proof is essentially identical to those developed for several other variants of hybrid logic presented in the literature (see, e.g. [9]).

Proposition 2.5 (Local satisfaction condition) *Let $\chi\colon \Delta \to \Delta^1$ be a signature morphism. Then $(W, M) \models^w \chi(\varphi)$ iff $(W, M)\restriction_\chi \models^w \varphi$, for all Kripke structures over Δ^1, all sentences φ over Δ.* [7]

Global satisfaction relation The global satisfaction relation is defined by

$$(W, M) \models \varphi \text{ iff for each possible world } w \in |W| \text{ we have } (W, M) \models^w \varphi$$

for all signatures Δ, all Kripke Δ-structures (W, M) and all Δ-sentences φ. The global consequence relation between sentences is defined by

$$\varphi \models \psi \text{ iff } (W, M) \models \varphi \text{ implies } (W, M) \models \psi \text{ for all Kripke structures } (W, M),$$

for all sentences φ and ψ over the same signature. The global consequence relation can be extended to sets of sentences in the usual way.

We adopt the terminology used in the algebraic specification literature. A pair (Δ, Φ) consisting of a signature Δ and a set of sentences Φ over Δ is called a *presentation*. We let $\Phi^\bullet$ denote $\{\varphi \in \mathtt{Sen}(\Delta) \mid \Phi \models \varphi\}$, the closure of Φ under the global consequence relation. A *presentation morphism* $\chi : (\Delta, \Phi) \to (\Delta^1, \Phi^1)$ consists of a signature morphism $\chi : \Delta \to \Delta^1$ such that $\Phi^1 \models \chi(\Phi)$. Any presentation (Δ, T) such that $T = T^\bullet$ is called a *theory* . A *theory morphism* is just a presentation morphism between theories.

Examples Fragments of HFOL have been studied extensively in the literature. We give a few examples.

---

[6] An expansion of (W, M) to $\Delta(X)$ is a Kripke structure (W', M') over $\Delta(X)$ that interprets all symbols in Δ in the same way as (W, M).

[7] By the definition of reducts, (W', M') and $(W', M')\restriction_\chi$ have the same possible worlds, which means that the statement of Proposition 4.4 is well-defined.

Example 2.6 (Rigid First-Order Hybrid Logic (RFOHL) [4]) This logic is obtained from HFOL by restricting the signatures $\Delta = (\Sigma^{\mathtt{n}}, \Sigma^{\mathtt{r}} \subseteq \Sigma)$ such that (a) $\Sigma^{\mathtt{n}}$ has only one binary modality, (b) Σ is single-sorted, (c) the unique sort is rigid, (d) there are no rigid function symbols except variables (regarded here as special constants), and (e) there are no rigid relation symbols.

Example 2.7 (Hybrid First-Order Logic with user-defined Sharing (HFOLS)) This logic has the same signatures and Kripke structures as HFOL. The sentences are obtained from atoms constructed with open terms only, that is, if $\Delta = (\Sigma^{\mathtt{n}}, \Sigma^{\mathtt{r}} \subseteq \Sigma)$, all (ground) equations over Δ are of the form $t_1 = t_2$, where $t_1, t_2 \in T_\Sigma$, and all (ground) relation over Δ are of the form $\varpi(t)$, where $(\varpi : \mathtt{ar}) \in P$ and $t \in T_{\Sigma,\mathtt{ar}}$. A version of HFOLS is the underlying logic of H system [6]. Other variants of HFOLS have been studied in [23,10,9].

Example 2.8 (Hybrid Propositional Logic (HPL)) This is the most common form of multi-modal hybrid logic (e.g. [1]). HPL is obtained from HFOL by restricting the signatures $\Delta = (\Sigma^{\mathtt{n}}, \Sigma^{\mathtt{r}} \subseteq \Sigma)$ such that $\Sigma^{\mathtt{r}}$ is empty and the set of sorts in Σ is empty. Notice that if $\Sigma = (S, F, P)$ and $S = \emptyset$ then P contains only propositional symbols.

Reachability Let (W, M) be a Kripke structure over a signature Δ.

- A possible world $w \in |W|$ is called *reachable* if it is the denotation of some nominal, that is, $w = W_k$ for some nominal $k \in F^{\mathtt{n}}$.
- Let $w \in |W|$ be a possible world and $s \in S$ a sort. An element $e \in M_{w,s}$ is called reachable if it is the denotation of some hybrid term, that is, $w = W_k$ and $e = M_{w,t}$ for some nominal $k \in F^{\mathtt{n}}$ and rigid hybrid term $t \in T_{@\Sigma,@_k s}$.
- (W, M) is reachable by an $S^{\mathtt{e}}$-sorted set C of nominals and rigid hybrid terms if (a) its set of possible worlds consists of denotations of nominals in C, and (b) its carrier sets for the rigid sorts consist of denotations of rigid hybrid terms from C.
- (W, M) is reachable if (W, M) is reachable by nominals and rigid hybrid terms.

The notion of reachability is connected to quantification, which is the reason for considering a Kripke structure reachable if its elements of rigid sorts are denotations of terms, thus disregarding elements of flexible sorts. This notion is semantic and it makes sense also for fragments of HFOL whose sentences do not contain rigid hybrid terms such as RFOHL, HFOLS or HPL. In institution theory, the notion of reachability was originally defined in [24] at an abstract level, and it played an important role in proving several proof-theoretic results [20,19,14] as well as model-theoretic properties [13,15,16,7].

3 Basic definitions and results

In this section, we establish the terminology and we state some foundational results necessary for the present study. We start by noticing that rigid quantification cannot refer to unreachable elements of flexible sorts.

Lemma 3.1 *Let (W, M) be a reachable Kripke structure over a signature Δ. Let (W, N) be a Kripke structure obtained from (W, M) by (a) replacing unreachable elements of flexible sorts by some new elements, (b) preserving the interpretation of function and relation symbols on the elements inherited from (W, M), and (c) interpreting function symbols arbitrarily on the new arguments. Then (W, M) and (W, N) are elementarily equivalent, in symbols, $(W, M) \equiv (W, N)$ (that is, $(W, M) \models \varphi$ iff $(W, N) \models \varphi$, for all $\varphi \in \mathtt{Sen}(\Delta)$).*

The proof of the lemma above is straightforward by induction on the structure of sentences. We recall Robinson consistency property as stated in institution theory (see, for example, [21]).

Definition 3.2 Consider the following square $\mathcal{S}$ of signature morphisms.

$$\begin{array}{ccc} \Delta^2 & \xrightarrow{\upsilon_2} & \Delta' \\ {\scriptstyle \chi_2}\uparrow & & \uparrow{\scriptstyle \upsilon_1} \\ \Delta & \xrightarrow[\chi_1]{} & \Delta^1 \end{array}$$

$\mathcal{S}$ is a *Robinson square*, if for every consistent theories $T^1 \subseteq \mathtt{Sen}(\Delta^1)$, $T^2 \subseteq \mathtt{Sen}(\Delta^2)$ and complete theory $T \subseteq \mathtt{Sen}(\Delta)$ such that χ_1, χ_2 are theory morphisms, it holds that $\upsilon_1(T^1) \cup \upsilon_2(T^2)$ is consistent.

As it was shown in [17], HFOL is compact, which means that interpolation is equivalent to Robinson consistency property.

Proposition 3.3 *The following are equivalent for a commutative square $\mathcal{S}$ of signature morphisms as depicted in the diagram of Definition 3.2:*

1) $\mathcal{S}$ is a Robinson square.

2) For every consistent theories $T^1 \subseteq \mathtt{Sen}(\Delta^1)$ and $T^2 \subseteq \mathtt{Sen}(\Delta^2)$ such that $\chi_1^{-1}(T^1) \cup \chi_2^{-1}(T^2)$ is consistent, the set $\upsilon_1(T^1) \cup \upsilon_2(T^2)$ is consistent.

3) $\mathcal{S}$ is a Craig Interpolation (CI) square , that is, for every $\Phi^1 \subseteq \mathtt{Sen}(\Delta^1)$ and $\Phi^2 \subseteq \mathtt{Sen}(\Delta^2)$ such that $\upsilon_1(\Phi^1) \models \upsilon_2(\Phi^2)$ there exists $\Phi \subseteq \mathtt{Sen}(\Delta)$ such that $\Phi^1 \models \chi_1(\Phi)$ and $\chi_2(\Phi) \models \Phi^2$.

The equivalence of the first two statements can be proved similarly to [21, Proposition 6], while the equivalence of first and last statement can be shown using ideas from [26, Corollary 3.1].

Example 3.4 Let $\Delta^1 \overset{\chi_1}{\leftarrow} \Delta \overset{\chi_2}{\rightarrow} \Delta^2$ be a span of signature morphisms such that

- (a) Δ has three nominals $\{k_1, k_2, k_3\}$, three flexible sorts $\{s_1, s_2, s_3\}$ and three flexible constants $\{c_1 :\to s_1, c_2 :\to s_2, c_3 :\to s_3\}$; (b) Δ^1 has two nominals $\{k, k_3\}$, one flexible sort $\{s\}$ and two flexible constants $\{c :\to s, c_3 :\to s\}$; (c) Δ^2 has two nominals $\{k, k_1\}$, two flexible sorts $\{s, s_2\}$ and three flexible constants $\{c_1 :\to s, c_2 :\to s_2, c_3 :\to s\}$;
- (a) on nominals $\chi_1(k_1) = \chi_1(k_2) = k$, $\chi_1(k_3) = k_3$, on sorts $\chi_1(s_1) = \chi_1(s_2) = \chi_1(s_3) = s$, on function symbols $\chi_1(c_1 :\to s_1) = \chi(c_2 :\to s_2) = c :\to s$, $\chi_1(c_3 :\to s_3) = c_3 :\to s$; (b) on nominals $\chi_2(k_1) = k_1$, $\chi_2(k_2) =$

$\chi_2(k_3) = k$, on sorts $\chi_2(s_1) = \chi_2(s_3) = s$, $\chi_2(s_2) = s_2$, on function symbols $\chi_2(c_1 :\to s_1) = c_1 :\to s$, $\chi_2(c_2 :\to s_2) = c_2 :\to s_2$, $\chi_2(c_3 :\to s_3) = c_3 :\to s$.

Let $\Delta^1 \overset{v_1}{\to} \Delta' \overset{v_2}{\leftarrow} \Delta^2$ be a pushout of the above span such that

- Δ' has one nominal $\{k\}$, one flexible sort $\{s\}$ and two flexible constants $\{c :\to s, c_3 :\to s\}$;
- (a) $v_1(k) = v_1(k_3) = k$, $v_1(c :\to s) = c :\to s$ and $v_1(c_3 :\to s) = c_3 :\to s$; (b) $v_2(k) = v_2(k_1) = k$, $v_2(c_1 :\to s) = c :\to s$, $v_2(c_2 :\to s_2) = c :\to s$ and $v_2(c_3 :\to s) = c_3 :\to s$.

According to the following lemma, interpolation doesn't hold in HFOL, in general.

Lemma 3.5 *The pushout described in Example 3.4 is not a CI square.*

Proof. Let $\Phi^1 := \{@_{k_3}(c = c_3)\}$ and $\Phi^2 := \{@_{k_1}(c_1 = c_3)\}$. Obviously, $v_1(\Phi^1) \models v_2(\Phi^2)$. Suppose towards a contradiction that there exists an interpolant Φ over Δ such that $\Phi^1 \models \chi_1(\Phi)$ and $\chi_2(\Phi) \models \Phi^2$.

Let (W^1, M^1) be the Kripke structure over Δ^1 defined as follows: W^1 consists of one possible world w, and M^1_w is the single-sorted algebra such that $M^1_{w,s} = \{d, e\}$ and $M^1_{w,c} = M^1_{w,c_3} = d$. We have $(W^1, M^1) \models \Phi^1$, and since $\Phi^1 \models \chi_1(\Phi)$, we get $(W^1, M^1) \models \chi_1(\Phi)$. By the satisfaction condition, $(W^1, M^1)\upharpoonright_{\chi_1} \models \Phi$. Let (V, N) be the Kripke structure over Δ obtained from $(W^1, M^1)\upharpoonright_{\chi_1}$ by changing the interpretation of $c_3 :\to s_3$ from d to e, which implies that $V_{w,c_3} = e$. There exists an isomorphism $h : (V, N) \to (W^1, M^1)\upharpoonright_{\chi_1}$ such that h_{w,s_1} and h_{w,s_2} are identities, while $h_{w,s_3}(d) = e$ and $h_{w,s_3}(e) = d$. It follows that $(V, N) \models \Phi$. There exists a χ_2-expansion (V^2, N^2) of (V, N). By the satisfaction condition, $(V^2, N^2) \models \chi_2(\Phi)$. Since $N^2_{w,c_1} = N_{w,c_1} = d$ and $N^2_{w,c_3} = N_{w,c_3} = e$, we have $(V^2, N^2) \not\models \Phi^2$, contradicting $\chi_2(\Phi) \models \Phi^2$. □

We are interested in characterizing a span of signature morphisms whose pushout is a CI square. For this purpose, it is necessary to restrict one of the arrows of the underlying span according to the following definition.

Definition 3.6 A signature morphism $\chi : \Delta \to \Delta^1$ *preserves flexible symbols* if

1) χ preserves flexible sorts, that is, $\chi(s) \in S^{\mathtt{f}}_1$ for all $s \in S^{\mathtt{f}}$, and
2) χ adds no new flexible operations on 'old' flexible sorts, that is, for all flexible sorts $s \in S^{\mathtt{f}}$ and all function symbols $\sigma_1 : \mathtt{ar}_1 \to \chi(s) \in F^{\mathtt{f}}_1$ there exists $\sigma : \mathtt{ar} \to s \in F^{\mathtt{f}}$ such that $\chi(\sigma : \mathtt{ar} \to s) = \sigma_1 : \mathtt{ar}_1 \to \chi(s)$.

If, in addition, χ is injective on flexible sorts and on flexible function and relation symbols that have at least one flexible sort $s \in S^{\mathtt{f}}$ in the arity then we say that χ *protects flexible symbols*.

If $\chi : \Delta \to \Delta^1$ is an inclusion that preserves flexible sorts and adds no new function symbols $\sigma : \mathtt{ar} \to s$ with $s \in S^{\mathtt{f}}$ on Δ then $\chi : \Delta \to \Delta^1$ protects flexible symbols. If Δ has no flexible sorts then $\chi : \Delta \to \Delta^1$ protects flexible symbols. In particular, if Δ is a HPL or RFOHL signature then $S^{\mathtt{f}} = \emptyset$, which means that χ protects flexible symbols. In applications, $\chi : \Delta \to \Delta^1$ from Definition 3.6

is appropriate for hiding information, which makes HFOL an instance of the abstract completeness result for structured specifications proved in [5].

Lemma 3.7 (Lifting Lemma) *Consider the following:*

1) a signature morphism $\chi : \Delta \to \Delta^1$ which is injective on sorts and protects flexible symbols;

2) a set C^1 of new nominals and new rigid constants for Δ^1;

3) $(V^1, N^1) \in |\mathtt{Mod}(\Delta^1(C^1))|$ reachable by C^1 and $(W, M) \in |\mathtt{Mod}(\Delta(C))|$ reachable by C such that $(V^1, N^1) \upharpoonright_{\chi^C} \equiv (W, M)$, where
- *C is the reduct of C^1 across χ, i.e. $C := \{c :\to s \mid c :\to \chi(s) \in C^1\}$, and*
- *$\chi^C : \Delta(C) \to \Delta^1(C^1)$ is the extension of χ that maps each constant $c :\to s \in C$ to $c :\to \chi(s) \in C^1$.*

Then $(W^1, M^1) \equiv (V^1, N^1)$ for some χ^C-expansion (W^1, M^1) of (W, M).

Proof. Let $(V^1, N^1) \in |\mathtt{Mod}(\Delta^1(C^1))|$ and $(W, M) \in |\mathtt{Mod}(\Delta(C))|$ be Kripke structures such that $(V^1, N^1) \upharpoonright_\chi \equiv (W, M)$. To keep notation consistent, we let (V, N) be the reduct $(V^1, N^1) \upharpoonright_\chi$ of (V^1, N^1), and we will construct an expansion (W^1, M^1) of (W, M) such that $(W^1, M^1) \equiv (V^1, N^1)$ in three steps.

1) We construct an isomorphism $h : (W, M) \to (V, R)$, where (V, R) is obtained from (V, N) by replacing all unreachable elements by the unreachable elements from (W, M).

   Firstly, we define h as a function, which implicitly means that we define the universe of (V, R). Let $k \in C_{\mathtt{n}}$ be a nominal, $v := V_k$ and $w := W_k$.
   CASE $s \in S^{\mathtt{r}}$: We define the set $R_{v,s} := N_{v,s}$, and the function $h_{w,s} : M_{w,s} \to R_{v,s}$ by $h_{w,s}(M_{w,c}) = N_{v,c}$ for all constants $c :\to s \in C$. Since both (W, M) and (V, N) are reachable by C and $(W, M) \equiv (V, N)$, the function $h_{w,s} : M_{w,s} \to R_{v,s}$ is bijective.
   CASE $s \in S^{\mathtt{f}}$: Let $R_{v,s}$ be the set obtained from $N_{v,s}$ by removing all unreachable elements and adding all unreachable elements from $M_{w,s}$. We define $h_{w,s} : M_{w,s} \to R_{v,s}$ by $h_{w,s}(M_{w,t}) = N_{v,t}$ for all rigid hybrid $\Delta(C)$-terms t of sort $@_k s$, and $h_{w,s}(e) = e$ for all unreachable elements $e \in M_{w,s}$. Since $(W, M) \equiv (V, N)$, the function $h_{w,s} : M_{w,s} \to R_{v,s}$ is bijective.

   Secondly, we interpret the function and relation symbols from $\Delta(C)$ in (V, R). Let $k \in C_{\mathtt{n}}$ be a nominal, $v := V_k$ and $w := W_k$.
   CASE $\sigma : \mathtt{ar} \to s \in F(C)$: We define $R_{v,\sigma} : R_{v,\mathtt{ar}} \to R_{v,s}$ by $R_{v,\sigma}(e) = h_{w,s}(M_{w,\sigma}(h^{-1}_{w,\mathtt{ar}}(e)))$ for all elements $e \in R_{v,\mathtt{ar}}$. Since $(W, M) \equiv (V, N)$, we have $R_{v,\sigma}(e) = N_{v,\sigma}(e)$ for all reachable elements $e \in N_{v,\mathtt{ar}} \cap R_{v,\mathtt{ar}}$.
   CASE $\varpi : \mathtt{ar} \in P$ We define $R_{v,\varpi} := h_{w,\mathtt{ar}}(M_{w,\varpi})$. Since $(W, M) \equiv (V, N)$, we have $e \in R_{v,\varpi}$ iff $e \in N_{v,\varpi}$ for all reachable elements $e \in N_{v,\mathtt{ar}} \cap R_{v,\mathtt{ar}}$.
   By construction, $h : (W, M) \to (V, R)$ is a homomorphism, and since it is bijective, $h : (W, M) \to (V, R)$ is an isomorphism.

2) We define an expansion (V^1, R^1) of (V, R) along χ such that $(V^1, R^1) \equiv (V^1, N^1)$. Roughly, (V^1, R^1) is obtained from (V^1, N^1) by replacing all unreachable elements of sorts in $\chi(S^{\mathtt{f}})$ with unreachable elements of flexible

sorts from (V,R). Concretely, (V^1,R^1) is obtained from (V^1,N^1) as follows:
CASE $s_1 \in \chi(S^{\mathtt{f}})$: $R^1_{v,s_1} := R_{v,\chi^{-1}(s_1)}$ for all $v \in |V^1|$, which is well-defined since χ is injective on sorts.
CASE $\sigma_1 : \mathtt{ar}_1 \to s_1 \in \chi(F^{\mathtt{f}})$, where $\mathtt{ar}_1$ contains at least one sort from $\chi(S^{\mathtt{f}})$: For all $v \in |V^1|$, $R^1_{v,\sigma_1} := R_{v,\chi^{-1}(\sigma_1)}$. Since χ protects flexible symbols, $\chi^{-1}(\sigma_1)$ is unique, which means that R^1_{v,σ_1} is well-defined. Also, we have $N^1_{v,\sigma_1}(e) = N_{v,\chi^{-1}(\sigma_1)}(e) = R_{v,\chi^{-1}(\sigma_1)}(e)$ for all reachable elements $e \in N^1_{v,\mathtt{ar}_1}$.
CASE $\pi_1 : \mathtt{ar}_1 \in \chi(P^{\mathtt{f}})$, where $\mathtt{ar}_1$ contains at least one sort from $\chi(S^{\mathtt{f}})$: For all $v \in |V^1|$, $R^1_{v,\pi_1} := R_{v,\chi^{-1}(\pi_1)}$. Since χ protects flexible symbols, $\chi^{-1}(\pi_1)$ is unique, which means that R^1_{v,π_1} is well-defined. Also, we have $e \in N^1_{v,\pi_1} = N_{v,\chi^{-1}(\pi_1)}$ iff $e \in R_{v,\chi^{-1}(\pi_1)} = R^1_{v,\pi_1}$ for all reachable elements $e \in N^1_{v,\mathtt{ar}_1}$.
CASE $\sigma_1 : \mathtt{ar}_1 \to s_1 \in F^{\mathtt{f}}_1 \setminus \chi(F^{\mathtt{f}})$, where $\mathtt{ar}_1$ has at least one sort from $\chi(S^{\mathtt{f}})$: For all $v \in |V^1|$, the function $R^1_{v,\sigma_1} : R^1_{v,\mathtt{ar}_1} \to R^1_{v,s_1}$ is defined by

- $R^1_{v,\sigma_1}(e) = N^1_{v,\sigma_1}(e)$ for all elements $e \in R^1_{v,\mathtt{ar}_1} \cap N^1_{v,\mathtt{ar}_1}$ and
- $R^1_{v,\sigma_1}(e)$ is an arbitrary value in R^1_{v,s_1} for all unreachable $e \in R^1_{v,\mathtt{ar}_1} \setminus N^1_{v,\mathtt{ar}_1}$.

CASE $\pi_1 : \mathtt{ar}_1 \in P^{\mathtt{f}}_1 \setminus \chi(P^{\mathtt{f}})$, where $\mathtt{ar}_1$ contains at least one sort from $\chi(S^{\mathtt{f}})$: For all possible worlds $v \in |V^1|$, let $R^1_{v,\pi_1} := \{e \in N^1_{v,\pi_1} \mid e \text{ is reachable}\}$.
Now, since χ protects flexible symbols, χ preserves flexible symbols, which means that for all possible worlds $v \in |V|$ and all sorts $s \in S$,

$$e \in N_{v,s} \text{ is unreachable iff } e \in N^1_{v,\chi(s)} \text{ is unreachable.}$$

It follows that the reachable sub-structures of (V^1,N^1) and (V^1,R^1) coincide. By Lemma 3.1, $(V^1,N^1) \equiv (V^1,R^1)$.

3) We define an isomorphism $h^1 : (W^1,M^1) \to (V^1,R^1)$ by expanding $h : (W,M) \to (V,R)$ along χ.
 Firstly, we define h^1 as a function. Let $h^1 : W \to V$ be $h : W \to V$, which is bijective. Assume k is a nominal in C and let $w = W_k$.
 CASE $s_1 \in \chi(S)$: Let $h^1_{w,s_1} := h_{w,\chi^{-1}(s_1)}$, which is bijective.
 CASE $s_1 \in S^1 \setminus \chi(S)$: $M^1_{w,s_1} := R^1_{w,s_1}$ and $h^1_{w,s_1} : M^1_{w,s_1} \to R^1_{w,s_1}$ is the identity.
 Secondly, we interpret the function and relation symbols from $\Delta^1(C^1)$ in (W^1,M^1). For any nominal or modality x in $\Delta^1(C^1)$, we define $W^1_x := (h^1)^{-1}(V^1_x)$. Take a nominal k in C^1, and let $w := W^1_k$ and $v := V^1_k$.
 CASE $\sigma_1 : \mathtt{ar}_1 \to s_1 \in F^1(C^1)$: We define $M^1_{w,\sigma_1} : M^1_{w,\mathtt{ar}_1} \to M^1_{w,s_1}$ by $M^1_{w,\sigma_1}(e) = (h^1_{w,s_1})^{-1}(R^1_{v,\sigma_1}(h^1_{w,\mathtt{ar}_1}(e)))$ for all elements $e \in M^1_{w,\mathtt{ar}_1}$.
 CASE $\pi_1 : \mathtt{ar}_1 \in P^1$: We define $M^1_{w,\pi_1} := (h^1_{w,\mathtt{ar}_1})^{-1}(R^1_{\pi_1})$.
 Since $h : (W,M) \to (V,R)$ is an isomorphism, $h^1 : (W^1,M^1) \to (V^1,R^1)$ is an isomorphism too.

It follows that $(W^1,M^1) \equiv (V^1,R^1)$. Since $(V^1,R^1) \equiv (V^1,N^1)$, we get $(W^1,M^1) \equiv (V^1,N^1)$. □

Definition 3.6 provides a general criterion for proving Robinson consistency property, while Lemma 3.7 is essential for completing the proof of Robinson consistency theorem.

4 Relativization

Relativization is a well-known method in classical model theory for defining substructures and their properties [11]. The substructures are usually characterized by some unary predicate and the technique is necessary in the absence of sorts when dealing with modular properties (such as putting together models defined over different signatures) which implicitly involve signature morphisms. For HFOL, the relativization is necessary to prove Robinson consistency property from omitting types property, since the signatures of nominals are single-sorted. It is worth mentioning that relativization is not necessary to prove Robinson consistency for many-sorted first-order logic.

Definition 4.1 The *relativized union* of any signatures Δ^1 and Δ^2 is a presentation $(\Delta^\diamond, \Phi^\diamond)$ defined as follows:

1) $\Delta^\diamond$ is the signature obtained from $\Delta^1 \coprod \Delta^2$ by adding two nominals o_1 and o_2, and two unary modalities $\pi_1 : 1$ and $\pi_2 : 1$,
2) $\Phi^\diamond \subseteq \mathtt{Sen}(\Delta^\diamond)$ consists of $\pi_1 \vee \pi_2$ and all sentences of the form $@_{k_i} \pi_i$, where $i \in \{1, 2\}$ and $k_i \in F_i^{\mathtt{n}} \cup \{o_i\}$.

Let $\mathtt{inj}_i : \Delta^i \to \Delta^1 \coprod \Delta^2$ be the canonical injection, for each $i \in \{1, 2\}$. Let $\theta : \Delta^1 \coprod \Delta^2 \hookrightarrow \Delta^\diamond$ be an inclusion. Here, we are interested more in the vertex $\Delta^\diamond$ and less in the arrows $(\mathtt{inj}_i; \theta)$, where $i \in \{1, 2\}$. The presentation $(\Delta^\diamond, \Phi^\diamond)$ is meant to define Kripke structures obtained from the union of a Kripke structure over Δ^1 and a Kripke structure over Δ^2. The new nominals o_1 and o_2 together with the sentences $@_{o_1} \pi_1$ and $@_{o_2} \pi_2$ ensure that the domains of π_1 and π_2 are not empty. For each nominal $k_1 \in F_1^{\mathtt{n}}$, the sentence $@_{k_1} \pi_1$ ensures that the interpretation of k_1 belongs to the denotation of π_1. A similar remark holds for any sentence $@_{k_2} \pi_2$ with $k_2 \in F_2^{\mathtt{n}}$. For the sake of simplifying the notation, we assume without loss of generality that Δ^1 and Δ^1 are disjoint, which means that $\Delta^1 \coprod \Delta^2 = \Delta^1 \cup \Delta^2$.

Definition 4.2 Let Δ^1 and Δ^2 be two disjoint signatures. For each $i \in \{1, 2\}$, the *relativized reduct* $\upharpoonright_{\pi_i} : \mathtt{Mod}(\Delta^\diamond) \to \mathtt{Mod}(\Delta^i)$ is defined as follows:

1) For each $(W, M) \in |\mathtt{Mod}(\Delta^\diamond, \Phi^\diamond)|$, the Kripke structure $(W, M) \upharpoonright_{\pi_i}$ denoted (W^i, M^i) is defined by (a) $|W^i| = W_{\pi_i}$, (b) $W_k^i = W_k$ for all nominals $k \in F_i^{\mathtt{n}}$, (c) $W_\varrho^i = W_\varrho \cap W_{\pi_i}$ for all unary modalities $(\varrho : \mathtt{n}) \in P_i^{\mathtt{n}}$, (d) $W_\lambda^i = \{(w, v) \in W_\lambda \mid w, v \in W_{\pi_i}\}$ for all modalities $(\lambda : \mathtt{n}\ \mathtt{n}) \in P_i^{\mathtt{n}}$, (e) $M^i = M|_{W^i}$ and $M_{w,x}^i = M_{w,x}$ for all possible worlds $w \in |W^i|$ and all sort/function/relation symbols x in Σ_i.
2) For each $h : (W, M) \to (W', M') \in \mathtt{Mod}(\Delta^\diamond, \Phi^\diamond)$, the homomorphism $h \upharpoonright_{\pi_i} : (W, M) \upharpoonright_{\pi_i} \to (W', M') \upharpoonright_{\pi_i}$ is defined by $(h \upharpoonright_{\pi_i})_w = h_w$ for all $w \in W_{\pi_i}$.

Definition 4.3 The *relativized translation* $\mathtt{rt}(\pi_i) : \mathtt{Sen}(\Delta^i) \to \mathtt{Sen}(\Delta^\diamond, \Phi^\diamond)$, where $i \in \{1, 2\}$, is defined by induction on the structure of sentences, simultaneously, for all disjoint signatures Δ^1 and Δ^2:

1) $\mathtt{rt}(\pi_i)(k) := \pi_i \Rightarrow k$ for all nominals $k \in F_i^{\mathtt{n}}$.
2) $\mathtt{rt}(\pi_i)(\varrho) := \pi_i \Rightarrow \varrho$ for all unary modalities $(\varrho : \mathtt{n}) \in P_i^{\mathtt{n}}$.
3) $\mathtt{rt}(\pi_i)(t_1 = t_2) := \pi_i \Rightarrow t_1 = t_2$ for all ground equations $t_1 = t_2$ over Δ^i.
4) $\mathtt{rt}(\pi_i)(\varpi(t_1, \ldots, t_n)) := \pi_i \Rightarrow \varpi(t_1, \ldots t_n)$ for all ground relations $\varpi(t_1, \ldots, t_n)$ over Δ^i.
5) $\mathtt{rt}(\pi_i)(@_k \gamma) := \pi_i \Rightarrow @_k \mathtt{rt}(\pi_i)(\gamma)$ for all nominals $k \in F_i^{\mathtt{n}}$ and all sentences $\gamma \in \mathtt{Sen}(\Delta^i)$.
6) $\mathtt{rt}(\pi_i)(\langle\lambda\rangle\gamma) := \pi_i \Rightarrow \langle\lambda\rangle\pi_i \wedge \mathtt{rt}(\pi_i)(\gamma)$ for all modalities $(\lambda : \mathtt{n}\ \mathtt{n}) \in P_i^{\mathtt{n}}$ and all sentences $\gamma \in \mathtt{Sen}(\Delta^i)$.
7) $\mathtt{rt}(\pi_i)(\neg\gamma) := \pi_i \Rightarrow \neg\mathtt{rt}(\pi_i)(\gamma)$ for all $\gamma \in \mathtt{Sen}(\Delta^i)$.
8) $\mathtt{rt}(\pi_i)(\gamma_1 \vee \gamma_2) := \mathtt{rt}(\pi_i)(\gamma_1) \vee \mathtt{rt}(\pi_i)(\gamma_2)$ for all $\gamma_1, \gamma_2 \in \mathtt{Sen}(\Delta^i)$.
9) $\mathtt{rt}(\pi_i)(\downarrow z \cdot \gamma) := \pi_i \Rightarrow \downarrow z \cdot \mathtt{rt}(\pi_i)(\gamma)$ for all sentences $\downarrow z \cdot \gamma \in \mathtt{Sen}(\Delta^i)$, where z a nominal variable. [8]
10) $\mathtt{rt}(\pi_i)(\exists x \cdot \gamma) := \pi_i \Rightarrow \exists x \cdot @_x \pi_i \wedge \mathtt{rt}(\pi_i)(\gamma)$ for all sentences $\exists x \cdot \gamma \in \mathtt{Sen}(\Delta^i)$ with x a nominal variable.
11) $\mathtt{rt}(\pi_i)(\exists y \cdot \gamma) := \pi_i \Rightarrow \exists y \cdot \mathtt{rt}(\pi_i)(\gamma)$ for all sentences $\exists y \cdot \gamma \in \mathtt{Sen}(\Delta^i)$ with y a rigid variable.

Simultaneous induction for all disjoint signatures is necessary for the case corresponding to quantified sentences. Unlike first-order logic, relativization is applied not only to quantifiers but it starts with atomic sentences. One can notice that locally the antecedent π_i of the implication is redundant, but globally it is not.

Proposition 4.4 (Satisfaction condition) *For all disjoint signatures Δ^1 and Δ^2, all Kripke structures $(W, M) \in |\mathtt{Mod}(\Delta^\diamond, \Phi^\diamond)|$ and all sentences $\gamma \in \mathtt{Sen}(\Delta^i)$, where $i \in \{1, 2\}$, the following* satisfaction conditions *hold:*

- *For all worlds $w \in W_{\pi_i}$, we have $(W, M) \models^w \mathtt{rt}(\pi_i)(\gamma)$ iff $(W, M)\upharpoonright_{\pi_i} \models^w \gamma$.*
- *For all worlds $w \in |W| \setminus W_{\pi_i}$, we have $(W, M) \models^w \mathtt{rt}(\pi_i)(\gamma)$.*
- *$(W, M) \models \mathtt{rt}(\pi_i)(\gamma)$ iff $(W, M)\upharpoonright_{\pi_i} \models \gamma$.*

The first two statements of Proposition 4.4 are straightforward by induction on the structure of sentences. The third statement, which corresponds to the global satisfaction condition, is a consequence of the first two statements.

Proposition 4.5 *For all disjoint signatures Δ^1 and Δ^2, and all Kripke structures (W^1, M^1) and (W^2, M^2) over Δ^1 and Δ^2, respectively, there exists $(W, M) \in |\mathtt{Mod}(\Delta^\diamond, \Phi^\diamond)|$ called the* relativized union *of (W^1, M^1) and (W^2, M^2) such that $(W, M)\upharpoonright_{\pi_1} = (W^1, M^1)$ and $(W, M)\upharpoonright_{\pi_2} = (W^2, M^2)$.*

The relativized union of Kripke structure is not unique.

---

[8] Notice that $\Delta^\diamond(z)$ is obtained from the relativized union of $\Delta^i(z)$ and Δ^j, where $i, j \in \{1, 2\}$ and $i \neq j$; therefore, $\mathtt{rt}(\pi)(\gamma)$ is well-defined.

5 Robinson consistency

Robinson consistency property is derived from Omitting Types Theorem, which was proved in [18] for HFOL. We start by defining the semantic opposite of a sentence. In first-order logic the semantic opposite of a sentence is its negation.

Definition 5.1 Given a sentence ψ over a signature Δ, we let (a) $+\psi$ denote the sentence $\forall z^\circ \cdot @_{z^\circ} \psi$, (b) $-\psi$ denote the sentence $\exists z^\circ \cdot @_{z^\circ} \neg\psi$, and (c) $\pm\psi$ range over $\{+\psi, -\psi\}$, where z° is a distinguished nominal variable for Δ.

The proof of the following lemma is straightforward.

Lemma 5.2 *For all Kripke structures (W, M) and all sentences ψ over a signature Δ, we have (a) $(W, M) \models \psi$ iff $(W, M) \models +\psi$ iff $(W, M) \models^w +\psi$ for some possible world $w \in |W|$, and (b) $(W, M) \not\models \psi$ iff $(W, M) \models -\psi$.*

By Lemma 5.2, the satisfaction of $+\psi$ does not depend on the possible world where the sentence $+\psi$ is evaluated. The same comment holds for $-\psi$ too.

5.1 Framework

We set the framework in which Robinson consistency property is proved. Let $(\Delta^1, \Phi^1) \overset{\chi_1}{\leftarrow} (\Delta, \Phi) \overset{\chi_2}{\rightarrow} (\Delta^2, \Phi^2)$ be a span of presentation morphisms such that (a) χ_2 is injective on sorts and protects flexible sorts, (b) Φ is maximally consistent over Δ, and (c) Φ^i is consistent over Δ^i for each $i \in \{1, 2\}$.

Assume a set of new rigid constants C for Δ such that $\mathtt{card}(C_s) = \alpha$ for all rigid sorts $s \in S^{\mathtt{r}}$, where $\alpha := max\{\mathtt{card}(\mathtt{Sen}(\Delta^1)), \mathtt{card}(\mathtt{Sen}(\Delta^2))\}$. For each $i \in \{1, 2\}$, let C^i be the set of new constants for Δ^i obtained by renaming the translation of the constants in C along χ_i and by adding a set of new constants $C^i_{s^i}$ of cardinality α for each rigid sort $s^i \in S^{\mathtt{r}}_i$ outside the image of χ_i:

$$C^i := \{c^i :\to \chi_i(s) \mid c :\to s \in C\} \cup (\textstyle\bigcup_{s^i \in S^{\mathtt{r}}_i \setminus \chi_i(S^{\mathtt{r}})} C^i_{s^i})$$

where (a) each constant $c^i :\to \chi_i(s)$ is the renaming of a constant $c :\to s \in C$, and (b) $C^i_{s^i}$ is a set of new constants of sort s^i for all rigid sorts $s^i \in S^{\mathtt{r}}_i \setminus \chi_i(S^{\mathtt{r}})$. Let $\chi_i^C : \Delta(C) \to \Delta^i(C^i)$ be the extension of $\chi_i : \Delta \to \Delta^i$ to $\Delta(C)$ which maps each constant $c :\to s \in C$ to its renaming $c^i :\to \chi_i(s) \in C^i$. Without loss of generality we assume that $\Delta^1(C^1)$ and $\Delta^2(C^2)$ are disjoint.

In Figure 1, $(\Delta^\diamond, \Phi^\diamond)$ is the relativized union of Δ^1 and Δ^2, while $(\Delta^\diamond(C^\diamond), \Phi^\diamond_C)$ is the relativized union of $\Delta^1(C^1)$ and $\Delta^2(C^2)$. The definitions of C, C^1 and C^2 are unique up to isomorphism and they are essential for the proof of Robinson consistency theorem.

Notation 5.3 (Semantics) *For each $(W, M) \in |\mathtt{Mod}(\Delta^\diamond, \Phi^\diamond)|$, we let*

- *(W^1, M^1) and (W^2, M^2) denote $(W, M)\upharpoonright_{\pi_1}$ and $(W, M)\upharpoonright_{\pi_2}$, respectively;*
- *(W^a, M^a) and (W^b, M^b) denote $(W^1, M^1)\upharpoonright_{\chi_1}$ and $(W^2, M^2)\upharpoonright_{\chi_2}$, respectively.*

Moreover, let (V, N) be a Kripke structure over $\Delta^\diamond(C^\diamond)$ and adopt a similar convention as above for its reducts.

Notation 5.4 (Syntax) *Let $i \in \{1, 2\}$ and $\psi \in \mathtt{Sen}(\Delta(C))$.*

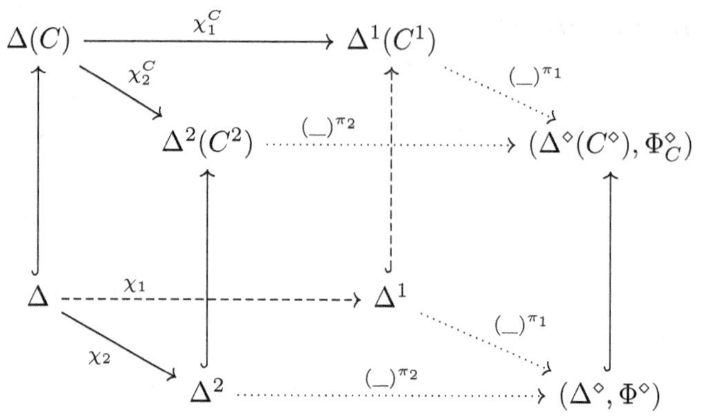

Fig. 1.

- *Let* ψ^i *and* ψ^{π_i} *denote* $\chi_i^C(\psi)$ *and* $\mathtt{rt}(\pi_i)(\chi_i^C(\psi))$, *respectively.*
- *Let* Φ^{π_i} *denote* $\mathtt{rt}(\pi_i)(\Phi^i)$.

5.2 Results

This section contains the main results, which are Robinson consistency theorem and its corollaries. The following lemma is crucial for subsequent developments and it is a consequence of Proposition 4.4 and Lemma 5.2.

Lemma 5.5 *For all* $(V, N) \in |\mathtt{Mod}(\Delta^\diamond(C^\diamond))|$ *and all* $\psi \in \mathtt{Sen}(\Delta(C))$,

$$(V, N) \models +(+\psi)^{\pi_1} \Leftrightarrow +(+\psi)^{\pi_2} \text{ iff } \Big((V^a, N^a) \models \psi \text{ iff } (V^b, N^b) \models \psi\Big).$$

Lemma 5.5 says that (V, N) globally satisfies $+(+\psi)^{\pi_1} \Leftrightarrow +(+\psi)^{\pi_2}$ for all $\Delta(D)$-sentences ψ iff (V^a, N^a) and (V^b, N^b) are elementarily equivalent.

Notation 5.6 *We define the following set of sentences over* $\Delta^\diamond(C^\diamond)$:

$$T_C := \Phi_C^\diamond \cup \Phi^{\pi_1} \cup \Phi^{\pi_2} \cup \{+(+\psi)^{\pi_1} \Leftrightarrow +(+\psi)^{\pi_2} \mid \psi \in \mathtt{Sen}(\Delta(C))\}.$$

Notice that T_C describes Kripke structures (V, N) over $\Delta^\diamond(C^\diamond)$ obtained from the relativized union of some Kripke structures $(V^1, N^1) \in |\mathtt{Mod}(\Delta^1(C^1), \Phi^1)|$ and $(V^2, N^2) \in |\mathtt{Mod}(\Delta^2(C^2), \Phi^2)|$ such that (V^a, N^a) and (V^b, N^b) are elementarily equivalent.

Proposition 5.7 T_C *is consistent.*

Proof. Let Ψ be a finite set of $\Delta(C)$-sentences. Let $D \subseteq C$ be the finite subset of all constants from C that occur in Ψ. We define $D^1 := \chi_1^C(D)$ and $D^2 := \chi_2^C(D)$. For each $i \in \{1, 2\}$, we let χ_i^D denote the restriction of χ_i^C to $\Delta(D)$.

We show that $\Phi_D^\diamond \cup \Phi^{\pi_1} \cup \Phi^{\pi_2} \cup \{+(+\psi)^{\pi_1} \Leftrightarrow +(+\psi)^{\pi_2} \mid \psi \in \Psi\}$ is consistent, where $(\Delta^\diamond(D^\diamond), \Phi_D^\diamond)$ is the relativized union of $\Delta^1(D^1)$ and $\Delta^2(D^2)$.

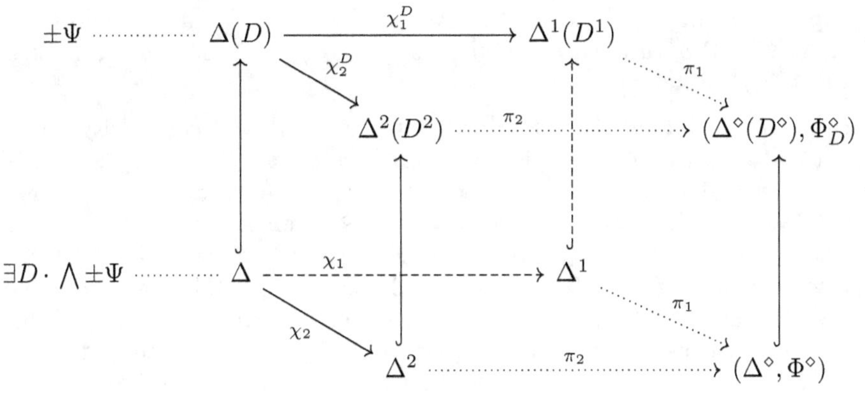

Fig. 2.

Since Φ^1 is consistent, $(W^1, M^1) \models \Phi^1$ for some Kripke structure (W^1, M^1) over Δ^1. Let (V^1, N^1) be an arbitrary expansion of (W^1, M^1) to $\Delta^1(D^1)$. Let $\pm\Psi := \{\pm\psi \mid \psi \in \Psi \text{ and } (V^a, N^a) \models \pm\psi\}$, where $(V^a, N^a) = (V^1, N^1)\!\upharpoonright_{\chi_1^D}$. Since $(V^a, N^a) \models \pm\Psi$, $(W^a, M^a) \models \exists D \cdot \bigwedge \pm\Psi$.[9] Since Φ^2 is consistent, $(W^2, M^2) \models \Phi^2$ for some Kripke structure (W^2, M^2) over Δ^2. Since $\chi_1(\Phi) \subseteq \Phi^1$ and $\chi_2(\Phi) \subseteq \Phi^2$, by satisfaction condition, $(W^a, M^a) \models \Phi$ and $(W^b, M^b) \models \Phi$. Since Φ is maximally consistent, $(W^a, M^a) \equiv (W^b, M^b)$. Since $(W^a, M^a) \models \exists D \cdot \bigwedge \pm\Psi$, $(W^b, M^b) \models \exists D \cdot \bigwedge \pm\Psi$. It follows that $(V^b, N^b) \models \pm\Psi$ for some expansion (V^b, N^b) of (W^b, M^b) to $\Delta(D)$. Since $\{\Delta(D) \hookleftarrow \Delta \overset{\chi_2}{\to} \Delta^2, \Delta(D) \overset{\chi_2^D}{\to} \Delta^2(D^2) \hookleftarrow \Delta^2\}$ is a pushout and $(V^b, N^b)\!\upharpoonright_\Delta = (W^b, M^b) = (W^2, M^2)\!\upharpoonright_{\chi_2}$, there exists an expansion (V^2, N^2) of (W^2, M^2) to $\Delta^2(D^2)$ such that $(V^2, N^2)\!\upharpoonright_{\chi_2^D} = (V^b, N^b)$. Let (V, N) be a relativized union of (V^1, N^1) and (V^2, N^2). Since $(V^1, N^1) \models \Phi^1$ and $(V^2, N^2) \models \Phi^2$, by Proposition 4.4, $(V, N) \models \Phi^\diamond_D \cup \Phi^{\pi_1} \cup \Phi^{\pi_2}$. By Lemma 5.5, $(V, N) \models \{+(+\psi)^{\pi_1} \Leftrightarrow +(+\psi)^{\pi_2} \mid \psi \in \Psi\}$.

Hence, by compactness, T_C is consistent. □

Recall that $\mathtt{n}$ denotes the sort of nominals.

Notation 5.8 (Nominal type) *We define a type in one nominal variable z:*

$$\Gamma_{\mathtt{n}} := \{@_z\, \pi_i \Rightarrow z \neq c^i \mid i = \overline{1,2} \text{ and } c :\to \mathtt{n} \in C\}$$

where c^i denotes $\chi_i(c)$ for all nominals $c :\to \mathtt{n} \in C$, and $z \neq c^i$ denotes $\neg @_z\, c^i$.

Notice that a Kripke structure (V, N) which satisfies $\pi_1 \vee \pi_2$ and omits $\Gamma_{\mathtt{n}}$ has the set of possible worlds reachable by the nominals in $C^\diamond$.

[9] Since D is not a set of variables, $\exists D \cdot \bigwedge \pm\Psi$ is not a sentence in our language, but there exists a Δ-sentence semantically equivalent to it.

Proposition 5.9 *T_C α-omits $\Gamma_{\mathtt{n}}$, that is, for each set of sentences $p \subseteq \mathtt{Sen}(\Delta^\diamond(C^\diamond, z))$ of cardinality strictly less than α such that $T_C \cup p$ is consistent, we have $T_C \cup p \not\models \Gamma_{\mathtt{n}}$.*

Proof. Let $p \subseteq \mathtt{Sen}(\Delta^\diamond(C^\diamond, z))$ be a set of sentences of cardinality strictly less than α such that $T_C \cup p$ is consistent. Let $D_{\mathtt{n}}$ be the set of all nominals $c \in C_{\mathtt{n}}$ such that either c^1 or c^2 occurs in p. Since $\mathtt{card}(p) < \alpha$, we have $\mathtt{card}(D_{\mathtt{n}}) < \alpha$. Let D be the set of constants obtained from C by removing all nominals from $C_{\mathtt{n}} \setminus D_{\mathtt{n}}$. We define $D^i := \chi_i^C(D)$ for each $i \in \{1,2\}$. It follows that $p \subseteq \mathtt{Sen}(\Delta^\diamond(D^\diamond, z))$, where $(\Delta^\diamond(D^\diamond), \Phi_D^\diamond)$ is the relativized union of $\Delta^1(D^1)$ and $\Delta^2(D^2)$. Let T_D be the set of all sentences from T_C which contains only constants from D^1 and D^2. Since $T_C \cup p$ is consistent, its subset $T_D \cup p$ is consistent too. Let (V, N) be a Kripke structure over $\Delta^\diamond(D^\diamond)$ such that $(V, N) \models T_D$ and let $v \in |V|$ such that $(V^{z \leftarrow v}, N) \models p$. Since $(V, N) \models \pi_1 \vee \pi_2$, we have $v \in \pi_1^V$ or $v \in \pi_2^V$. We assume that $v \in \pi_1^V$, as the case $v \in \pi_2^V$ is symmetrical. According to our conventions, $(V^1, N^1) = (V, N) \upharpoonright_{\pi_1}$, $(V^a, N^a) = (V, N) \upharpoonright_{\chi_1^D}$, $(V^2, N^2) = (V, N) \upharpoonright_{\pi_2}$ and $(V^b, N^b) = (V, N) \upharpoonright_{\chi_2^D}$, where χ_i^D denotes the restriction of χ_i^C to $\Delta(D)$ for each $i \in \{1,2\}$.

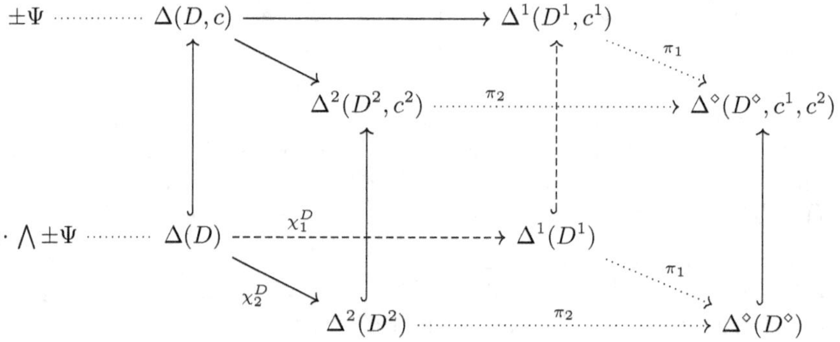

Fig. 3.

Since $\mathtt{card}(D_{\mathtt{n}}) < \alpha = \mathtt{card}(C_{\mathtt{n}})$, there exists a nominal $c \in C_{\mathtt{n}} \setminus D_{\mathtt{n}}$. Let $\Psi \subseteq \mathtt{Sen}(\Delta(D,c)) \setminus \mathtt{Sen}(\Delta(D))$ be a finite set of sentences. Then $T_D \cup p \cup \{(@_{c^1} \pi_1), (@_{c^2} \pi_2), (z = c^1)\} \cup \{+(+\psi)^{\pi_1} \Leftrightarrow +(+\psi)^{\pi_2} \mid \psi \in \Psi\}$ is consistent:

We define $\pm\Psi = \{\pm\psi \mid ((V^a)^{c \leftarrow v}, N^a) \models \pm\psi\}$. Since $((V^a)^{c \leftarrow v}, N^a) \models \pm\Psi$, $(V^a, N^a) \models \exists c \cdot \bigwedge \pm\Psi$. Since $(V, N) \models \{+(+\varphi)^{\pi_1} \Leftrightarrow +(+\varphi)^{\pi_2} \mid \varphi \in \mathtt{Sen}(\Delta(D))\}$, by Lemma 5.5, $(V^a, N^a) \equiv (V^b, N^b)$. It follows that $(V^b, N^b) \models \exists c \cdot \bigwedge \pm\Psi$. By semantics, $((V^b)^{c \leftarrow u}, N^b) \models \pm\Psi$ for some $u \in |V^b|$. We get $((V^a)^{c \leftarrow v}, N^a) \models \psi$ iff $((V^b)^{c \leftarrow u}, N^b) \models \psi$ for all $\psi \in \Psi$. By Lemma 5.5, $((V)^{(c^1,c^2) \leftarrow (v,u)}, N) \models \{+(+\psi)^{\pi_1} \Leftrightarrow +(+\psi)^{\pi_2} \mid \psi \in \Psi\}$. By satisfaction condition, $((V)^{(z,c^1,c^2) \leftarrow (v,v,u)}, N) \models \{+(+\psi)^{\pi_1} \Leftrightarrow +(+\psi)^{\pi_2} \mid \psi \in \Psi\}$. By satisfaction condition, since $(V, N) \models T_D$, $((V)^{(z,c^1,c^2) \leftarrow (v,v,u)}, N) \models T_D$. Since $(V^{z \leftarrow v}, N) \models p$, by satisfaction condition, $((V)^{(z,c^1,c^2)\leftarrow(v,v,u)}, N) \models p$. Since $v \in \pi_1^V$, $u \in \pi_2^V$ and the interpretations of z and c^1 are v, we obtain $((V)^{(z,c^1,c^2)\leftarrow(v,v,u)}, N) \models \{(@_{c^1}\, \pi_1), (@_{c^2}\, \pi_2), (z = c^1)\}$.

By compactness, $T_{D\cup\{c\}} \cup p \cup \{@_z\, \pi_1 \wedge z = c^1\}$ is consistent, where $T_{D\cup\{c\}}$ is the set of all sentences from T_C which contains only constants from $D^1 \cup D^2 \cup \{c^1 :\rightarrow \mathtt{n}, c^2 :\rightarrow \mathtt{n}\}$.

Now let $E \subset C$ be any proper subset which includes D. We define $E^i := \chi_i^C(E)$ for each $i \in \{1,2\}$. Assuming that $T_E \cup p \cup \{@_z\, \pi_1 \wedge z = c^1\}$ is consistent, we prove that $T_{E\cup\{k\}} \cup p \cup \{@_z\, \pi_1 \wedge z = c^1\}$ is consistent for any $k \in C_{\mathtt{n}} \setminus E_{\mathtt{n}}$. The proof is similar to the one above.

By compactness, $T_C \cup p \cup \{@_z\, \pi_1 \wedge z = c^1\}$ is consistent. Since $p \subseteq \mathtt{Sen}(\Delta^\diamond(C^\diamond, z))$ is an arbitrary set of cardinality strictly less than α consistent with T_C, it follows that T_C α-omits $\Gamma_{\mathtt{n}}$. □

Notation 5.10 (Rigid types) *Let z^1 be a variable of sort $s^1 \in S_1^{\mathtt{r}}$ and z^2 be a variable of sort $s^2 \in S_2^{\mathtt{r}}$. For each $i \in \{1,2\}$, we define a type in variable z^i:*

$$\Gamma_{s^i} := \{z^i \neq c^i \mid c^i :\rightarrow s^i \in C^i\}.$$

Notice that a Kripke structure (V,N) over $\Delta^\diamond(C^\diamond)$ which omits Γ_{s^i} has the carrier sets corresponding to the sort s^i reachable by the constants of sort s^i in C^i, where $i \in \{1,2\}$.

Proposition 5.11 *T_C α-omits both types Γ_{s^1} and Γ_{s^2}.*

Proof. We show that T_C omits Γ_{s^1}, since showing that T_C omits Γ_{s^2} is similar. Moreover, we focus on the case when $s^1 \in \chi_1(S^{\mathtt{r}})$, since the case $s^1 \notin \chi_1(S^{\mathtt{r}})$ is easy.

Let $p \subseteq \mathtt{Sen}(\Delta^\diamond(C^\diamond, z^1))$ be a set of sentences such that $\mathtt{card}(p) < \alpha$ and $T_C \cup p$ is consistent. We define the subset of constants $D \subseteq C$ as follows: (a) for all rigid sorts $s \in \chi_1^{-1}(s^1)$, the set D_s consists of all constants $c :\rightarrow s \in C$ such that either c^1 or c^2 occurs in p, and (b) for all rigid sorts $s \notin \chi_1^{-1}(s^1)$, we have $D_s := C_s$. For each $i \in \{1,2\}$, we define $D^i := \chi_i(D) \cup \{c :\rightarrow s \in C^i \mid s \in S_i^{\mathtt{r}} \setminus \chi_i(S^{\mathtt{r}})\}$. It follows that $p \subseteq \mathtt{Sen}(\Delta^\diamond(D^\diamond, z^1))$, where $(\Delta^\diamond(D^\diamond, z^1), \Phi_D^\diamond)$ is the relativized union of $\Delta^1(D^1)$ and $\Delta^2(D^2)$.

Let T_D be the set of all sentences from T_C which contains only constants from D^1 and D^2. Since $T_C \cup p$ is consistent, its subset $T_D \cup p$ is consistent too. Let (V,N) be a Kripke structure over $\Delta^\diamond(D^\diamond)$ such that $(V,N) \models T_D$ and $(V, N^{z^1\leftarrow e}) \models p$ for some possible world $v \in V_{\pi_1}$ and element $e \in N_{v,s^1}$. According to our conventions, $(V^1,N^1) = (V,N)\restriction_{\pi_1}$, $(V^a,N^a) = (V,N)\restriction_{\chi_1^D}$, $(V^2,N^2) = (V,N)\restriction_{\pi_2}$ and $(V^b,N^b) = (V,N)\restriction_{\chi_2^D}$, where χ_i^D denotes the restriction of χ_i^C to $\Delta(D)$ for each $i \in \{1,2\}$.

1) Let $s \in \chi_1^{-1}(s^1)$. Since $\mathtt{card}(D_s) < \alpha = \mathtt{card}(C_s)$, there exists $c \in C_s \setminus D_s$. Let $\Psi \subseteq \mathtt{Sen}(\Delta(D,c)) \setminus \mathtt{Sen}(\Delta(D))$ be a finite set of sentences. We show that $T_D \cup p \cup \{z^1 = c^1\} \cup \{+(+\psi)^{\pi_1} \Leftrightarrow +(+\psi)^{\pi_2} \mid \psi \in \Psi\}$ is consistent, where $c^1 :\rightarrow \chi_1(s)$ is the translation of $c :\rightarrow s$ along χ_1^C. The proof is similar to the first part of the proof of Proposition 5.9. By compactness, $T_{D\cup\{c\}} \cup p \cup \{z^1 =$

$c^1\}$ is consistent, where $T_{D\cup\{c\}}$ is the set of all sentences from T_C which contains only constants from $D^1 \cup \{c^1 :\to \chi_1(s)\}$ and $D^2 \cup \{c^2 :\to \chi_2(s)\}$.

2) Now let $E \subseteq C$ be an arbitrary subset of constants which includes D. We define $E^i := \chi_i^C(E) \cup \{c :\to s \in C^i \mid s \in S_i^{\mathtt{r}} \setminus \chi_i(S^{\mathtt{r}})\}$ for each $i \in \{1,2\}$. Assuming that $T_E \cup p \cup \{z^1 = c^1\}$ is consistent, we prove that $T_{E\cup\{d\}} \cup p \cup \{z^1 = c^1\}$ is consistent for any $d :\to s \in C \setminus E$. The proof is similar to the one above.

From (1) and (2), by compactness, $T_C \cup p \cup \{z^1 = c^1\}$ is consistent. Since $p \subseteq \mathtt{Sen}(\Delta^\diamond(C^\diamond, z^1))$ is an arbitrary set of cardinality strictly less than α an consistent with T_C, it follows that T_C α-omits Γ_{s^1}. □

All the preliminary results for proving Robinson consistency property are in place.

Theorem 5.12 (Robinson consistency) *Recall that χ_2 is injective on sorts. In addition, assume that χ_2 protects flexible symbols. Let $\Delta^1 \overset{\upsilon_1}{\to} \Delta' \overset{\upsilon_2}{\leftarrow} \Delta^2$ be the pushout of $\Delta^1 \overset{\chi_1}{\leftarrow} \Delta \overset{\chi_2}{\to} \Delta^2$. Then $\upsilon_1(\Phi^1) \cup \upsilon^2(\Phi^2)$ is consistent.*

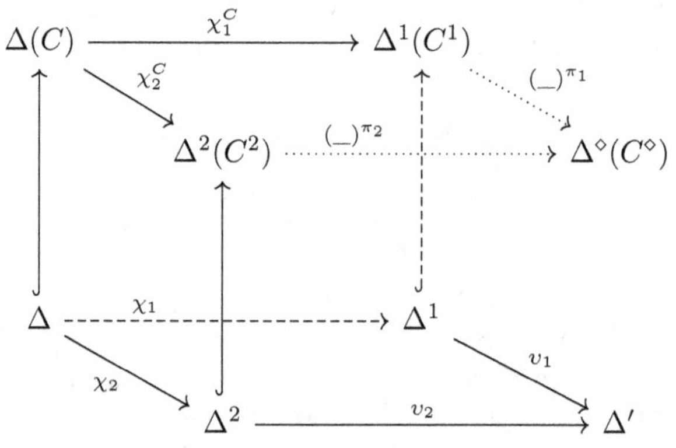

Fig. 4.

Proof. By Proposition 5.9, T_C α-omits $\Gamma_{\mathtt{n}}$. By Proposition 5.11, T_C α-omits Γ_{s^1} and Γ_{s^2} for all rigid sorts $s^1 \in S_1^{\mathtt{r}}$ and $s^2 \in S_2^{\mathtt{r}}$. By [18, Extended Omitting Types Theorem], there exists $(V,N) \in |\mathtt{Mod}(\Delta^\diamond(C^\diamond))|$ such that $(V,N) \models T_C$ and (V,N) omits $\Gamma_{\mathtt{n}}$, Γ_{s^1} and Γ_{s^2} for all rigid sorts $s^1 \in S_1^{\mathtt{r}}$ and $s^2 \in S_2^{\mathtt{r}}$. Since $(V,N) \models \Phi^{\pi_1} \cup \Phi^{\pi_2}$, by satisfaction condition, $(V^1,N^1) \models \Phi^1$ and $(V^2,N^2) \models \Phi^2$. Since $\Phi^1 \models \chi_1(\Phi)$ and $\Phi^2 \models \chi_2(\Phi)$, by satisfaction condition, $(V^a,N^a) \models \Phi$ and $(V^b,N^b) \models \Phi$. Since Φ is maximally consistent, $(V^a,N^a) \equiv (V^b,N^b)$. Since (V,N) omits $\Gamma_{\mathtt{n}}$ and Γ_{s^1} for all rigid sorts $s^1 \in S_1^{\mathtt{r}}$, (V^1,N^1) is reachable by C^1. Since (V,N) omits $\Gamma_{\mathtt{n}}$ and Γ_{s^2} for all rigid sorts $s^2 \in S_2^{\mathtt{r}}$, (V^2,N^2) is reachable by C^2. Since $(V^a,N^a) \equiv (V^b,N^b)$, (V^a,N^a) is reachable by C and (V^2,N^2) is reachable by C^2, by Lemma 3.7, $(U^2,R^2) \equiv (V^2,N^2)$ for some

χ_2^C-expansion (U^2, R^2) of (V^a, N^a). We define $(W^1, M^1) := (V^1, N^1) \upharpoonright_{\Delta^1}$ and $(W^2, M^2) := (U^2, R^2) \upharpoonright_{\Delta^2}$. Since $(V^1, N^1) \models \Phi^1$ and $(U^2, R^2) \models \Phi^2$, by satisfaction condition, $(W^1, M^1) \models \Phi^1$ and $(W^2, M^2) \models \Phi^2$. Since $\Delta^1 \overset{v_1}{\to} \Delta' \overset{v_2}{\leftarrow} \Delta^2$ is the pushout of $\Delta^1 \overset{\chi_1}{\leftarrow} \Delta \overset{\chi_2}{\to} \Delta^2$ and $(W^1, M^1) \upharpoonright_{\chi_1} = (W, M) = (W^2, M^2) \upharpoonright_{\chi_2}$, by [10, Example 3.5], there exists a unique Kripke structure $(W', M') \in |\texttt{Mod}(\Delta')|$ such that $(W', M') \upharpoonright_{v_1} = (W^1, M^1)$ and $(W', M') \upharpoonright_{v_2} = (W^2, M^2)$. By satisfaction condition, $(W', M') \models v_1(\Phi^1) \cup v^2(\Phi^2)$. □

In many-sorted first-order logic, if one of the signature morphisms in the span is injective on sorts then the corresponding pushout is a CI square [21]. Lemma 3.5 shows that in HFOL, this condition is also necessary. The following two examples focus on the second condition, the protection of flexible symbols.

Example 5.13 Let $\Delta^1 \overset{\chi_1}{\hookleftarrow} \Delta \overset{\chi_2}{\hookrightarrow} \Delta^2$ be a span of inclusions such that

- Δ has one nominal $\{k\}$, one flexible sort $\{s\}$ and one constant $\{c :\to s\}$;
- Δ^1 has nominals $\{k, k_1\}$, one rigid sort $\{s\}$ and one flexible constant $\{c :\to s\}$;
- Δ^2 has two nominals $\{k, k_2\}$, one flexible sort s, and two flexible constants $\{c :\to s, c_2 :\to s\}$.

Let $\Delta^1 \overset{v_1}{\hookrightarrow} \Delta' \overset{v_2}{\hookleftarrow} \Delta^2$ be a pushout of the above span such that

- Δ' consists of three nominals $\{k, k_1, k_2\}$, one rigid sort $\{s\}$, and two flexible constants $\{c :\to s, c_2 :\to s\}$.

In Example 5.13, the signature morphism χ_2 adds a new constant $c_2 :\to s$ on the flexible sort s, which means that flexible symbols are not protected. The following lemma shows that the pushout constructed above is not a CI square.

Lemma 5.14 *The pushout described in Example 5.13 is not a CI square.*

Proof. Let $\Phi_1 := \{\forall y \cdot c = y \in \texttt{Sen}(\Delta^1)\}$ and $\Phi_2 := \{c_2 = c \in \texttt{Sen}(\Delta^2)\}$. It is straightforward to show $\Phi_1 \models \Phi_2$. Suppose towards a contradiction that there exists $\Phi \subseteq \texttt{Sen}(\Delta)$ such that $\Phi_1 \models \Phi$ and $\Phi \models \Phi_2$.

Let (W^1, M^1) be the Kripke structure over Δ^1 that consists of one possible world w, which means that $W^1_k = W^1_{k_1} = w$ and M^1_w is the single-sorted algebra consisting of one element $M^1_{w,s} = \{e\}$, which means that $M^1_{w,c} = e$. Obviously, $(W^1, M^1) \models \forall x \cdot x = c$. By the satisfaction condition, $(W^1, M^1) \upharpoonright_\Delta \models \Phi$. Let (W, M) be the Kripke structure over Δ obtained from $(W^1, M^1) \upharpoonright_\Delta$ by adding a new flexible element d of sort s. Since d is an unreachable element, by Lemma 3.1, $(W, M) \equiv (W^1, M^1) \upharpoonright_\Delta$. It follows that $(W, M) \models \Phi$. Let (W^2, M^2) be the expansion of (W, M) to Δ^2 which interprets $c_2 :\to s$ as d. By the satisfaction condition, $(W^2, M^2) \models \Phi$. Since $\Phi \models \Phi_2$, we have $(W^2, M^2) \models \Phi_2$, which is a contradiction, as $M^2_{w,c} = e \neq d = M^2_{w,c_2}$. □

We give another example of pushout which is not a CI square.

Example 5.15 Let $\Delta^1 \overset{\chi_1}{\hookleftarrow} \Delta \overset{\chi_2}{\hookrightarrow} \Delta^2$ be a span of inclusions such that

- Δ has one nominal $\{k\}$, one flexible sort $\{Nat\}$ and two flexible function symbols $\{0 :\to Nat, succ : Nat \to Nat\}$;

- Δ^1 has two nominals $\{k, k_1\}$, one rigid sort $\{Nat\}$ and three flexible function symbols $\{0 :\to Nat, succ : Nat \to Nat, \_+\_ : Nat\ Nat \to Nat\}$;
- Δ^2 has two nominals $\{k, k_2\}$, two rigid sorts $\{Nat, List\}$, four flexible operations $\{0 :\to Nat, succ : Nat \to Nat, nil :\to List, \_|\_ : Nat\ List \to List\}$.

Let $\Delta^1 \overset{v_1}{\hookrightarrow} \Delta' \overset{v_2}{\hookleftarrow} \Delta^2$ be a pushout of the above span such that

- Δ' has three nominals $\{k, k_1, k_2\}$, two rigid sorts $\{Nat, List\}$ and five flexible function symbols $\{0 :\to Nat, succ : Nat \to Nat, \_+\_ : Nat\ Nat \to Nat, nil :\to List, cons : Nat\ List \to List\}$.

In Example 5.15, the signature morphism χ_2 does not preserve the flexible sort Nat.

Lemma 5.16 *The pushout described in Example 5.15 is not a CI square.*

Proof. Let Φ^1 be the set of Δ^1-sentences which consists of

- $\forall x : Nat \cdot succ(succ(x)) = x$,
- $\forall x : Nat \cdot 0 + x = x$, and
- $\forall x : Nat, y : Nat \cdot succ(y) + x = succ(y + x)$.

Let $\Phi^2 := \{\forall x : Nat \cdot succ(succ(x)) = x\}$. Suppose towards a contradiction that there exists a set of Δ-sentences Φ such that $\Phi^1 \models \Phi$ and $\Phi \models \Phi^2$.

Let (W^1, M^1) be a Kripke structure over Δ^1 that consists of two possible worlds $\{w_1, w_2\}$, both $M^1_{w_1}$ and $M^1_{w_2}$ are the quotient algebra $\mathbb{Z}_2$. Obviously, $(W^1, M^1) \models \Phi^1$. Since $\Phi^1 \models \Phi$, $(W^1, M^1) \models \Phi$. By the satisfaction condition, $(W^1, M^1)\upharpoonright_\Delta \models \Phi$. Let (W, M) obtained by adding a new element $\widehat{2}$ of sort Nat such that $M_{w_i,succ}(\widehat{2}) = \widehat{0}$ for each $i \in \{1, 2\}$. Since $\widehat{2}$ is an unreachable element, by Lemma 3.1, $(W, M) \equiv (W^1, M^1) \upharpoonright_\Delta$. Let (W^2, M^2) be the expansion of (W, M) to Δ^2 which interprets $List$ in both worlds as the set of all lists with elements from $\{\widehat{0}, \widehat{1}, \widehat{2}\}$. By the satisfaction condition, $(W^2, M^2) \models \Phi$. Since $M_{w_1,succ}(M_{w_1,succ}(\widehat{2})) = \widehat{1}$, we have $(W^2, M^2) \not\models \Phi^2$, which contradicts $\Phi \models \Phi^2$. □

6 Conclusions

Lemma 5.14 and Lemma 5.16 show that not only injectivity on sorts but also protection of flexible symbols is necessary for interpolation in HFOL.

Recall that RFOHL signatures form a subcategory of HFOL signatures. It is not difficult to check that the subcategory of RFOHL signatures is closed under pushouts. It follows that Theorem 5.12 is applicable to RFOHL. Since intersection-union square of signature morphism are, in particular, pushouts, our results cover the ones obtained in [3].

Similarly, HPL signatures form a subcategory of HFOL signatures. It is not difficult to check that the subcategory of HPL signatures is closed under pushouts. It follows that Theorem 5.12 is applicable to HPL. Since Theorem 5.12 is derived from Omitting Types Theorem, our results rely on quantification over possible worlds. Therefore, the present work does not cover the interpolation result from [2], which is applicable to hybrid propositional logic without quantification.

HFOL and HFOLS have the same signatures and Kripke structures. By [17, Lemma 2.20], HFOLS and HFOL have the same expressivity power. The relationship between HFOL and HFOLS is similar to the relationship between first-order logic and unnested first-order logic, which allows only terms of depth one [11]. Therefore, a square of HFOL signature morphisms is a CI square in HFOL iff it is a CI square in HFOLS.

Proposition 3.2 does not give an effective construction of an interpolant, which is an open problem.

References

[1] Areces, C. and P. Blackburn, *Bringing them all Together*, Journal of Logic and Computation **11** (2001), pp. 657–669.

[2] Areces, C., P. Blackburn and M. Marx, *Hybrid logics: characterization, interpolation and complexity*, Journal of Symbolic Logic **66** (2001), pp. 977–1010.

[3] Areces, C., P. Blackburn and M. Marx, *Repairing the interpolation theorem in quantified modal logic*, Ann. Pure Appl. Log. **124** (2003), pp. 287–299.

[4] Blackburn, P., M. A. Martins, M. Manzano and A. Huertas, *Rigid first-order hybrid logic*, in: R. Iemhoff, M. Moortgat and R. J. G. B. de Queiroz, editors, *Logic, Language, Information, and Computation - 26th International Workshop, WoLLIC 2019, Utrecht, The Netherlands, July 2-5, 2019, Proceedings*, Lecture Notes in Computer Science **11541** (2019), pp. 53–69.

[5] Borzyszkowski, T., *Logical systems for structured specifications*, Theor. Comput. Sci. **286** (2002), pp. 197–245.

[6] Codescu, M., *Hybridisation of institutions in HETS (tool paper)*, in: M. Roggenbach and A. Sokolova, editors, *8th Conference on Algebra and Coalgebra in Computer Science, CALCO 2019, June 3-6, 2019, London, United Kingdom*, LIPIcs **139** (2019), pp. 17:1–17:10.

[7] Ţutu, I. and J. L. Fiadeiro, *From conventional to institution-independent logic programming*, J. Log. Comput. **27** (2017), pp. 1679–1716.

[8] Diaconescu, R., "Institution-independent Model Theory," Studies in Universal Logic, Birkhäuser, Basel, 2008, 1 edition.

[9] Diaconescu, R., *Quasi-varieties and initial semantics for hybridized institutions*, Journal of Logic and Computation **26** (2016), pp. 855–891.

[10] Diaconescu, R. and A. Madeira, *Encoding Hybridised Institutions into First-Order Logic*, Mathematical Structures in Computer Science **26** (2016), pp. 745–788.

[11] Ebbinghaus, H., J. Flum and W. Thomas, "Mathematical logic (2. ed.)," Undergraduate Texts in Mathematics, Springer, 1994.

[12] Goguen, J. and R. Burstall, *Institutions: Abstract model theory for specification and programming*, Journal of the Association for Computing Machinery **39** (1992), pp. 95–146.

[13] Găină, D., *Interpolation in logics with constructors*, Theoretical Computer Science **474** (2013), pp. 46–59.

[14] Găină, D., *Birkhoff style calculi for hybrid logics*, Formal Asp. Comput. **29** (2017), pp. 805–832.

[15] Găină, D., *Downward Löwenheim-Skolem Theorem and interpolation in logics with constructors*, Journal of Logic and Computation **27** (2017), pp. 1717–1752.

[16] Găină, D., *Foundations of logic programming in hybrid logics with user-defined sharing*, Theor. Comput. Sci. **686** (2017), pp. 1–24.

[17] Găină, D., *Forcing and calculi for hybrid logics*, Journal of the Association for Computing Machinery **67** (2020), pp. 25:1–25:55.

[18] Găină, D., G. Badia and T. Kowalski, *Omitting types theorem in hybrid-dynamic logics with rigid symbols*, math **abs/2203.08720** (2022).

[19] Găină, D., K. Futatsugi and K. Ogata, *Constructor-based Logics*, J. UCS **18** (2012), pp. 2204–2233.

[20] Găină, D. and M. Petria, *Completeness by forcing*, Journal of Logic and Computation **20** (2010), pp. 1165–1186.

[21] Găină, D. and A. Popescu, *An institution-independent proof of the robinson consistency theorem*, Stud Logica **85** (2007), pp. 41–73.

[22] Lindström, P., *Omitting uncountable types and extensions of elementary logic*, Theoria - a Swedish Journal of Philosophy **44** (1978), pp. 152—156.

[23] Martins, M. A., A. Madeira, R. Diaconescu and L. S. Barbosa, *Hybridization of institutions*, in: A. Corradini, B. Klin and C. Cîrstea, editors, *Algebra and Coalgebra in Computer Science - 4th International Conference, CALCO 2011, Proceedings*, Lecture Notes in Computer Science **6859** (2011), pp. 283–297.

[24] Petria, M., *An Institutional Version of Gödel's Completeness Theorem*, in: T. Mossakowski, U. Montanari and M. Haveraaen, editors, *Algebra and Coalgebra in Computer Science, Second International Conference, CALCO 2007, Bergen, Norway, August 20-24, 2007, Proceedings*, Lecture Notes in Computer Science **4624** (2007), pp. 409–424.

[25] Robinson, A., *A result on consistency and its application to thetheory of definition*, Indagationes Mathematicae (Proceedings) **59** (1956), pp. 47–58.
URL `https://www.sciencedirect.com/science/article/pii/S138572585650008X`

[26] Tarlecki, A., *Bits and Pieces of the Theory of Institutions*, , **240**, Springer, 1986 pp. 334–360.

Direct elimination of additive-cuts in GL4ip: verified and extracted.

Rajeev Goré

Technical University of Vienna, Austria, and Polish Academy of Science, Poland

Ian Shillito

Australian National University, Australia

Abstract

Recently, van der Giessen and Iemhoff proved cut-admissibility for the sequent calculus GL4ip for propositional intuitionistic provability logic. To do so, they were forced to use an indirection via the GL3ip calculus as GL4ip resists all standard direct cut-admissibility techniques. This indirection leaves little hope for the extraction of a comprehensible cut-elimination procedure for GL4ip from their work.
We eliminate this indirection: we prove the admissibility of additive cut for GL4ip in a direct way by using a recently discovered proof technique which requires the existence of a terminating backward proof-search procedure in this calculus. By formalising our results in Coq we: (1) exhibit a successful direct proof technique for cut-admissibility for GL4ip ; (2) extract a syntactic cut-elimination procedure for GL4ip in Haskell ; and (3) use a local measure on sequents based on the shortlex order to show that the proof-search terminates. Once again, we see an unusual phenomenon in that terminating backward proof-search forms the basis for syntactic cut-elimination rather than for semantic cut-free completeness.

Keywords: Intuitionistic provability logic, Cut elimination, Backward proof-search, Interactive theorem proving, Proof theory.

Acknowledgement

Rajeev Goré supported by the FWF project P 33548. We acknowledge the support of the National Centre for Research and Development, Poland (NCBR), and the Luxembourg National Research Fund (FNR), under the PolLux/FNR-CORE project STV (POLLUX-VII/1/2019).

1 Introduction

Classical modal provability logics have gained a lot of attention because of the ability to interpret the formula $\Box A$ as "A is provable in Peano Arithmetic" [12]. As usual, the completeness of the traditional sequent calculus for provability logic with respect to the traditional Hilbert axiomatisation requires showing cut-admissibility. But cut-admissibility is usually not trivial because the standard double-induction on the size of the cut-formula and the height of the derivation do not suffice. To solve this problem, Valentini [18] introduced a third complex parameter called "width" in addition to these two traditional induction measures. The complications in his cut-admissibility argument in a set-based setting led to many claims and counter-claims, finally resolved thirty years later by Goré and Ramanayake [8] in a multiset setting.

Recently, van der Giessen and Iemhoff [19] showed that the proof-theory of intuitionistic provability logics is also complicated. They gave a cut-free sequent calculus GL3ip for intuitionistic provability logic extending the standard G3ip [17] calculus for intuitionistic logic with the following well known rule:

$$\frac{X, \Box X, \Box A \Rightarrow A}{W, \Box Y, \Box X \Rightarrow \Box A} \text{(GLR)}$$

Similarly to G3ip, the admissibility of the rules of weakening and contraction can easily be shown for GL3ip. However, the admissibility of cut encounters the same problems as for GL, leading van der Giessen and Iemhoff to successfully adapt the technique developed by Valentini, thus obtaining a direct proof of cut-admissibility for intuitionistic provability logic.

However, GL3ip cannot support a simple terminating backward proof-search strategy because its left-implication rule, inherited from G3ip and shown below, allows trivial cycles up the left premise as is well known:

$$\frac{X, A \to B \Rightarrow A \quad X, B \Rightarrow C}{X, A \to B \Rightarrow C} (\to\text{L})$$

To solve this problem and characterize a terminating proof-search procedure, they follow Dyckhoff [6] and Hudelmaier [10] and define the calculus GL4ip by both slightly modifying the rule (GLR) and mimicking G4ip by replacing ($\to$L) with a collection of rules sensitive to the form of the formula A in $A \to B$. To prove cut-admissibility they show that GL3ip and GL4ip are equivalent, in that they prove the same sequents.

They point out that although the calculus GL4ip enjoys terminating backward proof-search, the existence of a direct proof of cut-admissibility is doubtful: all standard methods fail, including Valentini's. While a direct and syntactic proof of cut-admissibility usually leads to a straightforward algorithm for cut-elimination, here the only potential cut-elimination algorithm for GL4ip is quite convoluted: (1) take a GL4ip proof containing cuts; (2) transform it to a GL3ip proof containing cuts ; (3) apply the cut-elimination procedure for GL3ip to obtain a cut-free GL3ip proof ; (4) transform the cut-free GL3ip proof into a cut-free GL4ip proof. In particular, the steps from (2) to (3), which rely on Valentini's complicated argument, and from (3) to (4), which involve intricate transformations, are anything but trivial. This indirection, coupled with the intricacies mentioned, can only lead to a painful and obscure algorithm for cut-elimination for GL4ip.

Naturally, the following question comes to mind: can we eliminate the indirection from GL4ip+(cut) to GL3ip+(cut) to GL3ip to GL4ip, and obtain a direct cut-elimination procedure for GL4ip? Moreover, can we guarantee that this cut-elimination proof is correct?

Here, we answer both questions positively by giving a direct syntactic proof of cut-admissibility for GL4ip. First, we show the admissibility of the structural rules by adapting the arguments from Dyckhoff and Negri [7]. Second, we define a proof-search procedure PSGL4ip on GL4ip. Furthermore, we develop a thorough termination argument by defining a local measure on sequents and a well-founded relation along which this measure decreases upwards in the proof-search. Finally, we directly prove cut-admissibility for GL4ip using the *mhd proof technique*, which makes use of the termination of PSGL4ip to attribute a *m*aximum *h*eight of *d*erivations to each sequent [3]. We use this number as an induction measure in an argument involving local and syntactic transformations, allowing us to exhibit and hence extract a cut-elimination procedure. All of our claims have been formally verified in the interactive theorem prover Coq (https://github.com/ianshil/CE_GL4ip.git). Using the automatic program extraction facilities of Coq, we extracted the formally verified computer program for cut-elimination associated to our formalisation.

2 Preliminaries

Let $\mathbb{V} = \{p, q, r \dots\}$ be an infinite set of propositional variables. Modal formulae are defined by the following grammar.

$$A ::= p \in \mathbb{V} \mid \bot \mid A \wedge A \mid A \vee A \mid A \to A \mid \Box A$$

We encode formulae as a type `(MPropF V)` over some parametric type `(V)` of propositional variables. A list of such formulae then has the type `list ( MPropF V)`. The usual operations on lists "append" and "cons" are respectively represented by `++` and `::` but Coq also allows us to write lists in infix notation using `;`. Thus the terms `A1 :: A2 :: A3` and `[A1] ++ [A2] ++ [A3]` and `[A1 ; A2 ; A3]` all encode the list A_1, A_2, A_3.

Definition 2.1 The *weight* $w(A)$ of a formula A is defined as follows:

$$\begin{array}{rcl} w(\bot) = w(p) & = & 1 \\ w(C \vee D) = w(C \to D) & = & w(C) + w(D) + 1 \\ w(C \wedge D) & = & w(C) + w(D) + 2 \\ w(\Box C) & = & w(C) + 1 \end{array}$$

We say that a formula A is a *boxed formula* if it has $\Box$ as its main connective. A boxed multiset contains only boxed formulae. For a set $X = \{A_1, \dots, A_n\}$, define $\boxtimes X = \{A_1, \Box A_1, \dots, A_n, \Box A_n\}$. We denote the set of subformulae of a formula A, including itself, by $\mathrm{Sub}(A)$. We abuse the notation to designate the set of subformulae of all formulae in the set X by $\mathrm{Sub}(X)$. We use the letters $A, B, C, \dots$ for formulae and $X, Y, Z, \dots$ for multisets of formulae.

The Hilbert calculus for the intuitionistic normal modal logic iK extends a Hilbert-calculus for intuitionistic propositional logic with the axiom $\Box(p \to q) \to (\Box p \to \Box q)$ and the inference rule of necessitation: from A infer $\Box A$. Intuitionistic Gödel-Löb logic iGL is obtained by the addition of the axiom $\Box(\Box p \to p) \to \Box p$ to iK. We write $A \in$ iK when A is a theorem of iK.

A *sequent* is a pair of a multiset of formulae and a formula, denoted $X \Rightarrow C$. For multisets X and Y, the multiset sum $X \uplus Y$ is the multiset whose multiplicity (at each formula) is a sum of the multiplicities of X and Y. We write X, Y to mean $X \uplus Y$. For a formula A, we write A, X and X, A to mean $\{A\} \uplus X$. From the formalisation perspective, a pair of a list of formulae and a formula has type `list (MPropF V) * (MPropF V)`, using the Coq notation `*` for forming pairs. The latter is the type we give to sequents in our formalisation, for which we use the macro `Seq`. Thus the sequent $A_1, A_2, A_3 \Rightarrow B$ is encoded by the term `[A_1 ; A_2 ; A_3] * B`, which itself can also be written as the pair `([A_1 ; A_2 ; A_3], B)`. Note that `[A_1 ; A_2 ; A_3] * B` is different from `[A_2 ; A_1 ; A_3] * B` since the order of the elements is crucial, so our lists do not capture multisets (yet).

A *sequent calculus* consists of a finite set of *sequent rule schemas*. Each rule schema consists of a conclusion sequent and some number of premise sequents. If a rule schema has no premise sequents, then it is called an initial sequent. The conclusion and premises are built in the usual way from propositional-variables, formula-variables and multiset-variables. A *rule instance* is obtained by uniformly instantiating every variable in the rule schema with a concrete object of that type. This is the standard definition from structural proof theory.

Definition 2.2 [Derivation/Proof] A *derivation* of a sequent s in the sequent calculus C is a finite tree of sequents such that (i) the root node is s; and (ii) each interior node and its direct children are the conclusion and premise(s) of a rule instance in C. A *proof* is a derivation where every leaf is an instance of an initial sequent.

In what follows, it should be clear from context whether the word "proof" refers to the object defined in Definition 2.2, or to the meta-level notion. We say that a sequent is *provable* in C if it has a proof in C. We elide the details of the encodings of sequent rules, collections of sequent rules and derivations as these can be found elsewhere [4]. For a sequent calculus C we define two predicates on sequents: `C_drv` for *derivability* in C, and `C_prv` for *provability* in C. Instances of these predicates are `GL4ip_prv`, `GL4ip_cut_prv` or `PSGL4ip_drv`. We note that our encodings primarily rely on the type `Type`, which bears computational content and is crucially compatible with the extraction function of Coq while `Prop` is not.

Definition 2.3 [Height] For any derivation δ, its *height* $h(\delta)$, is the maximum number of nodes on a path from root to leaf.

In this article we assume some familiarity with the notions of admissibility, invertibility, and height-preservation.

The sequent calculus GL4ip is given in Figure 1. When defining rules we put the label naming the rule on the left of the horizontal line, while the label appears on the right of the line in *instances* of rules.

$$(\bot\text{L})\ \frac{}{\bot, X \Rightarrow C} \qquad (\text{IdP})\ \frac{}{X, p \Rightarrow p}$$

$$(\wedge\text{L})\ \frac{X, A, B \Rightarrow C}{X, A \wedge B \Rightarrow C} \qquad (\wedge\text{R})\ \frac{X \Rightarrow A \qquad X \Rightarrow B}{X \Rightarrow A \wedge B}$$

$$(\vee\text{L})\ \frac{X, A \Rightarrow C \qquad X, B \Rightarrow C}{X, A \vee B \Rightarrow C} \qquad (\vee_i\text{R})\ \frac{X \Rightarrow A_i}{X \Rightarrow A_1 \vee A_2}\ (i \in \{1,2\})$$

$$(p{\to}\text{L})\ \frac{X, p, A \Rightarrow C}{X, p, p \to A \Rightarrow C} \qquad ({\to}\text{R})\ \frac{X, A \Rightarrow B}{X \Rightarrow A \to B}$$

$$(\Box{\to}\text{L})\ \frac{\boxtimes X, \Box A \Rightarrow A \qquad W, \Box X, B \Rightarrow C}{W, \Box X, \Box A \to B \Rightarrow C} \qquad (\text{GLR})\ \frac{\boxtimes X, \Box A \Rightarrow A}{W, \Box X \Rightarrow \Box A}$$

$$(\wedge{\to}\text{L})\ \frac{X, A \to (B \to C) \Rightarrow D}{X, (A \wedge B) \to C \Rightarrow D} \qquad (\vee{\to}\text{L})\ \frac{X, A \to C, B \to C \Rightarrow D}{X, (A \vee B) \to C \Rightarrow D}$$

$$({\to}{\to}\text{L})\ \frac{X, B \to C \Rightarrow A \to B \qquad X, C \Rightarrow D}{X, (A \to B) \to C \Rightarrow D}$$

Fig. 1. The sequent calculus GL4ip. Here, W does not contain any boxed formula.

In (IdP), a propositional variable instantiating the featured occurrences of p is principal. In a rule instance of ($\wedge$R), ($\wedge$L), ($\vee_i$R), ($\vee$L) or ($\to$R), the *principal formula* of that instance is defined as usual. In a rule instance of ($p\to$L), both a propositional variable instantiating p and the formula instantiating the featured $p \to A$ are principal formulae of that instance. In a rule instance of ($\wedge\to$L), ($\vee\to$L), ($\to\to$L) or ($\Box\to$L), the formula instantiating respectively $(A \wedge B) \to C$, $(A \vee B) \to C$, $(A \to B) \to C$ or $\Box A \to B$ is the principal formula of that instance. In a rule instance of (GLR), the formula $\Box A$ is called the *diagonal formula* [14].

Example 2.4 The following are examples of derivations in GL4ip. Note that while the first and second examples are derivations, the third is a proof.

$$p \Rightarrow q \to r \qquad \frac{\Rightarrow p}{\Rightarrow p \vee (p \to \bot)}\ (\vee_1\text{R}) \qquad \frac{\dfrac{}{\Box p, p, \Box p \Rightarrow p}\ (\text{IdP})}{\Box p \Rightarrow \Box p}\ (\text{GLR})$$

Example 2.5 A special example of a derivation in GL4ip is the following:

$$\frac{\Box A \to A, \Box(\Box A \to A), A, A, \Box A, \Box A, \Box A \Rightarrow A \qquad \Box(\Box A \to A), A, \Box A, \Box A \Rightarrow A}{\Box A \to A, \Box(\Box A \to A), A, \Box A, \Box A \Rightarrow A}\ (\Box{\to}\text{L})$$

The conclusion and left premise are identical modulo formula multiplicities, so the rule ($\Box\to$L) can be infinitely applied upwards on the left branch.

Finally, we consider the additive cut rule.

$$\frac{X \Rightarrow A \qquad A, X \Rightarrow C}{X \Rightarrow C}\ (\text{cut})$$

In the above, we call A the *cut-formula*. It is known that GL4ip+(cut) is sound and complete w.r.t. the Hilbert calculus iGL [19] as stated next.

Theorem 2.6 *For all A we have: $A \in$ iGL iff $\Rightarrow A$ is provable in* GL4ip+(cut).

3 A path to contraction for GL4ip

As mentioned above, our formalisation encodes sequents using lists and not multisets. Despite this distance between our formalisation and the pen-and-paper definition, list-sequents from the former mimic multiset-sequents from the latter. Below, `exch s se` encodes the fact that `se` is obtained from the sequent `s` by permuting two sub-lists in the list representing its antecedent.

Lemma 3.1 (Admissibility of exchange) *For all X_0, X_1, A, B and C, if $X_0, A, B, X_1 \Rightarrow C$ is provable in* GL4ip, *then so is $X_0, B, A, X_1 \Rightarrow C$.*

```
Lemma GL4ip_adm_exch : forall s, (GL4ip_prv s) ->
      (forall se, (exch s se) -> (GL4ip_prv se)).
```

Note that the admissibility of exchange is not an accident, nor is it hard-wired as an explicit rule in Coq. That is, our encoding of the multiset-based rules shown in Figure 1 is designed to entail exchange. For example, the conclusion $X, A \wedge B \Rightarrow C$ of the rule ($\wedge$L) rule is encoded as the list-sequent `(X0++(And A B)::X1, C)` which allows us to "slide" `(And A B)` to any point in the antecedent by appropriate choices of the lists `X0` and `X1`. The list-encoding requires a very pedantic analysis of the position of the occurrence of `(And A B)` in the antecedent of a rule instance. This is a major disadvantage of our approach: for example, the admissibility of exchange itself requires some 5000 lines of Coq code!

Given the above lemma, we allow ourselves to consider that the left-hand side of sequents is indeed a multiset. The remaining of this section extends the work of Dyckhoff and Negri [7] on G4ip to the sequent calculus GL4ip. Thus, the proofs they developed are embedded in our proofs and hence formalised. Most lemmata are proven by straightforward inductions on the structure of formulae or derivations, and the order in which we present them gives an account of the dependencies between them. We omit the Coq encodings for brevity.

Lemma 3.2 (Height-preserving admissibility of weakening) *For all X, A and C, if $X \Rightarrow C$ has a proof π in* GL4ip, *then $X, A \Rightarrow C$ has a proof π_0 in* GL4ip *such that $h(\pi_0) \leq h(\pi)$.*

Lemma 3.3 (Height-preserving invertibility of rules) *The rules $(\wedge R)$, $(\wedge L)$, $(\vee L)$, $(\rightarrow R)$, $(p \rightarrow L)$, $(\wedge \rightarrow L)$, $(\vee \rightarrow L)$ are height-preserving invertible.*

Lemma 3.4 *For all X and A, the sequent $A, X \Rightarrow A$ has a proof.*

We can show that the height-preserving invertibility of the rules ($\rightarrow\rightarrow$L) and ($\Box\rightarrow$L) holds for the right premise:

Lemma 3.5 (Height-preserving right-invertibility of rules) *For all X, A, B, D and C:*

(i) *If* $X, (A \to B) \to D \Rightarrow C$ *has a proof* π *in* GL4ip, *then* $X, D \Rightarrow C$ *has a proof* π_0 *in* GL4ip *such that* $h(\pi_0) \leq h(\pi)$.

(ii) *If* $X, \Box A \to B \Rightarrow C$ *has a proof* π *in* GL4ip, *then* $X, B \Rightarrow C$ *has a proof* π_0 *in* GL4ip *such that* $h(\pi_0) \leq h(\pi)$.

To obtain the key Lemma 3.7 for admissibility of contraction, pertaining to the rule ($\to\to$L), we need to show that the usual left-implication rule is admissible:

Lemma 3.6 *The rule* ($\to$L) *is admissible in* GL4ip:

$$\frac{X \Rightarrow A \qquad X, B \Rightarrow C}{X, A \to B \Rightarrow C}\ (\to L)$$

Lemma 3.7 *For all* X, A, B, D *and* C, *if* $X, (A \to B) \to D \Rightarrow C$ *is provable in* GL4ip, *then* $X, A, B \to D, B \to D \Rightarrow C$ *is provable in* GL4ip.

We finally obtain the admissibility of contraction for GL4ip:

Lemma 3.8 (Admissibility of contraction) *For all* X, A *and* C: *If* $A, A, X \Rightarrow C$ *is provable in* GL4ip, *then* $A, X \Rightarrow C$ *is provable in* GL4ip.

In the following section we introduce a second calculus PSGL4ip which embodies a terminating non-deterministic backward proof-search procedure for GL4ip. This will allow us to define the maximum height of derivations for a sequent with respect to this procedure. Later on this will constitute the secondary induction measure in the proof of admissibility of cut.

4 PSGL4ip: terminating backward proof-search

Given a sequent calculus C, one can define a backward proof-search procedure on C by imposing further constraints on the backward applicability of the rules of C. This procedure captures a subset of the set of all derivations of C, i.e. those which are built using the restricted version of the rules of C. Consequently, a backward proof-search procedure can be identified with the calculus PSC consisting of these restricted rules of C, under the condition that PSC allows to decide the provability of sequents in C.

We present such a sequent calculus for GL4ip. PSGL4ip restricts the rules of GL4ip in the following way.

(Ident) The rule (IdP) is replaced by the identity rule (Id) on formulae of any weight shown. Note that it is derivable in GL4ip as shown in Lemma 3.4.

$$\frac{}{A, X \Rightarrow A}\ (\text{Id})$$

(NoInit) The conclusion of no rule is permitted to be an instance of either (Id) or ($\bot$L).

Before commenting on the above, we note that it is straightforward to prove that GL4ip and PSGL4ip are equivalent in the following sense: a sequent is provable in one if it is provable in the other. So, according to the above general description, it suffices to prove that PSGL4ip can be used to decide the provability of sequents in GL4ip to show that the former deserves its prefix.

Conjointly, these restrictions aim at avoiding repetitions along a branch of a sequent which is either an identity or an instance of ($\bot$L), as in Example 2.5. Restriction (NoInit) disallows the destruction of a formula upwards in presence of a sequent which is obviously provable, while (Ident) allows to designate the latter as provable. In fact, by showing that no loop can appear in a branch of a PSGL4ip derivation, we concretely show that the only type of loop present in GL4ip are loops on provable sequents.

In the remainder of this section we proceed to show that no loop can exist in PSGL4ip. We do so by proving that each sequent has a derivation of maximum height in PSGL4ip. The existence of such derivations is ensured by the strict decreasing of a local measure on sequents upwards in the rules of PSGL4ip.

4.1 A well-founded order on $(\mathbb{N} \times \mathbb{N} \times \mathit{list}\,\mathbb{N})$

We define a well-founded order on triples $(n, m, l) \in (\mathbb{N} \times \mathbb{N} \times \mathit{list}\,\mathbb{N})$ where $\mathit{list}\,\mathbb{N}$ is the set of all lists of natural numbers.

In the following, we use $<$ to mean the usual ordering on natural numbers. Let us recall the general definition of a lexicographic order.

Definition 4.1 [Lexicographic order] Let $(A_1, <_1), \cdots, (A_n, <_n)$ be a collection of sets A_i with respective (strict total) orders $<_i$ on these sets. We define the lexicographic order $<_{lex}^{(A_1,<_1),\cdots,(A_n,<_n)}$ as follows. For two n-tuples $(a_1, \cdots, a_n)$ and $(a'_1, \cdots, a'_n)$ of the Cartesian product $A_1 \times \cdots \times A_n$, we write $(a_1, \cdots, a_n) <_{lex}^{(A_1,<_1),\cdots,(A_n,<_n)} (a'_1, \cdots, a'_n)$ if there is a $1 \leq j \leq n$ such that:

(i) $a_p = a'_p$, for all $1 \leq p < j$

(ii) $a_j <_j a'_j$

Note that if $<_i$ is a well-founded relation for all $1 \leq i \leq n$, then $<_{lex}^{(A_1,<_1),\ldots,(A_n,<_n)}$ is also well-founded [13]. If $(A_i, <_i) = (A_j, <_j)$ for all $1 \leq i, j \leq n$, then we note $(A_i, <_i)^n$ the sequence $(A_1, <_1), \ldots, (A_n, <_n)$. We define the shortlex order, also called *breadth-first* [11] or *length-lexicographic* order, over lists of natural numbers $\ll$:

Definition 4.2 [Shortlex order] The shortlex order over lists of natural numbers, noted $\ll$, is defined as follows. For two lists l_0 and l_1 of natural numbers, we say that $l_0 \ll l_1$ whenever one of the following conditions is satisfied:

(i) $\mathit{length}(l_0) < \mathit{length}(l_1)$;

(ii) $\mathit{length}(l_0) = \mathit{length}(l_1) = n$ and $l_0 <_{lex}^{(\mathbb{N},<)^n} l_1$;

Intuitively, the shortlex order is ordering lists according to their length and follows the lexicographic order whenever length does not discriminate.

Finally, we define the order $<^3$ on $(\mathbb{N} \times \mathbb{N} \times \mathit{list}\,\mathbb{N})$ as $<_{lex}^{(\mathbb{N},<),(\mathbb{N},<),(\mathit{list}(\mathbb{N}),\ll)}$. Given that $<$ and $\ll$ are well-founded orders, we get that $<^3$ also is.

4.2 A $(\mathbb{N} \times \mathbb{N} \times \mathit{list}\,\mathbb{N})$-measure on sequents

In what follows we use the term "measure" in an informal way. We proceed to attach to each sequent $X \Rightarrow C$ a measure $\Theta(X \Rightarrow C)$ which is a triple

$(\alpha(X \Rightarrow C), \beta(X \Rightarrow C), \gamma(X \Rightarrow C)) \in (\mathbb{N} \times \mathbb{N} \times \mathit{list}\,\mathbb{N})$. For simplicity, in the following paragraphs we consider a fixed sequent $X \Rightarrow C$ for which we define the triple, and thus erase the mention of the sequent in the measures.

First, we focus on γ. As $X \Rightarrow C$ is built from a finite multiset of formulae, it contains a *topmost* formula of maximal weight. Let n be that maximal weight. We can create a list of length n such that at each position m in the list (counting from right to left) for $1 \leq m \leq n$, we find the number of occurrences in $X \Rightarrow C$ of *topmost* formulae of weight m. Such a list gives the count of occurrences in $X \Rightarrow C$ of formulae of weight n in its leftmost (i.e. n-th) component, then of occurrences of formulae of weight $n-1$ in the next (i.e. $(n-1)$-th) component, and so on until we reach 1. We define γ to be this unique list. For example, $\gamma(p \wedge q, p \vee q \Rightarrow q \to p)$ is the list $[1, 2, 0, 0]$ because $p \wedge q$ is the formula of maximum weight 4, and it is the only formula with this weight occurring in the list, while both $p \vee q$ and $q \to p$ are of weight 3. Two things needs to be noted about such lists. First, if no topmost occurrence of a formula is of weight $1 \leq k \leq n$, then a 0 appears in position k in the list. This is the case for the weight 2 in the example. Second, as in general no formula is of weight 0 we do not need to dedicate a position for this particular weight in our list.

Why do we need such a list? With this list, the shortlex order becomes an adequate substitute to the Dershowitz-Manna order [5] considered in Dyckhoff's work on G4ip. We recall this order, given two multisets Γ_0 and Γ_1, by quoting van der Giessen and Iemhoff [19]: "$\Gamma_0 \ll \Gamma_1$ if and only if Γ_0 is the result of replacing one or more formulas in Γ_1 by zero or more formulas of lower degree". As our use of the symbol $\ll$ for the shortlex order suggests, the shortlex order can replace the order given above to order finite multisets of formulae.

A similar list was independently formalised in Coq by Daniel Schepler in the study of the calculus G4ip which he calls LJT [15], following Dyckhoff. However, he does not involve this list in a termination argument: instead, he uses it to show the equivalence of G4ip and the usual natural deduction system for intuitionistic logic.

Second, we turn to β. On the contrary to the measure defined by Bílková [1] and used by van der Giessen and Iemhoff, which attributes a natural number to a sequent *appearing in a proof-search tree which depends on the root*, we use a *local* notion of "number of usable boxes" as done by Goré et al. [9].

Definition 4.3 We define:

(i) the *usable boxes* $ub(X \Rightarrow C)$ of $X \Rightarrow C$ as:

$$ub(X \Rightarrow C) := \{\Box A \mid \Box A \in \mathrm{Sub}(X \cup C)\} \setminus \{\Box A \mid \Box A \in X\}$$

(ii) the number $\beta(X \Rightarrow C)$ of usable boxes of $X \Rightarrow C$ as $\beta(X \Rightarrow C) = \mathrm{Card}(ub(X \Rightarrow C))$, where $\mathrm{Card}(U)$ is the cardinality of the set U.

Thus, the notion of usable boxes of $X \Rightarrow C$ is the set of boxed subformulae of $X \Rightarrow C$ minus the topmost boxed formulae in X. Intuitively, this notion captures the set of boxed formulae of a sequent s which might be the diagonal formula of an instance of (GLR) in a derivation of s in PSGL4ip.

Third, we finally consider α. As X is a finite multiset of formulae, the checking of whether or not $X \Rightarrow C$ is an instance of the rule (Id) or ($\bot$L) is decidable. So, we can constructively define the following test function:

$$\alpha(X \Rightarrow C) = \begin{cases} 0 & \text{if } X \Rightarrow C \text{ is an instance of (Id) or } (\bot\text{L}) \\ 1 & \text{otherwise} \end{cases}$$

4.3 Every rule of PSGL4ip reduces Θ upwards

We proceed to prove that the measure Θ decreases upwards through the rules of PSGL4ip on the $<^3$ ordering.

Lemma 4.4 *For all sequents $s_0, s_1, ..., s_n$ and for all $1 \leq i \leq n$, if there is an instance of a rule r of* PSGL4ip *of the form below, then $\Theta(s_i) <^3 \Theta(s_0)$:*

$$\frac{s_1 \quad \ldots \quad s_n}{s_0}\ r$$

Note that contraction and weakening as rules allow Θ to *increase* upwards. While it is rather obvious for contraction, this statement for weakening is surprising. The key point here is to note that weakening allows the deletion of boxed formulae in the antecedent of sequents, leading to a potential increase in the number of usable boxes β: that is, weakening may remove some of the boxes that "block" some applications of (GLR) upwards and so the number of usable boxes increases.

4.4 The existence of a derivation of maximum height

For convenience, we define the order $\lessdot 3$ on sequents as follows:

$$s_0 \lessdot 3\, s_1 \text{ if and only if } \Theta(s_0) <^3 \Theta(s_1)$$

As $<^3$ is a well-founded order, it is obvious that $\lessdot 3$ is so as well. As a consequence we obtain a strong induction principle following the $\lessdot 3$ order.

Theorem 4.5 *For any property P on sequents, to prove the statement $\forall s P(s)$ it is sufficient to show that every sequent s_0 satisfies P under the assumption that all its $\lessdot 3$-predecessors satisfy P.*

```
Theorem less_than3_strong_inductionT:
forall (P : Seq -> Type),
(forall s0, (forall s1, ((s1 <3 s0) -> P s1)) -> P s0)
-> forall s, P s.
```

If we use this principle with the previous Lemma 4.4, we can easily prove the existence of a derivation in PSGL4ip of maximum height for all sequents.

Theorem 4.6 *Every sequent s has a* PSGL4ip *derivation of maximum height.*

```
Theorem PSGL4ip_termin :
 forall s, existsT2 (D: PSGL4ip_drv s), (is_mhd D).
```

Here, `D` is a *derivation*, the existence of which is guaranteed by the constructive existential quantifier **existsT2**. This quantifier not only requires us to construct a witnessing term but also to provide a proof that the witness is of the correct type. The function `is_mhd` returns the constructive Coq proposition `True` if and only if its argument, `D`, is a derivation of maximum height.

As the previous lemma implies the *constructive* existence of a derivation δ of maximum height in PSGL4ip for any sequent s, we are entitled to let mhd(s) denote the height of δ. As in the work of Goré et al. [9], we later use mhd(s) as the secondary induction measure used in the proof of admissibility of cut.

Before proving the only property we need from mhd(s), let us interpret the previous lemma from the point of view of the proof-search procedure underlying PSGL4ip. The existence of a derivation of maximum height for each sequent in PSGL4ip shows that in the backward application of rules of PSGL4ip on a sequent, i.e. the expansion of branches rooted in this sequent, a halting point has to be encountered. As a consequence, the expansion of every branch must meet a halting point: the proof-search procedure *terminates*.

While this is the essence of the content of the previous lemma, we effectively only use the fact that mhd(s) decreases upwards in the rules of PSGL4ip.

Lemma 4.7 *If r is a rule instance from* PSGL4ip *with conclusion s_0 and s_1 as one of the premises, then $mhd(s_1) < mhd(s_0)$.*

5 Cut-elimination for GL4ip

To reach cut-elimination, our main theorem, we first state and prove cut-admissibility in a purely syntactic way. More precisely, we proceed to prove that the *additive*-cut rule is admissible. The latter statement, stating that the provability of the sequents $X \Rightarrow A$ and $X, A \Rightarrow C$ entails the provability of $X \Rightarrow C$, is formalised in Coq in the following way:

```
Theorem GL4ip_cut_adm : forall A X0 X1 C,
 (GL4ip_prv (X0++X1,A)  *  GL4ip_prv (X0++A::X1,C)) ->
              GL4ip_prv (X0++X1,C).
```

Here, the term `(X0++X1,A)` encodes the sequent $X_0, X_1 \Rightarrow A$ as a pair, thus hiding a lower level occurrence of `*`. Then, given that `GL4ip_prv s` is in `Type` and not in `Prop`, we are required to use the constructor `*` for pairs at the higher level shown instead of `/\` which is the usual conjunction in `Prop`. So, the existence of *proofs* in GL4ip for the sequent `(X0++X1,A)` as well as for the sequent `(X0++A::X1,C)` asserted in the second line entail the existence of a *proof* in GL4ip for the sequent `(X0++X1,C)`. It is now clear that this statement formalises the following theorem:

Theorem 5.1 *The additive cut rule is admissible in* GL4ip.

Proof. Let d_1 (with last rule r_1) and d_2 (with last rule r_2) be proofs in GL4ip of $X \Rightarrow A$ and $A, X \Rightarrow C$ respectively, as shown below.

$$\dfrac{d_1}{X \Rightarrow A}\, r_1 \qquad \dfrac{d_2}{A, X \Rightarrow C}\, r_2$$

It suffices to show that there is a proof in GL4ip of $X \Rightarrow C$. We reason by strong primary induction (PI) on the weight of the cut-formula A, giving the primary inductive hypothesis (PIH). We also use a strong secondary induction (SI) on mhd of the conclusion of a cut, giving the secondary inductive hypothesis (SIH).

We make a first case distinction: does $X \Rightarrow C$ violate (NoInit)? If it is the case, then this sequent is an instance of (Id) or ($\bot$L). So, we use Lemma 3.4 or apply ($\bot$L) to obtain a proof of $X \Rightarrow C$. If $X \Rightarrow C$ satisfies (NoInit), then it is not an instance of (Id) or ($\bot$L). In this case we consider r_1. In total, there are thirteen cases to consider for r_1: one for each rule in GL4ip. However, we can gather some of the cases together and reduce the number of cases to eight. We separate them by using Roman numerals and showcase the most interesting ones.

(I) $\mathbf{r_1}$ =($\to$R) : Then r_1 has the following form where $A = B \to D$:

$$\dfrac{B, X \Rightarrow D}{X \Rightarrow B \to D}\ (\to\text{R})$$

We consider one sub-case.

(I-a) If r_2 is ($\to\to$L) where the cut formula is not principal in r_2, then it must have the following form where $(E \to F) \to G, X_0 = X$:

$$\dfrac{B \to D, F \to G, X_0 \Rightarrow E \to F \qquad B \to D, G, X_0 \Rightarrow C}{B \to D, (E \to F) \to G, X_0 \Rightarrow C}\ (\to\to\text{L})$$

Thus, we have that the sequents $X \Rightarrow C$ and $X \Rightarrow B \to D$ are respectively of the form $(E \to F) \to G, X_0 \Rightarrow C$ and $(E \to F) \to G, X_0 \Rightarrow B \to D$. Using the right-invertibility of ($\to\to$L), proven in Lemma 3.5, on $(E \to F) \to G, X_0 \Rightarrow B \to D$ we obtain a proof of the sequent $G, X_0 \Rightarrow B \to D$. Then, we make a case distinction on whether the sequent $F \to G, X_0 \Rightarrow E \to F$ is an instance of (Id) or ($\bot$L). If it is the case, then we proceed as follows:

$$\dfrac{F \to G, X_0 \Rightarrow E \to F \qquad \dfrac{G, X_0 \Rightarrow B \to D \qquad B \to D, G, X_0 \Rightarrow C}{G, X_0 \Rightarrow C}\ \text{SIH}}{(E \to F) \to G, X_0 \Rightarrow C}\ (\to\to\text{L})$$

Here the left branch is obviously provable either by invoking Lemma 3.4 or by applying ($\bot$L). If $F \to G, X_0 \Rightarrow E \to F$ is not an instance of these rules, then consider the following proof of this sequent, where Lemma 3.7 deconstructs the implication $(E \to F) \to G$, Lemma 3.8 contracts $F \to G$ and Lemma 3.3 is the invertibility of the rule ($\to$R).

$$\dfrac{\dfrac{\dfrac{\dfrac{(E \to F) \to G, X_0 \Rightarrow B \to D}{E, F \to G, F \to G, X_0 \Rightarrow B \to D}\ \text{Lem.3.7}}{E, F \to G, X_0 \Rightarrow B \to D}\ \text{Lem.3.8} \qquad \dfrac{B \to D, F \to G, X_0 \Rightarrow E \to F}{B \to D, E, F \to G, X_0 \Rightarrow F}\ \text{Lem.3.3}}{E, F \to G, X_0 \Rightarrow F}\ \text{SIH}}{F \to G, X_0 \Rightarrow E \to F}\ (\to\text{R})$$

The crucial point here is to see that the use of SIH is justified, i.e. that $\text{mhd}(E, F \to G, X_0 \Rightarrow F) < \text{mhd}((E \to F) \to G, X_0 \Rightarrow C)$. This is the case as we made sure that the rule applications ($\to\to$L) and ($\to$R) are both instances of rules of PSGL4ip because their respective conclusions $(E \to F) \to G, X_0 \Rightarrow C$ and $F \to G, X_0 \Rightarrow E \to F$ are not instances of (Id) or ($\bot$L). So, we get that $\text{mhd}(E, F \to G, X_0 \Rightarrow F) < \text{mhd}(F \to G, X_0 \Rightarrow E \to F) < \text{mhd}((E \to F) \to$

$G, X_0 \Rightarrow C)$ by Lemma 4.7 hence $\text{mhd}(E, F \to G, X_0 \Rightarrow F) < \text{mhd}((E \to F) \to G, X_0 \Rightarrow C)$ by transitivity of $<$. So, we are done. Note that the created cut could not be justified by usual induction on height, as Lemma 3.7 is not height-preserving.

(II) $\mathbf{r_1}$ =(GLR): Then A is the diagonal formula in r_1:

$$\dfrac{\boxtimes X_0, \Box B \Rightarrow B}{W, \Box X_0 \Rightarrow \Box B}\ (\text{GLR})$$

where $A = \Box B$ and $W, \Box X_0 = X$. Thus, we have that the sequents $X \Rightarrow C$ and $A, X \Rightarrow C$ are respectively of the form $W, \Box X_0 \Rightarrow C$ and $\Box B, W, \Box X_0 \Rightarrow C$. We now consider one case for r_2.

(II-a) If r_2 is $(\Box\!\to\!\text{L})$. Then r_2 is of the following form and where $\Box D \to E, W_0 = W$:

$$\dfrac{B, \Box B, \boxtimes X_0, \Box D \Rightarrow D \qquad E, W_0, \Box B, \Box X_0 \Rightarrow C}{\Box D \to E, W_0, \Box B, \Box X_0 \Rightarrow C}\ (\Box\!\to\!\text{L})$$

We proceed as follows.

$$\dfrac{\begin{matrix}\pi \\ \boxtimes X_0, \Box D \Rightarrow D\end{matrix} \qquad \dfrac{\dfrac{\dfrac{\boxtimes X_0, \Box B \Rightarrow B}{\Box X_0 \Rightarrow \Box B}\ (\text{GLR})}{E, W_0, \Box X_0 \Rightarrow \Box B}\ \text{Lem.3.2} \qquad E, W_0, \Box B, \Box X_0 \Rightarrow C}{E, W_0, \Box X_0 \Rightarrow C}\ \text{SIH}}{\Box D \to E, W_0, \Box X_0 \Rightarrow C}\ (\Box\!\to\!\text{L})$$

where π is:

$$\dfrac{\dfrac{\dfrac{\boxtimes X_0, \Box B \Rightarrow B}{\Box X_0 \Rightarrow \Box B}\ (\text{GLR})}{\boxtimes X_0, \Box D \Rightarrow \Box B}\ \text{Lem.3.2} \qquad \dfrac{\dfrac{\boxtimes X_0, \Box B \Rightarrow B}{\boxtimes X_0, \Box B, \Box D \Rightarrow B}\ \text{Lem.3.2} \qquad B, \Box B, \boxtimes X_0, \Box D \Rightarrow D}{\boxtimes X_0, \Box B, \Box D \Rightarrow D}\ \text{PIH}}{\boxtimes X_0, \Box D \Rightarrow D}\ \text{SIH}$$

Note that both uses of SIH are justified here as the assumption (NoInit) ensures that the last rule in this proof is effectively an instance of $(\Box\!\to\!\text{L})$ in PSGL4ip, hence $\text{mhd}(\boxtimes X_0, \Box D \Rightarrow D) < \text{mhd}(\Box D \to E, W_0, \Box X_0 \Rightarrow C)$ and $\text{mhd}(E, W_0, \Box X_0 \Rightarrow C) < \text{mhd}(\Box D \to E, W_0, \Box X_0 \Rightarrow C)$ by Lemma 4.7. Q.E.D.

Before turning to cut-elimination let us comment on the need to use additive cuts in the previous proof. To justify a cut through SIH, we need to link the sequent-conclusion of the initial cut to the sequent-conclusion of the newly created cut by a chain of rule applications which make mhd decrease upwards. Now, contraction and weakening can increase mhd upwards. So, in the mhd technique we cannot use contraction or weakening in the chain linking the two sequent-conclusion, forbidding us from considering multiplicative cuts. The use of additive cuts allows us to circumvent this difficulty. This sensitivity of the proof technique is surprising as both calculi admit weakening and contraction, making additive and multiplicative cuts equivalent.

It is commonly accepted that a purely syntactic proof of cut-admissibility provides a cut-elimination procedure: eliminate topmost cuts first. So, the above proof theoretically establishes that cuts are eliminable in the calculus

GL4ip extended with (cut). To effectively prove this statement in Coq we explicitly encode the additive cut rule as follows:

$$\frac{\texttt{(X0++X1, A)} \qquad \texttt{(X0++A::X1, C)}}{\texttt{(X0++X1, C)}}$$

With this rule in hand, we can encode the set of rules `GL4ip_cut_rules` as `GL4ip_rules` enhanced with (cut), i.e. the calculus GL4ip + (cut). We can finally turn to the elimination of additive cuts:

Theorem 5.2 *The additive cut rule is eliminable from* GL4ip + (*cut*).

```
Theorem GL4ip_cut_elimination : forall s,
 (GL4ip_cut_prv s) -> (GL4ip_prv s).
```

The above theorem shows that given a proof in GL4ip + (cut) of a sequent, i.e. `GL4ip_cut_prv s`, we can transform this proof directly to obtain a proof in GL4ip of the same sequent. Given that this theorem is in fact a constructive function based on elements defined on `Type`, we can use the extraction feature of Coq and obtain a cut-eliminating Haskell program.

6 Discussion

The *mhd proof technique* for cut-admissibility, based on terminating backward proof-search, was recently discovered by Brighton [3] and successfully applied to the provability logic GL [2,16] by Goré et al. [9]. The novelty of this technique consists in the binary induction measure it relies on: while the first component is the traditional "size of the cut formula", the second is the intriguing "maximum height of derivations". The latter is defined using a terminating backward proof-search procedure which allows to exhibit for a given sequent a derivation of maximum height, hence bounding the height of all the possible derivations of this sequent. The mhd technique is interesting for four reasons.

First, as shown by Goré et al. [9], the mhd technique gives simpler proofs in difficult cases such as GL and we do not need Valentini's extra measure of width but can utilise only two measures. This advantage carries over to iGL.

Second, it reverses the usual order of cut-admissibility and termination of backward proof-search. Indeed, we usually prove that cut is admissible, then design a proof-search procedure on the cut-free system and show its termination. This oddity is promising for a general treatment of cut admissibility via local transformations for calculi with a terminating backward proof-search.

Third, it is sensitive to the type of cut admitted. More precisely, this technique seems applicable only to *additive* cuts, in cases where weakening and contraction are admissible in the calculus. Intuitively, the mhd technique involves the backward application of rules on the conclusion of the initial cut. For termination, it must exclude (backward applications of) contraction and weakening as both can increase the termination measure upwards. But banishing these also banishes the use of multiplicative cuts of the form below:

$$\frac{X_0 \Rightarrow A \quad X_1 \Rightarrow C}{X_0, X_1 \Rightarrow C}$$

Fourth, many sequent calculi for non-classical logics enjoy terminating backward proof-search, and often, they are based upon G4ip. Is there a general theory of cut-admissibility hidden inside the mhd method for these calculi?

7 Conclusion

In the conclusion of a previous work [9], we hinted at the interest of using mhd as an induction measure to prove the admissibility of cut for a sequent calculus for intuitionistic GL based on Dyckhoff's terminating calculus G4ip. Here, we ventured down this alley and obtained a cut-admissibility result for GL4ip relying on the termination of backward proof-search. More than an alternative proof technique, the use of mhd in the case of GL4ip is to date the only known pathway to a direct proof of admissibility of cut: as admitted by van der Giessen and Iemhoff [19], all other available proof techniques fail.

So, in addition to using a local measure for proving termination of proof-search instead of Bíková's non-local measure [1], and formalising on the way most of Dyckhoff and Negri's results on G4ip, we consequently addressed van der Giessen and Iemhoff's issue by providing a formalised direct proof of cut-admissibility for GL4ip. Crucially, this direct syntactic proof allows to obtain an extractable simple cut-elimination procedure for GL4ip hardly obtainable from the indirection in van der Giessen and Iemhoff's work.

8 Further work

While the use of the termination of a backward proof-search procedure as a basis for cut-elimination is an intriguing and unconventional argument, it seems to have limitations. The calculi GLS, G4ip and GL4ip either contain no cycles or only contain *provable* cycles, i.e. cycles going through a provable sequent. Thus, the proof-search on these calculi only need to get rid of provable cycles. This is done by imposing restrictions on the application of rules which, when violated, entail the provability of the sequent under consideration. For example, if a sequent violates the restrictions of the PSGL4ip calculus, then we know that either it is an instance of ($\bot$L) or (Id), which entails its provability. So, for every rule application of GL4ip we have the crucial case distinction, which we make use of in the admissibility of cut: either it is an instance of PSGL4ip, which makes mhd decrease, or its conclusion is obviously provable. Now, if we face a calculus containing unprovable cycles, such as the standard ones for modal logic K4 or S4, then a terminating proof-search on this calculus need to involve restrictions which, when violated, do not entail the provability of the sequent violating them. Then, the case distinction mentioned above does not give us much when the sequent violates the restrictions of the proof-search: its provability is not obvious. We are currently investigating further adaptations of the technique to sequent calculi with unprovable cycles.

The Haskell program extractable from our formalisation should effectively eliminate cuts from GL4ip+(cut) proofs, as ensured from the extraction feature of Coq. However, we have neither tested it nor tried to optimize it. We intend to follow D'Abrera et al. [4] by exploring both of these alleys in future works.

References

[1] Bílková, M., "Interpolation in modal logics," Ph.D. thesis, Univerzita Karlova, Prague (2006).

[2] Boolos, G., "The Unprovability of Consistency: An Essay in Modal Logic," Cambridge University Press, 1979.

[3] Brighton, J., *Cut Elimination for GLS Using the Terminability of its Regress Process*, Journal of Philosophical Logic **45** (2016), pp. 147–153.

[4] D'Abrera, C., J. E. Dawson and R. Goré, *A formally verified cut-elimination procedure for linear nested sequents for tense logic*, in: A. Das and S. Negri, editors, *Automated Reasoning with Analytic Tableaux and Related Methods - 30th International Conference, TABLEAUX 2021, Birmingham, UK, September 6-9, 2021, Proceedings*, Lecture Notes in Computer Science **12842** (2021), pp. 281–298.
URL https://doi.org/10.1007/978-⊠3-⊠030-⊠86059-⊠2_17

[5] Dershowitz, N. and Z. Manna, *Proving termination with multiset orderings*, Commun. ACM **22** (1979), p. 465–476.
URL https://doi.org/10.1145/359138.359142

[6] Dyckhoff, R., *Contraction-free sequent calculi for intuitionistic logic*, The Journal of Symbolic Logic **57** (1992), pp. 795–807.
URL http://www.jstor.org/stable/2275431

[7] Dyckhoff, R. and S. Negri, *Admissibility of structural rules for contraction-free systems of intuitionistic logic*, The Journal of Symbolic Logic **65** (2000), pp. 1499–1518.
URL http://www.jstor.org/stable/2695061

[8] Goré, R. and R. Ramanayake, *Valentini's cut-elimination for provability logic resolved*, Rev. Symb. Log. **5** (2012), pp. 212–238.
URL https://doi.org/10.1017/S1755020311000323

[9] Goré, R., R. Ramanayake and I. Shillito, *Cut-Elimination for Provability Logic by Terminating Proof-Search: Formalised and Deconstructed Using Coq*, in: A. Das and S. Negri, editors, *Automated Reasoning with Analytic Tableaux and Related Methods - 30th International Conference, TABLEAUX 2021, Birmingham, UK, September 6-9, 2021, Proceedings*, Lecture Notes in Computer Science **12842** (2021), pp. 299–313.
URL https://doi.org/10.1007/978-⊠3-⊠030-⊠86059-⊠2_18

[10] Hudelmaier, J., *An O(n log n)-Space Decision Procedure for Intuitionistic Propositional Logic*, Journal of Logic and Computation **3** (1993), pp. 63–75.
URL https://doi.org/10.1093/logcom/3.1.63

[11] Larchey-Wendling, D. and R. Matthes, *Certification of breadth-first algorithms by extraction*, in: G. Hutton, editor, *Mathematics of Program Construction - 13th International Conference, MPC 2019, Porto, Portugal, October 7-9, 2019, Proceedings*, Lecture Notes in Computer Science **11825** (2019), pp. 45–75.
URL https://doi.org/10.1007/978-⊠3-⊠030-⊠33636-⊠3_3

[12] Leivant, D. M., "Absoluteness of intuitionistic logic," Ph.D. thesis, University of Amsterdam, Amsterdam (1975).

[13] Paulson, L. C., *Constructing recursion operators in intuitionistic type theory*, Journal of Symbolic Computation **2** (1986), pp. 325–355.
URL https://www.sciencedirect.com/science/article/pii/S0747717186800025

[14] Sambin, G. and S. Valentini, *The modal logic of provability: the sequential approach*, Journal of Philosophical Logic **11** (1982), p. 311–342.

[15] Schepler, D., *coq-sequent-calculus*, https://github.com/dschepler/coq-⊠sequent-⊠calculus/blob/master/LJTStar.v (2016).

[16] Solovay, R., *Provability interpretations of modal logic*, Israel Journal of Mathematics **25** (1976), pp. 287–304.

[17] Troelstra, A. S. and H. Schwichtenberg, "Basic Proof Theory," Cambridge Tracts in Theoretical Computer Science, Cambridge University Press, 2000, 2 edition.

[18] Valentini, S., *The modal logic of provability: Cut-elimination*, Journal of Philosophical Logic **12** (1983), p. 471–476.

[19] van der Giessen, I. and R. Iemhoff, *Sequent Calculi for Intuitionistic Gödel-Löb Logic*, Notre Dame Journal of Formal Logic **62** (2021), pp. 221 – 246.
URL https://doi.org/10.1215/00294527-⊠2021-⊠0011

Appendix

Proof. [of Lemma 4.4] We reason by case analysis on r:

(i) If r is (Id) or ($\bot$L), then we are done as there is no premise.

(ii) If r is ($\wedge$R), ($\wedge$L), ($\vee_1$R), ($\vee_2$R), ($\vee$L), ($\to$R), ($p\to$L), ($\wedge\to$L), ($\vee\to$L) or ($\to\to$L), then we have that $\gamma(s_0) \ll \gamma(s_1)$ and $\gamma(s_0) \ll \gamma(s_2)$ (if it exists), as shown by Dyckhoff and Negri [7]. It has to be noted that the use of the different weight for the conjunction is crucial for the case where r is the rule ($\wedge\to$L). Obviously, α can only decrease upwards in these rules, as no rule of PSGL4ip with premises can be applied to an initial sequent. Also, it is not hard to convince oneself that the number of usable boxes can only decrease in these rules as the boxed formulae on the left of the sequent are preserved upwards and the set of boxed subformulae is either stable or loses elements. So we can easily deduce that Θ decreases on $<^3$ from the conclusion to the premises of these rules.

(iii) If r is (GLR) then it must have the following form.

$$\frac{\boxtimes X, \Box B \Rightarrow B}{W, \Box X \Rightarrow \Box B}\ (\text{GLR})$$

Clearly, we have that $\{\Box A \mid \Box A \in \text{Sub}(\boxtimes X \cup \{\Box B\} \cup \{B\})\} \subseteq \{\Box A \mid \Box A \in \text{Sub}(W \cup \Box X \cup \{\Box B\})\}$. Also, given that we consider a derivation in PSGL4ip, we can note that (Id) is not applicable on $W, \Box X \Rightarrow \Box B$ by assumption, hence $\Box B \notin \Box X$. Consequently, we get $\{\Box A \mid \Box A \in W \cup \Box X\} \subset \{\Box A \mid \Box A \in \boxtimes X \cup \{\Box B\}\}$. An easy set-theoretic argument leads to $ub(\boxtimes X, \Box B \Rightarrow B) \subset ub(W, \Box X \Rightarrow \Box B)$. As a consequence we obtain $\beta(\boxtimes X, \Box B \Rightarrow B) < \beta(W, \Box X \Rightarrow \Box B)$, hence $\Theta(\boxtimes X, \Box B \Rightarrow B) <^3 \Theta(W, \Box X \Rightarrow \Box B)$.

(iv) If r is ($\Box\to$L) then it must have the following form.

$$\frac{\boxtimes X, \Box A \Rightarrow A \qquad W, \Box X, B \Rightarrow C}{W, \Box X, \Box A \to B \Rightarrow C}\ (\Box\to\text{L})$$

For the right premise we can straightforwardly see that $\gamma(W, \Box X, B \Rightarrow C) \ll \gamma(W, \Box X, \Box A \to B \Rightarrow C)$, and that both α and β either are stable or decrease upwards. So, we obtain $\Theta(W, \Box X, B \Rightarrow C) <^3 \Theta(W, \Box X, \Box A \to B \Rightarrow C)$. The case of the left premise is more complex but can be treated similarly to the (GLR) as follows. Note that $\{\Box D \mid \Box D \in \text{Sub}(\boxtimes X \cup \{\Box A\} \cup \{A\})\} \subseteq \{\Box D \mid \Box D \in \text{Sub}(W \cup \Box X \cup \{\Box A \to B\} \cup \{C\})\}$. We consider two cases.

In the first case, we have that $\Box A \notin \Box X$. Then as in (GLR) we obtain $\{\Box D \mid \Box D \in W \cup \Box X \cup \{\Box A \to B\}\} \subset \{\Box D \mid \Box D \in \boxtimes X \cup \{\Box A\}\}$ and consequently $\beta(\boxtimes X, \Box A \Rightarrow A) < \beta(W, \Box X, \Box A \to B \Rightarrow C)$. So, regardless of the value of $\alpha(\boxtimes X, \Box A \Rightarrow A)$, we obtain $\Theta(\boxtimes X, \Box A \Rightarrow A) <^3$

$\Theta(W, \Box A, \Box X, \Box A \to B \Rightarrow C)$.

In the second case, we have that $\Box A \in \Box X$. Then the rule application is of the following form:

$$\frac{\boxtimes X, \Box A, A, \Box A \Rightarrow A \qquad W, \Box A, \Box X, B \Rightarrow C}{W, \Box A, \Box X, \Box A \to B \Rightarrow C}\ (\Box{\to}\mathrm{L})$$

Clearly, we get $\alpha(\boxtimes X, \Box A, A, \Box A \Rightarrow A) = 0$ as it is an instance of an initial sequent, hence $\alpha(\boxtimes X, \Box A, A, \Box A \Rightarrow A) < \alpha(W, \Box A, \Box X, \Box A \to B \Rightarrow C)$. Consequently, we get $\Theta(\boxtimes X, \Box A, A, \Box A \Rightarrow A) <^3 \Theta(W, \Box A, \Box X, \Box A \to B \Rightarrow C)$.

Q.E.D.

Proof. [of Theorem 4.6] We use `less_than3_strong_inductionT`, the strong induction principle on $\mathrel{<\!3}$ from Theorem 4.5. As the applicability of the rules of PSGL4ip is decidable, we distinguish two cases:

(I) No PSGL4ip rule is applicable to s. Then the derivation of maximum height sought after is simply the derivation constituted of s solely, which is the only derivation for s.

(II) Some PSGL4ip rule is applicable to s. Either only initial rules are applicable, in which case the derivation of maximum height sought after is simply the derivation of height 1 constituted of the application of the applicable initial rule to s. Or, some other rules than the initial rules are applicable. Then consider the finite list $Prems(s)$ of all sequents s_{prem} such that there is an application of a PSGL4ip rule r with s as conclusion of r and s_{prem} as premise of r. Note that this list is effectively computable, as shown by the lemma `finite_premises_of_S` in our formalisation. By Lemma 4.4 we know that every element s_0 in the list $Prem(s)$ is such that $s_{prem} \mathrel{<\!3} s$. Consequently, the strong induction hypothesis allows us to consider the derivation of maximum height of all the sequents in $Prem(s)$. As $Prem(s)$ is finite, there must be an element s_{max} of $Prem(s)$ such that its derivation of maximum height is higher or of same height than the derivation of maximum height of all sequents in $Prem(s)$. It thus suffices to pick that s_{max}, use its derivation of maximum height, and apply the appropriate rule to obtain s as a conclusion: this is by choice the derivation of maximum height of s. Q.E.D.

Proof. [of Lemma 4.7] As $<$ and $=$ are decidable relations over natural numbers, we can reason by contradiction. So, suppose that $\mathrm{mhd}(s_1) \geq \mathrm{mhd}(s_0)$. Let δ_0 be the derivation of s_0 of maximal height and let δ_1 be the derivation of s_1 of maximal height as guaranteed by Theorem 4.6. If r is a rule instance from PSGL4ip with s_1 as one of the premises and with conclusion s_0, then δ_2 as shown below is a derivation of s_0 of height greater than $\mathrm{mhd}(s_1) + 1$:

$$\frac{\begin{array}{c}\delta_1\\[4pt] s_1\end{array} \qquad \cdots}{s_0}\ r$$

The maximality of δ_0 implies that the height of δ_0 is greater than the height of δ_2: thus $\text{mhd}(s_1) + 1 \leq \text{mhd}(s_0)$. As our initial assumption implies that $\text{mhd}(s_1) + 1 > \text{mhd}(s_0)$, we reached a contradiction. Q.E.D.

Proof. [of Theorem 5.1] As in the partial proof given in the main body of the article, we need to show the existence of a proof in GL4ip of $X \Rightarrow C$ while being given GL4ip proofs d_1 (with last rule r_1) and d_2 (with last rule r_2) of $X \Rightarrow A$ and $A, X \Rightarrow C$. Here again, we use the primary and secondary inductive hypothesis PIH and SIH.

We make a first case distinction: does $X \Rightarrow C$ violate (NoInit)? If it is the case, then this sequent is an instance of (Id) or ($\bot$L). So, we use Lemma 3.4 or apply ($\bot$L) to obtain a proof of $X \Rightarrow C$. If $X \Rightarrow C$ satisfies (NoInit), then it is not an instance of (Id) or ($\bot$L). In this case we consider r_1. In total, there are thirteen cases to consider for r_1: one for each rule in GL4ip. However, we can gather some of the cases together and reduce the number of cases to eight. We separate them by using Roman numerals.

(I) $\mathbf{r_1}$ =(IdP) : then we have that $A = p$. Consequently, $X \Rightarrow C$ is of the form $X_0, p \Rightarrow C$. Also, the conclusion of r_2 is of the form $X_0, p, p \Rightarrow C$. We can apply the contraction Lemma 3.8 to obtain a proof of $X_0, p \Rightarrow C$.

(II) $\mathbf{r_1}$ =($\bot$L): Then r_1 must have the following form.

$$\frac{}{X_0, \bot \Rightarrow A}\,(\bot\text{L})$$

where $X_0, \bot = X$. Thus, we have that the sequent $X \Rightarrow C$ is of the form $X_0, \bot \Rightarrow C$, and is an instance of $\bot$L. But this is in contradiction with (NoInit). So we are done.

(III) $\mathbf{r_1}$ ∈{($\wedge$L), ($\vee$L), ($p\to$L), ($\wedge\to$L), ($\vee\to$L)}: In all these cases, the cut formula is not principal in r_1 so it is preserved in the premise. Given that the rules considered are invertible, we simply take the conclusion of r_2 and use the corresponding invertibility lemma to destruct the principal formula of r_1. Then, we use SIH to cut on A in the obtained premises, and apply r_1 on the conclusion of the cut.

(IV) $\mathbf{r_1}$ ∈{($\wedge$R), ($\vee_1$R), ($\vee_2$R)}: In all these cases, the cut formula is principal in r_1 so it is deconstructed in the premise. Given that the corresponding left rules are invertible, we simply take the conclusion of r_2 and use the adequate invertibility lemma to destruct the cut formula. Then, we use PIH to cut on the obtained subformulae.

(V) $\mathbf{r_1}$ =($\to$R) : Then r_1 has the following form where $A = B \to D$:

$$\frac{B, X \Rightarrow D}{X \Rightarrow B \to D}\,(\to\text{R})$$

For the cases where $B \to D$ is principal in r_2 and $r_2 \neq (\Box\to\text{L})$, or where $r_2 \in \{(\text{IdP}), (\bot\text{L})\}$, we refer to Dyckhoff and Negri's proof [7] as the cuts produced in these cases involve the traditional induction hypothesis PIH. We are left with seven sub-cases.

(V-a) If r_2 is ($\to$R) then it must have the following form.

$$\frac{B \to D, E, X \Rightarrow F}{B \to D, X \Rightarrow E \to F}\,(\to\text{R})$$

where $E \to F = C$. We can use Lemma 3.2 on the proof of $X \Rightarrow B \to D$ to get a proof of $E, X \Rightarrow B \to D$. Proceed as follows.

$$\dfrac{\dfrac{E, X \Rightarrow B \to D \qquad B \to D, E, X \Rightarrow F}{E, X \Rightarrow F}\ \text{SIH}}{X \Rightarrow E \to F}\ (\to\text{R})$$

Note that the use of SIH is justified here as the last rule in this proof is effectively an instance of $(\to\text{R})$ in PSGL4ip, hence $\text{mhd}(E, X \Rightarrow F) < \text{mhd}(X \Rightarrow E \to F)$ by Lemma 4.7.

(V-b) If r_2 is $(\wedge\text{R})$ or $(\vee_i\text{R})$, then we simply use cut with the premise(s) of r_2 and the conclusion of r_1 using SIH.

(V-c) If r_2 is $(\wedge\text{L})$, $(\vee\text{L})$, $(p\to\text{L})$, $(\vee\to\text{R})$ or $(\wedge\to\text{R})$ where the cut formula is not principal in r_2, then we use the inversion lemma for r_2 on the conclusion of r_1, and then apply cut using SIH.

(V-d) If r_2 is $(\to\to\text{L})$ where the cut formula is not principal in r_2, then see case (I-a) in the partial proof given in the main body of the article.

(V-e) If r_2 is $(\Box\to\text{L})$ with the cut formula as principal formula, then it must have the following form, where $W, \Box X_0 = X$ and $\Box E = B$.

$$\dfrac{\boxtimes X_0, \Box E \Rightarrow E \qquad D, W, \Box X_0 \Rightarrow C}{\Box E \to D, W, \Box X_0 \Rightarrow C}\ (\Box\to\text{L})$$

Thus, we have that the sequents $X \Rightarrow C$ and $B, X \Rightarrow D$ are respectively of the form $W, \Box X_0 \Rightarrow C$ and $\Box E, W, \Box X_0 \Rightarrow D$. Then, we proceed as follows.

$$\dfrac{\dfrac{\dfrac{\boxtimes X_0, \Box E \Rightarrow E}{\Box X_0 \Rightarrow \Box E}\ \text{GLR}}{W, \Box X_0 \Rightarrow \Box E}\ \text{Lem.3.2} \qquad \dfrac{\Box E, W, \Box X_0 \Rightarrow D \qquad \dfrac{D, W, \Box X_0 \Rightarrow C}{D, \Box E, W, \Box X_0 \Rightarrow C}\ \text{Lem.3.2}}{\Box E, W, \Box X_0 \Rightarrow C}\ \text{PIH}}{W, \Box X_0 \Rightarrow C}\ \text{PIH}$$

(V-f) If r_2 is $(\Box\to\text{L})$ with a principal formula different from the cut formula, then it must have the following form where $\Box E \to F, W, \Box X_0 = X$.

$$\dfrac{\boxtimes X_0, \Box E \Rightarrow E \qquad F, B \to D, W, \Box X_0 \Rightarrow C}{B \to D, \Box E \to F, W, \Box X_0 \Rightarrow C}\ (\Box\to\text{L})$$

Thus, we have that $X \Rightarrow C$ and $X \Rightarrow B \to D$ are respectively of the form $\Box E \to F, W, \Box X_0 \Rightarrow C$ and $\Box E \to F, W, \Box X_0 \Rightarrow B \to D$. Using the right-invertibility of $(\Box\to\text{L})$, proven in Lemma 3.5, on $\Box E \to F, W, \Box X_0 \Rightarrow B \to D$ we obtain a proof of $F, W, \Box X_0 \Rightarrow B \to D$. Then, we proceed as follows.

$$\dfrac{\boxtimes X_0, \Box E \Rightarrow E \qquad \dfrac{F, W, \Box X_0 \Rightarrow B \to D \qquad F, B \to D, W, \Box X_0 \Rightarrow C}{F, W, \Box X_0 \Rightarrow C}\ \text{SIH}}{\Box E \to F, W, \Box X_0 \Rightarrow C}\ (\Box\to\text{L})$$

Note that the use of SIH is justified here as the assumption (NoInit) ensures that the last rule in this proof is effectively an instance of $(\Box\to\text{L})$ in PSGL4ip, hence $\text{mhd}(F, W, \Box X_0 \Rightarrow C) < \text{mhd}(\Box E \to F, W, \Box X_0 \Rightarrow C)$ by Lemma 4.7.

(V-g) If r_2 is (GLR) then it must have the following form.

$$\dfrac{\boxtimes X_0, \Box E \Rightarrow E}{W, B \to D, \Box X_0 \Rightarrow \Box E}\ (\text{GLR})$$

where $W, \Box X_0 = X$ and $\Box E = C$. In that case, note that the sequent $X \Rightarrow C$ is of the form $W, \Box X_0 \Rightarrow \Box E$. To obtain a proof of the latter, we apply the rule (GLR) on the premise of r_2 without weakening $B \to D$:

$$\frac{\boxtimes X_0, \Box E \Rightarrow E}{W, \Box X_0 \Rightarrow \Box E}\ (\text{GLR})$$

(VI) $\mathbf{r_1}$ =($\to\to$L): Then r_1 is as follows, where $(B \to D) \to E, X_0 = X$.

$$\frac{D \to E, X_0 \Rightarrow B \to D \qquad E, X_0 \Rightarrow A}{(B \to D) \to E, X_0 \Rightarrow A}\ (\to\to\text{L})$$

Thus, we have that the sequents $X \Rightarrow C$ and $A, X \Rightarrow C$ are respectively of the form $(B \to D) \to E, X_0 \Rightarrow C$ and $A, (B \to D) \to E, X_0 \Rightarrow C$. Using the right-invertibility of ($\to\to$L), proven in Lemma 3.5, on $A, (B \to D) \to E, X_0 \Rightarrow C$ we obtain a proof of the sequent $A, E, X_0 \Rightarrow C$. Then, we proceed as follows.

$$\frac{D \to E, X_0 \Rightarrow B \to D \qquad \dfrac{E, X_0 \Rightarrow A \qquad A, E, X_0 \Rightarrow C}{E, X_0 \Rightarrow C}\ \text{SIH}}{(B \to D) \to E, X_0 \Rightarrow C}\ (\to\to\text{L})$$

Note that the use of SIH is justified here as the assumption (NoInit) ensures that the last rule in this proof is effectively an instance of ($\to\to$L) in PSGL4ip, hence $\text{mhd}(E, X_0 \Rightarrow C) < \text{mhd}((B \to D) \to E, X_0 \Rightarrow C)$ by Lemma 4.7.

(VII) $\mathbf{r_1}$ =($\Box\to$L): We proceed as in (V-f).

(VIII) $\mathbf{r_1}$ =(GLR): Then A is the diagonal formula in r_1:

$$\frac{\boxtimes X_0, \Box B \Rightarrow B}{W, \Box X_0 \Rightarrow \Box B}\ (\text{GLR})$$

where $A = \Box B$ and $W, \Box X_0 = X$. Thus, we have that the sequents $X \Rightarrow C$ and $A, X \Rightarrow C$ are respectively of the form $W, \Box X_0 \Rightarrow C$ and $\Box B, W, \Box X_0 \Rightarrow C$. We now consider r_2.

(VIII-a) If r_2 is one of (IdP), ($\bot$L), ($\wedge$R), ($\wedge$L), ($\vee_1$R), ($\vee_2$R), ($\vee$L), ($\to$R), ($p\to$L), ($\wedge\to$L), ($\vee\to$L) and ($\to\to$L) then proceed similarly to the cases (I), (II), (III), (IV) and (VI), where the cut-formula is not principal in the rules considered by using SIH.

(VIII-b) If r_2 is ($\Box\to$L), then see case (II-a) in the main body of the article.

(VIII-c) If r_2 is (GLR). Then r_2 is of the following form where $\Box D = C$:

$$\frac{B, \Box B, \boxtimes X_0, \Box D \Rightarrow D}{W, \Box B, \Box X_0 \Rightarrow \Box D}\ (\text{GLR})$$

We proceed as follows where π is taken from the case (VIII-b):

$$\frac{\overset{\pi}{\boxtimes X_0, \Box D \Rightarrow D}}{W, \Box X_0 \Rightarrow \Box D}\ (\text{GLR})$$

Note that the use of SIH is justified here as the assumption (NoInit) ensures that the last rule in this proof is effectively an instance of (GLR) in PSGL4ip, hence $\text{mhd}(\boxtimes X_0, \Box D \Rightarrow D) < \text{mhd}(W, \Box X_0 \Rightarrow \Box D)$ by Lemma 4.7. Q.E.D.

Medvedev logic is the logic of finite distributive lattices without top element

Gianluca Grilletti [1]

Munich Center for Mathematical Philosophy
Geschwister-Scholl-Platz 1
D-80539 München

Abstract

Medvedev logic ML (also known as the logic of finite problems) is an intermediate logic firstly introduced by Yu.T. Medvedev in 1962. ML can be characterized as the logic of the class of intuitionistic Kripke frames corresponding to finite Boolean algebras (regarded as partially ordered structures) without their top element. Several fundamental questions about this logic still remain open to this day, most notably whether the set of its validities is decidable. In this work we provide an alternative characterization of ML in terms of finite distributive lattices without top element, in the same spirit as the characterization in terms of Boolean algebras.

Keywords: Medvedev logic, Finite distributive lattices, Meet-prime elements.

1 Introduction

Yu.T. Medvedev introduced in [10] the logic of finite problems ML, building on the informal interpretation of intuitionistic logic as "the calculus of a constructive solution to problems" proposed by Kolmogorov [6]. In [11] he also provided several results on the logic, enough to easily infer the following semantic characterization: ML is the logic of the intuitionistic Kripke frames obtained by removing the topmost element from a finite Boolean algebra—the so-called Medvedev frames.

Medvedev further developed this framework in [12], where he proposed a theory of "informational types and their transformations" based on the same models used for ML (see also [20] for further discussion). These works also led to other quite interesting logics, building on the same fundamental ideas. Among these we find the *logic of infinite problems* ML_1 proposed by D.P. Skvortsov [19], a family of logics based on "informational types" proposed by V.B. Shehtman

[1] Email: g.grilletti@lmu.de
Gefördert durch die Deutsche Forschungsgemeinschaft (DFG) - Projektnummer 446711878.
Funded by the German Research Foundation (DFG) - Project number 446711878.
We would like to thank I. Ciardelli and V. Punčochář for discussions on the result here presented. Moreover, we would like to thank the anonymous reviewers for the feedback provided.

and Skvortsov [18] and some tense modal logics built over Medvedev frames developed by W.H. Holliday [5].

Although the theory behind ML was a source of inspiration, not much is known about the properties of the original logic, and the few results available in the literature have been achieved over the span of five decades. In 1976 T. Prucnal showed that ML is structurally complete [13]. In 1979 L.L. Maksimova, Skvortsov and Shehtman showed that ML is not finitely axiomatizable [9]. In 1986 Maksimova showed that ML is maximal among the intermediate logics with the disjunction property [8]. In 1990 Shehtman showed that, not only the logic itself, but also its modal counterparts are not finitely axiomatizable [17]. In 2013, more than 20 years later, M. Łazarz showed that ML can also be characterized as the logic of the so-called *Kubiński frames* [7], obtained by removing the topmost element from a *Kubiński lattice.* And these, as far as the author knows, are the most salient results currently available.

It is surprising that, despite the amount of effort spent to study ML, several fundamental questions about the logic still remain open to this day. For example, the semantics characterization in terms of finite Boolean algebras readily implies that the set of non-valid formulas of the logic is recursively axiomatizable, but it is still not known whether the logic is decidable. A way to settle this issue in the positive would be to prove that ML (the logic of finite problems) coincides with ML_1 (the logic of infinite problems); however, whether this is the case is yet another open question. It is also worth noticing that some of the results about ML are obtained by studying the bounded-morphic images of Medvedev frames (compare with [9,17]), but we currently do not have a satisfying characterization of this class of bounded-morphic images.

In this paper we make a novel contribution to the study of Medvedev logic, by showing that ML can be characterized as the logic of the intuitionistic Kripke frames obtained by removing the topmost element from a finite *distributive lattice.* [2] This characterization is very similar in spirit to the original one presented by Medvedev in [11], as the frames considered in both cases are obtained by removing the topmost element from a class of finite algebraic structures. Moreover, both Medvedev frames and Kubiński frames are examples of distributive lattices without the topmost elements, which means that we found a family extending the ones considered in [7,11], again with ML as its logic.

This characterization is an addition to the study of ML, that might help to uncover new properties of the logic and to simplify the proofs of the results currently available. In fact, the class if finite distributive lattices without topmost element seems more stable under basic transformations than the class of Medvedev frames, and at the same time the structure of finite distributive lattices is well-understood and characterized. This characterization also opens venues for future research: following the approach of [19], we can consider the class of distributive lattices without topmost elements, leaving aside the

[2] As far as the author knows, the question whether the logic of this class of frames is ML was firstly proposed by V. Punčochář in 2018.

finiteness requirement. We expect the logic of this class to be ML_1, the logic of infinite problems. If this were the case, the new semantic approach to the study of ML and ML_1 based on distributive lattices could shed some light on the relation between the two logics. Another potential application of our result relates to inquisitive semantics [3], a formalism shown to have tight connections with Medvedev logic (see for example [2, Section 3.4]). This novel characterization of ML could reveal new properties of different inquisitive logics, in particular those explicitly built on algebraic structures, as for example the ones presented in [14,15,16]. [3]

The structure of the paper is as follows: In Section 2 we fix some notation and present the basic notions used in the rest of the paper. In Section 3 we prove the characterization of Medvedev logic in terms of finite distributive lattices, that is, our main result. In Section 4 we provide two examples showcasing the salient passages of the proof from Section 3.

2 Preliminaries

We assume the reader to be familiar with the basic notions on order theory, lattices, intermediate logics and intuitionistic Kripke frames. In this section we limit ourselves to fix some notations and recall the results used throughout the paper. For a basic introduction on order theory and lattices see [4]. And for a basic introduction on intermediate logics and intuitionistic Kripke frames see [1].

2.1 Distributive lattices

Throughout the paper we use the term *lattice* to indicate a *bounded lattice*. In particular, given a lattice $\langle \mathrm{L}, \leq \rangle$, we indicate with $\top$ and $\bot$ respectively its greatest and smallest elements. Given two elements $a, b \in \mathrm{L}$, we indicate with $a \wedge b$ and $a \vee b$ respectively the *meet* and the *join* of a and b. Likewise, given a finite set $S \subseteq \mathrm{L}$ we indicate with $\bigwedge S$ and $\bigvee S$ respectively the *meet* and the *join* of the elements of S (as customary, we define $\bigwedge \emptyset = \top$ and $\bigvee \emptyset = \bot$). With a slight abuse of notation, we indicate with L both the ordered structure $\langle \mathrm{L}, \leq \rangle$ and its underlying set of elements.

We call a lattice L *distributive* if, for every choice of $a, b, c \in \mathrm{L}$ the following two identities hold:

$$a \vee (b \wedge c) = (a \vee b) \wedge (a \vee c) \qquad a \wedge (b \vee c) = (a \wedge b) \vee (a \wedge c)$$

We indicate with $\mathbf{DL}$ the class of distributive lattices and with $\mathbf{DL}_{\mathrm{fin}}$ the class of *finite* distributive lattices.

In the following section we focus on two special families of elements of finite distributive lattices: *meet-prime elements* and *coatoms*. Given a distributive lattice L we indicate with $\mathfrak{M}_{\mathrm{L}}$ the set of its meet-prime elements, that is, the elements $p \in \mathrm{L} \backslash \{\top\}$ satisfying the condition that, for every $a, b \in \mathrm{L}$, if $a \wedge b \leq p$, then $a \leq p$ or $b \leq p$. And we indicate with $\mathfrak{C}_{\mathrm{L}}$ the set of coatoms of L, that is,

[3] Thanks to V. Punčochář for pointing out this very interesting connection.

the maximal elements of the set $\mathrm{L} \setminus \{\top\}$. Notice that coatoms are in particular meet-prime elements, so $\mathfrak{C}_\mathrm{L} \subseteq \mathfrak{M}_\mathrm{L}$. Moreover, a non-trivial finite lattice always contains coatoms, thus for $\mathrm{L} \in \mathbf{DL}_{\mathrm{fin}}$ we have $\mathfrak{C}_\mathrm{L} \neq \emptyset$.

Meet-prime elements play a special role in the study of distributive lattices. In this paper we use some well-known technical results from the literature. The first is the following lemma. [4]

Lemma 2.1 **([4, Lemma 5.11])** *Consider a lattice* $\mathrm{L} \in \mathbf{DL}$. *For an element* $a \in \mathrm{L} \setminus \{\top\}$ *the following conditions are equivalent:*

- *a is meet-prime;*
- *for any $b_1, b_2 \in \mathrm{L}$, if $a = b_1 \wedge b_2$ then $a = b_1$ or $a = b_2$;*
- *for any $b_1, \ldots, b_k \in \mathrm{L}$, if $a \geq b_1 \wedge \cdots \wedge b_k$ then $a \geq b_i$ for some $1 \leq i \leq k$.*

A fundamental result about finite distributive lattices—another essential ingredient for the results of this paper—is *Birkhoff's representation theorem.* Let us indicate with $\mathcal{P}$ the powerset functor.

Theorem 2.2 ([4, Theorem 5.12] Birkhoff's representation for $\mathbf{DL}_{\mathrm{fin}}$) *Consider a lattice* $\mathrm{L} \in \mathbf{DL}_{\mathrm{fin}}$. *For an element* $a \in \mathrm{L}$ *define the set*

$$\mathfrak{M}_a := \{\, p \in \mathfrak{M}_\mathrm{L} \mid p \geq a \,\}$$

Then for every $a, b \in \mathrm{L}$ it holds that $a = \bigwedge \mathfrak{M}_a$ and that $a \leq b$ iff $\mathfrak{M}_a \supseteq \mathfrak{M}_b$.

We refer to the set $\mathfrak{M}_a$ as the set of meet-prime elements *above a*.

2.2 Intuitionistic Kripke frames and bounded morphisms

An *intuitionistic Kripke frame* (henceforth, just *frame*) is a pair $\mathcal{F} = \langle F, \leq \rangle$ where F is a set (the points of the frame) and $\leq$ is a partial order (the accessibility relation). Frames are commonly used to provide a semantics for intuitionistic and intermediate logics. We do not spell out the definition of this semantics—which can be found in [1, Chapter 2]—since we do not need it to present the results of this paper. We limit ourselves to define some basic notions about frames and their logics and to recall some basic results.

Henceforth we indicate with AP a fixed infinite set of atomic propositions, and we focus exclusively on propositional formulas over AP, that is, formulas generated by the following grammar:

$$\phi \ ::= \ p \in \mathrm{AP} \mid \bot \mid \phi \wedge \phi \mid \phi \vee \phi \mid \phi \rightarrow \phi$$

Given a frame $\mathcal{F}$, the *logic of* $\mathcal{F}$ is the set $\mathrm{Log}(\mathcal{F})$ of formulas valid over $\mathcal{F}$. Similarly, given a class of frames $\mathcal{C}$, the *logic of the class* $\mathcal{C}$ is the collection of formulas valid over *every* frame of the class, that is, the set $\mathrm{Log}(\mathcal{C}) := \bigcap_{\mathcal{F} \in \mathcal{C}} \mathrm{Log}(\mathcal{F})$.

[4] The statements in [4, Lemma 5.11, Theorem 5.12] are formulated in terms of the set $\mathcal{J}(\mathrm{L})$ of *join-prime elements* of L. The formulation we employ here using meet-prime elements is obtained by applying the *duality principle for lattices* (Statement 1.20, *ibidem*).

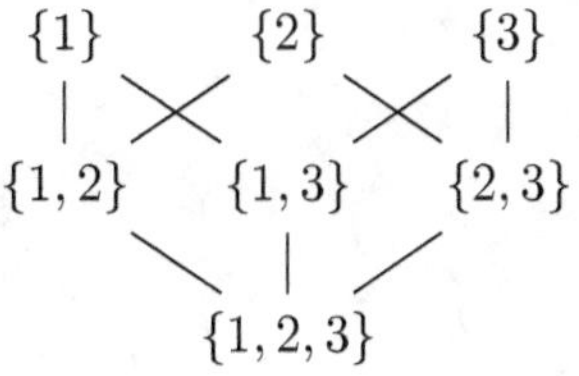

Figure 1. The Medvedev frame $\mathcal{M}_3$, represented using its Hasse diagram.

In this paper we focus on a particular family of frames, that is, *Medvedev frames*. For $n \in \mathbb{N}$ a positive natural number, the Medvedev frame of order n is defined as

$$\mathcal{M}_n \; := \; \langle \mathcal{P}_0(\{1, \ldots, n\}), \supseteq \rangle$$

where $\mathcal{P}_0(\{1, \ldots, n\})$ indicates the non-empty subsets of $\{1, \ldots, n\}$. Notice that these frames—modulo isomorphism—are obtained by removing the topmost element from a finite Boolean algebra (considered as a partial order). *Medvedev logic* ML is the logic of the family of Medvedev frames, or in the notation introduced above: $\mathsf{ML} = \mathrm{Log}(\{\mathcal{M}_n \mid n \geq 1\})$ (compare with [10,11]). We give a graphical representation of a Medvedev frame in Figure 1.

We also introduce *bounded morphisms* (also known as *reductions* or *p-morphisms*), a powerful tool to study the logics of frames and their families.

Definition 2.3 [Bounded morphism [1, Section 2.3]] Given two frames $\mathcal{F}$ and $\mathcal{F}'$ a *bounded morphism* from $\mathcal{F} = \langle F, \leq \rangle$ to $\mathcal{F}' = \langle F', \preceq \rangle$ is a function $f : F \to F'$ satisfying the following properties.

Forth condition For every $a, b \in F$, if $a \leq b$ then $f(a) \preceq f(b)$.

Back condition For every $a \in F$ and $b' \in F'$, if $f(a) \preceq b'$ then there exists $b \in F$ such that $a \leq b$ and $f(b) = b'$.

We indicate with $f : \mathcal{F} \twoheadrightarrow \mathcal{F}'$ that f is a *surjective* bounded morphism from $\mathcal{F}$ to $\mathcal{F}'$.

Bounded morphisms allow us to compare the logics of two frames, as the following result shows.

Lemma 2.4 ([1, Corollary 2.16]) *Let $\mathcal{F}$ and $\mathcal{F}'$ be two frames, and suppose there exists $f : \mathcal{F} \twoheadrightarrow \mathcal{F}'$. Then* $\mathrm{Log}(\mathcal{F}') \subseteq \mathrm{Log}(\mathcal{F})$.

This is all we need about frames and their logics to prove the results of this paper!

3 A novel class of frames for Medvedev logic

For ease of reading, in the rest of this section we indicate $\mathfrak{M}_{\mathrm{L}}$ and $\mathfrak{C}_{\mathrm{L}}$ simply as $\mathfrak{M}$ and $\mathfrak{C}$ (omitting the subscript). Given a finite distributive lattice L, we indicate with L^- the *partial order* obtained by removing the topmost element from L, that is, the set $\mathrm{L}^- = \mathrm{L} \setminus \{\top\}$ with the induced order. We indicate with $\mathbf{DL}^-_{\mathrm{fin}}$ the collection of all the structures obtained this way $\mathbf{DL}^-_{\mathrm{fin}} := \{\mathrm{L}^- \mid \mathrm{L} \in$

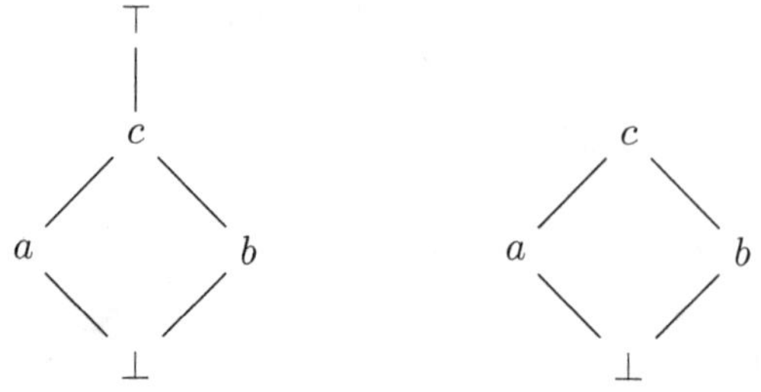

Figure 2. A distributive lattice L (on the left) and the corresponding partial order L^- (on the right), both represented using their Hasse diagrams. Recall that L^- is obtained by removing the topmost element $\top$.

$\mathbf{DL}_{\text{fin}}\}$. In Figure 2 we depict a distributive lattice L and its corresponding partial order L^-. We will use it as a running example through the paper.

The aim of this paper is to provide a proof of the following theorem.

Theorem 3.1 $\mathsf{ML} = \mathrm{Log}(\, \mathbf{DL}^-_{\text{fin}}\,)$.

That is, we want to show that Medvedev logic can be characterized as the logic of finite distributive lattices without their top element. The argument we use mainly relies on the following lemma, and in fact most of this section is devoted to its proof.

Lemma 3.2 *For every finite distributive lattice* L, *there exists a Medvedev frame* $\mathcal{M}_n$ *such that* $\mathcal{M}_n \twoheadrightarrow L^-$.

Let us first show how to use Lemma 3.2 to prove Theorem 3.1.

Proof [Proof of Theorem 3.1] Notice that $\langle \mathcal{P}(\{1,\dots,n\}), \supseteq\rangle$ is a distributive lattice (since it is a boolean algebra) with topmost element $\emptyset$. Thus $\langle \mathcal{P}(\{1,\dots,n\}), \supseteq\rangle^- = \mathcal{M}_n$. In particular $\{\, \mathcal{M}_n \mid n \geq 1\,\} \subseteq \mathbf{DL}^-_{\text{fin}}$, so

$$\mathrm{Log}(\, \mathbf{DL}^-_{\text{fin}}\,) \subseteq \mathrm{Log}(\, \{\, \mathcal{M}_n \mid n \geq 1\,\}\,) = \mathsf{ML}$$

As for the other inclusion, by Lemma 3.2 for every $L \in \mathbf{DL}_{\text{fin}}$ there exists a Medvedev frame $\mathcal{M}_n$ such that $\mathcal{M}_n \twoheadrightarrow L^-$. In particular by Lemma 2.4 we have that that $\mathrm{Log}(\mathcal{M}_n) \subseteq \mathrm{Log}(L^-)$, and so

$$\mathsf{ML} = \bigcap_{n\geq 1} \mathrm{Log}(\mathcal{M}_n) \subseteq \bigcap_{L\in\mathbf{DL}_{\text{fin}}} \mathrm{Log}(L^-) = \mathrm{Log}(\, \mathbf{DL}^-_{\text{fin}}\,)$$

which concludes the proof. □

We showed how to use Lemma 3.2 to prove Theorem 3.1, so now we are at the hard part: proving the lemma.

To prove Lemma 3.2, given a finite distributive lattice $L \in \mathbf{DL}_{\text{fin}}$ we need to provide two ingredients: a Medvedev frame $\mathcal{M}_n$ and a surjective bounded morphism from $\mathcal{M}_n$ to L^-. For ease of presentation, we work modulo isomorphism: we define a frame $\mathcal{M}$ *isomorphic* to a Medvedev frame and then we construct

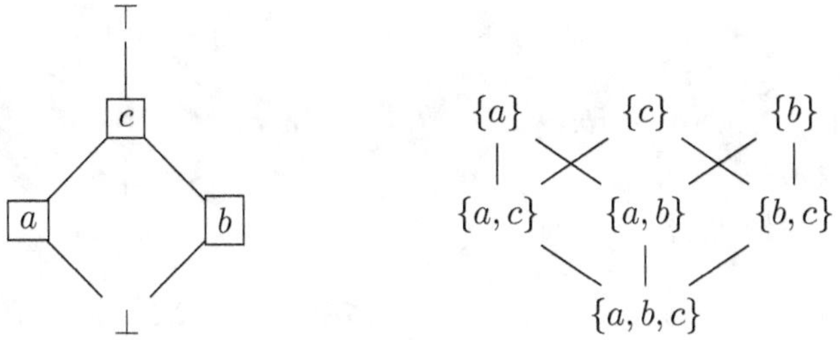

Figure 3. On the left, the lattice L from Figure 2 where we highlighted the elements of $\mathfrak{M}$. On the right, the corresponding frame $\mathcal{M}$.

a surjective bounded morphism $f : \mathcal{M} \twoheadrightarrow \mathrm{L}^-$. The desired bounded morphism can then be obtained by composing f with an automorphism between $\mathcal{M}$ and a Medvedev frame.

The frame we are going to use is $\mathcal{M} = \langle \mathcal{P}_0(\mathfrak{M}), \supseteq \rangle$, that is, the frame whose points are non-empty subsets of meet-prime elements of L ordered by reverse inclusion. This is trivially isomorphic to the Medvedev frame $\mathcal{M}_{|\mathfrak{M}|}$, where $|\mathfrak{M}|$ indicates the cardinality of the set $\mathfrak{M}$. In Figure 3 we depict the frame $\mathcal{M}$ corresponding to the running example of Figure 2.

The construction of the bounded morphism $f : \mathcal{M} \twoheadrightarrow \mathrm{L}^-$ requires the definition of two auxiliary functions $g, h : \mathcal{P}_0(\mathfrak{M}) \to \mathcal{P}(\mathfrak{M})$. We start with the former: given $N \subseteq \mathfrak{M}$ we define the map[5]

$$\begin{aligned} g(N) &= \{\, p \in \mathfrak{M} \mid \mathfrak{M}_p \subseteq N \,\} \\ &= \{\, p \in \mathfrak{M} \mid \forall p' \geq p.\, p' \in N \,\} \end{aligned}$$

So $g(N)$ collects all the meet-prime elements $p \in N$ for which *p and its successors* are contained in N. In Figure 4 we represent the function g for our running example. We collect some properties of g in the following proposition, after recalling some basic definitions. We call $U \subseteq \mathfrak{M}$ an *upset* (of $\mathfrak{M}$) if for every $p \in U$ and $q \in \mathfrak{M}$, if $p \leq q$ then $q \in U$. An upset is called *principal* if it contains a minimum element, that is, if it is of the form $\mathfrak{M}_q$ for some $q \in \mathfrak{M}$. Finally, we indicate with $\mathrm{Up}(\mathfrak{M})$ the collection of the upsets of $\mathfrak{M}$ and with $\mathrm{Up}_0(\mathfrak{M})$ the collection of *non-empty* subsets.

Proposition 3.3 *For every $N \subseteq \mathfrak{M}$, $g(N)$ is the greatest (under set-theoretic inclusion)* upset of $\mathfrak{M}$ *contained in N. In particular, the map g is monotone and $g \restriction_{\mathrm{Up}_0(\mathfrak{M})}$ is the identity function.*

Proof By definition, $g(N)$ is the union of all the principal upsets contained in N, which implies that $g(N)$ is the greatest *upset of* $\mathfrak{M}$ contained in N. Given this, it follows trivially that g is monotone and that $g(U) = U$ for an upset U. □

[5] Recall that $\mathfrak{M}_q := \{p \in \mathfrak{M} \,|\, p \geq q\}$, as defined in Theorem 2.2.

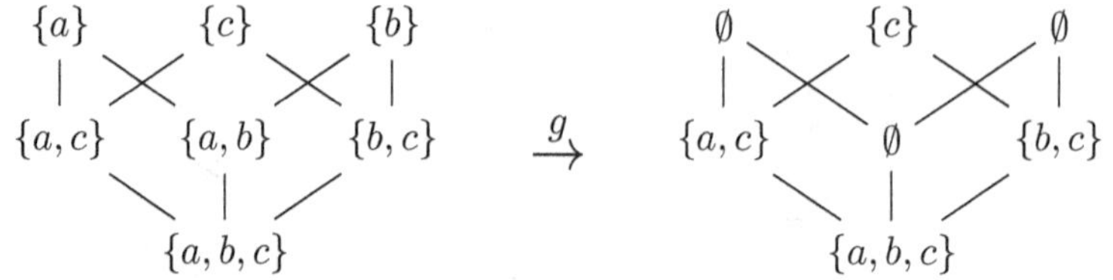

Figure 4. Representation of the function $g : \mathcal{P}_0(\mathfrak{M}) \to \mathcal{P}(\mathfrak{M})$ for $\mathcal{M}$ the running example from Figures 2 and 3. Recall that the points of the frame $\mathcal{M}$ (on the left) are the elements of the set $\mathcal{P}_0(\mathfrak{M})$. On the right, we indicate for each $N \in \mathcal{P}_0(\mathfrak{M})$ the corresponding image $g(N)$.

The function g allows to shift our focus from the collection of non-empty subsets $\mathcal{P}_0(\mathfrak{M})$ to the collection of *upsets* $\mathrm{Up}(\mathfrak{M})$. And this is particularly interesting since there is a natural surjective bounded morphism from the frame $\langle \mathrm{Up}_0(\mathfrak{M}), \supseteq \rangle$ to L^-:

$$\begin{aligned} \bigwedge : \mathrm{Up}_0(\mathfrak{M}) &\twoheadrightarrow \mathrm{L}^- \\ U &\mapsto \bigwedge U \end{aligned}$$

Recall that $\bigwedge U$ is the meet of all the elements in U, computed in the lattice L. Notice that this map is well-defined: since U is non-empty, for a meet-prime $p \in U$ we have $\bigwedge U \leq p < \top$. Moreover, the surjectiveness of this map is a direct consequence of Theorem 2.2.

So a naive approach to define a surjective bounded morphism $f : \mathcal{P}_0(\mathfrak{M}) \twoheadrightarrow \mathrm{L}^-$ would be to compose the maps $\bigwedge$ and g, that is, to consider the map $N \mapsto \bigwedge g(N)$. However, there is a complication: the set $g(N)$ might be *empty* (for example when N does not contain any maximal element of $\mathfrak{M}$) and the bounded morphism $\bigwedge : \mathrm{Up}_0(\mathfrak{M}) \to \mathrm{L}^-$ cannot be extended to $\mathrm{Up}(\mathfrak{M})$. To avoid this issue, we need another auxiliary function to *tweak* the set $g(N)$ before we take the meet $\bigwedge g(N)$. Recall that we indicate with $\mathfrak{C}$ the set of coatoms of the lattice L.

Definition 3.4 [link function] Given L a finite distributive lattice, we define a *link function* as a map $h : \mathfrak{M} \to \mathfrak{C}$ such that $h(p) \geq p$ for every $p \in \mathfrak{M}$.

So a link function is an increasing map from $\mathfrak{M}$ to $\mathfrak{C}$. In particular a link function restricted to the set $\mathfrak{C}$ is always the identity. Notice that every element of L^- is smaller than some coatom since L is finite, so under our assumptions link functions always exist.

In the rest of the section we work with a fixed arbitrary link function h. With a slight abuse of notation, we indicate with the same notation also the lifting of h to sets of meet-prime elements: $h(N) = \{ h(p) \mid p \in N \}$. In Figure 5 we represent the function h for our running example. Notice that (in contrast to the map g) for a non-empty subset $N \subseteq \mathfrak{M}$ the image $h(N)$ is not empty. Moreover, h is monotone over $\mathcal{P}_0(\mathfrak{M})$ by definition.

Proposition 3.5 *For $N, N' \in \mathcal{P}_0(\mathfrak{M})$ with $N \subseteq N'$, it holds that $h(N) \subseteq h(N')$.*

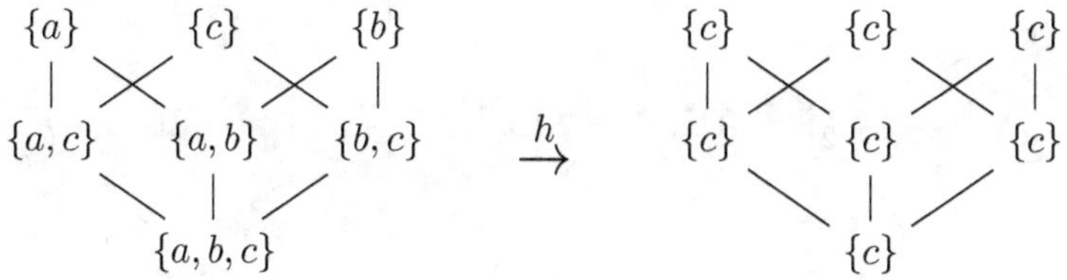

Figure 5. The function $h : \mathcal{P}_0(\mathfrak{M}) \to \mathcal{P}(\mathfrak{M})$ for the running example from Figures 2 and 3. The only link function for the lattice L of Figure 2 is the constant function with value c, that is, the function h used in this example. As in Figure 4, on the left we depict the points of the frame $\mathcal{M}$, that is, the elements of the set $\mathcal{P}_0(\mathfrak{M})$, and on the right their images under h. As the map g of Figure 4, the map h is monotone.

The map h is also well-behaved when restricted to upsets, and this allows to draw an interesting connection between g and h that will be useful later on.

Proposition 3.6 *For $U \in \mathrm{Up}(\mathfrak{M})$ it holds $h(U) \subseteq U$. In particular, $g(U) \cup h(U) = U$.*

Proof Firstly we prove that $h(U) \subseteq U$. Given an arbitrary $p \in U$, we have $h(p) \geq p$, and since U is an upset it follows that $h(p) \in U$. Since p was arbitrary, it follows that $h(U) \subseteq U$. Given this, the identity $g(U) \cup h(U) = U$ is a direct consequence of Proposition 3.3. □

Proposition 3.6 and the previous observations on the map $\bigwedge : \mathrm{Up}_0(\mathfrak{M}) \twoheadrightarrow \mathrm{L}^-$ suggest the following definition for the desired bounded morphism $f : \mathcal{P}_0(\mathfrak{M}) \twoheadrightarrow \mathrm{L}^-$.

$$f(N) := \bigwedge(\, g(N) \cup h(N)\,)$$

In Figure 6 we represent the function f for our running example.

We now have our two ingredients for the proof of Lemma 3.2, that is, the frame $\langle \mathcal{P}_0(\mathfrak{M}), \supseteq \rangle$ and the map f. What remains to be proved is that f is a *surjective bounded morphism* between the structures $\mathcal{P}_0(\mathfrak{M})$ and L^-.

Proof [Proof of Lemma 3.2] We want to show that f is well-defined, surjective and a bounded morphism. In particular, the latter amounts to proving that f satisfies the forth and back conditions from Definition 2.3. We show these results separately.

f is well-defined. Given a non-empty subset $N \in \mathcal{P}_0(\mathfrak{M})$, we already noticed that $h(N) \neq \emptyset$. In particular, this implies that for $c \in h(N)$ it holds

$$f(N) \;=\; \bigwedge(\, g(N) \cup h(N)\,) \;\leq\; c \;<\; \top$$

thus $f(N)$ is an element of L^-.

f is surjective. Let $b \in \mathrm{L}^-$ be an arbitrary element and recall that by Theorem 2.2 we have $b = \bigwedge \mathfrak{M}_b$. We claim that $f(\mathfrak{M}_b) = b$. Since $b < \top$ we have that $\mathfrak{M}_b \neq \emptyset$, and so $\mathfrak{M}_b \in \mathcal{P}_0(\mathfrak{M})$. By definition $\mathfrak{M}_b$ is an upset in $\mathcal{P}_0(\mathfrak{M})$, thus by

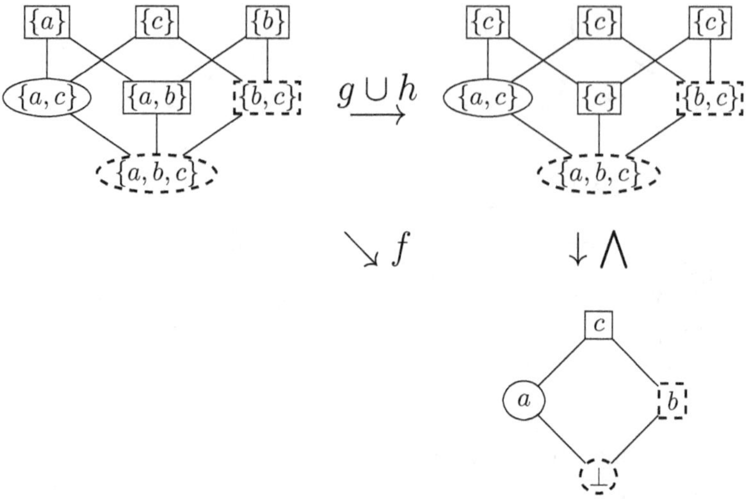

Figure 6. The function $f : \mathcal{P}_0(\mathfrak{M}) \to \mathrm{L}^-$ for the running example, computed as a composition of the functions $g \cup h$ and $\bigwedge$. The image of $g \cup h$ is depicted as in Figures 4 and 5, while the images of the functions $\bigwedge$ and f are coded by the drawing styles of the nodes (e.g., $f(\{b, c\}) = b$ and this is indicated in the picture by the matching rectangular dashed nodes).

Proposition 3.6 it follows that $g(\mathfrak{M}_b) \cup h(\mathfrak{M}_b) = \mathfrak{M}_b$, and so $f(\mathfrak{M}_b) = \bigwedge \mathfrak{M}_b = b$. Since b was an arbitrary element in L^-, this shows that f is surjective.

f respects the forth condition. That is, we want to show that for every $N, N' \in \mathcal{P}_0(\mathfrak{M})$, if $N \supseteq N'$ then $f(N) \leq f(N')$.[6]

Consider arbitrary $N, N' \in \mathcal{P}_0(\mathfrak{M})$ and assume that $N \supseteq N'$. By Propositions 3.3 and 3.5, we know that $g(N) \cup h(N) \supseteq g(N') \cup h(N')$. Thus we have:

$$f(N) \;=\; \bigwedge(\,g(N) \cup h(N)\,) \;\leq\; \bigwedge(\,g(N') \cup h(N')\,) \;=\; f(N')$$

Since N, N' were arbitrary, the forth condition follows.

f respects the back condition. That is, we want to show that for $N \in \mathcal{P}_0(\mathfrak{M})$ and $b \in \mathrm{L}^-$, if $f(N) \leq b$ then there exists $N' \in \mathcal{P}_0(\mathfrak{M})$ such that $N \supseteq N'$ and $f(N') = b$.

Consider arbitrary $N \in \mathcal{P}_0(\mathfrak{M})$ and $b \in \mathrm{L}^-$, and assume that $f(N) \leq b$. We aim to find a non-empty $N' \subseteq N$ such that $f(N') = b$. Recall that $b = \bigwedge \mathfrak{M}_b$ by Theorem 2.2. In particular, for every $p \in \mathfrak{M}_b$ we have that:

$$f(N) \;=\; \bigwedge(\,g(N) \cup h(N)\,) \;\leq\; b \;\leq\; p$$

By Lemma 2.1 and the previous inequality we obtain:

$$\forall p \in \mathfrak{M}_b. \quad \exists p' \in g(N) \cup h(N). \quad p' \leq p \tag{1}$$

6 Recall that we are considering $\mathcal{P}_0(\mathfrak{M})$ ordered by the relation $\supseteq$.

We proceed by considering the partition of the set $\mathfrak{M}_b = G \cup H$, where G contains all the $p \in \mathfrak{M}_b$ such that $p' \leq p$ for some $p' \in g(N)$, and $H := \mathfrak{M}_b \setminus G$. Using (1), we can give alternative characterizations of the sets G and H.

- **Claim 1:** $p \in G \iff \mathfrak{M}_p \subseteq N$. Firstly suppose that $p \in G$. By definition of G there exists $p' \in g(N)$ such that $p' \leq p$. This implies that $\mathfrak{M}_p \subseteq \mathfrak{M}_{p'} \subseteq N$, where the last containment follows from the definition of $g(N)$. Secondly, suppose that $\mathfrak{M}_p \subseteq N$. Then, by definition of $g(N)$, we have that $p \in g(N)$ and consequently $p \in G$ (since we can take $p' = p$).
- **Claim 2:** $H = \mathfrak{M}_b \cap h(N) \cap (\mathfrak{M} \setminus N)$. We prove the two inclusions $H \subseteq \mathfrak{M}_b \cap h(N) \cap (\mathfrak{M} \setminus N)$ and $\mathfrak{M}_b \cap h(N) \cap (\mathfrak{M} \setminus N) \subseteq H$ separately.

  $\underline{H \subseteq \mathfrak{M}_b \cap h(N) \cap (\mathfrak{M} \setminus N)}$. We have that $H \subseteq \mathfrak{M}_b$ by definition of H, so we only have to show that $H \subseteq h(N)$ and that $H \subseteq \mathfrak{M} \setminus N$.

  We first prove that $H \subseteq h(N)$. By Claim 1 we have that $g(N) \subseteq G$, and so by (1) and the definition of H it follows that:

  $$\forall c \in H. \quad \exists c' \in h(N) \setminus g(N). \quad c' \leq c$$

  In particular, since the elements of $h(N)$ are coatoms, the previous property boils down to $H \subseteq h(N) \setminus g(N) \subseteq \mathfrak{C}$, which in particular implies $H \subseteq h(N)$.

  Finally, we prove that $H \subseteq \mathfrak{M} \setminus N$. Assume towards a contradiction that there exists $c \in H \cap N$. Since c is a coatom, it follows that $\mathfrak{M}_c = \{c\} \subseteq N$. Thus by definition of $g(N)$ we have $c \in g(N)$, which contradicts $H \subseteq h(N) \setminus g(N)$. So we conclude that $H \cap N = \emptyset$, that is, $H \subseteq \mathfrak{M} \setminus N$.

  $\underline{\mathfrak{M}_b \cap h(N) \cap (\mathfrak{M} \setminus N) \subseteq H}$. Consider an arbitrary element $c \in \mathfrak{M}_b \cap h(N) \cap (\mathfrak{M} \setminus N)$. $c \in h(N)$ is a coatom, so we have $\mathfrak{M}_c = \{c\} \nsubseteq N$ which in turn implies $c \notin g(N)$. Since $g(N)$ us an upset it follows that there is no $p \in g(N)$ such that $c \geq p$, that is, $c \notin G$. In particular, we have that $c \in \mathfrak{M}_b \setminus G = H$. Since c was an arbitrary element of $\mathfrak{M}_b \cap h(N) \cap (\mathfrak{M} \setminus N)$, we conclude that $\mathfrak{M}_b \cap h(N) \cap (\mathfrak{M} \setminus N) \subseteq H$.

Now that we have these characterizations of the sets G and H, we are ready to define the set N' we are looking for. Fix an enumeration of the sets $G = \{p_1, \ldots, p_k\}$ and $H = \{c_1, \ldots, c_l\}$. By Claim 2, for every $j \in \{1, \ldots, l\}$ we can fix an element $p'_j \in N$ such that $h(p'_j) = c_j$. We define the set N' as follows:

$$N' := \left(\bigcup_{i=1}^{k} \mathfrak{M}_{p_i} \right) \cup \{p'_1, \ldots, p'_l\}$$

We need to show that $N' \subseteq N$ and that $f(N') = b$. We start with the former condition. By Claim 1, for $i \in \{1, \ldots, k\}$ we have $\mathfrak{M}_{p_i} \subseteq N$. And for $j \in \{1, \ldots, l\}$ we have $p'_j \in N$ by definition of p'_j. Combining these facts we obtain $N' \subseteq N$.

We now show that $f(N') = b$. We firstly prove that $f(N') \leq b$. Notice that $G = \{p_1, \ldots, p_k\} \subseteq g(N')$ and that $H = \{c_1, \ldots, c_l\} = \{h(p'_1), \ldots, h(p'_l)\} \subseteq$

$h(N')$. So in particular we have:

$$f(N') = \bigwedge(\, g(N') \cup h(N')\,) \leq \bigwedge(G \cup H) = \bigwedge \mathfrak{M}_b = b$$

As for the other inequality, consider an element $q \in g(N') \cup h(N')$. We first show that $q \geq b$. We consider two (non mutually exclusive) cases:

- If $q \in g(N')$, by definition of $g(N')$ we have that $\mathfrak{M}_q \subseteq N' \subseteq N$. We want to show that $q \in \cup_{i=1}^{k} \mathfrak{M}_{p_i}$. We can do it reasoning by contradiction: assume that $q = p'_j$ for some $j \in \{1, \ldots, l\}$. Since $c_j \geq p'_j$ and $\mathfrak{M}_q$ is an upset, it follows that $c_j \in \mathfrak{M}_q \subseteq N$, which contradicts Claim 2. So it follows that $q \neq p'_j$ for every $j \in \{1, \ldots, l\}$, and so $q \in \bigcup_{i=1}^{k} \mathfrak{M}_{p_i}$. In particular, this implies that there exists $i \in \{1, \ldots, k\}$ such that $q \geq p_i \geq b$.
- If $q \in h(N')$, by definition of $h(N')$ there exists $q' \in N'$ such that $h(q') = q$. If $q' \in \mathfrak{M}_{p_i}$ for some $i \in \{1, \ldots, k\}$ then $q = h(q') \geq h(p_i) \geq p_i \geq b$. Otherwise, if $q' = p'_j$ for some $j \in \{1, \ldots, l\}$ then $q = h(q') = h(p'_j) = c_j \geq b$. In both cases we have $q \geq b$.

So for every choice of $q \in g(N') \cup h(N')$ we have that $q \geq b$. From this it follows that:

$$f(N') = \bigwedge(\, g(N') \cup h(N')\,) \geq b$$

Thus we conclude that $f(N') = b$. Since $N \in \mathcal{P}_0(\mathfrak{M})$ and $b \in \mathrm{L}^-$ were arbitrary elements, this shows that the back condition holds for f, thus concluding the proof of the lemma. □

4 Examples

In this section we showcase the construction presented in the proof of Lemma 3.2 with two examples. For reasons of layout, we move the figures to the end of the manuscript.

We build the bounded morphisms f_1 and f_2 corresponding to distinct link functions h_1 and h_2 over the same lattice L, depicted in Figure 7. We firstly compute the function g, which is independent from the link functions. The values of g are depicted in Figure 8, following the representation given in Figure 4. Henceforth, we use the same notational conventions used in the running example from the previous section without mentioning it explicitly.

4.1 First example

Consider the link function $h_1 : \mathfrak{M} \to \mathfrak{C}$ defined as $h_1(a) = d$, $h_1(b) = d$, $h_1(d) = d$ and $h_1(e) = e$. We depict the lifting of the function h_1 in Figure 9 and the resulting bounded morphism $f_1 : \mathcal{M} \twoheadrightarrow \mathrm{L}^-$ in Figure 10.

4.2 Second example

Consider the link function $h_2 : \mathfrak{M} \to \mathfrak{C}$ defined as $h_2(a) = e$, $h_2(b) = d$, $h_2(d) = d$ and $h_2(e) = e$. We depict the lifting of function h_2 in Figure 11 and the resulting bounded morphism $f_2 : \mathcal{M} \twoheadrightarrow \mathrm{L}^-$ in Figure 12.

References

[1] Chagrov, A. and M. Zakharyaschev, "Modal Logic," Oxford University Press, 1997.
[2] Ciardelli, I., *Inquisitive semantics and intermediate logics* (2009), MSc Thesis, University of Amsterdam.
[3] Ciardelli, I. and F. Roelofsen, *Inquisitive logic*, Journal of Philosophical Logic **40** (2011), pp. 55–94.
[4] Davey, B. and H. Priestley, "Introduction to lattices and order," Cambridge University Press, Cambridge, 1990.
URL `http://www.worldcat.org/search?qt=worldcat_org_all&q=0521367662`
[5] Holliday, W., *On the modal logic of subset and superset: Tense logic over Medvedev frames*, Studia Logica **105** (2017), pp. 13–35.
URL `https://doi.org/10.1007/s11225-016-9680-1`
[6] Kolmogorov, A., *Zur deutung der intuitionistischen logik*, Matematische Zeitschrift **35** (1932), pp. 58–65.
[7] Lazarz, M., *Characterization of Medvedev's logic by means of Kubiński's frames*, Bulletin of the Section of Logic **42** (2013), pp. 83–90.
[8] Maksimova, L., *On maximal intermediate logics with the disjunction property*, Studia Logica **45** (1986), pp. 69–75.
[9] Maksimova, L., V. Shetman and D. Skvortsov, *The impossibility of a finite axiomatization of Medvedev's logic of finitary problems*, Soviet Mathematics Doklady **20** (1979), pp. 394–398.
[10] Medvedev, Y., *Finite problems*, Soviet Math. Dokl. **3** (1962 (Russian)), pp. 227–230.
[11] Medvedev, Y., *Interpretation of logical formulas by means of finite problems*, Soviet Math. Dokl. **7** (1966 (Russian)), pp. 857–860.
[12] Medvedev, Y., *Transformations of information and calculi that describe them: types of information and their possible transformations*, Semiotika i informatika **13** (1979 (Russian)), pp. 109–141.
[13] Prucnal, T., *Structural completeness of Medvedev's propositional calculus*, Reports on Mathematical Logic **6** (1976), pp. 103–105.
[14] Punčochář, V., *Algebras of Information States*, Journal of Logic and Computation **27** (2016), pp. 1643–1675.
URL `https://doi.org/10.1093/logcom/exw021`
[15] Punčochář, V., *Substructural inquisitive logics*, The Review of Symbolic Logic **12** (2019), pp. 296–330.
[16] Punčochář, V., *Inquisitive heyting algebras*, Studia Logica **109** (2021), pp. 995–1017.
URL `https://doi.org/10.1007/s11225-020-09936-9`
[17] Shehtman, V., *Modal counterparts of Medvedev logic of finite problems are not finitely axiomatizable*, Studia Logica: An International Journal for Symbolic Logic **49** (1990), pp. 365–385.
URL `http://www.jstor.org/stable/20015517`
[18] Shehtman, V. and D. Skvortsov, *Logics of some kripke frames connected with Medvedev notion of informational types*, Studia Logica: An International Journal for Symbolic Logic **45** (1986), pp. 101–118.
URL `http://www.jstor.org/stable/20015250`
[19] Skvortsov, D., *Logic of infinite problems and Kripke models on atomic semilattices of sets*, Doklady AN SSSR **245** (1979 (Russian)), pp. 798–801.
[20] Skvortsov, D., *On some propositional logics related to the concept of information types of Yu. T. Medvedev*, Semiotika i informatika **13** (1979 (Russian)), pp. 142–149.

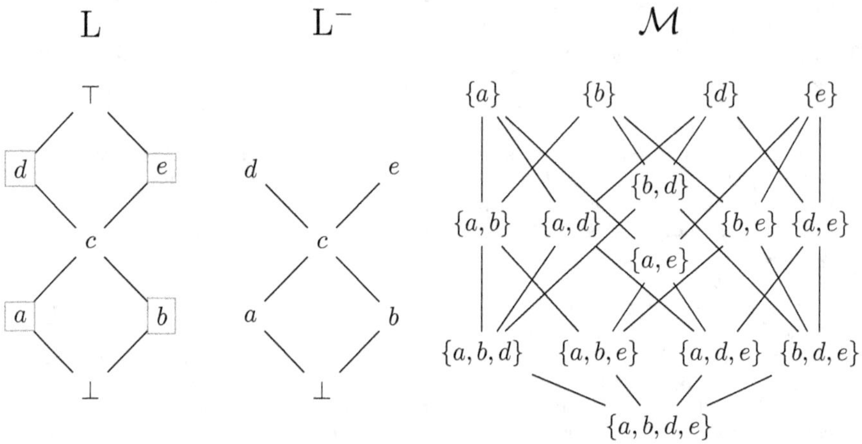

Figure 7. A distributive lattice L, the partial order L$^-$ and the associated Medvedev frame $\mathcal{M}$. As we did in Figure 3, we highlighted the elements of $\mathfrak{M}$.

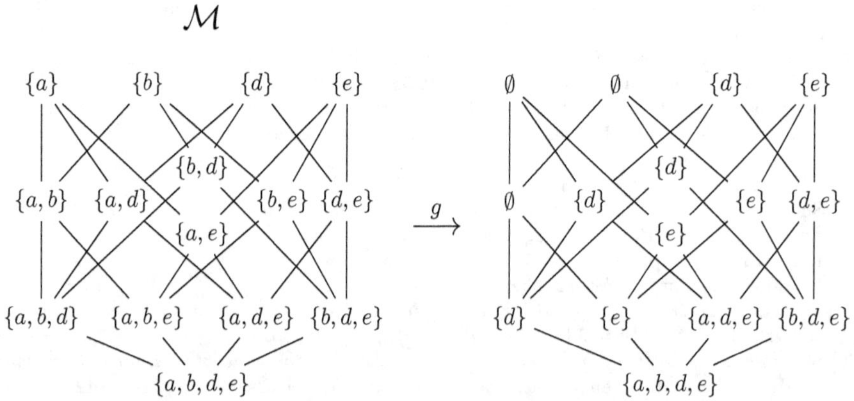

Figure 8. Representation of the function g from the proof of Lemma 3.2, for $\mathcal{M}$ the Medvedev frame in Figure 7.

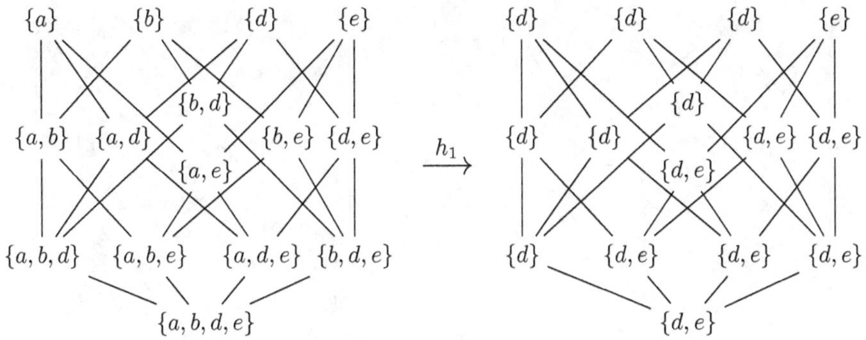

Figure 9. Representation of the function $h_1 : \mathcal{P}_0(\mathfrak{M}) \to \mathcal{P}_0(\mathfrak{C})$.

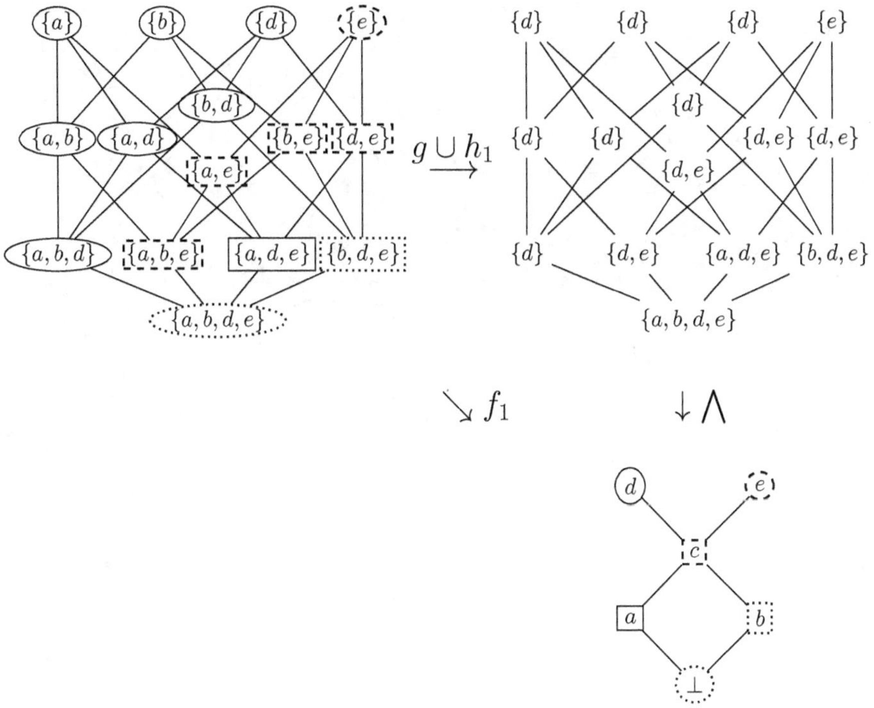

Figure 10. Representation of the bounded morphism $f_1 : \mathcal{P}_0(\mathfrak{M}) \twoheadrightarrow \mathrm{L}^-$.

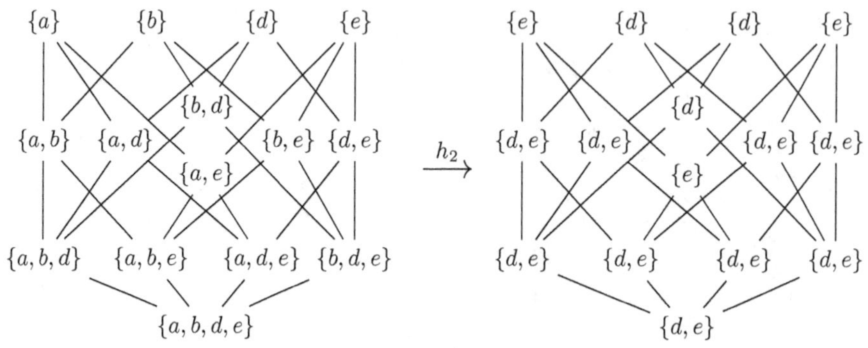

Figure 11. Representation of the function $h_2 : \mathcal{P}_0(\mathfrak{M}) \to \mathcal{P}_0(\mathfrak{C})$.

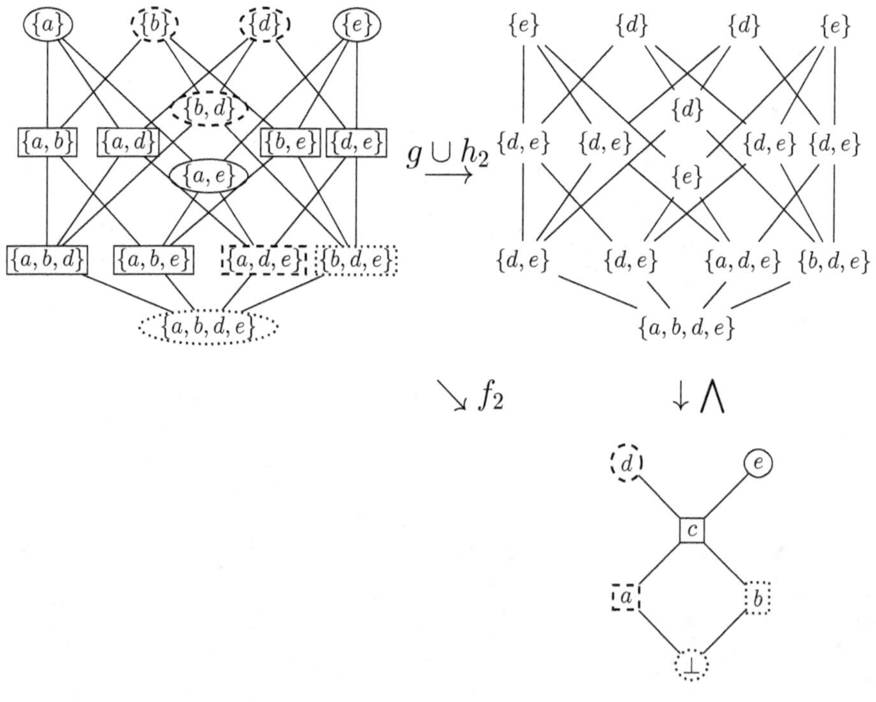

Figure 12. Representation of the bounded morphism $f_2 : \mathcal{P}_0(\mathfrak{M}) \twoheadrightarrow \mathrm{L}^-$.

Goldblatt-Thomason Theorems for Modal Intuitionistic Logics

Jim de Groot

The Australian National University
Canberra
Australia

Abstract

We prove a Goldblatt-Thomason theorem for dialgebraic intuitionistic logics, and instantiate it to Goldblatt-Thomason theorems for a wide variety of modal intuitionistic logics from the literature.

Keywords: Modal logic, intuitionistic logic, Goldblatt-Thomason theorem.

1 Introduction

A prominent question in the study of (modal) logics and their semantics is what classes of frames can be defined as the class of frames satisfying some set of formulae. Such a class is usually called *axiomatic* or *modally definable*. A milestone result partially answering this question in the realm of classical normal modal logic is from Goldblatt and Thomason and dates back to 1974 [16]. It states that an elementary class of Kripke frames is axiomatic if and only if it reflects ultrafilter extensions and is closed under p-morphic images, generated subframes and disjoint unions. The proof in [16] relies on Birkhoff's variety theorem [4] and makes use of the algebraic semantics of the logic. A model-theoretic proof was provided almost twenty years later by Van Benthem [1].

A similar result for (non-modal) intuitionistic logic was proven by Rodenburg [30] (see also [15]), where the interpreting structures are *intuitionistic* Kripke frames and models. This, of course, requires analogues of the notions of p-morphic images, generated subframes, disjoint unions and ultrafilter extensions. While the first three carry over straightforwardly from the setting of classical normal modal logic, ultrafilters need to be replaced by *prime filters*.

In recent years, Goldblatt-Thomason style theorems (which we will simply refer to as "Goldblatt-Thomason theorems") for many other logics have been proven, including for positive normal modal logic [8], graded modal logic [31], modal extensions of Łukasiewicz finitely-valued logics [35], LE-logics [10], and modal logics with a universal modality [32]. A general Goldblatt-Thomason theorem for coalgebraic logics for Set-coalgebras was given in [22].

In the present paper we prove Goldblatt-Thomason theorems for modal intuitionistic logics. These include the extensions of intuitionistic logic with a normal modality [36,37,38], a monotone one [14, Sec. 6], a neighbourhood modality [11], and a strict implication modality [25,26,12]. For each we obtain:

A class $\mathscr{K}$ of frames closed under prime filter extensions is axiomatic if and only if it reflects prime filter extensions and is closed under disjoint unions, regular subframes and p-morphic images.

Instead of proving each of these results individually, we prove a more general Goldblatt-Thomason theorem for *dialgebraic intuitionistic logics*, merging techniques from [15] and [22]. We then apply this to specific instances.

Dialgebraic logic slightly generalises coalgebraic logic and was recently introduced in [18]. It provides a framework where modal logics are developed parametric in the signature of the language and a functor $\mathcal{T} : \mathsf{C}' \to \mathsf{C}$, where C' is some subcategory of C. While coalgebraic logics are too restrictive to describe modal intuitionistic logics (see e.g. [24, Rem. 8], [18, Sec. 2]), the additional flexibility of dialgebraic logic does allow us to model a number of them.

The paper is structured as follows. In Sec. 2 we recall a semantics for the extension of intuitionistic logic with a normal modality $\Box$ from [38]. Using this as running example, in Sec. 3 we recall the basics of dialgebraic logic and prove the Goldblatt-Thomason theorem. In particular, this yields a new Goldblatt-Thomason theorem for the logic and semantics from Sec. 2. In Sec. 4 we instantiate the general theorem to several more modal intuitionistic logics from the literature to obtain new Goldblatt-Thomason theorems.

2 Normal Modal Intuitionistic Logic

For future reference, we recall the extension of intuitionistic logic with a unary meet-preserving modality from Wolter and Zakharyaschev [37,38].

Definition 2.1 Denote the language of intuitionistic logic by $\mathbf{L}$, with proposition letters from some countably infinite set Prop. That is, $\mathbf{L}$ is generated by the grammar

$$\varphi ::= \top \mid \bot \mid p \mid \varphi \wedge \varphi \mid \varphi \vee \varphi \mid \varphi \to \varphi,$$

where $p \in \text{Prop}$. Write $\mathbf{L}_\Box$ for its extension with a unary operator $\Box$. Further, let $\mathcal{L}$ be the intuitionistic propositional calculus, and let $\mathcal{L}_\Box$ be the logic that arises from extending an axiomatisation for $\mathcal{L}$ (that we assume includes uniform substitution) with the axioms and rule

$$\Box\top \leftrightarrow \top, \qquad \Box p \wedge \Box q \leftrightarrow \Box(p \wedge q), \qquad (p \leftrightarrow q)/(\Box p \leftrightarrow \Box q) \qquad (1)$$

We write Pos for the category of posets and order-preserving functions. In this paper, we define an *intuitionistic Kripke frame* as a poset and we write Krip for the full subcategory of Pos whose morphisms are p-morphisms [2, Sec. 2.1.1]. (Sometimes intuitionistic Kripke frames are defined to be preorders. For the results presented in this paper there is no discernible difference.)

Definition 2.2 A $\Box$-*frame* is a triple $(X, \leq, R)$ where $(X, \leq)$ is an intuitionistic Kripke frame and R is a relation on X satisfying $(\leq \circ R \circ \leq) = R$.

Adding a valuation $V : \text{Prop} \to \mathcal{U}p(X, \leq)$ $(= \{a \subseteq X \mid x \in a \text{ and } x \leq y \text{ implies } y \in a\})$ yields a $\Box$-*model*, in which we can interpret $\mathbf{L}_\Box$-formulae. Proposition letters are interpreted via the valuation, intuitionistic connectives are interpreted as usual in the underlying intuitionistic Kripke frame and a state x satisfies $\Box\varphi$ if all its R-successors satisfy φ.

While morphisms are not defined in [37,38], there is an obvious choice:

Definition 2.3 A $\Box$-*morphism* form $(X, \leq, R)$ to $(X', \leq', R')$ is a function $f : X \to X'$ such that for $E \in \{\leq, R\}$ and for all $x, y \in X$ and $z' \in X'$:

- If xEy then $f(x)E'f(y)$;
- If $f(x)E'z'$ then $\exists z \in X$ such that xEz and $f(z) = z'$.

We write $\mathsf{WZ}\Box$ for the category of $\Box$-frames and -morphisms.

The algebraic semantics of $\mathcal{L}_\Box$ is given as follows.

Definition 2.4 A *Heyting algebra with operators* (HAO) is a pair $(A, \Box)$ of a Heyting algebra A and a function $\Box : A \to A$ satisfying $\Box\top = \top$ and $\Box a \wedge \Box b = \Box(a \wedge b)$ for all $a, b \in A$. Together with $\Box$-preserving Heyting homomorphisms, these constitute the category HAO.

We briefly recall some categories, functors and natural transformations.

Definition 2.5 DL and HA denote the categories of distributive lattices and Heyting algebras. Let *up* be the contravariant functor $\mathsf{Pos} \to \mathsf{DL}$ that sends a poset to the distributive lattice of its upsets and an order-preserving function f to f^{-1}. Write $\mathit{pf} : \mathsf{DL} \to \mathsf{Pos}$ for the contravariant functor sending $A \in \mathsf{DL}$ to the set of prime filters of A ordered by inclusion, and a homomorphism to its inverse image. These restrict to $\mathit{up}' : \mathsf{Krip} \to \mathsf{HA}$ and $\mathit{pf}' : \mathsf{HA} \to \mathsf{Krip}$.

Let $\eta : \mathit{id}_{\mathsf{Pos}} \to \mathit{pf} \circ \mathit{up}$ and $\theta : \mathit{id}_{\mathsf{DL}} \to \mathit{up} \circ \mathit{pf}$ be the natural transformations defined by $\eta_{(X,\leq)}(x) = \{a \in \mathit{up}(X, \leq) \mid x \in a\}$ and $\theta_A(a) = \{\mathfrak{p} \in \mathit{pf}A \mid a \in \mathfrak{p}\}$. (These are the units of the dual adjunction between Pos and DL.) Furthermore, θ restricts to the natural transformation $\theta' : \mathit{id}_{\mathsf{HA}} \to \mathit{up}' \circ \mathit{pf}'$.

Every $\Box$-frame $(X, \leq, R)$ yields a HAO $(\mathit{up}'(X, \leq), \Box_R)$ (called its *complex algebra*), with $\Box_R(a) = \{x \in X \mid xRy \text{ implies } y \in a\}$. Conversely, every HAO $(A, \Box)$ gives rise to a $\Box$-frame $(\mathit{pf}'A, \subseteq, R_\Box)$, where $\mathfrak{p}R_\Box\mathfrak{q}$ iff for all $a \in A$, $\Box a \in \mathfrak{p}$ implies $a \in \mathfrak{q}$. Concatenating these constructions yields:

Definition 2.6 The *prime filter extension* of a $\Box$-frame $(X, \leq, R)$ is the frame $(X^{pe}, \subseteq, R^{pe})$, where X^{pe} is the set of prime filters on $(X, \leq)$ and R^{pe} is defined by $\mathfrak{p}R^{pe}\mathfrak{q}$ iff for all $a \in \mathit{up}'(X, \leq)$, $\Box_R(a) \in \mathfrak{p}$ implies $a \in \mathfrak{q}$.

3 A General Goldblatt-Thomason Theorem

We restrict the framework of dialgebraic logic [18] to an intuitionistic base. Within this, we prove a Goldblatt-Thomason theorem. Throughout this section, we show how general constructions specialise to the normal modal intuitionistic logic from Sec. 2. Our focus on an intuitionistic propositional base allows us to augment the framework of dialgebraic logic from [18] in the following ways:

- In [18] a logic is identified via an initial object in some category, which plays the role of the Lindenbaum-Tarski algebra. Here we define logics explicitly, by means of an axiomatisation.
- Whereas proposition letters in [18] are regarded as predicate liftings, here we elevate them to a special status. This has two reasons: first, it simplifies the connection to (frames and models for) modal intuitionistic logics from the literature; second, they facilitate the use of Birkhoff's variety theorem.
- We give dialgebraic definitions of subframes, p-morphic images and disjoint unions, and corresponding preservation results.
- We give prime filter extensions for models (not just for frames).

We work towards a Goldblatt-Thomason theorem as follows. First we recall the use of dialgebras as frames for modal extensions of intuitionistic logic (Sec. 3.1), and we prove some invariance properties (Sec. 3.2). Then we describe algebraic semantics and prime filter extensions dialgebraically (Sec. 3.3 and 3.4). This culminates in the Goldblatt-Thomason theorem in Sec. 3.5.

3.1 Languages and Frames

Dialgebras were introduced by Hagino in [19] to describe data types. Here we use them to describe frames for modal intuitionistic logics.

Definition 3.1 Let $\mathcal{F}, \mathcal{G} : \mathsf{C} \to \mathsf{D}$ be functors. An $(\mathcal{F}, \mathcal{G})$-*dialgebra* is a pair (X, γ) where $X \in \mathsf{C}$ and $\gamma : \mathcal{F}X \to \mathcal{G}X$ is a D-morphism. An $(\mathcal{F}, \mathcal{G})$-*dialgebra morphism* from (X, γ) to (X', γ') is a C-morphism $f : X \to X'$ such that $\mathcal{G}f \circ \gamma = \gamma' \circ \mathcal{F}f$. They constitute the category $\mathsf{Dialg}(\mathcal{F}, \mathcal{G})$. In diagrams:

$$\text{objects:} \quad \begin{array}{c} \mathcal{F}X \\ \downarrow{\scriptstyle\gamma} \\ \mathcal{G}X \end{array} \qquad \text{arrows:} \quad \begin{array}{ccc} \mathcal{F}X & \xrightarrow{\mathcal{F}f} & \mathcal{F}X' \\ {\scriptstyle\gamma}\downarrow & & \downarrow{\scriptstyle\gamma'} \\ \mathcal{G}X & \xrightarrow{\mathcal{G}f} & \mathcal{G}X' \end{array}$$

We will be concerned with two classes of dialgebras. First, $(i, \mathcal{T})$-dialgebras, where $i : \mathsf{Krip} \to \mathsf{Pos}$ is the inclusion functor and $\mathcal{T} : \mathsf{Krip} \to \mathsf{Pos}$ is any functor, serve as frame semantics for our dialgebraic intuitionistic logics. Second, dialgebras for functors $\mathsf{HA} \to \mathsf{DL}$ will be used as algebraic semantics.

Example 3.2 Let $\mathcal{P}_{up} : \mathsf{Krip} \to \mathsf{Pos}$ be the functor that sends an intuitionistic Kripke frame $(X, \leq)$ to its set of upsets ordered by reverse inclusion, and a p-morphism $f : (X, \leq) \to (X', \leq')$ to $\mathcal{P}_{up}f : \mathcal{P}_{up}(X, \leq) \to \mathcal{P}_{up}(X', \leq') : a \mapsto f[a]$. Then identifying a relation R on X with the map $\gamma_R : (X, \leq) \to \mathcal{P}_{up}(X, \leq) : x \mapsto \{y \in X \mid xRy\}$ yields an isomorphism $\mathsf{WZ}\Box \cong \mathsf{Dialg}(i, \mathcal{P}_{up})$ [18, Sec. 2].

Modalities for $\mathsf{Dialg}(i, \mathcal{T})$ are defined via predicate liftings [18, Def. 5.7].

Definition 3.3 An *n-ary predicate lifting* for a functor $\mathcal{T} : \mathsf{Krip} \to \mathsf{Pos}$ is a natural transformation

$$\lambda : (\mathcal{U}p \circ i)^n \to \mathcal{U}p \circ \mathcal{T}.$$

Here $\mathcal{U}p : \mathsf{Pos} \to \mathsf{Set}$ is the contravariant functor that sends a poset to its set of upsets, and $(\mathcal{U}p \circ i)^n(X, \leq)$ is the n-fold product of $\mathcal{U}p(i(X, \leq))$ in Set.

Definition 3.4 Let Prop be a countably infinite set of proposition letters. For a set Λ of predicate liftings, define the language $\mathbf{L}(\Lambda)$ by the grammar

$$\varphi ::= \top \mid \bot \mid p \mid \varphi \wedge \varphi \mid \varphi \vee \varphi \mid \varphi \to \varphi \mid \heartsuit^\lambda(\varphi_1, \ldots, \varphi_n),$$

where p ranges over Prop and $\lambda \in \Lambda$ is n-ary.

Definition 3.5 Let Λ be a set of predicate liftings for $\mathcal{T} : \mathsf{Krip} \to \mathsf{Pos}$. An $(i, \mathcal{T})$-*model* $\mathfrak{M}$ is an $(i, \mathcal{T})$-dialgebra $\mathfrak{X} = (X, \leq, \gamma)$ with a valuation $V : \mathrm{Prop} \to \mathcal{U}p(X, \leq)$. Truth of $\varphi \in \mathbf{L}(\Lambda)$ at $x \in X$ is defined by

$$\mathfrak{M}, x \Vdash \top \quad \text{always}, \qquad \mathfrak{M}, x \Vdash \bot \quad \text{never}, \qquad \mathfrak{M}, x \Vdash p \quad \text{iff} \quad x \in V(p)$$
$$\mathfrak{M}, x \Vdash \varphi \wedge \psi \quad \text{iff} \quad \mathfrak{M}, x \Vdash \varphi \text{ and } \mathfrak{M}, x \Vdash \psi$$
$$\mathfrak{M}, x \Vdash \varphi \vee \psi \quad \text{iff} \quad \mathfrak{M}, x \Vdash \varphi \text{ or } \mathfrak{M}, x \Vdash \psi$$
$$\mathfrak{M}, x \Vdash \varphi \to \psi \quad \text{iff} \quad x \leq y \text{ and } \mathfrak{M}, y \Vdash \varphi \text{ imply } \mathfrak{M}, y \Vdash \psi$$
$$\mathfrak{M}, x \Vdash \heartsuit^\lambda(\varphi_1, \ldots, \varphi_n) \quad \text{iff} \quad \gamma(x) \in \lambda_{(X,\leq)}(\llbracket \varphi_1 \rrbracket^{\mathfrak{M}}, \ldots, \llbracket \varphi_n \rrbracket^{\mathfrak{M}})$$

Here $\llbracket \varphi \rrbracket^{\mathfrak{M}} = \{x \in X \mid \mathfrak{M}, x \Vdash \varphi\}$. We write $\mathfrak{M} \Vdash \varphi$ if $\mathfrak{M}, x \Vdash \varphi$ for all $x \in X$ and $\mathfrak{X} \Vdash \varphi$ if $(\mathfrak{X}, V) \Vdash \varphi$ for all valuations V for $\mathfrak{X}$. If $\Phi \subseteq \mathbf{L}(\Lambda)$ then we say that Φ is *valid* on $\mathfrak{X}$, and write $\mathfrak{X} \Vdash \Phi$, if $\mathfrak{X} \Vdash \varphi$ for all $\varphi \in \Phi$. Also, let

$$\mathrm{Fr}\,\Phi = \{\mathfrak{X} \in \mathsf{Dialg}(i, \mathcal{T}) \mid \mathfrak{X} \Vdash \Phi\}.$$

We call a class $\mathscr{K} \subseteq \mathsf{Dialg}(i, \mathcal{T})$ *axiomatic* if $\mathscr{K} = \mathrm{Fr}\,\Phi$ for some $\Phi \subseteq \mathbf{L}(\Lambda)$.

Example 3.6 Since $\Box$-frames correspond to $(i, \mathcal{P}_{up})$-dialgebras, it is easy to see that $\Box$-models correspond to $(i, \mathcal{P}_{up})$-models. The modal operator $\Box$ can be induced by the predicate lifting $\lambda^\Box : \mathcal{U}p \circ i \to \mathcal{U}p \circ \mathcal{P}_{up}$ given by

$$\lambda^\Box_{(X,\leq)} : \mathcal{U}p(i(X, \leq)) \to \mathcal{U}p(\mathcal{P}_{up}(X, \leq)) : a \mapsto \{b \in \mathcal{P}_{up}(X, \leq) \mid b \subseteq a\}.$$

Indeed, if $\mathfrak{M} = (X, \leq, R, V)$ is a $\Box$-model and $(X, \leq, \gamma_R, V)$ the corresponding $(i, \mathcal{P}_{up})$-model then we have $x \Vdash \Box\varphi$ iff every R-successor of x satisfies φ, i.e. iff $\gamma_R(x) \subseteq \llbracket \varphi \rrbracket^{\mathfrak{M}}$. By definition the latter is equivalent to $\gamma_R(x) \in \lambda^\Box_{(X,\leq)}(\llbracket \varphi \rrbracket^{\mathfrak{M}})$.

Finally, we define morphisms between $(i, \mathcal{T})$-models.

Definition 3.7 An $(i, \mathcal{T})$-*model morphism* from $\mathfrak{M} = (\mathfrak{X}, V)$ to $\mathfrak{M}' = (\mathfrak{X}', V')$ is an $(i, \mathcal{T})$-dialgebra morphism $f : \mathfrak{X} \to \mathfrak{X}'$ such that $V = f^{-1} \circ V'$.

Proposition 3.8 *If $f : \mathfrak{M} \to \mathfrak{M}'$ is an $(i, \mathcal{T})$-model morphism, then for all states x of $\mathfrak{M}$ and $\varphi \in \mathbf{L}(\Lambda)$, we have $\mathfrak{M}, x \Vdash \varphi$ iff $\mathfrak{M}', f(x) \Vdash \varphi$.*

Proof. Let $\mathfrak{M} = (X, \leq, \gamma, V)$ and $\mathfrak{M}' = (X', \leq', \gamma', V')$. The proof proceeds by induction on the structure of φ. If $\varphi \in \mathrm{Prop}$ then the claim follows from the definition of an $(i, \mathcal{T})$-model morphism. The inductive cases for propositional connectives are routine, so we focus on the modal case. We restrict our attention to unary modalities, higher arities being similar. Compute:

$$
\begin{aligned}
&\mathfrak{M}, x \Vdash \heartsuit^{\lambda}\varphi && \\
&\quad\text{iff} \quad \gamma(x) \in \lambda_{(X,\leq)}([\![\varphi]\!]^{\mathfrak{M}}) && \text{(Def. 3.5)}\\
&\quad\text{iff} \quad \gamma(x) \in \lambda_{(X,\leq)}(f^{-1}([\![\varphi]\!]^{\mathfrak{M}'})) && \text{(Induction hypothesis)}\\
&\quad\text{iff} \quad \gamma(x) \in \lambda_{(X,\leq)}((if)^{-1}([\![\varphi]\!]^{\mathfrak{M}'})) && \text{(Because } if = f\text{)}\\
&\quad\text{iff} \quad \gamma(x) \in (\mathcal{T}f)^{-1}(\lambda_{(X',\leq')}([\![\varphi]\!]^{\mathfrak{M}'})) && \text{(Naturality of } \lambda\text{)}\\
&\quad\text{iff} \quad (\mathcal{T}f)(\gamma(x)) \in \lambda_{(X',\leq')}([\![\varphi]\!]^{\mathfrak{M}'}) && \\
&\quad\text{iff} \quad \gamma'((if)(x)) \in \lambda_{(X',\leq')}([\![\varphi]\!]^{\mathfrak{M}'}) && (f \text{ is a dialgebra morphism})\\
&\quad\text{iff} \quad \mathfrak{M}', f(x) \Vdash \heartsuit^{\lambda}\varphi && \text{(Def. 3.5 and } if = f\text{)}
\end{aligned}
$$

This proves the proposition. □

3.2 Disjoint Unions, Generated Subframes and p-Morphic Images

The category theoretic analogue of a disjoint union is a coproduct. For any $\mathcal{T}$: $\mathsf{Krip} \to \mathsf{Pos}$ the category $\mathsf{Dialg}(i, \mathcal{T})$ has coproducts because Krip has coproducts and i preserves them [6, Thm. 3.2.1]. So we define:

Definition 3.9 The *disjoint union* of a K-indexed family of $(i, \mathcal{T})$-dialgebras $\mathfrak{X}_k = (X_k, \leq_k, \gamma_k)$ is the coproduct $\coprod_{k \in K} \mathfrak{X}_k$ in $\mathsf{Dialg}(i, \mathcal{T})$.

Example 3.10 Let $(X_k, \leq_k, R_k)$ be a K-indexed set of □-frames, and $(X_k, \leq_k, \gamma_k)$ the corresponding $(i, \mathcal{P}_{up})$-dialgebras. The coproduct $\coprod_{k \in K}(X_k, \leq_k, \gamma_k)$ is given by $(X, \leq, \gamma)$, where $(X, \leq)$ is the coproduct of the intuitionistic Kripke frames $(X_k, \leq_k)$ (which is computed as in Set), and $\gamma : (X, \leq) \to \mathcal{P}_{up}(X, \leq)$ is given by $\gamma(x_k) = \gamma_k(x_k)$ (for $x_k \in X_k$). Transforming this back into a □-frame, we obtain $(X, \leq, R)$, with xRy iff there is a $k \in K$ with $x, y \in X_k$ and xR_ky. So this corresponds to the expected notion of disjoint union of □-frames.

Proposition 3.11 *Let $\mathfrak{X}_k = (X_k, \leq_k, \gamma_k)$ be a family of $(i, \mathcal{T})$-dialgebras indexed by some set K. Suppose $\mathfrak{X}_k \Vdash \varphi$ for all $k \in K$. Then $\coprod \mathfrak{X}_k \Vdash \varphi$.*

Proof. Let V be a valuation for $\coprod \mathfrak{X}_k$. Define the valuation V_k for $\mathfrak{X}_k$ by $V_k(p) = V(p) \cap X_k$. Then the coproduct inclusion maps $\kappa_k : (\mathfrak{X}_k, V_k) \to (\coprod \mathfrak{X}_k, V)$ are $(i, \mathcal{T})$-model morphisms, hence the assumption $\mathfrak{X}_k \Vdash \varphi$ for all $k \in K$ implies that $(\coprod \mathfrak{X}_k, V) \Vdash \varphi$. Since V was arbitrary, $\coprod \mathfrak{X}_k \Vdash \varphi$. □

Definition 3.12 Let $\mathfrak{X}' = (X', \leq', \gamma')$ and $\mathfrak{X} = (X, \leq, \gamma)$ be $(i, \mathcal{T})$-dialgebras.

(i) $\mathfrak{X}'$ is called a *generated subframe* of $\mathfrak{X}$ if there exists a p-morphism f : $\mathfrak{X}' \to \mathfrak{X}$ such that $f : (X', \leq') \to (X, \leq)$ is an embedding.

(ii) $\mathfrak{X}'$ is a *p-morphic image* of $\mathfrak{X}$ if there exists a surjective dialgebra morphism $\mathfrak{X} \to \mathfrak{X}'$.

Example 3.13 Guided by [5, Def. 2.5 and 3.13], we could define a generated sub-□-frame of a □-frame $(X, \leq, R)$ as a □-frame $(X', \leq', R')$ such that:

- $X' \subseteq X$ and $\leq' = (\leq \cap (X' \times X'))$ and $R' = (R \cap (X' \times X'))$;
- if $x \in X'$ and $x \leq y$ or xRy, then $y \in X'$.

With this definition, it can be shown that a $\Box$-frame $\mathfrak{X}'$ is isomorphic to a generated sub-$\Box$-frame of a $\Box$-frame $\mathfrak{X}$ if and only if the dialgebraic rendering of $\mathfrak{X}'$ is a generated subframe of the dialgebraic rendering of $\mathfrak{X}$ (as per Def. 3.12).

Proposition 3.14 *Let $\mathfrak{X}$ be an $(i, \mathcal{T})$-dialgebra such that $\mathfrak{X} \Vdash \varphi$.*

(i) *If $\mathfrak{X}'$ is a generated subframe of $\mathfrak{X}$ then $\mathfrak{X}' \Vdash \varphi$.*

(ii) *If $\mathfrak{X}'$ is a p-morphic image of $\mathfrak{X}$ then $\mathfrak{X}' \Vdash \varphi$.*

Proof. We prove the first item, the second item being similar. If $\mathfrak{X}' = (X', \leq', \gamma')$ is a generated subframe of $\mathfrak{X} = (X, \leq, \gamma)$ then there exists a $(i, \mathcal{T})$-dialgebra morphism $f : \mathfrak{X}' \to \mathfrak{X}$ that is an embedding of the underlying posets. Let V' be any valuation for $\mathfrak{X}'$. Define a valuation $V^{\uparrow}$ for $\mathfrak{X}$ by $V^{\uparrow}(p) = \{x \in X \mid \exists y \in V'(p) \text{ s.t. } f(y) \leq x\}$. Then the fact that f is an embedding implies that $V' = f^{-1}V^{\uparrow}$, and therefore $f : (\mathfrak{X}', V') \to (\mathfrak{X}, V^{\uparrow})$ is a dialgebra model morphism. The assumption that $\mathfrak{X} \Vdash \varphi$ together with Prop. 3.8 implies that $(\mathfrak{X}', V') \Vdash \varphi$. Since V' is arbitrary we find $\mathfrak{X}' \Vdash \varphi$. $\Box$

3.3 Axioms and Algebraic Semantics

In order to get intuition for the dialgebraic perspective of algebraic semantics, we observe that the category HAO is isomorphic to a category of dialgebras. In this case, we consider dialgebras for functors $\mathsf{HA} \to \mathsf{DL}$. Again, one of the functors is simply the inclusion functor, which we denote by $j : \mathsf{HA} \to \mathsf{DL}$.

Example 3.15 Let $\mathcal{K} : \mathsf{HA} \to \mathsf{DL}$ be the functor that sends a Heyting algebra A to the free distributive lattice generated by $\{\boxdot a \mid a \in A\}$ modulo $\boxdot\top = \top$ and $\boxdot a \wedge \boxdot b = \boxdot(a \wedge b)$, where a and b range over A. The action of $\mathcal{K}$ on a Heyting homomorphism $h : A \to A'$ is defined on generators by $\mathcal{K}h(\boxdot a) = \boxdot h(a)$. Then $\mathsf{HAO} \cong \mathsf{Dialg}(\mathcal{K}, j)$ [18, Exm. 3.3].

We denote generators by dotted boxes to distinguish them from the modality $\Box$. Observe that the relations defining $\mathcal{K}$ correspond to the axioms we want a normal box to satisfy. We investigate how to generalise this to the setting of some arbitrary set Λ of predicate liftings for a functor $\mathcal{T} : \mathsf{Krip} \to \mathsf{Pos}$.

Definition 3.16 A *rank-1 formula* in $\mathbf{L}(\Lambda)$ is a formula φ such that

- φ does not contain intuitionistic implication;
- each proposition letter appears in the scope of precisely one modal operator.

A *rank-1 axiom* is a formula of the form $\varphi \leftrightarrow \psi$, where φ, ψ are rank-1 formulae. It is called *sound* if it is valid in all $(i, \mathcal{T})$-dialgebras.

Let Ax be a collection of sound rank-1 axioms. Define the logic $\mathcal{L}(\Lambda, \mathrm{Ax})$ as the smallest set of $\mathbf{L}(\Lambda)$-formulae containing Ax and an axiomatisation for intuitionistic logic, which is closed under modus ponens, uniform substitution, and

$$\frac{\varphi_1 \leftrightarrow \psi_1 \quad \cdots \quad \varphi_n \leftrightarrow \psi_n}{\heartsuit^{\lambda}(\varphi_1, \ldots, \varphi_n) \leftrightarrow \heartsuit^{\lambda}(\psi_1, \ldots, \psi_n)} \quad \text{(congruence rule).}$$

Example 3.15 generalises as follows [18, Sec. 5].

Definition 3.17 Let Λ be a set of predicate liftings for $\mathcal{T}$ and Ax a set of sound rank-1 axioms for $\mathbf{L}(\Lambda)$. For a Heyting algebra A, define $\mathcal{L}^{(\Lambda,\mathrm{Ax})}A$ to be the free distributive lattice generated by $\{\dot{\heartsuit}^{\lambda}(a_1,\dots,a_n) \mid \lambda \in \Lambda, a_i \in A\}$ modulo the axioms in Ax, where each occurrence of $\heartsuit$ is replaced by the formal generator $\dot{\heartsuit}$, $\leftrightarrow$ is replaced by $=$, and the proposition letters range over the elements of A. (This is well defined since the axioms in Ax are rank-1 axioms, which result in equations constructed from elements of the form $\dot{\heartsuit}(a_1,\dots,a_n)$ and distributive lattice connectives.)

If $h : A \to A'$ is a Heyting homomorphism, define $\mathcal{L}^{(\Lambda,\mathrm{Ax})}h : \mathcal{L}^{(\Lambda,\mathrm{Ax})}A \to \mathcal{L}^{(\Lambda,\mathrm{Ax})}A'$ on generators by $\mathcal{L}^{(\Lambda,\mathrm{Ax})}h(\dot{\heartsuit}^{\lambda}(a_1,\dots,a_n)) = \dot{\heartsuit}^{\lambda}(h(a_1),\dots,h(a_n))$. Then $\mathcal{L}^{(\Lambda,\mathrm{Ax})} : \mathsf{HA} \to \mathsf{DL}$ defines a functor.

Again, we use a symbol with a dot in it to denote formal generators, and separate them from symbols in the language.

Example 3.18 Let $\Lambda = \{\lambda^{\Box}\}$, where $\lambda^{\Box}$ is the predicate lifting from Exm. 3.6, and write $\Box$ instead of $\heartsuit^{\lambda^{\Box}}$. Let Ax consist of the two axioms (not the rule) from (1), and note that these are both rank-1 axioms. Then the logic $\mathcal{L}(\Lambda,\mathrm{Ax})$ coincides with $\mathcal{L}_{\Box}$, and the functor obtained from the procedure in Def. 3.17 is naturally isomorphic to $\mathcal{K}$ from Exm. 3.15. (The only difference is the symbol used to represent the formal generators.)

The following observation allows us to use the Birkhoff variety theorem when proving the Goldblatt-Thomason theorem below.

Lemma 3.19 *Let $\mathcal{L}$ be obtained from predicate liftings and axioms via Def. 3.17. Then the category* $\mathsf{Dialg}(\mathcal{L},j)$ *is a variety of algebras.*

Proof. It is known that the category HA of Heyting algebras is a variety of algebras. We add to its signature an n-ary operation symbol for each n-ary predicate lifting in Λ, and to the set of equations defining HA the equations obtained from Ax by replacing $\leftrightarrow$ with equality and proposition letters with variables. □

We can evaluate $\mathbf{L}(\Lambda)$-formulae in a $(\mathcal{L}^{(\Lambda,\mathrm{Ax})},j)$-dialgebra (A,α) with an assignment of the proposition letters to elements of A. Intuitionistic connectives are interpreted as in the Heyting algebra A, and the interpretation of $\heartsuit^{\lambda}(\varphi_1,\dots,\varphi_n)$ is given by $\alpha(\dot{\heartsuit}^{\lambda}(\llparenthesis\varphi_1\rrparenthesis,\dots,\llparenthesis\varphi_n\rrparenthesis))$, where $\llparenthesis\varphi_i\rrparenthesis$ is the interpretation of φ_i. We say that φ is valid in (A,α), and write $(A,\alpha) \models \varphi$, if φ evaluates to $\top$ under every assignment of the proposition letters.

This evaluation is closely related to the interpretation of formulae in $(i,\mathcal{T})$-dialgebras: a formula φ is valid in some $(i,\mathcal{T})$-dialgebra if and only if it is valid in some related algebra, called the complex algebra.

Definition 3.20 Define $\rho : \mathcal{L}^{(\Lambda,\mathrm{Ax})} \circ \mathit{up}' \to \mathit{up} \circ \mathcal{T}$ on generators by

$$\rho_{(X,\leq)}(\dot{\heartsuit}^{\lambda}(a_1,\dots,a_n)) = \lambda_{(X,\leq)}(a_1,\dots,a_n).$$

Then ρ is a well defined transformation because Ax is assumed to be sound, and it is natural because predicate liftings are natural transformations.

It gives rise to a functor $(\cdot)^+ : \mathsf{Dialg}(i, \mathcal{T}) \to \mathsf{Dialg}(\mathcal{L}^{(\Lambda,\mathrm{Ax})}, j)$, which sends an $(i, \mathcal{T})$-dialgebra $(X, \leq, \gamma)$ to its *complex algebra* $(up'(X, \leq), \gamma^+)$, given by

$$\mathcal{L}^{(\Lambda,\mathrm{Ax})}(up'(X,\leq)) \xrightarrow{\rho(X,\leq)} up(\mathcal{T}(X,\leq)) \xrightarrow{up\gamma} up(i(X,\leq)) = j(up'(X,\leq)).$$
$$\underbrace{\qquad\qquad\qquad\qquad\qquad}_{\gamma^+}$$

The action of $(\cdot)^+$ on an $(i, \mathcal{T})$-dialgebra morphism f is given by $f^+ = up' f$.

Example 3.21 Let $(X, \leq, R)$ be a $\Box$-frame and $(X, \leq, \gamma)$ the corresponding $(i, \mathcal{P}_{up})$-dialgebra. The complex algebra of $(X, \leq, \gamma)$ is the $(\mathcal{K}, j)$-dialgebra $(up'(X, \leq), \gamma^+)$, where γ^+ is given by $\gamma^+(\boxdot a) = \gamma^{-1}(\lambda^\Box(a)) = \{x \in X \mid \gamma(x) \subseteq a\}$. Translating this to a HAO, we see that this corresponds precisely to the complex algebra of $(X, \leq, R)$ in the sense of Sec. 2.

Proposition 3.22 *Let $\mathfrak{X}$ be an $(i, \mathcal{T})$-dialgebra and $\varphi \in \mathbf{L}(\Lambda)$. Then we have*

$$\mathfrak{X} \Vdash \varphi \quad \textit{iff} \quad \mathfrak{X}^+ \models \varphi.$$

Proof. This follows from a routine induction on the structure of φ, where the base case follows from the fact that valuations for $\mathfrak{X}$ correspond bijectively to assignments of the proposition letters to elements of $\mathfrak{X}^+$. □

3.4 Prime Filter Extensions

The proof of the Goldblatt-Thomason theorem relies on Birkhoff's variety theorem and the connection between frame semantics and algebraic semantics of a logic. As we have seen above, every $\Box$-frame gives rise to a complex algebra, or, more generally, every $(i, \mathcal{T})$-dialgebra gives rise to a $(\mathcal{L}, j)$-dialgebra. To transfer the variety theorem from $(\mathcal{L}, j)$-dialgebras back to $(i, \mathcal{T})$-dialgebras, we need a functor $(\cdot)_+ : \mathsf{Dialg}(\mathcal{L}, j) \to \mathsf{Dialg}(i, \mathcal{T})$ such that for each $(i, \mathcal{T})$-dialgebra $\mathfrak{X}$,

$$(\mathfrak{X}^+)_+ \Vdash \varphi \quad \text{implies} \quad \mathfrak{X} \Vdash \varphi. \qquad (\star)$$

Assumption 3.23 *Throughout this subsection, let $\mathcal{T} : \mathsf{Krip} \to \mathsf{Pos}$ be a functor, Λ a set of predicate liftings for $\mathcal{T}$, and a set Ax of sound rank-1 axioms from $\mathbf{L}(\Lambda)$. Abbreviate $\mathcal{L} := \mathcal{L}^{(\Lambda,\mathrm{Ax})}$ and $\rho := \rho^{(\Lambda,\mathrm{Ax})}$.*

A functor $(\cdot)_+ : \mathsf{Dialg}(\mathcal{L}, j) \to \mathsf{Dialg}(i, \mathcal{T})$ arises from a natural transformation τ in the same way as ρ induced a functor from frames to complex algebras. To stress its dependence on the choice of τ, we denote it by $(\cdot)_\tau$ instead of $(\cdot)_+$.

Definition 3.24 Let $\tau : pf \circ \mathcal{L} \to \mathcal{T} \circ pf'$ be a natural transformation. Then we define the contravariant functor $(\cdot)_\tau : \mathsf{Dialg}(\mathcal{L}, j) \to \mathsf{Dialg}(i, \mathcal{T})$ on objects by sending a $(\mathcal{L}, j)$-dialgebra $\mathcal{H} = (H, \alpha)$ to the $(i, \mathcal{T})$-dialgebra $\mathcal{H}_\tau$ given by

$$i(pf'H) = pf(jH) \xrightarrow{pf\alpha} pf(\mathcal{L}H) \xrightarrow{\tau_H} \mathcal{T}(pf'H).$$

For a $(\mathcal{L}, j)$-dialgebra morphism $h : \mathcal{H} \to \mathcal{H}'$ we define $h_\tau = pf'h : \mathcal{H}'_\tau \to \mathcal{H}_\tau$. Naturality of τ ensures that this is well defined.

We call $(\mathfrak{X}^+)_\tau$ the τ-prime filter extension of an $(i, \mathcal{T})$-dialgebra $\mathfrak{X}$ if τ satisfies a sufficient condition that ensures that $(\star)$ holds (by Prop. 3.27). This condition relies on the following variation of the adjoint mate of ρ.

Definition 3.25 Let $\rho : \mathcal{L} \circ up' \to up \circ \mathcal{T}$. Then we write $\rho^\flat$ for the natural transformation defined as the composition

$$\mathcal{T} \circ pf' \xrightarrow{\eta_{\mathcal{T} \circ pf'}} pf \circ up \circ \mathcal{T} \circ pf' \xrightarrow{pf \rho_{pf'}} pf \circ \mathcal{L} \circ up' \circ pf' \xrightarrow{pf(\mathcal{L}\theta')} pf \circ \mathcal{L},$$

where η and θ are defined as in Def. 2.5.

Definition 3.26 Let τ be a natural transformation such that $\rho^\flat \circ \tau = id_{pf \circ \mathcal{L}}$.

(i) Define $\mathfrak{pe}_\tau := (\cdot)_\tau \circ (\cdot)^+ : \mathsf{Dialg}(i, \mathcal{T}) \to \mathsf{Dialg}(i, \mathcal{T})$. We call $\mathfrak{pe}_\tau \mathfrak{X}$ the *τ-prime filter extension* of $\mathfrak{X} \in \mathsf{Dialg}(i, \mathcal{T})$.

(ii) The *τ-prime filter extension* of a model $\mathfrak{M} = (\mathfrak{X}, V)$ is $\mathfrak{pe}_\tau \mathfrak{M} := (\mathfrak{pe}_\tau \mathfrak{X}, V^{pe})$, where $V^{pe}(p) = \{\mathfrak{q} \in \mathfrak{pe}_\tau \mathfrak{X} \mid V(p) \in \mathfrak{q}\}$ for all $p \in \mathrm{Prop}$.

Observe that the prime filter extension of an $(i, \mathcal{T})$-dialgebra $\mathfrak{X} = (X, \leq, \gamma)$ is of the form $\mathfrak{pe}_\tau \mathfrak{X} = (X^{pe}, \subseteq, \gamma^{pe})$, where X^{pe} denotes the set of prime filters of upsets of $(X, \leq)$ and γ^{pe} is computed using both ρ and τ.

We now show that τ-prime filter extensions satisfy $(\star)$.

Proposition 3.27 *Let τ be a natural transformation such that $\rho^\flat \circ \tau = id_{pf \circ \mathcal{L}}$, $\mathfrak{X} = (X, \leq, \gamma)$ an $(i, \mathcal{T})$-dialgebra, $\mathfrak{M} = (\mathfrak{X}, V)$ a model based on $\mathfrak{X}$, $\varphi \in \mathbf{L}(\Lambda)$.*

(i) *For all prime filters $\mathfrak{q} \in X^{pe}$ we have $\mathfrak{pe}_\tau \mathfrak{M}, \mathfrak{q} \Vdash \varphi$ iff $[\![\varphi]\!]^{\mathfrak{M}} \in \mathfrak{q}$.*

(ii) *For all states $x \in X$ we have $\mathfrak{M}, x \Vdash \varphi$ iff $\mathfrak{pe}_\tau \mathfrak{M}, \eta_{(X,\leq)}(x) \Vdash \varphi$.*

(iii) *If $\mathfrak{pe}_\tau \mathfrak{X} \Vdash \varphi$ then $\mathfrak{X} \Vdash \varphi$.*

Proof. The proof of the proposition is given in the appendix. □

Example 3.28 Returning to our example of $\Box$-frames, we wish to find a natural transformation $\tau^\Box$ such that $(\rho^\Box)^\flat \circ \tau^\Box = id_{pf \circ \mathcal{L}^\Box}$.

Before defining $\tau^\Box$, let us get an idea of what $(\rho^\Box)^\flat$ looks like. Let A be a Heyting algebra and $Q \in pf(\mathcal{L}^\Box A)$. Since Q is determined by elements of the form $\boxdot a$ it contains, where $a \in A$, we pay special attention to these elements. For $D \in \mathcal{P}_{up} \circ pf' A$ and $a \in A$ we have

$$\begin{aligned} \boxdot a \in (\rho^\Box_A)^\flat(D) \quad &\text{iff} \quad \rho_{pf'A}((pf(\mathcal{L}^\Box \theta'_A))(\boxdot a)) \in \eta_{\mathcal{T}(pf'A)}(D) \\ &\text{iff} \quad \rho_{pf'A}(\boxdot \theta'_A(a)) \in \eta_{\mathcal{T}(pf'A)}(D) \\ &\text{iff} \quad D \in \rho_{pf'A}(\boxdot \theta'_A(a)) \\ &\text{iff} \quad D \subseteq \theta'_A(a) \end{aligned}$$

Guided by this we define $\tau : pf \circ \mathcal{L}^\Box \to \mathcal{P}_{up} \circ pf'$ on components by

$$\tau^\Box_A : pf(\mathcal{L}^\Box A) \to \mathcal{P}_{up}(pf' A) : Q \mapsto \{\mathfrak{p} \in pf' A \mid \forall a \in A, \boxdot a \in Q \text{ implies } a \in \mathfrak{p}\}$$

With this definition we can prove the following lemma, the proof of which can be found in the appendix.

Lemma 3.29 *$\tau^{\Box}$ is a natural transformation such that $(\rho^{\Box})^{\flat} \circ \tau^{\Box} = id_{pf \circ \mathcal{L}^{\Box}}$.*

Now suppose $(A, \Box)$ is a HAO, and $\mathcal{A} = (A, \alpha)$ its corresponding $(\mathcal{L}^{\Box}, j)$-dialgebra (with α given by $\alpha(\boxdot a) = \Box a$). We have $\mathcal{A}_{\tau} = (pf' A, \subseteq, \gamma)$, where

$$\gamma(\mathfrak{q}) = \{\mathfrak{p} \in pf' A \mid \forall a \in A, \boxdot a \in \alpha^{-1}(\mathfrak{q}) \text{ implies } a \in \mathfrak{p}\}.$$

Note that $\boxdot a \in \alpha^{-1}(\mathfrak{q})$ iff $\Box a = \alpha(\boxdot a) \in \mathfrak{q}$. Therefore, translating γ to a relation R_{γ}, we obtain: $\mathfrak{q} R_{\gamma} \mathfrak{p}$ iff $\Box a \in \mathfrak{q}$ implies $a \in \mathfrak{p}$ for all $a \in A$.

It follows that the $(i, \mathcal{T})$-dialgebra corresponding to the prime filter extension of a $\Box$-frame $(X, \leq, R)$ (as in Sec. 2) coincides with the $\tau^{\Box}$-prime filter extension of the dialgebraic rendering of $\mathfrak{X}$. So, modulo dialgebraic translation, prime filter extensions and $\tau^{\Box}$-prime filter extensions of $\Box$-frames coincide.

3.5 The Goldblatt-Thomason Theorem

Finally, we put our theory to work and prove a Goldblatt-Thomason theorem for dialgebraic intuitionistic logics. We work with the same assumptions as in Assum. 3.23. Additionally, we assume that we have a natural transformation $\tau : pf \circ \mathcal{L} \to \mathcal{T} \circ pf'$ such that $\rho^{\flat} \circ \tau = id_{pf \circ \mathcal{L}}$. This allows us to use Def. 3.26.

Definition 3.30 If $\Phi \subseteq \mathbf{L}(\Lambda)$ and $\mathcal{A} \in \mathsf{Dialg}(\mathcal{L}, j)$ then we write $\mathcal{A} \models \Phi$ if $\mathcal{A} \models \varphi$ for all $\varphi \in \Phi$. Besides, we let $\mathrm{Alg}\,\Phi = \{\mathcal{A} \in \mathsf{Dialg}(\mathcal{L}, j) \mid \mathcal{A} \models \Phi\}$ be the collection of $(\mathcal{L}, j)$-dialgebras satisfying Φ. We say that a class $\mathscr{C} \subseteq \mathsf{Dialg}(\mathcal{L}, j)$ is *axiomatic* if $\mathscr{C} = \mathrm{Alg}\,\Phi$ for some collection Φ of $\mathbf{L}(\Lambda)$-formulae.

Lemma 3.31 *$\mathscr{C} \subseteq \mathsf{Dialg}(\mathcal{L}, j)$ is axiomatic iff it is a variety of algebras.*

Proof. If $\mathsf{A} = \{\mathcal{A} \in \mathsf{Dialg}(\mathcal{L}, j) \mid \mathcal{A} \models \Phi\}$, then it is precisely the variety of algebras satisfying $\varphi^x \leftrightarrow \top$, where $\varphi \in \Phi$ and φ^x is the formula we get from φ by replacing the proposition letters with variables from some set S of variables. Conversely, suppose A is a variety of algebras given by a set E of equations using variables in S. For each equation $\varphi = \psi$ in E, let $(\varphi \leftrightarrow \psi)^p$ be the formula we get from replacing the variables in $\varphi \leftrightarrow \psi$ with proposition letters. Then we have $\mathsf{A} = \mathrm{Alg}\{(\varphi \leftrightarrow \psi)^p \mid \varphi = \psi \in E\}$. □

For a class $\mathscr{K}$ of $(i, \mathcal{T})$-dialgebras, write $\mathscr{K}^+ = \{\mathfrak{X}^+ \mid \mathfrak{X} \in \mathscr{K}\}$ for the collection of corresponding complex algebras. Also, if $\mathscr{C}$ is a class of algebras, then we write $H\mathscr{C}$, $S\mathscr{C}$ and $P\mathscr{C}$ for its closure under homomorphic images, subalgebras and products, respectively.

Lemma 3.32 *A class $\mathscr{K} \subseteq \mathsf{Dialg}(i, \mathcal{T})$ is axiomatic if and only if*

$$\mathscr{K} = \{\mathfrak{X} \in \mathsf{Dialg}(i, \mathcal{T}) \mid \mathfrak{X}^+ \in HSP(\mathscr{K}^+)\}. \tag{2}$$

Proof. Suppose $\mathscr{K}$ is axiomatic, i.e. $\mathscr{K} = \mathrm{Fr}\,\Phi$. Then it follows from Prop. 3.22 and the fact that H, S and P preserve validity of formulae that (2) holds. Conversely, suppose (2) holds. Since $HSP(\mathscr{K}^+)$ is a variety, Birkhoff's variety theorem states that it is of the from $\mathrm{Alg}\,\Phi$. It follows that $\mathscr{K} = \mathrm{Fr}\,\Phi$.□

We now have all the ingredients to prove the Goldblatt-Thomason theorem.

Theorem 3.33 *Let $\mathscr{K} \subseteq \mathsf{Dialg}(i, \mathcal{T})$ be closed under τ-prime filter extensions. Then $\mathscr{K}$ is axiomatic if and only if $\mathscr{K}$ reflects τ-prime filter extensions and is closed under disjoint unions, generated subframes and p-morphic images.*

Proof. The implication from left to right follows from Sec. 3.2 and Prop. 3.27. For the converse, by Lem. 3.32 it suffices to prove that $\mathscr{K} = \{\mathfrak{X} \in \mathsf{Dialg}(i, \mathcal{T}) \mid \mathfrak{X}^+ \in HSP(\mathscr{K}^+)\}$. So let $\mathfrak{X} = (X, \gamma) \in \mathsf{Dialg}(i, \mathcal{T})$ and suppose $\mathfrak{X}^+ \in HSP(\mathscr{K}^+)$. Then there are $\mathfrak{Z}_i \in \mathscr{K}$ such that $\mathfrak{X}^+$ is the homomorphic image of a sub-dialgebra $\mathcal{A}$ of the product of the $\mathfrak{Z}_i^+$. In a diagram:

$$\mathfrak{X}^+ \xleftarrow{\text{surjective}} \mathcal{A} \xrightarrow{\text{injective}} \prod \mathfrak{Z}_i^+$$

Since $\prod \mathfrak{Z}_i^+ = (\coprod \mathfrak{Z}_i)^+$, dually this yields

$$(\mathfrak{X}^+)_\tau \xrightarrow{\text{gen. subframe}} \mathcal{A}_\tau \xleftarrow{\text{p-morphic image}} \left((\coprod \mathfrak{Z}_i)^+\right)_\tau$$

We have $\coprod \mathfrak{Z}_i \in \mathscr{K}$ because $\mathscr{K}$ is closed under coproducts, and $\left((\coprod \mathfrak{Z}_i)^+\right)_\tau \in \mathscr{K}$ because $\mathscr{K}$ is closed under prime filter extensions. Then $\mathcal{A}_\tau \in \mathscr{K}$ and $(\mathfrak{X}^+)_\tau \in \mathscr{K}$ because $\mathscr{K}$ is closed under p-morphic images and generated subframes. Finally, since $\mathscr{K}$ reflects prime filter extensions we find $\mathfrak{X} \in \mathscr{K}$. □

Circling back to □-frames, it follows from Lem. 3.29 and Thm. 3.33 that:

Theorem 3.34 *Suppose $\mathscr{K} \subseteq \mathsf{WZ}\Box$ is closed under prime filter extensions. Then $\mathscr{K}$ is axiomatic if and only if it reflects prime filter extensions and is closed under disjoint unions, generated subframes and p-morphic images.*

4 Applications

In each of the following subsection we recall a modal intuitionistic logic and model it dialgebraically. We use this to derive a notion of prime filter extension and we apply Thm. 3.33 to obtain a Goldblatt-Thomason theorem.

4.1 Goldblatt's Geometric Modality I

The extension of intuitionistic logic with a monotone modality, here denoted by $\triangle$, was first studied by Goldblatt in [14, Sec. 6]. It is closely related to its classical counterpart [9,20,21], except that the underlying propositional logic is intuitionistic. A dialgebraic perspective was given in [18, Sec. 8].

Let $\mathbf{L}_\triangle$ denote the language of intuitionistic logic extended with a unary operator $\triangle$, and write $\boldsymbol{\mathcal{L}}_\triangle$ for the logic obtained from extending intuitionistic logic $\boldsymbol{\mathcal{L}}$ with the axiom $\triangle(p \wedge q) \to \triangle p$ and the congruence rule for $\triangle$.

Definition 4.1 An *intuitionistic monotone frame* (or *IM-frame*) is a triple $(X, \leq, N)$ where $(X, \leq)$ is an intuitionistic Kripke frame and N is a function that assigns to each $x \in X$ a collection of upsets of $(X, \leq)$ such that:

- if $a \in N(x)$ and $a \subseteq b \in \mathcal{U}p(X, \leq)$, then $b \in N(x)$;
- if $x \leq y$ then $N(x) \subseteq N(y)$.

An *intuitionistic monotone frame morphism* (IMF-morphism) from $(X_1, \leq_1, N_1)$ to $(X_2, \leq_2, N_2)$ is a p-morphism $f : (X_1, \leq_1) \to (X_2, \leq_2)$ such that

$f^{-1}(a_2) \in N_1(x_1)$ iff $a_2 \in N_2(f(x_1))$ for all $x_1 \in X_1$ and $a_2 \in \mathcal{U}p(X_2, \leq_2)$. We write Mon for the category of intuitionistic monotone frames and morphisms.

An *intuitionistic monotone model* is a tuple $\mathfrak{M} = (X, \leq, N, V)$ such that $(X, \leq, N)$ is an intuitionistic monotone frame and $V : \text{Prop} \to \mathcal{U}p(X, \leq)$ is a valuation. The interpretation of $\mathbf{L}_\Delta$-formulae at a state x in $\mathfrak{M}$ is defined recursively, where the propositional cases are as usual and $\mathfrak{M}, x \Vdash \Delta\varphi$ iff $[\![\varphi]\!]^{\mathfrak{M}} \in N(x)$. We now take a dialgebraic perspective.

Definition 4.2 For an intuitionistic Kripke frame $(X, \leq)$, define

$$\mathcal{M}(X, \leq) = \{W \subseteq \mathcal{U}p(X, \leq) \mid \text{ if } a \in W \text{ and } a \subseteq b \in \mathcal{U}p(X, \leq) \text{ then } b \in W\}$$

ordered by inclusion. For a p-morphism $f : (X_1, \leq_1) \to (X_2, \leq_2)$, let

$$\mathcal{M}f : \mathcal{M}(X_1, \leq_1) \to \mathcal{M}(X_2, \leq_2) : W \mapsto \{a_2 \in \mathcal{U}p(X_2, \leq_2) \mid f^{-1}(a_2) \in W\}.$$

Then $\mathcal{M} : \mathsf{Krip} \to \mathsf{Pos}$ defines a functor.

Theorem 4.3 ([18], Thm. 8.3) *We have* $\mathsf{Mon} \cong \mathsf{Dialg}(i, \mathcal{M})$.

Translating the dialgebraic notion of disjoint union to IM-frames gives:

Definition 4.4 Let $\{(X_k, \leq_k, N_k) \mid k \in K\}$ be a K-indexed set of IM-frames. The disjoint union $\coprod_{k \in K}(X_k, \leq_k, N_k)$ is the frame $(X, \leq, N)$ where $(X, \leq)$ is the disjoint union of the intuitionistic Kripke frames $(X_k, \leq_k)$, and N is given by $a \in N(x_k)$ iff $a \cap X_k \in N_k(x_k)$ for all $a \in \mathcal{U}p(X, \leq)$ and $x_k \in X_k$.

Definition 4.5 An IM-frame $\mathfrak{X}'$ is a *generated subframe* of an IM-frame $\mathfrak{X}$ if there exists an IMF-morphism $\mathfrak{X}' \to \mathfrak{X}$ that is an embedding of posets, and $\mathfrak{X}'$ is a *p-morphic image* of $\mathfrak{X}$ if there is a surjective IMF-morphism $\mathfrak{X} \to \mathfrak{X}'$.

The modal operator Δ can be introduced by the predicate lifting $\lambda^\Delta : \mathcal{U}p \circ i \to \mathcal{U}p \circ \mathcal{M}$ given by

$$\lambda^\Delta_{(X,\leq)}(a) = \{W \in \mathcal{M}(X, \leq) \mid a \in W\}.$$

With $\text{Ax} = \{\Delta(a \wedge b) \wedge \Delta a \leftrightarrow \Delta(a \wedge b)\}$ we have $\mathcal{L}_\Delta = \mathcal{L}(\{\lambda^\Delta\}, \text{Ax})$. Its algebraic semantics is given by $(\mathcal{L}^\Delta, j)$-dialgebras, where $\mathcal{L}^\Delta : \mathsf{HA} \to \mathsf{DL}$ is the functor sending A to the free distributive lattice generated by $\{\blacktriangle a \mid a \in H\}$ modulo $\blacktriangle(a \wedge b) \leq \blacktriangle b$. The corresponding natural transformation $\rho^\Delta : \mathcal{L}^\Delta \circ up' \to up \circ \mathcal{M}$ is defined on generators by $\rho^\Delta_{(X,\leq)}(a) = \{W \in \mathcal{M}(X, \leq) \mid a \in W\}$.

Towards prime filter extensions and a Goldblatt-Thomason theorem we need to define a right inverse τ of $(\rho^\Delta)^\flat$. To garner inspiration we investigate what $(\rho^\Delta)^\flat_A : \mathcal{M}(pf'A) \to pf(\mathcal{L}^\Delta A)$ looks like for $A \in \mathsf{HA}$. We have

$$\blacktriangle a \in (\rho^\Delta)^\flat_A(W) \quad \text{iff} \quad \rho_{pf'A}(\blacktriangle\theta'_A(a)) \in \eta_{\mathcal{M} \circ pf'A}(W) \quad \text{iff} \quad \theta'_A(a) \in W$$

for all $W \in \mathcal{M}(pf'A)$ and $a \in A$. (Recall that $\theta'_A(a) = \{\mathfrak{q} \in pf'A \mid a \in \mathfrak{q}\}$.)

Definition 4.6 Let $A \in \mathsf{HA}$. We call $D \in up'(pf'A)$ *closed* if $D = \bigcap\{\theta'_A(a) \mid a \in A \text{ and } D \subseteq \theta'_A(a)\}$, and *open* if $D = \bigcup\{\theta'_A(a) \mid a \in A \text{ and } \theta'_A(a) \subseteq D\}$.

(Indeed, this coincides with closed and open upsets of $pf'A$, conceived of as an Esakia space [2, Sec. 2.3.3].) Upsets of the form $\theta'_A(a)$ are closed *and* open.

Definition 4.7 For a Heyting algebra A, define $\tau_A : pf \circ \mathcal{L}^{\triangle} A \to \mathcal{M} \circ pf'A$ as follows. Let $Q \in pf(\mathcal{L}^{\triangle} A)$ and $D \in \mathcal{U}p(pf'A)$, and define:

- If $D = \theta'_A(a)$ for some $a \in A$, then $\theta'_A(a) \in \tau_A(Q)$ if $\blacktriangle a \in Q$;
- If D is closed then $D \in \tau_A(Q)$ if for all $a \in A$, $D \subseteq \theta'_A(a)$ implies $\blacktriangle a \in Q$.
- For other D, $D \in \tau_A(Q)$ if there is a closed upset $C \subseteq D$ such that $C \in \tau_A(Q)$.

It is easy to see that $\tau_A^{\triangle}$ is an order-preserving function, i.e. a morphism in Pos. The next lemma states that $\tau^{\triangle}$ is a natural transformation. We postpone the unexciting proof to the appendix.

Lemma 4.8 *The transformation τ from Def. 4.7 is natural. Moreover, $(\rho^{\triangle})^{\flat}_A \circ \tau_A^{\triangle} = id(pf(\mathcal{L}^{\triangle} A))$ for every Heyting algebra A.*

Translating the dialgebraic definition of a prime filter extension to IM-frames gives a definition of prime filter extension for IM-frames. We emphasise that this definition relies on $\tau^{\triangle}$. In the next section we derive a different notion of prime filter extension for IM-frames, with its own Goldblatt-Thomason theorem.

Definition 4.9 The *$\tau^{\triangle}$-prime filter extension* of an IM-frame $(X, \leq, N)$ is the IM-frame $(X^{pe}, \subseteq, N^{pe})$, where N^{pe} is given as follows. Let $\triangle_N(a) = \{x \in X \mid a \in N(x)\}$, and for $\mathfrak{q} \in X^{pe}$ and $D \in \mathcal{U}p(X^{pe}, \subseteq)$ define:

- If $a \in up'(X, \leq)$, then $\theta'_A(a) \in N^{pe}(\mathfrak{q})$ if $\triangle_N a \in \mathfrak{q}$;
- If D is closed then $D \in N^{pe}(\mathfrak{q})$ if $\theta'_A(a) \in N^{pe}(\mathfrak{q})$ for all $\theta'_A(a)$ containing D;
- For any D, $D \in N^{pe}(\mathfrak{q})$ if there is a closed $C \subseteq D$ such that $C \in N^{pe}(\mathfrak{q})$.

Now Thm. 3.33 instantiates to:

Theorem 4.10 *Suppose $\mathcal{K}$ is a class of IM-frames closed under $\tau^{\triangle}$-prime filter extensions. Then $\mathcal{K}$ is axiomatic iff it reflects $\tau^{\triangle}$-prime filter extensions and is closed under disjoint unions, generated subframes and p-morphic images.*

4.2 Goldblatt's Geometric Modality II

We substantiate the claim that a logic may have several notions of prime filter extension by giving a different right-inverse of $(\rho^{\triangle})^{\flat}$ from Sec. 4.1. The setup is the same as in Sec. 4.1, so we proceed by defining a right-inverse of $(\rho^{\triangle})^{\flat}$.

Definition 4.11 For a Heyting algebra A, define $\sigma_A : pf \circ \mathcal{L}^{\triangle} A \to \mathcal{M} \circ pf'A$ by sending $Q \in pf(\mathcal{L}^{\triangle} A)$ to $\sigma_A(Q)$, where:

- For open upsets D, let $D \in \sigma_A(Q)$ if $\exists a \in A$ s.t. $\blacktriangle a \in Q$ and $\theta'_A(a) \subseteq D$;
- For any other upset D, let $D \in \sigma_A(Q)$ if all open supersets of D are in $\sigma_A(Q)$.

Similar to Lem. 4.8 we can prove the following.

Lemma 4.12 *$\sigma = (\sigma_A)_{A \in \mathsf{HA}} : pf \circ \mathcal{L}^{\triangle} \to \mathcal{M} \circ pf'$ is a natural transformation, and for every Heyting algebra A, we have $\rho^{\flat}_A \circ \sigma_A = id_{pf(\mathcal{L}^{\triangle} A)}$.*

Now σ yields a different notion of prime filter extension, the precise definition of which we leave to the reader. Thm. 3.33 yields a Goldblatt-Thomason theorem with respect to this different notion of prime filter extension.

Theorem 4.13 *Let $\mathscr{K}$ be a class of IM-frames closed under σ-prime filter extensions. Then $\mathscr{K}$ is axiomatic iff it reflects σ-prime filter extensions and is closed under disjoint unions, generated subframes and p-morphic images.*

4.3 Non-Normal Intuitionistic Modal Logic

Neighbourhood semantics is used to accommodate for non-normal modal operators [33,27,9,28]. Dalmonte, Grellois and Olivett recently put forward an intuitionistic analogue [11] to interpret the extension of intuitionistic logic with unary modalities $\Box$ and $\Diamond$ which a priori do not satisfy any interaction axioms.

The ordered sets underlying the neighbourhood semantics from [11] are allowed to be preorders. Conforming to our general framework, we shall assume them to be posets. However, as mentioned in the introduction, we can obtain exactly the same (dialgebraic) results when replacing posets with preorders.

We use $\wp$ to denote the (covariant) powerset functors on Set.

Definition 4.14 A *coupled intuitionistic neighbourhood frame* or *CIN-frame* is a tuple $(X, \leq, N_\Box, N_\Diamond)$ such that $(X, \leq)$ is an intuitionistic Kripke frame and $N_\Box, N_\Diamond$ are functions $X \to \wp\wp X$ such that for all $x, y \in X$:

$$x \leq y \quad \text{implies} \quad N_\Box(x) \subseteq N_\Box(y) \quad \text{and} \quad N_\Diamond(x) \supseteq N_\Diamond(y).$$

A CIN-morphism $f : (X, \leq, N_\Box, N_\Diamond) \to (X', \leq', N'_\Box, N'_\Diamond)$ is a p-morphism $f : (X, \leq) \to (X', \leq')$ where for all $N \in \{N_\Box, N_\Diamond\}$, $x \in X$, $a' \in \wp X'$, $f^{-1}(a') \in N(x)$ iff $a' \in N'(f(x))$. CIN denotes the category of CIN-frames and -morphisms.

The language $\mathbf{L}_{\Box\Diamond}$ extending the intuitionistic language with unary modalities $\Box$ and $\Diamond$ can be interpreted in models based on CIN-frames, where

$$x \Vdash \Box\varphi \quad \text{iff} \quad [\![\varphi]\!] \in N_\Box(x), \qquad x \Vdash \Diamond\varphi \quad \text{iff} \quad X \setminus [\![\varphi]\!] \notin N_\Diamond(x).$$

We now view this dialgebraically:

Definition 4.15 Define $\mathcal{N} : \mathsf{Krip} \to \mathsf{Pos}$ on objects $(X, \leq)$ by $\mathcal{N}(X, \leq) = (\wp\wp X, \subseteq) \times (\wp\wp X, \supseteq)$, and on morphisms $f : (X, \leq) \to (X', \leq')$ by

$$\mathcal{N}f(W_1, W_2) = \big(\{a'_1 \in \wp X' \mid f^{-1}(a'_1) \in W_1\}, \{a'_2 \in \wp X' \mid f^{-1}(a'_2) \in W_2\}\big).$$

Theorem 4.16 *We have* $\mathsf{CIN} \cong \mathsf{Dialg}(i, \mathcal{N})$.

Proof. The isomorphism on objects is obvious. The isomorphism on morphisms follows from a computation similar to that in the proof of Thm. 4.3. $\Box$

The modal operators $\Box, \Diamond$ are induced by $\lambda^\Box, \lambda^\Diamond : \mathcal{U}p \circ i \to \mathcal{U}p \circ \mathcal{N}$, where

$$\lambda^\Box_{(X,\leq)}(a) = \{(W_1, W_2) \in \mathcal{N}(X, \leq) \mid a \in W_1\}$$
$$\lambda^\Diamond_{(X,\leq)}(a) = \{(W_1, W_2) \in \mathcal{N}(X, \leq) \mid X \setminus a \notin W_2\}$$

Unravelling the definition of a disjoint union of (the dialgebraic renderings of) CIN-frames shows that it is computed similar to Def. 4.4. Generated subframes and p-morphic images are defined by means of CIN-morphisms.

Since $\Box$ and $\Diamond$ only satisfy the congruence rule, the algebraic semantics is given by dialgebras for the functor $\mathcal{L}^{\Box\Diamond} : \mathsf{HA} \to \mathsf{DL}$ that sends A to the free distributive lattice generated by $\{\boxdot a, \diamonddot a \mid a \in A\}$. The induced natural transformation $\rho^{\Box\Diamond} : \mathcal{L}^{\Box\Diamond} \circ \mathit{up}' \to \mathit{up} \circ \mathcal{N}$ is defined on components via $\rho^{\Box\Diamond}_{(X,\leq)}(\boxdot a) = \lambda^{\Box}_{(X,\leq)}(a)$ and $\rho^{\Box\Diamond}_{(X,\leq)}(\diamonddot a) = \lambda^{\Diamond}_{(X,\leq)}(a)$. Akin to Sec. 4.1 we find $\boxdot a \in (\rho^{\Box\Diamond}_A)^\flat(W_1, W_2)$ iff $\theta'_A(a) \in W_1$ and $\diamonddot a \in (\rho^{\Box\Diamond}_A)^\flat(W_1, W_2)$ iff $\mathit{pf}'A \setminus \theta'_A(a) \notin W_1$ for all $A \in \mathsf{HA}$, $(W_1, W_2) \in \mathcal{N}(\mathit{pf}'A)$ and $a \in A$.

Definition 4.17 For a Heyting algebra A, define

$$\tau_A : \mathit{pf}(\mathcal{L}^{\Box\Diamond} A) \to \mathcal{N}(\mathit{pf}'A) : Q \mapsto \big(\{\theta'_A(a) \mid \boxdot a \in Q\}, \{\mathit{pf}'A \backslash \theta'_A(a) \mid \diamonddot a \notin Q\}\big).$$

Then $\tau = (\tau_A)_{A \in \mathsf{HA}}$ defines a natural transformation $\mathit{pf} \circ \mathcal{L}^{\Box\Diamond} \to \mathcal{N} \circ \mathit{pf}'$. It follows from the definitions that $(\rho^{\Box\Diamond})^\flat \circ \tau = \mathit{id}_{\mathit{pf} \circ \mathcal{L}^{\Box\Diamond}}$. We get the following definition of τ-prime filter extensions and Goldblatt-Thomason theorem.

Definition 4.18 The τ-prime filter extension of a CIN-frame $\mathfrak{X} = (X, \leq, N_\Box, N_\Diamond)$ is given by $\mathfrak{pe}_\tau \mathfrak{X} = (X^{pe}, \subseteq, N^{pe}_\Box, N^{pe}_\Diamond)$, where for $\mathfrak{q} \in X^{pe}$ we have

$$N^{pe}_\Box(\mathfrak{q}) = \{\theta'_{\mathit{up}'(X,\leq)}(a) \in \wp X^{pe} \mid a \in \mathit{up}(X, \leq) \text{ and } \Box_N(a) \in \mathfrak{q}\}$$
$$N^{pe}_\Diamond(\mathfrak{q}) = \{X^{pe} \setminus \theta'_{\mathit{up}'(X,\leq)}(a) \in \wp X^{pe} \mid a \in \mathit{up}(X, \leq) \text{ and } \Diamond_N(a) \in \mathfrak{q}\}$$

Here $\Box_N(a) = \{x \in X \mid a \in N_\Box(x)\}$ and $\Diamond_N(a) = \{x \in X \mid X \setminus a \notin N_\Diamond(x)\}$.

Theorem 4.19 *Let $\mathcal{K}$ be a class of CIN-frames closed under τ-prime filter extensions. Then $\mathcal{K}$ is axiomatic iff it reflects τ-prime filter extensions and is closed under disjoint unions, generated subframes and p-morphic images.*

4.4 Heyting-Lewis Logic

Finally we discuss Heyting-Lewis logic, the extension of intuitionistic logic with a binary strict implication operator $⥽$ [25,26,12].

Definition 4.20 A *strict implication frame* is a tuple $(X, \leq, R_s)$, where $(X, \leq)$ is an intuitionistic Kripke frame and R_s is a relation on X such that $x \leq y R_s z$ implies $x R_s z$. Morphisms between them are functions that are p-morphisms with respect to both orders. Models are defined as expected, and $⥽$ is interpreted via

$$x \Vdash \varphi ⥽ \psi \quad \text{iff} \quad \text{for all } y \in X, \text{ if } x R_s y \text{ and } y \Vdash \varphi \text{ then } y \Vdash \psi.$$

Strict implication frames can be modelled as $(i, \mathcal{P}_s)$-dialgebras, where $\mathcal{P}_s : \mathsf{Krip} \to \mathsf{Pos}$ is the functor that sends $(X, \leq)$ to $(\wp X, \subseteq)$ ($\wp$ denotes the covariant powerset functor) and a p-morphism f to $\wp f$. The modality $⥽$ can then be defined via the binary predicate lifting $\lambda^{⥽}$, given on components by

$$\lambda^{⥽}_{(X,\leq)}(a, b) = \{c \in \mathcal{P}_s(X, \leq) \mid c \cap a \subseteq b\}.$$

Disjoint unions, generated subframes and p-morphic images are defined as for □-frames.

The algebraic semantics for this logic given in [12, Def. III.1] can be modelled dialgebraically in a similar way as we have seen above. Computation of the natural transformation $\rho^{\dashv}$ is, by now, routine. Examining the proof of the duality for Heyting-Lewis logic sketched in [12, Section III-D], we can compute a one-sided inverse τ to $(\rho^{\dashv})^{\flat}$. We suppress the details, but do give the resulting notion of prime filter extension:

Definition 4.21 The *prime filter extension* of a strict implication frame $(X, \leq, R_s)$ is given by the frame $(X^{pe}, \subseteq, R_s^{pe})$, with R_s^{pe} defined by

$$\mathfrak{p} R_s^{pe} \mathfrak{q} \quad \text{iff} \quad \forall a, b \in \mathit{up}(X, \leq), \text{if } a \dashv_R b \in \mathfrak{p} \text{ and } a \in \mathfrak{q} \text{ then } b \in \mathfrak{q}$$

where $a \dashv_R b = \{x \in X \mid R[x] \cap a \subseteq b\}$.

With this notion of prime filter extension, Thm. 3.33 instantiates to:

Theorem 4.22 *A class $\mathcal{K}$ of strict implication frames that is closed under prime filter extensions is axiomatic iff it reflects prime filter extensions and is closed under disjoint unions, generated subframes and p-morphic images.*

5 Conclusions

We have given a general way to obtain Goldblatt-Thomason theorems for modal intuitionistic logics, using the framework of dialgebraic logic. Subsequently, we applied the general result to several concrete modal intuitionistic logics. The results in this paper can be generalised in several directions.

More applications. The general Goldblatt-Thomason theorem can also be instantiated to ◇-frames and □◇-frames [38]. Using preorders instead of posets, we can obtain Goldblatt-Thomason theorems for ((strictly) condensed) H□-frames and H□◇ frames used by Božić and Došen [7].

More base logics. The framework of dialgebraic logic is not restricted to an intuitionistic base. Generalising the results from this paper, we can obtain a general Goldblatt-Thomason theorem that also covers modal bi-intuitionistic logics [17] and modal lattice logics [3]. Moreover, this would also cover coalgebraic logics over a classical and a positive propositional base. The results in this paper can be generalised to dialgebraic logics for different base logics. This would give rise to Goldblatt-Thomason

Other modal intuitionistic logics The results in the paper do not apply to the modal intuitionistic logics investigated by Fischer Servi [13], Plotkin and Sterling [29], and Simpson [34], because these formalisms are not covered by the dialgebraic approach. It would be interesting to see if similar techniques can be applied to these logics to still prove Goldblatt-Thomason theorems.

Acknowledgements. I am grateful to the anonymous reviewers for many constructive and helpful comments.

References

[1] Benthem, J. F. A. K. v., *Modal frame classes revisited*, Fundamenta Informaticae **18** (1993), pp. 307–317.

[2] Bezhanishvili, N., "Lattices of intermediate and cylindric modal logics," Ph.D. thesis, University of Amsterdam (2006).

[3] Bezhanishvili, N., A. Dmitrieva, J. de Groot and T. Moraschini, *Positive (modal) logic beyond distributivity* (2022), arxiv:2204.13401.

[4] Birkhoff, G., *On the structure of abstract algebras*, Mathematical Proceedings of the Cambridge Philosophical Society **31** (1935), pp. 433–454.

[5] Blackburn, P., M. d. Rijke and Y. Venema, "Modal Logic," Cambridge University Press, Cambridge, 2001.

[6] Blok, A., "Interaction, observation and denotation," Master's thesis, University of Amsterdam (2012).

[7] Božić, M. and K. Došen, *Models for normal intuitionistic modal logics*, Studia Logica **43** (1984), pp. 217–245.

[8] Celani, S. A. and R. Jansana, *Priestley duality, a Sahlqvist theorem and a Goldblatt-Thomason theorem for positive modal logic*, Logic Journal of the IGPL **7** (1999), pp. 683–715.

[9] Chellas, B. F., "Modal Logic: An Introduction," Cambridge University Press, Cambridge, 1980.

[10] Conradie, W., A. Palmigiano and A. Tzimoulis, *Goldblatt-thomason for LE-logics* (2018), arxiv:1809.08225.

[11] Dalmonte, T., C. Grellois and N. Olivetti, *Intuitionistic non-normal modal logics: A general framework*, Journal of Philosophical Logic **49** (2020), pp. 833–882.

[12] de Groot, J., T. Litak and D. Pattinson, *Gödel-McKinsey-Tarski and Blok-Esakia for Heyting-Lewis implication*, in: *Proc. LICS 2021*, 2021, pp. 1–15.

[13] Fischer Servi, G., *Semantics for a class of intuitionistic modal calculi*, in: D. Chiara and M. Luisa, editors, *Italian Studies in the Philosophy of Science* (1980), pp. 59–72.

[14] Goldblatt, R. I., "Mathematics of Modality," CSLI publications, Stanford, California, 1993.

[15] Goldblatt, R. I., *Axiomatic classes of intuitionistic models*, Journal of Universal Computer Science **11** (2005), pp. 1945–1962.

[16] Goldblatt, R. I. and S. K. Thomason, *Axiomatic classes in propositional modal logic*, in: J. Crossley, editor, *Algebra and Logic* (1974), pp. 163–173.

[17] Groot, J. d. and D. Pattinson, *Hennessy-Milner properties for (modal) bi-intuitionistic logic*, in: R. Iemhoff, M. Moortgat and R. de Queiroz, editors, *Proc. WoLLIC 2019* (2019), pp. 161–176.

[18] Groot, J. d. and D. Pattinson, *Modal intuitionistic logics as dialgebraic logics*, in: *Proc. LICS 2020* (2020), pp. 355—369.

[19] Hagino, T., "A categorical programming language," Ph.D. thesis, University of Edinburgh (1987), arxiv:2010.05167.

[20] Hansen, H. H., "Monotonic modal logics," Master's thesis, Institute for Logic, Language and Computation, University of Amsterdam (2003).

[21] Hansen, H. H. and C. Kupke, *A coalgebraic perspective on monotone modal logic*, Electronic Notes in Theoretical Computer Science **106** (2004), pp. 121–143.

[22] Kurz, A. and J. Rosický, *The Goldblatt-Thomason theorem for coalgebras*, in: T. Mossakowski, U. Montanari and M. Haveraaen, editors, *Proc. CALCO 2007* (2007), pp. 342–355.

[23] Kurz, A. and J. Rosický, *Strongly complete logics for coalgebras*, Logical Methods in Computer Science **8** (2012).

[24] Litak, T., *Constructive modalities with provability smack* (2017), arxiv:1708.05607.

[25] Litak, T. and A. Visser, *Lewis meets Brouwer: Constructive strict implication*, Indagationes Mathematicae **29** (2018), pp. 36–90.

[26] Litak, T. and A. Visser, *Lewisian fixed points I: two incomparable constructions* (2019), arxiv:1905.09450.

[27] Montague, R., *Universal grammar*, Theoria **36** (1970), pp. 373–398.

[28] Pacuit, E., "Neighborhood Semantics for Modal Logic," Springer, Cham, 2017, xii+154 pp.

[29] Plotkin, G. and C. Stirling, *A framework for intuitionistic modal logics: Extended abstract*, in: *Proc. TARK 1986* (1986), pp. 399–406.

[30] Rodenburg, P. H., "Intuitionistic Correspondence Theory," Ph.D. thesis, University of Amsterdam (1986).

[31] Sano, K. and M. Ma, *Goldblatt-Thomason-style theorems for graded modal language*, in: *Proc. AiML 2010* (2010), pp. 330–349.

[32] Sano, K. and J. Virtema, *Characterising modal definability of team-based logics via the universal modality*, Annals of Pure and Applied Logic **170** (2019), pp. 1100–1127.

[33] Scott, D., *Advice in modal logic*, in: K. Lambert, editor, *Philosophical Problems in Logic* (1970), pp. 143–173.

[34] Simpson, A. K., "The Proof Theory and Semantics of Intuitionistic Modal Logic," Ph.D. thesis, University of Edinburgh (1994).

[35] Teheux, B., *Modal definability based on Łukasiewicz validity relations*, Studia Logica **104** (2016), pp. 343–363.

[36] Wolter, F. and M. Zakharyaschev, *The relation between intuitionistic and classical modal logics*, Algebra and Logic **36** (1997), pp. 73–92.

[37] Wolter, F. and M. Zakharyaschev, *Intuitionistic modal logics as fragments of classical bimodal logics*, in: E. Orlowska, editor, *Logic at Work, Essays in honour of Helena Rasiowa*, Springer–Verlag, 1998 pp. 168–186.

[38] Wolter, F. and M. Zakharyaschev, *Intuitionistic modal logic*, in: A. Cantini, E. Casari and P. Minari, editors, *Logic and Foundations of Mathematics: Selected Contributed Papers of the Tenth International Congress of Logic, Methodology and Philosophy of Science* (1999), pp. 227–238.

Appendix
A Omitted proofs

We use the following lemma in the proof of Prop. 3.27.

Lemma A.1 *Let τ be a natural transformation such that $\rho^\flat \circ \tau = \mathrm{id}_{pf \circ \mathcal{L}}$, and $\mathcal{A} = (A, \alpha) \in \mathsf{Dialg}(\mathcal{L}, j)$. Then $\theta'_A : A \to \mathit{up}'(\mathit{pf}'H)$ defines a $(\mathcal{L}, j)$-dialgebra morphism from $\mathcal{A}$ to $(\mathcal{A}_\tau)^+$.*

Proof. This is similar to [23, Theorem 6.4(1)]. We repeat the argument here.

Let $\mathcal{A} = (A, \alpha)$ be a $(\mathcal{L}, j)$-dialgebra. Then $(\mathcal{A}_\tau)^+$ is given by the composition

$$\mathcal{L}(\mathit{up}'(\mathit{pf}'A)) \xrightarrow{\rho_{\mathit{pf}'A}} \mathit{up}(\mathcal{T}(\mathit{pf}'A)) \xrightarrow{\mathit{up}\tau_A} \mathit{up}(\mathit{pf}(\mathcal{L}A)) \xrightarrow{\mathit{up} \circ \mathit{pf}\alpha} \mathit{up}(\mathit{pf}(jA)) = j(\mathit{up}'(\mathit{pf}'A))$$

In order to show that θ'_A is a morphism from $\mathcal{A}$ to $(\mathcal{A}_\tau)^+$ we need to show that the outer shell of the following diagram commutes:

$$\begin{array}{c}
\mathcal{L}A \xrightarrow{\alpha} jA \\
\theta_{\mathcal{L}A} \searrow \quad \theta_{jA} \nearrow \\
\mathcal{L}\theta'_A \downarrow \qquad\qquad\qquad\qquad \downarrow j\theta'_A \\
\mathcal{L}(\mathit{up}'(\mathit{pf}'A)) \xrightarrow[\rho_{\mathit{pf}'A}]{} \mathit{up}(\mathcal{T}(\mathit{pf}'A)) \xrightarrow[\mathit{up}\tau_A]{} \mathit{up}(\mathit{pf}(\mathcal{L}A)) \xrightarrow[\mathit{up}(\mathit{pf}\alpha)]{} \mathit{up}(\mathit{pf}(jA)) = j(\mathit{up}'(\mathit{pf}'A))
\end{array}$$

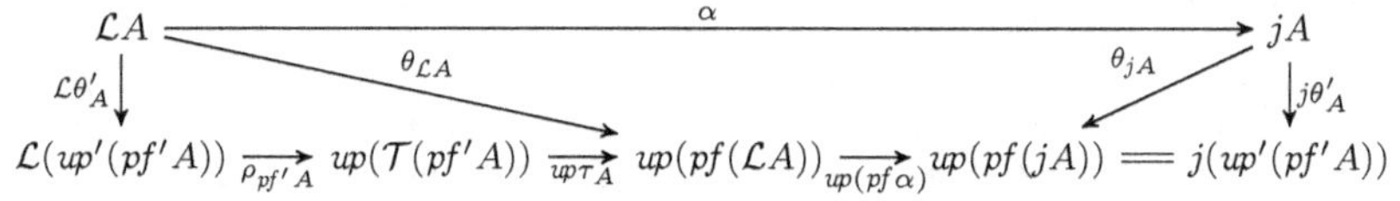

The right triangle commutes by definition. The middle square commutes by naturality of θ. So we are left to prove that $\theta_{\mathcal{L}A} = \mathit{up}\tau_A \circ \rho_{\mathit{pf}'A} \circ \mathcal{L}\theta'_A$.

Since $\rho^\flat \circ \tau = \mathrm{id}$, hence $\mathit{up}\tau \circ \mathit{up}\rho^\flat = \mathrm{id}_{\mathit{up}}$, it suffices to prove that $\mathit{up}\rho^\flat \circ \theta_{\mathcal{L}A} = \rho_{\mathit{pf}'A} \circ \mathcal{L}\theta'_A$. (The result then follows from composing both sides with $\mathit{up}\tau_A$ on the left.) This is precisely the outer shell of the diagram

$$\begin{array}{c}
\mathcal{L}A \xrightarrow{\mathcal{L}\theta'_A} \mathcal{L}(\mathit{up}'(\mathit{pf}'A)) \xrightarrow{\rho_{\mathit{pf}'A}} \mathit{up}(\mathcal{T}(\mathit{pf}'A)) \xrightarrow{\mathrm{id}} \\
\theta_{\mathcal{L}A} \downarrow \quad \theta_{\mathcal{L}(\mathit{up}'(\mathit{pf}'A))} \downarrow \quad \theta_{\mathit{up}(\mathcal{T}(\mathit{pf}'A))} \downarrow \\
\mathit{up}(\mathit{pf}(\mathcal{L}H)) \xrightarrow{\mathit{up}(\mathit{pf}(\mathcal{L}\theta'_A))} \mathit{up}(\mathit{pf}(\mathcal{L}(\mathit{up}'(\mathit{pf}'A)))) \xrightarrow{\mathit{up}(\mathit{pf}\rho_{\mathit{pf}'A})} \mathit{up}(\mathit{pf}(\mathit{up}(\mathcal{T}(\mathit{pf}'A)))) \xrightarrow{\mathit{up}\eta_{\mathcal{T}(\mathit{pf}'A)}} \mathit{up}(\mathcal{T}(\mathit{pf}'A)) \\
\underbrace{\qquad\qquad\qquad\qquad\qquad}_{\mathit{up}\rho^\flat_A}
\end{array}$$

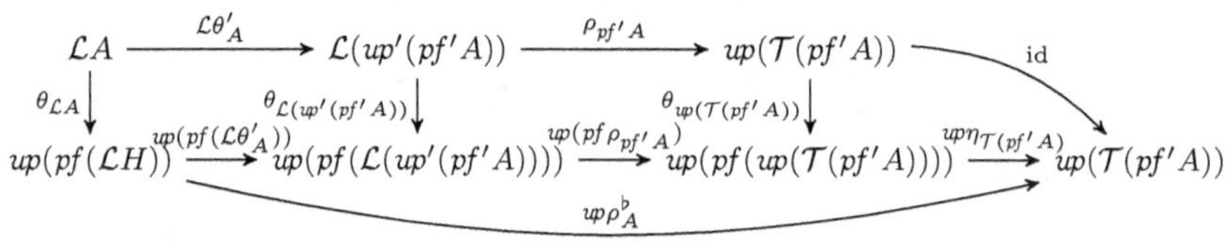

Here the bottom square commutes by definition of $\rho^\flat$. The other two squares commute by naturality of θ and the triangle on the right commutes because θ and η are the units of a dual adjunction. □

Proof of Proposition 3.27. Recall that $\theta'_{\mathit{up}'(X, \leq)}(\llbracket \varphi \rrbracket^{\mathfrak{M}}) = \{\mathfrak{p} \in \mathit{pf}'(\mathit{up}'(X, \leq)) \mid \llbracket \varphi \rrbracket^{\mathfrak{M}} \in \mathfrak{p}\}$. So the first item is equivalent to

$$\llbracket \varphi \rrbracket^{\mathfrak{pe}_\tau \mathfrak{M}} = \theta'_{\mathit{up}'(X, \leq)}(\llbracket \varphi \rrbracket^{\mathfrak{M}}),$$

where we view truth sets of formulae as elements in the relevant complex algebras (cf. Prop. 3.22). The proof proceeds by induction on the structure of φ. If $\varphi = q \in \mathsf{Prop}$ then the statement holds by definition of V^{pe}. The cases $\varphi = \top$ and $\varphi = \bot$ hold by definition of a prime filter.

If φ is of the form $\varphi_1 \star \varphi_2$, where $\star \in \{\wedge, \vee, \to\}$ then we use Lem. A to find

$$\begin{aligned}
[\![\varphi_1 \star \varphi_2]\!]^{\mathfrak{pe}_\tau \mathfrak{M}} &= [\![\varphi_1]\!]^{\mathfrak{pe}_\tau \mathfrak{M}} \star [\![\varphi_1]\!]^{\mathfrak{pe}_\tau \mathfrak{M}} \\
&= \theta'_{\mathit{up}'(X,\leq)}([\![\varphi_1]\!]^{\mathfrak{M}}) \star \theta'_{\mathit{up}'(X,\leq)}([\![\varphi_2]\!]^{\mathfrak{M}}) && \text{(IH)} \\
&= \theta'_{\mathit{up}'(X,\leq)}([\![\varphi_1]\!]^{\mathfrak{M}} \star [\![\varphi_2]\!]^{\mathfrak{M}}) \\
&= \theta'_{\mathit{up}'(X,\leq)}([\![\varphi_1 \star \varphi_2]\!]^{\mathfrak{M}})
\end{aligned}$$

The case where $\varphi = \heartsuit^\lambda(\varphi_1, \dots, \varphi_n)$ follows from a similar computation, using the fact that $\theta'_{\mathit{up}(X,\leq)}$ preserves operators of the form $\heartsuit^\lambda$.

Item (ii) follows from Item (i) and the definition of $\eta_{(X,\leq)}(x)$ via

$$\mathfrak{M}, x \Vdash \varphi \quad \text{iff} \quad x \in [\![\varphi]\!]^{\mathfrak{M}} \quad \text{iff} \quad [\![\varphi]\!]^{\mathfrak{M}} \in \eta_{(X,\leq)}(x) \quad \text{iff} \quad \mathfrak{pe}_\tau \mathfrak{M}, \eta_{(X,\leq)}(x) \Vdash \varphi.$$

For Item (iii), let V be any valuation for $\mathfrak{X}$ and $x \in X$. By assumption $(\mathfrak{pe}_\tau \mathfrak{X}, V^{pe}), \theta'_{(X,\leq)}(x) \Vdash \varphi$, so by Item (ii) $(\mathfrak{X}, V), x \Vdash \varphi$ and hence $\mathfrak{X} \Vdash \varphi$. $\square$

Proof of Lemma 3.29. Let A be a Heyting algebra. Recall that $\theta'_A(a) = \{\mathfrak{q} \in \mathit{pf}'A \mid a \in \mathfrak{q}\}$. Using this we can rewrite $\tau^{\Box}_A : \mathit{pf}(\mathcal{L}^{\Box} A) \to \mathcal{P}_{\mathit{up}}(\mathit{pf}'A)$ as

$$\tau^{\Box}_A(Q) = \bigcap \{\theta'_A(a) \mid a \in A, \boxdot a \in Q\}. \tag{$\star$}$$

Since $\theta'_A(a)$ is an upset of $\mathit{pf}'A$, $\tau^{\Box}_A(Q)$ is also an upset of $\mathit{pf}'A$, hence in $\mathcal{P}_{\mathit{up}}(\mathit{pf}'A)$. The elements of $\mathit{pf}(\mathcal{L}^{\Box} A)$ are ordered by inclusion. If $Q, Q' \in \mathit{pf}(\mathcal{L}^{\Box} A)$ and $Q \subseteq Q'$ then it follows immediately that $\tau^{\Box}_A(Q) \supseteq \tau^{\Box}_A(Q')$. Since $\mathcal{P}_{\mathit{up}}(\mathit{pf}'A)$ is ordered by reverse inclusion, so $\tau^{\Box}_A$ is a morphism of Pos.

For naturality, let $h : A \to B$ be a Heyting homomorphism. We need that

$$\begin{array}{ccc}
\mathit{pf}(\mathcal{L}^{\Box} A) & \xrightarrow{\tau^{\Box}_A} & \mathcal{P}_{\mathit{up}}(\mathit{pf}'A) \\
{\scriptstyle (\mathcal{L}^{\Box} h)^{-1}} \uparrow & & \uparrow {\scriptstyle \mathcal{P}_{\mathit{up}}(h^{-1})} \\
\mathit{pf}(\mathcal{L}^{\Box} B) & \xrightarrow{\tau^{\Box}_B} & \mathcal{P}_{\mathit{up}}(\mathit{pf}'B)
\end{array}$$

commutes. Let $Q \in \mathit{pf}(\mathcal{L}^{\Box} B)$, $\mathfrak{q} \in \mathit{pf}'A$, and suppose $\mathfrak{q} \in \tau^{\Box}_A(\mathcal{L}^{\Box} h)^{-1}(Q)$. To show $\mathfrak{q} \in \mathcal{P}_{\mathit{up}}(h^{-1})(\tau^{\Box}_B(Q))$ it suffices to find a prime filter $\mathfrak{p} \in \tau^{\Box}_B(Q)$ such that $h^{-1}(\mathfrak{p}) \subseteq \mathfrak{q}$, because $\tau^{\Box}_A(\mathcal{L}^{\Box} h)^{-1}(Q)$ is an upset of $\mathit{pf}'A$. Define $F = \{b \in B \mid \boxdot b \in Q\}$ and $I = \{b \in B \mid \exists a \in A \setminus \mathfrak{q} \text{ s.t. } b \leq c\}$. If $I \cap F \neq \emptyset$ then there exists $b \in B$ and $a \in A \setminus \mathfrak{q}$ such that $b \leq h(a)$. Since $\boxdot b \in Q$ this implies $\boxdot h(a) \in Q$ and hence $\boxdot a \in (\mathcal{L}^{\Box} h)^{-1}(Q)$. But then $a \in \mathfrak{q}$ because $\mathfrak{q} \in \tau^{\Box}_A((\mathcal{L}^{\Box} h)^{-1}(Q))$, a contradiction. So $I \cap F = \emptyset$. The prime filter lemma then gives a prime filter $\mathfrak{p}$ containing F and disjoint from I. This satisfies $\mathfrak{p} \in \tau^{\Box}_B(Q)$ and $h^{-1}(\mathfrak{p}) \subseteq \mathfrak{q}$ by design.

Conversely, suppose $\mathfrak{q} \in \mathcal{P}_{\mathit{up}}(h^{-1})(\tau^{\Box}_B(Q))$. Then there exists a $\mathfrak{p} \in \tau^{\Box}_B(Q)$ such that $\mathfrak{q} = h^{-1}(\mathfrak{p})$. We show that $\mathfrak{q} \in \tau^{\Box}_A(\mathcal{L}^{\Box} h)^{-1}(Q)$. Let $a \in A$ and suppose $\boxdot a \in (\mathcal{L}^{\Box} h)^{-1}(Q)$. Then $\boxdot h(a) = \mathcal{L}^{\Box} h(\boxdot a) \in Q$ so $h(a) \in \mathfrak{p}$. But this implies $a \in \mathfrak{q} = h^{-1}(\mathfrak{p})$. So by definition $\mathfrak{q} \in \tau^{\Box}_A(\mathcal{L}^{\Box} h)^{-1}(Q)$.

Finally, we show that $(\rho^{\boxdot}_A)^\flat \circ \tau^{\boxdot}_A = id_{pf \circ \mathcal{L}^\Box A}$ for $A \in \mathsf{HA}$. Let $Q \in pf(\mathcal{L}^\Box A)$. Since elements of $pf(\mathcal{L}^\Box A)$ are determined uniquely by the generators of the form $\boxdot a$ they contain, it suffices to show that $\boxdot a \in Q$ iff $\boxdot a \in (\rho^{\boxdot}_A)^\flat(\tau^{\boxdot}_A(Q))$. Because of the computation in Exm. 3.28 this is equivalent to showing $\boxdot a \in Q$ iff $\tau^{\boxdot}_A(Q) \subseteq \theta'_A(a)$. The direction from left to right follows from $(\star)$. For the converse, suppose $\boxdot a \notin Q$. Let $F = \{b \in A \mid \Box b \in Q\}$ and $I = \{c \in A \mid c \leq a\}$. Then F is a filter and I is an ideal of A, and $F \cap I = \emptyset$. By the prime filter lemma we obtain some $\mathfrak{q} \in pf' A$ extending F and disjoint from I. This implies that $\mathfrak{q} \in \tau^{\boxdot}_A(Q)$ while $\mathfrak{q} \notin \theta'_A(a)$, so that $\tau^{\boxdot}_A(Q) \not\subseteq \theta'_A(a)$. □

Proof of Lemma 4.8. Throughout this proof use the fact that $pf'A$ forms an Esakia space (which in particular is a Stone space), with a topology generated by sets of the form $\theta'_A(a)$ and their complements [2, Sec. 2.3.3]. Furthermore, we note that for any Heyting homomorphism $h : A \to B$ we have

$$\theta'_B(h(a)) = (h^{-1})^{-1}(\theta'_A(a)) \tag{$\dagger$}$$

We first prove naturality of $\tau^{\triangle}$. Let $h : A \to B$ be a Heyting homomorphism. We need to show that the following diagram commutes:

$$\begin{array}{ccc} pf(\mathcal{L}^\triangle A) & \xrightarrow{\tau^{\dot\triangle}_A} & \mathcal{P}_{up}(pf'A) \\ {\scriptstyle (\mathcal{L}^\triangle h)^{-1}}\uparrow & & \uparrow{\scriptstyle \mathcal{P}_{up}(h^{-1})} \\ pf(\mathcal{L}^\triangle B) & \xrightarrow{\tau^{\dot\triangle}_B} & \mathcal{P}_{up}(pf'B) \end{array}$$

Let $Q \in pf(\mathcal{L}^\triangle B)$ and $D \in \mathcal{U}p(pf'A)$. We go by the items of Def. 4.7.

- If $D = \theta'_A(a)$ for some $a \in A$ then

$$\begin{array}{rll} \theta'_A(a) \in \tau^{\dot\triangle}_A(\mathcal{L}^\triangle h)^{-1}(Q) & \text{iff} \ \ \dot\triangle a \in (\mathcal{L}^\triangle h)^{-1}(Q) & (\text{Def. } \tau^\triangle) \\ & \text{iff} \ \ (\mathcal{L}^\triangle h)(\dot\triangle a) \in Q & \\ & \text{iff} \ \ \dot\triangle h(a) \in Q & (\text{Def. of } \mathcal{L}^\triangle) \\ & \text{iff} \ \ \theta'_B(h(a)) \in \tau^{\dot\triangle}_B(Q) & (\text{Def. of } \tau^\triangle) \\ & \text{iff} \ \ (h^{-1})^{-1}(\theta'_A(a)) \in \tau^{\dot\triangle}_B(Q) & (\text{By } (\dagger)) \\ & \text{iff} \ \ \theta'_A(a) \in \mathcal{M}(h^{-1})(\tau^{\dot\triangle}_B(Q)) & (\text{Def. of } \mathcal{M}) \end{array}$$

- Suppose D is closed in $pf'A$. If $D \in \tau^{\dot\triangle}_A(\mathcal{L}^\triangle h)^{-1}(Q)$, then for all $a \in A$, $D \subseteq \theta'_A(a)$ implies $\dot\triangle a \in (\mathcal{L}^\triangle h)^{-1}(Q)$, i.e. $\dot\triangle h(a) \in Q$. In order to prove that $D \in \mathcal{M}(h^{-1})(\tau^{\dot\triangle}_B(Q))$, we need to show that $(h^{-1})^{-1}(D) \in \tau^{\dot\triangle}_B$. Since h^{-1} is an Esakia morphism (hence continuous), $(h^{-1})^{-1}(D)$ is closed in $pf'B$, so it suffices to show that for all $b \in B$, $(h^{-1})^{-1}(D) \subseteq \theta'_B(b)$ implies $\dot\triangle b \in Q$. Let $b \in B$ be such that $(h^{-1})^{-1}(D) \subseteq \theta'_B(b)$. Then since D is closed we have

$$\bigcap\{(h^{-1})^{-1}(\theta'_A(a)) \mid a \in A, D \subseteq \theta'_A(a)\} \subseteq \theta'_B(b)\}.$$

Using $(\dagger)$ and compactness of $pf'B$ we can find $a_1, \ldots, a_n \in A$ such that

$$\theta'_B(h(a_1 \wedge \cdots \wedge a_n)) = \theta'_B(h(a_1)) \cap \cdots \cap \theta'_B(h(a_n)) \subseteq \theta'_B(b).$$

As a consequence of Esakia duality it follows that $h(a_1 \wedge \cdots \wedge a_n) \leq b$. Since $D \subseteq \theta'_A(a_1 \wedge \cdots \wedge a_n)$, we have $\blacktriangle(a_1 \wedge \cdots \wedge a_n) \in (\mathcal{L}^\triangle h)^{-1}(Q)$, so $\blacktriangle(h(a_1 \wedge \cdots \wedge a_n)) \in Q$. Monotonicity of $\blacktriangle$ now implies $\blacktriangle b \in Q$.

Conversely, if $D \in \mathcal{M}(h^{-1})(\tau_B^\blacktriangle(Q))$ then a similar but easier argument shows that $D \in \tau_A^\blacktriangle(\mathcal{L}^\triangle h)^{-1}(Q)$.

- Finally, suppose D is any upset. If $D \in \tau_A^\blacktriangle(\mathcal{L}^\triangle h)^{-1}(Q)$ then there exists a closed upset C such that $C \subseteq D$ and $C \in \tau_A^\blacktriangle(\mathcal{L}^\triangle h)^{-1}(Q)$. This implies $C \in \mathcal{M}(h^{-1})(\tau_B^\blacktriangle(Q))$, so that $(h^{-1})^{-1}(C) \in \tau_B^\blacktriangle(Q)$. Since $(h^{-1})^{-1}(C)$ is closed again and $(h^{-1})^{-1}(C) \subseteq (h^{-1})^{-1}(D)$ we have $(h^{-1})^{-1}(D) \in \tau_B^\blacktriangle(Q)$, and therefore $D \in \mathcal{M}(h^{-1})(\tau_B^\blacktriangle(Q))$.

  Conversely, suppose $D \in \mathcal{M}(h^{-1})(\tau_B^\blacktriangle(Q))$. Then there exists a closed upset $C \in \tau_B^\blacktriangle(Q)$ such that $C \subseteq (h^{-1})^{-1}(D)$. Define $C' = h^{-1}[C]$ to be direct image of C under h^{-1}. Since h^{-1} is an Esakia morphism it sends closed upsets to closed upsets. Furthermore $C \subseteq (h^{-1})^{-1}(C')$ so $C' \in \mathcal{M}(h^{-1})(\tau_B^\blacktriangle(Q))$. This implies $C' \in \tau_A^\blacktriangle(\mathcal{L}^\triangle h)^{-1}(Q)$. By design $C' \subseteq D$, hence $D \in \tau_A^\blacktriangle(\mathcal{L}^\triangle h)^{-1}(Q)$.

Next we prove that $(\rho^\triangle)_A^\flat \circ \tau_A = id_{pf(\mathcal{L}^\triangle A)}$ for $A \in \mathsf{HA}$. It follows from the definitions of $\rho^\flat$ and τ that for any Heyting algebra A, $a \in A$ and prime filter $Q \in pf(\mathcal{L}^\triangle A)$ we have $\theta'_A(a) \in \rho_A^\flat(\tau_A(Q))$ iff $\blacktriangle a \in \tau_A(Q)$ iff $\theta'_A(a) \in Q$. Since elements of $pf(\mathcal{L}^\triangle A)$ are determined uniquely by the elements of the form $\blacktriangle a$ they contain, this proves the lemma. □

EXPTIME-hardness of higher-dimensional Minkowski spacetime

Robin Hirsch [1]

Department of Computer Science, University College London
Gower Street
London WC1E 6BT
United Kingdom

Brett McLean [2]

Department of Mathematics: Analysis, Logic and Discrete Mathematics, Ghent University
Building S8, Krijgslaan 281
9000 Ghent
Belgium

Abstract

We prove the EXPTIME-hardness of the validity problem for the basic temporal logic on Minkowski spacetime with more than one space dimension. We prove this result for both the lightspeed-or-slower and the slower-than-light accessibility relations (and for both the irreflexive and the reflexive versions of these relations). As an auxiliary result, we prove the EXPTIME-hardness of validity on any frame for which there exists an embedding of the infinite complete binary tree satisfying certain conditions. The proof is by a reduction from the two-player corridor-tiling game.

Keywords: Temporal logic, tense logic, Minkowski spacetime, EXPTIME-hard, corridor-tiling game

1 Introduction

Temporal logic has traditionally treated time as absolute, independent of other aspects of state, such as position. However, special and general relativity tell us that no absolute view of time reflects physical reality, and we must be content with understanding space and time jointly in the form of a unified spacetime. Thus the study of temporal logics of spacetime can be seen as a natural and foundational topic in temporal logic.

---

[1] r.hirsch@ucl.ac.uk

[2] brett.mclean@ugent.be
The second author was supported by the Research Foundation – Flanders (FWO) under the SNSF–FWO Lead Agency Grant 200021L 196176 (SNSF)/G0E2121N (FWO).

The frames we study in this paper are: Minkowski spacetime with *lightspeed-or-slower* accessibility and Minkowski spacetime with *slower-than-lightspeed* accessibility. Minkowski spacetime is the flat spacetime of special relativity, where light travels in straight lines. Let m be the number of space dimensions. Then Minkowski spacetime with (irreflexive) lightspeed-or-slower accessibility can be realised as the Kripke frame $(\mathbb{R}^{m+1}, <)$ where

$$(x_1, \ldots, x_m, t) < (y_1, \ldots, y_m, t') \iff t < t' \text{ and } \sum_i (x_i - y_i)^2 \leq (t' - t)^2.$$

(Here the order relation on the right-hand side is the usual ordering on the reals.) Minkowski spacetime with (irreflexive) slower-than-lightspeed accessibility can be realised as the frame $(\mathbb{R}^{m+1}, \prec)$ where

$$(x_1, \ldots, x_m, t) \prec (y_1, \ldots, y_m, t') \iff t < t' \text{ and } \sum_i (x_i - y_i)^2 < (t' - t)^2.$$

Depictions of these frames for $m = 2$ can be found in Figure 1.

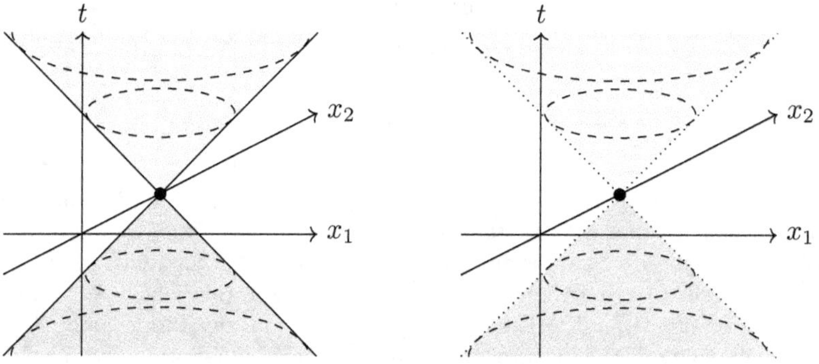

Figure 1. Minkowski spacetime frames $(\mathbb{R}^3, <)$ and $(\mathbb{R}^3, \prec)$

One may also consider reflexive versions $\leq$ and $\preceq$ of these accessibility relations, given by the reflexive closure of the respective irreflexive relations. This distinction will not play an important role in this paper.

The language we use is the most classical temporal language, the language of Prior's tense logic [6]. Formulas are built from a countable supply $p, q, \ldots$ of propositional variables using the standard propositional connectives $\neg$ and $\vee$ and the unary temporal operators $\mathbf{F}$ ('at some time in the future') and $\mathbf{P}$ ('at some time in the past'). We call the formulas of this language $(\mathbf{F}, \mathbf{P})$-formulas. Usual classical propositional abbreviations apply, and $\mathbf{G} := \neg\mathbf{F}\neg$ and $\mathbf{H} := \neg\mathbf{P}\neg$. A *modal* formula is built using propositional connectives and $\mathbf{F}$ (but no $\mathbf{P}$).

It is known that the modal validities of reflexive Minkowski spacetime with m spatial dimensions are axiomatised by **S4.2** (reflexive, transitive, confluent) [3], regardless of choice of $m \geq 1$, and regardless of whether speed-of-light is allowed; hence the logic is PSPACE-complete [7]. For irreflexive spacetime, with slower-than-light accessibility the validities again do not depend on the dimension, and are given by a logic called **OI.2** (transitive, confluent, serial, two-dense: $\mathbf{F}p \wedge \mathbf{F}q \rightarrow \mathbf{F}(\mathbf{F}p \wedge \mathbf{F}q)$) [9], which is also PSPACE-complete [7]. With irreflexive lightspeed-or-slower accessibility, the logics depend on the dimension, and there is a formula satisfiable with two spatial dimensions yet not with only one spatial dimension [3]. By a simple reduction from the reflexive cases it is clear that the modal logic of irreflexive lightspeed-or-slower accessibility is PSPACE-hard in all cases $m \geq 1$, though we could not find a proof in the literature that in all cases the logic is in PSPACE.

For temporal validities, (i.e. valid $(\mathbf{F}, \mathbf{P})$-formulas) over two-dimensional Minkowski spacetime with lightspeed-or-slower accessibility, that is, either $(\mathbb{R}^2, <)$ or $(\mathbb{R}^2, \leq)$, validity is PSPACE-complete [5], and the same is true for slower-than-light accessibility [4].[3] However, little is known about the logics of the higher-dimensional frames, except that they differ from those in two dimensions [5]. In the problem (5) at the conclusion of [5], Hirsch and Reynolds ask for the decidability of the $(\mathbb{R}^3, <)$ validities and conjecture undecidability.

In this paper, we make a small amount of progress towards identifying the complexity of the validity problem in higher dimensions, by proving an EXPTIME lower bound (with any of the four accessibility relations: lightspeed-or-slower or slower-than-light and reflexive or irreflexive). We do this by reducing the two-person corridor-tiling game to satisfiability in our higher-dimensional frames. We describe this tiling game, which is known to be EXPTIME-complete [2], in the following section. In Section 3 we carry out the reduction, giving a general EXPTIME-hardness result (Theorem 3.1), and in Section 4 we apply this theorem to Minkowski spacetime frames (Theorem 4.2). Section 5 lists open problems.

2 The two-person corridor-tiling game

In order that this paper be self contained, in this section we present the two-person corridor-tiling game. Our version of the game is that found in [1, §6.8].

A Wang tile type, T, is a square with its sides labelled by colours *left*(T), *right*(T), *up*(T) and *down*(T) [10]. From this presentation it is clear that tiles are 'fixed in orientation', that is, what is obtained by rotating a tile by 90° is considered distinct from the original tile type. Figure 2 shows the conventional way of depicting a Wang tile.

Game instances. An instance of the two-player corridor-tiling game is a pair $((T_0, \ldots, T_{s+1}), (I_1, \ldots, I_n))$, where $(T_0, \ldots, T_{s+1})$ is a sequence of tile types and $(I_1, \ldots, I_n)$ is a sequence of tiles drawn from $\{T_0, \ldots, T_{s+1}\}$.

The players and board. The game is played between two players—Eloise and Abelard—on an $n \times \omega$ grid—the 'corridor'. There are walls to the left of the first column and to the right of the nth column, which we consider to be columns 0 and $n+1$ respectively, and throughout the game both these columns are filled

[3] A contrasting result is that with *exactly lightspeed* accessibility, validity is undecidable [8].

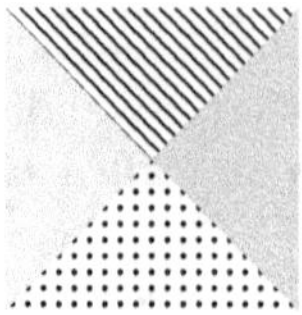

Figure 2. Depiction of a Wang tile

entirely with instances of tile type T_0. We assume this gives both walls the same colour, *white* say. At the start of the game the first row of the corridor is filled with the tiles $(I_1, \ldots, I_n)$ in that order. (This is the only purpose of the data $(I_1, \ldots, I_n)$, beyond determining the value of n.[4]) This initial position is depicted in Figure 3.

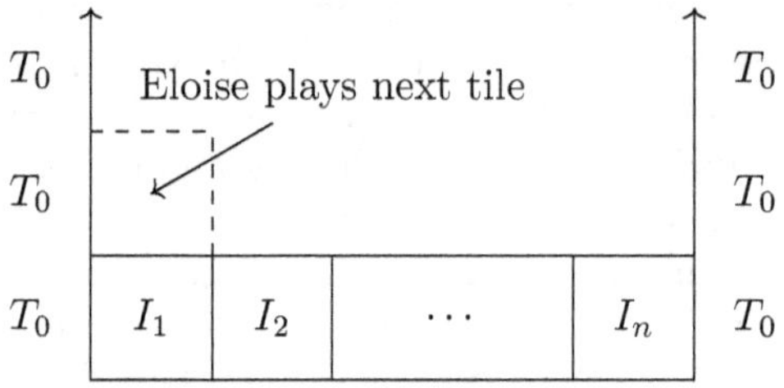

Figure 3. Game instance $((T_0, \ldots, T_{s+1}), (I_1, \ldots, I_n))$ at the start of play

The play. During play, the two players alternate, with Eloise playing first. To make a legal move, a player must chose a tile type and place one instance of that tile type on the board. The placement of the new tile is fixed: it is placed in the leftmost available position in the first incomplete row. It is required that the edge-colours of the placed tile match up with those of adjacent tiles already on the board.

Winning conditions. If at any point during the play, tile type T_{s+1} is placed in the first column then the game ends and Eloise wins. In all other cases, Abelard wins; there are two ways this can happen. Firstly, if T_{s+1} has not yet been placed in the first column, and the current player (Eloise or Abelard) is unable to make a move then the game ends and Abelard wins. The other possible way Abelard can win is if the play goes on infinitely long without T_{s+1} being placed in the first column, and where each player always makes a legal move.

Based on these rules, in order to have all relevant information about the state of play it is sufficient to know the *position* $(p, \bar{T}, i)$ consisting of:

(i) the player $p \in \{\exists, \forall\}$ whose turn it is,

[4] Note though that the presence of $(I_1, \ldots, I_n)$ in instance specification ensures the corridor width is polynomial—not exponential—in the size of the input, matching the stipulation in Chlebus' original presentation that n be encoded in unary.

(ii) a sequence $\bar{T}$ of n tiles, indicating the last tile placed in columns 1 to n,

(iii) a column index i, where $1 \leq i \leq n$, indicating the column where the next tile must be placed.

The game tree and strategy. Instances of the corridor-tiling game may analysed via their game tree. Let $\mathcal{I} = ((T_0, \ldots, T_{s+1}), (I_1, \ldots, I_n))$ be an instance. Let $p_0 = (\exists, (I_1, \ldots, I_n), 1)$ be the initial position in the game tree $\mathcal{T}(\mathcal{I})$. Each node of $\mathcal{T}(\mathcal{I})$ is labelled by a position, and the root of $\mathcal{T}(\mathcal{I})$ is labelled by the initial position p_0. In any position there are at most $s+2$ possible moves (as there are $s+2$ tile types). Each possible move determines the position of a child node. Each branch is either a win for Eloise or Abelard, not both.

A *strategy* for Eloise is a subtree of the game tree including the root, all children of included nodes where it is Abelard's turn, and at least one child of included nodes where it is Eloise's turn. Such a strategy is a winning strategy if all branches are wins for Eloise. Of course, whether such a winning strategy exists depends on the instance. The decision problem associated to the two-person corridor-tiling game is the set of instances for which Eloise has a winning strategy.

The two-person corridor-tiling game was first presented in [2], where it was called the *rectangle tiling game* and was proven to be EXPTIME-complete.

3 Reduction from corridor-tiling game

In this section we take an arbitrary instance $\mathcal{I} = ((T_0, \ldots, T_{s+1}), (I_1, \ldots, I_n))$ of the two-person corridor-tiling game and compute an associated temporal formula $\phi_{\mathcal{I}}$. Then we show that for any Kripke frame into which there is a certain kind of embedding of a complete binary tree with no leaves, $\mathcal{I}$ is a yes-instance if and only if $\phi_{\mathcal{I}}$ is satisfiable in the frame.

The rough idea is that we can use the image of the embedded tree as a scaffolding for drawing the game tree $\mathcal{T}(\mathcal{I})$ in the frame, simply by labelling (images of) nodes using propositional variables. (We do however re-encode $\mathcal{T}(\mathcal{I})$, with its finite branching factor, as a binary-branching tree, in order that the embedding be easier to produce.) Then $\phi_{\mathcal{I}}$ expresses that a given point is the root of a game tree for which Eloise has a winning strategy.

We first introduce some propositional formulas and definable modalities, to help express the formula $\phi_{\mathcal{I}}$.

In Minkowski spacetime we may express the universal modality $\blacksquare$ by

$$\blacksquare\varphi := \mathbf{GH}\varphi.$$

More generally, let $(V, <)$ be any frame that is both *transitive* and *confluent* ($x < y, z$ implies there is w with $y, z < w$). Then $\blacksquare$ is 'for all within the current weakly connected component' (of $(V, <)$, viewed as a directed graph), which is strong enough for our purposes.[5] Note that we will refer to elements of V as

[5] Technically this is not true if $<$ is empty on a (necessarily singleton) connected component. If this occurs in the frame then one can add a conjunct $\mathbf{F}\top$ to the $\phi_{\mathcal{I}}$ we describe to prevent

points.

We will use the following propositional variables.

(i) f to denote a 'forbidden set' F.

(ii) $index_i$, for $1 \leq i \leq n$, to indicate a node of the tree at which the next tile is to be played in column i. (If no $index_i$ holds at a point in V then the point is not a node.)

Later, in the definition (1), we use F to help define a modality $\Diamond$ (see also Figure 4).

The remaining variables give information only at nodes and have no significance elsewhere (where they may be assigned arbitrarily).

(iv) $col_i(T)$, for all $0 \leq i \leq n+1$ and all $T \in \{T_0, \ldots, T_{s+1}\}$, used to indicate that the latest tile placed in column i was of type T.

(v) *eloise*, true if it is Eloise's turn to play, false if Abelard's.

(vi) *win* to indicate that if play continues from the current position, Eloise has a winning strategy.

For the encoding of $\mathcal{T}(\mathcal{I})$ as a labelled binary tree $\mathcal{B}$, the maximum number of children of a node of $\mathcal{T}(\mathcal{I})$ is $s+2$; let $b = \lceil \log_2(s+2) \rceil$. By duplicating branches, we may assume that the branching factor of $\mathcal{T}(\mathcal{I})$ is always either 2^b or zero. We also assume $b \geq 3$.

Let Δ be a complete binary tree of depth b, whose nodes are strings over $\{0,1\}$ of length at most b, and with edges $(x, x0)$ and $(x, x1)$ when x is a string of length less than b. The root of Δ is the empty string ε. Each node of Δ has a depth equal to the length of the bit-string, so the root has depth 0 and the leaves have depth b.

For the *tree structure* of $\mathcal{B}$, first take a forest consisting of copies Δ_p of Δ indexed by nodes of $\mathcal{T}(\mathcal{I})$, so a typical node will be (p, x) where $p \in \mathcal{T}(\mathcal{I})$ and $x \in \Delta$. Now for every non-leaf node p of $\mathcal{T}(\mathcal{I})$ choose a bijection θ_p between the leaves of Δ_p and those nodes q of $\mathcal{T}(\mathcal{I})$ where there is a move from p to q in $\mathcal{T}(\mathcal{I})$ (each set has cardinality 2^b). Now, for each non-leaf node p of $\mathcal{T}(\mathcal{I})$ identify each leaf (p, x) of Δ_p with $(\theta_p(p, x), \varepsilon)$. Finally, remove any leaf nodes (precisely the nodes of depth b that were not identified). This defines the binary tree structure of $\mathcal{B}$. After this identification, the nodes are (p, x) where p is a node of $\mathcal{T}(\mathcal{I})$, and the node x of Δ has depth at most $b-1$. The children of (p, x) are $(p, x0), (p, x1)$ when the depth of x is less than $b-1$. When the depth of x is $b-1$ the children of (p, x) are (q, ε) where q can be one of two successor positions of p in $\mathcal{T}(\mathcal{I})$. For the *labelling* of $\mathcal{B}$, each node (p, x) is labelled by the node p of $\mathcal{T}(\mathcal{I})$ and the depth of x.

We thus also have propositions

(vii) $depth_j$, for $0 \leq j \leq b-1$, to indicate the depth of x in Δ.

---

the formula being satisfied at such a point.

Let

$$\beta(i,j) := index_i \wedge depth_j,$$

and

$$\beta := \bigvee_{1\le i\le n} index_i.$$

Thus β is the proposition 'is a node'.

Define a function $^+ : [1,n] \times [0,b-1] \to [1,n] \times [0,b-1]$ that updates the column-index and depth by either incrementing the depth if less than $b-1$ leaving column-index fixed, else updating the column-index and resetting depth to 0. That is, for $(i,j) \in [1,n] \times [0,b-1]$, let

$$(i,j)^+ = \begin{cases} (i,j+1) & j \le b-2 \\ (i+1,0) & j = b-1,\ i < n \\ (1,0) & j = b-1,\ i = n\,. \end{cases}$$

If (i,j) is the index and depth in an encoded version of the game where b levels encode one move, then $(i,j)^+$ is the index and depth of children in this encoded game. Since $b \ge 3$ we know that $(i,j)^{++} \ne (i,j)$ (this allows us to distinguish the children from the parent of any node).

We will now define a pair of modalities $\Diamond, \Box$ that will express 'possibly for a child' and 'necessarily for children' respectively, and thus will do most of the heavy lifting in the formula $\phi_{\mathcal{I}}$. It may be helpful to know that for the Minkowski frames, we will be embedding the binary tree into the (hyper)plane $t = 0$. To give some intuition, Figure 4 illustrates a point where $\Diamond\varphi$ holds.

We define $\Diamond$ and $\Box$ as follows.

$$\Diamond\varphi := \bigvee_{1\le i\le n,\ 0\le j\le b-1} \beta(i,j) \wedge \mathbf{F}(\mathbf{P}(\beta((i,j)^+) \wedge \varphi) \wedge \mathbf{H}\neg f) \tag{1}$$

$$\Box\varphi := \bigwedge_{1\le i\le n,\ 0\le j\le b-1} \beta(i,j) \to \mathbf{G}(\mathbf{H}\neg f \to \mathbf{H}(\beta((i,j)^+) \to \varphi))$$

Note that $\Box\varphi$ is indeed $\mathbf{K}_t$-equivalent to $\neg\Diamond\neg\varphi$, and that $\mathbf{K}_t \vdash (\Box\varphi \wedge \Box\psi) \to \Box(\varphi \wedge \psi)$.[6]

Now is a good time to state our main theorem.

Theorem 3.1 *Let $(V,<)$ be any transitive and confluent Kripke frame, and let $\mathcal{B}$ be the infinite complete binary tree. (We view the edges, $E(\mathcal{B})$, of $\mathcal{B}$ as directed from parent to child.) Suppose there is a map $' : \mathcal{B} \to V$ and a subset $F \subseteq V$ such that the following condition holds for all distinct $x, y \in \mathcal{B}$.*

$$\begin{gathered} (x,y) \in E(\mathcal{B}) \vee (y,x) \in E(\mathcal{B}) \\ \Longleftrightarrow \\ \exists z \in V(z > x' \wedge z > y' \wedge \forall w(w < z \implies w \notin F)) \end{gathered} \tag{2}$$

[6] Recall that $\mathbf{K}_t$ is the temporal analogue of modal $\mathbf{K}$, that is, the set of temporal formulas valid on all frames.

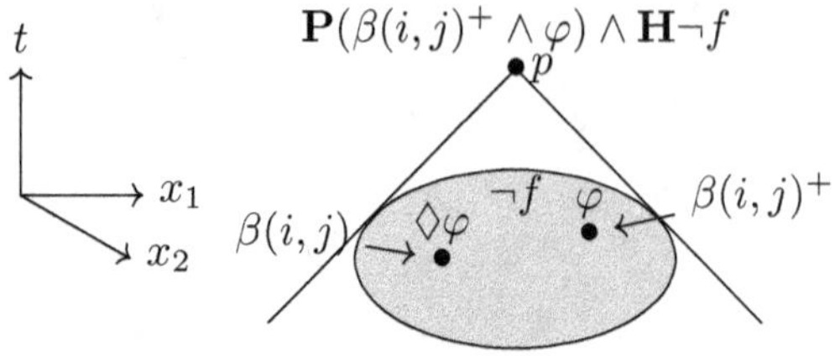

Figure 4. The formula $\Diamond\varphi$ holds at a point if some $\beta(i,j)$ holds there, and a point p in the future sees a point in its past where $\varphi \wedge \beta(i,j)^+$ holds, but no point satisfying f is in the past of p. In particular, f must not hold at any point in the shaded disc where the past of p intersects the spatial plane.

Then the satisfiability of $(\mathbf{F},\mathbf{P})$*-formulas over* $(V,<)$ *is EXPTIME-hard.*

Note that the map $'$ of Theorem 3.1 is necessarily injective, since if $x_1' = x_2'$ then condition (2) implies x_1 and x_2 have (excluding each other) the same neighbours in $\mathcal{B}$.

As mentioned previously, given an instance $\mathcal{I} = ((T_0,\ldots,T_{s+1}), (I_1,\ldots,I_n))$ of the two-person corridor-tiling game, we construct a formula $\phi_{\mathcal{I}}$ that is satisfiable in $(V,<)$ if and only if Eloise has a winning strategy for $\mathcal{T}(\mathcal{I})$. The formula $\phi_{\mathcal{I}}$ will state that the initial position is a winning position for Eloise.

The formula $\phi_{\mathcal{I}}$ is the conjunction of all the conditions I–XX we are about to lay out. We follow the structure of the presentation of the reduction to PDL satisfiability given on pages 398–401 of [1]. That reduction originates from [2]. Most of the conditions have the $\blacksquare$ modality in outermost position, so that what is inside this modality holds throughout the play. We omit the phrase 'Throughout the play...' from the textual descriptions of these conditions.

I Position and depth at start of play:

$$\mathit{eloise} \wedge \beta(1,0) \wedge \mathit{col}_0(T_0) \wedge \mathit{col}_1(I_1) \wedge \ldots \wedge \mathit{col}_n(I_n) \wedge \mathit{col}_{n+1}(T_0).$$

II Nodes have a unique index:

$$\blacksquare \bigwedge_{1\leq i,j\leq n;\ i\neq j} (\mathit{index}_i \to \neg \mathit{index}_j)$$

III Nodes have a unique depth:

$$\blacksquare \bigwedge_{0\leq i,j\leq b-1;\ i\neq j} (\mathit{depth}_i \to \neg \mathit{depth}_j)$$

IV The depths partition the same set as the indexes:

$$\blacksquare(\beta \leftrightarrow \bigvee_{0\leq j\leq b-1} \mathit{depth}_j)$$

V In every column i, at least one tile type was previously placed:

$$\blacksquare(\beta \to \bigwedge_{0 \le i \le n+1} (col_i(T_0) \vee \ldots \vee col_i(T_{s+1}))).$$

VI In every column i, at most one tile type was last placed:

$$\blacksquare(col_i(T_u) \to \neg col_i(T_v)) \qquad (0 \le i \le n+1 \text{ and } 0 \le u \ne v \le s+1).$$

VII Tile T_0 is already placed in columns 0 and $n+1$:

$$\blacksquare(\beta \to (col_0(T_0) \wedge col_{n+1}(T_0))).$$

With these preliminaries behind us, we can now describe the structure of the game tree.

VIII No change in game position when depth is less than $b-1$:

$$\blacksquare(\beta \wedge \neg depth_{b-1} \to (eloise \to \Box eloise) \wedge (\neg eloise \to \Box \neg eloise))$$

and for $1 \le i \le n$ and $0 \le u \le s+1$

$$\blacksquare(\beta \wedge \neg depth_{b-1} \to (col_i(T_u) \to \Box col_i(T_u) \wedge (\neg col_i(T_u) \to \Box \neg col_i(T_u))).$$

IX In columns where no tile is placed, nothing changes when a move is made: for all $1 \le i, j \le n$ with $j \ne i$ and all $0 \le u \le s+1$

$$\blacksquare(\beta(i, b-1) \to (col_j(T_u) \to \Box col_j(T_u)) \wedge (\neg col_j(T_u) \to \Box \neg col_j(T_u)))$$

X Players alternate:

$$\blacksquare(depth_{b-1} \to (eloise \to \Box \neg eloise) \wedge (\neg eloise \to \Box eloise)).$$

Next, both players make legal moves; that is, they only place tiles that correctly match adjacent tiles. It will be helpful to define the following ternary relation of 'compatibility' between tile types:

$$C(T', T, T'') \iff right(T') = left(T) \text{ and } down(T) = up(T'').$$

That is, $C(T', T, T'')$ holds if and only if the tile T can be placed to the right of tile T' and above tile T''. With the aid of this relation we can formulate the first constraint on tile placement as follows.

XI Adjacencies: for all $1 \le i \le n$ and all tile types T' and T'',

$$\blacksquare(\beta(i, b-1) \wedge col_{i-1}(T') \wedge col_i(T'') \to \Box \bigvee \{col_i(T) \mid C(T', T, T'')\}).$$

(Here, by convention, $\bigvee \emptyset = \bot$.)

The previous constraint only ensures matching to the left and downwards. We also need to ensure that tiles placed in column n match the white wall tile to their right.

XII Match white on right in column n:

$$\blacksquare(\beta \to \bigvee\{col_n(T) \mid right(T) = white\}).$$

XIII All possible Abelard moves are included: for all $1 \leq i < n$,

$$\blacksquare(\neg eloise \wedge depth_{b-1} \wedge col_i(T'') \wedge col_{i-1}(T') \to \bigwedge\{\Diamond col_i(T) \mid C(T', T, T'')\}),$$

(Here, by convention, $\bigwedge \emptyset = \top$).

This completes the description of the game tree. The remaining conditions ensure Eloise has a winning strategy.

XIV The initial position is a winning position for Eloise:

$$win.$$

XV Recursive conditions:

$$\blacksquare(win \to (col_1(T_{s+1}) \vee (\neg eloise \wedge \Diamond\top \wedge \Box win) \vee (eloise \wedge \Diamond win))).$$

To bound the length of the game, first define $N = 2n(s+2)^n$. If the game goes on for N plays, then the position has repeated,[7] which does not help Eloise: if she can win, she can do so in fewer than N moves. Let L be the number of binary digits necessary to write N. We introduce new propositions $q_1, \ldots, q_L$ to denote the sequence of bits from the least to the most significant digit of a *counter*.

XVI Counter is initially 0:

$$\neg q_L \wedge \ldots \wedge \neg q_1.$$

XVII Counter does not change when $1 \leq j \leq b-1$:

$$\blacksquare(\beta \wedge \neg depth_{b-1} \to \bigwedge_{k=1}^{L}((q_k \to \Box q_k) \wedge (\neg q_k \to \Box\neg q_k)).$$

XVIII Incrementation of counter when least-significant digit is 0:

$$\blacksquare(depth_{b-1} \wedge \neg q_1 \to \Box q_1 \wedge \bigwedge_{k=2}^{L}((q_k \to \Box q_k) \wedge (\neg q_k \to \Box\neg q_k))).$$

XIX Incrementation of counter when least-significant digit is 1: for $1 \leq k < L$,

$$\blacksquare(depth_{b-1} \wedge \neg q_{k+1} \wedge \bigwedge_{l=1}^{k} q_l \to$$
$$\Box(q_{k+1} \wedge \bigwedge_{l=1}^{k} \neg q_l) \wedge \bigwedge_{l=k+2}^{L}((q_l \to \Box q_l) \wedge (\neg q_l \to \Box\neg q_l))).$$

[7] Including the initial position, $2n(s+2)^n + 1$ positions have occurred.

XX If the counter reaches N then the game has gone on too long and Abelard wins:

$$\blacksquare(\beta \wedge (counter = N) \rightarrow \neg win).$$

This concludes the definition of $\phi_{\mathcal{I}}$. We are now in a position to prove Theorem 3.1.

Proof We argue that whenever $(V, <)$ satisfies conditions of the theorem (i.e. there is $', F$ satisfying (2)), then the derivation of $\phi_{\mathcal{I}}$ from $\mathcal{I}$ is a correct, polynomial-time reduction of the two-player corridor-tiling problem to satisfiability of $(\mathbf{F}, \mathbf{P})$-formulas over $(V, <)$. Since the two-player corridor-tiling problem is EXPTIME-hard [2], it follows that satisfiability of $(\mathbf{F}, \mathbf{P})$-formulas over $(V, <)$ is too.

First note that the length of $\phi_{\mathcal{I}}$ is polynomial in s and n, at which point it is clear that $\phi_{\mathcal{I}}$ can be calculated from $\mathcal{I}$ in time polynomial in the size of (the encoding) of $\mathcal{I}$. We now argue that the reduction is correct.

Formula is satisfiable implies Eloise has a winning strategy: Suppose v is a valuation and $r \in V$ with $(V, <), v, r \models \phi_{\mathcal{I}}$. Define a tree rooted at r by setting the children of a node x satisfying $\beta(i, j)$ to be

$$\{y \models \beta(i, j)^+ \mid \exists z : z > x \wedge z > y \wedge \forall w(w < z \implies w \not\models f)\}.$$

By the construction of $\phi_{\mathcal{I}}$, this determines a labelled tree $\mathcal{B}$ encoding a portion the game tree $\mathcal{T}(\mathcal{I})$ sufficient for defining a strategy. If this $\mathcal{B}$ is not technically a tree because there is some unwanted sharing of descendants, then unwind it into the tree of paths in this directed graph that depart from r. Note also that there is no particular reason for $\mathcal{B}$ to be binary. The formula $\phi_{\mathcal{I}}$ further ensures that the initial position is a position from which Eloise can force a win. Hence Eloise has a winning strategy.

Eloise has a winning strategy implies formula is satisfiable: Assume the condition in the statement of Theorem 3.1. Let $\mathcal{T}(\mathcal{I})$ be a labelled game tree for the instance $((T_0, \ldots, T_{s+1}), (I_1, \ldots, I_n))$. As we saw, we can encode $\mathcal{T}(\mathcal{I})$ as a labelled binary tree $\mathcal{B}$ in which each node has a position and depth, and by the hypothesis of the theorem, there is an embedding $'$ of $\mathcal{B}$ as $\mathcal{B}' \subseteq V$ together with a subset F of V satisfying condition (2).

Define a propositional valuation $v : \text{Prop} \rightarrow \wp(V)$ by:

- $v(f) = F$,
- $v(index_i) = \{b' \in \mathcal{B}' \mid b \text{ has column-index } i\}$,
- $v(depth_j) = \{b' \in \mathcal{B}' \mid b \text{ has depth } j\}$,
- $v(eloise)$ consists of points in $\mathcal{B}'$ where it is Eloise's turn,
- $v(col_i(T)) = \{b' \in \mathcal{B}' \mid T \text{ was last placed in column } i \text{ at } b\}$.

We may define $v(q_i)$ for $1 \leq i \leq L$ so that the q_is that hold at b' encode, as a binary number, the minimum of $2^L - 1$ and the number of edges on a path from b to the root r of $\mathcal{T}(\mathcal{I})$. Let win hold at all points in the range of $'$ where Eloise has a winning strategy and the counter is less than N, but nowhere else.

Let b be any node of the binary tree $\mathcal{B}$, say $b' \models \beta(i,j)$ (some i,j). Then b is adjacent to two child nodes c_1, c_2 and perhaps to its parent, but no other node, and $\beta(i,j)^+$ holds at c_1' and at c_2'. By (1) and (2), if φ holds at c_1' or at c_2' then $\Diamond\varphi$ holds at b'. Conversely, if $\Diamond\varphi$ holds at b' then by (1) and (2), we know that $\beta(i,j)^+ \wedge \varphi$ must hold where a node adjacent to b in $\mathcal{B}$ embeds. However, $\beta(i,j)^+$ does not hold where the parent of b embeds, so φ must hold at c_1' or at c_2'.

By the assumption that Eloise has a winning strategy, $r' \models win$; other conjuncts of $\phi_{\mathcal{I}}$ are now easily verified. Thus $(V,<), v, r' \models \phi_{\mathcal{I}}$, as required. □

4 Minkowski spacetime frames

In this section we apply Theorem 3.1 to our target frames, thereby proving EXPTIME lower bounds.

Lemma 4.1 *Let a, b, c, d be points on a circle C, in that order, as you go round the circle (clockwise or anticlockwise). Every disc containing a and c also contains b or d.*

Proof For contradiction, suppose D is a disc containing a and c but neither b nor d. Since $b \notin D$, but $a \in D$, the boundary of D must cross C between a and b. Similarly, the boundary of D meets C three more times, between b and c, between c and d, and between d and a. Since $a, c \in D$ and $b, d \notin D$, the four points in C on the boundary of D must be distinct. But two circles intersecting more than twice must be identical, contradicting $a, c \in D$ and $b, d \notin D$. □

Theorem 4.2 *Validity of temporal formulas over $(m+1)$-dimensional Minkowski spacetime is EXPTIME-hard for $m \geq 2$ with 'lightspeed-or-slower' or with 'slower-than-lightspeed' accessibility, either the reflexive or irreflexive versions.*

Proof Let $(V,<)$ be any of the frames in the Theorem; clearly the frame is transitive and confluent. To apply Theorem 3.1, we need to find an embedding $'$ from the complete binary tree $\mathcal{B}$ of depth ω into V together with a subset $F \subseteq V$, in such a way that condition (2) is satisfied. To do this, we will embed $\mathcal{B}$ in a two-dimensional spacelike plane P perpendicular to the time axis and select $F \subseteq P$. We could, for example, assume P is the plane defined by $x_3 = x_4 = \cdots = x_m = t = 0$. Points in P are determined by a pair (x,y) of real coordinates with respect to some fixed choice of orthonormal basis for P (equipped with the standard inner product). Intersections of past light cones with P are discs (closed or open depending on whether lightspeed is included in $<$). In the following, D will range over all discs of the appropriate type—closed if lightspeed is included, open otherwise.

To apply Theorem 3.1, we need to establish condition (2), which is equivalent to the following, for all $p \neq q \in \mathcal{B}$,

$$(p,q) \in E(\mathcal{B}) \vee (q,p) \in E(\mathcal{B}) \iff \exists D(p', q' \in D \wedge (D \cap F = \emptyset)) \qquad (3)$$

For p', q' let $p' \sim q' \iff \exists D(p', q' \in D \wedge (D \cap F = \emptyset))$. An embedding $'$ and forbidden set F satisfying (3) is shown in Figure 5.

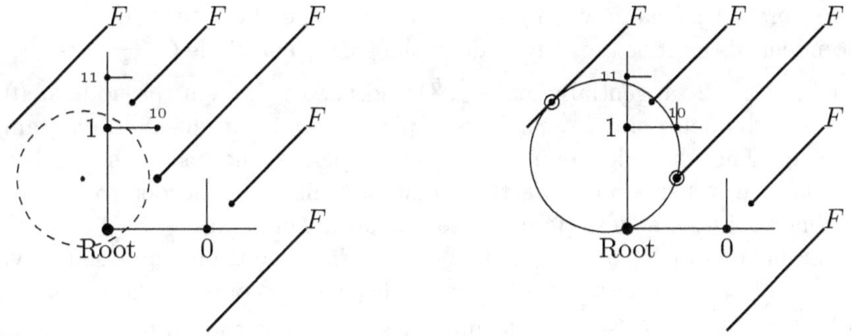

Figure 5. Embedding of complete binary tree $\mathcal{B}$, and forbidden set F in a plane, to scale. Left: parent $\sim$ child; a disc containing root and child, disjoint from F. Right: grandparent $\not\sim$ grandchild; a circle through $(0,0), (\frac{1}{2},\frac{1}{2})$, and $(-\frac{3}{4},\frac{5}{4})$; nodes 11, 10 and descendants are outside the circle, between the two circled forbidden nodes.

The nodes of $\mathcal{B}$ may be written as finite strings of bits, and the two children of string s are obtained by appending 0 or 1 to the end of s. For the embedding, map string $s = (b_0b_1 \ldots b_{k-1})$ to $\sum_{i<k,\, b_i=0} 2^{-i}(1,0) + \sum_{i<k,\, b_i=1} 2^{-i}(0,1)$, so the root (the empty string) maps to $(0,0)$, its two children 0 and 1 map to $(1,0)$ and $(0,1)$ respectively, its grandchildren 00, 01, 10, and 11 map to $(\frac{3}{2},0)$, $(1,\frac{1}{2})$, $(\frac{1}{2},1)$, and $(0,\frac{3}{2})$ respectively, and so on; see Figure 5.

The lines $y = x-2$ and $y = x+2$ are included in F. Also, if μ is the midpoint in P of (the images of) a pair of siblings, every point at or above/right of μ on the line $y = x + c$ passing through μ, is contained in F. Nothing else is included in F. These forbidden lines are shown as diagonal lines or half-lines in Figure 5.

For the left-to-right implication of (3), first suppose p is the root of $\mathcal{B}$. Consider the open disc $O_{\frac{2}{3}}(-\frac{1}{4},\frac{1}{2})$ shown on the left of Figure 5 (to scale), and first observe that the root and its child 1 map into the disc, since $\frac{\sqrt{5}}{4} < \frac{2}{3}$. Then observe that all forbidden points are outside the closure of the disc, since the nearest point on the line $y = x + 2$ is $(-\frac{7}{8},\frac{9}{8})$ at a distance $\frac{5\sqrt{2}}{8}$, and the nearest points in F on the lines $y = x + 1$ and $y = x$ are the endpoints $(\frac{1}{4},\frac{5}{4})$ and $(\frac{1}{2},\frac{1}{2})$ at distances $\frac{\sqrt{13}}{4}$ and $\frac{3}{4}$, respcctively. Similarly, the root and its other child, 0, map into a disc disjoint from F. Cases where p is not the root are treated shortly.

For the right-to-left implication of (3), suppose D avoids forbidden nodes, and $p' \neq q' \in D$. By convexity of D all points on the line segment from p' to q' belong to D, which is disjoint from F. The midpoint of (the images of) two siblings is in F, so p and q cannot be siblings. Moreover, if p and q are descendants of two distinct siblings then there is a point on the line segment between p' and q' that is forbidden, so this case cannot happen either. So either p is a descendant of q, or q is a descendant of p.

We claim that p and q cannot be more than one generation apart. Again, we first prove this claim when p is the root, which embeds to $(0,0)$; thus q is a descendent of the root (i.e. any node of the tree). The circle $C_{\frac{\sqrt{49+529}}{32}}(-\frac{7}{32},\frac{23}{32})$ meets $y=x+2$ tangentially at $(-\frac{3}{4},\frac{5}{4})$ and passes through the node at $(0,0)$ and the forbidden point $(\frac{1}{2},\frac{1}{2})$. This is pictured in the right-hand diagram of Figure 5. The forbidden point at $(\frac{1}{4},\frac{5}{4})$ happens to lie inside the circle; we make no use it here, but note this point only makes it harder to find discs avoiding forbidden nodes. Here we use the forbidden points $(-\frac{3}{4},\frac{5}{4})$ and $(\frac{1}{2},\frac{1}{2})$. All descendants of 11, at $(0,\frac{3}{2})$, or of 10, at $(\frac{1}{2},1)$, lie, from the point of view of $(0,0)$, between these two forbidden nodes and are outside the circle, since $\frac{\sqrt{49+625}}{32},\frac{\sqrt{529+81}}{32}>\frac{\sqrt{529+48}}{32}$. (Reading glasses may be needed to see that 10 is outside the circle in Figure 5.) Thus every disc containing $(0,0)$ and avoiding forbidden nodes, excludes 11, 10, and their descendants, else by convexity the disk would also contain a point c on $C_{\frac{\sqrt{49+529}}{32}}(-\frac{7}{32},\frac{23}{32})$, such that $(0,0)$, $(-\frac{3}{4},\frac{5}{4})$, c, $(\frac{1}{2},\frac{1}{2})$ is the clockwise order, contradicting Lemma 4.1. So q is none of 11, 10, or their descendants. By a symmetric argument, we see that q cannot be 00, 01, or their descendants. It follows that q must be a child, either 0 or 1, of the root, proving the right-to-left implication in (3), for the case where p is the root.

To generalise to arbitrary p, we now define a P-similarity. Recall that $P\subseteq V=\mathbb{R}^{m+1}$ is a two-dimensional spatial plane, the image of $'\colon\mathcal{B}\to V$ is included in P, and $F\subseteq V$. Let $\mathcal{B}'=\{p'\mid p\in\mathcal{B}\}$ and $E'=\{(p',q')\mid(p,q)\in E(\mathcal{B})\}$. A *$P$-similarity* $\sigma:P\to P$ is a similarity of P with respect to the Euclidean metric over $P\subseteq\mathbb{R}^{m+1}$, such that

- $p'\in\mathcal{B}'\implies\sigma(p')\in\mathcal{B}'$ ('nodes map to nodes'),
- $(p',q')\in E'\implies(\sigma(p'),\sigma(q'))\in E'$ ('edges map to edges'),
- $\sigma(P\setminus F)\subseteq P\setminus F$ ('permitted points map to permitted points').

for all $p,q\in\mathcal{B}$. Observe that if σ is a P-similarity and $p'\in D$ for some disc D disjoint from F, then $\sigma[D]$ is a disc disjoint from F containing $\sigma(p')$. The map $(x,y)\mapsto(1+\frac{x}{2},\frac{y}{2})$ is a P-similarity mapping $(0,0)$ to $(1,0)$ $(=0')$, and likewise for all $p\in\mathcal{B}$ there is a P-similarity mapping $(0,0)$ to p'. Hence the left-to-right implication of (3) holds for all $p\in\mathcal{B}$. The proof of the right-to-left implication of (3) goes through when p is not the root, since there is a P-similarity mapping $(0,0)$ to p' carrying the forbidden points $(-\frac{3}{4},\frac{5}{4})$ and $(\frac{1}{2},\frac{1}{2})$ to forbidden points. □

5 Open problems

(1) Our main result is the EXPTIME-hardness of the temporal validity problem over the Minkowski spacetimes $(\mathbb{R}^n,\leq),(\mathbb{R}^n,<),(\mathbb{R}^n,\prec),(\mathbb{R}^n,\preceq)$, for $n\geq 3$, but we have no upper bound on the complexity of these frames; we do not know if the logic of any of these frames is decidable, nor even if they are recursively enumerable. For $n=2$ (only one spatial dimension) we know that all four validity problems are PSPACE-complete [5,4]. Find more precise bounds on the complexity of the validities of $(\mathbb{R}^n,<)$, for

$n \geq 3$.

(2) There are purely modal formulas satisfiable in $(\mathbb{R}^3, <)$ but not in $(\mathbb{R}^2, <)$ [3], and there are temporal formulas satisfiable in $(\mathbb{R}^3, \leq)$ but not in $(\mathbb{R}^2, \leq)$ [5]. Are there temporal formulas that distinguish $(\mathbb{R}^3, <)$ from $(\mathbb{R}^4, <)$?

(3) Find sound and complete axioms for the temporal validities of any of these Minkowski frames. (These are open even with one spatial dimension.)

References

[1] Blackburn, P., M. d. Rijke and Y. Venema, "Modal Logic," Cambridge Tracts in Theoretical Computer Science, Cambridge University Press, 2001.

[2] Chlebus, B. S., *Domino-tiling games*, Journal of Computer and System Sciences **32** (1986), pp. 374–392.

[3] Goldblatt, R., *Diodorean modality in Minkowski spacetime*, Studia Logica **39** (1980), pp. 219–236.

[4] Hirsch, R. and B. McLean, *The temporal logic of two-dimensional Minkowski spacetime with slower-than-light accessibility is decidable*, Advances in Modal Logic **12** (2018), pp. 347–366.

[5] Hirsch, R. and M. Reynolds, *The temporal logic of two dimensional Minkowski spacetime is decidable*, The Journal of Symbolic Logic **83** (2018), pp. 829–867.

[6] Prior, A. N., "Time and Modality," John Locke lectures, Clarendon Press, 1957.

[7] Shapirovsky, I., *On PSPACE-decidability of transitive modal logic*, Advances in Modal Logic **5** (2005), pp. 269–287.

[8] Shapirovsky, I., *Simulation of two dimensions in unimodal logics*, Advances in Modal Logic **8** (2010), pp. 371–391.

[9] Shapirovsky, I. and V. B. Shehtman, *Chronological future modality in Minkowski spacetime*, Advances in Modal Logic **4** (2003), pp. 437–460.

[10] Wang, H., *Proving theorems by pattern recognition — II*, The Bell System Technical Journal **40** (1961), pp. 1–41.

Compatibility and accessibility: lattice representations for semantics of non-classical and modal logics

Wesley H. Holliday

University of California, Berkeley

Abstract

In this paper, we study three representations of lattices by means of a set with a binary relation of compatibility in the tradition of Ploščica. The standard representations of complete ortholattices and complete perfect Heyting algebras drop out as special cases of the first representation, while the second covers arbitrary complete lattices, as well as complete lattices equipped with a negation we call a protocomplementation. The third topological representation is a variant of that of Craig, Haviar, and Priestley. We then extend each of the three representations to lattices with a multiplicative unary modality; the representing structures, like so-called graph-based frames, add a second relation of accessibility interacting with compatibility. The three representations generalize possibility semantics for classical modal logics to non-classical modal logics, motivated by a recent application of modal orthologic to natural language semantics.

Keywords: lattices, representation theorems, ortholattices, orthologic, Heyting algebras, intuitionistic logic, Boolean algebras, modal logic, negation, graph-based frames, possibility semantics

1 Introduction

Semantics for non-classical and modal logics may be seen as arising from more basic algebraic representation theorems. For example, traditional semantics for intuitionistic logic, orthologic, and classical modal logic may be understood in terms of the following well-known representations:

- Any Heyting algebra H embeds into the lattice of downsets of a poset, and the embedding is an isomorphism if H is complete and its completely join-irreducible elements are join-dense (see, e.g., [22], [17, Prop. 1.1]).
- Any ortholattice L embeds into the lattice of $\perp$-closed sets of an orthoframe $(X, \perp)$ equipped with an orthocomplementation $\neg$ induced by the relation $\perp$, and the embedding is an isomorphism if L is complete (see [29] for the orthoframe description and [6, §§ 32-4], [19] for other descriptions).
- Any Boolean algebra B equipped with a multiplicative unary operation $\Box$ embeds into the powerset of a set W equipped with an operation $\Box_R$ induced

by a binary relation R on W, and the embedding is an isomorphism if B is complete and atomic and $\Box$ completely multiplicative (see [46,38,47]).

In each case, adding topology to the relevant relational structures allows one to characterize topologically the image of the relevant embedding [22,30,5,27,44].

Here we study representations that subsume and go beyond all of those mentioned above. In § 2, we explain how to go from a set together with a binary relation of "compatibility" to a complete lattice. In § 3, we study three ways of going back: one economical representation of certain complete lattices, including but not limited to Heyting and ortholattice cases; one less economical but fully general representation of complete lattices, including complete lattices equipped with a type of negation that we call a protocomplementation; and one representation of arbitrary lattices. In § 4, we extend the three representations to lattices with a multiplicative unary modality $\Box$, by adding a second relation of accessibility interacting with compatibility. We conclude in § 5.

After writing this paper, I discovered Ploščica's [43] representation of bounded lattices using certain compatibility frames [1] as in Definition 2.1 with a topology, as well as Craig et al.'s [16] modification of Ploŝĉica's approach. In § 3.3, we briefly cover a variant of this representation with a different topology. A referee also informed me that the addition of modal accessibility interacting with compatibility (Definition 4.2) appears in the *graph-based frames* of Conradie et al. [10]. We will return to this connection in § 4.

A Jupyter notebook with code to verify examples and investigate conjectures and questions is available at github.com/wesholliday/compat-frames.

2 From compatibility frames to lattices

2.1 Basic concepts

Our starting point is a certain way of going from a set with a binary relation to a complete lattice. For a comparison with other ways of realizing complete lattices using doubly ordered structures and polarities, see [34].

Definition 2.1 A *relational frame* is a pair $\mathcal{F} = (X, \lhd)$ where X is a nonempty set and $\lhd$ is a binary relation on X. A *compatibility frame* is a relational frame in which $\lhd$ is reflexive.

We read $x \lhd y$ as "x is compatible with y," also written $y \rhd x$. [2]

Convention 2.2 In diagrams, such as Fig. 1, an arrow with a triangle arrowhead from y to x indicates $y \rhd x$. Thus, we draw the directed graph $(X, \rhd)$ to represent the compatibility frame $(X, \lhd)$. Reflexive loops are not shown.

Recall that a unary operation on a lattice is a *closure operator* if c is inflationary ($x \leq c(x)$), idempotent ($c(c(x)) = c(x)$), and monotone ($x \leq y$ implies $c(x) \leq c(y)$). We will use the compatibility relation $\lhd$ to define a closure op-

[1] Also see [15] for *TiRS graphs*, which are compatibility frames with extra properties, which we do not require here (as is crucial for a number of our results and for Conjecture 3.13).

[2] In [34], we wrote $x \between y$ and $y \between^{-1} x$ instead of $x \lhd y$ and $y \rhd x$, respectively.

erator on $\wp(X)$, whose fixpoints give us a complete lattice as in the following classic result (see, e.g., [8, Thm. 5.2]).

Proposition 2.3 *Let X be a nonempty set and c a closure operator on $\wp(X)$. Then the fixpoints of c, i.e., those $A \subseteq X$ with $c(A) = A$, ordered by $\subseteq$ form a complete lattice with*

$$\bigwedge_{i\in I} A_i = \bigcap_{i\in I} A_i \text{ and } \bigvee_{i\in I} A_i = c(\bigcup_{i\in I} A_i).$$

Definition 2.4 Given a relational frame $(X, \lhd)$, define $c_\lhd : \wp(X) \to \wp(X)$ by

$$c_\lhd(A) = \{x \in X \mid \forall x' \lhd x\ \exists x'' \rhd x' : x'' \in A\}.$$

Thus, x is in $c_\lhd(A)$ iff every state compatible with x is compatible with some state in A. Given a compatibility frame, we are interested in the $c_\lhd$-fixpoints, i.e., those $A \subseteq X$ such that $c_\lhd(A) = A$. Looking at a diagram of a compatibility frame, one can check that $c_\lhd(A) = A$ by checking that the following holds:

- from any $x \in X \setminus A$, you can step forward along an arrow to a state x' that cannot step backward along an arrow into A.

Informally, "from x you can see a state that cannot be seen from A."

Example 2.5 Consider the cycle on three elements on the left of Fig. 1, regarded as a compatibility frame according to Convention 2.2: $\{y\}$ is a $c_\lhd$-fixpoint because z and x can both see x, which cannot be seen from $\{y\}$. Yet $\{y, z\}$ is not a $c_\lhd$-fixpoint, because x cannot see a state that cannot be seen from $\{y, z\}$, since both x and y can be seen from $\{y, z\}$.

We get the reverse verdicts on $\{y\}$ and $\{y, z\}$ in the acyclic (ignoring loops) but non-transitive frame on the right of Fig. 1: $\{y\}$ is *not* a $c_\lhd$-fixpoint, because now z cannot see a state that cannot be seen from $\{y\}$; but $\{y, z\}$ *is* a $c_\lhd$-fixpoint, because x can see a state, namely x, that cannot be seen from $\{y, z\}$.

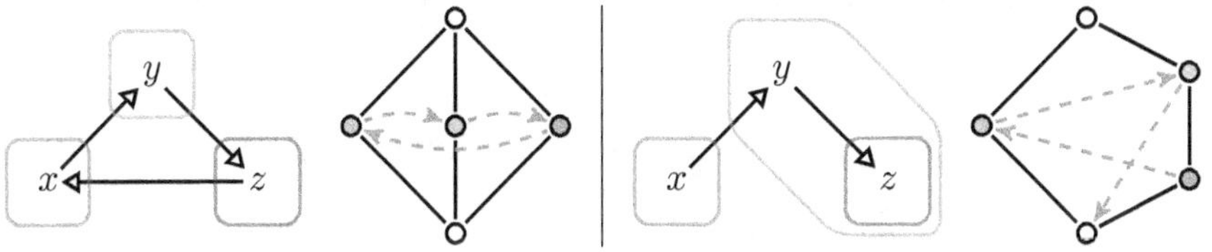

Fig. 1. Two compatibility frames drawn according to Convention 2.2 with their $c_\lhd$-fixpoints (except X and $\varnothing$) outlined, followed by their associated lattices.

Proposition 2.6 *For any relational frame, $c_\lhd$ is a closure operator on $\wp(X)$.*

Proof. That $Y \subseteq c_\lhd(Y)$ and that $Y \subseteq Z$ implies $c_\lhd(Y) \subseteq c_\lhd(Z)$ are obvious. To see $c_\lhd(c_\lhd(Y)) \subseteq c_\lhd(Y)$, suppose $x \in c_\lhd(c_\lhd(Y))$ and $x' \lhd x$. Hence there is an $x'' \rhd x'$ with $x'' \in c_\lhd(Y)$. This implies there is an $x''' \rhd x'$ with $x''' \in Y$. Thus, for any $x' \lhd x$ there is an $x''' \rhd x'$ with $x''' \in Y$. Therefore, $x \in c_\lhd(Y)$.□

Given Propositions 2.3 and 2.6, we have the following immediate corollary.

Corollary 2.7 *For any relational frame* $(X, \lhd)$, *the* $c_{\lhd}$*-fixpoints ordered by* $\subseteq$ *form a complete lattice* $\mathfrak{L}(X, \lhd)$ *with meet and join as in Proposition 2.3.*

Example 2.8 We see in Fig. 1 that the $\mathbf{M}_3$ lattice (ignoring the dashed arrows for now) arises from the cycle on three elements, while the $\mathbf{N}_5$ lattice arises from the acyclic but non-transitive frame on three elements.

We can relate Corollary 2.7 to possible world semantics for classical and intuitionistic logic as follows. Let $=_W$ be the identity relation on the set W.

Proposition 2.9

(i) *Given any set* W, *the pair* $(W, =_W)$ *is a compatibility frame, and* $\mathfrak{L}(W, =_W)$ *is the Boolean algebra of all subsets of* W.

(ii) *Given any preorder* $\leq$ *on a set* P, *the pair* $(P, \leq)$ *is a compatibility frame, and* $\mathfrak{L}(P, \leq)$ *is the Heyting algebra of all downsets of* $(P, \leq)$.

Part (ii) appears in [13, Prop. 4.1.1] but we include a proof for convenience.

Proof. Part (i) is obvious. For (ii), let A be a downset and $x \in P \setminus A$. Setting $x' = x$, we have $x' \lhd x$, and for all $x'' \rhd x'$, i.e., all $x'' \geq x'$, $x'' \notin A$, since A is a downset. Thus, $c_{\lhd}(A) = A$. Conversely, suppose $c_{\lhd}(A) = A$, $x \in A$, and $y \leq x$. Let $y' \lhd y$, so $y' \leq y$ and hence $y' \leq x$. Then setting $y'' = x$, we have $y' \lhd y'' \in A$. Since $c_{\lhd}(A) = A$, it follows that $y \in A$. Thus, A is a downset. □

Example 2.10 To illustrate part (ii), if we add to the non-transitive frame in Fig. 1 the arrow from x to z required by transitivity, then instead of realizing $\mathbf{N}_5$, we realize the four-element chain in Fig. 2, which is a Heyting algebra.

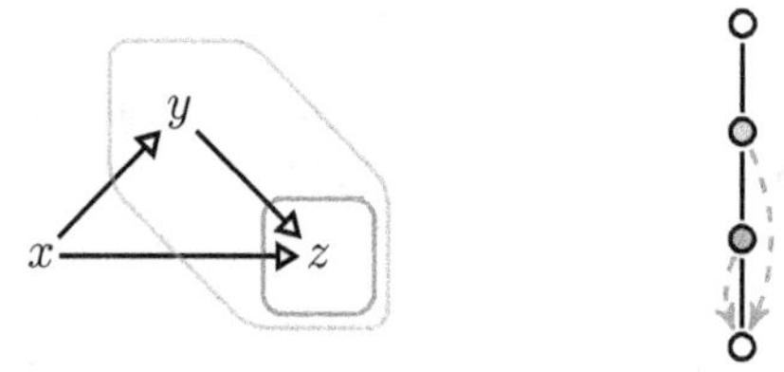

Fig. 2. A compatibility frame realizing a Heyting algebra.

Some appealing aspects of working with downsets of a preorder also apply to our fixpoints; let $^{\partial}$, $\oplus$, $\overline{\oplus}$, $\dot{\cup}$, and $\times$ be the dual, linear sum, vertical sum, disjoint union, and product operations [18, § 1.31, § 1.24, Ex. 1.18, Prop. 1.32].

Proposition 2.11 *For any relational frames* $\mathcal{F}$ *and* $\mathcal{G}$, (i) $\mathfrak{L}(\mathcal{F}^{\partial}) \cong \mathfrak{L}(\mathcal{F})^{\partial}$; (ii) $\mathfrak{L}(\mathcal{F} \oplus \mathcal{G}) \cong \mathfrak{L}(\mathcal{F}) \overline{\oplus} \mathfrak{L}(\mathcal{G})$; *and* (iii) $\mathfrak{L}(\mathcal{F} \dot{\cup} \mathcal{G}) \cong \mathfrak{L}(\mathcal{F}) \times \mathfrak{L}(\mathcal{G})$.

Proof. Let $\mathcal{F} = (X, \lhd)$, so $\mathcal{F}^{\partial} = (X, \rhd)$. For an isomorphism φ from $\mathfrak{L}(\mathcal{F}^{\partial})$ to $\mathfrak{L}(\mathcal{F})^{\partial}$, let $\varphi(A) = \{x \in X \mid \forall y \lhd x \; y \notin A\}$. From $\mathfrak{L}(\mathcal{F}) \overline{\oplus} \mathfrak{L}(\mathcal{G})$ to $\mathfrak{L}(\mathcal{F} \oplus \mathcal{G})$, $\varphi(A) = A$ if $A \in \mathfrak{L}(\mathcal{F})$, and $\varphi(A) = A \cup X$ if $A \in \mathfrak{L}(\mathcal{G})$. Part (iii) is also easy. □

We can also relate our approach to that of realizing complete Boolean algebras as in forcing [45] or possibility semantics [33] for classical logic as follows.

Proposition 2.12 *Given a preordered set $(P, \leq)$, define $\lhd$ on P by: $x \lhd y$ if $\exists z \in P$: $z \leq x$ and $z \leq y$. Then $(P, \lhd)$ is a compatibility frame, and $\mathfrak{L}(P, \lhd)$ is the Boolean algebra $\mathcal{RO}(P, \leq)$ of all regular open downsets of $(P, \leq)$.*

Proof. Observe that any $c_\lhd$-fixpoint is a $\leq$-downset, and for any $\leq$-downset A, (i) $\forall x' \leq x \, \exists x'' \leq x' : x'' \in A$ and (ii) $\forall x' \lhd x \, \exists x'' \rhd x' : x'' \in A$ are equivalent. It follows that A is a regular open downset iff A is a $c_\lhd$-fixpoint. □

Now $(X, \lhd)$ gives us not only $\mathfrak{L}(X, \lhd)$ but also an operation $\neg_\lhd$ on $\mathfrak{L}(X, \lhd)$.

Proposition 2.13 *For any relational frame $(X, \lhd)$ and $A \subseteq X$, the set $\neg_\lhd A = \{x \in X \mid \forall y \lhd x \;\; y \notin A\}$ is a $c_\lhd$-fixpoint.*

Proof. If $x \in X \setminus \neg_\lhd A$, then $\exists x' \lhd x$ with $x' \in A$, so $\forall x'' \rhd x'$, $x'' \notin \neg_\lhd A$. □

To characterize the $\neg_\lhd$ operation, let us recall some terminology.

Definition 2.14 Let L be a bounded lattice and $a \in L$. An $x \in L$ is a *semicomplement* of a if $a \wedge x = 0$, a *complement* of a if $a \wedge x = 0$ and $a \vee x = 1$, and a *pseudocomplement* of a if x is the maximum in L of $\{y \in X \mid a \wedge y = 0\}$.

A unary operation $\neg$ on L is a *semicomplementation* (resp. *complementation, pseudocomplementation*) if for all $a \in L$, $\neg a$ is a semicomplement (resp. complement, pseudocomplement) of a. It is *antitone* if for all $a, b \in L$, $a \leq b$ implies $\neg b \leq \neg a$, *involutive* if $\neg\neg a = a$ for all $a \in L$, and *anti-inflationary* if $a \not\leq \neg a$ for all nonzero $a \in L$. An *ortholattice* is a bounded lattice equipped with an involutive antitone complementation, called an *orthocomplementation.* A *p-algebra* is a bounded lattice equipped with a pseudocomplementation.

Finally, for a non-standard piece of terminology, we say $\neg$ is a *protocomplementation* if $\neg$ is an antitone semicomplementation such that $\neg 0 = 1$.

An antitone $\neg$ is anti-inflationary iff it is a semicomplementation. Also recall that the operation in a Heyting algebra H defined by $\neg a = a \to 0$ is a pseudocomplementation, so H may also be regarded as a p-algebra; and the complementation in a Boolean algebra B is an orthocomplementation, so B is an ortholattice. As for the operation $\neg_\lhd$, the following is easy to check.

Proposition 2.15 *For any compatibility (resp. relational) frame $(X, \lhd)$, $\neg_\lhd$ is a protocomplementation (resp. antitone and such that $\neg 0 = 1$) on $\mathfrak{L}(X, \lhd)$.*

In our diagrams of lattices arising from compatibility frames, the dashed arrows represent the operation $\neg_\lhd$. We omit arrows representing $\neg 0 = 1$ and $\neg 1 = 0$.

2.2 Frames for ortholattices

If we assume that $\lhd$ is *symmetric*, then we get a standard representation (as in [29] via "proximity frames") of ortholattices. The proof is straightforward.

Proposition 2.16 *For any compatibility frame $(X, \lhd)$, if $\lhd$ is symmetric, then $\neg_\lhd$ is an orthocomplementation on $\mathfrak{L}(X, \lhd)$.*

Example 2.17 Fig. 3 shows two symmetric compatibility frames and their associated ortholattices, $\mathbf{MO}_2$ or $\mathbf{M}_4$ (left) and the Benzene ring $\mathbf{O}_6$ (right).

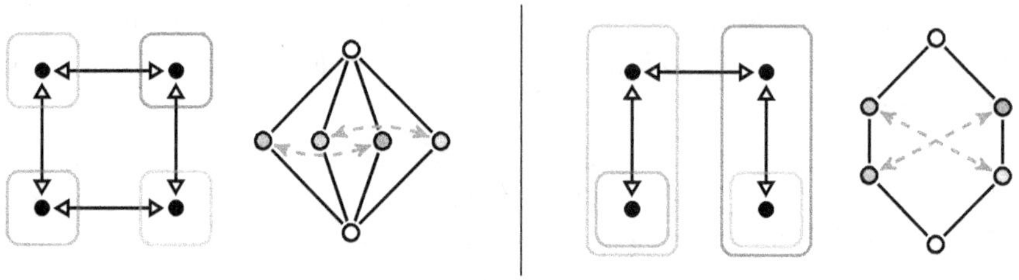

Fig. 3. Compatibility frames realizing ortholattices.

2.3 Frames for Heyting and Boolean algebras

We conclude this section with sufficient conditions on $\lhd$ for realizing Heyting and Boolean algebras, weaker than those in Proposition 2.9 (where $\lhd$ was a preorder or the identity relation). First, we define some auxiliary notions.

Definition 2.18 Given a compatibility frame $(X, \lhd)$ and $x, y \in X$:

(i) x *pre-refines* y, written $x \sqsubseteq_{pr} y$, if for all $z \in X$, $z \lhd x$ implies $z \lhd y$;

(ii) x *post-refines* y, written $x \sqsubseteq_{po} y$, if for all $z \in X$, $x \lhd z$ implies $y \lhd z$;

(iii) x *refines* y, written $x \sqsubseteq y$, if x pre-refines and post-refines y;

(iv) x is *compossible with* y if there is a $w \in X$ that refines x and pre-refines y.

Note that if $\lhd$ is symmetric, then pre-refinement and post-refinement are equivalent, and x is compossible with y just in case they have a common refinement.

Lemma 2.19 *For any compatibility frame $(X, \lhd)$, $\sqsubseteq_{pr}$ and $\sqsubseteq_{po}$ are preorders on X. Moreover, each $c_\lhd$-fixpoint is a $\sqsubseteq_{pr}$-downset.*

Proof. The preorder part is obvious. Next suppose A is a $c_\lhd$-fixpoint, $x \in A$, and $y \sqsubseteq_{pr} x$. Toward showing that $y \in A$, consider any $y' \lhd y$. Then since $y \sqsubseteq_{pr} x$, we have $y' \lhd x$, so taking $y'' = x$, we have shown that for every $y' \lhd y$ there is a $y'' \rhd y'$ with $y'' \in A$. Since A is a $c_\lhd$-fixpoint, it follows that $y \in A$. □

Definition 2.20 A *compossible compatibility frame* is a compatibility frame $(X, \lhd)$ in which for any $x, y \in X$, if $x \lhd y$, then x is compossible with y.

Theorem 2.21

(i) *If $(X, \lhd)$ is a compossible compatibility frame, then $\mathfrak{L}(X, \lhd)$ is a Heyting algebra with $\to$ defined by $A \to B = \{x \in X \mid \forall y \sqsubseteq_{pr} x\, (y \in A \Rightarrow y \in B)\}$, and $\neg_\lhd$ is the Heyting pseudocomplementation.*

(ii) *If $(X, \lhd)$ is a compossible symmetric compatibility frame, then $\mathfrak{L}(X, \lhd)$ is the Boolean algebra $\mathcal{RO}(X, \sqsubseteq)$, and $\neg_\lhd$ is the Boolean complementation.*

Proof. For part (i), recall that a *nucleus* on a Heyting algebra H is a closure operator that is also multiplicative, i.e., for all $x, y \in H$, we have $c(x) \wedge c(y) \leq c(x \wedge y)$. This follows from c being a closure operator such that for all $a, b \in H$, $a \wedge c(b) \leq c(a \wedge b)$. For then setting $a = c(x)$ and $b = y$, $c(x) \wedge c(y) \leq c(c(x) \wedge y) = c(y \wedge c(x))$. Then setting $a = y$ and $b = x$, we have $y \wedge c(x) \leq c(y \wedge x)$, in which case monotonicity and idempotence yield $c(y \wedge c(x)) \leq c(c(y \wedge x)) = c(y \wedge x) = c(x \wedge y)$. Thus, combining the two long strings of equations, $c(x) \wedge c(y) \leq c(x \wedge y)$.

Now it is well known that the fixpoints of a nucleus j on a (complete) Heyting algebra H form a (complete) Heyting algebra H_j under the restricted lattice order (see, e.g., [21, p. 71]), where for $a, b \in H_j$, $a \to_{H_j} b = a \to_H b$.

Let H be the Heyting algebra of all $\sqsubseteq_{pr}$-downsets. By the results above, to prove the first part of (i), it suffices to show that $c_\lhd$ restricted to H is a nucleus, for which it suffices to show that for all $A, B \in H$, $A \cap c_\lhd(B) \subseteq c_\lhd(A \cap B)$. Suppose $x \in A \cap c_\lhd(B)$ but $x \notin c_\lhd(A \cap B)$. Since $x \notin c_\lhd(A \cap B)$, there is a $y \lhd x$ such that $(\star)$ for all $y' \rhd y$, we have $y' \notin A \cap B$. Since $y \lhd x$, by compossibility there is a z that refines y and pre-refines x. Since z pre-refines x, $z \lhd x$. Then since $x \in c_\lhd(B)$, there is a $z' \rhd z$ with $z' \in B$. Since $z \lhd z'$, there is a w that refines z and pre-refines z'. Since w pre-refines z', from $z' \in B$ we have $w \in B$ by Lemma 2.19. Since w pre-refines z and z pre-refines x, w pre-refines x, so $x \in A$ implies $w \in A$. Thus, $w \in A \cap B$. Moreover, since w post-refines z and z post-refines y, w post-refines y. Hence $y \lhd w$. But then by $(\star)$, $w \notin A \cap B$. This is a contradiction. Finally, for the claim about $\neg_\lhd$, observe that for any $\sqsubseteq_{pr}$-downset A, we have $A \to \varnothing = \{x \in X \mid \forall y \sqsubseteq_{pr} x \; y \notin A\} = \neg_\lhd A$; for the second equality, the right-to-left inclusion uses that $y \sqsubseteq_{pr} x$ implies $y \lhd x$, while the left-to-right inclusion uses the assumption of compossibility.

For part (ii), to see that $\mathfrak{L}(X, \lhd)$ is isomorphic to $\mathcal{RO}(X, \sqsubseteq)$, by Proposition 2.12 it suffices to show that for all $x, y \in X$, we have $x \lhd y$ iff there is a $z \in X$ such that $z \sqsubseteq x$ and $z \sqsubseteq y$. From left to right, if $x \lhd y$, then since $(X, \lhd)$ is a compossible compatibility frame, there is a z that refines x and pre-refines y, which implies that z refines y by the symmetry of $\lhd$. From right to left, $z \sqsubseteq x$ implies $x \lhd z$, which with $z \sqsubseteq y$ implies $x \lhd y$. Finally, for the claim about $\neg_\lhd$, recall that $\neg A$ in $\mathcal{RO}(X, \sqsubseteq)$ is $\{x \in X \mid \forall y \sqsubseteq x \; y \notin A\}$, which is equal to $\neg_\lhd A$ by reasoning analogous to that at the end of the previous paragraph. □

3 From lattices to compatibility frames

3.1 Representation of special complete lattices via join-dense sets

In this section, we give an economical representation of certain complete lattices L using compatibility frames based on a join-dense set V of nonzero elements of L, so the frame representing L is smaller than L. In all our figures, the frame can be seen as obtained from the lattice via this representation.

Let $CJI(L)$ be the set of completely join-irreducible elements of L. Recall that a complete Heyting algebra is *perfect* if $CJI(L)$ is join-dense in L. The standard representations of complete perfect Heyting algebras and complete ortholattices drop out of the representation in this section as special cases.

3.1.1 Suitable compatibility relations

Before giving the crucial definition of the compatibility relation used in our representation (Definition 3.5(i) below), it is instructive to distill conditions on a compatibility relation sufficient for a successful representation.

Definition 3.1 Let L be a lattice and V a set of elements of L. A binary relation $\lhd$ on V is *suitable for* L if $\lhd$ is reflexive and:

(i) for $a \in \mathrm{V}$ and $b \in L$, if $a \nleq b$, then $\exists a' \lhd a\ \forall a'' \rhd a'\ a'' \nleq b$;

(ii) if $a \in \mathrm{V}$, B is a $c_{\lhd}$-fixpoint of $(\mathrm{V}, \lhd)$ with a join b in L, and $a \leq b$, then $a \in B$.

We call (i) and (ii) the first and second *suitability conditions*, respectively.

Proposition 3.2 *Let L be a lattice,* V *a join-dense set of elements of L, and $\lhd$ a reflexive relation on* V*. For $b \in L$, define $\varphi(b) = \{x \in \mathrm{V} \mid x \leq b\}$.*

(i) *If $\lhd$ satisfies the first suitability condition, then φ embeds L into $\mathfrak{L}(\mathrm{V}, \lhd)$.*

(ii) *If L is complete and $\lhd$ suitable, φ is an isomorphism from L to $\mathfrak{L}(\mathrm{V}, \lhd)$.*

Proof. For part (i), clearly φ is order-preserving: if $a \leq b$, then $\varphi(a) \subseteq \varphi(b)$. Moreover, φ is order-reflecting: since V is join-dense in L, we have $a = \bigvee A$ for some $A \subseteq \mathrm{V}$, so $a \nleq b$ implies that for some $a_0 \in A$, we have $a_0 \nleq b$, so $a_0 \in \varphi(a)$ but $a_0 \notin \varphi(b)$, and hence $\varphi(a) \nsubseteq \varphi(b)$.

Next we claim $\varphi(b)$ is a $c_{\lhd}$-fixpoint. Suppose for $a \in \mathrm{V}$ that $a \notin \varphi(b)$, so $a \nleq b$. Then by the first suitability condition, there is an $a' \lhd a$ such that for all $a'' \rhd a'$, $a'' \nleq b$ and hence $a'' \notin \varphi(b)$. Thus, $\varphi(b)$ is a $c_{\lhd}$-fixpoint.

For (ii), φ is surjective. Let B be a $c_{\lhd}$-fixpoint and $b = \bigvee B$. We claim $B = \varphi(b)$. For $B \subseteq \varphi(b)$, if $b_0 \in B$, then $b_0 \leq b$, so $b_0 \in \varphi(b)$. For $B \supseteq \varphi(b)$, suppose $a \in \varphi(B)$, so $a \leq b$. Then by the second suitability condition, $a \in B$.□

Our strategy for defining suitable compatibility relations on a join-dense set V of elements of L will be to assume that L comes equipped with an anti-inflationary operation $\neg$, e.g., as in the case of an ortholattice with orthcomplementation $\neg$ or a p-algebra or Heyting algebra with pseudocomplementation $\neg$ or a lattice to which we have added an anti-inflationary operation $\neg$ as in Fig. 1. We will then use $\neg$ to define a compatibility relation $\lhd^{\neg}_{\mathrm{V}}$ on V.

3.1.2 The first suitability condition

In the case of an ortholattice, our defined compatibility relation will be equivalent to $x \lhd y$ if $y \nleq \neg x$ (Lemma 3.7(i)). To see why this compatibility relation satisfies the first suitability condition, the following concept is useful.

Definition 3.3 Let L be a lattice equipped with a unary operation $\neg$ and V a set of elements of L. Given $a, b \in L$, we say that *a escapes b in* V *with* $\neg$ if there is some $c \in \mathrm{V}$ such that $a \nleq \neg c$ but $b \leq \neg c$.

Lemma 3.4 *Let L be an ortholattice,* V *a join-dense set of elements of L, and $a, b \in L$. If $a \nleq b$, then a escapes b in* V *with the orthocomplementation* $\neg$.

Proof. Suppose $a \nleq b$, so $\neg b \nleq \neg a$. Since V is join-dense, we have $\neg b = \bigvee C$ for some $C \subseteq \mathrm{V}$. Then $\neg b \nleq \neg a$ implies $c \nleq \neg a$ for some $c \in C$. From $c \nleq \neg a$ we have $a \nleq \neg c$, and from $c \in C$ we have $c \leq \neg b$ and hence $b \leq \neg c$. □

Using Lemma 3.4, it is easy to see that the relation $\lhd$ on ortholattices defined by $x \lhd y$ if $y \nleq \neg x$ satisfies the first suitability condition.

However, in lattices with $\neg$ that are not ortholattices, we can have $a \nleq b$ while a cannot escape b, as shown with two examples in Fig. 4.

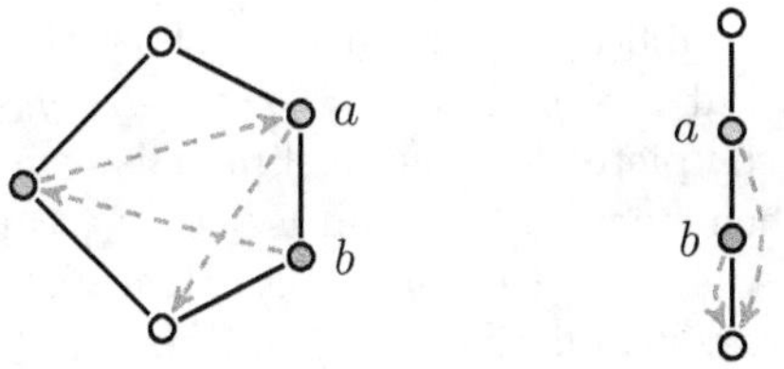

Fig. 4. Lattices with $\neg$ (dashed arrows) in which $a \not\leq b$ but a cannot escape b.

In order to deal with lattices in which $a \not\leq b$ does not imply that a can escape b, we introduce the key definition of this section.

Definition 3.5 Let L be a lattice, $\neg$ an anti-inflationary operation on L, and V a set of nonzero elements of L.

(i) Define $\lhd^{\neg}_{\mathrm{V}}$ on V by: $x \lhd^{\neg}_{\mathrm{V}} y$ if both $y \not\leq \neg x$ and for all $z \in L$, if $y \leq z$ but $x \not\leq z$, then x escapes z in V with $\neg$.

(ii) Given $a \in \mathrm{V}$ and $b \in L$, we say that a *compatibly escapes* b *in* V *with* $\neg$ if there is some $c \in \mathrm{V}$ such that $c \lhd^{\neg}_{\mathrm{V}} a$ and $b \leq \neg c$.

(iii) L *has compatible escape with* $\neg$ *in* V if for all $a \in \mathrm{V}$ and $b \in L$, if a escapes b in V with $\neg$, then a compatibly escapes b in V with $\neg$.

The following is easy to check.

Lemma 3.6 *Under the assumptions of Definition 3.5,* (i) $\lhd^{\neg}_{\mathrm{V}}$ *is reflexive,* (ii) $x \leq y$ *implies* $x \lhd^{\neg}_{\mathrm{V}} y$, *and* (iii) $x \lhd^{\neg}_{\mathrm{V}} y \leq y'$ *implies* $x \lhd^{\neg}_{\mathrm{V}} y'$.

It will turn out (Proposition 3.8) that L having compatible escape ensures that $\lhd^{\neg}_{\mathrm{V}}$ satisfies the first suitability condition. First, we show that ortholattices and Heyting algebras are alike in having compatible escape, and $\lhd^{\neg}_{\mathrm{V}}$ reduces to familiar relations in ortholattices and complete perfect Heyting algebras.

Proposition 3.7 *Let L be a lattice and V a join-dense set of nonzero elements.*

(i) *If L is an ortholattice, then $x \lhd^{\neg}_{\mathrm{V}} y$ iff $y \not\leq \neg x$ for $x, y \in \mathrm{V}$; hence L has compatible escape in V with $\neg$.*

(ii) *If L is a p-algebra, then L has compatible escape in V with $\neg$.*

(iii) *If L is a complete Heyting algebra and $\mathrm{V} = CJI(L)$, then $x \lhd^{\neg}_{\mathrm{V}} y$ iff $x \leq y$ for $x, y \in \mathrm{V}$.*

Proof. For part (i), by definition, $x \lhd^{\neg}_{\mathrm{V}} y$ implies $y \not\leq \neg x$. Conversely, suppose $y \not\leq \neg x$. To show $x \lhd^{\neg}_{\mathrm{V}} y$, suppose $y \leq z$ but $x \not\leq z$. From $x \not\leq z$, it follows by Lemma 3.4 that x escapes z in V with $\neg$. This shows $x \lhd^{\neg}_{\mathrm{V}} y$.

For (ii), suppose a escapes b in V, so for some $c \in \mathrm{V}$, $a \not\leq \neg c$ but $b \leq \neg c$. Since $a \not\leq \neg c$ and $\neg$ is pseudocomplementation, we have $a \wedge c \neq 0$. From $b \leq \neg c$, we have $b \leq \neg(a \wedge c)$. Since V is join-dense in L, there is a $d \in \mathrm{V}$ such that $d \leq a \wedge c$, so $\neg(a \wedge c) \leq \neg d$. Then since $b \leq \neg(a \wedge c)$, we have $b \leq \neg d$. Moreover, by Lemma 3.6(ii), $d \leq a$ implies $d \lhd^{\neg}_{\mathrm{V}} a$. Hence we have found a $d \lhd^{\neg}_{\mathrm{V}} a$ such that $b \leq \neg d$. Thus, a compatibly escapes b in V with $\neg$.

For (iii), given Lemma 3.6(ii), we need only show that $x \lhd^{\neg}_{\mathrm{V}} y$ implies $x \leq y$. Suppose $x \lhd^{\neg}_{\mathrm{V}} y$ but $x \not\leq y$. Let $w = \bigvee\{z \in L \mid y \leq z, x \not\leq z\}$. Since $x \in CJI(L)$ and L is a complete Heyting algebra, x is completely join-prime, which implies $x \not\leq w$. Then since $x \lhd^{\neg}_{\mathrm{V}} y$ and $y \leq w$, x escapes w, so for some $v \in \mathrm{V}$, $x \not\leq \neg v$ but $w \leq \neg v$ and thus $y \leq \neg v$. Hence $x \not\leq v$, for otherwise $\neg v \leq \neg x$ and so $y \leq \neg x$, contradicting $x \lhd^{\neg}_{\mathrm{V}} y$. From $x \not\leq w$ and $x \not\leq v$, we have $x \not\leq w \vee v$, while $y \leq w \vee v$. Since $w \leq \neg v$, we also have $w < w \vee v$. But together $y \leq w \vee v$, $x \not\leq w \vee v$, and $w < w \vee v$ contradict the definition of w.□

Crucially, there are other lattices with compatible escape besides ortholattices and p-algebras, as in $\mathbf{N}_5$ equipped with the $\neg$ operation in Fig. 4 (left).

Proposition 3.8 *Let L be a lattice, $\neg$ an anti-inflationary operation on L, and V a set of nonzero elements of L. If L has compatible escape in V with $\neg$, then $\lhd^{\neg}_{\mathrm{V}}$ satisfies the first suitability condition.*

Proof. For $a \in \mathrm{V}$ and $b \in L$, suppose $a \not\leq b$.

Case 1: a escapes b. Then since L has compatible escape in V with $\neg$, a compatibly escapes b using some $c \in \mathrm{V}$, so $c \lhd^{\neg}_{\mathrm{V}} a$ and $b \leq \neg c$. Let $a' = c$, so $a' \lhd^{\neg}_{\mathrm{V}} a$. Suppose $a'' \rhd^{\neg}_{\mathrm{V}} a'$, so $a'' \not\leq \neg c$. Then since $b \leq \neg c$, we have $a'' \not\leq b$.

Case 2: a does not escape b. Then let $a' = a$, so $a' \lhd^{\neg}_{\mathrm{V}} a$. Suppose $a'' \rhd^{\neg}_{\mathrm{V}} a'$. If $a'' \leq b$, then since $a \not\leq b$, from $a \lhd^{\neg}_{\mathrm{V}} a''$ it follows that a escapes b, contradicting the assumption of the case. Hence $a'' \not\leq b$. □

Corollary 3.9 *Let L be a lattice and V a join-dense set of nonzero elements.*

(i) *If L is an ortholattice, then $\lhd^{\neg}_{\mathrm{V}}$ satisfies the first suitability condition.*

(ii) *If L is a p-algebra, then $\lhd^{\neg}_{\mathrm{V}}$ satisfies the first suitability condition.*

3.1.3 The second suitability condition

We can treat the second suitability condition more quickly.

Proposition 3.10 (i) *If L is a complete ortholattice and V a join-dense set of nonzero elements of L, then $\lhd^{\neg}_{\mathrm{V}}$ satisfies the second suitability condition.*

(ii) *If L is a complete Heyting algebra and $\mathrm{V} = CJI(L)$ is join-dense in L, then $\lhd^{\neg}_{\mathrm{V}}$ satisfies the second suitability condition.*

Proof. Suppose L is an ortholattice, $a \in \mathrm{V}$, B is a $c_{\lhd^{\neg}_{\mathrm{V}}}$-fixpoint, and $a \leq b = \bigvee B$. Suppose $a' \lhd^{\neg}_{\mathrm{V}} a$, so $a \not\leq \neg a'$ and hence $b \not\leq \neg a'$. Then for some $b_0 \in B$, we have $b_0 \not\leq \neg a'$. Now let $a'' = b_0$, so $a'' \in B$. Since $a'' \not\leq \neg a'$, by Proposition 3.7(i) we have $a' \lhd^{\neg}_{\mathrm{V}} a''$. Given that B is a $c_{\lhd^{\neg}_{\mathrm{V}}}$-fixpoint, this shows $a \in B$.

Suppose L is a complete Heyting algebra, $a \in \mathrm{V}$, B is a $c_{\lhd^{\neg}_{\mathrm{V}}}$-fixpoint, and $a \leq \bigvee B$. Since L is a complete Heyting algebra and $\mathrm{V} = CJI(L)$, $a \in \mathrm{V}$ implies that a is completely join-prime. Hence $a \leq b_0$ for some $b_0 \in B$. Now consider any $a' \lhd^{\neg}_{\mathrm{V}} a$. Let $a'' = b_0$, so $a'' \in B$. Since $a' \lhd^{\neg}_{\mathrm{V}} a \leq a''$, by Lemma 3.6(iii) we have $a' \lhd^{\neg}_{\mathrm{V}} a''$. Given that B is a $c_{\lhd^{\neg}_{\mathrm{V}}}$-fixpoint, this shows $a \in B$.□

3.1.4 Representation theorem

Combining Propositions 3.2-3.10, we have the following representation theorem.

Theorem 3.11 *Let L be a complete lattice and V a join-dense set of nonzero elements of L.*

(i) *If L is an ortholattice, then L is isomorphic to $\mathfrak{L}(\mathrm{V}, \lhd_{\mathrm{V}}^{\neg})$.*[3]

(ii) *If L is a Heyting algebra and $\mathrm{V} = CJI(L)$, L is isomorphic to $\mathfrak{L}(\mathrm{V}, \lhd_{\mathrm{V}}^{\neg})$.*

(iii) *If $\neg$ is an anti-inflationary operation on L such that L has compatible escape in V with $\neg$, and $\lhd_{\mathrm{V}}^{\neg}$ satisfies the second suitability condition, then L is isomorphic to $\mathfrak{L}(\mathrm{V}, \lhd_{\mathrm{V}}^{\neg})$.*

Part (iii) applies to many non-ortholattice and non-Heyting examples, as in Fig. 1, but its precise scope is an open question.

Question 3.12 *For which complete lattices L is there a set V of nonzero elements and an anti-inflationary $\neg$ on L such that L is isomorphic to $\mathfrak{L}(\mathrm{V}, \lhd_{\mathrm{V}}^{\neg})$?*

The Jupyter notebook cited in § 1 verifies that every lattice L up to size 8 is such an L. It also verifies the following for every lattice up to size 16 (cf. Footnote 4).

Conjecture 3.13 *For any nondegenerate finite lattice L, there is a compatibility frame $(X, \lhd)$ with $|X| < |L|$ such that L is isomorphic to $\mathfrak{L}(X, \lhd)$.*

Finally, Appendix A.1 modifies (iii) to represent not only L but also $(L, \neg)$.

3.2 Representation of arbitrary complete lattices

In this section, we turn to the representation of arbitrary complete lattices. Instead of representing a lattice L using special elements of L as in § 3.1, here we use a potentially less economical representation in terms of pairs of elements, as in the birelational representations of complete lattices in [2,3,34,41].

Definition 3.14 Let L be a lattice and P a set of pairs of elements of L. Define a binary relation $\lhd$ on P by $(a, b) \lhd (c, d)$ if $c \not\leq b$. Then we say P is *separating* if for all $a, b \in L$:

(i) if $a \not\leq b$, then there is a $(c, d) \in P$ with $c \leq a$ and $c \not\leq b$;

(ii) for all $(c, d) \in P$, if $c \not\leq b$, then there is a $(c', d') \lhd (c, d)$ such that for all $(c'', d'') \rhd (c', d')$, we have $c'' \not\leq b$.

Proposition 3.15 *Let L be a lattice and P a separating set of pairs of elements of L. For $a \in L$, define $\varphi(a) = \{(x, y) \in P \mid x \leq a\}$. Then:*

(i) *φ is an embedding of L into $\mathfrak{L}(P, \lhd)$;*

(ii) *if L is complete, then φ is an isomorphism from L to $\mathfrak{L}(P, \lhd)$.*

Proof. For part (i), clearly φ is order-preserving: $a \leq b$ implies $\varphi(a) \subseteq \varphi(b)$. Condition (i) of Definition 3.14 ensures that φ is order-reflecting. Condition (ii) ensures that $\varphi(b)$ is a $c_{\lhd}$-fixpoint for each $b \in L$.

For part (ii), we claim φ is surjective. Given a $c_{\lhd}$-fixpoint A, define $a = \bigvee\{a_i \mid \exists b_i : (a_i, b_i) \in A\}$. We claim $A = \varphi(a)$. For $A \subseteq \varphi(a)$, suppose

[3] The orthocomplementation of L is represented as $\neg_{\lhd_{\mathrm{V}}^{\neg}}$ by Propositions A.1 and A.2.

$(a_i, b_i) \in A$. Then by definition of a, $a_i \leq a$, so $(a_i, b_i) \in \varphi(a)$. For $A \supseteq \varphi(a)$, suppose $(c, d) \in \varphi(a)$, so $c \leq a$. Since A is a $c_\lhd$-fixpoint, to show $(c, d) \in A$, it suffices to show that for every $(c,' d') \lhd (c, d)$ there is a $(c'', d'') \rhd (c', d')$ with $(c'', d'') \in A$. Suppose $(c,' d') \lhd (c, d)$, so $c \not\leq d'$, which with $c \leq a$ implies $a \not\leq d'$. Then for some $(a_i, b_i) \in A$, we have $a_i \not\leq d'$. Setting $(c'', d'') = (a_i, b_i)$, from $a_i \not\leq d'$ we have $(c', d') \lhd (c'', d'')$, and $(c'', d'') \in A$, so we are done. □

Every lattice has a separating set of pairs, and for ortholattices and Heyting algebras we can cut down the sets of pairs, as in Proposition 3.16. Compare part (i) to the representation of a complete lattice by join-dense and meet-dense sets in [39,40] or in the fundamental theorem of concept lattices [18, Thm. 3.9].

Proposition 3.16 *Let L be a lattice, V a join-dense set of elements of L, and Λ a meet-dense set of elements of L. Then:*

(i) *if P_1 is a subset of $P_0 = \{(a, b) \mid a \in \mathrm{V}, b \in \Lambda, a \not\leq b\}$ such that for each $a \in \mathrm{V}$ there is a $b \in \Lambda$ with $(a, b) \in P_1$, and for each $b \in \Lambda$ there is an $a \in \mathrm{V}$ with $(a, b) \in P_1$, then P_1 is separating;*[4]

(ii) *if L is an ortholattice, then $P_2 = \{(a, \neg a) \mid a \in \mathrm{V}, a \neq 0\}$ is separating;*

(iii) *if L is Heyting, then $P_3 = \{(a, a \to b) \mid a \in \mathrm{V}, b \in \Lambda, a \not\leq b\}$ is separating.*

Proof. In each case, that condition (i) of Definition 3.14 is satisfied is obvious, so we focus on condition (ii).

For P_1, suppose $(c, d) \in P_1$ and $c \not\leq b$. Then where $b = \bigwedge\{b_i \in \Lambda \mid i \in I\}$, there is some $b_i \in \Lambda$ such that $c \not\leq b_i$, but $b \leq b_i$, and some $a \in \mathrm{V}$ with $(a, b_i) \in P_1$. Let $(c', d') = (a, b_i)$, so $(c', d') \lhd (c, d)$. Now consider any $(c'', d'') \in P_1$ with $(c', d') \lhd (c'', d'')$. Then $c'' \not\leq d' = b_i$, so $c'' \not\leq b$.

For the ortholattice case with P_2, suppose $(c, d) = (c, \neg c) \in P_2$ and $c \not\leq b$. Hence $\neg b \not\leq \neg c$. Then where $\neg b = \bigvee\{x_i \in \mathrm{V} \mid i \in I\}$, for some $i \in I$ we have $x_i \not\leq \neg c$, but $x_i \leq \neg b$, so $c \not\leq \neg x_i$ and $b \leq \neg x_i$. Let $(c', d') = (x_i, \neg x_i)$. Since $c \not\leq \neg x_i$, we have $(c', d') \lhd (c, d)$. Now consider any $(c'', d'') \in P_2$ with $(c', d') \lhd (c'', d'')$. Then $c'' \not\leq d' = \neg x_i$, which with $b \leq \neg x_i$ implies $c'' \not\leq b$.

For the Heyting case with P_3, suppose $(c, d) = (c, c \to e) \in P_3$ and $c \not\leq b$. Then where $b = \bigwedge\{b_i \in \Lambda \mid i \in I\}$, there is some $b_i \in \Lambda$ such that $c \not\leq b_i$, but $b \leq b_i$. From $c \not\leq b_i$, we have $c \not\leq c \to b_i$. Then where $(c', d') = (c, c \to b_i)$, we have $(c', d') \in P_3$ and $(c', d') \lhd (c, d)$. Now consider any $(c'', d'') \in P_3$ such that $(c', d') \lhd (c'', d'')$, so $c'' \not\leq d' = c \to b_i$. Then $c'' \wedge c \not\leq b_i$, which with $b \leq b_i$ implies $c'' \wedge c \not\leq b$ and hence $c'' \not\leq b$. □

By Propositions 3.15 and 3.16, L embeds into $\mathfrak{L}(P_1, \lhd)$. As the image of the embedding is join-dense and meet-dense in $\mathfrak{L}(P_1, \lhd)$, it follows that $\mathfrak{L}(P_1, \lhd)$ is (up to isomorphism) the MacNeille completion of L (see [26, Thm. 2.2]), and similarly for $\mathfrak{L}(P_2, \lhd)$ and $\mathfrak{L}(P_3, \lhd)$ in the ortholattice[5] and Heyting cases.

---

[4] Here P_1 is an edge cover for the bipartite graph $(\mathrm{V}, \Lambda, P_0)$. If for every finite lattice L, the smallest edge cover for the bipartite graph of its join- and meet-irreducibles has cardinality less than $|L|$ (as we verified for all L with $|L| \leq 16$ using [23]), then Conjecture 3.13 is true.

[5] It is also easy to check that φ respects the orthocomplementation: $\varphi(\neg a) = \neg_\lhd \varphi(a)$.

Recall from Propositions 2.16 and 2.21 how symmetric, compossible, and symmetric compossible compatibility frames give rise to ortholattices, Heyting algebras, and Boolean algebras, respectively. We now prove a converse.

Proposition 3.17 *Let L be a lattice, V a join-dense set of elements of L, and Λ a meet-dense set of elements of L. Let P_1, P_2, and P_3 be defined from V and Λ as in Proposition 3.16. Then:*

(i) *$(P_1, \lhd)$ is a compatibility frame;*

(ii) *if L is an ortholattice, then $(P_2, \lhd)$ is a symmetric frame;*

(iii) *if L is a Heyting algebra, then $(P_3, \lhd)$ is a compossible frame;*

(iv) *if L is a Boolean algebra, then $(P_2, \lhd)$ is a symmetric compossible frame.*

Proof. For each part, $(a, b) \in P_i$ implies $a \not\leq b$, so $(a, b) \lhd (a, b)$. Hence $\lhd$ is reflexive. For part (ii), if $(a, \neg a) \lhd (b, \neg b)$, so $b \not\leq \neg a$, then $a \not\leq \neg b$, so $(b, \neg b) \lhd (a, \neg a)$. Hence $\lhd$ is symmetric.

For (iii), toward showing that $(P_3, \lhd)$ is a compossible compatibility frame, suppose $(a, a \to b) \lhd (c, c \to d)$. We must show there is an $(x, x \to y) \in P_3$ that refines $(a, a \to b)$ and pre-refines $(c, c \to d)$. Since $(a, a \to b) \lhd (c, c \to d)$, we have $c \not\leq a \to b$, so $c \wedge a \not\leq b$. Then where $c \wedge a = \bigvee\{x_i \in \mathrm{V} \mid i \in I\}$ and $b = \bigwedge\{b_k \in \Lambda \mid k \in K\}$, there are $i \in I$ and $k \in K$ such that $x_i \not\leq b_k$ and hence $x_i \not\leq x_i \to b_k$. We claim that $(x_i, x_i \to b_k)$ refines $(a, a \to b)$ and pre-refines $(c, c \to d)$. To see that $(x_i, x_i \to b_k)$ pre-refines $(a, a \to b)$ and $(c, c \to d)$, suppose $(w, w \to v) \lhd (x_i, x_i \to b_k)$, so $x_i \not\leq w \to v$. Then since $x_i \leq a$, we have $a \not\leq w \to v$, so $(w, w \to v) \lhd (a, a \to b)$; and since $x_i \leq c$, we have $c \not\leq w \to v$, so $(w, w \to v) \lhd (c, c \to d)$. To see that $(x_i, x_i \to b_k)$ post-refines $(a, a \to b)$, suppose $(x_i, x_i \to b_k) \lhd (w, w \to v)$, so $w \not\leq x_i \to b_k$. It follows that $w \wedge x_i \not\leq b_k$ and hence $w \wedge x_i \not\leq b$, which with $x_i \leq a$ implies $w \wedge a \not\leq b$, so $w \not\leq a \to b$ and hence $(a, a \to b) \lhd (w, w \to v)$.

For (iv), symmetry follows from part (ii). As for compossibility, if $(a, \neg a) \lhd (b, \neg b)$, so $b \not\leq \neg a$, then $a \wedge b \neq 0$. Hence there is some nonzero $c \in \mathrm{V}$ with $c \leq a \wedge b$, which implies that $(c, \neg c) \in P_2$ refines both $(a, \neg a)$ and $(b, \neg b)$. □

Combining Propositions 3.15-3.17, we have the following.

Theorem 3.18 *If L is a lattice (resp. ortholattice, Heyting algebra, Boolean algebra), then L embeds into the lattice of $c_\lhd$-fixpoints of a compatibility frame (resp. symmetric, compossible, symmetric compossible compatibility frame), and if L is complete, the embedding is an isomorphism.*

Thus, compossible compatibility frames yield a semantics for intermediate logics as general as complete Heyting algebras in the "semantic hierarchy" of [3].

Finally, we prove that we can also represent any complete lattice expanded with a protocomplementation using a compatibility frame.

Theorem 3.19 *For any bounded lattice L equipped with a protocomplementation $\neg$, the expansion $(L, \neg)$ embeds into the lattice of $c_\lhd$-fixpoints of a compatibility frame equipped with $\neg_\lhd$, and if L is complete, the embedding is an isomorphism.*

Proof. First, we claim $P = \{(a,b) \mid a,b \in L, a \neq 0, a \not\leq b, \neg a \leq b\}$ is separating. For part (i) of Definition 3.14, take $(c,d) = (a, \neg a)$. For (ii), suppose $(c,d) \in P$ and $c \not\leq b$. Let $(c',d') = (1,b)$. Since $b \neq 1$ and $\neg 1 = 0 \leq b$, $(1,b) \in P$, and since $c \not\leq b$, $(c',d') \lhd (c,d)$. Now consider any $(c'',d'') \in P$ with $(c',d') \lhd (c'',d'')$. Then $c'' \not\leq d' = b$, so (ii) holds. Thus, by Proposition 3.15, L embeds into $\mathfrak{L}(P,\lhd)$ via the map φ, which is an isomorphism if L is complete. Also observe that $\lhd$ is reflexive on P. It only remains to show $\varphi(\neg a) = \neg_{\lhd}\varphi(a)$. Suppose $(x,y) \in \varphi(\neg a)$, so $x \leq \neg a$, and $(x',y') \lhd (x,y)$. If $x' \leq a$, then $\neg a \leq \neg x'$, which with $x \leq \neg a$ implies $x \leq \neg x'$, which with $\neg x' \leq y'$ implies $x \leq y'$, contradicting $(x',y') \lhd (x,y)$. Thus, $x' \not\leq a$, so $(x',y') \notin \varphi(a)$. Hence $(x,y) \in \neg_{\lhd}\varphi(a)$. Conversely, let $(x,y) \in P \setminus \varphi(\neg a)$, so $x \not\leq \neg a$. Since $\neg 0 = 1$, it follows that $a \neq 0$, so $(a, \neg a) \in P$, and $(a,\neg a) \lhd (x,y)$, so $(x,y) \notin \neg_{\lhd}\varphi(a)$.□

Note that any bounded lattice can be equipped with the protocomplementation such that $\neg 0 = 1$ and $\neg a = 0$ for $a \neq 0$ (in which case the set P in the proof of Theorem 3.19 coincides with P_0 from Proposition 3.16 where V and Λ are the non-minimum and non-maximum elements of L, respectively), so Theorem 3.19 generalizes the part of Theorem 3.18 concerning bounded lattices.

3.3 Representation of arbitrary lattices

There is another way of representing any lattice L as a sublattice of the lattice of $c_{\lhd}$-fixpoints of a compatibility frame, which is now the canonical extension of L (see [25,14]) rather than its MacNeille completion.[6] The sublattice can then be characterized in a simple way topologically. This approach uses disjoint filter-ideal pairs and appears already in [16], building on [48,1], though we use a different topology in order to generalize the choice-free Stone duality of [4]. Given a lattice L, define $\mathsf{FI}(L) = (X, \lhd)$ as follows: X is the set of all pairs (F,I) such that F is a filter in L, I is an ideal in L, and $F \cap I = \varnothing$; and $(F,I) \lhd (F',I')$ iff $I \cap F' = \varnothing$. Given $a \in L$, let $\widehat{a} = \{(F,I) \in X \mid a \in F\}$. Let $\mathsf{S}(L)$ be $\mathsf{FI}(L)$ endowed with the topology generated by $\{\widehat{a} \mid a \in L\}$.

In Appendix A.2, we prove the following (without choice).

Theorem 3.20 *For any lattice (resp. bounded lattice) L, the map $a \mapsto \widehat{a}$ is* (i) *a lattice (resp. bounded lattice) embedding of L into $\mathfrak{L}(\mathsf{FI}(L))$ and* (ii) *an isomorphism from L to the sublattice of $\mathfrak{L}(\mathsf{FI}(L))$ consisting of $c_{\lhd}$-fixpoints that are compact open in the space $\mathsf{S}(L)$.*

In Appendix A.2 we also prove the following characterization of spaces equipped with a relation $\lhd$ that are isomorphic to $\mathsf{S}(L)$ for some L in a manner analogous to the characterization of UV-spaces in [4]. Let X be a topological space and $\lhd$ a reflexive relation on X. Let $\mathsf{COFix}(X,\lhd)$ be the set of all compact open sets of X that are also $c_{\lhd}$-fixpoints. Given $U, V \in \mathsf{COFix}(X,\lhd)$, $U \vee V = c_{\lhd}(U \cup V)$. Given $x \in X$, let

$$\mathsf{F}(x) = \{U \in \mathsf{COFix}(X,\lhd) \mid x \in U\}$$
$$\mathsf{I}(x) = \{U \in \mathsf{COFix}(X,\lhd) \mid \forall y \rhd x \; y \notin U\}.$$

[6] We do not have space to discuss $(L,\neg)$ under this approach, so we save this for the future.

Proposition 3.21 *For any space X and reflexive binary relation $\lhd$ on X, there is a lattice L such that $(X, \lhd)$ and $\mathsf{S}(L)$ are homeomorphic as spaces and isomorphic as relational frames iff the following conditions hold for all $x, y \in X$:* (i) *$x = y$ iff $(\mathsf{F}(x), \mathsf{I}(x)) = (\mathsf{F}(y), \mathsf{I}(y))$;* (ii) *$\mathsf{COFix}(X, \lhd)$ is closed under $\cap$ and $\vee$ and forms a basis for X;* (iii) *each disjoint filter-ideal pair from $\mathsf{COFix}(X, \lhd)$ is $(\mathsf{F}(x), \mathsf{I}(x))$ for some $x \in X$;* (iv) *$x \lhd y$ iff $\mathsf{I}(x) \cap \mathsf{F}(y) = \varnothing$.*

4 Compatibility and accessibility frames

In this section, we extend the three representations from §§ 3.1, 3.2, and 3.3 to lattices equipped with a modal operation $\Box$ (we defer other modalities not definable from $\Box$ to future work). First, we add an accessibility relation R to compatibility frames and require that the standard modal operation $\Box_R$ sends $c_{\lhd}$-fixpoints to $c_{\lhd}$-fixpoints. These frames are similar to the *graph-based frames* of [10, Definition 2], which have been applied in [9,13,12]. Conradie et al. [10, Theorem 1] use the filter-ideal frame $\mathsf{FI}(L)$ equipped with accessibility relations to prove completeness of the minimal non-distributive modal logic with respect to graph-based frames (compare our Theorem 4.10); in addition, they treat Sahlqvist correspondence theory for graph-based frames.

There are many related approaches to representing lattices with modalities in the literature (see, e.g., [42,24,11,31,32,28,20] and references therein). The approach below using just two binary relations on a single set is of special interest to us as a non-classical generalization of classical "possibility semantics" for modal logic [37,35,33,49,50]. Our motivation for such a non-classical generalization comes from a recent application of the approach below to modal ortholattices for natural language semantics [36].

All proofs in this section are deferred to Appendix A.3.

Definition 4.1 A *necessity lattice* is a pair $(L, \Box)$ where L is a lattice and $\Box$ is a unary operation on L that is multiplicative, i.e., $\Box(a \wedge b) = \Box a \wedge \Box b$ for all $a, b \in L$, and $\Box 1 = 1$ if L contains a maximum element 1. We say $\Box$ is *completely multiplicative* if for any $A \subseteq L$, if $\bigwedge A$ exists in L, then $\Box \bigwedge A = \bigwedge \{\Box a \mid a \in A\}$.

Definition 4.2 A *compatibility and accessibility* (*CA*) *frame* is a triple $(X, \lhd, R)$ such that $(X, \lhd)$ is a compatibility frame and R is a binary relation on X such that for any $A \subseteq X$, if A is a $c_{\lhd}$-fixpoint, then so is

$$\Box_R A = \{x \in X \mid R(x) \subseteq A\},$$

where $R(x) = \{y \in X \mid xRy\}$.

Stronger conditions on the interplay of $\lhd$ and R could be imposed (see [10, Def. 2] and Proposition 4.5 below) but Definition 4.2 suffices for the following.

Proposition 4.3 *For any CA frame $(X, \lhd, R)$, the pair $(\mathfrak{L}(X, \lhd), \Box_R)$ is a complete necessity lattice with $\Box_R$ completely multiplicative.*

Proof. That $\mathfrak{L}(X, \lhd)$ is a complete lattice is Corollary 2.7. Recall that meet is intersection. Then the complete multiplicativity of $\Box_R$ is obvious. $\Box$

Example 4.4 Fig. 5 shows a CA frame (left) where a dotted line from w to v means wRv. Observe that $\Box_R(\{x\}) = \{x\}$, $\Box_R(\{y,z\}) = \{z\}$, and $\Box_R(\{z\}) = \{z\}$. Thus, $\Box_R$ sends $c_\lhd$-fixpoints to $c_\lhd$-fixpoints. The $\Box_R$ operation on the lattice of $c_\lhd$-fixpoints (right) is represented by the double-shafted arrows.

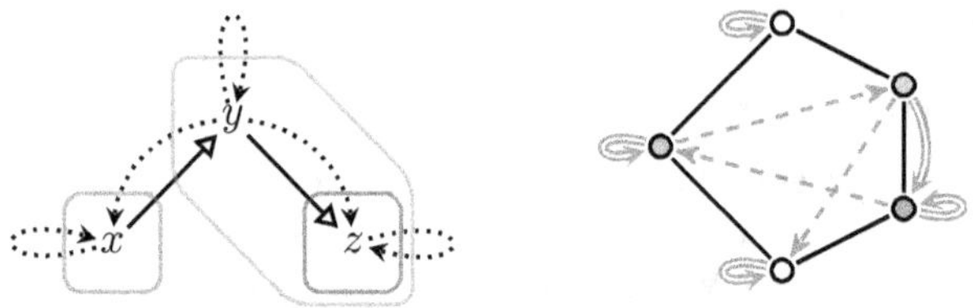

Fig. 5. CA frame and associated necessity lattice.

Under certain assumptions about $\lhd$, the condition that $\Box_R$ sends $c_\lhd$-fixpoints to $c_\lhd$-fixpoints corresponds to a first-order condition on R and $\lhd$. For example, when $\lhd$ is a preorder, in light of Proposition 2.9(ii), the condition that $\Box_R$ sends $c_\lhd$-fixpoints to $c_\lhd$-fixpoints corresponds to $\Box_R$ sending downsets to downsets and hence to the standard interaction condition for intuitonistic modal frames [7]: if $y \in R(x)$ and $x \lhd x'$, then there is a $y' \in R(x')$ with $y \lhd y'$. When $\lhd$ is symmetric, we get the first-order condition in the following proposition. For useful notation, define $z \lhd_R x \Leftrightarrow \exists y : z \lhd y \in R(x)$.

Proposition 4.5 *If $(X, \lhd)$ is a compatibility frame and R a binary relation on X, then $(X, \lhd, R)$ is a CA frame if the following condition holds: for all $x, z \in X$, if $z \lhd_R x$, then $\exists x' \lhd x\ \forall x'' \rhd x'\ \ z \lhd_R x''$.*

Moreover, if $(X, \lhd)$ is a symmetric compatibility frame, then $(X, \lhd, R)$ is a CA frame if and only if *the stated condition holds.*

In fact, the proof of Proposition 4.10 below shows that any necessity lattice can be represented using a CA frame satisfying the condition in Proposition 4.5, so in that sense we can work with such CA frames without loss of generality.

Next we turn to extending the representation theorems of § 3.

Proposition 4.6 *Let L be a complete lattice satisfying the hypotheses of Proposition 3.2, so L is isomorphic to $\mathfrak{L}(\mathrm{V}, \lhd)$ via $b \mapsto \varphi(b) = \{x \in \mathrm{V} \mid x \leq b\}$. Given a completely multiplicative operation $\Box$ on L, define R on V by xRy iff $y \leq \bigwedge\{a \in L \mid x \leq \Box a\}$. Then $(\mathrm{V}, \lhd, R)$ is a CA frame, and φ is an isomorphism from $(L, \Box)$ to $(\mathfrak{L}(\mathrm{V}, \lhd), \Box_R)$.*

Example 4.7 Fig. 6 shows a necessity ortholattice $(L, \Box)$ (right) along with the CA frame (left) that comes from the representation of $(L, \Box)$ via join-irreducible elements given by Theorem 3.11 and Proposition 4.6. It is argued in [36] that this necessity ortholattice captures some important logical entailments involving the epistemic modals 'must' (formalized as $\Box$) and 'might' (formalized as $\Diamond = \neg\Box\neg$) in natural language, including the phenomenon of "epistemic contradiction" whereby sentences of the form "p, but it might be that $\neg p$" are judged contradictory, even though "it might be that $\neg p$" does not entail $\neg p$.

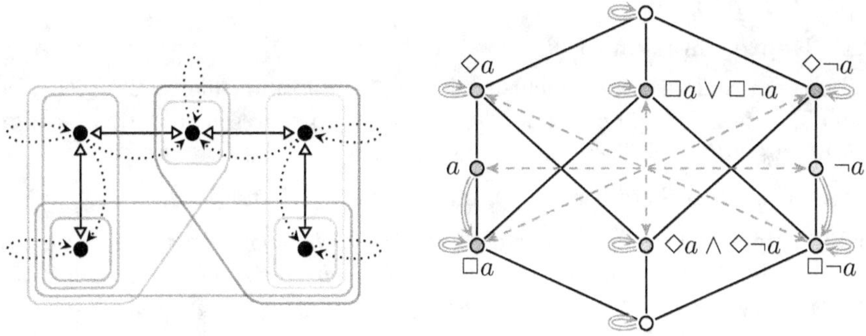

Fig. 6. CA frame realizing a necessity ortholattice (with $\Diamond$ defined by $\neg\Box\neg$).

We can extend our other representation results to the modal setting as well. Given $(L, \Box)$ and P a separating set of pairs of elements of L as in § 3.2, define a relation R on P by $(x, x')R(y, y')$ iff xRy as defined in Proposition 4.6.

Proposition 4.8 *If $(L, \Box)$ is a complete necessity lattice with $\Box$ completely multiplicative and P a separating set of pairs of elements of L, then $(P, \lhd, R)$ is a CA frame and $(L, \Box)$ is isomorphic to $(\mathfrak{L}(P, \lhd), \Box_R)$.*

Using Theorem 3.19 we can similarly represent complete necessity lattices equipped with a protocomplementation.

Proposition 4.9 *If $(L, \Box)$ is a complete necessity lattice with $\Box$ completely multiplicative and $\neg$ is a protocomplementation on L, then there is a CA frame $(P, \lhd, R)$ such that $(L, \Box, \neg)$ is isomorphic to $(\mathfrak{L}(P, \lhd), \Box_R, \neg_\lhd)$.*

Finally, define $\mathsf{FI}(L, \Box)$ just like $\mathsf{FI}(L)$ in § 3.3 but with the addition of a relation R with $(F, I)R(F', I')$ if for all $a \in L$, $\Box a \in F$ implies $a \in F'$.

Proposition 4.10 *For any necessity lattice $(L, \Box)$, $\mathsf{FI}(L, \Box)$ is a CA frame, and the map $a \mapsto \widehat{a}$ is* (i) *an embedding of $(L, \Box)$ into $(\mathfrak{L}(\mathsf{FI}(L)), \Box_R)$ and* (ii) *an isomorphism from $(L, \Box)$ to the subalgebra of $(\mathfrak{L}(\mathsf{FI}(L)), \Box_R)$ consisting of $c_\lhd$-fixpoints that are compact open in the space* $\mathsf{S}(L)$ (recall § 3.3).

5 Conclusion

We have investigated three representations of complete lattices by means of compatibility frames, as well as modal analogues thereof. For future work, we hope to make progress on Question 3.12 and Conjecture 3.13, as well as applications of the representations studied here to lattice-based logics. For modal logic in particular, for reasons in [36] we would like to understand the lattice of modal orthologics, for which we hope that CA frames will be useful.

Acknowledgements

I thank Peter Jipsen, Guillaume Massas, and the referees for *Advances in Modal Logic* for helpful feedback.

A Appendix

A.1 Representation of $(L, \neg)$

In Theorem 3.11(iii), the conclusion is that L is isomorphic to $\mathfrak{L}(\mathrm{V}, \lhd_{\mathrm{V}}^{\neg})$, not that $(L, \neg)$ is isomorphic to $(\mathfrak{L}(\mathrm{V}, \lhd_{\mathrm{V}}^{\neg}), \neg_{\lhd_{\mathrm{V}}^{\neg}})$. To represent $(L, \neg)$, we ask for a *third suitability condition* on $\lhd_{\mathrm{V}}^{\neg}$, namely that for $x \in \mathrm{V}$ and $y \in L$:

if $x \not\leq \neg y$, then there is a $y_0 \in \mathrm{V}$ such that $y_0 \leq y$ and $y_0 \lhd_{\mathrm{V}}^{\neg} x$.

Proposition A.1 *If L is a complete lattice, V is a join-dense set of nonzero elements, $\neg$ is a protocomplementation on L such that L has compatible escape in V with $\neg$, and $\lhd_{\mathrm{V}}^{\neg}$ satisfies the second and third suitability conditions, then $(L, \neg)$ is isomorphic to $(\mathfrak{L}(\mathrm{V}, \lhd_{\mathrm{V}}^{\neg}), \neg_{\lhd_{\mathrm{V}}^{\neg}})$.*

Proof. We need only add to the proof of Theorem 3.11(iii) that $\varphi(\neg a) = \neg_{\lhd_{\mathrm{V}}^{\neg}} \varphi(a)$. Suppose $x \in \varphi(\neg a)$, so $x \leq \neg a$, and $x' \lhd_{\mathrm{V}}^{\neg} x$, so $x \not\leq \neg x'$. If $x' \leq a$, then $\neg a \leq \neg x'$, which with $x \leq \neg a$ implies $x \leq \neg x'$, contradicting the previous sentence. Thus, $x' \not\leq a$, so $x' \notin \varphi(a)$. Hence $x \in \neg_{\lhd_{\mathrm{V}}^{\neg}} \varphi(a)$. Conversely, suppose $x \in \mathrm{V} \setminus \varphi(\neg a)$, so $x \not\leq \neg a$. Then by the third suitability condition, there is an $a_0 \in \mathrm{V}$ such that $a_0 \leq a$, so $a_0 \in \varphi(a)$, and $a_0 \lhd_{\mathrm{V}}^{\neg} x$, so $x \notin \neg_{\lhd_{\mathrm{V}}^{\neg}} \varphi(a)$. □

It remains to be seen how broadly Proposition A.1 applies to lattices expanded with a protocomplementation. In the Jupyter notebook cited in § 1, we show there are such expansions $(L, \neg)$ that cannot be represented by any compatibility frame with $|X| \leq |L|$. Of course, Proposition A.1 applies to all ortholattices.

Proposition A.2 *If $(L, \neg)$ is a complete ortholattice, and V is a join-dense set of nonzero elements, then $\lhd_{\mathrm{V}}^{\neg}$ satisfies the third suitability condition.*

Proof. If $x \not\leq \neg y$, then $x \not\leq \neg \bigvee\{z \in \mathrm{V} \mid z \leq y\} = \bigwedge\{\neg z \mid z \in \mathrm{V}, z \leq y\}$, so there is $y_0 \in \mathrm{V}$ with $y_0 \leq y$ and $x \not\leq \neg y_0$, so $y_0 \lhd_{\mathrm{V}}^{\neg} x$ by Proposition 3.7(i). □

Recall that using the less economical representation of § 3.2, any complete lattice expanded with a protocomplementation is representable (Theorem 3.19).

A.2 Proofs for § 3.3

Theorem 3.20 *For any lattice (resp. bounded lattice) L, the map $a \mapsto \widehat{a}$ is (i) a lattice (resp. bounded lattice) embedding of L into $\mathfrak{L}(\mathsf{FI}(L))$ and (ii) an isomorphism from L to the sublattice of $\mathfrak{L}(\mathsf{FI}(L))$ consisting of $c_{\lhd}$-fixpoints that are compact open in the space $\mathsf{S}(L)$.*

Proof. First observe that for any $a \in L$, $\widehat{a}$ is a $c_{\lhd}$-fixpoint. It suffices to show that if $(F, I) \notin \widehat{a}$, then there is an $(F', I') \lhd (F, I)$ such that for all $(F'', I'') \rhd (F', F')$, we have $(F'', I'') \notin \widehat{a}$. Suppose $(F, I) \notin \widehat{a}$, so $a \notin F$. Let $F' = F$ and $I' = {\downarrow} a$. Then $a \notin F$ implies $F' \cap I' = \varnothing$. Thus, $(F', I') \in \mathsf{FI}(L)$. Now consider any (F'', I'') such that $(F', I') \lhd (F'', I'')$, so $I' \cap F'' = \varnothing$. Then since $a \in I'$, we have $a \notin F''$, so $(F'', I'') \notin \widehat{a}$, as desired.

Next, the map $a \mapsto \widehat{a}$ is clearly injective: if $a \not\leq b$, then ${\uparrow} a \cap {\downarrow} b = \varnothing$, so $({\uparrow} a, {\downarrow} b) \in \mathsf{FI}(L)$, $({\uparrow} a, {\downarrow} b) \in \widehat{a}$, and $({\uparrow} a, {\downarrow} b) \notin \widehat{b}$. If L has bounds, then $\widehat{1} = X$ and $\widehat{0} = \varnothing$. The map also preserves $\wedge$: $\widehat{a \wedge b} = \{(F, I) \in X \mid a \wedge b \in F\} = \{(F, I) \in X \mid a, b \in F\} = \{(F, I) \in X \mid a \in F\} \cap \{(F, I) \in X \mid b \in F\} = \widehat{a} \cap \widehat{b} = \widehat{a} \wedge \widehat{b}$.

To complete part (i), we show $\widehat{a \vee b} = \widehat{a} \vee \widehat{b}$. Recall from Proposition 2.3 that $\widehat{a} \vee \widehat{b} = c_{\lhd}(\widehat{a} \cup \widehat{b})$. Suppose $(F, I) \in \widehat{a \vee b}$, so $a \vee b \in F$. Consider any $(F', I') \lhd (F, I)$, so $I' \cap F = \varnothing$ and hence $a \vee b \notin I'$. Then since I' is an ideal, $a \notin I'$ or $b \notin I'$. Without loss of generality, suppose $a \notin I'$. Then setting $F'' = \uparrow a$ and $I'' = \downarrow c$ for any $c \in I'$, we have $(F'', I'') \in \mathsf{FI}(L)$ and $I' \cap F'' = \varnothing$, so $(F', I') \lhd (F'', I'')$, and $(F'', I'') \in \widehat{a}$. Thus, we have shown that for any $(F', I') \lhd (F, I)$ there is an $(F'', I'') \rhd (F', I')$ with $(F'', I'') \in \widehat{a} \cup \widehat{b}$. Hence $(F, I) \in \widehat{a} \vee \widehat{b}$. Conversely, suppose $(F, I) \notin \widehat{a \vee b}$, so $a \vee b \notin F$. Then setting $F' = \uparrow c$ for any $c \in F$ and $I' = \downarrow (a \vee b)$, we have $(F', I') \in \mathsf{FI}(L)$ and $I' \cap F = \varnothing$, so $(F', I') \lhd (F, I)$. Now consider any (F'', I'') with $(F', I') \lhd (F'', I'')$, so $I' \cap F'' = \varnothing$ and hence $a \vee b \notin F''$. Then since F'' is a filter, $a \notin F''$ and $b \notin F''$, so $F'' \notin \widehat{a} \cup \widehat{b}$. Thus, we have shown there is an $(F', I') \lhd (F, I)$ such that for all $(F'', I'') \rhd (F', I')$, we have $(F'', I'') \notin \widehat{a} \cup \widehat{b}$. Hence $(F, I) \notin \widehat{a} \vee \widehat{b}$.

For part (ii), we first show that $\widehat{a}$ is compact open. Since $\widehat{b}$'s form a basis, we need only show that if $\widehat{a} \subseteq \bigcup \{\widehat{b_i} \mid i \in I\}$, then there is a finite subcover. Suppose $a \not\leq b_i$ for some $i \in I$. Then since $(\uparrow a, \downarrow b_i) \in \widehat{a}$, we have $(\uparrow a, \downarrow b_i) \in \widehat{b_j}$ for some $j \in I$, which implies $a \leq b_j$. Thus, $a \leq b_k$ for some $k \in I$, so $\widehat{a} \subseteq \widehat{b_k}$.

Finally, we show that φ is onto the set of compact open $c_{\lhd}$-fixpoints. Suppose U is compact open, so $U = \widehat{a_1} \cup \dots \cup \widehat{a_n}$ for some $a_1, \dots, a_n \in L$. Further suppose U is a $c_{\lhd}$-fixpoint, so $c_{\lhd}(U) = U$. Where $d = a_1 \vee \dots \vee a_n$, an obvious induction using part (i) and the fact that $c_{\lhd}(c_{\lhd}(A) \cup B) = c_{\lhd}(A \cup B)$ for any $A, B \subseteq X$ yields $\widehat{d} = c_{\lhd}(\widehat{a_1} \cup \dots \cup \widehat{a_n})$, so $\widehat{d} = c_{\lhd}(U) = U$. □

Proposition 3.21 *For any space X and reflexive binary relation $\lhd$ on X, there is a lattice L such that $(X, \lhd)$ and $\mathsf{S}(L)$ are homeomorphic as spaces and isomorphic as relational frames iff the following conditions hold for all $x, y \in X$:* (i) $x = y$ *iff* $(\mathsf{F}(x), \mathsf{I}(x)) = (\mathsf{F}(y), \mathsf{I}(y))$; (ii) $\mathsf{COFix}(X, \lhd)$ *is closed under $\cap$ and $\vee$ and forms a basis for X;* (iii) *each disjoint filter-ideal pair from* $\mathsf{COFix}(X, \lhd)$ *is* $(\mathsf{F}(x), \mathsf{I}(x))$ *for some $x \in X$;* (iv) $x \lhd y$ *iff* $\mathsf{I}(x) \cap \mathsf{F}(y) = \varnothing$.

Proof. Suppose there is such an L. It suffices to show $\mathsf{S}(L)$ satisfies (i)–(iv) in place of $(X, \lhd)$. That (ii) holds for $\mathsf{COFix}(\mathsf{S}(L))$ and $\mathsf{S}(L)$ follows from the proof of Theorem 3.20. Let φ be the isomorphism $a \mapsto \widehat{a}$ from L to $\mathsf{COFix}(\mathsf{S}(L))$ in Theorem 3.20, which induces a bijection $(F, I) \mapsto (\varphi[F], \varphi[I])$ between disjoint filter-ideal pairs of L and of $\mathsf{COFix}(\mathsf{S}(L))$. Parts (i), (iii), and (iv) follow from the fact that for any $x = (F, I) \in \mathsf{S}(L)$, $(\varphi[F], \varphi[I]) = (\mathsf{F}(x), \mathsf{I}(x))$. First, $\widehat{a} \in \varphi[F]$ iff $a \in F$ iff $x \in \widehat{a}$ iff $\widehat{a} \in \mathsf{F}(x)$. Second, $\widehat{a} \in \varphi[I]$ iff $a \in I$, and we claim that $a \in I$ iff $\widehat{a} \in \mathsf{I}(x)$, i.e., for all $(F', I') \rhd (F, I)$, $(F', I') \notin \widehat{a}$, i.e., $a \notin F'$. If $a \in I$ and $(F, I) \lhd (F', I')$, then $a \notin F'$ by definition of $\lhd$. Conversely, if $a \notin I$, let $F' = \uparrow a$ and $I' = I$; then $(F, I) \lhd (F', I')$ and $a \in F'$. Now for (i), given $x, y \in \mathsf{S}(L)$ with $x = (F, I)$ and $y = (F', I')$, we have $(F, I) = (F', I')$ iff $(\varphi[F], \varphi[I]) = (\varphi[F'], \varphi[I'])$ iff $(\mathsf{F}(x), \mathsf{I}(x)) = (\mathsf{F}(y), \mathsf{I}(y))$; similarly, for (iv), $(F, I) \lhd (F', I')$ iff $I \cap F' = \varnothing$ iff $\varphi[I] \cap \varphi[F'] = \varnothing$ iff $\mathsf{I}(x) \cap \mathsf{F}(y) = \varnothing$. Finally, for (iii), if $(\mathscr{F}, \mathscr{I})$ if a disjoint filter-ideal pair from $\mathsf{COFix}(\mathsf{S}(L))$, then setting $x = (\varphi^{-1}[\mathscr{F}], \varphi^{-1}[\mathscr{I}])$, we have $x \in \mathsf{S}(L)$ and $(\mathscr{F}, \mathscr{I}) = (\mathsf{F}(x), \mathsf{I}(x))$.

Assuming X satisfies the conditions, $\mathsf{COFix}(X, \lhd)$ is a lattice, and we define a map ϵ from $(X, \lhd)$ to $\mathsf{S}(\mathsf{COFix}(X, \lhd))$ by $\epsilon(x) = (\mathsf{F}(x), \mathsf{I}(x))$. The proof that ϵ is a homeomorphism using (i)–(iii) is analogous to the proof of Thm. 5.4(2) in [4]. That ϵ preserves and reflects $\lhd$ follows from (iv). □

A.3 Proofs for § 4

Proposition 4.5 *If $(X, \lhd)$ is a compatibility frame and R a binary relation on X, then $(X, \lhd, R)$ is a CA frame if the following condition holds: for all $x, z \in X$, if $z \lhd_R x$, then $\exists x' \lhd x\ \forall x'' \rhd x'\ \ z \lhd_R x''$.*

Moreover, if $(X, \lhd)$ is a symmetric compatibility frame, then $(X, \lhd, R)$ is a CA frame if and only if *the stated condition holds.*

Proof. For the first part, we must show that $\Box_R A$ is a $c_\lhd$-fixpoint for any $c_\lhd$-fixpoint A. That is, we must show that $x \in X \setminus \Box_R A \Rightarrow \exists x' \lhd x \forall x'' \rhd x'\ x'' \notin \Box_R A$. Suppose $x \notin \Box_R A$, so there is some $y \in R(x)$ with $y \notin A$. Then since A is a $c_\lhd$-fixpoint, there is a $z \lhd y$ such that $(\star)$ for all $z' \rhd z$, we have $z' \notin A$. Since $z \lhd y \in R(x)$, by the condition we have $\exists x' \lhd x \forall x'' \rhd x' \exists y' : z \lhd y' \in R(x'')$. Now $z \lhd y'$ implies $y' \notin A$ by $(\star)$, which with $y' \in R(x'')$ implies $x'' \notin \Box_R A$.

For the second part, assume $\lhd$ is symmetric and $\Box_R$ sends $c_\lhd$-fixpoints to $c_\lhd$-fixpoints. Toward proving the condition, suppose $z \lhd y \in R(x)$. Hence $y \notin \neg_\lhd c_\lhd(\{z\})$, so $x \notin \Box_R \neg_\lhd c_\lhd(\{z\})$. Since by assumption $\Box_R \neg_\lhd c_\lhd(\{z\})$ is a $c_\lhd$-fixpoint, it follows that there is an $x' \lhd x$ such that for all $x'' \rhd x'$, we have that $x'' \notin \Box_R \neg_\lhd c_\lhd(\{z\})$. Thus, there is a $w \in R(x'')$ such that $w \notin \neg_\lhd c_\lhd(\{z\})$, so for some $w' \lhd w$, we have $w' \in c_\lhd(\{z\})$, which means that for all $w'' \lhd w'$, there is a $w''' \rhd w''$ such that $w''' \in \{z\}$, i.e., for all $w'' \lhd w'$, $w'' \lhd z$. Since $\lhd$ is symmetric, from $w' \lhd w$, we have $w \lhd w'$, so setting $w'' = w$, we conclude $w \lhd z$, so $z \lhd w$. Thus, $z \lhd w \in R(x'')$, i.e., $z \lhd_R x''$. □

Proposition 4.6 *Let L be a complete lattice satisfying the hypotheses of Proposition 3.2, so L is isomorphic to $\mathfrak{L}(\mathrm{V}, \lhd)$ via $b \mapsto \varphi(b) = \{x \in \mathrm{V} \mid x \leq b\}$. Given a completely multiplicative operation $\Box$ on L, define R on V by xRy iff $y \leq \bigwedge\{a \in L \mid x \leq \Box a\}$. Then $(\mathrm{V}, \lhd, R)$ is a CA frame, and φ is an isomorphism from $(L, \Box)$ to $(\mathfrak{L}(\mathrm{V}, \lhd), \Box_R)$.*

Proof. First, recall the key fact provided by complete multiplicativity of $\Box$: if $x \nleq \Box b$, then $\bigwedge\{a \in L \mid x \leq \Box a\} \nleq b$. For if $\bigwedge\{a \in L \mid x \leq \Box a\} \leq b$, then $\Box \bigwedge\{a \in L \mid x \leq \Box a\} \leq \Box b$ and hence $\bigwedge\{\Box a \in L \mid x \leq \Box a\} \leq \Box b$, so $x \leq \Box b$.

Now we show that for all $b \in L$, $\varphi(\Box b) = \Box_R \varphi(b)$. Suppose $x \in \varphi(\Box b)$, so $x \leq \Box b$. Then for all $y \in R(x)$, we have $y \leq b$ and hence $y \in \varphi(b)$. Thus, $x \in \Box_R \varphi(b)$. Now suppose $x \notin \varphi(\Box b)$, so $x \nleq \Box b$. Hence $\bigwedge\{a \in L \mid x \leq \Box a\} \nleq b$ as above. Then since V is join-dense, there is a $y \in \mathrm{V}$ such that $y \leq \bigwedge\{a \in L \mid x \leq \Box a\}$ but $y \nleq b$. Hence xRy and $y \notin \varphi(b)$, so $x \notin \Box_R \varphi(b)$.

Finally, we prove that $(\mathrm{V}, \lhd, R)$ is indeed a CA frame: if B is a $c_\lhd$-fixpoint of $(\mathrm{V}, \lhd)$, so is $\Box_R B$. By the surjectivity of φ, $B = \varphi(b)$ for some $b \in B$. Then $\Box_R B = \Box_R \varphi(b) = \varphi(\Box b)$, and $\varphi(\Box b)$ is a $c_\lhd$-fixpoint, so we are done. □

Proposition 4.8 *If $(L, \Box)$ is a complete necessity lattice with $\Box$ completely multiplicative and P a separating set of pairs of elements of L, then $(P, \lhd, R)$*

is a CA frame and $(L,\Box)$ *is isomorphic to* $(\mathfrak{L}(P,\lhd),\Box_R)$.

Proof. We showed in the proof of Proposition 3.15 that $a \mapsto \varphi(a) = \{(x,y) \in P \mid x \leq a\}$ is an isomorphism from L to $\mathfrak{L}(P,\lhd)$. It only remains to show that $\varphi(\Box b) = \Box_R\varphi(b)$. Suppose $(x,y) \in \varphi(\Box b)$, so $x \leq \Box b$. Then $(x,y)R(x',y')$ implies $x' \leq b$ and hence $(x',y') \in \varphi(b)$. Thus, $(x,y) \in \Box_R\varphi(b)$. Conversely, suppose $(x,y) \notin (\Box b)$, so $x \not\leq \Box b$. Since $\Box$ is completely multiplicative, it follows that $\bigwedge\{a \in L \mid x \leq \Box a\} \not\leq b$. Then since P is separating, there is some $(c,d) \in P$ with $c \leq \bigwedge\{a \in L \mid x \leq \Box a\}$ but $c \not\leq b$. Hence $(x,y)R(c,d)$ but $c \notin \varphi(b)$, so $(x,y) \notin \Box_R\varphi(b)$. Now the proof that $(P,\lhd,R)$ is a CA frame is analogous to the last paragraph of the previous proof. □

Proposition 4.9 *If* $(L,\Box)$ *is a complete necessity lattice with* $\Box$ *completely multiplicative and* $\neg$ *is a protocomplementation on* L*, then there is a CA frame* $(P,\lhd,R)$ *such that* $(L,\Box,\neg)$ *is isomorphic to* $(\mathfrak{L}(P,\lhd),\Box_R,\neg_\lhd)$.

Proof. Where $P = \{(a,b) \mid a,b \in L, a \neq 0, a \not\leq b, \neg a \leq b\}$, we showed in the proof of Theorem 3.19 that P is separating. Hence by the proof of Proposition 4.8, $(L,\Box)$ is isomorphic to $(\mathfrak{L}(P,\lhd),\Box_R)$ via the map $a \mapsto \varphi(a) = \{(x,y) \in P \mid x \leq a\}$. We also showed in the proof of Theorem 3.19 that φ preserves the protocomplementation. Hence $(L,\Box,\neg)$ is isomorphic to $(\mathfrak{L}(P,\lhd),\Box_R,\neg_\lhd)$. □

Proposition 4.10 *For any necessity lattice* $(L,\Box)$, $\mathsf{FI}(L,\Box)$ *is a CA frame, and the map* $a \mapsto \widehat{a}$ *is* (i) *an embedding of* $(L,\Box)$ *into* $(\mathfrak{L}(\mathsf{FI}(L)),\Box_R)$ *and* (ii) *an isomorphism from* $(L,\Box)$ *to the subalgebra of* $(\mathfrak{L}(\mathsf{FI}(L)),\Box_R)$ *consisting of* $c_\lhd$*-fixpoints that are compact open in the space* $\mathsf{S}(L)$.

Proof. In the proof of Proposition 4.5, we showed that for any $(X,\lhd)$ and binary relation R on X, if the first-order condition in Proposition 4.5 holds, then $(X,\lhd,R)$ is a CA frame. We claim $\mathsf{FI}(L,\Box)$ satisfies the condition. Suppose $(G,H) \lhd (G',H') \in R((F,I))$, which implies $H \cap \{a \in L \mid \Box a \in F\} = \varnothing$. Then where $F' = F$ and I' is the ideal generated by $\{\Box a \mid a \in H\}$, we claim $F' \cap I' = \varnothing$, so $(F',I') \lhd (F,I)$. For if $b \in F' \cap I'$, then for some $a_1,\dots,a_n \in H$, $b \leq \Box a_1 \vee \dots \vee \Box a_n$, which implies $b \leq \Box(a_1 \vee \dots \vee a_n)$, so $\Box(a_1 \vee \dots \vee a_n) \in F$, whence $a_1 \vee \dots \vee a_n \notin H$, contradicting $a_1,\dots,a_n \in H$. Now suppose $(F',I') \lhd (F'',I'')$, so $I' \cap F'' = \varnothing$. Let J be the filter generated by $\{b \in L \mid \Box b \in F''\}$. We claim $J \cap H = \varnothing$. For if $a \in J$, then $b_1 \wedge \dots \wedge b_n \leq a$ for $\Box b_i \in F''$, which implies $\Box a \in F''$, so $\Box a \notin I'$, whence $a \notin H$. Thus, $(G,H) \lhd (J,H) \in R((F'',I''))$, which establishes the desired condition.

Now for (i)–(ii), we need only add to Theorem 3.20 that $\widehat{\Box a} = \Box_R\widehat{a}$. Suppose $(F,I) \in \widehat{\Box a}$, so $\Box a \in F$. Then if $(F,I)R(F',I')$, we have $a \in F'$ and hence $(F',I') \in \widehat{a}$. Thus, $(F,I) \in \Box_R\widehat{a}$. Conversely, suppose $(F,I) \notin \widehat{\Box a}$, so $\Box a \notin F$. Let F' be the filter generated by $\{b \in L \mid \Box b \in F\}$ and $I' = {\downarrow}a$. Since $\Box a \notin F$ and $\Box$ is multiplicative, it follows that $F' \cap I' = \varnothing$. Hence $(F',I') \in \mathsf{FI}(L,\Box)$, $(F',I') \notin \widehat{a}$, and by construction of F', $(F,I)R(F',I')$. Thus $(F,I) \notin \Box_R\widehat{a}$. □

References

[1] Allwein, G. and C. Hartonas, *Duality for bounded lattices* (1993), Indiana University Logic Group, Preprint Series, IULG-93-25 (1993).

[2] Allwein, G. and W. MacCaull, *A Kripke semantics for the logic of Gelfand quantales*, Studia Logica **68** (2001), pp. 173–228.

[3] Bezhanishvili, G. and W. H. Holliday, *A semantic hierarchy for intuitionistic logic*, Indagationes Mathematicae **30** (2019), pp. 403–469.

[4] Bezhanishvili, N. and W. H. Holliday, *Choice-free Stone duality*, The Journal of Symbolic Logic **85** (2020), pp. 109–148.

[5] Bimbó, K., *Functorial duality for ortholattices and De Morgan lattices*, Logica Universalis **1** (2007), pp. 311–333.

[6] Birkhoff, G., "Lattice Theory," American Mathematical Society, New York, 1940.

[7] Božic, M. and K. Došen, *Models for normal intuitionistic modal logics*, Studia Logica **43** (1984), pp. 217–245.

[8] Burris, S. N. and H. P. Sankappanavar, "A Course in Universal Algebra," Springer-Verlag, New York, 1981.

[9] Conradie, W., A. Craig, A. Palmigiano and N. Wijnberg, *Modelling competing theories*, in: *Proceedings of the 11th Conference of the European Society for Fuzzy Logic and Technology (EUSFLAT 2019)*, Atlantis Studies in Uncertainty Modelling **1**, 2019, pp. 721–739.

[10] Conradie, W., A. Craig, A. Palmigiano and N. M. Wijnberg, *Modelling informational entropy*, in: R. Iemhoff, M. Moortgat and R. Queiroz, editors, *Logic, Language, Information, and Computation. WoLLIC 2019*, Lectures Notes in Computer Science **11541**, 2019, pp. 140–160.

[11] Conradie, W., S. Frittella, A. Palmigiano, M. Piazzai, A. Tzimoulis and N. M. Wijnberg, *Categories: How I learned to stop worrying and love two sorts*, in: J. Väänänen, A. Hirvonen and R. de Queiroz, editors, *Logic, Language, Information, and Computation. WoLLIC 2016*, Lectures Notes in Computer Science **9803**, 2016, pp. 145–164.

[12] Conradie, W., A. Palmigiano, C. Robinson, A. Tzimoulis and N. Wijnberg, *Modelling socio-political competition*, Fuzzy Sets and Systems **407** (2021), pp. 115–141.

[13] Conradie, W., A. Palmigiano, C. Robinson and N. Wijnberg, *Non-distributive logics: from semantics to meaning*, in: A. Rezus, editor, *Contemporary Logic and Computing*, Landscapes in Logic, College Publications, 2020 pp. 38–86.

[14] Craig, A. and M. Haviar, *Reconciliation of approaches to the construction of canonical extensions of bounded lattices*, Mathematica Slovaca **64** (2014), pp. 1335–1356.

[15] Craig, A. P. K., M. J. Gouveia and M. Haviar, *TiRS graphs and TiRS frames: a new setting for duals of canonical extensions*, Algebra Universalis **74** (2015), pp. 123–138.

[16] Craig, A. P. K., M. Haviar and H. A. Priestley, *A fresh perspective on canonical extensions for bounded lattices*, Applied Categorical Structures **21** (2013), pp. 725–749.

[17] Davey, B. A., *On the lattice of subvarieties*, Houston Journal of Mathematics **5** (1979), pp. 183–192.

[18] Davey, B. A. and H. A. Priestley, "Introduction to Lattices and Order," Cambridge University Press, New York, 2002, 2nd edition.

[19] Dishkant, H., *Semantics of the minimal logic of quantum mechanics*, Studia Logica **30** (1972), pp. 23–30.

[20] Dmitrieva, A., "Positive modal logic beyond distributivity: duality, preservation and completeness," Master's thesis, University of Amsterdam (2021).

[21] Dragalin, A. G., "Mathematical Intuitionism: Introduction to Proof Theory," Translations of Mathematical Monographs, American Mathematical Society, Providence, Rhode Island, 1988.

[22] Esakia, L., "Heyting Algebras. Duality Theory," Springer, Cham, 2019, english translation by A. Evseev, eds. G. Bezhanishvili and W. H. Holliday.

[23] Gebhardt, V. and S. Tawn, *Constructing unlabelled lattices*, Journal of Algebra **545** (2020), pp. 213–236.

[24] Gehrke, M., *Generalized Kripke frames*, Studia Logica **84** (2006), pp. 241–275.
[25] Gehrke, M. and J. Harding, *Bounded lattice expansions*, Journal of Algebra **238** (2001), pp. 345–371.
[26] Gehrke, M., J. Harding and Y. Venema, *MacNeille completions and canonical extensions*, Transactions of the American Mathematical Society **358** (2005), pp. 573–590.
[27] Goldblatt, R., "Metamathematics of Modal Logic," Ph.D. thesis, Victoria University, Wellington (1974).
[28] Goldblatt, R., *Morphisms and duality for polarities and lattices with operators* (2019), arXiv:1902.09783 [math.LO].
[29] Goldblatt, R. I., *Semantic analysis of orthologic*, Journal of Philosophical Logic **3** (1974), pp. 19–35.
[30] Goldblatt, R. I., *The Stone space of an ortholattice*, Bulletin of the London Mathematical Society **7** (1975), pp. 45–48.
[31] Hartonas, C., *Discrete duality for lattices with modal operators*, Journal of Logic and Computation **29** (2018), pp. 71–89.
[32] Hartonas, C. and E. Orłowska, *Representation of lattices with modal operators in two-sorted frames*, Fundamenta Informaticae **166** (2019), pp. 29–56.
[33] Holliday, W. H., *Possibility semantics*, in: M. Fitting, editor, *Selected Topics from Contemporary Logics*, Landscapes in Logic, College Publications, 2021 pp. 363–476.
[34] Holliday, W. H., *Three roads to complete lattices: Orders, compatibility, polarity*, Algebra Universalis **82** (2021).
[35] Holliday, W. H., *Possibility frames and forcing for modal logic*, The Australasian Journal of Logic (Forthcoming).
URL `https://escholarship.org/uc/item/0tm6b30q`
[36] Holliday, W. H. and M. Mandelkern, *The orthologic of epistemic modals* (2021), arXiv:2203.02872 [cs.LO].
[37] Humberstone, I. L., *From worlds to possibilities*, Journal of Philosophical Logic **10** (1981), pp. 313–339.
[38] Jónsson, B. and A. Tarski, *Boolean algebras with operators. Part I.*, American Journal of Mathematics **73** (1951), pp. 891–939.
[39] Markowsky, G., *Some combinatorial aspects of lattice theory*, in: *Proc. Univ. of Houston Lattice Theory Conf.*, 1973, pp. 36–68.
[40] Markowsky, G., *The factorization and representation of lattices*, Transactions of the American Mathematical Society **203** (1975), pp. 185–200.
[41] Massas, G., *B-frame duality for complete lattices* (2020), manuscript.
[42] Orłowska, E. and D. Vakarelov, *Lattice-based modal algebras and modal logics*, in: *Logic, methodology and philosophy of science. Proceedings of the 12th international congress*, 2005, pp. 147–170.
[43] Ploščica, M., *A natural representation of bounded lattices*, Tatra Mountains Mathematical Publication **5** (1995), pp. 75–88.
[44] Sambin, G. and V. Vaccaro, *Topology and duality in modal logic*, Annals of Pure and Applied Logic **37** (1988), pp. 249–296.
[45] Takeuti, G. and W. M. Zaring, "Axiomatic Set Theory," Springer-Verlag, New York, 1973.
[46] Tarski, A., *Zur Grundlegung der Bool'schen Algebra. I*, Fundamenta Mathematicae **24** (1935), pp. 177–198.
[47] Thomason, S. K., *Categories of frames for modal logic*, The Journal of Symbolic Logic **40** (1975), pp. 439–442.
[48] Urquhart, A., *A topological representation theory for lattices*, Algebra Universalis **8** (1978), pp. 45–58.
[49] Yamamoto, K., *Results in modal correspondence theory for possibility semantics*, Journal of Logic and Computation **27** (2017), pp. 2411–2430.
[50] Zhao, Z., *Algorithmic correspondence and canonicity for possibility semantics*, Journal of Logic and Computation **31** (2021), pp. 523–572.

Verification of Multi-Agent Properties in Electronic Voting: A Case Study

Damian Kurpiewski[1,3], Wojciech Jamroga[1,2], Łukasz Maśko[1]
Łukasz Mikulski[3,1], Witold Pazderski[1]
Wojciech Penczek[1], and Teofil Sidoruk[1,4] *

[1] *Institute of Computer Science, Polish Academy of Sciences*
ul. Jana Kazimierza 5, 01-248 Warsaw, Poland

[2] *Interdisciplinary Centre for Security, Reliability, and Trust, SnT, University of Luxembourg*
29 Av. John F. Kennedy, 1855 Luxembourg, Luxembourg

[3] *Faculty of Mathematics and Computer Science, Nicolaus Copernicus Univeristy*
ul. Chopina 12/18, 87-100 Toruń, Poland

[4] *Faculty of Mathematics and Information Science, Warsaw University of Technology*
ul. Koszykowa 75, 00-662 Warsaw, Poland

Abstract

Formal verification of multi-agent systems is hard, both theoretically and in practice. In particular, studies that use a single verification technique typically show limited efficiency, and allow to verify only toy examples. Here, we propose some new techniques and combine them with several recently developed ones to see what progress can be achieved for a real-life scenario. Namely, we use fixpoint approximation, domination-based strategy search, partial order reduction, and parallelization to verify heterogeneous scalable models of the SELENE e-voting protocol. The experimental results show that the combination allows to verify requirements for much more sophisticated models than previously.

Keywords: multi-agent systems, formal verification, e-voting.

1 Introduction

Multi-agent systems (MAS) provide models and methodologies for analysis of systems that feature interaction of multiple autonomous components [64,69,71]. Formal specification and verification of such systems becomes essential due to

* The authors acknowledge the support of the National Centre for Research and Development, Poland (NCBR), and the Luxembourg National Research Fund (FNR), under the PolLux/FNR-CORE project STV (POLLUX-VII/1/2019). W. Penczek and T. Sidoruk acknowledge the support of CNRS/PAS under the project MOSART.

the dynamic development of AI solutions that enter practical applications [4,5]. In particular, it is crucial to assess requirements that refer to *strategic abilities* of agents and their groups, such as the ability of a passenger to leave an autonomous cab (preferably alive), or the inability of an intruder to take remote control of the cab.

Specification and verification of MAS. Properties of this kind can be conveniently specified in *modal logics of strategic ability*, of which alternating-time temporal logic **ATL*** [7,8] is probably the most popular. Logic-based methods for MAS are relatively well studied from the theoretical perspective [21,25,26,34], including theories of agents and agency [10,11,19,60,70] semantic issues [2,3,23,31,42,63], meta-logical properties [14,32], and the complexity of model checking [13,24,32,63,67]. There is even a number of model checking approaches and tools. Unfortunately, they only admit temporal properties [9,22,44,45,49], deal with the less practical case of perfect information strategies [6,17,46,47,48], treat imperfect information with limited interest and effectiveness [54], or have restricted verification capabilities [5,50,52]. No less importantly, attempts to verify actual requirements on realistic agent systems have been scarce.

In this paper, we combine and extend some of the recent advances in model checking of modal specifications for MAS [37,39,51], and apply them to see how far we can get with the verification of an existing e-voting protocol. Anonymous, coercion-resistant, and verifiable e-voting procedures have been proposed and studied for over 10 years now [16,27,43,62], including implementations and their use in real-life elections [1,15,20]. This makes e-voting a great case for testing verification algorithms and tools, being developed for MAS.

Contribution. Verification for modal logics of strategic ability is hard, both theoretically and in practice. Likely, no single technique suffices to deal with it alone. Here, we try the "all out" approach, and combine several techniques, developed recently by our team, to verify properties of the Selene e-voting protocol [61]. We use the algorithms of *fixpoint approximation* [37], *depth-first* and *domination-based strategy search* [51], as well as *partial order reduction* [40,41]. To apply the latter, we extend it to handle strategic-epistemic properties, and prove the correctness of the extension. We also propose and study a distributed variant of the depth-first and domination-based synthesis. We evaluate the power of the combined approach on a new model of Selene, consisting of voters, coercers, and the election infrastructure. While our model does not yet match the complexity of a real-life election, it goes beyond typically used examples.

Related work. Formal verification of voting protocols typically focuses on their cryptographic aspects. The multi-agent and social interaction in the models is limited, and the verification restricts to temporal and bisimulation-based properties. Examples include the automated analysis of Selene [12] and Electryo [72] using the Tamarin prover for security protocols. Theorem proving in first-order logic was also used to capture some socio-technical factors of Helios

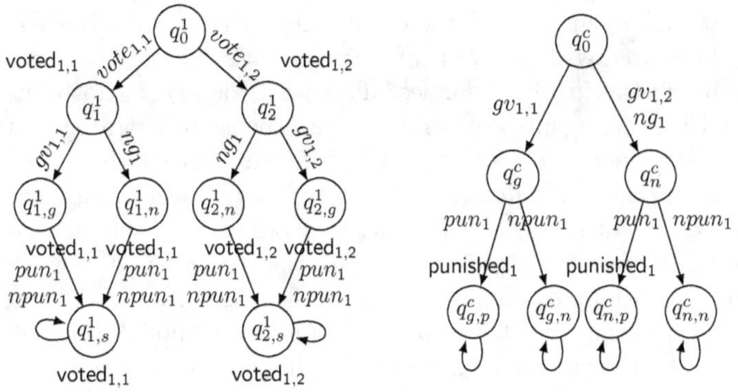

Fig. 1. ASV_1^2: agents *Voter₁* (left) and *Coercer* (right)

in [56]. Similarly, the interactive theorem prover Coq for higher-order logic was used in [29,33].

A slightly more detailed multi-agent model of the Prêt à Voter protocol was used in [35], but the verification concerned only simple temporal formulas of **CTL**. In [38], strategic abilities in Prêt à Voter were considered, but the strategies were hand-crafted rather than synthesized in the verification process. A preliminary formalization of receipt-freeness and coercion resistance using **ATL** was shown in [65], but no verification was proposed. Perhaps the closest work to the present one was our previous attempt to model and verify SELENE in [36], but the model used there was extremely simple, and the performance results were underwhelming.

2 Preliminaries

We first recall the models of asynchronous interaction in MAS, defined in [40] and inspired by [53,59].

2.1 Models of Asynchronous Interaction

Definition 2.1 [Asynchronous MAS] An *asynchronous multi-agent system (AMAS)* S consists of n agents $\mathcal{A} = \{1, \ldots, n\}$, each associated with a tuple $A_i = (L_i, Evt_i, R_i, T_i, PV_i, V_i)$ including a set of *local states* $L_i = \{l_i^1, l_i^2, \ldots, l_i^{n_i}\}$, a nonempty finite set of *events* $Evt_i = \{\alpha_i^1, \alpha_i^2, \ldots, \alpha_i^{m_i}\}$, and a *repertoire of choices* $R_i : L_i \to 2^{2^{Evt_i}}$. For each $l_i \in L_i$, $R_i(l_i) = \{E_1, \ldots, E_m\}$ is a nonempty list of nonempty choices available to i at l_i. If the agent chooses $E_j = \{\alpha_1, \alpha_2, \ldots\}$, then only an event in E_j can be executed at l_i within the agent's module. Moreover, $T_i : L_i \times Evt_i \rightharpoonup L_i$ is a (partial) *local transition function* such that $T_i(l_i, \alpha)$ is defined iff $\alpha \in \bigcup R_i(l_i)$.

Agents are endowed with mutually disjoint, finite and possibly empty sets of *local propositions* PV_i, and their *valuations* $V_i : L_i \to 2^{PV_i}$. $Evt = \bigcup_{i \in \mathcal{A}} Evt_i$ is the set of all events, and $Agent(\alpha) = \{i \in \mathcal{A} \mid \alpha \in Evt_i\}$ the set of agents who have access to event α. $PV = \bigcup_{i \in \mathcal{A}} PV_i$ is the set of all propositions.

Note that each agent "owns" the events affecting its state, but some of the events may be shared with other agents. Such events can only be executed synchronously by all the involved parties. This way, the agent can influence how the states of the other agents evolve. Moreover, the agent's strategic choices are restricted by its repertoire function. Assigning sets rather than single events in R_i, which subsequently determines the type of strategy functions in Section 2.2, is a deliberate decision, allowing to avoid certain semantic issues that fall outside the scope of this paper. We refer the reader to [40] for the details.

The following example demonstrates a simple AMAS, while also introducing some key concepts that will be expanded upon in our model of the real-world protocol SELENE (discussed in Section 3). In particular, there is a *coercer* agent, whose goal is to ensure that the voter(s) select a particular candidate. To that end, the coercer may threaten them with punishment, e.g. if they refuse to cooperate by not sharing the ballot, or openly defy by voting for another candidate.

Example 2.2 [Asynchronous Simple Voting] Consider a simple voting system ASV_n^k with $n+1$ agents (n voters and 1 coercer). Each $Voter_i$ agent can cast her vote for a candidate $\{1, \ldots, k\}$, and decide whether to share her vote receipt with the *Coercer* agent. The coercer can choose to punish the voter or refrain from it. A graphical representation of the agents for $n = 1, k = 2$ is shown in Fig. 1. We assume that the coercer only registers if the voter hands in a receipt for candidate 1 or not. The repertoire of the coercer is defined as $R_c(q_0^c) = \{\{gv_{1,1}, gv_{1,2}, ng_1\}\}$ and $R_c(q_g^c) = R_c(q_n^c) = \{\{pun_1\}, \{npun_1\}\}$, i.e., the coercer first *receives* the voter's decision regarding the receipt, and then *controls* whether the voter is punished or not. Analogously, the voter's repertoire is given by: $R_1(q_0^1) = \{\{vote_{1,1}\}, \{vote_{1,2}\}\}$, $R_1(q_j^1) = \{\{gv_{1,j}\}, \{ng_1\}\}$ for $j = 1, 2$, and $R_1(q_{1,g}^1) = R_1(q_{1,n}^1) = R_1(q_{2,g}^1) = R_1(q_{2,n}^1) = \{\{pun_1, npun_1\}\}$.

Notice that the coercer cannot determine which of the events $gv_{1,1}, gv_{1,2}, ng_1$ will occur; this is entirely under the voter's control. This way we model the situation where it is the decision of the voter to show her vote or not. Similarly, the voter cannot avoid punishment by choosing the strategy allowing only $npun_1$, because the choice $\{npun_1\}$ is *not* in the voter's repertoire. She can only execute $\{pun_1, npun_1\}$, and await the decision of the coercer.

The execution semantics is based on interleaving with synchronization on shared events. Note that for a shared event to be executed, it must be done jointly by all agents who have it in their repertoires.

Definition 2.3 [Interleaved interpreted systems] Let S be an AMAS with n agents, and let $I \subseteq L_1 \times \ldots \times L_n$. The *full model* $IIS(S, I)$ extends S with: (i) the set of initial states I; (ii) the set of global states $St \subseteq L_1 \times \ldots \times L_n$ that collects all the configurations of local states, reachable from I by T (see below); (iii) the (partial) *global transition function* $T : St \times Evt \rightharpoonup St$, defined by $T(g_1, \alpha) = g_2$ iff $T_i(g_1^i, \alpha) = g_2^i$ for all $i \in Agent(\alpha)$ and $g_1^i = g_2^i$ for all

$i \in \mathcal{A} \setminus Agent(\alpha)$;[1] (iv) the *global valuation* of propositions $V : St \to 2^{PV}$, defined as $V(l_1, \ldots, l_n) = \bigcup_{i \in \mathcal{A}} V_i(l_i)$.

We will sometimes write $g_1 \xrightarrow{\alpha} g_2$ instead of $T(g_1, \alpha) = g_2$.

Definition 2.4 [Enabled events] Event $\alpha \in Evt$ is *enabled* at $g \in St$ if $g \xrightarrow{\alpha} g'$ for some $g' \in St$, i.e., $T(g, \alpha) = g'$. The set of such events is denoted by *enabled*(g).

Let $A = \{a_1, \ldots, a_k\} \subseteq \mathcal{A} = \{1, \ldots, n\}$ and $\overrightarrow{E}_A = (E_{a_1}, \ldots, E_{a_k})$ for some $k \leq n$, such that $E_i \in R_i(g^i)$ for every $i \in A$. Event $\beta \in Evt$ is *enabled by the vector of choices* $\overrightarrow{E}_A$ *at* $g \in St$ iff, for every $i \in Agent(\beta) \cap A$, we have $\beta \in E_i$, and for every $i \in Agent(\beta) \setminus A$, it holds that $\beta \in \bigcup R_i(g^i)$. That is, the "owners" of β in A have selected choices that admit β, while all the other "owners" of β *might* select choices that do the same. We denote the set of such events by *enabled*$(g, \overrightarrow{E}_A)$. Clearly, *enabled*$(g, \overrightarrow{E}_A) \subseteq$ *enabled*(g).

Some combinations of choices enable no events. To account for this, the models of AMAS are augmented with "silent" ϵ-loops, added when no "real" event can occur.

Definition 2.5 [Undeadlocked IIS] Let S be an AMAS, and assume that no agent in S has ϵ in its alphabet of events. The *undeadlocked model of S*, denoted $IIS^{\epsilon}(S, I)$, extends the model $IIS(S, I)$ as follows:

- $Evt_{IIS^{\epsilon}(S,I)} = Evt_{IIS(S,I)} \cup \{\epsilon\}$, where $Agent(\epsilon) = \emptyset$;
- For each $g \in St$, we add the transition $g \xrightarrow{\epsilon} g$ iff there is a selection of some agents' choices $\overrightarrow{E}_A = (E_{a_1}, \ldots, E_{a_k})$, $E_i \in R_i(g^i)$, such that $enabled_{IIS(S,I)}(g, \overrightarrow{E}_A) = \emptyset$. Then, we also fix $enabled_{IIS^{\epsilon}(S,I)}(g, \overrightarrow{E}_A) = \{\epsilon\}$.

We use the term *model* to refer to subgraphs of $IIS(S, I)$ as well as $IIS^{\epsilon}(S, I)$.

Example 2.6 The undeadlocked model of ASV_1^2 is shown in Figure 2. Note that it contains no ϵ-transitions, since no choices of the voter and the coercer can cause a deadlock.

2.2 Reasoning About Strategies and Knowledge

Let PV be a set of propositions and $\mathcal{A}$ the set of all agents. The syntax of *alternating-time logic* **ATL*** [8,63] is given by:

$$\varphi ::= \mathsf{p} \mid \neg\varphi \mid \varphi \wedge \varphi \mid \langle\langle A \rangle\rangle\gamma, \qquad \gamma ::= \varphi \mid \neg\gamma \mid \gamma \wedge \gamma \mid \mathrm{X}\,\gamma \mid \gamma\,\mathrm{U}\,\gamma,$$

where $\mathsf{p} \in PV$, $A \subseteq \mathcal{A}$, X stands for "next", U for "until", and $\langle\langle A \rangle\rangle\gamma$ for "agent coalition A has a strategy to enforce γ". Temporal operators F ("eventually") and G ("always"), Boolean connectives, and constants are defined as usual.

[1] g^i denotes agent i's state in $g = (l_1, \ldots, l_n)$, i.e., $g^i = l_i$.

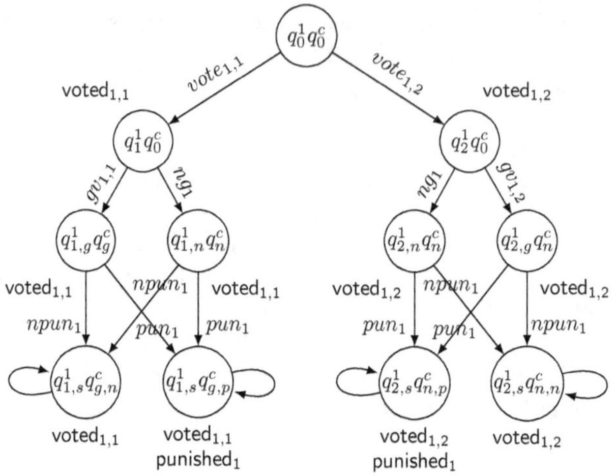

Fig. 2. The undeadlocked model $IIS^{\epsilon}(ASV_1^2)$

Example 2.7 *Coercion feasibility* against voter i can be expressed by formula $\langle\langle Coercer \rangle\rangle \mathrm{F}\,\mathsf{punished}_i$ (the coercer can ensure that the voter is eventually punished).

Strategic ability of agents. Following [40], a *positional imperfect information strategy (*ir*-strategy) for agent i* is defined by a function $\sigma_i\colon L_i \to 2^{Evt_i}$, such that $\sigma_i(l) \in R_i(l)$ for each $l \in L_i$. Note that σ_i is uniform by construction, as it is based on local, and not global states. The set of such strategies is denoted by Σ_i^{ir}. Joint strategies Σ_A^{ir} for $A = \{a_1, \ldots, a_k\} \subseteq \mathcal{A}$ are defined as usual, i.e., as tuples of strategies σ_i, one for each agent $i \in A$. By $\sigma_A(g) = (\sigma_{a_1}(g), \ldots, \sigma_{a_k}(g))$, we denote the joint choice of coalition A at global state g. An infinite sequence of global states and events $\pi = g_0\alpha_0 g_1 \alpha_1 g_2 \ldots$ is called a *path* if $g_j \xrightarrow{\alpha_j} g_{j+1}$ for every $j \geq 0$. The set of all paths in model M starting at state g is denoted by $\Pi_M(g)$.

Definition 2.8 [Standard outcome] Let $A \subseteq \mathcal{A}$. The *standard outcome* of strategy $\sigma_A \in \Sigma_A^{\mathrm{ir}}$ in state g of model M is the set $out_M^{\mathrm{Std}}(g, \sigma_A) \subseteq \Pi_M(g)$ such that $\pi = g_0\alpha_0 g_1 \alpha_1 \cdots \in out_M(g, \sigma_A)$ iff $g_0 = g$, and for each $m \geq 0$ we have that $\alpha_m \in enabled_M(g_m, \sigma_A(g_m))$.

Definition 2.9 [Reactive outcome] The *reactive outcome* is the set $out_M^{\mathrm{React}}(g, \sigma_A) \subseteq out_M^{\mathrm{Std}}(g, \sigma_A)$ such that $\pi = g_0\alpha_0 g_1 \alpha_1 \cdots \in out_M^{\mathrm{React}}(g, \sigma_A)$ iff $\alpha_m = \epsilon$ implies $enabled_M(g_m, \sigma_A(g_m)) = \{\epsilon\}$.

Intuitively, the standard outcome collects all the paths where agents in A follow σ_A, while the others freely choose from their repertoires. The reactive outcome includes only those outcome paths where the opponents cannot miscoordinate on shared events. Let $x \in \{\mathrm{Std}, \mathrm{React}\}$. The ir-semantics of $\langle\langle A \rangle\rangle \gamma$ in asynchronous MAS [8,63,40] is defined by the clause:

$M, g \models^x \langle\langle A \rangle\rangle \gamma$ iff there is a strategy $\sigma_A \in \Sigma_A^{\text{ir}}$ such that for all $\pi \in out_M^x(g, \sigma_A)$ we have $M, \pi \models^x \gamma$.

Adding knowledge operators. The following relations capture the notion of indistinguishability between states, needed to define semantics for the epistemic modality.

Definition 2.10 [Indistinguishable states] For each $i \in \mathcal{A}$, the relation $\sim_i = \{(g, g') \in St \times St \mid g^i = g'^i\}$ denotes that states g, g' are *indistinguishable* for agent i. The relation $\sim_J = \bigcap_{j \in J} \sim_j$ extends it to the distributed knowledge of a group of agents $J \subseteq \mathcal{A}$.

By **ATL*K**, we denote the extension of **ATL*** with knowledge operators K_i where $\mathrm{K}_i\psi$ means "agent i knows that ψ". Note that temporal and strategic operators cannot be nested inside ψ. The semantics of $\mathrm{K}_i\psi$ can be defined by the clause [26,68]:

$M, g \models \mathrm{K}_i\psi$ iff $M, g' \models \psi$ for every g' such that $g' \sim_i g$.

In e-voting, epistemic properties arise due to information exchanged by cryptographic protocols, but also published in plaintext, e.g., on the Web Bulletin Board.

Subjective ability. The asynchronous semantics of $\langle\langle A \rangle\rangle \gamma$ in [40] is based on the notion of "objective" ability, i.e., it suffices that there exists a strategy σ_A which enforces γ on outcome paths from the objective starting point g of the model. The more popular notion of "subjective" ability requires that σ_A succeeds on all outcome paths from the states that A might consider as possible starting points, cf. [14] for an in-depth discussion. The "subjective" semantics of strategic operators can be defined as:

$M, g \models_S^x \langle\langle A \rangle\rangle \gamma$ iff there is a strategy $\sigma_A \in \Sigma_A^{\text{ir}}$ such that, for each $\pi \in \bigcup_{i \in A} \bigcup_{g' \sim_i g} out_M^x(g', \sigma_A)$, we have $M, \pi \models_S^x \gamma$.

Example 2.11 Let $M = IIS^{\epsilon}(ASV_n^k, \{g_0\})$ where $g_0 = (q_0^1, \dots, q_0^n, q_0^c)$. That is, M is the undeadlocked model of the AMAS in Example 2.2 with g_0 as its sole initial state. Note that the Std and React semantics coincide on M, as it includes no ϵ-transitions. Clearly, $M, g_0 \models \langle\langle \mathit{Coercer} \rangle\rangle \mathrm{F}\, \mathsf{punished}_i$ for both the objective and subjective semantics. On the other hand, the stronger requirement $\langle\langle \mathit{Coercer} \rangle\rangle \mathrm{F}\, K_i \mathsf{punished}_i$ does not hold in (M, g_0).

In this paper, we focus on the (more popular) subjective semantics. Moreover, we only use formulas with no next step operators X and no nested strategic modalities, which is essential for the application of partial-order reduction. The corresponding "simple" subset of **ATL*** (resp. **ATL*K**) is denoted by **sATL*** (resp. **sATL*K**). The restriction is less prohibitive than it seems at a glance. First, the X operator is of little value for asynchronous systems. Secondly, nested strategic modalities would only allow us to express an agent's ability to endow another agent with ability (or deprive the other agent of ability). Such properties are sometimes interesting, e.g., one may want to require that $\langle\langle \mathit{Voter}_1 \rangle\rangle \mathrm{G} \neg \langle\langle \mathit{Coercer} \rangle\rangle \mathrm{F}\, \mathsf{punished}_1$ (the voter can keep the coercer unable to punish the voter). Still, simpler properties like $\langle\langle \mathit{Voter_1} \rangle\rangle \mathrm{G}\,\neg\mathsf{punished}_1$ and $\langle\langle \mathit{Voter_1} \rangle\rangle \mathrm{G} \bigwedge_{j=1,\dots,k} \neg K_{\mathit{Coercer}} \mathsf{voted}_{1,j}$ are usually of more immediate interest.

Finally, we remark that, for subjective ability, epistemic operators are definable with strategic operators, since $K_i\varphi \equiv \langle\langle i \rangle\rangle \bot \,\mathrm{U}\, \varphi$. Moreover, $M, g \models \langle\langle i \rangle\rangle\gamma$ always implies $M, g \models K_i\langle\langle i \rangle\rangle\gamma$ in the subjective semantics.

3 How to Specify Voters and Coercers

To keep the paper self-contained, in this section we provide a short description of the SELENE voting protocol and its formal specification.

3.1 Short Description of SELENE

SELENE [61] is an electronic voting protocol aimed to provide an effective mechanism for *voter verifiability* and *coercion resistance*. On the one hand, the voter receives a piece of evidence that allows to check if her vote has been registered correctly. On the other hand, she can present a fake vote evidence to the coercer, thus convincing him that she voted according to the coercer's request.

The protocol proceeds as follows. Before the election, the Election Authority (EA) sets up the system, generating the election keys used for the encryption and decryption of the votes and preparing the vote trackers, one per voter. The trackers are then encrypted and mixed to break any link between the voter and her tracker, and published on the Web Bulletin Board (WBB).

In the voting phase, each voter fills in, encrypts, and signs her vote, and sends it to the system. After several intermediate steps, a pair $(\mathit{Vote}_v, \mathit{tr}_v)$, consisting of the decrypted ballot and the tracker of v, is published on the WBB for each $v \in \mathit{Voters}$. At this stage, no voters know their trackers. All the votes are presented in plaintext on the WBB. Thus, the tally of the election is open to a public audit.

At the final stage, the voters receive their trackers by an independent channel (e.g., sms). If the voter has not been coerced, then she requests the special term α_v, which allows for obtaining the correct tracker tr_v. If she was coerced to fill her ballot in a certain way, she sends a description of the requested vote to the election server. After such a request, a fake term α'_v is sent to the voter, which can be presented to the coercer. The α'_v token, together with the public commitment of the voter, reveals a tracker pointing out to a vote compatible with the coercer's demand.

SELENE uses the ElGamal key encryption scheme, and relies on multiplicative homomorphism and non-interactive zero-knowledge proofs of knowledge that accompany all the transformations of data presented in WBB. In this work we abstract away from the cryptography and focus on the interaction between the involved agents.

3.2 Asynchronous MAS for SELENE

In our model of SELENE, we concentrate on the non-cryptographic interaction between the agents. In this sense, the present model is similar to [36], with three important differences.

First, the new modeling is fully modular and scalable. To this end, we use AMAS as input (rather than concurrent game structures or concurrent interpreted systems). Secondly, we have defined a flexible specification language for AMAS, based on asynchronous agent templates similar to those of UPPAAL [9], and implemented an interpreter for it. Besides modularity and scalability, this allowed us to adapt and use partial order reduction for our models. Thirdly, our specifications include much more details of the SELENE procedure than the skeletal model in [36].

3.3 Agents

There are 4 templates for agents in the modeling: the Election Authority (EA) that handles the generation and distribution of trackers and provides the Web Bulletin Board for voters and coercers, the Coercer, the standard Voter, and the Coerced Voter interacting with a coercer. The templates are written in a simple specification language created to provide the input for our algorithms. Each template consists of the name of the agent, the number of instances of that template in the model (e.g., the number of voters), the initial state, the list of transitions, and the agent's repertoire of choices (called "PROTOCOL" here). The Coerced Voter and Coercer templates for an election with 2 candidates are shown in Figures 3 and 4. The full specification is available at github.com/blackbat13/stv/blob/master/models/Selene.txt.

A template always begins with the keyword *Agent*, the agent's name, and the number of instances. The next line specifies the initial state, followed by the list of local transitions. Each transition starts with an event name, optionally preceded by the keyword *shared* if the event is shared with another agent. Then, the source and the target states are given. Optionally, the transition specification can also include a precondition (in the form of a simple Boolean formula) and/or a postcondition (via a list of updates specifying propositions and their new values). These conditions are only a technical shortcut that allows us to write clearer and shorter specifications. The keyword *aID* represents the ID of the current agent and is automatically replaced when preparing local models of agents. For example, the template VoterC[2] would produce two agents, $Voter_{C1}$ and $Voter_{C2}$. Then, each instance of the keyword *aID* will be replaced with $Voter_{C1}$ for the first coerced voter, and with $Voter_{C2}$ for the second one.

Consider the template in Figure 3. The first step of the coerced voter is her interaction with the coercer who can request the vote for a particular candidate (events *coerce1_VoterC1* and *coerce2_VoterC1*). The next step is the event *start_voting* synchronized with the *EA*. When the election has begun, the voter can create her commitment, fill in the vote, encrypt it, and send it to the EA. After that, she waits for the publication of votes, which is controlled by the EA. When the votes are published on the WBB, the voter can decide to compute the false alpha term and the false tracker using publicly available data, or she can just wait for her real tracker. The last few steps for the VoterC agent consist of checking the WBB, verifying the vote, and interacting with the

```
Agent VoterC[1]:
init start
shared coerce1_aID: start -> coerced [aID_required=1]
shared coerce2_aID: start -> coerced [aID_required=2]
select_vote1: coerced -> prepared [aID_vote=1, aID_prep_vote=1]
select_vote2: coerced -> prepared [aID_vote=2, aID_prep_vote=2]
shared is_ready: prepared -> ready
shared start_voting: ready -> voting
shared aID_vote: voting -> vote [Coercer1_aID_vote=?aID_vote, Coercer1_aID_revote=?aID_revote]
shared send_vote_aID: vote -> send
revote_vote_1: send -[aID_revote==1]> voting [aID_vote=?aID_required, aID_revote=2]
skip_revote_1: send -[aID_revote==1]> votingf
revote_vote_2: send -[aID_revote==2]> voting [aID_vote=?aID_required, aID_revote=3]
skip_revote_2: send -[aID_revote==2]> votingf
final_vote: send -[aID_revote==3]> votingf [aID_vote=?aID_prep_vote]
skip_final: send -[aID_revote==3]> votingf
shared send_fvote_aID: votingf -> sendf
shared finish_voting: sendf -> finish
shared send_tracker_aID: finish -> tracker
shared finish_sending_trackers: tracker -> trackers_sent
shared give1_aID: trackers_sent -> interact [Coercer1_aID_tracker=1]
shared give2_aID: trackers_sent -> interact [Coercer1_aID_tracker=2]
shared not_give_aID: trackers_sent -> interact [Coercer1_aID_tracker=0]
shared punish_aID: interact -> ckeck [aID_punish=true]
shared not_punish_aID: interact -> check [aID_punish=false]
shared check_tracker1_aID: check -> end
shared check_tracker2_aID: check -> end
PROTOCOL: [[coerce1_aID, coerce2_aID], [punish, not_punish]]
```

Fig. 3. Voter template

```
Agent Coercer[1]:
init coerce
shared coerce1_VoterC1: coerce -> coerce [aID_VoterC1_required=1]
shared coerce2_VoterC1: coerce -> coerce [aID_VoterC1_required=2]
shared start_voting: coerce -> voting
shared VoterC1_vote: voting -> voting
shared finish_voting: voting -> finish
shared finish_sending_trackers: finish -> trackers_sent
shared give1_VoterC1: trackers_sent -> trackers_sent
shared give2_VoterC1: trackers_sent -> trackers_sent
shared not_give_VoterC1: trackers_sent -> trackers_sent
to_check: trackers_sent -> check
shared check_tracker1_Coercer1: check -> check
shared check_tracker2_Coercer1: check -> check
to_interact: check -> interact
shared punish_VoterC1: interact -> interact
shared not_punish_VoterC1: interact -> interact
finish: interact -> end [aID_finish=1]
PROTOCOL: [[give1_VoterC1, give2_VoterC1, not_give_VoterC1]]
```

Fig. 4. Coercer template

coercer again. The voter can show one of her trackers (the false one or the real one) to the coercer, who then either punishes her or does not.

Additionally, the voter can do *revoting*, i.e., cast her vote multiple times, which is a well-known technique to counter in-house coercion by family members.

3.4 Specification of Properties

To specify interesting properties of the voting system, we use simple formulas of **ATL*K**. In the experiments, we will concentrate on the property of *coercion-vulnerability*, using the following formula:

$$\varphi_{vuln,i,k} \equiv \langle\!\langle \mathit{Coercer} \rangle\!\rangle \mathrm{G}\, ((\mathsf{end} \wedge \mathsf{revote}_{\mathsf{v1}} = \mathsf{k} \wedge \mathsf{voted}_{\mathsf{v1}} = \mathsf{i}) \rightarrow \mathrm{K}_{Coercer}\mathsf{voted}_{\mathsf{v1}} = \mathsf{i})$$

Formula $\varphi_{vuln,i}$ says that the coercer has a strategy so that, at the end of the election, if the voter has effectively voted for candidate i, the coercer knows about it. This can be seen as the opposite of *receipt-freeness* and *coercion-*

resistance formalizations in [43,65]. Note that the formula is parameterized by the name i of the preferred candidate of the coercer, as well as the number k of revoting rounds that the coercer is able to observe and learn the value of the cast vote. Proposition revote_{v1} corresponds to the number of revoting rounds performed by the first voter.

We will use the above models and formulas in our verification experiments in Section 6.

3.5 Model definition

Model file consists of definitions of agents templates and various properties. Empty lines and Lines starting with % are comments lines and are ignored by the parser. Properties that can be defined in the model file are described below. Each proposition starts with uppercase property name followed by a colon.

PERSISTENT List of propositions names that should be considered as persistent when generating the model. Persistent proposition retains its value after it was set. Non-persistent propositions are present only in the state that is the result of the transition that created this proposition.

REDUCTION List of propositions names used for the partial-order reduction method.

FORMULA ATL formula to be verified.

SHOW_EPISTEMIC Boolean value used only in the graphical interface. If set to *true* links of the epistemic relations will be displayed in the model view.

3.6 Agent definition

Agent definition consists of a set of non-empty lines specifying the local model of the agent. Each agent definition is a template that can be used to generate multiple agents of a given type. Individual definitions are explained below.

Header The first line should specify the name of the agent template and number of agents of this type, given in the following format: `Agent AgentName[count]:`. For example, in case of `Agent Voter[3]:` three agents will be generated: *Voter1*, *Voter2* and *Voter3*.

init First line of the agent specification, name of the initial state.

Local transition Transition is defined in format:
`actionName: state1 -[preCondition]> state2 [propositions]`.
Propositions are given in a form of a comma-separated list of variable definitions, i.e. `[prop1=true, prop2=false, prop3=2]`. Precondition can be any boolean formula that can be evaluated in Python.

Shared transition Shared transitions are defined similarly to local transitions, but are prefixed with *shared* keyword.

Dynamic names In order to simplify the specification dynamic names can be used. Every occurence of the keyword *aID* in the agent template specification will be replaced with the name of the computed agent.

4 Verification Algorithms

Model checking strategic ability under imperfect information is hard, both theoretically and in practice. In this section, we present two recently developed techniques that we use to tackle the high complexity of verification: fixpoint approximation and dominance-based depth-first strategy search (Sections 4.1 and 4.3). We also propose a novel approach based on distributed strategy synthesis for **sATL*K** (Section 4.4).

4.1 Verification by Fixpoint Approximation

The main idea of the fixpoint approximations method presented in [37] is that sometimes, instead of the exact model-checking, it suffices to provide a lower and an upper bound for the output. In particular, given a formula ϕ, we construct two translations $tr_L(\phi)$ and $tr_U(\phi)$, such that $tr_L(\phi) \Rightarrow \phi \Rightarrow tr_U(\phi)$. In other words, if the lower bound translation $tr_L(\phi)$ evaluates to true, then the original formula ϕ must hold in the model. Similarly, if the upper bound translation $tr_U(\phi)$ is verified as false, then ϕ is also false.

The approximations are built on fixpoint-definable properties. For the upper bound translation we just compute the given formula under the perfect information assumption. For the lower bound we rely on translations that map the formula of $\mathbf{ATL}^*_{\mathbf{ir}}$ to an appropriate variant of alternating μ-calculus, see [37] for more details.

4.2 Depth-First Strategy Search

Further, we have implemented a depth-first search algorithm for strategy synthesis. The typical recursive DFS-based approach needed to be adapted due to the presence of epistemic classes and the nondeterministic outcome of the coalition's actions. In case of nondeterminism due to multiple possible transitions, it is possible that decisions taken in one of the branches may determine some choices in the other branches. If such a decision leads to a locally winning strategy, it may need to be changed since decisions in nodes being members of the same epistemic class can influence the outcome of transitions in other branches. The proposed algorithm allows to backtrack and change locally winning strategy when no winning strategy is found in another branch.

Clearly, even for relatively small models with hundreds of states, the space of strategies is too big to perform a full search. To address this, we used our DFS algorithm as the backbone for implementing more refined methods, namely **Domino DFS** and two variants of parallel model checking, see Sections 4.3 and 4.4.

4.3 Domination-Based Strategy Search

The strategy space grows exponentially with respect to the number of states and transitions in the model (on top of the state space explosion). However, in reality it often suffices to only check a subset of possible strategies. This idea was used in the **Domino DFS** algorithm [51]. In that method, a notion of strategy dominance was proposed, according to which the dominated strategies are omitted during the search. Although in the worst case scenario all the strategies in the model need to be checked, the experimental results in [51] showed that in some cases the method achieves significant improvement in performance.

4.4 Parallel Implementation

The verification process itself has exponential nature with complexity of $O(k^n)$ for a single agent (where k is the number of possible actions and n is the number of states in the model). To reduce the search time, parallel computation can be used. As a basis and a reference, we used a sequential version of the recursive DFS-based strategy search. The biggest advantage of the recursive approach is its ability to perform multiple, independent searches at a time. Such a solution may be very effective for distributed computing architectures, when the search space can be divided independently and distributed between computing nodes with separate memory spaces. The downside is that all the nodes obtain copies of the same model, but then they all may compute independently looking for different solutions.

We have defined and implemented two different approaches to parallelizing depth-first strategy search.

Distributed Strategy Search: Simple Branching. The simple approach tries to concurrently execute a number of instances of sequential search, but assuming different strategy prefixes (i.e., such sequences of actions starting at a starting state, which correspond to deterministic paths in a graph – except for the last one, where there may be a branch). To achieve this, the algorithm first executes a breadth-first search to determine a set of potential different strategy prefixes up to a number equal to a value given as parameter. Due to the way in which the prefixes are selected, they correspond to single paths and they can be identified with single states. The algorithm tries to expand the prefixes by performing a BFS graph traversal.

Similarly to most BFS-based algorithms, a queue of pending prefixes is used. Initially, this queue contains an empty prefix denoting only one process to be executed. In each step of the algorithm, the first pending prefix is selected. The algorithm tries to expand this prefix by adding actions available in a state to which this prefix leads. For every deterministic action in such state, a new prefix is generated by appending this action to the current prefix. So obtained new prefix is added to the pending queue for further expansion. If the action is nondeterministic, the prefix obtained by appending this action to the current prefix is added to the set of resulting prefixes and is not expanded any more. The current prefix is then discarded. The loop stops when either there are no more pending prefixes or the number of prefixes in both the result set and the pending queue exceeds the given parameter. In the latter case, all the pending prefixes are copied to the resulting set.

After a set of prefixes is determined, the main process spawns a number of children processes. Each subprocess tries to find a winning strategy using the sequential algorithm, assuming that the strategy starts with actions from its assigned prefix. The main process waits for the results from its children; if any of them reports that a winning solution has been found, all the other children are terminated and the search ends.

Distributed Strategy Search: Flexible Version. Our second approach uses concurrent execution to examine parallel branches of a single strategy in a flexible way. To this end, it directly expands the single–threaded DFS method. In this approach, parallel threads cooperate in building the same subgraph of the model that corresponds to the outcome of the candidate strategy. The initial master thread spawns new worker threads whenever it finds a state in which selection of the same action leads to multiple states. A sequential algorithm in such case would try to recursively build a substrategy considering all the possible targets one by one. In the concurrent version, an additional worker thread is created for each possible outcome of the same action.

In order to ease the synchronization between threads, only the main thread is allowed to spawn new threads due to parallel branches. It is also assumed that only one parallel thread is allowed to select actions for states from any epistemic class. If any worker thread reaches a state in such a class, it must wait until the master thread fixes the action for this class. Synchronization is also needed when one thread meets a node which is already examined by another thread – since the nodes cooperate, it is not necessary to redo the same work, therefore the second node is suspended until the node is checked by the first one. The main thread is allowed to intercept the work already performed by a worker thread.

5 Taming State Space Explosion

The main obstacle in model checking of MAS logics is the prohibitive complexity of models, in particular due to the state-space explosion. Partial order reduction (POR) is a well-known technique for state space reduction, dating back more than three decades [30,57,66]. The idea is to restrict the set of all enabled transitions to a representative (i.e. provably sufficient) subset, based on some underlying notion of equivalence. Crucially, this occurs while generating the unfolding of the system, so the full model, which may be far too large, is never created.

POR has been defined for variants of **LTL** and **CTL**, including temporal-epistemic logics [53]. Furthermore, the reduction for **LTL** was recently adapted (notably, at the same computational cost) to the much more expressive logic **sATL*** [39,41], allowing to leverage the existing algorithms and tools for the verification of strategic abilities. In [40], POR for **sATL*** was shown to still work under an improved execution semantics for AMAS, which we also adopt in this paper. Note, however, that our formalization of coercion-vulnerability in Section 3.4 is a *strategic-epistemic property*, and hence those results do not cover this case. Moreover, the reductions in [39,40,41] are *not* correct if we allow for nested modalities. The question is: can we adapt the scheme so that it works if we only allow to nest *epistemic* operators?

In this section, we prove that the answer is affirmative. We also show that the reduction is correct for the subjective semantics of ability, whereas the previous works used the less intuitive (and less popular) objective semantics.

5.1 Conceptual Machinery

We first recall the concept of stuttering equivalence. Intuitively, two paths are stuttering equivalent if they can be divided into corresponding finite segments, each satisfying exactly the same propositions. If all states in corresponding segments are also indistinguishable for agents $i \in J$, then we say the paths are $J-$ stuttering equivalent.

Definition 5.1 [(J-)stuttering equivalence] Paths $\pi, \pi' \in \Pi_M(g)$ are *stuttering equivalent*, denoted $\pi \equiv_s \pi'$, if there exists a partition $B_0 = (\pi[0], \ldots, \pi[i_1 - 1])$, $B_1 = (\pi[i_1], \ldots, \pi[i_2 - 1])$, ... of the states of π, and an analogous partition $B'_0, B'_1, \ldots$ of the states of π', such that for each $j \geq 0 : B_j$ and B'_j are nonempty and finite, and $V(g) \cap PV = V(g') \cap PV$ for every $g \in B_j$ and $g' \in B'_j$.

If $\pi \equiv_s \pi'$, and additionally it holds that $\forall j > 0 \quad \forall g \in B_j, g' \in B'_j : g \sim_J g'$, then paths π and π' are *J-stuttering equivalent*, denoted $\pi \equiv_s^J \pi'$.

States g and g' are *stuttering path equivalent* (resp. *J-stuttering path equivalent*), denoted $g \equiv_s g'$ (resp. $g \equiv_s^J g'$), iff for every path π starting from g, there is a path π' starting from g' such that $\pi' \equiv_s \pi$ (resp. $\pi' \equiv_s^J \pi$), and for every path π' starting from g', there is a path π starting from g such that $\pi \equiv_s \pi'$ (resp. $\pi \equiv_s^J \pi'$).

Models M and $M' \subseteq M$ are *stuttering path equivalent* (resp. *J-stuttering path equivalent*), denoted $M \equiv_s M'$ (resp. $M \equiv_s^J M'$), iff they have the same initial states, and for each initial state $\iota_i \in I$ and each path $\pi \in \Pi_M(\iota_i)$, there is a path $\pi' \in \Pi_{M'}(\iota_i)$ such that $\pi \equiv_s \pi'$ (resp. $\pi \equiv_s^J \pi'$).

The POR algorithm uses the notions of invisible and independent events. Intuitively, an event is invisible iff it does not change the valuations of the propositions.[2] Two events are independent iff at least one is invisible and they are not in the same agent's repertoire. We designate a subset of agents $A \subseteq \mathcal{A}$ whose events are visible by definition.

Definition 5.2 [Invisibility and independence of events] Let $M = IIS^{\epsilon}(S, I)$, and $A \subseteq \mathcal{A}$. An event $\alpha \in Evt$ is *invisible* wrt. A and PV if $Agent(\alpha) \cap A = \emptyset$ and for each two global states $g, g' \in St$ we have that $g \xrightarrow{\alpha} g'$ implies $V(g) \cap PV = V(g') \cap PV$. The set of all invisible events for A, PV is denoted by $Invis_{A,PV}$, and its closure, i.e. the set of visible events, by $Vis_{A,PV} = Evt \setminus Invis_{A,PV}$.

The notion of *independence* $Ind_{A,PV} \subseteq Evt \times Evt$ is defined as: $Ind_{A,PV} = \{(\alpha, \alpha') \in Evt \times Evt \mid Agent(\alpha) \cap Agent(\alpha') = \emptyset\} \setminus (Vis_{A,PV} \times Vis_{A,PV})$. Events $\alpha, \alpha' \in Evt$ are called *dependent* if $(\alpha, \alpha') \notin Ind_{A,PV}$. If it is clear from the context, we omit the subscript PV.[3] Note that ϵ events are always

[2] This technical concept of invisibility is not connected to any agent's view, unlike in [55].

[3] The sets of agents' local propositions PV_i are explicitly disjoint in our model (cf. Definition

invisible and independent from others, since they do not modify propositions and we have that $Agent(\epsilon) = \emptyset$ (by Definition 2.5).

The reduced model (or *submodel*) $M' \subseteq M$ obtained with POR extends the same AMAS S as $M = IIS^{\epsilon}(S, I)$. In particular, we have $St' \subseteq St$, $I = I'$, T is an extension of T', and $V' = V|_{St'}$. Note that, for each $g \in St'$, it holds that $\Pi_{M'}(g) \subseteq \Pi_M(g)$.

M' is generated by modifying the standard DFS [28], so that for each g, the successor state g_1 such that $g \stackrel{\alpha}{\to} g_1$ is selected from $E(g) \cup \{\epsilon\}$ such that $E(g) \subseteq enabled(g) \setminus \{\epsilon\}$. That is, the algorithm always selects ϵ, plus a subset of the enabled events at g. This modified DFS is called for each initial state of the model, and we have $\Pi_{M'} = \bigcup_{g \in I} \Pi_{M'}(g)$. The conditions on the heuristic selection of $E(g)$ given below are inspired by [18,41,58].

C1 Along each path π in M that starts at g, each event that is dependent on an event in $E(g)$ cannot be executed in π without an event in $E(g)$ being executed first in π. Formally, $\forall \pi \in \Pi_M(g)$ such that $\pi = g_0 \alpha_0 g_1 \alpha_1 \ldots$ with $g_0 = g$, and $\forall b \in Evt$ such that $(b, c) \notin Ind_A$ for some $c \in E(g)$, if $\alpha_i = b$ for some $i \geq 0$, then $\alpha_j \in E(g)$ for some $j < i$.

C2 If $E(g) \neq enabled(g) \setminus \{\epsilon\}$, then $E(g) \subseteq Invis_A$.

C3 For every cycle in M' containing no ϵ-transitions, there is at least one node g in the cycle for which $E(g) = enabled(g) \setminus \{\epsilon\}$, i.e., all the successors of g are expanded.

Submodel $M' \subseteq M$ generated with this algorithm satisfies property $\mathbf{AE}_A$ [40]: $\forall \sigma_A \in \Sigma_A^{\mathrm{ir}} \quad \forall \iota_i \in I \quad \forall \pi \in out_M^x(\iota_i, \sigma_A) \quad \exists \pi' \in out_{M'}^x(\iota_i, \sigma_A) : \quad \pi \equiv_s \pi'$, where $x \in \{\mathrm{Std}, \mathrm{React}\}$.

5.2 POR for sATL*K

We will show that the reduction algorithm for **sATL*** [41,40] can be applied also to formulas that include the knowledge operator (in subformulas of the form $\mathrm{K}_i\varphi$), provided that $J \subseteq A$. That is, any agents in these epistemic subformulas are added to the set $A \subseteq \mathcal{A}$ that parametrises the relations of invisibility and independence.

Theorem 5.3 *Let S be an AMAS, $J \subseteq A \subseteq \mathcal{A}$, $M = IIS^{\epsilon}(S, I)$, and let $M' \subseteq M$ be the reduced model generated by DFS with the choice of $E(g')$ for $g' \in St'$ given by conditions* **C1-C3**. *Then, for any starting state $\iota_i \in I$ and any* **sATL*K** *formula φ over PV that refers only to coalitions $\hat{A} \subseteq A$, we have that $M, \iota_i \models \varphi$ iff $M', \iota_i \models \varphi$.*

Proof. First, note that conditions **C1-C3** remain unchanged from the reduction algorithm for **sATL*** [40]. Thus, by [40, Theorems A.8 and A.9],[4] we

2.1), allowing for simpler checking of event independence in the actual implementation of POR.

[4] The conference paper [40] does not include a technical appendix with proofs of Theorems A.8 and A.9, so we refer to its extended arXiv manuscript here.

have that:

(*) M and M' are stuttering path equivalent. For each path $\pi = g_0\alpha_0 g_1\alpha_1\ldots$ with $g_0 = \iota_i$ in M, there is a stuttering equivalent path $\pi' = g'_0\alpha'_0 g'_1\alpha'_1\ldots$ with $g'_0 = \iota_i$ in M' such that $Evt(\pi)|_{Vis_A} = Evt(\pi')|_{Vis_A}$, i.e., π and π' have the same maximal sequence of visible events for A.

(**) M and M' satisfy structural condition $\mathbf{AE}_A$.

That is, we have $M, \iota_i \models \varphi$ iff $M', \iota_i \models \varphi$ for all non-epistemic φ. To extend the reasoning to any **sATL*K** formula, we first show that the full and reduced model are also J-stuttering equivalent, which then allows to prove that epistemic subformulas are preserved in the reduced model M'. Finally, we show that these subformulas can be replaced with equivalent new propositions, effectively reducing the problem to the previously proven case for **sATL***.

Because $J \subseteq A$ (and so all transitions of the agents in group J are visible), it follows directly from **C2** that if $E(g) \neq enabled(g)\setminus\{\epsilon\}$, then $Agent(\alpha)\cap J = \emptyset$ for any event $\alpha \in E(g)$. This is a direct analogue of the extra condition **CJ** from [53]. Together with (*), this implies that the full and reduced model are also J-stuttering equivalent:

(***) $M \equiv_s^J M'$.

Consider any subformula $\varphi = \mathrm{K}_i\psi$. As per the syntax of **sATL*K**, temporal operators and strategic modalities cannot be nested inside K_i, so φ is a purely epistemic formula that only contains knowledge operator(s) and propositional variables with Boolean connectives. Now, we will show it follows from (***) that epistemic subformulas are preserved in the reduced model, i.e., for any state g such that $g \equiv_s^J g'$, we have $M, g \models \varphi$ iff $M', g' \models \varphi$:

($\Rightarrow$) Assume that $M, g \models \mathrm{K}_i\psi$. Let $St_\psi = \{g_\psi \in St \mid g \sim_i g_\psi\}$, and take g'_ψ such that $g' \sim_i g'_\psi$. We need to show that $M', g'_\psi \models \psi$. From $g \equiv_s^J g'$ and by transitivity of relation $\sim_i$, we have that $g'_\psi \in St_\psi$. So, clearly $M, g'_\psi \models \psi$. As $g_\psi \equiv_s^J g'_\psi$, it follows from the inductive assumption that $M', g'_\psi \models \psi$. Hence, $M', g' \models \mathrm{K}_i\psi$.

($\Leftarrow$) Assume that $M', g' \models \mathrm{K}_i\psi$. Let $St'_\psi = \{g'_\psi \in St' \mid g' \sim_i g'_\psi\}$, and take g_ψ such that $g \sim_i g_\psi$. We need to show that $M, g_\psi \models \psi$. Consider a path $\pi \in M$ that contains g_ψ. From (***), there is a path $\pi' \in M'$, which contains a state $g''_\psi \in St'$, such that $g_\psi \equiv_s^J g''_\psi$. By transitivity of $\sim_i$, we get that $g''_\psi \in St'_\psi$, and thus $M', g''_\psi \models \psi$. As $g_\psi \equiv_s^J g''_\psi$, it follows from the inductive assumption that $M, g_\psi \models \psi$. Hence, $M, g \models \mathrm{K}_i\psi$.

From the above we get that any epistemic subformula $K_i\psi$ holds in the reduced model M' iff it holds in the corresponding state of the (J-stuttering equivalent) full model M. Now, we introduce auxiliary propositional variables to replace epistemic subformulas, including nested ones.

Consider subformulas $\varphi_0, \varphi_1, \ldots$, where $\varphi_0 = \varphi$, and for all $i > 0$, φ_i is an epistemic subformula nested in φ_{i-1}. Note that in the reduced model M', one can add a set of new propositional variables $PV' = \bigcup_i\{\mathsf{sat}_{\varphi_i}\}$ to PV, and extend the valuation function accordingly, so that we have $V' : St' \to 2^{PV \cup PV'}$, and sat_{φ_i} is true in state $g' \in St'$ iff $M', g' \models \varphi_i$. That is, for each epistemic (sub)formula φ_i, a new proposition sat_{φ_i} is added, whose valuation in each state $g' \in St'$ corresponds to the satisfaction of φ_i in that state of M'. Then, for the formula $\varphi' = \mathsf{sat}_{\varphi_0}$, it clearly holds that $M', g' \models \varphi'$ iff $M', g' \models \varphi$. Furthermore, since epistemic subformulas only refer to agents $i \in J$ and we have that $J \subseteq A$, it follows that $Vis_{A,PV} = Vis_{A,PV \cup PV'}$ and $Ind_{A,PV} = Ind_{A,PV \cup PV'}$. That is, replacing epistemic subformulas with new propositions in this manner does not affect the visibility or independence of events wrt. A and PV. Hence, we also have $M', g' \models \varphi'$ iff $M, g \models \varphi$, from (*) and (**) and by [40, Theorems A.8 and A.9], as φ' is a **sATL*** formula. But from the construction of φ', we have $M', g' \models \varphi'$ iff $M', g' \models \varphi$, so it also holds that $M, g \models \varphi$ iff $M', g' \models \varphi$ for any **sATL*K** formula φ, both for standard outcomes [40, Theorem A.8] and under the opponent reactivity condition [40, Theorem A.9]. Thus, in particular we have that $M, \iota_i \models \varphi$ iff $M', \iota_i \models \varphi$, QED. □

5.3 POR for Subjective Semantics of Ability

Recall the subjective semantics of strategic ability in Section 2.2. We will show that the reduction scheme remains applicable in this setting, i.e., when the set of initial states is comprised of previously designated subset $I \subseteq St$, plus all states indistinguishable from those in I for coalition agents.

Theorem 5.4 *Let* $x \in \{\mathrm{Std}, \mathrm{React}\}$, $\hat{A} \subseteq A$ *and* $I_{\hat{A}} = I \cup \{g \in St \mid \exists_{\iota_i \in I} \exists_{j \in \hat{A}} : g \sim_j \iota_i\}$. *Let* $M = IIS(S, I_{\hat{A}})$, *and let* M' *be the reduction of* M *generated by POR using conditions* ***C1-C3*** *for the choice of ample sets. Then, for any initial state* $\iota_i \in I_{\hat{A}}$ *and any* **sATL*K** *formula* φ *that refers only to coalition* $\hat{A}$, *we have that* $M, \iota_i \models^x_S \varphi$ *iff* $M', \iota_i \models^x_S \varphi$.

Proof. [Proof sketch] For each $\iota_i \in I_{\hat{A}}$, take M_i constructed by DFS starting from ι_i (i.e., with a single initial state), and let M_i' be the reduction of M_i generated by POR.

Take any path $\pi \in \Pi_M$. Clearly, $\pi \in \Pi_{M_i}$ for some $i > 0$. By Theorem 5.3, we have $M_i \equiv^J_s M_i'$, so there is a J-stuttering equivalent path $\pi' \in \Pi_{M_i'}$. From the construction by DFS, $\Pi_{M'} = \bigcup_{i \in I_{\hat{A}}} \Pi_{M_i'}$. Hence, $\pi' \in \Pi_{M'}$, which implies that $M \equiv^J_s M'$. (*)

Take any joint strategy $\sigma_{\hat{A}}$. The subjective outcome of $\sigma_{\hat{A}}$ in M (resp. M') is the sum of objective outcomes of $\sigma_{\hat{A}}$ in M_i (resp. M_i'). But from $\mathbf{AE}_A$, $out^x_{M_i}(\iota_i, \sigma_{\hat{A}}) \equiv_s out^x_{M_i'}(\iota_i, \sigma_{\hat{A}})$. So, analogously to the reasoning for (*), it follows that $\bigcup_{i \in I_{\hat{A}}} out^x_{M_i}(\iota_i, \sigma_{\hat{A}}) \equiv_s \bigcup_{i \in I_{\hat{A}}} out^x_{M_i'}(\iota_i, \sigma_{\hat{A}})$. Hence, M and M' also satisfy $\mathbf{AE}_A$. (**)

Since (*) and (**), the thesis follows from Theorem 5.3, as in the case for objective semantics of strategic ability. □

6 Experiments and Results

In this section, we give a brief summary of the experimental results.

#A	#R	Full Model					Reduced Model					Result
		#st	#tr	Seq.	Par.	Appr.	#st	#tr	Seq.	Par.	Appr.	
4	3	3.63e4	7.46e4	0.003	0.009	1.121	2.60e4	5.99e4	0.001	0.002	0.184	True
4	5	5.62e4	1.15e5	0.004	0.003	0.345	4.01e4	9.26e4	0.002	0.002	0.283	True
4	10	1.06e5	2.18e5	0.009	0.005	0.691	7.55e4	1.74e5	0.004	0.002	0.563	True
5	3	1.55e6	5.91e6	0.158	0.004	14.78	1.09e6	4.65e6	0.112	0.021	12.99	True
6	3	7.61e7	4.98e8	0.524	0.051	41.24	5.34e7	3.82e8	0.427	0.042	37.35	True
7	3	model generation timeout					model generation timeout					-

Table 1
Verification of $\varphi_{vuln,i,k}$ for the first candidate ($i = 1$) and $k = \#R$ revotes

#Ag	#R	Full Model					Reduced Model					Result
		#st	#tr	Seq.	Par.	Appr.	#st	#tr	Seq.	Par.	Appr.	
4	3	3.63e4	7.46e4	0.003	0.010	1.103	2.60e4	5.99e4	0.002	0.003	0.166	True
4	5	5.62e4	1.15e5	0.004	0.005	0.348	4.01e4	9.26e4	0.003	0.003	0.280	True
4	10	1.06e5	2.18e5	0.008	0.009	0.700	7.55e4	1.74e5	0.005	0.004	0.567	True
5	3	1.55e6	5.91e6	0.160	0.055	14.03	1.09e6	4.65e6	0.112	0.053	12.49	True
6	3	7.61e7	4.98e8	0.602	0.083	42.44	5.34e7	3.82e8	0.501	0.057	38.20	True
7	3	model generation timeout					model generation timeout					-

Table 2
Verification of $\varphi_{vuln,i,k}$ for the last candidate ($i = \#C$) and $k = \#R$ revotes

#Ag	#R	Full Model					Reduced Model					Result
		#st	#tr	Seq.	Par.	Appr.	#st	#tr	Seq.	Par.	Appr.	
4	3	3.63e4	7.46e4	0.303	0.317	1.128	2.60e4	5.99e4	0.202	0.205	0.179	False
4	5	5.62e4	1.15e5	0.524	0.592	0.325	4.01e4	9.26e4	0.411	0.503	0.280	False
4	10	1.06e5	2.18e5	0.721	0.668	0.459	7.55e4	1.74e5	0.525	0.512	0.364	False
5	3	1.55e6	5.91e6	2.146	1.257	0.981	1.09e6	4.65e6	1.513	1.003	0.583	False
6	3	7.61e7	4.98e8	5.232	3.228	1.892	5.34e7	3.82e8	4.986	2.427	1.092	False
7	3	model generation timeout					model generation timeout					-

Table 3
Verification of $\varphi_{vuln,i,k}$ for the first candidate ($i = 1$) and $k = \#R - 1$ revotes

#Ag	#R	Full Model					Reduced Model					Result
		#st	#tr	Seq.	Par.	Appr.	#st	#tr	Seq.	Par.	Appr.	
4	3	3.63e4	7.46e4	0.302	0.311	0.180	2.60e4	5.99e4	0.201	0.213	0.126	False
4	5	5.62e4	1.15e5	0.519	0.584	0.310	4.01e4	9.26e4	0.410	0.475	0.283	False
4	10	1.06e5	2.18e5	0.742	0.627	0.462	7.55e4	1.74e5	0.558	0.544	0.370	False
5	3	1.55e6	5.91e6	2.160	1.358	0.942	1.09e6	4.65e6	1.621	1.009	0.519	False
6	3	7.61e7	4.98e8	5.504	3.516	1.903	5.34e7	3.82e8	5.110	2.380	1.112	False
7	3	model generation timeout					model generation timeout					-

Table 4
Verification of $\varphi_{vuln,i,k}$ for the last candidate ($i = \#C$) and $k = \#R - 1$ revotes

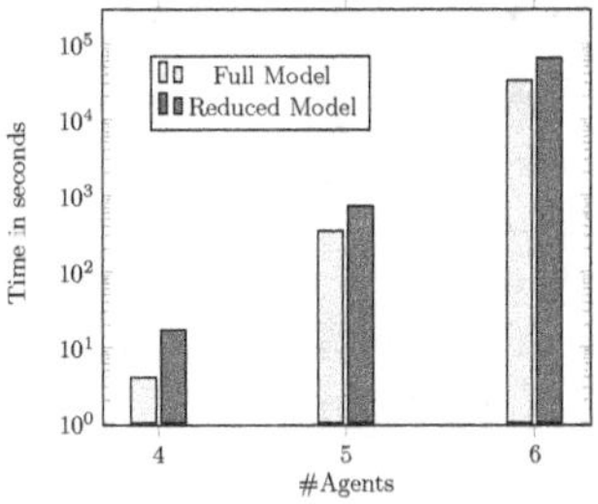

Fig. 5. Model generation times for 3 candidates 3 revotes

Models and formulas. The scalable class of models has been described in detail in Section 3. They can be configured using four parameters: the number of voters (V), the number of coerced voters (CV), the number of candidates (C) and the number of revoting turns (R). In all experiments we use configurations with one coerced voter and three candidates. We also use the *coercion-vulnerability formula*

$$\varphi_{vuln,i,k} \equiv \\ \langle\!\langle Coercer \rangle\!\rangle \mathrm{G}\,((\mathsf{end} \wedge \mathsf{revote_{v1}} = \mathsf{k} \wedge \mathsf{voted_{v1}} = \mathsf{i}) \rightarrow \mathrm{K}_{Coercer}\mathsf{voted_{v1}} = \mathsf{i})$$

saying that the coercer has a strategy so that, at the end of the election, if the voter has effectively voted for candidate i, then the coercer knows about it. This can be seen as the opposite of *receipt-freeness* and *coercion-resistance* formalizations in [43,65], see Section 3.4 for more details.

Configuration of the experiments. The test platform was a server equipped with ninety-six 2.40 GHz Intel Xeon Platinum 8260 CPUs, 991 GB RAM, and 64-bit Linux. All times are given in seconds. The algorithms have been implemented in C++. The code is available on github: github.com/blackbat13/ATLFormulaCheckerC.

Results. We present the verification results in the Tables 1-4 and model generation times in the Figure 5. All times are given in seconds and the timeout for verification was set to 1 hour and 20 hours for model generation. Model generation times are presented using a logarithmic scale. We present the size of the model and the verification times both for the full and the reduced model. We have compared time results for four verification methods, as described in Section 4: the sequential strategy synthesis (Seq.), the parallelized version (Par.) and the fixpoint approximation (Appr.).

Results: Domino DFS and flexible distributed algorithm. We omit the results for the **Domino DFS** algorithm, as it performed much slower than other algorithms. The reason is that in this scenario there is no redundancy in the strategy space, and hence no room for elimination of dominated strategies. In consequence, no gain is achieved compared to the standard DFS strategy synthesis.

The same applies to the flexible distributed algorithm proposed in Section 4.4. Apparently, cloning the model and synchronization between the threads grossly outweighs the benefits of using parallel computation.

Discussion of the results. As the results show, the simple parallel verification performs quite well in most cases. Only for relatively small models the sequential algorithm achieves better performance than the parallel algorithm, which is the result of a overhead associated with the parallelization. As various experiments have shown, the performance of the parallel algorithm is heavily dependent on the structure of the model. Depending on the amount of branching and loops in the model, it can result in a timeout even for small models. The reason for this behavior seems simple: the space of all strategies is too big to manage. As the strategy is generated from top to bottom, less branching means less configurations to check.

The fixpoint approximation algorithm performs well in cases where no strategy can be found, as it quickly reaches thc fixpoint. On the other hand, when the formula is satisfied, multiple loops maybe required before reaching the fixpoint, with each loop adding more states to the computation set.

Challenges and lessons learned. When conducting the experiments, we have encountered two main difficulties: high memory usage and slow model generation. The first one results in a memout for ordinary computers. The second one results in a timeout for more powerful machines with hundreds of gigabytes of RAM. Before moving to better computers, we have tried to overcome the memory usage problem by implementing a communication with an external database engine. The idea was to store parts of the model during the generation in the database. That, in theory, allows to shift some part of the memory requirements to other parts of the system. Unfortunately, this also heavily impacted generation times, resulting in approximately 10 times slower computation, which lead us to abandoning this idea.

To overcome the high model generation times one can also try to parallelize the generation of models. Our experiments have shown that the parallel algorithm, if implemented in a clever way, can greatly reduce the computation times. The model generation procedure works similarly to the DFS algorithm, which suggests that its parallel version can be as effective.

7 Conclusions

Modal logics for MAS, and the related verification problems, have been studied for many years. Unfortunately, they are characterized by high computational complexity, both in theory and in practice. On the other hand, it is necessary to try and verify real-life scenarios, with all their complexity, to make substantial progress in the field.

In this paper, we propose a hands-on case study in verification of a genuine protocol for secure voting. We use the "all out" approach, implementing multiple existing techniques (depth-first strategy search, domination-based strategy synthesis, fixpoint approximation) as well as proposing novel adaptations of others (partial-order reduction, distributed strategy search). Of those, partial order reduction, simple DFS, simple distributed DFS, and fixpoint approximation show very promising performance. Moreover, they produce significant gains in some of the tested configurations. On the other hand, the domination-based synthesis and flexible distributed turn out rather ill-fitted to the model checking task at hand. Overall, the experimental results are promising, and suggest that the time is ripe for the community to engage more in realistic application of the algorithms that we develop.

We also emphasize that the extension of partial order reduction, presented here, is a nontrivial technical result in its own right, as it is the first POR algorithm that handles the nesting of epistemic modalities in the context of strategic ability.

References

[1] Adida, B., *Helios: web-based open-audit voting*, in: *Proceedings of the 17th conference on Security symposium*, SS'08 (2008), pp. 335–348.

[2] Agotnes, T., *A note on syntactic characterization of incomplete information in ATEL*, in: *Procedings of Workshop on Knowledge and Games*, 2004, pp. 34–42.

[3] Agotnes, T., V. Goranko, W. Jamroga and M. Wooldridge, *Knowledge and ability*, in: H. van Ditmarsch, J. Halpern, W. van der Hoek and B. Kooi, editors, *Handbook of Epistemic Logic*, College Publications, 2015 pp. 543–589.

[4] Akintunde, M. E., E. Botoeva, P. Kouvaros and A. Lomuscio, *Formal verification of neural agents in non-deterministic environments*, in: *Proceedings of the 19th International Conference on Autonomous Agents and Multiagent Systems AAMAS*, 2020, pp. 25–33.

[5] Akintunde, M. E., E. Botoeva, P. Kouvaros and A. Lomuscio, *Verifying strategic abilities of neural-symbolic multi-agent systems*, in: *Proceedings of the 17th International Conference on Principles of Knowledge Representation and Reasoning KR*, 2020, pp. 22–32.

[6] Alur, R., L. de Alfaro, T. A. Henzinger, S. Krishnan, F. Mang, S. Qadeer, S. Rajamani and S. Tasiran, *MOCHA: Modularity in model checking*, Technical report, University of Berkeley (2000).

[7] Alur, R., T. A. Henzinger and O. Kupferman, *Alternating-time Temporal Logic*, in: *Proceedings of the 38th Annual Symposium on Foundations of Computer Science (FOCS)* (1997), pp. 100–109.

[8] Alur, R., T. A. Henzinger and O. Kupferman, *Alternating-time Temporal Logic*, Journal of the ACM **49** (2002), pp. 672–713.

[9] Behrmann, G., A. David and K. Larsen, *A tutorial on* UPPAAL, in: *Formal Methods for the Design of Real-Time Systems: SFM-RT*, number 3185 in LNCS (2004), pp. 200–236.

[10] Broersen, J., M. Dastani, Z. Huang and L. van der Torre, *The BOID architecture: conflicts between beliefs, obligations, intentions and desires*, in: J. Müller, E. Andre, S. Sen and C. Frasson, editors, *Proceedings of the Fifth International Conference on Autonomous Agents* (2001), pp. 9–16.

[11] Broersen, J., A. Herzig and N. Troquard, *Embedding Alternating-time Temporal Logic in Strategic STIT logic of agency*, Journal of Logic and Computation **16** (2006), pp. 559–578.

[12] Bruni, A., E. Drewsen and C. Schürmann, *Towards a mechanized proof of Selene receipt-freeness and vote-privacy*, in: *Proceedings of E-Vote-ID*, Lecture Notes in Computer Science **10615** (2017), pp. 110–126.

[13] Bulling, N., J. Dix and W. Jamroga, *Model checking logics of strategic ability: Complexity*, in: M. Dastani, K. Hindriks and J.-J. Meyer, editors, *Specification and Verification of Multi-Agent Systems*, Springer, 2010 pp. 125–159.

[14] Bulling, N. and W. Jamroga, *Comparing variants of strategic ability: How uncertainty and memory influence general properties of games*, Journal of Autonomous Agents and Multi-Agent Systems **28** (2014), pp. 474–518.

[15] Chaum, D., R. Carback, J. Clark, A. Essex, S. Popoveniuc, R. Rivest, P. Ryan, E. Shen, A. Sherman and P. Vora, *Scantegrity II: end-to-end verifiability by voters of optical scan elections through confirmation codes*, Trans. Info. For. Sec. **4** (2009), pp. 611–627.

[16] Chaum, D., P. Y. A. Ryan and S. A. Schneider, *A practical voter-verifiable election scheme*, in: *Proceedings of ESORICS*, 2005, pp. 118–139.

[17] Chen, T., V. Forejt, M. Kwiatkowska, D. Parker and A. Simaitis, *PRISM-games: A model checker for stochastic multi-player games*, in: *Proceedings of Tools and Algorithms for Construction and Analysis of Systems (TACAS)*, Lecture Notes in Computer Science **7795** (2013), pp. 185–191.

[18] Clarke, E. M., O. Grumberg and D. A. Peled, "Model Checking," The MIT Press, Cambridge, Massachusetts, 1999.

[19] Cohen, P. and H. Levesque, *Intention is choice with commitment*, Artificial Intelligence **42** (1990), pp. 213–261.

[20] Culnane, C., P. Ryan, S. Schneider and V. Teague, *vvote: A verifiable voting system*, ACM Trans. Inf. Syst. Secur. **18** (2015), pp. 3:1–3:30.

[21] Dastani, M., K. Hindriks and J. Meyer, editors, "Specification and Verification of Multi-Agent Systems," Springer, 2010.

[22] Dembiński, P., A. Janowska, P. Janowski, W. Penczek, A. Półrola, M. Szreter, B. Woźna and A. Zbrzezny, *VerICS: A tool for verifying timed automata and Estelle specifications*, in: *Proceedings of the of the 9th Int. Conf. on Tools and Algorithms for Construction and Analysis of Systems (TACAS'03)*, Lecture Notes in Computer Science **2619**, Springer, 2003 pp. 278–283.

[23] Dima, C., C. Enea and D. Guelev, *Model-checking an alternating-time temporal logic with knowledge, imperfect information, perfect recall and communicating coalitions*, in: *Proceedings of Games, Automata, Logics and Formal Verification (GandALF)*, 2010, pp. 103–117.

[24] Dima, C. and F. Tiplea, *Model-checking ATL under imperfect information and perfect recall semantics is undecidable*, CoRR **abs/1102.4225** (2011).

[25] Emerson, E. A., *Temporal and modal logic*, in: J. van Leeuwen, editor, *Handbook of Theoretical Computer Science, Volume B: Formal Models and Semantics*, Elsevier and MIT Press, 1990 pp. 995–1072.

[26] Fagin, R., J. Y. Halpern, Y. Moses and M. Y. Vardi, "Reasoning about Knowledge," MIT Press, 1995.

[27] Gardner, R. W., S. Garera and A. D. Rubin, *Coercion resistant end-to-end voting*, in: R. Dingledine and P. Golle, editors, *Financial Cryptography and Data Security*, Lecture Notes in Computer Science **5628**, Springer Berlin Heidelberg, 2009 pp. 344–361.

[28] Gerth, R., R. Kuiper, D. Peled and W. Penczek, *A partial order approach to branching time logic model checking*, Information and Computation **150** (1999), pp. 132–152.

[29] Ghale, M., R. Goré, D. Pattinson and M. Tiwari, *Modular formalisation and verification of STV algorithms*, in: *Proceedings of E-Vote-ID*, Lecture Notes in Computer Science **11143** (2018), pp. 51–66.

[30] Godefroid, P., "Partial-Order Methods for the Verification of Concurrent Systems: An Approach to the State-Explosion Problem," Springer-Verlag, Berlin, Heidelberg, 1996.

[31] Guelev, D. and C. Dima, *Epistemic ATL with perfect recall, past and strategy contexts*, in: *Proceedings of Computational Logic in Multi-Agent Systems (CLIMA)*, Lecture Notes in Computer Science **7486** (2012), pp. 77–93.

[32] Guelev, D., C. Dima and C. Enea, *An alternating-time temporal logic with knowledge, perfect recall and past: axiomatisation and model-checking*, Journal of Applied Non-Classical Logics **21** (2011), pp. 93–131.

[33] Haines, T., R. Goré and M. Tiwari, *Verified verifiers for verifying elections*, in: *Proceedings of CCS* (2019), pp. 685–702.

[34] Jamroga, W., "Logical Methods for Specification and Verification of Multi-Agent Systems," ICS PAS Publishing House, 2015.

[35] Jamroga, W., Y. Kim, D. Kurpiewski and P. Y. A. Ryan, *Towards model checking of voting protocols in uppaal*, in: *Proceedings of E-Vote-ID*, Lecture Notes in Computer Science **12455** (2020), pp. 129–146.

[36] Jamroga, W., M. Knapik and D. Kurpiewski, *Model checking the SELENE e-voting protocol in multi-agent logics*, in: *Proceedings of the 3rd International Joint Conference on Electronic Voting (E-VOTE-ID)*, Lecture Notes in Computer Science **11143** (2018), pp. 100–116.

[37] Jamroga, W., M. Knapik, D. Kurpiewski and Ł. Mikulski, *Approximate verification of strategic abilities under imperfect information*, Artificial Intelligence **277** (2019).

[38] Jamroga, W., D. Kurpiewski and V. Malvone, *Natural strategic abilities in voting protocols*, CoRR **abs/2007.12424** (2020).

[39] Jamroga, W., W. Penczek, P. Dembiński and A. Mazurkiewicz, *Towards partial order reductions for strategic ability*, in: *Proceedings of the 17th International Conference on Autonomous Agents and Multiagent Systems (AAMAS)* (2018), pp. 156–165.

[40] Jamroga, W., W. Penczek and T. Sidoruk, *Strategic abilities of asynchronous agents: Semantic side effects and how to tame them*, in: *Proceedings of KR 2021*, 2021, pp. 368–378.

[41] Jamroga, W., W. Penczek, T. Sidoruk, P. Dembiński and A. Mazurkiewicz, *Towards partial order reductions for strategic ability*, Journal of Artificial Intelligence Research **68** (2020), pp. 817–850.

[42] Jamroga, W. and W. van der Hoek, *Agents that know how to play*, Fundamenta Informaticae **63** (2004), pp. 185–219.

[43] Juels, A., D. Catalano and M. Jakobsson, *Coercion-resistant electronic elections*, in: *Proceedings of the 2005 ACM workshop on Privacy in the electronic society*, ACM, 2005, pp. 61–70.

[44] Kacprzak, M., A. Lomuscio and W. Penczek, *From bounded to unbounded model checking for temporal epistemic logic*, Fundamenta Informaticae **63** (2004), pp. 221–240.

[45] Kacprzak, M., W. Nabialek, A. Niewiadomski, W. Penczek, A. Pólrola, M. Szreter, B. Wozna and A. Zbrzezny, *VerICS 2007 - a model checker for knowledge and real-time*, Fundamenta Informaticae **85** (2008), pp. 313–328.

[46] Kacprzak, M. and W. Penczek, *Unbounded model checking for alternating-time temporal logic*, in: *3rd International Joint Conference on Autonomous Agents and Multiagent Systems (AAMAS 2004), 19-23 August 2004, New York, NY, USA* (2004), pp. 646–653.

[47] Kacprzak, M. and W. Penczek, *Fully symbolic unbounded model checking for alternating-time temporal logic*, Autonomous Agents and Multi-Agent Systems **11** (2005), pp. 69–89.

[48] Kanski, M., A. Niewiadomski, M. Kacprzak, W. Penczek and W. Nabialek, *SMT-based unbounded model checking for ATL*, in: A. Nouri, W. Wu, K. Barkaoui and Z. Li, editors, *Verification and Evaluation of Computer and Communication Systems - 15th International Conference, VECoS 2021, Virtual Event, November 22-23, 2021, Revised Selected Papers*, Lecture Notes in Computer Science **13187** (2021), pp. 43–58.

[49] Kant, G., A. Laarman, J. Meijer, J. van de Pol, S. Blom and T. van Dijk, *LTSmin: High-performance language-independent model checking*, in: *Tools and Algorithms for the Construction and Analysis of Systems. Proceedings of TACAS*, Lecture Notes in Computer Science **9035** (2015), pp. 692–707.

[50] Kurpiewski, D., W. Jamroga and M. Knapik, *STV: Model checking for strategies under imperfect information*, in: *Proceedings of the 18th International Conference on Autonomous Agents and Multiagent Systems AAMAS 2019* (2019), pp. 2372–2374.

[51] Kurpiewski, D., M. Knapik and W. Jamroga, *On domination and control in strategic ability*, in: *Proceedings of the 18th International Conference on Autonomous Agents and Multiagent Systems AAMAS 2019* (2019), pp. 197–205.

[52] Kurpiewski, D., W. Pazderski, W. Jamroga and Y. Kim, *STV+Reductions: Towards practical verification of strategic ability using model reductions*, in: *Proceedings of AAMAS* (2021), pp. 1770–1772.

[53] Lomuscio, A., W. Penczek and H. Qu, *Partial order reductions for model checking temporal-epistemic logics over interleaved multi-agent systems*, Fundamenta Informaticae **101** (2010), pp. 71–90.

[54] Lomuscio, A., H. Qu and F. Raimondi, *MCMAS: An open-source model checker for the verification of multi-agent systems*, International Journal on Software Tools for Technology Transfer **19** (2017), pp. 9–30.

[55] Malvone, V., A. Murano and L. Sorrentino, *Hiding actions in multi-player games*, in: *Proceedings of the 16th Conference on Autonomous Agents and MultiAgent Systems, AAMAS 2017, São Paulo, Brazil, May 8-12, 2017*, 2017, pp. 1205–1213.

[56] Martimiano, T., E. D. Santos, M. Olembo and J. Martina, *Ceremony analysis meets verifiable voting: Individual verifiability in Helios*, in: *SECURWARE*, 2015, pp. 195–204.

[57] Peled, D., *All from one, one for all: on model checking using representatives*, in: C. Courcoubetis, editor, *Computer Aided Verification* (1993), pp. 409–423.

[58] Peled, D., *Combining partial order reductions with on-the-fly model-checking*, in: *Proceedings of the 6th International Conference on Computer Aided Verification*, LNCS 818 (1994), pp. 377–390.

[59] Priese, L., *Automata and concurrency*, Theoretical Computer Science **25** (1983), pp. 221–265.

[60] Rao, A. and M. Georgeff, *Modeling rational agents within a BDI-architecture*, in: *Proceedings of the 2nd International Conference on Principles of Knowledge Representation and Reasoning*, 1991, pp. 473–484.

[61] Ryan, P., P. Rønne and V. Iovino, *Selene: Voting with transparent verifiability and coercion-mitigation*, in: *Financial Cryptography and Data Security: Proceedings of FC 2016. Revised Selected Papers*, Lecture Notes in Computer Science **9604** (2016), pp. 176–192.

[62] Ryan, P. Y. A., S. A. Schneider and V. Teague, *End-to-end verifiability in voting systems, from theory to practice*, IEEE Security & Privacy **13** (2015), pp. 59–62.

[63] Schobbens, P. Y., *Alternating-time logic with imperfect recall*, Electronic Notes in Theoretical Computer Science **85** (2004), pp. 82–93.

[64] Shoham, Y. and K. Leyton-Brown, "Multiagent Systems - Algorithmic, Game-Theoretic, and Logical Foundations," Cambridge University Press, 2009.

[65] Tabatabaei, M., W. Jamroga and P. Y. A. Ryan, *Expressing receipt-freeness and coercion-resistance in logics of strategic ability: Preliminary attempt*, in: *Proceedings of the 1st International Workshop on AI for Privacy and Security, PrAISe@ECAI 2016* (2016), pp. 1:1–1:8.

[66] Valmari, A., *Stubborn sets for reduced state space generation*, in: G. Rozenberg, editor, *Advances in Petri Nets 1990* (1991), pp. 491–515.

[67] van der Hoek, W., A. Lomuscio and M. Wooldridge, *On the complexity of practical ATL model checking*, in: *Proceedings of International Joint Conference on Autonomous Agents and Multiagent Systems (AAMAS)* (2006), pp. 201–208.

[68] van der Hoek, W. and M. Wooldridge, *Tractable multiagent planning for epistemic goals*, in: C. Castelfranchi and W. Johnson, editors, *Proceedings of the First International Joint Conference on Autonomous Agents and Multi-Agent Systems (AAMAS-02)* (2002), pp. 1167–1174.

[69] Weiss, G., editor, "Multiagent Systems. A Modern Approach to Distributed Artificial Intelligence," MIT Press: Cambridge, Mass, 1999.

[70] Wooldridge, M., "Reasoning about Rational Agents," MIT Press : Cambridge, Mass, 2000.

[71] Wooldridge, M., "An Introduction to Multi Agent Systems," John Wiley & Sons, 2002.

[72] Zollinger, M., P. Roenne and P. Ryan, *Mechanized proofs of verifiability and privacy in a paper-based e-voting scheme*, in: *Proceedings of 5th Workshop on Advances in Secure Electronic Voting*, 2020.

Choice-Free de Vries Duality

Guillaume Massas [1]

University of California, Berkeley

Abstract

De Vries duality generalizes Stone duality between Boolean algebras and Stone spaces to a duality between de Vries algebras (complete Boolean algebras equipped with a subordination relation satisfying some axioms) and compact Hausdorff spaces. This duality allows for an algebraic approach to region-based theories of space that differs from point-free topology. Building on the recent choice-free version of Stone duality developed by Bezhanishvili and Holliday, this paper establishes a choice-free duality between de Vries algebras and a category of de Vries spaces. We also investigate connections with the Vietoris functor on the category of compact Hausdorff spaces and with the category of compact regular frames in point-free topology, and we provide an alternative, choice-free topological semantics for the Symmetric Strict Implication Calculus of Bezhanishvili et al.

Keywords: Duality theory, de Vries algebras, point-free topology.

1 Introduction

Stone's [27] representation of Boolean algebras as clopen sets of compact, Hausdorff and zero-dimensional topological spaces has had a profound influence on the study of interactions between logic, algebra and topology. The realization that some properties of topological spaces could be retrieved by considering the algebraic properties of their lattices of open sets led to the development of *point-free* topology [16,18,24], in which open sets are taken as basic elements of a *frame* rather than defined as sets of points. Stone's representation theorem, and therefore Stone duality, relies on the Boolean Prime Ideal Theorem (BPI), a fragment of the Axiom of Choice. By contrast, the point-free approach has a more constructive flavor: even in the absence of the Axiom of Choice, the open set functor Ω mapping a topological space to its lattice of open sets has an adjoint functor *pt*, mapping a frame to its set of "points" endowed with a natural Stone-like topology. But the restriction of this adjunction to Stone spaces and compact zero-dimensional frames is only a duality under (BPI). In [7], a choice-free duality between Boolean algebras and a category of *UV*-spaces has been developed. It is based on the simple but powerful idea that the appeal to (BPI) could be eliminated by working with a partially-ordered

[1] gmassas@berkeley.edu

set of filters rather than a set of ultrafilters and by viewing these filters as partial approximations of a classical point. This approach has strong ties to both possibility semantics in modal logic [12,13,14] and the Vietoris functor on Stone spaces [28] and provides a *semi-pointfree* approach, i.e., both spatial *and* choice-free, to the representation of algebraic objects in *semi-constructive mathematics*, i.e., mathematics carried out in $ZF + DC$ [21,26].

In [8], de Vries generalized Stone duality to a duality between de Vries algebras (complete Boolean algebras equipped with a *subordination* relation) and compact Hausdorff spaces. Just like Stone, de Vries used (BPI) in his representation of complete compingent algebras as the regular open sets of a compact Hausdorff space. On the point-free side, Isbell [15] showed that the Ω-*pt* adjunction restricts to a duality between compact Hausdorff spaces and compact regular frames, also under the assumption of (BPI). This leaves open the question of whether a choice-free duality between these algebraic categories and a category of topological spaces can be defined.

In this paper, we show that this is indeed possible by generalizing the approach of [7]. Just like Bezhanishvili and Holliday, we work with a poset of filters rather than with a set of maximal filters, and we define our dual spaces both in terms of their topological properties and in terms of order-theoretic aspects of the induced specialization order. We also show how the spaces we define naturally relate to the Vietoris functor on compact Hausdorff spaces and compact regular frames. We take this as evidence of the naturality and fruitfulness of this semi-pointfree approach, in which the basic "points" of our spaces are the closed sets of the standard, non-constructive duality.

The paper is organized as follows. After reviewing some background on de Vries algebras, compact regular frames and UV-spaces (Section 2), we provide a choice-free representation of any de Vries algebra as the regular open sets of some topological space (Section 3). In Section 4, we characterize the choice-free duals of de Vries algebras, which we call dV-spaces. Section 5 deals with morphisms and ends with our main result, a choice-free dual equivalence between the category of de Vries algebras and the category of dV-spaces. In Section 6, we connect our duality to point-free topology and provide an alternative characterization of dV-spaces via the Vietoris functor on compact regular frames. Finally, Section 7 lists two applications of this duality: a choice-free analogue of Tychonoff's Theorem for compact Hausdorff spaces and a choice-free topological semantics for the system $\mathsf{S^2IC}$ introduced in [5].

2 Background

In this section, we briefly recall the de Vries and Isbell dualities for compact Hausdorff spaces as well as the choice-free Stone duality between Boolean algebras and UV-spaces presented in [7]. We start by fixing some notation that we will use throughout. Let L be a complete lattice and (X, τ) be a topological space.

(i) When no confusion arises, we write $\leq$ to designate the order on L. We designate the meet and join operations on L by $\wedge$ and $\vee$ respectively, and, whenever L is pseudo-complemented, we use $\neg$ for the pseudo-complement operation.

(ii) We will designate (a subset of) the set of all maximal filters on L by X_L and (a subset of) the set of all filters on L by S_L.

(iii) By a Stone-like topology on a set Y of filters of L, we mean the topology generated by the sets of the form $\widehat{a} = \{F \in Y \mid a \in F\}$, and we will usually designate such a topology by σ.

(iv) For any $U \subseteq X$, we write $-U$ for $X \setminus U$, $\overline{U}$ for the closure of U and $U^{\perp}$ for $-\overline{U}$. We write $\mathsf{CO}(X)$ for the set of compact open subsets of X and $\mathsf{RO}(X)$ for the Boolean algebra of regular open subsets of X, i.e., subsets U such that $U^{\perp\perp} = U$.

(v) The *specialization preorder* on (X, τ) is represented by the symbol $\leq$ when no confusion arises, and it is defined as $x \leq y$ iff $x \in U$ implies $y \in U$ for every $U \in \tau$.

(vi) The *up-set topology* on X is the topology generated by the set of all upward closed subsets in the specialization preorder. Given $U \subseteq X$, we let $\Uparrow U$ be the interior of U in the up-set topology, and $\downarrow U$ the closure of U. We write $\mathcal{RO}(X)$ for the Boolean algebra of order-regular open subsets of X, i.e., subsets U such that $\Uparrow\downarrow U = U$, and $\mathsf{CO}\mathcal{RO}(X)$ for $\mathsf{CO}(X) \cap \mathcal{RO}(X)$.

2.1 De Vries Algebras

De Vries algebras were introduced in [8] as an algebraic dual to compact Hausdorff spaces.

Definition 2.1 A *compingent algebra* is a pair $(B, \prec)$ such that B is a Boolean algebra with induced order $\leq$, and $\prec$ is a relation on $B \times B$ satisfying the following set of axioms:

(A1) $1 \prec 1$;

(A2) $a \prec b$ implies $a \leq b$;

(A3) $a \leq b \prec c \leq d$ implies $a \prec d$;

(A4) $a \prec b$ and $a \prec c$ together imply $a \prec b \wedge c$;

(A5) $a \prec b$ implies $\neg b \prec \neg a$;

(A6) $a \prec c$ implies that there is $b \in B$ such that $a \prec b \prec c$;

(A7) $a \neq 0$ implies that there is $b \neq 0 \in B$ such that $b \prec a$.

A *de Vries algebra* is a compingent algebra $V = (B, \prec)$ such that B is a complete Boolean algebra. It is *zero-dimensional* if for any $a \prec b \in V$ there is $c \in V$ such that $a \prec c \prec c \prec b$.

Compingent algebras constitute a specific kind of *contact algebras*, Boolean algebras equipped with a binary relation of subordination satisfying (A1)-(A5). One motivation for contact algebras is to develop a region-based theory of space [10,20], according to which regions of space form a Boolean algebra and a region

a is subordinated to a region b precisely if b completely surrounds a. For more on contact and subordination algebras, we refer the reader to [5,6,9,11].

Definition 2.2 Let $V = (B, \prec)$ be a de Vries algebra. For any filter F on B, let $\Uparrow F = \{a \in F \mid \exists b \in F : b \prec a\}$. A *concordant filter* on V is a filter F such that $\Uparrow F = F$. An *end* is a maximal concordant filter.

The dual space of a de Vries algebra V is obtained by taking the set X_V of all ends of V and endowing it with the Stone-like topology σ generated by all sets of the form $\{p \in X_V \mid a \in p\}$ for some $p \in V$. Conversely, the dual de Vries algebra of a compact Hausdorff space (X, τ) is the complete Boolean algebra $\mathsf{RO}(X)$ of regular open sets, with the subordination relation $\sqsubset$ given by $U \sqsubset V$ iff $\overline{U} \subseteq V$.

Theorem 2.3 ([8], Thm. I.4.3-5) *For any de Vries algebra $V = (B, \prec)$, (X_V, σ) is compact Hausdorff, and $(B, \prec)$ is isomorphic to $(\mathsf{RO}(X_V), \sqsubset)$. Conversely, for any compact Hausdorff space (X, τ), $(\mathsf{RO}(X), \sqsubset)$ is a de Vries algebra, and (X, τ) is homeomorphic to $(X_{(\mathsf{RO}(X),\sqsubset)}, \sigma)$.*

We now introduce the relevant notion of morphism between de Vries algebras.

Definition 2.4 Let $V_1 = (B_1, \prec_1)$ and $V_2 = (B_2, \prec_2)$ be de Vries algebras. A de Vries morphism from V_1 to V_2 is a function $h : B_1 \to B_2$ satisfying the following set of conditions:

(V1) $h(0) = 0$;

(V2) $h(a \wedge b) = h(a) \wedge h(b)$;

(V3) $a \prec_1 b$ implies $\neg h(\neg a) \prec_2 h(b)$;

(V4) $h(a) = \bigvee\{h(b) \mid b \prec_1 a\}$.

Given two de Vries morphisms $h : V_1 \to V_2$ and $k : V_2 \to V_3$, their composition $k \star h : V_1 \to V_3$ is defined as the map $a \mapsto \bigvee\{kh(b) : b \prec_1 a\}$.

One easily verifies that de Vries morphisms preserve both the order $\leq$ and the subordination relation $\prec$. Given a de Vries morphism $h : V_1 \to V_2$, the map $h^* : X_{V_2} \to X_{V_1}$ given by $h^*(p) = \Uparrow h^{-1}[p]$ for any end p on V_2 is a continuous function. Conversely, for any continuous function $f : (X_1, \tau_1) \to (X_2, \tau_2)$, the map $f_* : \mathsf{RO}(X) \to \mathsf{RO}(Y)$ given by $f_*(U) = (f^{-1}[U])^{\perp\perp}$ for any regular open set U is a de Vries morphism. This allowed de Vries to obtain the following:

Theorem 2.5 *The category* **deV** *of de Vries algebras and de Vries morphisms between them is dually equivalent to the category* **KHaus** *of compact Hausdorff spaces and continuous maps between them.*

2.2 Compact Regular Frames

Recall that a *frame* is a complete lattice L that satisfies the join-infinite distributive law, i.e., is such that $a \wedge \bigvee B = \bigvee\{a \wedge b \mid b \in B\}$ for any $a \in L$ and $B \subseteq L$. Frames are the central object of study of point-free topology, for which [16,18,24] are standard introductions. A frame L is *compact* if $1_L = \bigvee B$

for some $B \subseteq L$ implies that $1_L = \bigvee B'$ for some finite $B' \subseteq B$. A morphism between frames is a map preserving finite meets and arbitrary joins.

Definition 2.6 Let L be a frame and $a, b \in L$. Then a is said to be *rather below* b [24, Def. V.5.2], denoted $a \prec b$, if $b \vee \neg a = 1_L$. A *compact regular frame* is a compact frame L such that for any $a \in L$, $a = \bigvee\{b \in L \mid b \prec a\}$.

Given any topological space (X, τ), one can define its frame of open sets $\Omega(X)$. Conversely, given a frame L, one can define a Stone-like topology on the set of completely prime filters $pt(L)$. These constructions give rise to adjoint functors Ω and pt between the categories **Frm** of frames and frame morphisms and **Top** of topological spaces and continuous functions. Assuming (BPI), Isbell [15] showed that this adjunction restricts to a duality in the specific case of compact regular frames:

Theorem 2.7 *The category* **KRFrm** *of compact regular frames is dually equivalent to* **KHaus**.

As an immediate consequence of Theorems 2.5 and 2.7, the categories **deV** and **KRFrm** are equivalent. This equivalence has also been given a direct description in [2], which has the advantage of being choice-free. Given a frame L, an element $a \in L$ is *regular* if $\neg\neg a = a$. The *Booleanization* of L [1], denoted $B(L)$, is the subframe of all the regular elements of L. It is straightforward to verify that if L is a compact regular frame, $B(L)$ equipped with the rather below relation $\prec$ is a de Vries algebra. In order to go from de Vries algebras to frames, we need the following definition:

Definition 2.8 Let $V = (B, \prec)$ be a de Vries algebra. An ideal on B is *round* if for any $a \in I$, there is $b \in I$ such that $a \prec b$.

It is immediate to see that a proper ideal I on a de Vries algebra V is round if and only if its dual filter $I^\delta = \{\neg a \mid a \in I\}$ is concordant. Given a de Vries algebra V, its set of round ideals ordered by inclusion forms a compact regular frame $\mathfrak{R}(V)$. The equivalence between **KRFrm** and **deV** is then given by the following result:

Theorem 2.9 *Any compact regular frame L is isomorphic to $\mathfrak{R}(B(L))$. Conversely, any de Vries algebra V is isomorphic to $B(\mathfrak{R}(V))$, and the maps B and $\mathfrak{R}$ lift to an equivalence between* **KRFrm** *and* **deV**.

2.3 *UV*-spaces

We conclude this section by recalling the choice-free version of Stone duality presented in [7].

Definition 2.10 A topological space (X, τ) is a *UV-space* if it satisfies the following conditions:

(i) (X, τ) is compact and T_0;

(ii) $\mathsf{CO}\mathcal{RO}(X)$ is closed under $\cap$ and $-\downarrow$ and forms a basis for τ;

(iii) Any filter on $\mathsf{CO}\mathcal{RO}(X)$ is $\mathsf{CO}\mathcal{RO}(x) = \{U \in \mathsf{CO}\mathcal{RO}(X) \mid x \in U\}$ for some $x \in X$.

Given a Boolean algebra B, one considers the set S_B of all filters on B, equipped with the usual Stone-like topology σ. It can then be showed without appealing to (BPI) that UV-spaces are the duals of Boolean algebras:

Theorem 2.11 ([7], Thm. 5.4) *For any Boolean algebra B, (S_B, σ) is a UV-space, and B is isomorphic to $\mathsf{CORO}(S_B)$. Conversely, for any UV-space (X, τ), $\mathsf{CORO}(X)$ is a Boolean algebra, and (X, τ) is homeomorphic to $(S_{\mathsf{CORO}(X)}, \sigma)$.*

Moving on to morphisms, recall that a spectral map between two topological spaces is a map such that the preimage of any compact open set is compact open.

Definition 2.12 Given two UV-spaces (X, τ_1) and (Y, τ_2) with induced specialization orders $\leq_1$ and $\leq_2$, a UV-map from (X, τ_1) to (Y, τ_2) is a spectral map $f : X \to Y$ that is also a p-morphism with respect to $\leq_1$ and $\leq_2$, i.e., for any $x \in X$, $y \in Y$, if $f(x) \leq_2 y$, then there is $x' \geq_1 x$ such that $y = f(x')$.

Any UV-map $f : (X, \tau_1) \to (Y, \tau_2)$ gives rise to a Boolean algebra homomorphism $f_* : \mathsf{CORO}(Y) \to \mathsf{CORO}(X)$ given by $f_*(U) = f^{-1}[U]$ for any $U \in \mathcal{RO}(Y)$. Conversely, for any Boolean homomorphism $h : B_1 \to B_2$, the map $h^* : (S_{B_2}, \sigma_2) \to (S_{B_1}, \sigma_1)$ given by $h^*(F) = h^{-1}[F]$ for any filter F on B_2 is a UV-map. This yields the following result, which, unlike Stone duality, does not rely on the Axiom of Choice:

Theorem 2.13 *The category* **BA** *of Boolean algebras and Boolean homomorphisms between them is dually equivalent to the category* **UV** *of UV-spaces and UV-maps between them.*

3 A Choice-free Representation for de Vries Algebras

In this section, we complete the first step of the duality by obtaining a choice-free representation of any de Vries algebra as the regular open sets of some topological space. Our approach combines the techniques of Sections 2.1 and 2.3 in a natural way.

Definition 3.1 Let $V = (B, \prec)$ be a de Vries algebra. The *dual filter space* of V is the topological space (S_V, σ), where:

- S_V is the set of all concordant filters on V;
- σ is the Stone-like topology generated by $\{\widehat{a} = \{F \in S_V \mid a \in F\} \mid a \in V\}$.

The following two lemmas will help us investigate the structure of the space of concordant filters on a de Vries algebra.

Lemma 3.2 *Let $V = (B, \prec)$ be a de Vries algebra. Then:*

(i) *For any $a \neq 0$, $F = \{c \in V \mid a \prec c\}$ is a concordant filter.*

(ii) *If F and G are concordant filters and $c \wedge d \neq 0$ for any $c \in F, d \in G$, then the set $H = \{c \wedge d \mid c \in F, d \in G\}$ is a concordant filter.*

Proof. For part (i), by (A3), F is upward-closed, and by (A4), it is downward directed. To verify that $\Uparrow F = F$, note that if $a \prec c$, then by (A6) there is c' such that $a \prec c' \prec c$, so $c \in \Uparrow F$.

For part (ii), let $H = \{c \wedge d \mid c \in F, d \in G\}$. I claim that H is a concordant filter. It is routine to verify that H is a proper filter. To see that $\Uparrow H = H$, take $c \in F$ and $d \in G$. Since F and G are concordant there are $c' \prec c$ in F and $d' \prec d$ in G. Thus $c' \wedge d' \in H$ and $c' \wedge d' \prec c \wedge d$ by (A4), which means that $c \wedge d \in \Uparrow H$. This shows that $H \subseteq \Uparrow H$, and the converse is immediate from (A2). □

Lemma 3.3 *Let $V = (B, \prec)$ be a de Vries algebra, $a \in V$ and F a concordant filter on V. If $a \notin F$, then there is a concordant filter $G \supseteq F$ such that for any concordant filter $H \supseteq G$, $a \notin H$.*

Proof. Suppose $a \notin F$, and consider the set $G = \{c \wedge d \mid c \in F, \neg a \prec d\}$. I claim that G is a concordant filter. If $c \wedge d = 0$ for some $c \in F$ and d such that $\neg a \prec d$, then $c \leq \neg d \prec \neg\neg a = a$, which contradicts the assumption that $a \notin F$. Thus by Lemma 3.2 G is a concordant filter.

Now suppose H is a concordant filter such that $H \supseteq G$. If $a \in H$, then there is $d \in H$ such that $d \prec a$. But this implies that $\neg a \prec \neg d$, so $\neg d \in G \subseteq H$, a contradiction. □

Given a de Vries algebra V with dual space (S_V, σ), we now show that the map $a \mapsto \widehat{a}$ is a Boolean embedding of V into $\mathsf{RO}(S_V)$:

Lemma 3.4 *Let $V = (B, \prec)$ be a de Vries algebra with dual filter space (S_V, σ). Then for any $a, b \in V$:*

(i) $\widehat{a} \cap \widehat{b} = \widehat{a \wedge b}$;

(ii) *The set $\{\widehat{a} \mid a \in V\}$ is a basis for σ, and the specialization order on (S_V, σ) coincides with the inclusion order;*

(iii) $\widehat{a} \subseteq \widehat{b}$ *iff* $a \leq b$;

(iv) $\widehat{a}^{\perp} = \widehat{\neg a}$;

(v) $\Uparrow\!\downarrow \widehat{a} = \widehat{a} = \widehat{a}^{\perp\perp}$.

Proof. Part (i) is a consequence of the fact that the elements of S_V are filters, and part (ii) immediately follows from part (i). For part (iii), the right-to-left direction is immediate, and for the converse, since B is a Boolean algebra it is enough to show that for any $a \neq 0$, there is some concordant filter F such that $a \in F$. To see this, note that, by (A7), if $a \neq 0$ there is $b \neq 0$ such that $b \prec a$. Then $F = \{c \in V \mid b \prec c\}$ is a concordant filter by Lemma 3.2, and $a \in F$.

For part (iv), since the set $\{\widehat{a} \mid a \in V\}$ is a basis for σ by (ii), we have that for any $F \in S_V$, $F \in \overline{\widehat{a}}$ iff for any basic open $\widehat{b}$, $F \in \widehat{b}$ implies $\widehat{a} \cap \widehat{b} \neq \emptyset$. By (i) and (iii), this means that $F \in \overline{\widehat{a}}$ iff $b \wedge a \neq 0$ for all $b \in F$ iff $\neg a \notin F$ iff $F \notin \widehat{\neg a}$. Hence $\widehat{a}^{\perp} = \widehat{\neg a}$.

Finally, for part (v), $\widehat{a} = \widehat{a}^{\perp\perp}$ follows directly from (iv). To show that $\Uparrow\!\downarrow \widehat{a} = \widehat{a}$, note that the right-to-left inclusion is immediate since $\widehat{a}$ is upward-closed. Since the specialization order on (S_V, σ) coincides with the inclusion ordering, establishing the converse amounts to showing that for any $F \in S_V$, if $a \notin F$, then there is $G \supseteq F$ such that for all $H \supseteq G$, $a \notin H$. But this is precisely Lemma 3.3. □

Corollary 3.5 *Let $V = (B, \prec)$ be a de Vries algebra with dual filter space (S_V, σ). Then B is isomorphic to $\mathsf{RO}(S_V)$.*

Proof. Lemma 3.4 implies that the map $a \mapsto \widehat{a}$ is an injective Boolean homomorphism of B into $\mathsf{RO}(S_V)$. Therefore it only remains to show that every regular open subset of S_V is of the form $\widehat{a}$ for some $a \in V$. Let $U = \bigcup_{a \in A} \widehat{a}$ be a regular open set. Recall that $\bigvee A \in B$ since B is a complete Boolean algebra. I claim that $\overline{U} = \overline{\widehat{\bigvee A}}$. Since U is regular open, this will readily imply that $U = \widehat{\bigvee A}$. For the proof of the claim, recall that for any $F \in S_V$, $F \in \overline{\widehat{\bigvee A}}$ iff $\neg \bigvee A \notin F$. Similarly, $F \in \overline{U}$ iff for any $b \in F$ there is $a \in A$ such that $b \not\leq \neg a$. But the latter condition is equivalent to $b \not\leq \bigwedge\{\neg a \mid a \in A\}$, which is in turn equivalent to $\neg \bigvee A \notin F$. Hence $F \in \overline{U}$ iff $F \in \overline{\widehat{\bigvee A}}$ for any $F \in S_V$, which means that $\overline{U} = \overline{\widehat{\bigvee A}}$. This completes the proof that B is isomorphic to $\mathsf{RO}(S_V)$. □

We now turn to representing the subordination relation on a de Vries algebra. For any topological space (X, τ) and any $U, V \subseteq X$, let $U \ll V$ iff $\overline{U} \subseteq {\downarrow} V$.

Lemma 3.6 *Let $V = (B, \prec)$ be a de Vries algebra with dual filter space (S_V, σ). For any $a, b \in V$, $a \prec b$ iff $\widehat{a} \ll \widehat{b}$.*

Proof. For the first direction, suppose that $a \prec b$. Then if F is a concordant filter such that $\neg a \notin F$, by Lemma 3.2 $G = \{c \wedge d \mid c \in F, a \prec d\}$ is a concordant filter extending F and containing b. Now for any concordant filter F, $F \in \overline{\widehat{a}}$ iff $\neg a \notin F$. This shows that $\overline{\widehat{a}} \subseteq {\downarrow}\widehat{b}$. Conversely, assume that $a \not\prec b$. I claim that there is a concordant filter F such that $\neg a \notin F$ and $b \notin G$ for any concordant filter $G \supseteq F$. Let $F = \{c \wedge d \mid a \prec c, \neg b \prec d\}$. By Lemma 3.2, F is a concordant filter if $c \wedge d \neq 0$ for any $a \prec c$, $\neg b \prec d$. But if $c \wedge d = 0$, then $a \prec c \leq \neg d \prec \neg\neg b = b$, so $a \prec b$ by (A3), contradicting our assumption. Hence F is a concordant filter. Now if $\neg a \in F$ there must be some $e \in F$ such that $e \prec \neg a$. But this means that $a \prec \neg e$ and therefore $\neg e \in F$, a contradiction. Similarly for any concordant $G \supseteq F$, if $b \in G$ there must be some $e \in G$ such that $e \prec b$. But then $\neg b \prec \neg e$ so $\neg e \in F \subseteq G$, a contradiction. Therefore $F \in \overline{\widehat{a}} \setminus {\downarrow}\widehat{b}$. □

Putting Corollary 3.5 and Lemma 3.6 together yields the desired representation theorem.

Theorem 3.7 *Let $V = (B, \prec)$ be a de Vries algebra with dual filter space (S_V, σ). Then V is isomorphic to $(\mathsf{RO}(S_V), \ll)$.*

4 De Vries Spaces

In this section, we characterize the choice-free duals of de Vries algebras. In other words, we give an axiomatization of topological spaces of the form (S_V, σ) for some de Vries algebra V. In order to do so, we first need to introduce the following separation axioms:

Definition 4.1

(i) A topological space (X, τ) is *order-regular* if for any closed set B and any $x \notin {\Uparrow}B$, there are disjoint open sets U, V such that $x \in U$ and ${\Uparrow}B \subseteq V$.

(ii) A topological space (X, τ) is *order-normal* if for any closed set A and any regular closed set B such that A is disjoint from ${\Uparrow}B$, there are disjoint open sets U and V such that $A \subseteq {\downarrow}U$ and ${\Uparrow}B \subseteq V$.

Order-regularity and order-normality are straightforward variations of the usual regularity and normality separation axioms in general topology. Separation axioms for ordered topological spaces have been studied in the past [22,23,25], but here we are concerned with a very specific kind of ordered topological spaces, in which the order is determined by the topology. In the case of compact T_1 spaces, these separation properties are essentially equivalent to Hausdorffness:

Lemma 4.2 *Let (X, τ) be a compact T_1-space. The following are equivalent:*

(i) *(X, τ) is Hausdorff;*

(ii) *(X, τ) is order-regular;*

(iii) *(X, τ) is order-normal and order-regular.*

Proof. Recall that if (X, τ) is T_1, then the specialization preorder on X is just the identity relation. Thus a T_1 order-regular space is regular Hausdorff, which implies that it is also Hausdorff. As compact Hausdorff spaces are also regular, this shows that (i) and (ii) are equivalent. Moreover, (iii) clearly implies (ii), and compact Hausdorff spaces are also normal, which for T_1 spaces implies order-normality, showing that (i) implies (iii). □

As we will now see, for spaces in which the regular opens are also order-regular open, order-normality suffices to establish that they form a de Vries algebras when equipped with the relation $\ll$ defined above.

Lemma 4.3 *Let (X, τ) be an order-normal space such that $\mathsf{RO}(X) \subseteq \mathcal{RO}(X)$. For any $U, V \in \mathsf{RO}(X)$, let $U \ll V$ iff $\overline{U} \subseteq {\downarrow}V$. Then $(\mathsf{RO}(X), \ll)$ is a de Vries algebra.*

Proof. Since $\mathsf{RO}(X)$ is a complete Boolean algebra, we only need to verify axioms (A1)-(A7):

(A1) $X \ll X$. Immediate.

(A2) $U \ll V$ implies $U \subseteq V$. Suppose $\overline{U} \subseteq {\downarrow}V$. Taking complements, this yields $-{\downarrow}V \subseteq U^{\perp}$. Because every closed set is a downset, ${\downarrow}A \subseteq \overline{A}$

for any $A \subseteq X$, so $\downarrow - \downarrow V \subseteq \overline{U^\perp}$. Complementing again, we conclude that $U = U^{\perp\perp} \subseteq \Uparrow\downarrow V = V$.

(A3) $U_1 \subseteq U_2 \ll V_1 \subseteq V_2$ implies $U_1 \ll V_2$. We have the following chain on inclusions: $\overline{U_1} \subseteq \overline{U_2} \subseteq \downarrow V_1 \subseteq \downarrow V_2$.

(A4) $U \ll V_1$ and $U \ll V_2$ together imply $U \ll V_1 \cap V_2$. Suppose both $\overline{U} \subseteq \downarrow V_1$ and $\overline{U} \subseteq \downarrow V_2$. Then since $U, V_1, V_2 \in \mathcal{RO}(X)$, we have that $-\downarrow(U^\perp) \subseteq V_1$ and $-\downarrow(U^\perp) \subseteq V_2$, hence $\downarrow - \downarrow(U^\perp) \subseteq \downarrow(V_1 \cap V_2)$. Now since $U^\perp \in \mathcal{RO}(X)$, we have that $\downarrow - \downarrow(U^\perp) = -(U^\perp) = \overline{U}$, and therefore $\overline{U} \subseteq \downarrow(V_1 \cap V_2)$.

(A5) $U \ll V$ implies $V^\perp \ll U^\perp$. Suppose $\overline{U} \subseteq \downarrow V$. Then $\downarrow - \downarrow V \subseteq \downarrow(U^\perp)$. Taking complements, we have $-\downarrow(U^\perp) \subseteq \Uparrow\downarrow V = V$ since $V \in \mathcal{RO}(X)$. Now since $V \in \mathsf{RO}(X)$, $-V = \overline{V^\perp}$. Therefore, taking complements again, we have that $\overline{V^\perp} \subseteq \downarrow(U^\perp)$, hence $V^\perp \ll U^\perp$.

(A6) $U \ll V$ implies that there is W such that $U \ll W \ll V$. Suppose $\overline{U} \subseteq \downarrow V$, and consider the set $X \setminus \downarrow V = \Uparrow -V$. As $\overline{U}$ and $\Uparrow -V$ are disjoint and $-V$ is regular closed, by order-normality we get some disjoint open sets W_1, W_2 such that $\overline{U} \subseteq \downarrow W_1$ and $\Uparrow -V \subseteq W_2$. Note that this implies that $\overline{W_1} \cap \Uparrow -V = \emptyset$, and therefore $\overline{W_1} \subseteq \downarrow V$. Letting $W = W_1^{\perp\perp}$, we have that $\overline{U} \subseteq \downarrow W_1 \subseteq \downarrow W$, and $\overline{W} = \overline{W_1} \subseteq \downarrow V$.

(A7) If $U \neq \emptyset$ then there is $V \neq \emptyset$ such that $V \ll U$. Suppose $U \neq \emptyset$ and let $x \in U$. Consider $X \setminus \downarrow U = \Uparrow -U$. Note that $\downarrow x$ is disjoint from $\Uparrow -U$ and is closed, since $\downarrow x = \bigcap_{x \notin U, U \in \tau} -U$. By order-normality, we have disjoint open sets V_1 and V_2 such that $\downarrow x \subseteq \downarrow V_1$ and $\Uparrow -U \subseteq V_2$. Note that this implies that $V_1 \neq \emptyset$ and that $\overline{V_1} \subseteq \downarrow U$. Now letting $V = V_1^{\perp\perp}$, it follows that $V \neq \emptyset$ and $\overline{V} = \overline{V_1} \subseteq \downarrow U$.

Thus $(\mathsf{RO}(X), \ll)$ is a de Vries algebra. □

We are now in a position to define the choice-free duals of de Vries algebras:

Definition 4.4 A *de Vries space* (*dV*-space for short) is a topological space (X, τ) satisfying the following conditions:

(i) (X, τ) is T_0, compact and order-normal;

(ii) $\mathsf{RO}(X)$ is a basis for τ and $\mathsf{RO}(X) \subseteq \mathcal{RO}(X)$;

(iii) For every $x \in \mathsf{RO}(X)$, $\mathsf{RO}(x) = \{U \in \mathsf{RO}(X) \mid x \in U\}$ is a concordant filter on $\mathsf{RO}(X)$, and for every filter F on $\mathsf{RO}(X)$, there is $x \in X$ such that $\uparrow F = \mathsf{RO}(x)$.

Lemma 4.5 *Let $V = (B, \prec)$ be a de Vries algebra. Then (S_V, σ) is an order-regular dV-space.*

Proof. Condition (ii) follows from Lemma 3.4, and condition (iii) is immediate from Theorem 3.7, so we only have to check that (S_V, σ) is T_0, compact, order-normal and order-regular. It is routine to verify that (S_V, σ) is T_0. For compactness, note that $\uparrow\{1\} = \{1\} \in S_V$, so if $S_V \subseteq \bigcup_{a \in A} \widehat{a}$ for some $A \subseteq V$, it follows that $1 \in A$ and thus S_V has a finite subcover.

For order-regularity, let $B = \bigcap_{a \in A} -\widehat{a}$ be a closed set and $F \notin \uparrow B$. Then $F \in \downarrow -B = \bigcup_{a \in A} \downarrow \widehat{a}$, which means that there is $a \in A$ and $c \prec a$ such that $\neg c \notin F$. By ($A6$) there is some $c' \in V$ such that $c \prec c' \prec a$. Now $F \in -\widehat{\neg c} = \overline{\widehat{c}} \subseteq \downarrow \widehat{c'}$, and $-\widehat{\neg c'} = \overline{c'} \subseteq \downarrow \widehat{a}$, so $\uparrow B \subseteq \widehat{\neg c'}$. Thus $\widehat{c'}$ and $\widehat{\neg c'}$ are the required open sets.

Finally, for order-normality, fix a closed set $U = \bigcap_{a \in A} -\widehat{a}$ and a regular closed set B such that $\bigcap_{a \in A} -\widehat{a} \subseteq \downarrow -B$. Because B is regular closed it is of the form $-\widehat{b}$ for some $b \in V$. Now consider the concordant filter $F = \{c \in V \mid \neg b \prec c\}$. If there is $G \supseteq F$ such that $G \in \widehat{b}$, then there must be $c \in G$ such that $c \prec b$. But then $\neg c \in F \subseteq G$, and G is not a proper filter, a contradiction. Thus $F \notin \downarrow \widehat{b}$, which means that $F \in \bigcup_{a \in A} \widehat{a}$. Hence there is some $a \in A$ and some $c \in V$ such that $\neg b \prec c \prec a$, which in turn implies that $\neg a \prec \neg c \prec b$. This implies that $-\widehat{a} = \overline{\widehat{\neg a}} \subseteq \downarrow \widehat{\neg c}$, and $-\widehat{c} = \overline{\widehat{\neg c}} \subseteq \downarrow \widehat{b}$, and therefore we have two disjoint open sets, $\widehat{\neg c}$ and $\widehat{c}$, such that $\bigcap_{a \in A} -\widehat{a} \subseteq \downarrow \widehat{\neg c}$ and $\uparrow -\widehat{b} \subseteq \widehat{c}$. □

Theorem 4.6 *Let (X, τ) be a dV-space. Then (X, τ) is homeomorphic to $(S_{(\mathsf{RO}(X), \ll)}, \sigma)$.*

Proof. Let $f : X \to S_{(\mathsf{RO}(X), \ll)}$ be given by $f(x) = \mathsf{RO}(x)$. Then f is well-defined and surjective by condition (iii), and it is injective because X is T_0. Moreover, for any $U \in \mathsf{RO}(X)$, we have that $x \in U$ iff $U \in \mathsf{RO}(x)$ iff $U \in f(x)$ iff $f(x) \in \widehat{U}$. By Theorem 3.7 and since $\mathsf{RO}(X)$ is a basis for X, this is enough to conclude that f is open and continuous and therefore a homeomorphism. □

Note that, as a corollary to Lemma 4.5 and Theorem 4.6, we obtain that any dV-space is order-regular.

Let us conclude this section by characterizing UV-spaces as a special kind of dV-spaces. In order to do so, it is convenient to introduce first the following notion.

Definition 4.7 Let (X, τ) be a topological space. An open subset of (X, τ) is *well rounded* if for any closed set B such that $B \subseteq \downarrow U$, there are disjoint open sets V and W such that $B \subseteq \downarrow V$ and $-W \subseteq \downarrow U$.

Well-rounded subsets of a dV-space will play an important role later on when connecting our results with some standard notions of point-free topology. For now, let us note that a topological space in which every open is well-rounded is also order-regular and order-normal. Indeed, order-normality amounts to the requirement that every regular open set be well-rounded, and order-regularity follows from the fact that $\downarrow x$ is closed in every topological space. While not every open subset of a dV-space is well rounded, this is true for a special class of those, namely UV-spaces.

Lemma 4.8 *A topological space (X, τ) is a UV-space if and only if it is a dV-space such that $(\mathsf{RO}(X), \ll)$ is zero-dimensional.*

Proof. For the left-to-right direction, suppose (X, τ) is a UV-space. We may therefore view it as (S_B, σ) for some Boolean algebra B. This can be used to show that every open set (X, τ) is well-rounded. Indeed, let $U = \bigcap_{a \in A} -\widehat{a}$ and

$V = \bigcap_{c\in C} -\widehat{c}$ for some A, C, subsets of B such that $\bigcap_{a\in A} -\widehat{a} \subseteq {\downarrow}\bigcup_{c\in C}\widehat{c}$. Without loss of generality, we may assume that C is a proper ideal: if $F \in {\downarrow}\widehat{c'}$ for some $c' = c_1 \vee \ldots \vee c_n$ with $c_1, ..., c_n \in C$, then there must be some $i \leq n$ such that $\neg c_i \notin F$, and therefore $F \in {\downarrow}\widehat{c_i}$. So let $F = \{\neg c \mid c \in C\}$ be the dual filter of C. Clearly $F \notin {\downarrow}\bigcup_{c\in C}\widehat{c}$, so $A \cap F \neq \emptyset$. This means that there is some $a \in A$ such that $\neg a \in C$. Thus $U \subseteq -\widehat{a} \subseteq {\downarrow}\widehat{\neg a}$, and $\overline{\widehat{\neg a}} = {\downarrow}\widehat{\neg a} \subseteq {\downarrow}-V$. This shows that (X, τ) satisfies condition (i).

By [7, Prop. 4.3.1], $\mathsf{RO}(X) \subseteq \mathcal{RO}(X)$, so condition (ii) follows from condition (ii) of UV-spaces. Finally, condition (iii) follows from condition (iii) on UV-spaces once we show there is a one-to-one correspondence between concordant filters on $\mathsf{RO}(X)$ and proper filters on B, given by $F \mapsto \{a \in B \mid \widehat{a} \in F\}$. Recall first the observation that for any compact open set U in a UV-space, $\overline{U} = {\downarrow}U$ [7, Prop. 4.1]. This means that $\widehat{a} \ll \widehat{a}$ for any $W \in \mathsf{CO}\mathcal{RO}(X)$. Now assume $U \ll V$ for some $U, V \in \mathsf{RO}(X)$. By [7, Fact 8.2], we may write $U = \bigcup_{a\in A}\widehat{a}$ and $V = \bigcup_{c\in C}\widehat{c}$ for some ideals $A, C \subseteq B$. It is straightforward to see that $\overline{\bigcup_{a\in A}\widehat{a}} \subseteq {\downarrow}\bigcup_{c\in C}\widehat{c}$ implies that there is $c \in C$ such that $a \leq c$ for all $a \in A$, and thus that $\overline{U} \subseteq {\downarrow}\widehat{c}$ for some $c \in C$. Since $\widehat{c} \in \mathsf{CO}\mathcal{RO}(X)$, we also have that $\overline{\widehat{c}} \subseteq {\downarrow}\widehat{c} \subseteq {\downarrow}V$, hence $U \ll \widehat{c} \ll \widehat{c} \ll V$. This shows that $\mathsf{RO}(X)$ is zero-dimensional. Moreover, if F and G are distinct concordant filters on $\mathsf{RO}(X)$, without loss of generality there is $V \in F \setminus G$. But then there is some $U \in F$ such that $U \ll V$, hence $U \ll \widehat{c} \ll V$ for some $c \in B$. This shows that the map $F \mapsto \{c \in B \mid \widehat{c} \in F\}$ is injective. For surjectivity, note that given any proper filter G on B, $G' = \{U \in \mathsf{RO}(X) \mid \exists c \in G : \widehat{c} \ll U\}$ will be a preimage of G. This completes the proof that X is a dV-space such that $\mathsf{RO}(X, \ll)$ is a zero-dimensional de Vries algebra.

Conversely, suppose that (X, τ) is a dV-space such that $(\mathsf{RO}(X), \ll)$ is zero-dimensional. Let $B = \{U \in \mathsf{RO}(X) \mid U \prec U\}$. Clearly, B is a Boolean algebra, so we may consider its dual UV-space $UV(B)$. Since points in X are in one-to-one correspondence with concordant filters on $\mathsf{RO}(X)$, by the same argument as above, there is a one-to-one correspondence between X and $UV(B)$, given by $x \mapsto \{U \in B \mid x \in U\}$. As this map is easily seen to be a homeomorphism, it follows that (X, τ) is a UV-space. □

5 Morphisms

Having established the object part of our duality, the last step to obtain our duality result is to isolate the adequate notion of morphism between dV-spaces. It turns out to be a natural generalization of UV-maps:

Definition 5.1 Let (X, τ_1) and (Y, τ_2) be dV-spaces, and let $\leq_1$ and $\leq_2$ be the specialization orders induced by τ_1 and τ_2 respectively. A *de Vries map* (dV-map for short) $f : X \to Y$ is a continuous function that is also weakly dense, i.e., is such that for any $x \in X$, if $f(x) \leq_2 y$ for some $y \in Y$, then there is $x' \geq_1 x$ such that $y \leq_2 f(x')$.

Let **dVS** be the category of dV-spaces and dV-maps between them. It is straightforward to verify that if $f : (X, \tau_1) \to (Y, \tau_2)$ is weakly dense, then for any upward-closed $V \subseteq Y$, ${\downarrow}f^{-1}[V] = f^{-1}[{\downarrow}V]$. This implies in particular that the preimage of any order-regular open set under a weakly dense map is order-regular open. This fact plays a role in the proof of the following lemma:

Lemma 5.2 *Let $f : (X, \tau_1) \to (Y, \tau_2)$ be a dV-map between dV-spaces. Then $\Phi(f) : (\mathsf{RO}(Y), \ll_2) \to (\mathsf{RO}(X), \ll_1)$, given by $\Phi(f)(U) = (f^{-1}[U])^{\perp\perp}$ for any $U \in \mathsf{RO}(Y)$, is a de Vries morphism.*

Proof. We check the four conditions on de Vries morphisms in turn:

(V1) $\Phi(f)(\emptyset) = \emptyset$. Immediate.

(V2) $\Phi(f)(U \cap V) = \Phi(f)(U) \cap \Phi(f)(V)$. Simply compute that:

$$\begin{aligned}\Phi(f)[U] \cap \Phi(f)[V] &= (f^{-1}[U])^{\perp\perp} \cap (f^{-1}[V])^{\perp\perp} \\ &= (f^{-1}[U] \cap f^{-1}[V])^{\perp\perp} \\ &= \Phi(f)(U \cap V).\end{aligned}$$

(V3) $U \ll_2 V$ implies $(\Phi(f)(U^{\perp}))^{\perp} \ll_1 \Phi(f)(V)$. Suppose $\overline{U} \subseteq {\downarrow}V$. This means that $f^{-1}[\overline{U}] \subseteq f^{-1}[{\downarrow}V] = {\downarrow}f^{-1}[V]$, since f is weakly dense. Complementing, this gives us

$$-{\downarrow}f^{-1}[V] \subseteq f^{-1}[U^{\perp}] \subseteq \Phi(f)(U^{\perp}),$$

which, using the fact that $f^{-1}[V]$ is order-regular open, yields

$$-{\downarrow}(\Phi(f)(U^{\perp})) \subseteq f^{-1}[V] \subseteq \Phi(f)(V).$$

Taking order-closure and complements again, this yields

$$-{\downarrow}(\Phi(f)(V)) \subseteq \Phi(f)(U^{\perp}) = (\Phi(f)(U^{\perp}))^{\perp\perp},$$

and therefore

$$\overline{(\Phi(f)(U^{\perp}))^{\perp}} \subseteq {\downarrow}(\Phi(f)(V)).$$

(V4) $\Phi(f)(V) = (\bigcup\{\Phi(U) \mid U \ll_2 V\})^{\perp\perp}$. The right-to-left direction is immediate. For the converse, suppose that $f(x) \in V$, and let $x' \geq_1 x$. Then $f(x') \in V$, which implies that there is some $U \ll_2 V$ such that $f(x') \in {\downarrow}U$. This means that $f(x') \leq_2 y$ for some $y \in U$. Since f is weakly-dense, there is $z \geq_1 x'$ such that $f(z) \geq_2 y$, and therefore $z \in \Phi(f)(U)$. This shows that $f^{-1}[V] \subseteq (\bigcup\{\Phi(U) \mid U \ll_2 V\})^{\perp\perp}$, which clearly implies that $\Phi(f)(V) \subseteq (\bigcup\{\Phi(U) \mid U \ll_2 V\})^{\perp\perp}$.

Therefore $\Phi(f)$ is a de Vries morphism. □

It follows that we may define a contravariant functor Φ : **dVS** $\to$ **deV** by letting $\Phi(X, \tau) = (\mathsf{RO}(X), \ll)$ for any dV-space (X, τ) and mapping any $f : (X, \tau_1) \to (Y, \tau_2)$ to $\Phi(f)$ as in Lemma 5.2. It is straightforward to verify that Φ preserves composition and identity arrows. Going from de Vries algebras to dV-spaces requires the following result:

Lemma 5.3 *Let $h : V_1 \to V_2$ be a de Vries morphism. Then the function $\Lambda(h) : (S_{V_2}, \sigma_2) \to (S_{V_1}, \sigma_1)$, given by $\Lambda(h)(F) = {\Uparrow} h^{-1}[F]$ for any $F \in S_{V_2}$, is a dV-map.*

Proof. Let us first show that $\Lambda(h)$ is continuous. For any $a \in V_1$ we compute:

$$\begin{aligned}\Lambda(h)^{-1}[\widehat{a}] &= \{F \in S_{V_2} \mid \Lambda(h)(F) \in \widehat{a}\}\\ &= \{F \in S_{V_2} \mid a \in {\Uparrow} h^{-1}[F]\}\\ &= \{F \in S_{V_2} \mid \exists c \prec a : h(c) \in F\}\\ &= \bigcup_{c \prec a} \widehat{h(c)}.\end{aligned}$$

Now we check that $\Lambda(h)$ is weakly dense. Let $F \in S_{V_2}$ and $G \in S_{V_1}$ be such that ${\Uparrow} h^{-1}[F] \subseteq G$. I claim that

$$H = \{a \in V_2 \mid a \geq \neg h(\neg c) \wedge d \text{ for some } c \in G, d \in F\}$$

is a concordant filter. To see that this is a proper subset of V_2, note that if $h(\neg c) \in F$ for some $c \in G$, then there is $c' \prec c \in G$, which implies that $\neg c \prec \neg c'$ and thus that $\neg c' \in G$, a contradiction. To see that H is a filter, it is enough to verify that for any $c_1, c_2 \in G$, $\neg h(\neg c_1) \wedge \neg h(\neg c_2) \in H$. Since G is concordant, there is $c' \in G$ such that $c' \prec c_1 \wedge c_2$, which implies that

$$\neg h(\neg c) \prec \neg h(\neg(c_1 \wedge c_2)) \leq \neg h(\neg c_1) \wedge \neg h(\neg c_2),$$

and therefore $\neg h(\neg c_1) \wedge \neg h(\neg c_2) \in H$. A similar argument shows that ${\Uparrow} H = H$, which completes the proof of the claim.

By construction of H, $F \subseteq H$. Moreover, if $c \in G$, then there are $c_1, c_2 \in G$ such that $c_2 \prec c_1 \prec c$. Then $\neg h(\neg c_2) \prec h(c_1)$, which shows that $c \in \Lambda(h)[H]$, and therefore $G \subseteq \Lambda(h)[H]$. This completes the proof that $\Lambda(h)$ is a dV-map.□

We can therefore construct a functor $\Lambda : \mathbf{deV} \to \mathbf{dVS}$ by mapping any de Vries algebra V to $\Lambda(V) = (S_V, \sigma)$ and any de Vries morphism h to $\Lambda(h)$ as in Lemma 5.3. Again, it is straightforward to verify that Λ preserves composition and identity arrows. We conclude with the main result of this paper:

Theorem 5.4 *The functors Φ and Λ establish a dual equivalence between the categories* **deV** *and* **dVS**.

Proof. In light of Theorems 3.7 and 4.6, we only need to verify that:

(i) for any de Vries morphism $h : V_1 \to V_2$, $\Phi\Lambda(h)(\widehat{a}) = \widehat{h(a)}$ for any $a \in V_1$;

(ii) for any dV-map $f : (X, \tau_1) \to (Y, \tau_2)$, $\Lambda\Phi(f)(\mathsf{RO}(x)) = \mathsf{RO}(f(x))$ for any $x \in X$.

For (i), it is enough to compute that:

$$\begin{aligned}\Phi\Lambda(h)(\widehat{a}) &= ((\Lambda(h))^{-1}[\widehat{a}])^{\perp\perp}\\ &= (\bigcup\{\widehat{h(b)} \mid b \prec a\})^{\perp\perp}\\ &= \widehat{\bigvee\{h(b) \mid b \prec a\}}\\ &= \widehat{h(a)}.\end{aligned}$$

For (ii), we first compute that:

$$\begin{aligned}\Lambda\Phi(f)(\mathsf{RO}(x)) &= \Uparrow(\Phi(f))^{-1}[\mathsf{RO}(x)]\\ &= \Uparrow\{U \mid \Phi(f)(U) \in \mathsf{RO}(x)\}\\ &= \Uparrow\{U \mid (f^{-1}[U])^{\perp\perp} \in \mathsf{RO}(x)\}.\end{aligned}$$

Now if $V \in \mathsf{RO}(f(x))$, then there is $U \ll V$ such that $U \in \mathsf{RO}(f(x))$, and therefore $x \in f^{-1}[U] \subseteq (f^{-1}[U])^{\perp\perp}$, and hence $V \in \Lambda\Phi(\mathsf{RO}(x))$. For the converse direction, suppose that $x \in (f^{-1}[U])^{\perp\perp}$ and that $U \ll V$ for some $U, V \in \mathsf{RO}(Y)$. I claim that for any $y \geq_2 f(x)$, $y \in \overline{U}$. Since $\overline{U} \subseteq \downarrow V$, this implies that $f(x) \in \Uparrow\downarrow V$, and therefore that $v \in \mathsf{RO}(f(x))$. For the proof of the claim, note first that $x \in (f^{-1}[U])^{\perp\perp}$ implies that there is some regular open set $Z \in \mathsf{RO}(x)$ such that for any $x' \in Z$ and any open set W, $x' \in Z \cap W$ implies that $W \cap f^{-1}[U] \neq \emptyset$. Now fix some $y \in Y$ such that $f(x) \leq_2 y$. Since f is weakly dense, there is $x' \geq_1 x$ such that $y \leq_2 f(x')$. The claim is proved if $f(x') \in \overline{U}$. Assume towards a contradiction that this is not the case. Then $x' \in f^{-1}[U^{\perp}]$, which is open since f is continuous. But $x \leq_1 x'$ implies that $x' \in Z$, so $f^{-1}[U^{\perp}] \cap f^{-1}[U] \neq \emptyset$, a contradiction. This completes the proof.□

6 Point-Free and Hyperspace Approaches

In this section, we relate dV-spaces to compact regular frames. Because both the equivalence between de Vries algebras and compact regular frames on the one hand, and the duality between de Vries algebras and dV-spaces on the other hand, do not rely on the Axiom of Choice, we already know that there is a choice-free duality between compact regular frames and dV-spaces. In order to describe this duality more precisely, we first need the following lemma:

Lemma 6.1 *For any de Vries algebra $V = (B, \prec)$, there is an order isomorphism between the poset $w\mathcal{ORO}(\Lambda(V))$ of well-rounded $\mathcal{ORO}$ subsets of $\Lambda(V)$ and the round ideals on V.*

Proof. Let $\mathfrak{R}(V)$ be the frame of all round ideals of V and $w\mathcal{ORO}(\Lambda(V))$ the poset of all well-rounded $\mathcal{ORO}$ subsets of $\Lambda(V)$ ordered by inclusion. Define $\alpha : \mathfrak{R}(V) \to w\mathcal{ORO}(\Lambda(V))$ as $I \mapsto \bigcup_{b\in I} \widehat{b}$ and $\beta : w\mathcal{ORO}(\Lambda(V)) \to \mathfrak{R}(V)$ as $U \mapsto \{b \in B \mid \overline{\widehat{b}} \subseteq \downarrow U\}$. I claim that α and β are order preserving and inverses of one another.

First, let us verify that $\alpha(I)$ is a well-rounded $\mathcal{ORO}$ set for any round ideal I. Clearly, for any round ideal I, $\alpha(I)$ is open. To see that it is order-regular open, suppose $F \notin \alpha(I)$ for some concordant filter F, and consider the set $G = \{c \wedge \neg d \mid c \in F, d \in I\}$. I claim that $G \in \Lambda(V)$. Since I is round, $I^\delta = \{\neg d \mid d \in I\}$ is a concordant filter, so by Lemma 3.2 we only need to verify that $c \wedge \neg d \neq 0$ for any $c \in F, d \in I$. But this follows immediately from the assumption that $F \notin \alpha(I)$. Thus $G \in \Lambda(V)$, and clearly we have that $F \subseteq G$ and $G \notin \downarrow\alpha(I)$. Thus $F \notin \Uparrow\!\downarrow\alpha(I)$, which shows that $\alpha(I) \in \mathcal{RO}(\Lambda(V))$. Finally, let us check that $\alpha(I)$ is well-rounded. Suppose $W \subseteq \downarrow\alpha(I)$ is a closed set of the form $\bigcap_{a \in A} -\widehat{a}$ for some $A \subseteq B$. Note that I^δ is a concordant filter and clearly $I^\delta \notin \downarrow\alpha(I)$, so $A \cap I^\delta \neq \emptyset$. This means that $\neg a \in I$ for some $a \in A$. But then $\widehat{\neg a}$ and $\widehat{a}$ are the required open sets. This completes the proof that $\alpha(I) \in w\mathcal{ORO}(\Lambda(V))$.

Conversely, let us show that for any $w\mathcal{ORO}$ set U, $\beta(U)$ is a round ideal. Clearly, $\beta(U)$ is downward closed. Now suppose we have $a, b \in V$ such that $\overline{\widehat{a}}, \overline{\widehat{b}} \subseteq \downarrow U$. Then $\overline{\widehat{a} \cup \widehat{b}} = \overline{\widehat{a \vee b}} \subseteq \downarrow U$. Since U is well-rounded, there must be disjoint open sets W_1, W_2 such that $\overline{\widehat{a \cup b}} \subseteq \downarrow W_1$ and $-W_2 \subseteq \downarrow U$. By Theorem 3.7, $W_1^{\perp\perp} = \widehat{c}$ for some $c \in V$, and it is straightforward to verify that $\overline{\widehat{a \vee b}} \subseteq \downarrow\widehat{c}$ and $\overline{\widehat{c}} \subseteq \downarrow U$. This shows that $a \vee b \prec c$ and that $c \in \beta(U)$, establishing that $\beta(U)$ is a round ideal.

It is immediate to see that both maps are order preserving, so we only need to show that they are inverses of one another. Let I be a round ideal. If $b \in I$, then $b \prec a$ for some $a \in I$. But then $\overline{\widehat{b}} \subseteq \downarrow\widehat{a} \subseteq \downarrow\alpha(I)$, so $b \in \beta\alpha(I)$. Conversely, assume $b \notin I$, and let $F = \{c \wedge \neg d \mid b \prec c, d \in I\}$. If $c \wedge \neg d \leq \neg b$ for some $d \in I$ and c such that $b \prec c$, then $b \wedge \neg d \prec c \wedge \neg d \leq \neg b$, hence $b \wedge \neg d \leq b \wedge \neg d \wedge \neg b \leq 0$. But this implies that $b \leq d$ and thus that $b \in I$, contradicting our assumption. Thus $\neg b \notin F$. By Lemma 3.2, this shows that F is a concordant filter and moreover $F \in \overline{\widehat{b}}$ by Lemma 3.4 (iv). But clearly $F \notin \downarrow\alpha(I) = \bigcup_{d \in I} \downarrow\widehat{d}$. By contraposition, it follows that if $\overline{\widehat{b}} \subseteq \downarrow\alpha(I)$, then $b \in I$. This shows that $\beta\alpha(I) = I$ for any round ideal I.

Similarly, if $F \in U$ for $U \in w\mathcal{ORO}(\Lambda(V))$, then since U is open there must be some $a \in F$ such that $\widehat{a} \subseteq U$. Since F is concordant, there is $b \prec a$ for some $b \in F$. But then $F \in \widehat{b}$ and $\overline{\widehat{b}} \subseteq \downarrow\widehat{a} \subseteq \downarrow U$, so $F \in \alpha\beta(U)$. Conversely, suppose $F \in \alpha\beta(U)$. Then there is $a \in F$ such that $\overline{\widehat{a}} \subseteq \downarrow U$. Since $\overline{\widehat{a}} = -\widehat{\neg a}$ and for any concordant $G \supseteq F$, $\neg a \notin G$, it follows that $F \in \Uparrow\!\downarrow U = U$. This shows that $\alpha\beta(U) = U$, which completes the proof. □

As a consequence, the well-rounded $\mathcal{ORO}$ subsets of any dV-space form a compact regular frame, and we can lift this correspondence to a functor $w\mathcal{ORO} : \mathbf{deV} \to \mathbf{KRFrm}$. To go from compact regular frames to dV-spaces, it is enough to recall that the round ideals on a de Vries algebra V are precisely the duals of concordant filters on V. Thus given a compact regular frame L, we may simply define the topological space $\Xi(L) = (L^-, \delta)$, where $L^- = L \setminus 1_L$ and δ is the topology generated by sets of the form $\breve{a} = \{b \mid \neg a \prec b\}$ for any $a \in L$. Indeed, since L is isomorphic to $\mathfrak{R}(B(L))$, we may think of any $b \in L$ as a round ideal I_b on the de Vries algebra $(B(L), \prec)$ such that for any $b \in L$

and $c \in B(L)$, $c \prec b$ iff $\neg c \in I_b$. But since $B(L) = \{\neg a \mid a \in L\}$, we therefore have for any $a \in L$:

$$\begin{aligned}\breve{a} &= \{b \in L^- \mid \neg a \prec b\} \\ &= \{b \in L^- \mid \neg\neg a \in I_b\} \\ &= \{b \in L^- \mid \neg a \in (I_b)^\delta\} \\ &= \{b \in L^- \mid (I_b)^\delta \in \widehat{\neg a}\}.\end{aligned}$$

This shows that the correspondence $b \mapsto (I_b)^\delta$ is a homeomorphism between $\Xi(L)$ and $\Lambda(B(L))$. It follows that Ξ lifts to a contravariant functor from **KRFrm** to **dVS** and that we have the following theorem:

Theorem 6.2 *For any compact regular frame L, L is isomorphic to $w\mathcal{ORO}(\Xi(L))$. Conversely, any dV-space (X, τ) is homeomorphic to $\Xi(w\mathcal{ORO}(X))$. Moreover, $w\mathcal{ORO}$ and Ξ establish a duality between* **KRFrm** *and* **dVS**.

We may think of Theorem 6.2 as establishing a choice-free analogue of Isbell duality. In the presence of (BPI), any compact regular frame is spatial, meaning that any compact regular frame L is isomorphic to $\Omega(pt(L))$, or equivalently that any compact regular frame is the lattice of open sets of some compact Hausdorff space. In our choice-free case, we do not represent L as the open sets of a topological space (since doing so would imply Isbell duality), but only as the well-rounded order-regular open sets of a dV-space. We might however be interested in better understanding the relationship between the Isbell dual of a compact regular frame and its de Vries dual. The answer turns out to involve the upper Vietoris functor on compact regular frames.

Recall that the Vietoris hyperspace of a compact Hausdorff space (X, τ) is obtained by taking as points the closed subsets of X. That a Vietoris-like construction would play a role in our duality is far from surprising. De Vries had already remarked [8, Theorem I.3.12] that there was a dual order-isomorphism between the closed sets of a compact Hausdorff space and the concordant filters on its de Vries algebra of regular open sets. Moreover, assuming (BPI), the dual UV-space of a Boolean algebra B is homeomorphic to the upper Vietoris hyperspace of the dual Stone space of B [7, Theorem 7.7]. The upper Vietoris construction can also be defined on compact regular locales [7,16,19]:

Definition 6.3 Let L be a compact regular locale. The *upper Vietoris space* of L is the topological space $UV(L) = (L^-, \tau_\Box)$, where $\tau_\Box$ is the topology generated by the sets $\Box a = \{b \in L^- \mid a \vee b = 1_L\}$ for any $a \in L$.

Lemma 6.4 *For any locale L, $\Xi(L)$ is homeomorphic to $UV(L)$.*

Proof. Since $\Xi(L)$ and $UV(L)$ have the same domain, it is enough to show that the two topologies coincide. For any $a \in L$:

$$\begin{aligned}\breve{a} &= \{b \in L^- \mid \neg a \prec b\} \\ &= \{b \in L^- \mid \neg\neg a \vee b = 1_L\} \\ &= \Box\neg\neg a,\end{aligned}$$

which shows that $\delta \subseteq \tau_{\Box}$. Conversely, I claim that for any $a \in L$,

$$\Box a = \bigcup_{b \prec a} \check{b} = \{c \in L \mid \exists b \prec a : \neg b \prec c\}.$$

To see this, notice first that if $\neg b \prec c$ for some $b \prec a$, then $c \vee \neg\neg b = 1_L$ and $\neg\neg b \leq a$, which implies that $a \vee c = 1_L$. This shows the right-to-left inclusion. For the converse, suppose that $a \vee c = 1_L$. Since L is regular, $a = \bigvee\{b \in L \mid b \prec a\}$, and hence $1_L = \bigvee\{b \vee c \mid b \prec a\}$. Since L is also compact, this means that there are $b_1, ..., b_n$ such that $b_1 \vee ... \vee b_n \prec a$ and $c \vee b_1 \vee ... \vee b_n = 1_L$. Letting $b = \neg\neg(b_1 \vee ... \vee b_n)$, it follows that $b \prec a$ and that $\neg b \prec c$. This shows that $\Box a = \bigcup_{b \prec a} \check{b}$, and therefore that $\tau_{\Box} \subseteq \delta$. □

As an immediate corollary of the previous lemma, we obtain the following characterization of dV-spaces, which can be seen as a generalization of Theorem 7.7 in [7]:

Theorem 6.5 *A topological space is a dV-space if and only if it is homeomorphic to the upper Vietoris space of a compact regular locale.*

Let us conclude this section by noting that connections between de Vries duality and the Vietoris functor on compact Hausdorff spaces have already been studied in [3,4]. In particular, the authors define modal de Vries algebras and prove that they are the duals of coalgebras of the Vietoris functor. For lack of space, we leave as an open problem the relationship between modal de Vries algebras and dV-spaces.

7 Two Applications

We conclude by briefly mentioning two straightforward applications of the duality presented here. The first one is a choice-free version of Tychonoff's Theorem for compact Hausdorff spaces and the second one deals with the topological semantics of the strong implication calculus defined in [7].

7.1 The Choice-free Product of Compact Hausdorff Spaces

The following is a well-known result in point-free topology [16,17,24]:

Lemma 7.1 *The category* **KRFrm** *is closed under coproducts.*

By the duality obtained in the previous section, this means that the category of dV-spaces is closed under products. This means that a version of Tychonoff's Theorem for dV-spaces (the product in **dVS** of a family of dV-spaces is compact) holds in a choice-free setting. Moreover, this also motivates the following definition.

Definition 7.2 Let $\{(X_i, \tau_i)\}_{i \in I}$ be a family of compact Hausdorff spaces. The *choice free product* of this family is the dV-space $\Xi(\bigoplus_{i \in I} \Omega(X_i))$.

As an immediate consequence of the results in the previous section, we get the following choice-free Tychonoff Theorem for compact Hausdorff spaces:

Theorem 7.3 *The choice-free product of a family of compact Hausdorff spaces $\{(X_i, \tau_i)\}_{i\in I}$ is compact. Moreover, under (BPI), it is homeomorphic to the upper-Vietoris space of $\prod_{i\in I}(X_i, \tau_i)$.*

It is worth contrasting this result to one that can be obtained using Isbell duality. Since the category of compact regular frames is closed under coproducts, it can be proved without appealing to the Axiom of Choice that the coproduct of the frames of opens of any family $\{(X_i, \tau_i)\}_{i\in I}$ of compact Hausdorff spaces is a compact frame. Under (BPI), this frame is precisely the frame of opens of the product of $\{(X_i, \tau_i)\}_{i\in I}$ in the category of topological spaces. In the absence of (BPI) however, it may fail to be spatial. We may therefore see Theorem 7.3 as a *semi-pointfree* version of Tychonoff's Theorem, that is choice-free yet remains spatial.

7.2 Topological Completeness for the Symmetric Strong Implication Calculus

De Vries duality has been used in [5] to prove that the Symmetric Strong Implication Calculus $\mathsf{S^2IC}$ is sound and complete with respect to the class of compact Hausdorff spaces. This calculus is obtained by adding a binary relation $\rightsquigarrow$ to the language of classical propositional calculus, to be interpreted as a *strong implication* connective. Given a contact algebra $(B, \prec)$, one can interpret the strong implication connective by letting $a \rightsquigarrow b = 1_B$ if $a \prec b$ and $a \rightsquigarrow b = 0$ otherwise. This gives rise to a binary normal and additive operator $\Delta(a,b) := \neg(a \rightsquigarrow \neg b)$, meaning that one may think of the pair $(B, \rightsquigarrow)$ as a BAO. For details on the axiomatization of $\mathsf{S^2IC}$, we refer to [5]. In order to provide a choice-free topological semantics for $\mathsf{S^2IC}$, we introduce the following notion:

Definition 7.4 A *de Vries topological model* is a triple (X, τ, V) such that (X, τ) is a dV-space, and V is a valuation such that for any formulas φ, ψ of $\mathsf{S^2IC}$:

- If φ is propositional letter p, then $V(\varphi) \in \mathsf{RO}(X)$;
- $V(\neg\varphi) = V(\varphi)^{\perp}$ and $V(\varphi \wedge \psi) = V(\varphi) \cap V(\psi)$;
- $V(\varphi \rightsquigarrow \psi) = X$ if $\overline{V(\varphi)} \subseteq {\downarrow}V(\psi)$ and $V(\varphi \rightsquigarrow \psi) = \emptyset$ otherwise.

A formula φ is *valid* on a dV-space (X, τ) iff $V(\varphi) = X$ for any de Vries topological model (X, τ, V).

As a consequence of Theorem 5.4, we have the following result, which does not assume the Axiom of Choice:

Theorem 7.5 *The system $\mathsf{S^2IC}$ is sound and complete with respect to the class of all dV-spaces.*

Proof. By Theorem 5.10 and Remark 5.11 in [5], de Vries algebras provide a sound and complete algebraic semantics for $\mathsf{S^2IC}$, and this result can be obtained choice-free. Combining this result with Theorem 5.4, it follows that dV-spaces also provide a choice-free sound and complete semantics for $\mathsf{S^2IC}$. □

Since dV-spaces constitute a choice-free, filter-based representation of de Vries algebras, we may think of our choice-free de Vries duality as providing a possibility semantics for the logic of region-based theories of space, just as the choice-free Stone duality through UV-spaces serves as a foundation for possibility semantics for classical and modal propositional logic [12,13,14].

Acknowledgments

I thank Wes Holliday and an anonymous referee for helpful comments that greatly improved the clarity of the paper.

References

[1] Banaschewski, B. and A. Pultr, *Booleanization*, Cahiers de Topologie et Géométrie Différentielle Catégoriques **37** (1996), pp. 41–60.

[2] Bezhanishvili, G., *De Vries algebras and compact regular frames*, Applied Categorical Structures **20** (2012), pp. 569–582.

[3] Bezhanishvili, G., N. Bezhanishvili and J. Harding, *Modal compact Hausdorff spaces*, Journal of Logic and Computation **25** (2015), pp. 1–35.

[4] Bezhanishvili, G., N. Bezhanishvili and J. Harding, *Modal operators on compact regular frames and de Vries algebras*, Applied Categorical Structures **23** (2015), pp. 365–379.

[5] Bezhanishvili, G., N. Bezhanishvili, T. Santoli and Y. Venema, *A strict implication calculus for compact Hausdorff spaces*, Annals of Pure and Applied Logic **170** (2019), p. 102714.

[6] Bezhanishvili, G., N. Bezhanishvili, S. Sourabh and Y. Venema, *Irreducible equivalence relations, Gleason spaces, and de Vries duality*, Applied Categorical Structures **25** (2017), pp. 381–401.

[7] Bezhanishvili, N. and W. H. Holliday, *Choice-free Stone duality*, Journal of Symbolic Logic **85** (2020), pp. 109–148.

[8] De Vries, H., "Compact Spaces and Compactification: An Algebraic Approach," Ph.D. thesis, University of Amsterdam (1962).

[9] Dimov, G., E. Ivanova-Dimova and D. Vakarelov, *A generalization of the Stone duality theorem*, Topology and its Applications **221** (2017), pp. 237–261.

[10] Dimov, G. and D. Vakarelov, *Contact algebras and region-based theory of space: a proximity approach–I*, Fundamenta Informaticae **74** (2006), pp. 209–249.

[11] Fedorchuk, V. V., *Boolean δ-algebras and quasiopen mappings*, Siberian Mathematical Journal **14** (1973), pp. 759–767.

[12] Holliday, W. H., *Possibility semantics*, in: M. Fitting, editor, *Selected Topics from Contemporary Logics*, Landscapes in Logic **2**, College Publications, 2021 pp. 363–476. URL `https://escholarship.org/uc/item/9ts1b228`

[13] Holliday, W. H., *Possibility frames and forcing for modal logic* (Forthcoming in The Australasian Journal of Logic).
URL `https://escholarship.org/uc/item/0tm6b30q`

[14] Holliday, W. H. and T. Litak, *Complete additivity and modal incompleteness*, The Review of Symbolic Logic **12** (2019), pp. 487–535.

[15] Isbell, J. R., *Atomless parts of spaces*, Mathematica Scandinavica **31** (1972), pp. 5–32.

[16] Johnstone, P. J., "Stone Spaces," Number 3 in Cambridge Studies in Advanced Mathematics, Cambridge University Press, Cambridge, 1982.

[17] Johnstone, P. T., *Tychonoff's theorem without the axiom of choice*, Fund. Math **113** (1981), pp. 21–35.

[18] Johnstone, P. T., *The point of pointless topology*, Bulletin (New Series) of the American Mathematical Society **8** (1983), pp. 41–53.

[19] Johnstone, P. T., *Vietoris locales and localic semilattices*, in: *Continuous lattices and their applications*, CRC Press, 2020 pp. 155–180.

[20] Lando, T. and D. Scott, *A calculus of regions respecting both measure and topology*, Journal of Philosophical Logic **48** (2019), pp. 825–850.
[21] Massas, G., *A semi-constructive approach to the hyperreal line*, arXiv preprint arXiv:2201.10818 (2022).
[22] McCartan, S., *Separation axioms for topological ordered spaces*, Mathematical Proceedings of the Cambridge Philosophical Society **64** (1968), pp. 965–973.
[23] Nachbin, L., "Topology and order," Van Nostrand Mathematical Studies, 1965.
[24] Picado, J. and A. Pultr, "Frames and Locales: topology without points," Springer Science & Business Media, 2011.
[25] Priestley, H. A., *Representations of distributive lattices by means of ordered Stone spaces*, Bulletin of the London Mathematical Society **2** (1970), pp. 186–190.
[26] Schechter, E., "Handbook of Analysis and its Foundations," Academic Press, 1996.
[27] Stone, M. H., *The theory of representation for Boolean algebras*, Transactions of the American Mathematical Society **40** (1936), pp. 37–111.
[28] Vietoris, L., *Bereiche zweiter Ordnung*, Monatshefte für Mathematik und Physik **32** (1922), pp. 258–280.

Intuitionistic Modality and Beth Semantics

Satoru Niki [1]

Ruhr University Bochum,
Universitätsstraße 150,
44801 Bochum

Abstract

One of the standard methods to understand intuitionistic logic is to see it through its semantical counterpart. Kripke semantics in particular offers an insightful interpretation in terms of the growth of knowledge. This interpretation is extended to modal operators in the case of intuitionistic modal logics. In the framework of M. Božić and K. Došen, the four intuitionistic modalities (necessity, possibility, impossibility and non-necessity) are characterised in a uniform manner, suggesting that they share a type of assumption on how modal notions interact with the growth of knowledge. On the other hand, there is another intuitionistic semantics called Beth semantics, which supports a different perspective on the notion of the growth of knowledge. A natural question then is how the four modalities appear from this alternative perspective. The main observation of this paper is how the above-mentioned uniformity breaks down in Beth semantics, which hints that the modalities can be seen to be based on different conceptions of the growth of knowledge. In addition, we look at the Beth correspondence theory of some modal principles, which is then applied to obtain a Beth completeness of a paraconsistent system by R. Sylvan.

Keywords: Beth semantics, Intuitionistic Logic, Modal Logic, Negation, Paraconsistent Logic.

1 Introduction

Relational semantics is a valuable tool to provide intuitive interpretations to logical systems. An important example of this type of semantics is *Kripke semantics* for intuitionistic logic [22]. This semantics captures the validity of formulas using the pictures of the growth of knowledge (of an agent), depicted with a partially ordered set of worlds (or information states). A characteristic feature of intuitionistic Kripke semantics is the valuation of implication; in order to establish that an implication holds at a world, one has to look at

[1] This research was supported by funding from the German Academic Exchange Service (DAAD) during the production, and funding from the European Research Council (ERC) under the European Union's Horizon 2020 research and innovation programme, grant agreement ERC-2020-ADG, 101018280, ConLog, during the revision. The author would also like to thank the anonymous referees for their helpful comments.

not only the world in question, but also all later worlds. This feature plays an essential role in invalidating constructively unacceptable logical principles such as the law of excluded middle.

Kripke semantics is one of the standard semantics for intuitionistic logic. There is, however, another type of relational semantics that is worthy of attention. This semantics, introduced by E.W. Beth [3], is accordingly called *Beth semantics*. Like Kripke semantics, Beth semantics can also be understood to depict the growth of knowledge, but as the main differences, a model of Beth semantics (a) typically involves infinitely many worlds, (b) has a stronger constraint on the valuation of propositional variables, and (c) uses a different condition for the forcing of disjunction, which allows a disjunction to hold at a world without either of the disjuncts being so. The last point is analysed in depth by W.H. Holliday [20].

Beth semantics is known to be more general than Kripke semantics (see [4,31]), and so we may embed a Kripke model into a Beth model, whereas the converse is not always possible. On the other hand, as far as logic is concerned, the two semantics capture the same logic, namely intuitionistic logic. The situation can change, however, when we add an additional operator. For instance, it is observed in [24] that when a type of alternative negation called *empirical negation* [7,8] is considered, an identical forcing condition (falsity at the root of a model) ends up in two different logics, depending on which of the semantics is used. In other words, it is reflected in the logics that the two semantics offer different philosophical interpretations of the notion of 'growth of knowledge'.

Since empirical negation can be seen as a kind of modal operator in intuitionistic setting, this gives a motivation to investigate modal operators in Beth semantics. We note there already exists a related investigation by R. Goldblatt [19], which established that the lax operator [15] can be captured by the forcing condition of (classical) possibility operator in cover semantics, which is a generalisation of Beth semantics. Analogous observations are also established in [19] for bimodal systems **CK** [23] and **CS4** [1]. D. Rogozin [27] also applies the framework for the systems of intuitionistic epistemic logic by S. Artemov and T. Protopopescu [2].

Intuitionistic modal logics have been studied by various authors since Fitch [17] (see e.g. [18,28,34] for overviews of early approaches). Among these, we shall base our enquiry on the systems investigated by K. Došen and M. Božić [5,10,11,12,13]. In particular, we shall mainly focus on (i) the system $\mathbf{HK}\Box$ for necessity operator, and (ii) its negative counterpart, the system $\mathbf{HK}\Box'$ for non-necessity operator. We shall also consider operators for possibility and impossibility, and the corresponding systems $\mathbf{HK}\Diamond$ and $\mathbf{HK}\Diamond'$[2]. These systems can all be seen as intuitionistic analogues of the classical modal logic **K**. Kripke semantics for the logics are defined in a uniform way, by (a) using the same forcing condition as the classical case, and (b) employing a frame condition specifying the interaction of the intuitionistic ordering and the modal acces-

[2] See [14] for an investigation of further operators definable in these systems.

sibility relation. Our central observation will be that this uniformity breaks down once we move to Beth semantics.

The structure of this paper is as follows. Firstly, we shall look at the necessity operator: after introducing the axiomatisation and the corresponding Kripke semantics, we shall formulate the Beth semantics for **HK□** and show the soundness and completeness with respect to the axiomatic system. Secondly, we shall look at the non-necessity operator; we shall analogously formulate the Beth semantics for **HK□**′ using a different forcing condition from the Kripke one, and show the soundness and completeness. The third section treats the cases for **HK◇** and **HK◇**′. We shall point out that these cases are different from the first two cases in that they require a new condition for the soundness. This is followed by a step into the correspondence theory of a couple of modal principles in the Beth semantics of necessity and non-necessity. The result is then used in the final section to obtain a Beth semantics for a paraconsistent logic of R. Sylvan [29].

2 Beth semantics for intuitionistic necessity

In this section, we shall give a complete Beth semantics for Božić and Došen [5]'s system **HK□**, which is an intuitionistic analogue of the classical modal logic **K** (with □ primitive). We first specify the language to be used.

Definition 2.1 [$\mathcal{L}_\Box$] We shall use the following language $\mathcal{L}_\Box$:

$$A ::= p \mid \bot \mid A \wedge A \mid A \vee A \mid A \to A \mid \Box A.$$

We shall adopt $A \leftrightarrow B$ as an abbreviation for $(A \to B) \wedge (B \to A)$.

As an inessential difference from [5], we take $\bot$ as primitive and define a negation $\neg A$ as $A \to \bot$.

2.1 HK□: axiomatisation and Kripke semantics

Next, we give a Hilbert-style axiomatisation of **HK□**.

Definition 2.2 [**HK□**] We define the system **HK□** in $\mathcal{L}_\Box$ by the following axiom schemata and rules.

$(A\to(B\to C))\to((A\to B)\to(A\to C))$	(S)	$\bot \to A$	(EFQ)
$A \to (B \to A)$	(K)	$(\Box A \wedge \Box B) \to \Box(A \wedge B)$	(P1)
$(A\to B)\to((A\to C)\to(A\to B\wedge C))$	(CI)	$\Box(A \to A)$	(P2)
$A_1 \wedge A_2 \to A_i$	(CE)	$\dfrac{A \qquad A \to B}{B}$	(MP)
$A_i \to A_1 \vee A_2$	(DI)		
$(A\to C)\to((B\to C)\to(A\vee B\to C))$	(DE)	$\dfrac{A \to B}{\Box A \to \Box B}$	(RM)

where $i \in \{1, 2\}$. A formula A is called a *theorem* of **HK□** if there is a finite sequence $B_1, \ldots, B_n \equiv A$ of formulas such that each B_i is either an instance of one of the axiom schemata, or obtained from the preceding formulas by one of the rules. A *proof* of A from a set of formulas Γ in **HK□** is a finite sequence $B_1, \ldots, B_n \equiv A$, where each B_i is either an element of Γ, a theorem of **HK□**, or obtained from previous items in the sequence by (MP). We shall write $\Gamma \vdash_{\mathbf{HK}\Box} A$ to denote the provability. In particular, when $\Gamma = \emptyset$ we shall write $\vdash_{\mathbf{HK}\Box} A$. Similar conventions apply to later systems.

We have the next Kripke semantics corresponding to the proof system.

Definition 2.3 [Kripke semantics for **HK□**] A *Kripke frame* $\mathcal{F}$ is a triple $(K, \leq, R)$, where $(K, \leq)$ is a non-empty partially ordered set, and $R \subseteq K \times K$ such that $\leq R \subseteq R \leq$.[3] A *Kripke model* $\mathcal{M}$ then is a pair $(\mathcal{F}, \mathcal{V})$ where $\mathcal{V}$ is a mapping assigning to each propositional variable p a set $\mathcal{V}(p) \subseteq K$. We require each $\mathcal{V}(p)$ to be *upward closed*, i.e. $k \in \mathcal{V}(p)$ and $k' \geq k$ implies $k' \in \mathcal{V}(p)$. $\mathcal{V}$ is uniquely extended to the forcing $\Vdash_{kp}$ of formulas by the clauses below.

$$
\begin{aligned}
k \Vdash_{kp} \bot &\text{ iff } \textit{never}. \\
k \Vdash_{kp} p &\text{ iff } k \in \mathcal{V}(p). \\
k \Vdash_{kp} A \wedge B &\text{ iff } k \Vdash_{kp} A \text{ and } k \Vdash_{kp} B. \\
k \Vdash_{kp} A \vee B &\text{ iff } k \Vdash_{kp} A \text{ or } k \Vdash_{kp} B. \\
k \Vdash_{kp} A \to B &\text{ iff } \text{ for all } k' \geq k (k' \Vdash_{kp} A \text{ implies } k' \Vdash_{kp} B). \\
k \Vdash_{kp} \Box A &\text{ iff } \text{ for all } k' R^{-1} k (k' \Vdash_{kp} A).
\end{aligned}
$$

where $R^{-1} = \{(k', k) : kRk'\}$. We shall write $\mathcal{M} \vDash_{kp} A$ if $k \Vdash_{kp} A$ for any k in $\mathcal{M}$. We write $\mathcal{F} \vDash_{kp} A$ if $\mathcal{M} \vDash_{kp} A$ for any $\mathcal{M}$ with $\mathcal{F}$ as the frame. Finally, we write $\vDash_{kp} A$ if $\mathcal{M} \vDash_{kp} A$ for all $\mathcal{M}$.

The next propositions are established in [5, Lemma 2, Theorem 1].

Proposition 2.4 (Upward closure) *$k \Vdash_{kp} A$ and $k' \geq k$ implies $k' \Vdash_{kp} A$.*

Theorem 2.5 (Kripke soundness and completeness of HK□) $\vdash_{\mathbf{HK}\Box} A$ *if and only if* $\vDash_{kp} A$.

2.2 HK□: Beth semantics

We now set out to formulate a Beth semantics for **HK□**. An important point to note is that Beth semantics uses a stronger condition than the upward closure of valuation. This means we have to generalise the condition $\leq R \subseteq R \leq$, which is put in place in order to preserve upward closure.

Beth semantics can be based either on trees (e.g. [31]) or posets (e.g. [32]). In this paper we take the former approach as it appears simpler and perhaps more loyal to the intuitionistic picture. At the same time, the conditions we will see are not necessarily optimised for trees, in view of possible generalisations into posets.

We define a *tree* $\mathcal{T} = (T, \preceq)$ to be a poset s.t. there is the minimum element $g \in T$ and for each t, $\{t' : t' \preceq t\}$ is linearly ordered. A *path* of T is then a maximal linearly ordered subset of T.[4] We will use $\alpha, \beta, \ldots$ to denote a path

---

[3] where $R_1 R_2 = \{(x, z) : \exists y (x R_1 y \wedge y R_2 z)\}$.

[4] For a more constructive approach, we can also take a tree to be a certain collection of sequences of natural numbers [30, p.186].

in $\mathcal{T}$.

Definition 2.6 [Beth semantics for **HK□**] We define a *Beth frame* $\mathcal{F}$ for **HK□** as a triple $(B, \preceq, S)$, where $(B, \preceq)$ is a tree s.t. $\forall b \in B \exists b' \in B(b' \succ b)$ (i.e. it is a *spread*), and $S \subseteq B \times B$. S also has to satisfy the next condition:

$$\forall b, b' \in B(b \preceq S\ b' \Rightarrow \exists \alpha \ni b \forall c \in \alpha(c\ S \preceq b'))$$

A *Beth model* $\mathcal{M}$ for **HK□** is then a pair $(\mathcal{F}, \mathcal{V})$ such that $\mathcal{V}$ is a mapping assigning to each propositional variable p a set $\mathcal{V}(p)$ subject to the following condition (*covering property*):

$$b \in \mathcal{V}(p) \text{ if and only if } \forall \alpha \ni b \exists b' \in \alpha(b' \in \mathcal{V}(p)).$$

$\mathcal{V}$ is now extended to the forcing $\Vdash_{bp}$ by the clauses below.

$$\begin{aligned}
b \Vdash_{bp} \bot &\text{ iff } \textit{never}.\\
b \Vdash_{bp} p &\text{ iff } b \in \mathcal{V}(p).\\
b \Vdash_{bp} A \wedge B &\text{ iff } b \Vdash_{bp} A \text{ and } b \Vdash_{bp} B.\\
b \Vdash_{bp} A \vee B &\text{ iff } \forall \alpha \ni b \exists b' \in \alpha(b' \Vdash_{bp} A \text{ or } b' \Vdash_{bp} B).\\
b \Vdash_{bp} A \to B &\text{ iff } \forall b' \succeq b(b' \Vdash_{bp} A \text{ implies } b' \Vdash_{bp} B).\\
b \Vdash_{bp} \Box A &\text{ iff } \forall b' S^{-1} b(b' \Vdash_{bp} A).
\end{aligned}$$

We shall write $\mathcal{M} \vDash_{bp} A$ if $b \Vdash_{bp} A$ for any b in $\mathcal{M}$. We write $\mathcal{F} \vDash_{bp} A$ if $\mathcal{M} \vDash_{bp} A$ for any $\mathcal{M}$ with $\mathcal{F}$ as the frame, and $\vDash_{bp} A$ if $\mathcal{M} \vDash_{bp} A$ for all $\mathcal{M}$.

The conditions for disjunction are changed from those of Kripke semantics and satisfy the following properties.

Proposition 2.7 (Covering property)
(i) $b \Vdash_{bp} A$ *if and only if* $\forall \alpha \ni b \exists b' \in \alpha(b' \Vdash_{bp} A)$.
(ii) $b \Vdash_{bp} A$ *and* $b' \succeq b$ *implies* $b' \Vdash_{bp} A$.

Proof. We proceed by induction on the complexity of A, and treat (i) and (ii) simultaneously. Note that the cases for (ii) follows from the cases for (i), because if $b' \succeq b$, then any $\alpha \ni b'$ must pass through b as well. [5] Hence $b \Vdash_{bs} A$ implies $b' \Vdash_{bs} A$ from the case of A for (i).

As for (i), the left-to-right direction is always immediate. For the right-to-left direction, the case $A \equiv p$ follows from the covering property of $\mathcal{V}$. The case for conjunction is straightforward, and the case for disjunction is analogous to the case for non-necessity we shall look at in the next section. For $A \equiv B \to C$, if $\forall \alpha \ni b \exists b' \in \alpha(b' \Vdash_{bp} B \to C)$, then $c \succeq b$ implies $\forall \alpha \ni c \exists c' \in \alpha(c' \Vdash_{bp} B \to C)$. Hence if $c \Vdash_{bp} B$, then for any α s.t. $c \in \alpha$ there is $c' \in \alpha$ with $c' \Vdash_{bp} B \to C$. We have either $c \preceq c'$ or $c \succeq c'$, but in each case, the later world must (using (ii) for B in one of the cases) force C. So $\forall \alpha \ni c \exists c' \in \alpha(c' \Vdash_{bp} C)$. Thus by I.H. again $c \Vdash_{bp} C$. Therefore $b \Vdash_{bp} B \to C$.

[5] It is essential here that we are considering trees and not posets.

When $A \equiv \Box B$, then if $\forall \alpha \ni b \exists b' \in \alpha(b' \Vdash_{bp} \Box B)$, suppose bSc. Then $b \preceq S\ c$, and so there is $\alpha \ni b$ s.t. $\forall c' \in \alpha(c'\ S \preceq\ c)$, by the frame condition of S. Thus in particular, $b'\ S\preceq\ c$ for some b' s.t. $b' \Vdash_{bp} \Box B$. Hence there exists c' such that $b'Sc'$ and $c' \preceq c$, which implies $c' \Vdash_{bp} B$. So by (ii) for B, we infer $c \Vdash_{bp} B$. Therefore $b \Vdash_{bp} \Box B$. □

Theorem 2.8 (Beth soundness of HK□) *If* $\vdash_{\mathbf{HK}\Box} A$ *then* $\models_{bp} A$.

Proof. See Appendix. □

In order to prove the Beth completeness, we shall employ an embedding of Kripke models to Beth models. This method is similar to the one given in [31] for the Beth completeness of intuitionistic logic, except that we put an extra world to guarantee the existence of the root in the constructed Beth model.

Theorem 2.9 (Beth completeness of HK□) *If* $\models_{bp} A$ *then* $\vdash_{\mathbf{HK}\Box} A$.

Proof. See Appendix. □

3 Beth semantics for non-necessity

In this section, we continue the investigation of Beth semantics for intuitionistic modality. We now turn our attention to the non-necessity operator, denoted by $\Box'$ in [10]. Also, $\bot$ turns out to be definable in the system we will consider; we shall however include $\bot$ in the language for uniformity.

Definition 3.1 [$\mathcal{L}_{\Box'}$] We shall use the following language $\mathcal{L}_{\Box'}$:

$$A ::= p \mid \bot \mid A \wedge A \mid A \vee A \mid A \to A \mid \Box' A.$$

3.1 HK□′: axiomatisation and Kripke semantics

The next system gives [6] an axiomatisation of the system **HK□′** in [10].

Definition 3.2 [**HK□′**] We define the system **HK□′** in $\mathcal{L}_{\Box'}$ with the axiom schemata (S)–(EFQ), the rule (MP) in addition to the following axiom schemata and a rule.

$$\Box'(A \wedge B) \to (\Box' A \vee \Box' B) \quad \text{(N1)}$$
$$\Box'(A \to A) \to B \quad \text{(N2)}$$
$$\frac{A \to B}{\Box' B \to \Box' A} \quad \text{(RC)}$$

The notion of a proof is then defined as in **HK□**.

The following Kripke semantics is given to **HK□′**.

Definition 3.3 [Kripke semantics for **HK□′**] A *Kripke frame* $\mathcal{F}$ for **HK□′** is a triple $(K, \leq, R)$, where $(K, \leq)$ is as before, and $R \subseteq K \times K$ is such that $\geq R\ \subseteq\ R \leq$. A *Kripke model* $\mathcal{M}$ for **HK□′** then is a pair $(\mathcal{F}, \mathcal{V})$ where $\mathcal{V}$ is again an upward closed assignment of propositional variables to worlds. The forcing $\Vdash_{kn}$ is defined similarly to $\Vdash_{kp}$; for non-necessity, we use the next clause.

$$k \Vdash_{kn} \Box' A \text{ iff for some } k' R^{-1} k (k' \nVdash_{kn} A).$$

[6] Strictly speaking, there is a difference in that Došen's system again has $\neg$ as primitive.

We will use $\models_{kn}$ for the validity with respect to this semantics.

The next propositions are established in [10, Lemma 10, Theorem 3].

Proposition 3.4 (Upward closure) *$k \Vdash_{kn} A$ and $k' \geq k$ implies $k' \Vdash_{kn} A$.*

Theorem 3.5 (Kripke soundness and completeness of HK□′)
$\vdash_{\mathbf{HK}\Box'} A$ if and only if $\models_{kn} A$.

3.2 HK□′: Beth semantics

We now move on to the definition of Beth semantics for **HK□′**. In Kripke semantics, the clauses for necessity and non-necessity are in a sense dual of each other, as we have seen. In comparison, the clauses will turn out to be quite different in Beth semantics, a situation comparable to those of conjunction and disjunction. This is because the condition on the accessibility relation, even when generalised for Beth semantics, guarantees only the upward closure of forcing, and not the covering property. On the other hand, once we have a forcing condition for non-necessity that assures covering property, there is no longer a need for a condition on the accessibility relation: upward closure follows automatically from the covering property. As a result, Beth semantics for non-necessity will have simpler frames, but models are not necessarily so.

Definition 3.6 [Beth semantics for **HK□′**] We define a *Beth frame* $\mathcal{F}$ for **HK□′** as a triple $(B, \preceq, S)$, where $(B, \preceq)$ is as before, and $S \subseteq B \times B$ without any other conditions. A *Beth model* $\mathcal{M}$ for **HK□′** is a pair $(\mathcal{F}, \mathcal{V})$ such that $\mathcal{V}$ is an assignment satisfying the covering property. For the forcing $\Vdash_{bn}$, it has the same clauses as $\Vdash_{bp}$, except the clause for non-necessity, which is:

$$b \Vdash_{bn} \Box' A \text{ iff } \forall \alpha \ni b \exists b' \in \alpha \exists c S^{-1} b' (c \nVdash_{bn} A).$$

We shall write $\models_{bn}$ for the validity with respect to this semantics.

Remark 3.7 The Beth clause for disjunction may be interpreted as saying that one can assert a disjunction even when one can not assert either of the disjuncts. This perhaps better capture the informal usage of disjunction than the Kripke clause. In a similar manner, the Beth clause for non-necessity appears to be telling that one can assert “It is not necessary that A” even when all the states that can be referred to (accessed) support A. This possibility is however under the condition that a world which does not support A will eventually become accessible in all cases.

Let us check that the forcing condition establishes the covering property.

Proposition 3.8 (Covering property)
$b \Vdash_{bn} A$ if and only if $\forall \alpha \ni b \exists b' \in \alpha (b' \Vdash_{bn} A)$.

Proof. We look at the right-to-left direction for the case of non-necessity. When $A \equiv \Box' B$, then if $\forall \alpha \ni b \exists b' \in \alpha (b' \Vdash_{bn} \Box' B)$, it follows that

$$\forall \alpha \ni b \exists b' \in \alpha (\forall \beta \ni b' \exists c \in \beta \exists c' S^{-1} c (c' \nVdash_{bn} B)).$$

In particular, since $\alpha \ni b'$, we infer $\forall \alpha \ni b \exists b' \in \alpha \exists c' S^{-1} b' (c' \nVdash_{bn} B)$. Hence $b \Vdash_{bn} \Box' B$. □

Corollary 3.9 (Upward closure) *$b \Vdash_{bn} A$ and $b' \succeq b$ implies $b' \Vdash_{bn} A$.*

We leave the proofs of Beth soundness and completeness for **HK□′** in Appendix.

Theorem 3.10 (Beth soundness of HK□′) *If $\vdash_{\mathbf{HK}\Box'} A$ then $\models_{bn} A$.*

Theorem 3.11 (Beth completeness of HK□′) *If $\models_{bn} A$ then $\vdash_{\mathbf{HK}\Box'} A$.*

4 Beth semantics for possibility and impossibility

Having looked at the necessity and non-necessity operators, let us move on to consider their counterparts, possibility and impossibility operators [5,10]. At first glance, one may expect that the Beth semantics for these operators are identical to the semantics for non-necessity and necessity modulo a simple rewriting of the accessibility and forcing conditions. However, as we shall find out, there is a subtle difference for these operators, which is perhaps not visible in Kripke semantics. It thus points to a divergence of the two semantics for intuitionistic modal logics at this basic level.

Let us again start with specifying the languages.

Definition 4.1 $[\mathcal{L}_\Diamond, \mathcal{L}_{\Diamond'}]$ The languages $\mathcal{L}_\Diamond$ and $\mathcal{L}_{\Diamond'}$ are defined as follows.

$$A ::= p \mid \bot \mid A \wedge A \mid A \vee A \mid A \to A \mid \Diamond A.$$
$$A ::= p \mid \bot \mid A \wedge A \mid A \vee A \mid A \to A \mid \Diamond' A.$$

4.1 HK◇, HK◇′: axiomatisation and Kripke semantics

The proof systems **HK◇** and **HK◇′** quite resemble **HK□** and **HK□′**.

Definition 4.2 [**HK◇**, **HK◇′**] We define the system **HK◇** in $\mathcal{L}_\Diamond$ and **HK◇′** in $\mathcal{L}_{\Diamond'}$ with the axiom schemata (S)–(EFQ), rule (MP) in addition to the following axiom schemata and a rule.
For **HK◇**:

$$\Diamond(A \vee B) \to (\Diamond A \vee \Diamond B) \quad \text{(Q1)} \qquad \neg\Diamond\neg(A \to A) \quad \text{(Q2)} \qquad \frac{A \to B}{\Diamond A \to \Diamond B} \quad \text{(RM2)}$$

For **HK◇′**:

$$(\Diamond' A \wedge \Diamond' B) \to \Diamond'(A \vee B) \quad \text{(O1)} \qquad \Diamond'\neg(A \to A) \quad \text{(O2)} \qquad \frac{A \to B}{\Diamond' B \to \Diamond' A} \quad \text{(RC2)}$$

The notions of a proof are then defined as in **HK□**.

Definition 4.3 [Kripke semantics for **HK◇**, **HK◇′**] The Kripke semantics for **HK◇** and **HK◇′** are defined similarly to those of **HK□** and **HK□′**. The only difference for the frames is the frame condition for R, given respectively as $\geq R \subseteq R \geq$ and $\leq R \subseteq R \geq$. The forcing conditions $\Vdash_{kq}$ and $\Vdash_{ko}$ have the next conditions for the respective modality.

$$k \Vdash_{kq} \Diamond A \text{ iff for some } k' R^{-1} k (k' \Vdash_{kq} A).$$
$$k \Vdash_{ko} \Diamond' A \text{ iff for all } k' R^{-1} k (k' \nVdash_{ko} A).$$

We will use $\models_{kq}, \models_{ko}$ for the validity in the semantics, respectively.

Proposition 4.4 (Upward closure) *For $x \in \{q, o\}$, $k \models_{kx} A$ and $k' \geq k$ implies $k' \Vdash_{kx} A$.*

Proof. See [5, Lemma 16] and [10, Lemma 2]. □

Theorem 4.5 (Kripke soundness and completeness of HK◇, HK◇′)
(i) $\vdash_{\mathbf{HK}\Diamond} A$ *if and only if* $\models_{kq} A$.
(ii) $\vdash_{\mathbf{HK}\Diamond'} A$ *if and only if* $\models_{ko} A$.

Proof. See [5, Theorem 4] and [10, Theorem 1]. □

4.2 HK◇, HK◇′: Beth semantics

Definition 4.6 [Beth semantics for **HK◇**] Beth semantics for **HK◇** is mostly identical to that of **HK□′**. We need an extra condition that $\forall b, b' \in B(bSb' \Rightarrow \exists \alpha \ni b' \forall c \in \alpha(c \succeq b' \Rightarrow bSc))$. Possibility has the next condition in $\Vdash_{bq}$:

$$b \Vdash_{bq} \Diamond A \text{ iff } \forall \alpha \ni b \exists b' \in \alpha \exists c S^{-1} b' (c \Vdash_{bn} A).$$

We shall write $\models_{bq}$ for the validity with respect to this semantics.

Definition 4.7 [Beth semantics for **HK◇′**] We define Beth semantics for **HK◇′** in a similar way as that of **HK□**. We have the following condition on the accessibility relation:

$$\forall b, b' \in B(b \preceq S\ b' \Rightarrow \exists \alpha \ni b \forall c \in \alpha(c\ S \succeq b'))$$

In addition, we again require that $\forall b, b' \in B(bSb' \Rightarrow \exists \alpha \ni b' \forall c \in \alpha(c \succeq b' \Rightarrow bSc))$. The forcing $\Vdash_{bo}$ has the next clause for impossibility.

$$b \Vdash_{bo} \Diamond' A \text{ iff } \forall b' S^{-1} b (b' \nVdash_{bo} A).$$

We shall then denote the validity in the semantics by $\models_{bo}$.

Remark 4.8 As we shall see, the condition we added for the above two semantics is required to show the Beth soundness. The additional condition ensures that all Beth frames behave like the ones obtained from Kripke frames; this allows us to overcome the difference in the two semantics.

Proposition 4.9 (Covering property) *For $x \in \{q, o\}$, the following hold.*
(i) $b \Vdash_{bx} A$ *if and only if* $\forall \alpha \ni b \exists b' \in \alpha(b' \Vdash_{bx} A)$.
(ii)$b \Vdash_{bx} A$ *and* $b' \succeq b$ *implies* $b' \Vdash_{bx} A$.

Proof. Analogous to the cases for **HK□′** and **HK□**, respectively. □

Let us denote by $\models_{bq'}$ and $\models_{bo'}$ the validity with respect to Beth semantics for **HK◇** and **HK◇′** minus the condition that $\forall b, b' \in B(bSb' \Rightarrow \exists \alpha \ni b' \forall c \in \alpha(c \succeq b' \Rightarrow bSc))$. Then we observe that they fail one of the axiom schemata of the corresponding system.

Proposition 4.10
(i) $\not\models_{bq'} \Diamond(A \vee B) \to (\Diamond A \vee \Diamond B)$.
(ii) $\not\models_{bo'} (\Diamond' A \wedge \Diamond' B) \to \Diamond'(A \vee B)$.

Proof. Let $(B, \preceq)$ be a tree defined from two paths $\alpha_1 = (g, b_1, b_2, \ldots)$ and $\alpha_2 = (g, b'_1, b'_2, \ldots)$ branching at g.

For (i), we set a Beth frame $\mathcal{F} = (B, \preceq, S)$ with $S = \{(g, g)\}$. Then define a Beth model $\mathcal{M} = (\mathcal{F}, \mathcal{V})$ where $\mathcal{V}(p) = \{b_i : i \in \mathbb{N}\}$ and $\mathcal{V}(q) = \{b'_i : i \in \mathbb{N}\}$. Then it is straightforward to check that $\mathcal{M}$ satisfies the covering property, and so is well-defined. Now, since $g \Vdash_{bq'} p \vee q$ and gSg it holds that $g \Vdash_{bq'} \Diamond(p \vee q)$. On the other hand, for any $b \in \alpha_1$ we have $b \nVdash_{bq'} \Diamond p$ and $b \nVdash_{bq'} \Diamond q$, because the only world accessible by S is g, which neither forces p nor q. Hence $g \nVdash_{bq'} \Diamond p \vee \Diamond q$ and so $g \nVdash_{bq'} \Diamond(p \vee q) \to (\Diamond p \vee \Diamond q)$.

For (ii), we use the same tree and $\mathcal{V}$ but define $\mathcal{F}$ with $S = \{(b, g) : b \in B\}$. Then we have to check that $b \preceq S\ b' \Rightarrow \exists \alpha \ni b \forall c \in \alpha (c\ S \succeq\ b')$. This is not difficult to see, as the premise holds precisely for arbitrary b and $b' = g$, and taking any $\alpha \ni b$ allows us to reach the conclusion. Thus the model is again well-defined. Now we readily observe that $g \Vdash_{bo'} \Diamond' p \wedge \Diamond' q$, but as $g \Vdash_{bo'} p \vee q$ we have to conclude that $g \nVdash_{bo'} \Diamond'(p \vee q)$. Therefore $g \nVdash_{bo'} (\Diamond' p \wedge \Diamond' q) \to \Diamond'(p \vee q)$. □

The same thing does not happen for the models we defined earlier. For soundness and completeness, again we leave them to Appendix.

Theorem 4.11 (Beth soundness of HK◇, HK◇′)
(i) *If* $\vdash_{\mathbf{HK}\Diamond} A$ *then* $\models_{bq} A$.
(ii) *If* $\vdash_{\mathbf{HK}\Diamond'} A$ *then* $\models_{bo} A$.

Theorem 4.12 (Beth completeness of HK◇, HK◇′)
(i) *If* $\models_{bq} A$ *then* $\vdash_{\mathbf{HK}\Diamond} A$.
(ii) *If* $\models_{bo} A$ *then* $\vdash_{\mathbf{HK}\Diamond'} A$.

Remark 4.13 The acceptability of $\Diamond(A \vee B) \to (\Diamond A \vee \Diamond B)$ in intuitionistic modal logic has been questioned by some authors [1,33] on computational grounds. K. Kojima [21] points out that whether to admit the formula depends on the viewpoint one takes for a Kripke model. One viewpoint is that of an *external observer*, who can check all worlds in a model. For such an observer, assuming $\Diamond(A \vee B)$ at a world gives an accessible world in which the disjunction holds. Then he can look at the world in question to find out which of the disjuncts holds there, and thereby conclude $\Diamond A \vee \Diamond B$. Another viewpoint is that of an *internal observer*, who is confined to a world and only has incomplete information for other worlds. Such an agent cannot ascertain which of the disjuncts is the case in the accessible world, and so cannot assert $\Diamond A \vee \Diamond B$. Hence the formula is not acceptable for an internal observer.

5 Beth frame conditions for HK□ and HK□′

In this section, we shall consider a few modal principles for necessity and non-necessity, as a first step to see what kind of frame conditions correspond to them in Beth semantics. Correspondence theory for Beth semantics is already undertaken for some intermediate logics by B. de Beer [9]. He worked with the version of Beth semantics based on posets, and the frame conditions are quite complex compared with the frame conditions for Kripke semantics. Our version of Beth semantics is based on trees, which should simplify the matter to a certain extent: nonetheless, as we shall see, the conditions are still more complex[7] than those of intuitionistic Kripke semantics in [10,11,12,13].

We shall focus on relatively simple frame conditions in Kripke semantics, namely the reflexivity and symmetry for the relation $R\leq$. They correspond to different formulas in **HK□** and **HK□′**. We shall show how even in these simple cases, the frame conditions for the formulas diverge in Beth semantics.

In what follows, $S\preceq$ in some of the conditions can in fact be replaced with S; nonetheless we use the first relation to make the comparison clearer.

5.1 Correspondence for HK□

Let us start with the case for necessity. It is shown in [11] that the reflexivity and symmetry for $R\leq$ correspond to the validity of $\Box A \to A$ and $A \vee \Box\neg\Box A$, respectively. For the first schema, the corresponding Beth-frame condition is still relatively straightforward. The condition may be seen to generalise the notion of reflexivity to each path.

Proposition 5.1 *Let $\mathcal{F} = (B, \preceq, S)$ be a Beth frame for* **HK□**. *Then the following are equivalent.*
(i) $\mathcal{F} \vDash_{bp} \Box A \to A$ *for all* A.
(ii) $\forall\alpha \ni b\exists b' \in \alpha(b\ S\!\preceq b')$.

Proof. From (i) to (ii), we argue by contradiction. Suppose for some b, there is $\alpha \ni b$ s.t. for all $b' \in \alpha$, it is not the case that $b\ S\!\preceq b'$[8]. Then let $\mathcal{V}(p) = \{b' : \forall\beta \ni b'\exists c \in \beta(b\ S\!\preceq c)\}$. In order to see that $\mathcal{V}$ satisfies the covering property, assume $\forall\alpha \ni c\exists c' \in \alpha(c' \in \mathcal{V}(p))$. Then $\forall\alpha \ni c\exists c' \in \alpha\forall\beta \ni c'\exists d \in \beta(b\ S\!\preceq d)$. So taking $\beta = \alpha$ allows us to conclude $c \in \mathcal{V}(p)$. Now if bSb' then $\forall\beta \ni b'\exists c \in \beta(b\ S\!\preceq c)$. So $b' \Vdash_{bp} p$ and thus $b \Vdash_{bp} \Box p$. On the other hand, if $b \Vdash_{bp} p$ then $\forall\beta \ni b\exists c \in \beta(b\ S\!\preceq c)$, contradicting our initial supposition. Thus $b \nVdash_{bp} p$ and so $\mathcal{F} \nvDash_{bp} \Box p \to p$.

From (ii) to (i), if for some b and $b' \succeq b$ we have $b' \Vdash_{bp} \Box A$, then $\forall cS^{-1}b'(c \Vdash_{bp} A)$. By the assumption, in any $\alpha \ni b'$ there is $c \in \alpha$ such that $b'Sc'$ and $c' \preceq c$ for some c'. This means $c' \Vdash_{bp} A$ and so $c \Vdash A$. Thus $\forall\alpha \ni b'\exists c \in \alpha(c \Vdash_{bp} A)$. Hence by the covering property, $b' \Vdash_{bp} A$ and so $b \Vdash_{bp} \Box A \to A$ for all $b \in B$. □

The case for the second schema is a bit more involved because it contains a disjunction. Like the previous case, there is a flavour of symmetry in the condition, but one has to come back only to the original path, and not necessarily to the same world.

Proposition 5.2 *Let $\mathcal{F} = (B, \preceq, S)$ be a Beth frame for* **HK□**. *Then the following are equivalent.*
(i) $\mathcal{F} \vDash_{bp} A \vee \Box\neg\Box A$ *for all* A.
(ii) $\forall\alpha \ni b\exists b' \in \alpha\forall c(b'\ S\!\preceq c \Rightarrow \exists d \in \alpha(c\ S\!\preceq d))$.

---

[7] Note however that **CS4**-modalities in the setting of Goldblatt [19] have simple conditions.

[8] That is to say, $\neg(b\ S\!\preceq b')$.

Proof. From (i) to (ii), we argue by contraposition. Suppose for some b, there exists $\alpha \ni b$ such that $\forall b' \in \alpha \exists c(b' \; S\preceq \; c$ and $\forall d \in \alpha(\neg(c \; S\preceq \; d)))$. Then define a set $\Sigma = \{c : \exists b' \in \alpha(b' \; S\preceq \; c)$ and $\forall d \in \alpha(\neg(c \; S\preceq \; d))\}$ and let $\mathcal{V}(p) = \{d : \forall \gamma \ni d \exists d' \in \gamma \exists c \in \Sigma(c \; S\preceq \; d')\}$. We can check that $\mathcal{V}$ satisfies the covering property similarly to Proposition 5.1. Now for any $b' \in \alpha$, if $b' \Vdash_{bp} p$ then $\exists d' \in \alpha \exists c \in \Sigma(c \; S\preceq \; d')$. However any $c \in \Sigma$ cannot access any $d \in \alpha$ via $S\preceq$, a contradiction. So $b' \nVdash_{bp} p$. Also if $b' \Vdash_{bp} \Box\neg\Box p$, then we first infer from our initial supposition that there is c such that $b' \; S\preceq \; c$ and $\forall d \in \alpha(\neg(c \; S\preceq \; d))$. This means $c \in \Sigma$, which implies that $c \Vdash_{bp} \Box p$. But as $b' \; S\preceq \; c$, by the upward closure we have $c \Vdash_{bp} \neg\Box p$ as well, a contradiction. Hence $b' \nVdash_{bp} \Box\neg\Box p$ as well for any $b' \in \alpha$. Therefore $b \nVdash_{bp} p \vee \Box\neg\Box p$ and so $\mathcal{F} \nvDash_{bp} p \vee \Box\neg\Box p$.

From (ii) to (i), suppose for $\alpha \ni b$ we have $b' \nVdash_{bp} A$ for any $b' \in \alpha$. By assumption, there is $c \in \alpha$ s.t. for all c', $c \; S\preceq \; c'$ implies $c' \; S\preceq \; d$ for some $d \in \alpha$. We claim $c \Vdash_{bp} \Box\neg\Box A$. If $c \; S\preceq \; c'$, then $c' \nVdash_{bp} \Box A$ because $d \in \alpha$ accessible from c' via $S\preceq$ does not force A. It then follows that $c \Vdash_{bp} \Box\neg\Box A$. Therefore $b \Vdash_{bp} A \vee \Box\neg\Box A$ for all b. □

5.2 Correspondence for HK□′

We now move on to the cases for non-necessity. It does not seem to be treated explicitly by Došen, but it is apparent from the results in [13,29] for related systems that the reflexivity of $R\leq$ in Kripke semantics should correspond to an analogue of the *law of excluded middle* for non-necessity.

Proposition 5.3 *Let $\mathcal{F} = (K, \leq, R)$ be a Kripke frame for* **HK□′**. *Then the following are equivalent.*
(i) $\mathcal{F} \vDash_{kn} A \vee \Box' A$ *for all* A.
(ii) $k \; R\leq \; k$.

Proof. See Appendix. □

Now, the frame condition for Beth semantics can be calculated as follows. It can still be seen as a generalisation of reflexivity, but is more complex than the condition for necessity.

Proposition 5.4 *Let $\mathcal{F} = (B, \preceq, S)$ be a Beth frame for* **HK□′**. *Then the following are equivalent.*
(i) $\mathcal{F} \vDash_{bn} A \vee \Box' A$ *for all* A.
(ii) $\forall \alpha \ni b \exists b' \in \alpha \forall \beta \ni b' \exists c \in \beta \exists c' \in \alpha(c \; S\preceq \; c')$.

Proof. From (i) to (ii), we show the contrapositive. Suppose for some $b \in B$,

$$\exists \alpha \ni b \forall b' \in \alpha \exists \beta \ni b' \forall c \in \beta \forall c' \in \alpha(\neg(c \; S\preceq \; c'))$$

Then we let $\mathcal{M} = (\mathcal{F}, \mathcal{V})$ be a model such that $b' \in \mathcal{V}(p) \Leftrightarrow b' \notin \alpha$. We need to check that $\mathcal{V}$ satisfies the covering property. If $\forall \beta \ni b' \exists c \in \beta(c \in \mathcal{V}(p))$, then we have that $b' \in \alpha$ implies $\exists c \in \alpha(c \notin \alpha)$, a contradiction. Hence $b' \notin \alpha$, and so $b' \in \mathcal{V}(p)$. Thus the model satisfies the covering property.

Now for $b' \in \alpha$, on one hand if $b' \Vdash_{bn} p$ then $b' \notin \alpha$, a contradiction. Hence $b' \nVdash_{bn} p$. On the other hand, $b' \Vdash_{bn} \Box' p$ implies $\forall \beta \ni b' \exists c \in \beta \exists c' S^{-1} c(c' \in \alpha)$.

But by our supposition, $\exists\beta \ni b'\forall c \in \beta\forall c' \in \alpha(\neg(c\ S_{\preceq}\ c'))$, a contradiction. Thus $\forall b' \in \alpha(b' \nVdash_{bn} p$ and $b' \nVdash_{bn} \Box' p)$. So $b \nVdash_{bn} p \vee \Box' p$. Therefore $\mathcal{F} \nvDash_{bn} p \vee \Box' p$.

From (ii) to (i), assume that $\mathcal{F}$ satisfies (ii) and let $\mathcal{M}$ be a model and suppose $b \nVdash_{bn} A \vee \Box' A$. Then

$$\exists\alpha \ni b\forall b' \in \alpha(b' \nVdash_{bn} A \text{ and } b' \nVdash_{bn} \Box' A).$$

The latter conjunct means $\forall b' \in \alpha\exists\beta \ni b'\forall c \in \beta\forall c' S^{-1} c(c' \Vdash_{bn} A)$. But as $c' \in \alpha \Rightarrow c' \nVdash_{bn} A$ by the first conjunct, it follows that

$$\forall b' \in \alpha\exists\beta \ni b'\forall c \in \beta\forall c' S^{-1} c(c' \notin \alpha).$$

However, this contradicts the assumption that

$$\exists b' \in \alpha\forall\beta \ni b'\exists c \in \beta\exists c' \in \alpha(c\ S_{\preceq}\ c'),$$

because $d \preceq c'$ and $c' \in \alpha$ implies $d \in \alpha$. Thus $b \Vdash_{bn} A \vee \Box' A$ and as the model was arbitrary, we conclude $\mathcal{F} \vDash_{bn} A \vee \Box' A$. □

The second principle we shall consider is the analogue of *double negation elimination* $\Box'\Box' A \to A$ which corresponds Kripke-semantically to the symmetry of $R_{\leq}$.

Proposition 5.5 *Let $\mathcal{F} = (K, \leq, R)$ be a Kripke frame for* **HK**$\Box'$. *Then the following are equivalent.*
(i) $\mathcal{F} \vDash_{kn} \Box'\Box' A \to A$ *for all A.*
(ii) $k\ R_{\leq}\ k' \Rightarrow k'\ R_{\leq}\ k$.

Proof. From (i) to (ii), we argue by contradiction. If there are k, k' with $k\ R_{\leq}\ k'$ but not $k'\ R_{\leq}\ k$, then take $\mathcal{V}$ s.t. $\mathcal{V}(p) = \{l : k'\ R_{\leq}\ l\}$. Then $k \Vdash_{kn} \Box'\Box' p$ but $k \nVdash_{kn} p$. Hence $\mathcal{F} \nvDash_{kn} \Box'\Box' p \to p$. From (ii) to (i), if $k' \Vdash_{kn} \Box'\Box' A$ for some k and $k' \geq k$, then there is a world l accessible from k', all of whose accessible worlds force A. But by the frame condition, there must be a world l' s.t. lRl' and $l' \leq k'$. Thus $k' \Vdash_{kn} A$. Therefore $\mathcal{F} \vDash_{kn} \Box'\Box' A \to A$. □

The frame condition for Beth semantics is on the other hand rather involved.

Proposition 5.6 *Let $\mathcal{F} = (B, \preceq, S)$ be a Beth frame for* **HK**$\Box'$. *Then the following are equivalent.*
(i) $\mathcal{F} \vDash_{bn} \Box'\Box' A \to A$ *for all A.*
(ii) $\forall\alpha \ni b\exists\beta \ni b\forall b' \in \beta\forall c(b'\ S_{\preceq}\ c \Rightarrow \forall\gamma \ni c\exists c' \in \gamma\exists d \in \alpha(c'\ S_{\preceq}\ d))$.

Proof. From (i) to (ii), we show the contrapositive. Assume for some $b \in B$,

$$\exists\alpha \ni b\forall\beta \ni b\exists b' \in \beta\exists c(b'\ S_{\preceq}\ c \text{ and } \exists\gamma \ni c\forall c' \in \gamma\forall d \in \alpha(\neg(c'\ S_{\preceq}\ d))).$$

We then fix, for each $\beta \ni b$, worlds $b'_\beta \in \beta$, $c_\beta \succeq S^{-1}\ b'_\beta$ and a path $\gamma_\beta \ni c_\beta$ s.t. $\forall c' \in \gamma_\beta\forall d \in \alpha(\neg(c'\ S_{\preceq}\ d))$. We define a model $\mathcal{M} = (\mathcal{F}, \mathcal{V})$ by choosing $\mathcal{V}$ such that:

$$x \in \mathcal{V}(p) \Leftrightarrow \forall\delta \ni x\exists\beta \ni b\exists y \in \gamma_\beta\exists z \in \delta(y\ S_{\preceq}\ z).$$

In order to check that $\mathcal{V}$ satisfies the covering property, suppose $\forall\alpha \ni x\exists y \in \alpha(y \in \mathcal{V}(p))$ but $x \notin \mathcal{V}(p)$. Then there is $\delta \ni x$ such that for all $\beta \ni b$, $y \in \gamma_\beta$ and $z \in \delta$, it is not the case that $y\ S{\preceq}\ z$. But by our supposition, there is $x' \in \delta$ such that $x' \in \mathcal{V}(p)$. So for some $\beta_0 \ni b$, $y_0 \in \gamma_{\beta_0}$ and $z_0 \in \delta$, we have $y_0\ S{\preceq}\ z_0$. Hence we obtain a contradiction, which allows us to conclude $x \in \mathcal{V}(p)$, as desired.

Now, for each $\beta \ni b$, if $c'Sd$ for some $c' \in \gamma_\beta$, then $d \Vdash_{bn} p$ because for each $\delta \ni d$, d itself is an element accessible from c'. Hence $c_\beta \nVdash_{bn} \Box' p$ for each $\beta \ni b$. Therefore $b \Vdash_{bn} \Box'\Box' p$. On the other hand, by assumption, for any $\beta \ni b$ and $c' \in \gamma_\beta$, there is no $d \in \alpha$ such that $c'\ S{\preceq}\ d$. So $b \nVdash_{bn} p$ and thus $\mathcal{F} \nvDash_{bn} \Box'\Box' p \to p$.

From (ii) to (i), let $\mathcal{M} = (\mathcal{F}, \mathcal{V})$ be a model and suppose $b' \Vdash_{bn} \Box'\Box' A$ for some b and $b' \succeq b$. This is equivalent to saying that

$$\forall\beta \ni b'\exists c \in \beta\exists c'S^{-1}c\exists\gamma \ni c'\forall d \in \gamma\forall d'S^{-1}d(d' \Vdash_{bn} A).$$

Now take $\alpha \ni b'$. then by our assumption, there is $\beta \ni b'$ satisfying

$$\circledast : \forall c \in \beta\forall c'(c\ S{\preceq}\ c' \Rightarrow \forall\gamma \ni c'\exists d \in \gamma\exists d' \in \alpha(d\ S{\preceq}\ d')).$$

Also by the equivalence above, there is $c_0 \in \beta$, c_0' satisfying $c_0\ S{\preceq}\ c_0'$ and $\gamma \ni c_0'$ such that $\forall d \in \gamma\forall d'S^{-1}d(d' \Vdash_{bn} A)$. With respect to this γ, it holds from $\circledast$ that $\exists d \in \gamma\exists d' \in \alpha(d\ S{\preceq}\ d')$. Hence $d' \in \alpha$ and $d' \Vdash_{bn} A$ for such d'. Therefore $\forall\alpha \ni b'\exists d' \in \alpha(d' \Vdash_{bn} A)$. Then by the covering property, we infer $b' \Vdash_{bn} A$. Consequently $b \Vdash_{bn} \Box'\Box' A \to A$; so $\mathcal{F} \vDash_{bn} \Box'\Box' A \to A$ for all A. $\Box$

6 An application to paraconsistent logic

In what follows, we shall utilise the frame conditions we discovered to obtain the completeness of a paraconsistent logic called $\mathbf{CC}_\omega$. This system is formulated in [29] as an extension of N.C.A. da Costa's system $\mathbf{C}_\omega$ [6] by (RC). $\mathbf{CC}_\omega$ can be extended to both $\mathbf{TCC}_\omega$ and the logic **daC** of dual negation by G. priest [26]: see [24,25] for more details. Like empirical negation, the forcing conditions for negation in these systems define different systems depending on which semantics (Kripke/Beth) is used. Indeed, empirical negation itself can be obtained as an extension from these systems [8,16]. In order to better understand this phenomenon, it would be desirable to have a characterisation of its basis in terms of Beth semantics. Beth completeness of $\mathbf{CC}_\omega$ is therefore expected to provide us insights into the difference between the semantics.

Let us first look at the axiomatic system for $\mathbf{CC}_\omega$ and its Kripke semantics. For the sake of consistency, we will keep using the symbol $\Box'$ to denote the negation of the system.

Definition 6.1 [$\mathbf{CC}_\omega$] We define the system $\mathbf{CC}_\omega$ by adding to $\mathbf{HK}\Box'$ the next axiom schemata.

$$A \vee \Box' A \quad \text{(LEM)} \qquad\qquad \Box'\Box' A \to A \quad \text{(DNE)}$$

System	(a)	(b)	(c)
HK□	+	−	−
HK□′	−	+	−
HK◇	−	+	+
HK◇′	+	−	+

Table 1
differences among the semantics

It is known that (N1) becomes redundant in $\mathbf{CC}_\omega$. To see this, we note $\Box'(\Box' A \vee \Box' B) \to \Box'\Box' A \wedge \Box'\Box' B$ is derivable in **HK□′** without appealing to (N1). Then use (DNE), (RC) and (DNE) again to obtain $\Box'(A \wedge B) \to \Box' A \vee \Box' B$.

Definition 6.2 [Kripke semantics for $\mathbf{CC}_\omega$] A *Kripke frame* $\mathcal{F}$ for $\mathbf{CC}_\omega$ is a triple $(K, \leq, R)$ where $(K, \leq)$ is as before, and $R \subseteq K \times K$ is a reflexive and symmetric relation such that $\geq R \subseteq R$. (This corresponds to the 'condensed' relation $R_{\Box'}$ in [10, p.11].) Then the notions of Kripke models and validity are defined analogously to the ones in **HK□′**. We will use $\Vdash_{kc}$ etc. as the notations.

Theorem 6.3 (Kripke soundness and completeness of $\mathbf{CC}_\omega$) $\vdash_{\mathbf{CC}_\omega} A$ *if and only if* $\vDash_{kc} A$.

Proof. See [29, p.56]. □

Applying the results from the preceding subsection, we shall define the Beth semantics of $\mathbf{CC}_\omega$ in the following manner.

Definition 6.4 [Beth semantics for $\mathbf{CC}_\omega$] We define a *Beth frame* $\mathcal{F}$ for $\mathbf{CC}_\omega$ as the ones of **HK□′** which satisfies the following conditions.

- $\forall \alpha \ni b \exists b' \in \alpha \forall \beta \ni b' \exists c \in \beta \exists c' \in \alpha (c\ S \preceq\ c')$.
- $\forall \alpha \ni b \exists \beta \ni b \forall b' \in \beta \forall c (b'\ S \preceq\ c \Rightarrow \forall \gamma \ni c \exists c' \in \gamma \exists d \in \alpha (c'\ S \preceq\ d))$

Otherwise, the notions of Beth models and validity are defined as in the ones for **HK□′**. We shall use the notation $\Vdash_{bc}$ etc.

Then as for Beth completeness, we shall again argue via an embedding of Kripke models into Beth models: see Appendix for the details.

Theorem 6.5 (Beth soundness and completeness of $\mathbf{CC}_\omega$) $\vdash_{\mathbf{CC}_\omega} A$ *if and only if* $\vDash_{bc} A$.

7 Conclusion

In this paper, we investigated intuitionistic modal logics in terms of Beth semantics. In the Božić-Došen style of Kripke semantics for intuitionistic modal logic, modal operators are incorporated by means of a frame condition expressing the interaction between the intuitionistic ordering and the accessibility relation. We observed that the situation is rather different in Beth semantics. The semantics for the operators differ on whether (a) there is a frame condition ensuring upward closure; (b) the forcing condition is altered to satisfy the covering property; and (c) it has an extra condition to keep the soundness. Table 1 summarises the properties of each semantics according to these criteria.

The divergence suggests that we have to make different types of assumptions on the notion of growth of knowledge in Beth semantics, depending on which modality is considered. Beth semantics therefore appears more advantageous than Kripke semantics in capturing the particularity of each modal notion. In addition, the Beth forcing condition for non-necessity and possibility might be seen as more natural or preferable in that they allow one to assert a non-necessity (possibility) even when currently accessible worlds say otherwise. Furthermore, some people may see the condition in (a) and its Kripke counterpart as rather ad hoc; so it is perhaps more satisfying that the Beth semantics for non-necessity and possibility does not endorse it.

For future works, a natural direction is to explore more fully the correspondence theory of Beth semantics for each of the modal operators. In particular, the frame conditions for the axiom schemata extending $\mathbf{CC}_\omega$ are of interest for gaining more insights into the difference between Kripke and Beth semantics. Another important direction would be to study the precise relationship between the semantics of this paper and that of Goldblatt. Finally, we only considered one operator at a time, but it should offer new insights into Beth semantics to study the interaction of different operators in a system, as Božić and Došen already did for Kripke semantics.

Appendix

Proof of Theorem 2.8

Proof. By induction on the depth of derivation. Here we look at the cases for (P1), (P2) and (RM). For (P1), if $b' \Vdash_{bs} \Box A \wedge \Box B$ for $b' \succeq b$, then for all $cS^{-1}b'$, it holds that $c \Vdash_{bp} A \wedge B$. So, $b' \Vdash_{bp} \Box(A \wedge B)$. Thus $b \Vdash_{bp} (\Box A \wedge \Box B) \to \Box(A \wedge B)$. For (P2), we have $b' \Vdash A \to A$ for any $b'S^{-1}b$. So $b \Vdash_{bp} \Box(A \to A)$.

For (RM), if $b' \Vdash_{bp} \Box A$ for any b and $b' \succeq b$, then $\forall cS^{-1}b'(c \Vdash_{bp} A)$. Also, by I.H. $\vDash_{bp} A \to B$. Thus $\forall cS^{-1}b'(c \Vdash_{bp} B)$. Hence $b' \Vdash_{bp} \Box B$. Consequently $b \Vdash_{bp} \Box A \to \Box B$; so $\vDash_{bp} \Box A \to \Box B$. □

Proof of Theorem 2.9

Proof. We argue via Kripke completeness. If $\nvdash_{\mathbf{HK}\Box} A$, then by Theorem 2.5 there is a Kripke model $\mathcal{M}_k = ((K, \leq, R), \mathcal{V}_K)$ such that $k \nVdash_{kp} A$ for some $k \in K$. Then we construct a Beth model $\mathcal{M}_b = ((B, \preceq, S), \mathcal{V}_B)$ by the following clauses.

- $B = \{(k_1, \dots, k_n) : k_i \in K \text{ and } k_1 \leq \dots \leq k_n\} \cup \{g\}$.

(i) $(k_1, \dots, k_n) \preceq (k'_1, \dots, k'_m)$ iff $n \leq m$ and $k_i = k'_i$ for $i \leq n$.
(ii) $g \preceq b$ for all $b \in B$ and $b \preceq g \Rightarrow b = g$.

(i) $(k_1, \dots, k_n)S(k'_1, \dots, k'_m)$ iff $k_n R k'_m$.
(ii) $gS(k_1, \dots, k_n)$ iff $k' R k_n$ for some $k' \in K$.

(iii) $\neg(bSg)$ for all $b \in B$.

(i) $(k_1, \ldots, k_n) \in \mathcal{V}_B(p)$ iff $k_n \in \mathcal{V}_K(p)$.

(ii) $g \in \mathcal{V}_B(p)$ iff $k \in \mathcal{V}_K(p)$ for all $k \in K$.

Then it is straightforward to see that $\mathcal{M}_{\mathcal{B}}$ is a Beth model: in particular, if $\forall \alpha \ni b \exists b' \in \alpha(b' \in \mathcal{V}_B(p))$, then for $b = (k_1, \ldots, k_n)$ we look at

$$\alpha = (g, (k_1), \ldots, (k_1, \ldots, k_n), (k_1, \ldots, k_n, k_n), \ldots).$$

Then $k' \in \mathcal{V}_K(p)$ for some $k' \leq k_n$, hence $k_n \in \mathcal{V}_K(p)$ and so $(k_1, \ldots, k_n) \in \mathcal{V}_B(p)$. For $b = g$, for each $k \in K$ we look at $\alpha = (g, (k), (k, k), \ldots)$ to conclude $k \in \mathcal{V}_K(p)$; it then follows that $g \in \mathcal{V}_B(p)$.

We also have to check that S satisfies the condition that for all $b, b' \in B$:

$$b \preceq S\ b' \Rightarrow \exists \alpha \ni b \forall c \in \alpha (c\ S{\preceq}\ b').$$

If $b \preceq S\ b'$, then note that $b' \neq g$; so suppose $b' = (k'_1, \ldots, k'_m)$. We first consider the case when $b = g$. Then for some c, we have $b \preceq c$ and cSb'. Now, regardless of whether $c = g$, there is a world $(k''_1, \ldots, k''_l)$ such that $k''_l R k'_m$. Take $\alpha = (g, (k''_l), (k''_l, k''_l), \ldots)$. Then if $c' \in \alpha$, either $c' = g$ or c' ends with k''_l. In each case, we have $c'Sb'$. Therefore $\exists \alpha \ni b \forall c \in \alpha(c\ S{\preceq}\ b')$, as required. Next, for the case when $b = (k_1, \ldots, k_n)$, first we observe that there is a world $(k_1, \ldots, k_n, \ldots, k_l)$ such that $k_l R k'_m$ and consequently gSb'. For the path, we take $\alpha = (g, (k_1), \ldots, (k_1, \ldots, k_n), (k_1, \ldots, k_n, k_n), \ldots)$. Suppose now $c \in \alpha$. If $c = g$, then it is immediate from the above observation that $c\ S{\preceq}\ b'$. Otherwise, c ends with $k_{n'} \leq k_n$ which satisfies $k_{n'} \leq R\ k'_m$. Then by the frame condition on R, we infer $k_{n'}\ R{\leq}\ k'_m$. Consequently $c\ S{\preceq}\ k'_m$. Thus $\exists \alpha \ni b \forall c \in \alpha(c\ S{\preceq}\ b')$ in this case as well.

We next claim that the following equivalences hold between the two models.

$$k \Vdash_{kp} A \Longleftrightarrow (k_1, \ldots, k) \Vdash_{bp} A.$$
$$\mathcal{M}_k \vDash_{kp} A \Longleftrightarrow g \Vdash_{bp} A.$$

We shall establish this by induction on the complexity of A. The case $A \equiv p$ follows by above. The cases when $A \equiv B \wedge C, B \rightarrow C$ are straightforward. The case when $A \equiv B \vee C$ is similar to the case for non-necessity we shall see later.

When $A = \Box B$, then for the first equivalence, if $k \Vdash_{kp} \Box B$ then $\forall k' R^{-1} k (k' \Vdash_{kp} B)$. Then by the I.H. $\forall b S^{-1}(k_1, \ldots, k)(b \Vdash_{bp} B)$, because g is not accessible from any worlds. Hence $(k_1, \ldots, k) \Vdash_{bp} \Box B$. The converse direction is similarly shown. For the second equivalence, note that $\mathcal{M}_k \vDash_{kp} \Box B$ is equivalent to $\forall k (\exists k'(k'Rk) \Rightarrow k \Vdash_{kp} B)$. By the I.H. and the definition of S, this is further equivalent to $\forall b(gSb \Rightarrow b \Vdash_{bp} B)$, and so to $g \Vdash_{bp} \Box B$. (We are again using the fact that g is not accessible from any world.)

Now since $k \nVdash_{kp} A$, the above equivalence implies $(k) \nVdash_{bp} A$. Therefore $\nvDash_{bp} A$. □

Proof of Theorem 3.10

Proof. By induction on the depth of derivation. Here we look at the cases for (N1),(N2) and (RC). For (N1), if $b' \Vdash_{bn} \Box'(A \wedge B)$ for some b and $b' \succeq b$, then $\forall \alpha \ni b' \exists c \in \alpha \exists c' S^{-1} c(c' \nVdash_{bn} A \wedge B)$. So,

$$\forall \alpha \ni b' \exists c \in \alpha(\exists c' S^{-1} c(c' \nVdash_{bn} A) \text{ or } \exists c' S^{-1} c(c' \nVdash_{bn} B)).$$

Consequently it follows that

$$\forall \alpha {\in} b' \exists c {\in} \alpha(\forall \beta {\in} c \exists c' {\in} \beta \exists d S^{-1} c'(d \nVdash_{bn} A) \text{ or } \forall \beta {\in} c \exists c' {\in} \beta \exists d S^{-1} c'(d \nVdash_{bn} B)).$$

That is to say, $\forall \alpha \ni b' \exists c \in \alpha(c \Vdash_{bn} \Box' A \text{ or } c \Vdash_{bn} \Box' B)$. Thus $b' \Vdash_{bs} \Box' A \vee \Box' B$ and therefore $b \Vdash_{bn} \Box'(A \wedge B) \to (\Box' A \vee \Box' B)$.

For (N2), suppose $b' \Vdash_{bn} \Box'(A \to A)$ for some b and $b' \succeq b$. Consider any $\alpha \ni b'$. Then there has to be $c \in \alpha$ and $c' S^{-1} c$ such that $c' \nVdash_{bn} A \to A$, a contradiction. Thus $b \Vdash_{bn} \Box'(A \to A) \to B$.

For (RC), if $b' \Vdash_{bn} \Box' B$ for some b and $b' \succeq b$, then $\forall \alpha \ni b' \exists c \in \alpha \exists c' S^{-1} c(c' \nVdash_{bn} B)$. Also, by I.H. $\vDash_{bn} A \to B$. Thus $\forall \alpha \ni b' \exists c \in \alpha \exists c' S^{-1} c(c' \nVdash_{bn} A)$. Hence $b' \Vdash_{bn} \Box' A$. Consequently $b \Vdash_{bn} \Box' B \to \Box' A$; so $\vDash_{bn} \Box' B \to \Box' A$. □

Proof of Theorem 3.11

Proof. We argue via Kripke completeness. If $\nvdash_{\mathbf{HK}\Box'} A$, then by Theorem 3.5 there is an **HK□′**-Kripke model $\mathcal{M}_k = ((K, \leq, R), \mathcal{V}_K)$ such that $k \nVdash_{ks} A$ for some $k \in K$. Then we construct an **HK□′**-Beth model $\mathcal{M}_b = ((B, \preceq, S), \mathcal{V}_B)$ in almost the same way as Theorem 2.9, except that:

(i) $gS(k'_1, \ldots, k'_m)$ iff kRk'_m for all $k \in K$.

(ii) $(k_1, \ldots k_n)Sg$ iff $k_n R k$ for all $k \in K$.

(iii) (gSg) iff kRk' for all $k, k' \in K$.

Because there is no condition on S this time, it is immediate that $\mathcal{M}_k$ is a Beth model for **HK□′**.

We again need to check that the next equivalences hold between the models.

$$k \Vdash_{kn} A \Longleftrightarrow (k_1, \ldots, k) \Vdash_{bn} A.$$
$$\mathcal{M}_k \vDash_{kn} A \Longleftrightarrow g \Vdash_{bn} A.$$

When $A \equiv \Box' B$, if $k \Vdash_{kn} \Box' B$, then $\exists k' R_k^{-1} k(k' \nVdash_{kn} B)$. By I.H. this is equivalent to $\exists (k'_1, \ldots, k') S^{-1}(k_1, \ldots, k)((k'_1, \ldots, k') \nVdash_{bn} B)$. Thus we can infer that for any $\alpha \ni (k_1, \ldots, k)$ there is a node b in the path such that $\exists (k'_1 \ldots, k') S^{-1} b((k'_1, \ldots, k') \nVdash_{bn} B)$. Hence $(k_1, \ldots, k) \Vdash_{bn} \Box' B$.

For the converse direction, if $(k_1, \ldots, k) \Vdash_{bn} \Box' B$ then it follows that $\forall \alpha \ni (k_1, \ldots k) \exists b \in \alpha \exists b' S^{-1} b(b' \nVdash_{bn} B)$. We then choose the path $\alpha = (g, (k_1), \ldots, (k_1, \ldots, k), (k_1, \ldots, k, k), \ldots)$. Then there are four possibilities, depending on whether $b = g$ and $b' = g$. If $b = g$, then gSb' implies kRk_0 for some k_0 s.t. $k_0 \nVdash_{kn} B$, in both of the cases $b' = g$ and $b' \neq g$. If $b \neq g$, then $b = (k_1, \ldots, k')$ for some $k' \leq k$ such that $k'Rl$ for some l with $l \nVdash_{kn} B$, independent of whether $b' = g$. Then $k \geq R\ l$, so by the frame condition for Kripke frames we infer $k\ R\leq\ l$. Hence there is l' such that kRl' and $l' \nVdash_{kn} B$. Thus in all case $\exists k'R^{-1}k(k' \nVdash_{bn} B)$ and so $k \Vdash_{kn} \Box' B$. The case for $\mathcal{M}_k \vDash_{kn} \Box' B \Longleftrightarrow g \Vdash_{bn} \Box' B$ is analogous.

Now, since $k \nVdash_{kn} A$, the above equivalence implies $(k) \nVdash_{bn} A$. Thus $\nvDash_{bn} A$. □

Proof of Theorem 4.11

Proof. For (i), we need to check the cases for (Q1),(Q2) and (RM2). To see that (Q1) is valid, suppose $b' \Vdash_{bq} \Diamond(A \vee B)$ for some b and $b' \succeq b$. Then $\forall \alpha \ni b' \exists c \in \alpha \exists c' S^{-1} c(c' \Vdash_{bq} A \vee B)$. So for any $\beta \ni c'$ there is $d \in \beta$ such that $d \Vdash_{bq} A$ or $d \Vdash_{bq} B$. Note that we can assume $d \succeq c'$ without loss of generality because of the upward closure. Then by the frame condition added for **HK**$\Diamond$, it holds that cSd for such d with respect to some $\beta \ni c'$. Consequently $c \Vdash_{bq} \Diamond A$ or $c \Vdash_{bq} \Diamond B$. Therefore $b' \Vdash_{bq} \Diamond A \vee \Diamond B$; so $b \Vdash_{bq} \Diamond(A \vee B) \to (\Diamond A \vee \Diamond B)$. The cases for (Q2) and (RM2) are straightforward.

For (ii), we use the same construction of Beth model as **HK**$\Box'$. We need then to check first that the condition $\forall b, b' \in B(bSb' \Rightarrow \exists \alpha \ni b' \forall c \in \alpha(c \succeq b' \Rightarrow bSc))$ is satisfied. If $b' = (k_1, \ldots, k_m)$, then we can take the path $\alpha = (g, (k_1), \ldots, (k_1, \ldots, k_m), (k_1, \ldots, k_m, k_m), \ldots)$. If $b' = g$, then bSc for any $c \in B$, so we can take an arbitrary α. We furthermore need to show that the equivalences between Kripke and Beth forcings; These can be shown analogously to the cases for **HK**$\Box'$. □

Proof of Theorem 4.12

Proof. For (1), we need to check the cases for (O1),(O2) and (RC2). To see that (O1) is valid, suppose $b' \Vdash_{bo} \Diamond' A \wedge \Diamond' B$ for some b and $b' \succeq b$. Then $b'Sc$ implies $c \nVdash_{bo} A$ and $c \nVdash_{bo} B$. Now if $c' \Vdash_{bo} A \vee B$ for some $c'S^{-1}b'$, then by the added frame condition there is $\alpha \ni c'$ s.t. $b'Sd$ for all $d \in \alpha$ satisfying $d \succeq c'$. So no world in α can force A nor B, a contradiction. Thus $b' \Vdash_{bo} \Diamond'(A \vee B)$ and so $b \Vdash_{bo} (\Diamond' A \wedge \Diamond' B) \to \Diamond'(A \vee B)$. The cases for (O2) and (RC2) are simple.

For (ii), the condition $\forall b, b' \in B(bSb' \Rightarrow \exists \alpha \ni b' \forall c \in \alpha(c \succeq b' \Rightarrow bSc))$ can be checked as in (i), but notice that the case $b' = g$ does not apply. Then we can show $\forall b, b' \in B(b \preceq S\ b' \Rightarrow \exists \alpha \ni b \forall c \in \alpha(c\ S\succeq\ b'))$ analogously to the respective condition for **HK**$\Box$. Similarly for the equivalences of the forcings.□

Proof of Proposition 5.3

Proof. From (i) to (ii), we argue by contradiction. Suppose there is k s.t. $\neg(k\ R\leq\ k)$. Choose $\mathcal{V}$ s.t. $\mathcal{V}(p) = \{k' : k\ R\leq\ k'\}$. $\mathcal{V}(p)$ is upward closed, and $k \nVdash_{kn} p \vee \Box' p$. Hence $\mathcal{F} \nvDash_{kn} p \vee \Box' p$. From (ii) to (i), suppose $k \nVdash_{kn} \Box' A$.

Then all worlds accessible from k force A. In particular, by the frame condition there is k' s.t. kRk' and $k' \leq k$. So $k \Vdash_{kn} A$. □

Proof of Theorem 6.5

Proof. The soundness direction follows immediately from Theorem 3.10, Proposition 5.4 and 5.6.

For completeness, the outline is identical to the proof of Theorem 3.11. As we noted, the Kripke semantics for $\mathbf{CC}_\omega$ has the condition $\geq R \subseteq R$ rather than $\geq R \subseteq R\leq$ used for $\mathbf{HK}\Box'$. The latter condition is used in establishing the equivalence of valuation for non-necessity, but it is straightforward to check that the former condition works as well.

We need to check that the Beth model constructed out of a Kripke model satisfies the conditions corresponding to (LEM) and (DNE). For the condition of (LEM), if $b = (k_1, \ldots, k) \neq g$ and $\alpha \ni b$, take $b' = b$. Then given $\beta \ni b'$, we note kRk by reflexivity and so $b\ S\preceq\ b$ and $b \in \alpha$. Thus $\exists c \in \beta \exists c' \in \alpha(c\ S\preceq\ c')$, as desired. If $b = g$, then for $\alpha = (g, (k), \ldots)$ take $b' = (k)$; we can then argue in the same manner to obtain the same conclusion. For the condition of (DNE), given $b = (k_1, \ldots k) \neq g$ and $\alpha \ni b$, we take $\beta = (g, (k_1), \ldots, (k_1, \ldots, k), (k_1, \ldots, k, k), \ldots)$. Suppose $b' \in \beta$, $c \succeq c'S^{-1}b'$ and $\gamma \ni c$. If $b' \neq g$, then $b' = (k_1, \ldots, k')$ for $k' \leq k$. When $c' = (l_1, \ldots, l) \neq g$, we then have $k'Rl$ and so kRl from the frame condition that $\geq R \subseteq R$. So by symmetry lRk, which means $c'S \preceq b$. Consequently, (noting $c' \in \gamma$) we have $\exists c' \in \gamma \exists d \in \alpha(c'\ S\preceq\ d)$. When $c' = g$, then $k'Rl$ for all $l \in K$ and so is k by the frame condition. Thus by symmetry lRk for all $l \in K$, which implies $c'S \preceq b$ again. If $b' = g$ then $b'Sc'$ means either: there is $l \in K$ which is accessible from k (if $c' \neq g$), or every point is accessible from k (if $c' = g$). In both case we can use symmetry to conclude $c'S \preceq b$. Finally, if $b = g$, then for $\alpha = (g, (k), \ldots)$ take $\beta = (g, (k), (k, k), \ldots)$. The rest is analogous. □

References

[1] Alechina, N., M. Mendler, V. de Paiva and E. Ritter, *Categorical and Kripke semantics for constructive S4 modal logic*, in: *International Workshop on Computer Science Logic*, Springer, 2001, pp. 292–307.

[2] Artemov, S. and T. Protopopescu, *Intuitionistic epistemic logic*, The Review of Symbolic Logic **9** (2016), pp. 266–298.

[3] Beth, E. W., *semantic construction of intuitionistic logic*, (Mededelingen van de) Koninklijke Nederlandse Akademie van Wetenschappen (Afdeling Letteren) (Nieuwe Reeks) **19** (1956), pp. 357–388.

[4] Bezhanishvili, G. and W. H. Holliday, *A semantic hierarchy for intuitionistic logic*, Indagationes Mathematicae **30** (2019), pp. 403–469.

[5] Božić, M. and K. Došen, *Models for normal intuitionistic modal logics*, Studia Logica **43** (1984), pp. 217–245.

[6] da Costa, N. C. A., *On the theory of inconsistent formal systems*, Notre Dame Journal of Formal Logic **15** (1974), pp. 497–510.

[7] De, M., *Empirical negation*, Acta Analytica **28** (2013), pp. 49–69.

[8] De, M. and H. Omori, *More on empirical negation*, in: R. Goré, B. Kooi and A. Kurucz, editors, *Advances in Modal Logic*, 10 (2014), pp. 114–133.
[9] de Beer, B., "Frame Properties of Beth Models," B.S. thesis, Utrecht University (2012).
[10] Došen, K., *Negative modal operators in intuitionistic logic*, Publication de l'Instutute Mathematique, Nouv. Ser **35** (1984), pp. 3–14.
[11] Došen, K., *Models for stronger normal intuitionistic modal logics*, Studia Logica **44** (1985), pp. 39–70.
[12] Došen, K., *Negation as a modal operator*, Reports on Mathematical Logic **20** (1986), pp. 15–27.
[13] Došen, K., *Negation on the light of modal logic*, in: D. M. Gabbay and H. Wansing, editors, *What is Negation?*, Kluwer Academic Publishing, 1999 pp. 77–86.
[14] Drobyshevich, S., *On classical behavior of intuitionistic modalities*, Logic and Logical Philosophy **24** (2015), pp. 79–104.
[15] Fairtlough, M. and M. Mendler, *Propositional lax logic*, Information and Computation **137** (1997), pp. 1–33.
[16] Ferguson, T. M., *Extensions of Priest-da Costa Logic*, Studia Logica **102** (2014), pp. 145–174.
[17] Fitch, F. B., *Intuitionistic modal logic with quantifiers*, Portugaliae mathematica **7** (1948), pp. 113–118.
[18] Gabbay, D. M., "Fibring logics," Oxford Logic Guides **38**, Clarendon Press, 1998.
[19] Goldblatt, R., *Cover semantics for quantified lax logic*, Journal of Logic and Computation **21** (2011), pp. 1035–1063.
[20] Holliday, W. H., *Inquisitive intuitinostic logic*, in: S. N. Nicola Olivetti, Rineke Verbrugge and G. Sandu, editors, *Advances in Modal Logic*, 13, College Publications, 2020 pp. 329–348.
[21] Kojima, K., "Semantical study of intuitionistic modal logics," Ph.D. thesis, Kyoto University (2012).
[22] Kripke, S. A., *Semantical analysis of intuitionistic logic I*, in: *Studies in Logic and the Foundations of Mathematics*, 40, Elsevier, 1965 pp. 92–130.
[23] Mendler, M. and V. De Paiva, *Constructive CK for contexts*, Context Representation and Reasoning (CRR-2005) **13** (2005).
[24] Niki, S., *Empirical negation, co-negation and contraposition rule I: Semantical investigations*, Bulletin of the Section of Logic **49** (2020), pp. 231–253.
[25] Osorio, M. and J. A. C. Joo, *Equivalence among RC-type paraconsistent logics*, Logic Journal of the IGPL **25** (2017), pp. 239–252.
[26] Priest, G., *Dualising intuitionictic negation*, Principia: an international journal of epistemology **13** (2009), pp. 165–184.
[27] Rogozin, D., *Categorical and algebraic aspects of the intuitionistic modal logic* IEL^- *and its predicate extensions*, Journal of Logic and Computation **31** (2021), pp. 347–374.
[28] Simpson, A. K., "The proof theory and semantics of intuitionistic modal logic," Ph.D. thesis, University of Edinburgh (1994).
[29] Sylvan, R., *Variations on da Costa C systems and dual-intuitionistic logics I. analyses of* C_ω *and* CC_ω, Studia Logica **49** (1990), pp. 47–65.
[30] Troelstra, A. S. and D. van Dalen, "Constructivism in Mathematics: An Introduction I," Studies in Logic and the Foundations of Mathematics, Elsevier, 1988.
[31] Troelstra, A. S. and D. van Dalen, "Constructivism in Mathematics: An Introduction II," Studies in Logic and the Foundations of Mathematics **II**, Elsevier, 1988.
[32] van Dalen, D., *How to glue analysis models*, The Journal of symbolic logic **49** (1984), pp. 1339–1349.
[33] Wijesekera, D., *Constructive modal logics I*, Annals of Pure and Applied Logic **50** (1990), pp. 271–301.
[34] Wolter, F. and M. Zakharyaschev, *Intuitionistic modal logic*, in: *Logic and foundations of mathematics*, Springer, 1999 pp. 227–238.

Analytic Cut and Mints' Symmetric Interpolation Method for Bi-intuitionistic Tense Logic

Hiroakira Ono [1]

Japan Advanced Institute of Science and Technology
1-1 Asahidai,
Nomi, Ishikawa, 923-1292, Japan

Katsuhiko Sano [2]

Faculty of Humanities and Human Sciences, Hokkaido University
Nishi 7 Chome, Kita 10 Jo, Kita-ku,
Sapporo, Hokkaido, 060-0810, Japan

Abstract

This paper establishes the Craig interpolation theorem of bi-intuitionistic stable tense logic **BiSKt**, which is proposed by Stell et al. (2016). First, we define a sequent calculus G(**BiSKt**) with the cut rule for the logic and establish semantically that applications of the cut rule can be restricted to analytic ones, i.e., applications such that the cut formula is a subformula of the conclusion of the cut rule. Second, we apply a symmetric interpolation method, originally proposed by Mints (2001) for multi-succedent calculus for intuitionistic logic, to obtain the Craig interpolation theorem of the calculus G(**BiSKt**). Our argument also provides a simplification of Kowalski and Ono (2017)'s argument for the Craig interpolation theorem of bi-intuitionistic logic.

Keywords: Analytic Cut, Craig Interpolation, Subformula Property, Bi-intuitionistic Logic, Bi-intuitionistic Tense Logic.

1 Introduction

The Craig interpolation theorem for classical (first-order) logic states that if $\varphi \to \psi$ is a theorem then there exists a formula γ, which is called an *interpolant*, such that both $\varphi \to \gamma$ and $\gamma \to \psi$ are theorems and the set of all free variables (constant symbols or predicate symbols) occurring in γ is a subset of the set of all free variables (constant symbols or predicate symbols, respectively) occurring in both φ and ψ (we do not consider function symbols and equality

[1] ono@jaist.ac.jp

[2] v-sano@let.hokudai.ac.jp

here). Maehara [12] showed that an interpolant can be effectively calculated from a cut-free derivation of a sequent calculus **LK** of classical first-order logic by considering a partition of a sequent. This is now called Maehara's method.

Since then, Maehara's method has been applied to various non-classical logics, e.g., intuitionistic first-order logic [28,34] and propositional modal logics (see, e.g., [20,21]), when the target logic has a cut-free sequent calculus in terms of an ordinary notion of sequent (i.e., a pair of finite multisets or lists). For a single succedent sequent calculus of intuitionistic propositional logic **Int**, we restrict the form of a partition where to apply Maehara's method (see, e.g., [21, Theorem 3.7]). For a multi-succedent sequent calculus of **Int**, Mints [16] did not restrict the form of a partition but generalized the notion of Craig interpolant to calculate such a generalized interpolant inductively. Mints' method can be regarded as a generalization of Maehara's method. One remarkable point of Mints' method is that a generalized interpolant implies the existence of a Craig interpolant for the right rule of implication.

Since Maehara's method assumes a cut-free sequent calculus (based on the ordinary notion of sequent) for a logic, if the sequent calculus is not cut-free, it becomes challenging to establish the Craig interpolation theorem proof-theoretically. Well-known examples of the failure of cut-elimination theorem for proposed sequent calculi are modal logic **S5** [18], basic tense logic **Kt** [17], bi-intuitionistic logic [22], etc. For such cases, there are at least two proof-theoretic approaches to obtain the Craig interpolation theorem (if it holds). The first approach is to restrict applications of cut to ensure that such restricted applications are still compatible with Maehara's method. The second approach is to extend the notion of sequent to get a cut-free sequent calculus to apply Maehara's method. For **S5**, the reader is referred to [31,33,20] for the first approach and to [8] for the second approach which is based on the notion of hypersequent [1]. It is noted that we can restrict all applications of cut to analytic ones for **S5** [31,33] (for the relationship between a derivation with analytic cut of the sequent calculus for **S5** [18] and a cut-free derivation of a hypersequent calculus for **S5**, the reader is referred to [2]).

For bi-intuitionistic logic [23,24,25] (see [6] for one of the recent studies), the first approach is taken in [7]. Kowalski and the first author semantically established that applications of cut can be restricted to analytic ones and then showed that Maehara's method can be applied to a restricted form of a partition of a sequent. Maehara's partition argument for analytic cut (actually, analytic *mix* rule, see [7, Lemma 5.4]), however, becomes more involved than for **S5** [20, pp.245-6]. Later, the second approach is taken in [11] by Lyon et al. They employed nested sequent calculi and generalized the notion of interpolant to get the interpolation theorem. As far as the authors can see, this generalized notion of interpolant in [11] is exactly the same one as Mints' [16], but [11] has no reference to [16]. They also applied the same approach based on nested sequents to extensions of basic tense logic. As for the first approach to basic tense logic, the reader is referred to [27].

This paper takes the first approach above to bi-intuitionistic stable tense logic **BiSKt** [30], a combination of bi-intuitionistic logic and basic tense logic, where we choose a residuated pair of past possibility operator $\blacklozenge$ and future necessity operator $\Box$ as primitive symbols for tense logic. Semantically, it is a tense logic with a Kripke semantics where worlds in a frame are equipped with a pre-order $\leqslant$ as well as with an accessibility relation R which is *stable* with respect to this pre-order, i.e., $\leqslant \circ R \circ \leqslant \subseteq R$ where $\circ$ is the composition of two relations. This logic arose in the semantic context of hypergraphs because a special case of the pre-order can represent the incidence structure of a hypergraph (see [30] for the detail). A labelled tableau calculus [30] and Hilbert system [26] of the logic **BiSKt** have already been studied and it also enjoys the finite model property via filtration technique [26]. A sequent calculus for **BiSKt** and the Craig interpolation, however, have not been studied in the literature.

Our sequent calculus $\mathsf{G}(\mathbf{BiSKt})$ for **BiSKt** is a sequent calculus $\mathbf{LBJ}_1$ for bi-intuitionistic logic [7] expanded with two inference rules for tense operators, which are a reformulation of Nishimura [17]'s rules in terms of $\blacklozenge$ and $\Box$. It is noted that the left rule for coimplication and rules for $\blacklozenge$ and $\Box$ as well as the right rule for implication has context restrictions. We first establish semantically that every application of cut in a derivation of $\mathsf{G}(\mathbf{BiSKt})$ can be replaced with an analytic one. A key notion for this aim is: Ξ-*partial valuation*, which is also employed in [15,27] (the original idea in [15] is due to Mitio Takano). Then, we follow Mints' symmetric interpolation method to calculate an interpolant by induction on derivation, similarly to Maehara's method. A novelty of our argument is to demonstrate that Mints' symmetric interpolation method [16] works more properly with an analytic cut rule than the ordinary Maehara's method in [7]. For inference rules with context restrictions mentioned above, we can construct a Craig interpolant from a generalized interpolant.

We proceed as follows. Section 2 introduces the syntax and Kripke semantics of bi-intuitionistic stable tense logic **BiSKt**. In Section 3, we define three sequent calculi for **BiSKt** and prove the soundness of a sequent calculus $\mathsf{G}(\mathbf{BiSKt})$ for **BiSKt** with no restriction on rule applications (Theorem 3.3). In Section 4, we establish that a sequent calculus $\mathsf{G}^a(\mathbf{BiSKt})$, whose rule applications are always analytic, is semantically complete for Kripke semantics (Theorem 4.7) and conclude that $\mathsf{G}(\mathbf{BiSKt})$ enjoys the subformula property (Corollary 4.8). In Section 5, we introduce Mints' notion of interpolant [16] (a generalization of a Craig interpolant) to show the Craig interpolation theorem for $\mathsf{G}(\mathbf{BiSKt})$ (Theorem 5.19). Section 5 is the most important contribution of this paper. Section 6 concludes the paper.

2 Syntax and Kripke Semantics

We introduce the syntax and semantics for bi-intuitionistic stable tense logic [30,26]. Let Prop be a countably infinite set of propositional variables. Our syntax $\mathcal{L}$ for bi-intuitionistic stable tense logic consists of all logical connectives of bi-intuitionistic logic, i.e., two constant symbols $\bot$ and $\top$, disjunction $\vee$, conjunction $\wedge$, implication $\rightarrow$, coimplication $\prec$, as well as two modal operators $\{ \blacklozenge, \Box \}$. The set of all formulas in $\mathcal{L}$ is defined in a standard way as follows:

$$\varphi ::= \top \mid \bot \mid p \mid \varphi \wedge \varphi \mid \varphi \vee \varphi \mid \varphi \rightarrow \varphi \mid \varphi \prec \varphi \mid \blacklozenge\varphi \mid \Box\varphi \quad (p \in \mathsf{Prop}).$$

Given a finite set Γ of formulas, we define $\bigwedge\Gamma$ and $\bigvee\Gamma$ as the conjunction and disjunction of all formulas in Γ, respectively, where $\bigwedge\varnothing := \top$ and $\bigvee\varnothing := \bot$. Given any formula φ, we define $\mathsf{Sub}(\varphi)$ as the set of all subformulas of φ. Moreover, for any set (or multiset) Γ of formulas, we define $\mathsf{Sub}(\Gamma) = \bigcup_{\varphi\in\Gamma}\mathsf{Sub}(\varphi)$. We say that a set (or multiset) Γ is *subformula closed* if $\mathsf{Sub}(\Gamma) \subseteq \Gamma$. We define the set $\mathsf{Prop}(\Gamma)$ of propositional variables occurring in Γ as $\mathsf{Sub}(\Gamma) \cap \mathsf{Prop}$. We often write $\mathsf{Prop}(\varphi)$ instead of $\mathsf{Prop}(\{\varphi\})$. We use $\varphi[\psi_1/p_1, \ldots, \psi_n/p_n]$ as the result of uniformly substituting each propositional variable p_i in φ with a formula ψ_i simultaneously for all $1 \leqslant i \leqslant n$, where all p_is are pairwise distinct. For a set Γ of formulas, we also naturally define $\Gamma[\psi_1/p_1, \ldots, \psi_n/p_n]$ as $\{\varphi[\psi_1/p_1, \ldots, \psi_n/p_n] \mid \varphi \in \Gamma\}$.

Definition 2.1 We say that $F = (W, \leqslant, R)$ is a *frame* if W is a nonempty set, $\leqslant$ is a preorder on W, and R is a *stable* binary relation on W, i.e., R satisfies the following condition: $\leqslant \circ R \circ \leqslant \subseteq R$, where "$\circ$" is the relational composition.

Since $\leqslant$ is reflexive, it is easy to see that R is stable iff $\leqslant \circ R \circ \leqslant = R$. This condition is also employed in [35] for interpreting $\Box$ on the intuitionistic setting.

Definition 2.2 Given a frame $F = (W, \leqslant, R)$, $X \subseteq W$ is said to be $\leqslant$*-closed* (or *up-closed*) if $u \in X$ and $u \leqslant v$ jointly imply $v \in X$ for all $u, v \in W$. A *valuation* on a frame $F = (W, \leqslant, R)$ is a mapping V from Prop to the set of all $\leqslant$-closed sets on W. A pair $M = (F, V)$ is a *model* if $F = (W, \leqslant, R)$ is a frame and V is a valuation. Given a model $M = (W, \leqslant, R, V)$, a state $u \in W$ and a formula φ, the *satisfaction relation* $M, u \models \varphi$ is defined inductively as follows:

$$\begin{array}{lll}
M, u \models p & \text{iff} & u \in V(p), \\
M, u \models \top, & & \\
M, u \not\models \bot, & & \\
M, u \models \varphi \vee \psi & \text{iff} & M, u \models \varphi \text{ or } M, u \models \psi, \\
M, u \models \varphi \wedge \psi & \text{iff} & M, u \models \varphi \text{ and } M, u \models \psi, \\
M, u \models \varphi \rightarrow \psi & \text{iff} & \text{For all } v \in W\ ((u \leqslant v \text{ and } M, v \models \varphi) \text{ imply } M, v \models \psi), \\
M, u \models \varphi \prec \psi & \text{iff} & \text{For some } v \in W\ (v \leqslant u \text{ and } M, v \models \varphi \text{ and } M, v \not\models \psi), \\
M, u \models \blacklozenge\varphi & \text{iff} & \text{For some } v \in W\ (vRu \text{ and } M, v \models \varphi), \\
M, u \models \Box\varphi & \text{iff} & \text{For all } v \in W\ (uRv \text{ implies } M, v \models \varphi).
\end{array}$$

Proposition 2.3 *For every model $M = (W, \leqslant, R, V)$ and formula φ, the truth set $[\![\varphi]\!]_M := \{u \in W \mid M, u \models \varphi\}$ is $\leq$-closed.*

3 Sequent Calculi

In what follows in this section, we use Γ, Δ, etc. to denote finite multisets of formulas. A *sequent* is a pair (Γ, Δ) of finite multisets and it is denoted

Table 1
Sequent Calculi $\mathsf{G}(\mathbf{BiSKt})$, $\mathsf{G}^*(\mathbf{BiSKt})$, and $\mathsf{G}^a(\mathbf{BiSKt})$

Sequent Calculus $\mathsf{G}(\mathbf{BiSKt})$

Initial Sequents:

$$\Rightarrow \top \qquad \varphi \Rightarrow \varphi \qquad \bot \Rightarrow$$

Structural Rules:

$$\frac{\Gamma \Rightarrow \Delta}{\Gamma \Rightarrow \Delta, \varphi}\ (\Rightarrow w) \qquad \frac{\Gamma \Rightarrow \Delta}{\varphi, \Gamma \Rightarrow \Delta}\ (w \Rightarrow)$$

$$\frac{\Gamma \Rightarrow \Delta, \varphi, \varphi}{\Gamma \Rightarrow \Delta, \varphi}\ (\Rightarrow c) \qquad \frac{\varphi, \varphi, \Gamma \Rightarrow \Delta}{\varphi, \Gamma \Rightarrow \Delta}\ (c \Rightarrow)$$

Logical Rules:

$$\frac{\Gamma \Rightarrow \Delta, \varphi_1 \quad \Gamma \Rightarrow \Delta, \varphi_2}{\Gamma \Rightarrow \Delta, \varphi_1 \wedge \varphi_2}\ (\Rightarrow \wedge) \qquad \frac{\varphi_i, \Gamma \Rightarrow \Delta}{\varphi_1 \wedge \varphi_2, \Gamma \Rightarrow \Delta}\ (\wedge \Rightarrow)$$

$$\frac{\Gamma \Rightarrow \Delta, \varphi_i}{\Gamma \Rightarrow \Delta, \varphi_1 \vee \varphi_2}\ (\Rightarrow \vee) \qquad \frac{\varphi_1, \Gamma \Rightarrow \Delta \quad \varphi_2, \Gamma \Rightarrow \Delta}{\varphi_1 \vee \varphi_2, \Gamma \Rightarrow \Delta}\ (\vee \Rightarrow)$$

$$\frac{\varphi, \Gamma \Rightarrow \psi}{\Gamma \Rightarrow \varphi \to \psi}\ (\Rightarrow\to) \qquad \frac{\Gamma \Rightarrow \Delta, \varphi \quad \psi, \Pi \Rightarrow \Sigma}{\varphi \to \psi, \Gamma, \Pi \Rightarrow \Delta, \Sigma}\ (\to\Rightarrow)$$

$$\frac{\Gamma \Rightarrow \Delta, \varphi \quad \psi, \Pi \Rightarrow \Sigma}{\Gamma, \Pi \Rightarrow \Delta, \Sigma, \varphi \prec \psi}\ (\Rightarrow \prec) \qquad \frac{\varphi \Rightarrow \Delta, \psi}{\varphi \prec \psi \Rightarrow \Delta}\ (\prec \Rightarrow)$$

Modal Rules:

$$\frac{\blacklozenge \Theta, \Gamma \Rightarrow \varphi}{\Theta, \Box \Gamma \Rightarrow \Box \varphi}\ (\Box) \qquad \frac{\varphi \Rightarrow \Delta, \Box \Theta}{\blacklozenge \varphi \Rightarrow \blacklozenge \Delta, \Theta}\ (\blacklozenge)$$

Cut Rule:

$$\frac{\Gamma \Rightarrow \Delta, \varphi \quad \varphi, \Pi \Rightarrow \Sigma}{\Gamma, \Pi \Rightarrow \Delta, \Sigma}\ (Cut)$$

$\mathsf{G}^*(\mathbf{BiSKt})$: the same as $\mathsf{G}(\mathbf{BiSKt})$ except the following analytic cut rule.

$$\frac{\Gamma \Rightarrow \Delta, \varphi \quad \varphi, \Pi \Rightarrow \Sigma}{\Gamma, \Pi \Rightarrow \Delta, \Sigma}\ (Cut)^a, \text{ where } \varphi \in \mathsf{Sub}(\Gamma, \Pi, \Delta, \Sigma).$$

$\mathsf{G}^a(\mathbf{BiSKt})$: the same as $\mathsf{G}^*(\mathbf{BiSKt})$ except the following modal rules.

$$\frac{\blacklozenge \Theta, \Gamma \Rightarrow \varphi}{\Theta, \Box \Gamma \Rightarrow \Box \varphi}\ (\Box)^a, \text{ where } \blacklozenge \Theta \subseteq \mathsf{Sub}(\Gamma, \varphi).$$

$$\frac{\varphi \Rightarrow \Delta, \Box \Theta}{\blacklozenge \varphi \Rightarrow \blacklozenge \Delta, \Theta}\ (\blacklozenge)^a, \text{ where } \Box \Theta \subseteq \mathsf{Sub}(\varphi, \Delta).$$

by $\Gamma \Rightarrow \Delta$, where Γ is an *antecedent*, Δ is a *succedent* (or a *consequent*) and $\Gamma \Rightarrow \Delta$ is read as "if all the formulas of Γ hold, then some formula in Δ holds".

For any sequent calculus in Table 1, we define the notion of derivation (from initial sequents of the calculus) and derivable sequent in the calculus as usual. The bi-intuitionistic fragment of G(**BiSKt**) is the same as the system $\mathbf{LBJ}_1$ in [7]. We use the same modal rules as in [27] for basic tense logic **Kt** (over classical logic) but these are a reformulation in terms of $\blacklozenge$ and $\Box$ of Nishimura's rules [17] for tense operators G and H over classical logic. The calculus G*(**BiSKt**) is the same as G(**BiSKt**) except (Cut) is replaced with its analytic variant $(Cut)^a$ but the rules $(\Box)$ and $(\blacklozenge)$ do not satisfy the subformula property, i.e., each formula in the premise(s) of a rule may not be a subformula of the conclusion of the rule (we use this calculus for establishing the Craig interpolation theorem). Finally, G^a(**BiSKt**) is a fully analytic calculus all of whose rules enjoy the subformula property, where the side syntactic conditions for $(\blacklozenge)$ and $(\Box)$ were originally proposed by Takano [31].

Remark 3.1 For the bi-intuitionistic fragment of G(**BiSKt**) (i.e., $\mathbf{LBJ}_1$ of [7]), since all rules except (Cut) are analytic, the restriction of applications of (Cut) to analytic ones ensures the subformula property of the whole calculus and vice versa. However, it is not the case for the full calculus G(**BiSKt**). To establish the subformula property, we also need to restrict applications of the modal rules $\Box$ and $\blacklozenge$ to analytic ones. Since the Craig interpolation theorem is concerned with propositional variables, however, G*(**BiSKt**) will do the job in Section 5.

Definition 3.2 A sequent $\Gamma \Rightarrow \Delta$ is *valid* in a model $M = (W, \leqslant, R, V)$ if, for every $u \in W$, whenever $M, u \models \gamma$ for all $\gamma \in \Gamma$, then $M, u \models \delta$ for some $\delta \in \Delta$.

Theorem 3.3 *If a sequent* $\Gamma \Rightarrow \Delta$ *is derivable in* G(**BiSKt**), *then it is valid in all models.*

Proof. We only prove that $(\Box)$ preserves validity on a model $M = (W, \leqslant, R, V)$. Suppose that $\blacklozenge\Theta, \Gamma \Rightarrow \varphi$ is valid on M. To show that $\Theta, \Box\Gamma \Rightarrow \Box\varphi$ is valid on M, let us fix any state $u \in W$ such that $M, u \models \bigwedge(\Theta, \Box\Gamma)$. Fix any state $v \in W$ such that uRv. Our goal is to show $M, v \models \varphi$. By uRv and $M, u \models \bigwedge\Theta$, we get $M, v \models \bigwedge\blacklozenge\Theta$. It follows also from uRv and $M, u \models \bigwedge\Box\Gamma$ that $M, v \models \bigwedge\Gamma$. By our initial supposition, we obtain $M, v \models \varphi$, as desired. □

The bi-intuitionistic fragment $\mathbf{LBJ}_1$ in [7] is known to be not cut-free (cf. [7, Theorem 2.3]) where a counterexample is a sequent $p \Rightarrow q, r \to (p \prec q) \wedge r$, which was pointed out by Uustalu for Dragalin-style sequent calculus of bi-intuitionistic logic (see [22, Section 2]):

$$\dfrac{\dfrac{p \Rightarrow p \quad q \Rightarrow q}{p \Rightarrow q, p \prec q}(\Rightarrow \prec) \qquad \dfrac{\dfrac{\dfrac{\dfrac{p \prec q \Rightarrow p \prec q}{r, p \prec q \Rightarrow p \prec q}(w \Rightarrow) \quad \dfrac{r \Rightarrow r}{r, p \prec q \Rightarrow r}(w \Rightarrow)}{r, p \prec q \Rightarrow (p \prec q) \wedge r}(\Rightarrow \wedge)}{p \prec q \Rightarrow r \to ((p \prec q) \wedge r)}(\Rightarrow \to)}{p \Rightarrow q, r \to ((p \prec q) \wedge r)}(Cut)$$

Therefore we cannot also eliminate (Cut) from the sequent calculi $\mathsf{G}(\mathbf{BiSKt})$. Even if we remove the coimplication from our syntax, the resulting system is not cut-free, either. The sequent $p, \blacklozenge\Box(p \to \bot) \Rightarrow$ is derivable in $\mathsf{G}(\mathbf{BiSKt})$ with the help of (Cut) as in the following, but the application of (Cut) is indispensable for the purpose:

$$\dfrac{\dfrac{\Box(p \to \bot) \Rightarrow \Box(p \to \bot)}{\blacklozenge\Box(p \to \bot) \Rightarrow p \to \bot}\,(\blacklozenge) \qquad \dfrac{p \Rightarrow p \quad \bot \Rightarrow}{p \to \bot, p \Rightarrow}\,(\to\Rightarrow)}{p, \blacklozenge\Box(p \to \bot) \Rightarrow}\,(Cut)$$

(this kind of phenomena are well-known for a sequent calculus of modal logic **S5**, see, e.g., [20, p.222]). It is noted in the above derivation that the cut formula $p \to \bot$ is a subformula of the conclusion of (Cut) and moreover $\Box(p \to \bot)$ is also a subformula of the conclusion of the rule $(\blacklozenge)$. Therefore, all the applications of the inference rules in the derivation are *analytic*, i.e., they satisfy the subformula property locally. A similar argument also holds for the above derivation of Uustalu's sequent $p \Rightarrow q, r \to (p \prec q) \wedge r$. These motivate us to consider if a derivation of $\mathsf{G}(\mathbf{BiSKt})$ implies the existence of a derivation of $\mathsf{G}^a(\mathbf{BiSKt})$.

4 Subformula Property

In this section, we follow Takano [32,33]'s semantic approach to establish the subformula property of $\mathsf{G}(\mathbf{BiSKt})$. That is, by Theorem 3.3, it suffices for our purpose to show that the fully analytic variant $\mathsf{G}^a(\mathbf{BiSKt})$ is semantically complete for the semantics (see Theorem 4.7). A key notion for our proof is: *Ξ-partial valuation* [15,27]. In what follows in this section, we use Γ, Δ, Σ, etc. to denote finite sets of formulas.

Definition 4.1 [Ξ-partial valuation] Let Ξ be a subformula closed finite set. We say that a pair (Γ, Δ) of finite sets of formulas is a *Ξ-partial valuation* in $\mathsf{G}^a(\mathbf{BiSKt})$ if the following three conditions are satisfied: (i) $\Gamma \Rightarrow \Delta$ is underivable in $\mathsf{G}^a(\mathbf{BiSKt})$, (ii) $\Gamma \cup \Delta = \mathsf{Sub}(\Gamma \cup \Delta)$, and (iii) $\mathsf{Sub}(\Gamma \cup \Delta) \subseteq \Xi$.

A Ξ-partial valuation can be constructed from an unprovable sequent in $\mathsf{G}^a(\mathbf{BiSKt})$ by the next lemma.

Lemma 4.2 *Let $\Gamma \Rightarrow \Delta$ be underivable in $\mathsf{G}^a(\mathbf{BiSKt})$. For any subformula closed finite set Ξ such that $\mathsf{Sub}(\Gamma, \Delta) \subseteq \Xi$, there exists a Ξ-partial valuation (Γ^+, Δ^+) in $\mathsf{G}^a(\mathbf{BiSKt})$ such that* (i) $\Gamma \subseteq \Gamma^+$, (ii) $\Delta \subseteq \Delta^+$, *and* (iii) $\Gamma^+ \cup \Delta^+ = \mathsf{Sub}(\Gamma, \Delta)$.

Proof. We need to use $(Cut)^a$ in the proof. Suppose that $\Gamma \Rightarrow \Delta$ is underivable in $\mathsf{G}^a(\mathbf{BiSKt})$ and that $\mathsf{Sub}(\Gamma, \Delta) \subseteq \Xi$. Let $\varphi_1, \ldots, \varphi_n$ be an enumeration of all formulas in $\mathsf{Sub}(\Gamma, \Delta)$. In what follows, we inductively construct a sequence $(\Gamma_l, \Delta_l)_{1 \leqslant l \leqslant n}$ such that $\Gamma_l \Rightarrow \Delta_l$ is underivable in $\mathsf{G}^a(\mathbf{BiSKt})$, $\Gamma_l \subseteq \Gamma_{l+1}$ and $\Delta_l \subseteq \Delta_{l+1}$ for all $1 \leqslant l < n$.

For the base step, we define $(\Gamma_0, \Delta_0) := (\Gamma, \Delta)$, where $\Gamma \Rightarrow \Delta$ is clearly underivable in $\mathsf{G}^a(\mathbf{BiSKt})$. For the inductive step, let us suppose that we have constructed pairs $(\Gamma_l, \Delta_l)_{1 \leqslant l \leqslant k}$ such that each corresponding sequent $\Gamma_l \Rightarrow \Delta_l$ is underivable in $\mathsf{G}^a(\mathbf{BiSKt})$, $\Gamma_l \subseteq \Gamma_{l+1}$ and $\Delta_l \subseteq \Delta_{l+1}$ for all $1 \leqslant l < k$. We show that either $\Gamma_k \Rightarrow \Delta_k, \varphi_k$ or $\varphi_k, \Gamma_k \Rightarrow \Delta_k$ is underivable in $\mathsf{G}^a(\mathbf{BiSKt})$. Suppose otherwise, i.e., both sequents are derivable in $\mathsf{G}^a(\mathbf{BiSKt})$. Then we obtain the following derivation:

$$\dfrac{\dfrac{\Gamma_k \Rightarrow \Delta_k, \varphi_k \quad \varphi_k, \Gamma_k \Rightarrow \Delta_k}{\Gamma_k, \Gamma_k \Rightarrow \Delta_k, \Delta_k}\ (Cut)^a}{\Gamma_k \Rightarrow \Delta_k}\ (c \Rightarrow, \Rightarrow c)\quad ,$$

which implies a contradiction with the unprovability of $\Gamma_k \Rightarrow \Delta_k$. We note that the side condition of $(Cut)^a$ in the above derivation is satisfied because $\varphi_k \in \mathsf{Sub}(\Gamma, \Delta) \subseteq \mathsf{Sub}(\Gamma_k, \Delta_k)$. So, we define one of underivable sequents as $(\Gamma_{k+1}, \Delta_{k+1})$.

Finally, we define $(\Gamma^+, \Delta^+) := (\Gamma_{n+1}, \Delta_{n+1})$, which is easily shown to be a Ξ-partial valuation in $\mathsf{G}^a(\mathbf{BiSKt})$ satisfying the conditions from (i) to (iii) of the statement. □

Definition 4.3 Let Ξ be a subformula closed finite set. We define the *derived model* $M^\Xi = (W^\Xi, \leqslant^\Xi, R^\Xi, V^\Xi)$ *from* Ξ by:

- $W^\Xi := \{ (\Pi, \Sigma) \mid (\Pi, \Sigma)$ is a Ξ-partial valuation in $\mathsf{G}^a(\mathbf{BiSKt}) \}$.
- $(\Gamma, \Delta) \leqslant^\Xi (\Pi, \Sigma)$ iff $\Gamma \subseteq \Pi$ and $\Sigma \subseteq \Delta$.
- $(\Gamma, \Delta) R^\Xi (\Pi, \Sigma)$ iff the following two conditions hold:
  (i) if $\Box\psi \in \Gamma$ then $\psi \in \Pi$, for all formulas ψ,
  (ii) if $\blacklozenge\psi \in \Sigma$ then $\psi \in \Delta$, for all formulas ψ.
- $(\Gamma, \Delta) \in V^\Xi(p)$ iff $p \in \Gamma$.

Lemma 4.4 *Let* Ξ *be a subformula closed finite set. Then,* M^Ξ *is a model.*

Proof. It is easy to see that $\leqslant^\Xi$ is reflexive and transitive and that V^Ξ is $\leqslant$-closed. So, we show that R^Ξ is stable, i.e., $\leqslant^\Xi \circ R^\Xi \circ \leqslant^\Xi \subseteq R^\Xi$. Suppose that $(\Gamma, \Delta) \leqslant^\Xi (\Gamma_1, \Delta_1) R^\Xi (\Gamma_2, \Delta_2) \leqslant^\Xi (\Pi, \Sigma)$. To show our goal of $(\Gamma, \Delta) R^\Xi (\Pi, \Sigma)$, we need to verify two conditions (i) and (ii) for R^Ξ of Definition 4.3. However, these are easy to establish. □

Lemma 4.5 *Let* $(\Gamma, \Delta) \in W^\Xi$. *Then, all of the following hold.*

(i) *If* $\varphi \wedge \psi \in \Gamma$, *then* $\varphi \in \Gamma$ *and* $\psi \in \Gamma$.

(ii) *If* $\varphi \wedge \psi \in \Delta$, *then* $\varphi \in \Delta$ *or* $\psi \in \Delta$.

(iii) *If* $\varphi \vee \psi \in \Gamma$, *then* $\varphi \in \Gamma$ *or* $\psi \in \Gamma$.

(iv) *If* $\varphi \vee \psi \in \Delta$, *then* $\varphi \in \Delta$ *and* $\psi \in \Delta$.

(v) *If* $\varphi \to \psi \in \Gamma$, *then* $((\Gamma, \Delta) \leqslant^\Xi (\Pi, \Sigma)$ *and* $\varphi \in \Pi$ *imply* $\psi \in \Pi)$ *for all* $(\Pi, \Sigma) \in W^\Xi$.

(vi) *If* $\varphi \to \psi \in \Delta$, *then* $((\Gamma, \Delta) \leqslant^\Xi (\Pi, \Sigma)$ *and* $\varphi \in \Pi$ *and* $\psi \in \Sigma)$ *for some* $(\Pi, \Sigma) \in W^\Xi$.

(vii) *If* $\varphi \prec \psi \in \Gamma$, *then* $((\Pi, \Sigma) \leqslant^{\Xi} (\Gamma, \Delta)$ *and* $\varphi \in \Pi$ *and* $\psi \in \Sigma)$ *for some* $(\Pi, \Sigma) \in W^{\Xi}$.

(viii) *If* $\varphi \prec \psi \in \Delta$, *then* $((\Pi, \Sigma) \leqslant^{\Xi} (\Gamma, \Delta)$ *and* $\varphi \in \Pi$ *imply* $\psi \in \Pi)$ *for all* $(\Pi, \Sigma) \in W^{\Xi}$.

(ix) *If* $\Box\varphi \in \Gamma$, *then* $((\Gamma, \Delta)R^{\Xi}(\Pi, \Sigma)$ *implies* $\varphi \in \Pi)$ *for all* $(\Pi, \Sigma) \in W^{\Xi}$.

(x) *If* $\Box\varphi \in \Delta$, *then* $((\Gamma, \Delta)R^{\Xi}(\Pi, \Sigma)$ *and* $\varphi \in \Sigma)$ *for some* $(\Pi, \Sigma) \in W^{\Xi}$.

(xi) *If* $\blacklozenge\varphi \in \Gamma$, *then* $((\Pi, \Sigma)R^{\Xi}(\Gamma, \Delta)$ *and* $\varphi \in \Pi)$ *for some* $(\Pi, \Sigma) \in W^{\Xi}$.

(xii) *If* $\blacklozenge\varphi \in \Delta$, *then* $((\Pi, \Sigma)R^{\Xi}(\Gamma, \Delta)$ *implies* $\varphi \in \Sigma)$ *for all* $(\Pi, \Sigma) \in W^{\Xi}$.

Proof. It is easy to establish items from (i) to (iv), and items (ix) and (xii) are immediate from the definition of R^{Ξ}. Moreover, we can prove items (v) and (vi) similarly to (vii) and (viii), respectively, by duality. So, we prove the remaining items below.

(vii) Suppose that $\varphi \prec \psi \in \Gamma$. We show that $\varphi \Rightarrow \Delta, \psi$ is underivable in $\mathsf{G}^a(\mathbf{BiSKt})$. Suppose otherwise. Then, we obtain the following derivation:

$$\dfrac{\varphi \Rightarrow \Delta, \psi}{\varphi \prec \psi \Rightarrow \Delta}\ (\prec\Rightarrow),$$

which implies the derivability of $\Gamma \Rightarrow \Delta$, a contradiction. So, by Lemma 4.2, there exists $(\Pi, \Sigma) \in W^{\Xi}$ such that $\varphi \in \Pi$, $\{\psi\} \cup \Delta \subseteq \Sigma$ and $\Pi \cup \Sigma = \mathsf{Sub}(\varphi, \Delta, \psi)$. To finish our proof, it suffices to show that $(\Pi, \Sigma) \leqslant^{\Xi} (\Gamma, \Delta)$. Since $\Delta \subseteq \Sigma$ is easy, we show that $\Pi \subseteq \Gamma$. Suppose that $\pi \in \Pi$. We show that $\pi \in \Gamma$. We observe that $\pi \in \Pi \subseteq \Pi \cup \Sigma = \mathsf{Sub}(\varphi, \Delta, \psi) \subseteq \mathsf{Sub}(\Gamma, \Delta) = \Gamma \cup \Delta$. Suppose for contradiction that $\pi \notin \Gamma$, i.e., $\pi \in \Delta \subseteq \Sigma$. This implies the derivability of $\Pi \Rightarrow \Sigma$ in $\mathsf{G}^a(\mathbf{BiSKt})$. A contradiction. We conclude $\pi \in \Gamma$, as desired.

(viii) Assume that $\varphi \prec \psi \in \Delta$. Fix any $(\Pi, \Sigma) \in W^{\Xi}$ such that $(\Pi, \Sigma) \leqslant^{\Xi} (\Gamma, \Delta)$ and $\varphi \in \Pi$. We show $\psi \in \Pi$. Since $\Delta \subseteq \Sigma$, we have $\varphi \prec \psi \in \Sigma$. This implies that $\psi \in \mathsf{Sub}(\Pi, \Sigma) = \Pi \cup \Sigma$. To obtain our goal, it suffices to show that $\psi \notin \Sigma$. So, suppose for contradiction that $\psi \in \Sigma$. Since $\varphi \Rightarrow \varphi \prec \psi, \psi$ is derivable in $\mathsf{G}^a(\mathbf{BiSKt})$, this implies that $\Pi \Rightarrow \Sigma$ is derivable in $\mathsf{G}^a(\mathbf{BiSKt})$, a contradiction. Therefore, $\psi \notin \Sigma$ hence $\psi \in \Pi$.

(x) Assume that $\Box\varphi \in \Delta$. Let us define Π and Θ as follows:

$$\Pi := \{\psi \mid \Box\psi \in \Gamma\},\ \Theta := \{\psi \mid \psi \in \Gamma \text{ and } \blacklozenge\psi \in \mathsf{Sub}(\Pi, \varphi)\},$$

where it is noted that $\mathsf{Sub}(\Pi, \varphi) \subseteq \mathsf{Sub}(\Gamma, \Delta)$. We show that $\blacklozenge\Theta, \Pi \Rightarrow \varphi$ is underivable in $\mathsf{G}^a(\mathbf{BiSKt})$. Suppose not. Then, we can consider the following derivation:

$$\dfrac{\dfrac{\blacklozenge\Theta, \Pi \Rightarrow \varphi}{\Theta, \Box\Pi \Rightarrow \Box\varphi}\ (\Box)^a}{\Gamma \Rightarrow \Delta}\ (w\Rightarrow, \Rightarrow w),$$

where $\blacklozenge\Theta \subseteq \mathsf{Sub}(\Pi,\varphi)$ holds by definition of Θ and note that $\Theta\cup\Box\Pi \subseteq \Gamma$ and $\Box\varphi \in \Delta$. But, this is a contradiction with the underivability of $\Gamma \Rightarrow \Delta$. By Lemma 4.2, there exists $(\Pi^+,\Sigma^+) \in W^{\Xi}$ such that $\blacklozenge\Theta\cup\Pi \subseteq \Pi^+$, $\varphi \in \Sigma^+$, and $\Pi^+\cup\Sigma^+ = \mathsf{Sub}(\blacklozenge\Theta,\Pi,\varphi)$. Let us establish $(\Gamma,\Delta)R^{\Xi}(\Pi^+,\Sigma^+)$. We verify two conditions (i) and (ii) for R^{Ξ} of Definition 4.3. Condition (i) is easy to verify, so we focus on condition (ii). Suppose that $\blacklozenge\sigma \in \Sigma^+$. Our goal is to show $\sigma \in \Delta$. We first observe that $\blacklozenge\sigma \in \Sigma^+ \subseteq \Pi^+ \cup \Sigma^+ = \mathsf{Sub}(\blacklozenge\Theta,\Pi,\varphi) = \mathsf{Sub}(\blacklozenge\Theta)\cup\mathsf{Sub}(\Pi,\varphi) \subseteq \mathsf{Sub}(\Gamma,\Delta)$. It follows that $\sigma \in \mathsf{Sub}(\Gamma,\Delta) = \Gamma\cup\Delta$. To show our goal of $\sigma\in\Delta$, suppose for contradiction that $\sigma \notin \Delta$. It follows from $\sigma \in \Gamma\cup\Delta$ that $\sigma\in\Gamma$. From our supposition of $\blacklozenge\sigma\in\Sigma^+$, $\blacklozenge\sigma \in \mathsf{Sub}(\Pi,\varphi)$ holds. So, by our definition of Θ, we have $\sigma\in\Theta$ hence $\blacklozenge\sigma \in \blacklozenge\Theta \subseteq \Pi^+$. Since $\blacklozenge\sigma\in\Sigma^+$, we showed that $\Pi^+ \Rightarrow \Sigma^+$ is derivable in $\mathsf{G}^a(\mathbf{BiSKt})$, but this is a contradiction with the underivability of $\Pi^+ \Rightarrow \Sigma^+$ in $\mathsf{G}^a(\mathbf{BiSKt})$. Therefore, $\sigma\in\Delta$, as desired.

(xi) Assume that $\blacklozenge\varphi \in \Gamma$. Define Σ and Θ as follows:

$$\Sigma := \{\psi \mid \blacklozenge\psi \in \Delta\},\ \Theta := \{\psi \mid \psi\in\Delta \text{ and } \Box\psi \in \mathsf{Sub}(\varphi,\Sigma)\},$$

where we note that $\mathsf{Sub}(\varphi,\Sigma) \subseteq \mathsf{Sub}(\Gamma,\Delta)$. We show that $\varphi \Rightarrow \Sigma,\Box\Theta$ is underivable in $\mathsf{G}^a(\mathbf{BiSKt})$. Suppose not. Then, we can obtain the following derivation:

$$\dfrac{\dfrac{\varphi \Rightarrow \Sigma,\Box\Theta}{\blacklozenge\varphi \Rightarrow \blacklozenge\Sigma,\Theta}\,(\blacklozenge)^a}{\Gamma \Rightarrow \Delta}\,(w\Rightarrow,\Rightarrow w),$$

where we note that $\Box\Theta \subseteq \mathsf{Sub}(\varphi,\Sigma)$ and $\blacklozenge\varphi\in\Gamma$ and $\blacklozenge\Sigma\cup\Theta\subseteq\Delta$. But, this is a contradiction with the underivability of $\Gamma \Rightarrow \Delta$, So, by Lemma 4.2, there exists $(\Pi^+,\Sigma^+) \in W^{\Xi}$ such that: $\varphi\in\Pi^+$, $\Sigma\cup\Box\Theta \subseteq \Sigma^+$, and $\Pi^+\cup\Sigma^+ = \mathsf{Sub}(\varphi,\Sigma,\Box\Theta)$. Finally, let us establish $(\Pi^+,\Sigma^+)R^{\Xi}(\Gamma,\Delta)$. We check two conditions one by one. As for condition (ii), we proceed as follows. Assume that $\blacklozenge\sigma\in\Delta$. Our goal is to show $\sigma\in\Sigma^+$. By assumption, $\sigma\in\Sigma$. We deduce from $\Sigma\subseteq\Sigma^+$ that $\sigma\in\Sigma^+$, as desired.

Let us move to condition (i). Assume that $\Box\gamma\in\Pi^+$. We show $\gamma\in\Gamma$. Observe that $\Box\gamma \in \Pi^+\cup\Sigma^+ = \mathsf{Sub}(\varphi,\Sigma,\Box\Theta) \subseteq \mathsf{Sub}(\varphi,\Sigma) \subseteq \mathsf{Sub}(\Gamma,\Delta)$ by $\Box\Theta\subseteq\mathsf{Sub}(\varphi,\Sigma)$. This implies that $\gamma\in\mathsf{Sub}(\Gamma,\Delta) = \Gamma\cup\Delta$. Suppose for contradiction that $\gamma\notin\Gamma$. Then $\gamma\in\Delta$. By $\Box\gamma\in\mathsf{Sub}(\varphi,\Sigma)$, now we obtain $\gamma\in\Theta$ by definition of Θ, which implies $\Box\gamma\in\Box\Theta\subseteq\Sigma^+$. Together with $\Box\gamma\in\Pi^+$, we obtain the derivability of $\Pi^+\Rightarrow\Sigma^+$, a contradiction.

□

It is immediate from Lemma 4.5 to obtain the following.

Lemma 4.6 *Let Ξ be a subformula closed finite set. Then, for all $(\Gamma,\Delta)\in W^{\Xi}$ and for all $\chi\in\Gamma\cup\Delta$, the following hold:*

(i) *If* $\chi \in \Gamma$, *then* $M^{\Xi}, (\Gamma, \Delta) \models \chi$, (ii) *If* $\chi \in \Delta$, *then* $M^{\Xi}, (\Gamma, \Delta) \not\models \chi$.

Theorem 4.7 *If a sequent* $\Gamma \Rightarrow \Delta$ *is valid in all finite models, then it is derivable in* $\mathsf{G}^a(\mathbf{BiSKt})$.

Proof. We prove the contrapositive implication. Suppose that a sequent $\Gamma \Rightarrow \Delta$ is underivable in $\mathsf{G}^a(\mathbf{BiSKt})$. Define $\Xi := \mathsf{Sub}(\Gamma, \Delta)$, which is a subformula closed finite set. It follows from Lemma 4.2 that there exists a Ξ-partial valuation $(\Gamma^+, \Delta^+) \in W^{\Xi}$ such that $\Gamma \subseteq \Gamma^+ \subseteq \Xi$, $\Delta \subseteq \Delta^+ \subseteq \Xi$ and $\Gamma^+ \cup \Delta^+ = \mathsf{Sub}(\Gamma, \Delta)$. By Lemma 4.6, we can conclude that $\Gamma \Rightarrow \Delta$ is not valid in a finite model M^{Ξ}, where it is noted that the domain W^{Ξ} of M^{Ξ} is finite. □

Corollary 4.8 *The following are all equivalent: for any sequent* $\Gamma \Rightarrow \Delta$, (i) *it is valid in all models,* (ii) *it is valid in all finite models,* (iii) *it is derivable in* $\mathsf{G}^a(\mathbf{BiSKt})$, (iv) *it is derivable in* $\mathsf{G}^*(\mathbf{BiSKt})$, (v) *it is derivable in* $\mathsf{G}(\mathbf{BiSKt})$. *Therefore,* $\mathsf{G}(\mathbf{BiSKt})$ *enjoys the finite model property and the subformula property. Moreover,* $\mathsf{G}(\mathbf{BiSKt})$ *is decidable.*

Proof. The direction from (ii) to (iii) is due to Theorem 4.7 and the direction from (v) to (i) is due to Theorem 3.3. The remaining directions from (i) to (ii), from (iii) to (iv), and from (iv) to (v), are immediate. □

Corollary 4.8 also provides an alternative proof of the finite model property of **BiSKt** (see [26, Theorem 5]), which was originally established via filtration technique in [26].

5 Craig Interpolation Theorem

This section establishes the Craig Interpolation Theorem of **BiSKt** by Mints' symmetric interpolation method [16], which is a generalization of Maehara's method [12]. For this purpose, it suffices to make use of $\mathsf{G}^*(\mathbf{BiSKt})$ from Table 1, instead of the fully analytic calculus $\mathsf{G}^a(\mathbf{BiSKt})$. This is because the Craig interpolation theorem is concerned with propositional variables and we do not need to consider analytic applications of modal rules of ($\blacklozenge$) and ($\Box$). Also, we do not restrict partitions of a sequent to a particular form, e.g., a *normal partition* as in [7]. Instead, we use the same unrestricted notion of partition of a sequent as for classical propositional logic (this is why the method is called *symmetric* in [16]) and follow Mints [16] to generalize the notion of Craig interpolant.

5.1 Mints' Notion of Interpolant

In what follows, we use S, T, U, S_i, etc. to denote sequents. Given any sequent $S = \Gamma \Rightarrow \Delta$, the following abbreviation is used: $\mathsf{Ant}(S) := \Gamma$ and $\mathsf{Suc}(S) := \Delta$. When we list sequents, we use the semicolon ";" (instead of the comma) to separate sequents such as $S_1; S_2; S_3$. Given any finite list $S_1; \ldots; S_n$ of sequents, we say that a sequent S is *derivable* in a system from $S_1; \ldots; S_n$ if there is a finite tree generated by inference rules of the system from initial sequents of the system and sequents among $S_1; \ldots; S_n$. It is not difficult to see that a sequent

S is derivable in a system iff S is derivable from the empty list in the system. When S is a sequent $\Gamma \Rightarrow \Delta$, we define $S[\psi_1/p_1, \ldots, \psi_n/p_n]$ as the sequent $\Gamma[\psi_1/p_1, \ldots, \psi_n/p_n] \Rightarrow \Delta[\psi_1/p_1, \ldots, \psi_n/p_n]$.

The following simple notation introduced in [16] makes the essence of an argument in this section more explicit.

Definition 5.1 Given sequents $S := \Gamma \Rightarrow \Delta$ and $S' := \Gamma' \Rightarrow \Delta'$, the notation SS' is defined as $SS' := \Gamma, \Gamma' \Rightarrow \Delta, \Delta'$.

Definition 5.2 ([16, Definition 1]) Let p_1, ..., p_l be distinct propositional variables and $S_1; \ldots; S_k$ a finite list of sequents such that $\mathsf{Ant}(S_i) \cup \mathsf{Suc}(S_i) \subseteq \{p_1, \ldots, p_l\}$ and all elements in $\mathsf{Ant}(S_i)$ and $\mathsf{Suc}(S_i)$ are distinct $(1 \leqslant i \leqslant k)$. We say that the list $S_1; \ldots; S_k$ is *closed* if the empty sequent $\Rightarrow$ is derivable from the list by applying (Cut) and contraction rules alone.

For example, both $(p \Rightarrow\ ;\ \Rightarrow p)$ and $((p \Rightarrow q); (p, q \Rightarrow\ ); (\ \Rightarrow p))$ are closed.

Lemma 5.3 *Let $S_1; \ldots; S_m; S_{m+1}; \ldots; S_k$ $(m \geqslant 1)$, T and U be sequents. Suppose that the empty sequent $\Rightarrow$ is derivable from $S_1; \ldots; S_m; S_{m+1}; \ldots; S_k$ only by the rule of cut and contraction rules. Then, TU is derivable from $S_1T; \ldots; S_mT; S_{m+1}U; \ldots; S_kU$ by the rule of cut and contraction rules.*

Proof. It suffices to consider the following transformation of derivations:

$$\begin{array}{c} S_1 \quad \cdots \quad S_m \quad S_{m+1} \quad \cdots \quad S_k \\ \vdots\ (Cut), (c) \\ \Rightarrow \end{array} \quad \rightsquigarrow \quad \begin{array}{c} S_1T \quad \cdots \quad S_mT \quad S_{m+1}U \quad \cdots \quad S_kU \\ \vdots\ (Cut), (c) \\ \underbrace{T \cdots T}_{m}\underbrace{U \cdots U}_{k-m} \\ \vdots\ (c) \\ TU \end{array}.$$

□

It is noted that $S_1; \ldots; S_m; S_{m+1}; \ldots; S_k$ of Lemma 5.3 may not be atomic sequents and that T or U could be an empty sequent.

Definition 5.4 A *partition* of a sequent S is an arbitrary pair $(S_1; S_2)$ of sequents such that $S = S_1S_2$. A partition $(\Gamma_1 \Rightarrow \Delta_1; \Gamma_2 \Rightarrow \Delta_2)$ of a sequent S is also denoted by $(\Gamma_1 : \Delta_1); (\Gamma_2 : \Delta_2)$ (cf. [7]).

Definition 5.5 Let $S = \Gamma \Rightarrow \Delta$ be a sequent. The *formulaic translation* $\mathtt{f}(S)$ and *dual formulaic translation* $\mathtt{d}(S)$of S is defined as:

$$\mathtt{f}(S) := \bigwedge \Gamma \to \bigvee \Delta \quad \text{and} \quad \mathtt{d}(S) := \bigwedge \Gamma \prec \bigvee \Delta.$$

It is noted that $(\mathtt{f}(S) \Rightarrow)S$ is derivable, because the sequent $\bigwedge \Gamma \to \bigvee \Delta, \Gamma \Rightarrow \Delta$ is derivable. Similarly, $(\Rightarrow \mathtt{d}(S))S$ is derivable, because the sequent $\Gamma \Rightarrow \Delta, \bigwedge \Gamma \prec \bigvee \Delta$ is derivable.

Definition 5.6 We say that γ is a *Craig interpolant* for a partition $(S; S')$ if both $S(\Rightarrow \gamma)$ and $S'(\gamma \Rightarrow)$ are derivable and $\mathsf{Prop}(\gamma) \subseteq \mathsf{Prop}(S) \cap \mathsf{Prop}(S')$.

In other words, γ is a Craig interpolant for $(\Gamma_1 : \Delta_1); (\Gamma_2 : \Delta_2)$ if both $\Gamma_1 \Rightarrow \Delta_1, \gamma$ and $\gamma, \Gamma_2 \Rightarrow \Delta_2$ are derivable and $\mathsf{Prop}(\gamma) \subseteq \mathsf{Prop}(\Gamma_1, \Delta_1) \cap \mathsf{Prop}(\Gamma_2, \Delta_2)$.

Definition 5.7 (Interpolant [16, Definition 3]) Let S and S' be arbitrary sequents, $\gamma_1, \ldots, \gamma_l$ be formulas, $p_1, \ldots, p_l$ be distinct propositional variables, and $T_1; \ldots; T_k$ ($k \geqslant 2$) be sequents such that $\mathsf{Ant}(T_i) \cup \mathsf{Suc}(T_i) \subseteq \{p_1, \ldots, p_l\}$ and all elements in $\mathsf{Ant}(T_i)$ and $\mathsf{Suc}(T_i)$ are distinct. We say that $(\gamma_1, \ldots, \gamma_l, (T_1; \ldots; T_k))$ is an *interpolant* for a partition $(S; S')$ if the following are satisfied:

(i) $\mathsf{Prop}(\gamma_i) \subseteq \mathsf{Prop}(S) \cap \mathsf{Prop}(S')$ for all i such that $1 \leqslant i \leqslant l$,

(ii) $(T_1; \ldots; T_k)$ is closed,

(iii) there exists m such that all sequents $ST_1^*; \ldots; ST_m^*; S'T_{m+1}^*; \ldots; S'T_k^*$ are derivable, where $T_i^* := T_i[\gamma_1/p_1, \ldots, \gamma_l/p_l]$.

The notion of interpolant in Definition 5.7 is a generalization of the notion of Craig interpolant (Definition 5.6). This is because $(\gamma, (\Rightarrow p; p \Rightarrow))$ is an interpolant for $(\Gamma_1 \Rightarrow \Delta_1; \Gamma_2 \Rightarrow \Delta_2)$ iff $\mathsf{Prop}(\gamma) \subseteq \mathsf{Prop}(\Gamma_1, \Delta_1) \cap \mathsf{Prop}(\Gamma_2, \Delta_2)$, $(\Rightarrow p; p \Rightarrow)$ is closed (this is trivial) and both $\Gamma_1 \Rightarrow \Delta_1, \gamma$ and $\gamma, \Gamma_2 \Rightarrow \Delta_2$ are derivable.

Let us say that a partition $(\Gamma_1, \Delta_1); (\Gamma_2, \Delta_2)$ is *normal* if Δ_1 or Γ_2 is empty ([7, p.11]). The following two lemmas tell us that a Craig interpolant can be constructed from an interpolant for a normal partition. Lemma 5.8 is needed for calculating interpolants for rules $(\Rightarrow\to)$ and $(\Box)$ and Lemma 5.9 is for rules $(\prec\Rightarrow)$ and $(\blacklozenge)$. The following lemma generalizes [16, Lemma 1(b)].

Lemma 5.8 *From an interpolant for* $(\Gamma_1 : \varnothing); (\Gamma_2 : \Delta)$, *a Craig interpolant for* $(\Gamma_1 : \varnothing); (\Gamma_2 : \Delta)$ *can be constructed.*

Proof. Let $(\gamma_1, \ldots, \gamma_l, (T_1; \ldots; T_k))$ be an interpolant for $(\Gamma_1 : \varnothing); (\Gamma_2 : \Delta)$. We can find an m such that $1 \leqslant m \leqslant k$ and all of the following sequents are derivable:

$$(\Gamma_1 \Rightarrow)T_1^*; \ldots; (\Gamma_1 \Rightarrow)T_m^*; (\Gamma_2 \Rightarrow \Delta)T_{m+1}^*; \ldots; (\Gamma_2 \Rightarrow \Delta)T_k^*,$$

where recall that $T^* := T[\gamma_1/p_1, \ldots, \gamma_l/p_l]$. It is noted that $(T_1; \ldots; T_k)$ is closed and $\mathsf{Prop}(\gamma_i) \subseteq \mathsf{Prop}(\Gamma_1) \cap \mathsf{Prop}(\Gamma_2, \Delta)$. Let us define $\gamma := \bigwedge_{1 \leqslant i \leqslant m} \mathtt{f}(T_i^*)$. We show that γ is a Craig interpolant for $(\Gamma_1 : \varnothing); (\Gamma_2 : \Delta)$. The variable condition is easily verified. So, we focus on checking the derivability condition. First, let us check the derivability of $\Gamma_1 \Rightarrow \gamma$ is derivable. The empty succedent of $(\Gamma_1 : \varnothing)$ becomes crucial here. It suffices to show that $\Gamma_1 \Rightarrow \mathtt{f}(T_i^*)$ for every $1 \leqslant i \leqslant m$. By recalling $\mathtt{f}(T_i^*) := \bigwedge \mathsf{Ant}(T_i^*) \to \bigvee \mathsf{Suc}(T_i^*)$, this is shown as follows:

$$\frac{\Gamma_1, \bigwedge \mathsf{Ant}(T_i^*) \Rightarrow \bigvee \mathsf{Suc}(T_i^*)}{\Gamma_1 \Rightarrow \bigwedge \mathsf{Ant}(T_i^*) \to \bigvee \mathsf{Suc}(T_i^*)} \ (\Rightarrow\to),$$

where the upper sequent is derivable since $(\Gamma_1 \Rightarrow)T_i^*$ is derivable by assumption.

Second, let us establish that $\gamma, \Gamma_2 \Rightarrow \Delta$ is derivable. It suffices to derive $\mathtt{f}(T_1^*), \ldots, \mathtt{f}(T_m^*), \Gamma_2 \Rightarrow \Delta$. Since $(T_1; \ldots; T_k)$ is closed, we can derive the empty sequent $\Rightarrow$ from $T_1^*, \ldots, T_k^*$ by applying the rule of cut and contraction rules *alone*. It follows from Lemma 5.3 that $\mathtt{f}(T_1^*), \ldots, \mathtt{f}(T_m^*), \Gamma_2 \Rightarrow \Delta$ is derivable from $(\mathtt{f}(T_1^*), \ldots, \mathtt{f}(T_m^*) \Rightarrow)T_i^*$ $(1 \leqslant i \leqslant m)$ and $(\Gamma_2 \Rightarrow \Delta)T_j^*$ $(m+1 \leqslant j \leqslant k)$ by the rule of cut and contraction rules alone. Each $(\Gamma_2 \Rightarrow \Delta)T_j^*$ is derivable by assumption. Since $(\mathtt{f}(T_i^*) \Rightarrow)T_i^*$ is derivable, all the sequents $(\mathtt{f}(T_1^*), \ldots, \mathtt{f}(T_m^*) \Rightarrow)T_i^*$ are also derivable by weakening rules. This finishes establishing that $\mathtt{f}(T_1^*), \ldots, \mathtt{f}(T_m^*), \Gamma_2 \Rightarrow \Delta$ is derivable.

Therefore, γ is a Craig interpolant for $(\Gamma_1 : \varnothing); (\Gamma_2 : \Delta)$. □

While we use a formulaic translation $\mathsf{f}(S)$ of a sequent S in the proof of Lemma 5.8, we need to use its dual variant $\mathsf{d}(S)$ in the following proof.

Lemma 5.9 *From an interpolant for* $(\Gamma : \Delta_1); (\varnothing : \Delta_2)$, *a Craig interpolant for* $(\Gamma : \Delta_1); (\varnothing : \Delta_2)$ *can be constructed.*

Proof. Let $(\gamma_1, \ldots, \gamma_l, (T_1; \ldots; T_k))$ be an interpolant for $(\Gamma : \Delta_1); (\varnothing : \Delta_2)$. We can find an m such that $1 \leqslant m \leqslant k$ and all of the following sequents are derivable:

$$(\Gamma \Rightarrow \Delta_1)T_1^*; \ldots; (\Gamma \Rightarrow \Delta_1)T_m^*; (\Rightarrow \Delta_2)T_{m+1}^*; \ldots; (\Rightarrow \Delta_2)T_k^*,$$

where we recall that $T^* := T[\gamma_1/p_1, \ldots, \gamma_l/p_l]$. It is noted that every γ_i satisfies $\mathsf{Prop}(\gamma_i) \subseteq \mathsf{Prop}(\Gamma, \Delta_1) \cap \mathsf{Prop}(\Delta_2)$. Define

$$\rho := \bigvee_{m+1 \leqslant j \leqslant k} \mathsf{d}(T_j^*).$$

We show that ρ is a Craig interpolant for $(\Gamma : \Delta_1); (\varnothing : \Delta_2)$. The variable condition is easily verified. So, we focus on checking the derivability condition in what follows. First, let us check the derivability of $\rho \Rightarrow \Delta_2$. The empty antecedent of $(\varnothing : \Delta_2)$ becomes crucial here. It suffices to show that $\mathsf{d}(T_j^*) \Rightarrow \Delta_2$ for every $m+1 \leqslant j \leqslant k$. By recalling $\mathsf{d}(T_j^*) := \bigwedge \mathsf{Ant}(T_j^*) \prec \bigvee \mathsf{Suc}(T_j^*)$, this is shown as follows:

$$\frac{\bigwedge \mathsf{Ant}(T_j^*) \Rightarrow \Delta_2, \bigvee \mathsf{Suc}(T_j^*)}{\bigwedge \mathsf{Ant}(T_j^*) \prec \bigvee \mathsf{Suc}(T_j^*) \Rightarrow \Delta_2} (\prec \Rightarrow),$$

where the upper sequent is derivable since $(\Rightarrow \Delta_2)T_j^*$ is derivable by assumption.

Second, let us establish that $\Gamma \Rightarrow \Delta_1, \rho$ is derivable. It suffices to derive $\Gamma \Rightarrow \Delta_1, \mathsf{d}(T_{m+1}^*), \ldots, \mathsf{d}(T_k^*)$. Since $(T_1; \ldots; T_k)$ is closed, we can derive the empty sequent $\Rightarrow$ from $T_1^*, \ldots, T_k^*$ by applying the rule of cut and contraction rules *alone*. It follows from Lemma 5.3 that $\Gamma \Rightarrow \Delta_1, \mathsf{d}(T_{m+1}^*), \ldots, \mathsf{d}(T_k^*)$ is derivable from $(\Gamma \Rightarrow \Delta_1)T_i^*$ $(1 \leqslant i \leqslant m)$ and $(\Rightarrow \mathsf{d}(T_{m+1}^*), \ldots, \mathsf{d}(T_k^*))T_j^*$ $(m+1 \leqslant j \leqslant k)$ by the rule of cut and contraction rules. Note that each $\Gamma \Rightarrow \Delta_1, \mathsf{d}(T_i^*)$ is derivable by assumption. Since $(\Rightarrow \mathsf{d}(T_j^*))T_j^*$ is derivable, all the sequents $(\Rightarrow \mathsf{d}(T_{m+1}^*), \ldots, \mathsf{d}(T_k^*))T_j^*$ are also derivable by weakening rules. This finishes establishing that $\Gamma \Rightarrow \Delta_1, \mathsf{d}(T_{m+1}^*), \ldots, \mathsf{d}(T_{k+1}^*)$ is derivable.

Therefore, ρ is a Craig interpolant for $(\Gamma : \Delta_1); (\varnothing : \Delta_2)$. □

5.2 Transfer of Interpolant Across Inference Rules

Lemma 5.10 *Every partition of an initial sequent of Table 1 has an interpolant.*

Proof. We only check the initial sequent of the form $\varphi \Rightarrow \varphi$ here (because the other two cases are easy). There are four possible partitions as follows:

(i) $(\varphi : \varphi); (\varnothing : \varnothing)$. An interpolant is $(\bot, (\Rightarrow p; p \Rightarrow))$,

(ii) $(\varphi : \varnothing); (\varnothing : \varphi)$. An interpolant is $(\varphi, (\Rightarrow p; p \Rightarrow))$,

(iii) $(\varnothing : \varphi); (\varphi : \varnothing)$. An interpolant is $(\varphi, (p \Rightarrow; \Rightarrow p))$,

(iv) $(\varnothing : \varnothing); (\varphi : \varphi)$. An interpolant is $(\top, (\Rightarrow p; p \Rightarrow))$,

where item 3 is most important because there is no Craig interpolant for a partition $(\varnothing : p); (p : \varnothing)$ in the multi-succedent calculus of intuitionistic propositional logic (see [16, p.226]). □

In what follows, we show that interpolation transfers across applications of all inference rules of G*(**BiSKt**) in the following sense (cf. [7, p.12]).

Definition 5.11 We say that *interpolation transfers across applications of an inference rule* if from the assumption that an interpolant exists for each partition of the upper sequent(s) of the rule it follows that an interpolant exists for each partition of the lower sequent of the rule.

Lemma 5.12 *Interpolation transfers across applications of weakening and contraction rules,* $(\wedge \Rightarrow)$ *and* $(\Rightarrow \vee)$.

Proof. For these one premise rules, the same interpolant can be used from a partition of the premise to the corresponding partition of the conclusion. □

We go back to the remaining one premise rules later. To deal with two premise rules, we introduce the following notion of *composition* of lists of sequents from [16, p.231] (but we use the different notation).

Definition 5.13 Given two finite lists $(S_1; \ldots; S_m)$ and $(T_1; \ldots; T_n)$ of sequents, the composition $(S_1; \ldots; S_m) \circ (T_1; \ldots; T_n)$ of them is defined as:

$$(S_1; \ldots; S_m) \circ (T_1; \ldots; T_n) := (S_i T_j | 1 \leqslant i \leqslant m \text{ and } 1 \leqslant j \leqslant n),$$

where we use the lexicographic order for the composition (but the order will not matter below).

It is noted that the resulting composition has $m \cdot n$ sequents.

The following lemma witnesses that our combination of analytic cut and Mints' notion of interpolant work neatly (compare the following proof of Lemma 5.14 with [7, Lemma 5.4] which states that a Craig interpolant transfers across (essential) applications of the analytic mix rule, i.e., an extended version of (Cut)).

Lemma 5.14 *Interpolation transfers across applications of* $(Cut)^a$.

Proof. Let us write the application of the rule as:

$$\frac{\Gamma_1, \Gamma_2 \Rightarrow \Delta_1, \Delta_2, \varphi \quad \varphi, \Pi_1, \Pi_2 \Rightarrow \Sigma_1, \Sigma_2}{\Gamma_1, \Gamma_2, \Pi_1, \Pi_2 \Rightarrow \Delta_1, \Delta_2, \Sigma_1, \Sigma_2} \ (Cut)^a,$$

where $\varphi \in \mathsf{Sub}(\Gamma_1, \Gamma_2, \Pi_1, \Pi_2, \Delta_1, \Delta_2, \Sigma_1, \Sigma_2)$. Let $(\Gamma_1, \Pi_1 : \Delta_1, \Sigma_1); (\Gamma_2, \Pi_2 : \Delta_2, \Sigma_2)$ be a partition. We divide our argument into the following two cases: (i) $\varphi \in \mathsf{Sub}(\Gamma_1, \Pi_1, \Delta_1, \Sigma_1)$; (ii) $\varphi \in \mathsf{Sub}(\Gamma_2, \Pi_2, \Delta_2, \Sigma_2)$. We consider case (i) alone because case (ii) is shown similarly. By induction hypothesis, there is an interpolant $(\gamma_1, \ldots, \gamma_a, (T_1; \ldots; T_m; T_{m+1}; \ldots; T_b))$ for $(\Gamma_1 : \Delta_1, \varphi); (\Gamma_2 : \Delta_2)$. That is,

- $\mathsf{Prop}(\gamma_i) \subseteq \mathsf{Prop}(\Gamma_1, \Delta_1, \varphi) \cap \mathsf{Prop}(\Gamma_2, \Delta_2)$,
- $(T_1; \ldots; T_m; T_{m+1}; \ldots; T_b)$ is closed,
- all of the following sequents are derivable:

$$(\Gamma_1 \Rightarrow \Delta_1, \varphi)T_1^*; \ldots; (\Gamma_1 \Rightarrow \Delta_1, \varphi)T_m^*; (\Gamma_2 \Rightarrow \Delta_2)T_{m+1}^*; \ldots; (\Gamma_2 \Rightarrow \Delta_2)T_b^*$$

Again, by induction hypothesis, let $(\rho_1, \ldots, \rho_c, (U_1; \ldots; U_n; U_{n+1}; \ldots; U_d))$ be an interpolant for the partition $(\varphi, \Pi_1 : \Sigma_1); (\Pi_2 : \Sigma_2)$. That is,

- $\mathsf{Prop}(\rho_j) \subseteq \mathsf{Prop}(\varphi, \Pi_1, \Sigma_1) \cap \mathsf{Prop}(\Pi_2, \Sigma_2)$,
- $(U_1; \ldots; U_n; U_{n+1}; \ldots; U_d)$ is closed,
- all of the following sequents are derivable:

$$(\varphi, \Pi_1 \Rightarrow \Sigma_1)U_1^\star; \ldots; (\varphi, \Pi_1 \Rightarrow \Sigma_1)U_n^\star; (\Pi_2 \Rightarrow \Sigma_2)U_{n+1}^\star; \ldots; (\Pi_2 \Rightarrow \Sigma_2)U_d^\star.$$

We can assume without loss of generality that $(T_1; \ldots; T_b)$ and $(U_1; \ldots; U_d)$ have no common propositional variable. In what follows, we prove that our interpolant for the original partition is:

$$(\gamma_1, \ldots, \gamma_a, \rho_1, \ldots, \rho_c, ((T_1; \ldots; T_m) \circ (U_1; \ldots; U_n); (T_{m+1}; \ldots; T_b; U_{n+1}; \ldots; U_d))).$$

Let us verify the required three conditions. First, we verify the variable condition. For γ_i, we proceed as follows. By $\varphi \in \mathsf{Sub}(\Gamma_1, \Pi_1, \Delta_1, \Sigma_1)$,

$$\mathsf{Prop}(\gamma_i) \subseteq \mathsf{Prop}(\Gamma_1, \Delta_1, \varphi) \cap \mathsf{Prop}(\Gamma_2, \Delta_2) \subseteq \mathsf{Prop}(\Gamma_1, \Pi_1, \Delta_1, \Sigma_1) \cap \mathsf{Prop}(\Gamma_2, \Delta_2)$$

As for ρ_j, it is shown from $\varphi \in \mathsf{Sub}(\Gamma_1, \Pi_1, \Delta_1, \Sigma_1)$ as:

$$\mathsf{Prop}(\rho_j) \subseteq \mathsf{Prop}(\varphi, \Pi_1, \Sigma_1) \cap \mathsf{Prop}(\Pi_2, \Sigma_2) \subseteq \mathsf{Prop}(\Gamma_1, \Pi_1, \Delta_1, \Sigma_1) \cap \mathsf{Prop}(\Pi_2, \Sigma_2).$$

Second, we proceed as follows for the closure condition. By Lemma 5.3 and the closedness of $(U_1; \ldots; U_d)$, we can derive T_j from $(T_j U_1; \ldots; T_j U_n)$ and $(U_{n+1}; \ldots; U_d)$ by (Cut) and contraction rules, for every j such that $1 \leqslant j \leqslant m$. Then, we can derive $\Rightarrow$ from just obtained $T_1; \ldots; T_m$ and $(T_{m+1}; \ldots; T_b)$ since $(T_1; \ldots; T_b)$ is closed. To sum up, the empty sequent $\Rightarrow$ is derivable from $(T_1; \ldots; T_m) \circ (U_1; \ldots; U_n)$; $(T_1; \ldots; T_b)$; $(U_{n+1}; \ldots; U_d)$ by definition of the composition $\circ$.

For the derivability condition, we first show that $(\Gamma_1, \Pi_1 \Rightarrow \Delta_1, \Sigma_1)(T_i U_j)^{**}$ ("$*\star$" is the concatenation or union of two substitutions, recall that $(T_1; \ldots; T_b)$ and $(U_1; \ldots; U_d)$ have no common propositional variable. So, $(T_i U_j)^{**} = T_i^* U_j^\star$) is derivable as follows:

$$\frac{(\Gamma_1 \Rightarrow \Delta_1, \varphi)T_i^* \quad (\varphi, \Pi_1 \Rightarrow \Sigma_1)U_j^\star}{(\Gamma_1, \Pi_1 \Rightarrow \Delta_1, \Sigma_1)(T_i U_j)^{**}} \ (Cut)^a,$$

where the upper sequents are derivable by assumption. Second, sequents $(\Gamma_2, \Pi_2 \Rightarrow \Delta_2, \Sigma_2)T_i^*$ and $(\Gamma_2, \Pi_2 \Rightarrow \Delta_2, \Sigma_2)U_j^\star$ are derivable in terms of weakening rules from the derivability of sequents $(\Gamma_2 \Rightarrow \Delta_2)T_i^*$ and $(\Pi_2 \Rightarrow \Sigma_2)U_j^\star$, respectively. This finishes establishing the derivability condition. □

We can prove Lemma 5.15 similarly to Lemma 5.14.

Lemma 5.15 *Interpolation transfers across applications of inference rules* $(\rightarrow\Rightarrow)$, $(\Rightarrow \prec)$, $(\Rightarrow \wedge)$ *and* $(\Rightarrow \vee)$.

Let us discuss the remaining one premise rules below.

Lemma 5.16 *Interpolation transfers across applications of* $(\Rightarrow\rightarrow)$ *and* $(\prec \Rightarrow)$.

Proof. First, let us consider the following application of the rule $(\prec \Rightarrow)$:

$$\frac{\varphi \Rightarrow \psi, \Delta_1, \Delta_2}{\varphi \prec \psi \Rightarrow \Delta_1, \Delta_2} \ (\prec \Rightarrow)$$

All possible partitions are $(\varphi \prec \psi : \Delta_1); (\varnothing : \Delta_2)$ and $(\varnothing : \Delta_1); (\varphi \prec \psi : \Delta_2)$. First, we consider the former partition. By induction hypothesis and Lemma 5.9, there is a *Craig interpolant* ρ of $(\varphi : \psi, \Delta_1); (\varnothing : \Delta_2)$. We show that $(\rho, (\Rightarrow q; q \Rightarrow))$ is an interpolant. The variable condition and closure conditions are trivial. The derivability of $\rho \Rightarrow \Delta_2$ is immediate. Moreover, from the derivability of $\varphi \Rightarrow \psi, \Delta_1, \rho$, we conclude by $(\prec \Rightarrow)$ that $\varphi \prec \psi \Rightarrow \Delta_1, \rho$ is derivable. Next, we consider the partition $(\varnothing : \Delta_1); (\varphi \prec \psi : \Delta_2)$. By induction hypothesis and Lemma 5.9, there is a Craig interpolant ρ of $(\varphi : \psi, \Delta_2); (\varnothing : \Delta_1)$. We show that $(\rho, (q \Rightarrow; \Rightarrow q))$ is an interpolant for $(\varnothing : \Delta_1); (\varphi \prec \psi : \Delta_2)$ (be careful about the positions of q). All the conditions are verified similarly.

Second, we can apply a similar argument for the rule $(\Rightarrow\rightarrow)$ but we need Lemma 5.8 instead of Lemma 5.9. □

Lemma 5.17 *Interpolation transfers across applications of* $(\Box)$ *and* $(\blacklozenge)$.

Proof. First, we check the following application of the rule $(\blacklozenge)$:

$$\frac{\varphi \Rightarrow \Delta_1, \Delta_2, \Box\Theta_1, \Box\Theta_2}{\blacklozenge\varphi \Rightarrow \blacklozenge\Delta_1, \blacklozenge\Delta_2, \Theta_1, \Theta_2} \ (\blacklozenge).$$

All possible partitions have the following two forms:

$$(\blacklozenge\varphi : \blacklozenge\Delta_1, \Theta_1); (\varnothing : \blacklozenge\Delta_2, \Theta_2) \text{ or } (\varnothing : \blacklozenge\Delta_1, \Theta_1); (\blacklozenge\varphi : \blacklozenge\Delta_2, \Theta_2).$$

We focus on the partition of the form $(\varnothing : \blacklozenge\Delta_1, \Theta_1); (\blacklozenge\varphi : \blacklozenge\Delta_2, \Theta_2)$ below, because we can find an interpolant for the other partition by a similar argument. Let $((\gamma_1, \ldots, \gamma_a), T_1; \ldots; T_b)$ be an interpolant of $(\varphi : \Delta_2, \Box\Theta_2); (\varnothing : \Delta_1, \Box\Theta_1)$. Then, there exists a *Craig interpolant* for $(\varphi : \Delta_2, \Box\Theta_2); (\varnothing : \Delta_1, \Box\Theta_1)$ by Lemma 5.9. Then, we can find a formula γ such that $\varphi \Rightarrow \Delta_2, \gamma, \Box\Theta_2$ and $\gamma \Rightarrow \Delta_1, \Box\Theta_1$ are derivable, and $\mathsf{Prop}(\gamma) \subseteq \mathsf{Prop}(\varphi, \Delta_2, \Box\Theta_2) \cap \mathsf{Prop}(\Delta_1, \Box\Theta_1)$. By the following applications of the rule $(\blacklozenge)$:

$$\frac{\varphi \Rightarrow \Delta_2, \gamma, \Box\Theta_2}{\blacklozenge\varphi \Rightarrow \blacklozenge\Delta_2, \blacklozenge\gamma, \Theta_2}\ (\blacklozenge) \qquad \frac{\gamma \Rightarrow \Delta_1, \Box\Theta_1}{\blacklozenge\gamma \Rightarrow \blacklozenge\Delta_1, \Theta_1}\ (\blacklozenge),$$

we can conclude that $(\blacklozenge\gamma; (q \Rightarrow; \Rightarrow q))$ is our desired interpolant for the partition $(\varnothing : \blacklozenge\Delta_1, \Theta_1); (\blacklozenge\varphi : \blacklozenge\Delta_2, \Theta_2)$ because $\mathsf{Prop}(\blacklozenge\gamma) \subseteq \mathsf{Prop}(\blacklozenge\varphi, \blacklozenge\Delta_2, \Theta_2) \cap \mathsf{Prop}(\blacklozenge\Delta_1, \Theta_1)$ holds easily.

Second, we can also verify the case of $(\Box)$ similarly as above except we need Lemma 5.8 instead of Lemma 5.9. □

Lemma 5.18 *If* $\Gamma \Rightarrow \Delta$ *is derivable in* $\mathsf{G}^*(\mathbf{BiSKt})$, *then every partition* $(S; S')$ *for* $\Gamma \Rightarrow \Delta$ *has an interpolant in* $\mathsf{G}(\mathbf{BiSKt})$.

Proof. By induction on derivation of $\Gamma \Rightarrow \Delta$ in $\mathsf{G}^*(\mathbf{BiSKt})$. For the base case, we can use Lemma 5.10 and for the inductive step, it suffices to apply Lemmas 5.12, 5.14, 5.15, 5.16, and 5.17. □

Theorem 5.19 (Craig Interpolation) *If* $\Rightarrow \varphi \rightarrow \psi$ *is derivable in* $\mathsf{G}(\mathbf{BiSKt})$ *then there exists a formula* γ *such that* $\Rightarrow \varphi \rightarrow \gamma$ *and* $\Rightarrow \gamma \rightarrow \psi$ *are derivable in* $\mathsf{G}(\mathbf{BiSKt})$ *and* $\mathsf{Prop}(\gamma) \subseteq \mathsf{Prop}(\varphi) \cap \mathsf{Prop}(\psi)$.

Proof. Suppose that $\Rightarrow \varphi \rightarrow \psi$ is derivable in $\mathsf{G}(\mathbf{BiSKt})$. By

$$\frac{\overline{\Rightarrow \varphi \rightarrow \psi} \qquad \dfrac{\varphi \Rightarrow \varphi \quad \psi \Rightarrow \psi}{\varphi \rightarrow \psi, \varphi \Rightarrow \psi}\ (\rightarrow\Rightarrow)}{\varphi \Rightarrow \psi}\ (Cut),$$

we get the derivability of $\varphi \Rightarrow \psi$ in $\mathsf{G}(\mathbf{BiSKt})$. By Corollary 4.8, $\varphi \Rightarrow \psi$ is derivable in $\mathsf{G}^*(\mathbf{BiSKt})$. By Lemma 5.18, there is an interpolant for $(\varphi : \varnothing); (\varnothing : \psi)$ in $\mathsf{G}(\mathbf{BiSKt})$. By Lemma 5.8 (or Lemma 5.9), there is a *Craig interpolant* γ of $(\varphi : \varnothing); (\varnothing : \psi)$ in $\mathsf{G}(\mathbf{BiSKt})$. Therefore, $\Rightarrow \varphi \rightarrow \gamma$ and $\Rightarrow \gamma \rightarrow \psi$ are derivable in $\mathsf{G}(\mathbf{BiSKt})$ and the variable condition is also satisfied. □

6 Conclusion

This paper established the Craig interpolation theorem for bi-intuitionistic stable tense logic [30]. A novelty of this paper is in revealing that Mints' symmetric interpolation method [16] works properly with analytic cuts. Since our argument is modular, it also provides a simplification of Kowalski and the first author's argument in [7] for the Craig interpolation theorem of bi-intuitionistic logic in two respects. First, the restriction of applications of (Cut) to analytic ones for bi-intuitionistic logic is proved more directly than in [7]. Second,

Mints' symmetric interpolation method [16] enables us to prove the interpolation theorem in a simpler manner than in [7].

We comment on possible directions of further research. First, as we noted in the proof of Lemma 5.10, there is no Craig interpolant for a partition $(\varnothing : p); (p : \varnothing)$ in the multi-succedent calculus of intuitionistic propositional logic (see [16, p.226]). On the other hand, Lemmas 5.8 and 5.9 tell us that we can obtain Craig interpolants from interpolants for particular forms of partitions. From this respect, it would be nice to have a comprehensive study of the relationship between the Mints interpolation for analytic cut and the Craig interpolation for analytic mix.

Second, our calculation of an interpolant for the rule ($\blacklozenge$) (Lemma 5.17) depends on Lemma 5.9, in particular, the existence of coimplication $\prec$. So, it would be interesting to consider if the Craig interpolation theorem holds for the fragment of **G**(**BiSKt**) without coimplication, given that the subformula property of the fragment holds by our argument in Section 4. It is noted that this fragment is different from *intuitionistic* tense logic of Ewald [3] (see also [4,9] for the recent studies).

Third, to emphasize the effectiveness of a combination of analytic cut and Mints' symmetric interpolation method, we may apply our argument in this paper also to other non-classical logics, e.g., axiomatic extensions of bi-intuitionistic stable tense logic [26], bi-intuitionistic tense logic **BiKt** [5] (having future possibility operator $\Diamond$ and past necessity operator $\blacksquare$ in addition to $\blacklozenge$ and $\Box$), intuitionistic modal logic $\mathbf{L}_4$ [19] (the strongest **S5**-type intuitionistic modal logic in [19]), and bi-intuitionistic stable tense logic with universal modality [29]. Luppi [10] established the Craig interpolation theorem of $\mathbf{L}_4$ algebraically, but a proof-theoretic argument has not been provided. By a slight modification of Lemma 5.8, we have already confirmed that our argument is still applicable to $\mathbf{L}_4$. A semantic connection between **BiSKt** and **BiKt** [5] stated in [30, pp.517-8] allows us to apply our argument also to **BiKt** (see Appendix A for the detail). It may be also interesting to consider if we can generalize our argument of this paper to cover substructural modal or tense logics, though our argument depends on the existence of structural rules (e.g., in the proofs of Lemmas 5.3, 5.8 and 5.9).

Finally, algebras of the bi-intuitionistic stable tense logic are double Heyting algebras (cf. [7, pp.259-60]), which are distributive, equipped with a residuated pair of unary modal operators. It is well-known that, for normal modal logics based on classical logic, Craig interpolation corresponds to an algebraic property called superamalgamability [14,13]. It is an open problem to find the corresponding algebraic notion to Mints' symmetric interpolation method. [3]

[3] We would like to thank the three reviewers for their suggestions and comments to our draft. The work of the second author was partially supported by JSPS KAKENHI Grant-in-Aid for Scientific Research (B) Grant Number JP22H00597 and (C) Grant Number JP19K12113.

Appendix

A An Application to Bi-intuitionistic Tense Logic BiKt

The syntax of bi-intuitionistic tense logic **BiKt** by Goré et al. [5] is an expansion of the syntax for bi-intuitionistic stable tense logic **BiSKt** with future possibility operator $\Diamond$ and past necessity operator $\blacksquare$. Kripke semantics for **BiKt** [5, Section 6] is slightly different from the one for **BiSKt** as follows. Let us say that $(W, \leqslant, R, S)$ is a **BiKt**-*frame* if $(W, \leqslant)$ is a preorder and R and S are binary relations on W such that the following two condition hold:

$(F\Box)$ $R \circ \leqslant \subseteq \leqslant \circ R$,

$(F\Diamond)$ $S^{-1} \circ \leqslant \subseteq \leqslant \circ S^{-1}$,

where S^{-1} is the converse relation of S. A **BiKt**-*model* is a pair of a **BiKt**-frame and a valuation assigning $\leqslant$-closed sets to propositional variables. The satisfaction relation is the same as for the bi-intuitionistic stable tense logic except:

$$
\begin{array}{lll}
M, u \models \blacklozenge\varphi & \text{iff} & \text{For some } v \in W \ (vRu \text{ and } M, v \models \varphi), \\
M, u \models \Box\varphi & \text{iff} & \text{For all } v \in W \ (u(\leqslant \circ R)v \text{ implies } M, v \models \varphi), \\
M, u \models \Diamond\varphi & \text{iff} & \text{For some } v \in W \ (vS^{-1}u \text{ and } M, v \models \varphi), \\
M, u \models \blacksquare\varphi & \text{iff} & \text{For all } v \in W \ (u(\leqslant \circ S^{-1})v \text{ implies } M, v \models \varphi).
\end{array}
$$

We define the notion of validity of a sequent as for **BiSKt**. We use $[\![\varphi]\!]_M$ to mean $\{ u \in W \mid M, u \models \varphi \}$. We can prove that $[\![\varphi]\!]$ is $\leqslant$-closed by induction. As noted in [5], however, we need the conditions $(F\Box)$ and $(F\Diamond)$ for showing that $[\![\blacklozenge\varphi]\!]$ and $[\![\Diamond\varphi]\!]$ are $\leqslant$-closed, respectively.

To define sequent calculi for **BiKt**, it suffices to consider the following two additional rules for $\Diamond$ and $\blacksquare$ and their analytic variants:

$$
\frac{\varphi \Rightarrow \Delta, \blacksquare\Theta}{\Diamond\varphi \Rightarrow \Diamond\Delta, \Theta}\ (\Diamond) \qquad \frac{\Diamond\Theta, \Gamma \Rightarrow \varphi}{\Theta, \blacksquare\Gamma \Rightarrow \blacksquare\varphi}\ (\blacksquare)
$$

where the side conditions for analytic applications of $(\blacklozenge)$ and $(\Box)$ are similarly defined as in Table 1. In what follows, we use $\mathsf{G}(\mathbf{BiKt})$, $\mathsf{G}^*(\mathbf{BiKt})$, and $\mathsf{G}^a(\mathbf{BiKt})$ as the full calculus, the calculus where the rule of cut are analytic, and the analytic calculus, respectively (those are also similarly defined as in Table 1). Once we establish for **BiKt** a similar result to Corollary 4.8, we can apply the same argument to obtain the Craig interpolation theorem for **BiKt**, since the shape of rules $(\blacklozenge)$ and $(\Box)$ are the same as $(\Diamond)$ and $(\blacksquare)$. So, in what follows, we comment on the soundness and the subformula property of $\mathsf{G}(\mathbf{BiKt})$.

Theorem A.1 *If* $\Gamma \Rightarrow \Delta$ *is derivable in* $\mathsf{G}(\mathbf{BiKt})$ *then it is valid in all* **BiKt**-*models.*

Proof. We only prove that the rule $(\blacksquare)$ preserves the validity on a **BiKt**-model $M := (W, \leqslant, R, S, V)$. Suppose that $\Diamond\Theta, \Gamma \Rightarrow \varphi$ is valid on M and fix any state $u \in W$ such that $M, u \models \bigwedge(\Theta, \blacksquare\Gamma)$. To show $M, u \models \blacksquare\varphi$, let us also fix any state v such that $u(\leqslant \circ S^{-1})v$, i.e., $u \leqslant w$ and $wS^{-1}v$ for some $w \in W$. We fix such a state w. Our goal is to show $M, v \models \varphi$. By the supposition of $M, u \models \bigwedge(\Theta, \blacksquare\Gamma)$ and $u(\leqslant \circ S^{-1})v$, we obtain $M, v \models \bigwedge\Gamma$. By $u \leqslant w$ and the same supposition, we get $M, w \models \bigwedge\Theta$, since the truth set for $\bigwedge\Theta$ is $\leqslant$-closed. It follows from $wS^{-1}v$ (i.e., vSw) that $M, v \models \bigwedge\Diamond\Theta$ hence $M, v \models \bigwedge(\Diamond\Theta, \Gamma)$. Since $\Diamond\Theta, \Gamma \Rightarrow \varphi$ is valid on M, we conclude that $M, v \models \varphi$, as desired. □

The following proposition, which is extracted from [30, pp.517-8], is useful for relating the semantic completeness of G^a(**BiSKt**) with that of G^a(**BiKt**) below.

Proposition A.2 *Let* $(W, \leqslant)$ *be a preorder and* $R, S \subseteq W \times W$.

(i) *Let R be stable. Then, R satisfies the condition $(F\Box)$ and* $\leqslant \circ R = R$.

(ii) *Let S^{-1} be stable. Then, S satisfies the condition $(F\Diamond)$ and* $\leqslant \circ S^{-1} = S^{-1}$.

It is noted that the condition $\leqslant \circ R = R$ implies that the satisfaction relation of $\Box\varphi$ for **BiSKt** is equivalent to that for **BiKt**. So, if we focus on the fragment with $\blacklozenge$ and $\Box$ but not with $\Diamond$ and $\blacksquare$, this proposition implies that if $\Gamma \Rightarrow \Delta$ is valid on the semantics for **BiKt** then it is also valid on the semantics for **BiSKt**. Therefore, our semantic completeness argument for G^a(**BiSKt**) can be naturally extended to an argument for G^a(**BiKt**) as in the following proof.

Theorem A.3 *If* $\Gamma \Rightarrow \Delta$ *is valid in all finite* **BiKt**-*models then it is derivable in* G^a(**BiKt**).

Proof. Suppose that $\Gamma \Rightarrow \Delta$ is not derivable in G^a(**BiKt**). Put $\Xi := \mathsf{Sub}(\Gamma, \Delta)$. Because Lemma 4.2 still holds for G^a(**BiKt**) (its proof needs $(Cut)^a$ alone), we can find a Ξ-partial valuation (Γ^+, Δ^+) such that $\Gamma \subseteq \Gamma^+$ and $\Delta \subseteq \Delta^+$. We define the derived **BiKt**-model $M^\Xi := (W^\Xi, \leqslant^\Xi, R^\Xi, S^\Xi, V^\Xi)$ from Ξ in the same way as in Definition 4.3 except that $(\Gamma, \Delta)S^\Xi(\Pi, \Sigma)$ iff the following two conditions hold:

(i) if $\blacksquare\psi \in \Pi$ then $\psi \in \Gamma$, for all formulas ψ,

(ii) if $\Diamond\psi \in \Delta$ then $\psi \in \Sigma$, for all formulas ψ.

We can prove that $(S^\Xi)^{-1}$ is stable, i.e., $\leqslant^\Xi \circ (S^\Xi)^{-1} \circ \leqslant^\Xi \subseteq (S^\Xi)^{-1}$. By Proposition A.2, it suffices for us to check the corresponding items from (ix) to (xii) of Lemma 4.5 to $\Diamond$ and $\blacksquare$ (we can just replace $\blacklozenge$, $\Box$ and R^Ξ of items from (ix) to (xii) of Lemma 4.5 with $\Diamond$, $\blacksquare$ and $(S^\Xi)^{-1}$, respectively). These items are shown similarly to those for G^a(**BiSKt**). Then, we can establish the corresponding lemma to Lemma 4.6 and so conclude that $\Gamma \Rightarrow \Delta$ is not valid in a finite **BiKt**-model M^Ξ. □

To sum up all arguments in this section, we can obtain the following.

Theorem A.4 *For any sequent* $\Gamma \Rightarrow \Delta$, *the following are all equivalent:* (i) *it is valid in all* **BiKt**-*models,* (ii) *it is valid in all finite* **BiKt**-*models,* (iii) *it is derivable in* G^a(**BiKt**), (iv) *it is derivable in* G^*(**BiKt**), (v) *it is derivable* *in* G(**BiKt**)*. Therefore,* G(**BiKt**) *enjoys the finite model property, the subformula property and the decidability. Moreover,* G(**BiKt**) *enjoys the Craig interpolation theorem.*

References

[1] Avron, A., *The method of hypersequents in the proof theory of propositional non-classical logics*, in: W. Hodges, editor, *Logic: Foundations to Applications*, Oxford, 1996 pp. 1–32.

[2] Ciabattoni, A., T. Lang and R. Ramanayake, *Bounded-analytic sequent calculi and embeddings for hypersequent logics*, The Journal of Symbolic Logic **86** (2021), pp. 635–668.

[3] Ewald, W. B., *Intuitionistic tense and modal logic*, Journal of Symbolic Logic **51** (1986), pp. 166–179.

[4] Figallo, A. V. and G. Pelaitay, *An algebraic axiomatization of the Ewald's intuitionistic tense logic*, Soft Computing **18** (2014), pp. 1873–1883.

[5] Goré, R., L. Postniece and A. Tiu, *Cut-elimination and proof search for bi-intuitionistic tense logic*, in: L. D. Beklemishev, V. Goranko and V. B. Shehtman, editors, *Advances in Modal Logic 8, papers from the eighth conference on "Advances in Modal Logic," held in Moscow, Russia, 24-27 August 2010* (2010), pp. 156–177.

[6] Goré, R. and I. Shillito, *Bi-intuitionistic logics: a new instance of an old problem*, in: N. Olivetti, R. Verbrugge, S. Negri and G. Sandu, editors, *13th Conference on Advances in Modal Logic, AiML 2020, Helsinki, Finland, August 24-28, 2020*, College Publications, 2020 pp. 269–288.

[7] Kowalski, T. and H. Ono, *Analytic cut and interpolation for bi-intuisionistic logic*, The Review of Symbolic Logic **10(2)** (2017), pp. 259–283.

[8] Kuznets, R., *Craig interpolation via hypersequents*, in: *Concepts of Proof in Mathematics, Philosophy, and Computer Science*, De Gruyter, 2016 pp. 193–214.

[9] Liang, F. and Z. Lin, *On the decidability of intuitionistic tense logic without disjunction*, in: *IJCAI'20: Proceedings of the Twenty-Ninth International Joint Conference on Artificial Intelligence*, 2021, pp. 1798–1804.

[10] Luppi, C., *On the interpolation property of some intuitionistic modal logics*, Archive for Mathematical Logic **35** (1996), pp. 173–189.

[11] Lyon, T., A. Tiu, R. Goré and R. Clouston, *Syntactic Interpolation for Tense Logics and Bi-Intuitionistic Logic via Nested Sequents*, in: M. Fernández and A. Muscholl, editors, *28th EACSL Annual Conference on Computer Science Logic* (*CSL 2020*), Leibniz International Proceedings in Informatics (LIPIcs) **152** (2020), pp. 28:1–28:16.

[12] Maehara, S., *Craig no interpolation theorem* (in Japanese), Sugaku **12** (1961), pp. 235–237.

[13] Maksimova, L., *Amalgamation and interpolation in normal modal logics*, Studia Logica **50** (1991), pp. 457–471.

[14] Maksimova, L. L., *Interpolation theorems in modal logics and amalgamable varieties of topological boolean algebras*, Algebra and Logic **18** (1979), pp. 348–370.

[15] Maruyama, A., S. Tojo and H. Ono, *Decidability of temporal epistemic logics for multi-agent models*, Proceedings of the ICLP'01 Workshop on Computational Logic in Multi-Agent Systems (CLIMA-01) (2001), pp. 31–40.

[16] Mints, G., *Interpolation theorems for intuitionistic predicate logic*, Annals of Pure and Applied Logic **113** (2001), pp. 225–242.

[17] Nishimura, H., *A study of some tense logics by Gentzen's sequential method*, Publications of the Research Institute for Mathematical Sciences **16** (1980), pp. 343–353.

[18] Ohnishi, M. and K. Matsumoto, *Gentzen method in modal calculi II*, Osaka Journal of Mathematics **11** (1959), pp. 115–120.

[19] Ono, H., *On some intuitionistic modal logics*, Publications of the Research Institute for Mathematical Sciences **13** (1977), pp. 687–722.

[20] Ono, H., *Proof-theoretic methods in nonclassical logic – an introduction*, Mathematical Society of Japan Memoirs **2** (1998), pp. 207–254.

[21] Ono, H., "Proof Theory and Algebra in Logic," Springer Singapore, 2019.

[22] Pinto, L. and T. Uustalu, *Proof search and counter-model construction for bi-intuitionistic propositional logic with labelled sequents*, in: *Lecture Notes in Computer Science*, Springer Berlin Heidelberg, 2009 pp. 295–309.

[23] Rauszer, C., *A formalization of the propositional calculus of H-B logic*, Studia Logica **33** (1974), pp. 23–34.

[24] Rauszer, C., *Semi-Boolean algebras and their applications to intuitionistic logic with dual operations*, Fundamenta Mathematicae **LXXXIII** (1974), pp. 219–249.

[25] Rauszer, C., "An algebraic and Kripke-style approach to a certain extension of intuitionistic logic," Dissertationes Mathematicae **CLXVII**, PWN Polish Scientific Publishers, Warszawa, 1980.

[26] Sano, K. and J. G. Stell, *Strong completeness and the finite model property for bi-intuitionistic stable tense logics*, Electronic Proceedings in Theoretical Computer Science **243** (2017), pp. 105–121.

[27] Sano, K. and S. Yamasaki, *Subformula property and Craig interpolation theorem of sequent calculi for tense logics*, in: N. Olivetti and R. Verbrugge, editors, *Short Papers of Advances in Modal Logic* (*AiML 2020*), 2020, pp. 97–101.

[28] Schütte, K., *Der Interpolationssatz der intuitionistischen Prädikatenlogik*, Mathematische Annalen **148** (1962), pp. 192–200.

[29] Sindoni, G., K. Sano and J. G. Stell, *Expressing discrete spatial relations under granularity*, Journal of Logical and Algebraic Methods in Programming **122** (2021), p. 100682.

[30] Stell, J. G., R. A. Schmidt and D. Rydeheard, *A bi-intuitionistic modal logic: Foundations and automation*, Journal of Logical and Algebraic Methods in Programming **85** (2016), pp. 500–519.

[31] Takano, M., *Subformula property as a substitute for cut-elimination in modal propositional logics*, Mathematica Japonica **37** (1992), pp. 1129–1145.

[32] Takano, M., *A modified subformula property for the modal logics K5 and K5D*, Bulletin of the Section of Logic **30** (2001), pp. 115–122.

[33] Takano, M., *A semantical analysis of cut-free calculi for modal logics*, Reports of mathematical logic **53** (2018), pp. 43–65.

[34] Takeuti, G., "Proof Theory," Studies in Logic and the Foundations of Mathematics **81**, North-Holland Publishing Company, Amsterdam, 1975.

[35] Wolter, F. and M. Zakharyaschev, *Intuitionistic modal logic*, in: A. Cantini, editor, *Logic and Foundations of Mathematics*, Kluwer Academic Publishers, 1999 pp. 227–238.

Labelled sequent calculi for logics of strict implication

Eugenio Orlandelli [1]

Università di Bologna
Via Zamboni 38, 40126 Bologna, Italia

Matteo Tesi [2]

Scuola Normale Superiore
Piazza dei Cavalieri 7, 56126 Pisa, Italia

Abstract

We study the proof theory of C.I. Lewis' logics of strict conditional **S1-S5** and we propose the first modular and uniform presentation of C.I. Lewis' systems. In particular, for each logic **Sn** we present a labelled sequent calculus **G3Sn** and we discuss its structural properties: every rule is height-preserving invertible and the structural rules of weakening, contraction and cut are admissible. Completeness of **G3Sn** is established both indirectly via the embedding in the axiomatic system **Sn** and directly via the extraction of a countermodel out of a failed proof search. Finally, the sequent calculus **G3S1** is employed to obtain a syntactic proof of decidability of **S1**.

Keywords: Strict implication, non-normal modalities, S1, sequent calculi cut elimination.

1 Introduction

Clarence Irving Lewis proposed the first axiomatic systems of propositional modal logic [7,8]. In particular, due to his dissatisfaction towards the material conception of classical implication, he devised a new logical operator, namely *strict implication* ($\prec$). He introduced five systems from **S1** to **S5** [8]. The modal logics **S4** and **S5** have been intensively studied, whereas **S2**, **S3** and, above all, **S1** did not receive much attention.

It can be argued that this depended on the fact that the latter are non-normal modal logics, as the rule of necessitation does not hold unrestrictedly. The semantics of the systems **S2** and **S3** was obtained via a slight modification of the standard Kripke semantics, by considering models with non-normal (or queer) worlds [6].

[1] eugenio.orlandelli@unibo.it

[2] matteo.tesi@sns.it

On the contrary, Cresswell [3] proposed a semantics for **S1** which combines features of neighborhood and relational models. [3] Due to the rather complex formulation of the semantics **S1** was long considered as an uninteresting system, but see [2]. In our opinion, this position is not justified, insofar as the system exhibits some interesting metalogical properties. In particular, the system is decidable and the modal operator defined from the strict implication has some hyperintensional features: unrestricted substitution of materially equivalent formulas does not hold, cf. [1,5,12].

In the present work we shall focus on the proof theory of these systems. In a previous paper by one of the two authors labelled sequent calculi were introduced for the logics **S2**, **S3** and some related systems [14]. However, a modular and uniform treatment is still lacking, due to the impossibility to encompass the system **S1**.

We propose the first modular and analytic approach to the proof theory of the original systems by C. I. Lewis (related systems are omitted for brevity). As in the tradition of labelled systems, sequent calculi are obtained by converting the truth conditions for the logical operators in corresponding rules. The rules introduced in Table 1 for $\prec$ correspond more directly to a simplification of Cresswell's semantics where the neighborhood function is replaced by a bi-neighborhood as it is done in [4] for non-normal modal logics. In bi-neighborhood semantics worlds are mapped to pairs of disjoint sets of worlds, providing 'independent positive and negative evidence (or support) for a proposition' [4, p. 161]. Neverthteless, this paper sticks to Cresswell's semantics for simplicity.

The calculi satisfy good structural properties, namely admissibility of the rules of weakening, contraction and cut, as well as invertibility of all rules.

Completeness is first established by showing the embedding of Lewis' axiomatic calculi into the corresponding labelled sequent calculi. The admissibility of the rule of substitution of strict equivalents requires to prove a non trivial lemma, see Lemma 4.9. We then establish a more direct form of completeness via the extraction of a countermodel out of a failed proof search and we discuss the relation between the **S1** neighborhood semantics and the bi-neighborhood framework.

Our proof-theoretic approach enables us to investigate the system **S1** by purely syntactic means which are uniform with respect to the ones traditionally employed for **S2** – **S5**. In particular, we exploit the calculus **G3S1** to obtain the first purely syntactic proof of decidability of the logic **S1** via terminating proof search. The decidability of the system **S1** had already been established by semantic means. In particular, the proof used the filtration method to prove the finite model property, see [2,3]. We are not aware of a syntactic proof of decidability for **S1**. This depends on the fact that axiomatic presentations are not amenable to proof search due to their substantial lack of analyticity.

---

[3] A semantics for **S1** based on Rantala models has been given in [15], and a relational semantics has been given in [13].

The structure of the paper is as follows. Section 2 recalls the logics **S1** – **S5**. Then, Section 3 introduces the labelled calculi **G3S1** – **G3S5** and Section 4 studies their structural properties. Section 5 gives a direct and modular proof of completeness for **G3Sn**. Finally Section 6 proves the decidability of **G3S1**.

2 Logics of strict implication

Language

The language of strict implications is defined by the following grammar:

$$A \;::=\; p \mid \bot \mid A \wedge A \mid A \vee A \mid A \supset A \mid A \dashv\!\!\!3\, A \qquad (\mathcal{L}^{\dashv\!3})$$

where $p \in \mathcal{P}$ for a denumerable set of sentential variables $\mathcal{P}$.

Parentheses are used as customary ($\dashv\!3$ binds lighter than other operators). Capital roman letters will be used for arbitrary formulas and lower-case ones for sentential variables. We use $\equiv$ to denote syntactic identity. The symbol $\top$ is short for $\bot \supset \bot$ and $\neg A$ is short for $A \supset \bot$. The unary modalities $\Box$ and $\Diamond$ can be defined as: $\Box A \;\equiv\; \top \dashv\!3\, A$ and $\Diamond A \;\equiv \neg(\top \dashv\!3\, \neg A)$
We use $\mathcal{L}^{\Box}$ to denote the standard modal language—i.e., $\mathcal{L}^{\dashv\!3}$ with $\Box$ and $\Diamond$ in place of $\dashv\!3$. The formula $A \dashv\!3\, B$ can be defined in $\mathcal{L}^{\Box}$ either as $\Box(A \supset B)$ or as $\neg\Diamond(A \wedge \neg B)$. Observe that languages $\mathcal{L}^{\dashv\!3}$ and language $\mathcal{L}^{\Box}$ are not minimal since we have the usual classical and modal interdefinabilities—e.g., C.I. Lewis [8] considered a language with only $\neg$, $\wedge$ and $\Diamond$ as primitives.

We will use $A[B//C]$ for the formula obtained from A by replacing some occurrences of C with occurrences of B.

Axiomatic systems

We present here C.I. Lewis [8] axiomatisation of the logics **S1** – **S5**. As already anticipated Lewis considered a language with only $\neg$, $\wedge$ and $\Diamond$ as primitives. For simplicity we assume to have the definition of the other symbols as implicit axioms. We simplify Lewis' axiomatisations by dropping the redundant axiom $A \dashv\!3\, \neg\neg A$—see [9]—and by considering axiom schemes instead of having as primitive a rule of uniform substitution of material equivalents.

Definition 2.1 [Lewis' axiomatisation of **S1**]

- Axioms:
  - (i) $(A \wedge B) \dashv\!3\, (B \wedge A)$
  - (ii) $(A \wedge B) \dashv\!3\, A$
  - (iii) $A \dashv\!3\, (A \wedge A)$
  - (iv) $((A \wedge B) \wedge C) \dashv\!3\, (A \wedge (B \wedge C))$
  - (v) $((A \dashv\!3\, B) \wedge (B \dashv\!3\, C)) \dashv\!3\, (A \dashv\!3\, C)$
  - (vi) $(A \wedge (A \dashv\!3\, B)) \dashv\!3\, B$
- Rules:
  - (i) $\dfrac{A \quad (B \dashv\!3\, C) \wedge (C \dashv\!3\, B)}{A[B//C]}\, SSE$
  - (ii) $\dfrac{A \quad B}{A \wedge B}\, Adj$
  - (iii) $\dfrac{A \dashv\!3\, B \quad A}{B}\, MP_{\dashv 3}$

Definition 2.2 [Axiomatisation of **S2**–**S5**] **S2** $=$ **S1** $\oplus\, \Diamond(A \wedge B) \dashv\!3\, \Diamond A$; **S3** $=$ **S2** $\oplus\, (A \dashv\!3\, B) \dashv\!3\, (\Box A \dashv\!3\, \Box B)$; **S4** $=$ **S1** $\oplus\, \Box A \dashv\!3\, \Box\Box A$; **S5** $=$ **S4** $\oplus\, A \dashv\!3\, \Box\Diamond A$.

Semantics

As it is well-known, standard relational semantics can be used for the normal conditional logics **S4** and **S5**. A modification thereof has been used by Kripke [6] to give a semantics for the non-normal conditional logics **S2** and **S3**: we must add so-called *queer* (or *non-normal*) *worlds* where $\Diamond A$ is always true and $\Box A$ is always false. Finally, a semantics for **S1** has been introduced by Cresswell in [3] and generalised to logics weaker than **S1** in [2]. This semantics is interesting because it needs both an accessibility relation and a neighborhood functions to define strict implication (as well as $\Box$ and $\Diamond$): the accessibility relation is used in normal worlds and the neighborhood function is used in queer ones.

Formally an **S1**-*frame* is quadruple $\mathcal{F} = \langle \mathcal{W}, \mathcal{N}, \mathcal{R}, \mathcal{I} \rangle$ where: (i) $\mathcal{W}$ is a non-empty set of *worlds*; (ii) $\mathcal{N}$ is a subset of $\mathcal{W}$, of so-called *normal worlds* (worlds in $\mathcal{W} \backslash \mathcal{N}$ will be called *queer worlds*); (iii) $\mathcal{R} \subseteq \mathcal{W} \times \mathcal{W}$ is a reflexive *accessibility relation* on $\mathcal{W}$; (iv) $\mathcal{I} : \mathcal{W} \longrightarrow \mathcal{P}(\mathcal{P}(\mathcal{W}))$ is a *neighborhood functions* mapping worlds to sets of sets of worlds with the side conditions that if $\alpha \in \mathcal{I}(w)$ then $w \in \alpha$—i.e., $\mathcal{I}$ is reflexive— and that if $X, Y \in I(w)$ then $X \cup Y \neq \mathcal{W}$.

By adding conditions on $\mathcal{N}, \mathcal{R}$, and $\mathcal{I}$ we can define a class of frames for the other Lewis' systems. In particular: (i) an **S2**-*frame* is an **S1**-frame where $\mathcal{I}$ is such that it maps each world to $\emptyset$; (ii) an **S3**-*frame* is a transitive **S2**-frame—i.e., if $w\mathcal{R}v$ and $v\mathcal{R}u$ then $w\mathcal{R}u$; (iii) an **S4**-*frame* is an **S3**-frame where $\mathcal{N} = \mathcal{W}$; (iv) an **S5**-*frame* is a symmetric **S4**-frame—i.e., if $w\mathcal{R}v$ then $v\mathcal{R}w$. Some observations are in order. **S2**-frames can be equivalently defined by simply dropping $\mathcal{I}$ from **S1** fames, thus obtaining Kripke semantics for non-normal logics [6]. **S4** can be defined by dropping $\mathcal{N}$ and $\mathcal{I}$ from **S1**-frames, thus obtaining standard relational semantics for normal modalities.

A *model* $\mathcal{M}$ is a frame augmented with a *valuation function* $\mathcal{V} : \mathcal{W} \longrightarrow \mathcal{P}(\mathcal{W})$ mapping each sentential variable to the set of worlds where it holds. We say that $\mathcal{M}$ is an **Sn**-*model* if its underlying frame is an **Sn**-frame.

We are now ready to define *truth* of a formula A at a world w of a model $\mathcal{M}$, $\models_w^{\mathcal{M}} A$ or simply $\models_w A$ when $\mathcal{M}$ is clear from the context. The definition is standard for sentential variables and for the extensional operators—e.g., $\models_w p$ iff $w \in \mathcal{V}(p)$ and $\models_w A \wedge B$ iff $\models_w A$ and $\models_w B$. The only interesting case is that of strict implication where we have:

$$\models_w A \prec B \quad \text{iff} \quad \begin{cases} \forall v \in \mathcal{W}, w\mathcal{R}v \text{ and } \models_v A \text{ imply } \models_v B, & \text{if } w \in \mathcal{N} \\ [\![A \supset B]\!]_{\mathcal{M}} \in \mathcal{I}(w), & \text{else} \end{cases}$$

where $[\![A]\!]_{\mathcal{M}}$ is the truth set of A in $\mathcal{M}$: $[\![A]\!]_{\mathcal{M}} = \{w : \models_w^{\mathcal{M}} A\}$. Equivalently, we have that $\models_w A \prec B$ for $w \in \mathcal{W} \backslash \mathcal{N}$ iff $\exists \alpha \in I(w)$ such that, for all $v \in \mathcal{W}$, ($\not\models_v A$ or $\models_v B$) if and only if $v \in \alpha$.

We now introduce two abbreviations. $\alpha \Vdash B$ expresses $\forall u(u \in \alpha \Rightarrow \models_u B)$ —i.e., every world in α satisfies the formula B. $\alpha \lhd B$ is the *covering relation* which asserts $\forall u(\models_u B \Rightarrow u \in \alpha)$—i.e., every world which satisfies B is in α. The latter can also be equivalently formulated as $\forall u \neg(u \notin \alpha \ \& \ \models_u B)$. The expression $[\![A \supset B]\!]_{\mathcal{M}} \in \mathcal{I}(w)$ can be rewritten as:

$$\exists \alpha \in \mathcal{I}(w)(\alpha \Vdash A \supset B \ \& \ \alpha \lhd A \supset B)$$

Observe that for **S2**- and **S3**-models the clause for queer worlds says that $A \prec B$ cannot be true therein, and for **S4**- and **S5**-models it can be dropped.

A formula A is said to be: (i) *True in a model* $\mathcal{M}$, $\models^{\mathcal{M}} A$, if it true in every normal point of that model; (ii) **Sn**-*valid*, $\mathbf{Sn} \models A$, if it is true in all **Sn**-models; (iii) *An* **Sn**-*consequence* of a set of formulas X, $X \models_{\mathbf{Sn}} A$, if A is true in all normal world of each **Sn**-model where all formulas in X are true.

Theorem 2.3 (Characterisation, [3]) *The axiomatic calculus* **Sn** *is sound and compete for validity w.r.t. the class of all* **Sn**-*frames.*

3 Labelled sequent calculi

We are now going to introduce labelled sequent calculi for the logics of strict implication **S1**-**S5**. Labelled calculi for normal modal logics have been introduced in [10] and for the non-normal ones in [4,11]. Labelled calculi for the non-normal logics **S2** and **S3**, as well as for some of their extensions, based on the language $\mathcal{L}^{\Box}$ have been studied in [14]. Here we consider also **S1** and we work with a language with $\prec$ instead of $\Box$ as primitive.

To define the language of sequent calculi we consider two denumerable sets of labels: a set $\mathbb{W}$ of *world labels*, for which we use the metavariables $w, v, u, \ldots$, and a set $\mathbb{I}$ of *neighbours label*, denoted by $\alpha, \beta, \gamma, \ldots$. Moreover, we add the following atomic predicates $R, N, Q, \in$, and $\notin$ that are syntactic counterparts of the elements of **S1**-frames. The formulas of the labelled language $\mathcal{L}^{ll}$ are the following (where $w, v \in \mathbb{W}$, $\alpha \in \mathbb{I}$ and $A \in \mathcal{L}^{\prec}$): (i) *relational atoms* wRv; (ii) *normality atoms* Nw; (iii) *queer atoms* Qw; (iv) *neighbour atoms* $\alpha \in Iw$; (v) *inclusion atoms* $w \in \alpha$; (vi) *exclusion atoms* $w \notin \alpha$; (vii) *labelled formulas* $w : A$; (viii) *forcing formulas* $\alpha \Vdash A$; and (ix) *covering formulas* $\alpha \lhd A$.

Definition 3.1 The *label* of a formula E in $\mathcal{L}^{ll}$ of form $u : A$ (resp. $\alpha \Vdash A$ or $\alpha \lhd A$) is u (resp. α) and is denoted by $l(E)$. The *pure part* of a labelled formula E is obtained removing from E the label and the forcing and is denoted by $p(E)$. The notion of *weight* is defined for labels and pure parts of formulas. For every $u \in \mathbb{W}$, $\mathtt{w}(u) = 0$, and for every $a \in \mathbb{I}$, $\mathtt{w}(\alpha) = 1$. The weight of a pure formula A, $\mathtt{w}(A)$ is defined as follows: $\mathtt{w}(p) = \mathtt{w}(\bot) = 1$, $\mathtt{w}(A \circ B) = max(\{\mathtt{w}(A), \mathtt{w}(B)\}) + 1$, where $\circ \in \{\wedge, \vee, \supset\}$, $\mathtt{w}(A \prec B) = max(\{\mathtt{w}(A), \mathtt{w}(B)\}) + 2$. The *degree* of a labelled, forcing, or covering formula E is an ordered pair $\mathtt{deg}(E) = (\mathtt{w}(p(E)), \mathtt{w}(l(E)))$. Additionally, we stipulate $\mathtt{deg}(wRu) = \mathtt{deg}(Nu) = \mathtt{deg}(Qu) = \mathtt{deg}(\alpha \in Iu) = \mathtt{deg}(u \in \alpha) = \mathtt{deg}(u \notin \alpha) = (0, 1)$. *Degrees* of $\mathcal{L}^{ll}$-formulas are ordered lexicographically.

A *sequent* is an expression $\Gamma \Rightarrow \Delta$ where Γ is a finite multiset of $\mathcal{L}^{ll}$-formulas and Δ is a finite multiset of labelled, forcing, and covering formulas only. Substitutions of labels in an $\mathcal{L}^{ll}$-formula E, $E[v/u]$ and $E[\alpha/\beta]$, are defined as expected and it is extended to multisets by applying it componentwise.

The rules of the calculi **G3S1**–**G3S5** are given in Table 1: **G3S1** contains all initial sequent and all propositional, conditional, and relational rules. **G3S2** = **G3S1** plus rule *S*2. **G3S3** = **G3S2** plus rule *Trans.* **G3S4** = **G3S3** plus rule *Norm.* **G3S5** = **G3S4** plus rule *Sym.* Observe that the calculus

G3S2 (**G3S4**) is equivalent to the simpler calculus obtained by dropping rules $L/R \prec_Q$ (and removing normality atoms from rules $L/R \prec_N$) and all relational rules but Ref_R from the calculus **G3S1** (**G3S3**).

A **G3Sn**-*derivation* of a sequent $\Gamma \Rightarrow \Delta$ is a tree of sequents, whose leaves are initial sequents, whose root is $\Gamma \Rightarrow \Delta$, and which grows according to the rules of **G3Sn**. The *height* of a **G3Sn**-derivation is the number of nodes of a branch of maximal length. We say that $\Gamma \Rightarrow \Delta$ is **G3Sn**-derivable (with height n), and we write **G3Sn** $\vdash^{(n)} \Gamma \Rightarrow \Delta$, if there is a **G3Sn**-derivation (of height at most n) of $\Gamma \Rightarrow \Delta$. A rule is said to be *(height-preserving) admissible* in **G3Sn**, if, whenever its premisses are **G3Sn**-derivable (with height at most n), also its conclusion is **G3Sn**-derivable (with height at most n). In each rule depicted in Table 1, Γ and Δ are called *contexts*, the formulas occurring in the conclusion are called *principal*, and those occurring in the premisses only are called *active*.

Lemma 3.2 *The sequent* $E, \Gamma \Rightarrow \Delta, E$ *is* **G3Sn**-*derivable for every* Γ, Δ, E.

Proof By induction on the degree of the formula E: the rules are applied root-first since in each branch we reach a sequent with a formula occurring both in the antecedent and in the succedent and having lesser degree than E.□

4 Structural properties

Lemma 4.1 (Substitution) **G3Sn** $\vdash^n \Gamma \Rightarrow \Delta$ *implies* **G3Sn** $\vdash^n \Gamma[v/u] \Rightarrow \Delta[v/u]$ *and* **G3Sn** $\vdash^n \Gamma[\alpha/\beta] \Rightarrow \Delta[\alpha/\beta]$.

Proof A standard induction on the height of the derivation $\mathcal{D}$ of the sequent $\Gamma \Rightarrow \Delta$. We apply to $\mathcal{D}$ the inductive hypothesis either twice or once—depending on whether the last rule instance *Rule* in $\mathcal{D}$ has a variable condition that clashes with the substitution or not— and then we conclude by applying an instance of *Rule*. □

Theorem 4.2 (Weakening) *For every multiset* Π *and* Σ, **G3Sn** $\vdash^n \Gamma \Rightarrow \Delta$ *implies* **G3Sn** $\vdash^n \Pi, \Gamma \Rightarrow \Delta, \Sigma$.

Proof By induction on the height of the derivation $\mathcal{D}$ of $\Gamma \Rightarrow \Delta$, possibly applying an (hp-admissible) instance of substitution if the last rule instance in $\mathcal{D}$ has a variable condition. □

Lemma 4.3 *If* A *is an axiom of the axiomatic system* **Sn** *then the sequent* $Nw \Rightarrow w : A$ *is* **G3Sn**-*derivable.*

Proof The proof is straightforward by a root-first application of the rules of the calculi, possibly using the admissiblity of weakening. We limit ourselves to considering axiom (v).

Table 1
Rules of the calculi **G3S1–G3S5**

Initial Sequents

$$\dfrac{}{w:p,\Gamma\Rightarrow\Delta,w:p}\,Ax \qquad \dfrac{}{w:\bot,\Gamma\Rightarrow\Delta}\,L\bot$$

$$\dfrac{}{Nw,Qw,\Gamma\Rightarrow\Delta}\,Ax_N \qquad \dfrac{}{w\in\alpha,w\notin\alpha,\Gamma\Rightarrow\Delta}\,Ax_\in$$

Propositional Rules

$$\dfrac{w:A,w:B,\Gamma\Rightarrow\Delta}{w:A\wedge B,\Gamma\Rightarrow\Delta}\,L\wedge \qquad \dfrac{\Gamma\Rightarrow\Delta,w:A \qquad \Gamma\Rightarrow\Delta,w:B}{\Gamma\Rightarrow\Delta,w:A\wedge B}\,R\wedge \qquad \dfrac{w:A,\Gamma\Rightarrow\Delta \qquad w:B,\Gamma\Rightarrow\Delta}{w:A\vee B,\Gamma\Rightarrow\Delta}\,L\vee$$

$$\dfrac{\Gamma\Rightarrow\Delta,w:A,w:B}{\Gamma\Rightarrow\Delta,w:A\vee B}\,R\vee \qquad \dfrac{\Gamma\Rightarrow\Delta,w:A \qquad w:B,\Gamma\Rightarrow\Delta}{w:A\supset B,\Gamma\Rightarrow\Delta}\,L\supset \qquad \dfrac{w:A,\Gamma\Rightarrow\Delta,w:B}{\Gamma\Rightarrow\Delta,w:A\supset B}\,R\supset$$

Conditional Rules

$$\dfrac{Nw,wRv,w:A\strictif B,\Gamma\Rightarrow\Delta,v:A \qquad v:B,Nw,wRv,w:A\strictif B,\Gamma\Rightarrow\Delta}{Nw,wRv,w:A\strictif B,\Gamma\Rightarrow\Delta}\,L\strictif_N$$

$$\dfrac{u:A,wRu,Nw,\Gamma\Rightarrow\Delta,u:B}{Nw,\Gamma\Rightarrow\Delta,w:A\strictif B}\,R\strictif_N,\ u\text{ fresh} \qquad \dfrac{\alpha\in Iw,\alpha\Vdash A\supset B,\alpha\lhd A\supset B,Qw,\Gamma\Rightarrow\Delta}{Qw,w:A\strictif B,\Gamma\Rightarrow\Delta}\,L\strictif_Q,\ \alpha\text{ fresh}$$

$$\dfrac{Qw,\alpha\in Iw,\Gamma\Rightarrow\Delta,w:A\strictif B,\alpha\Vdash A\supset B \qquad Qw,\alpha\in Iw,\Gamma\Rightarrow\Delta,w:A\strictif B,\alpha\lhd A\supset B}{Qw,\alpha\in Iw,\Gamma\Rightarrow\Delta,w:A\strictif B}\,R\strictif_Q$$

Relational rules

$$\dfrac{v:A,v\in\alpha,\alpha\Vdash A,\Gamma\Rightarrow\Delta}{v\in\alpha,\alpha\Vdash A,\Gamma\Rightarrow\Delta}\,L\Vdash \qquad \dfrac{u\in\alpha,\Gamma\Rightarrow\Delta,u:A}{\Gamma\Rightarrow\Delta,\alpha\Vdash A}\,R\Vdash,\ u\text{ fresh}$$

$$\dfrac{v\notin\alpha,\alpha\lhd A,\Gamma\Rightarrow\Delta,v:A}{v\notin\alpha,\alpha\lhd A,\Gamma\Rightarrow\Delta}\,L\lhd \qquad \dfrac{u\notin\alpha,u:A,\Gamma\Rightarrow\Delta}{\Gamma\Rightarrow\Delta,\alpha\lhd A}\,R\lhd,\ u\text{ fresh} \qquad \dfrac{wRw,\Gamma\Rightarrow\Delta}{\Gamma\Rightarrow\Delta}\,Ref_R$$

$$\dfrac{u\notin\alpha,u\notin\beta,\alpha\in Iw,\beta\in Iw,\Gamma\Rightarrow\Delta}{\alpha\in Iw,\beta\in Iw,\Gamma\Rightarrow\Delta}\,S1,\ u\text{ fresh} \qquad \dfrac{w\in\alpha,\alpha\in Iw,\Gamma\Rightarrow\Delta}{\alpha\in Iw,\Gamma\Rightarrow\Delta}\,Ref_I$$

$$\dfrac{Nw,\Gamma\Rightarrow\Delta \qquad Qw,\Gamma\Rightarrow\Delta}{\Gamma\Rightarrow\Delta}\,Norm$$

Additional rules

$$\dfrac{}{\alpha\in Iw,\Gamma\Rightarrow\Delta}\,S2 \qquad \dfrac{wRu,wRv,vRu,\Gamma\Rightarrow\Delta}{wRv,vRu,\Gamma\Rightarrow\Delta}\,Trans$$

$$\dfrac{Nw,\Gamma\Rightarrow\Delta}{\Gamma\Rightarrow\Delta}\,Norm \qquad \dfrac{uRw,wRu,\Gamma\Rightarrow\Delta}{wRu,\Gamma\Rightarrow\Delta}\,Sym$$

$$\dfrac{\dfrac{\dfrac{Nw,Nu,wRu,u:A\strictif B,u:B\strictif C\Rightarrow u:A\strictif C \qquad \dfrac{\dfrac{\dfrac{\dfrac{\dfrac{\dfrac{\dfrac{\dfrac{}{[...],v:B,v:A\Rightarrow v:B,v:C,[...]}\,\text{Lem.3.2}}{[...],v:A\Rightarrow v:B,v:B\supset C,[...]}\,R\supset}{[...]\Rightarrow v:A\supset B,v:B\supset C,[...]}\,R\supset}{[...],v\notin\alpha,v\notin\beta,\alpha\lhd A\supset B,\beta\lhd B\supset C\Rightarrow u:A\strictif C,v:A\supset B}\,L\lhd}{[...],v\notin\alpha,v\notin\beta,\alpha\in Iu,\alpha\lhd A\supset B,\beta\in Iu,\beta\lhd B\supset C\Rightarrow u:A\strictif C}\,L\lhd}{[...],\alpha\in Iu,\alpha\lhd A\supset B,\beta\in Iu,\beta\Vdash B\supset C,\beta\lhd B\supset C\Rightarrow u:A\strictif C}\,S1}{[...],Qu,\alpha\in Iu,\alpha\Vdash A\supset B,\alpha\lhd A\supset B,u:B\strictif C\Rightarrow u:A\strictif C}\,L\strictif_Q}{Qu,u:A\strictif B,u:B\strictif C\Rightarrow u:A\strictif C}\,L\strictif_Q}{Nw,wRu,u:A\strictif B,u:B\strictif C\Rightarrow u:A\strictif C}\,Norm}{Nw,wRu,u:(A\strictif B)\wedge(B\strictif C)\Rightarrow u:A\strictif C}\,L\wedge}{Nw\Rightarrow w:((A\strictif B)\wedge(B\strictif C))\strictif(A\strictif C)}\,R\strictif_N$$

The leftmost top-sequent is provable via applications of rules R-$\strictif_N$ and L-$\strictif_N$.□

Lemma 4.4 *Each rule of* **G3Sn** *is height-preserving invertible.*

Proof For the rules with repetition of the principal formulas in the premiss hp-invertibility follows from Theorem 4.2. For the other rules, if we are inverting w.r.t. the principal formula of the last rule instance in $\mathcal{D}$, there is nothing to prove. Else, we reason by induction on the height of $\mathcal{D}$, possibly applying Lemma 4.1. The base case is trivial, because only atomic formulas are active in initial sequent. To illustrate, assume we are inverting rule $R \multimap_N$ and the last rule instance in $\mathcal{D}$ is the following instance of $R \Vdash$:

$$\frac{Nw, u \in \alpha, \Gamma \Rightarrow \Delta', w : B \multimap C, u : A}{Nw, \Gamma \Rightarrow \Delta', \alpha \Vdash A, w : B \multimap C} \; R \Vdash, u \text{ fresh}$$

We transform $\mathcal{D}$ into the following derivation having at most the same height:

$$\frac{\dfrac{\dfrac{Nw, u \in \alpha, \Gamma \Rightarrow \Delta', w : B \multimap C, u : A}{Nw, u' \in \alpha, \Gamma \Rightarrow \Delta', w : B \multimap C, u' : A} \; Lem.4.1}{Nw, w' : B, wRw', u' \in \alpha, \Gamma \Rightarrow \Delta', u' : A, w' : C} \; IH}{Nw, w' : B, wRw', \Gamma \Rightarrow \Delta', \alpha \Vdash A, w' : C} \; R \Vdash$$

where the substitutions are needed if $w' \equiv u$. □

Theorem 4.5 (Contraction) **G3Sn** $\vdash^n \Pi, \Pi, \Gamma \Rightarrow \Delta, \Sigma, \Sigma$ *implies* **G3Sn** $\vdash^n \Pi, \Gamma \Rightarrow \Delta, \Sigma$.

Proof By induction on the height of the derivation $\mathcal{D}$ of $\Pi, \Pi, \Gamma \Rightarrow \Delta, \Sigma, \Sigma$. If $\Pi, \Pi, \Gamma \Rightarrow \Delta, \Sigma, \Sigma$ is an initial sequent, then the proof is immediate. If the principal formula of last rule applied is not in Π or Σ, then the conclusion follows from an application of the induction hypothesis to the premise(s) and then of the rule. If the principal formula of the last rule applied is in Π or Σ, we exploit the invertibility of the corresponding rule. We give a concrete example of this qualitative analysis.

Let us assume the conclusion of $\mathcal{D}$ is $w : A \multimap B, w : A \multimap B, \Pi', \Pi', \Gamma' \Rightarrow \Delta, \Sigma, \Sigma$. we have two cases depending on whether *Rule* is an instance of $L \multimap_N$ or of $L \multimap_Q$. In the former case we can proceed as when no instance of $w : A \multimap B$ is principal since $L \multimap_N$ is a rule with repetition of the principal formulas. In the latter case we transform

$$\frac{\alpha \in Iw, \alpha \Vdash A \supset B, \alpha \lhd A \supset B, Qw, w : A \multimap B, \Pi', \Pi', \Gamma'' \Rightarrow \Delta, \Sigma, \Sigma}{Qw, w : A \multimap B, w : A \multimap B, \Pi', \Pi', \Gamma'' \Rightarrow \Delta, \Sigma, \Sigma} \; L \multimap_Q, \alpha \text{ fresh}$$

into the following derivation of at most the same height:

$$\frac{\dfrac{\dfrac{\dfrac{\alpha \in Iw, \alpha \Vdash A \supset B, \alpha \lhd A \supset B, Qw, w : A \multimap B, \Pi', \Pi', \Gamma'' \Rightarrow \Delta, \Sigma, \Sigma}{\beta \in Iw, \beta \Vdash A \supset B, \beta \lhd A \supset B, \alpha \in Iw, \alpha \Vdash A \supset B, \alpha \lhd A \supset B, Qw, \Pi', \Pi', \Gamma'' \Rightarrow \Delta, \Sigma, \Sigma} \; Lem.4.4}{\alpha \in Iw, \alpha \Vdash A \supset B, \alpha \lhd A \supset B, \alpha \in Iw, \alpha \Vdash A \supset B, \alpha \lhd A \supset B, Qw, \Pi', \Pi', \Gamma'' \Rightarrow \Delta, \Sigma, \Sigma} \; Lem.4.1}{\alpha \in Iw, \alpha \Vdash A \supset B, \alpha \lhd A \supset B, Qw, \Pi', \Gamma'' \Rightarrow \Delta, \Sigma} \; IH}{Qw, w : A \multimap B, \Pi', \Gamma'' \Rightarrow \Delta, \Sigma} \; L \multimap_Q$$

where both α and β do not occur in the conclusion. □

Theorem 4.6 (Cut) *Let E be a relational, forcing, or covering formula. The following rule of cut is admissible in* **G3Sn***:*

$$\frac{\Gamma \Rightarrow \Delta, E \qquad E, \Pi \Rightarrow \Sigma}{\Pi, \Gamma \Rightarrow \Delta, \Sigma} \; Cut$$

Proof We consider an uppermost instance of *Cut* and we proceed by induction on the degree of the cut-formula with a sub-induction on the cut-height of $\mathcal{D}$—i.e., the sum of the heights of the derivations $\mathcal{D}_1$ and $\mathcal{D}_2$ of the two premisses. The theorem then follows by induction on the number of cuts in the derivation.

As usual it is convenient to divide the proof in three exhaustive cases: in case (i) one premiss has a derivation of height 1; in case (ii) the cut-formula is not principal in the last step of at least one of the two premisses; in case (iii) the cut-formula is principal in the last step of both premisses.

The proof of cases (i) and (ii) as well as the sub-cases of case (iii) where the principal operator of the cut-formula is in $\wedge, \vee, \rightarrow$, are standard and can thus be omitted. The proof of the sub-cases of (iii) when the cut-formula has shape $\alpha \Vdash A$ or $\alpha \lhd A$ can be found in [11]. Hence, we have to consider only the sub-cases of (iii) where the cut-formula has shape $w : B \strictif C$ and either the multiset Nw, wRv or $Qw, \alpha \in Iw$ occurs in Γ.

In the first case suppose $\mathcal{D}$ is a follows (for u not in the conclusion):

$$\dfrac{\dfrac{\vdots\mathcal{D}_{11} \atop u : B, wRu, Nw, \Gamma' \Rightarrow \Delta, u : C}{Nw, \Gamma' \Rightarrow \Delta, w : B \strictif C}\,R{\strictif_N} \qquad \dfrac{{\vdots\mathcal{D}_{21} \atop Nw, wRv, w : B \strictif C, \Pi' \Rightarrow \Sigma, v : B} \qquad {\vdots\mathcal{D}_{22} \atop v : C, Nw, wRv, w : B \strictif C, \Pi' \Rightarrow \Sigma}}{w : B \strictif C, Nw, wRv, \Pi' \Rightarrow \Sigma}\,L{\strictif_N}}{Nw, Nw, wRv, \Pi', \Gamma' \Rightarrow \Delta', \Sigma}\,Cut$$

We transform it into the following derivation ($[\Gamma]^n$ stands for n copies of Γ, and, for the sake of space, we omit the premisses of dotted inferences):

$$\dfrac{\dfrac{\dfrac{\vdots\mathcal{D}_1 \quad \vdots\mathcal{D}_{21}}{[Nw]^2, wRv, \Pi', \Gamma' \Rightarrow \Delta, \Sigma, v : B}\,Cut_j \qquad \dfrac{\vdots\mathcal{D}_{11}}{v : B, wRv, Nw, \Gamma' \Rightarrow \Delta, v : C}\,Lem.4.1}{[Nw]^3, [wRv]^2, \Pi', [\Gamma']^2 \Rightarrow [\Delta]^2, \Sigma, v : C}\,Cut_i \qquad \dfrac{\vdots\mathcal{D}_1 \quad \vdots\mathcal{D}_{22}}{v : C, [Nw]^2, wRv, \Pi', \Gamma' \Rightarrow \Delta, \Sigma}\,Cut_j}{\dfrac{[Nw]^5, [wRv]^3, [\Pi']^2, [\Gamma']^3 \Rightarrow [\Delta]^3, [\Sigma]^2}{Nw, wRv, \Pi, \Gamma \Rightarrow \Delta, \Sigma}\,Lem.4.5}\,Cut_i$$

where instances of *Cut* with subscript i (j) are admissible by the (sub-)induction hypothesis.

Finally, if $\mathcal{D}$ is a follows (for β not in the conclusion):

$$\dfrac{\dfrac{{\vdots\mathcal{D}_{11} \atop Qw, \beta \in Iw, \Gamma' \Rightarrow \Delta, w : B \strictif C, \beta \Vdash B \supset C} \qquad {\vdots\mathcal{D}_{12} \atop Qw, \beta \in Iw, \Gamma' \Rightarrow \Delta, w : B \strictif C, \beta \lhd B \supset C}}{Qw, \beta \in Iw, \Gamma' \Rightarrow \Delta, w : B \strictif C}\,R{\strictif_Q} \qquad \dfrac{\vdots\mathcal{D}_{21} \atop \alpha \in Iw, \alpha \Vdash B \supset C, \alpha \lhd B \supset C, Qw, \Pi' \Rightarrow \Sigma}{Qw, w : B \strictif C, \Pi' \Rightarrow \Sigma}\,L{\strictif_Q}}{Qw, \alpha \in Iw, \Pi', \Gamma' \Rightarrow \Delta, \Sigma}\,Cut$$

we transform it into the following derivation ($\mathcal{D}_{1i}[\star]$ stands for the derivation $\mathcal{D}_{1i}$ with α in place of β by an instance of Lemma 4.1, and D stands for $B \supset C$):

$$\dfrac{\dfrac{\vdots\mathcal{D}_{12}[\star] \quad \vdots\mathcal{D}_2}{[Qw]^2, \alpha \in Iw, \Pi', \Gamma' \Rightarrow \Delta, \Sigma, \alpha \lhd D}\,Cut_j \qquad \dfrac{\dfrac{\vdots\mathcal{D}_{11}[\star] \quad \vdots\mathcal{D}_2}{[Qw]^2, \alpha \in Iw, \Pi', \Gamma' \Rightarrow \Delta, \Sigma, \alpha \Vdash D}\,Cut_j \qquad {\vdots\mathcal{D}_{21} \atop \alpha \Vdash D, \alpha \lhd D, \alpha \in Iw, Qw, \Pi' \Rightarrow \Sigma}}{\alpha \lhd D, [Qw]^3, [\alpha \in Iw]^2, [\Pi']^2, \Gamma' \Rightarrow \Delta, [\Sigma]^2}\,Cut_i}{\dfrac{[Qw]^5, [\alpha \in Iw]^3, [\Pi']^3, [\Gamma']^2 \Rightarrow [\Delta]^2, [\Sigma]^3}{Nw, \alpha \in Iw, \Pi, \Gamma \Rightarrow \Delta, \Sigma}\,Thm.4.5}\,Cut_i$$

□

Corollary 4.7 *The rule* $MP_{\strictif}$ *is* **G3Sn***-admissible:*

$$\dfrac{Nw \Rightarrow w : A \strictif B \qquad Nw \Rightarrow w : A}{Nw \Rightarrow w : B}\,Det.$$

Proof By applying Lemma 4.4 to $Nw \Rightarrow w : A \strictif B$ we obtain the derivability of $u : A, wRu, Nw \Rightarrow u : B$ for some fresh label u. By an instance of Lemma 4.1 this becomes $w : A, wRw, Nw \Rightarrow w : B$ and, by a *Cut* with $Nw \Rightarrow w : A$, we obtain $wRw, Nw, Nw \Rightarrow w : B$. Finally, we apply an instance of Rule Ref_R and one of Theorem 4.5 to conclude that $Nw \Rightarrow w : B$ is derivable. □

Corollary 4.8 **G3Sn**-*derivations are analytic, i.e. every label occurring in a derivation either occurs in its conclusion or it is an* eigenvariable *and every formula is a subformula of the formulas in the conclusion.*

Proof See [14, Lemma 3.17]. □

Lemma 4.9 *For every formula A, B and C, if the sequents $w : A \Rightarrow w : B$ and $w : B \Rightarrow w : A$ are derivable in* **G3Sn**, *then the sequents:*

$$w : C \Rightarrow w : C[A//B] \text{ and } w : C[A//B] \Rightarrow w : C$$

are provable in **G3Sn**.

Proof The proof runs by induction on the weight of the formula C. We assume that $C \not\equiv A$, otherwise the proof is trivial. If C is a sentential variable p or $\bot$, the claim is trivial. If C is a conjunction, a disjunction or a formula of the shape $D \supset E$, then the proof easily follows by applying the induction hypothesis. We discuss the case in which C is of the form $D \strictif E$. Since $C \not\equiv A$, we have $(D \strictif E)[A//B] \equiv D[A//B] \strictif E[A//B]$.

We first show that $Nw, w : D \strictif E \Rightarrow w : D[A//B] \strictif E[A//B]$ is derivable.

$$\dfrac{\dfrac{[...], w : D \strictif E, u : D[A//B] \Rightarrow u : E[A//B], u : D \qquad [...], w : D \strictif E, u : D[A//B], u : E \Rightarrow u : E[A//B]}{Nw, wRu, w : D \strictif E, u : D[A//B] \Rightarrow u : E[A//B]}\,\text{L}{\strictif}_N}{Nw, w : D \strictif E \Rightarrow w : D[A//B] \strictif E[A//B]}\,\text{R}{\strictif}_N$$

The derivability of the topmost sequents follows from the induction hypothesis and weakening. The sequent $Qw, w : D \strictif E \Rightarrow w : D[A//B] \strictif E[A//B]$ is derivable too.

$$\dfrac{\dfrac{\dfrac{\dfrac{[...], o \in \alpha, o : D \supset E \Rightarrow o : D[A//B] \supset E[A//B]}{[...], o \in \alpha, \alpha \Vdash D \supset E \Rightarrow o : D[A//B] \supset E[A//B]}\,\text{L}{\Vdash}}{[...], \alpha \Vdash D \supset E \Rightarrow \alpha \Vdash D[A//B] \supset E[A//B]}\,\text{R}{\Vdash} \qquad \dfrac{\dfrac{[...], u : D[A//B] \supset E[A//B], \alpha \lhd D \supset E \Rightarrow u : D \supset E}{[...], u \notin \alpha, u : D[A//B] \supset E[A//B], \alpha \lhd D \supset E \Rightarrow}\,\text{L}{\lhd}}{[...], \alpha \lhd D \supset E \Rightarrow \alpha \lhd D[A//B] \supset E[A//B]}\,\text{R}{\lhd}}{Qw, \alpha \in Iw, \alpha \lhd D \supset E, \alpha \Vdash D \supset E \Rightarrow w : D[A//B] \strictif E[A//B]}\,\text{R}{\strictif}_Q}{Qw, w : D \strictif E \Rightarrow w : D[A//B] \strictif E[A//B]}\,\text{L}{\strictif}_Q$$

The derivability of the topmost sequents follows from application of the rules L$\supset$, R$\supset$, the induction hypothesis and weakening. The desired conclusion follows from an application of $Norm$.

We now discuss the other part of the claim, i.e. $w : D[A//B] \strictif E[A//B] \Rightarrow w : D \strictif E$. We first show that $Nw, w : D[A//B] \strictif E[A//B] \Rightarrow w : D \strictif E$:

$$\dfrac{\dfrac{[...], u : D \Rightarrow u : D[A//B], u : E \qquad [...], , u : E[A//B], u : D \Rightarrow u : E}{[...], w : D[A//B] \strictif E[A//B], u : D \Rightarrow u : E}\,\text{L}{\strictif}_N}{Nw, w : D[A//B] \strictif E[A//B] \Rightarrow w : D \strictif E}\,\text{R}{\strictif}_N$$

Again, the derivability of the topmost sequents follows from the induction hypothesis and weakening. For the other direction we proceed as follows (we omit to display redundant repetition of formulas):

$$\dfrac{\dfrac{\dfrac{[...], o : D[A//B] \supset E[A//B] \Rightarrow o : D \supset E, [...]}{[...], o \in \alpha, \alpha \Vdash D[A//B] \supset E[A//B] \Rightarrow \alpha \Vdash D \supset E, [...]}\,\text{L}{\Vdash} \qquad \dfrac{\dfrac{[...], u : D \supset E \Rightarrow u : D[A//B] \supset E[A//B], [...]}{[...], u \notin \alpha, u : D \supset E, \alpha \lhd D[A//B] \supset E[A//B] \Rightarrow [...]}\,\text{L}{\lhd}}{[...], \alpha \lhd D[A//B] \supset E[A//B] \Rightarrow \alpha \lhd D \supset E, [...]}\,\text{R}{\lhd}}{Qw, \alpha \in Iw, \alpha \Vdash D[A//B] \supset E[A//B], \alpha \lhd D[A//B] \supset E[A//B] \Rightarrow w : D \strictif E}\,\text{R}{\strictif}_Q}{Qw, w : D[A//B] \strictif E[A//B] \Rightarrow w : D \strictif E}\,\text{L}{\strictif}_Q$$

The topmost sequents are derivable via applications of the rules R$\supset$, L$\supset$, the induction hypothesis and admissibility of weakening.

□

We shall now prove the admissibility of the rule of substitution of strict equivalents.

Corollary 4.10 *The rule of substitution of strict equivalents is* **G3Sn**-*admissible:*

$$\frac{Nw \Rightarrow w : A \qquad Nw \Rightarrow w : (B \prec C) \wedge (C \prec B)}{Nw \Rightarrow w : A[B//C]} \; SSE$$

Proof We assume that we have a proof of the sequents $Nw \Rightarrow w : A$ and $Nw \Rightarrow w : (B \prec C) \wedge (C \prec B)$. By invertibility of the rule R$\wedge$ we get the derivations of $Nw \Rightarrow w : B \prec C$ and $Nw \Rightarrow w : C \prec B$.

We apply again the invertibility of the rule R$\prec$ we get $Nw, wRu, u : B \Rightarrow u : C$ and $Nw, wRu, u : C \Rightarrow u : B$. By inspection of the rules, the normality atoms and the relational atoms displayed in such sequents are never active in a derivation, so we can remove them.

Therefore the sequents $u : C \Rightarrow u : B$ and $u : B \Rightarrow u : C$ are derivable and we can apply Lemma 4.9 which yields $w : A \Rightarrow w : A[B//C]$. Finally, a cut gives the desired result. □

We are now in the position to state and prove the embedding of the axiomatic calculi **Sn** into **G3Sn**.

Theorem 4.11 *If* **Sn** $\vdash A$*, then* **G3Sn** $\vdash Nw \Rightarrow w : A$.

Proof The proof runs by induction on the height of the derivation in the axiomatic calculi **Sn**. The axioms are derivable by Lemma 4.3. The rule *Adj* is admissible by rule R$\wedge$. The admissibility of $MP_{\prec}$ is a consequence of the Corollary 4.7, and that of *SSE* follows from Theorem 4.10. □

5 Characterisation

We will now propose an alternative and more direct form of completeness which is obtained by extracting a countermodel out of a failed proof search. We start by defining the notion of validity of labelled sequents [11].

Definition 5.1 Given a set $\mathbb{W}'$ of world labels w, a set $\mathbb{I}'$ of neighborhood labels α and an **Sn** model $\mathcal{M} = \langle \mathcal{W}, \mathcal{N}, \mathcal{R}, \mathcal{I}, \mathcal{V} \rangle$, an *SN* realisation (ρ, σ) is a pair of functions mapping each $w \in \mathbb{W}'$ into $\rho(x) \in \mathcal{W}$ and mapping each $\alpha \in \mathbb{I}'$ into $\sigma(\alpha) \in \mathcal{I}w$ for some $w \in \mathcal{W}$. We introduce the notion $\mathcal{M}$ *satisfies a formula E under an* $\mathbb{W}'\mathbb{I}'$-*realisation* (σ, ρ) and denote it by $\mathcal{M} \vDash_{\rho,\sigma} E$, where we assume that the labels in E occur in $\mathbb{W}', \mathbb{I}'$. The definition extends by cases on the form of E, we give some examples:

- $\mathcal{M} \vDash_{\rho,\sigma} w \in \alpha$ if $\rho(w) \in \sigma(\alpha)$.
- $\mathcal{M} \vDash_{\rho,\sigma} w : A$ if $\vDash_{\rho(w)} A$
- $\mathcal{M} \vDash_{\rho,\sigma} \alpha \Vdash A$ if for all u in $\sigma(\alpha)$, $\rho(u) \vDash A$

- $\mathcal{M} \vDash_{\rho,\sigma} \alpha \lhd A$ if for all u s. t. $\mathcal{M} \vDash_{\rho,\sigma} u : A$, $\rho(u) \in \sigma(\alpha)$.

Given a sequent $\Gamma \Rightarrow \Delta$, let $\mathbb{W}', \mathbb{I}'$ be the sets of world and neighborhood labels occurring in $\Gamma \cup \Delta$, and let (ρ, σ) be an $\mathbb{W}', \mathbb{I}'$-realisation; we define $\mathcal{M} \vDash_{\rho,\sigma} \Gamma \Rightarrow \Delta$ to hold if, whenever $\mathcal{M} \vDash_{\rho,\sigma} E$ for all formulas $E \in \Gamma$, then $\mathcal{M} \vDash_{\rho,\sigma} F$ for some formula $F \in \Delta$. We further define $\mathcal{M}$-validity by:

$$\mathcal{M} \vDash \Gamma \Rightarrow \Delta \text{ iff } \mathcal{M} \vDash_{\rho,\sigma} \Gamma \Rightarrow \Delta \text{ for every } SN\text{-realisation } (\rho, \sigma).$$

A sequent $\Gamma \Rightarrow \Delta$ is **Sn**-*valid* if $\mathcal{M} \vDash \Gamma \Rightarrow \Delta$ for every **Sn**-model $\mathcal{M}$.

Theorem 5.2 (Soundness) *If* **G3Sn** $\vdash \Gamma \Rightarrow \Delta$, *then* $\Gamma \Rightarrow \Delta$ *is* **Sn**-*valid.*

Proof By induction on the height of the derivations in the calculus **G3Sn**. □

We introduce the notion of saturated sequent in a derivation. For every branch in a derivation we write $\downarrow \Gamma$ ($\downarrow \Delta$) to denote the union of the antecedents (succedents) in the branch from the endsequent up to the sequent $\Gamma \Rightarrow \Delta$.

Definition 5.3 A branch in a proof search in the system **G3S1** from the endsequent up to the sequent $\Gamma \Rightarrow \Delta$ is *saturated* if, for every rule R, if the principal formulas of R occur in the branch, the formulas introduced by one of the premises of R also occur in the branch. In detail, a saturated branch up to $\Gamma \Rightarrow \Delta$ has to satisfy the following conditions (we omit some of them): (Ax) There is no sentential variable p such that $w : p \in \Gamma \cap \Delta$. ($\text{Ax}_C$) There are no α, w such that $w \in \alpha, w \notin \alpha, \in \Gamma$. ($\text{Ax}_N$) There is no w such that $Nw, Qw, \in \Gamma$. (L$\bot$) It is not the case that $w : \bot \in \Gamma$. (L-$\prec_Q$) If Qw and $w : A \prec B \in \downarrow \Gamma$, then $\alpha \in Iw$, $\alpha \lhd A \supset B$ and $\alpha \Vdash A \supset B$ are in $\downarrow \Gamma$ for some α. (R-$\prec_Q$) If Qw, $\alpha \in Iw$ are in $\downarrow \Gamma$ and $w : A \prec B \in \downarrow \Delta$, then $\alpha \lhd A \supset B \in \downarrow \Delta$ or $\alpha \Vdash A \supset B \in \downarrow \Delta$. The notion of saturated sequent is extended to the systems **G3Sn** by adding conditions relative to the additional rules.

Given a sequent $\Gamma \Rightarrow \Delta$ we build a *proof search tree* by applying all possible rules of the calculus. To avoid repetitions, we fix a counter. At stage 1 we apply rule L$\wedge$, at stage 2 the rule R$\wedge$ and so forth. There are 18 + m different stages (where m is the number of relational and additional rules depending on the system). At stage 18 + m + 1 we repeat stage 1. If the construction ends we obtain a derivation or a finite tree in which a branch is saturated, otherwise we obtain an infinite tree. By König's Lemma there is an infinite branch which is saturated from which we can extract a countermodel.

Theorem 5.4 *Given a saturated branch $\mathcal{B}$ in a proof search tree for a sequent $\Pi \Rightarrow \Sigma$ up to the sequent $\Gamma \Rightarrow \Delta$ built according to the rules of system* **G3Sn**, *we can extract a countermodel $\mathcal{M}$ for the endsequent based on an* **Sn**-*frame.*

Proof Given a saturated branch $\mathcal{B}$ up to $\Gamma \Rightarrow \Delta$ in a proof search tree for the endsequent $\Pi \Rightarrow \Sigma$ we define the following countermodel: $\langle \mathcal{W}, \mathcal{N}, \mathcal{R}, \mathcal{I}, \mathcal{V} \rangle$ such that:

- $\mathcal{W}$ is the set of all world labels occurring in $\downarrow \Gamma$.
- $\mathcal{N}$ is the set of all labels w such that $Nw \in \downarrow \Gamma$.
- $w\mathcal{R}u$ if and only if wRu occurs in $\downarrow \Gamma$.

- $\mathcal{I}(w)$ is the set of all the neighbors α such that $\alpha \in Iw$ occurs in $\downarrow \Gamma$ and every α consists of all the worlds w such that $w \in \alpha$ occur in $\downarrow \Gamma$.
- $\mathcal{V}(p)$ is the set of all worlds w such that $w : p$ occurs in $\downarrow \Gamma$.

Notice that $\mathcal{V}$ is well defined by the saturation conditions Ax, Ax_C, Ax_N, and $L\bot$. For every system **G3Sn**, the frame $\langle \mathcal{W}, \mathcal{N}, \mathcal{R}, \mathcal{I} \rangle$ satisfies the properties of **Sn**-frames by the saturation conditions regarding relational and additional rules. We define the realization (ρ, σ) such that $\rho(w) \equiv w$ and $\sigma(\alpha) \equiv \alpha$. We claim that:

(i) If $w : A$ is in $\downarrow \Gamma$, then $\mathcal{M} \vDash_{\rho,\sigma} w : A$.

(ii) If $w : A$ is in $\downarrow \Delta$, then $\mathcal{M} \nvDash_{\rho,\sigma} w : A$.

The proof is by simultaneous induction on the degree of A. We focus on the case of strict implication.

(a) If $w : A \prec B$ is in $\downarrow \Gamma$, then by the saturation condition there is either Qw or Nw in $\downarrow \Gamma$. In the first case, again by the saturation condition, there are $\alpha \in Iw$, $\alpha \lhd A \supset B$ and $\alpha \Vdash A \supset B$ in $\downarrow \Gamma$. By definition of $\mathcal{M}$ and induction hypothesis we have $\alpha \in \mathcal{I}(w)$, $\mathcal{M} \vDash_{\rho,\sigma} \alpha \lhd A \supset B$ and $\mathcal{M} \vDash_{\rho,\sigma} \alpha \Vdash A \supset B$, therefore $\mathcal{M} \vDash_{\rho,\sigma} w : A \prec B$. In the second case, we distinguish two subcases. If there is no label u such that wRu occurs in $\downarrow \Gamma$, then the claim trivially follows. Otherwise for every u sucht that wRu occurs in $\downarrow \Gamma$, by the saturation condition either $u : A$ is in $\downarrow \Delta$ or $u : B$ is in $\downarrow \Gamma$. By induction hypothesis we get $\mathcal{M} \nvDash_{\rho,\sigma} u : A$ or $\mathcal{M} \vDash_{\rho,\sigma} u : B$. Therefore we get $\mathcal{M} \vDash_{\rho,\sigma} w : A \prec B$.

(b) If $w : A \prec B$ is in $\downarrow \Delta$, then by the saturation condition there is either Qw or Nw in $\downarrow \Gamma$. In the first case, by the saturation condition, for every $\alpha \in \mathcal{I}(w)$, there is $\alpha \lhd A \supset B$ or $\alpha \Vdash A \supset B$ in $\downarrow \Delta$. In both cases by induction hypothesis it follows $\mathcal{M} \nvDash_{\rho,\sigma} w : A \prec B$. In the second case, by saturation there are $wRu, u : A \in \downarrow \Gamma$ and $u : B$ in $\downarrow \Delta$. By induction hypothesis we get $\mathcal{M} \vDash_{\rho,\sigma} u : A$ and $\mathcal{M} \nvDash_{\rho,\sigma} u : B$, which yields $\mathcal{M} \vDash_{\rho,\sigma} w : A \prec B$.

□

Corollary 5.5 (Completeness) *For every formula A:*

$$\textbf{Sn} \vDash A \textit{ if and only if } \textbf{G3Sn} \vdash Nw \Rightarrow w : A$$

Proof The direction from right to left is the content of the soundness theorem. For the other direction we prove the contrapositive. Suppose that **G3Sn** $\nvdash Nw \Rightarrow w : A$, hence there is a saturated branch and we can extract a **Sn**-countermodel for $Nw \Rightarrow w : A$, which gives **Sn** $\nvDash A$. □

We observe that our completeness proof builds a natural bridge from the neighborhood semantics for **S1** to a bi-neighborhood one, see [4]. Indeed, to build the countermodel out of a failed proof search we are actually considering the complement of every neighbor α, but the rules of our calculus naturally build a pair of disjoint sets for every modal operator as in the case of bi-neighborhood semantics.

6 Syntactic decidability of S1

In this section we use the labelled system **G3S1** to obtain a syntactic decidability result for **S1** via terminating proof search. The decidability proofs in the literature for **S1** are obtained via semantic methods showing that the system satisfies the finite model property [2].

To establish decidability we need to show that the search for a derivation can be interrupted at a certain point and that we can extract a finite countermodel if the search fails. Termination of the proof search and completeness entail decidability. However, the extraction of a countermodel can be regarded as a *desideratum* as it yields a constructive proof of the finite model property. First we prove some preliminary lemmata.

Lemma 6.1 *The rules Ref_R, Ref_I, $S1$, $Norm$, $L\Vdash$, $L\lhd$, $L\text{-}\!\!\exists_N$ and $R\text{-}\!\!\exists_Q$ need not be instantiated more than once on the same principal formula(s) in every branch in a proof search.*

Proof By height-preserving admissibility of contraction. □

In order to obtain a termination result we need to show that the number of labels introduced in a proof search is finite. Therefore we need to establish some bounds on the application of the *dynamic* rules—i.e., the rules which introduce new labels, which are $R\Vdash$, $S1$, $R\lhd$, $L\text{-}\!\!\exists_Q$ and $R\text{-}\!\!\exists_N$. We now introduce some definitions which allow us to check the relations between world labels and neighborhood labels.

Definition 6.2 In a branch $\mathcal{B}$ of a proof search tree of the sequent $Nw \Rightarrow w : A$ we define the relation $\rightarrow_{\mathcal{B}}$ of *immediate successor* (for $u, v \in \mathbb{W}$ and $\alpha \in \mathbb{I}$): (i) $u \rightarrow_{\mathcal{B}} \alpha$ if $\alpha \in Iu$ occurs in $\mathcal{B}$; (ii) $\alpha \rightarrow_{\mathcal{B}} u$ if in $\mathcal{B}$ there is either $u \in \alpha$ or $u \notin \alpha$; (iii) $u \rightarrow_{\mathcal{B}} v$ uRv is in $\mathcal{B}$.

Fact. 1 The transitive closure of $\rightarrow_{\mathcal{B}}$ defines a tree which, as it is easy to check, does not contain cycles *modulo* the reflexive ones.
Fact. 2 The immediate successors of a world label in a saturated brach of a proof search tree are either all neighborhood labels or world labels, but not both.

Theorem 6.3 *Each label in a branch $\mathcal{B}$ of a proof search tree of an endsequent $Nw \Rightarrow w : A$ has only a finite number of immediate successors.*

Proof The immediate successors of a label can be introduced only by applications of the dynamic rules $R\Vdash$, $S1$, $R\lhd$, $L\text{-}\!\!\exists_Q$ or $R\text{-}\!\!\exists_N$. The subformulas of the formula A which occur in a proof search are finite by Corollary 4.8, therefore if there were infinite immediate successors there would be more than one application of one of the above mentioned rules to the same principal labelled formulas. We show that every derivation can be transformed in a derivation of the same height in which every branch contains at most one application of such rules to the same principal labelled formulas. We detail the case of $L\text{-}\!\!\exists_Q$ as an example; notice that by Fact 2 we can assume that the formula is again principal in an application of $L\text{-}\!\!\exists_Q$ and not of $L\text{-}\!\!\exists_N$.

$$\frac{Qu, \beta \in Iu, \beta \Vdash A \supset B, \beta \lhd A \supset B, \alpha \in Iu, \alpha \Vdash A \supset B, \alpha \lhd A \supset B, \Gamma \Rightarrow \Delta}{Qu, u : A \dashv\!\exists\, B, \alpha \in Iu, \alpha \Vdash A \supset B, \alpha \lhd A \supset B, \Gamma \Rightarrow \Delta} \text{L-}\dashv\!\exists_Q$$

$$\vdots\, \mathcal{D}$$

$$\frac{Qu, \alpha \in Iu, \alpha \Vdash A \supset B, \alpha \lhd A \supset B, \Gamma \Rightarrow \Delta}{Qu, u : A \dashv\!\exists\, B, \Gamma \Rightarrow \Delta} \text{L-}\dashv\!\exists_Q$$

We transform the derivation as follows:

$$\dfrac{\dfrac{\dfrac{Qu, \beta \in Iu, \beta \Vdash A \supset B, \beta \lhd A \supset B, \alpha \in Iu, \alpha \Vdash A \supset B, \alpha \lhd A \supset B, \Gamma \Rightarrow \Delta}{Qu, \alpha \in Iu, \alpha \Vdash A \supset B, \alpha \lhd A \supset B, \alpha \in Iu, \alpha \Vdash A \supset B, \alpha \lhd A \supset B, \Gamma \Rightarrow \Delta} \text{Lem.4.1}}{Qu, \alpha \in Iu, \alpha \Vdash A \supset B, \alpha \lhd A \supset B, \Gamma \Rightarrow \Delta} \text{Thm.4.5}}{Qu, \alpha \in Iu, \alpha \Vdash A \supset B, \alpha \lhd A \supset B, u : A \dashv\!\exists\, B, \Gamma \Rightarrow \Delta} \text{Thm.4.2}$$

$$\vdots\, \mathcal{D}$$

$$\frac{Qu, \alpha \in Iu, \alpha \Vdash A \supset B, \alpha \lhd A \supset B, \Gamma \Rightarrow \Delta}{Qu, u : A \dashv\!\exists\, B, \Gamma \Rightarrow \Delta} \text{L-}\dashv\!\exists_Q$$

The application of the hp-admissible rules of substitution, contraction and weakening does not introduce new applications of L-$\dashv\!\exists$ (this is easily checked).□

As a consequence, the tree defined by $\rightarrow_{\mathcal{B}}$ is finitely branching. The second part of the proof of termination consists in showing that in every branch the length of a chain of labels is finite. This depends on the fact that the relation defined by $\rightarrow_{\mathcal{B}}$ in a proof search is intransitive. In particular, a label *sees* only its immediate successors and itself (by reflexivity). Therefore the length of a branch is determined by the number of modal operators occurring in a formula.

Theorem 6.4 *Every chain of labels in a branch in a proof search for the sequent $Nw \Rightarrow w : A$ is finite.*

Proof Given a chain of labels in a branch in a proof search for the sequent $Nw \Rightarrow w : A$ and a label u in the chain, every immediate successor of u is introduced by the application of one of the dynamical rules R-$\dashv\!\exists_N$, L-$\dashv\!\exists_Q$, R$\lhd$, R$\Vdash$ or $S1$ to a formula B labelled by u. By inspection, these rules can be applied whenever B contains at least one modal operator. However, since every label sees only itself and its immediate successors, every label introduced by the analysis of $u : B$ will label only formulas of lesser degree. Since by definition the degree of each formula is finite, the chain is finite. □

Theorem 6.5 *The proof search for a sequent $Nw \Rightarrow w : A$ in the system* **G3S1** *terminates.*

Proof The proof is immediate because in every branch the number of labels generated is finite. □

Corollary 6.6 *The relation* **G3S1** $\vdash Nw \Rightarrow w : A$ *is decidable.*

Proof By Theorem 5.4 we can extract a countermodel out of a saturated branch, so we get the finite model property and the decidability of the system.□

7 Conclusion

We introduced a modular and uniform approach to the proof theory of the strict implication logics by C. I. Lewis. By converting the truth conditions of the semantics into suitably formulated rules of the calculus, we obtained labelled systems with good structural properties. Furthermore, the analyticity of the systems enabled us to obtain a syntactic proof of the decidability of the system **S1**. We conjecture that the upper bound given by the proof search procedure is not optimal, but we have not investigated this and other related complexity issues yet.

There are some possible themes for future research. First, it would be interesting to see whether it is possible to obtain calculi for systems related to **S1**, cf. [2]. Second, the semantics of **S1**, or a modification theoreof, might be employed to model hyperintensional features—see [1,5,12]—in the context of epistemic logics. Finally, it might be interesting to see whether the semantics for **S1** can be simplified and if it can be adapted to the bi-neighborhood framework.

References

[1] Berto, F. and D. Nolan, *Hyperintensionality*, in: E. N. Zalta, editor, *The Stanford Encyclopedia of Philosophy*, Metaphysics Research Lab, Stanford University, 2021, Summer 2021 edition .
URL
`https://plato.stanford.edu/archives/sum2021/entries/hyperintensionality/`

[2] Chellas, B. F. and K. Segerberg, *Modal logics in the vicinity of* **S1**, Notre Dame Journal of Formal Logic **37** (1996), pp. 1 – 24.
URL `https://doi.org/10.1305/ndjfl/1040067312`

[3] Cresswell, M., *The completeness of* **S1** *and some related systems*, Notre Dame Journal of Formal Logic **13** (1972), pp. 485 – 496.
URL `https://doi.org/10.1305/ndjfl/1093890710`

[4] Dalmonte, T., N. Olivetti and S. Negri, *Non-normal modal logics: bi-neighbourhood semantics and its labelled calculi*, in: *Advances in Modal Logic 2018*, College Publications, London, 2018 pp. 159–178.

[5] Drobyshevich, S. and D. Skurt, *Neighbourhood semantics for FDE-based modal logics*, Studia Logica (2021), pp. 1273–1309.
URL `https://doi.org/10.1007/s11225-021-09948-z`

[6] Kripke, S. A., *Semantical analysis of modal logic II. Non-normal modal propositional calculi*, in: J. W. Addison, L. Enkin and A. Tarski, editors, *The Theory of Models*, Studies in Logic and the Foundations of Mathematics, North-Holland, 1965 pp. 206–220.

[7] Lewis, C. I., *Implication and the algebra of logic*, Mind **21** (1912), pp. 522–531.
URL `https://doi.org/10.1093/mind/XXI.84.522`

[8] Lewis, C. I. and C. H. Langford, "Symbolic Logic," Century Co, New York, 1932.

[9] McKinsey, J. C. C., *A reduction in number of the postulates for C. I. Lewis' system of strict implication*, Bulletin of the American Mathematical Society **40** (1934), pp. 425 – 427.
URL `https://doi.org/10.1090/S0002-9904-1934-05881-6`

[10] Negri, S., *Proof analysis in modal logic*, Journal of Philosophical Logic **34** (2005), pp. 507–544.
URL `https://doi.org/10.1007/s10992-005-2267-3`

[11] Negri, S., *Proof theory for non-normal modal logics: The neighbourhood formalism and basic results*, IfCoLog Journal of Logics and their Applications **4** (2017), pp. 1241–1286.

[12] Sedlar, I., *Hyperintensional logics for everyone*, Synthese (2021), pp. 933–956. URL `https://doi.org/10.1007/s11229-018-02076-7`

[13] Sylvan, R., *Relational semantics for all Lewis, Lemmon and Feys' modal logics, most notably for systems between* $S0.3^{\circ}$ *and* $S1$, The Journal of Non-Classical Logics **6** (1989), pp. 19–40.

[14] Tesi, M., *Labelled sequent calculi for Lewis' non-normal propositional modal logics*, Studia Logica **109(4)** (2021), pp. 725–757. URL `https://doi.org/10.1007/s11225-020-09924-z`

[15] Thijsse, E. and H. Wansing, *A fugue on the themes of awareness logic and correspondence*, Journal of Applied Non-Classical Logics **6** (1996), pp. 127–136. URL `https://doi.org/10.1080/11663081.1996.10510874`

Graded modal logic with a single modality

Mattia Panettiere

Vrije Universiteit, Amsterdam

Apostolos Tzimoulis

Vrije Universiteit, Amsterdam

Abstract

Graded modal logic is an extension of classical modal logic with modalities $\Diamond_n$, for $n \in \mathbb{N}$, that allows to count the number of successors of a state in a Kripke model. In this article we study the logics obtained by restricting the language to a single modality $\Diamond_n$ for a fixed natural number n, where $\Diamond_n\varphi$ is satisfied on a point w of a Kripke model exactly when w has at least n successors satisfying φ. We compare the logics $\mathcal{L}_n$ and $\mathcal{L}_m$ for $n \neq m$. We provide concrete axiomatizations in cases $n = 2$ and $n = 3$ and provide a method for generating axiomatizations for every n.

Keywords: Graded modal logic, Kripke models, completeness, monotone modal logic.

1 Introduction

Graded modal logic is an extension of classical modal logic with graded modalities $\Diamond_n (n \in \mathbb{N}^+)$ that allows to count the number of successors of a given state in a Kripke model. Intuitively, the formula $\Diamond_n A$ is satisfied at a point w of a Kripke frame if and only if w has at least n successors satisfying A.

Graded modal logic was originally introduced in Goble [10]. Kaplan [12] studied graded modal logic as an extension of **S5**. The completeness of graded modal logic and its extensions was investigated in [9,7,2]. Van der Hoek [15] and Cerrato [3] used filtrations to obtain the finite model property and decidability of graded modal logic. Van der Hoek [15] also studied the expressibility, definability and correspondence theory. Bisimulations for graded modal logic were introduced in [8], and used to provide an alternative proof of the finite model property, and show that a first-order formula is invariant under graded bisimulation iff it is equivalent to a graded modal formula. Aceto, Ingolfsdottir and Sack [1] showed that resource bisimulation and graded bisimulation coincide over image-finite Kripke frames. Finally, various notions of epistemic and dynamic graded modal logics have been investigated in [16] and [13].

Even though the modality $\Diamond_1$ corresponds to the standard classical modal logic connective, and therefore retains all its properties, the modalities $\Diamond_n$

for $n \geq 2$ do not. In particular, the modalities $\Diamond_n$ are monotone, i.e. they satisfy the rule $\vdash \varphi \rightarrow \psi / \vdash \Diamond_n\varphi \rightarrow \Diamond_n\psi$, and satisfy $\Diamond_n\bot \leftrightarrow \bot$, but are not *additive*, that is, the implication $\Diamond_n(p \vee q) \rightarrow (\Diamond_n p \vee \Diamond_n q)$ fails for $n \geq 2$. Modal logics with monotone modalities have been extensively studied [4,11,14]. However, not much work has been done regarding the connections between monotonic modal logics and graded modal logic. In [6], building on the proof-theoretic and algebraic analysis of non-normal modal logics of [5], a line of research studying these connection was initiated, where an elementary but not modally definable class of neighbourhood frames was shown to exactly correspond to graded Kripke frames, and the notion of graded bisimulation was recasted through the lens of neighbourhood bisimulations.

This article adds to the study of connections between monotonic modal logic and graded modal logic, albeit towards a different direction. Specifically, the standard axiomatization of graded modal logic relies on the *interaction* of the different graded modalities, and captures the properties of addition of natural numbers. However, when viewed as monotone modalities, each graded modality can also be studied in isolation. Accordingly, for every $n \in \mathbb{N}^+$, we introduce the logic $\mathcal{L}_n$, whose language contains a single modal operation, $\Diamond$, and whose theory is defined as the set of validities on Kripke frames, where $\Diamond$ is interpreted as the graded modality $\Diamond_n$ described in the first paragraph. We show the relationship between these logics, and that they have the finite model property and are decidable. Moreover, we investigate their axiomatizations.

This article is structured as follows: In Section 2, we present the basic definition of the logic $\mathcal{L}_n$ for each $n \in \mathbb{N}$ via its semantics. In Section 3 we show that if $n \neq m$ the logics $\mathcal{L}_n$ and $\mathcal{L}_m$ are distinct, we identify the relationship between them, and we observe that they are decidable and enjoy the strong finite model property. In Sections 4 we discuss possible axiomatizations and their completeness. In particular, we introduce axiomatizations of $\mathcal{L}_2$ and $\mathcal{L}_3$, and we present a method to generate axioms for $\mathcal{L}_n$, showing that all logics $\mathcal{L}_n$ are finitely axiomatizable. Finally, in Section 5 we suggest avenues for future research.

2 Preliminaries

In this section, we introduce the languages and semantics for the logics, and define some key notions that will be useful throughout this paper.

For every natural number $n \geq 2$, the language for graded modal logic restricted to the n-th modality will be the same Φ, generated by the following grammar:

$$\varphi ::= p \mid \neg\varphi \mid \varphi \wedge \varphi \mid \Diamond\varphi,$$

where p ranges over a countable collection of propositional variables AtProp. We define $\rightarrow$ and $\vee$, as usual.

Given some $n \geq 2$, the semantics of the language is given in terms of Kripke frames $\mathbb{X} = (X, R)$. For any valuation function $v : \mathsf{AtProp} \rightarrow \mathcal{P}(X)$ and any Kripke frame $\mathbb{X}$, $M = (\mathbb{X}, v)$ is a model for the graded modal logic restricted to the n-th modality. Truth in a model M at a state $x \in X$ is defined inductively as follows:

$$\begin{array}{lll} M, x \Vdash_n p & \text{iff} & x \in v(p) \\ M, x \Vdash_n \varphi \wedge \psi & \text{iff} & M, x \Vdash_n \varphi \text{ and } M, x \Vdash_n \psi \\ M, x \Vdash_n \neg\varphi & \text{iff} & M, x \nVdash_n \varphi \\ M, x \Vdash_n \Diamond\varphi & \text{iff} & |\{y \in R[x] : M, y \Vdash_n \varphi\}| \geq n, \end{array}$$

where $R[z]$ with $z \in X$ indicates the direct image of $\{z\}$ through R. We write $M \Vdash_n \varphi$ if $M, x \Vdash_n \varphi$ for each $x \in X$ and we write $\mathbb{X} \Vdash_n \varphi$ if $(\mathbb{X}, v) \Vdash_n \varphi$ for every valuation v. Finally, we define

$$\mathcal{L}_n := \{\varphi \in \Phi \mid \forall \mathbb{X} \quad \mathbb{X} \Vdash_n \varphi\}.$$

In what follows, to help the reader identify the intended interpretation of formulas, we will sometimes slightly abuse notation, and use the modality $\Diamond_n$ instead of $\Diamond$, when the formula is to be interpreted in $\mathcal{L}_n$. Seeing this formally, given Φ_G, the language of graded modal logic defined as

$$\Phi_G \ni \varphi ::= p \mid \neg\varphi \mid \varphi \wedge \varphi \mid \Diamond_n\varphi, \quad n \in \mathbb{N}^+,$$

we can define an embedding $\epsilon : \mathbb{N}^+ \times \Phi \to \Phi_G$ recursively by letting $\epsilon(n, \Diamond\varphi) = \Diamond_n \epsilon(n, \varphi)$.

Even though the language and semantics of $\mathcal{L}_n$ cannot express the classical normal modality [1], it turns out that the notion that a point w of a Kripke frame has at least one successor satisfying a formula φ can sometimes be captured with a help of an auxiliary formula. In particular, consider the following formula

$$\Diamond_1^\psi \varphi := \Diamond(\varphi \vee \psi) \wedge \neg\Diamond\psi \tag{1}$$

It is easy to see that if $M, w \models_n \Diamond_1^\psi \varphi$, then there exists some u, such that wRu, and $M, u \models_n \varphi$. Formulas of this form will be key in the construction of the canonical model used in the proofs of completeness

Another important convention that we will follow in this article is the following: We will reserve small letters from the Greek alphabet to denote sequences of mutually contradictory formulas. In particular, when we write $\alpha_1, \ldots, \alpha_n$ we understand that $\alpha_i \wedge \alpha_j \to \bot$ is provable in classical logic for $i \neq j$. For example, $\alpha_i = p_i \wedge \neg(\bigvee_{j \neq i} p_j)$.

Finally, throughout this paper we write $\mathbb{N}^+$ to denote the set of positive natural numbers; we will also slightly abuse notation and identify $n \in \mathbb{N}^+$ with the set $\{1, \ldots, n\}$.

3 Basic properties

In this section we discuss the basic properties of the logics $\mathcal{L}_n$ and we compare their validities.

[1] For instance, a model with one point that is reflexive and and a model with a point which is not reflexive are indistinguishable for all the logics $\mathcal{L}_n$ for $n \geq 2$.

3.1 Decidability and Finite model property

A formula $\varphi \in \Phi$ is a validity in $\mathcal{L}_n$ if and only if $\epsilon(n, \varphi)$ is a validity in graded modal logic. Since graded modal logic is decidable, it immediately follows that $\mathcal{L}_n$ is decidable. For the strong finite model property, the filtration construction in [15, Section 6.1] works in this case verbatim, given that $\epsilon[\{n\}, \Phi] \subseteq \Phi_G$.

3.2 Comparing the logics $\mathcal{L}_n$

Lemma 3.1 *If $n < m$, there exists a formula ζ_n, such that $\zeta_n \in \mathcal{L}_n$ but $\zeta_n \notin \mathcal{L}_m$.*

Proof. Consider the formula

$$\zeta_n := \left(\bigwedge_{i=1}^{n} \Diamond_1^{q_i} \alpha_i \right) \to \Diamond \left(\bigvee_{i=1}^{n} \alpha_i \right).$$

Given any model M, if $M, x \vDash_n \bigwedge_{i=1}^{n} \Diamond_1^{q_i} \alpha_i$, then for each $i \in n$, there exists a y_i, such that xRy_i and $M, y_i \vDash_n \alpha_i$. Since α_i are mutually contradictory formulas, the y_i are all distinct, and hence $M, x \vDash_n \Diamond (\bigvee_{i=1}^{n} \alpha_i)$.

On the other hand, let $m > n$ and consider the model

$$M = (m + n, R, v),$$

where $R = \{(1, k) \mid 2 \leq k \leq m+n\}$, $v(q_i) = m \setminus 1$, and $v(\alpha_i) = \{m+i\}$. Then $M, 1 \vDash_m \Diamond_1^{q_i} \alpha_i$, since $m \setminus 1 \cup \{m+i\} \subseteq R[1]$. However, there are only n points that satisfy $\bigvee_{i=1}^{n} \alpha_i$, and therefore $M, 1 \nvDash_m \Diamond (\bigvee_{i=1}^{n} \alpha_i)$. □

Lemma 3.2 *Assume $n < m$ such that $m - 1 = (n-1) \cdot k + r$ where $r < n - 1$. Then, $\mathcal{L}_m \subseteq \mathcal{L}_n$ if and only if $r < k$.*

Proof. First, let's assume that $r < k$. To show that $\mathcal{L}_m \subseteq \mathcal{L}_n$ it is enough to show that for every formula $\varphi \in \Phi$ and every model M, there exists a model M' such that $M, w \vDash_n \varphi$ if and only if $M, w' \vDash_m \varphi$. Given a model $M = (X, R, v)$, we define $M' = (X \times k, R', v')$, where $(x, i)R'(y, j)$ if and only if xRy, and $v'(p) = v(p) \times k$. We will show that for any formula $\varphi \in \Phi$,

$$M, w \vDash_n \varphi \quad \Longleftrightarrow \quad M', (w, j) \vDash_m \varphi$$

by induction on the complexity of φ. All cases are immediate, except for the case where $\varphi = \Diamond\psi$. Let's assume that $M, w \nvDash_n \Diamond\psi$. Then there are $z \leq n-1$ successors of w satisfying ψ, so, by the induction hypothesis, exactly $z \cdot k \leq (n-1) \cdot k \leq m-1$ successors of (w, j) satisfy ψ, hence $M, (w, j) \nvDash_m \Diamond\psi$. Now assume that $M, (w, j) \nvDash_m \Diamond\psi$. By definition of M', it follows that (w, j) has $z \cdot k$ successors satisfying ψ where $z \cdot k \leq m - 1$. By induction hypothesis, w has z successors satisfying φ. Since $z \cdot k \leq m - 1$ and $r < k$ we have

$$z \cdot k \leq m - 1 = (n-1) \cdot k + r < (n-1) \cdot k + k = n \cdot k,$$

which implies that $z < n$. Hence $M, w \nvDash_n \Diamond\psi$.

Now let us assume that $k \leq r$. We consider the formula

$$\theta_n := \bigwedge_{i=1}^{n}\bigwedge_{j=1}^{n-1}\left(\Diamond_1^{q_i^j}\alpha_i^j \wedge \bigwedge_{i=1}^{n}\neg\Diamond(\bigvee_{j=1}^{n-1}\alpha_i^j)\right) \rightarrow \bigvee_{s:n\rightarrow n-1}\neg\Diamond\bigvee_{i=1}^{n}\alpha_i^{s(i)}.$$

Let us show that $\theta_n \notin \mathcal{L}_n$. Recall that α_i^j are mutually contradictory. Let

$$M = (n \times n \sqcup \{x_1, \dots x_{n-1}\}, R, v)$$

where $R[(1,1)] = n \times n \sqcup \{x_1, \dots x_{n-1}\}$, $v(q_i^j) = \{x_1 \dots, x_{n-1}\}$, and $v(\alpha_i^j) = \{(i, j+1)\}$. It is routine to check that $M, (1,1) \not\models_n \theta_n$.

Let's show now that $\theta_n \in \mathcal{L}_m$. Let M be a model. If w satisfies the antededent of the implication, then, for each i, there exists some j (let's call it $s(i)$), such that w has at most k successors satisfying α_i^j. Indeed, otherwise for each j, w has at least $k+1$ successors satisfying α_i^j and hence it has at least $(n-1)\cdot(k+1)$ successors satisfying $\bigvee_{j=1}^{n-1}\alpha_i^j$ by the fact that the α_i^j are mutually contradictory. Since

$$(n-1)\cdot(k+1) = (n-1)\cdot k + (n-1) > k(n-1) + r = m-1,$$

it follows that $M, w \models_m \Diamond\bigvee_{j=1}^{n-1}\alpha_i^j$, a contradiction. Now, since w has at most k successors satisfying $\alpha_i^{s(i)}$ for every i, it follows that w has at most $n \cdot k$ successors satisfying $\bigvee_{i=1}^{n}\alpha_i^{s(i)}$. Since

$$n \cdot k = (n-1)\cdot k + k \leq (n-1)\cdot k + r = m-1,$$

it follows that $M, w \models_m \neg\Diamond\bigvee_{i=1}^{n}\alpha_i^{s(i)}$. Hence $M, w \models_m \theta_n$. □

Summarizing the above results, we obtain a complete description of the relation between the logics $\mathcal{L}_n$:

Theorem 3.3 *Let $n < m$ such that $m-1 = (n-1)\cdot k + r$ where $r < n-1$. Then, if $r < k$ it follows that $\mathcal{L}_m \subsetneq \mathcal{L}_n$. If $k \leq r$, then there exists $\zeta_n, \theta_n \in \Phi$, such that $\zeta_n \in \mathcal{L}_n$ while $\zeta_n \notin \mathcal{L}_m$ and $\theta_n \in \mathcal{L}_m$ while $\theta_n \notin \mathcal{L}_n$.*

4 Axiomatizations and completeness

In this section we will discuss the completeness of the logics $\mathcal{L}_n$ with respect to the proposed axiomatizations. Before going into the various cases, we will present the axioms that are present in the axiomatization of every $\mathcal{L}_n$, as well as a key result that will allow us to construct enough distinct points in the canonical model.

For every logic $\mathcal{L}_n$, their corresponding axiomatization GP_n will include all the propositional tautologies (or an axiomatization of them), the formula

$$(\bot) \quad \Diamond\bot \rightarrow \bot,$$

and it will be closed under modus ponens, uniform substitution, and the monotonicity rule

$$\frac{\vdash p \to q}{\vdash \Box p \to \Box q} \quad \text{(M)}.$$

We call this basic system GP_0, which we will augment with further axioms (depending on n) in the following sections.

Lemma 4.1 *Let $(\mathbb{B}, \Diamond)$ be a Boolean algebra with a monotone operation satisfying $\Diamond\bot = \bot$. Let u be an ultrafitler on $\mathbb{B}$ and let*

$$Z_u = \{a \in \mathbb{B} \mid \forall b \in \mathbb{B}(\Diamond(a \vee b) \in u \Rightarrow \Diamond b \in u)\}.$$

Then Z_u is an ideal on $\mathbb{B}$ such that $\Diamond a \in u$ implies that $a \notin Z_u$.

Proof. Clearly $\Diamond(c \vee \bot) = \Diamond(c)$ so $\bot \in Z_u$. Assume that $b \in Z_u$ and $a \leq b$. Then if $\Diamond(c \vee a) \in u$, by monotonicity it follows that $\Diamond(c \vee b) \in u$, which implies $\Diamond c \in u$, so $a \in Z_u$. Finally assume that $a, b \in Z_u$, and $\Diamond(c \vee (a \vee b)) \in u$. Then since $a \in Z_u$, it follows that $\Diamond(c \vee b) \in u$ and since $b \in Z_u$ it follows that $\Diamond c \in u$. Finally, assume that $\Diamond a \in u$. Then $\Diamond\bot \notin u$, while $\Diamond(a \vee \bot) \in u$. Hence $a \notin Z_u$. □

Remark 4.2 Notice that $Z_u = \{a \in \mathbb{B} \mid \forall b \in \mathbb{B}\ (\Diamond_1^b a \notin u)\}$.

4.1 Case n=2

The system GP_2 is obtained by adding the following axiom-schema to GP_0:

$$\text{(G2)} \quad [\Diamond_1^{q_1}(\alpha_1) \wedge \Diamond_1^{q_2}(\alpha_2)] \to \Diamond(\alpha_1 \vee \alpha_2);$$

where, as discussed in Section 2, α_1 and α_2 contradict each other. This axiom intuitively states if w has at least one successor satisfying α_1 (witnessed using q_1) and at least one successor satisfying α_2 (witnessed using q_2), then w has at least two successors satisfying $\alpha_1 \vee \alpha_2$. It is easy to check, also given this explanation, that G2 is sound.

Completeness. To show completeness we will construct a canonical model using, as usual, the ultrafilters of the free Boolean algebra generated by the axiomatic system. Let $\mathbb{B}$ be a Boolean algebra with a monotone operation satisfying the axioms and rules of GP_2. Let u be an ultrafilter on $\mathbb{B}$. Let us define $e_u : \mathbb{B} \to \{0, 1, 2\}$ as follows:

$$e_u(a) = \begin{cases} 2 & \text{if } \Diamond a \in u, \\ 1 & \text{if } \Diamond a \notin u \text{ and } (\exists b \in \mathbb{B})(\Diamond b \notin u \text{ and } \Diamond(a \vee b) \in u), \\ 0 & \text{otherwise.} \end{cases}$$

Intuitively, this function roughly represents the number of permissible successors of u satisfying the statement a. Indeed, if $\Diamond a \in u$, u must have at least 2 successors satisfying a, while if $e_u(a) = 1$, u must have exactly one successor satisfying a. In particular notice that $e_u(a) = 1$, if $\Diamond a \notin u$ and $\Diamond_1^b a \in u$, for some $b \in \mathbb{B}$.

Corollary 4.3 *The set* $Z_u = \{a \in \mathbb{B} \mid e_u(a) = 0\}$ *is an ideal.*

Proof. Follows immediately from Lemma 4.1 and the definition of e_u. □

In the remainder of the paper, we write $\mathbf{n}$ to denote the set $\{1, 2, \ldots, n\}$ given any natural number n.

Definition 4.4 Let $\mathbb{B}$ be the free Boolean algebra generated by the axiomatic system above. The *canonical frame* of GP_2 is the Kripke frame $\mathbb{X}_{\mathbb{B}} = (\mathcal{U}(\mathbb{B}) \times \mathbf{2}, R)$, where $\mathcal{U}(\mathbb{B})$ denotes the collection of ultrafilters on $\mathbb{B}$, and R is such that for any $u, w \in \mathcal{U}(\mathbb{B})$, and $i, j \in \mathbf{2}$,

$$(u, j)R(w, i) \qquad \text{iff} \qquad e(w, u) \geq i,$$

where $e(w, u) = \min\{e_u(a) \mid a \in w\}$. The *canonical model* of GP_2 is the Kripke model $M_{\mathbb{B}} = (\mathbb{X}_{\mathbb{B}}, v)$ such that, for any $u \in \mathcal{U}(\mathbb{B})$ and $p \in \mathsf{AtProp}$,

$$(u, i) \in v(p) \qquad \text{iff} \qquad p \in u.$$

Lemma 4.5 (Truth lemma for GP_2) *For any* Φ*-formula* φ, $u \in \mathcal{U}(\mathbb{B})$, *and* $i \in 2$,

$$M_{\mathbb{B}}, (u, i) \models_2 \varphi \qquad \textit{iff} \qquad \varphi \in u.$$

Proof. We proceed by induction on the complexity of φ. All the cases are trivial, except the one in which $\varphi = \Diamond\psi$ for some $\psi \in \Phi$.

Assume that $M_{\mathbb{B}}, (u, i) \models_2 \Diamond\psi$. Then, there exist $(w, j), (r, k) \in \mathcal{U}(\mathbb{B}) \times \mathbf{2}$ such that $(u, i)R(w, j)$, $(u, i)R(r, k)$, and $\psi \in w \cap r$ by the induction hypothesis. Consider two cases:

If $w = r$, then we can assume without loss of generality that $j = 1$ and $k = 2$. So $e(w, u) \geq 2$, and therefore for all $a \in w$, $e_u(a) = 2$ holds; hence $e_u(\psi) = 2$, i.e., $\Diamond\psi \in u$.

Now suppose that $w \neq r$ and $e(w, u) = e(r, u) = 1$, as otherwise, if $e(w, u) = 2$ or $e(r, u) = 2$, we proceed as above. Since $w \neq r$, there is some $\theta \in w \setminus r$, and therefore $\neg\theta \in r \setminus w$. Since $e(w, u) = e(r, u) = 1$, we can assume without loss of generality that $e_u(\psi \wedge \theta) = e_u(\psi \wedge \neg\theta) = 1$. By definition of e_u, there are a and b such that

$$\neg\Diamond a, \quad \neg\Diamond b, \quad \Diamond((\psi \wedge \theta) \vee a), \quad \Diamond((\psi \wedge \neg\theta) \vee b) \quad \in u.$$

Hence by (G2) and modus ponens $\Diamond((\psi \wedge \theta) \vee (\psi \wedge \neg\theta)) \in u$, i.e., $\Diamond\psi \in u$.

For the converse direction assume that $\Diamond\psi \in u$. If there is an ultrafilter w such that $e(w, u) = 2$ and $\psi \in w$, then we are done since both $(w, 1)$ and $(w, 2)$ are R-successors of (u, i). Suppose that there is no such ultrafilter. Since, Z_u is an ideal by Corollary 4.3, and $\psi \notin Z_u$ since $e_u(\psi) = 2$, by the prime ideal theorem (PIT), there exists some ultrafilter w such that $\psi \in w$ and $w \cap Z_u = \varnothing$; thus $(u, i)R(w, 1)$ as $e(w, u) \geq 1$. Having ruled out ultrafilters containing ψ and such that $e(w, u) = 2$, it must be $e(w, u) = 1$. Therefore, there exists some $\zeta \in w$ such that $e_u(\zeta) = 1$, implying $e_u(\psi \wedge \zeta) = 1$, so $\Diamond(\psi \wedge \zeta) \notin u$. By hypothesis $\Diamond\psi \in u$, thus $e_u(\psi \wedge \neg\zeta) \geq 1$. Using PIT again, there exists some ultrafilter w' such that $\psi \wedge \neg\zeta \in w'$ and $w' \cap Z_u = \varnothing$; hence $(u,i)R(w',1)$ as $e_u(w,u) \geq 1$. Since $\psi \in w \cap w'$ and $\psi \wedge \neg\zeta \notin w$, it follows that $w \neq w'$. Therefore, $M_{\mathbb{B}}, (u,i) \vDash_2 \varphi$ holds. This concludes the proof. □

From the lemma above, using the standard argument, the following theorem holds.

Theorem 4.6 *The system GP_2 is strongly complete with respect to the logic $\mathcal{L}_2$.*

4.2 Case n=3

The system GP_3 is obtained by adding the following axiom-schemata to GP_0:

(G3$_1$) $[\Diamond_1^{q_1}(\alpha_1) \wedge \Diamond_1^{q_2}(\alpha_2) \wedge \Diamond_1^{q_3}(\alpha_3)] \rightarrow \Diamond(\alpha_1 \vee \alpha_2 \vee \alpha_3)$,

(G3$_2$) $[\Diamond_1^{q_1}(\alpha_2) \wedge \Diamond_1^{q_2}(\beta_2) \wedge \Diamond(\alpha_1 \vee \beta_1) \wedge \neg\Diamond(\alpha_1 \vee \alpha_2)] \rightarrow \Diamond(\beta_1 \vee \beta_2)$;

where, as discussed in Section 2, the α_i contradict each other, and likewise the β_i. The axiom (G3$_1$) states that if w has at least one successor satisfying each of α_1, α_2 and α_3, then w must satisfy $\Diamond(\alpha_1 \vee \alpha_2 \vee \alpha_3)$. The axiom (G3$_2$) intuitively expresses the idea that if w has at most one successor satisfying α_1 (which is captured by the fact that w doesn't satisfy $\Diamond(\alpha_1 \vee \beta_1)$), while satisfying $\Diamond(\alpha_1 \vee \beta_1)$, then w must have at least 2 successors satisfying β_1. It is routine to verify the soundness of these axioms.

Completeness. Let $\mathbb{B}$ be a Boolean algebra with a monotone operation satisfying the axioms of GP_3. Let u be an ultrafilter on $\mathbb{B}$. Let us define $e_u : \mathbb{B} \rightarrow \{0,1,2,3\}$ as follows:

$$e_u(a) = \begin{cases} 3 & \text{if } \Diamond a \in u, \\ 2 & \text{if } e_u(a) \neq 3 \text{ and } (\forall b, c \in \mathbb{B})\big(\Diamond c \notin u \text{ and } \Diamond(b \vee c) \in u, \\ & \qquad\qquad a \wedge b = \bot \implies \Diamond(a \vee b) \in u\big) \\ 1 & \text{if } e_u(a) \notin \{3,2\} \text{ and } (\exists b \in \mathbb{B})(\Diamond b \notin u \text{ and } \Diamond(a \vee b) \in u), \\ 0 & \text{otherwise.} \end{cases}$$

Notice that we can write the condition for 2 as

$$\forall b, c \in \mathbb{B}\ ((\Diamond_1^c b \in u \text{ and } a \wedge b = \bot) \Rightarrow \Diamond(a \vee b) \in u).$$

Corollary 4.7 *The set $Z_u = \{a \in \mathbb{B} \mid e_u(a) = 0\}$ is an ideal.*

Proof. Follows immediately from Lemma 4.1 and the definition of e_u. □

Definition 4.8 Let $\mathbb{B}$ be the free Boolean algebra of GP_3. The *canonical frame* of GP_3 is the Kripke frame $\mathbb{X}_{\mathbb{B}} = (\mathcal{U}(\mathbb{B}) \times \mathbf{3}, R)$, where $\mathcal{U}(\mathbb{B})$ denotes the collection of ultrafilters of $\mathbb{B}$, and R is such that for any $u, w \in \mathcal{U}(\mathbb{B})$, and $i, j \in \mathbf{3}$,

$$(u,j)R(w,i) \quad \text{iff} \quad e(w,u) \geq i,$$

where $e(w,u) = \min\{e_u(a) \mid a \in w\}$. The *canonical model* of GP_3 is the Kripke model $M_{\mathbb{B}} = (\mathbb{X}_{\mathbb{B}}, v)$ such that, for any $u \in \mathcal{U}(\mathbb{V})$ and $p \in \mathsf{AtProp}$,

$$u \in v(p) \quad \text{iff} \quad p \in u.$$

Lemma 4.9 (Truth lemma for GP_3) *For any Φ-formula φ, $u \in \mathcal{U}(\mathbb{B})$, and $i \in 3$:*

$$M_{\mathbb{B}}, (u, i) \models_3 \varphi \quad \textit{iff} \quad \varphi \in u.$$

Proof. We proceed by induction on the complexity of φ. The only non-trivial case is when $\varphi = \Diamond\psi$ for some $\psi \in \Phi$.

Assume $M_{\mathbb{B}}, (u, i) \models_3 \varphi$. Then there are $(w_1, j_1), (w_2, j_2), (w_3, j_3) \in \mathcal{U}(\mathbb{B}) \times \mathbf{3}$ such that $(u, i)R(w_1, j_1)$, $(u, i)R(w_2, j_2)$, $(u, i)R(w_3, j_3)$, and $\psi \in w_1 \cap w_2 \cap w_3$. Without loss of generality, we assume $j_1 \leq j_2 \leq j_3$. There are three possible cases:

(1) $w_1 = w_2 = w_3$ and $j_1 = 1$, $j_2 = 2$, and $j_3 = 3$. In this case, $e(w, u) = 3$, thus $e_u(\psi) = 3$, i.e., $\varphi = \Diamond\psi \in u$.

(2) $w_1 \neq w_2 = w_3$, $j_1 = j_2 = 1$, and $j_3 = 2$. In this case, there is $\theta \in w_1 \setminus w_2$, hence $\neg\theta \in w_2 \setminus w_1$. If $e_u(\psi \wedge \theta) = 3$, then by monotonicity $e_u(\Diamond\psi) = 3$ and we proceed as the case above. So let us suppose that $e_u(\psi \wedge \theta) = 2$. Since $\neg\theta \in w_2$, it must be that $e_u(\psi \wedge \neg\theta) \geq 1$, i.e. $\Diamond_1^b(\psi \wedge \neg\theta) \in u$. By the definition of $e_u(\cdot)$, since $e_u(\psi \wedge \theta) = 2$ it follows that if $\Diamond_1^b c \in u$, then $\Diamond((\psi \wedge \theta) \vee c) \in u$. Hence, $\Diamond((\psi \wedge \theta) \vee (\psi \wedge \neg\theta)) \in u$, i.e. $\Diamond\psi \in u$.

(3) $w_1 \neq w_2 \neq w_3 \neq w_1$, and $j_1 = j_2 = j_3 = 1$. Clearly, there are

$$\theta_1 \in w_1 \setminus (w_2 \cup w_3), \qquad \theta_2 \in w_2 \setminus (w_1 \cup w_3), \qquad \theta_3 \in w_3 \setminus (w_1 \cup w_2),$$

such that $e_u(\theta_1) = e_u(\theta_2) = e_u(\theta_3) = 1$, and $\theta_1, \theta_2, \theta_3 \leq \psi$[2], which are w.l.o.g. contradictory. By the definition of $e_u(\cdot)$, there are $\zeta_1, \zeta_2, \zeta_3$ such that

$$\begin{array}{ccc} \neg\Diamond\zeta_1, & \neg\Diamond\zeta_2, & \neg\Diamond\zeta_3, \\ \Diamond(\theta_1 \vee \zeta_1) \in u, & \Diamond(\theta_2 \vee \zeta_2) \in u, & \Diamond(\theta_3 \vee \zeta_3) \in u. \end{array}$$

By axiom ($G3_1$) and modus ponens, $\Diamond(\theta_1 \vee \theta_2 \vee \theta_3) \in u$. By monotonicity of $\Diamond$ (axiom (M)), as $\theta_1 \vee \theta_2 \vee \theta_3 \leq \psi$, then $\Diamond\psi \in u$.

For the converse direction, assume $\Diamond\psi \in u$. There are three possible cases:

(1) There is an ultrafilter w such that $e(w, u) = 3$ and $\psi \in w$. In this case we are done since $(w, 1)$, $(w, 2)$, and $(w, 3)$ are (distinct) successors of (u, i) (for any $i \in 3$), i.e. $M_{\mathbb{B}}, (u, i) \models_3 \Diamond\psi$.

(2) There is an ultrafilter w such that $e(w, u) = 2$ and $\psi \in w$. Since $e(w, u) = 2$, there is $\theta \in w$ such that $e_u(\theta) = 2$; hence $e_u(\psi \wedge \theta) = 2$ and, since $\Diamond\psi \in u$, $e_u(\psi \wedge \neg\theta) \geq 1$. By the prime ideal theorem, there exists an ultrafilter w' such that $\psi \wedge \neg\theta \in w'$ and $w' \cap Z_u = \varnothing$, i.e. $e(w', u) \geq 1$, and thus $(u, 1)R(w', 1)$. Since $e(w, u) = 2$, we have that $(u, 1)R(w, 1)$ and $(u, 1)R(w, 2)$. Because $\psi \wedge \theta \in w$, $w \neq w'$. Hence there are at least three different successors of (u, i) that satisfy ψ, i.e. $M_{\mathbb{B}}, (u, i) \models_3 \Diamond\psi$.

---

[2] In general, given two distinct filters f_1 and f_2 of some lattice L, and an element $x \in f_1 \cap f_2$, there is an element $y \in f_1 \setminus f_2$ such that $y \leq x$. Indeed, without loss of generality, there is $z \in f_1 \setminus f_2$. As f_1 is a filter, $y := x \wedge z \in f_1$. Clearly $y \leq x$, and $y \notin f_2$, as otherwise also x would be in f_2.

(3) Every ultrafilter w that contains ψ is such that $e(w,u) \leq 1$. By the PIT, there is some ultrafilter w_1 such that $\psi \in w_1$ and $w_1 \cap Z_u = \varnothing$ (as $e_u(\psi) = 3$), hence $e(w_1,u) = 1$, yielding $(u,i)R(w_1,1)$. It follows that there exists $\theta \in w$ such that $e_u(\theta) = 1$, and so $e(\psi \wedge \theta) = 1$. Therefore, there are b and c such that $\Diamond_1^c b \in u$, $(\psi \wedge \theta) \wedge b = \bot$, and $\Diamond((\psi \wedge \theta) \vee b) \notin u$. Hence $\neg\Diamond((\psi \wedge \theta) \vee b) \in u$, and hence, by axiom (G3$_2$), for any $\Diamond_1^e d \in u$, with $d \wedge \psi \wedge \neg\theta = \bot$, it follows that $\Diamond(\psi \wedge \neg\theta \vee d) \in u$, i.e. $e_u(\psi \wedge \neg\theta) \geq 2$. Hence there exists ζ such that $e_u(\psi \wedge \neg\theta \wedge \zeta) = 1$ and an ultrafilter w_2, such that $\psi \wedge \neg\theta \wedge \zeta \in w_2$ and $w_2 \cap Z_u = \varnothing$, i.e. $(u,i)R(w_2,1)$. Now, if $\Diamond((\psi \wedge \theta) \vee (\psi \wedge \neg\theta \wedge \zeta)) \in u$, axiom (G3$_2$) implies (arguing as above) that $e_u(\psi \wedge \neg\theta \wedge \zeta)) \geq 2$, a contradiction. Hence $\neg\Diamond((\psi \wedge \theta) \vee (\psi \wedge \neg\theta \wedge \zeta)) \in u$ and, as $\Diamond\psi \in u$, it follows that $e_u(\psi \wedge \neg\theta \wedge \neg\zeta) \geq 1$. By the PIT there is an ultrafilter w_3 containing $\psi \wedge \neg\theta \wedge \neg\zeta$ and such that $w_2 \cap Z_u = \varnothing$, i.e. $(u,i)R(w_3,1)$. Clearly w_1, w_2, w_3 are all distinct and hence we have $M_{\mathbb{B}}, (u,i) \models_3 \Diamond\psi$.

This concludes the proof. □

From Lemma 4.9, using the standard argument, the following theorem holds.

Theorem 4.10 *The system GP_3 is strongly complete with respect to the logic $\mathcal{L}_3$.*

4.3 General case

Through the remainder of this section we will fix a natural number n and its corresponding logic $\mathcal{L}_n$. For the general case, we will not provide an explicit axiomatization but connect axioms with inequalities of positive natural numbers.

The logic. We intend to introduce axioms that encode, in the same way as the axioms of GP_2 and GP_3 did, several implications regarding inequalities of natural numbers. In particular, we want to express for each $m_i, m_j \leq n$, $I = \{1,2,\ldots,m_i\}$, $J = \{1,\ldots,m_j\}$, and positive natural numbers $x_i^j \in \mathbb{N}$ (with $i \in I$ and $j \in J$),

$$\bigwedge_{j\in J}(\sum_{i\in I} x_i^j < n) \to \bigvee_{k\in K\subset J}(\sum_{h\in H\subset I} x_h^k < n). \tag{2}$$

Clearly, as there can be just a finite number of sets I and J since their size is bounded, there is a finite amount of such implications, and each implication has a finite number of inequalities on both sides.

Definition 4.11 Let $GP_n \subseteq \Phi$ be the collection of Φ-formulas that contains GP_0 and for each *true* implication as in (2), an axiom

$$\left(\Diamond_1^{q_i^j}(\alpha_i^j) \wedge \bigwedge_{j\in J} \neg\Diamond \bigvee_{i\in I} \alpha_i^j\right) \to \bigvee_{k\in K\subset J} \neg\Diamond \bigvee_{h\in H\subset I} \alpha_h^k.$$

As discussed above, the number of possible such inequalities is finite and hence also the axiomatization proposed here is finite. Furthermore, knowing whether the implication of inequalities is true or not is a decidable procedure: these are statements in Presburger arithmetic which is a decidable theory.

Completeness. Let $\mathbb{B}$ be a Boolean algebra with a monotone operation satisfying the axioms of GP_n. Let u be an ultrafilter on $\mathbb{B}$. Similar to the previous cases, we want to define a function $e_u : \mathbb{B} \to \mathbb{N}$ which satisfies the following conditions for every $\varphi \in \Phi$:

(i) $e_u(\varphi) \geq n$ if $\Diamond(\varphi) \in u$,

(ii) $e_u(\varphi) < n$ if $\Diamond(\varphi) \notin u$,

(iii) $e_u(\varphi) + e_u(\psi) = e_u(\varphi \wedge \psi) + e_u(\varphi \vee \psi)$, for every $\psi \in \Phi$,

(iv) $e_u(\varphi) = 0$ whenever for every $\psi \in \Phi$, $\Diamond(\psi \vee \varphi) \to \Diamond(\psi) \in u$.

Lemma 4.12 *Such an e_u exists for each ultrafilter $u \in \mathcal{U}(\mathbb{B})$.*

Proof. These restrictions on e_u define a system of equations, Δ, that has a solution if and only if $\Delta \cup P$ is satisfiable, where P is the set of the axioms of Presburger arithmetic. This set, in turn, is satisfiable if and only if every finite subset $Y \subseteq \Delta \cup P$ of it is satisfiable, by the compactness of first-order logic. Finally Y is satisfiable if and only if $X = Y \cap \Delta$ has a solution. Hence, let Z be a finite subset of Δ. The system Z contains as parameters formulas of Φ. Let's call the set of these formulas Φ_Z We will strengthen the system Z, by adding extra condition: We stipulate that for $\varphi \in \Phi_Z$, $e_u(\varphi) = 0$ whenever for every $\psi \in \Phi_Z$, $\Diamond(\psi \vee \varphi) \to \Diamond(\psi) \in u$. Let's call this new system X. Notice that $\Phi_Z = \Phi_X$. Clearly if X has a solution, so does Z. So let's show that X has a solution.

Let $\mathbb{B}_X$ be the finite Boolean algebra generated by Φ_X. Since $\mathbb{B}_X$ is finite, it is routine to verify that the subsystem X has a solution if and only if there exists an assignment $s : A \to \mathbb{N}$ on the atoms A of $\mathbb{B}_X$ such that $\sum_{a \in C \subseteq A} s(a) \geq n$ if and only if $\Diamond(\bigvee_{a \in C \subseteq A} a) \in u$ for every $C \subseteq A$.

First notice that, by monotonicity of $\Diamond$, for every $a_1, \ldots, a_m$,

$$e_u(\bigvee_{i=1}^{m} a_i) = 0 \quad \text{iff} \quad (\forall i \in m) s(a_i) = 0.$$

Indeed, for $\varphi \in \Phi_X$ if $\Diamond(a_i \vee \varphi) \in u$ then $\Diamond(\bigvee_{i=1}^{m} a_i \vee \varphi) \in u$; therefore $\Diamond\varphi \in u$. For the other direction assume that $\Diamond((a_1 \vee \cdots \vee a_m) \vee \varphi) \in u$. Then $\Diamond(a_2 \vee \cdots \vee a_m \vee \varphi) \in u$. Continuing recursively on m, we conclude that $\Diamond\varphi \in u$.

Now let us show that such an s exists. If no such s exists, then this is a true statement about inequalities

$$\bigwedge_{j \in J} (\sum_{i \in I} x_i^j < n) \to \bigvee_{k \in K \subset J} (\sum_{h \in H \subset I} x_h^k < n),$$

where J corresponds to the set of inequalities in X of the form $e_u(\varphi) < n$ and K to the set of inequalities in X of the form $e_u(\varphi) \geq n$. But then GP_n contains an axiom of the form

$$\left(\Diamond_1^{q_i^j}(\alpha_i^j) \wedge \bigwedge_{j \in J} \neg\Diamond \bigvee_{i \in I} \alpha_i^j\right) \rightarrow \bigvee_{k \in K \subset J} \neg\Diamond \bigvee_{h \in H \subset I} \alpha_h^k.$$

This is a contradiction, since by definition on the conditions of e_u $\neg\Diamond(\bigvee_{i \in I} \alpha_i^j) \in u$ and $\Diamond(\bigvee_{h \in H} \alpha_h^k) \in u$ for every $k \in K$. Therefore s exists, and so e_u also exists. □

By Lemma 4.12, for each ultrafilter u, there is some map $e_u : \mathbb{B} \to \mathbb{N}$ satisfying the conditions above. By the axiom of choice, we choose one of such e_u for every ultrafilter and define

$$e(w, u) = \min\{e_u(a) \mid a \in w\}.$$

Notice that for each e_u, also the inverse of condition (iii) holds: that is if there exists ψ such that $\Diamond(\varphi \vee \psi) \in u$ and $\Diamond(\psi) \notin u$, then $e_u(\varphi \vee \psi) \geq n$, while $e_u(\psi) < n$. So

$$s(\varphi) = s(\varphi \vee \psi) - s(\psi) + s(\varphi \wedge \psi) > 0.$$

Corollary 4.13 *The set* $Z_u = \{a \in \mathbb{B} \mid e_u(a) = 0\}$ *is an ideal.*

Proof. Immediately by Lemma 4.1. □

Definition 4.14 Let $\mathbb{B}$ be the free Boolean algebra of GP_n. The *canonical frame* of GP_n is the Kripke frame $\mathbb{X}_{\mathbb{B}} = (\mathcal{U}(\mathbb{B}) \times \mathbf{n}, R)$, where $\mathcal{U}(\mathbb{B})$ denotes the collection of ultrafilters of $\mathbb{B}$, and R is such that for any $u, w \in \mathcal{U}(\mathbb{B})$, and $i, j \in \mathbf{n}$,

$$(u, j)R(w, i) \qquad \text{iff} \qquad e(w, u) \geq i.$$

The *canonical model* of GP_n is the Kripke model $M_{\mathbb{B}} = (\mathbb{X}_{\mathbb{B}}, v)$ such that, for any $u \in \mathcal{U}(\mathbb{V})$ and $p \in \mathsf{AtProp}$,

$$u \in v(p) \qquad \text{iff} \qquad p \in u.$$

Lemma 4.15 (Truth lemma for GP_n) *For any* Φ*-formula* φ, $u \in \mathcal{U}(\mathbb{B})$, *and* $j \in n$,

$$M_{\mathbb{B}}, (u, j) \vDash_n \varphi \qquad \textit{iff} \qquad \varphi \in u.$$

Proof. We prove the statement by induction on the complexity of φ. We check only the case $\varphi = \Diamond\psi$ for some $\phi \in \Phi$, the other cases being trivial.

Assume that $M, (u, j) \vDash_n \Diamond\psi$. Then there exist k different ultrafilters $w_1, \ldots, w_k \in \mathcal{U}(\mathbb{B})$, and $m_1, \ldots, m_k \in \mathbf{n}$ such that $m_1 + \cdots + m_k \geq n$ and $(u, j)R(w_i, m_i)$ and such that $M, (w_i, m_i) \vDash_n \psi$, and so $\psi \in w_i$ for each $i \in k$ by the induction hypothesis. Since all the w_i are distinct ultrafilters, there exist $\theta_1, \ldots, \theta_k \in \mathbb{B}$ such that for all $i \in \{1, \ldots, k\}$,

$$\theta_i \in w_i \qquad \text{and} \qquad \neg\theta_j \in w_i \text{ for all } j \in \{1, \ldots, k\} \setminus \{i\}.$$

Without loss of generality we can assume $\theta_1, \ldots, \theta_k$ are mutually disjoint[3] and by the argument in Footnote 2, we can also assume that $\theta_1, \ldots, \theta_k \leq \psi$. For each i, $e_u(\psi \wedge \theta_i) \geq m_i$ since $\mathrm{e}_u(w_i, u) \geq m_i$. Hence,

$$e_u(\psi) = e_u(\psi \wedge (\theta_1 \vee \cdots \vee \theta_k)) \geq \sum_{1 \leq i \leq k} m_i \geq n.$$

By the conditions that e_u satisfies, it follows that $\Diamond\psi \in u$.

For the other direction, assume that $\Diamond\psi \in u$, and so $e_u(\psi) \geq n$. We will recursively define a sequence of distinct ultrafilters $w_1, \ldots, w_k$, such that $0 < e(w_i, u) = m_i$, $m_1 + \cdots + m_k \geq n$, and $\psi \in w_i$ for all i. For the base case, by the PIT, there exists an ultrafilter w_1 disjoint from Z_u such that $\psi \in w_1$. Since it's disjoint from Z_u, $e(w_1, u) > 0$. Assume now that we already have ℓ distinct ultrafilters $w_1, \ldots, w_\ell$ and $m_1 + \cdots + m_\ell < n$ such that $\psi \in \bigcap_{i=1}^{\ell} w_i$, and such that $m_1, \ldots, m_\ell > 0$. Since the ultrafilters are distinct we have that there exist mutually disjoint $\theta_i \in w_i$ for $1 \leq i \leq \ell$, such that $\neg\theta_i \in w_1, \ldots, w_{i-1}, w_{i+1}, \ldots, w_\ell$, $\theta_i \leq \psi$ and $e_u(\psi \wedge \theta_i) = m_i$. Then,

$$\sum_{1=1}^{\ell} e_u(\psi \wedge \theta_i) = m_1 + \cdots + m_\ell < n,$$

Hence, since $\theta_1, \ldots, \theta_\ell$ are mutually disjoint,

$$\begin{array}{rcll}
e_u(\psi) & = & e_u\left((\psi \wedge \bigvee_{i=1}^{\ell} \theta_i) \vee (\psi \wedge \neg \bigvee_{i=1}^{\ell} \theta_i)\right) & \mathbb{B} \text{ Boolean} \\
& = & e_u\left(\bigvee_{i=1}^{\ell} (\psi \wedge \theta_i) \vee (\psi \wedge \neg \bigvee_{i=1}^{\ell} \theta_i)\right) & \mathbb{B} \text{ distributive} \\
& = & e_u\left(\bigvee_{i=1}^{\ell} (\psi \wedge \theta_i)\right) + e_u\left(\psi \wedge \neg \bigvee_{i=1}^{\ell} \theta_i\right) & \text{property of } e_u \\
& = & \sum_{1=1}^{\ell} e_u(\psi \wedge \theta_i) + e_u\left(\psi \wedge \neg \bigvee_{i=1}^{\ell} \theta_i\right) & \theta_1, \ldots, \theta_\ell \text{ disjoint} \\
& = & \sum_{1=1}^{\ell} m_i + e_u\left(\psi \wedge \neg \bigvee_{i=1}^{\ell} \theta_i\right) & \text{definition of } m_i \\
\text{iff} & & e_u\left(\psi \wedge \neg \bigvee_{i=1}^{\ell} \theta_i\right) = e_u(\psi) - \sum_{1=1}^{\ell} m_i & \\
\text{implies} & & e_u\left(\psi \wedge \neg \bigvee_{i=1}^{\ell} \theta_i\right) \geq n - \sum_{1=1}^{\ell} m_i & \Diamond\psi \in u \\
\text{implies} & & e_u\left(\psi \wedge \neg \bigvee_{i=1}^{\ell} \theta_i\right) > 0 & \sum_{1=1}^{\ell} m_i < n
\end{array}$$

Thus, by the PIT, there exists some ultrafilter, $w_{\ell+1}$ that contains $\psi \wedge \neg(\theta_1 \vee \cdots \vee \theta_\ell)$ (and so distinct from $w_1, \ldots, w_\ell$) that is disjoint from Z_u, and hence $e(w_{\ell+1}, u) > 0$.

Finally, given that this process will terminate after a finite number of steps we will obtain a sequence of distinct ultrafilters $w_1, \ldots, w_k$ such that

---

[3] From any $\theta_1, \ldots \theta_k$ such as the ones above, one could consider for each i,

$$\theta_i' := \theta_i \wedge \bigwedge_{j \neq i} \neg\theta_j \in w_i.$$

Clearly, for each $i, j \in \{1, \ldots, k\}$, $\theta_i' \wedge \theta_j' = \bot$ whenever $i \neq j$.

$0 < e(w_i, u) = m_i$, $m_1 + \cdots + m_k \geq n$, and $\psi \in w_i$ for all i. Therefore $(u, j)R(w_i, k_i)$, for $k_i \leq m_i$, i.e. $M, (u, j) \vDash_n \Diamond\psi$. □

The following theorem follows from Lemma 4.15 using standard completeness arguments.

Theorem 4.16 (AC) *The system GP_n is strongly complete with respect to the logic $\mathcal{L}_n$.*

5 Conclusions

Contributions. In this article we studied the family of the monotonic modal logics $\mathcal{L}_n$ for any $n \geq 2$, each with a single modality which is interpreted semantically as a graded modality $\Diamond_n$. We observed that all these logics are decidable and have the strong finite model property. We compared these logics with each other showing that, if $m - n$ is small, the logics $\mathcal{L}_n$ and $\mathcal{L}_m$ might be incomparable, while if $m - n$ is large enough, then $\mathcal{L}_m \subsetneq \mathcal{L}_n$. We also provided concrete complete axiomatizations for $\mathcal{L}_2$ and $\mathcal{L}_3$, and we showed that each $\mathcal{L}_n$ for $n \geq 4$ is finitely axiomatizable, by showing that each axiom corresponds to a "rule" in a finite set.

The maps e_u. The construction of the canonical model for classical graded modal logic (see e.g. [7]) pivots on the map $e : \mathcal{U}(\mathcal{B}) \times \mathcal{U}(\mathcal{B}) \to n$ (which we also used in Definition 4.14). However, thanks to the fact that the language of graded modal logic is much more expressive, in [7], the map e_u can be obtained immediately as $e_u(\varphi) = \sup\{k \in \mathbb{N} \mid \Diamond_k\varphi \in u\}$, for every formula $\varphi \in \Phi_G$. In the case of $\mathcal{L}_n$, since the language is restricted, defining the map e_u becomes much more intricate and complicated. We showed that, if $n = 2$ or $n = 3$, the language is expressive enough to define e_u in a unique and uniform way. However, this is no longer possible already for $n = 4$. To see this, consider the frame $(\{u\} \sqcup 4, u \times 4)$, and the valuation $v(p) = \{1\}$, $v(q) = \{2\}$ and $v(r) = \{3, 4\}$. Then under any permutation of $\{p, q, r\}$, the theory of u remains unchanged, meaning that u cannot "tell apart" p, q, r. If a uniform way of defining e_u existed, then $e_u(p) = e_u(q) = e_u(r)$, but this is impossible, since exactly one of them needs value 2, and the other two should have value 1. Therefore, when defining e_u for $n \geq 4$, arbitrary choices need to be made.

Future directions. Even though, as discussed above, some steps towards complete axiomatization for $\mathcal{L}_n$ for $n \geq 4$ were taken, the question of identifying concrete axiomatic systems for $\mathcal{L}_n$ remains open. From the discussion in the paragraph above, it is clear that such axiomatizations need to be more complex than the ones presented for $n = 2$ and $n = 3$.

Acknowledgements. We would like to thank Hans van Ditmarsch, for raising this very interesting question on the theory of the logics $\mathcal{L}_n$, and for further fruitful exchanges on this topic.

References

[1] Aceto, L., A. Ingolfsdottir and J. Sack, *Resource bisimilarity and graded bisimilarity coincide*, Information Processing Letters **111** (2010), pp. 68–76.

[2] Cerrato, C., *General canonical models for graded normal logics (graded modalities IV)*, Studia Logica **49** (1990), pp. 241–252.

[3] Cerrato, C., *Decidability by filtrations for graded normal logics (graded modalities V)*, Studia Logica **53** (1994), pp. 61–74.

[4] Chellas, B. F., "Modal logic: an introduction," Cambridge University Press, 1980.

[5] Chen, J., G. Greco, A. Palmigiano and A. Tzimoulis, *Non-normal modal logics and conditional logics: Semantic analysis and proof theory*, Information and Computation (2021), p. 104756.
URL `https://www.sciencedirect.com/science/article/pii/S0890540121000717`

[6] Chen, J., H. van Ditmarsch, G. Greco and A. Tzimoulis, *Neighbourhood semantics for graded modal logic*, Bulletin of the Section of Logic **50** (2021), pp. 373–395.

[7] De Caro, F., *Graded modalities, II (canonical models)*, Studia Logica **47** (1988), pp. 1–10.

[8] de Rijke, M., *A note on graded modal logic*, Studia Logica **64** (2000), pp. 271–283.

[9] Fine, K., *In so many possible worlds.*, Notre Dame Journal of Formal Logic **13** (1972), pp. 516 – 520.
URL `https://doi.org/10.1305/ndjfl/1093890715`

[10] Goble, L. F., *Grades of modality*, Logique et Analyse **13** (1970), pp. 323–334.
URL `http://www.jstor.org/stable/44083605`

[11] Hansen, H. H., "Monotonic modal logics," ILLC Report Nr: PP-2003-24, University of Amsterdam, 2003.

[12] Kaplan, D., *S5 with multiple possibility*, Journal of Symbolic Logic **35** (1970), p. 355.

[13] Ma, M. and H. van Ditmarsch, *Dynamic graded epistemic logic.*, The Review of Symbolic Logic **12** (2019), pp. 663–684.

[14] Pacuit, E., "Neighborhood semantics for modal logic," Short Textbooks in Logic, Springer, 2017.

[15] van der Hoek, W., *On the semantics of graded modalities*, Journal of Applied Non-Classical Logics **2** (1992), pp. 81–123.

[16] van der Hoek, W. and J.-J. C. Meyer, *Graded modalities in epistemic logic*, in: *International Symposium on Logical Foundations of Computer Science*, Springer, 1992, pp. 503–514.

An analytic proof system for common knowledge logic over S5

Jan Rooduijn [1]

ILLC, University of Amsterdam

Lukas Zenger [2]

Institute of Computer Science, University of Bern

Abstract

In this paper we present an analytic proof system for multi-modal logic with common-knowledge over S5 (called S5-CKL). The system is an annotated cyclic calculus manipulating two-sided Gentzen sequents and extending a known system for multi-modal S5. First a direct argument is used to show that the system is sound. Using a canonical model construction, we then show that the system is analytically complete. In particular, the use of the cut-rule is restricted to analytic cuts. Exploiting this analyticity, we then reduce the provability problem of a given sequent to the problem of solving a certain parity game. As a consequence we obtain an optimal decision procedure for proof search and thereby for the validity problem of S5-CKL.

Keywords: Common knowledge, S5 multi-modal logic, analytic proof systems, cyclic proofs, proof search games.

1 Introduction

Common knowledge is an important notion of group knowledge with applications ranging from philosophy to computer science. A proposition p is *common knowledge* in a group of agents if all agents know that p, all agents know that all agents know that p, and so on. In a formal setting, common knowledge is often studied using a logic called *common knowledge logic* (CKL, for short). The logic CKL is an extension of multi-agent modal logic by a fixed point operator ⊞ meant to express common knowledge. As such it is a fragment of the alternation-free modal μ-calculus [4]. CKL was introduced in 1990 by Halpern and Moses [12]. For an introduction to logics of common knowledge in general we refer the reader to [22].

[1] J.M.W. Rooduijn is supported by a grant from the Dutch Research Council NWO, project nr. 617.001.857.

[2] L. Zenger is supported by the Swiss National Science Foundation Grant 200021L_196176 Proof and Model Theory of Intuitionistic Temporal Logic.

Most commonly CKL is axiomatised using a Hilbert-style system for multi-modal logic, extended by fixed point axioms expressing that $\boxplus$ is a greatest fixed point (see for example [22]). However, axiomatisiations of this kind suffer from the usual drawback of Hilbert-style systems: the presence of the *modus ponens* rule frustrates proof-theoretic analysis.

The common solution to this problem is to construct a sequent system with restricted applications of the *cut* rule, but this has been proven difficult for the logic CKL. When the base modal logic is required to satisfy certain frame conditions (motivated for example by the study of distributed systems), the difficulty often further increases. It is for example an open question whether there exists a finite and *analytic* sequent calculus for the logic CKL interpreted over S5-frames (this logic will be called S5-CKL). As usual, we call a calculus analytic whenever every valid sequent admits a finite proof containing only formulas in some sense relevant to the endsequent. In the context of modal fixed point logics one usually counts as relevant the formulas in the endsequent's *Fischer-Ladner closure* (originally defined in [10]). Notable sequent calculi for S5-CKL have been constructed for example by Alberucci & Jäger [5] and Hill & Poggiolesi [14], but although both are finite, neither are analytic. A description of these systems, as well as a comparison between them and the present paper, is postponed to Section 7 below.

More related work exists in the area of tableau-based decision procedures. In [3], Ajspur et al. give such a procedure for S5-CKL and several of its extensions. While their procedure is analytic, the fact that it requires multiple passes makes it unclear how one could extract an analogous sequent-style proof system. A single pass tableau-based decision procedure for CKL is given by Abate et. al in [1], but only for the interpretation of CKL over the class of all frames.

In the first part of this paper we give a positive answer to the question of whether S5-CKL admits a finite and analytic sequent calculus. To that end we present the cyclic sequent calculus $\mathsf{sCKL_f}$. Instead of an induction rule as in [5] and [14], $\mathsf{sCKL_f}$ uses cycles to characterise $\boxplus$ as a greatest fixed point. These cycles are handled by a focus mechanism, a technique originally proposed by Lange & Stirling in [16]. Roughly, a cyclic branch of some proof is deemed valid whenever some form of progress is made. The soundness argument then exploits the fact that this form of progress cannot be made infinitely often. In our case we show the soundness of $\mathsf{sCKL_f}$ by a minimal countermodel approach that is sometimes found in the literature (see *e.g.* [21]). For completeness we use a canonical model construction that is similar to the construction in [20]. Importantly, in the completeness proof the cut rule is only applied to formulas in the Fischer-Ladner closure of the endsequent. As every other rule of $\mathsf{sCKL_f}$ enjoys the subformula property (or its analogue for the Fischer-Ladner closure), we obtain that the calculus is analytic. The approach of using cyclic proof systems for common knowledge logic has also been been taken by Ricardo Wehbe in his PhD Thesis [23], but, like Abate et. al, he does not consider the restriction to S5-frames.

In the second part of the paper we show that $\mathsf{sCKL_f}$ is suitable for proof search. To that end we translate the calculus into a parity game which is played by a player called Prover, who tries to show that a given sequent is derivable in $\mathsf{sCKL_f}$, and a player called Refuter who tries to show the opposite. We then show that a sequent is derivable in $\mathsf{sCKL_f}$ if and only if Prover has a memoryless winning strategy in the corresponding game. Next, we use a result proven by Calude et al. in [8] to establish the existence of an efficient algorithm which computes the winner of each game. Finally, by combining these two results we establish the existence of an algorithm deciding whether a given sequent is derivable in $\mathsf{sCKL_f}$, which runs in optimal time.

The paper is structured as follows: in Section 2 we introduce the basic definitions and notations for S5-CKL, including its syntax and semantics. In Section 3 we define the sequent calculus $\mathsf{sCKL_f}$. Sections 4 and 5 are devoted to prove soundness and completeness, respectively. In Section 6 we define the aforementioned proof search game and establish that there exists and optimal proof search algorithm. Finally, in Section 7 we compare our work to related work of different authors and sketch some ideas for further research.

2 Basic definitions

The language of CKL consists of a finite set of atomic agents $\mathcal{A} = \{1, \dots n\}$ and a countably infinite set of atomic propositions P.

Definition 2.1 *Formulas* φ, ψ of CKL are inductively defined as follows:

$$\varphi, \psi := p \mid \neg\varphi \mid \varphi \wedge \psi \mid \Box_i\varphi \mid \boxplus\varphi$$

where $p \in \mathsf{P}$ and $i \in \mathcal{A}$. The set of CKL-formulas is denoted by Fm.

We will use the abbreviation $\Box\varphi :\equiv \bigwedge_{i\in\mathcal{A}} \Box_i\varphi$. Furthermore, we define $\Box^0\varphi := \varphi$ and $\Box^{n+1}\varphi := \Box\Box^n\varphi$. The expression $\Box_i^n\varphi$ is defined analogously.

Definition 2.2 An *epistemic Kripke model* is a tuple $\mathbb{S} = (S, \{R_i \mid i \in \mathcal{A}\}, V\}$ where

- S is a non-empty set;
- R_i is an equivalence relation on S for each $i \in \mathcal{A}$;
- V is a function $S \to \mathcal{P}(P)$.

Elements in S are called *states.* The binary relations R_i are called *transition relations* and V is called a *valuation.* Instead of $(s,t) \in R_i$ we usually write sR_it.

Formulas of CKL are evaluated in epistemic Kripke models as follows:

Definition 2.3 Let $\mathbb{S} = (S, \{R_i \mid i \in \mathcal{A}\}, V\}$ be an epistemic Kripke model and $s \in S$ be a state. The relation $\Vdash \subseteq S \times Fm$ is inductively defined by:

$$\begin{array}{lll}
\mathbb{S}, s \Vdash p & \Leftrightarrow & p \in V(s) \\
\mathbb{S}, s \Vdash \neg\varphi & \Leftrightarrow & \mathbb{S}, s \nVdash \varphi \\
\mathbb{S}, s \Vdash \varphi \wedge \psi & \Leftrightarrow & \mathbb{S}, s \Vdash \varphi \text{ and } \mathbb{S}, s \Vdash \psi \\
\mathbb{S}, s \Vdash \Box_i\varphi & \Leftrightarrow & \text{for all } t \in S \text{ with } sR_it\text{, it holds that } \mathbb{S}, t \Vdash \varphi \\
\mathbb{S}, s \Vdash \boxplus\varphi & \Leftrightarrow & \text{for all } t \in S \text{ with } sR^*t\text{, it holds that } \mathbb{S}, t \Vdash \varphi
\end{array}$$

where R^* is the transitive closure of the relation $\bigcup_{i \in \mathcal{A}} R_i$.

If $\mathbb{S}, s \Vdash \varphi$, then we say that φ is *true* or *holds* at state s of the epistemic Kripke model $\mathbb{S}$. A formula φ is *satisfiable* if there exists an epistemic Kripke model $\mathbb{S}$ and a state s such that $\mathbb{S}, s \Vdash \varphi$, and *unsatisfiable* if not satisfiable. A formula φ is *valid* if $\mathbb{S}, s \Vdash \varphi$ for every epistemic Kripke model $\mathbb{S}$ and every state s in $\mathbb{S}$, and *invalid* if not valid.

Definition 2.4 The *Fischer-Ladner closure* of a formula φ is the smallest set of formulas $Cl(\varphi)$ which contains φ and is closed under the following conditions:

- $\neg\psi \in Cl(\varphi)$ implies $\psi \in Cl(\varphi)$;
- $\psi_1 \wedge \psi_2 \in Cl(\varphi)$ implies $\psi_k \in Cl(\varphi)$ for each $k \in \{1, 2\}$;
- $\Box_i\psi \in Cl(\varphi)$ implies $\psi \in Cl(\varphi)$;
- $\boxplus\psi \in Cl(\varphi)$ implies $\psi \in Cl(\varphi)$ and $\{\Box_i\boxplus\psi \mid i \in \mathcal{A}\} \subseteq Cl(\varphi)$.

We will usually denote $Cl(\varphi)$ simply as the *closure* of φ. The definition of the closure of a formula is extended to the definition of the *closure* of a set of formulas A as follows:

$$Cl(A) := \bigcup\{Cl(\varphi) \mid \varphi \in A\}.$$

A set A of formulas is called *closed* whenever $Cl(A) = A$.

3 An annotated sequent system

An *annotated formula* is a pair (φ, a), usually written φ^a, where φ is a formula and a is either u (designating that the formula is *unfocussed*) or f (designating that the formula is *in focus*). A *sequent* is a pair (Γ, Δ), usually written $\Gamma \Rightarrow \Delta$, where Γ and Δ are finite sets of annotated formulas. In later proofs we will often denote sequents using the Greek letter σ, possibly with subscript.

We will only consider sequents whose formulas are annotated in a very specific way. Namely, if $\Gamma \Rightarrow \Delta$ is a sequent, we require that every formula in Γ is unfocussed, and at most one formula in Δ is in focus. Moreover, if Δ contains a formula that is in focus, then it must be of the form $\boxplus\varphi$ or of the form $\Box_i\boxplus\varphi$. To emphasise that we are only considering sequents of this restricted form, we will also refer to them as CKL-*sequents*.

For Γ a finite set of annotated formulas, we define the following abbreviations:

$$\Gamma^u := \{\varphi^u \mid \varphi^a \in \Gamma\}, \qquad \Box_i\Gamma := \{\Box_i\varphi^a \mid \varphi^a \in \Gamma\}$$
$$\Gamma^- := \{\varphi \mid \varphi^a \in \Gamma\}, \qquad \Box_i^{-1}\Gamma := \{\varphi^a \mid \Box_i\varphi^a \in \Gamma\}$$

The following definition presents our proof system. Note that basic modal part, *i.e.* the part without the rules $\boxplus_L$, $\boxplus_R$, U, F, is based on a standard system for the modal logic $\mathsf{S5}$. This system was originally presented by Ohnishi and Matsumoto in 1957 [19].

Definition 3.1 The sequent calculus $\mathsf{sCKL_f}$ manipulates CKL-sequents by the following axioms and rules.

$$\mathsf{id}\ \frac{}{\varphi^u \Rightarrow \varphi^a} \qquad \mathsf{w}_L\ \frac{\Gamma \Rightarrow \Delta}{\Gamma, \varphi^u \Rightarrow \Delta} \qquad \mathsf{w}_R\ \frac{\Gamma \Rightarrow \Delta}{\Gamma \Rightarrow \varphi^a, \Delta}$$

$$\neg_L\ \frac{\Gamma \Rightarrow \varphi^u, \Delta}{\Gamma, \neg\varphi^u \Rightarrow \Delta} \qquad \neg_R\ \frac{\Gamma, \varphi^u \Rightarrow \Delta}{\Gamma \Rightarrow \neg\varphi^u, \Delta}$$

$$\wedge_L\ \frac{\Gamma, \varphi^u, \psi^u \Rightarrow \Delta}{\Gamma, (\varphi \wedge \psi)^u \Rightarrow \Delta} \qquad \wedge_R\ \frac{\Gamma \Rightarrow \varphi^u, \Delta \qquad \Gamma \Rightarrow \psi^u, \Delta}{\Gamma \Rightarrow (\varphi \wedge \psi)^u, \Delta}$$

$$\boxplus_L\ \frac{\Gamma, \varphi^u, \{\Box_i \boxplus \varphi^u\}_{i=1}^n \Rightarrow \Delta}{\Gamma, \boxplus\varphi^u \Rightarrow \Delta} \qquad \boxplus_R\ \frac{\Gamma \Rightarrow \varphi^u, \Delta \qquad \{\Gamma \Rightarrow \Box_i \boxplus \varphi^a, \Delta\}_{i=1}^n}{\Gamma \Rightarrow \boxplus\varphi^a, \Delta}$$

$$\Box_{\mathsf{T}}\ \frac{\Gamma, \varphi^u \Rightarrow \Delta}{\Gamma, \Box_i\varphi^u \Rightarrow \Delta} \qquad \Box_{\mathsf{S5}}\ \frac{\Box_i\Gamma \Rightarrow \varphi^a, \Box_i\Delta}{\Box_i\Gamma \Rightarrow \Box_i\varphi^a, \Box_i\Delta}$$

$$\mathsf{U}\ \frac{\Gamma \Rightarrow \Delta^u}{\Gamma \Rightarrow \Delta} \qquad \mathsf{F}\ \frac{\Gamma \Rightarrow \varphi^f, \Delta^u}{\Gamma \Rightarrow \varphi^u, \Delta^u} \qquad \mathsf{cut}\ \frac{\Gamma \Rightarrow \varphi^u, \Delta \qquad \Gamma, \varphi^u \Rightarrow \Delta}{\Gamma \Rightarrow \Delta}$$

As usual, we will call the main formula introduced in the conclusion of a rule the *principal formula* of that rule. In the case of the rule $\Box_{\mathsf{S5}}$, this is the formula $\Box_i\varphi^a$. The rules U and cut have no principal formula.

Remark 3.2 To readers familiar with cyclic proof systems for modal fixed point logics, we offer the following motivation for the focus annotations. The purpose is to capture *traces*, in the sense of [18]. Because we are working in a very simple fragment of the modal μ-calculus, our ν-traces are relatively elementary. In particular, since we are in the alternation-free fragment, our ν-traces do not pass through μ-unfoldings. Moreover, our ν-traces do not pass trough disjunctions, or, in terms of two-sided sequents, they do not pass through conjunctions on the left-hand side of the sequent. It is because of this latter property, characteristic of the *completely additive* fragment of the μ-calculus (see [9]), that the traces do not *split*, whence it suffices to have at most one formula in focus.

The fact that our sequent calculus only manipulates CKL-sequents imposes restrictions on how its rules can be applied. This is illustrated by the following lemma.

Lemma 3.3 *In any application of the $\boxplus_R$-rule such that the principal formula is in focus, the leftmost premiss has no formula in focus.*

Proof. Suppose that in the rule application

$$\boxplus_R\ \frac{\Gamma \Rightarrow \varphi^u, \Delta \qquad \{\Gamma \Rightarrow \Box_i \boxplus \varphi^f, \Delta\}_{i=1}^n}{\Gamma \Rightarrow \boxplus\varphi^f, \Delta}$$

the leftmost premiss has a formula in focus. Then this formula must belong to Δ. Specifically, for the conclusion is a CKL-sequent, we have $\boxplus\varphi^f \in \Delta$. However, that means that every other premiss has two formulas in focus, contradicting the fact that they must also be CKL-sequents. $\square$

A *derivation* in $\mathsf{sCKL_f}$ is a finite tree whose nodes are labelled by CKL-sequents and which is generated by the rules of the calculus $\mathsf{sCKL_f}$. Given a derivation π in $\mathsf{sCKL_f}$, an *upward path* ρ in π is a finite sequence of nodes $\rho = \rho(0), \rho(1), ..., \rho(n)$ of π such that for each $0 \leq i < n$ the node $\rho(i+1)$ is a child of the node $\rho(i)$. Observe that we do not require upward paths to start in the root of the derivation.

Definition 3.4 An upward path ρ in an $\mathsf{sCKL_f}$-derivation is said to be *successful* if the following holds:

(i) Every sequent $\Gamma \Rightarrow \Delta$ on the path ρ has a formula in focus, *i.e.* Δ contains an annotated formula of the form φ^f.

(ii) The path ρ passes through at least one application of $\boxplus_R$, where the principal formula is in focus.

Observe that on a successful path there are no applications of the focus rules U and F. Given a derivation π in $\mathsf{sCKL_f}$ and a leaf l in π we call l an *axiomatic leaf* if l is the conclusion of an application of the rule id. If l is not axiomatic, we call it a *non-axiomatic leaf.* A *repetition* in π is a pair of nodes $\langle u, v\rangle$ such that u is a proper ancestor of v and both u and v are labelled by the same sequent. A repetition $\langle u, v\rangle$ is called *successful* if the path from u to v is successful.

Definition 3.5 A derivation π in $\mathsf{sCKL_f}$ is a *proof* if every leaf l of π is either axiomatic or there exists a node $c(l)$ such that $\langle c(l), l\rangle$ is a successful repetition.

If π is a proof with root sequent $\Gamma \Rightarrow \Delta$, then π is said to be an $\mathsf{sCKL_f}$-*proof of* $\Gamma \Rightarrow \Delta$ and $\Gamma \Rightarrow \Delta$ is called $\mathsf{sCKL_f}$-*provable.*

4 Soundness

A sequent $\Gamma \Rightarrow \Delta$ is said to be *satisfied* at a state s of an epistemic Kripke model $\mathbb{S}$ - denoted by $\mathbb{S}, s \Vdash \Gamma \Rightarrow \Delta$ - if it holds that either $\mathbb{S}, s \nVdash \varphi$ for some $\varphi^u \in \Gamma$, or $\mathbb{S}, s \Vdash \psi$ for some $\psi^a \in \Delta$. A sequent is called *valid* if it is satisfied at every state of every epistemic Kripke model. Note that the focus annotations play no meaningful role in the above definitions.

The proof of the following lemma, which states that every rule of $\mathsf{sCKL_f}$ is individually sound, is standard and therefore omitted.

Lemma 4.1 *Let*

$$\mathsf{r}\,\frac{\sigma_1 \quad \cdots \quad \sigma_n}{\sigma}$$

be a rule application of $\mathsf{sCKL_f}$. *If* σ *is invalid, then so is one of the premisses.*

In order to prove that the system $\mathsf{sCKL_f}$ is sound as a whole, we will first prove a strengthening of Lemma 4.1 which takes the annotations into account.

Let σ be a sequent that has a formula in focus, *i.e.* for $j \in \{0,1\}$ the right-hand side Δ of σ contains a formula of the form $\Box_i^j \boxplus \psi^f$. We denote by $\sigma(n)$ the sequent obtained by adding the formula $\Box_i^j \Box^n \psi^u$ to Δ (recall the definition of $\Box^n$ below Definition 2.1). For any invalid sequent σ that has a formula in focus, we define:

$$\mu(\sigma) := \min\{n \in \omega : \sigma(n) \text{ is invalid}\}.$$

Observe that the function μ is well-defined.

Lemma 4.2 *Let*

$$\mathsf{r}\,\frac{\sigma_1 \quad \cdots \quad \sigma_n}{\sigma}$$

be any rule application of $\mathsf{sCKL_f}$*. If σ is invalid, then there is an invalid premiss σ_i such that, if σ and σ_i both have a formula in focus, then:*

$$\mu(\sigma_i) \leq \mu(\sigma), \tag{1}$$

and, if moreover $\mathsf{r} = \boxplus_R$ *and the principal formula is in focus, then the inequality (1) is strict.*

Proof. Note that the statement becomes vacuous if r is id, since id derives only valid sequents. Moreover, if either σ, or all of the σ_i, have no formula in focus, then the statement reduces to Lemma 4.1. This covers the cases where r is among $\{\mathsf{U}, \mathsf{F}\}$, because those rules require either the conclusion or the sole premiss to have no formula in focus.

Now suppose that $\mathsf{r} \notin \{\mathsf{U}, \mathsf{F}, \mathsf{id}\}$ and both σ and at least one of the σ_i have a formula in focus. We first consider the case where the formula that is in focus in σ is *not* the principal formula of the rule application (this includes the case where $\mathsf{r} = \mathsf{cut}$). Direct inspection of the rules shows that in this case every premiss σ_i has a formula in focus and that

$$\mathsf{r}\,\frac{\sigma_1(\mu(\sigma)) \quad \cdots \quad \sigma_n(\mu(\sigma))}{\sigma(\mu(\sigma))}$$

is a valid rule application. The required result then follows from Lemma 4.1.

The only cases left are those in which the principal formula is in focus, which can only be the case if $\mathsf{r} \in \{\mathsf{w}_R, \Box_{\mathsf{S5}}, \boxplus_R\}$. The case $\mathsf{r} = \mathsf{w}_R$ is immediate, as the sole premiss will have no formula in focus whenever the principal formula is in focus. We treat the two remaining cases separately.

▷ $\Box_{\mathsf{S5}}$: Then σ is of the form:

$$\Box_i \Gamma \Rightarrow \Box_i \boxplus \psi^f, \Box_i \Delta.$$

Let $n := \mu(\sigma)$. By the definition of μ, there is an epistemic Kripke model $\mathbb{S}$, and a state s of $\mathbb{S}$ such that $\mathbb{S}, s \not\Vdash \sigma(n)$. In particular, it holds that

$$\mathbb{S}, s \not\Vdash \Box_i \Box^n \psi.$$

It follows that there is a state t in $\mathbb{S}$ such that sR_it and $\mathbb{S}, t \not\Vdash \Box^n\psi$. Clearly this also means that $\mathbb{S}, t \not\Vdash \boxplus\psi$. We claim that, in fact,

$$\mathbb{S}, t \not\Vdash \Box_i\Gamma \Rightarrow \boxplus\psi^f, \Box^n\psi^u, \Box_i\Delta,$$

which gives the required result.

By the fact that R_i is transitive, it holds for all φ such that $\mathbb{S}, s \Vdash \Box_i\varphi$, that $\mathbb{S}, t \Vdash \Box_i\varphi$. It follows that $\mathbb{S}, t \Vdash \Box_i\varphi$ for each $\Box_i\varphi^u \in \Box_i\Gamma$. Moreover, suppose that $\Box_i\psi^a \in \Box_i\Delta$. Then $\mathbb{S}, s \not\Vdash \Box_i\psi$. Thus there is a state r in $\mathbb{S}$ such that sR_ir and $\mathbb{S}, r \not\Vdash \psi$. By symmetry and transitivity, we get tR_is, whence $\mathbb{S}, t \not\Vdash \Box_i\psi$, as required.

▷ $\boxplus_R$: As in the previous case, let $n := \mu(\sigma)$ and let $\mathbb{S}, s$ be such that $\mathbb{S}, s \not\Vdash \sigma(n)$. Then $\mathbb{S}, s \not\Vdash \boxplus\varphi$, where $\boxplus\varphi^f$ is the principal formula.

If $n = 0$, then $\mathbb{S}, s \not\Vdash \varphi$ and thus the leftmost premiss is invalid and forms a witness to the statement, as it has no formula in focus. If $n > 0$, then $\mathbb{S}, s \not\Vdash \Box_i\Box^{n-1}\varphi$, for some $i \in \mathcal{A}$. This means that there is an invalid premiss σ_k with $\mu(\sigma_k) = n - 1$, as required. □

We are now ready to prove the soundness theorem.

Theorem 4.3 *If there is an* sCKL$_f$*-proof with root* σ*, then* σ *is valid.*

Proof. Suppose, towards a contradiction, that an invalid sequent σ is the root of some sCKL$_f$-proof π. Repeatedly applying Lemma 4.2, we obtain an upward path

$$\rho = \sigma_0, \sigma_1, \ldots, \sigma_n$$

through π such that $\sigma_0 = \sigma$ and σ_n labels a leaf of π. Since σ_n is invalid by construction, this leaf cannot be axiomatic. Therefore, there there must be some $k < n$ such that $\langle\sigma_k, \sigma_n\rangle$ is a successful repeat. Observe that this implies that $\sigma_k = \sigma_n$. However, by the fact that we constructed this path using Lemma 4.2, it holds that $\mu(\sigma_k) < \mu(\sigma_n)$, a contradiction. □

5 Completeness

Let Σ be a finite and closed set of formulas and let $\Gamma \Rightarrow \Delta$ be a sequent. We say that $\Gamma \Rightarrow \Delta$ is a Σ-*sequent* if $\Gamma^- \cup \Delta^- \subseteq \Sigma$. A sequent $\Gamma \Rightarrow \Delta$ will be called Σ-*provable* whenever there is an sCKL$_f$-proof of $\Gamma \Rightarrow \Delta$ that contains only Σ-sequents. Finally, we say of a Σ-sequent $\Gamma \Rightarrow \Delta$ that it is *saturated* whenever it is Σ-unprovable and $\Gamma^- \cup \Delta^- = \Sigma$.

By the presence of the cut rule, the following lemma is immediate.

Lemma 5.1 *For any* Σ*-unprovable* Σ*-sequent* $\Gamma \Rightarrow \Delta$*, there is a saturated* Σ*-sequent* $\overline{\Gamma} \Rightarrow \overline{\Delta}$ *such that* $\Gamma \subseteq \overline{\Gamma}$ *and* $\Delta \subseteq \overline{\Delta}$.

Similarly, the following follows by the presence of the rule T.

Lemma 5.2 *If* $\Gamma \Rightarrow \Delta$ *is saturated and* $\Box_i\varphi \in \Gamma^-$*, then* $\varphi \in \Gamma^-$.

The following is a standard definition for the canonical model of S5-CKL (although the canonical model is usually defined with respect to a Hilbert-style proof system).

Definition 5.3 Let Σ be a non-empty, finite and closed set of formulas. The *canonical model* $\mathbb{S}^\Sigma$ of Σ is given by:

$$S^\Sigma := \{\Gamma^- \mid \Gamma \Rightarrow \Delta \text{ is a saturated } \Sigma\text{-sequent}\}$$
$$AR_i^\Sigma B :\Leftrightarrow \Box_i\Box_i^{-1}A = \Box_i\Box_i^{-1}B$$
$$V^\Sigma(A) := \{p \in \mathsf{P} \mid p \in A\}$$

It is immediate to verify that for every non-empty, finite and closed set Σ, its canonical model $\mathbb{S}^\Sigma$ is an epistemic Kripke model.

We are now ready to prove the Truth Lemma.

Lemma 5.4 (Truth Lemma) *For every* $\varphi \in \Sigma$*:* $\mathbb{S}^\Sigma, A \Vdash \varphi$ *if and only if* $\varphi \in A$.

Proof. We prove this by induction on φ. We only treat the cases $\varphi = \Box_i\psi$ and $\varphi = \boxplus\psi$. The other cases are standard.

- ▷ $\varphi = \Box_i\psi$.

  In case $\varphi \in A$, we must show that for every B with $AR_i^\Sigma B$, it holds that $\mathbb{S}^\Sigma, B \Vdash \psi$. By the induction hypothesis it suffices to show that $\psi \in B$. First note that, by the definition of R_i^Σ, we have $\Box_i\psi \in B$. Lemma 5.2 then gives $\psi \in B$.

  Now suppose that $\varphi \notin A$. By definition there is a saturated Σ-sequent $\Gamma \Rightarrow \Delta$ such that $\Gamma^- = A$. Note that by saturation, we have $\varphi^a \in \Delta$ for some $a \in \{u, f\}$. Let $\Delta_0 = \Delta \setminus \{\varphi^a\}$. We claim that the Σ-sequent

  $$\Box_i\Box_i^{-1}\Gamma \Rightarrow \psi^a, \Box_i\Box_i^{-1}\Delta_0 \tag{2}$$

  is Σ-unprovable. Indeed, consider the inference

  $$\Box_{\mathsf{S5}}\ \frac{\Box_i\Box_i^{-1}\Gamma \Rightarrow \psi^a, \Box_i\Box_i^{-1}\Delta_0}{\Box_i\Box_i^{-1}\Gamma \Rightarrow \Box_i\psi^a, \Box_i\Box_i^{-1}\Delta_0}$$

  If the premiss were Σ-provable, then so would be the conclusion. However, this cannot be the case because $\Gamma \Rightarrow \Delta$ can be obtained from the conclusion by a series of weakenings.

  By Lemma 5.1, there is a saturated Σ-sequent $\overline{\Gamma} \Rightarrow \overline{\Delta}$ extending the sequent depicted in (2). Define the set $B := (\overline{\Gamma})^-$. Since $\psi \notin B$, the induction hypothesis gives $\mathbb{S}^\Sigma, B \nVdash \psi$. Finally, we claim that $AR_i^\Sigma B$ and thus $\mathbb{S}^\Sigma, A \nVdash \Box_i\psi$. Indeed, we clearly have $\Box_i\Box_i^{-1}A \subseteq \Box_i\Box_i^{-1}B$. For the other direction, first note that $\Box_i\psi \notin B$, for otherwise we would have $\psi \in B$. It follows for any $\Box_i\chi \in \Sigma$ that $\Box_i\chi \in B$ entails $\Box_i\chi \notin \Delta^-$, whence $\Box_i\chi \in A$, as required.

- ▷ $\varphi = \boxplus\psi$. As before, we will first consider the case where $\boxplus\psi \in A$. Let $\Gamma \Rightarrow \Delta$ be a saturated Σ-sequent such that $A = \Gamma^-$. Then it holds by saturation that $\psi^u \in \Gamma$ and $\Box_i \boxtimes \psi^u \in \Gamma$ for all $1 \leq i \leq n$. By the induction hypothesis, we have $\mathbb{S}^\Sigma, A \Vdash \psi$. Moreover, for every $B \in S^\Sigma$ such that $AR_i^\Sigma B$, it holds by the definition of R_i^Σ and Lemma 5.2 that $\boxtimes\psi \in B$. By repeating this argument we obtain that $\mathbb{S}^\Sigma, A \Vdash \boxtimes\psi$.

In case $\varphi \notin A$, we again consider a saturated Σ-sequent $\Gamma \Rightarrow \Delta$ such that $A = \Gamma^-$. By the presence of the rules U and F, we may assume without loss of generality that $\varphi^f \in \Delta$. Now suppose, towards a contradiction, that $\mathbb{S}^\Sigma, A \Vdash \varphi$.

For every $A(R^\Sigma)^* B$ it holds that $\mathbb{S}^\Sigma, B \Vdash \psi$. In particular, it follows that $\mathbb{S}^\Sigma, A \Vdash \psi$, whence, by the induction hypothesis, we have $\psi^u \in \Gamma$.

As in the previous case, we take $\Delta_0 = \Delta \setminus \{\varphi^f\}$. Consider the following derivation:

$$\boxtimes R \frac{\begin{array}{c}\pi \\ \Gamma \Rightarrow \psi^u, \Delta_0\end{array} \quad \begin{array}{c}\pi_1 \\ \Gamma \Rightarrow \Box_1\boxtimes\psi^f, \Delta_0\end{array} \quad \cdots \quad \begin{array}{c}\pi_n \\ \Gamma \Rightarrow \Box_n\boxtimes\psi^f, \Delta_0\end{array}}{\Gamma \Rightarrow \boxtimes\psi^f, \Delta_0}$$

where π is a series of weakenings followed by an application of id to derive $\psi^u \Rightarrow \psi^u$, and each π_i is constructed as follows:

$$\begin{array}{l}
\boxtimes R \dfrac{\begin{array}{c}\pi' \\ \sigma'\end{array} \quad \begin{array}{c}\pi'_1 \\ \sigma'_1\end{array} \quad \begin{array}{c}\cdots \\ \cdots\end{array} \quad \begin{array}{c}\pi'_n \\ \sigma'_n\end{array}}{\Box_i\Box_i^{-1}\Gamma \Rightarrow \boxtimes\psi^f, \Box_i\Box_i^{-1}\Delta_0} \\
\Box\mathsf{S5} \dfrac{}{\Box_i\Box_i^{-1}\Gamma \Rightarrow \Box_i\boxtimes\psi^f, \Box_i\Box_i^{-1}\Delta_0} \\
\mathsf{w}_L \dfrac{}{\vdots} \\
\mathsf{w}_L \dfrac{}{\Gamma \Rightarrow \Box_i\boxtimes\psi^f, \Box_i\Box_i^{-1}\Delta_0} \\
\mathsf{w}_R \dfrac{}{\vdots} \\
\mathsf{w}_R \dfrac{}{\Gamma \Rightarrow \Box_i\boxtimes\psi^f, \Delta_0}
\end{array}$$

In the above derivation the sequent σ' is given by

$$\sigma' = \Box_i\Box_i^{-1}\Gamma \Rightarrow \psi^u, \Box_i\Box_i^{-1}\Delta_0$$

and the derivation π' is obtained from the Σ-provability of the sequent $\Box_i\Box_i^{-1}\Gamma \Rightarrow \psi^u, \Box_i\Box_i^{-1}\Delta_0$. Indeed, if it were not Σ-provable, then by applying Lemma 5.1 and the induction hypothesis, we would obtain a state B such that $\mathbb{S}^\Sigma, B \nVdash \psi$. By the same argument as in the previous case, we would moreover have $AR_i^\Sigma B$, contradicting the assumption that $A \Vdash \boxtimes\psi$

Furthermore, each sequent σ'_k in the derivation π_i is given by

$$\sigma'_k = \Box_i\Box_i^{-1}\Gamma \Rightarrow \Box_k\boxtimes\psi^f, \Box_i\Box_i^{-1}\Delta_0$$

and each derivation π'_k is constructed by repeatedly applying cut to add formulas from Σ until every leaf is either saturated or Σ-provable. To the leaves that are Σ-provable we append their respective proofs, and to the saturated sequents we apply the same process as we have now applied to $\Gamma \Rightarrow \Box_i \boxtimes \psi^f, \Delta_0$. Crucially, for every such saturated leaf $\Gamma' \Rightarrow \Box_k \boxtimes \psi^f, \Delta'$ we have that $AR_i^{\Sigma} C$, where $C := (\Gamma')^-$. Our assumption $\mathbb{S}^{\Sigma}, A \Vdash \boxtimes\psi$ therefore entails that $\mathbb{S}^{\Sigma}, C \Vdash \boxtimes\psi$, allowing us to repeat the same argument.

By the pidgeonhole principle, at some point one of these saturated leaves must be identical to a saturated leaf reached earlier in the construction. Note that in this case the upward path from the earlier saturated leaf to the later one is successful. We then terminate the construction of this branch. Since every branch is terminated at some point, we end up with a Σ-proof of $\Gamma \Rightarrow \Delta$, a contradiction. □

The following theorem can now be proven by a standard argument.

Theorem 5.5 *If a Σ-sequent σ is valid, then it is Σ-provable in* $\mathsf{sCKL_f}$.

6 A proof search game

Deciding whether a CKL-formula is valid is known to be EXPTIME-complete [13]. In this section we are going to show that our proof system admits optimal proof search, *i.e.* there exists an algorithm that decides whether a formula is provable in $\mathsf{sCKL_f}$ in time exponential in the size of the formula. To that end we are going to take a game-theoretic perspective on our proof system. For each sequent σ we will define a parity game $\mathcal{G}_\sigma$ which is played by two players called *Prover* and *Refuter*. Roughly, the goal of Prover is to show that σ has a proof in $\mathsf{sCKL_f}$ while Refuter tries to show the opposite. We will show that a sequent σ has a $\mathsf{sCKL_f}$-proof if and only if Prover has a memoryless winning strategy in the game $\mathcal{G}_\sigma$. In order to obtain our complexity-theoretic result, we will then refer to an algorithm given in [8] which decides for each game $\mathcal{G}_\sigma$ whether Prover has a memoryless winning strategy in time polynomial in the size of the input i.e. the size of $\mathcal{G}_\sigma$. Finally, as the size of $\mathcal{G}_\sigma$ is exponential in the size of σ, we obtain an optimal proof search result.

Throughout this section Σ always denotes a finite and closed set of formulas. For the basic definitions around parity games we refer the reader to [11]. The game $\mathcal{G}_\sigma$ has two types of positions: CKL-sequents and rule instances.

Definition 6.1 A *rule instance* (or *instance* for short) is a triple $(\sigma, \mathsf{r}, \langle \sigma_1, \ldots, \sigma_n \rangle)$ such that

$$\mathsf{r}\,\frac{\sigma_1 \quad \cdots \quad \sigma_n}{\sigma}$$

is a valid rule application in $\mathsf{sCKL_f}$.

If i is a rule instance, we write $\mathsf{conc}(i)$ for the first element of i, i.e. for the conclusion of i. A Σ-*instance* is a rule instance involving only Σ-sequents. For a Σ-sequent σ the game $\mathcal{G}_\sigma$ is defined as follows:

Definition 6.2 Let σ be a Σ-sequent. The *proof search game* $\mathcal{G}_\sigma$ associated to σ takes positions in $S \cup I$, where S is the set of Σ-sequents and I is the set of Σ-instances. The ownership function and admissible moves are as described in the following table:

Position	Owner	Admissible moves
σ	Prover	$\{i \in I \mid \mathsf{conc}(i) = \sigma\}$
$(\sigma, \mathsf{r}, \langle \sigma_1, \ldots, \sigma_n \rangle)$	Refuter	$\{\sigma_i \mid 1 \leq i \leq n\}$

The positions are given the following priorities:

(i) Every position of the form $\Gamma \Rightarrow \Delta^u$ has priority 3;

(ii) Every position of the form $(\sigma, \boxplus_R, \langle \sigma_0, \ldots, \sigma_n \rangle)$ where the principal formula is in focus has priority 2;

(iii) Every other position has priority 1.

A position is called a *dead end* if its owner has no admissible moves in this position available. A *match* in $\mathcal{G}_\sigma$ is a sequence of positions starting in σ, such that any two consecutive positions are related by an admissible move. A match is either finite and ends in a dead end or infinite. The *winning conditions* for a match are as follows: Prover wins every finite match in which the dead end belongs to Refuter and she wins every infinite match in which the highest priority encountered infinitely often is even. Refuter wins every finite match in which the dead end belongs to Prover and he wins every infinite match in which the highest priority visited infinitely often is odd.

Observe that the only positions that are dead ends are rule instances of id. Therefore Prover wins every finite match. In the following we are going to use standard terminology in the theory of parity games such as *strategy*, *memoryless strategy*, *winning strategy* and *strategy tree*. For definitions of these concepts we again refer the reader to [11].

Proposition 6.3 *Let σ be a Σ-sequent. If σ is Σ-provable, then Prover has a memoryless winning strategy in $\mathcal{G}_\sigma$.*

The basic proof idea is to read off a winning strategy for Prover from a Σ-proof π for σ. This can be done by identifying each play in $\mathcal{G}_\sigma$ with a finite or infinite path through π. As Prover can always choose which rule to apply to a given sequent and every play starts in σ (which labels the root of π), we can ensure that every possible play in $\mathcal{G}_\sigma$ - when Prover uses this strategy - corresponds to some path through π. The fact that π is a proof then guarantees that Prover wins every play. Finally, since in each parity game exactly one player has a memoryless winning strategy [11], the existence of a winning strategy for Prover implies the existence of a memoryless winning strategy for her. In order to present a detailed proof, we require some preliminary work.

Recall that in a proof π there exists for every non-axiomatic leaf l a node $c(l)$ such that $\langle c(l), l \rangle$ is a successful repetition. As there might exist several candidates for the node $c(l)$, we fix for each non-axiomatic leaf l one candidate $c(l)$ which we call the *companion* of l.

Definition 6.4 Let π be a $\mathsf{sCKL_f}$-proof. A *path ρ through π* is a (possibly infinite) sequence $\rho = \rho(0), \rho(1), \ldots$ of nodes in π which satisfies the following properties:

(i) $\rho(0)$ is the root of π.

(ii) If $\rho(i)$ is an axiomatic leaf, then ρ is finite and ends at $\rho(i)$.

(iii) If $\rho(i)$ is an non-axiomatic leaf l with companion $c(l)$, then $\rho(i+1)$ is a child node of $c(l)$.

(iv) Otherwise, $\rho(i+1)$ is a child node of $\rho(i)$.

If condition (iii) applies, then we say that ρ *passes through the non-axiomatic leaf* l. Observe that paths never pass through axiomatic leaves. Notice that there is a difference between paths as defined here and *upward paths* as defined is section 3. Namely, an upward path is always finite, has to end at latest at a leaf of π (axiomatic or non-axiomatic) and might start at any node. In particular, an upward path cannot pass through a non-axiomatic leaf and continue at its companion. A path as defined here has to start at the root and in case it reaches a non-axiomatic leaf, it has to continue at its companion. Moreover, a finite path can only end at an axiomatic leaf. Notice that a finite path might still pass through some non-axiomatic leaves first before eventually reaching an axiomatic leaf. Furthermore, notice that paths are defined with respect to $\mathsf{sCKL_f}$-proofs and not with respect to arbitrary derivations like upward paths. The reason for this is simply to avoid the case that some non-axiomatic leaf is not part of a successful repetition.

The following lemma states a first basic result about infinite paths through $\mathsf{sCKL_f}$-proofs. The proof of the lemma follows immediately from the fact that $\mathsf{sCKL_f}$-proofs are finite.

Lemma 6.5 *Suppose π is an $\mathsf{sCKL_f}$-proof and ρ is an infinite path through π. Then there exists a non-axiomatic leaf l through which ρ passes infinitely often.*

Given sets X, Y and a function $f : X \longrightarrow Y$ we denote by $dom(f)$ the *domain* of f and by $ran(f)$ the *range* of f.

Definition 6.6 A *finite tree with back edges* is a pair (T, f) consisting of a finite tree T and a (partial) function $f : T \longrightarrow T$, such that every $u \in dom(f)$ is a leaf of T and the node $f(u)$ is a proper ancestor of u.

Observe that cyclic proofs can be considered as finite trees with back edges, that satisfy the property that if $u \in dom(f)$, then u is a non-axiomatic leaf and $f(u)$ is the companion of u. The definition of path through a cyclic proof is generalised to path through a finite tree with back edges in the obvious way.

Definition 6.7 Let (T, f) be a finite tree with back edges. Define the *one-step dependency order* $\preceq_1$ on $ran(f)$ as follows:

$$u \preceq_1 v :\Leftrightarrow u \text{ occurs on the upward path from } v \text{ to } v' \text{ for some } v' \in f^{-1}(v)$$

Define the *dependency order* $\preceq$ on $ran(f)$ as the transitive closure of $\preceq_1$.

Observe that the dependency order $\preceq$ is reflexive and transitive. Furthermore it is also anti-symmetric and so $\preceq$ is a partial order on $ran(f)$. Observe that $u \preceq v$ implies that there exists an upward path from v to u. Let (T, f)

be a finite tree with back edges and let ρ be an infinite path through (T, f). Denote by $Inf(\rho)$ the set of nodes of T that occur infinitely often in ρ.

Lemma 6.8 *Let (T, f) be a finite tree with back edges and let ρ be an infinite path through (T, f). Then the set $Inf(\rho) \cap ran(f)$ has a $\preceq$-greatest element.*

Proof. Observe that the set $Inf(\rho) \cap ran(f)$ is finite since T is a finite tree. Furthermore, observe that $Inf(\rho) \cap ran(f)$ is non-empty, as ρ must pass through some leaf $l \in dom(f)$ infinitely often (see Lemma 6.5). It therefore suffices to prove that all $\preceq$-maximal elements in $Inf(\rho) \cap ran(f)$ are identical. To that end let $u \in Inf(\rho) \cap ran(f)$ be a $\preceq$-maximal element. We claim that all nodes in $Inf(\rho)$ belong to T_u, where T_u denotes the subtree of T which is rooted at u. Notice that it suffices to prove the claim for each $f(l)$ for which $l \in Inf(\rho) \cap dom(f) \cap T_u$. Therefore let l be an arbitrary such leaf. Then there exists an upward path ρ_l starting at the root of T and ending in l, such that both $f(l)$ and u occur on ρ_l. Moreover, since $l \in Inf(\rho)$ we have that $f(l) \in Inf(\rho) \cap ran(f)$. As u is a $\preceq$-maximal element of $Inf(\rho) \cap ran(f)$ it follows that $u \not\prec f(l)$. Therefore $f(l)$ belongs to T_u. Since u was chosen arbitrarily, it follows that $Inf(\rho) \cap ran(f)$ has a $\preceq$-greatest element. □

Corollary 6.9 *Let π be a $\mathsf{sCKL_f}$-proof and ρ be an infinite path through π. Let $l_1, \ldots, l_k$ be the non-axiomatic leaves through which ρ passes infinitely often. There exists $1 \leq i \leq k$ such that there exists an upward path from $c(l_i)$ to $c(l_j)$ for each $1 \leq j \leq k$.*

Recall that a match m in a proof search game is a finite or infinite sequence of positions $m(0), m(1), m(2), \ldots$. Observe that each even position $m(2i)$ is owned by Prover and each odd position $m(2i+1)$ by Refuter. For an initial segment $m(0), \ldots m(i)$ of m we say that its *length* is i.

Definition 6.10 Let σ be a Σ-sequent and let π be a Σ-proof of σ. Let ρ be an infinite path through π and let m be an infinite match in $\mathcal{G}_\sigma$.

- An initial segment $m(0), \ldots, m(i)$ of m *corresponds* to an initial segment $\rho(0), \ldots, \rho(j)$ of ρ if $i = 2j$ and for each $0 \leq k \leq i$ with $k = 2l$ it holds that $m(k)$ is the sequent that labels $\rho(l)$.
- The match m and the path ρ are called *corresponding* if every initial segment of m with even length corresponds to an initial segment of ρ.

Proof. [of Proposition 6.3] Suppose that σ is Σ-provable. So there exists a proof π of σ in which every occurring sequent is a Σ-sequent. We denote the root of π by r_π. We simultaneously define a strategy $\mathcal{S}$ for Prover in the game $\mathcal{G}_\sigma$ and show how to map each initial segment of a match of even length in which Prover uses $\mathcal{S}$ onto an initial segment of a path through π. The strategy $\mathcal{S}$ is a function which maps initial segments of matches $\langle m(0), \ldots, m(2i) \rangle$ of even length onto rule instances. Therefore strategy $\mathcal{S}$ uses a *memory.*

For the base case observe that each match in $\mathcal{G}_\sigma$ begins in σ. Therefore $\langle m(0) \rangle$ for $m(0) = \sigma$ is an initial segment of every match in $\mathcal{G}_\sigma$. Similarly, every path through π starts in r_π which is labelled by σ. Thus $\langle\rho(0)\rangle$ for $\rho(0) = r_\pi$ is an initial segment of every path through π. Observe that $\langle m(0)\rangle$ corresponds to $\langle\rho(0)\rangle$.

For the inductive step suppose that we have already mapped the initial segment

$$m_i = \langle m(0), \ldots, m(2i)\rangle$$

of a match onto the initial segment

$$\rho_i = \langle\rho(0), \ldots, \rho(i)\rangle$$

of a path, such that m_i corresponds to ρ_i, where $i \geq 0$. Let $j \in I$ be the Σ-instance

$$j = (m(2i), \mathsf{r}, \langle\sigma'_1, \ldots, \sigma'_k\rangle),$$

which generated $\rho(i)$ in π when read top down. Then define

$$\mathcal{S}(m_i) = j.$$

Now suppose that Refuter extends the match by choosing premiss σ'_l. Then let $m(2i+1) = j$ and let $m(2i+2) = \sigma'_l$ and extend the initial segment m_i to

$$m_{i+1} = \langle m(0), \ldots, m(2i), m(2i+1), m(2i+2)\rangle$$

Furthermore, let $\rho(i+1)$ be the child of $\rho(i)$ which is labelled by σ'_l and extend ρ_i to

$$\rho(i+1) = \langle\rho(0), \ldots, \rho(i), \rho(i+1)\rangle$$

Observe that $m(i+1)$ corresponds to $\rho(i+1)$.

Finally, in order to turn $\mathcal{S}$ into a total function which maps *every* initial segment of a match with even length onto a rule instance (and not just those that correspond to initial segments of paths), we add the following clause. Fix a Σ-instance $j' \in I$. For any initial segment $m'_i = \langle m(0)', \ldots, m(2i)'\rangle$ of a match which is not covered in the above construction we define

$$\mathcal{S}(m'_i) = j'$$

Observe that $\mathcal{S}$ is a well-defined strategy for Prover, which has the property that if m is a match of $\mathcal{G}_\sigma$ in which Prover uses strategy $\mathcal{S}$, then there exists by construction a path ρ through π such that every initial segment of m of even length corresponds to some initial segment of ρ. Therefore m corresponds to ρ.

We show that $\mathcal{S}$ is a winning strategy for Prover. To that end let m be a match in $\mathcal{G}_\sigma$ in which Prover uses strategy $\mathcal{S}$ and let ρ be the path through π which corresponds to m. In case m is finite Prover wins by default and we have nothing to show. So suppose m is infinite. Then ρ is also infinite. By Lemma 6.5 and Corollary 6.9 there exists a non-axiomatic leaf l_0 with companion $c(l_0)$ such that the following holds:

(i) ρ passes through l_0 infinitely often.

(ii) If $l_0, l_1, ..., l_k$ are the non-axiomatic leaves through which ρ passes infinitely often, then there is an upward path from $c(l_0)$ to $c(l_i)$ for each $1 \leq i \leq k$.

Consider the subtree π_{l_0} of π rooted at $c(l_0)$. Observe that for each upward path from $c(l_0)$ to one of the non-axiomatic leaves l_i there is always a formula in focus (for a proof of this claim, see for example Proposition 2 in [20]). Next, observe that ρ contains a final segment in which each of $l_0, l_1, ..., l_k$ occur infinitely often and no other non-axiomatic leaf occurs. Therefore on this final segment ρ only passes through nodes of π_{l_0} which occur on the upward paths between $c(l_0)$ and l_i and so there is a formula in focus in each step on that final segment. Hence, the match m passes, after finitely many moves, only through positions with priority 1 or 2. As π is a proof, there exists a rule application of $\boxplus_R$ between $c(l_0)$ and l_0 where the principal formula is in focus. This means that, since ρ passes infinitely often through l_0, priority 2 is encountered infinitely often. Hence, the highest priority encountered infinitely often is even and Prover wins the match. We conclude that $\mathcal{S}$ is a winning strategy for Prover. Finally, since in a given parity game exactly one of the two players has a memoryless winning strategy [11], the existence of a winning strategy for Prover implies the existence of a *memoryless* winning strategy for her. □

Let us now consider the converse direction of Proposition 6.3.

Proposition 6.11 *Let σ be a Σ-sequent. If Prover has a memoryless winning strategy in $\mathcal{G}_\sigma$, then σ is Σ-provable.*

Proof. Suppose that Prover has a memoryless winning strategy in $\mathcal{G}_\sigma$. Let T be the corresponding strategy tree. Let π be the finite subtree of T obtained by pruning every infinite branch of T at the first repetition. Observe that the root of π is labelled by σ and π is generated by rules of $\mathsf{sCKL_f}$. Therefore π is a derivation. In order to show that π is indeed a $\mathsf{sCKL_f}$-proof, let l be an arbitrary leaf of π. If l is also a leaf of T, then l is axiomatic as T is the strategy tree of a winning strategy for Prover. Otherwise, l was generated by pruning an infinite branch of T. In that case there exists a node $c(l)$, such that $\langle c(l), l\rangle$ is a repetition. Therefore it remains to show $\langle c(l), l\rangle$ is successful. Suppose towards a contradiction that $\langle c(l), l\rangle$ is not successful. Then on the upward path ρ from $c(l)$ to l either some sequent does not have a formula in focus or ρ does not pass through an application of $\boxplus_R$ where the principal formula is in focus. This implies that there is a position occurring in ρ which has priority 3, or every position in ρ has priority 1. Since T is the strategy tree of a *memoryless* strategy, there exists an infinite branch in T that has a final segment which is an infinite concatenation $\rho \cdot \rho \cdot \rho \cdots$. On this path the highest priority occurring infinitely often is odd, contradicting the assumption that T is the strategy tree of a winning strategy for Prover. Therefore each repetition is successful and so we conclude that π is a $\mathsf{sCKL_f}$-proof of σ. □

Observe that the above constructed proof is *uniform* in the following sense:

Definition 6.12 A Σ-proof π is *uniform* if there exists a function $f : S \longrightarrow I$ such that whenever a sequent $\sigma \in S$ occurs in π, it occurs as the conclusion of the rule instance $f(\sigma)$.

Observe that in a uniform proof the first repetition in each branch is successful. We conclude:

Theorem 6.13 *The following are equivalent for any sequent σ:*

(i) *σ is $\mathsf{sCKL_f}$-provable*

(ii) *Prover has a memoryless winning strategy in $\mathcal{G}_\sigma$.*

(iii) *σ has a uniform $\mathsf{sCKL_f}$-proof.*

Let us now turn towards complexity. The *size* $c(\varphi)$ of a formula φ is the number of subformulas of φ. The *size* $c(\Gamma \Rightarrow \Delta)$ of a sequent $\Gamma \Rightarrow \Delta$ is defined as follows:

$$c(\Gamma \Rightarrow \Delta) := \sum_{\varphi \in \Gamma^- \uplus \Delta^-} c(\varphi)$$

where $\uplus$ denotes the disjoint union. Observe that if Σ is the closure of $\Gamma^- \cup \Delta^-$, then $|\Sigma|$ is linear in $c(\Gamma \Rightarrow \Delta)$.

Lemma 6.14 *Given a Σ-sequent σ, the number of positions in $\mathcal{G}_\sigma$ is polynomially bounded by $|\mathcal{P}(\Sigma)|$.*

Proof. Observe that each unannotated Σ-sequent is an ordered pair of subsets of Σ. Therefore there are $|\mathcal{P}(\Sigma)|^2$ many unannotated Σ-sequents. By taking the focus annotations into account we obtain at most $|\mathcal{P}(\Sigma)|^3$ many Σ-sequents. Hence $|S| \leq |\mathcal{P}(\Sigma)|^3$. Next, observe that for each Σ-sequent σ' and each rule r, there are at most $|\Sigma|$ many ways of applying r to σ'. We therefore obtain the following upper bound:

$$|I| \leq 14 \cdot |\mathcal{P}(\Sigma)|^3 \cdot |\Sigma| \leq 14 \cdot |\mathcal{P}(\Sigma)|^4$$

Together, the set of positions of $\mathcal{G}_\sigma$ is bounded by $14 \cdot |\mathcal{P}(\Sigma)|^4 + |\mathcal{P}(\Sigma)|^3$ which is polynomial in $|\mathcal{P}(\Sigma)|$. □

In order to get a polynomial bound for deciding the winner of a given proof search game we can now make use of one of the many existing algorithms for solving parity games. For instance the following result by Calude et al. [8].

Theorem 6.15 ([8, Theorem 2.9]) *There is an algorithm which finds the winner of a parity game in time $\mathcal{O}(n^{log(m)+6})$ for a parity game with n positions and priorities in $\{1, 2, ..., m\}$. Furthermore, the algorithm can compute a memoryless winning strategy for the winner in time $\mathcal{O}(n^{log(m)+7} \cdot log(n))$.*

Let σ be a sequent and let Σ be its closure. By Lemma 6.14 the number n of positions in $\mathcal{G}_\sigma$ is polynomial in the size of $|\mathcal{P}(\Sigma)|$. Since the number of different priorities in our games is constant, Theorem 6.15 implies that there is an algorithm deciding the winner of $\mathcal{G}_\sigma$ in time polynomial in $|\mathcal{P}(\Sigma)|$ and so exponential in $c(\sigma)$. By the same argument the above mentioned algorithm also computes a memoryless winning strategy in time exponential in $c(\sigma)$.

Corollary 6.16 *For any* CKL*-sequent* σ*, there is an algorithm deciding whether* σ *is provable that runs in time exponential in* $c(\sigma)$.

7 Related research and future work

In this section we discuss the relation of the present paper with earlier research. In passing we also propose some directions for further research.

7.1 Explicit induction

In [5], Alberucci & Jäger present another proof system for S5-CKL. In contrast to our system $\mathsf{sCKL_f}$, their system $\mathsf{S5}_n(\mathsf{C})$ does not allow cyclic proofs. Rather, it uses an *explicit induction* rule, which can be thought of as the Gentzen-style translation of the well-known *induction axiom*: $⊞(p \to \Box p) \to (p \to ⊞p)$.

Like us, Alberucci & Jäger obtain a partial cut-elimination result. However, given some endsequent $\Gamma \Rightarrow \Delta$, they do not manage to restrict cuts to $Cl(\Gamma \cup \Delta)$, but only to a larger set that they call the *disjunctive-conjunctive* closure of $Cl(\Gamma \cup \Delta)$. One could try to sharpen their cut-elimination result by translating our $\mathsf{sCKL_f}$-proofs into $\mathsf{S5}_n(\mathsf{C})$-proofs. Such a translation from cyclic proofs into inductive proofs occurs more often in the literature (see *e.g.* [16] and [2]).

A first problem with this approach is caused by the fact that our language is not sufficiently expressive to capture the *strengthened induction* rule from [2]. Indeed, an adaptation of the rule $\mathsf{ind_s}$ to our system would have to take the following form:

$$\mathsf{ind_s}\ \frac{\Gamma \Rightarrow \varphi, \Delta \qquad \{\Gamma \Rightarrow \Box_i \nu x.\Box(\overline{\Gamma \Rightarrow \Delta} \vee x) \wedge \varphi\}_{i=1}^{n}}{\Gamma \Rightarrow ⊞\varphi, \Delta}$$

where $\overline{\Gamma \Rightarrow \Delta}$ is defined as

$$\bigwedge_{\varphi \in \Gamma} \varphi \wedge \bigwedge_{\psi \in \Delta} \neg \psi$$

But the μ-calculus formulas in the right premisses are not expressible in the language of CKL. This problem is circumvented by Brünnler & Lange in [7] by augmenting the language with *annotations*. However, if we were to extend our language analogously, it would be unclear how to translate sequents from this extended language into the ordinary sequents of the system $\mathsf{S5}_n(\mathsf{C})$.

One could resort to a reformulation of $\mathsf{S5}_n(\mathsf{C})$ in terms of the augmented language, but then a second problem arises. Namely, to make the translation one has to show that the strengthened induction rule is derivable in $\mathsf{S5}_n(\mathsf{C})$. But for this one needs to apply cut to formulas of the form $\overline{\Gamma \Rightarrow \Delta}$, where Γ and Δ only contain formulas in the ordinary language. Interestingly, this seems to give a partial cut-elimination result very similar to the one obtained by Alberucci & Jäger. This suggests that inductive proofs for S5-CKL might require strictly more complex cut formulas than cyclic proofs. We leave it as future work to investigate this conjecture.

Finally, we wish to point out that the sequent given in [5, p.10] as example of a sequent that does not admit a cut-free proof in $\mathsf{S5}_n(\mathsf{C})$, namely the sequent $\Box_1(p \wedge ⊞q) \wedge \Box_2(q \wedge ⊞p) \Rightarrow ⊞(p \vee q)$, does have a cut-free proof in our system.

7.2 Generalised sequent calculi

Another proposal for a proof system for S5-CKL is made by Hill & Poggiolesi in [14]. Their system HS5C does not manipulate ordinary sequents, but a generalised form of sequents called *indexed hypersequents.* Indexed hypersequents are akin to the more well-known formalisms of nested sequents (see *e.g.* [6]) or labelled sequents (see *e.g.* [17]) .

Like us, Hill & Poggiolesi start with a non-analytic system, and then use semantic methods to show that proofs can be restricted to a certain shape. While this restriction does not result in fully analytic proofs, Hill & Poggiolesi claim that it nevertheless makes their system suitable for proof search. No complexity bound for proof search is given.

Using a syntactic cut-elimination procedure, Hill & Poggiolesi show that the system HS5C is cut-free. In particular, its non-analyticity does not arise from unavoidable applications of the cut-rule. Rather, the system HS5C features a non-analytic explicit induction rule similar to that of the system $\mathsf{S5}_n(\mathsf{C})$ discussed in the previous subsection. It is this induction rule whose non-analyticity cannot be avoided.

We consider it a very interesting avenue for further research to see if the explicit induction rule in HS5C can be replaced by cyclic proofs. The main obstacle to this approach is the fact that even an analytic proof of a given endsequent may still contain infinitely many distinct labels. In terms of our approach, this would obstruct the appeal to the pidgeonhole principle in the proof of the Truth Lemma.

7.3 Focus games

In [15], Lange introduces a two-player game for checking satisfiability of Converse-PDL (CPDL, for short). This is an extension of Propositional Dynamic Logic (or, PDL) introduced by Fischer and Ladner in [10]. The logic CPDL introduces the possibility to reason about the backwards application of programs. The satisfiability game introduced in [15] is played by a player $\exists$ whose goal is to show that a given formula of CPDL is satisfiable and a player $\forall$ whose goal is to show that the formula is unsatisfiable. As in our case, the game uses a focus mechanism to capture traces. Furthermore, as our proof system, the satisfiability game is cyclic, *i.e.* plays are finite and end when either a leaf is encountered or a certain condition for cycles (similar to our condition for successful repetitions) is met. For details about the game we refer the reader to [15]. Lange claims that the satisfiability game is sound and complete, *i.e.* player $\exists$ has a winning strategy in the game for a sequent Φ if and only if Φ is satisfiable. Unfortunately, however, the authors of the present paper discovered a counterexample to the soundness of Lange's system. Namely, consider the sequent $\Phi = \{[a]p, \langle a\rangle[\bar{a}]\langle a\rangle\neg p\}$, where a is an atomic program, p an atomic proposition and $\bar{a}$ is the converse program of a. It is readily checked that Φ is unsatisfiable in any Kripke model. However, player $\exists$ has a winning strategy in the satisfiability game for Φ. This can be seen by observing that no matter in what order rules are applied to Φ, neither winning condition 1 nor 2 of $\forall$ can ever be met. For winning condition 2, this is immediate, as Φ does not contain a formula of the form $\langle \alpha^* \rangle \varphi$. For winning condition 1, it suffices to notice that no play starting in Φ can ever reach a sequent containing the formulas p and $\neg p$, as by eliminating a modal operator in the left formula, one has to introduce a modal operator in the right formula and vice versa. Therefore, no play can ever be won by $\forall$, implying that $\exists$ has a winning strategy.

Due to the close connection between CPDL and S5-CKL, we conjecture that the method of the present paper could be adapted to CPDL. If true, one could extract a focus game for validity for CPDL in the same way as described in the previous section. Such a game could then be dualised into a focus game for satisfiability, thus fixing the original system proposed in [15]. The realisation of this task is left for future research.

Acknowledgements

The authors would like to thank Johannes Marti and the anonymous reviewers for helpful comments on a final draft of this paper.

References

[1] Abate, P., R. Goré and F. Widmann, *Cut-free single-pass tableaux for the logic of common knowledge*, in: *In Workshop on Agents and Deduction at TABLEAUX*, 2007.

[2] Afshari, B. and G. E. Leigh, *Cut-free completeness for modal mu-calculus*, in: *2017 32nd Annual ACM/IEEE Symposium on Logic in Computer Science (LICS)*, IEEE, 2017, pp. 1–12.

[3] Ajspur, M., V. Goranko and D. Shkatov, *Tableau-based decision procedure for the multiagent epistemic logic with all coalitional operators for common and distributed knowledge*, Log. J. IGPL **21** (2013), pp. 407–437.

[4] Alberucci, L., "The Modal μ-Calculus and the Logic of Common Knowledge," Ph.D. thesis, Universität Bern, Institut für Informatik und angewandte Mathematik (2002).

[5] Alberucci, L. and G. Jäger, *About cut elimination for logics of common knowledge*, Annals of Pure and Applied Logic **133** (2005), pp. 73–99.

[6] Brünnler, K., "Nested sequents," habilitation, University of Bern (2010).

[7] Brünnler, K. and M. Lange, *Cut-free sequent systems for temporal logic*, The Journal of Logic and Algebraic Programming **76** (2008), pp. 216–225.

[8] Calude, C. S., S. Jain, B. Khoussainov, W. Li and F. Stephan, *Deciding parity games in quasipolynomial time*, in: *Proceedings of the 49th Annual ACM SIGACT Symposium on Theory of Computing*, STOC 2017 (2017), p. 252–263.

[9] Carreiro, F. and Y. Venema, *PDL inside the μ-calculus: a syntactic and an automata-theoretic characterization*, Advances in Modal Logic **10** (2014), pp. 74–93.

[10] Fischer, M. J. and R. E. Ladner, *Propositional dynamic logic of regular programs*, J. Comput. Syst. Sci. **18** (1979), pp. 194–211.

[11] Grädel, E., W. Thomas and T. Wilke, editors, "Automata, Logics, and Infinite Games: A Guide to Current Research [outcome of a Dagstuhl seminar, February 2001]," Lecture Notes in Computer Science **2500**, Springer, 2002.

[12] Halpern, J. Y. and Y. Moses, *Knowledge and common knowledge in a distributed environment*, J. ACM **37** (1990), p. 549–587.

[13] Halpern, J. Y. and Y. Moses, *A guide to completeness and complexity for modal logics of knowledge and belief*, Artificial intelligence **54** (1992), pp. 319–379.

[14] Hill, B. and F. Poggiolesi, *Common knowledge: a finitary calculus with a syntactic cut-elimination procedure*, Logique et Analyse (2015), pp. 279–306.

[15] Lange, M., *Satisfiability and completeness of converse-PDL replayed*, in: A. Günter, R. Kruse and B. Neumann, editors, *KI 2003: Advances in Artificial Intelligence, 26th Annual German Conference on AI, KI 2003, Hamburg, Germany, September 15-18, 2003, Proceedings*, Lecture Notes in Computer Science **2821** (2003), pp. 79–92.

[16] Lange, M. and C. Stirling, *Focus games for satisfiability and completeness of temporal logic*, in: *Proceedings 16th Annual IEEE Symposium on Logic in Computer Science*, IEEE, 2001, pp. 357–365.

[17] Negri, S., *Proof theory for modal logic*, Philosophy Compass **6** (2011), pp. 523–538.

[18] Niwiński, D. and I. Walukiewicz, *Games for the μ-calculus*, Theoretical Computer Science **163** (1996), pp. 99–116.

[19] Ohnishi, M. and K. Matsumoto, *Gentzen method in modal calculi*, Osaka Mathematical Journal **9** (1957), pp. 113–130.

[20] Rooduijn, J. M. W., *Cyclic hypersequent calculi for some modal logics with the master modality*, in: A. Das and S. Negri, editors, *Automated Reasoning with Analytic Tableaux and Related Methods* (2021), pp. 354–370.

[21] Stirling, C., *A tableau proof system with names for modal mu-calculus*, in: A. Voronkov and M. V. Korovina, editors, *HOWARD-60: A Festschrift on the Occasion of Howard Barringer's 60th Birthday*, EPiC Series in Computing **42**, EasyChair, 2014 pp. 306–318.

[22] van Ditmarsch, H., W. van der Hoek and B. Kooi, "Dynamic Epistemic Logic," Springer Publishing Company, Incorporated, 2007, 1st edition.

[23] Wehbe, R., "Annotated Systems for Common Knowledge," Ph.D. thesis, Universität Bern, Institut für Informatik und angewandte Mathematik (2010).

Medvedev's logic and products of converse well orders

Denis I. Saveliev [1]

Institute for Information Transmission Problems RAS

Ilya Shapirovsky [2]

New Mexico State University

Abstract

We show that modal (and hence, intuitionistic) Medvedev's logic is the intersection of the logics of finite direct powers of ordinals with the converse ordering taken without the top element, and that the latter logics have the finite model property. Then we provide other semantic characterizations of Medvedev's logic and related systems in terms of various natural substructures of products of ordinals.

Keywords: modal logic, Noetherian order, Medvedev's logic, finite model property, direct product of ordinals

1 Introduction

In this paper we study modal logics of Noetherian (in other terms, converse well-founded) partially ordered sets which are substructures of direct products of converse well-ordered sets, in particular, the products without an upper part.

Well-known examples of this kind are Grz.2 (the Grzegorczyk modal logic extended with the axiom of weak directedness) and modal Medvedev's logic. If follows from [6] that Grz.2 is the logic of all Boolean cubes $(2, \geq)^n$ with $n < \omega$. Such cubes without the top element are called *Medvedev's frames*; the intuitionistic logic of all Medvedev's frames is the well-known *Medvedev's logic of finite problems* [7,8]. *Modal Medvedev's logic* Mdv is the modal logic of these structures [9,12]. Also, Grz.2 and Mdv can be characterized as the logics of finite subsets of ω ordered by the converse inclusion:

$$\mathrm{Grz.2} = \mathrm{Log}(P_\omega(\omega), \supseteq) = \mathrm{Log}\{(2, \geq)^n\} : n < \omega\}, \quad (1)$$
$$\mathrm{Mdv} = \mathrm{Log}(P_\omega(\omega)\setminus\{\emptyset\}, \supseteq) = \mathrm{Log}\{(2, \geq)^n \setminus \{\mathrm{top}\} : n < \omega\}. \quad (2)$$

[1] d.i.saveliev@iitp.ru

[2] ilshapir@nmsu.edu

In spite of the similarity in the above semantic characterizations, logical properties of Grz.2 and Mdv are different. In particular, Grz.2 is given by a finite set of axioms, so in view of its completeness with respect to a class of finite frames, it is decidable. It is well known that both modal and intuitionistic Medvedev logics are not finitely axiomatizable [6,9]. Whether the modal or intuitionistic logics of the structures $\{(2, \geq)^n \setminus \{\text{top}\} : n < \omega\}$ are recursively axiomatizable is an old-standing open problem.

In Sections 3 and 4, we consider modal logics of similar structures where instead of 2 any ordinal is allowed. For a finite n, we define $\Gamma(n)$ and $\Delta(n)$ as the logics of the frames $(\omega, \geq)^n$ and $(\omega, \geq)^n \setminus \{\text{top}\}$, respectively. It is straightforward that these logics have the finite model property. In Theorem 3.3, we show that the frames $(\alpha, \geq)^n$ and $(\alpha, \geq)^n \setminus \{\text{top}\}$ have the same logics for any infinite α; consequently (Corollary 3.4),

$$\Gamma(n) = \text{Log}\{\textstyle\prod_{i<n}(\alpha_i, \geq) : \alpha_i \text{ is an ordinal}\},$$

$$\Delta(n) = \text{Log}\{\textstyle\prod_{i<n}(\alpha_i, \geq) \setminus \{\text{top}\} : \alpha_i \text{ is an ordinal}\}.$$

We observe that $\bigcap_{n<\omega} \Gamma(n) = \text{Grz.2}$ (this fact is straightforward from (1)) and that $\bigcap_{n<\omega} \Delta(n) = \text{Mdv}$ (Theorem 4.5).

As well as Mdv, each of the logics $\Gamma(n)$, $\Delta(n)$ is characterized by a recursive set of finite frames, so these logics are co-recursively enumerable. It is known [12] that Mdv.2 = Grz.2. We show that $\Delta(n).2 = \Gamma(n)$ for all $n < \omega$ (Theorem 4.7); consequently, if $\Delta(n)$ is recursively axiomatizable, then so is $\Gamma(n)$. However, decidability of $\Gamma(n)$ and $\Delta(n)$ is an open problem for $n > 1$.

In Section 5, we study the restrictions of the direct powers $(\omega, \geq)^n$ to several $A \subseteq \omega^n$ and the modal logics of these restrictions, which thus generalize the above logics $\Gamma(n)$ and $\Delta(n)$; we also consider Noetherian subframes of infinite powers of converse ordinals.

In Section 6, we discuss other semantic interpretations of the logics $\Gamma(n)$ and $\Delta(n)$.

2 Preliminaries

We assume the reader's familiarity with the basic notions in modal logic, which can be found, e.g., in [1] or [2]; we recall only some standard concepts.

Modal formulas are built from a countable set of propositional variables $p, q, \dots$ by using $\bot$ and $\to$ (chosen as the primitive Boolean connectives) and the unary modal operator $\Diamond$. Other connectives are standard abbreviations; in particular, $\Box$ abbreviates $\neg\Diamond\neg$. By a *(modal) logic* we mean a normal propositional uni-modal logic.

A Kripke frame $\mathfrak{F}' = (W', R')$ is a *weak subframe* of a Kripke frame $\mathfrak{F} = (W, R)$ iff $\emptyset \neq W' \subseteq W$ and $R' \subseteq R$. A Kripke model $\mathfrak{M}' = (\mathfrak{F}', \theta')$ is a *weak submodel* of a Kripke model $\mathfrak{M} = (\mathfrak{F}, \theta)$ iff $\mathfrak{F}'$ is a weak subframe of $\mathfrak{F}$ and $\theta'(p) = \theta(p) \cap W'$ for all propositional variables p. If $R' = R \cap (W' \times W')$, then $\mathfrak{F}'$ and $\mathfrak{M}'$ are called the *restrictions of* $\mathfrak{F}$ and *of* $\mathfrak{M}$ *to* W', respectively. If moreover, W' is *upward closed w.r.t.* R (in other words, is an *upper cone*, i.e., $x \in W'$ and xRy implies $y \in W'$), then these restrictions are called a *generated subframe* and a *generated submodel*, respectively. If $w \in W$ and W' is the least upward closed set that contains w, then the resulting substructures are said to be *point-generated.* A frame is *rooted* if it is generated by one of its points.

Log($\mathfrak{F}$) denotes the modal logic of a frame $\mathfrak{F}$ (or of a class of frames). If L is a modal logic, L.2 denotes its extension with the formula $\Diamond\Box p \to \Box\Diamond p$, and L.3 with the formula $\Diamond p \wedge \Diamond q \to \Diamond(p \wedge \Diamond q) \vee \Diamond(q \wedge \Diamond p)$. Recall that $(W, R) \models \Diamond\Box p \to \Box\Diamond p$ iff $R^{-1} \circ R \subseteq R \circ R^{-1}$; in particular, if (W, R) is a finite partial order with a least element, the latter means that there is a top element. For a non-strict partial order $\mathfrak{F}$, the formula $\mathfrak{F} \models \Diamond p \wedge \Diamond q \to \Diamond(p \wedge \Diamond q) \vee \Diamond(q \wedge \Diamond p)$ means that every point-generated subframe of $\mathfrak{F}$ is linear. Grz denotes the logic of the class of all Noetherian (in other words, converse well-founded) non-strict partial orders, and is called the *Grzegorczyk logic.* The following facts are well known (see, e.g., [2]): Grz is the logic of all (finite) Noetherian non-strict partial orders; Grz.2 is the logic of the class of all (finite) Noetherian non-strict partial orders with a greatest element; Grz.3 is the logic of all (finite) linear Noetherian non-strict partial orders.

The following proposition is standard, see, e.g., [2, Section 8.5].

Proposition 2.1 (Generated Subframe Lemma)

(i) *If $\mathfrak{F}'$ is a generated subframe of $\mathfrak{F}$, then* Log($\mathfrak{F}$) $\subseteq$ Log($\mathfrak{F}'$).

(ii) *Let $\{\mathfrak{F}_i : i \in I\}$ be a family of generated subframes of a frame $\mathfrak{F}$, and let every x in $\mathfrak{F}$ belong to some $\mathfrak{F}_i$. Then* $\operatorname{Log}\mathfrak{F} = \operatorname{Log}\{\mathfrak{F}_i : i \in I\}$.

The following two propositions are also well known (see, e.g., [1, Proposition 2.14 and Lemma 3.20]).

Proposition 2.2 *If $\mathfrak{G}$ is a p-morphic image of $\mathfrak{F}$, then* Log($\mathfrak{F}$) $\subseteq$ Log($\mathfrak{G}$).

Proposition 2.3 *Let $\mathfrak{F}$ be transitive, $\mathfrak{G}$ finite rooted. If* Log($\mathfrak{F}$) $\subseteq$ Log($\mathfrak{G}$), *then $\mathfrak{G}$ is a p-morphic image of a point-generated subframe of $\mathfrak{F}$.*

Given frames $\mathfrak{F}_i = (W_i, R_i)$, $i \in I$, their *direct product* is the frame $\prod_I \mathfrak{F}_i = (W, R)$ where $W = \prod_{i\in I} W_i$, the Cartesian product of the sets W_i, and R is defined point-wise: xRy iff $x_i R_i y_i$ for all $i \in I$. Given a frame $\mathfrak{F}$, we write $\mathfrak{F}^n$ for its nth direct power.

Proposition 2.4 *If $(\mathfrak{F}_i)_{i\in I}$ is a non-empty family of frames and we have $\forall x \exists y\, (xR_iy)$ in every $\mathfrak{F}_i$, then for every $i \in I$ the ith projection is a p-morphism of $\prod_I \mathfrak{F}_i$ onto $\mathfrak{F}_i$.*

Proof. Immediate from the definition. □

Proposition 2.5 *If $(\mathfrak{F}_i)_{i\in I}$ and $(\mathfrak{G}_i)_{i\in I}$ are non-empty families of frames such that for every $i \in I$ there exists a p-morphism of $\mathfrak{F}_i$ onto $\mathfrak{G}_i$, then there exists a p-morphism of $\prod_I \mathfrak{F}_i$ onto $\prod_I \mathfrak{G}_i$.*

Proof. For each $i \in I$, pick a p-morphism π_i of $\mathfrak{F}_i$ onto $\mathfrak{G}_i$. For each f in $\prod_I \mathfrak{F}_i$ put $\pi(f)(i) := \pi_i(f(i))$. It is straightforward that π is the required p-morphism. □

Definition 2.6 Let $\mathfrak{M}'$ be a weak submodel of $\mathfrak{M}$, and let Ψ be a set of modal formulas. $\mathfrak{M}'$ is a *selective filtration of* $\mathfrak{M}$ *through* Ψ iff the following holds: whenever $\mathfrak{M}, w \models \Diamond\psi$ for some $w \in W'$ and $\Diamond\psi \in \Psi$, then there exists $v \in W'$ such that $wR'v$ and $\mathfrak{M}, v \models \psi$ (i.e., v witnesses that $\Diamond\psi$ holds at w in $\mathfrak{M}$).

The following fact was known since 1970s.

Proposition 2.7 (Selective Filtration Lemma) *Let Ψ be a set of formulas closed under taking subformulas, and let $\mathfrak{M}'$ be a selective filtration of $\mathfrak{M}$ through Ψ. Then for all $w \in W'$ and $\varphi \in \Psi$, we have*

$$\mathfrak{M}', w \models \varphi \Leftrightarrow \mathfrak{M}, w \models \varphi.$$

Proof. Induction on φ. □

Through the paper, we regularly consider a frame $\mathfrak{F}$ with the removed top element (provided the relation of the frame is an order and the top element exists), for which we reserve the notation $\mathfrak{F} \setminus \{\text{top}\}$. In particular, if α is an ordinal, $(\alpha, \geq)^n \setminus \{\text{top}\}$ denotes the direct power $(\alpha, \geq)^n$ without the n-sequence of 0's.

3 Products of converse ordinals

Below we consider frames $(\alpha, \geq)$ where α is an ordinal and $\geq$ its converse ordering, and their direct products $\prod_{i<n}(\alpha_i, \geq)$ with finite n.

It follows from Proposition 2.4 that the logics of such frames decrease by inclusion with increasing ordinals α. We are going to show that, in fact, these logics do not depend on a particular α whenever α is infinite; moreover, they are characterized by upper finite cones of their frames (and thus have the finite model property).

For a formula φ and a Kripke model $\mathfrak{M} = (W, R, \theta)$, let $\|\varphi\|_{\mathfrak{M}} = \{x \in W : \mathfrak{M}, x \models \varphi\}$; in particular, $\|p\|_{\mathfrak{M}} = \theta(p)$.

Lemma 3.1 *Let $\mathfrak{M} = (W, \geq, \theta)$ be a Kripke model such that $\leq$ is a well-founded partial order on W. Let Ψ be a set of formulas closed under taking subformulas, and let V be a non-empty subset of W such that*

$$\{x \in W : \exists\varphi \in \Psi \; (x \textit{ is } \leq\textit{-minimal in } \|\varphi\|_{\mathfrak{M}})\} \subseteq V.$$

Then the restriction of $\mathfrak{M}$ to V is a selective filtration of $\mathfrak{M}$ through Ψ.

Proof. Suppose $\mathfrak{M}, w \models \Diamond\psi$ for some $w \in V$ and $\Diamond\psi \in \Psi$. Then the set $\{v \in W : w \geq v\} \cap \|\psi\|_{\mathfrak{M}}$ is non-empty. Let v be a $\leq$-minimal element of this set. Then v is a $\leq$-minimal element of $\|\psi\|_{\mathfrak{M}}$, and hence $v \in V$. By the construction, $w \geq v$ and $\mathfrak{M}, v \models \psi$. □

Remark 3.2 Lemma 3.1 remains true if $\leq$ is a well-founded pre-order (in this case, x is said to be $\leq$-*minimal* iff for all $y \in W$, $y \leq x$ implies $x \leq y$.)

Theorem 3.3 *For all $\alpha \geq \omega$ and $n < \omega$, we have:*

(i) $\text{Log}((\alpha, \geq)^n) = \text{Log}((\omega, \geq)^n) = \text{Log}\{(m, \geq)^n : m < \omega\}$,

(ii) $\mathrm{Log}((\alpha,\geq)^n \setminus \{\mathrm{top}\}) = \mathrm{Log}((\omega,\geq)^n \setminus \{\mathrm{top}\}) = \mathrm{Log}\{(m,\geq)^n \setminus \{\mathrm{top}\} : m < \omega\}$.

Proof. (i). For all $m < \omega$, $(m,\geq)^n$ is a generated subframe of $(\omega,\geq)^n$, and the latter is a generated subframe of $(\alpha,\geq)^n$ since $\omega \leq \alpha$. It follows that

$$\mathrm{Log}((\alpha,\geq)^n) \subseteq \mathrm{Log}((\omega,\geq)^n) \subseteq \mathrm{Log}\{(m,\geq)^n : m < \omega\}.$$

So it remains to show that $\mathrm{Log}\{(m,\geq)^n : m < \omega\} \subseteq \mathrm{Log}(\alpha,\geq)^n$. To this end, suppose that a modal formula φ is satisfiable in $(\alpha,\geq)^n$ and show that φ is satisfiable in $(m,\geq)^n$ for some $m < \omega$.

Assume there is a model $\mathfrak{M}$ on the frame $(\alpha,\geq)^n$ satisfying φ at some point, in other words, $\|\varphi\|_{\mathfrak{M}} \neq \emptyset$. Let Ψ consist of all subformulas of φ. For each $\psi \in \Psi$, we let

$$U_\psi := \{s \in \alpha^n : s \text{ is } \leq\text{-minimal in } \|\psi\|_{\mathfrak{M}}\} \text{ and } U := \bigcup_{\psi\in\Psi} U_\psi\,.$$

Every antichain in $(\alpha,\leq)^n$ is finite (by generalized Dickson's lemma, see, e.g., [5, Theorem 2.10]), so every U_ψ is finite. Hence, since Ψ is finite, U is finite as well.

Further, consider the projections of U for all $i < n$ and their product:

$$U_i := \{\beta < \alpha : \exists x \in U\ x(i) = \beta\} \text{ and } V := \prod_{i<n} U_i\,.$$

Since $\|\varphi\|_{\mathfrak{M}} \neq \emptyset$, and so $U_\varphi \neq \emptyset$, the set U contains some s such that $\mathfrak{M}, s \models \varphi$. Clearly, $U \subseteq V$. By Lemma 3.1, the restriction $\mathfrak{M}'$ of $\mathfrak{M}$ to V is a selective filtration of $\mathfrak{M}$ through Ψ. By the Selective Filtration Lemma (Proposition 2.7), we get $\mathfrak{M}', s \models \varphi$.

This proves that φ is satisfiable in $(V,\geq)$. Finally, for each $i < n$, we let

$$m_i := |U_i| \text{ and } m := \max_{i<n} m_i\,.$$

Clearly, $(V,\geq)$ is isomorphic to $(\prod_{i<n} m_i, \geq)$, and the latter is a generated subframe of $(m,\geq)^n$. It follows that φ is satisfiable in $(m,\geq)^n$, as required.

(ii). The proof is analogous with the only modification at the step when we define U_i. Namely, let $\mathfrak{M}$ be a model on the frame $(\alpha,\geq)^n \setminus \{\mathrm{top}\}$ such that $\|\varphi\|_{\mathfrak{M}} \neq \emptyset$, and let the set U be defined as above. Now for each $i < n$, we let

$$U_i := \{\beta < \alpha : \exists x \in U\ x(i) = \beta\} \cup \{0\}.$$

We define the numbers m_i and m as above and observe that the frame of the restriction of $\mathfrak{M}$ to the set $\prod_{i<n} U_i \setminus \{\mathrm{top}\}$ is isomorphic to the frame $(\prod_{i<n} m_i, \geq) \setminus \{\mathrm{top}\}$. The latter is a generated subframe of $(m,\geq)^n \setminus \{\mathrm{top}\}$. By the same reasoning as above, φ is satisfiable in $(m,\geq)^n \setminus \{\mathrm{top}\}$, as required.

The proof is complete. □

We introduce the following notation for these logics:

$$\Gamma(n) := \mathrm{Log}((\omega, \geq)^n),$$
$$\Delta(n) := \mathrm{Log}((\omega, \geq)^n \setminus \{\mathrm{top}\}).$$

Corollary 3.4 *For all finite n, we have:*

(i) $\Gamma(n) = \mathrm{Log}\{\prod_{i<n}(\alpha_i, \geq) : \alpha_i$ *is an ordinal*$\}$,

(ii) $\Delta(n) = \mathrm{Log}\{\prod_{i<n}(\alpha_i, \geq) \setminus \{\mathrm{top}\} : \alpha_i$ *is an ordinal*$\}$.

Proof. Follows from Theorem 3.3 and Proposition 2.4. □

Corollary 3.5 *For all finite n,* $\mathrm{Log}(\mathfrak{F}) = \mathrm{Grz.3}$ *implies* $\mathrm{Log}(\mathfrak{F}^n) = \Gamma(n)$ *and* $\mathrm{Log}(\mathfrak{F}^n \setminus \{\mathrm{top}\}) = \Delta(n)$.

Proof. For every finite m, there is a point-generated subframe of $\mathfrak{F}$ isomorphic to $(m, \leq)$, and so a point-generated subframe of $\mathfrak{F}^n$ isomorphic to $(m, \leq)^n$. Hence, $\mathrm{Log}(\mathfrak{F}^n) \subseteq \Gamma(n)$ by Theorem 3.3. For the converse inclusion, observe that any point-generated subframe of $\mathfrak{F}^n$ is the direct product of n converse well-orders, and so $\mathrm{Log}(\mathfrak{F}^n) \supseteq \Gamma(n)$ by Corollary 3.4. The same reasoning proves that $\mathrm{Log}(\mathfrak{F}^n \setminus \{\mathrm{top}\}) = \Delta(n)$. □

Remark 3.6 One can generalize the above corollary for the direct product of n frames $\mathfrak{F}_i$ with $\mathrm{Log}(\mathfrak{F}_i) = \mathrm{Grz.3}$ for all $i < n$.

4 The logics $\Gamma(n)$, $\Delta(n)$, and Medvedev's logic

In this section, we study the connection between the logics $\Gamma(n)$ and Grz.2 and between the logics $\Delta(n)$ and Medvedev's logic.

4.1 The sequences $\Gamma(n)$ and $\Delta(n)$

Let Mdv denote (modal) *Medvedev's logic*, which is the logic of all non-empty finite subsets of ω endowed with their converse inclusion order:

$$\mathrm{Mdv} := \mathrm{Log}(\mathcal{P}_\omega(\omega) \setminus \{\emptyset\}, \supseteq).$$

The following fact follows from [6] (see also [13], [12]):

Proposition 4.1 $\mathrm{Log}(\mathcal{P}_\omega(\omega), \supseteq) = \mathrm{Grz.2}$.

Lemma 4.2 *For any $n < \omega$, there are modal formulas $\varphi^{\max}_{\leq n}$ and $\varphi^{\mathrm{ram}}_{\leq n}$ such that for any finite partial order $\mathfrak{F} = (W, \leq)$ with a least element,*

(i) $\mathfrak{F} \models \varphi^{\max}_{\leq n}$ *iff there exist $\leq n$ maximal points above each $x \in W$,*

(ii) $\mathfrak{F} \models \varphi^{\mathrm{ram}}_{\leq n}$ *iff there exist $\leq n$ immediate successors of each $x \in W$.*

Proof. (i). Put

$$\varphi^{\max}_{\leq n} := \bigwedge_{i<n} \Diamond \Box \Big(p_i \wedge \bigwedge_{i \neq j < n} \neg p_j\Big) \to \Box \Diamond \bigvee_{i<n} p_i .$$

If $\mathfrak{F}$ has at most n maximal points and the premise of $\varphi^{\max}_{\leq n}$ holds in a point w in a model on $\mathfrak{F}$, then at each maximal point one of p_i, $i < n$, is true; hence the conclusion $\varphi^{\max}_{\leq n}$ holds at w.

If $\mathfrak{F}$ has at least $n+1$ maximal points $w_0, \ldots, w_n$, put $\theta(p_i) = \{w_i\}$ for all $i < n$ and consider the model $(\mathfrak{F}, \theta)$; at the least element of $\mathfrak{F}$ (which exists by our assumption), the premise of $\varphi^{\max}_{\leq n}$ holds, and the conclusion is false, since all p_i, $i < n$, are false at w_n.

(ii). This statement is due to [3] (see also [2, Proposition 2.41]): there are intuitionistic formulas expressing the property on finite posets; the formulas $\varphi^{\text{ram}}_{\leq n}$ are their Gödel–Tarski translations. □

Lemma 4.3 *For any n, $0 < n < \omega$,*

(i) $\varphi^{\max}_{\leq 1}$ *belongs to* $\Gamma(n)$*,*

(ii) $\varphi^{\max}_{\leq n}$ *belongs to* $\Delta(n)$ *but not to* $\Delta(n+1)$*,*

(iii) $\varphi^{\text{ram}}_{\leq n}$ *belongs to* $\Delta(n)$ *and* $\Gamma(n)$ *but not to* $\Delta(n+1)$ *and* $\Gamma(n+1)$*.*

Proof. Clear. □

Proposition 4.4 *For all $n < \omega$,*

(i) $\Delta(1) = \Gamma(1) = \text{Grz.3}$*,*

(ii) $\text{Grz} \subset \Delta(n) \subset \Gamma(n)$ *if* $n \geq 2$*,*

(iii) $\Delta(n+1) \subset \Delta(n)$ *and* $\Gamma(n+1) \subset \Gamma(n)$*,*

(iv) $\Delta(n) \not\subseteq \Gamma(n+1)$ *and* $\Gamma(n) \not\subseteq \Delta(2)$*.*

Proof. Grz is valid in any Noetherian poset, so we have $\text{Grz} \subseteq \Delta(n)$ in (ii). The logic Grz.3 is the logic of all Noetherian linearly ordered sets. This proves (i).

For $\Delta(n) \subseteq \Gamma(n)$, note that $(\omega^n, \geq)$ is a p-morphic image of the frame $\mathfrak{F} = (\omega, \geq)^n \setminus \{\text{top}\}$; e.g., the map π defined by letting $(\pi s)(i) = \max(s(i), 1)$, for all s in $\mathfrak{F}$ and $i < n$, is a p-morphism of $\mathfrak{F}$ onto $(\omega \setminus 1, \geq)^n$, and the latter frame is obviously isomorphic to $(\omega, \geq)^n$. Also $(\omega, \geq)^n$ is isomorphic to a generated subframe of $(\omega, \geq)^{n+1}$, e.g., to the subframe consisting of elements $s \in \omega^{n+1}$ with $s(0) = 0$, and similarly for the frames without their top elements, thus proving $\Delta(n+1) \subseteq \Delta(n)$ and $\Gamma(n+1) \subseteq \Gamma(n)$ in (iii).

Furthermore, by Lemma 4.2(i),(ii), the formula $\varphi^{\max}_{\leq 1}$ is in $\Gamma(n) \setminus \Delta(n)$ whenever $n \geq 2$, whence it follows $\Delta(n) \neq \Gamma(n)$ in (ii) and $\Gamma(n) \not\subseteq \Delta(2)$ in (iv). Also by Lemma 4.2(ii), the formula $\varphi^{\max}_{\leq n}$ is in $\Delta(n) \setminus \Delta(n+1)$ and $\Gamma(n) \setminus \Gamma(n+1)$ whenever $n \geq 1$, whence it follows $\Delta(n+1) \neq \Delta(n)$ and $\Gamma(n+1) \neq \Gamma(n)$, and so $\text{Grz} \neq \Delta(n)$, thus completing the proof of (ii) and (iii). Finally, by Lemma 4.2(iii), the formula $\varphi^{\text{ram}}_{\leq n}$ is in $\Delta(n) \setminus \Gamma(n+1)$ whenever $n \geq 1$, which completes the proof of (iv) and also provides another way to see $\text{Grz} \neq \Delta(n)$.

The proposition is proved. □

Theorem 4.5 $\bigcap_{n<\omega} \Delta(n) = \text{Mdv}$ *and* $\bigcap_{n<\omega} \Gamma(n) = \text{Grz.2}$*.*

Proof. The inclusion $\bigcap_{n<\omega} \Delta(n) \subseteq \text{Mdv}$ is immediate from the fact that every frame $(2, \geq)^n \setminus \{\text{top}\}$ is a point-generated subframe of $(\omega, \geq)^n \setminus \{\text{top}\}$. By the same argument and Proposition 4.1, we get $\bigcap_{n<\omega} \Gamma(n) \subseteq \text{Grz.2}$ as well.

Let us check the inclusion $\text{Mdv} \subseteq \bigcap_{n<\omega} \Delta(n)$. For this, it suffices to construct, for each $n < \omega$, a p-morphism σ_n of $(\mathcal{P}_\omega(\omega) \setminus \{\emptyset\}, \supseteq)$ onto $(\omega, \geq)^n \setminus \{\text{top}\}$.

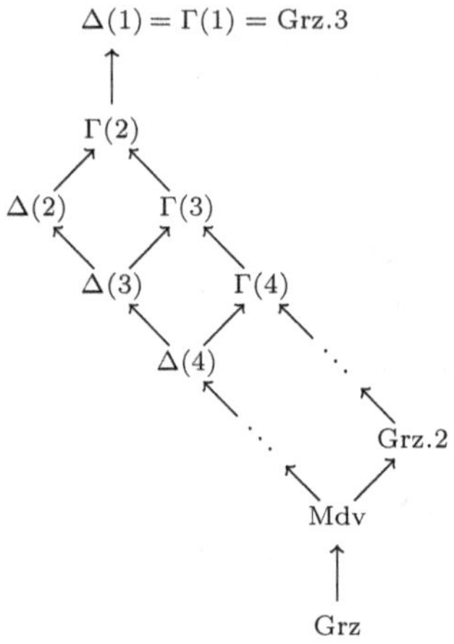

Fig. 1. Inclusions between logics $\Gamma(n)$ and $\Delta(n)$

Define it as follows. Partition ω into n infinite disjoint subsets X_i, $i < n$, and let $\sigma_n(A) = (|A \cap X_0|, \ldots, |A \cap X_{n-1}|)$, for all $A \in \mathcal{P}_\omega(\omega) \setminus \{\emptyset\}$. It is easy to see that σ_n is a surjective p-morphism. Indeed, $A \supseteq B$ clearly implies $|A \cap X_i| \geq |B \cap X_i|$ for all $i < n$, and thus $\sigma_n(A) \geq \sigma_n(B)$; also if $\sigma_n(A) = (|A \cap X_0|, \ldots, |A \cap X_{n-1}|) \geq (b_0, \ldots, b_{n-1})$ then letting B_i consisting of the first b_i elements of $A \cap X_i$, we have $A \supseteq B$ and $\sigma_n(B) = (b_0, \ldots, b_{n-1})$; and the surjectivity is obvious by the same reason.

Finally, we have $\mathrm{Grz.2} \subseteq \bigcap_{n<\omega} \Gamma(n)$ since every $(\omega, \geq)^n$ is a Noetherian poset with a top element.

The proof is complete. □

The diagram of strict inclusions between the considered logics is shown on Figure 1.

4.2 Remarks on axiomatization and decidability

By Lemma 4.3 (and Proposition 4.4), we have

$$\mathrm{Grz.2} + \varphi^{\mathrm{ram}}_{\leq n} \subseteq \Gamma(n) \ \text{ and } \ \mathrm{Grz} + \varphi^{\max}_{\leq n} + \varphi^{\mathrm{ram}}_{\leq n} \subseteq \Delta(n).$$

By Theorem 3.3, each of the logics $\Gamma(n)$, $\Delta(n)$ is characterized by a recursive set of finite frames, so these logics are co-recursively enumerable; hence, they are decidable if they are recursively axiomatizable.

Question 4.6 *Are the logics* $\Gamma(n)$, $\Delta(n)$, $2 \leq n < \omega$, *finitely axiomatizable? recursively axiomatizable?*

Theorem 4.7

(i) $\mathrm{Mdv.2} = \mathrm{Grz.2}$,

(ii) $\Delta(n).2 = \Gamma(n)$ *for all* $n < \omega$.

Proof. (i) This fact is known, see Propositions 13 and 9 in [12].

(ii) As $\Delta(n) \subseteq \Gamma(n)$, we have $\Delta(n).2 \subseteq \Gamma(n).2 = \Gamma(n)$.

In [10], it was shown that if a transitive logic L has the finite model property, then so does L.2. Hence $\Delta(n).2$ is the logic of the class of finite rooted $\Delta(n)$-frames with a greatest element.

Assume that a modal formula φ is refutable in a model $\mathfrak{M}$ on one of such frames $\mathfrak{F}$. Since $\mathfrak{F}$ validates $\Delta(n)$, it follows from Proposition 2.3 that $\mathfrak{M}$ is a p-morphic image of a model $\mathfrak{N}$ based on a point-generated subframe of $(\omega, \geq)^n \setminus \{\text{top}\}$. This p-morphism maps all maximal points of $\mathfrak{N}$ to the greatest point of $\mathfrak{M}$, and so the valuations of variables coincide at these points. Add a top point to $\mathfrak{N}$ with the same valuation and denote the new model by $\mathfrak{N}'$. Clearly, $\mathfrak{N}$ is a selective filtration of $\mathfrak{N}'$ through the set of all modal formulas. The formula φ is refutable in $\mathfrak{N}$ since $\mathfrak{M}$ is a p-morphic image of $\mathfrak{N}$. By the Selective Filtration Lemma (Proposition 2.7), φ is refutable in $\mathfrak{N}'$. Now it remains to observe that the frame of $\mathfrak{N}'$ is a $\Gamma(n)$-frame. This proves the inclusion $\Gamma(n) \subseteq \Delta(n).2$. □

Corollary 4.8 *If $\Delta(n)$ is finitely (recursively) axiomatizable, then so is $\Gamma(n)$.*

5 Substructures of finite powers

In this section, we study the restrictions of the direct powers $(\omega, \geq)^n$ to several $A \subseteq \omega^n$ and the modal logics of these restrictions, which thus generalize the logics $\Gamma(n)$ and $\Delta(n)$ considered above.

5.1 The logics $\Gamma(n, m)$

For an ordinal α, the *αth level* of a Noetherian frame $\mathfrak{F} = (X, \geq)$ consists of all points of $\mathfrak{F}$ having the rank α in the well-founded frame $(X, \leq)$.

The following fact is obvious.

Lemma 5.1 *The kth level of $(\omega, \geq)^n$ consists precisely of sequences s such that $\sum_{i<n} s(i) = k$, so its size is the number of (weak) compositions $\binom{k+n-1}{n-1}$.*

Let $\mathfrak{P}_{n,m}$ denote the restriction of the frame $(\omega, \geq)^n$ to its levels $\geq m$; so the universe of $\mathfrak{P}_{n,m}$ is $\{s \in \omega^n : \sum_{i<n} s(i) \geq m\}$. Four first frames for the case of $n = 2$ are depicted on Figure 2.

For $1 \leq n < \omega$, $0 \leq m < \omega$, let

$$\Gamma(n, m) := \text{Log}(\mathfrak{P}_{n,m}).$$

Let also

$$\begin{aligned} \Gamma(\omega, m) &:= \text{Log}\{\mathfrak{P}_{n,m} : n < \omega\}, \\ \Gamma(n, \omega) &:= \text{Log}\{\mathfrak{P}_{n,m} : m < \omega\}, \\ \Gamma(\omega, \omega) &:= \text{Log}\{\mathfrak{P}_{n,m} : m, n < \omega\}. \end{aligned}$$

Thus $\Gamma(\omega, m) = \bigcap_{n<\omega} \Gamma(n, m)$, $\Gamma(n, \omega) = \bigcap_{m<\omega} \Gamma(n, m)$, and $\Gamma(\omega, \omega) = \bigcap_{n,m<\omega} \Gamma(n, m)$. For sake of completeness we may let $\Gamma(0, 0) :=$ the logic of a trivial frame consisting of a reflexive singleton, axiomatized by $p \leftrightarrow \Diamond p$.

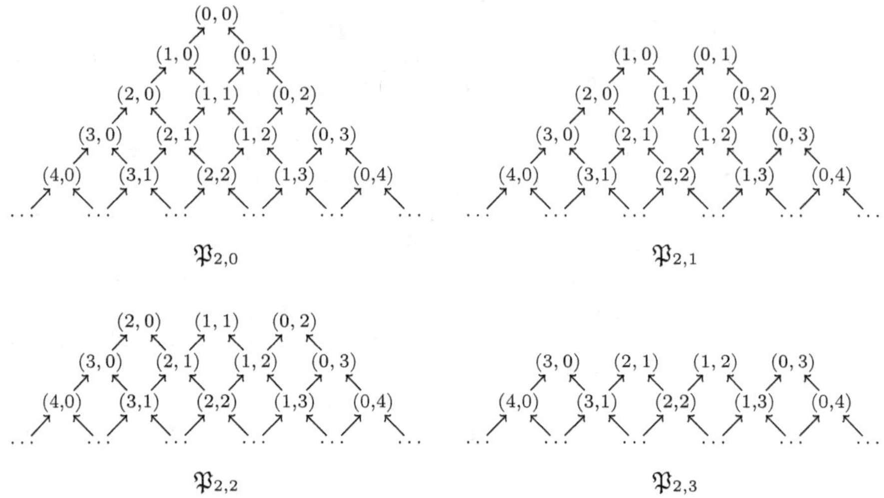

Fig. 2. Frames $\mathfrak{P}_{2,m}$

We restate the previous definitions and facts proved above:

(i) $\Gamma(1, m) = \text{Grz.3}$ for all $m \leq \omega$,

(ii) $\Gamma(n, 0) = \Gamma(n)$ and $\Gamma(n, 1) = \Delta(n)$,

(iii) $\Gamma(\omega, 0) = \text{Grz.2}$ and $\Gamma(\omega, 1) = \text{Mdv}$.

The following theorem generalizes Proposition 4.4.

Theorem 5.2

(i) $\text{Grz} \subseteq \Gamma(\omega, \omega)$,

(ii) $\Gamma(n, m) \supseteq \Gamma(n', m')$ *if* $n \leq n' \leq \omega$ *and* $m \leq m' \leq \omega$,

(iii) $\Gamma(n, \omega) \not\subseteq \Gamma(n + 1, 0)$,

(iv) $\Gamma(n, m) \not\subseteq \Gamma(n', m')$ *if* $\binom{m+n-1}{n-1} < \binom{m'+n'-1}{n'-1}$.

Proof. Item (i) is clear, and for (ii), it suffices to consider only the cases of $n' = n + 1$ and $m' = m + 1$.

We have $\Gamma(n, m) \supseteq \Gamma(n + 1, m)$ since the frame $\mathfrak{P}_{n,m}$ is isomorphic to the generated subframe of $\mathfrak{P}_{n+1,m}$ consisting of s with $s(0) = 0$. To show $\Gamma(n, m) \supseteq \Gamma(n, m+1)$, we shall construct a p-morphism of $\mathfrak{P}_{n,m+1}$ onto $\mathfrak{P}_{n,m}$.

To simplify our construction, we first replace $\mathfrak{P}_{n,m}$ with the substructure $\mathfrak{Q}_{n,m}$ of $\mathfrak{P}_{n,m+1}$ consisting of s with $s(0) > 0$. The shift ρ defined by letting $(\rho s)(0) = s(0) + 1$, and $(\rho s)(i) = s(i)$ if $0 < i < n$, is an isomorphism of $\mathfrak{P}_{n,m}$ onto $\mathfrak{Q}_{n,m}$.

Now we construct a map π that gives, for all $m < \omega$, a p-morphism of $\mathfrak{P}_{n,m+1}$ onto $\mathfrak{Q}_{n,m} \subseteq \mathfrak{P}_{n,m+1}$. For any $s \in \omega^n \setminus \{(0, \ldots, 0)\}$, define $\pi(s)$ as follows: if $s(0) > 0$, let $\pi s = s$; otherwise, i.e., if $s(0) = 0$, let

$$(\pi s)(i) := \begin{cases} 1 & \text{if } i = 0, \\ s(i) - 1 & \text{if } i = \min \operatorname{supp}(s), \\ s(i) & \text{otherwise,} \end{cases}$$

where $\operatorname{supp}(s) := \{i < n : s(i) \neq 0\}$, the *support* of s. For brevity, let also $i_s := \min \operatorname{supp}(s)$. Note that $s \geq t$ implies $\operatorname{supp}(s) \supseteq \operatorname{supp}(t)$ and so $i_s \leq i_t$.

Let us verify that π is as required. Obviously, π maps $\mathfrak{P}_{n,m+1}$ onto $\mathfrak{Q}_{n,m}$ and leaves all the points of $\mathfrak{Q}_{n,m}$ fixed. It is easy to see that π preserves the levels: $\sum_{i<n}(\pi s)(i) = \sum_{i<n} s(i)$. We must show that π is a p-morphism, i.e., a homomorphism with the lifting property. This is a routine check, which however we write down for the sceptic reader.

Pick $s \geq t$ and show $\pi s \geq \pi t$. For the case of $t(0) > 0$, we have $s(0) > 0$ and so $\pi s = s \geq t = \pi t$. Consider now the case of $t(0) = 0$. We have then $(\pi t)(0) = 1$. But if $s(0) > 0$ then $(\pi s)(0) = s(0) \geq 1$, and if $s(0) = 0$ then $(\pi s)(0) = 1$, thus we have $(\pi t)(0) \leq (\pi s)(0)$ in any case. Further, if $s(0) > 0$ then $(\pi s)(i) = s(i) \geq t(i) \geq (\pi t)(i)$ for all $i \in n \setminus \{0\}$. If $s(0) = 0$, the same relationship $(\pi s)(i) = s(i) \geq t(i) \geq (\pi t)(i)$ holds for all $i \in n \setminus \{0, i_s\}$; moreover, $i_s \leq i_t$. But if $i_s = i_t$ then $(\pi s)(i_s) = s(i_s) - 1 \geq t(i_s) - 1 = (\pi t)(i_s)$, and if $i_s < i_t$ then $(\pi t)(i_s) = 0$ since $0 < i_s < i_t$; thus we have $(\pi s)(i_s) \geq (\pi t)(i_s)$ in any case.

Pick now $\pi s \geq t'$ with $t' \in \operatorname{ran}(\pi)$, i.e., with $\pi t' = t'$, and find t with $s \geq t$ and $\pi t = t'$. If $s(0) > 0$ then $\pi s = s$, so it suffices to let $t = t'$. If $s(0) = 0$, define t by letting $t(0) = 0$, $t(i_s) = t'(i_s) + 1$, and $t(i) = t'(i)$ if $i \in n \setminus \{i_s\}$. Note that $i_t = i_s$. We have: $s(0) = t(0) = 0$, so $(\pi s)(0) = (\pi t)(0) = 1$, and $t'(0) = 1$ (since $1 \leq t'(0) \leq (\pi s)(0) = 1$); also $s(i_s) = (\pi s)(i_s) + 1 \geq t(i_s) + 1 = t(i_s)$ and $(\pi t)(i_s) = t(i_s) - 1 = t'(i_s) + 1 - 1 = t'(i_s)$; finally, if $i \in n \setminus \{0, i_s\}$ then $s(i) = (\pi s)(i) \geq t'(i) = t(i)$.

So we have proved that π is a p-morphism of $\mathfrak{P}_{n,m+1}$ onto $\mathfrak{Q}_{n,m}$.

Finally, for (iii), note that $\varphi^{\mathrm{ram}}_{\leq n} \in \Gamma(n, \omega) \setminus \Gamma(n+1, 0)$, and for (iv), that $\varphi^{\max}_{\leq N} \in \Gamma(n, m) \setminus \Gamma(n', m')$ where $N = \binom{m+n-1}{n-1}$ by Lemma 5.1. □

Corollary 5.3

(i) $\Gamma(n, m) \supset \Gamma(n', m)$ *if* $n < n'$,

(ii) $\Gamma(n, m) \supset \Gamma(n, m')$ *if* $m < m'$.

Proof. Theorem 5.2(ii)–(iv). □

The diagram of (non-strict) inclusions between the logics $\Gamma(n, m)$ is shown on Figure 3.

Question 5.4 *What about the inclusions of the logics that are not under the scope of Theorem 5.2? E.g., is* $\Gamma(3, 1) \subseteq \Gamma(2, 2)$*?*

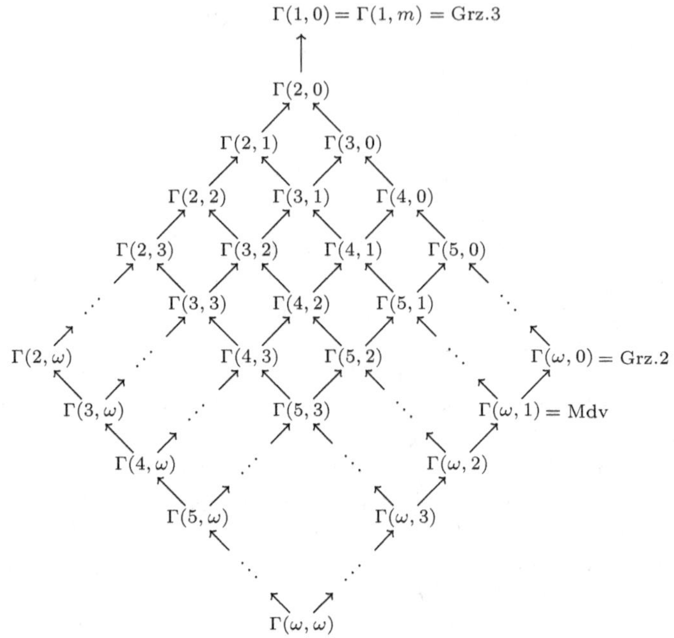

Fig. 3. Inclusions between logics $\Gamma(n, m)$

As we have Mdv $= \Gamma(\omega, 1) \supseteq \Gamma(\omega, 2) \supseteq \Gamma(\omega, 3) \supseteq \ldots \supseteq \Gamma(\omega, \omega) \supseteq$ Grz, these logics can be regarded as "sub-Medvedev"; the next proposition below confirms this view. Given cardinals $\kappa < \lambda$, let $\mathcal{P}_{\lambda,\kappa}(X) := \mathcal{P}_\lambda(X) \setminus \mathcal{P}_\kappa(X) = \{A \subseteq X : \kappa \leq |A| < \lambda\}$.

Proposition 5.5 *For all* $m < \omega$,

$$\Gamma(\omega, m) = \mathrm{Log}(\mathcal{P}_{\omega,m}(\omega), \supseteq) = \mathrm{Log}\{(\mathcal{P}_{n+1,m}(n), \supseteq) : n < \omega\}.$$

Proof. The second equality is clear since the finite frames $(\mathcal{P}_{n+1,m}(n), \supseteq)$ are point-generated subframes of $(\mathcal{P}_{\omega,m}(\omega), \supseteq)$ and cover it.

Also each $(\mathcal{P}_{n+1,m}(n), \supseteq)$ is isomorphic to a point-generated subframe of $\mathfrak{P}_{n,m}$ (via characteristic functions), whence we get $\subseteq$ in the first equality. To prove that $\supseteq$ holds as well, one use the maps σ_n constructed in the proof of Theorem 4.5. Recall that, for a partition of ω into n infinite subsets X_i, $i < n$, we let $\sigma_n(A)(i) := |A \cap X_i|$ for all $i < n$ and $A \in \mathcal{P}_\omega(\omega)$. Then σ_n is a p-morphism of $(\mathcal{P}_\omega(\omega), \supseteq)$ onto $(\omega, \geq)^n$; moreover, it is easy to see that it preserves the levels, i.e., the ranks of points (where, of course, the kth level of $(\mathcal{P}_\omega(\omega), \supseteq)$ consists of sets of size k), hence its restriction to $\mathcal{P}_{\omega,m}(\omega)$ is a p-morphism onto $\mathfrak{P}_{m,n}$. □

Related systems, the intuitionistic logics of $(\mathcal{P}(\omega)\backslash\mathcal{P}(m), \supseteq)$, $0 < m < \omega$, were considered in [13]; none of them are finitely axiomatizable. We conjecture that the logics $\Gamma(\omega, m)$ are not finitely axiomatizable as well.

5.2 The logics of monotone sequences

Given $n < \omega$, let

$$D^{\leq}(n) = \{s \in \omega^n : s(0) \leq \ldots \leq s(n-1)\},$$
$$D^{\geq}(n) = \{s \in \omega^n : s(0) \geq \ldots \geq s(n-1)\}.$$

Note that $D^{\leq}(2) = \leq$ and $D^{\geq}(2) = \geq$ (where $\leq$ and $\geq$ are on ω).

It is easy to see that for any n, the frames $(D^{\leq}(n), \geq)$ and $(D^{\geq}(n), \geq)$ are isomorphic under the natural isomorphism taking $(s(0), \ldots, s(n-1))$ to $(s(n-1), \ldots, s(0))$, hence it suffices to consider only one of them.

Theorem 5.6 $\mathrm{Log}(D^{\leq}(n), \geq) = \Gamma(n)$.

Proof. To see the inclusion $\Gamma(n) \subseteq \mathrm{Log}(D^{\leq}(n), \geq)$, define $\pi : \omega^n \to D^{\leq}(n)$ by letting for all $i < n$,

$$(\pi s)(i) := \max_{j \leq i} s(j).$$

Clearly, π is surjective, and it leaves all points of $D^{\leq}(n)$ fixed: $\pi s = s$ for all $s \in D^{\leq}(n)$. Also $s \geq t$ clearly implies $\pi s \geq \pi t$. And if $\pi s \geq t'$ for some $t' \in D^{\leq}(n)$, letting $t(i) = t'(i)$ if $s(i) = \pi s(i)$, and $t(i) = 0$ otherwise, we have $s \geq t$ and $\pi t = t'$. Thus π is a p-morphism of $(\omega^n, \geq)$ onto $(D^{\leq}(n), \geq)$, which proves the inclusion above.

To prove the converse inclusion $\mathrm{Log}(D^{\leq}(n), \geq) \subseteq \Gamma(n)$, for each $k < \omega$, let $S_k \subseteq D^{\leq}(n)$ be the subframe generated by the point $(k, k \cdot 2, \ldots, k \cdot n)$, and let $C_k \subseteq S_k$ be the "n-dimensional cube" of size k^n consisting of all points between the points $(0, k, k \cdot 2, \ldots, k \cdot (n-1))$ and $(k, k \cdot 2, k \cdot 3, \ldots, k \cdot n)$:

$$C_k := \{s \in \omega^n : k \cdot i \leq s(i) \leq k \cdot (i+1) \text{ for all } i < n\}.$$

Define $\pi_k : S_k \to C_k$ by letting for all $i < n$,

$$(\pi_k s)(i) := \max(s(i), k \cdot i).$$

It is easy to see that π_k leaves all points of C_k fixed, and it is a p-morphism of $(S_k, \geq)$ onto $(C_k, \geq)$, whence it follows $\mathrm{Log}(S_k) \subseteq \mathrm{Log}(C_k)$. But $(C_k, \geq)$ are isomorphic to $(n^k, \geq)$, whence $\bigcap_{k<\omega} \mathrm{Log}(C_k, \geq) = \Gamma(n)$ by Theorem 3.3, and $(S_k, \geq)$ are point-generated subframes of $(D^{\leq}(n), \geq)$ with $\bigcup_{k<\omega} S_k = D^{\leq}(n)$, whence $\bigcap_{k<\omega} \mathrm{Log}(S_k, \geq) = \mathrm{Log}(D^{\leq}(n), \geq)$ by Proposition 2.1(ii). Therefore,

$$\mathrm{Log}(D^{\leq}(n), \geq) = \bigcap_{k<\omega} \mathrm{Log}(S_k, \geq) \subseteq \bigcap_{k<\omega} \mathrm{Log}(C_k, \geq) = \Gamma(n),$$

as required. □

A similar result can be obtained for the logics $\Delta(n)$; for this, however, it does not suffice to remove only the top element from the frame of monotone sequences as the residual has its own top (e.g., in $D^{\leq}(2)$, under removing $(0,0)$ the new top $(0,1)$ appears). The precise statement of such a result on $\Delta(n)$ is hence more complicated, and we postpone this for the further work.

5.3 Noetherian substructures of infinite powers

The structures consisting of all finite subsets and of all finite sequences can be naturally identified with certain substructures of infinite powers. Hence, Mdv and, more generally, the logics $\Gamma(\omega, m)$ considered above, can be also regarded as instances of logics of such substructures.

Consider the frame consisting of all eventually zero sequences endowed with the point-wise order; it can be regarded as a subdirect infinite power (while all sequences form the usual direct power). Given ordinals α, β, define

$$E(\alpha, \beta) := \{f \in \alpha^\beta : |\mathrm{supp}(f)| < \omega\} \text{ where } \mathrm{supp}(f) := \{i < \beta : f(i) \neq 0\}.$$

As above, we consider the point-wise order on $E(\alpha, \beta)$: for $f, g \in \alpha^\beta$, we let $f \leq g$ iff $f(i) \leq g(i)$ for all $i < \beta$; so the top element of $(E(\alpha, \beta), \geq)$ is the β-sequence of 0's.

Theorem 5.7 *For all infinite ordinals α, β,*

$$\mathrm{Log}((E(\alpha, \beta), \geq) \setminus \{\mathrm{top}\}) = \mathrm{Mdv} \text{ and } \mathrm{Log}(E(\alpha, \beta), \geq) = \mathrm{Grz.2}.$$

Proof. Given $f \in \alpha^\beta$, the subframe of the frame $(E(\alpha, \beta), \geq)$ generated by the point f is isomorphic to $(\alpha^{|\mathrm{supp}(f)|}, \geq)$, and likewise for the frame without the top element. As $\alpha, \beta \geq \omega$, Theorem 3.3 gives the conclusion. □

6 Other characterizations of $\Gamma(n)$ and $\Delta(n)$

The logics $\Gamma(n)$ and $\Delta(n)$ admit different semantic interpretations.

1. By the *standard translation* argument (see, e.g., [1, Section 2.4]), the logics $\Gamma(n)$ and $\Delta(n)$ are fragments of the *n-adic* (i.e., relation variables are n-ary) second order logic over natural numbers with the standard ordering. Propositional variables p are interpreted as n-ary predicates on ω, the order on tuples in $(\omega, \geq)^n$ is interpreted via conjunctions $\bigwedge_{i<n} x_i \geq y_i$.

2. The logics $\Gamma(2)$ and $\Delta(2)$ can be considered in the context of interval temporal logic (see, e.g., [4]). Let W be the set of closed segments $[m, n]$ of integer numbers containing a fixed integer (e.g., 0):

$$W := \{[m, n] : m \leq 0 \leq n\} \quad \text{where } [m, n] := \{k \in \mathbb{Z} : m \leq k \leq n\}.$$

It is immediate that $\mathrm{Log}(W, \supseteq)$ is $\Gamma(2)$ and $\mathrm{Log}(W \setminus \{[0, 0]\}, \supseteq)$ is $\Delta(2)$ according to the fact that $(W, \supseteq)$ is isomorphic to $(\omega, \geq)^2$; the isomorphism is defined by letting $(m, n) \mapsto [-m, n]$.

3. For the first time, the logics $\Gamma(n)$ and $\Delta(n)$ has been appeared in studies of modal logics of model-theoretic relations undertaken in a recent paper [11]. Referring the reader to that paper for related concepts and results, we prove the following characterization of the logics $\Gamma(2^n)$ and $\Delta(2^n)$, announced there.

Theorem 6.1 *Let $n < \omega$, and let τ be a signature consisting of n unary predicates and possibly some constants. The robust modal logic of the class of models of τ with the submodel relation is $\Gamma(2^n)$ if τ has at least one constant, and $\Delta(2^n)$ otherwise.*

Proof. Let $\sqsupseteq$ denote the *submodel relation* on models of a given signature: $\mathfrak{A} \sqsupseteq \mathfrak{B}$ iff $\mathfrak{B}$ is a submodel of the model $\mathfrak{B}$. Let also $\simeq$ be the isomorphism (equivalence) relation and $[\mathfrak{A}]_{\simeq}$ the equivalence class $\{\mathfrak{A}' : \mathfrak{A}' \simeq \mathfrak{A}\}$, i.e., the isomorphism type of $\mathfrak{A}$. We define $[\mathfrak{A}]_{\simeq} \sqsupseteq [\mathfrak{B}]_{\simeq}$ iff there exist $\mathfrak{A}' \simeq \mathfrak{A}$ and $\mathfrak{B}' \simeq \mathfrak{B}$ such that $\mathfrak{A}' \sqsupseteq \mathfrak{B}'$. By [11, Theorem 29], the robust modal logic of the submodel relation on the class $\mathfrak{K}$ of all models of a given signature τ is $\mathrm{Log}\{(\mathrm{Sub}_{\simeq}(\mathfrak{A}), \sqsupseteq_{\simeq}) : \mathfrak{A} \in \mathfrak{K}\}$ where $\mathrm{Sub}_{\simeq}(\mathfrak{A})$ is $\{[\mathfrak{B}]_{\simeq} : \mathfrak{A} \sqsupseteq \mathfrak{B}\}$.

Fix an enumeration $P_0, \ldots, P_{n-1}$ of the unary predicates in τ. Let $\mathfrak{A}$ be a model of τ with the universe A, and let W be the set of cardinals $\leq |A|$. For each $B \subseteq A$ define a map $s_B : \mathcal{P}(n) \to W$ by letting, for all $X \subseteq n$,

$$s_B(X) := |\{a \in B : \mathfrak{A} \models P_i(a) \text{ iff } i \in X\}|.$$

Note that the submodels of $\mathfrak{A}$ given by subsets B and B' are isomorphic iff $s_B(X) = s_{B'}(X)$ for all $X \subseteq n$; we let $s_B \sim s_{B'}$ iff this is the case. Hence we can define a map $f : \mathrm{Sub}_{\simeq}(\mathfrak{A}) \to W^{\mathcal{P}(n)}$ by letting $f([B]_{\simeq}) := [s_B]_{\sim}$, and this map is injective.

Moreover, note that each model $\mathfrak{B}$ of τ is a submodel of a model $\mathfrak{A}$ of τ such that $s_{\mathfrak{A}}(n)$ has the greatest possible value $|A|$. Thus $\mathrm{Sub}_{\simeq}(\mathfrak{B})$ forms a point-generated subframe of $\mathrm{Sub}_{\simeq}(\mathfrak{A})$. Therefore, by Proposition 2.1, it suffices to handle the case of such models $\mathfrak{A}$. In this case, f is a bijection between $\mathrm{Sub}_{\simeq}(\mathfrak{A})$ and $W^{\mathcal{P}(n)}$ providing τ contains at least one constant; the top element relates to the least submodel of $\mathfrak{A}$ consisting of its constants (and the bottom element, of course, to the whole model $\mathfrak{A}$). Moreover, f is an isomorphism of $(\mathrm{Sub}_{\simeq}(\mathfrak{A}), \sqsupseteq_{\simeq})$ and $(W, \geq)^{2^n}$ where $\leq$ is the usual ordering of cardinals. For the case without constants, the corresponding structure is $(W, \geq)^{2^n} \setminus \{\mathrm{top}\}$; maximal elements relate to the singleton submodels of $\mathfrak{A}$. Since $\leq$ on W is a well-order (under the axiom of choice), the statement now follows from Theorem 3.3. □

Acknowledgements

The authors are grateful to the three anonymous reviewers for their questions and suggestions on an earlier version of the paper.

References

[1] Blackburn, P., M. de Rijke and Y. Venema, "Modal Logic," Cambridge Tracts in Theoretical Computer Science **53**, Cambridge University Press, 2002.

[2] Chagrov, A. and M. Zakharyaschev, "Modal Logic," Oxford Logic Guides **35**, Oxford University Press, 1997.

[3] Gabbay, D. M. and D. H. J. De Jongh, *A sequence of decidable finitely axiomatizable intermediate logics with the disjunction property*, The Journal of Symbolic Logic **39** (1974), pp. 67–78.

[4] Goranko, V., A. Montanari and G. Sciavicco, *A road map of interval temporal logics and duration calculi*, Journal of Applied Non-Classical Logics **14** (2004), pp. 54 – 9.

[5] Harzheim, E., "Ordered Sets," Advances in Mathematics, Springer US, 2006.

[6] Maksimova, L. L., V. B. Shehtman and D. P. Skvortsov, *Impossibility of finite axiomatization of Medvedev's logic of finite problems*, Dokl. Akad. Nauk SSSR **245** (1979), pp. 1051–1054.
[7] Medvedev, J. T., *Finite problems*, Sov. Math., Dokl. **3** (1962), pp. 227–230.
[8] Medvedev, J. T., *Interpretation of logical formulas by means of finite problems*, Sov. Math., Dokl. **7** (1966), pp. 857–860.
[9] Prucnal, T., *On two problems of Harvey Friedman*, Studia Logica **38** (1979), pp. 247–262.
[10] Rybakov, V. V., *A modal analog for Glivenko's theorem and its applications*, Notre Dame J. Formal Logic **33** (1992), pp. 244–248.
[11] Saveliev, D. I. and I. B. Shapirovsky, *On modal logics of model-theoretic relations*, Studia Logica **108** (2020), pp. 989–1017.
[12] Shehtman, V. B., *Modal counterparts of Medvedev logic of finite problems are not finitely axiomatizable*, Studia Logica **49** (1990), pp. 365–385.
[13] Shehtman, V. B. and D. P. Skvortsov, *Logics of some Kripke frames connected with Medvedev notion of informational types*, Studia Logica **45** (1986), pp. 101–118.

Relevant Reasoners in a Classical World

Igor Sedlár

Czech Academy of Sciences
Institute of Computer Science
Prague, The Czech Republic

Pietro Vigiani

Scuola Normale Superiore
Department of Philosophy
Pisa, Italy

Abstract

We develop a framework for epistemic logic that combines relevant modal logic with classical propositional logic. In our framework the agent is modelled as reasoning in accordance with a relevant modal logic while the propositional fragment of our logics is classical. In order to achieve this feature, we modify the relational semantics for relevant modal logics so that validity in a model is defined as satisfaction throughout a set of designated states that, as far as propositional connectives are concerned, behave like classical possible worlds. The main technical result of the paper is a modular completeness theorem parametrized by the relevant modal logic formalizing the agent's reasoning.

Keywords: Epistemic logic, logical omniscience, non-normal worlds, relevant logic.

1 Introduction

The *logical omniscience problem* consists in a discrepancy between properties of modal operators in normal modal epistemic logics on one hand and intuitions concerning epistemic attitudes of real-life agents on the other hand. More specifically, if $\varphi_1 \wedge \ldots \wedge \varphi_n \to \psi$ is valid in a normal modal logic, for some $n \geq 0$, then so is $\Box\varphi_1 \wedge \ldots \wedge \Box\varphi_n \to \Box\psi$, meaning informally that epistemic attitudes of moderately idealised agents represented by $\Box$ are closed under local consequence from arbitrary finite (possibly empty) sets of assumptions. As an assumption about agents with bounded memory, reasoning capacity etc., this is clearly unrealistic.

A number of alternative approaches to epistemic logic that avoid the logical omniscience problem have been proposed. One group of approaches, the *non standard states approaches*, consists in enriching Kripke models with so-called non-standard states in which the behaviour of logical connectives may differ from their behaviour in standard states (possible worlds). Closure of epistemic attitudes under logical consequence is avoided by (i) defining consequence in terms of standard states, but (ii) allowing the epistemic accessibility relation to connect standard states with non-standard ones. One of the first approaches of this kind is [13]; see also [21] for a generalization.

In the most general versions of the non-standard states approach, agents' epistemic attitudes are represented as lacking any kind of logical regimentation. It seems more realistic to assume closure under consequence of some weak *non-classical logic* instead of Scylla of logical anarchism and Charybdis of closure under normal modal logic. In this vein, Levesque [11] has advocated an epistemic logic based on the logic **FDE** of First Degree Entailment, an implication-free fragment of the relevant logic **E**; see also [10] for a version allowing nesting of modal operators. In this logic, closure of epistemic attitudes under some problematic principles involving negation is avoided; for instance, agents are allowed to have inconsistent yet non-trivial beliefs. Validity is defined in terms of standard possible worlds and so the propositional fragment of Levesque's logic is classical. The general motivation behind this approach is that Boolean connectives seem to correctly represent the logic of agent-independent facts, while agents' reasoning is regimented by a weaker non-classical logic. This approach will be followed in this paper.

Levesque acknowledges that an extension of his formalism with a non-standard implication connective would be desirable to properly formalize epistemic attitudes towards implicational statements. Fagin et al. [7] study an extension of **FDE** with epistemic modal operators and material implication, thus providing some means to formalize attitudes towards implicational statements, but the well known discrepancy between properties of material implication and the intuitive properties of implicational statements renders their formalization problematic.

Relevant modal logic including a relevant implication connective is used as a basis for epistemic logic in [3,4]. As is usual in relevant logic, this framework does not take standard states to be classical possible worlds. Hence, it is not an option if one intends to extend classical propositional logic with epistemic modal operators regimented by a relevant logic. Such an option is provided in [16,17] where combinations of classical propositional logic with relevant modal logic are studied. The semantics of [17] is two-sorted and the Hilbert-style axiomatizations provided in the paper use a meta-rule of inference, which is inconvenient if the underlying relevant logic is undecidable. The semantics of [16] is one-sorted, thus more elegant, but the Hilbert-style axiomatizations provided in that paper have two peculiar features: proofs are defined non-standardly as *pairs* of finite sequences of formulas, and the completeness proof relies on the presence of extensional truth constants $\top, \bot$ in the language.

In this paper we provide a framework for epistemic logic based on relevant modal logic that avoids the problematic features of the previous approaches. The crucial features of our approach are that (i) closure of epistemic attitudes under normal modal consequence is avoided in all forms (that is, taking into account zero, one or more assumptions) but (ii) epistemic attitudes are represented as being regimented by relevant modal logic while (iii) the non-modal fragments of our epistemic logics remain classical; (iv) our use of full relevant logic, including a relevant implication connective, enables a more realistic representation of attitudes towards implicational statements; and (v) our main technical result, a modular completeness theorem parametrized by the relevant modal logic governing the agents' reasoning, is stated in terms of a standard Hilbert-style proof system, and to obtain the result we do not need to assume the presence of extensional truth constants in the language.

The paper is structured as follows. In Section 2 we give some background information on relevant modal logic and in Section 3 we outline the general strategy behind our framework. The framework itself is introduced in Section 4 and the axiomatization results are obtained in Section 5. The concluding section summarizes the paper and discusses some attractive topics for future research. Proofs of some of our results are given in the appendix.

2 Relevant modal logics

This section gives some background information on relevant modal logic. We build on the approach of Fuhrmann [8,9], with the difference that relevant modal logics discussed here are *bimodal*, with a pair of modal operators $\Box$ and $\Box_L$. While $\Box$ will be seen as an epistemic modal operator, the auxiliary operator $\Box_L$ is introduced to capture "validity in relevant logic" in a specific technical sense that will be discussed later.

Definition 2.1 The *modal language* $\mathcal{L}$ contains a countable set Pr of propositional variables and operators $\wedge, \vee, \rightarrow$ (binary), and $\neg, \Box, \Box_L$ (unary). The set of formulas of this language is defined as usual and denoted as $Fm_{\mathcal{L}}$. We define $\varphi \leftrightarrow \psi := (\varphi \rightarrow \psi) \wedge (\psi \rightarrow \varphi)$.

The propositional operators $\wedge, \vee, \neg$ and $\rightarrow$ are read as usual. The modal operator $\Box$ is read (with a contextually fixed agent in mind) as "the agent believes that ...", while the operator $\Box_L$ is auxiliary. We note that we stick to a language with one epistemic modal operator $\Box$, instead of multiple $\Box_i$ for agents $i \in G$, only for the sake of simplicity and that the extension of our framework to the multi-agent case is trivial.

Before defining the semantics for the modal language, we introduce some simplifying notation. Let $(S_1, \leq_1)$ and $(S_2, \leq_2)$ be two partially ordered sets. If $k_1, \dots, k_n, k_{n+1} \subseteq \{\downarrow, \uparrow\}$, then an n-ary function f from $(S_1, \leq_1)$ to $(S_2, \leq_2)$ is said to be of type $k_1 \dots k_n \mapsto k_{n+1}$ iff

$$\bigwedge_{i \leq n} (s_i Z_i t_i) \implies f(s_1, \dots, s_n) Z_{n+1} f(t_1, \dots, t_n)$$

where $Z_i = \leq$ in case $k_i = \uparrow$ and $Z_i = \geq$ in case $k_i = \downarrow$. We denote as $S_1(k_1 \dots k_n, S_2(k_{n+1}))$ the set of n-ary functions from S_1 to S_2 of type $k_1 \dots k_n \mapsto k_{n+1}$. As a special case, n-ary relations on $(S, \leq)$ are n-ary operations from $(S, \leq)$ to $T = (\{\mathit{true}, \mathit{false}\}, \sqsubseteq)$, where it is assumed that $\mathit{false} \sqsubseteq \mathit{true}$.

For example, $S(\uparrow, T(\uparrow))$ denotes the set of all subsets of S that are closed upwards under $\leq$; $S(\downarrow\uparrow, T(\uparrow))$ denotes the set of binary relations on S that are anti-monotonic in the first position and monotonic in the second position; and $S(\uparrow, S(\downarrow))$ denotes the set of anti-monotonic unary functions on S. We will usually omit $T(\uparrow)$; hence $S(\uparrow)$ means $S(\uparrow, T(\uparrow))$. If B is a binary relation on S, then $B(s)$ denotes the set $\{t \mid Bst\}$, and if $X \subseteq S$, then $B(X) := \bigcup_{s \in X} B(s)$.

Definition 2.2 A *frame* is a tuple $F = (S, \leq, R, *, Q, Q_L)$ where $(S, \leq)$ is a partially ordered set, $R \in S(\downarrow\downarrow\uparrow)$, $* \in S(\uparrow, S(\downarrow))$ and $Q, Q_L \in S(\downarrow\uparrow)$. A *model* based on a frame F is $M = (F, V)$ where $V : Pr \to S(\uparrow)$.

We will consider later on structures that expand frames by additional relations on S and we will apply the terminology defined for frames to these structures.

Definition 2.3 For each frame F, we define the following operations on 2^S:

$$\begin{aligned}
X \wedge^F Y &= X \cap Y \\
X \vee^F Y &= X \cup Y \\
\neg^F X &= \{s \mid s^* \notin X\} \\
X \circ^F Y &= \{u \mid \exists s, t(s \in X \;\&\; t \in Y \;\&\; Rstu)\} \\
X \to^F Y &= \{s \mid \{s\} \circ^F X \subseteq Y\} \\
\Box^F X &= \{s \mid \forall t(Qst \Rightarrow t \in X)\} \\
\Box_L^F X &= \{s \mid \forall t(Q_L st \Rightarrow t \in X)\}
\end{aligned}$$

These operations are related to the standard satisfaction clauses of relevant (modal) logic. The tonicity conditions incorporated into the definition of a frame ensure that $S(\uparrow)$ is closed under c^F for $c \in \{\wedge, \vee, \to, \neg, \Box, \Box_L\}$.

Definition 2.4 For each $M = (F, V)$, the *M-interpretation* is a function $[\![\,]\!]_M : Fm_{\mathcal{L}} \to S(\uparrow)$ such that $[\![p]\!]_M = V(p)$ and

$$[\![c(\varphi_1, \ldots, \varphi_n)]\!]_M = c^F\left([\![\varphi_1]\!]_M, \ldots, [\![\varphi_n]\!]_M\right)$$

for all $c \in \{\wedge, \vee, \to, \neg, \Box, \Box_L\}$.

Since $S(\uparrow)$ is closed under each c^F and $V(p) \in S(\uparrow)$ by definition, it follows that $[\![\varphi]\!]_M \in S(\uparrow)$ for all $\varphi \in Fm_{\mathcal{L}}$. We will often write $(M, s) \models \varphi$ instead of $s \in [\![\varphi]\!]_M$ and $(M, X) \models \Gamma$ instead of $(M, s) \models \varphi$ for all $\varphi \in \Gamma$ and $s \in X$ ($s \models \varphi$, $X \models \Gamma$ when M is clear from context); and we will not distinguish between singletons and their elements when using this notation.

As usual in relevant logic, elements of S are seen as bodies of information, or *situations*, roughly in the sense of [1], partially ordered by the amount of information they support. Situations are not closed under the usual laws of classical logic, and the relations used to define the operations corresponding to $\neg$ and $\to$ are introduced to achieve this. First, situations may be incomplete or inconsistent, i.e. for some $s \in S$ we may have $s \in [\![\varphi]\!]_M \cap [\![\neg\varphi]\!]_M$ (meaning that $s \not\leq s^*$) or $s \notin [\![\varphi]\!]_M \cup [\![\neg\varphi]\!]_M$ (meaning that $s^* \not\leq s$). Informally, s^* is seen as the maximal situation that is *compatible* with s. Second, note that $s \in X \to Y$ iff, for all t, u, if $Rstu$ and $t \in X$, then $u \in Y$. Interpretation of implication in terms of a ternary relation on situations enables the failure of the so-called "paradoxes of strict implication"; note that it may be the case that $s \in [\![\varphi]\!]_M$ and $s \notin [\![\varphi \to \varphi]\!]_M$ if there are t, u such that $Rstu$ and $t \not\leq u$. Informally, the ternary relation R is seen as representing *combination* of information supported by situations; $Rstu$ can be seen as representing the fact that the body of information that results from combining the information supported by s with the information supported by t is contained in the information supported by u. [1] Not much is assumed about R in the general setting; for instance, we do not assume that information combination is commutative ($Rstu \Rightarrow Rtsu$), associative ($Rstuv \Rightarrow Rs(tu)v$; see the explanation of the notation before Figure 1) or reflexive ($Rsss$). However, these and similar properties of R can be assumed when one considers stronger relevant logics; see Figure 1. The modal accessibility relation Q represents information about the beliefs of our contextually fixed agent. In particular, Qst expresses that all information that the agent believes according to s is supported by t.

We now define an expansion of frames that is used in relational semantics for relevant logics. The key feature is a semantic deduction theorem according to which $\varphi \to \psi$ is valid iff ψ follows from φ in the frame.

Definition 2.5 An *L-frame* is a structure $\boldsymbol{F} = (F, L)$, where F is a frame and $L \in S(\uparrow)$ such that

$$\forall s \exists x (x \in L \ \& \ Rxss) \tag{1}$$
$$s \in L \ \& \ Rstu \Rightarrow t \leq u \tag{2}$$

A *model* based on an L-frame $\boldsymbol{F}$ (an L-model) is a tuple $\boldsymbol{M} = (\boldsymbol{F}, V)$ where $V : Pr \to S(\uparrow)$. A formula φ is *valid* in an L-model iff $L \subseteq [\![\varphi]\!]_{\boldsymbol{M}}$; notation $\boldsymbol{M} \models \varphi$. A formula φ is valid in a class of L-frames iff it is valid in each L-model based on an L-frame in the class.

Lemma 2.6 *For all L-models $\boldsymbol{M}$, $\varphi \to \psi$ is valid in $\boldsymbol{M}$ iff $[\![\varphi]\!]_{\boldsymbol{M}} \subseteq [\![\psi]\!]_{\boldsymbol{M}}$.*

Proof. Frame conditions (1), (2) and the fact that $[\![\varphi]\!]_{\boldsymbol{M}} \in S(\uparrow)$ for all φ. □

The set of formulas valid in all L-models is denoted as **BM.C**. We use the notation of [8,9], where **BM.C** denotes the smallest *conjunctively regular* modal extension of **BM**. A modal logic is said to be conjunctively regular if its modal operators distribute over conjunctions. The logic **BM** is one of the weakest propositional relevant logics; see [8, I.3–I.4].

---

[1] For instance, Dunn and Restall point out that "perhaps the best reading [of $Rstu$] is to say that the combination of the pieces of information s and t (not necessarily the union) is a piece of information in u" [5, p. 67].

Definition 2.7 The *axiom system* BM.C consists of the following axioms

(a1)	$p \to p$	(a7)	$q \to (p \vee q)$
(a2)	$\neg(p \wedge q) \to (\neg p \vee \neg q)$	(a8)	$((p \to q) \wedge (p \to r)) \to (p \to (q \wedge r))$
(a3)	$(\neg p \wedge \neg q) \to \neg(p \vee q)$	(a9)	$((p \to r) \wedge (q \to r)) \to ((p \vee q) \to r)$
(a4)	$(p \wedge q) \to p$	(a10)	$(p \wedge (q \vee r)) \to ((p \wedge q) \vee (p \wedge r))$
(a5)	$(p \wedge q) \to q$	(a11)	$(\Box p \wedge \Box q) \to \Box(p \wedge q)$
(a6)	$p \to (p \vee q)$	(a12)	$(\Box_L p \wedge \Box_L q) \to \Box_L(p \wedge q)$

plus the rules of Uniform substitution (US) and Modus ponens (MP) and

$$(\text{Adj})\ \frac{\varphi \qquad \psi}{\varphi \wedge \psi} \qquad (\text{Aff})\ \frac{\varphi' \to \varphi \qquad \psi \to \psi'}{(\varphi \to \psi) \to (\varphi' \to \psi')} \qquad (\text{Con})\ \frac{\varphi \to \psi}{\neg\psi \to \neg\varphi}$$

$$(\Box_L\text{-Mon})\ \frac{\varphi \to \psi}{\Box_L\varphi \to \Box_L\psi} \qquad (\Box\text{-Mon})\ \frac{\varphi \to \psi}{\Box\varphi \to \Box\psi}$$

The set of theorems of BM.C, Th(BM.C), is defined in the usual way. The axiom system BM.C given here differs from Fuhrmann's axiom system for the conjunctively regular modal extension of **BM** as follows: instead of axiom schemata we use axioms and add (US); we add a conjunctive regularity axiom and a monotonicity rule for the second modal box operator $\Box_L$. Note that we do not assume any interplay between $\Box$ and $\Box_L$.

Some frame conditions assumed in the semantics for relevant modal logics stronger than **BM.C** are listed in Figure 1, where we define $Rstuv := \exists x(Rstx \ \& \ Rsuv)$, $Rs(tu)v := \exists x(Rsxv \ \& \ Rtux)$, $RQstu := \exists x(Rstx \ \& \ Qxu)$ and $QRstu := \exists x(Qsx \ \& \ Rxtu)$.

For example, the logic **DW.C** is defined in terms of frames satisfying (Cp), the logic **TW.C** adds (B) and (CB), **T.C** adds (WB), (X), (RD) and (W), **E.C** adds the rule (ER), **R.C** adds (C) and the logic **RM.C** adds (M). For a propositional relevant logic **PL**, the logic **PL.R** adds *implicational regularity* ($\Box$K) to **PL.C** and **PL.K** further adds (Nec). If $\mathbf{X} \in \{\mathbf{C}, \mathbf{R}, \mathbf{K}\}$, then **PL.XT** adds ($\Box$T) to **PL.X**, and similarly for **D**, **4** and **5**. For a more extensive list, including especially more variation on the propositional level, see [8,9,15].

If a relevant modal logic **L** is defined as a set of formulas valid in all L-frames satisfying a selection of frame conditions from Figure 1, then we will denote the set of L-frames satisfying those frame conditions as *L-frames for* **L**, or simply **L**-frames, and the selection of the frame conditions as **L**-*conditions*. For any relevant modal logic **L**, the axiom system L is obtained by adding to BM.C the axioms and rules corresponding to the **L**-conditions according to Figure 1.

Theorem 2.8 *For all* **L**, $\mathbf{L} = Th(\mathsf{L})$.

Proof. Virtually the same argument as the one in [8,9]. □

3 Classical epistemic logic based on relevant modal logic

It is clear that relevant modal logics, as presented above, can be used to model reasoning about agents that are not logically omniscient with respect to classical

Frame condition		**Corresponding axiom/rule**
(DN)	$s^{**} = s$	$p \leftrightarrow \neg\neg p$
(Cp)	$Rstu \Rightarrow Rsu^*t^*$	$(p \rightarrow q) \rightarrow (\neg q \rightarrow \neg p)$
(WB)	$Rstu \Rightarrow Rs(st)u$	$((p \rightarrow q) \wedge (q \rightarrow r)) \rightarrow (p \rightarrow r)$
(X)	$s \in L \Rightarrow s^* \leq s$	$p \vee \neg p$
(Rd)	Rss^*s	$(p \rightarrow \neg p) \rightarrow \neg p$
(B)	$Rstuv \Rightarrow Rs(tu)v$	$(p \rightarrow q) \rightarrow ((r \rightarrow p) \rightarrow (r \rightarrow q))$
(CB)	$Rstuv \Rightarrow Rt(su)v$	$(p \rightarrow q) \rightarrow ((q \rightarrow r) \rightarrow (p \rightarrow r))$
(W)	$Rstu \Rightarrow Rsttu$	$(p \rightarrow (p \rightarrow q)) \rightarrow (p \rightarrow q)$
(C)	$Rstuv \Rightarrow Rsutv$	$(p \rightarrow (q \rightarrow r)) \rightarrow (q \rightarrow (p \rightarrow r))$
(M)	$Rstu \Rightarrow (s \leq u \vee t \leq u)$	$p \rightarrow (p \rightarrow p)$
(ER)	$\exists x(x \in L \ \& \ Rsxs)$	$\dfrac{\varphi}{(\varphi \rightarrow \psi) \rightarrow \psi}$
(Nec)	$(x \in L \ \& \ Qxs) \Rightarrow s \in L$	$\dfrac{\varphi}{\Box\varphi}$
($\Box$K)	$RQstu \Rightarrow \exists x(Qtx \ \& \ QRsxu)$	$\Box(p \rightarrow q) \rightarrow (\Box p \rightarrow \Box q)$
($\Box$T)	Qss	$\Box p \rightarrow p$
($\Box$D)	$\exists x(Qsx^* \ \& \ Qs^*x)$	$\Box\neg p \rightarrow \neg\Box p$
($\Box$4)	$(Qst \ \& \ Qtu) \Rightarrow Qsu$	$\Box p \rightarrow \Box\Box p$
($\Box$5)	$(Qs^*u \ \& \ Qst) \Rightarrow Qt^*u$	$\neg\Box p \rightarrow \Box\neg\Box p$

Fig. 1. Some prominent frame conditions with the corresponding axioms and rules.

logic, but whose beliefs are still logically regimented. For instance, $\Box(p \wedge \neg p) \rightarrow \Box q$ is invalid in all **L** mentioned above, but $\Box(p \wedge q) \rightarrow \Box p$ is valid. Interestingly, the degree to which agents are represented as being omniscient with respect to the relevant modal logic at hand is smaller than in the case of classical modal logic. As the Regularity axiom (a11) and the Monotonicity rule ($\Box$-Mon) entail, each L is closed under

$$\frac{\varphi_1 \wedge \ldots \wedge \varphi_n \rightarrow \psi}{\Box\varphi_1 \wedge \ldots \wedge \Box\varphi_n \rightarrow \Box\psi} \tag{CR}$$

for $n > 0$, but not each **L** is closed under the Necessitation rule (Nec). This means that even though the beliefs of agents are assumed to be closed under valid implications with non-empty antecedents, agents are not assumed to believe all valid formulas.

Moreover, in general **L** is not closed under the implicational version of (CR)

$$\frac{\varphi_1 \rightarrow (\ldots(\varphi_n \rightarrow \psi)\ldots)}{\Box\varphi_1 \rightarrow (\ldots(\Box\varphi_n \rightarrow \Box\psi)\ldots)}. \tag{IR}$$

Relational semantics for relevant modal logics yield conjunctively regular but not necessarily implicatively regular modal logics. For instance, even in logics with the implicational version of contraposition (Cp) as an axiom, the formula $\Box(p \rightarrow q) \rightarrow (\Box\neg q \rightarrow \Box\neg p)$ is not necesarily valid. In this respect relevant modal logic is more fine-grained than classical modal logic, where the distinction between (CR) and (IR) collapses due to the provability of

$(\varphi \to (\psi \to \chi)) \leftrightarrow ((\varphi \wedge \psi) \to \chi)$ [2]. The difference between (CR) and (IR) expresses the assumption that while beliefs of agents are represented as "automatically" closed under conjunction introduction, they are not seen as closed under implication elimination. As Sequoiah-Grayson [20] points out, this can be understood as meaning that while agents are assumed to automatically *aggregate* their beliefs, they are not assumed to automatically *combine* them. While this view constitutes a prima facie motivation for (CR), belief aggregation has been criticised in the epistemic logic literature, e.g. by [6], who argue that beliefs tend to come in non-interacting clusters, or frames of mind. Accommodating this idea in our framework leads to a shift from relational to neighborhood semantics, and is left for further work.

The informational interpretations of relevant logics make them good candidates for logics that regiment epistemic attitudes of realistic, non-omniscient agents. One may wonder if such an employment of relevant logics is in conflict with using classical propositional logic. In fact, most frameworks dealing with logical omniscience build on classical propositional logic and add various extra requirements to normal modal logic on the epistemic level; see [2,6,11] for instance. It can be argued that while classical consequence models truth preservation, relevant consequence models preservation of informational support in situations and their combinations. The latter is naturally associated with models of how agents reason.

In the rest of this paper we will develop a framework for epistemic logic based on these considerations. In our framework, the agent is modelled as a *relevant reasoner in a classical world:* the agent reasons in accordance with a relevant modal logic, but the propositional fragment is classical. Our framework extends the framework of [10,11] by including relevant implication and allowing for a greater variability on the propositional level.

More specifically, for each relevant modal logic L we develop a "classical" modal logic CL with two main features. First, the propositional fragment of each CL is classical propositional logic. Second, the set of theorems of CL is not closed under (CR), but only under the so-called *Relevant reasoning meta-rule*

$$\frac{\vdash_{\mathsf{L}} \varphi_1 \wedge \ldots \wedge \varphi_n \to \psi}{\vdash_{\mathsf{CL}} \Box\varphi_1 \wedge \ldots \wedge \Box\varphi_n \to \Box\psi} \qquad \text{(RR)}$$

for $n > 0$. In order to ensure that the propositional fragment of CL is classical propositional logic, CPC, we modify the relational semantics for relevant modal logics presented above so that validity in a model is defined as satisfaction throughout a set of designated states that, as far as propositional connectives are concerned, behave like classical possible worlds.

The auxiliary modal operator $\Box_L$ is crucial in achieving the second feature of our framework, closure under (RR). In particular, we will show that

$$\vdash_{\mathsf{L}} \varphi \to \psi \implies \vdash_{\mathsf{CL}} \Box_L(\varphi \to \psi) \implies \vdash_{\mathsf{CL}} \varphi \to \psi$$

[2] This will be the case also for any L containing the K-axiom ($\Box$K), as well as for classical logic, where $(\varphi \to (\psi \to \chi)) \leftrightarrow ((\varphi \wedge \psi) \to \chi)$ is provable.

In fact, we will show that $\vdash_{\mathsf{L}} \varphi \iff \vdash_{\mathsf{CL}} \Box_L \varphi$. In this sense, $\Box_L$ captures validity in L.

Closure under (RR) is the key feature of our framework. We stress that while it is satisfied, the standard logical omniscience problem is avoided in our framework since CL is generally *not* closed under (CR) nor under (Nec). This follows from the fact that while validity is defined as satisfaction in all standard states (in our case, possible worlds), the epistemic accessibility relation Q may connect standard states with non-standard states.

4 Possible worlds in relevant models

In this section we introduce the central semantic framework of the paper. Our semantics is a modification of the relational semantics for relevant modal logics introduced in Section 2 where the set of logical states L is replaced with a set of states W that, as far as propositional connectives are concerned, behave like classical possible worlds. For technical reasons, our frames need to be *bounded* in a specific sense. Bounded models for relevant modal logics were introduced by Seki [18,19].

Definition 4.1 A *bounded* frame is a frame F where $(S, \leq)$ is a bounded poset, i.e. there are elements $0, 1 \in S$ such that for all $s \in S$ $0 \leq s \leq 1$, and where, for all $s, t \in S$, the following are satisfied ($Q_{(L)} \in \{Q, Q_L\}$):

$$1^* = 0 \text{ and } 0^* = 1 \tag{3}$$
$$Q_{(L)}00 \tag{4}$$
$$Q_{(L)}1s \Rightarrow s = 1 \tag{5}$$
$$R010 \tag{6}$$
$$R1st \Rightarrow (s = 0 \text{ or } t = 1) \tag{7}$$

A *bounded model* is a model $M = (F, V)$ where F is a bounded frame and, for all $p \in Pr$, $0 \notin V(p)$ and $1 \in V(p)$.

Lemma 4.2 *If M is a bounded model, then*

(i) $(M, 1) \models \varphi$ *for all* φ*;*

(ii) $(M, 0) \not\models \varphi$ *for all* φ*.*

Proof. By simultaneous induction on the complexity of φ. The base case holds by definition of V in bounded models. The cases of $\varphi := \psi \wedge \chi$, $\varphi := \psi \vee \chi$ are trivial. When $\varphi := \neg\psi$, $1 \models \neg\psi$ iff $1^* \not\models \psi$ iff by (3) $0 \not\models \psi$, which holds by the induction hypothesis (IH); $0 \not\models \neg\psi$ iff $0^* \models \psi$ iff by (3) $1 \models \psi$, which holds by IH. When $\varphi := \psi \to \chi$, $1 \models \psi \to \chi$ iff $\forall s, t \in S(R1st,\ s \models \psi \Rightarrow t \models \chi)$. If $R1st$, then by (7) we distinguish two cases: if $s = 0$ then by IH $0 \not\models \psi$, and if $t = 1$, then $t \models \chi$ by IH. In both cases $1 \models \varphi \to \psi$. $0 \not\models \psi \to \chi$ iff $\exists s, t \in S$ such that $R0st$, $s \models \psi$ and $t \not\models \chi$; by (6) in particular $R010$, so $1 \models \psi$ and $0 \not\models \chi$ by IH. When $\varphi := \Box\psi$, $1 \models \Box\varphi$ iff $\forall s \in S$, $Q1s \Rightarrow s \models \psi$; by (5), $Q1s$ entails $s = 1$ and by IH $1 \models \psi$. $0 \not\models \Box\psi$ iff $\exists s \in S$ such that $Q0s$ and $s \not\models \psi$; by (4) $Q00$ and $0 \not\models \psi$ by IH. The case $\varphi := \Box_L\psi$ is analogous. $\square$

Definition 4.3 Let M be a bounded model. An element $w \in S$ is a *possible world* iff, for all $s, t \in S$:

$$w^* = w \tag{8}$$
$$Rwww \tag{9}$$
$$Rwst \Rightarrow (s = 0 \text{ or } w \leq t) \tag{10}$$
$$Rwst \Rightarrow (t = 1 \text{ or } s \leq w^*) \tag{11}$$

Definition 4.4 A *W-frame* is a structure $\boldsymbol{F} = (F, W)$ where F is a bounded frame, $W \subseteq S$ is a set of possible worlds and the following conditions are satisfied:

$$(\forall w \in W)(\forall s, t, u)(Q_L wu \ \& \ Rust \Rightarrow s \leq t) \tag{12}$$
$$(\forall s)(\exists w \in W)(\exists u)(Q_L wu \ \& \ Russ) \tag{13}$$

A *W-model* based on $\boldsymbol{F}$ is $\boldsymbol{M} = (\boldsymbol{F}, V)$ where $V : Pr \rightarrow S(\uparrow)$ such that $1 \in V(p)$ for all p and $0 \notin V(p)$ for all $p \in Pr$.

Conditions (12)-(13) enable W-frames to simulate validity in L-models. In W-frames, the set of states $Q_L(W) = \{u \mid \exists w(w \in W \ \& \ Q_L wu)\}$ "plays the role" of L in L-models: comparing (1)-(2) with (12)-(13), we observe that the set of states Q_L-accessible from W has the crucial properties of L in L-models. In fact, if (F, W) is a W-frame, then $(F, Q_L(W))$ is an L-frame. Moreover, we note that we have to explicitly mention the bounds $0, 1$ in (10) and (11) since $Rwst \Rightarrow w \leq t$ and $Rwst \Rightarrow s \leq w^*$ do not hold in the canonical model (see Section 5 and Footnote 6).

Definition 4.5 For each W-model $\boldsymbol{M}$, the $\boldsymbol{M}$*- interpretation* $[\![\,]\!]_{\boldsymbol{M}}$ is defined as in Definition 2.4. A formula φ is *valid* in a W-model $\boldsymbol{M}$ iff $W \subseteq [\![\varphi]\!]_{\boldsymbol{M}}$. A formula φ is valid in a class of W-frames iff it is valid in each W-model based on a W-frame belonging to the class.

Lemma 4.6 *$\Box_L(\varphi \rightarrow \psi)$ is valid in a W-model $\boldsymbol{M}$ iff $[\![\varphi]\!] \subseteq [\![\psi]\!]$.*

Proof. (12) and (13), together with the fact that $[\![\varphi]\!]_{\boldsymbol{M}} \in S(\uparrow)$, for all φ. □

Proposition 4.7 *Let $\boldsymbol{M}$ be any W-model and w a possible world. Then:*

(i) $(\boldsymbol{M}, w) \models \neg\varphi$ *iff* $(\boldsymbol{M}, w) \not\models \varphi$

(ii) $(\boldsymbol{M}, w) \models \varphi \rightarrow \psi$ *iff* $(\boldsymbol{M}, w) \not\models \varphi$ *or* $(\boldsymbol{M}, w) \models \psi$.

Proof. The first claim follows easily from $w = w^*$. The left-to-right implication of the second claim follows from $Rwww$. The right-to-left implication is established as follows. If $w \models \psi$, then $w \models \varphi \rightarrow \psi$ since $Rwst$ implies $w \leq t$. If $w \not\models \varphi$ and $Rwst$ with $s \models \varphi$, then we reason as follows. If $t = 1$, then $t \models \psi$ by Lemma 4.2. If $t \neq 1$, then $s \leq w = w^*$ by (11) and so $w \models \varphi$, which is a contradiction. □

Note that even though $\neg$ and $\rightarrow$ behave like Boolean negation and material implication, respectively, when evaluated in possible worlds, their semantic interpretation is uniform across the model, namely, it is given by the semantic operations $\neg^{\boldsymbol{F}}$ and $\rightarrow^{\boldsymbol{F}}$. The difference in their behaviour is given by the specific properties of the states of evaluation. Hence, we do not assume that the meaning of the symbols $\neg$ and $\rightarrow$ is context-dependent in our setting.[3]

Definition 4.8 A *C-variant* of an L-frame condition Φ from Figure 1 is a first-order formula that results from Φ by replacing each occurrence of $s \in L$ by $\exists w(w \in W \ \& \ Q_L ws)$, where w is a fresh variable.

CL *frame conditions* are the C-variants of the **L** frame conditions, for all **L**. A **CL**-frame is any W-frame that satisfies the **CL** frame conditions. A **CL**-model is a model based on a **CL**-frame.

It follows from the definition that if L does not occur in frame condition Φ, then Φ is identical to its C-variant.

We have already noted that each W-model can be seen as an L-model where $L = Q_L(W)$. Conversely, each L-model can be transformed into an "equivalent" W-model satisfying the "right" frame conditions. This fact will be used in the completeness proof in the next section.

Proposition 4.9 *For each L-model* $\boldsymbol{M}$ *for* **L** *with a set of states S there is a W-model* $\boldsymbol{M'}$ *for* **CL** *with the set of states* $S' \supsetneq S$ *such that, for all* $\varphi \in Fm_{\mathcal{L}}$:

(i) *for all* $s \in S$, $(\boldsymbol{M}, s) \models \varphi$ *iff* $(\boldsymbol{M'}, s) \models \varphi$;

(ii) *if* $\boldsymbol{M} \not\models \varphi$, *then* $\boldsymbol{M'} \not\models \Box_L \varphi$.

Proof. See the appendix. □

We note that, in comparison with [8,9], we do not consider logics arising by using two specific frame conditions, namely, the weakening frame condition (K) $Rstu \Rightarrow s \leq u$ and (M3) $s \in L \ \& \ t \leq u \Rightarrow t \leq s$. The reason for avoiding these is that (i) both are rather strong from the relevant logic perspective and (ii) including them would force us to use a substantially more complicated version of the construction used in the proof of Proposition 4.9.

5 Axiomatization

In this section we establish the main technical result of the paper, namely, a modular axiomatization result for logics of the form **CL**.

Definition 5.1 Let L be the axiom system for one of the relevant modal logics discussed in Section 2. We define CL as the axiom system comprising

(i) CPC with (MP) and (US) where substitutions are functions from Pr to $Fm_{\mathcal{L}}$;

(ii) for all axioms φ of L, an axiom $\Box_L \varphi$, and for all inference rules $\dfrac{\varphi_1 \ldots \varphi_n}{\psi}$ of L, the rule $\dfrac{\Box_L \varphi \ldots \Box_L \varphi_n}{\Box_L \psi}$;

[3] We thank Peter Verdée and Pierre Saint-Germier for pushing us on this point.

(iii) The Bridge Rule (BR) $\dfrac{\Box_L(\varphi \to \psi)}{\varphi \to \psi}$.

Lemma 5.2 (Soundness) *For all* L *and all* φ, *if* $\varphi \in Th(\mathsf{CL})$, *then* $\varphi \in \mathbf{CL}$.

Proof. Induction on the length of proofs. The base case is established by showing that all axioms of CL are valid in all **CL**-frames. (i) All axioms of CPC are valid in all W-frames thanks to Prop. 4.7. (ii) The fact that $\Box_L\varphi$ is valid in all **CL**-frames for each axiom φ of L can be shown using Lemma 4.6 and, where applicable, using the C-variants of the **L** frame conditions. In most cases, this boils down to standard arguments [8,9,15].[4] The one case where the C-variant differs from the original frame condition is established as follows. Assume that the C-variant of the frame condition (X), namely (C-X) $\exists w(w \in W \ \ \& \ \ Q_Lws) \Rightarrow s^* \leq s$, holds in a W-frame $\boldsymbol{F}$. To show that $\Box_L(p \vee \neg p)$ is valid in any $\boldsymbol{M}$ based on $\boldsymbol{F}$, take $w \in W$ and assume that Q_Lws. By (C-X), $s^* \leq s$. Thus, if $s \not\models p$, then $s^* \not\models p$ and so $s \models \neg p$.

To establish the induction step, we have to show that each rule of inference of CL preserves validity in **CL**-frames. (i) (MP) preserves validity in W-frames thanks to Proposition 4.7; and (US) clearly preserves validity in all frames.

(ii) The fact that the $\Box_L$-version of (Adj) preserves validity in W-frames is established using the satisfaction clause for $\Box_L$ and $\wedge$. The fact that $\Box_L$-versions of (Aff), (Con), ($\Box$-Mon) and ($\Box_L$-Mon) preserve validity in W-frames is established easily using Lemma 4.6 and satisfaction clauses for the operators involved. The fact that the $\Box_L$-version of (ER) preserves validity in W-frames satisfying (C-ER) $\forall s \exists wt(w \in W \ \ \& \ \ Q_Lwt \ \ \& \ \ Rsts)$ is established as follows. Assume that $\Box_L\varphi$ is valid in some $\boldsymbol{M}$ based on $\boldsymbol{F}$ satisfying (C-ER). Take some s and assume $s \models \varphi \to \psi$; we have to show that $s \models \psi$. Using (C-ER), there are $w \in W$ and $t \in S$ such that Q_Lwt and $Rsts$. Hence, $t \models \varphi$ and so $s \models \psi$. The fact that the $\Box_L$-version of (Nec) preserves validity in W-frames satisfying (C-Nec) $s \in Q_L(W) \ \ \& \ \ Qst \Rightarrow t \in Q_L(W)$ is established as follows. Assume that $\Box_L\varphi$ is valid in some $\boldsymbol{M}$ based on $\boldsymbol{F}$ satisfying (C-Nec). Take some $w \in W$ and s, t such that Q_Lws and Qst. By (C-Nec), there is $u \in W$ such that Q_Lut. It follows from this that $t \models \varphi$. Hence, $w \models \Box_L\Box\varphi$.

(iii) The Bridge Rule preserves validity in W-frames by Lemma 4.6 and Proposition 4.7. □

Lemma 5.3 *For all* L, $\vdash_{\mathsf{L}} \varphi$ *iff* $\vdash_{\mathsf{CL}} \Box_L\varphi$.

Proof. The fact that $\vdash_{\mathsf{L}} \varphi$ implies $\vdash_{\mathsf{CL}} \Box_L\varphi$ can be established by induction on the length of L-proofs. The base case holds since $\Box_L\varphi$ is an axiom of CL for all axioms φ of L. The induction step is established using the fact that, by definition of CL, if $\dfrac{\varphi_1 \cdots \varphi_n}{\psi}$ is an instance of an inference rule of L, then $\dfrac{\Box_L\varphi \ldots \Box_L\varphi_n}{\Box_L\psi}$ is an instance of an inference rule of CL.

[4] We note in relation to axiom ($\Box$D) that Fuhrmann's frame condition $\forall s \exists x(Qs^*x \ \ \& \ \ Qsx^*)$ corresponds to $\Box p \to \neg\Box\neg p$, not to $\Box\neg p \to \neg\Box p$, as stated in [8,9]. The frame condition suffices to show that $\Box\neg p \to \neg\Box p$ is valid only if it is assumed that $x \leq x^{**}$.

Conversely, if $\nvdash_{\mathsf{L}} \varphi$, then $\boldsymbol{M} \not\models \varphi$ for some $\mathbf{L}$-model $\boldsymbol{M}$. By Prop. 4.9 there is a $\mathbf{CL}$-model $\boldsymbol{M'}$ such that $\boldsymbol{M'} \not\models \Box_L\varphi$ and so, by Lemma 5.2, $\nvdash_{\mathsf{CL}} \Box_L\varphi$. $\Box$

Lemma 5.3 clarifies the role of the operator $\Box_L$ in our framework. We stress that $\Box_L$ is not an epistemic operator expressing attitudes of agents.[5] It follows from Lemma 5.3 and the presence of the Bridge Rule that $\vdash_{\mathsf{L}} \varphi \to \psi$ entails $\vdash_{\mathsf{CL}} \varphi \to \psi$. In fact, we can prove a stronger claim.

Proposition 5.4 *The following hold for all* L *such that the rule (Nec) is not an inference rule of* L*:*

(i) $\vdash_{\mathsf{L}} \varphi$ *entails* $\vdash_{\mathsf{CL}} \varphi$*;*

(ii) $\vdash_{\mathsf{CL}} \Box_L\varphi$ *entails* $\vdash_{\mathsf{CL}} \varphi$.

Proof. (i) The claim is established by induction on the length of L-proofs. All implicational axioms of L are provable in CL by Lemma 5.3 and the Bridge Rule; the axiom (X) $p \lor \neg p$ is provable in any CL since the propositional fragment of each L is included in CPC (this argument can be used also to show that the "purely propositional" axioms of L are provable in CL). The cases of the induction step corresponding to rules with implicational conclusions are established using Lemma 5.3 and (BR) as before (the arguments do not need to use the induction hypothesis). The case corresponding to (Adj) is established using the fact that $\vdash_{\mathsf{CPC}} \varphi \to (\psi \to (\varphi \land \psi))$ for all φ, ψ. We note that (Nec) is problematic: φ is not necessarily an implication, so we can not use Lemma 5.3 and (BR); using the induction hypothesis gives us only $\vdash_{\mathsf{CL}} \varphi$ and using only Lemma 5.3 gives us $\vdash_{\mathsf{CL}} \Box_L\varphi$, from which we can infer only that $\vdash_{\mathsf{CL}} \Box_L\Box\varphi$ using the $\Box_L$-version of (Nec). (ii) follows from (i) and Lemma 5.3. $\Box$

Note that the converses of (i) and (ii) from the previous proposition do not hold.

Proposition 5.5 (Relevant Reasoning) *Each* CL *is closed under the Relevant reasoning meta-rule* (RR)

$$\frac{\vdash_{\mathsf{L}} \varphi_1 \land \ldots \land \varphi_n \to \psi}{\vdash_{\mathsf{CL}} \Box\varphi_1 \land \ldots \land \Box\varphi_n \to \Box\psi}$$

for all $n > 0$.

Proof. If $\vdash_{\mathsf{L}} \bigwedge_{i\leq n} \varphi_i \to \psi$, then $\vdash_{\mathsf{L}} \bigwedge_{i\leq n} \Box\varphi_i \to \Box\psi$ using monotonicity and regularity of $\Box$ in L, and so $\vdash_{\mathsf{CL}} \Box_L(\bigwedge_{i\leq n} \Box\varphi_i \to \Box\psi)$ by Lemma 5.3. But then $\vdash_{\mathsf{CL}} \bigwedge_{i\leq n} \Box\varphi_i \to \Box\psi$ follows using the Bridge Rule. $\Box$

Let $(\mathsf{C})\mathsf{L} \in \{\mathsf{L}, \mathsf{CL}\}$. A $(\mathsf{C})\mathsf{L}$-*theory* is any set Γ of formulas such that (i) $\varphi \in \Gamma$ and $\psi \in \Gamma$ only if $\varphi \land \psi \in \Gamma$; and (ii) if $\varphi \in \Gamma$ and $\vdash_{(\mathsf{C})\mathsf{L}} \varphi \to \psi$, then $\psi \in \Gamma$. A $(\mathsf{C})\mathsf{L}$-theory Γ is *prime* iff (iii) $\varphi \lor \psi \in \Gamma$ only if $\varphi \in \Gamma$ or $\psi \in \Gamma$; it is *proper* iff $\Gamma \neq Fm$.

[5] Nevertheless, formulas of the form $\Box_L(\varphi \to \psi)$ can be seen as expressing a form of information constraint; see Lemma 4.6.

A pair of sets of formulas (Γ, Δ) is (C)L-*independent* iff there are no finite non-empty sets $\Gamma' \subseteq \Gamma$ and $\Delta' \subseteq \Delta$ such that

$$\vdash_{(\mathsf{C})\mathsf{L}} \bigwedge \Gamma' \to \bigvee \Delta' .$$

Lemma 5.6 (Extension Lemma) *For all* L*:*

(i) *If* (Γ, Δ) *is* L*-independent, then there is a prime* L*-theory* Σ *such that* $\Gamma \subseteq \Sigma$ *and* $\Delta \cap \Sigma = \emptyset$.

(ii) *If* (Γ, Δ) *is* CL*-independent and both* Γ *and* Δ *are non-empty, then there is a non-empty proper prime* CL*-theory* Σ *such that* $\Gamma \subseteq \Sigma$ *and* $\Delta \cap \Sigma = \emptyset$.

Proof. (i) If $\Gamma = \emptyset$, then let $\Sigma := \emptyset$. If $\Gamma \neq \emptyset$ and $\Delta = \emptyset$, then let $\Sigma := Fm$. If both Γ and Δ are non-empty, then use the standard "prime extension" argument (cf. Theorem 5.17 in [14], for example).

(ii) This is established similarly as the first claim (note that Σ needs to be proper if $\Delta \neq \emptyset$). In fact, this is the well-known Lindenbaum Lemma. □

A (C)L-theory Γ is *maximal* (C)L-*consistent* iff it is a proper (C)L-theory such that each (C)L-theory $\Delta \supsetneq \Gamma$ is not proper. It is easily shown that a CL-theory is non-empty, proper and prime iff it is maximal CL-consistent. It is also easily shown that if Γ is a non-empty proper prime CL-theory, then $Th(\mathsf{CL}) \subseteq \Gamma$.

Definition 5.7 The *canonical* CL-*frame* is the structure

$$\boldsymbol{F}^{\mathsf{CL}} = (S^{\mathsf{CL}}, \leq^{\mathsf{CL}}, W^{\mathsf{CL}}, R^{\mathsf{CL}}, *^{\mathsf{CL}}, Q^{\mathsf{CL}}, Q_L^{\mathsf{CL}})$$

where

- S^{CL} is the set of all prime L-theories;
- $\leq^{\mathsf{CL}}$ is set inclusion;
- W^{CL} is the set of all non-empty proper prime CL-theories;
- $R^{\mathsf{CL}}stu$ iff $\forall \varphi, \psi$, if $\varphi \to \psi \in s$ and $\varphi \in t$, then $\psi \in u$;
- $s^{*^{\mathsf{CL}}} = \{\varphi \mid \neg\varphi \notin s\}$;
- $Q^{\mathsf{CL}}st$ iff, $\forall \varphi$, $\Box\varphi \in s$ only if $\varphi \in t$.
- $Q_L^{\mathsf{CL}}st$ iff, $\forall \varphi$, $\Box_L\varphi \in s$ only if $\varphi \in t$;

The *canonical* CL-*model* is $\boldsymbol{M}^{\mathsf{CL}} = (\boldsymbol{F}^{\mathsf{CL}}, V^{\mathsf{CL}})$ where $V^{\mathsf{CL}} : Pr \to 2^{S^{\mathsf{CL}}}$ such that $V^{\mathsf{CL}}(p) = \{s \in S^{\mathsf{CL}} \mid p \in s\}$.

Canonical models are well defined since, for each L, any prime CL-theory is a prime L-theory (note that if $\vdash_{\mathsf{L}} \varphi \to \psi$, then $\vdash_{\mathsf{CL}} \Box_L(\varphi \to \psi)$ by Lemma 5.3 and then $\vdash_{\mathsf{CL}} \varphi \to \psi$ using the Bridge Rule).

Note that $\boldsymbol{M}^{\mathsf{CL}}$ is defined almost exactly as the canonical L-model $\boldsymbol{M}^{\mathsf{L}}$ (cf. [8,9]); the only difference is that in $\boldsymbol{M}^{\mathsf{L}}$ we have $L^{\mathsf{L}} = \{s \mid Th(\mathsf{L}) \subseteq s\}$ instead of W^{CL}. It follows that $\boldsymbol{M}^{\mathsf{CL}}$ automatically satisfies all the L-frame conditions Φ that are identical to their C-variant and are satisfied by $\boldsymbol{M}^{\mathsf{L}}$.

Lemma 5.8 *For all* L*, the structure* $\boldsymbol{M}^{\mathsf{CL}}$ *is a* **CL***-model.*

Proof. First we have to show that $\boldsymbol{M}^{\mathsf{CL}}$ is a W-model, that is, (i) relations $R^{\mathsf{CL}}, *^{\mathsf{CL}}, Q^{\mathsf{CL}}$ and Q_L^{CL} satisfy the required tonicity conditions, (ii) $\boldsymbol{M}^{\mathsf{CL}}$ is bounded, (iii) W is a set of possible worlds and (iv) conditions (12)-(13) are satisfied. Then we have to show that (v) $\boldsymbol{M}^{\mathsf{CL}}$ satisfies the **CL** frame conditions. In the remainder of the proof, we will mostly omit the superscript CL.

Claim (i) follows easily from the definitions. Claim (ii) is established as follows. Let $0 := \emptyset$ and $1 := Fm$; both are prime L-theories. Then 0 is the least element (with respect to set inclusion) of S and 1 is the greatest element. Next, $1^* = \{\varphi \mid \neg\varphi \notin 1\} = \emptyset = 0$ and $0^* = \{\varphi \mid \neg\varphi \notin 0\} = Fm = 1$. Next, clearly $Q_{(L)}00$ for all s. Next, if $Q_{(L)}1s$ and $s \neq 1$, then there is $\varphi \notin s$ and so $\Box_{(L)}\varphi \notin 1$, which is a contradiction; hence, $s = 1$. Next, $R010$ by the definition of R. Finally, if $R1st$, $s \neq 0$ and $t \neq 1$, then there is $\varphi \in s$ and $\psi \notin t$ such that $\varphi \to \psi \in 1$, which contradicts the definition of R.

(iii) We know that non-empty proper prime CL-theories are maximal CL-consistent theories, i.e. $\varphi \in s$ iff $\neg\varphi \notin s$. Hence $w^* = \{\varphi \mid \neg\varphi \notin w\} = w$. The fact that $Rwww$ follows from the fact that $\vdash_{\mathsf{CPC}} (\varphi \wedge (\varphi \to \psi)) \to \psi$. Now assume that $Rwst$ and $s \neq 0$; we have to prove that $w \subseteq t$. Thus assume $\varphi \in w$ and $\psi \in s$. Since $\vdash_{\mathsf{CPC}} \varphi \to (\psi \to \varphi)$, we have $\psi \to \varphi \in w$, and so $\varphi \in w$ by the definition of R. Since φ is arbitrary, we established $w \subseteq t$. Finally, assume $Rwst$ and $t \neq 1$; we have to prove that $s \subseteq w^*$. Hence, assume that $\neg\varphi \in w$, $\psi \notin t$ and, towards a contradiction, that $\varphi \in s$. Since $\vdash_{\mathsf{CPC}} \neg\varphi \to (\varphi \to \psi)$, we have $\varphi \to \psi \in w$ and so $\psi \in t$ by the definition of R. This is a contradiction; hence, if $\varphi \in s$, then $\neg\varphi \notin w$, meaning in general that $s \subseteq w^*$.[6]

(iv) Note that (12) follows from the fact that $\Box_L(\varphi \to \varphi)$ is a theorem of CL. Now we prove that the canonical frame satisfies (13). Take any $s \in S$; we will prove that there are $w \in W$ and $t \in S$ such that Q_Lwt and $Rtss$. If $s = 1$, then Q_Lws for all $w \in W$ and $Rsss$ and so we are done. If $s = 0$, then Q_Lw1 for all $w \in W$ and $R1ss$ and we are done. Finally, assume that $s \neq 1$ and $s \neq 0$. First, it is easily shown that the pair $\big(Th(\mathsf{L}), \{\varphi \to \psi \mid \varphi \in s \ \ \& \ \ \psi \notin s\}\big)$ is L-independent and so, by the Extension Lemma 5.6, there is a non-empty prime L-theory t such that $Rtss$. If $t = 1$, then also Q_Lwt for all $w \in W$ and we are done. If $t \neq 1$, then we can show that the pair

$$(Th(\mathsf{CL}), \{\Box_L\varphi \mid \varphi \notin t\})$$

is CL-independent. If it were not, then

- $\vdash_{\mathsf{CL}} \bigvee_{i<n} \Box_L\varphi_i$ for some $n > 0$, and so
- $\vdash_{\mathsf{CL}} \Box_L \bigvee_{i<n} \varphi_i$ by the properties of $\to$ in CL (namely, $\vdash_{\mathsf{CL}} \Box_L\varphi \vee \Box_L\psi \to \Box_L(\varphi \vee \psi)$), hence
- $\vdash_{\mathsf{L}} \bigvee_{i<n} \varphi_i$ by Lemma 5.3, so

[6] Note that $Rwst \Rightarrow w \subseteq t$ does not hold since $Rw0t$ for all w and t. Similarly, $Rwst \Rightarrow s \subseteq w^*$ does not hold since $Rws1$ for all w and s.

- $\bigvee_{i<n} \varphi_i \in t$ by the construction of u, and so
- $\varphi_i \in t$ for some $i < n$ since u is prime.

This is a contradiction, so the pair is CL-independent. Note also that both sets in the pair are non-empty (since, recall, $t \neq 1$). It follows using the Extension Lemma 5.6 that there is a non-empty maximal consistent CL-theory w such that $Q_L wt$.

(v) Assume first that Φ is a L-frame condition that is identical to its C-variant. As noted above, $\boldsymbol{F}^{\mathsf{CL}}$ has Φ iff $\boldsymbol{F}^{\mathsf{L}}$ has Φ. However, the fact that that $\boldsymbol{F}^{\mathsf{L}}$ has Φ follows from the well-know result that all L considered in this paper are canonical [8,9,15]. We will omit the details.

Assume now that Φ is a L-frame condition such that the C-variant Φ' is not identical to Φ. We reason by cases. Take (C-X) $s \in Q_L(W) \Rightarrow s^* \leq s$. If L contains axiom $p \vee \neg p$, then CL contains axiom $\Box_L(p \vee \neg p)$. Hence, if $Q_L ws$ for some $w \in W$, then $\varphi \vee \neg\varphi \in s$ for all φ. Now assume that $\psi \in s^*$. Hence, $\neg\psi \notin s$ and so $\psi \in s$. In general, $s^* \subseteq s$ as we wanted to show.

Now take (C-ER), $\forall s \exists w, t(w \in W \;\&\; Q_L wt \;\&\; Rsts)$. Fix any $s \in S$. We first construct a prime L-theory t such that $Rsts$. In order to do this, we show that the pair

$$(Th(\mathsf{L}), \{\varphi \mid \exists\psi(\varphi \to \psi \in s \;\&\; \psi \notin s)\})$$

is L-independent if (ER) is an admissible rule of L. If the pair were not L-independent, then

- $\vdash_{\mathsf{L}} \bigvee_{i<n} \varphi_i$ for some $n > 0$, which entails that
- $\vdash_{\mathsf{L}} (\bigvee_{i<n} \varphi_i \to \bigvee_{i<n} \psi_i) \to \bigvee_{i<n} \psi_i$ using (ER) where, for each $i < n$, ψ_i is a formula such that $\varphi_i \to \psi_i \in s$ and $\psi_i \notin s$; but this entails that
- $\bigvee_{i<n} \psi_i \in s$ since $\bigwedge_{i<n}(\varphi_i \to \psi_i) \in s$ and so $(\bigvee_{i<n} \varphi_i \to \bigvee_{i<n} \psi_i) \in s$. But this means that
- $\psi_i \in s$ for some $i < n$.

This is a contradiction and so the pair is L-independent. Consequently, there is a prime L-theory t such that $Th(\mathsf{L}) \subseteq t$ and $Rsts$, using the Extension Lemma 5.6. As a second step of the argument, we show that there is a non-empty proper prime CL-theory w such that $Q_L wt$. If $t = 1$, then $Q_L wt$ for all $w \in W$ and we are done. If $t \neq 1$, then we can show that the pair

$$(Th(\mathsf{CL}), \{\Box_L \varphi \mid \varphi \notin t\})$$

is CL-independent. (The argument is similar to the one establishing (13).) Note also that both sets in the pair are non-empty (since it is assumed that $t \neq 1$) and so, using the Extension Lemma 5.6, there is a non-empty proper prime CL-theory w such that $Q_L wt$. (C-Nec) is dealt with in a similar fashion.

□

Lemma 5.9 (Truth Lemma) *For all* L *and all* φ, $(\boldsymbol{M}^{\mathsf{CL}}, s) \models \varphi$ *iff* $\varphi \in s$.

Proof. Induction on the complexity of φ. The base holds by definition and the cases of the induction step where the main connective is propositional are established as usual in relevant logic [14,15]. The cases where the main connective is $\Box$ or $\Box_L$ are established as usual in relevant modal logic [8,9] using monotonicity and regularity of the box operators (in L since the claim of the lemma concerns an arbitrary prime L-theory). $\Box$

Theorem 5.10 *For all* L, $Th(\mathsf{CL}) = \mathbf{CL}$.

Proof. Soundness is established as Lemma 5.2. Completeness follows from Lemmas 5.6, 5.8 and 5.9. $\Box$

6 Conclusion

In this paper we studied a framework for epistemic logic that avoids the logical omniscience problem by introducing non-standard states, but where the epistemic attitudes of agents are regimented by a relevant logic. Unlike the classic non-standard-states approaches of Levesque [11] and Lakemeyer [10], the relevant logic regimenting attitudes is not the $\wedge, \vee, \neg$-fragment of $\mathbf{E}$, but any relevant modal logic from a wide variety of logics previously studied by Fuhrmann [8,9], for example. Hence, our approach provides a more realistic relevant formalization of attitudes towards implicational statements, which is an improvement with respect to the approach of Fagin et al. [7] as well. In comparison to earlier work on relevant epistemic logic [3,4,16,17], our framework combines relevant modal logic with classical propositional logic and it uses natural Hilbert-style proof theory without unnecessary linguistic assumptions. The main technical result of the paper is a modular completeness theorem, parametrized by the relevant modal logic governing the agents' reasoning.

There is a number of issues to pursue in the future. Firstly, it would be interesting to consider a version of our framework without the assumption of conjunctive regularity $\Box\varphi \wedge \Box\psi \to \Box(\varphi \wedge \psi)$. Such a framework needs to be based on neighborhood semantics. Secondly, it would be interesting to look at extensions of our logics with operators representing group epistemic notions such as common and distributed belief, or dynamic phenomena such as public announcements. A combination of our framework with the approach of [12] is an option to consider. Thirdly, it is tempting to consider a first-order version of the present framework.

Acknowledgement. Igor Sedlár acknowledges the support of the long-term strategic development financing of the Institute of Computer Science (RVO:67985807) and the support of the Czech Science Foundation project GA22-01137S. We thank Nicholas Ferenz for valuable comments on an earlier version of the paper. We are grateful to three anonymous reviewers for AiML for a number of useful suggestions. The paper was presented at the workshop *Relevant Logic Today* organized by Peter Verdée at UCLouvain in May 2022. We are grateful to the audience at the workshop for helpful discussion.

References

[1] Barwise, J. and J. Perry, "Situations and Attitudes," MIT Press, 1983.

[2] Berto, F. and P. Hawke, *Knowability relative to information*, Mind **130** (2018), pp. 1–33.

[3] Bílková, M., O. Majer and M. Peliš, *Epistemic logics for sceptical agents*, Journal of Logic and Computation **26** (2016), pp. 1815–1841.

[4] Bílková, M., O. Majer, M. Peliš and G. Restall, *Relevant agents*, in: L. Beklemishev, V. Goranko and V. Shehtman, editors, *Advances in Modal Logic, Volume 8* (2010), pp. 22–38.

[5] Dunn, J. M. and G. Restall, **6**, Kluwer, 2002, 2nd edition pp. 1 – 128.

[6] Fagin, R. and J. Y. Halpern, *Belief, awareness, and limited reasoning*, Artificial Intelligence **34** (1987), pp. 39 – 76.

[7] Fagin, R., J. Y. Halpern and M. Vardi, *A nonstandard approach to the logical omniscience problem*, Artificial Intelligence **79** (1995), pp. 203–240.

[8] Fuhrmann, A., "Relevant Logics, Modal Logics and Theory Change," PhD Thesis, Australian National University (1988).

[9] Fuhrmann, A., *Models for relevant modal logics*, Studia Logica **49** (1990), pp. 501–514.

[10] Lakemeyer, G., *Tractable meta-reasoning in propositional logics of belief.*, in: *IJCAI 1987*, Morgan Kaufmann Publishers Inc., 1987 pp. 401–408.

[11] Levesque, H., *A logic of implicit and explicit belief*, in: *Proceedings of AAAI 1984*, 1984, pp. 198–202.

[12] Punčochář, V. and I. Sedlár, *Relevant epistemic logic with public announcements and common knowledge*, in: P. Baroni, C. Benzmüller and Y. N. Wáng, editors, *Logic and Argumentation. Proc. of the 4th International Conference on Logic and Argumentation (CLAR 2021)*, Lecture Notes in Computer Science **13040**, pp. 342–361.

[13] Rantala, V., *Impossible worlds semantics and logical omniscience*, Acta Philosophica Fennica **35** (1982), pp. 106–115.

[14] Restall, G., "An Introduction to Substrucutral Logics," Routledge, London, 2000.

[15] Routley, R., V. Plumwood, R. K. Meyer and R. T. Brady, **1**, Ridgeview, 1982.

[16] Sedlár, I., *Substructural epistemic logics*, Journal of Applied Non-Classical Logics **25** (2015), pp. 256–285.

[17] Sedlár, I., *Epistemic extensions of modal distributive substructural logics*, Journal of Logic and Computation **26** (2016), pp. 1787–1813.

[18] Seki, T., *General frames for relevant modal logics*, Notre Dame Journal of Formal Logic **44** (2003), pp. 93–109.

[19] Seki, T., *A sahlqvist theorem for relevant modal logics*, Studia Logica **73** (2003), pp. 383–411.

[20] Sequoiah-Grayson, S., *A logic of affordances*, in: M. Blicha and I. Sedlár, editors, *The Logica Yearbook 2020*, 2021, pp. 219–236.

[21] Wansing, H., *A general possible worlds framework for reasoning about knowledge and belief*, Studia Logica **49** (1990), pp. 523–539.

A Technical appendix

This technical appendix contains the proof of Proposition 4.9.

Proposition 4.9. *For each L-model $\boldsymbol{M}$ for* **L** *with a set of states S there is a W-model $\boldsymbol{M'}$ for* **CL** *with the set of states $S' \supsetneq S$ such that, for all $\varphi \in Fm_{\mathcal{L}}$:*

(i) *for all $s \in S$, $(\boldsymbol{M}, s) \models \varphi$ iff $(\boldsymbol{M'}, s) \models \varphi$;*

(ii) *if $\boldsymbol{M} \not\models \varphi$, then $\boldsymbol{M'} \not\models \Box_L \varphi$.*

Proof. Let $\boldsymbol{M} = (S, \leq, L, R, *, Q, Q_L, V)$ be a L-model. The structure $\boldsymbol{M}^+ = (S^+, \leq^+, W^+, R^+, *^+, Q^+, Q_L^+, V^+)$ is defined as follows:

$S^+ = S \cup \{w, 0, 1\}$

$$\leq^+ = \leq \cup \{(w,w)\} \cup \{(s,1) \mid s \in S^+\} \cup \{(0,s) \mid s \in S^+\}$$
$$W = \{w\}$$
$$R^+ = R \cup \{(w,w,w)\} \cup \{(0,s,t),(s,0,t),(s,t,1) \mid s,t \in S^+\}$$
$$*^+ = * \cup \{(w,w)\} \cup \{(0,1),(1,0)\}$$
$$Q^+ = Q \cup \{(w,w)\} \cup \{(s,1) \mid s \in S^+\} \cup \{(0,s) \mid s \in S^+\}$$
$$Q_L^+ = Q_L \cup \{(w,w)\} \cup \{(w,s) \mid s \in L\} \cup \{(s,1) \mid s \in S^+\} \cup \{(0,s) \mid s \in S^+\}$$
$$V^+(p) = V(p) \cup \{1\} \text{ for all } p$$

We prove first that for all $\boldsymbol{M}$, the structure $\boldsymbol{M}^+$ is a W-model (Claim A.1), then we prove that for all $s \in S$ and all $\varphi \in Fm_{\mathcal{L}}$, $(\boldsymbol{M}, s) \models \varphi$ iff $(\boldsymbol{M}^+, s) \models \psi$ (Claim A.2). It follows from the construction of $\boldsymbol{M}^+$ and Claim A.2 that if $\boldsymbol{M} \not\models \varphi$, then $\boldsymbol{M}^+ \not\models \Box_L\varphi$ (since $(\boldsymbol{M}, s) \not\models \varphi$ for some $s \in L$ implies that $(\boldsymbol{M}^+, w) \not\models \Box_L\varphi$). Finally, we prove that if $\boldsymbol{M}$ is an **L**-model, then $\boldsymbol{M}^+$ is a **CL**-model (Claim A.3).

Claim A.1 *For all $\boldsymbol{M}$, the structure $\boldsymbol{M}^+$ is a W-model.*

Firstly, we have to show that $\boldsymbol{M}^+$ is a model, i.e. $\leq^+$ is a partial order and R^+, $*^+$, Q^+ and Q_L^+ satisfy the required tonicity conditions. A simple analysis of cases shows that $\leq^+$ is a partial order. To show that $R^+ \in S^+(\downarrow\downarrow\uparrow)$, assume for instance that R^+stu and $v \leq s$. If $\{s,t,u\} \subseteq S$, then either $v \in S$ or $v = 0$; in both cases R^+vtu. If $s \notin S$ and $s = 0$, then $v = 0$ and we are done since R^+0tu for all t, u. If $s = w$, then either $s = t = u$ and so either $v = w$ or $v = 0$ and we are done, or $t = 0$ and we are done, or $u = 1$ and we are done. Finally, if $s = 1$, then either $t = 0$ or $u = 1$, but then we are done in both cases. The cases $t \notin S$ and $u \notin S$ are dealt with similarly. The fact that the other relations satisfy the required tonicity conditions is established by a similar tedious examination of cases.

Secondly, we have to show that $\boldsymbol{M}^+$ is a bounded model and that w is a possible world. This is easily checked. Thirdly, we have to show that the frame conditions (12) and (13) are satisfied. Note that Q_L^+wu iff $u \in L \cup \{w, 1\}$. We deal with (12) first. If $u \in L$ and R^+ust, then either $s, t \in S$ and we are done thanks to (1), or $s = 0$ or $t = 1$, and then we are done thanks to the properties of $\leq^+$. If $u = w$, then by (10) either $s = 0$ in which case we are done or $w \leq^+ t$. If $w = t$, then either $s = w$ or $s = 0$ and we are done; if $t = 1$, then we are done no matter what s is. Now we check (13). If $s \in S$, then there is $u \in L$ such that $Russ$ by (11); hence R^+uss, but we also know that Q_L^+wu and so we are done. If $s \notin S$, then R^+wss by easy inspection of cases, but we also know that Q_L^+ww, and so we are done. This concludes the proof of Claim A.1.

Claim A.2 *For all $s \in S$ and all $\varphi \in Fm_{\mathcal{L}}$, $(\boldsymbol{M}, s) \models \varphi$ iff $(\boldsymbol{M}^+, s) \models \varphi$.*

The proof is by induction on the complexity of φ. The base case and the induction steps for $\wedge, \vee$ are trivial. $(\boldsymbol{M}, s) \models \neg\varphi$ iff $(\boldsymbol{M}, s^*) \not\models \varphi$ iff $(\boldsymbol{M}^+, s^*) \not\models \varphi$ (by IH) iff $(\boldsymbol{M}^+, s^{*^+}) \not\models \varphi$ (by the fact, easily confirmed by inspection of the definition, that if $s \in S$, then $s^* = s^{*^+}$) iff $(\boldsymbol{M}^+, s) \models \neg\varphi$. If $(\boldsymbol{M}, s) \not\models \varphi \to \psi$, then there are $t, u \in S$ such that $Rstu$ and $(\boldsymbol{M}^+, t) \models \varphi$ and $(\boldsymbol{M}^+, u) \not\models \psi$ by IH. The rest follows from $R \subseteq R^+$. Conversely, if $(\boldsymbol{M}^+, s) \not\models \varphi \to \psi$, then there are $t, u \in S^+$ such that R^+stu, $(\boldsymbol{M}^+, t) \models \varphi$ and $(\boldsymbol{M}^+, u) \not\models \psi$. By inspection of the definition of R^+, either $t, u \in S$, or $t = 0$, or $u = 1$. Since we already know that $\boldsymbol{M}^+$ is a bounded model, the latter two options are ruled out by Lemma 4.2. Hence, $t, u \in S$ but in this case $Rstu$ and so $(\boldsymbol{M}, s) \not\models \varphi \to \psi$ by IH. If $(\boldsymbol{M}, s) \not\models \Box_L \varphi$, then $(\boldsymbol{M}^+, s) \not\models \Box_L \varphi$ by $Q_L \subseteq Q_L^+$ and IH. Conversely, if $(\boldsymbol{M}^+, s) \not\models \Box_L \varphi$, then there is t such that $Q_L^+ st$ and $(\boldsymbol{M}^+, t) \not\models \varphi$. By inspection of the definition of Q_L^+ we see that either $t = 1$, which contradicts Lemma 4.2, or $t \in S$, in which case $Q_L st$ and so we are done by IH. The case of $\Box$ is analogous. This concludes the proof of Claim A.2.

Claim A.3 *if $\boldsymbol{M}$ is an* **L***-model, then $\boldsymbol{M}^+$ is a* **CL***-model.*

The proofs for (B), (CB), (W), (C), (WB) and (M) follow the same strategy, hence we show the details for (B) only. Assume that R^+stx and R^+xuv; we have to prove that there is y such that R^+tuy and R^+syv. Let $T = \{s, t, u, v, x\}$. First, if $T \subseteq S$, then $Rstx$ & $Rxuv$ and so we are done, since (B) holds in $\boldsymbol{M}$ and $R \subseteq R^+$. Second, if $1 \in T$ or $0 \in T$, then we distinguish three cases:

(i) If $0 \in \{s, t, u\}$ or $v = 1$, then we are done. (For instance, if $s = 0$, then R^+s1v and R^+tu1; the other cases are similar.)

(ii) If $x = 0$, then by (7) either $s = 0$ or $t = 0$ and we are in (i). If $x = 1$, then by (7) either $u = 0$ or $v = 1$; in both cases we are in (i). We will use (7) without explicit reference below.

(iii) If $s = 1$, then $t = 0$ (i) or $x = 1$ (ii). If $t = 1$, then $s = 0$ (i) or $x = 1$ (ii). If $u = 1$, then $x = 0$ or $v = 1$ (ii). If $v = 0$, then $x = 0$ (ii) or $u = 0$ (i).

Third, if $T \subseteq S \cup \{w\}$, then we are either in case (i) or $T = \{w\}$. In the latter case, set $y = w$ and we are done. These three groups of cases exhaust all possibilities and so $\boldsymbol{M}^+$ has to satisfy (B) if $\boldsymbol{M}$ does.

(DN) and (Rd) are preserved since they hold for $s \in \{w, 0, 1\}$ irrespectively of the properties of $\boldsymbol{M}$. (Cp) Assume that R^+stu. If $T = \{s, t, u\} \subseteq S$, then $\boldsymbol{M}^+$ satisfies the fame condition if $\boldsymbol{M}$ does. If $s = 0$ or $t = 0$ or $u = 1$ then $R^+su^{*^+}t^{*^+}$ holds by definition of $*^+$ and R^+. The cases where $s = 1$ or $t = 1$ or $u = 0$ reduce to the previous cases. If $T \subseteq S \cup \{w\}$, then either $T \subseteq S$ or $T = \{w\}$; we are done in both cases. To prove that if $\boldsymbol{M}$ satisfies (X), then $\boldsymbol{M}^+$ satisfies (C-X) $s \in Q_L(W) \Rightarrow s^* \leq s$, assume $s \in Q_L^+(w)$, i.e. $s = w$, $s = 1$ or $s \in L$. If $s = 1$, $0 \leq^+ 1$; if $s = w$ $w \leq^+ w$; and if $s \in L$ then $s^* \leq s$ by assumption and so $s^{*^+} \leq^+ s$.

To show that $\boldsymbol{M}^+$ satisfies (C-ER) $\forall s \exists x (x \in Q_L^+(w)$ & $R^+sxs)$ if $\boldsymbol{M}$ satisfies (ER), we reason as follows. If $s \in \{w, 0, 1\}$, then R^+sws, and we know that Q_L^+ww. If $s \in S$, then by (ER) there is $x \in L$ such that $Rsxs$, and so Q_L^+wx and R^+sxs.

To show that $\boldsymbol{M}^+$ satisfies (C-Nec) $\forall s \forall x (Q_L^+wx$ & $Q^+xs) \Rightarrow Q_L^+ws$ if $\boldsymbol{M}$ satisfies (Nec), we reason as follows. Assume that Q_L^+wx & Q^+xs. WE have to prove Q_L^+ws. If $T = \{x, s\} \subseteq S$, then $x \in L$ and Qxs. Using (Nec), we obtain $s \in L$, which entails that $Q^+_L ws$. If $0 \in T$ or $1 \in T$, then we reason as follows. If $x = 0$, then Q^+wx entails that $w = 0$, which is a contradiction. If $x = 1$, then Q^+xs entails $s = 1$ and then $Q^+_L ws$ by definition of Q^+_L, and we are done. If $s = 0$, then Q^+xs entails $x = 0$ which we already know to lead to a contradiction. The only remaining possibility is that $T \subseteq S \cup \{w\}$. We have already checked the case $T \subseteq S$. If, on the other hand, $w \in T$, then we reason as follows. If $x = w$, then Q^+xs entails that either $s = w$ or $s = 1$. In both cases $Q^+_L ws$. If $s = w$, then $Q^+_L ws$ as before. This exhausts all possibilities and so we are done.

To show that $\boldsymbol{M}^+$ satisfies ($\Box$K) if $\boldsymbol{M}$ does, assume that R^+stx and Q^+xu; we have to prove that there are y, z such that Q^+ty, Q^+sz and R^+zyu. First, if $T = \{s,t,u,x\} \subseteq S$, then $Rstx \;\;\&\;\; Qxu$, and so we are done using the assumption that $\boldsymbol{M}$ satisfies ($\Box$K) and $R \subseteq R^+$, $Q \subseteq Q^+$. Second, if $0 \in T$ or $1 \in T$, then we reason as follows. (i) If $s = 0$ or $t = 0$ or $u = 1$, then we can easily find $y, z \in \{0,1\}$ such that Q^+ty, Q^+sz and R^+zyu; (ii) if $x = 0$, then $s = 0$ or $t = 0$, which reduce to case (i), and if $x = 1$, then $u = 1$, which also reduces to case (i); (iii) if $s = 1$, then $t = 0$ (i) or $x = 1$ (ii); if $t = 1$, then $s = 0$ (i) or $x = 1$ (ii); and if $u = 0$, then $x = 0$ (ii). Third, if $T \subseteq S \cup \{w\}$, then either $T \subseteq S$, which is the first case, or $w \in T$; but then it can be shown that $T = \{w\}$ from which it follows easily that Q^+tw, Q^+sw and R^+wwu. This exhausts all possibilities and so we are done.

If $\boldsymbol{M}$ satisfies ($\Box$T), then so does $\boldsymbol{M}^+$ since Q^+ss if $s \in \{w,0,1\}$. ($\Box$D) is dealt with similarly; if $s \in \{w,0,1\}$ then a suitable $x \in \{w,0,1\}$ is easily found. To show that ($\Box$4) is satisfied in $\boldsymbol{M}^+$ if it is satisfied in $\boldsymbol{M}$, we reason as follows. Assume that Q^+st and Q^+tu; we have to show that Q^+su. First, if $T = \{s,t,u\} \subseteq S$, then Q^+su follows from the assumption that ($\Box$4) holds in $\boldsymbol{M}$ and $Q \subseteq Q^+$. Second, if $0 \in T$ or $1 \in T$, then we reason by cases as follows: (i) if $s = 0$ or $u = 1$, then trivially Q^+su. (ii) If $s = 1$, then $t = 1$ and so $u = 1$, which brings us back to case (i); if $u = 0$, then $t = 0$ and so $s = 0$, which brings us back to (i). Third, if $T \subseteq S \cup \{w\}$, then either $T \subseteq S$, in which case we are done, or $T = \{w\}$, in which case of course Q^+su. This exhausts all possibilities and so we are done. Preservation of ($\Box$5) is established in a similar way. This concludes the proof of Claim A.3 and Proposition 4.9. $\Box$

An Epistemic Interpretation of Tensor Disjunction

Haoyu Wang Yanjing Wang

Department of Philosophy, Peking University

Yunsong Wang

ILLC, University of Amsterdam

Abstract

This paper aims to give an epistemic interpretation to the *tensor disjunction* in dependence logic, through a rather surprising connection to the so-called *weak disjunction* in Medvedev's early work on intermediate logic under the Brouwer-Heyting-Kolmogorov (BHK)-interpretation. We expose this connection in the setting of inquisitive logic with tensor $\mathbf{InqB}^{\otimes}$ [7], but from an epistemic perspective. More specifically, we translate the propositional formulas of $\mathbf{InqB}^{\otimes}$ into modal formulas in a powerful epistemic language of knowing how following the proposal by [21,18]. We give a complete axiomatization of the logic of our full language based on Fine's axiomatization of S5 modal logic with propositional quantifiers. Finally we generalize the tensor operator with parameters k and n, which intuitively captures the epistemic situation that one has n potential answers to n questions and knows that at least k of them must be correct. The original tensor disjunction is the special case when $k = 1$ and $n = 2$. We show that adding the generalized tensor operators do not increase the expressive power of our logic, inquisitive logic and propositional dependence logic, though most of these generalized tensors are not uniformly definable in these logics, except in ours.

1 Introduction

As a rapidly growing field of research, *Dependence Logic* studies reasoning patterns expressed by logical languages extended with (in)dependence atoms (cf. e.g., [11] for a survey). The intuitive meaning of the atomic formula is best fleshed out formally by the *team semantics* capturing the (in)dependence between variables, where a *team* can be viewed as a collection of assignments or possible worlds. In defining the truth conditions for logical connectives, one guideline is to keep the property of *flatness*, i.e., for any formula α without the (in)dependence atoms, it is true w.r.t. a team X ($X \vDash \alpha$) if it is true on each singleton team $\{w\}$ such that $w \in X$. To some extent, flatness preserves the intuition of the classical logical connectives. In particular, the semantics of the distinct tensor disjunction $\otimes$ in dependence logic can be viewed as a natural lifting of the world-based semantics for classical disjunction to teams, viewed as *sets* of possible worlds:

$$X \vDash \alpha \otimes \beta \text{ iff there are } U, V \subseteq X \text{ such that } X \subseteq U \cup V,\ U \vDash \alpha \text{ and } V \vDash \beta$$

Note that a disjunction $\alpha \vee \beta$ is classically true on each world in a set X of possible worlds if and only if there are two subsets jointly covering the whole space of possible worlds such that one subset satisfies α homogeneously and the other satisfies β homogeneously. This lifting may also give the impression that $\otimes$ can be read more or less as a classical disjunction. However, it is not so straightforward. For example, the truth of the propositional dependence formula $(=(p,q) \otimes =(p,q))$ over a team is not equivalent to $=(p,q)$. In fact, $(=(p,q) \otimes =(p,q))$ is valid technically but $=(p,q)$, which says that the truth value of q depends on the truth value of p, is clearly *not* valid. A natural question arises: how to understand $\otimes$ intuitively and precisely?[1] Our work proposes a possible epistemic understanding of $\otimes$ (and its generalizations) from a Brouwer-Heyting-Kolmogorov (BHK)-like perspective to be explained below.

The initial idea is based on an unexpected connection between the tensor disjunction and the so-called *weak disjunction* in Medvedev's early work [14] on the *problem semantics* of intuitionistic logic, following Kolmogorov's problem-solving interpretation [13]. This connection is best exposed in the setting of *inquisitive logic* with tensor disjunction discussed in [7], since inquisitive logic has intimate connections with both the propositional dependence logic [25,4] and Medvedev's logic [9]. More specifically, various versions of propositional dependence logic can be viewed as disguised inquisitive logic, e.g., the dependence atom $=(p,q)$ becomes $(p \vee \neg p) \rightarrow (q \vee \neg q)$ in inquisitive logic [23,25,6]. On the other hand, Medvedev's logic is the substitution-closed core of inquisitive logic **InqB** that also admits a BHK-like interpretation via resolutions [3,9].[2] Another advantage of using inquisitive logic as the "medium" is that we can put classical, intuitionistic, and tensor disjunctions in the same picture to reveal their differences. The last missing piece for an intuitive reading of tensor is an epistemic interpretation that can incorporate the BHK-interpretation. Wang proposed to capture intuitionistic truth using a modality Kh to express *knowing how to prove/solve* [21], which reflects Heyting's often-overlooked early view of intuitionistic logic as an epistemic logic [12]. This also led to an *alternative* epistemic interpretation of inquisitive logic [18], where a state supports a formula α is rendered as *it is known how α is resolved* when viewing the state as a set of possible worlds capturing the epistemic uncertainty. This can give us alternative epistemic readings of inquisitive formulas, e.g., the excluded middle $p \vee \neg p$ in inquisitive logic is first rendered as $\mathsf{Kh}(p \vee \neg p)$ (knowing how $p \vee \neg p$ is true), then it can be reduced to $\mathsf{Kh}p \vee \mathsf{Kh}\neg p$ (knowing how p is true or knowing

[1] In [17], it is suggested that the (in)dependence formulas can be viewed as *types* of teams, i.e., each formula specifies a property of the team. The truth conditions of connectives also have their roots in the game semantics for classical and IF logic [17].

[2] In recent literature, inquisitive logic is also viewed as an extension of classical logic [4].

how $\neg p$ is true), and finally it is equivalent to the intuitively invalid $\mathsf{K}p \vee \mathsf{K}\neg p$ [18]. We will also see these reductions later in this work.

Now we are ready to give the epistemic interpretation of tensor disjunction. According to Medvedev's problem semantics [14], the *weak disjunction* $\alpha \sqcup \beta$ captures a composite problem where the solutions are pairs of *potential* solutions to the problems of α and β respectively such that *at least one* solution in each pair is correct.[3] From our epistemic perspective, Medvedev's truth concept for a formula γ means it is known how to solve γ. In particular, a weak disjunction $\alpha \sqcup \beta$ is true w.r.t. a set of possible worlds (i.e., a team or a *state* in inquisitive logic) iff there are two solutions r_1 and r_2 such that it is known that one of r_1 and r_2 is a correct solution to the corresponding problems. In the setting of inquisitive logic, instead of problems, we take formulas as statements or questions, which may have *resolutions* instead of solutions. We will show such a truth condition amounts to exactly the team semantics for tensor.

We first summarize what we are going to do in the paper before diving into the technical details. After introducing inquisitive logic with tensor disjunction $\mathbf{InqB}^{\otimes}$ in Section 2, we first propose in Section 3 a logical language of know-that and know-how, with extra machinery of announcements and propositional quantifiers, interpreted over epistemic models that are essentially states/teams in the literature. The semantics of the know-how operator Kh is given by using the $\exists x\mathsf{K}$ schema as in know-wh logics [20], based on a BHK-like interpretation following Medvedev's idea. The intention is to capture the alternative epistemic meaning of an $\mathbf{InqB}^{\otimes}$ formula α as knowing how to resolve α. In Section 4, we show that valid know-how formulas correspond exactly the theorems in $\mathbf{InqB}^{\otimes}$. Moreover, we also show that the announcements and propositional quantifiers facilitate a recursive process to "open up" the know-how formulas, in particular to decode the $\otimes$, and eventually translate them into classical ones free of the know-how operator. Based on such a process we give a complete axiomatization of our full dynamic epistemic logic in Section 5. Finally, in Section 6 we generalize the idea of the tensor, from our epistemic interpretation, to obtain a spectrum of n-ary disjunctions $\otimes^k_n$, which captures the interesting epistemic situation of knowing n potential answers to n questions and being sure at least k of them must be correct. We show that adding the generalized tensor operators does not increase the expressive power of our logic, the inquisitive logic and propositional dependence logic, though most of these generalized tensors are not uniformly definable in these logics. In contrast, we can uniformly define the generalize tensors in our epistemic language.

2 Inquisitive logic with tensor

Following [7], we introduce the language and semantics of *Inquisitive Logic with Tensor Disjunction* ($\mathbf{InqB}^{\otimes}$). In contrast with [7], we use the symbol $\vee$ for the inquisitive disjunction and adopt the model-based semantics as in [5]. Throughout the paper, we fix a countable set $\mathbf{P}$ of proposition letters.

---

[3] See [2], for the corresponding Kripke semantics of weak disjunction.

Definition 2.1 (Language $\mathbf{PL}^{\otimes}$) *The language of propositional logic with tensor disjunction ($\mathbf{PL}^{\otimes}$) is defined as follows:*

$$\alpha ::= p \mid \bot \mid (\alpha \wedge \alpha) \mid (\alpha \vee \alpha) \mid (\alpha \to \alpha) \mid (\alpha \otimes \alpha)$$

where $p \in \mathbf{P}$. We write $\neg\alpha$ for $\alpha \to \bot$, $\top$ and $\alpha \leftrightarrow \beta$ are defined as usual.

Definition 2.2 (Model and state) *A model is a pair $\mathcal{M} = \langle W, V \rangle$ where:*

- *W is a non-empty set of possible worlds;*[4]
- *$V : \mathbf{P} \to \wp(W)$ is a valuation function.*

A state *(or, say, a* team*) s in $\mathcal{M}$ is a subset of W.*

We will also view these models as ***epistemic models*** for our dynamic epistemic language to be introduced in Section 3.

Given $\mathcal{M}$, we refer to its components as $W_{\mathcal{M}}$ and $V_{\mathcal{M}}$. We write $w \in \mathcal{M}$ in case that $w \in W_{\mathcal{M}}$, and $\mathcal{M}' \subseteq \mathcal{M}$ in case that $W'_{\mathcal{M}} \subseteq W_{\mathcal{M}}$. The semantics is defined through the *support relation* between *states* (in models) and formulas.

Definition 2.3 (Support [7]) *The* support *relation $\Vdash$ is defined inductively:*

$\mathcal{M}, s \Vdash p$	*iff*	$\forall w \in s, w \in V(p)$
$\mathcal{M}, s \Vdash \bot$	*iff*	$s = \varnothing$
$\mathcal{M}, s \Vdash (\alpha \wedge \beta)$	*iff*	$\mathcal{M}, s \Vdash \alpha$ *and* $\mathcal{M}, s \Vdash \beta$
$\mathcal{M}, s \Vdash (\alpha \vee \beta)$	*iff*	$\mathcal{M}, s \Vdash \alpha$ *or* $\mathcal{M}, s \Vdash \beta$
$\mathcal{M}, s \Vdash (\alpha \to \beta)$	*iff*	$\forall t \subseteq s$: *if* $\mathcal{M}, t \Vdash \alpha$ *then* $\mathcal{M}, t \Vdash \beta$
$\mathcal{M}, s \Vdash (\alpha \otimes \beta)$	*iff*	*there exist two sets* $t \subseteq s$ *and* $t' \subseteq s$ *such that* $\mathcal{M}, t \Vdash \alpha$, $\mathcal{M}, t' \Vdash \beta$, *and* $t \cup t' = s$.

A formula α is valid *if it is supported by any state in any model.*

Here are some simple properties.

Proposition 2.4 (Downward closure) *For any $\alpha \in \mathbf{PL}^{\otimes}$, if $\mathcal{M}, s \Vdash \alpha$ then $\mathcal{M}, t \Vdash \alpha$ for any $t \subseteq s$. Moreover, $\mathcal{M}, \varnothing \Vdash \alpha$ for all $\alpha \in \mathbf{PL}^{\otimes}$.*

Proposition 2.5 *For any $\alpha \in \mathbf{PL}^{\otimes}$, $\mathcal{M}, s \Vdash \alpha$ implies $\mathcal{M}', s \Vdash \alpha$ for any $\mathcal{M}' \subseteq \mathcal{M}$ such that $s \subseteq \mathcal{M}'$. Conversely, if $\mathcal{M}', s \Vdash \alpha$ then $\mathcal{M}, s \Vdash \alpha$ given $\mathcal{M}' \subseteq \mathcal{M}$. Namely, the support relation only depends on the state.*

Definition 2.6 *Inquisitive Logic with Tensor Disjunction ($\mathbf{InqB}^{\otimes}$) is the set of valid $\mathbf{PL}^{\otimes}$ formulas under the support relation.*

3 A dynamic epistemic language

Definition 3.1 (Language PALKhΠ) *The language of* Public Announcement Logic with Know-how Operator and Propositional Quantifier *is:*[5]

$$\varphi ::= p \mid \bot \mid (\varphi \wedge \varphi) \mid (\varphi \vee \varphi) \mid (\varphi \otimes \varphi) \mid (\varphi \to \varphi) \mid \mathsf{K}\varphi \mid \mathsf{Kh}\alpha \mid \forall p \varphi \mid [\varphi]\varphi$$

*where $p \in \mathbf{P}$ and $\alpha \in \mathbf{PL}^{\otimes}$. We write $\neg\varphi$ for $\varphi \to \bot$, $\widehat{\mathsf{K}}$ for $\neg\mathsf{K}\neg$, $\exists p$ for $\neg\forall p \neg$ for all $p \in \mathbf{P}$ and $\langle\varphi\rangle$ for $\neg[\varphi]\neg$ for all $\varphi \in$ **PALKhΠ**.*

---

[4] In [8], the world set W could be empty. The distinction is not technically significant.

[5] Π in the name **PALKhΠ** denotes propositional quantifiers as in the literature [10].

Intuitively, $\mathsf{K}\varphi$ expresses "the agent *knows that* φ", $\mathsf{Kh}\alpha$ says that "the agent *knows how* to resolve α" or simply "the agent *knows how* α is true", $\forall p\varphi$ says that "for any proposition p, φ holds" and $[\varphi]\psi$ means that "after announcing φ, ψ holds". Note that Kh only allows $\mathbf{PL}^{\otimes}$-formula α in its scope. For instance, we can express $\mathsf{K}\neg\mathsf{Kh}\alpha$ but not $\mathsf{Kh}\mathsf{K}\alpha$ in **PALKhΠ**. We write $\varphi[\psi/\chi]$ for any formula obtained by replacing one or several occurrences of ψ with χ in φ.

We view the models in Definition 2.2 as *epistemic models* where the implicit epistemic relation is the total relation. The semantics of **PALKhΠ** is given on such models, with the notions of *resolution space* and *resolution* as below.

Definition 3.2 (Resolution space) *S is a function assigning each $\alpha \in \mathbf{PL}^{\otimes}$ its (non-empty) set of potential resolutions:*

$$\begin{array}{rlrl} S(p) &= \{p\},\ \textit{for}\ p \in \mathbf{P} & S(\bot) &= \{\bot\} \\ S(\alpha \vee \beta) &= (S(\alpha) \times \{0\}) \cup (S(\beta) \times \{1\}) & S(\alpha \to \beta) &= S(\beta)^{S(\alpha)} \\ S(\alpha \wedge \beta) &= S(\alpha) \times S(\beta) & S(\alpha \otimes \beta) &= S(\alpha) \times S(\beta) \end{array}$$

Resolution spaces reflect the BHK-interpretation, e.g., a possible resolution of an implication is a function transforming resolutions of the antecedent into resolutions of the consequent.[6] Note that resolution spaces for atomic propositions are singletons, based on the assumption in inquisitive semantics that atomic propositions are statements without inquisitiveness. The set of *actual* resolutions of each formula on each world in a given model is a (possibly empty) subset of the corresponding resolution space, as defined below.

Definition 3.3 (Resolution in model) *Given $\mathcal{M}$, $R : W_{\mathcal{M}} \times \mathbf{PL}^{\otimes} \to \bigcup_{\alpha \in \mathbf{PL}^{\otimes}} S(\alpha)$ gives the (actual) resolutions for each formula on each world:*

$$R(w,\bot) = \varnothing \qquad R(w,p) = \begin{cases} \{p\} & \textit{if } w \in V_{\mathcal{M}}(p) \\ \varnothing & \textit{otherwise} \end{cases}$$

$$\begin{aligned} R(w,\alpha \vee \beta) &= (R(w,\alpha) \times \{0\}) \cup (R(w,\beta) \times \{1\}) \\ R(w,\alpha \wedge \beta) &= R(w,\alpha) \times R(w,\beta) \\ R(w,\alpha \to \beta) &= \{f \in S(\beta)^{S(\alpha)} \mid f[R(w,\alpha)] \subseteq R(w,\beta)\} \\ R(w,\alpha \otimes \beta) &= (R(w,\alpha) \times S(\beta)) \cup (S(\alpha) \times R(w,\beta)) \end{aligned}$$

Important notation: *For $U \subseteq W_{\mathcal{M}}$, we write $R(U,\alpha)$ for $\bigcap_{w \in U} R(w,\alpha)$.*

While $S(\bot) = \{\bot\}$ is non-empty, it never has any actual resolution on specific worlds. For any $p \in \mathbf{P}$, p has itself as its resolution iff it is true on w. For any implication $\alpha \to \beta \in \mathbf{PL}^{\otimes}$, each of its resolution on w is a function in $S(\alpha \to \beta)$ which maps an actual resolution of α to an actual resolution of β on w. Following the idea of the weak disjunction introduced in [14], each resolution for $\alpha \otimes \beta \in \mathbf{PL}^{\otimes}$ on w is a pair of resolutions in $S(\alpha \otimes \beta)$, such that *at least one* in the pair is actual on w for the corresponding formula.

[6] See [18] for a more detailed explaination (without $\otimes$). The definition of resolutions is based on Medvedev's problem semantics [14]. A similar definition can be found in [15].

Note that by definition, $R(w, \alpha) \subseteq S(w, \alpha) \neq \varnothing$.

Proposition 3.4 *For any $\alpha \in \mathbf{PL}^{\otimes}$, $S(\alpha) \neq \varnothing$ and $S(\alpha)$ is finite.*

Below is a useful observation on the resolution of negations ($\neg\alpha := \alpha \to \bot$).

Proposition 3.5 ([18]) *For any $\mathcal{M}, w$, any α, $R(w, \neg\alpha)$ is either $\varnothing$ or a fixed singleton set independent from w, and $R(w, \neg\alpha) = \varnothing$ iff $R(w, \alpha) \neq \varnothing$.*

Let $\mathbf{P}(\alpha)$ be the set of propositional letters occurring in α and let $V^{\alpha}_{\mathcal{M}}(w)$ be the collection of $p \in \mathbf{P}(\alpha)$ that are true on w in $\mathcal{M}$. Proposition 3.6 says that $R(w, \alpha)$ only depends on the relevant valuation on w itself.

Proposition 3.6 *For any $\mathcal{M}, w$ and $\mathcal{N}, v$, for all $\alpha \in \mathbf{PL}^{\otimes}$, if $V^{\alpha}_{\mathcal{M}}(w) = V^{\alpha}_{\mathcal{N}}(v)$, then $R(w, \alpha) = R(v, \alpha)$.*

Now we define the satisfaction relation of **PALKhΠ** on *pointed models*, i.e, a model with a designated world, in contrast with the state-based support-semantics. Note that the connectives *outside* the scope of Kh are classical, e.g., $\otimes$ just functions as a classical disjunction. K is the standard epistemic modality of know-that. The semantics for $\mathsf{Kh}\alpha$ following the $\exists x\mathsf{K}$ schema in [20,19] via resolutions, and is intended to capture the know-how interpretation of $\mathbf{InqB}^{\otimes}$. $\forall p$ is a propositional quantifier over the full power set of $W_{\mathcal{M}}$. The semantics of the dynamic operator $[\psi]$ is as in public announcement logic [16].

Definition 3.7 (Semantics) *For $\varphi, \psi \in$ **PALKhΠ**, $\alpha \in \mathbf{PL}^{\otimes}$ and $\mathcal{M}, w$ where $\mathcal{M} = \langle W, V\rangle$, the semantics is defined as below ($\bigcirc \in \{\vee, \otimes\}$):*

$\mathcal{M}, w \nvDash \bot$		
$\mathcal{M}, w \vDash p$	$\iff$	$w \in V(p)$
$\mathcal{M}, w \vDash (\varphi \bigcirc \psi)$	$\iff$	$\mathcal{M}, w \vDash \varphi$ *or* $\mathcal{M}, w \vDash \psi$
$\mathcal{M}, w \vDash (\varphi \wedge \psi)$	$\iff$	$\mathcal{M}, w \vDash \varphi$ *and* $\mathcal{M}, w \vDash \psi$
$\mathcal{M}, w \vDash (\varphi \to \psi)$	$\iff$	$\mathcal{M}, w \vDash \varphi$ *implies* $\mathcal{M}, w \vDash \psi$
$\mathcal{M}, w \vDash \mathsf{K}\varphi$	$\iff$	*for any* $v \in \mathcal{M}, \mathcal{M}, v \vDash \varphi$
$\mathcal{M}, w \vDash \mathsf{Kh}\alpha$	$\iff$	*there is an* $x \in S(\alpha)$ *s.t. for any* $v \in \mathcal{M}, x \in R(v, \alpha)$
$\mathcal{M}, w \vDash \forall p\varphi$	$\iff$	*for any* $U \in \wp(W), \mathcal{M}[p \mapsto U], w \vDash \varphi$
$\mathcal{M}, w \vDash [\psi]\varphi$	$\iff$	$\mathcal{M}, w \vDash \psi$ *implies* $\mathcal{M}|_{[\![\psi]\!]}, w \vDash \varphi$

where:

- *Given $U \in \wp(W_{\mathcal{M}})$ and $p \in \mathbf{P}$, recall that $\mathcal{M}[p \mapsto U] = \langle W, V'\rangle$, where the assignment V' assigns U to p and coincides with V on all other atoms; and*
- *$[\![\psi]\!]_{\mathcal{M}} = \{w \in W_{\mathcal{M}} \mid \mathcal{M}, w \vDash \psi\}$ and $\mathcal{M}|_X$ is the submodel of $\mathcal{M}$ by restricting to $\varnothing \neq X \subseteq W_{\mathcal{M}}$. Thus $\mathcal{M}|_{[\![\psi]\!]_{\mathcal{M}}}$ is the submodel restricted to the worlds satisfying ψ in $\mathcal{M}$. We also write $\mathcal{M}|_{[\![\psi]\!]_{\mathcal{M}}}$ as $\mathcal{M}|_{\psi}$ for brevity.*

Validity and entailment are defined as usual.

In [18], to handle the implication in the know-how scope, we have a dynamic operator $\Box$ such that $\Box\varphi$ says that "given any information updates, φ holds". This can be expressed by $\forall p[p]\varphi$ in our language, given that p is not free in φ.

We write $\mathcal{M} \vDash \varphi$ iff $\mathcal{M}, w \vDash \varphi$ for all $w \in W_{\mathcal{M}}$. Clearly, $\mathcal{M}, w \vDash \mathsf{Kh}\alpha$ iff $\mathcal{M} \vDash \mathsf{Kh}\alpha$ and $\mathcal{M}, w \vDash \mathsf{K}\varphi$ iff $\mathcal{M} \vDash \varphi$. The rephrased truth condition of Kh

below says that $\mathsf{Kh}\alpha$ holds on a (pointed) model as long as there is a uniform resolution for α on that model, where we define $R(U, \alpha)$ as $\bigcap_{w \in U} R(w, \alpha)$.

$$\boxed{\mathcal{M} \vDash \mathsf{Kh}\alpha \iff \mathcal{M}, w \vDash \mathsf{Kh}\alpha \iff R(W_{\mathcal{M}}, \alpha) \neq \varnothing}$$

As in [18], we can give a uniform alternative truth condition for $\mathbf{PL}^{\otimes}$-formulas via the existence of actual resolutions.

Proposition 3.8 *For any $\alpha \in \mathbf{PL}^{\otimes}$ and $\mathcal{M}, w$, $\mathcal{M}, w \vDash \alpha \iff R(w, \alpha) \neq \varnothing$.*

Proof. We prove by induction on the structure of α. We only show the cases for $\otimes$. The other cases can be found in [18].

$$\begin{aligned}
\mathcal{M}, w \vDash (\alpha \otimes \beta) &\iff \mathcal{M}, w \vDash \alpha \text{ or } \mathcal{M}, w \vDash \beta \\
&\iff R(w, \alpha) \neq \varnothing \text{ or } R(w, \beta) \neq \varnothing \\
&\iff \text{ there is an } x \in R(w, \alpha) \text{ or there is a } y \in R(w, \beta) \\
&\iff \exists \langle x, y' \rangle \in R(w, \alpha \otimes \beta) \text{ s.t. } x \in R(w, \alpha) \wedge y' \in S(\beta) \\
&\qquad \text{or } \exists \langle x', y \rangle \in R(w, \alpha \otimes \beta) \text{ s.t. } x' \in S(\alpha) \wedge y \in R(w, \beta) \\
&\iff R(w, \alpha \otimes \beta) \neq \varnothing
\end{aligned}$$

□

From Proposition 3.8, we see that in propositional formulas, both $\vee$ and $\otimes$ collapse to the classical disjunction outside the scope of Kh. Yet $\otimes$ is weaker than $\vee$ in the way that we can construct a resolution of $\alpha \otimes \beta$ from that of $\alpha \vee \beta$. It also follows from Proposition 3.8 that for any $\alpha \in \mathbf{PL}^{\otimes}$, $\mathcal{M}, w \vDash \mathsf{K}\alpha$ iff for each $v \in \mathcal{M}$, there is *some* resolution for α on v, in the shape of $\mathsf{K}\exists x$. In contrast, $\mathcal{M}, w \vDash \mathsf{Kh}\alpha$ iff there is a *uniform* resolution for α on $\mathcal{M}$ in the shape of $\exists x \mathsf{K}$. The following is then immediate.

Proposition 3.9 *$\mathsf{Kh}\alpha \to \mathsf{K}\alpha$ is valid for all $\alpha \in \mathbf{PL}^{\otimes}$.*

Since each $p \in \mathbf{P}$ only has one possible resolution, when each point has a resolution for p, the model has a uniform one. Thus we have Proposition 3.10

Proposition 3.10 *$\mathsf{Kh}p \leftrightarrow \mathsf{K}p$ is valid for all $p \in \mathbf{P}$.*

Based on the semantics, we can have more intuitive readings of the formulas in inquisitive logic, e.g., $\mathsf{Kh}(p \vee \neg p)$ is equivalent to $\mathsf{Kh}p \vee \mathsf{Kh}\neg p$ and $\mathsf{K}p \vee \mathsf{K}\neg p$. This also explains the failure of *excluded middle* in inquisitive logic (cf. [18] for more discussions).

The rule of *replacement of equals* is not valid in general, for instance, although $(p \vee \neg p) \leftrightarrow (p \to p)$ is valid, $\mathsf{Kh}(p \vee \neg p) \leftrightarrow \mathsf{Kh}(p \to p)$ is not. However, if we only allow substitution to happen outside the scope of Kh operators, the rule becomes valid. It is not hard to verify the following:

Proposition 3.11 *For any $\varphi, \psi, \chi \in \mathbf{PALKh\Pi}$, the validity of $\varphi \leftrightarrow \psi$ implies the validity of $\chi[\varphi/\psi] \leftrightarrow \chi$, given that the substitution does not happen in the scope of Kh.*

4 Expressivity

Let **PALΠ** be the Kh-free fragment of **PALKhΠ**, **ELΠ** be the $[\varphi]$-free fragment of **PALΠ** and **EL** be the $\forall p$-free fragment of **ELΠ**. In Subsection 4.1, we show Kh and $[\varphi]$ can be eliminated, thus making **PALKhΠ**, **PALΠ** and **ELΠ** equally expressive. In Subsection 4.2, we show that the valid Kh formulas of **PALKhΠ** correspond to $\mathbf{InqB}^{\otimes}$ precisely.

4.1 Reduction

We introduce the reduction schemata to eliminate the Kh modality, which will also be used in the proof system to be introduced later. First, we have the following observation.

Proposition 4.1 *For any $\alpha_1, \alpha_2 \in \mathbf{PL}^{\otimes}$ where p_1 and p_2 do not occur free, for any pointed model $\mathcal{M}, w$, $\mathcal{M}, w \vDash \exists p_1 \exists p_2 \mathsf{K}((p_1 \otimes p_2) \wedge [p_1]\mathsf{Kh}\alpha_1 \wedge [p_2]\mathsf{Kh}\alpha_2)$ iff there are $U_1, U_2 \subseteq W_{\mathcal{M}}$ s.t. $U_1 \cup U_2 = W_{\mathcal{M}}$ and $U_i \neq \varnothing$ implies $R(U_i, \alpha_i) \neq \varnothing$ for $i = 1, 2$.*

Proof. Given a $U \subseteq W_{\mathcal{M}}$, for any $w \in \mathcal{M}$, $\mathcal{M}[p \mapsto U], w \vDash p \iff w \in U$ $(\star)$. For brevity, we write $\exists U$ for there exists $U \in W_{\mathcal{M}}$. Recall that $\mathcal{M}|_U$ denotes the submodel of $\mathcal{M}$ restricted to U, if U is non-empty (otherwise undefined).

$$\begin{aligned}
& \mathcal{M}, w \vDash \exists p_1 \exists p_2 \mathsf{K}((p_1 \otimes p_2) \wedge [p_1]\mathsf{Kh}\alpha_1 \wedge [p_2]\mathsf{Kh}\alpha_2) \\
\iff & \exists U_1 \exists U_2, \mathcal{M}[p_1, p_2 \mapsto U_1, U_2], w \vDash \mathsf{K}((p_1 \otimes p_2) \wedge \textstyle\bigwedge_{i=1}^{2} [p_i]\mathsf{Kh}\alpha_i) \\
\iff & \exists U_1 \exists U_2, \forall v \in W_{\mathcal{M}}, \mathcal{M}[p_1, p_2 \mapsto U_1, U_2], v \vDash (p_1 \otimes p_2) \wedge \textstyle\bigwedge_{i=1}^{2} [p_i]\mathsf{Kh}\alpha_i \\
& (\text{by } (\star) \text{ and the fact that } [\varphi]\psi \text{ holds trivially if } \varphi \text{ is false}) \\
\iff & \exists U_1 \exists U_2, \forall v \in W_{\mathcal{M}}, \mathcal{M}[p_1, p_2 \mapsto U_1, U_2], v \vDash p_1 \vee p_2 \\
& \text{and } v \in U_i \text{ implies } \mathcal{M}[p_1, p_2 \mapsto U_1, U_2], v \vDash [p_i]\mathsf{Kh}\alpha_i \text{ for } i = 1, 2 \\
\iff & \exists U_1 \exists U_2, U_1 \cup U_2 = W_{\mathcal{M}} \text{ and } \forall v \in W_{\mathcal{M}}, v \in U_i \text{ implies} \\
& \mathcal{M}[p_1, p_2 \mapsto U_1, U_2]|_{p_i}, v \vDash \mathsf{Kh}\alpha_i \text{ for } i = 1, 2 \\
& (\text{since } p_1 \text{ and } p_2 \text{ do not occur free in } \alpha_1 \text{ and } \alpha_2, \text{ we have:}) \\
\iff & \exists U_1 \exists U_2, U_1 \cup U_2 = W_{\mathcal{M}} \text{ and } \forall v \in W_{\mathcal{M}}, v \in U_i \text{ implies} \\
& \mathcal{M}|_{U_i}, v \vDash \mathsf{Kh}\alpha_i \text{ for } i = 1, 2 \\
\iff & \exists U_1 \exists U_2, U_1 \cup U_2 = W_{\mathcal{M}} \text{ and } U_i \neq \varnothing \text{ implies } R(U_i, \alpha_i) \neq \varnothing \text{ for } i = 1, 2
\end{aligned}$$

□

Together with Proposition 3.10 and 3.11, Proposition 4.2 helps us to first eliminate the Kh modality without changing the expressive power, i.e., each **PALKhΠ**-formula is equivalent to a **PALΠ**-formula.

Proposition 4.2 *The following formulas and schemata are valid:*

$$\begin{array}{llll}
\mathtt{KKhp}: & \mathsf{K}p \to \mathsf{Kh}p & \mathtt{Kh}_{\bot}: & \mathsf{Kh}\bot \leftrightarrow \bot \\
\mathtt{Kh}_{\vee}: & \mathsf{Kh}(\alpha \vee \beta) \leftrightarrow \mathsf{Kh}\alpha \vee \mathsf{Kh}\beta & \mathtt{Kh}_{\wedge}: & \mathsf{Kh}(\alpha \wedge \beta) \leftrightarrow \mathsf{Kh}\alpha \wedge \mathsf{Kh}\beta \\
\mathtt{Kh}_{\to}: & \multicolumn{3}{l}{\mathsf{Kh}(\alpha \to \beta) \leftrightarrow \mathsf{K}\forall p[p](\mathsf{Kh}\alpha \to \mathsf{Kh}\beta), \quad p \text{ does not occur free in } \alpha \text{ or } \beta} \\
\mathtt{Kh}_{\otimes}: & \multicolumn{3}{l}{\mathsf{Kh}(\alpha \otimes \beta) \leftrightarrow \exists p_1 \exists p_2 \mathsf{K}((p_1 \otimes p_2) \wedge [p_1]\mathsf{Kh}\alpha \wedge [p_2]\mathsf{Kh}\beta),} \\
& \multicolumn{3}{r}{\text{where } p_1, p_2 \text{ do not occur free in } \alpha \text{ or } \beta}
\end{array}$$

Proof. We only show $\mathtt{Kh}_{\otimes}$ and $\mathtt{Kh}_{\to}$. The other cases can be found in [18].

For $\mathtt{Kh}_{\otimes}$:

$\Longrightarrow$: Suppose $\mathcal{M}, w \vDash \mathsf{Kh}(\alpha \otimes \beta)$, then by the semantics, there is some $\langle x, y\rangle \in R(W_{\mathcal{M}}, \alpha \otimes \beta)$. Let $U = \{u \in W_{\mathcal{M}} \mid x \in R(u, \alpha)\}$ and $V = \{v \in W_{\mathcal{M}} \mid x \in R(v, \beta)\}$. It is not hard to see that $U \cup V = W_{\mathcal{M}}$, $U \neq \varnothing$ implies $R(U, \alpha) \neq \varnothing$ and $V \neq \varnothing$ implies $R(V, \beta) \neq \varnothing$. By Proposition 4.1, $\mathcal{M}, w \vDash \exists p_1 \exists p_2 \mathsf{K}((p_1 \otimes p_2) \wedge [p_1]\mathsf{Kh}\alpha \wedge [p_2]\mathsf{Kh}\beta)$.

$\Longleftarrow$: Suppose $\mathcal{M}, w \vDash \exists p_1 \exists p_2 \mathsf{K}((p_1 \otimes p_2) \wedge [p_1]\mathsf{Kh}\alpha \wedge [p_2]\mathsf{Kh}\beta)$, by Proposition 4.1, there are U_1, U_2 satisfying the desired property. If $U_1 \neq \varnothing$ and $U_2 \neq \varnothing$, pick $\langle x, y\rangle$ as the witness for $R(W_{\mathcal{M}}, \alpha \otimes \beta)$ s.t. $x \in R(U_1, \alpha)$ and $y \in R(U_2, \beta)$. If $U_1 = \varnothing$ then $U_2 \neq \varnothing$ since $W_{\mathcal{M}}$ is non-empty, then we pick $\langle x, y\rangle$ such that $y \in R(U_2, \beta)$ and $x \in S(\alpha)$. Similar for the case when $U_2 = \varnothing$. This suffices to show $\mathcal{M}, w \vDash \mathsf{Kh}(\alpha \otimes \beta)$.

For $\mathsf{Kh}_{\rightarrow}$: In [18], we showed the validity of $\mathsf{Kh}(\alpha \rightarrow \beta) \leftrightarrow \mathsf{K}\Box(\mathsf{Kh}\alpha \rightarrow \mathsf{Kh}\beta)$ where $\Box$ is the informational update operator such that $\mathcal{M}, w \vDash \Box\varphi \iff$ for any $\mathcal{M}' \subseteq \mathcal{M}$ s.t. $w \in \mathcal{M}'$, $\mathcal{M}', w \vDash \varphi$. Note that $\Box\varphi$ can be defined by $\forall p[p]\varphi$ where p does not occur free in φ. The rest is the same as in [18]. $\square$

By Proposition 4.3 we can further eliminate the announcement operator in a formula without Kh. [7]

Proposition 4.3 *The following formulas and schemata are valid:*

$$
\begin{array}{ll}
[\,]_{\mathsf{P}} & [\chi]p \leftrightarrow (\chi \rightarrow p),\ p \in \mathbf{P} \cup \{\bot\} \\
[\,]_{\bigcirc} & [\chi](\varphi \bigcirc \psi) \leftrightarrow [\chi]\varphi \bigcirc [\chi]\psi, \bigcirc \in \{\wedge, \vee, \otimes, \rightarrow\} \\
[\,]_{\mathsf{K}} & [\chi]\mathsf{K}\varphi \leftrightarrow (\chi \rightarrow \mathsf{K}([\chi]\varphi)) \\
[\,]_{\forall} & [\chi]\forall p\varphi \leftrightarrow \forall p[\chi]\varphi, p \textit{ is not in } \chi
\end{array}
$$

Proof. We only show $[\,]_{\vee}$ and $[\,]_{\forall}$ as examples.

For $[\,]_{\vee}$: $\mathcal{M}, w \vDash [\chi](\varphi \vee \psi)$ iff $\mathcal{M}, w \vDash \chi$ implies $\mathcal{M}|_{\chi}, w \vDash \varphi \vee \psi$ iff $\mathcal{M}, w \vDash \chi$ implies $(\mathcal{M}|_{\chi}, w \vDash \varphi$ or $\mathcal{M}|_{\chi}, w \vDash \psi)$ iff $(\mathcal{M}, w \vDash \chi$ implies $\mathcal{M}|_{\chi}, w \vDash \varphi)$ or $(\mathcal{M}, w \vDash \chi$ implies $\mathcal{M}|_{\chi}, w \vDash \psi)$ iff $\mathcal{M}, w \vDash [\chi]\varphi$ or $\mathcal{M}, w \vDash [\chi]\psi$ iff $\mathcal{M}, w \vDash [\chi]\varphi \vee [\chi]\psi$.

For $[\,]_{\forall}$: $\mathcal{M}, w \vDash [\chi]\forall p\varphi$ iff $\mathcal{M}, w \vDash \chi$ implies $\mathcal{M}|_{\chi}, w \vDash \forall p\varphi$ iff $\mathcal{M}, w \vDash \chi$ implies for all $U' \subseteq W_{\mathcal{M}|_{\chi}}$, $(\mathcal{M}|_{\chi})[p \mapsto U'], w \vDash \varphi$. If $U \subseteq W_{\mathcal{M}}$ then $U' = U \cap W_{\mathcal{M}|_{\chi}} \subseteq W_{\mathcal{M}|_{\chi}}$. Conversely, if $U' \subseteq W_{\mathcal{M}|_{\chi}}$ then for some $U \subseteq W_{\mathcal{M}}$, $U' = U \cap W_{\mathcal{M}|_{\chi}}$. Hence, $\mathcal{M}, w \vDash [\chi]\forall p\varphi$ iff $\mathcal{M}, w \vDash \chi$ implies for all $U \subseteq W_{\mathcal{M}}$, $(\mathcal{M}[p \mapsto U])|_{\chi}, w \vDash \varphi$ iff for all $U \subseteq W_{\mathcal{M}}$, $\mathcal{M}, w \vDash \chi$ implies $(\mathcal{M}[p \mapsto U])|_{\chi}, w \vDash \varphi$. Since p is not in χ, it is easy to see that $\mathcal{M}, w \vDash \chi$ iff $\mathcal{M}[p \mapsto U], w \vDash \chi$. Hence, $\mathcal{M}, w \vDash [\chi]\forall p\varphi$ iff for all $U \subseteq W_{\mathcal{M}}$, $\mathcal{M}[p \mapsto U], w \vDash \chi$ implies $(\mathcal{M}[p \mapsto U])|_{\chi}, w \vDash \varphi$ iff for all $U \subseteq W_{\mathcal{M}}$, $\mathcal{M}[p \mapsto U], w \vDash [\chi]\varphi$ iff $\mathcal{M}, w \vDash \forall p[\chi]\varphi$. $\square$

Without loss of generality, we can always rename the bound variable in case it occurs in χ. Then for any Kh-free formula φ, by repeatedly applying

[7] An alternative set of reduction formulas for the announcement operator is presented in Lemma 12 of [1] on top of reduction axioms in [16]. See [22] for more detailed discussions on reduction axioms for **PAL**.

Proposition 4.3, we can get rid of all $[\cdot]$ operators and find an equivalent **ELΠ**-formula for each **PALΠ**-formula. We will give a formal presentation of this result in Theorem 5.8 as a natural consequence of Theorem 5.2 (Soundness).

4.2 $\mathbf{Inq}^{\otimes}\mathbf{Kh} = \mathbf{InqB}^{\otimes}$

Now we show that $\mathbf{Inq}^{\otimes}\mathbf{Kh} = \{\alpha \in \mathbf{PL}^{\otimes} \mid \vDash \mathsf{Kh}\alpha\}$ is exactly $\mathbf{InqB}^{\otimes}$.

Lemma 4.4 *For any $\alpha \in \mathbf{PL}^{\otimes}$, $\mathcal{M}, w \vDash \mathsf{Kh}\alpha$ iff $\mathcal{M}, W_{\mathcal{M}} \Vdash \alpha$. As a consequence, for any non-empty state s in $\mathcal{M}$, $\mathcal{M}, s \Vdash \alpha$ iff $\mathcal{M}|_s \vDash \mathsf{Kh}\alpha$.*

Proof. Note that $\mathcal{M}, w \vDash \mathsf{Kh}\alpha$ iff $\mathcal{M} \vDash \mathsf{Kh}\alpha$ by the semantics, so we simply show $\mathcal{M} \vDash \mathsf{Kh}\alpha$ iff $\mathcal{M}, W_{\mathcal{M}} \Vdash \alpha$ inductively on the structure of α. We only prove the case of $\otimes$ since the rest are the same as in [18]. By Proposition 4.1 and 4.2, $\mathcal{M} \vDash \mathsf{Kh}(\alpha \otimes \beta)$ amounts to $\exists U, V$ s.t. $U \cup V = W_{\mathcal{M}}$, $U \neq \varnothing$ implies $R(U, \alpha) \neq \varnothing$ and $V \neq \varnothing$ implies $R(V, \beta) \neq \varnothing$. We show this is exactly $\mathcal{M}, W_{\mathcal{M}} \Vdash \alpha \otimes \beta$.

$\Longrightarrow$: If both U and V are non-empty, then $\mathcal{M} \vDash \mathsf{Kh}(\alpha \otimes \beta)$ amounts to $\mathcal{M}|_U \vDash \mathsf{Kh}\alpha$ and $\mathcal{M}|_V \vDash \mathsf{Kh}\beta$. By IH, it is equivalent to $\mathcal{M}|_U, U \Vdash \alpha$ and $\mathcal{M}|_V, V \Vdash \beta$, which implies $\mathcal{M}, W_{\mathcal{M}} \Vdash \alpha \otimes \beta$ since $U \cup V = W_{\mathcal{M}}$. If one of U and V is empty, suppose w.l.o.g. $U = \varnothing$, then we can also show $\mathcal{M}, V \Vdash \beta$ (as before), and $\mathcal{M}, U \Vdash \alpha$, for the empty state support all formulas by Proposition 2.4. Thus $\mathcal{M}, W_{\mathcal{M}} \Vdash \alpha \otimes \beta$.

$\Longleftarrow$: Suppose $\mathcal{M}, W_{\mathcal{M}} \Vdash \alpha \otimes \beta$, then there are substates t and t' such that $t \cup t' = W_{\mathcal{M}}$, $\mathcal{M}, t \Vdash \alpha$ and $\mathcal{M}, t' \Vdash \beta$. Take $U = t$ and $V = t'$, by IH, we have $U \cup V = W_{\mathcal{M}}$, $U \neq \varnothing$ implies $\mathcal{M}|_U \vDash \mathsf{Kh}\alpha$ and $V \neq \varnothing$ implies $\mathcal{M}|_V \vDash \mathsf{Kh}\beta$. Hence, $U \neq \varnothing$ implies $R(U, \alpha) \neq \varnothing$ and $V \neq \varnothing$ implies $R(V, \beta) \neq \varnothing$. By Proposition 4.1 and 4.2, we have $\mathcal{M}, W_{\mathcal{M}} \vDash \mathsf{Kh}(\alpha \otimes \beta)$.

For the consequence, $\mathcal{M}|_s, w \vDash \mathsf{Kh}\alpha$ iff $\mathcal{M}|_s, s \Vdash \alpha$ iff $\mathcal{M}, s \Vdash \alpha$, and the last step is due to Proposition 2.5. □

Remark 4.5 Note that the proof for the $\otimes$ case above actually established the equivalence between our semantics inspired by Medvedev's weak disjunction and the team/support semantics in dependence/inquisitive logics. As mentioned in the introduction, the formula $=(p, q) \otimes =(p, q)$ in propositional dependence logic is equivalent to the following formula in inquisitive logic

$$((p \vee \neg p) \to (q \vee \neg q)) \otimes ((p \vee \neg p) \to (q \vee \neg q)).$$

In our setting, it says that there is a pair of dependence functions $\langle f_1, f_2 \rangle$ s.t. you know that one of these functions captures how q depends on p actually. Note that such a pair of functions always exists: every state can be split into two substates such that one collects all the worlds where q is true, and the other one collects the rest. Then in the substate where q is homogeneously true we can define a constant function essentially assigning any resolution of $p \vee \neg p$ to the fixed resolution q. Similarly for the other substate where q is homogeneously false. Therefore this formula is valid.

Based on the lemma above, we can establish the relation between $\mathbf{InqB}^{\otimes}$ and $\mathbf{Inq}^{\otimes}\mathbf{Kh}$, where $\mathsf{Kh}\Gamma = \{\mathsf{Kh}\alpha \mid \alpha \in \Gamma\}$.

Theorem 4.6 *Given any* $\{\alpha\} \cup \Gamma \subseteq \mathbf{PL}^{\otimes}$, $\Gamma \Vdash \alpha$ *iff* $\mathsf{Kh}\Gamma \vDash \mathsf{Kh}\alpha$. *As a consequence when* $\Gamma = \varnothing$, $\mathbf{InqB}^{\otimes} = \mathbf{Inq}^{\otimes}\mathbf{Kh}$.

Proof. Suppose $\Gamma \Vdash \alpha$ and $\mathcal{M}, w \vDash \mathsf{Kh}\Gamma$. Now we have $\mathcal{M}, W_{\mathcal{M}} \Vdash \Gamma$ by Lemma 4.4 thus $\mathcal{M}, W_{\mathcal{M}} \Vdash \alpha$, therefore $\mathcal{M}, w \vDash \alpha$. For the other way around, if $\mathsf{Kh}\Gamma \vDash \mathsf{Kh}\alpha$ and $\mathcal{M}, s \Vdash \Gamma$, then $\mathcal{M}|_s \vDash \mathsf{Kh}\Gamma$ by Lemma 4.4, thus $\mathcal{M}|_s \vDash \mathsf{Kh}\alpha$. By Lemma 4.4 again, $\mathcal{M}, s \Vdash \alpha$. □

5 Axiomatization of PALKhΠ

In this section, we introduce the proof system $\mathsf{SPALKh\Pi}^+$ as below. The axioms can help us to "open up" the Kh-formulas step by step, and eventually eliminate all the Kh operator and the announcement operators.

System $\mathsf{SPALKh\Pi}^+$

Axioms

TAUT	Propositional tautologies
Rd⊗	$(\varphi \otimes \psi) \leftrightarrow (\varphi \vee \psi)$
$\mathtt{DIST}_{\mathsf{K}}$	$\mathsf{K}(\varphi \to \psi) \to (\mathsf{K}\varphi \to \mathsf{K}\psi)$
$[]_{\mathrm{P}}$	$[\chi]p \leftrightarrow (\chi \to p)$, $p \in \mathbf{P} \cup \{\bot\}$
$[]_{\bigcirc}$	$[\chi](\varphi \bigcirc \psi) \leftrightarrow [\chi]\varphi \bigcirc [\chi]\psi$
$[]_{\mathsf{K}}$	$[\chi]\mathsf{K}\varphi \leftrightarrow \chi \to \mathsf{K}[\chi]\varphi$
$[]_{\forall}$	$[\chi]\forall p\varphi \leftrightarrow \forall p[\chi]\varphi$, p is not in χ
$\mathtt{DIST}_{\forall}$	$\forall p(\varphi \to \psi) \to (\forall p\varphi \to \forall p\psi)$
$\mathtt{SUB}_{\forall}$	$\forall p\varphi \to \varphi[\psi/p]$, ψ is free for p in φ
SU	$\exists p(p \wedge \forall q(q \to \mathsf{K}(p \to q)))$
BC	$\forall p\mathsf{K}\varphi \to \mathsf{K}\forall p\varphi$
KhK	$\mathsf{Kh}\alpha \to \mathsf{K}\alpha$
KKhp	$\mathsf{K}p \to \mathsf{Kh}p$
$\mathtt{Kh}_{\bot}$	$\mathsf{Kh}\bot \leftrightarrow \bot$
$\mathtt{Kh}_{\vee}$	$\mathsf{Kh}(\alpha \vee \beta) \leftrightarrow \mathsf{Kh}\alpha \vee \mathsf{Kh}\beta$
$\mathtt{Kh}_{\wedge}$	$\mathsf{Kh}(\alpha \wedge \beta) \leftrightarrow \mathsf{Kh}\alpha \wedge \mathsf{Kh}\beta$
$\mathtt{Kh}_{\to}$	$\mathsf{Kh}(\alpha \to \beta) \leftrightarrow \mathsf{K}\forall p[p](\mathsf{Kh}\alpha \to \mathsf{Kh}\beta)$
$\mathtt{Kh}_{\otimes}$	$\mathsf{Kh}(\alpha \otimes \beta) \leftrightarrow \exists p_1 \exists p_2 \mathsf{K}((p_1 \otimes p_2) \wedge [p_1]\mathsf{Kh}\alpha \wedge [p_2]\mathsf{Kh}\beta)$
$\mathtt{T}_{\mathsf{K}}$	$\mathsf{K}\varphi \to \varphi$
4_{K}	$\mathsf{K}\varphi \to \mathsf{K}\mathsf{K}\varphi$
5_{K}	$\neg\mathsf{K}\varphi \to \mathsf{K}\neg\mathsf{K}\varphi$
4_{Kh}	$\mathsf{Kh}\alpha \to \mathsf{K}\mathsf{Kh}\alpha$
5_{Kh}	$\neg\mathsf{Kh}\alpha \to \mathsf{K}\neg\mathsf{Kh}\alpha$

Rules

MP	$\dfrac{\varphi, \varphi \to \psi}{\psi}$
$\mathtt{NEC}_{\mathsf{K}}$	$\dfrac{\vdash \varphi}{\vdash \mathsf{K}\varphi}$
$\mathtt{GEN}_{\forall}$	$\dfrac{\vdash \varphi \to \psi}{\vdash \varphi \to \forall p\psi}$ p not free in φ
rRE	$\dfrac{\vdash \varphi \leftrightarrow \psi}{\vdash \chi[\varphi/\psi] \leftrightarrow \chi}$, given that the substitution does not happen in the scope of Kh

where $p \in \mathbf{P}$, $\alpha, \beta \in \mathbf{PL}^{\otimes}$, $\varphi, \psi, \chi \in \mathbf{PALKh\Pi}$, $\bigcirc \in \{\wedge, \vee, \otimes, \to\}$; p, p_1, p_2 *do not occur free in* α *and* β *in* $\mathtt{Kh}_{\to}$ *and* $\mathtt{Kh}_{\otimes}$.

Together with **rRE**, Rd⊗ states the fact that ⊗ behaves exactly like ∨ when it occurs outside Kh. S5 axiom schemata/rules for the know-that modality K together with TAUT, $\mathtt{DIST}_{\forall}$, $\mathtt{SUB}_{\forall}$, SU and rule $\mathtt{GEN}_{\forall}$ form a complete axiomatization $\mathsf{S5\Pi}^+$ of S5 logic with propositional quantifiers [10], where SU states the existence of *atoms*, crucial to capture the powerset domain for the propositional quantifier. Axioms $[]_{\mathrm{P}}$, $[]_{\bigcirc}$, $[]_{\mathsf{K}}$ and $[]_{\forall}$ are reduction axioms for the announcement operator $[\cdot]$ [16,1].[8] KKhp, $\mathtt{Kh}_{\bot}$, $\mathtt{Kh}_{\vee}$, $\mathtt{Kh}_{\wedge}$, $\mathtt{Kh}_{\to}$ and $\mathtt{Kh}_{\otimes}$ are the reduction axioms decoding the $\mathbf{PL}^{\otimes}$ formulas, whose usages are shown in

[8] The original form of $[]_{\forall}$ in [1] is $[\chi]\forall p\varphi \leftrightarrow (\chi \to \forall p[\chi]\varphi)$ (p is not in χ).

Lemma 5.3. Barcan formula BC, introspection schemata 4_{K}, 4_{Kh} and 5_{Kh} can be proved from the rest of the system. In particular, 4_{Kh} requires an inductive proof on the structure of α. We include them for their intuitive meanings.

Remark 5.1 Compared to the proof system SDELKh in [18] for the standard propositional inquisitive logic, there are a few notable differences:

- In order to capture tensor, we need a more powerful language **PALKhΠ** than the language **DELKh** in [18]. Since we have the public announcement operator $[\cdot]$ and propositional quantifier $\forall p$ in **PALKhΠ**, the informational update operator $\Box$ in **DELKh** can be expressed by $\forall p[p]$. The axioms and rules of $[\cdot]$ and $\forall p$ can handle $\Box$ implicitly. In Section 6, we will see our language can uniformly define various generalised versions of tensor as well.
- In the proof system SDELKh of [18], there is a set of axiom schemata $\{\mathsf{EU}_k \mid k \in \mathbb{N}\}$, which captures the idea (roughly) that given a definable finite set of worlds, we can have an updated submodel with it as the set of possible worlds. These axioms are also necessary in the process of eliminating $\Box$ in [18] and have an intimate connection with the axioms $\mathtt{ND}_k$ in inquisitive logic. However, these axioms are no longer needed here, as the same function of postulating the existence of certain updated models can be realized by concrete announcements in our language **PALKhΠ**.

The next subsection explores reductions of Kh systematically.

5.1 Provable equivalence

In Section 4.1, we showed that **PALKhΠ** is expressively equivalent to **ELΠ**. Now we can show that each **PALKhΠ**-formula φ is *provably equivalent* to an **ELΠ**-formula φ' (Lemma 5.7) in $\mathsf{SPALKh\Pi}^+$. Meanwhile we provide a translation from φ to φ'.

Theorem 5.2 (Soundness) $\mathsf{SPALKh\Pi}^+$ *is sound over the class of all models.*

Proof. The validity of $[\,]_{\mathsf{p}}$, $[\,]_{\bigcirc}$, $[\,]_{\mathsf{K}}$ and $[\,]_{\forall}$ are given in Proposition 4.3. $\mathtt{DIST}_{\forall}$, $\mathtt{SUB}_{\forall}$, SU and rule $\mathtt{GEN}_{\forall}$ are given in [10]. KKhp, $\mathtt{Kh}_{\bot}$, $\mathtt{Kh}_{\vee}$, $\mathtt{Kh}_{\wedge}$, $\mathtt{Kh}_{\to}$, and $\mathtt{Kh}_{\otimes}$ are shown to be valid in Proposition 4.2. The validity of rRE is given in Proposition 3.11. The rest are trivial. □

To prove the completeness we first prove Lemmata 5.3 and 5.6 with two sets of reduction axioms for Kh and $[\cdot]$ respectively. Recall that **PALΠ** is the Kh-free fragment of **PALKhΠ**, and **ELΠ** is the $[\cdot]$-free fragment of **PALΠ**.

Lemma 5.3 *Each* **PALKhΠ**-*formula is provably equivalent to a* Kh-*free* **PALΠ** *formula in* $\mathsf{SPALKh\Pi}^+$.

Proof. We use rRE and axioms $\mathtt{Kh}_{\bot}, \mathtt{Kh}_{\wedge}, \mathtt{Kh}_{\vee}, \mathtt{Kh}_{\to}$, $\mathtt{Kh}_{\otimes}$ repeatedly to reduce $\mathsf{Kh}\alpha$ to some formula with $\mathsf{Kh}p$ only. With $\vdash \mathsf{Kh}p \leftrightarrow \mathsf{K}p$ from KhK and KKhp, we can eliminate all Kh modalities. □

To eliminate the announcement operator, we need a notion of complexity.

Definition 5.4 (Announcement rank) *For each* $\varphi \in$ **PALΠ**, *we define its announcement rank* $\mathbf{ar}(\varphi)$ *inductively as follows:*

- *If* $\varphi = p$ *or* $\varphi = \bot$, *then* $\mathbf{ar}(\varphi) = 0$.
- *If* $\varphi = \psi_1 \bigcirc \psi_2$ *where* $\bigcirc = \wedge, \vee, \otimes$ *or* $\to$, *then* $\mathbf{ar}(\varphi) = \max\{\mathbf{ar}(\psi_1), \mathbf{ar}(\psi_2)\}$.
- *If* $\varphi = \mathsf{K}\psi$, *then* $\mathbf{ar}(\varphi) = \mathbf{ar}(\psi)$.
- *If* $\varphi = \forall p \psi$, $p \in \mathbf{P}$, *then* $\mathbf{ar}(\varphi) = \mathbf{ar}(\psi)$.
- *If* $\varphi = [\chi]\psi$, *then* $\mathbf{ar}(\varphi) = \mathbf{ar}(\psi) + \mathbf{ar}(\chi) + 1$.

Lemma 5.5 *Each* **PALΠ**-*formula of the form* $[\chi]\psi$ *is provably equivalent to a* **PALΠ**-*formula* φ *in* $\mathsf{SPALKh\Pi^+}$ *such that* $\mathbf{ar}(\varphi) < \mathbf{ar}([\chi]\psi)$.

Proof. We prove by induction on ψ:

(i) If $\psi = p$ or $\psi = \bot$, then by axiom $[\,]_\mathtt{p}$, $[\chi]\psi \leftrightarrow (\chi \to \psi)$ and $\mathbf{ar}(\chi \to \psi) = \max\{\mathbf{ar}(\chi), \mathbf{ar}(\psi)\} < \mathbf{ar}([\chi]\psi)$. Hence $\varphi = \chi \to \psi$ is what we need.

(ii) If $\psi = \psi_1 \bigcirc \psi_2$ where $\bigcirc = \wedge, \vee, \otimes, \to$, then by $[\,]_\bigcirc$, $[\chi]\psi \leftrightarrow [\chi]\psi_1 \bigcirc [\chi]\psi_2$. By IH, there are $\varphi_1 \leftrightarrow [\chi]\psi_1$ and $\varphi_2 \leftrightarrow [\chi]\psi_2$ such that $\mathbf{ar}(\varphi_1) < \mathbf{ar}([\chi]\psi_1)$ and $\mathbf{ar}(\varphi_1) < \mathbf{ar}([\chi]\psi_1)$. Hence, $\varphi = \varphi_1 \bigcirc \varphi_2$ is what we need.

(iii) If $\psi = \mathsf{K}\psi'$, then by $[\,]_\mathsf{K}$, $[\chi]\psi \leftrightarrow (\chi \to \mathsf{K}[\chi]\psi')$. By IH, there is a φ' such that $\varphi' \leftrightarrow [\chi]\psi'$ and that $\mathbf{ar}(\varphi') < \mathbf{ar}([\chi]\psi')$. Then $\mathbf{ar}(\chi \to \mathsf{K}\varphi') = \max\{\mathbf{ar}(\chi), \mathbf{ar}(\mathsf{K}\varphi')\} = \max\{\mathbf{ar}(\chi), \mathbf{ar}(\varphi')\} < \max\{\mathbf{ar}(\chi), \mathbf{ar}([\chi]\psi')\} = \mathbf{ar}([\chi]\mathsf{K}\psi')$. Hence, $\varphi = \chi \to \mathsf{K}\varphi'$ is what we need.

(iv) If $\psi = \forall p \psi'$ where $p \in \mathbf{P}$, we consider two subcases. 1) If p is not in χ, we use $[\,]_\forall$ and the proof is similar to the above cases. 2) If p is in χ, replace p with the first letter $q \in \mathbf{P}$ which is not in χ (such relettering can be done in the system), and then go to case 1).

(v) If $\psi = [\chi']\psi'$, by IH, there is a φ' such that $\varphi' \leftrightarrow [\chi']\psi'$ and that $\mathbf{ar}(\varphi') < \mathbf{ar}([\chi']\psi')$. So $[\chi][\chi']\psi' \leftrightarrow [\chi]\varphi'$ and $\mathbf{ar}([\chi]\varphi') < \mathbf{ar}([\chi][\chi']\psi')$. Hence $\varphi = [\chi]\varphi'$ is what we need.

□

Given the above lemma, we can eliminate the announcement operators eventually. The idea is that each formula in the shape of $[\chi]\varphi$ has a finite announcement rank, and can be reduced to an equivalent formula φ' with a lower rank. In case φ' still has subformulas with the announcement operators, we can replace each of these subformulas with a provably equivalent one with a lower rank. Note that by definition, a subformula's rank is no greater than the whole formula. Eventually, by repeating this process, we can decrease the rank to zero and obtain an equivalent formula without the announcement operators.

Lemma 5.6 *Each* **PALΠ**-*formula is provably equivalent to an* **ELΠ**-*formula in* $\mathsf{SPALKh\Pi^+}$.

Combining Lemmata 5.3 and 5.6 we immediately have.

Lemma 5.7 *Each* **PALKhΠ**-*formula is provably equivalent to an* **ELΠ**-*formula in* $\mathsf{SPALKh\Pi^+}$.

Theorem 5.8 follows naturally from Lemma 5.7 and Theorem 5.2.

Theorem 5.8 **PALKhΠ** *is equally expressive as* **ELΠ** *over all models.*

Note that **ELΠ** is more expressive than **EL** [10].

5.2 Completeness

With Lemma 5.7 and Theorem 5.8, the completeness of System $\mathsf{SPALKh\Pi}^+$ can be reduced to that of $\mathsf{S5\Pi}^+$, which is given in [10]. $\mathsf{S5\Pi}^+$ is a variation of second-order modal logic, containing all the axiom schemata/rules of S5 as well as those concerning propositional quantifiers in $\mathsf{SPALKh\Pi}^+$.

Theorem 5.9 (Completeness of $\mathsf{S5\Pi}^+$ [10]) $\mathsf{S5\Pi}^+$ *is a complete axiomatization with regard to the class of models.*

Theorem 5.10 (Completeness) *System* $\mathsf{SPALKh\Pi}^+$ *is complete over the class of all models.*

Proof. We first use Lemma 5.3 and Lemma 5.6 to translate each **PALKhΠ**-formula φ into an equivalent **ELΠ**-formula φ' and then use the completeness of $\mathsf{S5\Pi}^+$. Note that $\vdash \varphi$ below means φ is in $\mathsf{SPALKh\Pi}^+$.

$$\models \varphi \;\underset{\text{Theorem 5.8}}{\overset{\text{expressive equivalence}}{\Longleftrightarrow}}\; \models \varphi' \;\underset{\text{Theorem 5.9}}{\overset{\text{completeness of } \mathsf{S5\Pi}^+}{\Longleftrightarrow}}\; \vdash_{\mathsf{S5\Pi}^+} \varphi'$$

$$\overset{\mathsf{S5\Pi}^+ \subseteq \mathsf{SPALKh\Pi}^+}{\Longrightarrow}\; \vdash \varphi' \;\underset{\text{Lemma 5.7}}{\overset{\text{provable equivalence}}{\Longleftrightarrow}}\; \vdash \varphi$$

□

6 Generalization of Tensor Disjunction

Inspired by our epistemic interpretation, we generalize the binary $\otimes$ to n-ary operators for any $n \geq 2$ with another parameter $k \leq n$. Due to the lack of space, most of the proofs are omitted, except for a few crucial ones.

6.1 Generalizing the tensor operator

Consider the following scenario: You completed an exam with n questions, for which you need at least k correct answers to pass. Now you only know you have passed the exam. What is your epistemic state about your answers? For any $n \geq 2$ and $1 \leq k \leq n$, we now define an n-ary connective $\otimes_n^k$ capturing that you are sure at least k of your n answers must be correct, but may not know which ones are correct. The original tensor actually captures the special case when $k = 1$ and $n = 2$.

Definition 6.1 (Language $\mathbf{PL}^{\otimes_n^k}$) *The* propositional language with general tensor *($\mathbf{PL}^{\otimes_n^k}$) is defined as follows:*

$$\alpha ::= p \mid \bot \mid (\alpha \wedge \alpha) \mid (\alpha \vee \alpha) \mid (\alpha \to \alpha) \mid \otimes_n^k(\underbrace{\alpha, \cdots, \alpha}_{n})$$

where $p \in \mathbf{P}$ and $n \geq 2$, $1 \leq k \leq n$.

Definition 6.2 (Language $\mathbf{PALKh\Pi_G}$) *The* Public Announcement Logic with Know-how and General Tensor *($\mathbf{PALKh\Pi_G}$) is as follows:*

$$\varphi ::= p \mid \bot \mid (\varphi \wedge \varphi) \mid (\varphi \vee \varphi) \mid (\varphi \to \varphi) \mid \otimes^k_n(\underbrace{\varphi, \cdots, \varphi}_{n}) \mid \mathsf{K}\varphi \mid \mathsf{Kh}\alpha \mid \forall p\varphi \mid [\varphi]\varphi$$

where $p \in \mathbf{P}$ *and* $\alpha \in \mathbf{PL}^{\otimes^k_n}$.

Now, we introduce the semantics of new connectives $\otimes^k_n$ via resolutions.

Definition 6.3 *For any positive integer* $n \geq 2$ *and* $1 \leq k \leq n$, *we define the resolution space and resolution of* $\otimes^k_n$ *as follow:*

$$S(\otimes^k_n(\alpha_1, \cdots, \alpha_n)) = S(\alpha_1) \times \cdots \times S(\alpha_n)$$
$$R(w, \otimes^k_n(\alpha_1, \cdots, \alpha_n)) = \{(r_1, \cdots, r_n) \mid k \leq |\{i \in [1, n] \mid r_i \in R(w, \alpha_i)\}|\}$$

The truth condition for Kh is as before in Definition 3.7. In particular, $\mathcal{M}, w \vDash \mathsf{Kh} \otimes^k_n (\alpha_1, \cdots, \alpha_n)$ iff $R(W_{\mathcal{M}}, \otimes^k_n(\alpha_1, \cdots, \alpha_n)) \neq \varnothing$.

By Definition 3.7 and 6.3, it is not hard to see the following.

Proposition 6.4 $\mathcal{M}, w \vDash \mathsf{Kh} \otimes^k_n (\alpha_1, \cdots, \alpha_n)$ *if and only if there is an n-tuple* $\langle r_1, \cdots, r_n \rangle$ *such that for any* $v \in W_{\mathcal{M}}$, $|\{i \mid r_i \in R(v, \alpha_i)\}| \geq k$, *i.e., there are at least* k *indexes* $i \in [1, n]$ *such that* $r_i \in R(v, \alpha_i)$.

Based on the above proposition, the truth condition for $\otimes^1_2$ is exactly as the one for the standard $\otimes$ defined earlier. Note that $\otimes^k_n$ can also appear outside the scope of Kh in our language $\mathbf{PALKh\Pi_G}$ and we define its semantics below.

Definition 6.5 (Semantics)

$$\mathcal{M}, w \vDash \otimes^k_n(\varphi_1, \cdots, \varphi_n) \iff \mathcal{M}, w \vDash \bigvee_{\substack{I \subseteq \{1,2,\cdots,n\} \\ |I|=k}} \bigwedge_{i \in I} \varphi_i$$

The semantics is guided by Proposition 3.8, with the desired property below.

Proposition 6.6 *For any* $\alpha \in \mathbf{PL}^{\otimes^k_n}$ *and* $\mathcal{M}, w$, $\mathcal{M}, w \vDash \alpha$ *iff* $R(w, \alpha) \neq \varnothing$.

Next, we show how to reduce the general tensors in $\mathbf{PALKh\Pi_G}$.

Proposition 6.7 *The following schemata are valid:*

$$\mathtt{Rd}\otimes^{\mathtt{k}}_{\mathtt{n}} \quad \otimes^k_n (\varphi_1, \cdots, \varphi_n) \leftrightarrow \bigvee_{\substack{I \subseteq \{1,2,\cdots,n\} \\ |I|=k}} \bigwedge_{i \in I} \varphi_i$$

$$\mathtt{Kh}_{\otimes^{\mathtt{k}}_{\mathtt{n}}} \quad \mathsf{Kh} \otimes^k_n (\alpha_1, \cdots, \alpha_n) \leftrightarrow \exists p_1 \cdots \exists p_n \mathsf{K}(\otimes^k_n(p_1, \cdots, p_n) \wedge \bigwedge_{i=1}^{n} [p_i]\mathsf{Kh}\alpha_i)$$

(where all the p_i *do not occur free in all the* α_i*)*

Proof. $\mathtt{Rd}\otimes^{\mathtt{k}}_{\mathtt{n}}$ is valid by the truth condition of $\otimes^k_n$ in Definition 6.5.

For $\mathtt{Kh}_{\otimes^{\mathtt{k}}_{\mathtt{n}}}$:

$\Longrightarrow$: By Proposition 6.4 $\mathcal{M}, w \vDash \mathsf{Kh} \otimes_n^k (\alpha_1, \cdots, \alpha_n)$ iff there is an n-tuple $\langle r_1, \cdots, r_n \rangle$ s.t. for any $v \in W_\mathcal{M}$, there are at least k indexes $i \in [1, n]$ such that $r_i \in R(v, \alpha_i)$. Let $U_i = \{v \in W_\mathcal{M} \mid r_i \in R(v, \alpha_i)\}$, then consider $\mathcal{M}[\bar{p} \mapsto \bar{U}] = \langle W, V' \rangle$ such that V' assigns U_i to p_i for $i \in \{1, \dots, n\}$ and coincides with V on all other atoms. Then, for any $v \in W_\mathcal{M}$, there are at least k indexes $i \in [1, n]$ s.t. $\mathcal{M}[\bar{p} \mapsto \bar{U}], v \vDash p_i$, so $\mathcal{M}[\bar{p} \mapsto \bar{U}], v \vDash \otimes_n^k(p_1, \cdots, p_n)$. And since for any $v \in U_i$ we have $r_i \in R(v, \alpha_i)$, so $R(U_i, \alpha_i) \neq \varnothing$. Hence, for any $v \in W_\mathcal{M}$, $\mathcal{M}[\bar{p} \mapsto \bar{U}], v \vDash [p_i]\mathsf{Kh}\alpha_i$. In sum, $\mathcal{M}[\bar{p} \mapsto \bar{U}], w \vDash \mathsf{K}(\otimes_n^k(p_1, \cdots, p_n) \wedge \bigwedge_{i=1}^n [p_i]\mathsf{Kh}\alpha_i)$, which is equivalent to $\mathcal{M}, w \vDash \exists p_1 \cdots \exists p_n \mathsf{K}(\otimes_n^k(p_1, \cdots, p_n) \wedge \bigwedge_{i=1}^n [p_i]\mathsf{Kh}\alpha_i)$.

$\Longleftarrow$: Suppose $\mathcal{M}, w \vDash \exists p_1 \cdots \exists p_n \mathsf{K}(\otimes_n^k(p_1, \cdots, p_n) \wedge \bigwedge_{i=1}^n [p_i]\mathsf{Kh}\alpha_i)$, then there are $U_i \subseteq W_\mathcal{M}$ such that $\mathcal{M}[\bar{p} \mapsto \bar{U}], w \vDash \mathsf{K}(\otimes_n^k(p_1, \cdots, p_n) \wedge \bigwedge_{i=1}^n [p_i]\mathsf{Kh}\alpha_i)$, which is equivalent to $\mathcal{M}[\bar{p} \mapsto \bar{U}], w \vDash (\mathsf{K} \otimes_n^k (p_1, \cdots, p_n)) \wedge \bigwedge_{i=1}^n \mathsf{K}[p_i]\mathsf{Kh}\alpha_i$.

For the first conjunct: $\mathcal{M}[\bar{p} \mapsto \bar{U}], w \vDash \mathsf{K} \otimes_n^k (p_1, \cdots, p_n)$ means that for any $v \in W_\mathcal{M}$ we have $\mathcal{M}[\bar{p} \mapsto \bar{U}], v \vDash \otimes_n^k(p_1, \cdots, p_n)$. So at least k of p_i is true in v, which means v belongs to at least k of the corresponding U_i. For the second conjunct: $\mathcal{M}[\bar{p} \mapsto \bar{U}], w \vDash \bigwedge_{i=1}^n \mathsf{K}[p_i]\mathsf{Kh}\alpha_i$ means that for any $v \in W_\mathcal{M}$ and $i \in [1, n]$, $v \in U_i$ implies that $R(U_i, \alpha_i) \neq \varnothing$. So, if $U_i \neq \varnothing$, pick a r_i from $R(U_i, \alpha_i)$. If $U_i = \varnothing$, pick an arbitrary r_i from $S(\alpha_i)$. Hence, we have for any $i \in [1, n]$, $U_i \neq \varnothing$ implies $r_i \in R(U_i, \alpha_i)$.

Combining the meaning of the two conjuncts, we know that for any $v \in W_\mathcal{M}$, there are at least k indexes $i \in [1, n]$ such that $v \in U_i$ and $U_i \neq \varnothing$ implies $r_i \in R(U_i, \alpha_i)$ for any $i \in [1, n]$. Hence, $\langle r_1, \cdots, r_n \rangle$ is a n-tuple satisfying the desired property, by Proposition 6.4, we have $\mathcal{M}, w \vDash \mathsf{Kh} \otimes_n^k (\alpha_1, \cdots, \alpha_n)$. □

By using the reduction axioms above, *all* the general tensors can be eliminated semantically, and thus $\mathbf{PALKh\Pi_G}$ and $\mathbf{PALKh\Pi}$ are equally expressive.

Let $\mathsf{SPALKh\Pi_G^+}$ be $\mathsf{SPALKh\Pi^+}$ extended with $\mathtt{Rd}\otimes_\mathtt{n}^\mathtt{k}$ and $\mathtt{Kh}_{\otimes_\mathtt{n}^\mathtt{k}}$ for any $n \geq 2$ and $1 \leq k \leq n$. Similar to Theorem 5.10, it is straightforward to show:

Theorem 6.8 (Soundness and completeness) *Proof system* $\mathsf{SPALKh\Pi_G^+}$ *is sound and complete over the class of all models.*

6.2 Support semantics for $\otimes_n^k$

We can now go back to define the support semantics for $\otimes_n^k$.

Definition 6.9 (Support for $\otimes_n^k$) $\mathcal{M}, s \Vdash \otimes_n^k(\alpha_1, \cdots, \alpha_n)$ *iff there exist n subsets $t_1, \cdots, t_n$ of s such that for any $i \in [1, n]$, $\mathcal{M}, t_i \Vdash \alpha_i$ and for any $w \in s$, there are at least k indexes $i \in [1, n]$ such that $w \in t_i$.*

The support semantics for other connectives stays the same as in Definition 2.3. Let $\mathbf{InqB}^{\otimes_\mathbf{n}^\mathbf{k}}$ be the set of valid $\mathbf{PL}^{\otimes_n^k}$ formulas by the support semantics. We can show $\mathbf{Inq}^{\otimes_\mathbf{n}^\mathbf{k}}\mathbf{Kh} = \{\alpha \in \mathbf{PL}^{\otimes_n^k} \mid \vDash \mathsf{Kh}\alpha\}$ is exactly $\mathbf{InqB}^{\otimes_\mathbf{n}^\mathbf{k}}$, based on the following generalization of Lemma 4.4.

Proposition 6.10 *For any* $\alpha \in \mathbf{PL}^{\otimes_n^k}$, $\mathcal{M}, w \vDash \mathsf{Kh}\alpha \iff \mathcal{M}, W_\mathcal{M} \Vdash \alpha$.

Proof. Based on Lemma 4.4, we only consider the case of $\otimes_n^k(\alpha_1, \cdots, \alpha_n)$ and write $\exists U$ for $\exists U \subseteq W_{\mathcal{M}}$ for brevity, similarly for $\exists t$.

$$\mathcal{M}, w \vDash \mathsf{Kh}(\otimes_n^k(\alpha_1, \cdots, \alpha_n))$$
$$\iff \mathcal{M}, w \vDash \exists p_1 \cdots \exists p_n \mathsf{K}(\otimes_n^k(p_1, \cdots, p_n) \wedge \bigwedge_{i=1}^{n} [p_i]\mathsf{Kh}\alpha_i) \text{ (by Proposition 6.7).}$$
$$\iff \exists U_1, \cdots, U_n, \forall v \in W_{\mathcal{M}}, \text{ there are at least } k \text{ indexes } i \in [1, n] \text{ s.t. } v \in U_i \text{ and for any } i, U_i \neq \varnothing \text{ implies } R(U_i, \alpha_i) \neq \varnothing. \text{(similar to Proposition 4.1)}$$
$$\iff \exists t_1, \cdots, t_n, \forall i \in [1, n], t_i \Vdash \alpha_i \text{ and } \forall v \in W_{\mathcal{M}}, k \leq |\{i \in [1, n] \mid v \in U_i\}|$$
$$\iff \mathcal{M}, W_{\mathcal{M}} \Vdash \otimes_n^k(\alpha_1, \cdots, \alpha_n).$$

□

As shown in [24], adding tensor does not increase the expressive power of inquisitive logic. In fact, adding all the general tensors also does not increase the expressive power of inquisitive logic.

First, we extend the definition of realization in [6] to our new connectives.

Definition 6.11 (Realizations)

- $RL(p) = \{p\}$ *for* $p \in \mathbf{P}$
- $RL(\bot) = \{\bot\}$
- $RL(\alpha \vee \beta) = RL(\alpha) \cup RL(\beta)$
- $RL(\alpha \wedge \beta) = \{\rho \wedge \sigma \mid \rho \in RL(\alpha) \text{ and } \sigma \in RL(\beta)\}$
- $RL(\alpha \to \beta) = \{\bigwedge_{\rho \in RL(\alpha)} (\rho \to f(\rho)) \mid f : RL(\alpha) \to RL(\beta)\}$
- $RL(\otimes_n^k(\alpha_1, \cdots, \alpha_n)) = \{\neg \bigwedge_{\substack{I \subseteq \{1,2,\cdots,n\} \\ |I|=k}} \neg \bigwedge_{i \in I} \rho_i \mid \text{ for all } i, \rho_i \in RL(\alpha_i)\}$

Then we can generalize the inquisitive *normal form* in [6,9].

Proposition 6.12 *For any* $\alpha \in \mathbf{PL}^{\otimes_n^k}$, $s \Vdash \alpha$ *iff* $s \Vdash \bigvee_{\rho \in RL(\alpha)} \rho$.

Theorem 6.13 $\mathbf{PL}$ *and* $\mathbf{PL}^{\otimes_n^k}$ *are equally expressive w.r.t. the support semantics.*

Proof. By Proposition 6.12, for any $\alpha \in \mathbf{PL}^{\otimes_n^k}$, α is equivalent to a disjunction of some ρ without general tensors. □

In [25], it is shown that the variants of propositional dependence logics **PD**, $\mathbf{PD}^{\vee}$, **PID**, **InqB** are all equally expressive. Similarly, adding general tensors to the languages of these logics will also not increase the expressive power.

Corollary 6.14 *Adding general tensors to the langugaes of* **PD**, $\mathbf{PD}^{\vee}$, **PID** *or* **InqB** *does not increase their expressive power.*

6.3 Interdefinability

In [7], it is proved that although adding $\otimes$ to **PL** does not increase the expressive power, $\otimes$ is not *uniformly definable* in **PL**, i.e., $\varphi \otimes \psi$ cannot be expressed by a uniform formula "template" with φ and ψ as its only parameters. It is also natural to ask whether general tensors are uniformly definable by the standard binary tensor $\otimes$.

First, on the positive side, we show that $\otimes^n_n$ is a trivial conjunction and $\otimes^1_n$ can be uniformly defined by $\otimes^1_2$. Moreover, by fixing some components as $\top$ or $\bot$, some general tensors can be uniformly defined by others with smaller parameters.

Proposition 6.15 *For any $\alpha_1, \cdots, \alpha_n \in \mathbf{PL}^{\otimes^k_n}$, the following hold:*

- *For any $n \geq 2$ and any state s, $s \Vdash \otimes^n_n(\alpha_1, \cdots, \alpha_n) \iff s \Vdash \bigwedge_{i=1}^n \alpha_i$.*
- *For any $n \geq 3$, $s \Vdash \otimes^1_n(\alpha_1, \cdots, \alpha_n) \iff s \Vdash \otimes^1_2(\otimes^1_{n-1}(\alpha_1, \cdots, \alpha_{n-1}), \alpha_n)$.*
- *For any $n \geq 3$, $1 \leq k \leq n$ and any state s, $s \Vdash \otimes^k_n(\alpha_1, \cdots, \alpha_{n-1}, \top) \iff s \Vdash \otimes^{k-1}_{n-1}(\alpha_1, \cdots, \alpha_{n-1})$.*
- *For any $n \geq 3$, $1 \leq k \leq n-1$ and any state s, $s \Vdash \otimes^k_n(\alpha_1, \cdots, \alpha_{n-1}, \bot) \iff s \Vdash \otimes^k_{n-1}(\alpha_1, \cdots, \alpha_{n-1})$.*

Inspired by the proof in [7], we can show that $\otimes^2_3$ is not uniformly definable by the connectives $\{\bot, \wedge, \rightarrow, \vee, \otimes^1_2\}$ in $\mathbf{PL}^{\otimes}$.

To show the negative results, we need the following definitions about uniform definability from [24].

Definition 6.16 (Context) *A context for a propositional logic* $\mathbf{L}$ *is an* $\mathbf{L}$*-formula $\varphi(p_1, \cdots, p_n)$ with distinguished atoms $p_1, \cdots, p_n$, and it is also allowed to contain other atoms besides $p_1, \cdots, p_n$. For any* $\mathbf{L}$*-formulas $\psi_1, \cdots, \psi_n$, we write $\varphi(\psi_1, \cdots, \psi_n)$ for the formula $\varphi(\psi_1/p_1, \cdots, \psi_n/p_n)$.*

Definition 6.17 (Uniform definability) *In a language* $\mathbf{L}$*, we say that an n-ary connective $\odot$ is* uniformly definable *if there exists a context $\zeta(p_1, \cdots, p_n)$ such that for all $\chi_1, \cdots, \chi_n \in \mathbf{L}$: $\odot(\chi_1, \cdots, \chi_n)$ is equivalent to $\zeta(\chi_1, \cdots, \chi_n)$.*

As the first negative result, we show $\otimes^2_3$ is not uniformly definable in $\mathbf{PL}^{\otimes}$. We consider relative equivalence with respect to a state s.

Definition 6.18 (Relative equivalence [7]) *Let s be a state in $\mathcal{M}$ and $\varphi, \psi \in \mathbf{PL}^{\otimes^k_n}$. We say that φ and ψ are relatively equivalent in s, $\varphi \equiv_s \psi$ iff for all states $t \subseteq s$, $t \Vdash \varphi \iff t \Vdash \psi$.*

It is easy to see that if φ and ψ are equivalent then they are relatively equivalent in any state s.

Consider $\psi = p_1 \vee p_2 \vee p_3 \vee p_4$ and $s = \{w_{12}, w_{13}, w_{14}, w_{23}, w_{24}, w_{34}\}$ where only p_i, p_j are true in w_{ij} and all the other propositional letters are false. Now, we show that with respect to this state s, $\otimes^2_3$ cannot be uniformly defined by any context in $\mathbf{PL}^{\otimes}$.

Lemma 6.19 *For any context $\varphi(p_0)$, with $\varphi \in \mathbf{PL}^{\otimes}$ not containing p_1, p_2, p_3, p_4, $\varphi(\psi/p_0)$ would be equivalent to $\bot, \psi, \otimes^1_2(\psi, \psi)$ or $\top$ in s.*

Proof. First we notice for any state t, $t \Vdash \bot \Rightarrow t \Vdash \psi \Rightarrow t \Vdash \otimes^1_2(\psi, \psi) \Rightarrow t \Vdash \top$ $(\star)$. Then we prove by induction on φ. For short, we write φ^* for $\varphi(\psi/p_0)$:

- For $\varphi = \bot$ or $\varphi = p$ with $p \neq p_0$: Since we assume that p_1, p_2, p_3, p_4 are not in φ, so p is different from them. Hence, it is obvious that $\varphi^* \equiv_s \bot$.
- For $\varphi = p_0$: It is obvious that $\varphi^* \equiv_s \psi$.
- For $\varphi = \varphi_1 \wedge \varphi_2$: so $\varphi^* = \varphi_1^* \wedge \varphi_2^*$. By IH, φ_1^* and φ_2^* are both equivalent to $\bot, \psi, \otimes_2^1(\psi, \psi)$ or $\top$ in s. Since $(t \Vdash \chi_1 \Rightarrow t \Vdash \chi_2)$ implies $(t \Vdash \chi_1 \wedge \chi_2 \iff t \Vdash \chi_1)$, by $(\star)$, obviously φ^* is also equivalent to $\bot, \psi, \otimes_2^1(\psi, \psi)$ or $\top$ in s.
- For $\varphi = \varphi_1 \vee \varphi_2$: similar to the case of conjunction.
- For $\varphi = \varphi_1 \to \varphi_2$: so $\varphi^* = \varphi_1^* \to \varphi_2^*$, and for any state t, $t \Vdash \varphi_1^* \to \varphi_2^*$ iff for any $t' \subseteq t$, $t' \Vdash \varphi_1^*$ implies $t' \Vdash \varphi_2^*$. By $(\star)$, we could know that:
  · $\bot \to \bot$, $\bot \to \psi$, $\bot \to \otimes_2^1(\psi, \psi)$, $\bot \to \top$, $\psi \to \psi$, $\psi \to \otimes_2^1(\psi, \psi)$, $\psi \to \top$, $\otimes_2^1(\psi, \psi) \to \otimes_2^1(\psi, \psi)$, $\otimes_2^1(\psi, \psi) \to \top$, $\top \to \top$ are all equivalent to $\top$ in s.
  · $\psi \to \bot$, $\otimes_2^1(\psi, \psi) \to \bot$ and $\top \to \bot$ are all equivalent to $\bot$ in s.
  · $\otimes_2^1(\psi, \psi) \to \psi$ and $\top \to \psi$ are equivalent to ψ in s.
  · $\top \to \otimes_2^1(\psi, \psi)$ is equivalent to $\otimes_2^1(\psi, \psi)$ in s.
  Hence, φ^* is equivalent to $\bot, \psi, \otimes_2^1(\psi, \psi)$ or $\top$ in s.
- For $\varphi = \otimes_2^1(\varphi_1, \varphi_2)$: so $\varphi^* = \otimes_2^1(\varphi_1^*, \varphi_2^*)$. We consider the following cases:
  · $\varphi_1^* \equiv_s \top$. Then $\otimes_2^1(\varphi_1^*, \varphi_2^*) \equiv_s \top$.
  · $\varphi_1^* \equiv_s \bot$. Then $\otimes_2^1(\varphi_1^*, \varphi_2^*) \equiv_s \varphi_2^*$.
  · $\varphi_1^* \equiv_s \psi$. If $\varphi_2^* \equiv_s \top$ or $\varphi_2^* \equiv_s \bot$, it would be the same as former cases. Then we need to discuss two sub-cases:
    * $\varphi_2^* \equiv_s \psi$. Then $\otimes_2^1(\varphi_1^*, \varphi_2^*) \equiv_s \otimes_2^1(\psi, \psi)$.
    * $\varphi_2^* \equiv_s \otimes_2^1(\psi, \psi)$. Then $t \Vdash \varphi^* \iff$ there are $t_1, t_2 \subseteq t$ and $t_1 \cup t_2 = t$ such that $t_1 \Vdash \psi$ and $t_2 \Vdash \otimes_2^1(\psi, \psi) \iff$ there are $t_1, t_2 \subseteq t$, $t_1 \cup t_2 = t$ and $p_{i_1}, p_{i_2}, p_{i_3}$ such that p_{i_1} is true in any $w \in t_1$ and for any $w \in t_2$, p_{i_2} or p_{i_3} is true in $w \iff$ there are $p_{i_1}, p_{i_2}, p_{i_3}$ such that for any $w \in t$, p_{i_1}, p_{i_2} or p_{i_3} is true in w. However, there are only four propositional letters p_1, p_2, p_3, p_4 and in each $w \in s$, two of these propositional letters are true. So consider p_1, p_2 and p_3, we will know that for any $w \in t \subseteq s$, at least one of p_1, p_2 and p_3 is true in w. Hence, $\otimes_2^1(\psi, \otimes_2^1(\psi, \psi)) \equiv_s \top$.
  · $\varphi_1^* \equiv_s \otimes_2^1(\psi, \psi)$. Then if $\varphi_2^* \equiv_s \top$, $\varphi_2^* \equiv_s \bot$ or $\varphi_2^* \equiv_s \psi$, it would be the same as former cases. And if $\varphi_2^* \equiv_s \otimes_2^1(\psi, \psi)$, the proof is similar to the previous case and the result is that $\otimes_2^1(\otimes_2^1(\psi, \psi), \otimes_2^1(\psi, \psi)) \equiv_s \top$. □

Lemma 6.20 *$\otimes_3^2$ is not uniformly definable in* $\mathbf{PL}^{\otimes}$.

Proof. If $\otimes_3^2$ is uniformly definable in $\mathbf{PL}^{\otimes}$, there will be a context $\varphi(p)$ such that for any $\chi \in \mathbf{PL}^{\otimes}$: $\varphi(\chi)$ is equivalent to $\otimes_3^2(\chi, \chi, \chi)$.

However, as we proved in Lemma 6.19, for any context $\varphi(p_0) \in \mathbf{PL}^{\otimes}$, $\varphi(\psi/p_0)$ would be relatively equivalent to $\bot$, ψ, $\otimes_2^1(\psi, \psi)$ or $\top$ in s. But it is obvious that $\otimes_3^2(\psi, \psi, \psi)$ is not relatively equivalent to $\bot$, ψ, $\otimes_2^1(\psi, \psi)$ or $\top$ in s. Hence, $\otimes_3^2(\psi, \psi, \psi)$ and $\varphi(\psi/p_0)$ are not relatively equivalent in s, and hence not equivalent in general, which gives rise to a contradiction. □

Theorem 6.21 *All the $\otimes_n^k$ are **not** uniformly definable in $\mathbf{PL}^{\otimes}$ except $\otimes_n^1$ and $\otimes_n^n$, i.e., for any $2 \leq k \leq n-1$, $\otimes_n^k$ is not uniformly definable.*

Proof. Note that $n \geq 2$ by definition. When $2 \leq k \leq n-1$ (thus $n \geq 3$), by Proposition 6.15, $\otimes_3^2$ can be uniformly defined by $\otimes_n^k$, so $\otimes_3^2$ is not uniformly definable in $\mathbf{PL}^{\otimes}$ implies that $\otimes_n^k$ is not uniformly definable in $\mathbf{PL}^{\otimes}$. Based on the first two items of Proposition 6.15, we have the desired result. □

7 Conclusions and future work

In this paper, we proposed an epistemic interpretation of the tensor disjunction in dependence logic. The interpretation is inspired by the notion of weak disjunction in Medvedev's early work in terms of the BHK-like semantics. The connection between the two disjunctions is exposed in inquisitive logic with tensor disjunction ($\mathbf{InqB}^{\otimes}$) studied in the literature. We introduce a powerful dynamic epistemic language which can turn each formula in the language of $\mathbf{InqB}^{\otimes}$ into a know-how formula, which can be further reduced into a know-how-free formula. Along the way we need to use announcement operators and the propositional quantifiers to capture the epistemic meaning of the tensor disjunction. We give the axiomatization of our full logic, and generalize the tensor disjunction to a family of n-ary operators parameterized by a $k \leq n$, which capture the intuitive epistemic situations that knowing a list of n possible answers to n questions such that one knows at least k of them are correct.

We have seen that the propositional quantifiers are playing an important role in our framework, i.e., in defining the tensor and its generalizations. However, technically speaking, it might be an overkill since the expressive power of **InqB** and $\mathbf{InqB}^{\otimes}$ are the same. It remains to see whether we can use a simpler machinery to capture the epistemic meaning of tensor discussion without using the full power of the propositional quantifiers.

Besides further technical questions regarding our logic, the generalized tensor clearly has a life of its own, and invites further investigations. Its obvious combinatorial features may find applications in cryptographic protocols and game theory. To see the connection with the latter, we end the paper with the following interesting scenario where $\otimes_3^2$ makes perfect sense. Consider a badminton match between two teams. Each team has one good player with two other less capable ones. We can measure the abilities of the players by numbers which will determine the result of the matches in the most natural way. For team A, it is $6, 2, 2$ for the three players, and for team B it is $5, 3, 3$. The battle between the two teams consists of three single matches, and the rule of the game does not prevent one player from playing two matches if not in a row, although the second time the player will lose $1/3$ of his or her ability due to tiredness. Now, with some reflection, we can see team B has a unique arrangement of the playing players to make sure they can win at least two out of the three matches no matter how team A will do. Do you know which one?

Acknowledgements The authors would like to thank the participants of the ESSLLI2021-affiliated workshop *Logics of Dependence and Independence* for their questions and remarks, and thank the anonymous reviewers of AiML2022 for their insightful comments, which improved the presentation of the paper a lot. In particular, Yanjing Wang thanks Fan Yang for pointers on tensor disjunction in the literature of dependence logic. The work is supported by NSSF grant 19BZX135.

References

[1] Belardinelli, F., H. van Ditmarsch and W. van der Hoek, *Second-order propositional announcement logic*, in: *Proceedings of AAMAS 2016* (2016), p. 635–643.

[2] Chagrov, A. and M. Zakharyaschev, "Modal Logic," Clarendon Press, 1997.

[3] Ciardelli, I., *A first-order inquisitive semantics*, in: *17th Amsterdam Colloquium*, Lecture Notes in Computer Science **6042** (2009), pp. 234–243.

[4] Ciardelli, I., *Dependency as Question Entailment*, in: *Dependence Logic*, Springer International Publishing, Cham, 2016 pp. 129–181.

[5] Ciardelli, I., "Questions in logic," Ph.D. thesis, Institute for Logic, Language and Computation, University of Amsterdam (2016).

[6] Ciardelli, I., *Questions as information types*, Synthese **195** (2018), pp. 321–365.

[7] Ciardelli, I. and F. Barbero, *Undefinability in inquisitive logic with tensor*, in: *Proceedings of LORI VII*, Lecture Notes in Computer Science **11813** (2019), pp. 29–42.

[8] Ciardelli, I., R. Iemhoff and F. Yang, *Questions and dependency in intuitionistic logic*, Notre Dame Journal of Formal Logic **61** (2020), pp. 75–115.

[9] Ciardelli, I. and F. Roelofsen, *Inquisitive logic*, Journal of Philosophical Logic **40** (2011), pp. 55–94.

[10] Fine, K., *Propositional quantifiers in modal logic*, Theoria **36** (1970), pp. 336–346.

[11] Galliani, P., *Dependence Logic*, in: E. N. Zalta, editor, *The Stanford Encyclopedia of Philosophy*, Metaphysics Research Lab, Stanford University, Summer 2021 edition .

[12] Heyting, A., *La conception intuitionniste de la logique*, Les études philosophiques **11** (1956), pp. 226–233.

[13] Kolmogorov, A., *Zur deutung der intuitionistischen logik*, Mathematische Zeitschrift (1932), pp. 58–65.

[14] Medvedev, Y. T., *Interpretation of logical formulas by means of finite problems*, Dokl. Akad. Nauk SSSR **169** (1966), pp. 20–23.

[15] Miglioli, P., U. Moscato, M. Ornaghi, S. Quazza and G. Usberti, *Some results on intermediate constructive logics*, Notre Dame Journal of Formal Logic **30** (1989), pp. 543–562.

[16] Plaza, J., *Logics of public communications*, Synthese **158** (2007), pp. 165–179.

[17] Väänänen, J. A., "Dependence Logic - A New Approach to Independence Friendly Logic," London Mathematical Society student texts **70**, Cambridge University Press, 2007.

[18] Wang, H., Y. Wang and Y. Wang, *Inquisitive logic as an epistemic logic of knowing how*, Annals of Pure and Applied Logic (2022), p. 103145, in press.

[19] Wang, Y., *A New Modal Framework for Epistemic Logic*, Proceedings of TARK 2017 **251** (2017), pp. 515–534.

[20] Wang, Y., *Beyond Knowing That: A New Generation of Epistemic Logics*, in: *Jaakko Hintikka on Knowledge and Game-Theoretical Semantics*, Outstanding Contributions to Logic **12**, Springer Nature, 2018 pp. 499–533.

[21] Wang, Y., *Knowing how to understand intuitionistic logic* (2021), manuscript.

[22] Wang, Y. and Q. Cao, *On axiomatizations of public announcement logic*, Synthese **190** (2013), pp. 103–134.

[23] Yang, F., "On Extensions and Variants of Dependence Logic," Ph.D. thesis, University of Helsinki (2014).

[24] Yang, F., *Uniform definability in propositional dependence logic*, The Review of Symbolic Logic **10** (2017), p. 65–79.

[25] Yang, F. and J. Väänänen, *Propositional logics of dependence*, Annals of Pure and Applied Logic **167** (2016), pp. 557–589.

www.ingramcontent.com/pod-product-compliance
Lightning Source LLC
Chambersburg PA
CBHW071932220426
43662CB00009B/891
* 9 7 8 1 8 4 8 9 0 4 1 3 2 *